Tools for Your Success!

MyMathLab®

MyMathLab® is a series of text-specific, easily customizable online courses for Prentice Hall textbooks in mathematics and statistics. Powered by CourseCompass™ (Pearson Education's online teaching and learning environment) and MathXL® (our online homework, tutorial, and assessment system), MyMathLab gives you the tools you need to deliver all or a portion of your course online, whether your students are in a lab setting or working from home. MyMathLab provides a rich and flexible set of course materials, featuring free-response exercises that are algorithmically generated for unlimited practice and mastery. Students can also use online tools, such as video lectures, animations, and a multimedia textbook, to independently improve their understanding and performance. Instructors can use MyMathLab's homework and test managers to select and assign online exercises correlated directly to the textbook, and they can also create and assign their own online exercises and import TestGen tests for added flexibility. MyMathLab's online gradebook—designed specifically for mathematics and statistics—automatically tracks students' homework and test results and gives the instructor control over how to calculate final grades. Instructors can also add offline (paper-and-pencil) grades to the gradebook. MyMathLab is available to qualified adopters. For more information, visit our website at www.mymathlab.com or contact your Prentice Hall sales representative. (MyMathLab must be set up and assigned by your instructor.)

MathXL®

MathXL® is a powerful online homework, tutorial, and assessment system that accompanies Prentice Hall textbooks in mathematics or statistics. With MathXL, instructors can create, edit, and assign online homework and tests using algorithmically generated exercises correlated at the objective level to the textbook. They can also create and assign their own online exercises and import TestGen tests for added flexibility. All student work is tracked in MathXL's online gradebook. Students can take chapter tests in MathXL and receive personalized study plans based on their test results. The study plan diagnoses weaknesses and links students directly to tutorial exercises for the objectives they need to study and retest. Students can also access supplemental animations and video clips directly from selected exercises. MathXL is available to qualified adopters. For more information, visit our website at www.mathxl.com, or contact your Prentice Hall sales representative. (MathXL must be set up and assigned by your instructor.)

More Tools for Your Success!

STUDENT STUDY PACK

Includes:

- the *Student Solutions Manual*

- access to the *Prentice Hall Tutor Center*

- the *CD Lecture Series Videos*

Student Solutions Manual

Provides full, worked-out solutions to:

- the odd-numbered section exercises

- all (even and odd) exercises in the Cumulative Review Exercises (within the section exercises), Mid-Chapter Tests, Chapter Review Exercises, Chapter Practice Tests, and Cumulative Review Tests

Chapter Test Prep Video CD

Provides step-by-step video solutions to each problem in each *Chapter Practice Test* in the textbook.

CD Lecture Series Videos

Each section includes:

- a 20-minute mini-lecture

- several worked-out section exercises (identified by a CD icon)

Prentice Hall Tutor Center

Staffed by developmental math faculty, the Tutor Center provides live tutorial support via phone, fax, or email.

ELEMENTARY
and
INTERMEDIATE
ALGEBRA

Ir College Students

Third Edition

Allen R. Angel

Monroe Community College

with assistance from
Richard Semmler

Northern Virginia Community College

Aimee L. Calhoun
Donna R. Petrie

Monroe Community College

PEARSON

Prentice
Hall

er Saddle River, New Jersey 07458

Library of Congress Cataloging-in-Publication Data

Angel, Allen R.

Elementary and intermediate algebra for college students—3rd ed.

   Allen R. Angel; with assistance from Richard Semmler, Aimee L. Calhoun, Donna R. Petrie

   p.   cm,

Includes index.

ISBN 0-13-233722-3 (SE)—ISBN 0-13-614445-4 (AIE)

1. Algebra—Textbooks.   I.   Calhoun, Aimee.   II.   Petrie, Donna, R.   III   Title.

QA152.3.A57 2008

512.9—dc22   2006052505

Executive Editor: *Paul Murphy*
Project Manager: *Dawn Nuttall*
Editorial Director, Mathematics: *Christine Hoag*
Production Editor: *Lynn Savino Wendel*
Executive Managing Editor: *Kathleen Schiaparelli*
Senior Managing Editor: *Linda Mihatov Behrens*
Media Project Manager, Developmental Math: *Audra J. Walsh*
Media Production Editor: *Ashley Booth*
Assistant Managing Editor, Science and Math Supplements: *Karen Bosch*
Manufacturing Buyer: *Maura Zaldivar*
Manufacturing Manager: *Alexis Heydt-Long*
Director of Marketing: *Patrice Jones*
Senior Marketing Manager: *Kate Valentine*
Marketing Assistant: *Jennifer de Leeuwerk*
Editorial Assistant/Print Supplements Editor: *Abigail Rethore*
Editor in Chief, Development: *Carol Trueheart*
Art Director: *John Christiana*
Interior Designer: *Studio Indigo*
Cover Designer: *Michael J. Fruhbeis*
Art Editor: *Thomas Benfatti*
Creative Director: *Juan R. López*
Director of Creative Services: *Paul Belfanti*
Director, Image Resource Center: *Melinda Patelli*
Manager, Rights and Permissions: *Zina Arabia*
Manager, Visual Research: *Beth Brenzel*
Image Permission Coordinator: *Craig A. Jones*
Photo Researcher: *Teri Stratford*
Compositor: *Prepare, Inc.*
Art Studios: *Precision Graphics and Laserwords*

© 2008, 2004, 2000 Pearson Education, Inc.
Pearson Prentice Hall
Pearson Education, Inc.
Upper Saddle River, NJ 07458

10 9 8 7 6 5 4 3

ISBN 0-13-233722-3

Pearson Education, LTD., *London*
Pearson Education Australia PTY, Limited, *Sydney*
Pearson Education Singapore, Pte. Ltd
Pearson Education North Asia Ltd, *Hong Kong*
Pearson Education Canada, Ltd., *Toronto*
Pearson Educación de Mexico, S.A. de C.V.
Pearson Education –Japan, *Tokyo*
Pearson Education Malaysia, Pte. Ltd

To my brother,
Jerry Angel

Contents

Preface

This book was written for college students and other adults who have never been exposed to algebra or those who have been exposed but need a refresher course. My primary goal was to write a book that students can read, understand, and enjoy. To achieve this goal I have used short sentences, clear explanations, and many detailed, worked-out examples. I have tried to make the book relevant to college students by using practical applications of algebra throughout the text.

Features of the Text

Full-Color Format
Color is used pedagogically in the following ways:

- Important definitions and procedures are color screened.
- Color screening or color type is used to make other important items stand out.
- Artwork is enhanced and clarified with use of multiple colors.
- The full-color format allows for easy identification of important features by students.
- The full-color format makes the text more appealing and interesting to students.

Readability
One of the most important features of the text is its readability. The book is very readable, even for those with weak reading skills. Short, clear sentences are used and more easily recognized, and easy-to-understand language is used whenever possible.

Accuracy
Accuracy in a mathematics text is essential. To ensure accuracy in this book, mathematicians from around the country have read the pages carefully for typographical errors and have checked all the answers.

Connections
Many of our students do not thoroughly grasp new concepts the first time they are presented. In this text we encourage students to make connections. That is, we introduce a concept, then later in the text briefly reintroduce it and build upon it. Often an important concept is used in many sections of the text. Students are reminded where the material was seen before, or where it will be used again. This also serves to emphasize the importance of the concept. Important concepts are also reinforced throughout the text in the Cumulative Review Exercises and Cumulative Review Tests.

Chapter Opening Application
Each chapter begins with a real-life application related to the material covered in the chapter. By the time students complete the chapter, they should have the knowledge to work the problem.

Goals of this Chapter
This feature on the chapter opener page gives students a preview of the chapter and also indicates where this material will be used again in other chapters of the book. This material helps students see the connections among various topics in the book and the connection to real-world situations.

The Use of Icons
At the begining of each exercise set the icon for MathXL®, *Math* $\boxed{XL}$, and for MyMathLab, **MyMathLab** , are illustrated. Both of these icons will be explained shortly.

Keyed Section Objectives
Each section opens with a list of skills that the student should learn in that section. The objectives are then keyed to the appropriate portions of the sections with blue numbers such as **1**.

Problem Solving
Polya's five-step problem-solving procedure is discussed in Section 1.2. Throughout the book, problem solving and Polya's problem-solving procedure are emphasized.

Practical Applications
Practical applications of algebra are stressed throughout the text. Students need to learn how to translate application problems into algebraic symbols. The problem-solving approach used throughout this text gives students ample practice in setting up and solving application problems. The use of practical applications motivates students.

Detailed, Worked-Out Examples
A wealth of examples have been worked out in a step-by-step, detailed manner. Important steps are highlighted in color, and no steps are omitted until after the student has seen a sufficient number of similar examples.

Now Try Exercises
In each section, after each example, students are asked to work an exercise that parallels the example given in the text. These Now Try Exercises make the students *active*, rather than passive, learners and they reinforce the concepts as students work the exercises. Through these exercises, students have the opportunity to immediately apply what they have learned. After each example, Now Try Exercises are indicated in green type such as ▶ **Now Try Exercise 27**. They are also indicated in green type in the exercise sets, such as 27.

Study Skills Section
Many students taking this course have poor study skills in mathematics. Section 1.1,

the first section of this text, discusses the study skills need-ed to be successful in mathematics. This section should be very beneficial for your students and should help them to achieve success in mathematics.

Helpful Hints
The Helpful Hint boxes offer useful suggestions for problem solving and other varied topics. They are set off in a special manner so that students will be sure to read them.

Helpful Hints—Study Tip
The Helpful Hint–Study Tip boxes offer valuable information on items related to studying and learning the material.

Avoiding Common Errors
Errors that students often make are illustrated. The reasons why certain procedures are wrong are explained, and the correct procedure for working the problem is illustrated. These Avoiding Common Errors boxes will help prevent your students from making those errors we see so often.

Using Your Calculator
The Using Your Calculator boxes, placed at appropriate locations in the text, are written to reinforce the algebraic topics presented in the section and to give the student pertinent information on using a scientific calculator to solve algebraic problems.

Using Your Graphing Calculator
Using Your Graphing Calculator boxes are placed at appropriate locations throughout the text. They reinforce the algebraic topics taught and sometimes offer alternate methods of working problems. This book is designed to give the instructor the option of using or not using a graphing calculator in his or her course. Some of the Using Your Graphing Calculator boxes contain graphing calculator exercises, whose answers appear in the answer section of the book. The illustrations shown in the Using Your Graphing Calculator boxes are from a Texas Instruments TI-84 Plus calculator. The Using Your Graphing Calculator boxes are written assuming that the student has no prior graphing calculator experience.

Exercise Sets

The exercise sets are broken into three main categories: Concept/Writing Exercises, Practice the Skills, and Problem Solving. Many exercise sets also contain Challenge Problems and/or Group Activities. Each exercise set is graded in difficulty. The early problems help develop the student's confidence, and then students are eased gradually into the more difficult problems. A sufficient number and variety of examples are given in each section for the student to successfully complete even the more difficult exercises. The number of exercises in each section is more than ample for student assignments and practice.

Concept/Writing Exercises
Most exercise sets include exercises that require students to write out the answers in words. These exercises improve students'

understanding and comprehension of the material. Many of these exercises involve problem solving and conceptualization and help develop better reasoning and critical thinking skills. Writing exercises are indicated by the symbol ✎.

Problem Solving Exercises
These exercises have been added to help students become better thinkers and problem solvers. Many of these exercises involve real-life applications of algebra. It is important for students to be able to apply what they learn to real-life situations. Many problem-solving exercises help with this.

Challenge Problems
These exercises, which are part of many exercise sets, provide a variety of problems. Many were written to stimulate student thinking. Others provide additional applications of algebra or present material from future sections of the book so that students can see and learn the material on their own before it is covered in class. Others are more challenging than those in the regular exercise set.

CD Lecture Exercises
The exercises that are worked out in detail on the CD Lecture Videos are marked with the CD icon, ◉. This will prove helpful for your students.

Cumulative Review Exercises
All exercise sets (after the first two) contain questions from previous sections in the chapter and from previous chapters. These Cumulative Review Exercises will reinforce topics that were previously covered and help students retain the earlier material, while they are learning the new material. For the students' benefit, Cumulative Review Exercises are keyed to the section where the material is covered, using brackets, such as [3.4].

Group Activities
Many exercise sets have group activity exercises that lead to interesting group discussions. Many students learn well in a cooperative learning atmosphere, and these exercises will get students talking mathematics to one another.

Mid-Chapter Tests
In the middle of each chapter is a new Mid-Chapter Test. Students should take each Mid-Chapter Test to make sure they understand the material presented in the chapter up to that point. In the student answers, brackets such as [2.3] are used to indicate the section where the material was first presented.

Chapter Summary
At the end of each chapter is a newly formatted, comprehensive chapter summary that includes important chapter facts and examples illustrating these important facts.

Chapter Review Exercises
At the end of each chapter are review exercises that cover all types of exercises presented in the chapter. The review exercises are keyed, using color numbers and brackets, such as [1.5], to the sections where the material was first introduced.

Chapter Practice Tests

The comprehensive end-of-chapter practice test will enable the students to see how well they are prepared for the actual class test. The section where the material was first introduced is indicated in brackets in the student answers.

Cumulative Review Tests

These tests, which appear at the end of each chapter after the first, test the students' knowledge of material from the beginning of the book to the end of that chapter. Students can use these tests for review, as well as for preparation for the final exam. These exams, like the Cumulative Review Exercises, will serve to reinforce topics taught earlier. In the answer section, after each answer, the section where that material was covered is given using brackets.

Answers

The *odd answers* are provided for the exercise sets. *All answers* are provided for the Using Your Graphing Calculator Exercises, Cumulative Review Exercises, Mid-Chapter Tests, Chapters Review Exercises, Chapter Practice Tests, and Cumulative Review Tests. Answers are not provided for the Group Activity exercises since we want students to reach agreement by themselves on the answers to these exercises.

National Standards

Recommendations of the *Curriculum and Evaluation Standards for School Mathematics*, prepared by the National Council of Teachers of Mathematics (NCTM), and *Beyond Crossroads: Implementing Mathematics Standards in the First Two Years of College*, prepared by the American Mathematical Association of Two Year Colleges (AMATYC), are incorporated into this edition.

Prerequisite

This text assumes no prior knowledge of algebra. However, a working knowledge of arithmetic skills is important. Fractions are reviewed early in the text, and decimals and percent are reviewed in Appendix A.

Modes of Instruction

The format and readability of this book lends itself to many different modes of instruction. The constant reinforcement of concepts will result in greater understanding and retention of the material by your students.

The features of the text and the large variety of supplements available make this text suitable for many types of instructional modes, including

- lecture
- distance learning
- self-paced instruction
- modified lecture
- cooperative or group study
- learning laboratory

Changes in the Third Edition

When I wrote the third edition I considered many letters and reviews I got from students and faculty alike. I would like to thank all of you who made suggestions for improving the third edition. I would also like to thank the many instructors and students who wrote to inform me of how much they enjoyed, appreciated, and learned from the text. Some of the changes made in the third edition of the text include the following:

- Chapter 3, "Applications of Algebra," has been shortened from five sections to four sections and has been rewritten to make the material more understandable for students. Sections 3.1 and 3.4 have been reorganized to make the material clearer and flow more smoothly. The chapter also has many new and exciting real-life applications.

- The material on formulas has been moved from Chapter 3 to Chapter 2, where it now reinforces the material on solving equations.

- More emphasis is placed on solving equations containing decimal numbers and fractions in Chapter 2.

- Addition and subtraction of fractions and decimals have been enhanced and improved in Chapter 1.

- Material on rounding decimal numbers has been added to the decimal number appendix.

- A *Chapter Test Prep Video CD* now comes with the book. This video CD shows the worked-out solution to each exercise in the Chapter Practice Test for each chapter. This is yet another aid to improve student learning and understanding.

- *Every* example in the book now has a corresponding Now Try Exercise associated with it. Students are encouraged to work the exercise immediately after they finish studying the respective example. This gives students an opportunity to reinforce the concepts or topics covered in the example.

- A new feature called *Mid-Chapter Test* has been added to the middle of each chapter. These tests are designed to see how well students understand the topics covered in the first part of the chapter. If a student misses a question, the student should review the appropriate material. The section where the material was introduced is given in brackets next to the answer in the back of the book.

- More *Helpful Hints* and *Avoiding Common Errors* boxes have been added where appropriate.

- The Chapter Summary has been rewritten to include examples of important facts and concepts covered in the chapter. The left-hand column gives the fact or concept, and the right-hand column gives an example of the fact or concept. The new chapter summary should be an aid to students in reviewing the chapter and preparing for a test.

- New and exciting examples and exercises have been added throughout the book.

  ○ New exercises were added to many exercise sets to ensure that every example in the book now has exercises that correspond to that given example.

  ○ In some sections, more difficult exercises have been added at the end, or easier exercises have been added at the beginning of the exercise set so that there is a continuous increase in the level of difficulty of the exercises.

  ○ Every effort has been made to include applications that are of interest to students.

  ○ Variables other than x and y are used more often in examples and exercises.

- *Goals of This Chapter* have replaced *A Look Ahead*. The information provided gives students an overview of what they will see and are expected to learn in the chapter.

- *Using Your Graphing Calculator* boxes now show keystrokes for the TI-84 Plus calculator. Note that the same keystrokes are appropriate for the TI-83 Plus calculator.

- The *Mathematics in Action* feature has been removed to conserve space.

- More art and photos have been added to the text to make the material either more understandable or more interesting for students.

- The basic colors used in the text have been softened to make the text more attractive and easier for students to read.

Supplements for the Third Edition

FOR INSTRUCTORS

Printed Supplements

Annotated Instructor's Edition (0-13-614445-4)

Contains all the content found in the student edition, plus the following:

- Answers to exercises are printed on the same text page with graphing answers in a special Graphing Answer Section in the back of the text.

- *Teaching Tips* throughout the text are placed at key points in the margins.

- **NEW!** An extra *Instructor Example* is now provided in the margin next to each student example. These extra examples are meant to be used as additional examples in the classroom.

Instructor's Solutions Manual (0-13-233725-8)

This manual contains complete solutions to every exercise in the text.

Instructor's Resource Manual with Tests (0-13-614775-5)

- **NEW!** Now includes a Mini-Lecture for every section of the text.

- Provides several test forms, both free response and multiple choice, for every chapter, as well as cumulative tests and final exams.

- Answers to all items also included.

Media Supplements

TestGen (0-13-233726-6)

TestGen enables instructors to build, edit, print, and administer tests using a computerized bank of questions developed to cover all the objectives of the text. TestGen is algorithmically based, allowing instructors to create multiple but equivalent versions of the same question or test with the click of a button. Instructors can also modify test bank questions or add new questions. Tests can be printed or administered online. The software is available on a dual-platform Windows/Macintosh CD-ROM.

CD Lecture Series Videos—Lab Pack (0-13-241511-9)

For each section, there are about 20 minutes of lecture along with several of the section exercises worked out. The exercises that are worked out are identified in the student edition by the icon ⊙.

MyMathLab   NEW! MyMathLab Instructor Version
             (0-13-147898-2)

MyMathLab is a series of text-specific, easily customizable online courses for Prentice Hall textbooks in mathematics and statistics. Powered by CourseCompass™ (Pearson Education's online teaching and learning environment) and MathXL® (our online homework, tutorial, and assessment system), MyMathLab gives you the tools you need to deliver all or a portion of your course online, whether your students are in a lab setting or working from home.

MathXL   NEW! MathXL® Instructor Version
           (0-13-147895-8)

MathXL® is a powerful online homework, tutorial, and assessment system that accompanies Prentice Hall textbooks in mathematics or statistics. With MathXL, instructors can create, edit, and assign online homework and tests using algorithmically generated exercises correlated at the objective level to the textbook.

FOR STUDENTS

NEW! Student Study Pack (0-13-234885-3)

Includes

- The *Student Solutions Manual*
- Access to the *Prentice Hall Tutor Center*
- The *CD Lecture Series Videos*

Printed Supplements

Student Solutions Manual (0-13-614777-1)

Provides full, worked-out solutions to

- The odd-numbered section exercises
- All (even and odd) exercises in the Cumulative Review Exercises (within the section exercises), Mid-Chapter Tests, Chapter Review Exercises, Chapter Practice Tests, and Cumulative Review Tests

Media Supplements

NEW! Chapter Test Prep Video CD (0-13-159449-4)

Provides step-by-step video solutions to each problem in each *Chapter Practice Test* in the textbook. Packaged with a new text, inside the back cover.

NEW! MathXL® Tutorials on CD (0-13-233729-0)

This interactive tutorial CD-ROM provides algorithmically generated practice exercises that are correlated at the objective level to the exercises in the textbook. Every practice exercise is accompanied by an example and a guided solution designed to involve students in the solution process. Selected exercises may also include a video clip to help students visualize concepts. The software provides helpful feedback for incorrect answers and can generate printed summaries of students' progress.

Prentice Hall Tutor Center: www.prenhall.com/tutorcenter (0-13-064604-0)

Staffed by developmental math faculty, the Tutor Center provides live tutorial support via phone, fax, or e-mail. Tutors are available Sunday through Thursday 5 P.M. EST to midnight, 5 days a week, 7 hours a day. The Tutor Center may be accessed through a registration number that may be bundled with a new text or purchased separately with a used book. Comes automatically within *MyMathLab*.

InterAct Math Tutorial Web site: www.interactmath.com

Get practice and tutorial help online! This interactive tutorial Web site provides algorithmically generated practice exercises that correlate directly to the exercises in the textbook. Students can retry an exercise as many times as they like with new values each time for unlimited practice and mastery. Every exercise is accompanied by an interactive guided solution that provides helpful feedback for incorrect answers, and students can also view a worked-out sample problem that steps them through an exercise similar to the one they're working on.

Acknowledgments

Writing a textbook is a long and time-consuming project. Many people deserve thanks for encouraging and assisting me with this project. Most importantly, my special thanks goes to my wife Kathy and sons, Robert and Steven. Without their constant encouragement and understanding, this project would not have become a reality. I would also like to thank my daughter-in-law, Kathy, for her support.

I would like to thank Richard Semmler of Northern Virginia Community College. Richard has worked with me throughout this project, and with many of my other projects throughout the years. He has helped me in too many ways to list, and he has always been there to help when I needed assistance with my books. Richard, I truly thank you.

I would also like to thank both Aimee Calhoun and Donna Petrie of Monroe Community College for assisting me with this book. Both Aimee and Donna were extremely conscientious and helpful in numerous ways.

I want to thank Rafiq Ladhani and his team at Edutorial for accuracy reviewing the pages and checking all answers.

I would like to thank several people at Prentice Hall, including Paul Murphy, Executive Editor; Dawn Nuttall, Project Manager; Thomas Benfatti, Art Editor; John Christiana, Art Director; and Lynn Savino Wendel, Production Editor, for their many valuable suggestions and conscientiousness with this project.

I want to thank those who worked with me on the print supplements for this book.

- Student and Instructor's Solutions Manuals: Kevin Bodden and Randy Gallaher, *Lewis and Clark Community College, IL*
- Instructor's Resource Manual: Kevin Bodden and Randy Gallaher, *Lewis and Clark Community College, IL*

I would like to thank the following reviewers of the last two editions for their thoughtful comments and suggestions:

Laura Adkins, *Missouri Southern State College, MO*
Darla Aguilar, *Pima Community College, AZ*
Arthur Altshiller, *Los Angeles Valley College, CA*
Frances Alvarado, *University of Texas–Pan American, TX*
Jose Alvarado, *University of Texas–Pan American, TX*
Jacob Amidon, *Cayuga Community College, NY*
Ben Anderson, *Darton College, GA*
Sheila Anderson, *Housatonic Community College, CT*
Peter Arvanites, *State University of New York–Rockland Community College, NY*
Jannette Avery, *Monroe Community College, NY*
Mary Lou Baker, *Columbia State Community College, TN*
Jon Becker, *Indiana University, IN*
Sharon Berrian, *Northwest Shoals Community College, AL*
Paul Boisvert, *Oakton Community College, IL*

Dianne Bolen, *Northeast Mississippi Community College, MS*

Julie Bonds, *Sonoma State University, CA*

Beverly Broomell, *Suffolk County Community College, NY*

Clark Brown, *Mojave Community College, AZ*

Connie Buller, *Metropolitan Community College, NE*

Lavon Burton, *Abilene Christian University, TX*

Marla Dresch Butler, *Gavilan Community College, CA*

Marc Campbell, *Daytona Beach Community College, FL*

Mitzi Chaffer, *Central Michigan University, MI*

Terry Cheng, *Irvine Valley College, CA*

Julie Chesser, *Owens Community College, OH*

Kim Christensen, *Maple Woods Community College, MO*

Barry Cogan, *Macomb Community College, MI*

Pat C. Cook, *Weatherford College, TX*

Ted Corley, *Arizona State University and Glendale Community College, AZ*

Charles Curtis, *Missouri Southern State College, MO*

Joseph de Guzman, *Riverside City College (Norco), CA*

Lisa DeLong Cuneo, *Pennsylvania State University–Dubois, PA*

Stephan Delong, *Tidewater Community College, VA*

Deborah Doucette, *Erie Community College (North), NY*

William Echols, *Houston Community College, TX*

Gary Egan, *Monroe Community College, NY*

Mark W. Ernsthausen, *Monroe Community College, NY*

Elizabeth Farber, *Bucks County Community College, PA*

Dale Felkins, *Arkansas Technical University, AR*

Warrene Ferry, *Jones County Junior College, MS*

Christine Fogal, *Monroe Community College, NY*

Reginald Fulwood, *Palm Beach Community College, FL*

Gary Glaze, *Spokane Falls Community College, WA*

James Griffiths, *San Jacinto College, TX*

Susan Grody, *Broward Community College, FL*

Kathy Gross, *Cayuga Community College, NY*

Abdollah Hajikandi, *State University of New York–Buffalo, NY*

Olga Cynthia Harrison, *Baton Rouge Community College, LA*

Mary Beth Headlee, *Manatee Community College, FL*

Richard Hobbs, *Mission College, CA*

Joe Howe, *St. Charles Community College, MO*

Laura L. Hoye, *Trident Technical College, SC*

Barbara Hughes, *San Jacinto Community College (Central), TX*

Kelly Jahns, *Spokane Community College, WA*

Mary Johnson, *Inver Hills Community College, MN*

Judy Kasabian, *El Camino College, CA*

Jane Keller, *Metropolitan Community College, NE*

Mike Kirby, *Tidewater Community College, VA*

Maryanne Kirkpatrick, *Laramie County Community College, WY*

Marcia Kleinz, *Atlantic Cape Community College, NJ*

William Krant, *Palo Alto College, TX*

Gayle L. Krzemien, *Pikes Peak Community College, CO*

Shannon Lavey, *Cayuga Community College, NY*

Mitchel Levy, *Broward Community College, FL*

Mitzi Logan, *Pitt Community College, NC*

Jason Mahar, *Monroe Community College, NY*

Kimberley A. Martello, *Monroe Community College, NY*

Constance Meade, *College of Southern Idaho, ID*

Lynette Meslinsky, *Erie Community College, NY*

Shywanda Moore, *Meridian Community College, MS*

Elizabeth Morrison, *Valencia Community College, FL*

Catherine Moushon, *Elgin Community College, IL*

Elsie Newman, *Owens Community College, OH*

Charlotte Newsom, *Tidewater Community College, VA*

Kathy Nickell, *College of DuPage, IL*

Charles Odion, *Houston Community College, TX*

Jean Olsen, *Pikes Peak Community College, CO*

Shelle Patterson, *Moberly Area Community College, MO*

Patricia Pifko, *Housatonic Community College, CT*

Jeanne Pirie, *Erie Community College (North), NY*

Dennis Reissig, *Suffolk County Community College, NY*

Linda Retterath, *Mission College, CA*

Dale Rohm, *University of Wisconsin–Stevens Point, WI*

Behnaz Rouhani, *Athens Technical College, GA*

Troy Rux, *Spokane Falls Community College, WA*

Hassan Saffari, *Prestonburg Community College, KY*

Brian Sanders, *Modesto Junior College, CA*

Glenn R. Sandifer, *San Jacinto Community College (Central), TX*

Rebecca Schantz, *Prairie State College, IL*

Cristela Sifuentez, *University of Texas–Pan American, TX*

Rick Silvey, *St. Mary College, KS*

Julia Simms, *Southern Illinois University–Edwardsville, IL*

Linda Smoke, *Central Michigan University, MI*

Jed Soifer, *Atlantic Cape Community College, NJ*

Richard C. Stewart, *Monroe Community College, NY*

Elizabeth Suco, *Miami–Dade Community College, FL*

Harold Tanner, *Orangeburg–Calhoun Technical College, SC*

Fereja Tahir, *Illinois Central College, IL*

Dale Thielker, *Ranken Technical College, MO*

Burnette Thompson, Jr., *Houston Community College, TX*

Mary Vachon, *San Joaquin Delta College, CA*

Andrea Vorwark, *Maple Woods Community College, MO*

Ken Wagman, *Gavilan Community College, CA*

Patrick Ward, *Illinois Central College, IL*

Robert E. White, *Allan Hancock College, CA*

Cindy Wilson, *Henderson State University, AZ*

Ronald Yates, *Community College of Southern Nevada, NV*

To the Student

Algebra is a course that cannot be learned by observation. To learn algebra you must become an active participant. You must read the text, pay attention in class, and, most importantly, you must work the exercises. The more exercises you work, the better.

The text was written with you in mind. Short, clear sentences are used, and many examples are given to illustrate specific points. The text stresses useful applications of algebra. Hopefully, as you progress through the course, you will come to realize that algebra is not just another math course that you are required to take, but a course that offers a wealth of useful information and applications.

This text makes full use of color. The different colors are used to highlight important information. Important procedures, definitions, and formulas are placed within colored boxes.

The boxes marked **Helpful Hints** should be studied carefully, for they stress important information. The boxes marked **Avoiding Common Errors** should also be studied carefully. These boxes point out errors that students commonly make, and provide the correct procedures for doing these problems.

After each example you will see a Now Try Exercise reference, such as ▶ **Now Try Exercise 27**. The exercise indicated is very similar to the example given in the book. You may wish to try the indicated exercise after you read the example to make sure you truly understand the example. In the exercise set, the Now Try exercises are written in green, such as 27.

In the exercise sets, the exercises marked with a pencil, ✎, indicate writing exercises—that is, exercises that require a written answer. The exercises marked with a CD, 💿, indicate that these exercises are worked out on the CD Lecture Videos.

Ask your professor early in the course to explain the policy on when the calculator may be used. Pay particular attention to the 🖩 **Using Your Calculator** boxes. You should also read the 🖩 **Using Your Graphing Calculator** boxes even if you are not using a graphing calculator in class. You may find the information presented here helps you better understand the algebraic concepts.

Other questions you should ask your professor early in the course include: What supplements are available for use? Where can help be obtained when the professor is not available? Supplements that may be available include: the Student Solutions Manual; the CD Lecture Series Videos; the Chapter Test Prep Video CD; MathXL® *Math* XL ; MyMathLab *MyMathLab* ; and the Prentice Hall Mathematics Tutor Center. All these items are discussed under the heading of Supplements in Section 1.1, as well as in the Preface.

You may wish to form a study group with other students in your class. Many students find that working in small groups provides an excellent way to learn the material. By discussing and explaining the concepts and exercises to one another, you reinforce your own understanding. Once guidelines and procedures are determined by your group, make sure to follow them.

One of the first things you should do is to read Section 1.1, Study Skills for Success in Mathematics. Read this section slowly and carefully, and pay particular attention to the advice and information given. Occasionally, refer back to this section. This could be the most important section of the book. Carefully read the material on doing your homework and on attending class.

At the end of all Exercise Sets (after the first two) are **Cumulative Review Exercises**. You should work these problems on a regular basis, even if they are not assigned. These problems are from earlier sections and chapters of the text, and they will refresh your memory and reinforce those topics. If you have a problem when working these exercises, read the appropriate section of the text or study your notes that correspond to that material. The section of the text where the Cumulative Review Exercise was introduced is indicated in brackets, [ ], to the left of the exercise. After reviewing the material, if you still have a problem, make an appointment to see your professor. Working the Cumulative Review Exercises throughout the semester will also help prepare you to take your final exam.

Near the middle of each chapter is a **Mid-Chapter Test**. You should take each Mid-Chapter Test to make sure you understand the material up to that point. The section where the material was first introduced is given in brackets after the answer in the answer section of the book.

At the end of each chapter are a **Chapter Summary**, **Chapter Review Exercises**, a **Chapter Practice Test**, and a **Cumulative Review Test.** Before each examination you should review this material carefully and take the Chapter Practice Test (you may want to review the *Chapter Test Prep Video CD* also). If you do well on the Chater Practice Test, you should do well on the class test. The questions in the Review Exercises are marked to indicate the section in which that material was first introduced. If you have a problem with a Review Exercise question, reread the section indicated. You may also wish to take the Cumulative Review Test that appears at the end of every chapter.

In the back of the text there is an **answer section** that contains the answers to the *odd-numbered* exercises, including the Challenge Problems. Answers to *all* Using Your Graphing Calculator Exercises, Cumulative Review Exercises, Mid-Chapter Tests, Chapter Review Exercises, Chapter Practice Tests, and Cumulative Review Tests are provided. Answers to the Group Activity exercises are not provided, for we wish students to reach agreement by themselves on answers to these exercises. The answers should be used only to check your work. For the Mid-Chapter Tests, Chapter Practice Tests, and Cumulative Review Tests, after each answer the section number where that type of exercise was covered is provided.

I have tried to make this text as clear and error free as possible. No text is perfect, however. If you find an error in the text, or an example or section that you believe can be improved, I would greatly appreciate hearing from you. If you enjoy the text, I would also appreciate hearing from you. You can submit comments at *http://247.prenhall.com*.

Allen R. Angel

1 Real Numbers

IN ADDITION TO YEARLY PREMIUMS, certain medical insurance policies require a co-payment for each visit made to a doctor's office. Others require that the individual pay a certain dollar amount in medical expenses each year, after which the insurance company pays a large percentage of the remaining costs. On pages 10 and 11, we use a problem-solving technique to determine the portion of a medical bill a person is responsible for and how much of that bill the insurance company will pay.

1.1 Study Skills for Success in Mathematics

1 Recognize the goals of this text.

2 Learn proper study skills.

3 Prepare for and take exams.

4 Learn to manage time.

5 Purchase a calculator.

This section is extremely important. Take the time to read it carefully and follow the advice given. For many of you this section may be the most important section of the text.

Most of you taking this course fall into one of three categories: (1) those who did not take algebra in high school, (2) those who took algebra in high school but did not understand the material, or (3) those who successfully completed algebra in high school but have been out of school for some time and need to take the course again. Whichever the case, you will need to acquire study skills for mathematics courses.

Before we discuss study skills, we will present the goals of this text. These goals may help you realize why certain topics are covered in the text and why they are covered as they are.

1 Recognize the Goals of This Text

The goals of this text include:

1. Presenting traditional algebra topics
2. Preparing you for more advanced mathematics courses
3. Building your confidence in, and your enjoyment of, mathematics
4. Improving your reasoning and critical thinking skills
5. Increasing your understanding of how important mathematics is in solving real-life problems
6. Encouraging you to think mathematically, so that you will feel comfortable translating real-life problems into mathematical equations, and then solving the problems.

In addition to teaching you the mathematical content, our goals are to teach you to be more *mathematically literate*, which is also called *quantitatively literate*. We wish to teach you to *communicate mathematically*, to teach you *to understand and interpret data* in a variety of formats, to teach you measurement and geometric concepts, to teach you to *reason more logically*, and to teach you to be able to represent real world applications mathematically, which is called *modeling*. Throughout the book we will strive to increase your mathematical understanding to help you become more successful in mathematics, in your future job, and throughout life.

Another goal of this text is to help you become more familiar with technology that may help you in this and future mathematics courses, and throughout life. To do so, we use Calculator Boxes throughout the book for both scientific and graphing calculators.

We also realize that some of you may have some mathematics anxiety. We have written the book to try to help you overcome that anxiety by building your confidence in mathematics.

It is important to realize that this course is the foundation for more advanced mathematics courses. A thorough understanding of algebra will make it easier for you to succeed in later mathematics courses and in life.

2 Learn Proper Study Skills

Have a Positive Attitude

You may be thinking to yourself, "I hate math," or "I wish I did not have to take this class." You may have heard of "math anxiety" and feel you fit this category. The first thing to do to be successful in this course is to change your attitude to a more positive one. You must be willing to give this course, and yourself, a fair chance.

Based on past experiences in mathematics, you may feel that this is difficult. However, mathematics is something you need to work at. Many of you are more mature now than when you took previous mathematics courses. Your maturity and desire to learn are extremely important and can make a tremendous difference in your ability to succeed in mathematics. I believe you can be successful in this course, but you also need to believe it.

Prepare for and Attend Class

To be prepared for class, you need to do your homework assignments completely. If you have difficulty with the homework, or some of the concepts, write down questions to ask your instructor. If you were given a reading assignment, read the appropriate material carefully before class.

After the material is explained in class, read the corresponding sections of the text slowly and carefully, word by word.

You should plan to attend every class. Generally, the more absences you have, the lower your grade will be. Every time you miss a class, you miss important information. If you must miss a class, contact your instructor ahead of time, and get the reading assignment and homework. If possible, before the next class, try to copy a friend's notes to help you understand the material you missed.

In algebra and other mathematics courses, the material you learn is cumulative. The new material is built on material that was presented previously. You must understand each section before moving on to the next section, and each chapter before moving on to the next chapter. Therefore, do not let yourself fall behind. Seek help as soon as you need it—do not wait! You will greatly increase your chance of success in this course by following the study skills presented in this section.

While in class, pay attention to what your instructor is saying. If you don't understand something, ask your instructor to repeat the material. If you don't ask questions, your instructor will not know that you have a problem understanding the material.

In class, take careful notes. Write numbers and letters clearly, so that you can read them later. Make sure your x's do not look like y's and vice versa. It is not necessary to write down every word your instructor says. Copy the major points and the examples that do not appear in the text. You should not be taking notes so frantically that you lose track of what your instructor is saying.

Read the Text

Mathematics textbooks should be read slowly and carefully, word by word. If you don't understand what you are reading, reread the material. When you come across a new concept or definition, you may wish to underline or highlight it, so that it stands out. Then it will be easier to find later. When you come across an example, read and follow it line by line. Don't just skim it. Also, work the **Now Try Exercises** that appear in the text following each example. The Now Try Exercises are designed so that you have the opportunity to immediately apply new ideas. Make notes of anything you don't understand to ask your instructor.

This textbook has special features to help you. I suggest that you pay particular attention to these highlighted features, including the **Avoiding Common Errors** boxes, the **Helpful Hint** boxes, and important procedures and definitions identified by color. The Avoiding Common Errors boxes point out the most common errors made by students. Read and study this material very carefully and make sure that you understand what is explained. If you avoid making these common errors, your chances of success in this and other mathematics classes will be increased greatly. The Helpful Hints offer many valuable techniques for working certain problems. They may also present some very useful information or show an alternative way to work a problem.

Do the Homework

Two very important commitments that you must make to be successful in this course are attending class and doing your homework regularly. Your assignments must be worked conscientiously and completely. Do your homework as soon as possible, so the material presented in class will be fresh in your mind. It is through doing homework that you truly learn the material. While working homework you will become aware of the types of problems that you need further help with. If you do not work the assigned exercises, you will not know what questions to ask in class.

When you do your homework, make sure that you write it neatly and carefully. Pay particular attention to copying signs and exponents correctly.

Don't forget to check the answers to your homework assignments. This book contains the answers to the odd-numbered exercises in the back of the book. In addition, the answers to all the Cumulative Review Exercises, Mid-Chapter Tests, Chapter Review Exercises, Chapter Practice Tests, and Cumulative Review Tests are in the back of the book. The section number where the material is first introduced is provided next to the exercises for the Cumulative Review Exercises and Chapter Review Exercises. The section number where the material is first introduced is provided with the answers in the back of the book for the Mid-Chapter Tests, Chapter Practice Tests, and Cumulative Review Tests. Answers to the Group Activity Exercises are not provided because we want you to arrive at the answers as a group.

Ask questions in class about homework problems you don't understand. You should not feel comfortable until you understand all the concepts needed to work every assigned problem successfully.

Study for Class

Study in the proper atmosphere, in an area where you will not be constantly disturbed, so that your attention can be devoted to what you are reading. The area where you study should be well ventilated and well lit. You should have sufficient desk space to spread out all your materials. Your chair should be comfortable. You should try to minimize distractions while you are studying. You should not study for hours on end. Short study breaks are a good idea.

Before you begin studying, make sure that you have all the materials you need (pencils, markers, calculator, etc.). You may wish to highlight the important points covered in class or in the book.

It is recommended that students study and do homework for at least two hours for each hour of class time. Some students require more time than others. It is important to spread your studying time out over the entire week rather than studying during one large block of time.

When studying, you should not only understand how to work a problem but also know *why* you follow the specific steps you do to work the problem. If you do not have an understanding of why you follow the specific process, you will not be able to transfer the process to solve similar problems.

This book has Mid-Chapter Tests in the middle of each chapter. These exercises reinforce material presented in the first half of the chapter. They will also help you determine if you need to go back and review the topics covered in the first half of the chapter. For any of the Mid-Chapter Test questions that you get incorrect, turn to the section provided with the answers in the back of the book and review that section. This book also has Cumulative Review Exercises at the end of every section after Section 1.2. These exercises reinforce material presented earlier in the course, and you will be less likely to forget the material if you review it repeatedly throughout the course. The exercises will also help prepare you for the final exam. Even if these exercises are not assigned for homework, I urge you to work them as part of your studying process.

3 Prepare for and Take Exams

If you study a little bit each day you should not need to cram the night before an exam. Begin your studying early. If you wait until the last minute, you may not have time to seek the help you may need if you find you cannot work a problem.

To prepare for an exam:

1. Read your class notes.
2. Review your homework assignments.
3. Study formulas, definitions, and procedures you will need for the exam.
4. Read the Avoiding Common Errors boxes and Helpful Hint boxes carefully.

5. Read the summary at the end of each chapter.

6. Work the Chapter Review Exercises at the end of each chapter. If you have difficulties, restudy those sections. If you still have trouble, seek help.

7. Work the Mid-Chapter Test and the Chapter Practice Test.

8. Rework quizzes previously given if the material covered in the quizzes will be included on the test.

9. If your exam is a cumulative exam, work the Cumulative Review Test.

Prepare for Midterm and Final Exam

When studying for a comprehensive midterm or final exam follow the procedures discussed for preparing for an exam. However, also:

1. Study all your previous tests and quizzes carefully. Make sure that you have learned to work the problems that you may have previously missed.

2. Work the Cumulative Review Tests at the end of each chapter. These tests cover the material from the beginning of the book to the end of that chapter.

3. If your instructor has given you a worksheet or practice exam, make sure that you complete it. Ask questions about any problems you do not understand.

4. Begin your studying process early so that you can seek all the help you need in a timely manner.

Take an Exam

Make sure you get sufficient sleep the night before the test. Arrive at the exam site early so that you have a few minutes to relax before the exam. If you rush into the exam, you will start out nervous and anxious. After you are given the exam, you should do the following:

1. Carefully write down any formulas or ideas that you want to remember.

2. Look over the entire exam quickly to get an idea of its length. Also make sure that no pages are missing.

3. Read the test directions carefully.

4. Read each question carefully. Show all of your work. Answer each question completely, and make sure that you have answered the specific question asked.

5. Work the questions you understand best first; then go back and work those you are not sure of. Do not spend too much time on any one problem or you may not be able to complete the exam. Be prepared to spend more time on problems worth more points.

6. Attempt each problem. You may get at least partial credit even if you do not obtain the correct answer. If you make no attempt at answering the question, you will lose full credit.

7. Work carefully step by step. Copy all signs and exponents correctly when working from step to step, and make sure to copy the original question from the test correctly.

8. Write clearly so that your instructor can read your work. If your instructor cannot read your work, you may lose credit. When appropriate, make sure that your final answer stands out by placing a box around it.

9. If you have time, check your work and your answers.

10. Do not be concerned if others finish the test before you or if you are the last to finish. Use any extra time to check your work.

Stay calm when taking your test. Do not get upset if you come across a problem you can't figure out right away. Go on to something else and come back to that problem later.

4 Learn to Manage Time

As mentioned earlier, it is recommended that students study and do homework for at least two hours for each hour of class time. Finding the necessary time to study is not always easy. The following are some suggestions that you may find helpful.

1. Plan ahead. Determine when you will study and do your homework. Do not schedule other activities for these periods. Try to space these periods evenly over the week.

2. Be organized, so that you will not have to waste time looking for your books, your pencil, your calculator, or your notes.

3. If you are allowed to use a calculator, use it for tedious calculations.

4. When you stop studying, clearly mark where you stopped in the text.

5. Try not to take on added responsibilities. You must set your priorities. If your education is a top priority, as it should be, you may have to reduce time spent on other activities.

6. If time is a problem, do not overburden yourself with too many courses.

Use Supplements

This text comes with a large variety of supplements. Find out from your instructor early in the semester which supplements are available and might be beneficial for you to use. Supplements should not replace reading the text, but should be used to enhance your understanding of the material. If you miss a class, you may want to review the video on the topic you missed before attending the next class.

The supplements available are: the Student Solutions Manual which works out the odd section exercises as well as all the end-of-chapter exercises; the CD Lecture Series Videos, which show about 20 minutes of lecture per section and include the worked out solutions to the exercises marked with this icon ⊙; the Chapter Test Prep Video CD, which works out every problem in every Chapter Practice Test; *Math_XL* MathXL®, a powerful online tutorial and homework system, which is also available on CD; *MyMathLab* MyMathLab, the online course which houses MathXL plus a variety of other supplements; and the Prentice Hall Mathematics Tutor Center, which provides live tutorial support via phone, fax or email.

Seek Help

Be sure to get help as soon as you need it! Do not wait! In mathematics, one day's material is usually based on the previous day's material. So, if you don't understand the material today, you may not be able to understand the material tomorrow.

Where should you seek help? There are often a number of resources on campus. Try to make a friend in the class with whom you can study. Often, you can help one another. You may wish to form a study group with other students in your class. Discussing the concepts and homework with your peers will reinforce your own understanding of the material.

You should know your instructor's office hours, and you should not hesitate to seek help from your instructor when you need it. Make sure you read the assigned material and attempt the homework before meeting with your instructor. Come prepared with specific questions to ask.

There are often other sources of help available. Many colleges have a mathematics lab or a mathematics learning center where tutors are available. Ask your instructor early in the semester where and when tutoring is available. Arrange for a tutor as soon as you need one.

5 Purchase a Calculator

You may wish to purchase a scientific or graphing calculator for this course. Ask your instructor if he or she recommends a particular calculator for this or a future mathematics class. Also ask your instructor if you may use a calculator in class, on homework, and on tests. If so, you should use your calculator whenever possible to save time.

If a calculator contains a ⌑LOG⌑ key or ⌑SIN⌑ key, it is a scientific calculator. You *cannot* use the square root key ⌑√⌑ to identify scientific calculators since both scientific calculators and nonscientific calculators may have this key. You should pay particular attention to the **Using Your Calculator** boxes in this book. The boxes explain how to use your calculator to solve problems. If you are using a graphing calculator, pay particular attention to the **Using Your Graphing Calculator** boxes. You may also need to use the reference manual that comes with your calculator at various times.

A Final Word

You can be successful at mathematics if you attend class regularly, pay attention in class, study your text carefully, do your homework daily, review regularly, and seek help as soon as you need it. Good luck in your course.

EXERCISE SET 1.1

Do you know:

1. your professor's name and office hours?
2. your professor's office location and telephone number?
3. where and when you can obtain help if your professor is not available?
4. the name and phone number of a friend in your class?
5. what supplements are available to assist you in learning?
6. if your instructor is recommending the use of a particular calculator?
7. when you can use your calculator in this course?
8. if your instructor is requiring the use of MyMathLab?

If you do not know the answer to questions 1–8, you should find out as soon as possible.

9. What are your goals for this course?
10. What are your reasons for taking this course?
11. List the things you need to do to prepare properly for class.

12. Are you beginning this course with a positive attitude? It is important that you do!
13. For each hour of class time, how many hours outside of class are recommended for studying and doing homework?
14. Explain how a mathematics text should be read.
15. Two very important commitments that you must make to be successful in this course are **(a)** doing homework regularly and completely and **(b)** attending class regularly. Explain why these commitments are necessary.
16. When studying, you should not only understand how to work a problem, but also why you follow the specific steps you do. Why is this important?
17. Have you given any thought to studying with a friend or a group of friends? Can you see any advantages in doing so? Can you see any disadvantages in doing so?
18. Write a summary of the steps you should follow when taking an exam.

🔘 indicates an exercise worked out on the CD Lecture Series Videos.

🖊 indicates a writing exercise. That is, an exercise that requires a written answer.

1.2 Problem Solving

1 Learn the five-step problem-solving procedure.

2 Solve problems involving bar, line, and circle graphs.

3 Solve problems involving statistics.

1 Learn the Five-Step Problem-Solving Procedure

One of the main reasons we study mathematics is that we can use it to solve many real-life problems. Throughout the book, we will be problem solving. To solve most real-life problems mathematically, we need to be able to express the problem in mathematical symbols. This is an important part of the problem-solving procedure that we will present shortly. In Chapter 3, we will also spend a great deal of time explaining how to express real-life applications mathematically.

We will now give the general five-step **problem-solving procedure** that was developed by George Polya and presented in his book *How to Solve It*. You can approach any problem by following this general procedure.

George Polya

Guidelines for Problem Solving

1. **Understand the problem.**
   - Read the problem *carefully* at least twice. In the first reading, get a general overview of the problem. In the second reading, determine **(a)** exactly what you are being asked to find and **(b)** what information the problem provides.
   - Make a list of the given facts. Determine which are pertinent to solving the problem.
   - Determine whether you can substitute smaller or simpler numbers to make the problem more understandable.
   - If it will help you organize the information, list the information in a table.
   - If possible, make a sketch to illustrate the problem. Label the information given.

2. **Translate the problem to mathematical language.**
   - This will generally involve expressing the problem in terms of an algebraic expression or equation. (We will explain how to express application problems as equations in Chapter 3.)
   - Determine whether there is a formula that can be used to solve the problem.

3. **Carry out the mathematical calculations necessary to solve the problem.**

4. **Check the answer obtained in step 3.**
   - Ask yourself, "Does the answer make sense?" "Is the answer reasonable?" If the answer is not reasonable, recheck your method for solving the problem and your calculations.
   - Check the solution in the original problem if possible.

5. **Make sure you have answered the question.**
   - State the answer clearly.

In step 2 we use the words *algebraic expression*. An **algebraic expression**, sometimes just referred to as an **expression**, is a general term for any collection of numbers, letters (called variables), grouping symbols such as parentheses ( ) or brackets [ ], and **operations** (such as addition, subtraction, multiplication, and division). In this section we will not be using variables, so we will discuss their use later.

Examples of Expressions

$$3 + 4, \quad 6(12 \div 3), \quad (2)(7)$$

The following examples show how to apply the guidelines for problem solving. We will sometimes provide the steps in the examples to illustrate the five-step procedure. However, in some problems it may not be possible or necessary to list every step in the procedure. In some of the examples, we use decimal numbers and percents. If you need to review procedures for adding, subtracting, multiplying, or dividing decimal numbers, or if you need a review of percents, read Appendix A before going on.

EXAMPLE 1 ▶ Buying Games Georgia May is deciding which would be less expensive, buying her son's birthday presents on eBay or buying them at a local toy store. Founded in 1995, eBay is The World's Online Marketplace® for the sale of goods and services by a diverse community of individuals and small businesses. The eBay community includes more than 100 million registered members from around the world. The local toy store is only minutes from Georgia's house. Therefore, the cost of gasoline for her car will not factor into her decision. On eBay, the three games Georgia would like to purchase cost $5.99, $9.95, and $19.95. Shipping costs for the games would total $11.10. There would be no sales tax on this purchase. At the local toy store, the total cost for the same three games would be $57.89 plus 8.25% sales tax.

a) Which would be less expensive for Georgia, purchasing the games on eBay or at the local toy store?

b) How much would Georgia save by making the less expensive purchase?

Solution **a) Understand the problem** A careful reading of the problem shows that the task is to determine if it would be less expensive for Georgia to purchase the games on eBay or at the local toy store. Make a list of all the information given and determine which information is needed to solve the problem.

Information Given	Pertinent to Solving the Problem?
eBay includes more than 100 million registered members	no
$5.99, $9.95, and $19.95 cost of the games on eBay	yes
$11.10 shipping costs on eBay	yes
no sales tax on the eBay purchase	yes
cost of gasoline not a factor in Georgia's decision	yes
$57.89 cost of the games at the local toy store	yes
8.25% sales tax at the local toy store	yes

To determine if purchasing the games on eBay or at the local toy store would be less expensive, it is not necessary to know that eBay includes more than 100 million registered members. Solving this problem involves calculating the total cost of the games on eBay, including the shipping costs, and calculating the total cost of the games at the local toy store, including the sales tax. To calculate the sales tax, you need to determine 8.25% of the cost of the games at the local toy store. When performing calculations, numbers given in percent are changed to decimal numbers.

Translate the problem into mathematical language

total cost of games on eBay = cost of each individual game + shipping costs

total cost of games at local toy store = total cost of games + 8.25% sales tax

Carry out the calculations

total cost of games on eBay = $5.99 + $9.95 + $19.95 + $11.10 = $46.99

total cost of games at local toy store = $57.89 + 0.0825($57.89)

= $57.89 + $4.78 = $62.67

Check the answer The total costs of $46.99 and $62.67 are reasonable based on the information given.

Answer the question asked It would be less expensive for Georgia to purchase the games on eBay.

b) Understand To determine how much Georgia would save by making the less expensive purchase, you need to subtract the total cost of the games on eBay from the total cost of the games at the local toy store.

Translate amount Georgia would save = total cost at toy store − total cost on eBay

Carry Out $62.67 − $46.99 = $15.68

Check The answer $15.68 seems reasonable.

Answer Georgia would save $15.68 by purchasing the games on eBay.

▶ **Now Try Exercise 27**

EXAMPLE 2 ▶ **Processor Speed** In November 2004, the fastest Intel processor, the Pentium 4, could perform about 3.8 billion operations per second (3.8 gigahertz, symbolized 3.8 GHz). How many operations could it perform in 0.4 second?

Solution   Understand   We are given the name of the processor, a speed of about 3.8 billion (3,800,000,000) operations per second, and 0.4 second. To determine the answer to this problem, the name of the processor, Pentium 4, is not needed.

To obtain the answer, will we need to multiply or divide? Often a fairly simple problem seems more difficult because of the numbers involved. When very large or very small numbers make the problem seem confusing, try substituting commonly used numbers in the problem to determine how to solve the problem. Suppose the problem said that the processor can perform 6 operations per second. How many operations can it perform in 2 seconds? The answer to this question should be more obvious. It is 6×2 or 12. Since we multiplied to obtain this answer, we also will need to multiply to obtain the answer to the given problem.

Translate

number of operations in 0.4 second $= 0.4$(number of operations per second)

Carry Out $= 0.4(3,800,000,000)$

                                              $= 1,520,000,000$ *From a calculator*

Check   The answer, 1,520,000,000 operations, is less than the 3,800,000,000 operations per second, which makes sense because the processor is operating for less than a second.

Answer   In 0.4 second, the processor can perform about 1,520,000,000 operations.

▶ **Now Try Exercise 21**

EXAMPLE 3 ▶ **Medical Insurance** Brook Matthews's medical insurance policy is similar to that of many workers. Her policy requires that she pay the first $100 of medical expenses each calendar year (called a deductible). After the deductible is paid, she pays 20% of the medical expenses (called a co-payment) and the insurance company pays 80%. (There is a maximum co-payment of $600 that she must pay each year. After that, the insurance company pays 100% of the fee schedule.) On January 1, Brook sprained her ankle playing tennis. She went to the doctor's office for an examination and X rays. The total bill of $325 was sent to the insurance company.

a) How much of the bill will Brook be responsible for?

b) How much will the insurance company be responsible for?

Solution   **a)** Understand   First we list all the *relevant* given information.

Given Information

   $100 deductible
   20% co-payment after deductible
   80% paid by insurance company after deductible
   $325 doctor bill

All the other information is not needed to solve the problem. Brook will be responsible for the first $100 and 20% of the remaining balance. The insurance company will be responsible for 80% of the balance after the deductible. Before we can find what Brook owes, we need to first find the balance of the bill after the deductible. The balance of the bill after the deductible is $325 - $100 = $225.

Translate

Brook's responsibility = deductible + 20% of balance of bill after the deductible

Carry Out

$$\text{Brook's responsibility} = 100 + 20\%(225)$$
$$= 100 + 0.20(225)$$
$$= 100 + 45$$
$$= 145$$

Check and Answer   The answer appears reasonable. Brook will be responsible for $145.

b) The insurance company will be responsible for 80% of the balance after the deductible.

$$\text{insurance company's responsibility} = 80\% \text{ of balance after deductible}$$
$$= 0.80(225)$$
$$= 180$$

Thus, the insurance company is responsible for $180. This checks because the sum of Brook's responsibility and the insurance company's responsibility is equal to the doctor's bill.

$$\$145 + \$180 = \$325$$

We could have also found the answer to part **(b)** by subtracting Brook's responsibility from the total amount of the bill, but to give you more practice with percents we decided to show the solution as we did.

▸ **Now Try Exercise 33**

2 Solve Problems Involving Bar, Line, and Circle Graphs

Problem solving often involves understanding and reading graphs and sets of data (or numbers). We will be using bar, line, and circle (or pie) graphs and sets of data, throughout the book. We will illustrate a number of such graphs in this section and explain how to interpret them. To work Example 4, you must interpret a bar graph and work with data.

EXAMPLE 4 ▸ **Walking It Off** Experts suggest that people walk 10,000 steps daily. Depending on stride length, each mile ranges between 2000 and 2500 steps. **Figure 1.1** is a bar graph that shows the number of steps it takes to burn off calories from a garden salad with fat-free dressing, a 12-ounce can of soda, a doughnut, and a cheeseburger.

  a) Using the bar graph in **Figure 1.1**, estimate the number of steps it takes to burn off calories from a cheeseburger.

  b) If Cliff Jackson can walk a mile in 2000 steps, how many miles will he have to walk in order to burn off the calories from the cheeseburger he ate for lunch?

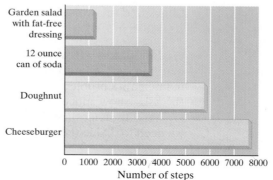

FIGURE 1.1   *Source*: USA Today

Solution   **a)** Using the bar to the right of Cheeseburger in **Figure 1.1**, we estimate that the number of steps it takes to burn off calories from a cheeseburger is about 7600. Notice that we are using the front face of the bar to get our estimate. Whenever we are working with a three-dimensional bar graph, as is the case here, we will use the front face to determine the reading.

b) Understand   Since it takes Cliff 2000 steps to walk a mile, it follows that he would need to take 4000 steps to walk 2 miles, 6000 steps to walk 3 miles, and so on. To determine how many miles Cliff will have to walk in order to burn off the calories from the cheeseburger, we need to divide as follows.

Translate $\text{miles to walk} = \dfrac{\text{number of steps to burn off calories}}{2000}$

Carry Out $\text{miles to walk} = \dfrac{7600}{2000} = 3.8$

Check and Answer   The answer appears reasonable. Cliff will need to walk 3.8 miles in order to burn off the calories from the cheeseburger that he ate for lunch.

▸ **Now Try Exercise 35**

In Example 5, we will use the symbol $\approx$, which is read "**is approximately equal to.**" If, for example, the answer to a problem is 34.12432, we may write the answer as ≈ 34.1.

EXAMPLE 5 ▸ **The Super Bowl** The **line graph** in **Figure 1.2** shows the cost of a 30-second commercial during Super Bowls from 1997 to 2005. The advertising prices are set by the TV network.

a) Estimate the cost of 30-second commercials in 1997 and 2004.

b) How much more was the cost for a 30-second commercial in 2004 than in 1997?

c) How many times greater was the cost of a 30-second commercial in 2004 than in 1997?

Solution   **a)** When reading a line graph where the line has some thickness, as in **Figure 1.2**, we will use the center of the line to make our estimate. By observing the dashed lines on the graph, we can estimate that the cost of a 30-second commercial was about $1.2 million (or $1,200,000) in 1997 and about $2.3 million (or $2,300,000) in 2004.

b) We use the problem-solving procedure to answer the question.

Understand   To determine how much more the cost of a 30-second commercial was in 2004 than in 1997, we need to subtract.

Translate   difference in cost = cost in 2004 − cost in 1997

Carry Out   = $2,300,000 − $1,200,000 = $1,100,000

Check and Answer   The answer appears reasonable. The cost was $1,100,000 more in 2004 than in 1997.

c) Understand   If you examine parts (b) and (c), they may appear to ask the same question, but they do not. In Section 1.1, we indicated that it is important to read a mathematics problem carefully, word by word. The two parts are different in that part (b) asks "how much more was the cost" whereas part (c) asks "how many <u>times</u> greater." To determine the number of times greater the cost was in 2004 than in 1997, we need to divide the cost in 2004 by the cost in 1997, as shown below.

Translate $\text{number of times greater} = \dfrac{\text{cost in 2004}}{\text{cost in 1997}}$

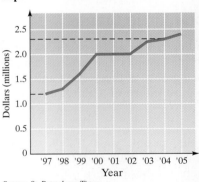

Super Bowl Commercial Cost

Dollars (millions) vs. Year

Source: St. Petersburg Times

FIGURE 1.2

Carry Out

$$\text{number of times greater} = \frac{2,300,000}{1,200,000} \approx 1.92$$

Check and Answer   By observing the graph, we see that the answer is reasonable. The cost of a 30-second commercial during the Super Bowl in 2004 was about 1.92 times the cost in 1997.

▶ **Now Try Exercise 37**

EXAMPLE 6 ▶ **Stay-at-Home Parents** The following information was provided in a press release on November 30, 2004, by the U.S. Census Bureau. **Figure 1.3** is a circle graph that shows the reasons why married mothers with children under the age of 15 have stayed out of the labor force for the past year. Use **Figure 1.3** to determine the number of married mothers with children under the age of 15 who have stayed out of the labor force for the following reasons: to care for home and family, ill/disabled, retired, going to school, could not find work, and other.

Reasons for Married Mothers with Children Under Age 15 Staying Out of the Labor Force for the Past Year

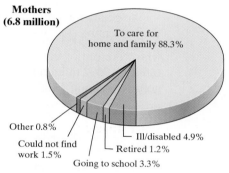

Source: U.S. Census Bureau, Current Population Survey, Annual Social and Economic Supplement, 2003.

FIGURE 1.3

Solution   Understand   The circle graph in **Figure 1.3** indicates that there were 6.8 million married mothers with children under the age of 15 who were out of the labor force. Of these mothers, 88.3% were out of the labor force to care for home and family, 4.9% were ill/disabled, 1.2% were retired, 3.3% were going to school, 1.5% could not find work, and 0.8% were out of the labor force for other reasons. The sum of these percents is 100%. (Notice that 88.3% of the total area of the circle indicates to care for home and family, 4.9% indicates ill/disabled, 1.2% indicates retired, 3.3% indicates going to school, 1.5% indicates could not find work, and 0.8% indicates other.) To determine the number of married mothers with children under the age of 15 out of the labor force to care for home and family, we need to find 88.3% of the total number of married mothers with children under the age of 15 who were out of the labor force. To do this, we multiply as follows.

Translate

$$\begin{pmatrix} \text{to care for home} \\ \text{and family} \end{pmatrix} = \begin{pmatrix} \text{percent out of the labor force} \\ \text{to care for home and family} \end{pmatrix}\begin{pmatrix} \text{total number out} \\ \text{of the labor force} \end{pmatrix}$$

Carry Out

$$\text{to care for home and family} = 0.883(6.8 \text{ million})$$

$$= 6.0044 \text{ million}$$

Thus, 6.0044 million married mothers with children under the age of 15 were out of the labor force to care for home and family.

To find the number of married mothers with children under the age of 15 who were out of the labor force for being ill/disabled, we do a similar calculation.

$$\text{ill/disabled} = 0.049(6.8 \text{ million})$$
$$= 0.3332 \text{ million}$$

We do similar calculations to find the number of married mothers with children under the age of 15 who were retired, going to school, could not find work, or for other reasons were out of the labor force.

$$\text{retired} = 0.012(6.8 \text{ million})$$
$$= 0.0816 \text{ million}$$
$$\text{going to school} = 0.033(6.8 \text{ million})$$
$$= 0.2244 \text{ million}$$
$$\text{could not find work} = 0.015(6.8 \text{ million})$$
$$= 0.102 \text{ million}$$
$$\text{other reasons} = 0.008(6.8 \text{ million})$$
$$= 0.0544 \text{ million}$$

Check   If we add the six amounts, we obtain the 6.8 million total. Therefore, our answer is correct.

$$6.0044 \text{ million} + 0.3332 \text{ million} + 0.0816 \text{ million} + 0.2244 \text{ million}$$
$$+ 0.102 \text{ million} + 0.0544 \text{ million} = 6.8 \text{ million}$$

Answer   In 2003, the number of married mothers with children under the age of 15 who were out of the labor force were as follows: 6.0044 million to care for home and family; 0.3332 million were ill/disabled; 0.0816 million were retired; 0.2244 million were going to school; 0.102 million could not find work; for other reasons, 0.0544 million.

▸ **Now Try Exercise 39**

3 Solve Problems Involving Statistics

Because understanding statistics is so important in our society, we will now discuss certain statistical topics and use them in solving problems.

The *mean* and *median* are two **measures of central tendency**, which are also referred to as *averages*. An average is a value that is representative of a set of data (or numbers). If you take a statistics course you will study these averages in more detail, and you will be introduced to other averages.

The **mean** of a set of data is determined by adding all the values and dividing the sum by the number of values. For example, to find the mean of 6, 9, 3, 12, 12, we do the following.

$$\text{mean} = \frac{6 + 9 + 3 + 12 + 12}{5} = \frac{42}{5} = 8.4$$

We divided the sum by 5 since there are five values. The mean is the most commonly used average and it is generally what is thought of when we use the word *average*.

Another average is the median. The **median** is the value in the middle of a set of **ranked data**. The data may be ranked from smallest to largest or largest to smallest. To find the median of 6, 9, 3, 12, 12, we can rank the data from smallest to largest as follows.

3, 6, 9, 12, 12
↑
Middle value

The value in the middle of the ranked set of data is 9. Therefore, the median is 9. Note that half the values will be above the median and half will be below the median.

If there is an even number of pieces of data, the median is halfway between the two middle pieces. For example, to find the median of 3, 12, 5, 12, 17, 9, we can rank the data as follows.

3, 5, 9, 12, 12, 17
↑
Middle values

Since there are six pieces of data (an even number), we find the value halfway between the two middle pieces, the 9 and the 12. To find the median, we add these values and divide the sum by 2.

$$\text{median} = \frac{9 + 12}{2} = \frac{21}{2} = 10.5$$

Thus, the median is 10.5. Note that half the values are above and half are below 10.5.

EXAMPLE 7 ▸ **The Mean Grade** Alfonso Ramirez's first six exam grades are 90, 87, 76, 84, 78, and 62.

a) Find the mean for Alfonso's six grades.

b) If one more exam is to be given, what is the minimum grade that Alfonso can receive to obtain at least a B average (a mean average of 80 or better)?

c) Is it possible for Alfonso to obtain an A average (90 or better)? Explain.

Solution   **a)** To obtain the mean, we add the six grades and divide by 6.

$$\text{mean} = \frac{90 + 87 + 76 + 84 + 78 + 62}{6} = \frac{477}{6} = 79.5$$

b) We will show the problem-solving steps for this part of the example.

Understand   The answer to this part may be found in a number of ways. For the mean average of seven exams to be 80, the total points for the seven exams must be 7(80) or 560. Can you explain why? The minimum grade needed can be found by subtracting the sum of the first six grades from 560.

Translate   minimum grade needed = 560 − sum of first six exam grades

Carry Out
$$\begin{aligned} &= 560 - (90 + 87 + 76 + 84 + 78 + 62) \\ &= 560 - 477 \\ &= 83 \end{aligned}$$

Check   We can check to see that a seventh grade of 83 gives a mean of 80 as follows.

$$\text{mean} = \frac{90 + 87 + 76 + 84 + 78 + 62 + 83}{7} = \frac{560}{7} = 80$$

Answer   A seventh grade of 83 or higher will result in at least a B average.

c) We can use the same reasoning as in part **(b)**. For a 90 average, the total points that Alfonso will need to attain is 90(7) = 630. Since his total points are 477, he will need 630 − 477 or 153 points to obtain an A average. Since the maximum number of points available on most exams is 100, Alfonso would not be able to obtain an A in the course.

▸ **Now Try Exercise 41**

EXERCISE SET 1.2   *Math XL*   *MyMathLab*

MathXL®   MyMathLab

Concept/Writing Exercises

✎ **1.** Outline the five-step problem-solving procedure.

✎ **2.** What is an expression?

✎ **3.** What does the symbol ≈ mean?

✎ **4.** If a problem is difficult to solve because the numbers in the problem are very large or very small, what can you do that may help make the problem easier to solve?

✎ **5.** Explain how to find the median of a set of data.

6. What measure of central tendency do we generally think of as "the average"?

✎ **7.** Explain how to find the mean of a set of data.

✎ **8.** Consider the set of data 2, 3, 5, 6, 30. Without doing any calculations, can you determine whether the mean or the median is greater? Explain your answer.

9. To get a grade of B, a student must have a mean average of 80. Walter Kirby has a mean average of 79 for 10 quizzes. He approaches his teacher and asks for a B, reasoning that he missed a B by only one point. What is wrong with his reasoning?

10. Consider the set of data 3, 3, 3, 4, 4, 4. If one 4 is changed to 5, which of the following will change, the mean and/or the median? Explain.

Practice the Skills

In this exercise set, use a calculator as needed to save time.

11. Test Grades Jenna Webber's test grades are 78, 97, 59, 74, and 74. For Jenna's grades, determine the **(a)** mean and **(b)** median.

12. Bowling Scores William Krant's bowling scores for five games were 161, 131, 187, 163, and 145. For William's games, determine the **(a)** mean and **(b)** median.

13. Electric Bills The Malones' electric bills for January through June, 2006, were $96.56, $108.78, $87.23, $85.90, $79.55, and $65.88. For these bills, determine the **(a)** mean and **(b)** median.

14. Grocery Bills Antoinette Payne's monthly grocery bills for the first five months of 2006 were $204.83, $153.85, $210.03, $119.76, and $128.38. For Antoinette's grocery bills, determine the **(a)** mean and **(b)** median.

15. Dry Summers The following figure shows the 10 driest summers in the Southeast from 1895 through 2004. Determine

the **(a)** mean and **(b)** median inches of rainfall for the 10 years shown.

Driest Summers in the Southeast

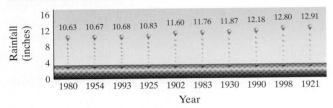

Source: Gloria Forthun, Southeast Regional Climate Center
Records are from 1895 through 2004. The Southeast is Va., N.C., S.C., Ga., Fla., and Ala.

16. Homes for Sale Eight homes are for sale in a community. The sale prices are $124,100, $175,900, $142,300, $164,800, $146,000, $210,000, $112,200, and $153,600. Determine the **(a)** mean and **(b)** median sale price of the eight homes.

Problem Solving

17. Commissions Barbara Riedell earns a 5% commission on appliances she sells. Her sales last week totaled $9400. Find her week's earnings.

18. Empire State Building May 1, 1931, was the opening day of the Empire State Building. It stands 1454 feet or 443 meters high. Use this information to determine the approximate number of feet in a meter.

19. Sales Tax a) The sales tax in Jefferson County is 8%. What was the sales tax that Scott Reed paid on a used car that cost $16,700 before tax?

b) What is the total cost of the car including tax?

20. Checking Account The balance in Debbie Ogilvie's checking account is $312.60. She purchased five DVDs at $17.11 each including tax. If she pays by check, what is the new balance in her checking account?

21. Computer Processor Suppose a computer processor can perform about 2.3 billion operations per second. How many operations can it perform in 0.7 second?

22. Buying a Computer Pat Sullivan wants to purchase a computer that sells for $950. He can either pay the total amount at the time of purchase or agree to pay the store $200 down and $33 per month for 24 months.

a) If he pays the down payment and monthly charge, how much will he pay for the computer?

b) How much money can he save by paying the total amount at the time of purchase?

23. Energy Values The following table gives the approximate energy values of some foods and the approximate energy consumption of some activities, in kilojoules (kJ). Determine how long it would take for you to use up the energy from the following.

a) a hamburger by running

b) a chocolate milkshake by walking

c) a glass of skim milk by cycling

Energy Value, Food	(kJ)	Energy Consumption, Activity	(kJ/min)
Chocolate milkshake	2200	Walking	25
Fried egg	460	Cycling	35
Hamburger	1550	Swimming	50
Strawberry shortcake	1440	Running	80
Glass of skim milk	350		

See Exercise 23.b)

A green numbered exercise, such as **21.** indicates a Now Try Exercise.

24. Jet Ski The rental cost of a jet ski from Don's Ski Rental is $20 per half-hour, and the rental cost from A. J.'s Ski Rental is $50 per hour. Suppose you plan to rent a jet ski for 3 hours.

a) Which is the better deal?

b) How much will you save?

25. Gas Mileage When the odometer in Tribet LaPierre's car reads 16,741.3, he fills his gas tank. The next time he fills his tank it takes 10.5 gallons, and his odometer reads 16,935.4. Determine the number of miles per gallon that his car gets.

26. Income Taxes The federal income tax rate schedule for a *joint return* in 2005 is illustrated in the following table.

Adjusted Gross Income	Taxes
$0–$14,600	10% of income
$14,600–$59,400	$1460.00 + 15% in excess of $14,600
$59,400–$119,950	$8180.00 + 25% in excess of $59,400
$119,950–$182,800	$23,317.50 + 28% in excess of $119,950
$182,800–$326,450	$40,915.50 + 33% in excess of $182,800
$326,450 and up	$88,320.00 + 35% in excess of $326,450

Source: www.irs.gov

a) If the Donovins' adjusted gross income in 2005 was $34,612 determine their taxes.

b) If the Ortegas' 2005 adjusted gross income was $75,610 determine their taxes.

27. Buying Tires Eric Weiss purchased four tires through the Internet. He paid $62.30 plus $6.20 shipping and handling per tire. There was no sales tax on this purchase. When he received the tires, Eric had to pay $8.00 per tire for mounting and balancing. At a local tire store, his total cost for the four tires with mounting and balancing would have been $425 plus 8% sales tax. How much did Eric save by purchasing the tires through the Internet?

28. Baseball Salaries Manny Ramirez, of the Boston Red Sox, was the highest paid professional baseball player in 2004, earning $22.5 million. Pedro Martinez, also of the Boston Red Sox, was the highest paid pitcher, earning $17.5 million. In 2004, Ramirez batted 568 times and Martinez pitched 217 innings. Determine approximately how much more Martinez received per inning pitched than Ramirez did per at bat.

29. Balance Consider the figure shown. Assuming the green and red blocks have the same weight, where should a single green block, , be placed to make the scale balanced? Explain how you determined your answer.

30. Taxi Ride A taxicab charges $2 upon a customer's entering the taxi, then 30 cents for each $\frac{1}{4}$ mile traveled and 20 cents for each 30 seconds stopped in traffic. David Lopez takes a taxi ride for a distance of 3 miles where the taxi spends 90 seconds stopped in traffic. Determine David's cost of the taxi ride.

31. Leaky Faucet A faucet that leaks 1 ounce of water per minute wastes 11.25 gallons in a day.

a) How many gallons of water are wasted in a (non-leap) year?

b) If water costs $5.20 per 1000 gallons, how much additional money per year is being spent on the water bill?

32. Conversions a) What is 1 mile per hour equal to in feet per hour? One mile contains 5280 feet.

b) What is 1 mile per hour equal to in feet per second?

c) What is 60 miles per hour equal to in feet per second?

33. Medical Insurance Mel LeBar's medical insurance policy requires that he pay a $150 deductible each calendar year. After the deductible is paid, he pays 20% of the medical expenses and the insurance company pays 80%. On January 1, Mel's daughter accidentally closed the car door on his finger. He went to the doctor's office for an examination and X rays. The total bill of $365 was sent to the insurance company. If Mel had not as yet paid any of the deductible,

a) how much of the bill will Mel be responsible for?

b) how much will the insurance company be responsible for?

34. Insurance Drivers under the age of 25 who pass a driver education course generally have their auto insurance premium decreased by 10%. Most insurers will offer this 10% deduction until the driver reaches 25. A particular driver education course costs $70. Don Beville, who just turned 18, has auto insurance that costs $630 per year.

a) Excluding the cost of the driver education course, how much would Don save in auto insurance premiums, from the age of 18 until the age of 25, by taking the driver education course?

b) What would be his net savings after the cost of the course?

35. Math Scores The bar graph below shows some 2003 scores from the Program for International Student Assessment, a test for 15-year-olds. The scale was constructed so that an average score is 500.

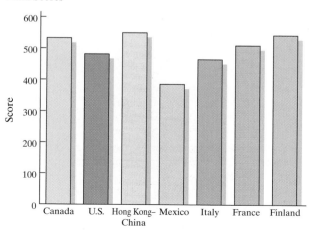

Source: New York Times

a) Which country had the highest math score shown? Estimate the score.

b) Which country had the lowest math scores shown? Estimate the score.

c) Estimate the difference between the scores for Hong Kong–China and Mexico.

36. Commercial Airlines The bar graph shows the total scheduled passengers on U.S. commercial airlines from 1999 to 2004 with the forecasted totals through 2016. From the bar graph, estimate

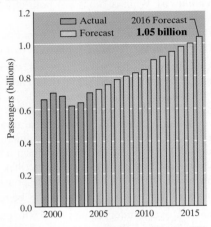

Total Scheduled Passengers on U.S. Commercial Airlines

Note: Data are for fiscal years beginning in October; 2004 figure is an estimate

Source: Federal Aviation Administration

a) the total number of scheduled passengers on U.S. commercial airlines in 2000.

b) the forecasted total number of scheduled passengers on U.S. commercial airlines in 2015.

c) how many times greater is the forecasted total number of scheduled passengers on U.S. commercial airlines in 2015 than in 2000?

37. Motorcycle Sales The line graph below shows the U.S. sales of new motorcycles from 1992 to 2004.

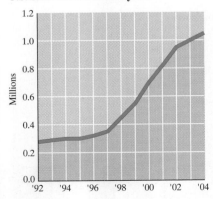

U.S. Sales of New Motorcycles

Note: 2004 figure is an estimate.

Source: Motorcycle Industry Council

a) Estimate the number of new motorcycles sold in the U.S. in 1992 and in 2004.

b) How many more new motorcycles were sold in the U.S. in 2004 than in 1992?

c) How many times greater was the number of new motorcycles sold in 2004 than in 1992?

38. NBA Draft The line graph below shows the number of high school students selected in the NBA draft's first round from 1995 to 2004.

a) During which consecutive years did the number of high school students selected in the NBA draft's first round remain constant?

During which consecutive years did the number of high school students selected in the NBA draft's first round

b) increase the most?

c) decrease the most?

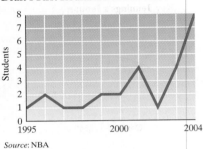

High School Students Selected in the NBA Draft's First Round from 1995–2004

Source: NBA

39. Adoption The following information was provided by *www.census.gov*. In 2004, approximately 1.7 million U.S. households contained adopted children. The circle graph below shows the percent of these households that had one adopted child, two adopted children, and three or more adopted children. Estimate the number of U.S. households in 2004 that had

a) one adopted child.

b) two adopted children.

c) three or more adopted children.

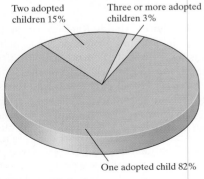

U.S. Households with Adopted Children (1.7 million households)

Two adopted children 15%

Three or more adopted children 3%

One adopted child 82%

Source: www.census.gov

40. Jeopardy! As of this writing, Ken Jennings, a software engineer from Salt Lake City, Utah, is the record holder for the most money ever won on a television game show. The circle graph on the next page shows the outcome of the 160 Daily

Doubles that Ken attempted on his *Jeopardy!* run. Use the circle graph to determine

a) the number of Daily Doubles Ken answered correctly.

b) the number of Daily Doubles Ken answered incorrectly.

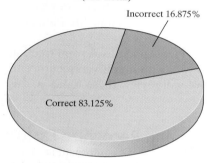

Ken Jennings's Jeopardy Daily Doubles (160 Total)

Incorrect 16.875%

Correct 83.125%

Source: www.tvgameshows.net/jenningswinnings.htm

41. Test Grades A mean of 60 on all exams is needed to pass a course. On his first five exams Lamond Paine's grades are 50, 59, 67, 80, and 56.

a) What is the minimum grade that Lamond can receive on the sixth exam to pass the course?

b) An average of 70 is needed to get a C in the course. Is it possible for Lamond to get a C? If so, what is the minimum grade that Lamond can receive on the sixth exam?

42. Test Grades A mean of 80 on all exams is needed to earn a B in a course. On her first four exams, Heather Feldman's grades are 95, 88, 82, and 85.

a) What is the minimum grade that Heather can receive on the fifth exam to earn a B in the course?

b) A mean of 90 is needed to earn an A in the course. Is it possible for Heather to get an A? If so, what is the minimum grade that Heather can receive on the fifth exam?

43. Level of Education The following bar graph shows the median annual earnings by level of education in 2003 in the United States. Which degree has median annual earnings that are approximately 2.3 times the median annual earnings of someone with less than a high school diploma?

Median Annual Earnings by Level of Education

Level of Education	Median Annual Earnings
Professional degree	$95,700
Doctorate	$79,400
Master's degree	$59,500
Bachelor's degree	$49,900
Associate degree	$37,600
Some college, no degree	$35,700
High school diploma	$30,800
Less than HS diploma	$21,600

Sources: College Board,
U.S. Census Bureau (2003 data)

44. Exams Mike Ambrosino's mean average on six exams is 78. Find the sum of his scores.

45. Construct Data Construct a set of five pieces of data with a mean of 70 and no two values the same.

Challenge Problem

46. Reading Meters The figure shows how to read an electric (or gas) meter.

Step 1. Start with the dial on the right. If the arrow falls between two numbers, use the smaller number on the dial (except when the dial is between 9 and 0, then use the 9). Notice the arrows above the meters indicate the direction the dial is moving (clockwise, then counterclockwise).

Step 2. If the pointer is directly on a number, check the dial to the *right* to make sure it has passed 0 and is headed toward 1. If the dial to the right has not passed 0, use the next lower number. The number on the meters on the right is 16064.

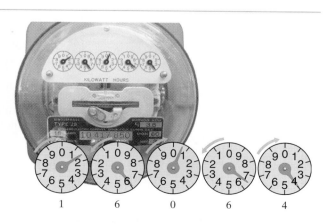

Source: Southern California Edison, Understanding Your Electricity Bill

Suppose your previous month's reading was as shown on the previous page, and this month's meter reading is as shown below.

a) Determine this month's meter reading.

b) Determine your electrical cost for this month by first subtracting last month's meter reading from this month's meter reading (measured in kilowatt hours), and then multiplying the difference by the cost per kilowatt hour of electricity. Assume electricity costs 24.3 cents per kilowatt hour.

1.3 Fractions

1 Learn multiplication symbols and recognize factors.

2 Simplify fractions.

3 Multiply fractions.

4 Divide fractions.

5 Add and subtract fractions.

6 Change mixed numbers to fractions and vice versa.

Students taking algebra for the first time often ask, "What is the difference between arithmetic and algebra?" When doing arithmetic, all the quantities used in the calculations are known. In algebra, however, one or more of the quantities are unknown and must be found. Consider the following:

Mr. Piersma has 1 gallon of paint. In order to paint his bedroom, he needs 3 gallons of paint. How many additional gallons does he need?

Although elementary, this is an example of an algebraic problem. The unknown quantity is the number of additional gallons of paint needed. Mr. Piersma needs 2 more gallons of paint.

An understanding of decimal numbers (see Appendix A) and fractions is essential to success in algebra. You will need to know how to simplify a fraction and how to add, subtract, multiply, and divide fractions. We will review these topics in this section. We will also explain the meaning of factors.

1 Learn Multiplication Symbols and Recognize Factors

In algebra we often use letters called **variables** to represent numbers. Letters commonly used as variables are x, y, and z, but other letters can be used as variables. Variables are usually shown in italics. So that we do not confuse the variable x with the multiplication sign, we often use different notation to indicate multiplication.

Multiplication Symbols

If a and b represent any two mathematical quantities, then each of the following may be used to indicate the product of a and b ("a times b").

$$ab \quad a \cdot b \quad a(b) \quad (a)b \quad (a)(b)$$

EXAMPLES

3 times 4 may be written:	3 times x may be written:	x times y may be written:
	$3x$	xy
$3(4)$	$3(x)$	$x(y)$
$(3)4$	$(3)x$	$(x)y$
$(3)(4)$	$(3)(x)$	$(x)(y)$

Now we will introduce the term *factors*, which we shall be using throughout the text. Below we define factors.

Factors

The numbers or variables that are multiplied in a multiplication problem are called **factors**.

If $a \cdot b = c$, then a and b are *factors* of c.

For example, in $3 \cdot 5 = 15$, the numbers 3 and 5 are factors of the product 15. In $2 \cdot 15 = 30$, the numbers 2 and 15 are factors of the product 30. Note that 30 has many other factors. Since $5 \cdot 6 = 30$, the numbers 5 and 6 are also factors of 30. Since $3x$ means 3 times x, both the 3 and the x are factors of $3x$.

2 Simplify Fractions

Now we have the necessary information to discuss **fractions**. The top number of a fraction is called the **numerator**, and the bottom number is called the **denominator**. In the fraction $\frac{3}{5}$, the 3 is the numerator and the 5 is the denominator.

Helpful Hint

Consider the fraction $\frac{3}{5}$. There are equivalent methods of expressing this fraction, as illustrated below.

$$\frac{3}{5} = 3 \div 5 = 5\overline{)3}$$

In general, $\frac{a}{b} = a \div b = b\overline{)a}$.

Now we will discuss how to simplify a fraction.

A fraction is **simplified** (or **reduced to its lowest terms**) when the numerator and denominator have no common factors other than 1. To simplify a fraction, follow these steps.

To Simplify a Fraction

1. Find the largest number that will divide (without remainder) both the numerator and the denominator. This number is called the **greatest common factor** (GCF).

2. Then divide both the numerator and the denominator by the greatest common factor.

If you do not remember how to find the greatest common factor of two or more numbers, read Appendix B.

EXAMPLE 1 ▶ Simplify   **a)** $\frac{10}{25}$ **b)** $\frac{6}{18}$.

Solution

a) The largest number that divides both 10 and 25 is 5. Therefore, 5 is the greatest common factor. Divide both the numerator and the denominator by 5 to simplify the fraction.

$$\frac{10}{25} = \frac{10 \div 5}{25 \div 5} = \frac{2}{5}$$

b) Both 6 and 18 can be divided by 1, 2, 3, and 6. The largest of these numbers, 6, is the greatest common factor. Divide both the numerator and the denominator by 6.

$$\frac{6}{18} = \frac{6 \div 6}{18 \div 6} = \frac{1}{3}$$

Note in Example **1b)** that both the numerator and denominator could have been written with a factor of 6. Then the common factor 6 could be divided out.

$$\frac{6}{18} = \frac{1 \cdot \cancel{6}}{3 \cdot \cancel{6}} = \frac{1}{3}$$

▶ **Now Try Exercise 23**

When you work with fractions you should simplify your answers.

3 Multiply Fractions

To multiply two or more fractions, multiply their numerators together and multiply their denominators together.

To Multiply Fractions

$$\frac{a}{b} \cdot \frac{c}{d} = \frac{ac}{bd}$$

EXAMPLE 2 ▶ Multiply $\frac{3}{13}$ by $\frac{5}{11}$.

Solution $\frac{3}{13} \cdot \frac{5}{11} = \frac{3 \cdot 5}{13 \cdot 11} = \frac{15}{143}$

▶ **Now Try Exercise 51**

Before multiplying fractions, to help avoid having to simplify an answer, we often divide both a numerator and a denominator by a common factor.

EXAMPLE 3 ▶ Multiply   **a)** $\frac{8}{17} \cdot \frac{5}{16}$ **b)** $\frac{27}{40} \cdot \frac{16}{9}$.

Solution

a) Since the numerator 8 and the denominator 16 can both be divided by the common factor 8, we divide out the 8 first. Then we multiply.

$$\frac{8}{17} \cdot \frac{5}{16} = \frac{\overset{1}{8}}{17} \cdot \frac{5}{\underset{2}{16}} = \frac{1 \cdot 5}{17 \cdot 2} = \frac{5}{34}$$

b)
$$\frac{27}{40} \cdot \frac{16}{9} = \frac{\overset{3}{27}}{40} \cdot \frac{16}{\underset{1}{9}}$$ *Divide both 27 and 9 by 9.*

$$= \frac{\overset{3}{27}}{\underset{5}{40}} \cdot \frac{\overset{2}{16}}{\underset{1}{9}}$$ *Divide both 40 and 16 by 8.*

$$= \frac{3 \cdot 2}{5 \cdot 1} = \frac{6}{5}$$

▶ **Now Try Exercise 53**

The numbers 0, 1, 2, 3, 4, ... are called **whole numbers**. The three dots after the 4 which is called an *ellipsis*, indicate that the whole numbers continue indefinitely in the same manner. Thus, the numbers 468 and 5043 are also whole numbers. Whole numbers will be discussed further in Section 1.4. To multiply a whole number by a fraction, write the whole number with a denominator of 1 and then multiply.

EXAMPLE 4 ▶ **Lawn Mower Engine** Some engines run on a mixture of gas and oil. A particular lawn mower engine requires a mixture of $\frac{5}{64}$ gallon of oil for each gallon of gasoline used. A lawn care company wishes to make a mixture for this engine using 12 gallons of gasoline. How much oil must be used?

Solution   We must multiply 12 by $\frac{5}{64}$ to determine the amount of oil that must be used. First we write 12 as $\frac{12}{1}$, then we divide both 12 and 64 by their greatest common factor, 4, as follows.

$$12 \cdot \frac{5}{64} = \frac{12}{1} \cdot \frac{5}{64} = \frac{\overset{3}{12}}{1} \cdot \frac{5}{\underset{16}{64}} = \frac{3 \cdot 5}{1 \cdot 16} = \frac{15}{16}$$

Thus, $\dfrac{15}{16}$ gallon of oil must be added to the 12 gallons of gasoline to make the proper mixture.

▶ **Now Try Exercise 101**

4 Divide Fractions

To divide one fraction by another, invert the divisor (the second fraction if written with ÷) and proceed as in multiplication.

To Divide Fractions

$$\frac{a}{b} \div \frac{c}{d} = \frac{a}{b} \cdot \frac{d}{c} = \frac{ad}{bc}$$

Sometimes, rather than being asked to obtain the answer to a problem by adding, subtracting, multiplying, or dividing, you may be asked to evaluate an expression. To **evaluate** an expression means to obtain the answer to the problem using the operations given.

EXAMPLE 5 ▶ Evaluate   **a)** $\dfrac{3}{5} \div \dfrac{5}{6}$ **b)** $\dfrac{3}{8} \div 12$.

Solution

a) $\dfrac{3}{5} \div \dfrac{5}{6} = \dfrac{3}{5} \cdot \dfrac{6}{5} = \dfrac{3 \cdot 6}{5 \cdot 5} = \dfrac{18}{25}$

b) Write 12 as $\dfrac{12}{1}$. Then invert the divisor and multiply.

$$\frac{3}{8} \div 12 = \frac{3}{8} \div \frac{12}{1} = \frac{\overset{1}{\cancel{3}}}{8} \cdot \frac{1}{\underset{4}{\cancel{12}}} = \frac{1}{32}$$

▶ **Now Try Exercise 59**

5 Add and Subtract Fractions

Fractions that have the same (or a common) *denominator can be added or subtracted.* To add (or subtract) fractions with the same denominator, add (or subtract) the numerators and keep the common denominator.

To Add and Subtract Fractions

$$\frac{a}{c} + \frac{b}{c} = \frac{a + b}{c} \quad \text{or} \quad \frac{a}{c} - \frac{b}{c} = \frac{a - b}{c}$$

EXAMPLE 6 ▶   **a)** Add $\dfrac{6}{15} + \dfrac{2}{15}$.   **b)** Subtract $\dfrac{8}{13} - \dfrac{5}{13}$.

Solution   **a)** $\dfrac{6}{15} + \dfrac{2}{15} = \dfrac{6 + 2}{15} = \dfrac{8}{15}$ **b)** $\dfrac{8}{13} - \dfrac{5}{13} = \dfrac{8 - 5}{13} = \dfrac{3}{13}$

▶ **Now Try Exercise 67**

To add (or subtract) fractions with unlike denominators, we must first rewrite each fraction with the same, or a common, denominator. The smallest number that is divisible by two or more denominators is called the **least common denominator** or **LCD**. *If you have forgotten how to find the least common denominator review Appendix B now.*

EXAMPLE 7 ▶ Add $\frac{1}{2} + \frac{1}{5}$.

Solution   We cannot add these fractions until we rewrite them with a common denominator. Since the lowest number that both 2 and 5 divide into (without remainder) is 10, we will first rewrite both fractions with the least common denominator of 10.

$$\frac{1}{2} = \frac{1}{2} \cdot \frac{5}{5} = \frac{5}{10} \quad \text{and} \quad \frac{1}{5} = \frac{1}{5} \cdot \frac{2}{2} = \frac{2}{10}$$

Now we add.

$$\frac{1}{2} + \frac{1}{5} = \frac{5}{10} + \frac{2}{10} = \frac{7}{10}$$

▶ **Now Try Exercise 75**

Note that multiplying both the numerator and denominator by the same number is the same as multiplying by 1. Thus, the value of the fraction does not change.

EXAMPLE 8 ▶ How much larger is $\frac{3}{4}$ inch than $\frac{2}{3}$ inch?

Solution   To find the difference, we need to subtract $\frac{2}{3}$ inch from $\frac{3}{4}$ inch.

$$\frac{3}{4} - \frac{2}{3}$$

The least common denominator is 12. Therefore, we rewrite both fractions with a denominator of 12.

$$\frac{3}{4} = \frac{3}{4} \cdot \frac{3}{3} = \frac{9}{12} \quad \text{and} \quad \frac{2}{3} = \frac{2}{3} \cdot \frac{4}{4} = \frac{8}{12}$$

Now we subtract.

$$\frac{3}{4} - \frac{2}{3} = \frac{9}{12} - \frac{8}{12} = \frac{1}{12}$$

Thus, $\frac{3}{4}$ inch is $\frac{1}{12}$ inch greater than $\frac{2}{3}$ inch.

▶ **Now Try Exercise 87**

Avoiding Common Errors

It is important to remember that dividing out a common factor in the numerator of one fraction and the denominator of a different fraction can be performed only when multiplying fractions. *This process cannot be performed when adding or subtracting fractions.*

CORRECT	INCORRECT
MULTIPLICATION PROBLEMS	ADDITION PROBLEMS

$$\frac{\overset{1}{\cancel{3}}}{5} \cdot \frac{1}{\underset{1}{\cancel{3}}}$$

$$\frac{\overset{2}{\cancel{8}} \cdot 3}{\underset{1}{\cancel{4}}}$$

---

6 Change Mixed Numbers to Fractions and Vice Versa

Consider the number $5\frac{2}{3}$. This is an example of a **mixed number**. A mixed number consists of a whole number followed by a fraction. The mixed number $5\frac{2}{3}$ means $5 + \frac{2}{3}$. We can change $5\frac{2}{3}$ to a fraction as follows.

$$5\frac{2}{3} = \boxed{5} + \frac{2}{3} = \boxed{\frac{15}{3}} + \frac{2}{3} = \frac{15 + 2}{3} = \frac{17}{3}$$

Notice that we expressed the whole number, 5, as a fraction with a denominator of 3, then added the fractions.

EXAMPLE 9 ▸ Change $7\frac{3}{8}$ to a fraction.

Solution
$$7\frac{3}{8} = \boxed{7} + \frac{3}{8} = \boxed{\frac{56}{8}} + \frac{3}{8} = \frac{56 + 3}{8} = \frac{59}{8}$$

▸ **Now Try Exercise 37**

Now consider the fraction $\frac{17}{3}$. This fraction can be converted to a mixed number as follows.

$$\frac{17}{3} = \boxed{\frac{15}{3}} + \frac{2}{3} = \boxed{5} + \frac{2}{3} = 5\frac{2}{3}$$

Notice we wrote $\frac{17}{3}$ as a sum of two fractions, each with the denominator of 3. The first fraction being added, $\frac{15}{3}$, is the equivalent of the largest integer that is less than $\frac{17}{3}$.

EXAMPLE 10 ▸ Change $\frac{43}{6}$ to a mixed number.

Solution
$$\frac{43}{6} = \boxed{\frac{42}{6}} + \frac{1}{6} = \boxed{7} + \frac{1}{6} = 7\frac{1}{6}$$

▸ **Now Try Exercise 43**

Helpful Hint

Notice in Example 10 the fraction $\frac{43}{6}$ is simplified because the greatest common divisor of the numerator and denominator is 1. Do not confuse simplifying a fraction with changing a fraction with a value greater than 1 to a mixed number. The fraction $\frac{43}{6}$ can be converted to the mixed number $7\frac{1}{6}$. However, $\frac{43}{6}$ is a simplified fraction.

Now we will work examples that contain mixed numbers.

EXAMPLE 11 ▸ **Plumbing** To repair a plumbing leak, a coupling $\frac{1}{2}$ inch long is glued to a piece of plastic pipe. After the gluing, the piece of plastic pipe showing is $2\frac{9}{16}$ inches, as shown in **Figure 1.4**. How long is the combined length?

Solution Understand and Translate We need to add $2\frac{9}{16}$ inches and $\frac{1}{2}$ inch to obtain the combined lengths. We will place the two numbers one above the other. After we write both fractions with a common denominator, we add the numbers.

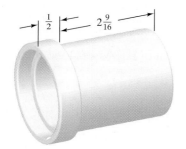

FIGURE 1.4

Carry Out

$$
\begin{array}{r}
2\dfrac{9}{16} \\
+\ \dfrac{1}{2} \\
\hline
\end{array}
\quad \rightarrow \quad
\begin{array}{r}
2\dfrac{9}{16} \\
+\ \dfrac{8}{16} \\
\hline
2\dfrac{17}{16}
\end{array}
$$

Since $2\frac{17}{16} = 2 + \frac{17}{16} = 2 + 1\frac{1}{16} = 3\frac{1}{16}$, the sum is $3\frac{1}{16}$.

Check and Answer The answer appears reasonable. Thus, the total length is $3\frac{1}{16}$ inches.

▸ **Now Try Exercise 105**

Jonathan's Growth

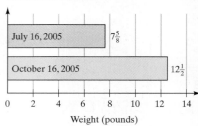

FIGURE 1.5

EXAMPLE 12 ▶ **Gaining Weight** The graph in **Figure 1.5** shows the weight of the Seeburgers' baby boy, Jonathan, on July 16, 2005, and on October 16, 2005. How much weight had Jonathan gained during this time period?

Solution   Understand and Translate   To find the increase, we need to subtract Jonathan's weight on July 16, 2005, from his weight on October 16, 2005. We will subtract vertically.

Carry Out

$$12\frac{1}{2} \qquad \rightarrow \qquad 12\frac{4}{8}$$
$$-\ 7\frac{5}{8} \qquad\qquad -\ 7\frac{5}{8}$$

Since we wish to subtract $\frac{5}{8}$ from $\frac{4}{8}$, and $\frac{5}{8}$ is greater than $\frac{4}{8}$, we write $12\frac{4}{8}$ as $11\frac{12}{8}$.

To get $11\frac{12}{8}$, we take 1 unit from the number 12 and write it as $\frac{8}{8}$. This gives $11 + 1 + \frac{4}{8} = 11 + \frac{8}{8} + \frac{4}{8} = 11 + \frac{12}{8} = 11\frac{12}{8}$. Now we subtract as follows.

$$12\frac{1}{2} \qquad \rightarrow \qquad 12\frac{4}{8} \qquad \rightarrow \qquad 11\frac{12}{8}$$
$$-\ 7\frac{5}{8} \qquad\qquad -\ 7\frac{5}{8} \qquad\qquad -\ 7\frac{5}{8}$$
$$\qquad\qquad\qquad\qquad\qquad\qquad\qquad\qquad\qquad 4\frac{7}{8}$$

Check and Answer   By examining the graph, we see that the answer is reasonable. Thus, Jonathan gained $4\frac{7}{8}$ pounds during this time period.

▶ **Now Try Exercise 97**

Although it is not necessary to change mixed numbers to fractions when adding or subtracting mixed numbers, it is necessary to change mixed numbers to fractions when multiplying or dividing mixed numbers. We illustrate this procedure in Example 13.

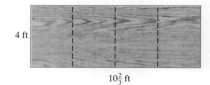

4 ft

$10\frac{2}{3}$ ft

FIGURE 1.6

EXAMPLE 13 ▶ **Cutting Wood** A carpenter is cutting a rectangular piece of wood 4 feet wide by $10\frac{2}{3}$ feet long into four equal strips, as shown in **Figure 1.6**. Find the dimensions of each strip.

Solution   Understand and Translate   We know from **Figure 1.6** that one side will have a width of 4 feet. To find the length of the strips, we need to divide $10\frac{2}{3}$ by 4.

Carry Out

$$10\frac{2}{3} \div 4 = \frac{32}{3} \div \frac{4}{1} = \frac{\overset{8}{\cancel{32}}}{3} \cdot \frac{1}{\underset{1}{\cancel{4}}} = \frac{8}{3} = 2\frac{2}{3}$$

Check and Answer   If you multiply $2\frac{2}{3}$ by 4 you obtain the original length, $10\frac{2}{3}$.

Thus, the calculation is correct. The dimensions of each strip will be 4 feet by $2\frac{2}{3}$ feet.

▶ **Now Try Exercise 99**

EXERCISE SET 1.3

MathXL®   MyMathLab

Concept/Writing Exercises

1. a) What are variables?
   b) What letters are often used to represent variables?
2. What are factors?
3. In a fraction, what is the name of the a) top number and b) bottom number?
4. Show five different ways that "5 times x" may be written.
5. Explain how to simplify a fraction.

6. a) What are the three dots in $4, 5, 6, 7, \ldots$ called?
   b) What do the three dots following the 7 mean?
7. a) What is the least common denominator of two or more fractions?
   b) Write two fractions and then give the LCD of the two fractions you wrote.
8. What is the LCD of the fractions $\frac{3}{8}$ and $\frac{7}{10}$? Explain.

In Exercises 9 and 10, which part a) or b) shows simplifying a fraction? Explain.

9. a) $\dfrac{\overset{1}{\cancel{2}}}{5} \cdot \dfrac{1}{\underset{2}{\cancel{4}}}$ b) $\dfrac{\overset{1}{\cancel{2}}}{\underset{2}{\cancel{4}}}$

10. a) $\dfrac{\overset{1}{\cancel{4}}}{\underset{3}{\cancel{12}}}$ b) $\dfrac{7}{\underset{3}{\cancel{12}}} \cdot \dfrac{\overset{1}{\cancel{4}}}{5}$

In Exercises 11 and 12, one of the procedures a) or b) is incorrect. Determine which is the incorrect procedure and explain why.

11. a) $\dfrac{\overset{1}{\cancel{4}}}{5} + \dfrac{3}{\underset{2}{\cancel{8}}}$ b) $\dfrac{\overset{1}{\cancel{4}}}{5} \cdot \dfrac{3}{\underset{2}{\cancel{8}}}$

12. a) $\dfrac{4}{\underset{3}{\cancel{15}}} \cdot \dfrac{\overset{1}{\cancel{5}}}{7}$ b) $\dfrac{4}{\underset{3}{\cancel{15}}} + \dfrac{\overset{1}{\cancel{5}}}{7}$

In Exercises 13 and 14, indicate any parts where a common factor can be divided out as a first step in evaluating each expression. Explain your answer.

13. a) $\dfrac{4}{5} + \dfrac{1}{4}$ b) $\dfrac{4}{5} - \dfrac{1}{4}$ c) $\dfrac{4}{5} \cdot \dfrac{1}{4}$ d) $\dfrac{4}{5} \div \dfrac{1}{4}$

14. a) $6 + \dfrac{5}{12}$ b) $6 \cdot \dfrac{5}{12}$ c) $6 - \dfrac{5}{12}$ d) $6 \div \dfrac{5}{12}$

15. Explain how to multiply fractions.
16. Explain how to divide fractions.
17. Explain how to add or subtract fractions.
18. Give an example of a mixed number.
19. Is the fraction $\dfrac{24}{5}$ simplified? Explain your answer.
20. Is the fraction $\dfrac{20}{3}$ simplified? Explain your answer.
21. Can the fraction $\dfrac{13}{5}$ be written as a mixed number? Explain your answer.
22. Can the fraction $\dfrac{8}{3}$ be written as a mixed number? Explain your answer.

Practice the Skills

Simplify each fraction. If a fraction is already simplified, so state.

23. $\dfrac{10}{15}$ 24. $\dfrac{40}{10}$ 25. $\dfrac{3}{12}$ 26. $\dfrac{19}{25}$ 27. $\dfrac{36}{76}$ 28. $\dfrac{16}{72}$

29. $\dfrac{9}{21}$ 30. $\dfrac{60}{105}$ 31. $\dfrac{18}{49}$ 32. $\dfrac{100}{144}$ 33. $\dfrac{12}{25}$ 34. $\dfrac{42}{138}$

Convert each mixed number to a fraction.

35. $2\dfrac{13}{15}$ 36. $5\dfrac{1}{3}$ 37. $7\dfrac{2}{3}$ 38. $4\dfrac{3}{4}$ 39. $3\dfrac{5}{18}$ 40. $6\dfrac{2}{9}$ 41. $9\dfrac{6}{17}$ 42. $3\dfrac{3}{32}$

Write each fraction as a mixed number.

43. $\dfrac{7}{4}$ 44. $\dfrac{18}{7}$ 45. $\dfrac{13}{4}$ 46. $\dfrac{9}{2}$ 47. $\dfrac{32}{7}$ 48. $\dfrac{110}{20}$ 49. $\dfrac{86}{14}$ 50. $\dfrac{72}{14}$

Find each product or quotient. Simplify the answer.

51. $\dfrac{2}{3} \cdot \dfrac{4}{5}$ 52. $\dfrac{6}{13} \cdot \dfrac{7}{17}$ 53. $\dfrac{5}{12} \cdot \dfrac{4}{15}$ 54. $\dfrac{36}{48} \cdot \dfrac{16}{45}$

55. $\dfrac{3}{4} \div \dfrac{1}{2}$ 56. $\dfrac{15}{16} \cdot \dfrac{4}{3}$ 57. $\dfrac{3}{8} \div \dfrac{3}{4}$ 58. $\dfrac{3}{8} \cdot \dfrac{10}{11}$

59. $\dfrac{10}{3} \div \dfrac{5}{9}$ 60. $\dfrac{5}{9} \div 30$ 61. $\dfrac{15}{4} \cdot \dfrac{2}{3}$ 62. $\dfrac{5}{12} \div \dfrac{4}{3}$

63. $5\dfrac{3}{8} \div 1\dfrac{1}{4}$ 64. $\left(2\dfrac{1}{5}\right)\left(\dfrac{7}{8}\right)$ 65. $\dfrac{28}{13} \cdot \dfrac{2}{7}$ 66. $4\dfrac{4}{5} \div \dfrac{8}{15}$

Add or subtract. Simplify each answer.

67. $\dfrac{3}{8} + \dfrac{2}{8}$

68. $\dfrac{18}{36} - \dfrac{1}{36}$

69. $\dfrac{3}{14} - \dfrac{1}{14}$

70. $\dfrac{1}{4} + \dfrac{3}{4}$

71. $\dfrac{4}{5} + \dfrac{6}{15}$

72. $\dfrac{7}{8} - \dfrac{5}{6}$

73. $\dfrac{8}{17} + \dfrac{2}{34}$

74. $\dfrac{3}{7} + \dfrac{17}{35}$

75. $\dfrac{1}{3} + \dfrac{1}{4}$

76. $\dfrac{1}{6} - \dfrac{1}{18}$

77. $\dfrac{7}{12} - \dfrac{2}{9}$

78. $\dfrac{3}{7} + \dfrac{5}{12}$

79. $3\dfrac{1}{8} - \dfrac{5}{12}$

80. $5\dfrac{6}{7} + 4\dfrac{5}{8}$

81. $6\dfrac{1}{3} - 3\dfrac{1}{2}$

82. $2\dfrac{3}{8} + 3\dfrac{3}{4}$

83. $9\dfrac{2}{5} - 6\dfrac{1}{2}$

84. $4\dfrac{5}{9} - \dfrac{7}{8}$

85. $5\dfrac{9}{10} + 3\dfrac{1}{3}$

86. $8\dfrac{2}{7} - 3\dfrac{1}{3}$

87. How much larger is $\dfrac{5}{6}$ mile than $\dfrac{3}{8}$ mile?

88. How much larger is $\dfrac{1}{5}$ meter than $\dfrac{1}{7}$ meter?

Problem Solving

89. Height Gain The following graph shows Rebecca Bersagel's height, in inches, on her 8th and 12th birthdays. How much had Rebecca grown in the 4 years?

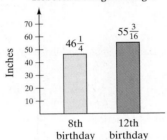

Rebecca Bersagel's Height

90. Road Paving The following graph shows the progress of the Davenport Paving Company in paving the Memorial Highway. How much of the highway was paved from June through August?

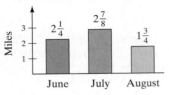

Highway Paved in Selected Months

In many problems you will need to subtract a fraction from 1, where 1 represents "the whole" or the "total amount." Exercises 91–94 are answered by subtracting the given fraction from 1.

91. Putting Success During the 2004 season, $\dfrac{31}{50}$ of all putts from six feet were made by PGA Tour players. What fraction of all putts from six feet were not made?

Source: USA Today

92. Global Warming The probability that an event does not occur may be found by subtracting the probability that the event does occur from 1. If the probability that global warming is occurring is $\dfrac{7}{9}$, find the probability that global warming is not occurring.

93. High School Diploma Use the following circle graph to determine the fraction of people ages 25 and older with no high school diploma who were unemployed in January, 2005.

Employment Status of People Ages 25 and Older with No High School Diploma in January, 2005

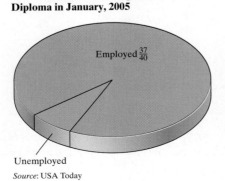

Employed $\frac{37}{40}$

Unemployed

Source: USA Today

94. Home Heating The following circle graph shows the fraction of U.S. homes that used electricity to heat their homes in 2003. Determine the fraction of U.S. homes that did not use electricity.

How Americans Heat Their Homes, 2003

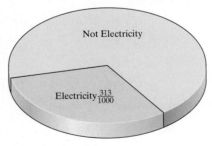

Not Electricity

Electricity $\frac{313}{1000}$

Source: www.census.gov

In Exercises 95–110, answer the questions asked.

95. Albino's Growth An albino python at Cypress Gardens, Florida, was 3 feet, $3\dfrac{1}{4}$ inches when born. Its present length is 15 feet, $2\dfrac{1}{2}$ inches. How much has it grown since birth?

96. Cream Pie A Boston cream pie weighs $1\dfrac{5}{16}$ pounds. If the pie is to be divided equally among 6 people, how much will each person get?

97. Running Denise started a running program in January when she could run a mile in $10\frac{1}{2}$ minutes. After 6 months, Denise could run a mile in $8\frac{2}{5}$ minutes. By how many minutes did Denise improve in 6 months?

98. Baking Turkey The instructions on a turkey indicate that a 12- to 16-pound turkey should bake at 325°F for about 22 minutes per pound. Josephine Nickola is planning to bake a $13\frac{1}{2}$-pound turkey. Approximately how long should the turkey be baked?

99. Wood Cut Debbie Anderson cuts a piece of wood measuring $3\frac{1}{8}$ inches into two equal pieces. How long is each piece?

100. Pants Inseam The inseam on a new pair of pants is 30 inches. If Don O'Neal's inseam is $28\frac{3}{8}$ inches, by how much will the pants need to be shortened?

101. Drug Amount A nurse must give $\frac{1}{16}$ milligram of a drug for each kilogram of patient weight. If Mr. Krisanda weighs (or has a mass of) 80 kilograms, find the amount of the drug Mr. Krisanda should be given.

102. Chopped Onions A recipe for pot roast calls for $\frac{1}{4}$ cup chopped onions for each pound of beef. For $5\frac{1}{2}$ pounds of beef, how many cups of chopped onions are needed?

103. Shampoo A bottle of shampoo contains 15 fluid ounces. If Tierra Bentley uses $\frac{3}{8}$ of an ounce each time she washes her hair, how many times can Tierra wash her hair using this bottle?

104. Fencing Matt Mesaros wants to fence in his backyard as shown. The three sides to be fenced measure $16\frac{2}{3}$ yards, $22\frac{2}{3}$ yards, and $14\frac{1}{8}$ yards.

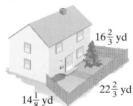

$16\frac{2}{3}$ yd

$22\frac{2}{3}$ yd

$14\frac{1}{8}$ yd

a) How much fence will Matt need?

b) If Matt buys 60 yards of fence, how much will be left over?

105. Windows An insulated window for a house is made up of two pieces of glass, each $\frac{1}{4}$-inch thick, with a 1-inch space between them. What is the total thickness of this window?

106. Truck Weight A flatbed tow truck weighing $4\frac{1}{2}$ tons is carrying two cars. One car weighs $1\frac{1}{6}$ tons, the other weighs $1\frac{3}{4}$ tons. What is the total weight of the tow truck and the two cars?

107. Cutting Wood A 28-inch length of wood is to be cut into $4\frac{2}{3}$ inch strips. How many whole strips can be made? Disregard loss of wood due to cuts made.

108. Fasten Bolts A mechanic wishes to use a bolt to fasten a piece of wood $4\frac{1}{2}$ inches thick to a metal tube $2\frac{1}{3}$ inches thick. If the thickness of the nut is $\frac{1}{8}$ inch, find the length of the shaft of the bolt so that the nut fits flush with the end of the bolt (see the figure below).

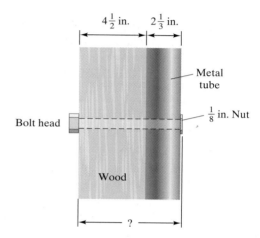

$4\frac{1}{2}$ in. $2\frac{1}{3}$ in.

Metal tube

$\frac{1}{8}$ in. Nut

Bolt head

Wood

?

109. Computer Cabinet Karen Kraatz is considering purchasing a mail order computer. The catalog describes the computer as $7\frac{1}{2}$ inches high and the monitor as $14\frac{3}{8}$ inches high. Karen is hoping to place the monitor on top of the computer and to place the computer and monitor together in the opening where the computer is shown.

$22\frac{1}{2}''$

$26\frac{1}{2}''$

$2\frac{1}{2}''$

$1\frac{1}{4}''$

a) Will there be sufficient room to do this?

b) If so, how much extra space will she have?

c) Find the total height of the computer desk

110. Soda If five 2-liter bottles of soda are split evenly among 30 people, how many liters of soda will each person get?

Challenge Problems

111. Add or subtract the following fractions using the rule discussed in this section. Your answer should be a single fraction, and it should contain the symbols given in the exercise.

a) $\dfrac{*}{a} + \dfrac{?}{a}$

b) $\dfrac{\odot}{?} - \dfrac{\square}{?}$

c) $\dfrac{\triangle}{\square} + \dfrac{4}{\square}$

d) $\dfrac{x}{3} - \dfrac{2}{3}$

e) $\dfrac{12}{x} - \dfrac{4}{x}$

112. Multiply the following fractions using the rule discussed in this section. Your answer should be a single fraction and it should contain the symbols given in the exercise.

a) $\dfrac{\triangle}{a} \cdot \dfrac{\square}{b}$

b) $\dfrac{6}{3} \cdot \dfrac{\triangle}{\square}$

c) $\dfrac{x}{a} \cdot \dfrac{y}{b}$

d) $\dfrac{3}{8} \cdot \dfrac{4}{y}$

e) $\dfrac{3}{x} \cdot \dfrac{x}{y}$

113. Drug Dosage An allopurinal pill comes in 300-milligram doses. Dr. Muechler wants a patient to get 450 milligrams each day by cutting the pills in half and taking $\frac{1}{2}$ pill three times a day. If he wants to prescribe enough pills for a 6-month period (assume 30 days per month), how many pills should he prescribe?

Group Activity

Discuss and answer Exercise 114 as a group.

114. Potatoes The following table gives the amount of each ingredient recommended to make 2, 4, and 8 servings of instant mashed potatoes.

Servings	2	4	8
Water	$\frac{2}{3}$ cup	$1\frac{1}{3}$ cups	$2\frac{2}{3}$ cups
Milk	2 tbsp	$\frac{1}{3}$ cup	$\frac{2}{3}$ cup
Butter*	1 tbsp	2 tbsp	4 tbsp
Salt†	$\frac{1}{4}$ tsp	$\frac{1}{2}$ tsp	1 tsp
Potato flakes	$\frac{2}{3}$ cup	$1\frac{1}{3}$ cup	$2\frac{2}{3}$ cups

*or margarine
†Less salt can be used if desired.

Determine the amount of potato flakes and milk needed to make 6 servings by the different methods described. When working with milk, 16 tbsp = 1 cup.

a) Group member 1: Determine the amount of potato flakes and milk needed to make 6 servings by multiplying the amount for 2 servings by 3.

b) Group member 2: Determine the amounts by adding the amounts for 2 servings to the amounts for 4 servings.

c) Group member 3: Determine the amounts by finding the average (mean) of 4 and 8 servings.

d) As a group, determine the amount by subtracting the amount for 2 servings from the amount for 8 servings.

e) As a group, compare your answers from parts **a)** through **d)**. Are they all the same? If not, can you explain why? (This might be a little tricky.)

Cumulative Review Exercises

[1.1] **115.** What is your instructor's name and office hours?

[1.2] **116.** What is the mean of 9, 8, 15, 32, 16?

117. What is the median of 9, 8, 15, 32, 16?

[1.3] **118.** What are variables?

1.4 The Real Number System

1 Identify sets of numbers.

2 Know the structure of the real numbers.

We will be talking about and using various types of numbers throughout the text. This section introduces you to some of those numbers and to the structure of the real number system.

1 Identify Sets of Numbers

A **set** is a collection of **elements** listed within braces. The set $\{a, b, c, d, e\}$ consists of five elements, namely a, b, c, d, and e. A set that contains no elements is called an **empty set** (or **null set**). The symbols { } or ∅ are used to represent the empty set.

There are many different sets of numbers. Two important sets are the natural numbers and the whole numbers. The whole numbers were introduced earlier.

Natural numbers: $\{1, 2, 3, 4, 5, \dots\}$

Whole numbers: $\{0, 1, 2, 3, 4, 5, \dots\}$

An aid in understanding sets of numbers is the real number line (**Fig. 1.7**).

FIGURE 1.7

The real number line continues indefinitely in both directions. The numbers to the right of 0 are positive and those to the left of 0 are negative. Zero is neither positive nor negative (**Fig. 1.8**).

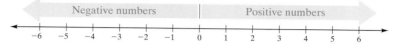

FIGURE 1.8

Figure 1.9 illustrates the natural numbers marked on a number line. The natural numbers are also called the **counting numbers** or the **positive integers**.

FIGURE 1.9

Another important set of numbers is the integers.

$$\text{Integers:} \quad \{\ldots, \underbrace{-5, -4, -3, -2, -1}_{\textit{Negative integers}}, 0, \underbrace{1, 2, 3, 4, 5, \ldots}_{\textit{Positive integers}}\}$$

The integers consist of the negative integers, 0, and the positive integers. The integers are marked on the number line in **Figure 1.10**.

FIGURE 1.10

Can you think of any numbers that are not integers? You probably thought of "fractions" or "decimal numbers." Fractions and certain decimal numbers belong to the set of rational numbers. The set of **rational numbers** consists of all the numbers that can be expressed as a quotient (or a ratio) of two integers, with the denominator not 0.

Rational numbers: {quotient of two integers, denominator not 0}

All integers are rational numbers since they can be written with a denominator of 1. For example, $3 = \dfrac{3}{1}$, $-12 = \dfrac{-12}{1}$, and $0 = \dfrac{0}{1}$. All fractions containing integers in the numerator and denominator (with the denominator not 0) are rational numbers. For example, the fraction $\dfrac{3}{5}$ is a quotient of two integers and is a rational number.

When a fraction that is a ratio of two integers is converted to a decimal number by dividing the numerator by the denominator, the quotient will always be either a *terminating decimal number*, such as 0.3 and 3.25, or a *repeating decimal number* such as 0.3333 . . . and 5.2727 . . . . The three dots at the end of a number like 0.3333 . . . indicate that the numbers continue to repeat in the same manner indefinitely. All terminating decimal numbers and all repeating decimal numbers are rational numbers which can be expressed as a quotient of two integers. For example, $0.3 = \dfrac{3}{10}$, $3.25 = \dfrac{325}{100}$, $0.3333 \ldots = \dfrac{1}{3}$ and $5.2727 \ldots = \dfrac{522}{99}$. Some rational numbers are illustrated on the number line in **Figure 1.11**.

FIGURE 1.11

Most of the numbers that we use are rational numbers. However, some numbers are not rational. Numbers such as the square root of 2, written $\sqrt{2}$, are not rational numbers. Any number that can be represented on the number line that is not a rational number is called an **irrational number**. Irrational numbers are non-terminating, non-repeating decimal numbers. For example, $\sqrt{2}$ cannot be expressed exactly as a decimal number. Irrational numbers can only be *approximated* by decimal numbers. $\sqrt{2}$ is *approximately* 1.41. Thus, we may write $\sqrt{2} \approx 1.41$. Some irrational numbers are illustrated on the number line in **Figure 1.12**. Rational and irrational numbers will be discussed further in later chapters.

FIGURE 1.12

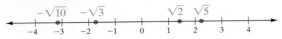

2 Know the Structure of the Real Numbers

Notice that many different types of numbers can be illustrated on a number line. Any number that can be represented on the number line is a **real number**.

Real numbers: {all numbers that can be represented on a real number line}

The symbol $\mathbb{R}$ is used to represent the set of real numbers. All the numbers mentioned thus far are real numbers. The natural numbers, the whole numbers, the integers, the rational numbers, and the irrational numbers are all real numbers. There are some types of numbers that are not real numbers, but these numbers are not discussed in this chapter. **Figure 1.13** illustrates the relationships between the various sets of numbers within the set of real numbers.

In **Figure 1.13a**, we can see that when we combine the rational numbers and the irrational numbers we get the real numbers. When we combine the integers with the noninteger rational numbers (such as $\frac{1}{2}$ and 0.42), we get the rational numbers. When we combine the whole numbers and the negative integers, we get the integers.

Consider the natural number 5. If we follow the branches in **Figure 1.13a** upward, we see that the number 5 is also a whole number, an integer, a rational number, and a real number. Now consider the number $\frac{1}{2}$. It belongs to the noninteger rational numbers. If we follow the branches upward, we can see that $\frac{1}{2}$ is also a rational number and a real number.

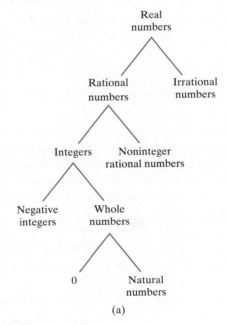

(a)

Real Numbers	
Rational numbers	Irrational numbers
(Integers and noninteger rational numbers)	(Certain* square roots and other special numbers)
-12 4 0	$\sqrt{2}$ $\sqrt{5}$
$\frac{3}{8}$ $\frac{1}{3}$ -1.24	π $\sqrt{12}$
$-1\frac{3}{5}$ -2.463	

*Other higher roots like $\sqrt[3]{2}$ and $\sqrt[4]{5}$ are also irrational numbers.

(b)

FIGURE 1.13

EXAMPLE 1 ▸ Consider the following set of numbers.

$$\left\{\sqrt{11}, -\frac{3}{8}, -0.6, -4, 7\frac{1}{2}, 12, 0, \sqrt{5}, -48, 3.9, -\sqrt{7}\right\}$$

List the elements of the set that are

a) natural numbers.   **b)** whole numbers.   **c)** integers.

d) rational numbers.   **e)** irrational numbers.   **f)** real numbers.

Solution   We will list the elements from left to right as they appear in the set. However, the elements may be listed in any order.

a) 12   **b)** 12, 0   **c)** −4, 12, 0, −48

d) $-\frac{3}{8}, -0.6, -4, 7\frac{1}{2}, 12, 0, -48, 3.9$ **e)** $\sqrt{11}, \sqrt{5}, -\sqrt{7}$

f) $\sqrt{11}, -\frac{3}{8}, -0.6, -4, 7\frac{1}{2}, 12, 0, \sqrt{5}, -48, 3.9, -\sqrt{7}$

▸ **Now Try Exercise 51**

EXERCISE SET 1.4   *Math XL* **MyMathLab**
MathXL®   MyMathLab

Concept/Writing Exercises

1. What is a set?

2. What is a set that contains no elements called?

3. Describe a set that is an empty set.

4. Give two symbols used to represent the empty set.

5. How do the sets of natural numbers and whole numbers differ?

6. What are two other names for the set of natural numbers?

7. Explain why the natural number 7 is also a
    a) whole number.
    b) rational number.
    c) real number.

8. **a)** What is a rational number?
    b) Explain why every integer is a rational number.

9. Is 0 a member of the set of
    a) integers?
    b) positive integers?
    c) negative integers?
    d) rational numbers?

10. Write a paragraph or two describing the structure of the real number system. Explain how whole numbers, counting numbers, integers, rational numbers, irrational numbers, and real numbers are related.

Practice the Skills

In Exercises 11–15, list each set of numbers.

11. Integers

12. Counting numbers

13. Whole numbers

14. Positive integers

15. Negative integers

16. Natural numbers

In Exercises 17–48, indicate whether each statement is true or false.

17. 0 is a whole number.

18. −1 is a negative integer.

19. −7.3 is a real number.

20. $\frac{3}{5}$ is an integer.

21. 0.6 is an integer.

22. 0 is a natural number.

23. $\sqrt{2}$ is a rational number.

24. $\sqrt{3}$ is a real number.

25. $-\frac{1}{5}$ is a rational number.

26. $-2\frac{1}{3}$ is a rational number.

27. 0 is a rational number.

28. 9.2 is a rational number.

29. $4\frac{5}{8}$ is an irrational number.

30. 0 is not a positive number.

31. $-\sqrt{5}$ is an irrational number.

32. Every counting number is a rational number.

33. The symbol $\varnothing$ is used to represent the empty set.

34. Every integer is negative.

35. Every real number is a rational number.

36. Every negative integer is a real number.

37. Every rational number is a real number.

38. When zero is added to the set of counting numbers, the set of whole numbers is formed.

39. Some real numbers are not rational numbers.

40. Some irrational numbers are not real numbers.

41. Every negative number is a negative integer.

42. All real numbers can be represented on a number line.

43. The symbol $\mathbb{R}$ is used to represent the set of real numbers.

44. Any number to the left of zero on a number line is a negative number.

45. Every number greater than zero is a positive integer.

46. Irrational numbers cannot be represented on a number line.

47. When the negative integers, the positive integers, and 0 are combined, the integers are formed.

48. The natural numbers, counting numbers, and positive integers are different names for the same set of numbers.

49. **Hotels** Some of the older hotels in Europe have elevators that list negative numbers for floors below the lobby level. For example, a floor might be designated as −2. In the United States and in many countries, floor number 13 is omitted because of superstition. Considering the numbers −2 and 13, list those that are

a) positive integers.

b) rational numbers.

c) real numbers.

d) whole numbers.

50. **Address** We generally think of house addresses as being integers greater than 0. Have you ever seen a house number that was not an integer greater than 0? There are quite a few

such addresses in cities and towns across America. The house numbers 0 and $2\frac{1}{2}$ appear on Legare Street in Charleston, SC. Considering the numbers 0 and $2\frac{1}{2}$, list those that are

a) integers

b) rational numbers

c) real numbers

51. Consider the following set of numbers.

$$\left\{-\frac{5}{7}, 0, -2, 3, 6\frac{1}{4}, \sqrt{7}, -\sqrt{3}, 1.63, 77\right\}$$

List the numbers that are

a) positive integers.

b) whole numbers.

c) integers.

d) rational numbers.

e) irrational numbers.

f) real numbers.

52. Consider the following set of numbers.

$$\left\{-6, 7, 12.4, -\frac{9}{5}, -2\frac{1}{4}, \sqrt{3}, 0, 9, \sqrt{7}, 0.35\right\}$$

List the numbers that are

a) positive integers.

b) whole numbers.

c) integers.

d) rational numbers.

e) irrational numbers.

f) real numbers.

Problem Solving

Give three examples of numbers that satisfy the given conditions.

53. An integer but not a negative integer.

54. A real number but not an integer.

55. An irrational number and a negative number.

56. A real number and a rational number.

57. A rational number but not a natural number.

58. An integer and a rational number.

59. A negative integer and a rational number.

60. A negative integer and a real number.

61. A real number but not a positive rational number.

62. A rational number but not a negative number.

63. A real number but not an irrational number.

64. A negative number but not a negative integer.

Three dots inside a set indicate that the set continues in the same manner. For example, {1, 2, 3, . . . , 84} is the set of natural numbers from 1 up to and including 84. In Exercises 65 and 66, determine the number of elements in each set.

65. {8, 9, 10, 11, . . . , 94}

66. {−4, −3, −2, −1, 0, 1, . . . , 64}

Challenge Problems

The diagrams in Exercises 67 and 68 are called **Venn diagrams** *(named after the English mathematician John Venn). Venn diagrams are used to illustrate sets. For example, in the diagrams, circle A contains all the elements in set A, and circle B contains all the elements in set B. For each diagram, determine a) set A, b) set B, c) the set of elements that belong to both set A and set B, and d) the set of elements that belong to either set A or set B.*

67.

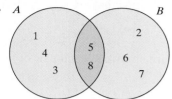

68.

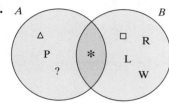

69. Consider the sets $A = \{1, 2, 3, 4\}$ and $B = \{1, 2, 3, 4, \ldots\}$.

    a) Explain the difference between set A and set B.

    b) How many elements are in set A?

    c) How many elements are in set B?

    d) Set A is an example of a *finite set*. Can you guess the name given to a set like set B?

70. How many decimal numbers are there

    a) between 1.0 and 2.0;

    b) between 1.4 and 1.5? Explain your answer.

71. How many fractions are there

    a) between 1 and 2;

    b) between $\frac{1}{3}$ and $\frac{1}{5}$? Explain your answer.

Group Activity

Discuss and answer Exercise 72 as a group.

72. Set A **union** set B, symbolized $A \cup B$, consists of the set of elements that belong to set A or set B (or both sets). Set A **intersection** set B, symbolized $A \cap B$, consists of the set of elements that both set A and set B have in common. Note that the elements that belong to both sets are listed only once in the union of the sets.
Consider the pairs of sets below.

       Group member 1: $A = \{2, 3, 4, 6, 8, 9\}$ $B = \{1, 2, 3, 5, 7, 8\}$

       Group member 2: $A = \{a, b, c, d, g, i, j\}$ $B = \{b, c, d, h, m, p\}$

       Group member 3: $A = \{red, blue, green, yellow\}$ $B = \{pink, orange, purple\}$

   a) Group member 1: Find the union and intersection of the sets marked Group member 1.

   b) Group member 2: Find the union and intersection of the sets marked Group member 2.

   c) Group member 3: Find the union and intersection of the sets marked Group member 3.

   d) Now as a group, check each other's work. Correct any mistakes.

   e) As a group, using group member 1's sets, construct a Venn diagram like those shown in Exercises 67 and 68.

Cumulative Review Exercises

[1.3]   **73.** Convert $5\frac{2}{5}$ to a fraction.

       74. Write $\frac{16}{3}$ as a mixed number.

75. Subtract $\frac{7}{8} - \frac{2}{3}$.

76. Divide $\frac{3}{5} \div 6\frac{3}{4}$.

1.5 Inequalities

1 Determine which is the greater of two numbers.

2 Find the absolute value of a number.

1 Determine Which Is the Greater of Two Numbers

The number line, which shows numbers increasing from left to right, can be used to explain inequalities (see **Fig. 1.14**). When comparing two numbers, **the number to the right on the number line is the greater number, and the number to the left is the lesser number**. The symbol $>$ is used to represent the words "is greater than." The symbol $<$ is used to represent the words "is less than."

FIGURE 1.14

The statement that the number 3 is greater than the number 2 is written $3 > 2$. Notice that 3 is to the right of 2 on the number line in **Figure 1.14** on page 35. The statement that the number 0 is greater than the number -1 is written $0 > -1$. Notice that 0 is to the right of -1 on the number line.

Instead of stating that 3 is greater than 2, we could state that 2 is less than 3, written $2 < 3$. Notice that 2 is to the left of 3 on the number line. The statement that the number -1 is less than the number 0 is written $-1 < 0$. Notice that -1 is to the left of 0 on the number line.

EXAMPLE 1 ▶ Insert either $>$ or $<$ in the shaded area between each pair of numbers to make a true statement.

a) -4 ▦ -2 **b)** $-\dfrac{3}{2}$ ▦ 2.5 **c)** $\dfrac{1}{2}$ ▦ $\dfrac{1}{4}$ **d)** -2 ▦ 4

Solution   The points given are shown on the number line (**Fig. 1.15**).

FIGURE 1.15

a) $-4 < -2$; notice that -4 is to the left of -2.

b) $-\dfrac{3}{2} < 2.5$; notice that $-\dfrac{3}{2}$ is to the left of 2.5.

c) $\dfrac{1}{2} > \dfrac{1}{4}$; notice that $\dfrac{1}{2}$ is to the right of $\dfrac{1}{4}$.

d) $-2 < 4$; notice that -2 is to the left of 4.

▶ **Now Try Exercise 35**

EXAMPLE 2 ▶ Insert either $>$ or $<$ in the shaded area between each pair of numbers to make a true statement.

a) -3 ▦ 3 **b)** -3 ▦ -4 **c)** -4 ▦ 0 **d)** -1.08 ▦ -1.8

Solution   The numbers given are shown on the number line (**Fig. 1.16**).

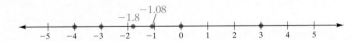

FIGURE 1.16

a) $-3 < 3$; notice that -3 is to the left of 3.

b) $-3 > -4$; notice that -3 is to the right of -4.

c) $-4 < 0$; notice that -4 is to the left of 0.

d) $-1.08 > -1.8$; notice that -1.08 is to the right of -1.8.

▶ **Now Try Exercise 43**

2 Find the Absolute Value of a Number

The concept of absolute value can be explained with the help of the number line shown in **Figure 1.17** on page 37. The **absolute value** of a number can be considered the distance between the number and 0 on a number line. Thus, the absolute value of 3, written $|3|$, is 3 since it is 3 units from 0 on a number line. Similarly, the absolute value of -3, written $|-3|$, is also 3 since -3 is 3 units from 0.

$$|3| = 3 \quad \text{and} \quad |-3| = 3$$

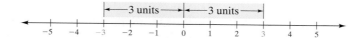

FIGURE 1.17

Since the absolute value of a number measures the distance (without regard to direction) of a number from 0 on the number line, *the absolute value of every number will be either positive or zero.*

Number	Absolute Value of Number
6	$\lvert 6 \rvert = 6$
-6	$\lvert -6 \rvert = 6$
0	$\lvert 0 \rvert = 0$
$-\dfrac{1}{2}$	$\left\lvert -\dfrac{1}{2} \right\rvert = \dfrac{1}{2}$

The negative of the absolute value of a nonzero number will always be a negative number. For example,

$$-\lvert 2 \rvert = -(2) = -2 \quad \text{and} \quad -\lvert -3 \rvert = -(3) = -3$$

EXAMPLE 3 ▶ Insert either $>$, $<$, or $=$ in each shaded area to make a true statement.

a) $\lvert 3 \rvert$ ▨ 3 **b)** $\lvert -2 \rvert$ ▨ $\lvert 2 \rvert$ **c)** -2 ▨ $\lvert -4 \rvert$

d) $\lvert -2 \rvert$ ▨ $\lvert -4 \rvert$ **e)** $\left\lvert -\dfrac{2}{5} \right\rvert$ ▨ $\lvert -0.42 \rvert$

Solution

a) $\lvert 3 \rvert = 3$.   **b)** $\lvert -2 \rvert = \lvert 2 \rvert$, since both $\lvert -2 \rvert$ and $\lvert 2 \rvert$ equal 2.

c) $-2 < \lvert -4 \rvert$, since $\lvert -4 \rvert = 4$.

d) $\lvert -2 \rvert < \lvert -4 \rvert$, since $\lvert -2 \rvert = 2$ and $\lvert -4 \rvert = 4$.

e) When an absolute value contains a fraction, we can compare it to an absolute value containing a decimal number by rewriting the fraction as a decimal number and comparing the absolute values of the decimal numbers.

$$\left\lvert -\frac{2}{5} \right\rvert = \lvert -0.40 \rvert = 0.40$$

$$\lvert -0.42 \rvert = 0.42$$

Therefore, $\left\lvert -\dfrac{2}{5} \right\rvert < \lvert -0.42 \rvert$ because $0.40 < 0.42$.

▶ **Now Try Exercise 57**

We will use absolute value in Sections 1.6 and 1.7 to add and subtract real numbers. The concept of absolute value is very important in higher-level mathematics courses. If you take a course in intermediate algebra, you will learn a more formal definition of absolute value.

EXERCISE SET 1.5

Math XL   MyMathLab
MathXL®   MyMathLab

Concept/Writing Exercises

1. a) Draw a number line.

  b) Mark the numbers -2 and -4 on your number line.

  c) Is -2 less than -4 or is -2 greater than -4? Explain.

  In parts **d)** and **e)** write a correct statement using -2 and -4 and the symbol given.

  d) $<$

  e) $>$

2. What is the absolute value of a number?

3. a) Explain why the absolute value of 4, $|4|$, is 4.

  b) Explain why the absolute value of -4, $|-4|$, is 4.

  c) Explain why the absolute value of 0, $|0|$, is 0.

4. Are there any real numbers whose absolute value is not a positive number? Explain your answer.

5. Suppose a and b represent any two real numbers. If $a > b$ is true, will $b < a$ also be true? Explain and give some examples using specific numbers for a and b.

6. Will $|a| - |a| = 0$ always be true for any real number a? Explain.

7. Suppose a and b represent any two real numbers. Suppose $a < b$ is true. Will $|a| < |b|$ also be true? Explain and give an example to support your answer.

8. Suppose a and b represent any two real numbers. Suppose $a > b$ is true. Will $|a| > |b|$ also be true? Explain and give an example to support your answer.

9. Suppose a and b represent any two real numbers. Suppose $|a| < |b|$ is true. Will $a < b$ also be true? Explain and give an example to support your answer.

10. Suppose a and b represent any two real numbers. Suppose $|a| > |b|$ is true. Will $a > b$ also be true? Explain and give an example to support your answer.

Practice the Skills

Evaluate.

11. $|7|$

12. $|-6|$

13. $|-15|$

14. $|0|$

15. $-|0|$

16. $|54|$

17. $-|-5|$

18. $-|92|$

19. $-|21|$

20. $-|-34|$

Insert either $<$ or $>$ in each shaded area to make a true statement.

21. $6 \quad\blacksquare\quad 2$

22. $4 \quad\blacksquare\quad -2$

23. $-4 \quad\blacksquare\quad 0$

24. $-6 \quad\blacksquare\quad -4$

25. $\dfrac{1}{2} \quad\blacksquare\quad -\dfrac{2}{3}$

26. $\dfrac{3}{5} \quad\blacksquare\quad \dfrac{4}{5}$

27. $0.7 \quad\blacksquare\quad 0.8$

28. $-0.2 \quad\blacksquare\quad -0.4$

29. $-\dfrac{1}{2} \quad\blacksquare\quad -1$

30. $-0.1 \quad\blacksquare\quad -0.9$

31. $-5 \quad\blacksquare\quad 5$

32. $-\dfrac{3}{4} \quad\blacksquare\quad -1$

33. $-2.1 \quad\blacksquare\quad -2$

34. $-1.83 \quad\blacksquare\quad -1.82$

35. $\dfrac{4}{5} \quad\blacksquare\quad -\dfrac{4}{5}$

36. $-9 \quad\blacksquare\quad -12$

37. $-\dfrac{3}{8} \quad\blacksquare\quad \dfrac{3}{8}$

38. $-4.09 \quad\blacksquare\quad -5.3$

39. $0.49 \quad\blacksquare\quad 0.43$

40. $-1.0 \quad\blacksquare\quad -0.7$

41. $-0.086 \quad\blacksquare\quad -0.095$

42. $\dfrac{1}{2} \quad\blacksquare\quad -\dfrac{1}{2}$

43. $0.001 \quad\blacksquare\quad 0.002$

44. $-0.006 \quad\blacksquare\quad -0.007$

45. $\dfrac{5}{8} \quad\blacksquare\quad 0.6$

46. $2.7 \quad\blacksquare\quad \dfrac{10}{3}$

47. $-\dfrac{4}{3} \quad\blacksquare\quad -\dfrac{1}{3}$

48. $\dfrac{9}{2} \quad\blacksquare\quad \dfrac{7}{2}$

49. $-0.8 \quad\blacksquare\quad -\dfrac{3}{5}$

50. $-0.7 \quad\blacksquare\quad -0.2$

51. $0.3 \quad\blacksquare\quad \dfrac{1}{3}$

52. $\dfrac{9}{20} \quad\blacksquare\quad 0.42$

53. $-\dfrac{17}{30} \quad\blacksquare\quad -\dfrac{16}{20}$

54. $\dfrac{13}{15} \quad\blacksquare\quad \dfrac{8}{9}$

55. $-(-6) \quad\blacksquare\quad -(-5)$

56. $-\left(-\dfrac{12}{13}\right) \quad\blacksquare\quad \dfrac{7}{8}$

Insert either $<$, $>$, or $=$ in each shaded area to make a true statement.

57. $5 \quad\blacksquare\quad |-2|$

58. $|-12| \quad\blacksquare\quad |-13|$

59. $\dfrac{3}{4} \quad\blacksquare\quad |-4|$

60. $|-4| \quad\blacksquare\quad -3$

61. $|0| \quad\blacksquare\quad |-4|$

62. $|-2.1| \quad\blacksquare\quad |-1.8|$

63. $4 \quad\blacksquare\quad \left|-\dfrac{9}{2}\right|$

64. $|-5| \quad\blacksquare\quad -|-6|$

65. $\left|-\dfrac{4}{5}\right| \quad\blacksquare\quad \left|-\dfrac{5}{4}\right|$

66. $\left|\dfrac{2}{5}\right| \quad\blacksquare\quad |-0.40|$

67. $|-4.6| \quad\blacksquare\quad \left|-\dfrac{23}{5}\right|$

68. $\left|-\dfrac{8}{3}\right| \quad\blacksquare\quad |-3.5|$

Insert either >, <, *or* = *in each shaded area to make a true statement.*

69. $\frac{2}{3}+\frac{2}{3}+\frac{2}{3}+\frac{2}{3}$ ▨ $4\cdot\frac{2}{3}$

70. $\frac{3}{4}+\frac{3}{4}$ ▨ $\frac{3}{4}\cdot\frac{3}{4}$

71. $\frac{1}{2}\cdot\frac{1}{2}$ ▨ $\frac{1}{2}\div\frac{1}{2}$

72. $5\div\frac{2}{3}$ ▨ $\frac{2}{3}\div 5$

73. $\frac{5}{8}-\frac{1}{2}$ ▨ $\frac{5}{8}\div\frac{1}{2}$

74. $3\frac{1}{5}+\frac{1}{3}$ ▨ $3\frac{1}{5}\cdot\frac{1}{3}$

Arrange the numbers from smallest to largest.

75. $0.46, \frac{4}{9}, |-5|, -|-1|, \frac{3}{7}$

76. $-\frac{3}{4}, -|0.6|, -\frac{5}{9}, -1.74, |-1.9|$

77. $\frac{2}{3}, 0.6, |-2.6|, \frac{19}{25}, \frac{5}{12}$

78. $-|-5|, |-9|, \left|-\frac{12}{5}\right|, 2.7, \frac{7}{12}$

Problem Solving

79. What numbers are 4 units from 0 on a number line?

80. What numbers are 100 units from 0 on the number line?

In Exercises 81–88, give three real numbers that satisfy all the stated criteria. If no real numbers satisfy the criteria, so state and explain why.

81. less than 4 and greater than 6
82. greater than 4 and less than 6
83. less than −2 and greater than −6
84. greater than −5 and greater than −9
85. greater than −3 and greater than 3
86. less than −3 and less than 3
87. greater than $|-2|$ and less than $|-6|$
88. greater than $|-3|$ and less than $|3|$
89. a) Consider the word *between*. What does this word mean?
   b) List three real numbers between 4 and 6
   c) Is the number 4 between the numbers 4 and 6? Explain.
   d) Is the number 5 between the numbers 4 and 6? Explain.
   e) Is it true or false that the real numbers between 4 and 6 are the real numbers that are both greater than 4 and less than 6? Explain.
90. Property Crime Rates The following line graph shows the U.S. property crime rates per 1000 households from 1973 to 2003.

Propery Crime Rates

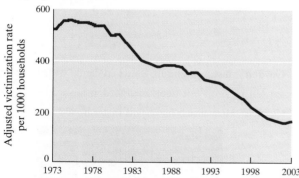

Note: Property crimes include burglary, theft, and motor vehicle theft.

Source: www.ojp.usdoj.gov

Estimate the year(s) when the property crime rate was
a) first less than 400 per 1000 households.
b) first less than 200 per 1000 households.
c) greater than 400 per 1000 households and less than 600 per 1000 households.

91. Peanuts The following bar graph shows the percent of the recommended daily allowance of certain nutrients in 1 ounce of dry-roasted, salted peanuts. Use the bar graph to determine which nutrients provide

Nutrients in Peanuts

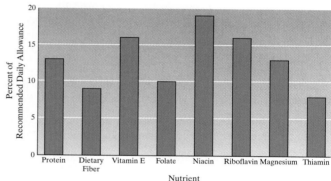

Source: Nutritional Wellness

a) less than 10% of the recommended daily allowance.
b) greater than 15% of the recommended daily allowance.

Challenge Problems

92. A number greater than 0 and less than 1 (or between 0 and 1) is multiplied by itself. Will the product be less than, equal to, or greater than the original number selected? Explain why this is always true.

93. A number between 0 and 1 is divided by itself. Will the quotient be less than, equal to, or greater than the original number selected? Explain why this is always true.

94. What two numbers can be substituted for x to make $|x| = 3$ a true statement?

95. Are there any values for x that would make $|x| = -3$ a true statement? Explain.

96. a) To what is $|x|$ equal if x represents a real number greater than or equal to 0?

b) To what is $|x|$ equal if x represents a real number less than 0?

c) Fill in the following shaded areas to make a true statement.

$$|x| = \begin{cases} \rule{0.5cm}{0.3cm} & , x \geq 0 \\ \rule{0.5cm}{0.3cm} & , x < 0 \end{cases}$$

Group Activity

Discuss and answer Exercise 97 as a group.

97. a) Group member 1: Draw a number line and mark points on the line to represent the following numbers.

$$|-2|, \quad -|3|, \quad -\left|\frac{1}{3}\right|$$

b) Group member 2: Do the same as in part **a)**, but on your number line mark points for the following numbers.

$$|-4|, \quad -|2|, \quad \left|-\frac{3}{5}\right|$$

c) Group member 3: Do the same as in parts **a)** and **b)**, but mark points for the following numbers.

$$|0|, \quad \left|\frac{16}{5}\right|, \quad -|-3|$$

d) As a group, construct one number line that contains all the points listed in parts **a)**, **b)**, and **c)**.

Cumulative Review Exercises

[1.3]   **98.** Add $2\frac{3}{5} + 3\frac{1}{3}$.

[1.4]   **99.** List the set of integers.

   100. List the set of whole numbers.

   101. Consider the following set of numbers.

$$\{5, -2, 0, \frac{1}{3}, \sqrt{3}, -\frac{5}{9}, 2.3\}$$

List the numbers that are

a) natural numbers.

b) whole numbers.

c) integers.

d) rational numbers.

e) irrational numbers.

f) real numbers.

Mid-Chapter Test: 1.1–1.5

To find out how well you understand the chapter material to this point, take this brief test. The answers, and the section where the material was initially discussed, are given in the back of the book. Review any questions that you answered incorrectly.

1. For each hour of class time, how many hours outside of class are recommended for studying and doing homework?

2. Phone Bills   Jason Wisely's monthly phone bills for the first six months of 2005 were $78.83, $96.57, $62.23, $88.79, $101.75, and $55.62. For Jason's phone bills, determine the **a)** mean and **b)** median.

3. Checking Account   The balance in Elizabeth Mater's checking account is $652.70. She deposited $230.75 and then purchased three books at $19.62 each, including tax. If she pays by check, what is the new balance in her checking account?

4. Boat Rental   The rental cost of a boat from Natwora's Boat Rental is $7.50 per 15 minutes, and the rental cost from Gurney's Boat Rental is $18 per half hour. Suppose you plan to rent a boat for 4 hours.

a) Which is the better deal?

b) How much will you save?

5. Water Rate   The water rate in Livingston County is $1.85 per 1000 gallons of water used. What is the water bill for a resident of Livingston County who uses 33,700 gallons of water?

Perform the indicated operation.

6. $\frac{3}{7} \cdot \frac{7}{18}$

7. $\frac{9}{16} \div \frac{12}{13}$

8. $\frac{5}{8} + \frac{3}{5}$

9. $6\frac{1}{4} - 3\frac{1}{5}$

10. Garden   Justin Calhoun wants to fence in his rectangular garden to keep out deer. His garden measures $14\frac{2}{3}$ feet by $12\frac{1}{2}$ feet. How much fence will Justin need?

In Exercises 11–15, indicate whether each statement is true or false.

11. 0 is a natural number.

12. −8.6 is a real number.

13. $3\frac{2}{3}$ is an irrational number.

14. Every integer is a real number.

15. Every whole number is a counting number.

16. Evaluate $-\left|-\dfrac{7}{10}\right|$.

Insert either $<$, $>$, or $=$ in each shaded area to make a true statement.

17. $-0.005 \quad\blacksquare\quad -0.006$

18. $\dfrac{7}{8} \quad\blacksquare\quad \dfrac{5}{6}$

19. $|-9| \quad\blacksquare\quad |-12|$

20. $\left|-\dfrac{3}{8}\right| \quad\blacksquare\quad |-0.375|$

1.6 Addition of Real Numbers

1. Add real numbers using a number line.

2. Add fractions.

3. Identify opposites.

4. Add using absolute values.

5. Add using calculators.

There are many practical uses for negative numbers. A submarine diving below sea level, a bank account that has been overdrawn, a business spending more than it earns, and a temperature below zero are some examples.

The four basic **operations** of arithmetic are addition, subtraction, multiplication, and division. In the next few sections, we will explain how to add, subtract, multiply, and divide numbers. We will consider both positive and negative numbers. In this section, we discuss the operation of addition.

1 Add Real Numbers Using a Number Line

To add numbers, we make use of a number line. Represent the first number to be added (first *addend*) by an arrow starting at 0. The arrow is drawn to the right if the number is positive. If the number is negative, the arrow is drawn to the left. From the tip of the first arrow, draw a second arrow to represent the second addend. The second arrow is drawn to the right or left, as just explained. The sum of the two numbers is found at the tip of the second arrow. Note that with the exception of 0, *any number without a sign in front of it is positive*. For example, 3 means +3 and 5 means +5.

EXAMPLE 1 ▶ Evaluate $3 + (-4)$ using a number line.

Solution   *Always begin at 0.* Since the first addend, the 3, is positive, the first arrow starts at 0 and is drawn 3 units to the right (**Fig. 1.18**).

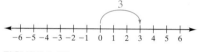

FIGURE 1.18

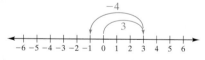

FIGURE 1.19

The second arrow starts at 3 and is drawn 4 units to the left, since the second addend is negative (**Fig. 1.19**). The tip of the second arrow is at −1. Thus,

$$3 + (-4) = -1$$

▶ **Now Try Exercise 27**

EXAMPLE 2 ▶ Evaluate $-4 + 2$ using a number line.

Solution   Begin at 0. Since the first addend is negative, −4, the first arrow is drawn 4 units to the left. From there, since 2 is positive, the second arrow is drawn 2 units to the right. The second arrow ends at −2 (**Fig. 1.20**).

FIGURE 1.20

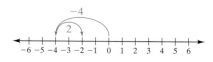

$$-4 + 2 = -2$$

▶ **Now Try Exercise 37**

EXAMPLE 3 ▶ Evaluate $-3 + (-2)$ using a number line.

Solution   Start at 0. Since both numbers being added are negative, both arrows will be drawn to the left (**Fig. 1.21**).

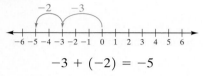

FIGURE 1.21

$$-3 + (-2) = -5$$

▶ **Now Try Exercise 39**

In Example 3, we can think of the expression $-3 + (-2)$ as combining a *loss* of 3 and a *loss* of 2 for a total *loss* of 5, or -5.

EXAMPLE 4 ▶ Add $5 + (-5)$ using a number line.

Solution   The first arrow starts at 0 and is drawn 5 units to the right. The second arrow starts at 5 and is drawn 5 units to the left. The tip of the second arrow is at 0. Thus, $5 + (-5) = 0$ (**Fig. 1.22**).

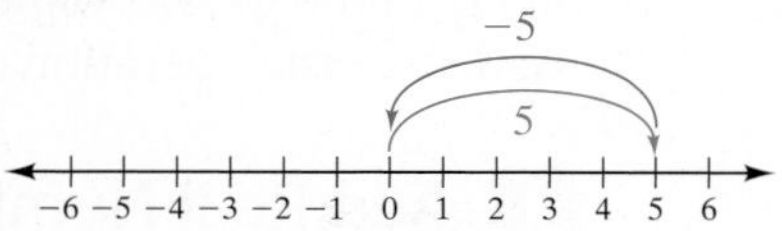

FIGURE 1.22

$$5 + (-5) = 0$$

▶ **Now Try Exercise 31**

EXAMPLE 5 ▶ **Below-Zero Temperatures** At the beginning of the five o'clock news broadcast on a winter day in Rochester, New York, the chief meteorologist reported that the temperature was three degrees below zero Fahrenheit. During the weather segment twenty minutes later, the chief meteorologist stated that the temperature had dropped four degrees since the beginning of the news broadcast. Find the temperature at 5:20 P.M.

Solution   A vertical number line (**Fig. 1.23**) may help you visualize this problem.

$$-3 + (-4) = -7°F$$

▶ **Now Try Exercise 117**

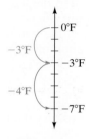

FIGURE 1.23

2 Add Fractions

We introduced fractions, and explained how to add positive fractions, in Section 1.3. To add fractions, where one or more of the fractions is negative, we use the same general procedure discussed in Section 1.3. Whenever the denominators are not the same, we must obtain a common denominator. Once we obtain the common denominator, we obtain the answer by adding the numerators while keeping the common denominator.

For example, suppose after obtaining a common denominator, we have $-\dfrac{19}{29} + \dfrac{13}{29}$.

To obtain the numerator of the answer, we may add $-19 + 13$ on the number line to obtain -6. The denominator of the answer is the common denominator, 29. Thus, the answer is $-\dfrac{6}{29}$. We show these calculations as follows.

$$-\frac{19}{29} + \frac{13}{29} = \frac{-19 + 13}{29} = \frac{-6}{29} = -\frac{6}{29}$$

Let's look at one more example. Suppose after obtaining a common denominator, we have $-\dfrac{7}{40} + \left(-\dfrac{5}{40}\right)$. To obtain the numerator of the answer, we may add $-7 + (-5)$ on the number line to obtain -12. The denominator of the answer is 40.

Thus, the answer before being simplified is $-\dfrac{12}{40}$. The final answer simplifies to $-\dfrac{3}{10}$. We show these calculations as follows.

$$-\frac{7}{40} + \left(-\frac{5}{40}\right) = \frac{-7 + (-5)}{40} = \frac{-12}{40} = -\frac{3}{10}$$

EXAMPLE 6 ▶ Add $\dfrac{7}{16} + \left(-\dfrac{2}{3}\right)$.

Solution   First, we determine that the least common denominator is 48. Changing each fraction to a fraction with a denominator of 48 yields

$$\frac{7}{16} \cdot \frac{3}{3} + \left(-\frac{2}{3}\right) \cdot \frac{16}{16}$$

or $\dfrac{21}{48} + \left(-\dfrac{32}{48}\right)$

Now the fractions being added have a common denominator. To add these fractions, we keep the common denominator and add the numerators to get

$$\frac{7}{16} + \left(-\frac{2}{3}\right) = \frac{21}{48} + \left(-\frac{32}{48}\right) = \frac{21 + (-32)}{48}$$

Now we add $21 + (-32)$ on a number line to get the numerator of the fraction, -11; see **Figure 1.24**.

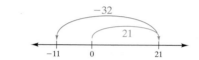

FIGURE 1.24

Thus, $\dfrac{7}{16} + \left(-\dfrac{2}{3}\right) = \dfrac{21}{48} + \left(-\dfrac{32}{48}\right) = \dfrac{21 + (-32)}{48} = -\dfrac{11}{48}$.

▶ **Now Try Exercise 77**

EXAMPLE 7 ▶ Add $-\dfrac{7}{8} + \left(-\dfrac{3}{40}\right)$.

Solution   The least common denominator (or LCD) is 40. Rewriting the first fraction with the LCD gives the following.

$$-\frac{7}{8} + \left(-\frac{3}{40}\right) = -\frac{7}{8} \cdot \frac{5}{5} + \left(-\frac{3}{40}\right)$$

$$= -\frac{35}{40} + \left(-\frac{3}{40}\right) = \frac{-35 + (-3)}{40}$$

Now we add $-35 + (-3)$ to get the numerator of the fraction, -38; see **Figure 1.25**.

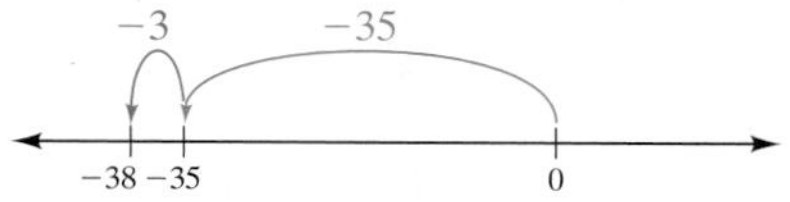

FIGURE 1.25

Thus, $-\dfrac{7}{8} + \left(-\dfrac{3}{40}\right) = \dfrac{-35}{40} + \left(\dfrac{-3}{40}\right) = \dfrac{-35 + (-3)}{40} = -\dfrac{38}{40} = -\dfrac{19}{20}$.

▶ **Now Try Exercise 85**

Helpful Hint

In Example 6, **Figure 1.24**, we illustrated that starting at 0 and moving 21 units to the right, then moving 32 units to the left, ends at a negative 11. Thus, $21 + (-32) = -11$.

When the numbers are large, you are not expected to actually mark and count the units. For example, you will not need to count 32 units to the left from 21 to obtain the answer -11.

In objective 3, we will show you how to obtain $21 + (-32) = -11$ without having to draw number lines. We present adding on a number line here to help you understand the concept of addition of signed numbers, and to help you in determining, without doing any calculations, whether the sum of two signed numbers will be a positive number, a negative number, or zero.

3   Identify Opposites

Now let's consider **opposites**, or **additive inverses**.

Opposites (or Additive Inverses)

Any two numbers whose sum is zero are said to be **opposites** (or **additive inverses**) of each other. In general, if we let a represent any real number, then its opposite is $-a$ and $a + (-a) = 0$.

In Example 4, the sum of 5 and -5 is 0. Thus, -5 is the opposite of 5 and 5 is the opposite of -5.

EXAMPLE 8 ▶ Find the opposite of each number. **a)** 3   **b)** -4 **c)** $-\dfrac{7}{8}$

Solution

a) The opposite of 3 is -3, since $3 + (-3) = 0$.

b) The opposite of -4 is 4, since $-4 + 4 = 0$.

c) The opposite of $-\dfrac{7}{8}$ is $\dfrac{7}{8}$, since $-\dfrac{7}{8} + \dfrac{7}{8} = 0$.

▶ Now Try Exercise 15

4   Add Using Absolute Values

Now that we have had some practice adding signed numbers on a number line, we give a rule (in two parts) for using absolute value to add signed numbers. Remember that the absolute value of a nonzero number will always be positive. The first part of the rule follows.

Adding Real Numbers with the Same Sign

To add real numbers with the same sign (either both positive or both negative), add their absolute values. The sum has the same sign as the numbers being added.

EXAMPLE 9 ▶ Add $4 + 8$.

Solution   Since both numbers have the same sign, both positive, we add their absolute values: $|4| + |8| = 4 + 8 = 12$. Since both numbers being added are positive, the sum is positive. Thus, $4 + 8 = 12$.

▶ Now Try Exercise 49

EXAMPLE 10 ▶ Add $-6 + (-9)$.

Solution   Since both numbers have the same sign, both negative, we add their absolute values: $|-6| + |-9| = 6 + 9 = 15$. Since both numbers being added are negative, their sum is negative. Thus, $-6 + (-9) = -15$.

▶ Now Try Exercise 51

The sum of two positive numbers will always be positive and the sum of two negative numbers will always be negative.

Adding Two Signed Numbers with Different Signs

To add two signed numbers with different signs (one positive and the other negative), subtract the smaller absolute value from the larger absolute value. The answer has the sign of the number with the larger absolute value.

EXAMPLE 11 ▶ Add $10 + (-6)$.

Solution   The two numbers being added have different signs, so we subtract the smaller absolute value from the larger: $|10| - |-6| = 10 - 6 = 4$. Since $|10|$ is greater than $|-6|$ and the sign of 10 is positive, the sum is positive. Thus, $10 + (-6) = 4$.

▶ **Now Try Exercise 53**

EXAMPLE 12 ▶ Add $12 + (-18)$.

Solution   The numbers being added have different signs, so we subtract the smaller absolute value from the larger: $|-18| - |12| = 18 - 12 = 6$. Since $|-18|$ is greater than $|12|$ and the sign of -18 is negative, the sum is negative. Thus, $12 + (-18) = -6$.

▶ **Now Try Exercise 55**

EXAMPLE 13 ▶ Add $-21 + 20$.

Solution   The two numbers being added have different signs, so we subtract the smaller absolute value from the larger: $|-21| - |20| = 21 - 20 = 1$. Since $|-21|$ is greater than $|20|$, the sum is negative. Therefore, $-21 + 20 = -1$.

▶ **Now Try Exercise 61**

Now let's look at some additional examples that contain fractions and decimal numbers.

EXAMPLE 14 ▶ Add $-\dfrac{3}{5} + \dfrac{4}{7}$.

Solution   We can write each fraction with the least common denominator, 35.

$$-\frac{3}{5} + \frac{4}{7} = -\frac{3}{5} \cdot \frac{7}{7} + \frac{4}{7} \cdot \frac{5}{5}$$

$$= \frac{-21}{35} + \frac{20}{35} = \frac{-21 + 20}{35}$$

Since $|-21|$ is greater than $|20|$, the final answer will be negative. In Example 13, we found that $-21 + 20 = -1$. Thus, we can write

$$\frac{-21}{35} + \frac{20}{35} = \frac{-21 + 20}{35} = \frac{-1}{35} = -\frac{1}{35}$$

Thus, $-\dfrac{3}{5} + \dfrac{4}{7} = -\dfrac{1}{35}$.

▶ **Now Try Exercise 87**

Examples 15 and 16 contain decimal numbers. If you have forgotten how to perform the basic operations of addition, subtraction, multiplication, and division of decimal numbers, read Appendix A now.

EXAMPLE 15 ▶ Add $-37.45 + (-26.98)$.

Solution   Since both numbers have the same sign, both negative, we add their absolute values: $|-37.45| + |-26.98| = 37.45 + 26.98$.

$$\begin{array}{r} 37.45 \\ +\ 26.98 \\ \hline 64.43 \end{array}$$

Since two negative numbers are being added, the sum is negative. Therefore, $-37.45 + (-26.98) = -64.43$.

▶ **Now Try Exercise 67**

EXAMPLE 16 ▸ Net Profit or Loss The B.J. Donaldson Printing Company had a loss of $4005.69 for the first 6 months of the year and a profit of $29,645.78 for the second 6 months of the year. Find the net profit or loss for the year.

Solution Understand and Translate This problem can be represented as $-4005.69 + 29,645.78$. Since the numbers have different signs, we subtract the smaller absolute value from the larger.

Carry Out $|29,645.78| - |-4005.69| = 29,645.78 - 4005.69$

$$
\begin{array}{r}
29,645.78 \\
-\ \ \ 4005.69 \\
\hline
25,640.09
\end{array}
$$

Since $|29,645.78|$ is greater than $|-4005.69|$ and the sign of 29,645.78 is positive, the sum is positive. Thus, $-4005.69 + 29,645.78 = 25,640.09$.

Check and Answer   The answer is reasonable. Thus, the net profit for the year was $25,640.09.

▸ **Now Try Exercise 121**

The sum of two signed numbers with different signs may be either positive or negative. The sign of the sum will be the same as the sign of the number with the larger absolute value.

Helpful Hint

Architects often make a scale model of a building before starting construction of the building. This "model" helps them visualize the project and often helps them avoid problems.

   Mathematicians also construct models. A mathematical *model* may be a physical representation of a mathematical concept. It may be as simple as using tiles or chips to represent specific numbers. For example, below we use a model to help explain addition of real numbers. This may help some of you understand the concepts better.

   We let a red chip represent +1 and a green chip represent −1.

   ● = +1   ● = −1

If we add +1 and −1, or a red and a green chip, we get 0.
Now consider the addition problem $3 + (-5)$. We can represent this as

   $\underbrace{● \ ● \ ●}_{3} + \underbrace{● \ ● \ ● \ ● \ ●}_{-5}$

If we remove 3 red chips and 3 green chips, or three zeros, we are left with 2 green chips, which represents a sum of −2. Thus, $3 + (-5) = -2$,

   $\cancel{●} \ \cancel{●} \ \cancel{●} + \cancel{●} \ \cancel{●} \ \cancel{●} \ ● \ ●$

Now consider the problem $-4 + (-2)$. We can represent this as

   $\underbrace{● \ ● \ ● \ ●}_{-4} + \underbrace{● \ ●}_{-2}$

Since we end up with 6 green chips, and each green chip represents −1, the sum is −6. Therefore, $-4 + (-2) = -6$.

5 Add Using Calculators

Throughout the book we will provide information about calculators in the Using Your Calculator Boxes. Some will be for scientific calculators; some will be for graphing calculators, also called graphers; and some will be for both. To the left are pictures of a scientific calculator (on left) and a graphing calculator (on right). Graphing calculators can do everything that a scientific calculator can do and more. Ask your instructor if he or she is recommending a particular calculator for this course. If you plan on taking additional mathematics courses, you may want to consider purchasing the calculator that will be used in those courses.

It is important that you understand the procedures for adding, subtracting, multiplying, and dividing real numbers *without* using a calculator. *You should not need to rely on a calculator to work problems.* If you are permitted to use a calculator, you can, however, use the calculator to help save time on difficult calculations. If you have an understanding of the basic concepts, you should be able to tell if you have made an error entering information on the calculator if the answer shown does not seem reasonable.

Following is the first of many Using Your Calculator Boxes and the first of many Using Your Graphing Calculator Boxes. Note that *no new material* will be presented in the boxes. The boxes are provided to help you use your calculator, if you are using one in this course. Do not be concerned if you do not know what all the calculator keys do. Your instructor will tell you which calculator keys you will need to know how to use.

In Using Your Graphing Calculator Boxes, when we show keystrokes or graphing screens (called windows), they will be for the Texas Instruments TI-84 Plus calculator. The same instructions will apply for the Texas Instruments TI-83 Plus calculator, although we will not indicate this in the Using Your Graphing Calculator Boxes. You should read your graphing calculator manual for more detailed instructions.

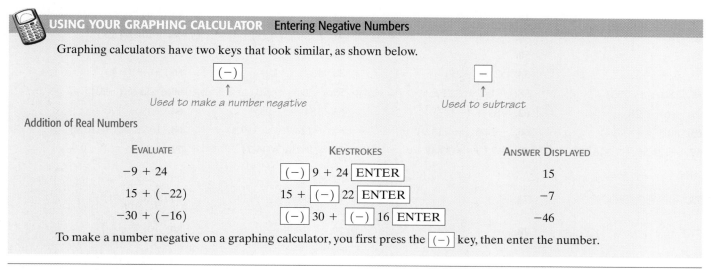

USING YOUR CALCULATOR Entering Negative Numbers

Most scientific calculators contain a $\boxed{+/-}$ key, which is used to enter a negative number. To enter the number -5, press $5\boxed{+/-}$ and a -5 will be displayed. Now we show how to evaluate some addition problems on a scientific calculator.

Addition of Real Numbers

EVALUATE	KEYSTROKES*	ANSWER DISPLAYED
$-9 + 24$	$9\boxed{+/-}\boxed{+}24\boxed{=}$	15
$15 + (-22)$	$15\boxed{+}22\boxed{+/-}\boxed{=}$	-7
$-30 + (-16)$	$30\boxed{+/-}\boxed{+}16\boxed{+/-}\boxed{=}$	-46

*With some scientific calculators the negative sign is entered before the number, as is done with graphing calculators.

USING YOUR GRAPHING CALCULATOR Entering Negative Numbers

Graphing calculators have two keys that look similar, as shown below.

$\boxed{(-)}$ ↑ *Used to make a number negative* $\boxed{-}$ ↑ *Used to subtract*

Addition of Real Numbers

EVALUATE	KEYSTROKES	ANSWER DISPLAYED
$-9 + 24$	$\boxed{(-)}9 + 24\boxed{\text{ENTER}}$	15
$15 + (-22)$	$15 + \boxed{(-)}22\boxed{\text{ENTER}}$	-7
$-30 + (-16)$	$\boxed{(-)}30 + \boxed{(-)}16\boxed{\text{ENTER}}$	-46

To make a number negative on a graphing calculator, you first press the $\boxed{(-)}$ key, then enter the number.

EXERCISE SET 1.6

 MathXL®   MyMathLab

Concept/Writing Exercises

1. What are the four basic operations of arithmetic?

2. a) What are opposites or additive inverses?
  b) Give an example of two numbers that are opposites.

3. a) Are the numbers $-\frac{2}{3}$ and $\frac{3}{2}$ opposites? Explain.

b) If the numbers in part **a)** are not opposites, what number is the opposite of $-\frac{2}{3}$?

4. If we add a positive number and a negative number, will the sum be positive, negative, or can it be either? Explain.

5. If we add two negative numbers, will the sum be a positive number or a negative number? Explain.

6. a) If we add −24,692 and 30,519, will the sum be positive or negative? Without performing any calculations, explain how you determined your answer.

   b) Repeat part a) for the numbers 24,692 and −30,519.

   c) Repeat part a) for the numbers −24,692 and −30,519.

7. Explain how to add two numbers with like signs.

8. Explain how to add two numbers with unlike signs.

9. Mr. Dabskic charged $162 worth of groceries on his charge card. He later made a payment of $85.

   a) Explain why the balance remaining on his card may be found by the addition −162 + 85.

   b) Find the sum of −162 + 85.

   c) In part b), you should have obtained a sum of −77. Explain why this −77 indicates that Mr. Dabskic owes $77 on his credit card.

10. Mrs. Goldstein owed $163 on her credit card. She charged another item costing $56.

   a) Explain why the new balance on her credit card may be found by the addition −163 + (−56).

   b) Find the sum of −163 + (−56).

   c) In part b), you should have obtained a sum of −219. Explain why this −219 indicates that Mrs. Goldstein owes $219 on her credit card.

In Exercises 11 and 12, are the calculations shown correct? If not, explain why not.

11. $\dfrac{-5}{12} + \dfrac{9}{12} = \dfrac{-5+9}{12} = \dfrac{4}{12} = \dfrac{1}{3}$

12. $\dfrac{-6}{70} + \left(\dfrac{-9}{70}\right) = \dfrac{-6+(-9)}{70} = \dfrac{-15}{70} = -\dfrac{3}{14}$

Practice the Skills

Write the opposite of each number.

13. 9	14. −7	15. −28	16. 3
17. 0	18. $-3\dfrac{1}{2}$	19. $\dfrac{5}{3}$	20. $-\dfrac{1}{4}$
21. $2\dfrac{3}{5}$	22. −1	23. 3.72	24. −0.721

Add.

25. 5 + 6	26. −8 + 2	27. 4 + (−3)	28. 9 + (−12)
29. −4 + (−2)	30. −3 + (−5)	31. 6 + (−6)	32. −8 + 8
33. −4 + 4	34. −6 + 6	35. −8 + (−2)	36. 6 + (−5)
37. −7 + 3	38. −6 + 9	39. −8 + (−5)	40. 0 + (−3)
41. 0 + 0	42. 0 + (−0)	43. −6 + 0	44. −9 + 13
45. 18 + (−9)	46. −7 + 7	47. −33 + (−31)	48. −27 + (−9)
49. 7 + 9	50. 12 + 3	51. −8 + (−4)	52. −25 + (−36)
53. 6 + (−3)	54. 52 + (−25)	55. 13 + (−19)	56. 34 + (−40)
57. 180 + (−200)	58. −452 + 312	59. −11 + (−20)	60. −33 + (−92)
61. −67 + 28	62. 183 + (−183)	63. 184 + (−93)	64. −19 + 176
65. 80.5 + (−90.4)	66. −24.6 + (−13.9)	67. −124.7 + (−19.3)	68. 106.3 + (−110.9)
69. −123.56 + (−18.35)	70. −72.79 + 33.47	71. −99.36 + 45.71	72. −84.15 + (−29.98)

Add.

73. $\dfrac{3}{5} + \dfrac{1}{7}$	74. $\dfrac{5}{8} + \dfrac{3}{5}$	75. $\dfrac{5}{12} + \dfrac{6}{7}$	76. $\dfrac{2}{9} + \dfrac{3}{10}$
77. $-\dfrac{8}{11} + \dfrac{4}{5}$	78. $-\dfrac{4}{9} + \dfrac{5}{27}$	79. $-\dfrac{7}{10} + \dfrac{11}{90}$	80. $\dfrac{8}{9} + \left(-\dfrac{1}{3}\right)$
81. $-\dfrac{7}{30} + \left(-\dfrac{5}{6}\right)$	82. $-\dfrac{7}{9} + \left(-\dfrac{1}{5}\right)$	83. $\dfrac{9}{25} + \left(-\dfrac{3}{50}\right)$	84. $-\dfrac{1}{15} + \left(-\dfrac{5}{6}\right)$
85. $-\dfrac{4}{5} + \left(-\dfrac{5}{75}\right)$	86. $\dfrac{5}{36} + \left(-\dfrac{5}{24}\right)$	87. $-\dfrac{9}{24} + \dfrac{5}{7}$	88. $-\dfrac{9}{40} + \dfrac{4}{15}$
89. $-\dfrac{5}{12} + \left(-\dfrac{3}{10}\right)$	90. $\dfrac{7}{16} + \left(-\dfrac{5}{24}\right)$	91. $-\dfrac{13}{14} + \left(-\dfrac{7}{42}\right)$	92. $-\dfrac{11}{27} + \left(-\dfrac{7}{18}\right)$

In Exercises 93–108, a) determine by observation whether the sum will be a positive number, zero, or a negative number; b) find the sum using your calculator; and c) examine your answer to part b) to see whether it is reasonable and makes sense.

93. 587 + (−197)	94. −140 + (−629)	95. −84 + (−289)	96. −647 + 352
97. −947 + 495	98. 762 + (−762)	99. −496 + (−804)	100. −354 + 1090

101. $-375 + 263$

102. $1127 + (-84)$

103. $-1833 + (-2047)$

104. $-426 + 572$

105. $3124 + (-2013)$

106. $-9095 + (-647)$

107. $-1025 + (-1025)$

108. $7513 + (-4361)$

Indicate whether each statement is true or false.

109. The sum of two negative numbers is always a negative number.

110. The sum of a negative number and a positive number is sometimes a negative number.

111. The sum of two positive numbers is never a negative number.

112. The sum of a positive number and a negative number is always a negative number.

113. The sum of a positive number and a negative number is always a positive number.

114. The sum of a number and its opposite is always equal to zero.

Problem Solving

Write an expression that can be used to solve each problem and then solve.

115. Credit Card David Nurkiewicz owed $94 on his bank credit card. He charged another item costing $183. Find the amount that David owed the bank.

116. Charge Card Mrs. Chu charged $142 worth of goods on her charge card. Find her balance after she made a payment of $87.

117. Football A football team lost 18 yards on one play and then lost 3 yards on the following play. What was the total loss in yardage?

118. Overdrawn Checking Account Mrs. Jahn is unaware that her checking account has been overdrawn by $56. While shopping, she writes a check for $162. Find the total amount by which Mrs. Jahn has overdrawn her account.

119. Drilling for Water A company is drilling a well. During the first week they drilled 27 feet, and during the second week they drilled another 34 feet before they struck water. How deep is the well?

120. Coffee Bar The Frenches opened a coffee bar. Their income and expenses for their first three months of operation are shown in the following graph.

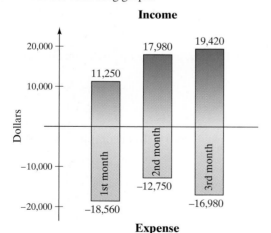

Income

a) Find the net profit or loss (the sum of income and expenses) for the first month.

b) Find the net profit or loss for the second month.

c) Find the net profit or loss for the third month.

121. High Mountain The Web site *www.guinnessworldrecords.com* lists Mauna Kea in Hawaii as the tallest mountain in the world when measured from its base to its peak. The base of Mauna Kea is 19,684 feet below sea level. The total height of the mountain from its base to its peak is 33,480 feet. How high is the peak of Mauna Kea above sea level?

122. Net Profit or Loss The Crafty Scrapbook Company had a loss of $3000 for the first 4 months of the year and a profit of $37,400 for the last 8 months of the year. Find the net profit or loss for the year.

123. Surplus or Deficit The following graph shows the surplus or deficit for Preferred Care, Inc. in Rochester, New York, for the years 1994 through 2004.

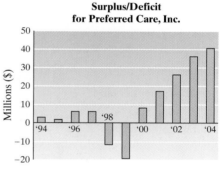

Source: Rochester Democrat & Chronicle

a) Estimate the surplus or deficit for Preferred Care, Inc. in 1998.

b) Estimate the surplus or deficit for Preferred Care Inc. for each of the years 1999, 2000, and 2001. Then estimate the surplus or deficit from 1999 through 2001 by adding these three estimates.

124. Stocks The following chart shows percent changes from the first quarter of 2004 through the first quarter of 2005 for stocks based in the Buffalo–Niagara, New York region.

Percent change from previous quarter for stocks based in the Buffalo–Niagara, New York region

1st quarter 2004	−3.2%
2nd quarter 2004	Unchanged
3rd quarter 2004	1.2%
4th quarter 2004	12.2%
1st quarter 2005	−2.6%

Source: The Buffalo News

Determine the percent change for stocks based in the Buffalo–Niagara, New York, region from the first quarter of 2004 through the first quarter of 2005 by adding the individual percents.

Challenge Problems

Evaluate each exercise by adding the numbers from left to right. We will discuss problems like this shortly.

125. $(-4) + (-6) + (-12)$ **126.** $5 + (-7) + (-8)$ **127.** $29 + (-46) + 37$

128. $4 + (-5) + 6 + (-8)$ **129.** $(-12) + (-10) + 25 + (-3)$ **130.** $(-4) + (-2) + (-15) + (-27)$

131. $\frac{1}{2} + \left(-\frac{1}{3}\right) + \frac{1}{5}$ **132.** $-\frac{3}{8} + \left(-\frac{2}{9}\right) + \left(-\frac{1}{2}\right)$

Find the following sums. Explain how you determined your answer. (Hint: Pair small numbers with large numbers from the ends inward.)

133. $1 + 2 + 3 + \cdots + 10$ **134.** $1 + 2 + 3 + \cdots + 20$

Cumulative Review Exercises

[1.3] **135.** Multiply $\left(\frac{4}{7}\right)\left(2\frac{3}{8}\right)$.   **136.** Subtract $3 - \frac{5}{16}$.

[1.4] **137.** True or False: Every number less than zero is a negative integer.

[1.5] *Insert either $<$, $>$, or $=$ in each shaded area to make a true statement.*

138. $|-3|$ ▨ 2 **139.** 8 ▨ $|-12|$

1.7 Subtraction of Real Numbers

1 Subtract numbers.

2 Subtract numbers mentally.

3 Evaluate expressions containing more than two numbers.

1 Subtract Numbers

Any subtraction problem can be rewritten as an addition problem using the additive inverse.

To Subtract Real Numbers

In general, if a and b represent any two real numbers, then

$$a - b = a + (-b)$$

This rule says that to subtract b from a, add the opposite or additive inverse of b to a.

EXAMPLE 1 ▶ Evaluate $9 - (+4)$.

Solution   We are subtracting a positive 4 from 9. To accomplish this, we add the opposite of +4, which is −4, to 9.

$$9 - (+4) = 9 + (-4) = 5$$

Subtract   Positive 4   Add   Negative 4

We evaluated $9 + (-4)$ using the procedures for *adding* real numbers presented in Section 1.6.

▶ **Now Try Exercise 13**

Often in a subtraction problem, when the number being subtracted is a positive number, the + sign preceding the number being subtracted is not shown. For example, in the subtraction $9 - 4$,

$$9 - \boxed{4} \text{ means } 9 - \boxed{(+4)}$$

Thus, to evaluate $9 - 4$, we must add the opposite of 4, which is -4, to 9.

$$9 - 4 = 9 + (-4) = 5$$

Subtract Positive 4 Add Negative 4

This procedure is illustrated in Example 2.

EXAMPLE 2 ▶ Evaluate $5 - 3$.

Solution We must subtract a positive 3 from 5. To change this problem to an addition problem, we add the opposite of 3, which is -3, to 5.

Subtraction problem   Addition problem

$$5 - 3 = 5 + (-3) = 2$$

Subtract Positive 3 Add Negative 3

▶ **Now Try Exercise 15**

EXAMPLE 3 ▶ Evaluate.

a) $4 - 9$ **b)** $-4 - 2$

Solution

a) Add the opposite of 9, which is -9, to 4.

$$4 - 9 = 4 + (-9) = -5$$

b) Add the opposite of 2, which is -2, to -4.

$$-4 - 2 = -4 + (-2) = -6$$

▶ **Now Try Exercise 17**

In Example 4, we subtract numbers that contain decimal points.

EXAMPLE 4 ▶ Evaluate $16.32 - 18.75$.

Solution Add the opposite of 18.75, which is -18.75, to 16.32.

$$16.32 - 18.75 = 16.32 + (-18.75) = -2.43$$

▶ **Now Try Exercise 53**

In Examples 5 and 6, we will show how to subtract a negative number.

EXAMPLE 5 ▶ Evaluate $4 - (-2)$.

Solution We are asked to subtract a negative 2 from 4. To do this, we add the opposite of -2, which is 2, to 4.

$$4 - (-2) = 4 + 2 = 6$$

Subtract Negative 2 Add Positive 2

▶ **Now Try Exercise 19**

Helpful Hint

By examining Example 5, we see that

$$4 - (-2) = 4 + 2$$

Two negative   Plus
signs together

Whenever we subtract a negative number, we can replace the two negative signs with a plus sign.

EXAMPLE 6 ▶ Evaluate.

a) $7 - (-5)$ **b)** $-15 - (-12)$

Solution

a) Since we are subtracting a negative number, adding the opposite of -5, which is 5, to 7 will result in the two negative signs being replaced by a plus sign.

$$7 - (-5) = 7 + 5 = 12$$

b) $-15 - (-12) = -15 + 12 = -3$

▶ **Now Try Exercise 29**

Helpful Hint

We will now indicate how we may illustrate subtraction using colored chips. Remember from the preceding section that a red chip represents $+1$ and a green chip -1.

● $= +1$ ● $= -1$

Consider the subtraction problem $2 - 5$. If we change this to an addition problem, we get $2 + (-5)$. We can then add, as was done in the preceding section. The figure below shows that $2 + (-5) = -3$.

Now consider $-2 - 5$. This means $-2 + (-5)$, which can be represented as follows:

Thus, $-2 - 5 = -7$.
Now consider the problem $-3 - (-5)$. This can be rewritten as $-3 + 5$, which can be represented as follows:

Thus, $-3 - (-5) = 2$.
Some students still have difficulty understanding why when you subtract a negative number you obtain a positive number. Let's look at the problem $3 - (-2)$. This time we will look at it from a slightly different point of view. Let's start with 3:

● ● ●

From this we wish to subtract a negative 2. To the $+3$ shown above we will add two zeros by adding two $+1 - 1$ combinations. Remember, $+1$ and -1 sum to 0.

    +3   0   0

Now we can subtract or "take away" the two -1's as shown:

From this we see that we are left with $3 + 2$ or 5. Thus, $3 - (-2) = 5$.

EXAMPLE 7 ▶ Subtract 12 from 3.

Solution

$$3 - 12 = 3 + (-12) = -9$$

▶ **Now Try Exercise 57**

Helpful Hint

Example 7 asked us to "subtract 12 from 3." Some of you may have expected this to be written as $12 - 3$ since you may be accustomed to getting a positive answer. However, the correct way of writing this is $3 - 12$. Notice that the number following the word "from" is our starting point. That is where the calculation begins. For example:

Subtract 2 from 7 means $7 - 2$.   From 7, subtract 2 means $7 - 2$.

Subtract 5 from -1 means $-1 - 5$.   From -1, subtract 5 means $-1 - 5$.

Subtract -4 from -2 means $-2 - (-4)$.   From -2, subtract -4 means $-2 - (-4)$.

Subtract -3 from 6 means $6 - (-3)$.   From 6, subtract -3 means $6 - (-3)$.

Subtract a from b means $b - a$.   From a, subtract b means $a - b$.

EXAMPLE 8 ▶ Subtract 5 from 5.

Solution

$$5 - 5 = 5 + (-5) = 0$$

▶ **Now Try Exercise 59**

EXAMPLE 9 ▶ Subtract -6.48 from 4.25.

Solution

$$4.25 - (-6.48) = 4.25 + 6.48 = 10.73$$

▶ **Now Try Exercise 61**

Now we will perform subtraction problems that contain fractions.

EXAMPLE 10 ▶ Subtract $\dfrac{5}{9} - \dfrac{13}{15}$.

Solution   Begin by changing the subtraction problem to an addition problem.

$$\frac{5}{9} - \frac{13}{15} = \frac{5}{9} + \left(-\frac{13}{15}\right)$$

Now rewrite the fractions with the LCD, 45, and add the fractions as was done in the last section.

$$\frac{5}{9} + \left(-\frac{13}{15}\right) = \frac{5}{9} \cdot \frac{5}{5} + \left(-\frac{13}{15}\right) \cdot \frac{3}{3}$$

$$= \frac{25}{45} + \left(-\frac{39}{45}\right) = \frac{25 + (-39)}{45} = \frac{-14}{45} = -\frac{14}{45}$$

Thus, $\dfrac{5}{9} - \dfrac{13}{15} = -\dfrac{14}{45}$.

▶ **Now Try Exercise 85**

EXAMPLE 11 ▶ Subtract $-\dfrac{7}{18}$ from $-\dfrac{9}{15}$.

Solution   This problem is written $-\dfrac{9}{15} - \left(-\dfrac{7}{18}\right)$.

We can simplify this as follows.

$$-\frac{9}{15} - \left(-\frac{7}{18}\right) = -\frac{9}{15} + \frac{7}{18}.$$

The LCD of 15 and 18 is 90. Rewriting the fractions with a common denominator gives

$$-\frac{9}{15}\cdot\frac{6}{6} + \frac{7}{18}\cdot\frac{5}{5} = -\frac{54}{90} + \frac{35}{90} = \frac{-54 + 35}{90}$$

$$= \frac{-19}{90} = -\frac{19}{90}.$$

▶ **Now Try Exercise 87**

Let us now look at some applications that involve subtraction.

EXAMPLE 12 ▶ **A Gift Card Balance** Brigitte Martineau's gift card indicates a balance of $86.23 before she makes purchases totaling $127.49. Find the amount that Brigitte owes the cashier after she uses her gift card.

Solution   Understand and Translate   We can obtain the amount that Brigitte owes by subtracting 127.49 from 86.23.

Carry Out $86.23 - 127.49 = 86.23 + (-127.49) = -41.26$

Check and Answer   The negative indicates a deficit, which is what we expect. Therefore, Brigitte owes the cashier $41.26.

▶ **Now Try Exercise 131**

EXAMPLE 13 ▶ **Temperature Difference** On January 4, 2005, the high temperature for the day in Laredo, Texas, was 88°F. On the same day, the low temperature in Wolf Point, Montana, was −29°F. Find the difference in their temperatures.

Solution   Understand and Translate   The word *difference* in the example title indicates subtraction. We can obtain the difference in their temperatures by subtracting as follows.

Carry Out $88 - (-29) = 88 + 29 = 117$

Check and Answer   Therefore, the high temperature in Laredo is 117°F greater than the low temperature in Wolf Point.

▶ **Now Try Exercise 135**

Laredo, Texas

EXAMPLE 14 ▶ **Measuring Snow** A kindergarten class in Richfield, Minnesota, has a snow gauge placed outside its window that is left untouched for two days. Suppose that on the first day $6\frac{3}{8}$ inches of snow falls. On the second day, no snow falls, but $1\frac{1}{2}$ inches of the first day's snowfall melts. How much snow remains after the second day?

Solution   Understand and Translate   From the first amount, $6\frac{3}{8}$ inches, we must subtract $1\frac{1}{2}$ inches.

Carry Out On page 26, we subtracted mixed numbers using the mixed numbers themselves. Addition and subtraction of mixed numbers can also be done by changing the mixed numbers to improper fractions. We illustrate the procedure here. We begin by changing the subtraction problem to an addition problem. We then change the mixed numbers to fractions, and then rewrite each fraction with the LCD, 8.

$$6\frac{3}{8} - 1\frac{1}{2} = 6\frac{3}{8} + \left(-1\frac{1}{2}\right)$$

$$= \frac{51}{8} + \left(-\frac{3}{2}\right)$$

$$= \frac{51}{8} + \left(-\frac{3}{2}\right) \cdot \frac{4}{4}$$

$$= \frac{51}{8} + \left(-\frac{12}{8}\right)$$

$$= \frac{51 + (-12)}{8} = \frac{39}{8} \quad \text{or} \quad 4\frac{7}{8}$$

Check and Answer Thus, after the second day there were $4\frac{7}{8}$ inches of snow remaining. Based upon the numbers given in the problem, the answer seems reasonable.

▶ **Now Try Exercise 133**

EXAMPLE 15 ▶ Evaluate.

a) $15 + (-4)$ **b)** $-16 - 3$ **c)** $19 + (-14)$
d) $7 - (-9)$ **e)** $-9 - (-3)$ **f)** $8 - 13$

Solution Parts **a)** and **c)** are addition problems, whereas the other parts are subtraction problems. We can rewrite each subtraction problem as an addition problem to evaluate.

a) $15 + (-4) = 11$ **b)** $-16 - 3 = -16 + (-3) = -19$
c) $19 + (-14) = 5$ **d)** $7 - (-9) = 7 + 9 = 16$
e) $-9 - (-3) = -9 + 3 = -6$ **f)** $8 - 13 = 8 + (-13) = -5$

▶ **Now Try Exercise 31**

2 Subtract Numbers Mentally

In the previous examples, we rewrote subtraction problems as addition problems. We did this because we know how to add real numbers. After this chapter, when we work out a subtraction problem, we will not show this step. *You need to practice and thoroughly understand how to add and subtract real numbers. You should understand this material so well that, when asked to evaluate an expression like $-4 - 6$, you will be able to compute the answer mentally. You should understand that $-4 - 6$ means the same as $-4 + (-6)$, but you should not need to write this using addition to find the value of the expression, -10.*

Let's evaluate a few subtraction problems without showing the process of changing the subtraction to addition.

EXAMPLE 16 ▶ Evaluate.

a) $-7 - 5$ **b)** $4 - 12$ **c)** $18 - 25$ **d)** $-20 - 12$

Solution

a) $-7 - 5 = -12$ **b)** $4 - 12 = -8$ **c)** $18 - 25 = -7$ **d)** $-20 - 12 = -32$

▶ **Now Try Exercise 33**

In Example 16 **a)**, we may have reasoned that $-7 - 5$ meant $-7 + (-5)$, which is -12, but we did not need to show it.

EXAMPLE 17 ▶ Evaluate $-\dfrac{3}{5}-\dfrac{7}{8}$.

Solution   The least common denominator is 40. Write each fraction with the LCD, 40.

$$-\frac{3}{5}\cdot\frac{8}{8}-\frac{7}{8}\cdot\frac{5}{5}=-\frac{24}{40}-\frac{35}{40}=\frac{-24-35}{40}=-\frac{59}{40}=-1\frac{19}{40}$$

▶ Now Try Exercise 77

Notice in Example 17, when we had $-\dfrac{24}{40}-\dfrac{35}{40}$, we could have written $\dfrac{-24+(-35)}{40}$, but at this time we elected to write it as $\dfrac{-24-35}{40}$. Since $-24-35$ is -59, the answer is $-\dfrac{59}{40}$ or $-1\dfrac{19}{40}$.

3 Evaluate Expressions Containing More Than Two Numbers

In evaluating expressions involving more than one addition and subtraction, work from left to right unless parentheses or other grouping symbols appear.

EXAMPLE 18 ▶ Evaluate.

a) $9-12+3$ **b)** $-7-15-6$ **c)** $-5+1-8$

Solution   We work from left to right.

a) $\underbrace{9-12}+3$ **b)** $\underbrace{-7-15}-6$ **c)** $\underbrace{-5+1}-8$
 $=-3\quad+3$ $=-22\quad-6$ $=-4\quad-8$
 $=0$ $=-28$ $=-12$

▶ Now Try Exercise 119

After this section you will generally not see an expression like $3+(-4)$. Instead, the expression will be written as $3-4$. Recall that $3-4$ means $3+(-4)$ by our definition of subtraction. **Whenever we see an expression of the form $a+(-b)$, we can write the expression as $a-b$.** For example, $12+(-15)$ can be written as $12-15$ and $-6+(-9)$ can be written as $-6-9$.

As discussed earlier, **whenever we see an expression of the form $a-(-b)$, we can rewrite it as $a+b$.** For example, $6-(-13)$ can be rewritten as $6+13$ and $-12-(-9)$ can be rewritten as $-12+9$.

> **Rewriting Expressions**
>
> In general, for any real numbers a and b,
> $$a+(-b)=a-b,\text{ and}$$
> $$a-(-b)=a+b$$

Using the given information, the expression $9+(-12)-(-8)$ may be simplified to $9-12+8$.

EXAMPLE 19 ▶

a) Evaluate $-5-(-9)+(-12)+(-3)$.

b) Simplify the expression in part **a)**.

c) Evaluate the simplified expression in part **b)**.

Solution

a) We work from left to right. The shading indicates the additions being performed to get to the next step.

$$-5 - (-9) + (-12) + (-3) = \boxed{-5 + 9} + (-12) + (-3)$$
$$= \boxed{4 + (-12)} + (-3)$$
$$= -8 + (-3)$$
$$= -11$$

b) The expression simplifies as follows:

$$-5 - (-9) + (-12) + (-3) = -5 + 9 - 12 - 3$$

c) Evaluate the simplified expression from left to right. Begin by adding $-5 + 9$ to obtain 4.

$$-5 + 9 - 12 - 3 = \boxed{4 - 12} - 3$$
$$= -8 - 3$$
$$= -11$$

When you come across an expression like the one in Example 19 **a)**, you should simplify it as we did in part **b)** and then evaluate the simplified expression.

▶ **Now Try Exercise 129**

USING YOUR CALCULATOR   Subtraction on a Scientific Calculator

In the Using Your Calculator box on page 47, we indicated that the $\boxed{+/-}$ key is usually pressed after a number is entered to make the number negative. Following are some examples of subtraction on a scientific calculator.

EVALUATE	KEYSTROKES	ANSWER DISPLAYED
$-5 - 8$	$5 \boxed{+/-} \boxed{-} 8 \boxed{=}$	-13
$2 - (-7)$	$2 \boxed{-} 7 \boxed{+/-} \boxed{=}$	9

Subtraction on a Graphing Calculator

In the Using Your Graphing Calculator box on page 47, we mentioned that on a graphing calculator we press the $\boxed{(-)}$ key before the number is entered to make the number negative. The $\boxed{-}$ key on a graphing calculator is used to perform subtraction. Following are some examples of subtraction on a graphing calculator.

EVALUATE	KEYSTROKE	DISPLAY
$-5 - 8$	$\boxed{(-)} 5 \boxed{-} 8 \boxed{\text{ENTER}}$	-13
$2 - (-7)$	$2 \boxed{-} \boxed{(-)} 7 \boxed{\text{ENTER}}$	9

EXERCISE SET 1.7   Math XL   MyMathLab
MathXL®   MyMathLab

Concept/Writing Exercises

1. Write an expression that illustrates 7 subtracted from 2.

2. Write an expression that illustrates -6 subtracted from -4.

3. Write an expression that illustrates ? subtracted from ☺.

4. Write an expression that illustrates ✳ subtracted from □.

5. a) Explain how to subtract a number b from a number a.

 b) Write an expression using addition that can be used to subtract 14 from 5.

 c) Evaluate the expression you determined in part **b)**.

6. a) Express the subtraction $a - (+b)$ in a simplified form.

 b) Simplify the expression $3 - (+9)$.

 c) Evaluate the simplified expression obtained in part **b)**.

7. a) Express the subtraction $a - (-b)$ in a simplified form.

 b) Write a simplified expression that can be used to evaluate $-4 - (-12)$.

 c) Evaluate the simplified expression obtained in part **b)**.

8. When we add three or more numbers without parentheses, how do we evaluate the expression?

9. a) Simplify $3 - (-6) + (-5)$ by eliminating two signs next to one another and replacing them with a single sign. (See Example 19b).) Explain how you determined your answer.

 b) Evaluate the simplified expression obtained in part **a)**.

10. a) Simplify $-12 + (-5) - (-4)$. Explain how you determined your answer.

 b) Evaluate the simplified expression obtained in part **a)**.

In Exercises 11 and 12, are the following calculations correct? If not, explain why.

11. $\dfrac{4}{9} - \dfrac{3}{7} = \dfrac{28}{63} - \dfrac{27}{63} = \dfrac{28 - 27}{63} = \dfrac{1}{63}$

12. $-\dfrac{5}{12} - \dfrac{7}{9} = -\dfrac{15}{36} - \dfrac{28}{36} = \dfrac{-15 - 28}{36} = -\dfrac{43}{36}$

Practice the Skills

Evaluate.

13. $8 - (+2)$ **14.** $17 - (+8)$ **15.** $12 - 5$ **16.** $9 - 4$

17. $8 - 9$ **18.** $-6 - 3$ **19.** $9 - (-3)$ **20.** $17 - (-5)$

21. $-6 - 6$ **22.** $-4 - (-3)$ **23.** $0 - 7$ **24.** $9 - (-9)$

25. $8 - 8$ **26.** $10 - 10$ **27.** $-3 - 1$ **28.** $-5.7 - (-3.1)$

29. $-8 - (-5)$ **30.** $4 - 9$ **31.** $6 - (-3)$ **32.** $6 - 10$

33. $-9 - 11$ **34.** $37 - 40$ **35.** $0 - (-9.8)$ **36.** $-6.3 - 4.7$

37. $-4.8 - (-5.1)$ **38.** $-4 - (-4)$ **39.** $14 - 7$ **40.** $9 - 9$

41. $-8 - (-12)$ **42.** $-6 - (-2)$ **43.** $18 - (-4)$ **44.** $-25 - 16$

45. $-9 - 2$ **46.** $-35 - (-8)$ **47.** $-90.7 - 40.3$ **48.** $-52.6 - 37.9$

49. $-45 - 37$ **50.** $-50 - (-40)$ **51.** $70 - (-70)$ **52.** $130 - (-90)$

53. $42.3 - 49.7$ **54.** $81.3 - 92.5$ **55.** $-7.85 - (-3.92)$ **56.** $-12.43 - (-9.57)$

57. Subtract 15 from 4.   **58.** Subtract 7 from 1.   **59.** Subtract 21 from 21.   **60.** Subtract 13 from 13.

61. Subtract -12.4 from -6.3.   **62.** Subtract 17.3 from -9.8.   **63.** Subtract -7.9 from 10.3.

64. Subtract -23 from -23.   **65.** Subtract 24 from 13.   **66.** Subtract -11.7 from -5.2.

67. Subtract 7.8 from -10.3.   **68.** Subtract -9.6 from 3.4.

Evaluate.

69. $\dfrac{5}{9} - \dfrac{3}{8}$ **70.** $\dfrac{4}{5} - \dfrac{5}{6}$ **71.** $\dfrac{8}{15} - \dfrac{7}{45}$ **72.** $\dfrac{5}{12} - \dfrac{7}{8}$

73. $-\dfrac{7}{10} - \dfrac{5}{12}$ **74.** $-\dfrac{1}{4} - \dfrac{2}{3}$ **75.** $-\dfrac{4}{15} - \dfrac{3}{20}$ **76.** $-\dfrac{5}{4} - \dfrac{7}{11}$

77. $-\dfrac{7}{12} - \dfrac{5}{40}$ **78.** $-\dfrac{5}{6} - \dfrac{3}{32}$ **79.** $\dfrac{3}{8} - \dfrac{6}{48}$ **80.** $\dfrac{17}{18} - \dfrac{13}{20}$

81. $-\dfrac{4}{9} - \left(-\dfrac{3}{5}\right)$ **82.** $\dfrac{5}{20} - \left(-\dfrac{1}{8}\right)$ **83.** $\dfrac{3}{16} - \left(-\dfrac{5}{8}\right)$ **84.** $-\dfrac{5}{12} - \left(-\dfrac{3}{8}\right)$

85. Subtract $\dfrac{7}{9}$ from $\dfrac{4}{7}$.   **86.** Subtract $\dfrac{7}{15}$ from $\dfrac{5}{8}$.   **87.** Subtract $-\dfrac{3}{10}$ from $-\dfrac{5}{12}$.   **88.** Subtract $-\dfrac{5}{16}$ from $-\dfrac{9}{10}$.

In Exercises 89–106, a) determine by observation whether the difference will be a positive number, zero, or a negative number; b) find the difference using your calculator; and c) examine your answer to part b) to see whether it is reasonable and makes sense.

89. $378 - 279$ **90.** $483 - 569$ **91.** $-482 - 137$ **92.** $178 - (-377)$

93. $843 - (-745)$ **94.** $864 - (-762)$ **95.** $-408 - (-604)$ **96.** $-623 - 111$

97. $-1024 - (-576)$ **98.** $-104.7 - 27.6$ **99.** $165.7 - 49.6$ **100.** $-40.2 - (-12.6)$

101. Subtract 364 from 295.   **102.** Subtract -433 from -932.   **103.** Subtract 647 from -1023.

104. Subtract 2432 from -4120.   **105.** Subtract -7.62 from -7.62.   **106.** Subtract 36.7 from -103.2.

Evaluate.

107. $7 + 5 - (+8)$ **108.** $15 - (+9) - (+5)$ **109.** $-6 + (-6) + 6$ **110.** $9 - 4 + (-2)$

111. $-13 - (+5) + 3$ **112.** $7 - (+4) - (-3)$ **113.** $-9 - (-3) + 4$ **114.** $15 + (-7) - (-3)$

115. $5 - (-9) + (-1)$ **116.** $12 + (-5) - (-4)$ **117.** $17 + (-8) - (+14)$ **118.** $-7 + 6 - 3$

119. $-36 - 5 + 9$ **120.** $45 - 3 - 7$ **121.** $-2 + 7 - 9$ **122.** $-2 - 7 - 13$

123. $25 - 19 + 3$ **124.** $-4 - 1 + 5$ **125.** $-4 - 6 + 5 - 7$ **126.** $-9 - 3 - (-4) + 5$

127. $17 + (-3) - 9 - (-7)$ **128.** $32 + 5 - 7 - 12$

129. $-9 + (-7) + (-5) - (-3)$ **130.** $6 - 9 - (-3) + 12$

Problem Solving

131. Lands' End The Lands' End catalog department had 300 ladies' blue cardigan sweaters in stock on December 1. By December 9, the department had taken orders for 343 of the sweaters.

a) How many sweaters were on back order?

b) If the catalog department wanted 100 sweaters in addition to those already ordered, how many sweaters would it need to back order?

132. Leadville, Co According to the *Guinness Book of World Records*, the city with the greatest elevation in the United States is Leadville, Colorado at 10,152 feet. The city with the lowest elevation in the United States, at 184 feet below sea level, is Calipatria, California. What is the difference in the elevation of these cities?

Leadville, Colorado

133. Measuring Rainfall At Kim Christensen's house, a rain gauge is placed in the yard and is left untouched for 2 days. Suppose that on the first day, $2\frac{1}{4}$ inches of rain falls. On the second day, no rain falls, but $\frac{3}{8}$ inch of the first day's rainfall evaporates. How much water remains in the gauge after the second day?

134. Death Valley A medical supply package is dropped into Death Valley, California, from a helicopter 1605.7 feet above sea level. The package lands at a location in Death Valley 267.4 feet below sea level. What vertical distance did the package travel?

135. Temperature Change The greatest change in temperature ever recorded within a 24-hour period occurred at Browning, Montana, on January 23, 1916. The temperature fell from 44°F to −56°F. How much did the temperature drop?

136. Going Home Two college students are driving on an expressway, going home for spring break. Shawntoya travels 58.5 miles in 1 hour. Marcelino travels 67.3 miles in 1 hour.

a) If Shawntoya and Marcelino start at the same parking lot and travel in opposite directions, how far apart will they be in 1 hour?

b) If Shawntoya and Marcelino start at the same parking lot and travel in the same direction, how far apart will they be in 1 hour?

137. Golf The chart below shows some final scores at the Masters golf tournament, held in Augusta, Georgia, in 2006.

The Masters, 2006

Golfer	Score (above or below par)
P. Mickelson	−7
T. Clark	−5
T. Woods	−4
V. Singh	−3
S. Cink	−2
D. Howell	+2
L. Donald	+8
O. Browne	+9

Source: www.masters.org

a) If par for the Masters is 288 strokes, determine P. Mickelson's score in 2006.

b) What was the difference in the strokes between T. Clark and S. Cink in 2006?

138. Inseam Christine Henry purchases a new pair of pants whose inseam is $32\frac{1}{2}$ inches. If $2\frac{3}{4}$ inches are cut from the inseam of the pants, what will be the new inseam of the pants?

Challenge Problems

Find each sum.

139. $1 - 2 + 3 - 4 + 5 - 6 + 7 - 8 + 9 - 10$

140. $1 - 2 + 3 - 4 + 5 - 6 + \cdots + 99 - 100$

141. Consider a number line.

a) What is the distance, in units, between −11 and −3?

b) Write a subtraction problem to represent this distance (the distance is to be positive).

142. Stock Amy Tait buys a stock for $50. Will the stock be worth more if it decreases by 10% and then increases by 10%, or if it increases by 10% and then decreases by 10%, or will the value be the same either way?

143. Rolling Ball A ball rolls off a table and follows the path indicated in the figure. Suppose the maximum height reached by the ball on each bounce is 1 foot less than on the previous bounce.

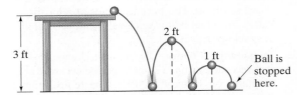

a) Determine the total vertical distance traveled by the ball.

b) If we consider the ball moving in a downward direction as negative, and the ball moving in an upward direction as positive, what was the net vertical distance traveled (from its starting point) by the ball?

Cumulative Review Exercises

[1.4] **144.** List the set of counting numbers.

145. Explain the relationship between the set of rational numbers, the set of irrational numbers, and the set of real numbers.

[1.5] *Insert either* >, <, *or* = *in each shaded area to make the statement true.*

146. $|-3|$ ▊ -5

147. $-|-9|$ ▊ $-|-5|$

[1.7] **148.** Subtract $\dfrac{7}{8}$ from $\dfrac{5}{6}$.

1.8 Multiplication and Division of Real Numbers

1 Multiply numbers.

2 Divide numbers.

3 Remove negative signs from denominators.

4 Evaluate divisions involving 0.

1 Multiply Numbers

The following rules are used in determining the sign of the product when two numbers are multiplied.

> ### The Sign of the Product of Two Real Numbers
> 1. The product of two numbers with **like** signs is a **positive** number.
> 2. The product of two numbers with **unlike** signs is a **negative** number.

By this rule, the product of two positive numbers or two negative numbers will be a positive number. The product of a positive number and a negative number will be a negative number.

EXAMPLE 1 ▶ Evaluate.

a) $4(-5)$ **b)** $(-6)(7)$ **c)** $(-9)(-3)$

Solution

a) Since the numbers have unlike signs, the product is negative.
$$4(-5) = -20$$

b) Since the numbers have unlike signs, the product is negative.
$$(-6)(7) = -42$$

c) Since the numbers have like signs, both negative, the product is positive.
$$(-9)(-3) = 27$$

▶ **Now Try Exercise 17**

EXAMPLE 2 ▶ Evaluate.

a) $(-8)(5)$ **b)** $(-4)(-8)$ **c)** $0(6)$

d) $0(-2)$ **e)** $4.2(-9.7)$ **f)** $(-3.62)(-6.18)$

Solution

a) $(-8)(5) = -40$ **b)** $(-4)(-8) = 32$ **c)** $0(6) = 0$

d) $0(-2) = 0$ **e)** $4.2(-9.7) = -40.74$ **f)** $(-3.62)(-6.18) = 22.3716$

Note that zero multiplied by any real number equals zero.
▶ **Now Try Exercise 29**

Helpful Hint

At this point some students begin confusing problems like $-2 - 3$ with $(-2)(-3)$ and problems like $2 - 3$ with $2(-3)$. If you do not understand the difference between problems like $-2 - 3$ and $(-2)(-3)$, make an appointment to see your instructor as soon as possible.

Subtraction Problems   Multiplication Problems

$-2 - 3 = -5$ $(-2)(-3) = 6$
$2 - 3 = -1$ $(2)(-3) = -6$

EXAMPLE 3 ▶ Evaluate.

a) $\left(\dfrac{-1}{8}\right)\left(\dfrac{-3}{5}\right)$ **b)** $\left(\dfrac{3}{20}\right)\left(\dfrac{-3}{10}\right)$

Solution

a) $\left(\dfrac{-1}{8}\right)\left(\dfrac{-3}{5}\right) = \dfrac{(-1)(-3)}{8(5)} = \dfrac{3}{40}$ **b)** $\left(\dfrac{3}{20}\right)\left(\dfrac{-3}{10}\right) = \dfrac{3(-3)}{20(10)} = -\dfrac{9}{200}$

▶ **Now Try Exercise 41**

Sometimes you may be asked to perform more than one multiplication in a given problem. When this happens, the sign of the final product can be determined by counting the number of *negative* numbers being multiplied. *The product of an even number of negative numbers will always be positive. The product of an odd number of negative numbers will always be negative.* Can you explain why?

EXAMPLE 4 ▶ Evaluate.

a) $(-5)(-3)(1)(-4)$ **b)** $(-2)(-4)(-1)(3)(-4)$

Solution

a) Since there are three negative numbers (an odd number of negative numbers), the product will be negative, as illustrated.

$$(-5)(-3)(1)(-4) = (15)(1)(-4)$$
$$= (15)(-4)$$
$$= -60$$

b) Since there are four negative numbers (an even number of negative numbers), the product will be positive, as illustrated.

$$(-2)(-4)(-1)(3)(-4) = (8)(-1)(3)(-4)$$
$$= (-8)(3)(-4)$$
$$= (-24)(-4)$$
$$= 96$$

▶ **Now Try Exercise 35**

2 Divide Numbers

The rules for dividing numbers are very similar to those used in multiplying numbers.

> ### The Sign of the Quotient of Two Real Numbers
> 1. The quotient of two numbers with **like** signs is a **positive** number.
> 2. The quotient of two numbers with **unlike** signs is a **negative** number.

Therefore, the quotient of two positive numbers or two negative numbers will be a positive number. The quotient of a positive number and a negative number will be a negative number.

EXAMPLE 5 ▶ Evaluate.

a) $\dfrac{10}{-5}$ **b)** $\dfrac{-45}{5}$ **c)** $\dfrac{-36}{-6}$

Solution

a) Since the numbers have unlike signs, the quotient is negative.

$$\frac{10}{-5} = -2$$

b) Since the numbers have unlike signs, the quotient is negative.

$$\frac{-45}{5} = -9$$

c) Since the numbers have like signs, both negative, the quotient is positive.

$$\frac{-36}{-6} = 6$$

▶ **Now Try Exercise 55**

EXAMPLE 6 ▶ Evaluate.

a) $-16 \div (-2)$ **b)** $\dfrac{-2}{3} \div \dfrac{-5}{7}$

Solution

a) Since the numbers have like signs, both negative, the quotient is positive.

$$\frac{-16}{-2} = 8$$

b) Invert the *divisor*, $\dfrac{-5}{7}$, and then multiply.

$$\frac{-2}{3} \div \frac{-5}{7} = \left(\frac{-2}{3}\right)\left(\frac{7}{-5}\right) = \frac{-14}{-15} = \frac{14}{15}$$

▶ **Now Try Exercise 77**

EXAMPLE 7 ▶ Evaluate, round your answer to the nearest hundredth when appropriate.

a) $-18.86 \div 4.1$ **b)** $\dfrac{-27.2}{-2.6}$

Solution

a) Since the numbers have unlike signs, the quotient is negative.

$$\frac{-18.86}{4.1} = -4.6$$

b) Since the numbers have like signs, both negative, the quotient is positive.

$$\frac{-27.2}{-2.6} \approx 10.46$$

The answer in part **b)** was rounded to two decimal places, or hundredths. If you have forgotten how to round decimal numbers, review Appendix A now.

▶ **Now Try Exercise 69**

Helpful Hint

For multiplication and division of two real numbers:

$$\left.\begin{array}{c} (+)(+) = + \\ (-)(-) = + \end{array}\right\} \quad \left.\begin{array}{c} \dfrac{(+)}{(+)} = + \\ \dfrac{(-)}{(-)} = + \end{array}\right\} \quad \textit{Like signs give positive products and quotients.}$$

$$\left.\begin{array}{c} (+)(-) = - \\ (-)(+) = - \end{array}\right\} \quad \left.\begin{array}{c} \dfrac{(+)}{(-)} = - \\ \dfrac{(-)}{(+)} = - \end{array}\right\} \quad \textit{Unlike signs give negative products and quotients.}$$

3 Remove Negative Signs from Denominators

We now know that the quotient of a positive number and a negative number is a negative number. The fractions $-\dfrac{3}{4}, \dfrac{-3}{4},$ and $\dfrac{3}{-4}$ all represent the same negative number, negative three-fourths.

The Quotient of a Positive Number and a Negative Number

If a and b represent any real numbers, $b \neq 0$, then

$$\frac{a}{-b} = \frac{-a}{b} = -\frac{a}{b}$$

In mathematics we generally do not write a fraction with a negative sign in the denominator. When a negative sign appears in a denominator, we can move it to the numerator or place it in front of the fraction. For example, the fraction $\dfrac{5}{-7}$ should be written as either $-\dfrac{5}{7}$ or $\dfrac{-5}{7}$. Fractions also can be written using a slash, /. For example, the fraction $-\dfrac{5}{7}$ may be written $-5/7$ or $-(5/7)$.

EXAMPLE 8 ▶ Evaluate $\dfrac{3}{7} \div \left(\dfrac{-12}{35}\right)$.

Solution
$$\frac{3}{7} \div \left(\frac{-12}{35}\right) = \frac{\overset{1}{\cancel{3}}}{\underset{1}{\cancel{7}}} \cdot \left(\frac{\overset{5}{\cancel{35}}}{\underset{4}{\cancel{-12}}}\right) = \frac{1(5)}{1(-4)} = \frac{5}{-4} = -\frac{5}{4}$$

▶ **Now Try Exercise 73**

The operations on real numbers are summarized in **Table 1.1**.

TABLE 1.1   Summary of Operations on Real Numbers

Signs of Numbers	Addition	Subtraction	Multiplication	Division
Both Numbers Are Positive	Sum Is Always Positive	Difference May Be Either Positive or Negative	Product Is Always Positive	Quotient Is Always Positive
Examples				
6 and 2	$6 + 2 = 8$	$6 - 2 = 4$	$6 \cdot 2 = 12$	$6 \div 2 = 3$
2 and 6	$2 + 6 = 8$	$2 - 6 = -4$	$2 \cdot 6 = 12$	$2 \div 6 = \frac{1}{3}$
One Number Is Positive and the Other Number Is Negative	Sum May Be Either Positive or Negative	Difference May Be Either Positive or Negative	Product Is Always Negative	Quotient Is Always Negative
Examples				
6 and -2	$6 + (-2) = 4$	$6 - (-2) = 8$	$6(-2) = -12$	$6 \div (-2) = -3$
-6 and 2	$-6 + 2 = -4$	$-6 - 2 = -8$	$-6(2) = -12$	$-6 \div 2 = -3$
Both Numbers Are Negative	Sum Is Always Negative	Difference May Be Either Positive or Negative	Product Is Always Positive	Quotient Is Always Positive
Examples				
-6 and -2	$-6 + (-2) = -8$	$-6 - (-2) = -4$	$-6(-2) = 12$	$-6 \div (-2) = 3$
-2 and -6	$-2 + (-6) = -8$	$-2 - (-6) = 4$	$-2(-6) = 12$	$-2 \div (-6) = \frac{1}{3}$

4 Evaluate Divisions Involving 0

Now let's look at divisions involving the number 0. What is $\frac{0}{1}$ equal to? Note that $\frac{6}{3} = 2$ because $3 \cdot 2 = 6$. We can follow the same procedure to determine the value of $\frac{0}{1}$. Suppose that $\frac{0}{1}$ is equal to some number, which we will designate by $\boxed{?}$.

$$\text{If} \quad \frac{0}{1} = \boxed{?} \quad \text{then} \quad 1 \cdot \boxed{?} = 0$$

Since only $1 \cdot 0 = 0$, the $\boxed{?}$ must be 0. Thus, $\frac{0}{1} = 0$. Using the same technique, we can show that zero divided by any nonzero number is zero.

Zero Divided by a Nonzero Number

If a represents any real number except 0, then

$$0 \div a = \frac{0}{a} = 0$$

Now what is $\frac{1}{0}$ equal to?

$$\text{If} \quad \frac{1}{0} = \boxed{?} \quad \text{then} \quad 0 \cdot \boxed{?} = 1$$

But since 0 multiplied by any number will be 0, there is no value that can replace $\boxed{?}$. We say that $\frac{1}{0}$ is **undefined**. Using the same technique, we can show that any real number, except 0, divided by 0 is undefined.

Division by Zero

If a represents any real number except 0, then

$$a \div 0 \quad \text{or} \quad \frac{a}{0} \quad \text{is **undefined**}$$

What is $\frac{0}{0}$ equal to?

$$\text{If} \quad \frac{0}{0} = \boxed{?} \quad \text{then} \quad 0 \cdot \boxed{?} = 0$$

Since the product of any number and 0 is 0, the $\boxed{?}$ can be replaced by any real number. Therefore, the quotient $\frac{0}{0}$ cannot be determined, and so there is no answer. Thus, we will not use it in this course.*

Summary of Division Involving 0

If a represents any real number except 0, then

$$\frac{0}{a} = 0 \quad \text{and} \quad \frac{a}{0} \text{ is undefined}$$

EXAMPLE 9 ▶ Indicate whether each quotient is 0 or undefined.

a) $\dfrac{0}{2}$ **b)** $\dfrac{5}{0}$ **c)** $\dfrac{0}{-4}$ **d)** $\dfrac{-2}{0}$

Solution The answer to parts **a)** and **c)** is 0. The answer to parts **b)** and **d)** is undefined.

▶ **Now Try Exercise 95**

USING YOUR CALCULATOR **Multiplication and Division on a Scientific Calculator**

Below we show how numbers may be multiplied and divided on many calculators.

EVALUATE	KEYSTROKES	ANSWER DISPLAYED
$6(-23)$	6 $\boxed{\times}$ 23 $\boxed{+/-}$ $\boxed{=}$	-138
$\dfrac{-240}{-16}$	240 $\boxed{+/-}$ $\boxed{\div}$ 16 $\boxed{+/-}$ $\boxed{=}$	15

Multiplication and Division on a Graphing Calculator

EVALUATE	KEYSTROKES	ANSWER DISPLAYED
$6(-23)$	6 $\boxed{\times}$ $\boxed{(-)}$ 23 $\boxed{\text{ENTER}}$	-138
$\dfrac{-240}{-16}$	$\boxed{(-)}$ 240 $\boxed{\div}$ $\boxed{(-)}$ 16 $\boxed{\text{ENTER}}$	15

Since a positive number multiplied by a negative number will be negative, to obtain the product of $6(-23)$, you can multiply, $(6)(23)$ and write a negative sign before the answer. Since a negative number divided by a negative number is positive, $\dfrac{-240}{-16}$ could have been found by dividing $\dfrac{240}{16}$.

---

*At this level, some professors prefer to call $\frac{0}{0}$ *indeterminate* while others prefer to call $\frac{0}{0}$ *undefined*. In higher-level mathematics courses, $\frac{0}{0}$ is sometimes referred to as the *indeterminate form*.

EXERCISE SET 1.8

Concept/Writing Exercises

1. State the rules used to determine the sign of the product of two real numbers.

2. State the rules used to determine the sign of the quotient of two real numbers.

3. When multiplying three or more real numbers, explain how to determine the sign of the product of the numbers.

4. What is the product of 0 and any real number?

5. **a)** What is $\dfrac{0}{a}$ equal to where a is any nonzero real number?

   b) What is $\dfrac{a}{0}$ equal to where a is any nonzero real number?

6. How do we generally rewrite a fraction of the form $\dfrac{a}{-b}$, where a and b represent any positive real numbers?

7. **a)** Explain the difference between $3 - 5$ and $3(-5)$.

   b) Evaluate $3 - 5$ and $3(-5)$.

8. **a)** Explain the difference between $-4 - 2$ and $(-4)(-2)$.

   b) Evaluate $-4 - 2$ and $(-4)(-2)$.

9. **a)** Explain the difference between $x - y$ and $x(-y)$ where x and y represent any real numbers.

   If x is 5 and y is -2, find the value of

   b) $x - y$,   **c)** $x(-y)$, and   **d)** $-x - y$.

10. If x is -9 and y is 7, find the value of

   a) xy,   **b)** $x(-y)$,

   c) $x - y$, and   **d)** $-x - y$.

Determine the sign of each product. Explain how you determined the sign.

11. $(8)(4)(-5)$

12. $(-9)(-12)(20)$

13. $(-102)(-16)(24)(19)$

14. $(1054)(-92)(-16)(-37)$

15. $(-40)(-16)(30)(50)(-13)$

16. $(-1)(3)(-462)(-196)(-312)$

Practice the Skills

Find each product.

17. $(-5)(-4)$

18. $-4(2)$

19. $6(-3)$

20. $6(-2)$

21. $(-8)(-10)$

22. $(-3)(2)$

23. $-2.1(6)$

24. $-1(8.7)$

25. $6(7)$

26. $-9(-4)$

27. $9(-9)$

28. $(7)(-8)$

29. $(-5)(-6)$

30. $0(-5)$

31. $(-9)(0)(-6)$

32. $5(-4)(2)$

33. $(21)(-1)(4)$

34. $2(8)(-1)(-3)$

35. $-1(-3)(3)(-8)$

36. $(2)(-4)(-5)(-1)$

37. $(-4)(5)(-7)(1)$

38. $(-3)(2)(5)(3)$

39. $(-1)(3)(0)(-7)$

40. $(-6)(6)(4)(-4)$

Find each product.

41. $\left(\dfrac{-1}{2}\right)\left(\dfrac{3}{5}\right)$

42. $\left(\dfrac{1}{3}\right)\left(\dfrac{-3}{5}\right)$

43. $\left(\dfrac{-5}{9}\right)\left(\dfrac{-7}{15}\right)$

44. $\left(\dfrac{4}{5}\right)\left(\dfrac{-3}{10}\right)$

45. $\left(\dfrac{6}{-3}\right)\left(\dfrac{4}{-2}\right)$

46. $\left(\dfrac{9}{-10}\right)\left(\dfrac{6}{-7}\right)$

47. $\left(\dfrac{3}{4}\right)\left(\dfrac{-2}{15}\right)$

48. $\left(\dfrac{9}{10}\right)\left(\dfrac{7}{-8}\right)$

Find each quotient.

49. $\dfrac{42}{6}$

50. $25 \div 5$

51. $-16 \div (-4)$

52. $\dfrac{-18}{9}$

53. $\dfrac{-36}{-9}$

54. $\dfrac{30}{-6}$

55. $\dfrac{36}{-2}$

56. $\dfrac{-15}{-1}$

57. $\dfrac{-19.8}{-2}$

58. $-15.6/(-3)$

59. $40/(-4)$

60. $\dfrac{63}{-7}$

61. $\dfrac{-66}{11}$

62. $\dfrac{-25}{-5}$

63. $\dfrac{48}{-6}$

64. $\dfrac{-10}{10}$

65. $-64.8 \div (-4)$

66. $-86.4/(-2)$

67. Divide 0 by 4.

68. Divide 26 by -13.

69. Divide 30.8 by -5.6.

70. Divide -67.64 by 7.6.

71. Divide -30 by -5.

72. Divide -36 by -6.

Find each quotient.

73. $\dfrac{3}{12} \div \left(\dfrac{-5}{8}\right)$

74. $4 \div \left(\dfrac{-6}{13}\right)$

75. $\dfrac{-5}{12} \div (-3)$

76. $\dfrac{6}{15} \div \left(\dfrac{7}{30}\right)$

77. $\dfrac{-15}{21} \div \left(\dfrac{-15}{21}\right)$

78. $\dfrac{-4}{9} \div \left(\dfrac{-6}{7}\right)$

79. $(-12) \div \dfrac{5}{12}$

80. $\dfrac{-16}{3} \div \left(\dfrac{5}{-9}\right)$

Evaluate.

81. $-4(8)$

82. $\dfrac{-18}{-2}$

83. $\dfrac{-100}{-5}$

84. $-50 \div (-10)$

85. $-7(2)$ **86.** $6.4(-8)$ **87.** $27.9 \div (-3)$ **88.** Divide 130 by -10.

89. $-100 \div 5$ **90.** $4(-2)(-1)(-5)$ **91.** Divide -90 by -90.   **92.** $(6)(1)(-3)(4)$

Indicate whether each quotient is 0 or undefined.

93. $0 \div 8.6$ **94.** $\dfrac{-2.7}{0}$ **95.** $\dfrac{5}{0}$ **96.** $\dfrac{0}{1}$

97. $0 \div (-7)$ **98.** $\dfrac{6}{0}$ **99.** 8 divided by 0   **100.** 0 divided by 12

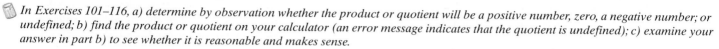 *In Exercises 101–116, a) determine by observation whether the product or quotient will be a positive number, zero, a negative number; or undefined; b) find the product or quotient on your calculator (an error message indicates that the quotient is undefined); c) examine your answer in part b) to see whether it is reasonable and makes sense.*

101. $92(-38)$ **102.** $-168 \div 42$ **103.** $-240/15$ **104.** $0/12$

105. $243 \div (-27)$ **106.** $(323)(-115)$ **107.** $(-49)(-126)$ **108.** $(1530)(0)$

109. $0 \div 5335$ **110.** $-86.4 \div (-36)$ **111.** $8.2 \div 0$ **112.** $-37.74 \div 0$

113. $8 \div (2.5)$ **114.** $(1.1)(9.72)(6.3)$ **115.** $(-3.0)(4.2)(-18)$ **116.** $-288.86/1.43$

Indicate whether each statement is true or false.

117. The product of two negative numbers is a negative number.

118. The product of a positive number and a negative number is a negative number.

119. The quotient of two numbers with unlike signs is a positive number.

120. The quotient of two negative numbers is a positive number.

121. The product of an even number of negative numbers is a positive number.

122. Zero divided by 1 is 1.

123. The product of an odd number of negative numbers is a negative number.

124. Six divided by 0 is 0.

125. Zero divided by 1 is undefined.

126. The product of 0 and any real number is 0.

127. Five divided by 0 is undefined.

128. Division by 0 does not result in a real number.

Problem Solving

129. **Football** A high school football team is penalized three times, each time with a loss of 15 yards, or -15 yards. Find the total loss due to penalties.

130. **Submarine Dive** A submarine is at a depth of -160 feet (160 feet below sea level). It dives to 3 times that depth. Find its new depth.

131. **Credit Card** Leona De Vito's balance on her credit card is $-\$520$ (she owes \$520). She pays back $\frac{1}{5}$ of this balance.
   a) How much did she pay back?
   b) What is her new balance?

132. **Money Owed** Brian Philip owes his Dad \$500. After he makes four payments of \$40 each, how much will he still owe?

133. **Garage Sale** Four sisters made a total of \$775.40 at a garage sale. After they each give their husbands \$50 and split the remaining amount equally, how much will each woman receive?

134. **Wind Chill** On Monday in Minneapolis the wind chill was $-30°F$. On Tuesday the wind chill was only $\frac{1}{3}$ of what it was on Monday. What was the wind chill on Tuesday?

135. **Test Score** Because of incorrect work, Josue Nunez lost 4 points on each of the five questions on his math test.
   a) How many points did Josue lose altogether?
   b) If the maximum score possible was 100%, what is Josue's test score?

136. **Reciprocals** Most calculators have a *reciprocal key*, $\boxed{1/x}$.
   a) Press a number key between 1 and 9 on your calculator. Then press the $\boxed{1/x}$ key. Indicate what happened when you pressed this key.
   b) Press the $\boxed{1/x}$ key a second time. What happens now?
   c) What do you think will happen if you enter 0 and then press the $\boxed{1/x}$ key? Explain your answer.
   d) Enter 0 and then press the $\boxed{1/x}$ key to see whether your answer to part **c)** was correct. If not, explain why.

137. **Heart Rate** The Johns Hopkins Medical Letter states that to find a person's *target heart rate* in beats per minute, follow this procedure. Subtract the person's age from 220, then multiply this difference by 60% and 75%. The difference multiplied by 60% gives the lower limit and the difference multiplied by 75% gives the upper limit.
   a) Find the target heart rate range of a 50-year-old.
   b) Find your own target heart rate.

Challenge Problems

We will learn in the next section that $2^3 = 2\cdot2\cdot2$ and $x^n = \underbrace{x\cdot x\cdot x\cdots\cdot x}_{(n\ factors\ of\ x)}.$

Use this information to evaluate each expression.

138. 3^4 　　　　**139.** $(-5)^3$ 　　　　**140.** $\left(\dfrac{2}{3}\right)^3$ 　　　　**141.** 1^{100} 　　　　**142.** $(-1)^{81}$

143. Will the product of $(-1)(-2)(-3)(-4)\cdots(-10)$ be a positive number or a negative number? Explain how you determined your answer.

144. Will the product of $(1)(-2)(3)(-4)(5)(-6)\cdots(33)(-34)$ be a positive number or a negative number? Explain how you determined your answer.

Group Activity

Discuss and answer Exercise 145 as a group, according to the instructions.

145. a) Each member of the group is to do this procedure separately. At this time do not share your number with the other members of your group.
1. Choose a number between 2 and 10.
2. Multiply your number by 9.
3. Add the two digits in the product together.
4. Subtract 5 from the sum.
5. Now choose the corresponding letter of the alphabet that corresponds with the difference found. For example, 1 is a, 2 is b, 3 is c, and so on.

6. Choose a *one-word* country that starts with that letter.
7. Now choose a *one-word* animal that starts with the last letter of the country selected.
8. Finally, choose a color that starts with the last letter of the animal chosen.

b) Now share your final answer with the other members of your group. Did you all get the same answer?

c) Most people will obtain the answer *orange*. As a group, write a paragraph or two explaining why.

Cumulative Review Exercises

[1.5] **146.** Insert either $<$, $>$, or $=$ in the shaded area to make a true statement.
$$|-3.6|\ \_\_\_\ |-2.7|$$

[1.6] **147.** Add $-\dfrac{7}{12} + \left(-\dfrac{1}{10}\right)$.

[1.7] **148.** Subtract -18 from -20.
149. Evaluate $6 - 3 - 4 - 2$
150. Evaluate $5 - (-2) + 3 - 7$

1.9 Exponents, Parentheses, and the Order of Operations

1 Learn the meaning of exponents.
2 Evaluate expressions containing exponents.
3 Learn the difference between $-x^2$ and $(-x)^2$.
4 Learn the order of operations.
5 Learn the use of parentheses.
6 Evaluate expressions containing variables.

1 Learn the Meaning of Exponents

To understand certain topics in algebra, you must understand exponents. Exponents are introduced in this section and are discussed in more detail in Chapter 4.

In the expression 4^2, the 4 is called the **base**, and the 2 is called the **exponent**. The number 4^2 is read "4 squared" or "4 to the second power" and means

$$\underbrace{4\cdot4}_{2\ factors\ of\ 4} = 4^2$$

The number 4^3 is read "4 cubed" or "4 to the third power" and means

$$\underbrace{4\cdot4\cdot4}_{3\ factors\ of\ 4} = 4^3$$

In general, the number b to the nth power, written b^n, means

$$\underbrace{b\cdot b\cdot b\cdot\ \cdots\ \cdot b}_{n\ factors\ of\ b} = b^n$$

Thus, $b^4 = b\cdot b\cdot b\cdot b$ or $bbbb$ and $x^3 = x\cdot x\cdot x$ or xxx.

2 Evaluate Expressions Containing Exponents

Let's evaluate some expressions that contain exponents.

EXAMPLE 1 ▶ Evaluate.   **a)** 3^2 **b)** 2^5 **c)** 1^5 **d)** $(-6)^2$ **e)** $(-2)^3$ **f)** $\left(\dfrac{2}{3}\right)^2$

Solution

a) $3^2 = 3 \cdot 3 = 9$

b) $2^5 = 2 \cdot 2 \cdot 2 \cdot 2 \cdot 2 = 32$

c) $1^5 = 1 \cdot 1 \cdot 1 \cdot 1 \cdot 1 = 1$ (1 raised to any power equals 1; why?)

d) $(-6)^2 = (-6)(-6) = 36$

e) $(-2)^3 = (-2)(-2)(-2) = -8$

f) $\left(\dfrac{2}{3}\right)^2 = \left(\dfrac{2}{3}\right)\left(\dfrac{2}{3}\right) = \dfrac{4}{9}$

▶ **Now Try Exercise 29**

It is not necessary to write exponents of 1. For example, when writing xxy, we write x^2y and not x^2y^1. **Whenever we see a variable or number without an exponent, we always assume that the variable or number has an exponent of 1.**

EXAMPLES OF EXPONENTIAL NOTATION

a) $xyxx = x^3y$

b) $xyzzy = xy^2z^2$

c) $3aabbb = 3a^2b^3$

d) $5xyyyy = 5xy^4$

e) $4 \cdot 4rrs = 4^2r^2s$

f) $5 \cdot 5 \cdot 5mmn = 5^3m^2n$

Notice in parts **a)** and **b)** that the order of the factors does not matter.

Helpful Hint

Note that $x + x + x + x + x + x = 6x$ and $x \cdot x \cdot x \cdot x \cdot x \cdot x = x^6$. Be careful that you do not get addition and multiplication confused.

3 Learn the Difference Between $-x^2$ and $(-x)^2$

An exponent refers only to the number or variable that directly precedes it unless parentheses are used to indicate otherwise. For example, in the expression $3x^2$, only the x is squared. In the expression $-x^2$, only the x is squared. We can write $-x^2$ as $-1x^2$ because any real number may be multiplied by 1 without affecting its value.

$$-x^2 = -1x^2$$

By looking at $-1x^2$ we can see that only the x is squared, not the -1. If the entire expression $-x$ were to be squared, we would need to use parentheses and write $(-x)^2$. Note the difference in the following two examples:

$$-x^2 = -(x)(x)$$
$$(-x)^2 = (-x)(-x)$$

Consider the expressions -3^2 and $(-3)^2$. How do they differ?

$$-3^2 = -(3)(3) = -9$$
$$(-3)^2 = (-3)(-3) = 9$$

Helpful Hint

The expression $-x^2$ is read "negative x squared," or "the opposite of x squared." The expression $(-x)^2$ is read "negative x, quantity squared."

EXAMPLE 2 ▶ Evaluate.

a) -5^2 **b)** $(-5)^2$ **c)** -2^3 **d)** $(-2)^3$

Solution

a) $-5^2 = -(5)(5) = -25$ **b)** $(-5)^2 = (-5)(-5) = 25$

c) $-2^3 = -(2)(2)(2) = -8$ **d)** $(-2)^3 = (-2)(-2)(-2) = -8$

▶ **Now Try Exercise 21**

EXAMPLE 3 ▶ Evaluate.

a) -2^4 **b)** $(-2)^4$

Solution

a) $-2^4 = -(2)(2)(2)(2) = -16$ **b)** $(-2)^4 = (-2)(-2)(-2)(-2) = 16$

▶ **Now Try Exercise 33**

USING YOUR CALCULATOR   Use of $\boxed{x^2}$, $\boxed{y^x}$, and $\boxed{\wedge}$ Keys

The $\boxed{x^2}$ key is used to square a value. For example, to evaluate 5^2, we would do the following.

	KEYSTROKES	ANSWER DISPLAYED
Scientific Calculator	5 $\boxed{x^2}$	25
Graphing Calculator	5 $\boxed{x^2}$ $\boxed{\text{ENTER}}$	25

To evaluate $(-5)^2$ on a calculator, we would do the following.

	KEYSTROKES	ANSWER DISPLAYED
*Scientific Calculator	5 $\boxed{+/-}$ $\boxed{x^2}$	25
Graphing Calculator	$\boxed{(}$ $\boxed{(-)}$ 5 $\boxed{)}$ $\boxed{x^2}$ $\boxed{\text{ENTER}}$	25

To raise a value to a power greater than 2, we use the $\boxed{y^x}$** or $\boxed{\wedge}$ key. To use these keys you enter the number, then press either the $\boxed{y^x}$ or $\boxed{\wedge}$ key, then enter the exponent. Following we show how to evaluate 2^5 and $(-2)^5$.

	EVALUATE	KEYSTROKES	ANSWER DISPLAYED
Scientific Calculator	2^5	2 $\boxed{y^x}$ 5 $\boxed{=}$	32
*Scientific Calculator	$(-2)^5$	2 $\boxed{+/-}$ $\boxed{y^x}$ 5 $\boxed{=}$	-32
Graphing Calculator	2^5	2 $\boxed{\wedge}$ 5 $\boxed{\text{ENTER}}$	32
Graphing Calculator	$(-2)^5$	$\boxed{(}$ $\boxed{(-)}$ 2 $\boxed{)}$ $\boxed{\wedge}$ 5 $\boxed{\text{ENTER}}$	-32

Possibly the easiest way to raise negative numbers to a power may be to raise the positive number to the power and then write a negative sign before the final answer if needed. *A negative number raised to an odd power will be negative, and a negative number raised to an even power will be positive.* Can you explain why this is true?

*Some newer scientific calculators have a $\boxed{(-)}$ key. To evaluate a negative expression raised to a power on these calculators, follow the instructions for the graphing calculator.
**Some calculators use a $\boxed{x^y}$ key instead of a $\boxed{y^x}$ key.

4 Learn the Order of Operations

Now that we have introduced exponents we can present the **order of operations**. Can you evaluate $2 + 3 \cdot 4$? Is it 20? Or is it 14? To answer this, you must know the order of operations to follow when evaluating a mathematical expression. You will often have to evaluate expressions containing multiple operations.

> ### Order of Operations To Evaluate Mathematical Expressions, Use the Following Order
>
> 1. First, evaluate the information within **parentheses** ( ), brackets [ ], or braces { }. These are **grouping symbols**, for they group information together. A fraction bar, −, also serves as a grouping symbol. If the expression contains nested parentheses (one pair of parentheses within another pair), evaluate the information in the innermost parentheses first.
>
> 2. Next, evaluate all **exponents**.
>
> 3. Next, evaluate all **multiplications** or **divisions** in the order in which they occur, working from left to right.
>
> 4. Finally, evaluate all **additions** or **subtractions** in the order in which they occur, working from left to right.

Some students remember the word PEMDAS or the phrase "Please Excuse My Dear Aunt Sally" to help them remember the order of operations. PEMDAS helps them remember the order: **P**arentheses, **E**xponents, **M**ultiplication, **D**ivision, **A**ddition, **S**ubtraction. Remember, this does not imply multiplication before division or addition before subtraction.

We can now evaluate $2 + 3 \cdot 4$. Since multiplications are performed before additions,

$$2 + 3 \cdot 4 \quad \text{means} \quad 2 + (3 \cdot 4) = 2 + 12 = 14$$

 **USING YOUR CALCULATOR**

We now know that $2 + 3 \cdot 4$ means $2 + (3 \cdot 4)$ and has a value of 14. What will a calculator display if you key in the following?

$$2 \boxed{+} 3 \boxed{\times} 4 \boxed{=}$$

The answer depends on your calculator. *Scientific and graphing calculators* will evaluate an expression following the rules just stated.

	KEYSTROKES	ANSWER DISPLAYED
Scientific Calculator	$2 \boxed{+} 3 \boxed{\times} 4 \boxed{=}$	14
Graphing Calculator	$2 \boxed{+} 3 \boxed{\times} 4 \boxed{\text{ENTER}}$	14

Nonscientific calculators will perform operations in the order they are entered.

		ANSWER DISPLAYED
Nonscientific Calculator	$2 \boxed{+} 3 \boxed{\times} 4 \boxed{=}$	20

Remember that in algebra, unless otherwise instructed by parentheses, we always perform multiplications and divisions before additions and subtractions. In this course you should be using either a scientific or graphing calculator.

5 Learn the Use of Parentheses

Parentheses or brackets may be used (1) to change the order of operations to be followed in evaluating an algebraic expression or (2) to help clarify the understanding of an expression.

To evaluate the expression $2 + 3 \cdot 4$, we would normally perform the multiplication, $3 \cdot 4$, first. If we wished to have the addition performed before the multiplication, we could indicate this by placing parentheses around $2 + 3$:

$$(2 + 3) \cdot 4 = 5 \cdot 4 = 20$$

Consider the expression $1 \cdot 3 + 2 \cdot 4$. According to the order of operations, multiplications are to be performed before additions. We can rewrite this expression as $(1 \cdot 3) + (2 \cdot 4)$. Note that the order of operations was not changed. The parentheses were used only to help clarify the order to be followed.

Sometimes it may be necessary to use more than one set of parentheses to indicate the order to be followed in an expression. As indicated in step 1 of the Order of Operations box, when one set of parentheses is within another set of parentheses, we call this **nested parentheses**. For example, the expression $6(2 + 3(4 + 1))$ has nested parentheses. Often, to make an expression with nested parentheses easier to follow, brackets, [ ], or braces, { }, are used in place of multiple parentheses. Thus, we could write the expression $6(2 + 3(4 + 1))$ as $6[2 + 3(4 + 1)]$. Whenever we are given an expression with nested parentheses, we always evaluate the numbers in the *innermost parentheses first*. Color shading is used in the following examples to indicate the order in which the expression is evaluated.

$$6[2 + 3(4 + 1)] = 6[2 + 3(5)] = 6[2 + 15] = 6[17] = 102$$
$$4[3(6 - 4) \div 6] = 4[3(2) \div 6] = 4[6 \div 6] = 4[1] = 4$$
$$\{2 + [(8 \div 4)^2 - 1]\}^2 = [2 + (2^2 - 1)]^2 = [2 + (4 - 1)]^2 = (2 + 3)^2 = 5^2 = 25$$

Now we will work some examples, but before we do, read the following Helpful Hint.

Helpful Hint

If parentheses are not used to change the order of operations, multiplications and divisions are always performed before additions and subtractions. When a problem has only multiplications and divisions, work from left to right. Similarly, when a problem has only additions and subtractions, work from left to right.

EXAMPLE 4 ▶ Evaluate $6 + 3 \cdot 5^2 - 4$.

Solution   Colored shading is used to indicate the order in which the expression is to be evaluated.

$$6 + 3 \cdot 5^2 - 4 \quad \text{Exponent}$$
$$= 6 + 3 \cdot 25 - 4 \quad \text{Multiply.}$$
$$= 6 + 75 - 4 \quad \text{Add.}$$
$$= 81 - 4$$
$$= 77$$

▶ **Now Try Exercise 61**

EXAMPLE 5 ▶ Evaluate $-7 + 2[-6 + (36 \div 3^2)]$.

Solution

$$-7 + 2[-6 + (36 \div 3^2)] \quad \text{Exponent}$$
$$= -7 + 2[-6 + (36 \div 9)] \quad \text{Divide.}$$
$$= -7 + 2[-6 + 4] \quad \text{Add.}$$
$$= -7 + 2[-2] \quad \text{Multiply.}$$
$$= -7 - 4$$
$$= -11$$

▶ **Now Try Exercise 75**

EXAMPLE 6 ▶ Evaluate $(8 \div 2) + 7(5 - 2)^2$.

Solution

$$\boxed{(8 \div 2)} + 7(\boxed{5 - 2})^2 \qquad \textit{Parentheses}$$

$$= 4 + 7\boxed{(3)^2} \qquad \textit{Exponent}$$

$$= 4 + \boxed{7 \cdot 9} \qquad \textit{Multiply.}$$

$$= 4 + 63$$

$$= 67$$

▶ **Now Try Exercise 81**

EXAMPLE 7 ▶ Evaluate $-8 - 81 \div 9 \cdot 2^2 + 7$.

Solution

$$-8 - 81 \div 9 \cdot \boxed{2^2} + 7 \qquad \textit{Exponent}$$

$$= -8 - \boxed{81 \div 9} \cdot 4 + 7 \qquad \textit{Divide.}$$

$$= -8 - \boxed{9 \cdot 4} + 7 \qquad \textit{Multiply.}$$

$$= \boxed{-8 - 36} + 7 \qquad \textit{Subtract.}$$

$$= -44 + 7$$

$$= -37$$

▶ **Now Try Exercise 79**

EXAMPLE 8 ▶ Evaluate.

a) $-4^2 + 6 \div 3$ **b)** $(-4)^2 + 6 \div 3$

Solution

a)

$$- \boxed{4^2} + 6 \div 3 \qquad \textit{Exponent}$$

$$= -16 + \boxed{6 \div 3} \qquad \textit{Divide.}$$

$$= -16 + 2$$

$$= -14$$

b)

$$\boxed{(-4)^2} + 6 \div 3 \qquad \textit{Exponent}$$

$$= 16 + \boxed{6 \div 3} \qquad \textit{Divide.}$$

$$= 16 + 2$$

$$= 18$$

▶ **Now Try Exercise 85**

EXAMPLE 9 ▶ Evaluate $\dfrac{3}{8} - \dfrac{2}{5} \cdot \dfrac{1}{12}$.

Solution First perform the multiplication.

$$\frac{3}{8} - \left(\frac{\overset{1}{2}}{5} \cdot \frac{1}{\underset{6}{12}} \right) \qquad \textit{Multiply.}$$

$$= \frac{3}{8} - \frac{1}{30} \qquad \textit{Subtract.}$$

$$= \frac{45}{120} - \frac{4}{120}$$

$$= \frac{41}{120}$$

▶ **Now Try Exercise 91**

EXAMPLE 10 ▶ Write the following statements as mathematical expressions using parentheses and brackets and then evaluate: Multiply 12 by 3. Add 8 to this product. Subtract 7 from this sum. Divide this difference by 6.

Solution

$$12 \cdot 3 \qquad \textit{Multiply 12 by 3.}$$
$$(12 \cdot 3) + 8 \qquad \textit{Add 8.}$$
$$[(12 \cdot 3) + 8] - 7 \qquad \textit{Subtract 7.}$$
$$\{[(12 \cdot 3) + 8] - 7\} \div 6 \qquad \textit{Divide the difference by 6.}$$

Now evaluate.

$$\{[(\; 12 \cdot 3 \;) + 8] - 7\} \div 6$$
$$= \{[36 + 8] - 7\} \div 6$$
$$= \{44 - 7\} \div 6$$
$$= 37 \div 6$$
$$= \frac{37}{6}$$

▶ **Now Try Exercise 135**

As shown in Example 10, sometimes brackets, [ ], and braces, { }, are used in place of parentheses to help avoid confusion. If only parentheses had been used, the preceding expression would appear as $(((12 \cdot 3) + 8) - 7) \div 6$.

USING YOUR CALCULATOR   Using Parentheses

When evaluating an expression on a calculator where the order of operations is to be changed, you will need to use parentheses. If you are not sure whether they are needed, it will not hurt to add them. Consider $\dfrac{8}{4 - 2}$. Since we wish to divide 8 by the difference $4 - 2$, we need to use parentheses.

EVALUATE	KEYSTROKES	ANSWER DISPLAYED
$\dfrac{8}{4 - 2}$	8 ÷ (4 − 2) = *	4

What would you obtain if you evaluated 8 ÷ 4 − 2 = on a scientific calculator? Why would you get that result?

To evaluate $\left(\dfrac{2}{5}\right)^2$ on a scientific calculator, we press the following keys.

EVALUATE	KEYSTROKES	ANSWER DISPLAYED
$\left(\dfrac{2}{5}\right)^2$	2 ÷ 5 = x^2 **	.16
	or (2 ÷ 5) x^2	.16

What would you obtain if you evaluated 2 ÷ 5 x^2 on a scientific calculator? Why?

* If using a graphing calculator, replace = with ENTER . Everything else remains the same.
** On a graphing calculator, replace = with ENTER and press ENTER after x^2 .

6 Evaluate Expressions Containing Variables

Now we will evaluate some expressions for given values of the variables.

EXAMPLE 11 ▶ Evaluate $5x - 4$ when $x = 3$.

Solution   Substitute 3 for x in the expression.

$$5x - 4 = 5(3) - 4 = 15 - 4 = 11$$

▶ **Now Try Exercise 115**

EXAMPLE 12 ▶ Evaluate **a)** x^2 **b)** $-x^2$ and **c)** $(-x)^2$ when $x = 3$.

Solution   Substitute 3 for x.

a) $x^2 = 3^2 = 3(3) = 9$

c) $(-x)^2 = (-3)^2 = (-3)(-3) = 9$

b) $-x^2 = -3^2 = -(3)(3) = -9$

▶ **Now Try Exercise 105**

EXAMPLE 13 ▶ Evaluate **a)** y^2 **b)** $-y^2$ and **c)** $(-y)^2$ when $y = -4$.

Solution   Substitute -4 for y.

a) $y^2 = (-4)^2 = (-4)(-4) = 16$

c) $(-y)^2 = [-(-4)]^2 = (4)^2 = 16$

b) $-y^2 = -(-4)^2 = -(-4)(-4) = -16$

▶ **Now Try Exercise 107**

   Note that $-x^2$ will always be a negative number for any nonzero value of x, and $(-x)^2$ will always be a positive number for any nonzero value of x. Can you explain why? See Exercise 6 on page 77.

Avoiding Common Errors

The expression $-x^2$ means $-(x^2)$. When asked to evaluate $-x^2$ for any real number x, many students will incorrectly treat $-x^2$ as $(-x)^2$. For example, to evaluate $-x^2$ when $x = 5$,

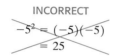

CORRECT	INCORRECT
$-5^2 = -(5^2) = -(5)(5)$	$-5^2 = (-5)(-5)$
$= -25$	$= 25$

EXAMPLE 14 ▶ Evaluate $2x^2 + 4x + 1$ when $x = \dfrac{1}{4}$.

Solution   Substitute $\dfrac{1}{4}$ for each x in the expression, then evaluate using the order of operations.

$$2x^2 + 4x + 1 = 2\left(\frac{1}{4}\right)^2 + 4\left(\frac{1}{4}\right) + 1 \qquad \textit{Substitute.}$$

$$= 2\left(\frac{1}{16}\right) + 4\left(\frac{1}{4}\right) + 1 \qquad \textit{Exponent}$$

$$= \frac{1}{8} + 1 + 1 \qquad \textit{Multiply.}$$

$$= \frac{1}{8} + 2 \qquad \textit{Add.}$$

$$= 2\frac{1}{8}$$

▶ **Now Try Exercise 123**

EXAMPLE 15 ▶ Evaluate $-y^2 + 3(x + 2) - 5$ when $x = -3$ and $y = -2$.

Solution   Substitute -3 for each x and -2 for each y, then evaluate using the order of operations.

$$-y^2 + 3(x + 2) - 5 = -(-2)^2 + 3(-3 + 2) - 5 \qquad \textit{Substitute.}$$

$$= -(-2)^2 + 3(-1) - 5 \qquad \textit{Parentheses}$$

$$= -(4) + 3(-1) - 5 \qquad \textit{Exponent}$$

$$= -4 - 3 - 5 \qquad \textit{Multiply.}$$

$$= -7 - 5 \qquad \textit{Subtract, left to right.}$$

$$= -12$$

▶ **Now Try Exercise 131**

USING YOUR CALCULATOR   Evaluating Expressions on a Scientific Calculator

Later in this course you will need to evaluate an expression like $3x^2 - 2x + 5$ for various values of x. Below we show how to evaluate such expressions.

EVALUATE

a) $3x^2 - 2x + 5$, for $x = 4$

$3(4)^2 - 2(4) + 5$

b) $3x^2 - 2x + 5$, for $x = -6$

$3(-6)^2 - 2(-6) + 5$

c) $-x^2 - 3x - 5$, for $x = -2$

$-(-2)^2 - 3(-2) - 5$

KEYSTROKES

$3\boxed{\times}4\boxed{x^2}\boxed{-}2\boxed{\times}4\boxed{+}5\boxed{=}45$

$3\boxed{\times}6\boxed{+/-}\boxed{x^2}\boxed{-}2\boxed{\times}6\boxed{+/-}\boxed{+}5\boxed{=}125$

$1\boxed{+/-}\boxed{\times}2\boxed{+/-}\boxed{x^2}\boxed{-}3\boxed{\times}2\boxed{+/-}\boxed{-}5\boxed{=}-3$

Remember in part **c)** that $-x^2 = -1x^2$.

USING YOUR GRAPHING CALCULATOR   Evaluating Expressions on a Graphing Calculator

All graphing calculators have a procedure for evaluating expressions. You generally need to enter the value to be used for the variable and the expression to be evaluated. After the $\boxed{\text{ENTER}}$ key is pressed, the graphing calculator displays the answer. The procedure varies from calculator to calculator. Below we show how an expression is evaluated on a TI-84 Plus. On this calculator the store key, $\boxed{\text{STO} \blacktriangleright}$, is used to store a value. Stored values and expressions are separated using a colon, which is obtained by pressing $\boxed{\text{ALPHA}}$ followed by $\boxed{\cdot}$.

EVALUATE

$3x^2 - 2x + 5$ for $x = -6$

KEYSTROKES ON TI-84 PLUS

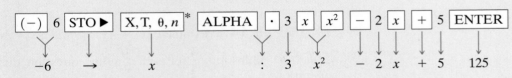

Display shown:

Notice that to obtain an x^2 on the display, we press the $\boxed{\text{X, T, θ, n}}$ key to select the variable x, then we press the $\boxed{x^2}$ key, which is used to square the variable or number selected.

A nice feature of graphing calculators is that to evaluate an expression for different values of a variable you do not have to re-enter the expression each time. For example, on the TI-84 Plus if you press $\boxed{2^{nd}}$ followed by $\boxed{\text{ENTER}}$ it displays the expression again. You then change the value stored for x to the new value and press $\boxed{\text{ENTER}}$ to evaluate the expression with the new value of the variable.

Each brand of calculator uses different keys and procedures. We have given just a quick overview. Please read the manual that comes with your graphing calculator for a complete explanation of how to evaluate expressions.

*This key can be used to generate any of these letters ($θ$ is a Greek letter). From this point on, in displays of keystrokes, to generate an x we will just show $\boxed{x}$ rather than $\boxed{\text{X, T, θ, n}}$.

EXERCISE SET 1.9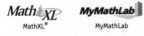

Concept/Writing Exercises

1. In the expression a^b, what is the a called and what is the b called?

2. Explain the meaning of

   a) 3^2

   b) 5^4

   c) x^n

3. a) What is the exponent on a number or variable that has no exponent illustrated?

   b) In the expression $5x^3y^2z$, what is the exponent on the 5, the x, the y, and the z?

4. Write a simplified expression for the following.

   a) $x + x + x + x + x$

   b) $x \cdot x \cdot x \cdot x \cdot x$

5. Write a simplified expression for the following.
  a) $y + y + y + y$
  b) $y \cdot y \cdot y \cdot y$

6. a) Explain why $-x^2$ will always be a negative number for any nonzero real number selected for x.
  b) Explain why $(-x)^2$ will always be a positive number for any nonzero real number selected for x.

7. List the order of operations to be followed when evaluating a mathematical expression.

8. When an expression has only additions and subtractions or only multiplications or divisions, how is it evaluated?

9. If you evaluate $4 + 5 \times 2$ on a calculator and obtain an answer of 18, is the calculator a scientific calculator? Explain.

10. List two reasons why parentheses are used in an expression.

11. Determine the results obtained on a scientific calculator if the following keys are pressed.
  a) $15 \boxed{-} 10 \boxed{\div} 5 \boxed{=}$
  b) $\boxed{(} 15 \boxed{-} 10 \boxed{)} \boxed{\div} 5 \boxed{=}$
  c) Which keystrokes, **a)** or **b)**, are used to evaluate $\dfrac{15 - 10}{5}$? Explain.

12. Determine the results obtained on a scientific calculator if the following keys are pressed.
  a) $20 \boxed{\div} 5 \boxed{-} 3 \boxed{=}$
  b) $20 \boxed{\div} \boxed{(} 5 \boxed{-} 3 \boxed{)} \boxed{=}$
  c) Which keystrokes, **a)** or **b)**, are used to evaluate $\dfrac{20}{5 - 3}$? Explain.

In Exercises 13 and 14, **a)** *write the step-by-step procedure you would use to evaluate the expression, and* **b)** *evaluate the expression.*

13. $[10 - (16 \div 4)]^2 - 6^3$

14. $[(8 \cdot 3) - 4^2]^2 - 5$

In Exercises 15 and 16, **a)** *write the step-by-step procedure you would use to evaluate the expression for the given value of the variable, and* **b)** *evaluate the expression for the given value of the variable.*

15. $-4x^2 + 3x - 6$ when $x = 5$

16. $-5x^2 - 2x + 8$ when $x = -2$

Practice the Skills

Evaluate.

17. 5^2

18. 2^3

19. 1^7

20. 4^1

21. -8^2

22. 7^3

23. $(-3)^2$

24. -6^3

25. $(-1)^3$

26. 2^5

27. -10^2

28. 5^3

29. $(-9)^2$

30. $(-3)^3$

31. 3^3

32. -7^2

33. $(-4)^4$

34. -4^4

35. -2^4

36. $3^2(4)^2$

37. $\left(\dfrac{3}{4}\right)^2$

38. $\left(\dfrac{5}{8}\right)^3$

39. $\left(-\dfrac{1}{2}\right)^5$

40. $\left(-\dfrac{2}{3}\right)^4$

41. $5^2 \cdot 3^2$

42. $(-1)^4(2)^4$

43. $4^3 \cdot 3^2$

44. $(-2)^3(-1)^9$

In Exercises 45–56, **a)** *determine by observation whether the answer should be positive or negative and explain your answer;* **b)** *evaluate the expression on your calculator; and* **c)** *determine whether your answer in part* **b)** *is reasonable and makes sense.*

45. 7^3

46. 4^6

47. 6^4

48. -2^5

49. $(-3)^5$

50. 10^3

51. $(-5)^4$

52. $(1.3)^3$

53. $-(-9)^2$

54. $(-3.3)^3$

55. $-\left(\dfrac{3}{8}\right)^2$

56. $\left(-\dfrac{3}{4}\right)^3$

Evaluate.

57. $3 + 2 \cdot 6$

58. $7 - 5^2 + 2$

59. $6 - 6 + 8$

60. $(8^2 \div 4) - (20 - 4)$

61. $-7 + 2 \cdot 6^2 - 8$

62. $6 + 2 \cdot 3^2 - 10$

63. $-3^3 + 27$

64. $(-2)^3 + 8 \div 4$

65. $(4 - 3) \cdot (5 - 1)^2$

66. $-10 - 6 - 3 - 2$

67. $3 \cdot 7 + 4 \cdot 2$

68. $4^2 - 3 \cdot 4 - 6$

69. $5 - 2(7 + 5)$

70. $8 + 3(6 + 4)$

71. $-32 - 5(7 - 10)^2$

72. $-40 - 3(4 - 8)^2$

73. $\dfrac{3}{4} + 2\left(\dfrac{1}{5}\right)^2$

74. $-\dfrac{2}{3} - 3\left(\dfrac{3}{4}\right)^2$

75. $-4 + 3[-1 + (12 \div 2^2)]$

76. $-2 + 4[-3 + (48 \div 4^2)]$

77. $(6 \div 3)^3 + 4^2 \div 8$

78. $4 + (4^2 - 13)^4 - 3$

79. $-7 - 48 \div 6 \cdot 2^2 + 5$

80. $-7 - 56 \div 7 \cdot 2^2 + 4$

81. $(9 \div 3) + 4(7 - 2)^2$

82. $(12 \div 4) + 5(6 - 4)^2$

83. $[4 + ((5 - 2)^2 \div 3)^2]^2$

84. $(20 \div 5 \cdot 5 \div 5 - 5)^2$

85. $(-3)^2 + 8 \div 2$

86. $-3^2 + 8 \div 2$

87. $2[1.55 + 5(3.7)] - 3.35$

88. $(8.4 + 3.1)^2 - (3.64 - 1.2)$

89. $\left(\dfrac{2}{5} + \dfrac{3}{8}\right) - \dfrac{3}{20}$

90. $\left(\dfrac{5}{6} \cdot \dfrac{4}{5}\right) + \left(\dfrac{2}{3} \cdot \dfrac{5}{8}\right)$

91. $\dfrac{3}{4} - 4 \cdot \dfrac{5}{40}$

92. $\dfrac{1}{8} - \dfrac{1}{4} \cdot \dfrac{3}{2} + \dfrac{3}{5}$

93. $\dfrac{4}{5} + \dfrac{3}{4} \div \dfrac{1}{2} - \dfrac{2}{3}$

94. $\dfrac{12 - (4 - 6)^2}{6 + 4^2 \div 2^2}$

95. $\dfrac{-4 - [2(9 \div 3) - 5]}{6^2 - 3^2 \cdot 7}$

96. $\dfrac{[(7 - 3)^2 - 4]^2}{9 - 16 \div 8 - 4}$

97. $\dfrac{-[4 - (6 - 12)^2]}{[(9 \div 3) + 4]^2 + 2^2}$

98. $\dfrac{[(5 - (3 - 7)) - 2]^2}{2[(16 \div 2^2) - (8 \cdot 4)]}$

99. $\{5 - 2[4 - (6 \div 2)]^2\}^2$

100. $\{-6 - [3(16 \div 4^2)]\}^2$

101. $-\{4 - [-3 - (2 - 5)]^2\}$

102. $3\{4[(3 - 4)^2 - 3]^3 - 1\}$

103. $\{4 - 3[2 - (9 \div 3)]^2\}^2$

104. $2\{5[(4 - 6)^3 - 1]^2 - 3\}$

*Evaluate **a)** x^2, **b)** $-x^2$, and **c)** $(-x)^2$ for the following values of x.*

105. 5

106. 8

107. -2

108. -5

109. 6

110. 7

111. $-\dfrac{1}{3}$

112. $\dfrac{3}{4}$

Evaluate each expression for the given value of the variable or variables.

113. $x + 6; x = -2$

114. $2x - 4x + 5; x = 3$

115. $-7z - 3; z = 6$

116. $3(x - 2); x = 5$

117. $a^2 - 6; a = -3$

118. $b^2 - 8; b = 5$

119. $3p^2 - 6p - 4; p = 2$

120. $2r^2 - 5r + 3; r = 1$

121. $-4x^2 - 2x + 1; x = -1$

122. $-t^2 - 4t + 5; t = -4$

123. $-x^2 - 2x + 5; x = \dfrac{1}{2}$

124. $2x^2 - 4x - 10; x = \dfrac{3}{4}$

125. $4(3x + 1)^2 - 6x; x = 5$

126. $3n^2(2n - 1) + 5; n = -4$

127. $r^2 - s^2; r = -2, s = -3$

128. $p^2 - q^2; p = 5, q = -3$

129. $5(x - 6y) + 3x - 7y; x = 1, y = -5$

130. $4(x + y)^2 + 2(x + y) + 3; x = 2, y = 4$

131. $3(x - 4)^2 - (3y - 4)^2; x = -1, y = -2$

132. $6x^2 + 3xy - y^2; x = 2, y = -3$

Problem Solving

Write the following statements as mathematical expressions using parentheses and brackets, and then evaluate.

133. Multiply 6 by 3. From this product, subtract 4. From this difference, subtract 2.

134. Add 4 to 9. Divide this sum by 2. Add 10 to this quotient.

135. Multiply 10 by 4. Add 9 to this product. Subtract 6 from this sum. Divide this difference by 7.

136. Multiply 6 by 3. To this product, add 27. Divide this sum by 8. Multiply this quotient by 10.

137. Add $\dfrac{4}{5}$ to $\dfrac{3}{7}$ Multiply this sum by $\dfrac{2}{3}$.

138. Multiply $\dfrac{3}{8}$ by $\dfrac{4}{5}$. To this product, add $\dfrac{7}{120}$. From this sum, subtract $\dfrac{1}{60}$.

139. For what value or values of x does $-(x^2) = -x^2$?

140. For what value or values of x does $x = x^2$?

141. Road Trip If a car travels at 65 miles per hour, the distance it travels in t hours is $65t$. Determine how far a car traveling at 65 miles per hour travels in 2.5 hours.

142. Sales Tax If the sales tax on an item is 8%, the sales tax on an item costing d dollars can be found by the expression $0.08d$. Determine the sales tax on a scrapbook that costs $19.99.

143. Projectile Height An object is projected upward with an initial velocity of 48 feet per second from the top of a 70-foot building. The height of the object above the ground at any time t, in seconds, can be found by the expression $-16t^2 + 48t + 70$. Determine the height of the object after 2 seconds.

144. Car Cost If the sales tax on an item is 8%, then the total cost of an item c, including sales tax, can be found by the expression $c + 0.08c$. Find the total cost of a car that costs $17,000.

145. In the Using Your Calculator box on page 74, we showed that to evaluate $\left(\dfrac{2}{5}\right)^2$ we press the following keys:

$$2 \;\boxed{\div}\; 5 \;\boxed{=}\; \boxed{x^2}$$

We obtained an answer of .16. Indicate what a scientific (or graphing) calculator would display if the following keys are pressed. (On a graphing calculator, press $\boxed{\text{ENTER}}$ instead of $\boxed{=}$ and end part **b)** with $\boxed{\text{ENTER}}$). Explain your reason for each answer. Check your answer on your calculator.

a) $2 \;\boxed{\div}\; 5 \;\boxed{x^2}\; \boxed{=}$

b) $\boxed{(}\; 2 \;\boxed{\div}\; 5 \;\boxed{)}\; \boxed{x^2}$

146. We will discuss using zero as an exponent in Section 6.1. On your calculator find the value of 4^0 by using your $\boxed{y^x}$, $\boxed{x^y}$, or $\boxed{\wedge}$ key and record its value. Evaluate a few other numbers raised to the zero power. Can you make any conclusions about a real number (other than 0) raised to the zero power?

Challenge Problems

147. Grass Growth The rate of growth of grass in inches per week depends on a number of factors, including rainfall and temperature. For a certain region of the country, the growth per week can be approximated by the expression $0.2R^2 + 0.003RT + 0.0001T^2$, where R is the weekly rainfall, in inches, and T is the average weekly temperature, in degrees Fahrenheit. Find the amount of growth of grass for a week in which the rainfall is 2 inches and the average temperature is 70°F.

Insert one pair of parentheses to make each statement true.

148. $14 + 6 \div 2 \times 4 = 40$

149. $12 - 4 - 6 + 10 = 24$

150. $24 \div 6 \div 2 + 2 = 1$

Group Activity

*Discuss and answer Exercises 151–154 as a group, according to the instructions. Each question has four parts. For parts **a)**, **b)**, and **c)**, simplify the expression and write the answer in exponential form. Use the knowledge gained in parts **a)**–**c)** to answer part **d)**. (General rules that may be used to solve exercises like these will be discussed in Chapter 6.)*

  a) *Group member 1: Do part **a)** of each exercise.*

  b) *Group member 2: Do part **b)** of each exercise.*

  c) *Group member 3: Do part **c)** of each exercise.*

  d) *As a group, answer part **d)** of each exercise. You may need to make up other examples like parts **a)**–**c)** to help you answer part **d)**.*

151. a) $2^2 \cdot 2^3$ **b)** $3^2 \cdot 3^3$ **c)** $2^3 \cdot 2^4$ **d)** $x^m \cdot x^n$

152. a) $\dfrac{2^3}{2^2}$ **b)** $\dfrac{3^4}{3^2}$ **c)** $\dfrac{4^5}{4^3}$ **d)** $\dfrac{x^m}{x^n}$

153. a) $\left(2^3\right)^2$ **b)** $\left(3^3\right)^2$ **c)** $\left(4^2\right)^2$ **d)** $\left(x^m\right)^n$

154. a) $(2x)^2$ **b)** $(3x)^2$ **c)** $(4x)^3$ **d)** $(ax)^m$

Cumulative Review Exercises

[1.2] **155. Dogs** The graph shows the number of dogs in various houses selected at random in a neighborhood.

Dogs in Selected Houses

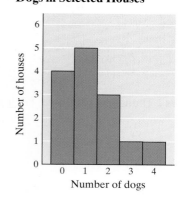

  a) How many houses have two dogs?

  b) Make a chart showing the number of houses that have no dogs, one dog, two dogs, and so on.

  c) How many dogs in total are there in all the houses?

  d) Determine the mean number of dogs in all the houses surveyed.

156. Taxi Cost Yellow Cab charges $2.40 for the first $\dfrac{1}{2}$ mile plus 20 cents for each additional $\dfrac{1}{8}$ mile or part thereof. Find the cost of a 3-mile trip.

[1.6] **157.** Add $-\dfrac{7}{12} + \dfrac{4}{9}$.

[1.8] **158.** Divide $\left(\dfrac{-5}{7}\right) \div \left(\dfrac{-3}{14}\right)$.

1.10 Properties of the Real Number System

1 Learn the commutative property.

2 Learn the associative property.

3 Learn the distributive property.

4 Learn the identity properties.

5 Learn the inverse properties.

Here, we introduce various properties of the real number system. We will use these properties throughout the text.

1 Learn the Commutative Property

The **commutative property of addition** states that the order in which any two real numbers are added does not matter.

> **Commutative Property of Addition**
>
> If a and b represent any two real numbers, then
> $$a + b = b + a$$

Notice that the commutative property involves a change in *order*. For example,

$$4 + 3 = 3 + 4$$
$$7 = 7$$

The **commutative property of multiplication** states that the order in which any two real numbers are multiplied does not matter.

> **Commutative Property of Multiplication**
>
> If a and b represent any two real numbers, then
> $$a \cdot b = b \cdot a$$

For example,

$$6 \cdot 3 = 3 \cdot 6$$
$$18 = 18$$

The commutative property **does not hold** *for subtraction or division.* For example, $4 - 6 \neq 6 - 4$ and $6 \div 3 \neq 3 \div 6$.

2 Learn the Associative Property

The **associative property of addition** states that, in the addition of three or more numbers, parentheses may be placed around any two adjacent numbers without changing the results.

> **Associative Property of Addition**
>
> If a, b, and c represent any three real numbers, then
> $$(a + b) + c = a + (b + c)$$

Notice that the associative property involves a change of *grouping*. For example,

$$(3 + 4) + 5 = 3 + (4 + 5)$$
$$7 + 5 = 3 + 9$$
$$12 = 12$$

In this example, the 3 and 4 are grouped together on the left, and the 4 and 5 are grouped together on the right.

The **associative property of multiplication**, states that, in the multiplication of three or more numbers, parentheses may be placed around any two adjacent numbers without changing the results.

Associative Property of Multiplication

If a, b, and c represent any three real numbers, then

$$(a \cdot b) \cdot c = a \cdot (b \cdot c)$$

For example,

$$(6 \cdot 2) \cdot 4 = 6 \cdot (2 \cdot 4)$$
$$12 \cdot 4 = 6 \cdot 8$$
$$48 = 48$$

Since the associative property involves a change of grouping, when the associative property is used, the content within the parentheses changes.

The associative property **does not hold** *for subtraction or division.* For example, $(4 - 1) - 3 \neq 4 - (1 - 3)$ and $(8 \div 4) \div 2 \neq 8 \div (4 \div 2)$.

Often when we add numbers we group the numbers so that we can add them easily. For example, when we add $70 + 50 + 30$ we may first add the $70 + 30$ to get 100. We are able to do this because of the commutative and associative properties. Notice that

$$
\begin{aligned}
(70 + 50) + 30 &= 70 + (50 + 30) &&\textit{Associative property of addition}\\
&= 70 + (30 + 50) &&\textit{Commutative property of addition}\\
&= (70 + 30) + 50 &&\textit{Associative property of addition}\\
&= 100 + 50 &&\textit{Addition facts}\\
&= 150
\end{aligned}
$$

Notice in the second step that the same numbers remained in parentheses but the order of the numbers changed, $50 + 30$ to $30 + 50$. Since this step involved a change in order (and not grouping), this is the commutative property of addition.

3 Learn the Distributive Property

A very important property of the real numbers is the **distributive property of multiplication over addition**. We often shorten the name to the **distributive property**.

Distributive Property

If a, b, and c represent any three real numbers, then

$$a(b + c) = ab + ac$$

For example, if we let $a = 2$, $b = 3$, and $c = 4$, then

$$2(3 + 4) = (2 \cdot 3) + (2 \cdot 4)$$
$$2 \cdot 7 = 6 + 8$$
$$14 = 14$$

Therefore, we may either add first and then multiply, or multiply first and then add. Another example of the distributive property is

$$2(x + 3) = 2 \cdot x + 2 \cdot 3 = 2x + 6$$

The distributive property can be expanded in the following manner:

$$a(b + c + d + \cdots + n) = ab + ac + ad + \cdots + an$$

For example, $3(x + y + 5) = 3x + 3y + 15$.

The distributive property will be discussed in more detail in Chapter 2.

Helpful Hint

The *commutative property* changes *order*.

The *associative property* changes *grouping*.

The *distributive property* involves *two operations*, usually multiplication and addition.

EXAMPLE 1 ▶ Name each property illustrated.

a) $4 + (-2) = -2 + 4$

b) $5(r + s) = 5 \cdot r + 5 \cdot s = 5r + 5s$

c) $x \cdot y = y \cdot x$

d) $(-12 + 3) + 4 = -12 + (3 + 4)$

Solution

a) Commutative property of addition

b) Distributive property

c) Commutative property of multiplication

d) Associative property of addition

▶ **Now Try Exercise 29**

Helpful Hint

Do not confuse the distributive property with the associative property of multiplication. Make sure you understand the difference.

Distributive Property	Associative Property of Multiplication
$3(4 + x) = 3 \cdot 4 + 3 \cdot x$	$3(4 \cdot x) = (3 \cdot 4)x$
$= 12 + 3x$	$= 12x$

For the distributive property to be used, there must be two *terms* within parentheses, separated by a plus or minus sign as in $3(4 \boxed{+} x)$.

4 Learn the Identity Properties

Now we will discuss the **identity properties**. When the number 0 is added to any real number, the real number is unchanged. For example, $5 + 0 = 5$ and $0 + 5 = 5$. For this reason we call 0 the **identity element of addition**. When any real number is multiplied by 1, the real number is unchanged. For example, $7 \cdot 1 = 7$ and $1 \cdot 7 = 7$. For this reason we call 1 the **identity element of multiplication**. The identity property for addition states that when 0 is added to any real number, the sum is the real number you started with. The identity property for multiplication states that when any real number is multiplied by 1, the product is the real number you started with. These properties are summarized below.

Identity Properties

If a represents any real number, then

$$a + 0 = a \quad \text{and} \quad 0 + a = a \qquad \textit{Identity property of addition}$$

and

$$a \cdot 1 = a \quad \text{and} \quad 1 \cdot a = a \qquad \textit{Identity property of multiplication}$$

We often use the identity properties without realizing we are using them. For example, when we reduce $\dfrac{15}{50}$, we may do the following:

$$\frac{15}{50} = \frac{3 \cdot 5}{10 \cdot 5} = \frac{3}{10} \cdot \frac{5}{5} = \frac{3}{10} \cdot 1 = \frac{3}{10}$$

When we showed that $\dfrac{3}{10} \cdot 1 = \dfrac{3}{10}$, we used the identity property of multiplication.

5 Learn the Inverse Properties

The last properties we will discuss in this chapter are the **inverse properties**. On page 44, we indicated that numbers like 3 and -3 were opposites or *additive inverses* because $3 + (-3) = 0$ and $-3 + 3 = 0$. Any two numbers whose sum is 0 are called *additive inverses* of each other. In general, for any real number a its additive inverse is $-a$.

We also have multiplicative inverses. Any two numbers whose product is 1 are called *multiplicative inverses* of each other. For example, because $4 \cdot \dfrac{1}{4} = 1$ and $\dfrac{1}{4} \cdot 4 = 1$, the numbers 4 and $\dfrac{1}{4}$ are *multiplicative inverses* (or *reciprocals*) of each other. In general, for any real number a, its multiplicative inverse is $\dfrac{1}{a}$. The inverse properties are summarized below.

> ### Inverse Properties
>
> If a represents any real number, then
>
> $$a + (-a) = 0 \quad \text{and} \quad -a + a = 0 \qquad \textit{Inverse property of addition}$$
>
> and
>
> $$a \cdot \frac{1}{a} = 1 \quad \text{and} \quad \frac{1}{a} \cdot a = 1 \, (a \neq 0) \qquad \textit{Inverse property of multiplication}$$

We often use the inverse properties without realizing we are using them. For example, to evaluate the expression $6x + 2$, when $x = \dfrac{1}{6}$ we may do the following:

$$6x + 2 = 6\left(\frac{1}{6}\right) + 2 = 1 + 2 = 3$$

When we multiplied $6\left(\dfrac{1}{6}\right)$ and replaced it with 1, we used the inverse property of multiplication. We will be using both the identity and inverse properties throughout the book, although we may not specifically refer to them by name.

EXAMPLE 2 ▶ Name each property illustrated.

a) $2(x + 6) = (2 \cdot x) + (2 \cdot 6) = 2x + 12$ b) $3x \cdot 1 = 3x$

c) $(3 \cdot 6) \cdot 5 = 3 \cdot (6 \cdot 5)$ d) $y \cdot \dfrac{1}{y} = 1$

e) $2a + (-2a) = 0$ f) $3y + 0 = 3y$

Solution

a) Distributive property
b) Identity property of multiplication
c) Associative property of multiplication
d) Inverse property of multiplication
e) Inverse property of addition
f) Identity property of addition

▶ Now Try Exercise 35

EXAMPLE 3 ▸ In parts **a)**–**f)**, the name of a property is given followed by part of an equation. Complete the equation, to the right of the equal sign, to illustrate the given property.

a) Associative property of multiplication   **b)** Inverse property of addition

$$(5 \cdot 4) \cdot 7 =$$ $$3c + (-3c) =$$

c) Identity property of multiplication   **d)** Distributive property

$$6y \cdot 1 =$$ $$3(x + 5) =$$

e) Identity property of addition   **f)** Inverse property of multiplication

$$2a + 0 =$$ $$b \cdot \frac{1}{b} =$$

Solution

a) $5 \cdot (4 \cdot 7)$ **b)** 0 **c)** $6y$ **d)** $3x + 15$ **e)** $2a$ **f)** 1

▸ **Now Try Exercise 55**

EXERCISE SET 1.10   *Math XL*   *MyMathLab*
MathXL®   MyMathLab

Concept/Writing Exercises

1. Explain the commutative property of addition and give an example of it.

2. Explain the commutative property of multiplication and give an example of it.

3. Explain the associative property of addition and give an example of it.

4. Explain the associative property of multiplication and give an example of it.

5. a) Explain the difference between $x + (y + z)$ and $x(y + z)$.

    b) Find the value of $x + (y + z)$ when $x = 4$, $y = 5$, and $z = 6$.

    c) Find the value of $x(y + z)$ when $x = 4$, $y = 5$, and $z = 6$.

6. Explain the distributive property and give an example of it.

7. Explain how you can tell the difference between the associative property of multiplication and the distributive property.

8. a) Write the associative property of addition using $x + (y + z)$.

    b) Write the distributive property using $x(y + z)$.

9. What number is the additive identity element?

10. What number is the multiplicative identity element?

Practice the Skills

In Exercises 11–22, for the given expression, determine **a)** *the additive inverse, and* **b)** *the multiplicative inverse.*

11. 6 **12.** 5 **13.** -3

14. -7 **15.** x **16.** z

17. 1.6 **18.** -0.125 **19.** $\dfrac{1}{5}$

20. $\dfrac{1}{8}$ **21.** $-\dfrac{5}{6}$ **22.** $-\dfrac{2}{9}$

Practice the Skills

Name each property illustrated.

23. $6(x + 7) = 6x + 42$ **24.** $3 + y = y + 3$

25. $(x + 3) + 5 = x + (3 + 5)$ **26.** $1(x + 3) = (1)(x) + (1)(3) = x + 3$

27. $5 \cdot y = y \cdot 5$ **28.** $-4x + 4x = 0$

29. $p \cdot (q \cdot r) = (p \cdot q) \cdot r$ **30.** $2(x + 4) = 2x + 8$

31. $4(d + 3) = 4d + 12$ **32.** $3 + (4 + t) = (3 + 4) + t$

33. $3z \cdot 1 = 3z$

34. $0 + 3y = 3y$

35. $2y \cdot \dfrac{1}{2y} = 1$

36. $x \cdot y = y \cdot x$

In Exercises 37–58, the name of a property is given followed by part of an equation. Complete the equation, to the right of the equal sign, to illustrate the given property.

37. commutative property of addition

$-4 + 1 =$

38. inverse property of addition

$(-7a) + 7a =$

39. associative property of multiplication

$-6 \cdot (4 \cdot 2) =$

40. associative property of addition

$-5 + (6 + 8) =$

41. distributive property

$-2(x + y) =$

42. distributive property

$4(x + 3) =$

43. commutative property of multiplication

$x \cdot y = \quad y \cdot x$

44. identity property of multiplication

$(1)\left(-\dfrac{1}{3}b\right) =$

45. commutative property of addition

$4x + 3y =$

46. associative property of multiplication

$-9 \cdot (3 \cdot 8) =$

47. associative property of addition

$(a + b) + 3 =$

48. commutative property of multiplication

$(x + 2)3 =$

49. associative property of addition

$(3x + 4) + 6 =$

50. commutative property of addition

$3(x + y) =$

51. commutative property of multiplication

$3(m + n) =$

52. associative property of multiplication

$(3x)y =$

53. distributive property

$4(x + y + 3)$

54. distributive property

$3(x + y + 2) =$

55. inverse property of addition

$3n + (-3n) =$

56. identity property of addition

$0 + 2x =$

57. identity property of multiplication

$\left(\dfrac{5}{2}n\right)(1) =$

58. inverse property of multiplication

$\left(\dfrac{x}{2}\right)\left(\dfrac{2}{x}\right) =$

Problem Solving

Indicate whether the given processes are commutative. That is, does changing the order in which the actions are done result in the same final outcome? Explain each answer.

59. Putting sugar and then cream in coffee; putting cream and then sugar in coffee.

60. Applying suntan lotion and then sunning yourself; sunning yourself and then applying suntan lotion.

61. Putting your contacts in and then washing your hair; washing your hair and then putting your contacts in.

62. Putting on your sweater and then your coat; putting on your coat and then your sweater.

63. Writing on the chalkboard and then erasing the chalkboard; erasing the chalkboard and then writing on the chalkboard.

64. Getting your hands dirty and then washing your hands; washing your hands and then getting your hands dirty.

In Exercises 65–70, indicate whether the given processes are associative. For a process to be associative, the final outcome must be the same when the first two actions are performed first or when the last two actions are performed first. Explain each answer.

65. Cleaning the kitchen sink, dusting the bedroom furniture, and doing the laundry.

66. In a store, buying cereal, soap, and dog food.

67. Turning on a DVD player, inserting a DVD, and watching the DVD.

68. Putting on a shirt, a tie, and a sweater.

69. Starting a car, moving the shift lever to drive, and then stepping on the gas.

70. Putting cereal, milk, and sugar in a bowl.

71. The commutative property of addition is $a + b = b + a$. Explain why $(3 + 4) + x = x + (3 + 4)$ also illustrates the commutative property of addition.

72. The commutative property of multiplication is $a \cdot b = b \cdot a$. Explain why $(3 + 4) \cdot x = x \cdot (3 + 4)$ also illustrates the commutative property of multiplication.

Challenge Problems

73. Consider $x + (3 + 5) = x + (5 + 3)$. Does this illustrate the commutative property of addition or the associative property of addition? Explain.

74. Consider $x + (3 + 5) = (3 + 5) + x$. Does this illustrate the commutative property of addition or the associative property of addition? Explain.

75. Consider $x + (3 + 5) = (x + 3) + 5$. Does this illustrate the commutative property of addition? Explain.

76. The commutative property of multiplication is $a \cdot b = b \cdot a$. Explain why $(3 + 4) \cdot (5 + 6) = (5 + 6) \cdot (3 + 4)$ also illustrates the commutative property of multiplication.

Cumulative Review Exercises

[1.3] **77.** Add $2\frac{3}{5} + \frac{2}{3}$.

78. Subtract $3\frac{5}{8} - 2\frac{3}{16}$.

[1.6] **79.** Add $102.7 + (-113.9)$.

80. Write the opposite of $\frac{7}{8}$.

Chapter 1 Summary

IMPORTANT FACTS AND CONCEPTS	EXAMPLES

Section 1.2

Guidelines for Problem Solving

1. **Understand the problem.**
2. **Translate the problem to mathematical language.**
3. **Carry out the mathematical calculations necessary to solve the problem.**
4. **Check the answer obtained in step 3.**
5. **Make sure you have answered the question.**

See page 8 for more details on problem solving.

Jacob Thomas can pay either $725 cash for a desk or pay $300 down and $20 a month for 24 months. How much money can he save by paying cash?

Solution: Understand We need to determine how much he would pay if he paid $300 down and $20 a month for 24 months. This amount would then be compared to $725.

Translate

$$\begin{pmatrix} \text{amount if} \\ \text{paid over 24} \\ \text{months} \end{pmatrix} = \$300 + \begin{pmatrix} \text{additional} \\ \text{amount paid} \\ \text{over 24 months} \end{pmatrix}$$

Carry Out
$$= \$300 + \$20(24)$$
$$= \$300 + \$480$$
$$= \$780$$

The difference in the amounts is $780 - $725 = $55.

Check and Answer

The answer appears reasonable. Jacob can save $55 by paying the total amount at the time of purchase.

IMPORTANT FACTS AND CONCEPTS	EXAMPLES

Section 1.2 (continued)

An **algebraic expression** is a general term for any collection of numbers, variables, grouping symbols, and operations.	$8(24 \div 4), 3x - 5, a^2 - 6$
Bar graph	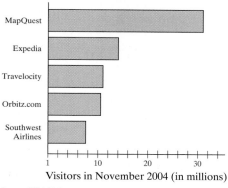 **Logging onto Travel Web Sites in November, 2004** *Source*: USA Today
Line graph	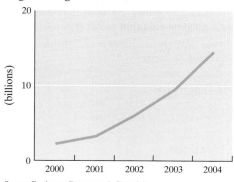 **Digital Images Not Printed** *Source:* Rochester Democrat & Chronicle
Circle graph	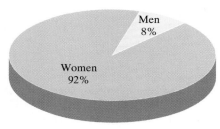 **Registered Nurses in the U.S., 2003** *Source:* www.census.gov
The symbol $\approx$ is read **"is approximately equal to."**	$57.91536 \approx 57.92$
Measures of central tendency or **averages** are values that are representative of a set of data.	The mean and median are averages.
The **mean** of a set of data is determined by adding all the values and dividing the sum by the number of values.	The mean of Adam Michael's test grades of 82, 95, 76, 92, and 88 is $\dfrac{82 + 95 + 76 + 92 + 88}{5} = \dfrac{433}{5} = 86.6.$
The **median** is the value in the middle of a set of **ranked data**.	The median of Adam's five test grades given above is 88 (circled). $76, 82, \textcircled{88}, 92, 95$

IMPORTANT FACTS AND CONCEPTS	EXAMPLES

Section 1.3

Multiplication Symbols If a and b represent any two mathematical quantities, then each of the following may be used to indicate the product of a and b ("a times b"). $\quad ab \quad a \cdot b \quad a(b) \quad (a)b \quad (a)(b)$	5 times z may be written: $\quad 5z, \quad 5 \cdot z, \quad 5(z), \quad (5)z \quad$ or $\quad (5)(z)$
The numbers or variables that are multiplied in a multiplication problem are called **factors**. $\quad$ If $a \cdot b = c$, then a and b are *factors* of c.	In $7 \cdot 9 = 63$, the numbers 7 and 9 are factors of 63.
The numbers $0, 1, 2, 3, 4, \ldots$ are called **whole numbers**.	$12, 105$, and 0 are whole numbers.
The top number of a fraction is called the **numerator**, and the bottom number is called the **denominator**.	In the fraction $\dfrac{6}{11}$, the 6 is the numerator and the 11 is the denominator.
To Simplify a Fraction 1. Find the largest number that will divide (without remainder) both the numerator and the denominator. This number is called the **greatest common factor** (GCF). 2. Then divide both the numerator and the denominator by the greatest common factor.	The GCF of 36 and 48 is 12. Therefore, $$\frac{36}{48} = \frac{36 \div 12}{48 \div 12} = \frac{3}{4}$$
To Multiply Fractions $$\frac{a}{b} \cdot \frac{c}{d} = \frac{ac}{bd}$$	$$\frac{3}{4} \cdot \frac{3}{5} = \frac{9}{20}$$
To Divide Fractions $$\frac{a}{b} \div \frac{c}{d} = \frac{a}{b} \cdot \frac{d}{c} = \frac{ad}{bc}$$	$$\frac{9}{7} \div \frac{4}{5} = \frac{9}{7} \cdot \frac{5}{4} = \frac{45}{28}$$
To Add and Subtract Fractions with Like Denominators $$\frac{a}{c} + \frac{b}{c} = \frac{a+b}{c} \quad \text{or} \quad \frac{a}{c} - \frac{b}{c} = \frac{a-b}{c}$$	1. $\dfrac{3}{13} + \dfrac{7}{13} = \dfrac{3+7}{13} = \dfrac{10}{13}$ 2. $\dfrac{4}{5} - \dfrac{1}{5} = \dfrac{4-1}{5} = \dfrac{3}{5}$
The smallest number that is divisible by two or more denominators is called the **least common denominator** or **LCD**. **To add (or subtract) fractions with unlike denominators,** first rewrite each fraction with a common denominator.	In the fractions $\dfrac{7}{9}$ and $\dfrac{2}{5}$, 45 is the LCD. Subtract $\dfrac{7}{9} - \dfrac{2}{5}$. $$\frac{7}{9} - \frac{2}{5} = \frac{7}{9} \cdot \frac{5}{5} - \frac{2}{5} \cdot \frac{9}{9}$$ $$= \frac{35}{45} - \frac{18}{45} = \frac{17}{45}$$
A **mixed number** consists of a whole number followed by a fraction. Change a mixed number to a fraction. Change a fraction to a mixed number.	$7\dfrac{4}{5}$ is a mixed number. $$6\frac{2}{3} = 6 + \frac{2}{3} = \frac{18}{3} + \frac{2}{3} = \frac{20}{3}$$ $$\frac{52}{7} = \frac{49}{7} + \frac{3}{7} = 7 + \frac{3}{7} = 7\frac{3}{7}$$

IMPORTANT FACTS AND CONCEPTS	EXAMPLES
Section 1.4	

A **set** is a collection of **elements** listed within braces.	The set {2, 4, 6, 8} consists of four elements, namely 2, 4, 6, and 8.
A set that contains no elements is called an **empty set** (or **null set**).	The set of dogs that can fly is an empty set.
Natural numbers: {1, 2, 3, 4, 5, ...} The natural numbers are also called the **positive integers** or the **counting numbers**.	23, 16, and 1231 are natural numbers.
Whole numbers: {0, 1, 2, 3, 4, 5, ...}	35, 0, and 257 are whole numbers.
Integers: {..., −3, −2, −1, 0, 1, 2, 3 ...} *Negative integers* *Positive integers*	−101, 0, and 236 are integers.
Rational numbers: {quotient of two integers, denominator not 0}	$-3, 2.8, 9\frac{1}{2}, 0, 15, -\frac{9}{13}$, and -3.6 are rational numbers.
Irrational numbers: {numbers that can be represented on the number line that are not rational numbers}	$\sqrt{13}$ and $-\sqrt{7}$ are irrational numbers.
Real numbers: {all numbers that can be represented on a number line}	$5.7, -\frac{3}{8}, -16, 0, \sqrt{5}, 3\frac{1}{7}, -2.1, -\sqrt{2}$, and 31 are real numbers.

Section 1.5	

The symbol > is used to represent the words "is greater than." The symbol < is used to represent the words "is less than."	**1.** 3.8 > 3.08 **2.** −7 < −1
The **absolute value** of a number can be considered the distance between the number and 0 on a number line. The absolute value of every number will be either *positive* or *zero*.	**1.** $\lvert 12 \rvert = 12$ **2.** $\lvert 0 \rvert = 0$ **3.** $\lvert -100 \rvert = 100$

Section 1.6	

Add Using a Number Line: Represent the first number to be added (first *addend*) by an arrow starting at 0 on the number line. The arrow is drawn to the right if the number is positive, and to the left if the number is negative. From the tip of the first arrow, draw a second arrow to represent the second addend. The second arrow is drawn to the right or left, as just explained. The sum of the two numbers is found at the tip of the second arrow.	Add 5 + (−7). 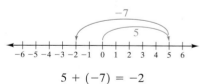 $5 + (-7) = -2$
Any two numbers whose sum is zero are said to be **opposites** (or **additive inverses**) of each other.	**1.** The opposite of 8 is −8. **2.** The opposite of $-\frac{5}{6}$ is $\frac{5}{6}$.
Add Using Absolute Value: **To add real numbers with the same sign**, add their absolute values. The sum has the same sign as the numbers being added.	Add −4 + (−9). $$\lvert -4 \rvert + \lvert -9 \rvert = 4 + 9 = 13$$ Since both numbers being added are negative, the sum is negative. Thus, $-4 + (-9) = -13$.
To add two signed numbers with different signs, subtract the smaller absolute value from the larger absolute value. The answer has the sign of the number with the larger absolute value.	Add −25 + 10. $$\lvert -25 \rvert - \lvert 10 \rvert = 25 - 10 = 15$$ Since $\lvert -25 \rvert$ is greater than $\lvert 10 \rvert$, the sum is negative. Thus, $-25 + 10 = -15$.

IMPORTANT FACTS AND CONCEPTS	EXAMPLES

Section 1.7

To Subtract Real Numbers

In general, if a and b represent any two real numbers, then

$$a - b = a + (-b)$$

In evaluating expressions involving more than one addition and subtraction, work from left to right unless parentheses or other grouping symbols appear.

1. $8 - (+4) = 8 + (-4) = 4$
2. $-7 - 5 = -7 + (-5) = -12$
3. $-12 - (-6) = -12 + 6 = -6$
4. $-8 + 5 - 11 = -3 - 11 = -14$

Section 1.8

The Sign of the Product of Two Real Numbers
1. The product of two numbers with **like** signs is a **positive** number.
2. The product of two numbers with **unlike** signs is a **negative** number.

1. $-8(-7) = 56$

2. $5(-6) = -30$

The Sign of the Quotient of Two Real Numbers
1. The quotient of two numbers with **like** signs is a **positive** number.
2. The quotient of two numbers with **unlike** signs is a **negative** number.

1. $\dfrac{-81}{-9} = 9$

2. $\dfrac{-35}{7} = -5$

The Quotient of a Positive Number and a Negative Number

If a and b represent any real numbers, $b \neq 0$, then

$$\frac{a}{-b} = \frac{-a}{b} = -\frac{a}{b}$$

$$\frac{3}{-8} = \frac{-3}{8} = -\frac{3}{8}$$

Summary of Division Involving 0

If a represents any real number except 0, then

$$\frac{0}{a} = 0 \qquad \frac{a}{0} \text{ is undefined}$$

1. $\dfrac{0}{12} = 0$

2. $\dfrac{5}{0}$ is undefined

Section 1.9

In general, the number b to the nth power, written b^n, means

$$\underbrace{b \cdot b \cdot b \cdots b}_{n \text{ factors of } b} = b^n$$

$$x^5 = x \cdot x \cdot x \cdot x \cdot x$$

Order of Operations
1. First, evaluate the information within **parentheses** ( ), brackets [ ], or braces { }. These are **grouping symbols**. A fraction bar, —, also serves as a grouping symbol. If the expression contains nested parentheses, evaluate the information in the innermost parentheses first.
2. Next, evaluate all **exponents**.
3. Next, evaluate all **multiplications** or **divisions** in order from left to right.
4. Finally, evaluate all **additions** or **subtractions** in order from left to right.

$$-16 - 3(6 - 8)^2$$
$$= -16 - 3(-2)^2$$
$$= -16 - 3(4)$$
$$= -16 - 12$$
$$= -28$$

IMPORTANT FACTS AND CONCEPTS	EXAMPLES
Section 1.10	

Commutative Property of Addition If a and b represent any two real numbers, then $$a + b = b + a$$	$9 + 6 = 6 + 9$ $15 = 15$
Commutative Property of Multiplication If a and b represent any two real numbers, then $$a \cdot b = b \cdot a$$	$6 \cdot 7 = 7 \cdot 6$ $42 = 42$
Associative Property of Addition If $a, b,$ and c represent any three real numbers, then $$(a + b) + c = a + (b + c)$$	$(1 + 2) + 3 = 1 + (2 + 3)$ $3 + 3 = 1 + 5$ $6 = 6$
Associative Property of Multiplication If $a, b,$ and c represent any three real numbers, then $$(a \cdot b) \cdot c = a \cdot (b \cdot c)$$	$(5 \cdot 4) \cdot 3 = 5 \cdot (4 \cdot 3)$ $20 \cdot 3 = 5 \cdot 12$ $60 = 60$
Distributive Property If $a, b,$ and c represent any three real numbers, then $$a(b + c) = ab + ac$$	$4(y + 9) = 4 \cdot y + 4 \cdot 9 = 4y + 36$
Identity Properties If a represents any real number, then **1.** $a + 0 = a$ and $0 + a = a$ *Identity property of addition* and **2.** $a \cdot 1 = a$ and $1 \cdot a = a$ *Identity property of multiplication*	**1.** $15 + 0 = 0 + 15 = 15$ **2.** $8 \cdot 1 = 1 \cdot 8 = 8$
Inverse Properties If a represents any real number, then **1.** $a + (-a) = 0$ and $-a + a = 0$ *Inverse property of addition* and **2.** $a \cdot \dfrac{1}{a} = 1$ and $\dfrac{1}{a} \cdot a = 1 (a \neq 0)$ *Inverse property of multiplication*	**1.** $9 + (-9) = -9 + 9 = 0$ **2.** $6 \cdot \dfrac{1}{6} = \dfrac{1}{6} \cdot 6 = 1$

Chapter 1 Review Exercises

[1.2] *Solve.*

1. **Final Exam Review** Eraj Basnayake bought doughnuts for his math classes for their final exam review sessions. His first class ate 32 doughnuts. His second class ate 29 doughnuts. His third class ate 36 doughnuts, and his fourth class ate 31 doughnuts. If he purchased 13 dozen doughnuts, how many did he have left over?

2. **Inflation** Assume that the rate of inflation is 5% per year for the next two years. What will be the cost of goods two years from now, adjusted for inflation, if the goods cost $500.00 today?

3. **Sales Tax** The sales tax in Cattaraugus County is 8.25%.
   a) What was the sales tax that Andrew Jacob paid on a laptop computer that cost $899.99 before tax?
   b) What is the total cost of the laptop computer including tax?

4. **Fax Machine** Neeta Gandhi wants to purchase a fax machine that sells for $300. She can either pay the total amount at the time of purchase, or she can agree to pay the store $30 down and $25 a month for 12 months. How much money can she save by paying the total amount at the time of purchase?

5. **Test Grades** On Angie Smajstrla's first five exams her grades were 75, 79, 86, 88, and 64. Find the **a)** mean and **b)** median of her grades.

6. **Little League** The number of points scored by a Little League football team in their last six games were 21, 3, 17, 10, 9, and 6. Find the **a)** mean and **b)** median of the points.

7. Commute Times In March, 2005, the U.S. Census Bureau released a report that included the following bar graph. The bar graph shows some of the longest average commute-to-work times in the United States. Estimate the average one-way commute time in
  a) New York.
  b) Illinois.

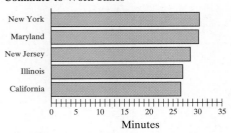

Some of the Longest Average
Commute-to-Work Times

Source: US Census Bureau

Number of Schools to Which Freshmen at Four-Year Colleges Applied

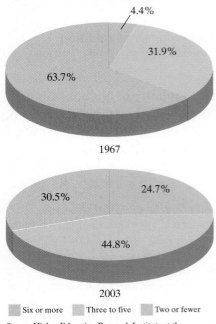

1967

2003

Six or more   Three to five   Two or fewer

Source: Higher Education Research Institute at the University of California-Los Angeles

8. College Applications The graphs on the right show the number of schools to which freshmen at four-year colleges applied in 1967 and 2003.
  a) If there were 1.6 million freshmen at four-year colleges in 1967, determine the number of these freshmen who applied to two or fewer colleges.
  b) If there were 2.1 million freshmen at four-year colleges in 2003, determine the number of these freshmen who applied to six or more colleges.

[1.3] *Perform each indicated operation. Simplify your answers.*

9. $\dfrac{3}{5} \cdot \dfrac{5}{6}$

10. $3\dfrac{5}{7} + 2\dfrac{1}{3}$

11. $\dfrac{5}{12} \div \dfrac{3}{5}$

12. $\dfrac{5}{6} + \dfrac{1}{3}$

13. $3\dfrac{1}{6} - 1\dfrac{1}{4}$

14. $7\dfrac{3}{8} \div \dfrac{5}{12}$

[1.4] **15.** List the set of natural numbers.

16. List the set of whole numbers.

17. List the set of integers.

18. Describe the set of rational numbers.

19. Consider the following set of numbers.

$$\left\{ 3, -5, -12, 0, \dfrac{1}{2}, -0.62, \sqrt{7}, 426, -3\dfrac{1}{4} \right\}$$

List the numbers that are
  a) positive integers.
  b) whole numbers.
  c) integers.
  d) rational numbers.
  e) irrational numbers.
  f) real numbers.

20. Consider the following set of numbers.

$$\left\{ -2.3, -8, -9, 1\dfrac{1}{2}, \sqrt{2}, -\sqrt{2}, 1, -\dfrac{3}{17} \right\}$$

List the numbers that are
  a) natural numbers.
  b) whole numbers.
  c) negative integers.
  d) integers.
  e) rational numbers.
  f) irrational numbers.
  g) real numbers.

[1.5] *Insert either* $<$, $>$, *or* $=$ *in each shaded area to make a true statement.*

21. -7 ▢ -5

22. -2.6 ▢ -3.6

23. 0.50 ▢ 0.509

24. 4.6 ▢ 4.06

25. -6.3 ▢ -6.03

26. 5 ▢ $|-3|$

27. $\left| -\dfrac{9}{2} \right|$ ▢ $|-4.5|$

28. $|-10|$ ▢ $|-7|$

[1.6–1.7] *Evaluate.*

29. $-9 + (-5)$

30. $-6 + 6$

31. $0 + (-3)$

32. $-10 + 4$

33. $-8 - (-2)$

34. $-2 - (-4)$

35. $4 - (-4)$

36. $12 - 12$

37. $2 - 7$

38. $7 - (-7)$

39. $0 - (-4)$

40. $-7 - 5$

41. $\dfrac{4}{3} - \dfrac{3}{4}$ **42.** $\dfrac{1}{2} + \dfrac{3}{5}$ **43.** $\dfrac{5}{9} - \dfrac{3}{4}$ **44.** $-\dfrac{5}{7} + \dfrac{3}{8}$

45. $-\dfrac{5}{12} - \dfrac{5}{6}$ **46.** $-\dfrac{6}{7} + \dfrac{5}{12}$ **47.** $\dfrac{2}{9} - \dfrac{3}{10}$ **48.** $\dfrac{5}{12} - \left(-\dfrac{3}{5}\right)$

Evaluate.

49. $9 - 4 + 3$ **50.** $-8 - 9 + 4$ **51.** $-5 - 4 - 3$ **52.** $-2 + (-3) - 2$

53. $7 - (+4) - (-3)$ **54.** $6 - (-2) + 3$

[1.8] *Evaluate.*

55. $7(-9)$ **56.** $(-8.2)(-3.1)$ **57.** $(-4)(-5)(-6)$ **58.** $\left(\dfrac{3}{5}\right)\left(\dfrac{-2}{7}\right)$

59. $\left(\dfrac{10}{11}\right)\left(\dfrac{3}{-5}\right)$ **60.** $\left(\dfrac{-5}{8}\right)\left(\dfrac{-3}{7}\right)$ **61.** $0\left(\dfrac{4}{9}\right)$ **62.** $(-4)(-6)(-2)(-3)$

Evaluate.

63. $15 \div (-3)$ **64.** $12 \div (-2)$ **65.** $-14.72 \div 4.6$ **66.** $-37.41 \div (-8.7)$

67. $-88 \div (-11)$ **68.** $-4 \div \left(\dfrac{-4}{9}\right)$ **69.** $\dfrac{28}{-3} \div \left(\dfrac{9}{-2}\right)$ **70.** $\dfrac{14}{3} \div \left(\dfrac{-6}{5}\right)$

Indicate whether each quotient is 0 or undefined.

71. $0 \div 5$ **72.** $0 \div (-6)$ **73.** $-12 \div 0$

74. $-4 \div 0$ **75.** $\dfrac{8.3}{0}$ **76.** $\dfrac{0}{-9.8}$

[1.6–1.8, 1.9] *Evaluate.*

77. $-5(3 - 8)$ **78.** $2(4 - 8)$ **79.** $(3 - 6) + 4$

80. $(-4 + 3) - (2 - 6)$ **81.** $[6 + 3(-2)] - 6$ **82.** $(-5 - 3)(4)$

83. $[12 + (-4)] + (6 - 8)$ **84.** $9[3 + (-4)] + 5$ **85.** $-4(-3) + [4 \div (-2)]$

86. $(-3 \cdot 4) \div (-2 \cdot 6)$ **87.** $(-3)(-4) + 6 - 3$ **88.** $[-2(3) + 6] - 4$

[1.9] *Evaluate.*

89. -6^2 **90.** $(-6)^2$ **91.** 2^4 **92.** $(-3)^3$

93. $(-1)^9$ **94.** $(-2)^5$ **95.** $\left(\dfrac{-4}{5}\right)^2$ **96.** $\left(\dfrac{2}{5}\right)^3$

97. $5^3 \cdot (-2)^2$ **98.** $(-2)^4\left(\dfrac{1}{2}\right)^2$ **99.** $\left(-\dfrac{2}{3}\right)^2 \cdot 3^3$ **100.** $(-4)^3(-2)^2$

Evaluate.

101. $-5 + 3 \cdot 4$ **102.** $4 \cdot 6 + 4 \cdot 2$ **103.** $(3.7 - 4.1)^2 + 6.2$

104. $10 - 36 \div 4 \cdot 3$ **105.** $6 - 3^2 \cdot 5$ **106.** $[6.9 - (3 \cdot 5)] + 5.8$

107. $\dfrac{6^2 - 4 \cdot 3^2}{-[6 - (3 - 4)]}$ **108.** $\dfrac{4 + 5^2 \div 5}{6 - (-3 + 2)}$ **109.** $3[9 - (4^2 + 3)] \cdot 2$

110. $(-3^2 + 4^2) + (3^2 \div 3)$ **111.** $2^3 \div 4 + 6 \cdot 3$ **112.** $(4 \div 2)^4 + 4^2 \div 2^2$

113. $(8 - 2^2)^2 - 4 \cdot 3 + 10$ **114.** $4^3 \div 4^2 - 5(2 - 7) \div 5$ **115.** $-\{-4[27 \div 3^2 - 2(4 - 2)]\}$

116. $2\{4^3 - 6[4 - (2 - 4)] - 3\}$

Evaluate each expression for the given values.

117. $3x - 7;\ x = 4$ **118.** $6 - 4x;\ x = -5$

119. $2x^2 - 5x + 3;\ x = 6$ **120.** $5y^2 + 3y - 2;\ y = -1$

121. $-x^2 + 2x - 3;\ x = -2$ **122.** $-x^2 + 2x - 3;\ x = 2$

123. $-3x^2 - 5x + 5;\ x = 1$ **124.** $-x^2 - 8x - 12y;\ x = -3,\ y = -2$

[1.6–1.9] **a)** *Use a calculator to evaluate each expression, and* **b)** *check to see whether your answer is reasonable.*

125. $278 + (-493)$ **126.** $324 - (-29.6)$ **127.** $\dfrac{-17.28}{6}$

128. $(-62)(-1.9)$ **129.** $(-4)^8$ **130.** $-(4.2)^3$

[1.10] *Name each indicated property.*

131. $(7 + 4) + 9 = 7 + (4 + 9)$ **132.** $-5(a + 2) = -5a - 10$

133. $6x + 3x = 3x + 6x$ **134.** $(x + 4)3 = 3(x + 4)$

135. $4(x + 3) = 4x + 12$ **136.** $(x + 7) + 4 = x + (7 + 4)$

137. $8b \cdot 1 = 8b$ **138.** $-8y + 8y = 0$

Chapter 1 Practice Test

To find out how well you understand the chapter material, take this practice test. The answers, and the section where the material was initially discussed, are given in the back of the book. Each problem is also fully worked out on the **Chapter Test Prep Video CD**. *Review any questions that you answered incorrectly.*

1. **Shopping** While shopping, Mia Nguyen purchases two half-gallons of milk for $1.30 each, one Boston cream pie for $4.75, and three 2-liter bottles of soda for $1.10 each.

   a) What is her total bill before tax?

   b) If there is a 7% sales tax on the bottles of soda, how much is the sales tax?

   c) How much is her total bill including tax?

   d) How much change will she receive from a $50 bill?

2. **TV Commercials** The following line graph shows the cost for a 30-second commercial during the same time slot on the same channel for 12 consecutive years. How many times greater was the cost for a 30-second commercial in the twelfth year than in the first year?

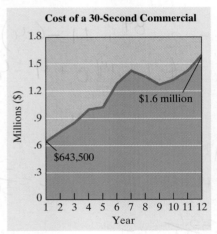

3. **Radio** The following graph shows the median number of listeners for various radio stations during a specific time.

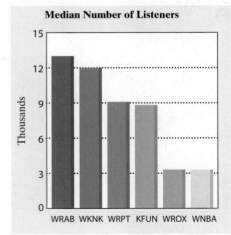

   a) Determine the median number of listeners who listened to WRAB during this time.

   b) The median number of listeners who listened to KFUN during this time was approximately 8.8 thousand. Explain what this means.

4. Consider the following set of numbers.

$$\left\{-6, 42, -3\tfrac{1}{2}, 0, 6.52, \sqrt{5}, \tfrac{5}{9}, -7, -1\right\}$$

   List the numbers that are

   a) natural numbers.

   b) whole numbers.

   c) integers.

   d) rational numbers.

   e) irrational numbers.

   f) real numbers.

Insert either $<$, $>$, or $=$ in each shaded area to make a true statement.

5. -9.9 ▢ -9.09

6. $|-3|$ ▢ $|-2|$

Evaluate.

7. $-7 + (-8)$

8. $-6 - 5$

9. $15 - 12 - 17$

10. $(-4 + 6) - 3(-2)$

11. $(-4)(-3)(2)(-1)$

12. $\left(\dfrac{-2}{9}\right) \div \left(\dfrac{-7}{8}\right)$

13. $\left(-18 \cdot \dfrac{1}{2}\right) \div 3$

14. $-\dfrac{3}{8} - \dfrac{4}{7}$

15. $-6(-2 - 3) \div 5 \cdot 2$

16. $\left(-\dfrac{2}{3}\right)^5$

17. $[6 + ((9 - 3)^2 \div 18)^2]^2$

18. Explain why $-x^2$ will always be a negative value for any nonzero real number selected for x.

Evaluate the expression for the given values.

19. $5x^2 - 8; \; x = -3$

20. $-x^2 - 6x + 3; \; x = -2$

21. $6x - 3y^2 + 4; \; x = 3, y = -2$

22. $-x^2 + xy + y^2; \; x = 1, y = -2$

Name each indicated property.

23. $x + 3 = 3 + x$

24. $4(x + 9) = 4x + 36$

25. $(2 + x) + 4 = 2 + (x + 4)$

2 Solving Linear Equations

GOALS OF THIS CHAPTER

The major emphasis of this chapter is to teach you how to solve linear equations. *To be successful in solving linear equations, you need to have a thorough understanding of adding, subtracting, multiplying, and dividing real numbers.* This material was discussed in Chapter 1.

The first four sections of this chapter will give you the building blocks you will need for solving linear equations. Section 2.5 combines the material previously presented to teach you how to solve a variety of linear equations. In the last few sections of this chapter, you will learn about formulas, ratios, and proportions.

You will be using principles learned in this chapter throughout the book and in real life.

PROPORTIONS ARE VERY USEFUL in everyday living. In Exercise 65 on page 162, we use a proportion to determine how many cups of onions are needed for a recipe.

2.1 Combining Like Terms

1 Identify terms.

2 Identify like terms.

3 Combine like terms.

4 Use the distributive property.

5 Remove parentheses when they are preceded by a plus or minus sign.

6 Simplify an expression.

1 Identify Terms

In Section 1.3, we indicated that letters called **variables** are used to represent numbers. A variable can represent a variety of different numbers.

As was indicated in Chapter 1, an **expression** (sometimes referred to as an **algebraic expression**) is a collection of numbers, variables, grouping symbols, and operation symbols.

Examples of Expressions

$$7, \quad x^2 - 6, \quad 4x - 3, \quad 2(x + 5) + 6, \quad \frac{x + 3}{4}$$

When an algebraic expression consists of several parts, the parts that are *added* are called the **terms** of the expression. Consider the expression $2x - 3y - 5$. The expression can be written as $2x + (-3y) + (-5)$, and so the expression $2x - 3y - 5$ has three terms: $2x$, $-3y$, and -5. The expression $3x^2 + 2xy + 5(x + y)$ also has three terms: $3x^2$, $2xy$, and $5(x + y)$.

When listing the terms of an expression, it is not necessary to list the $+$ sign at the beginning of a term.

Expression	Terms
$-2x + 3y - 8$	$-2x, 3y, -8$
$3y^2 - 2x + \frac{1}{2}$	$3y^2, -2x, \frac{1}{2}$
$7 + x + 4 - 5x$	$7, x, 4, -5x$
$3(x - 1) - 4x + 2$	$3(x - 1), -4x, 2$
$\frac{x + 4}{3} - 5x + 3$	$\frac{x + 4}{3}, -5x, 3$

The numerical part of a term is called its **numerical coefficient** or simply its **coefficient**. In the term $6x$, the 6 is the numerical coefficient. Note that $6x$ means the variable x is multiplied by 6.

Term	Numerical Coefficient
$5x$	5
$-\frac{1}{2}x$	$-\frac{1}{2}$
$4(x - 3)$	4
$\frac{2x}{3}$	$\frac{2}{3}$, since $\frac{2x}{3}$ means $\frac{2}{3}x$
$\frac{x + 4}{3}$	$\frac{1}{3}$, since $\frac{x + 4}{3}$ means $\frac{1}{3}(x + 4)$

Whenever a term appears without a numerical coefficient, we assume that the numerical coefficient is 1.

Examples

x means $1x$	$-x$ means $-1x$
x^2 means $1x^2$	$-x^2$ means $-1x^2$
xy means $1xy$	$-xy$ means $-1xy$
$(x + 2)$ means $1(x + 2)$	$-(x + 2)$ means $-1(x + 2)$

If an expression has a term that is a number (without a variable), we refer to that number as a **constant term**, or simply a **constant**. In the expression $x^2 + 3x - 4$, the -4 is a constant term, or a constant.

2 Identify Like Terms

Like terms (also called similar terms) are terms that have the same variables with the same exponents, respectively. Constants, such as 4 and -6, are like terms. Some examples of like terms and unlike terms follow. Note that if two terms are like terms, only their numerical coefficients may differ.

Like Terms	Unlike Terms	
$3x, \quad -4x$	$3x, \quad 2$	*(One term has a variable, the other is a constant.)*
$4y, \quad 6y$	$3x, \quad 4y$	*(Variables differ)*
$5, \quad -6$	$x, \quad 3$	*(One term has a variable, the other is a constant.)*
$3(x + 1), \quad -2(x + 1)$	$2x, \quad 3xy$	*(Variables differ)*
$3x^2, \quad 4x^2$	$3x, \quad 4x^2$	*(Exponents differ)*
$5ab, \quad 2ab$	$4a, \quad 2ab$	*(Variables differ)*

EXAMPLE 1 ▸ Identify any like terms.

a) $2x + 3x + 4$ **b)** $2x + 3y + 2$ **c)** $x + 3 + y - \dfrac{1}{2}$ **d)** $x + 3x^2 - 4x^2$

e) $5x - x + 6$ **f)** $3 - 2x + 4x - 6$ **g)** $12 + x^2 - x + 7$

Solution

a) $2x$ and $3x$ are like terms.

b) There are no like terms.

c) 3 and $-\dfrac{1}{2}$ are like terms.

d) $3x^2$ and $-4x^2$ are like terms.

e) $5x$ and $-x$ (or $-1x$) are like terms.

f) 3 and -6 are like terms; $-2x$ and $4x$ are like terms.

g) 12 and 7 are like terms.

▸ **Now Try Exercise 5**

3 Combine Like Terms

We often need to simplify expressions by combining like terms. **To combine like terms** means to add or subtract the like terms in an expression. To combine like terms, we can use the procedure that follows.

To Combine Like Terms

1. Determine which terms are like terms.
2. Add or subtract the coefficients of the like terms.
3. Multiply the number found in step 2 by the common variable(s).

Examples 2 through 7 illustrate this procedure.

EXAMPLE 2 ▶ Combine like terms: $5x + 4x$.

Solution $5x$ and $4x$ are like terms with the common variable x. Since $5 + 4 = 9$, then $5x + 4x = 9x$.

▶ **Now Try Exercise 9**

EXAMPLE 3 ▶ Combine like terms: $\dfrac{3}{5}x - \dfrac{2}{3}x$.

Solution   Since $\dfrac{3}{5} - \dfrac{2}{3} = \dfrac{9}{15} - \dfrac{10}{15} = -\dfrac{1}{15}$, then $\dfrac{3}{5}x - \dfrac{2}{3}x = -\dfrac{1}{15}x$.

▶ **Now Try Exercise 15**

EXAMPLE 4 ▶ Combine like terms: $6.47b - 8.39b$.

Solution   Since $6.47 - 8.39 = -1.92$, then $6.47b - 8.39b = -1.92b$.

▶ **Now Try Exercise 35**

EXAMPLE 5 ▶ Combine like terms: $3x + x + 5$.

Solution   The $3x$ and x are like terms.

$$3x + x + 5 = 3x + 1x + 5 = 4x + 5$$

▶ **Now Try Exercise 19**

Because of the commutative property of addition, the order of the terms in the answer is not critical. Thus, $5 + 4x$ is also an acceptable answer to Example 5. When writing answers, we generally list the terms containing variables in alphabetical order from left to right, and list the constant term last.

The commutative and associative properties of addition will be used to rearrange the terms in Examples 6 and 7.

EXAMPLE 6 ▶ Combine like terms: $3b + 6a - 5 - 2a$.

Solution   The only like terms are $6a$ and $-2a$.

$$\begin{aligned} 3b + 6a - 5 - 2a &= 6a - 2a + 3b - 5 &&\textit{Rearrange terms.} \\ &= 4a + 3b - 5 &&\textit{Combine like terms.} \end{aligned}$$

▶ **Now Try Exercise 21**

EXAMPLE 7 ▶ Combine like terms: $-2x^2 + 3y - 4x^2 + 3 - y + 5$.

Solution

$-2x^2$	and $-4x^2$	are like terms.
$3y$	and $-y$	are like terms.
3	and 5	are like terms.

Grouping the like terms together gives

$$\begin{aligned} -2x^2 + 3y - 4x^2 + 3 - y + 5 &= -2x^2 - 4x^2 + 3y - y + 3 + 5 \\ &= \qquad -6x^2 \quad + \quad 2y \quad + \quad 8 \end{aligned}$$

▶ **Now Try Exercise 29**

4 Use the Distributive Property

We introduced the distributive property in Section 1.10. Because this property is so important, we will study it again. But before we do, let's briefly review the subtraction of real numbers. Recall from Section 1.7 that

$$6 - 3 = 6 + (-3)$$

In general,

Subtraction of Real Numbers

For any real numbers a and b,

$$a - b = a + (-b)$$

We will use the fact that $a + (-b)$ means $a - b$ in discussing the distributive property.

Distributive Property

For any real numbers a, b, and c,

$$a(b + c) = ab + ac$$

EXAMPLE 8 ▶ Use the distributive property to remove parentheses.

a) $2(x + 4)$ **b)** $-5(p + 3)$

Solution

a) $2(x + 4) = 2x + 2(4) = 2x + 8$

b) $-5(p + 3) = -5p + (-5)(3) = -5p + (-15) = -5p - 15$

    Note in part **b)** that, instead of leaving the answer $-5p + (-15)$, we wrote it as $-5p - 15$, which is the proper form of the answer.

▶ **Now Try Exercise 61**

EXAMPLE 9 ▶ Use the distributive property to remove parentheses.

a) $3(x - 2)$ **b)** $-2(4x - 3)$

Solution

a) By the definition of subtraction, we may write $x - 2$ as $x + (-2)$.

$$3(x - 2) = 3[x + (-2)] = 3x + 3(-2)$$
$$= 3x + (-6)$$
$$= 3x - 6$$

b) $-2(4x - 3) = -2[4x + (-3)] = -2(4x) + (-2)(-3) = -8x + 6$

▶ **Now Try Exercise 63**

    The distributive property is used often in algebra, so you need to understand it well. You should understand it so well that you will be able to simplify an expression using the distributive property without having to write down all the steps that we listed in working Examples 8 and 9. Study closely the Helpful Hint that follows.

Helpful Hint

With a little practice, you will be able to eliminate some of the intermediate steps when you use the distributive property. When using the distributive property, there are eight possibilities with regard to signs. Study and understand the eight possibilities that follow.

Positive Coefficient

a) $2(x) = 2x$

$2(x+3) = 2x+6$

$2(+3) = +6$

b) $2(x) = 2x$

$2(x-3) = 2x-6$

$2(-3) = -6$

c) $2(-x) = -2x$

$2(-x+3) = -2x+6$

$2(+3) = +6$

d) $2(-x) = -2x$

$2(-x-3) = -2x-6$

$2(-3) = -6$

Negative Coefficient

e) $(-2)(x) = -2x$

$-2(x+3) = -2x-6$

$(-2)(+3) = -6$

f) $(-2)(x) = -2x$

$-2(x-3) = -2x+6$

$(-2)(-3) = +6$

g) $(-2)(-x) = 2x$

$-2(-x+3) = 2x-6$

$(-2)(+3) = -6$

h) $(-2)(-x) = 2x$

$-2(-x-3) = 2x+6$

$(-2)(-3) = +6$

The distributive property can be expanded as follows:

$$a(b + c + d + \cdots + n) = ab + ac + ad + \cdots + an$$

Examples of the Expanded Distributive Property

$$3(x + y + z) = 3x + 3y + 3z$$
$$2(x + y - 3) = 2x + 2y - 6$$

EXAMPLE 10 ▶ Use the distributive property to remove parentheses.

a) $4(x - 3)$ **b)** $-6(5y - 1)$ **c)** $-\dfrac{1}{2}(4r + 5)$ **d)** $-7(2x + 4y - 9z)$

Solution

a) $4(x - 3) = 4x - 12$ **b)** $-6(5y - 1) = -30y + 6$

c) $-\dfrac{1}{2}(4r + 5) = -2r - \dfrac{5}{2}$ **d)** $-7(2x + 4y - 9z) = -14x - 28y + 63z$

▶ **Now Try Exercise 79**

The distributive property can also be used from the right, as in Example 11.

EXAMPLE 11 ▶ Use the distributive property to remove parentheses from the expression $(2x - 8y)4$.

Solution   We distribute the 4 on the right side of the parentheses over the terms within the parentheses.

$$(2x - 8y)4 = 2x(4) - 8y(4)$$
$$= 8x - 32y$$

▶ **Now Try Exercise 83**

Example 11 could have been rewritten as $4(2x - 8y)$ by the commutative property of multiplication, and then the 4 could have been distributed from the left to obtain the same answer, $8x - 32y$.

Helpful Hint

Students sometimes try to use the distributive property when it cannot be used. For the distributive property to be used, there must be a + or − between the terms *within parentheses* and the terms within parentheses must be *multiplied* by some number or expression. Study the following correct simplifications carefully.

$$4(2xy) = 8xy \qquad \text{\textit{(distributive property is not used)}}$$
$$4(2x + y) = 8x + 4y \qquad \text{\textit{(distributive property is used)}}$$
$$(2x + y) - 4 = 2x + y - 4 \qquad \text{\textit{(distributive property is not used)}}$$
$$(2x + y)(-4) = -8x - 4y \qquad \text{\textit{(distributive property is used)}}$$

5 Remove Parentheses When They Are Preceded by a Plus or Minus Sign

In the expression $(4x + 3)$, how do we remove parentheses? Recall that the coefficient of a term is assumed to be 1 if none is shown. Therefore, we may write

$$(4x + 3) = 1(4x + 3)$$
$$= 1(4x) + (1)(3)$$
$$= 4x + 3$$

Note that $(4x + 3) = 4x + 3$. **When no sign or a plus sign precedes parentheses, the parentheses may be removed without having to change the expression inside the parentheses.**

Examples
$$(x + 3) = x + 3$$
$$(2x - 3) = 2x - 3$$
$$+(2x - 5) = 2x - 5$$
$$+(x + 2y - 6) = x + 2y - 6$$

Now consider the expression $-(4x + 3)$. How do we remove parentheses in this expression? Here, the coefficient in front of the parentheses is −1, so each term within the parentheses is multiplied by −1.

$$-(4x + 3) = -1(4x + 3)$$
$$= -1(4x) + (-1)(3)$$
$$= -4x + (-3)$$
$$= -4x - 3$$

Thus, $-(4x + 3) = -4x - 3$. **When a minus sign precedes parentheses, the signs of all the terms within the parentheses are changed when the parentheses are removed.**

Examples
$$-(x + 4) = -x - 4$$
$$-(-2x + 3) = 2x - 3$$
$$-(5x - y + 3) = -5x + y - 3$$
$$-(-4c - 3d - 5) = 4c + 3d + 5$$

6 Simplify an Expression

Combining what we learned in the preceding discussions, we have the following procedure for **simplifying an expression.**

To Simplify an Expression

1. Use the distributive property to remove any parentheses.
2. Combine like terms.

EXAMPLE 12 ▶ Simplify $6 - (2x + 3)$.

Solution

$$6 - (2x + 3) = 6 - 2x - 3 \qquad \text{Use the distributive property.}$$
$$= -2x + 3 \qquad \text{Combine like terms.}$$

Note: $3 - 2x$ is the same as $-2x + 3$; however, we generally write the term containing the variable first.

▶ **Now Try Exercise 89**

EXAMPLE 13 ▶ Simplify $-\left(\dfrac{2}{3}x - \dfrac{1}{4}\right) + 3x$.

Solution

$$-\left(\frac{2}{3}x - \frac{1}{4}\right) + 3x = -\frac{2}{3}x + \frac{1}{4} + 3x \qquad \text{Distributive property}$$
$$= -\frac{2}{3}x + 3x + \frac{1}{4} \qquad \text{Rearrange terms.}$$
$$= -\frac{2}{3}x + \frac{9}{3}x + \frac{1}{4} \qquad \text{Write x terms with the LCD, 3.}$$
$$= \frac{7}{3}x + \frac{1}{4} \qquad \text{Combine like terms.}$$

▶ **Now Try Exercise 101**

Notice in Example 13 that $\frac{7}{3}x$ and $\frac{1}{4}$ could not be combined because they are not like terms.

EXAMPLE 14 ▶ Simplify $\dfrac{3}{4}x + \dfrac{1}{3}(5x - 2)$.

Solution

$$\frac{3}{4}x + \frac{1}{3}(5x - 2) = \frac{3}{4}x + \frac{1}{3}(5x) + \frac{1}{3}(-2) \qquad \text{Distributive property}$$
$$= \frac{3}{4}x + \frac{5}{3}x - \frac{2}{3}$$
$$= \frac{9}{12}x + \frac{20}{12}x - \frac{2}{3} \qquad \text{Write x terms with the LCD, 12.}$$
$$= \frac{29}{12}x - \frac{2}{3} \qquad \text{Combine like terms.}$$

▶ **Now Try Exercise 103**

EXAMPLE 15 ▶ Simplify $3(2a - 5) - 3(b - 6) - 4a$.

Solution

$$3(2a - 5) - 3(b - 6) - 4a = 6a - 15 - 3b + 18 - 4a \qquad \text{Distributive property}$$
$$= 6a - 4a - 3b - 15 + 18 \qquad \text{Rearrange terms.}$$
$$= 2a - 3b + 3 \qquad \text{Combine like terms.}$$

▶ **Now Try Exercise 107**

Helpful Hint

Keep in mind the difference between the concepts of *term* and *factor*. When two or more expressions are **multiplied**, each expression is a **factor** of the product. For example, since $4 \cdot 3 = 12$, the 4 and the 3 are factors of 12. Since $3 \cdot x = 3x$, the 3 and the x are factors of $3x$. Similarly, in the expression $5xyz$, the 5, x, y, and z are all factors.

In an expression, the parts that are **added** are the **terms** of the expression. For example, the expression $2x^2 + 3x - 4$, has three terms, $2x^2$, $3x$, and -4. Note that the terms of an expression may have factors. For example, in the term $2x^2$, the 2 and the x^2 are factors because they are multiplied.

EXERCISE SET 2.1

Math XL MathXL®   *MyMathLab* MyMathLab

Concept/Writing Exercises

1. a) What are the terms of an expression?

  b) What are the terms of $3x - 4y - 5$?

  c) What are the terms of $6xy + 3x - y - 9$?

2. a) What is the name given to the numerical part of a term? List the coefficient of the following terms.

  b) $4x$ **c)** x **d)** $-x$

  e) $\dfrac{3x}{5}$ **f)** $\dfrac{4}{7}(3t - 5)$

3. Consider the expression $2x - 5$.

  a) What is the x called?

  b) What is the -5 called?

  c) What is the 2 called?

4. a) What are like terms? Determine whether the following are like terms. If not, explain why.

  b) $3x, 4y$ **c)** $7, -2$

  d) $5x^2, 2x$ **e)** $4x, -5xy$

5. Determine whether the following are like terms. If not, explain why.

  a) $5x, -7x$ **b)** $7y, 2$

  c) $-3(t + 2), 6(t + 2)$ **d)** $4pq, -9pq$

6. a) When no sign or a plus sign precedes an expression within parentheses, explain how to remove parentheses.

  b) Write $+(x - 8)$ without parentheses.

7. a) When a minus sign precedes an expression within parentheses, explain how to remove parentheses.

  b) Write $-(x - 8)$ without parentheses.

8. a) What are the factors of an expression?

  b) Explain why 3 and x are factors of $3x$.

  c) Explain why 5, x, and y are all factors of the expression $5xy$.

Practice the Skills

Combine like terms when possible. If not possible, rewrite the expression as is.

9. $6x + 3x$

10. $4x - 5x$

11. $3x + 6$

12. $4x + 3y$

13. $y + 3 + 4y$

14. $4x - 7x + 4$

15. $\dfrac{3}{4}a - \dfrac{6}{11}a$

16. $\dfrac{3}{4}p - \dfrac{2}{7}p$

17. $2 - 6x + 5$

18. $-7 - 4m - 6$

19. $-2w - 3w + 5$

20. $-8y - 4y - 7$

21. $-x + 2 - x - 2$

22. $8x - 2y - 1 - 3x$

23. $3 + 6x - 3 - 6x$

24. $y - 2y + 5$

25. $5 + 2x - 4x + 6$

26. $5s - 3s - 2s$

27. $4r - 6 - 6r - 2$

28. $-6t + 5 + 2t - 9$

29. $3x^2 - 9y^2 + 7x^2 - 5 - y^2 - 2$

30. $-4x^2 - 6y - 3x^2 + 6 - y - 1$

31. $-2x + 4x - 3$

32. $4 - x + 4x - 8$

33. $b + 4 + \dfrac{3}{5}$

34. $\dfrac{3}{4}x + 2 + x$

35. $5.1n + 6.42 - 4.3n$

36. $2x^2 + 3y^2 + 4x + 5y^2$

37. $\dfrac{1}{2}a + 3b + 1$

38. $x + \dfrac{1}{2}y - \dfrac{3}{8}y$

39. $13.4x + 1.2x + 8.3$

40. $-4x^2 - 3.1 - 5.2$

41. $-x^2 + 2x^2 + y$

42. $1 + x^2 + 6 - 3x^2$

43. $2x - 7y - 5x + 2y$

44. $3x - 7 - 9 + 4x$

45. $4 - 3n^2 + 9 - 2n$

46. $9x + y - 2 - 4x$

47. $-19.36 + 40.02x + 12.25 - 18.3x$

48. $52x - 52x - 63.5 - 63.5$

49. $\dfrac{3}{5}x - 3 - \dfrac{7}{4}x - 2$

50. $\dfrac{1}{2}y - 4 + \dfrac{3}{4}x - \dfrac{1}{5}y$

51. $5w^3 + 2w^2 + w + 3$

52. $4p^2 - 3p^2 + 2p - 5p$

53. $2z - 5z^3 - 2z^3 - z^2$

54. $5ab - 3ab$

55. $6x^2 - 6xy + 3y^2$

56. $x^2 - 3xy - 2xy + 6$

57. $4a^2 - 3ab + 6ab + b^2$

58. $4b^2 - 8bc + 5bc + c^2$

Use the distributive property to remove parentheses.

59. $5(x + 2)$

60. $2(-y + 5)$

61. $5(x + 4)$

62. $-2(y + 8)$

63. $3(x - 6)$

64. $-2(x - 4)$

65. $-\dfrac{1}{2}(2x - 4)$

66. $-4(x + 6)$

67. $1(-4 + x)$

68. $\frac{2}{3}(m - 9)$

69. $\frac{4}{5}(s - 5)$

70. $5(x - y + 5)$

71. $-0.3(3x + 5)$

72. $-(x - 3)$

73. $-\frac{1}{3}(3r - 12)$

74. $-2(x + y - z)$

75. $0.7(2x + 0.5)$

76. $-(x + 4y)$

77. $-(-x + y)$

78. $(3x + 4y - 6)$

79. $-(2x + 4y - 8)$

80. $-3(2a + 3b - 7)$

81. $1.1(3.1x - 5.2y + 2.8)$

82. $-4(-2m - 3n + 8)$

83. $(2x - 9y)5$

84. $(8b - 1)7$

85. $(x + 3y - 9)$

86. $(-p + 2q - 3)$

87. $-3(-x + 2y + 4)$

88. $2.3(1.6x + 5.1y - 4.1)$

Simplify.

89. $5 - (3x + 4)$

90. $7 - (2x - 9)$

91. $-2(3 - x) + 7$

92. $-(3x - 3) + 5$

93. $6x + 2(4x + 9)$

94. $3(x + y) + 2y$

95. $2(x - y) + 2x + 3$

96. $6 + (x - 5) + 3x$

97. $4(2c - 3) - 3(c - 4)$

98. $4 + (2y + 2) + y$

99. $8x - (x - 3)$

100. $-(x - 5) - 3x + 4$

101. $-\left(\frac{3}{4}x - \frac{1}{3}\right) + 2x$

102. $-\left(\frac{7}{8}x - \frac{1}{2}\right) - 3x$

103. $\frac{2}{3}x + \frac{1}{2}(5x - 4)$

104. $\frac{4}{5}x + \frac{1}{7}(3x - 1)$

105. $-(3s + 4) - (s + 2)$

106. $6 - 2(x + 3) + 5x$

107. $4(x - 1) + 2(3 - x) - 4$

108. $4(3b - 2) - 5(c - 4) - 6b$

109. $4(m + 3) - 4m - 12$

110. $-3(a + 2b) + 3(a + 2b)$

111. $0.4 - (y + 5) + 0.6 - 2$

112. $4 - (2 - x) + 3x$

113. $4 + (3x - 4) - 5$

114. $2y - 6(y - 2) + 3$

115. $4(x + 2) - 3(x - 4) - 5$

116. $6 - (a - 5) - (2b + 1)$

117. $-0.2(6 - x) - 4(y + 0.4)$

118. $-5(2y - 8) - 3(1 + x) - 7$

119. $-6x + 7y - (3 + x) + (x + 3)$

120. $3(t - 2) - 2(t + 4) - 6$

121. $\frac{1}{2}(x + 3) + \frac{1}{3}(3x + 6)$

122. $\frac{2}{3}(r - 2) - \frac{1}{2}(r + 4)$

Problem Solving

If $\square + \square + \square + \odot + \odot$ *can be represented as* $3\square + 2\odot$, *write an expression to represent each of the following.*

123. $\square + \ominus + \ominus + \square + \ominus$

124. $\otimes + \odot + \otimes + \odot + \odot + \odot$

125. $x + y + \triangle + \triangle + x + y + y$

126. $2 + x + 2 + \ominus + \ominus + 2 + y$

Combine like terms.

127. $3\triangle + 5\square - \triangle - 3\square$

128. $8\odot - 4\square - 2\square - 3\odot$

In Exercises 129 and 130, consider the following. The positive factors of 6 are 1, 2, 3, and 6 since

$$1 \cdot 6 = 6$$
$$2 \cdot 3 = 6$$
$$\uparrow \ \uparrow$$
$$\textit{factors}$$

129. List all the positive factors of 18.

130. List all the positive factors of 24.

Challenge Problems

Simplify.

131. $4x^2 + 5y^2 + 6(3x^2 - 5y^2) - 4x + 3$

132. $2x^2 - 4x + 8x^2 - 3(x + 2) - x^2 - 2$

133. $2[3 + 4(x - 5)] - [2 - (x - 3)]$

134. $\frac{1}{4}\left[3 - 2(y + 1)\right] - \frac{1}{3}\left[2 - (y - 6)\right]$

Cumulative Review Exercises

[1.5] *Evaluate.*

135. $|-7|$

136. $-|-16|$

[1.7] **137.** Evaluate $-4 - 3 - (-6)$.

[1.9] **138.** Write a paragraph explaining the order of operations.

139. Evaluate $-x^2 + 5x - 6$ when $x = -1$.

2.2 The Addition Property of Equality

1 Identify linear equations.

2 Check solutions to equations.

3 Identify equivalent equations.

4 Use the addition property to solve equations.

5 Solve equations by doing some steps mentally.

1 Identify Linear Equations

A statement that shows two algebraic expressions are equal is called an **equation**. For example, $4x + 3 = 2x - 4$ is an equation. In this chapter, we learn to solve **linear equations** in one variable.

> **Linear Equation**
>
> A **linear equation** in one variable is an equation that can be written in the form
>
> $$ax + b = c$$
>
> where a, b, and c are real numbers and $a \neq 0$.

Examples of Linear Equations

$$x + 4 = 7$$
$$2x - 4 = 6$$

2 Check Solutions to Equations

The **solution to an equation** is the number or numbers that when substituted for the variable or variables make the equation a true statement. For example, the solution to $x + 4 = 7$ is 3. We will shortly learn how to find the solution to an equation, or to **solve an equation**. But before we do this we will learn how to *check* the solution to an equation.

The solution to an equation may be **checked** by substituting the value that is believed to be the solution for the variable in the original equation. If the substitution results in a true statement, your solution is correct. If the substitution results in a false statement, then either your solution or your check is incorrect, and you need to go back and find your error. Try to check all your solutions. Checking the solutions will improve your arithmetic and algebra skills.

When we show the check of a solution we shall use the notation, $\stackrel{?}{=}$. This notation is used when we are questioning whether a statement is true. For example, if we use

$$2 + 3 \stackrel{?}{=} 2(3) - 1$$

we are asking "Does $2 + 3 = 2(3) - 1$?"

To check whether 3 is the solution to $x + 4 = 7$, we substitute 3 for each x in the equation.

Check:
$$x = 3$$
$$x + 4 = 7$$
$$3 + 4 \stackrel{?}{=} 7$$
$$7 = 7 \quad \textit{True}$$

Since the check results in a true statement, 3 is a solution.

EXAMPLE 1 ▶ Consider the equation $2x - 4 = 6$. Determine whether 3 is a solution.

Solution   To determine whether 3 is a solution to the equation, we substitute 3 for x.

Check:
$$x = 3$$
$$2x - 4 = 6$$
$$2(3) - 4 \stackrel{?}{=} 6$$
$$6 - 4 \stackrel{?}{=} 6$$
$$2 = 6 \quad \textit{False}$$

Since we obtained a false statement, 3 is not a solution.

▶ **Now Try Exercise 13**

Now check to see if 5 is a solution to the equation in Example 1. Your check should show that 5 is a solution.

We can use the same procedures to check more complex equations, as shown in Examples 2 and 3.

EXAMPLE 2 ▶ Determine whether 18 is a solution to the following equation.
$$3x - 2(x + 3) = 12$$

Solution   To determine whether 18 is a solution, we substitute 18 for each x in the equation. If the substitution results in a true statement, then 18 is a solution.

Check:
$$x = 18$$
$$3x - 2(x + 3) = 12$$
$$3(18) - 2(18 + 3) \stackrel{?}{=} 12$$
$$3(18) - 2(21) \stackrel{?}{=} 12$$
$$54 - 42 \stackrel{?}{=} 12$$
$$12 = 12 \quad \textit{True}$$

Since we obtained a true statement, 18 is a solution.

▶ **Now Try Exercise 17**

EXAMPLE 3 ▶ Determine whether $-\dfrac{3}{2}$ is a solution to the following equation.
$$3(n + 3) = 6 + n$$

Solution   In this equation n is the variable. Substitute $-\dfrac{3}{2}$ for each n in the equation.

Check:
$$n = -\frac{3}{2}$$
$$3(n + 3) = 6 + n$$
$$3\left(-\frac{3}{2} + 3\right) \stackrel{?}{=} 6 + \left(-\frac{3}{2}\right)$$
$$3\left(-\frac{3}{2} + \frac{6}{2}\right) \stackrel{?}{=} \frac{12}{2} - \frac{3}{2}$$
$$3\left(\frac{3}{2}\right) \stackrel{?}{=} \frac{9}{2}$$
$$\frac{9}{2} = \frac{9}{2} \quad \textit{True}$$

Thus, $-\dfrac{3}{2}$ is a solution.

▶ **Now Try Exercise 23**

USING YOUR CALCULATOR Checking Solutions

Calculators can be used to check solutions to equations. For example, to check whether $\dfrac{-10}{3}$ is a solution to the equation $2x + 3 = 5(x + 3) - 2$, we perform the following steps.

1. Substitute $\dfrac{-10}{3}$ for each x as shown below.

$$2x + 3 = 5(x + 3) - 2$$

$$2\left(\dfrac{-10}{3}\right) + 3 \overset{?}{=} 5\left(\dfrac{-10}{3} + 3\right) - 2$$

2. Evaluate each side of the equation separately using your calculator. If you obtain the same value on both sides, your solution checks.

Scientific Calculator

To evaluate the left side of the equation, $2\left(\dfrac{-10}{3}\right) + 3$, press the following keys:

2 $\boxed{\times}$ $\boxed{(}$ 10 $\boxed{+/-}$ $\boxed{\div}$ 3 $\boxed{)}$ $\boxed{+}$ 3 $\boxed{=}$ -3.6666667

To evaluate the right side of the equation, $5\left(\dfrac{-10}{3} + 3\right) - 2$, press the following keys:

5 $\boxed{\times}$ $\boxed{(}$ 10 $\boxed{+/-}$ $\boxed{\div}$ 3 $\boxed{+}$ 3 $\boxed{)}$ $\boxed{-}$ 2 $\boxed{=}$ -3.6666667

Since both sides give the same value, the solution checks. Note that because calculators differ in their electronics, sometimes the last digit of a calculation will differ.

Graphing Calculator

To evaluate the left side of equation: 2 $\boxed{(}$ $\boxed{(-)}$ 10 $\boxed{\div}$ 3 $\boxed{)}$ $\boxed{+}$ 3 $\boxed{\text{ENTER}}$ -3.666666667

Right side of the equation: 5 $\boxed{(}$ $\boxed{(-)}$ 10 $\boxed{\div}$ 3 $\boxed{+}$ 3 $\boxed{)}$ $\boxed{-}$ 2 $\boxed{\text{ENTER}}$ -3.666666667

Since both sides give the same value, the solution checks.

3 Identify Equivalent Equations

FIGURE 2.1

Now that we know how to check a solution to an equation we will discuss solving equations. Complete procedures for solving equations will be given shortly. For now, you need to understand that **to solve an equation, it is necessary to get the variable alone on one side of the equal sign. That is, we want to get an equation of the form $x =$ some number (or $1x =$ some number). When we get an equation in this form, we say that we isolate the variable.** To isolate the variable, we make use of two properties: the addition and multiplication properties of equality. Look first at **Figure 2.1**.

Think of an equation as a balanced statement whose left side is balanced by its right side. When solving an equation, we must make sure that the equation remains balanced at all times. That is, both sides must always remain equal. **We ensure that an equation always remains equal by doing the same thing to both sides of the equation.** For example, if we add a number to the left side of the equation, we must add exactly the same number to the right side. If we multiply the right side of the equation by some number, we must multiply the left side by the same number.

When we add the same number to both sides of an equation or multiply both sides of an equation by the same nonzero number, we do not change the solution to the equation, just the form of the equation. Two or more equations with the same solution are called **equivalent equations**. The equations $2x - 4 = 2$, $2x = 6$, and $x = 3$ are equivalent, since the solution to each is 3.

Check: $x = 3$

$2x - 4 = 2$	$2x = 6$	$x = 3$
$2(3) - 4 \stackrel{?}{=} 2$	$2(3) \stackrel{?}{=} 6$	$3 = 3$ *True*
$6 - 4 \stackrel{?}{=} 2$	$6 = 6$ *True*	
$2 = 2$ *True*		

When solving an equation, we use the addition and multiplication properties to express a given equation as simpler equivalent equations until we obtain the solution.

4 Use the Addition Property to Solve Equations

Now we are ready to define the **addition property of equality**.

> ### Addition Property of Equality
> If $a = b$, then $a + c = b + c$ for any real numbers a, b, and c.

This property means that the same number can be added to both sides of an equation without changing the solution. **The addition property is used to solve equations of the form $x + a = b$.** To isolate the variable x in equations of this form, add the opposite or additive inverse of a, $-a$, to both sides of the equation.

To isolate the variable when solving equations of the form $x + a = b$, **we use the addition property to eliminate the number on the same side of the equal sign as the variable.** Study the following examples carefully.

Equation	To Solve, Use the Addition Property to Eliminate the Number
$x - 4 = -3$	-4
$x + 5 = 9$	5
$-3 = k + 7$	7
$-5 = x - 4$	-4
$-6.25 = y + 12.78$	12.78

Now let's work some examples.

EXAMPLE 4 ▶ Solve the equation $x - 4 = -3$.

Solution   To isolate the variable, x, we must eliminate the -4 from the left side of the equation. To do this we add 4, the opposite of -4, to *both sides* of the equation.

$$x - 4 = -3$$
$$x - 4 + 4 = -3 + 4 \qquad \textit{Add 4 to both sides.}$$
$$x + 0 = 1$$
$$x = 1$$

Note how the process helps to isolate x.

Check:
$$x - 4 = -3$$
$$1 - 4 \stackrel{?}{=} -3$$
$$-3 = -3 \qquad \textit{True}$$

▶ **Now Try Exercise 29**

In Example 5, we will not show the check. Space limitations prevent us from showing all checks. However, *you should check all of your answers.*

EXAMPLE 5 ▶ Solve the equation $x + 5 = 9$.

Solution   To solve this equation, we must isolate the variable, x. Therefore, we must eliminate the 5 from the left side of the equation. To do this, we add -5, the opposite of 5, to *both sides* of the equation.

$$x + 5 = 9$$
$$x + 5 + (-5) = 9 + (-5) \quad \text{\textit{Add} } -5 \text{ \textit{to both sides.}}$$
$$x + 0 = 4$$
$$x = 4$$

▶ **Now Try Exercise 31**

In Example 5, we added -5 to both sides of the equation. From Section 1.7 we know that $5 + (-5) = 5 - 5$. Thus, we can see that adding a negative 5 to both sides of the equation is equivalent to subtracting a 5 from both sides of the equation. According to the addition property, the same number may be *added* to both sides of an equation. **Since subtraction is defined in terms of addition, the addition property also allows us to *subtract* the same number from both sides of the equation.** Thus, Example 5 could have also been worked as follows:

$$x + 5 = 9$$
$$x + 5 - 5 = 9 - 5 \quad \text{\textit{Subtract 5 from both sides.}}$$
$$x + 0 = 4$$
$$x = 4$$

In this text, unless there is a specific reason to do otherwise, rather than adding a negative number to both sides of the equation, we will subtract a number from both sides of the equation.

EXAMPLE 6 ▶ Solve the equation $-3 = k + 7$.

Solution   We must isolate the variable, k, which is on the right side of the equal sign.

$$-3 = k + 7$$
$$-3 - 7 = k + 7 - 7 \quad \text{\textit{Subtract 7 from both sides.}}$$
$$-10 = k + 0$$
$$-10 = k$$

Check:
$$-3 = k + 7$$
$$-3 \stackrel{?}{=} -10 + 7$$
$$-3 = -3 \quad \text{\textit{True}}$$

▶ **Now Try Exercise 27**

Helpful Hint

Remember that our goal in solving an equation is to get the variable alone on one side of the equation. To do this, we add or subtract **the number on the same side of the equation as the variable** to or from both sides of the equation.

Equation	Must Eliminate	Number to Add (or Subtract) to (or from) Both Sides of the Equation	Correct Results	Solution
$x - 5 = 8$	-5	add 5	$x - 5 + 5 = 8 + 5$	$x = 13$
$x - 3 = -12$	-3	add 3	$x - 3 + 3 = -12 + 3$	$x = -9$
$2 = x - 7$	-7	add 7	$2 + 7 = x - 7 + 7$	$9 = x$ or $x = 9$
$x + 12 = -5$	$+12$	subtract 12	$x + 12 - 12 = -5 - 12$	$x = -17$
$6 = x + 4$	$+4$	subtract 4	$6 - 4 = x + 4 - 4$	$2 = x$ or $x = 2$
$13 = x + 9$	$+9$	subtract 9	$13 - 9 = x + 9 - 9$	$4 = x$ or $x = 4$

Notice that under the *Correct Results* column, when the equation is simplified by combining terms, the x will become isolated because the sum of a number and its opposite is 0, and $x + 0$ equals x.

EXAMPLE 7 ▶ Solve the equation $-5 = x - 4$.

Solution   The variable, x, is on the right side of the equation. To isolate the x, we must eliminate the -4 from the right side of the equation. This can be accomplished by adding 4 to both sides of the equation.

$$-5 = x - 4$$
$$-5 \boxed{+\,4} = x - 4 \boxed{+\,4} \qquad \textit{Add 4 to both sides.}$$
$$-1 = x + 0$$
$$-1 = x$$

Thus, the solution is -1.

▶ **Now Try Exercise 37**

EXAMPLE 8 ▶ Solve the equation $-6.25 = y + 12.78$.

Solution   The variable, y, is on the right side of the equation. Subtract 12.78 from both sides of the equation to isolate the variable.

$$-6.25 = y + 12.78$$
$$-6.25 \boxed{-\,12.78} = y + 12.78 \boxed{-\,12.78} \qquad \textit{Subtract 12.78 from both sides.}$$
$$-19.03 = y + 0$$
$$-19.03 = y$$

Thus, the solution is -19.03.

▶ **Now Try Exercise 67**

Avoiding Common Errors

When solving an equation, our goal is to isolate the variable on one side of the equal sign. Consider the equation $x + 3 = -4$. How do we solve it?

CORRECT	INCORRECT
Remove the 3 from the left side of the equation.	Remove the -4 from the right side of the equation.

CORRECT

$$x + 3 = -4$$
$$x + 3 \boxed{-\,3} = -4 \boxed{-\,3}$$
$$x = -7$$

Variable is now isolated.

INCORRECT

$$x + 3 = -4$$
$$x + 3 \boxed{+\,4} = -4 \boxed{+\,4}$$
$$x + 7 = 0$$

*Variable is **not** isolated.*

Remember, use the addition property to *remove the number that is on the same side of the equal sign as the variable.*

---

5   Solve Equations by Doing Some Steps Mentally

Consider the following two problems.

a)
$$x \boxed{-\,5} = 12$$
$$x - 5 + 5 = 12 + 5$$
$$x + 0 = 12 \boxed{+\,5}$$
$$x = 17$$

b)
$$15 = x \boxed{+\,3}$$
$$15 - 3 = x + 3 - 3$$
$$15 \boxed{-\,3} = x + 0$$
$$12 = x$$

Note how the number on the same side of the equal sign as the variable is transferred to the opposite side of the equal sign when the addition property is used. Also note that the sign of the number changes when transferred from one side of the equal sign to the other.

When you feel comfortable using the addition property of equality, you may wish to do some of the steps mentally to reduce some of the written work. For example, the preceding two problems may be shortened as follows:

Shortened Form

a)

$$x - 5 = 12$$
$$x - 5 + 5 = 12 + 5 \quad \longleftarrow \quad \boxed{\text{Do this step mentally.}}$$
$$x = 12 + 5$$
$$x = 17$$

$$x - 5 = 12$$
$$x = 12 + 5$$
$$x = 17$$

Shortened Form

b)

$$15 = x + 3$$
$$15 - 3 = x + 3 - 3 \quad \longleftarrow \quad \boxed{\text{Do this step mentally.}}$$
$$15 - 3 = x$$
$$12 = x$$

$$15 = x + 3$$
$$15 - 3 = x$$
$$12 = x$$

EXERCISE SET 2.2   Math XL   MyMathLab

MathXL®   MyMathLab

Concept/Writing Exercises

1. What is an equation?

2. a) What is meant by the "solution to an equation"?

   b) What does it mean to "solve an equation"?

3. Explain how the solution to an equation may be checked.

4. Explain the addition property of equality.

5. What are equivalent equations?

6. To solve an equation we "isolate the variable."

   a) Explain what this means.

   b) Explain how to isolate the variable in the equations discussed in this section.

7. When solving the equation $6 = x + 2$, would you subtract 6 from both sides of the equation or subtract 2 from both sides of the equation? Explain.

8. When solving the equation $x - 4 = 6$, would you add 4 to both sides of the equation or subtract 6 from both sides of the equation? Explain.

9. Give an example of a linear equation in one variable.

10. Explain why the addition property allows us to subtract the same quantity from both sides of an equation.

11. Explain why the following three equations are equivalent.

$$2x + 3 = 5, \qquad 2x = 2, \qquad x = 1$$

12. To solve the equation $x - \square = \triangle$ for x, do we add $\square$ to both sides of the equation or do we subtract $\triangle$ from both sides of the equation? Explain.

Practice the Skills

13. Is $x = 2$ a solution of $4x - 3 = 5$?

14. Is $x = -6$ a solution of $2x + 1 = x - 5$?

15. Is $x = -3$ a solution of $2x - 5 = 5(x + 2)$?

16. Is $x = 1$ a solution of $2(x - 3) = -3(x + 1)$?

17. Is $p = -15$ a solution of $2p - 5(p + 7) = 10$?

18. Is $k = -2$ a solution of $5k - 6(k - 1) = 8$?

19. Is $x = 3.4$ a solution of $3(x + 2) - 3(x - 1) = 9$?

20. Is $x = \frac{3}{4}$ a solution of $x + 5 = 5x + 2$?

21. Is $x = \frac{1}{2}$ a solution of $4x - 4 = 2x - 2$?

22. Is $x = \frac{1}{3}$ a solution of $7x + 3 = 2x + 5$?

23. Is $x = \frac{11}{2}$ a solution of $3(x + 2) = 5(x - 1)$?

24. Is $h = 3$ a solution of $-(h - 5) - (h - 6) = 3h - 4$?

Solve each equation and check your solution.

25. $x + 2 = 7$

26. $x - 4 = 13$

27. $-6 = x + 1$

28. $-5 = x + 4$

29. $x - 4 = -8$

30. $x - 16 = -12$

31. $x + 9 = 52$

32. $x + 8 = 17$

33. $-6 + w = 9$

34. $3 = 7 + t$

35. $27 = x + 16$

36. $50 = x - 25$

37. $-18 = x - 14$

38. $-4 = x - 3$

39. $9 + x = 4$

40. $x + 29 = -29$

41. $4 + x = -9$

42. $9 = x - 3$

43. $7 + r = -23$

44. $a - 5 = -9$

45. $8 = 8 + v$	**46.** $9 + x = 12$	**47.** $7 + x = -50$	**48.** $-17 = 8 + x$
49. $12 = 16 + x$	**50.** $62 = z - 15$	**51.** $15 + x = -5$	**52.** $-20 = 4 + x$
53. $-15 + x = -15$	**54.** $8 = 8 + x$	**55.** $5 = x - 12$	**56.** $-12 = 20 + c$
57. $-50 = x - 24$	**58.** $-29 + x = -15$	**59.** $43 = 15 + p$	**60.** $-25 = 74 + x$
61. $40.2 + x = -5.9$	**62.** $-27.23 + x = 9.77$	**63.** $-37 + x = 9.5$	**64.** $7.2 + x = 7.2$
65. $x - 8.77 = -17$	**66.** $6.1 + x = 10.2$	**67.** $9.32 = x + 3.75$	**68.** $-5.62 = y + 11.39$

Problem Solving

69. Do you think the equation $x + 1 = x + 2$ has a real number as a solution? Explain. (We will discuss equations like this in Section 2.5.)

70. Do you think the equation $x + 4 = x + 4$ has more than one real number as a solution? If so, how many solutions does it have? Explain. (We will discuss equations like this in Section 2.5.)

Challenge Problems

We can solve equations that contain unknown symbols. Solve each equation for the symbol indicated by adding (or subtracting) a symbol to (or from) both sides of the equation. Explain each answer. (Remember that to solve the equation you want to isolate the symbol you are solving for on one side of the equation.)

71. $x - \triangle = \square$, for x

72. $\square + \odot = \triangle$, for $\odot$

73. $\odot = \square + \triangle$, for $\square$

74. $\square = \triangle + \odot$, for $\odot$

Group Activity

Discuss and answer Exercise 75 as a group.

75. Consider the equation $2(x + 3) = 2x + 6$.

a) Group member 1: Determine whether 4 is a solution to the equation.

b) Group member 2: Determine whether -2 is a solution to the equation.

c) Group member 3: Determine whether 0.3 is a solution to the equation.

d) Each group member: Select a number not used in parts a)–c) and determine whether that number is a solution to the equation.

e) As a group, write what you think is the solution to the equation $2(x + 3) = 2x + 6$ and write a paragraph explaining your answer.

Cumulative Review Exercises

[1.6] *Add.*

76. $-\dfrac{7}{15} + \dfrac{5}{6}$

77. $-\dfrac{11}{12} + \left(-\dfrac{3}{8}\right)$

[2.1] *Simplify.*

78. $4x + 3(x - 2) - 5x - 7$

79. $-(2t + 4) + 3(4t - 5) - 3t$

2.3 The Multiplication Property of Equality

1 Identify reciprocals.

2 Use the multiplication property to solve equations.

3 Solve equations of the form $-x = a$.

4 Do some steps mentally when solving equations.

1 Identify Reciprocals

In Section 1.10, we introduced the **reciprocal** (or multiplative inverse) of a number. Recall that two numbers are reciprocals of each other when their product is 1. Some examples of numbers and their reciprocals follow.

Number	Reciprocal	Product
2	$\dfrac{1}{2}$	$(2)\left(\dfrac{1}{2}\right) = 1$
$-\dfrac{3}{5}$	$-\dfrac{5}{3}$	$\left(-\dfrac{3}{5}\right)\left(-\dfrac{5}{3}\right) = 1$
-1	-1	$(-1)(-1) = 1$

The reciprocal of a positive number is a positive number and the reciprocal of a negative number is a negative number. Note that 0 has no reciprocal. Why?

In general, if a represents any nonzero number, its reciprocal is $\frac{1}{a}$. For example, the reciprocal of 3 is $\frac{1}{3}$ and the reciprocal of -2 is $\frac{1}{-2}$ or $-\frac{1}{2}$. The reciprocal of $-\frac{3}{5}$ is $\frac{1}{-\frac{3}{5}}$, which can be written as $1 \div \left(-\frac{3}{5}\right)$. Simplifying, we get $\left(\frac{1}{1}\right)\left(-\frac{5}{3}\right) = -\frac{5}{3}$.

Thus, the reciprocal of $-\frac{3}{5}$ is $-\frac{5}{3}$.

2 Use the Multiplication Property to Solve Equations

In Section 2.2, we used the addition property of equality to solve equations of the form $x + a = b$, where a and b represent real numbers. In this section, we use the multiplication property of equality to solve equations of the form $ax = b$, where a and b represent real numbers.

It is important that you recognize the difference between equations like $x + 2 = 8$ and $2x = 8$. In $x + 2 = 8$, the 2 is a *term* that is being added to x, so we use the addition property to solve the equation. In $2x = 8$, the 2 is a *factor* of $2x$. The 2 is the coefficient multiplying the x, so we use the multiplication property to solve the equation. The multiplication property of equality is used to solve linear equations where the coefficient of the x-term is a number other than 1.

Now we present the *multiplication property of equality*.

Multiplication Property of Equality

If $a = b$, then $a \cdot c = b \cdot c$ for any real numbers a, b, and c.

The multiplication property means that both sides of an equation can be multiplied by the same nonzero number without changing the solution. **The multiplication property can be used to solve equations of the form $ax = b$.** We can isolate the variable in equations of this form by multiplying both sides of the equation by the reciprocal of a, which is $\frac{1}{a}$. Doing so makes the numerical coefficient of the variable, x, become 1, which can be omitted when we write the variable.

Equation	To Solve, Use the Multiplication Property to Change
$4x = 9$	$4x$ to $1x$
$-5x = 20$	$-5x$ to $1x$
$15 = \frac{1}{2}x$	$\frac{1}{2}x$ to $1x$
$7 = -9x$	$-9x$ to $1x$

Now let's work some examples.

EXAMPLE 1 ▶ Solve the equation $9x = 63$.

Solution To isolate the variable, x, we must change the $9x$ on the left side of the equal sign to $1x$. To do this, we multiply both sides of the equation by the reciprocal of 9, which is $\frac{1}{9}$.

$$9x = 63$$

$$\frac{1}{9} \cdot 9x = \frac{1}{9} \cdot 63 \qquad \textit{Multiply both sides by } \frac{1}{9}.$$

$$\frac{1}{\cancel{9}} \cdot \cancel{9}x = \frac{1}{\cancel{9}} \cdot \cancel{63}^{7} \qquad \textit{Divide out the common factors.}$$

$$1x = 7$$

$$x = 7$$

▶ **Now Try Exercise 9**

Notice in Example 1 that $1x$ is replaced by x in the last step. Usually we do this step mentally.

EXAMPLE 2 ▶ Solve the equation $\dfrac{x}{2} = 4$.

Solution   Since dividing by 2 is the same as multiplying by $\dfrac{1}{2}$, the equation $\dfrac{x}{2} = 4$ is the same as $\dfrac{1}{2}x = 4$. We will therefore multiply both sides of the equation by the reciprocal of $\dfrac{1}{2}$, which is 2.

$$\frac{x}{2} = 4$$

$$\overset{1}{2}\left(\frac{x}{\underset{1}{2}}\right) = 2 \cdot 4 \qquad \textit{Multiply both sides by 2.}$$

$$x = 2 \cdot 4$$

$$x = 8$$

▶ **Now Try Exercise 11**

EXAMPLE 3 ▶ Solve the equation $\dfrac{2}{3}x = 6$.

Solution   The reciprocal of $\dfrac{2}{3}$ is $\dfrac{3}{2}$. We multiply both sides of the equation by $\dfrac{3}{2}$.

$$\frac{2}{3}x = 6$$

$$\frac{3}{2} \cdot \frac{2}{3}x = \frac{3}{2} \cdot 6 \qquad \textit{Multiply both sides by } \frac{3}{2}.$$

$$1x = 9$$

$$x = 9$$

We will show a check of this solution.

Check:
$$\frac{2}{3}x = 6$$

$$\frac{2}{3}(9) \overset{?}{=} 6$$

$$6 = 6 \qquad \textit{True}$$

▶ **Now Try Exercise 49**

In Example 1, we multiplied both sides of the equation $9x = 63$ by $\dfrac{1}{9}$ to isolate the variable. We could have also isolated the variable by dividing both sides of the equation by 9, as follows:

$$9x = 63$$

$$\frac{\overset{1}{9}x}{\underset{1}{9}} = \frac{\overset{7}{63}}{\underset{1}{9}} \qquad \textit{Divide both sides by 9.}$$

$$x = 7$$

We can do this because dividing by 9 is equivalent to multiplying by $\dfrac{1}{9}$. **Since division can be defined in terms of multiplication** $\left(\dfrac{a}{b} \text{ means } a \cdot \dfrac{1}{b}\right)$, **the multiplication property also allows us to divide both sides of an equation by the same nonzero number.** This process is illustrated in Examples 4 through 6.

EXAMPLE 4 ▸ Solve the equation $8w = 3$.

Solution   In this equation, w is the variable. To solve the equation we divide both sides of the equation by 8.

$$8w = 3$$

$$\frac{8w}{8} = \frac{3}{8} \qquad \text{\textit{Divide both sides by 8.}}$$

$$w = \frac{3}{8}$$

▸ **Now Try Exercise 33**

EXAMPLE 5 ▸ Solve the equation $-15 = -3z$.

Solution   In this equation, the variable, z, is on the right side of the equal sign. To isolate z, we divide both sides of the equation by -3.

$$-15 = -3z$$

$$\frac{-15}{-3} = \frac{-3z}{-3} \qquad \text{\textit{Divide both sides by −3.}}$$

$$5 = z$$

▸ **Now Try Exercise 21**

EXAMPLE 6 ▸ Solve the equation $0.24x = 1.20$.

Solution   We begin by dividing both sides of the equation by 0.24 to isolate the variable x.

$$0.24x = 1.20$$

$$\frac{0.24x}{0.24} = \frac{1.20}{0.24} \qquad \text{\textit{Divide both sides by 0.24.}}$$

$$1x = 5$$

$$x = 5$$

▸ **Now Try Exercise 35**

Working problems involving decimal numbers on a calculator will probably save you time.

Helpful Hint

When solving an equation of the form $ax = b$, we can isolate the variable by

1. multiplying both sides of the equation by the reciprocal of a, $\frac{1}{a}$, as was done in Examples 1, 2, and 3, or

2. dividing both sides of the equation by a, as was done in Examples 4, 5, and 6.

Either method may be used to isolate the variable. However, if the equation contains a fraction, or fractions, you will arrive at a solution more quickly by multiplying by the reciprocal of a. This is illustrated in Examples 7 and 8.

EXAMPLE 7 ▸ Solve the equation $-2x = \frac{3}{5}$.

Solution   Since this equation contains a fraction, we will isolate the variable by multiplying both sides of the equation by $-\frac{1}{2}$, which is the reciprocal of -2.

$$-2x = \frac{3}{5}$$

$$\left(-\frac{1}{2}\right)(-2x) = \left(-\frac{1}{2}\right)\left(\frac{3}{5}\right) \quad \textit{Multiply both sides by } -\frac{1}{2}.$$

$$1x = \left(-\frac{1}{2}\right)\left(\frac{3}{5}\right)$$

$$x = -\frac{3}{10}$$

▸ **Now Try Exercise 39**

In Example 7, if you wished to solve the equation by dividing both sides of the equation by -2, you would have to divide the fraction $\frac{3}{5}$ by -2.

EXAMPLE 8 ▸ Solve the equation $-6 = -\frac{3}{5}x$.

Solution   Since this equation contains a fraction, we will isolate the variable by multiplying both sides of the equation by the reciprocal of $-\frac{3}{5}$, which is $-\frac{5}{3}$.

$$-6 = -\frac{3}{5}x$$

$$\left(-\frac{5}{3}\right)(-6) = \left(-\frac{5}{3}\right)\left(-\frac{3}{5}x\right) \quad \textit{Multiply both sides by } -\frac{5}{3}.$$

$$10 = 1x$$

$$10 = x$$

▸ **Now Try Exercise 57**

In Example 8, the equation was written as $-6 = -\frac{3}{5}x$. This equation is equivalent to the equations $-6 = \frac{-3}{5}x$ and $-6 = \frac{3}{-5}x$. Can you explain why? All three equations have the same solution, 10.

3 Solve Equations of the Form $-x = a$

When solving an equation, we may obtain an equation like $-x = 7$. This is not a solution since $-x = 7$ means $-1x = 7$. The solution to an equation is of the form $x = $ some number. When an equation is of the form $-x = 7$, we can solve for x by multiplying both sides of the equation by -1, as illustrated in the following example.

EXAMPLE 9 ▸ Solve the equation $-x = 7$.

Solution $-x = 7$ means that $-1x = 7$. We are solving for x, not $-x$. We can multiply both sides of the equation by -1 to isolate x on the left side of the equation.

$$-x = 7$$

$$-1x = 7$$

$$(-1)(-1x) = (-1)(7) \quad \textit{Multiply both sides by } -1.$$

$$1x = -7$$

$$x = -7$$

Check:
$$-x = 7$$
$$-(-7) \stackrel{?}{=} 7$$
$$7 = 7 \quad \textit{True}$$

Thus, the solution is -7.

▸ **Now Try Exercise 23**

Example 9 may also be solved by dividing both sides of the equation by −1. Try this now and see that you get the same solution. Whenever we have the opposite (or negative) of a variable equal to a quantity, as in Example 9, we can solve for the variable by multiplying (or dividing) both sides of the equation by −1.

EXAMPLE 10 ▶ Solve the equation $-x = -5$.

Solution

$$-x = -5$$
$$-1x = -5$$
$$(-1)(-1x) = (-1)(-5) \quad \text{Multiply both sides by } -1.$$
$$1x = 5$$
$$x = 5$$

▶ **Now Try Exercise 25**

Helpful Hint

For any real number a, if $-x = a$, then $x = -a$.

Examples

$$-x = 7 \qquad\qquad -x = -2$$
$$x = -7 \qquad\qquad x = -(-2)$$
$$\qquad\qquad\qquad\qquad x = 2$$

4 Do Some Steps Mentally When Solving Equations

When you feel comfortable using the multiplication property, you may wish to do some of the steps mentally to reduce some of the written work. Now we present two examples worked out in detail, along with their shortened form.

EXAMPLE 11 ▶ Solve the equation $-3x = -21$.

Solution

$$-3x = -21$$

$$\frac{-3x}{-3} = \frac{-21}{-3} \quad \longleftarrow \boxed{\text{Do this step mentally.}}$$

$$x = \frac{-21}{-3}$$

$$x = 7$$

SHORTENED FORM

$$-3x = -21$$

$$x = \frac{-21}{-3}$$

$$x = 7$$

▶ **Now Try Exercise 61**

EXAMPLE 12 ▶ Solve the equation $\frac{1}{5}x = 20$.

Solution

$$\frac{1}{5}x = 20$$

$$5\left(\frac{1}{5}x\right) = 5(20) \quad \longleftarrow \boxed{\text{Do this step mentally.}}$$

$$x = 5(20)$$

$$x = 100$$

SHORTENED FORM

$$\frac{1}{5}x = 20$$

$$x = 5(20)$$

$$x = 100$$

▶ **Now Try Exercise 63**

In Section 2.2, we discussed the addition property and in this section we discussed the multiplication property. It is important that you understand the difference between the two. The following Helpful Hint should be studied carefully.

Helpful Hint

The **addition property** is used to solve equations of the form $x + a = b$. The *addition property* is used when a number is *added to or subtracted from* a variable.

$$x + 3 = -6$$
$$x + 3 - 3 = -6 - 3$$
$$x = -9$$

$$x - 5 = -2$$
$$x - 5 + 5 = -2 + 5$$
$$x = 3$$

The **multiplication property** is used to solve equations of the form $ax = b$. It is used when a variable is *multiplied* or *divided by* a number.

$$3x = 6$$
$$\frac{3x}{3} = \frac{6}{3}$$
$$x = 2$$

$$\frac{x}{2} = 4$$
$$2\left(\frac{x}{2}\right) = 2(4)$$
$$x = 8$$

$$\frac{2}{5}x = 12$$
$$\left(\frac{5}{2}\right)\left(\frac{2}{5}x\right) = \left(\frac{5}{2}\right)(12)$$
$$x = 30$$

EXERCISE SET 2.3   *Math XL*   **MyMathLab**
MathXL®   MyMathLab

Concept/Writing Exercises

1. Explain the multiplication property of equality.

2. Explain why the multiplication property allows us to divide both sides of an equation by a nonzero quantity.

3. a) If $-x = a$, where a represents any real number, what does x equal?

   b) If $-x = 5$, what is x?

   c) If $-x = -5$, what is x?

4. When solving the equation $-2x = 5$, would you divide both sides of the equation by -2 or by 5? Explain.

5. When solving the equation $3x = 5$, would you divide both sides of the equation by 3 or by 5? Explain.

6. When solving the equation $4 = \dfrac{x}{3}$, what would you do to isolate the variable? Explain.

7. When solving the equation $\dfrac{x}{2} = 3$, what would you do to isolate the variable? Explain.

8. When solving the equation $ax = b$ for x, would you divide both sides of the equation by a or b? Explain.

Practice the Skills

Solve each equation and check your solution.

9. $4x = 12$

10. $5x = 50$

11. $\dfrac{x}{3} = 7$

12. $\dfrac{y}{5} = 3$

13. $-4x = 12$

14. $8 = 16y$

15. $\dfrac{x}{4} = -2$

16. $\dfrac{x}{3} = -3$

17. $\dfrac{x}{5} = 1$

18. $-7x = 49$

19. $-27n = 81$

20. $\dfrac{x}{8} = -3$

21. $-7 = 3r$

22. $16 = -4y$

23. $-x = 13$

24. $-x = 9$

25. $-x = -8$

26. $-x = -15$

27. $-\dfrac{w}{3} = -13$

28. $-4 = \dfrac{c}{7}$

29. $4 = -12x$

30. $12y = -15$

31. $-\dfrac{x}{3} = -2$

32. $-\dfrac{a}{8} = -7$

33. $43t = 26$

34. $-24x = -18$

35. $-4.2x = -8.4$

36. $-3.88 = 1.94y$

37. $3x = \dfrac{3}{5}$

38. $7x = -7$

39. $5x = -\dfrac{3}{8}$

40. $-2b = -\dfrac{4}{5}$

41. $15 = -\dfrac{x}{4}$

42. $\dfrac{c}{9} = 0$

43. $-\dfrac{b}{4} = -60$

44. $-x = -\dfrac{5}{9}$

45. $\dfrac{x}{5} = -7$

46. $-3r = 0$

47. $5 = \dfrac{x}{4}$

48. $-3 = \dfrac{x}{-5}$

49. $\dfrac{3}{5}d = -30$

50. $\dfrac{2}{7}x = 7$

51. $\dfrac{y}{-2} = 0$

52. $-6x = \dfrac{5}{2}$

53. $\dfrac{-7}{8}w = 0$

54. $-x = \dfrac{5}{8}$

55. $\dfrac{1}{5}x = 4.5$

56. $-\dfrac{1}{4}x = \dfrac{3}{4}$

57. $-4 = -\dfrac{2}{3}z$

58. $-9 = \dfrac{-5}{3}n$

59. $-1.4x = 28.28$

60. $-0.42x = -2.142$

Solve each equation by doing some steps mentally. Check your solution.

61. $-8x = -56$

62. $-9x = -45$

63. $\dfrac{2}{3}x = 6$

64. $\dfrac{1}{3}x = 15$

Problem Solving

65. a) Explain the difference between $5 + x = 10$ and $5x = 10$.

  b) Solve $5 + x = 10$.

  c) Solve $5x = 10$.

66. a) Explain the difference between $3 + x = 6$ and $3x = 6$.

  b) Solve $3 + x = 6$.

  c) Solve $3x = 6$.

67. Consider the equation $\dfrac{2}{3}x = 4$. This equation could be solved by multiplying both sides of the equation by $\dfrac{3}{2}$, the reciprocal of $\dfrac{2}{3}$, or by dividing both sides of the equation by $\dfrac{2}{3}$. Which method do you feel would be easier? Explain your answer. Find the solution to the equation.

68. Consider the equation $4x = \dfrac{3}{5}$. Would it be easier to solve this equation by dividing both sides of the equation by 4 or by multiplying both sides of the equation by $\dfrac{1}{4}$, the reciprocal of 4? Explain your answer. Find the solution to the problem.

69. Consider the equation $\dfrac{3}{7}x = \dfrac{4}{5}$. Would it be easier to solve this equation by dividing both sides of the equation by $\dfrac{3}{7}$ or by multiplying both sides of the equation by $\dfrac{7}{3}$, the reciprocal of $\dfrac{3}{7}$? Explain your answer. Find the solution to the equation.

Challenge Problems

70. Consider the equation $\square \odot = \triangle$.

  a) To solve for $\odot$, what symbol do we need to isolate?

  b) How would you isolate the symbol you specified in part **a)**?

  c) Solve the equation for $\odot$.

71. Consider the equation $\odot = \triangle\square$.

  a) To solve for $\square$, what symbol do we need to isolate?

  b) How would you isolate the symbol you specified in part **a)**?

  c) Solve the equation for $\square$.

72. Consider the equation $\# = \dfrac{\odot}{\triangle}$.

  a) To solve for $\odot$, what symbol do we need to isolate?

  b) How would you isolate the symbol you specified in part **a)**?

  c) Solve the equation for $\odot$.

Cumulative Review Exercises

[1.7] **73.** Subtract -4 from -8.

[1.8] **74.** Evaluate $(-3)(-2)(5)(-1)$.

[1.9] **75.** Evaluate $4^2 - 2^3 \cdot 6 \div 3 + 6$.

[1.10] **76.** Name the property illustrated.
$2 + (4 + y) = (2 + 4) + y$

[2.2] **77.** Solve the equation $-48 = x + 9$.

2.4 Solving Linear Equations with a Variable on Only One Side of the Equation

1 Solve linear equations with a variable on only one side of the equal sign.

2 Solve equations containing decimal numbers or fractions.

1 Solve Linear Equations with a Variable on Only One Side of the Equal Sign

In this section, we discuss how to solve linear equations using *both* the addition and multiplication properties of equality when a variable appears on only one side of the equal sign. In Section 2.5, we will discuss how to solve linear equations using both properties when a variable appears on both sides of the equal sign.

The general procedure we use to solve equations is to "isolate the variable." That is, get the variable alone on one side of the equal sign.

No one method is the "best" to solve all linear equations. But the following general procedure can be used to solve linear equations when the variable appears on only one side of the equation.

> ## To Solve Linear Equations with a Variable on Only One Side of the Equal Sign
>
> 1. If the equation contains fractions, multiply **both** sides of the equation by the least common denominator (LCD). This will eliminate the fractions from the equation.
> 2. Use the distributive property to remove parentheses.
> 3. Combine like terms on the same side of the equal sign.
> 4. Use the addition property to obtain an equation with the term containing the variable on one side of the equal sign and a constant on the other side. This will result in an equation of the form $ax = b$.
> 5. Use the multiplication property to isolate the variable. This will give a solution of the form $x = \dfrac{b}{a}$ $\left(\text{or } 1x = \dfrac{b}{a}\right)$.
> 6. Check the solution in the original equation.

When solving an equation, you should always check your solution, as is indicated in step 6. To conserve space, we will not show all checks.

When solving an equation, remember that our goal is to isolate the variable on one side of the equation.

Consider the equation $2x + 4 = 10$ which contains no fractions or parentheses, and no like terms on the same side of the equal sign. Therefore, we start with step 4, using the addition property. Remember that the addition property allows us to add (or subtract) the same quantity to (or from) both sides of an equation without changing its solution. Here we subtract 4 from both sides of the equation to get the $2x$ by itself on one side of the equal sign.

Equation

$2x + 4 = 10$

$2x + 4 \;\boxed{-\; 4} = 10 \;\boxed{-\; 4}$ *Addition property*

$2x + 0 = 6$

or $2x = 6$ *x-term is now isolated*

Notice how the term containing the variable, $2x$, is now by itself on one side of the equal sign. We can now say that we have *isolated the term containing the variable* or have *isolated the variable term*. Now we use the multiplication property, step 5, to isolate the variable, x. Remember that the multiplication property allows us to multiply or divide both sides of the equation by the same nonzero number without changing its solution. Here we divide both sides of the equation by 2, the coefficient of the term containing the variable, to obtain the solution, 3.

$2x = 6$

$\dfrac{\overset{1}{\cancel{2}}x}{\underset{1}{\cancel{2}}} = \dfrac{\overset{3}{\cancel{6}}}{\underset{1}{\cancel{2}}}$ *Multiplication property*

$1x = 3$

$x = 3$ *x is now isolated*

The solution to the equation $2x + 4 = 10$ is 3. Now let's work some examples.

EXAMPLE 1 ▶ Solve the equation $5x - 7 = 13$.

Solution We will follow the procedure outlined for solving equations. Since the equation contains no fractions nor parentheses, and since there are no like terms to be combined, we start with step 4.

$$5x - 7 = 13$$

Step 4

$$5x - 7 \boxed{+ 7} = 13 \boxed{+ 7}$$ *Add 7 to both sides.*

$$5x = 20$$

Step 5

$$\frac{5x}{\boxed{5}} = \frac{20}{\boxed{5}}$$ *Divide both sides by 5.*

$$x = 4$$

Step 6 Check:

$$5x - 7 = 13$$

$$5(4) - 7 \overset{?}{=} 13$$

$$20 - 7 \overset{?}{=} 13$$

$$13 = 13 \qquad \textit{True}$$

Since the check is true, the solution is 4. Note that after completing step 4, we obtain $5x = 20$, which is an equation of the form $ax = b$. After completing step 5, we obtain the answer in the form $x =$ some real number.

▶ **Now Try Exercise 15**

Helpful Hint

When solving an equation that does not contain fractions, **the addition property (step 4) is to be used before the multiplication property (step 5)**. If you use the multiplication property before the addition property, it is still possible to obtain the correct answer. However, you will usually have to do more work, and you may end up working with fractions. What would happen if you tried to solve Example 1 using the multiplication property before the addition property?

EXAMPLE 2 ▶ Solve the equation $-2r - 6 = -3$.

Solution

$$-2r - 6 = -3$$

Step 4

$$-2r - 6 \boxed{+ 6} = -3 \boxed{+ 6}$$ *Add 6 to both sides.*

$$-2r = 3$$

Step 5

$$\frac{-2r}{\boxed{-2}} = \frac{3}{\boxed{-2}}$$ *Divide both sides by -2.*

$$r = -\frac{3}{2}$$

Step 6 Check:

$$-2r - 6 = -3$$

$$-2\left(-\frac{3}{2}\right) - 6 \overset{?}{=} -3$$

$$3 - 6 \overset{?}{=} -3$$

$$-3 = -3 \qquad \textit{True}$$

The solution is $-\dfrac{3}{2}$.

▶ **Now Try Exercise 23**

Note that checks are always made with the *original* equation. In some of the following examples, the check will be omitted to save space. You should check all of your answers.

EXAMPLE 3 ▶ Solve the equation $16 = 4x + 6 - 2x$.

Solution   Again we must isolate the variable, x. Since the right side of the equation has two like terms containing the variable, x, we will first combine these like terms.

$$16 = 4x + 6 - 2x$$

Step 3 $16 = 2x + 6$ *Like terms were combined.*

Step 4 $16 - 6 = 2x + 6 - 6$ *Subtract 6 from both sides.*

$$10 = 2x$$

Step 5 $\dfrac{10}{2} = \dfrac{2x}{2}$ *Divide both sides by 2.*

$$5 = x$$

▶ **Now Try Exercise 37**

The preceding solution can be condensed as follows.

$$16 = 4x + 6 - 2x$$

$16 = 2x + 6$ *Like terms were combined.*

$10 = 2x$ *6 was subtracted from both sides.*

$5 = x$ *Both sides were divided by 2.*

EXAMPLE 4 ▶ Solve the equation $5x - 2(x + 4) = 3$.

Solution

$$5x - 2(x + 4) = 3$$

Step 2 $5x - 2x - 8 = 3$ *Distributive property was used.*

Step 3 $3x - 8 = 3$ *Like terms were combined.*

Step 4 $3x - 8 + 8 = 3 + 8$ *Add 8 to both sides.*

$$3x = 11$$

Step 5 $\dfrac{3x}{3} = \dfrac{11}{3}$ *Divide both sides by 3.*

$$x = \dfrac{11}{3}$$

▶ **Now Try Exercise 65**

The solution to Example 4 can be condensed as follows:

$$5x - 2(x + 4) = 3$$

$5x - 2x - 8 = 3$ *Distributive property was used.*

$3x - 8 = 3$ *Like terms were combined.*

$3x = 11$ *8 was added to both sides.*

$x = \dfrac{11}{3}$ *Both sides were divided by 3.*

EXAMPLE 5 ▶ Solve the equation $3p - (2p + 5) = 7$.

Solution $3p - (2p + 5) = 7$

$3p - 2p - 5 = 7$ *Distributive property was used.*

$p - 5 = 7$ *Like terms were combined.*

$p = 12$ *5 was added to both sides.*

▶ **Now Try Exercise 69**

2 Solve Equations Containing Decimal Numbers or Fractions

In Chapter 3, we will be solving many equations that contain decimal numbers. To solve such equations, we may follow the same procedure as outlined earlier. Example 6 illustrates two methods to solve an equation that contains decimal numbers.

EXAMPLE 6 ▶ Solve the equation $x + 1.24 - 0.07x = 4.96$.

Solution We will work this example using two methods. In method 1, we work with decimal numbers throughout the solving process. In method 2, we multiply both sides of the equation by a power of 10 to change the decimal numbers to whole numbers.

Method 1
$$x + 1.24 - 0.07x = 4.96$$

$$0.93x + 1.24 = 4.96 \qquad \textit{Like terms were combined,}$$
$$\textit{1x} - \textit{0.07x} = \textit{0.93x.}$$

$$0.93x + 1.24 - 1.24 = 4.96 - 1.24 \qquad \textit{Subtract 1.24 from both sides.}$$

$$0.93x = 3.72$$

$$\frac{0.93x}{0.93} = \frac{3.72}{0.93} \qquad \textit{Divide both sides by 0.93.}$$

$$x = 4$$

Method 2 Some students prefer to eliminate the decimal numbers from the equation by multiplying both sides of the equation by 10 if the decimal numbers are given in tenths, by 100 if the decimal numbers are given in hundredths, and so on. In Example 6, since the decimal numbers are in hundredths, you can eliminate the decimals from the equation by multiplying both sides of the equation by 100. This alternate method would give the following.

$$x + 1.24 - 0.07x = 4.96$$

$$100(x + 1.24 - 0.07x) = 100(4.96) \qquad \textit{Multiply both sides of equation by 100.}$$

$$100(x) + 100(1.24) - 100(0.07x) = 496 \qquad \textit{Distributive property was used.}$$

$$100x + 124 - 7x = 496$$

$$93x + 124 = 496 \qquad \textit{Like terms were combined.}$$

$$93x = 372 \qquad \textit{124 was subtracted from both sides.}$$

$$x = 4 \qquad \textit{Both sides were divided by 93.}$$

Study both methods provided to see which method you prefer.

▶ **Now Try Exercise 41**

Now let's look at solving equations that contain fractions. There will be various times throughout the course when we will need to solve equations containing fractions. Often, the first step in solving equations containing fractions is to multiply both sides of the equation by the LCD to eliminate the fractions from the equations. Examples 7–9 illustrate this procedure. Example 7 will contain the expression $\frac{1}{5}(x + 1)$, Since the $\frac{1}{5}$ is multiplied by $(x + 1)$, the expression $\frac{1}{5}(x + 1)$ is a single term whose factors are $\frac{1}{5}$ and $(x + 1)$. When multiplying a term by a constant or variable, we multiply only one of the factors of the term by the number or variable. Therefore, for example,

$$5\left[\frac{1}{5}(x + 1)\right] = 5\left(\frac{1}{5}\right)(x + 1) = 1(x + 1) = x + 1$$

EXAMPLE 7 ▶ Solve $\frac{1}{5}(x + 1) = 1$.

Solution   The LCD of the fraction is 5. We will begin by multiplying both sides of the equation by the LCD. This step will eliminate fractions from the equation.

$$\frac{1}{5}(x + 1) = 1$$

Step 1 $$5\left[\frac{1}{5}(x + 1)\right] = 5 \cdot 1 \quad \textit{Multiply both sides by the LCD, 5.}$$

$$5\left(\frac{1}{5}\right)(x + 1) = 5$$

$$x + 1 = 5$$

Step 4 $$x = 4 \quad \textit{1 was subtracted from both sides.}$$

Step 6   Check: $$\frac{1}{5}(x + 1) = 1$$

$$\frac{1}{5}(4 + 1) \overset{?}{=} 1$$

$$\frac{1}{5}(5) \overset{?}{=} 1$$

$$1 = 1 \quad \textit{True}$$

The solution is 4.

▶ **Now Try Exercise 45**

Example 7 could also be written as $\frac{x + 1}{5} = 1$. To solve this equation, we would begin by multiplying both sides of the equation by the LCD, 5, as follows.

$$\frac{x + 1}{5} = 1$$

$$5\left(\frac{x + 1}{5}\right) = 5 \cdot 1$$

$$x + 1 = 5$$

$$x = 4$$

EXAMPLE 8 ▶ Solve the equation $\frac{d}{2} + 3d = 14$.

Solution   Step 1 tells us to multiply both sides of the equation by the LCD, 2. This step will eliminate fractions from the equation.

Step 1 $$2\left(\frac{d}{2} + 3d\right) = 2 \cdot 14 \quad \textit{Multiply both sides by the LCD, 2.}$$

Step 2 $$2\left(\frac{d}{2}\right) + 2 \cdot 3d = 2 \cdot 14 \quad \textit{Distributive property}$$

$$d + 6d = 28$$

Step 3 $$7d = 28 \quad \textit{Like terms were combined.}$$

Step 5 $$d = 4 \quad \textit{Both sides were divided by 7.}$$

Step 6   Check: $$\frac{d}{2} + 3d = 14$$

$$\frac{4}{2} + 3(4) \overset{?}{=} 14$$

$$2 + 12 \overset{?}{=} 14$$

$$14 = 14 \quad \textit{True}$$

▶ **Now Try Exercise 89**

EXAMPLE 9 ▸ Solve the equation $\dfrac{1}{5}x - \dfrac{3}{8}x = \dfrac{1}{10}$.

Solution   The LCD of 5, 8, and 10 is 40. Multiply both sides of the equation by 40 to eliminate fractions from the equation.

$$\frac{1}{5}x - \frac{3}{8}x = \frac{1}{10}$$

Step 1 $40\left(\dfrac{1}{5}x - \dfrac{3}{8}x\right) = 40\left(\dfrac{1}{10}\right)$ *Multiply both sides by the LCD, 40.*

Step 2 $40\left(\dfrac{1}{5}x\right) - 40\left(\dfrac{3}{8}x\right) = 40\left(\dfrac{1}{10}\right)$ *Distributive property*

$$8x - 15x = 4$$

Step 3 $-7x = 4$ *Like terms were combined.*

Step 5 $x = -\dfrac{4}{7}$ *Both sides were divided by −7.*

Step 6   Check: $\dfrac{1}{5}x - \dfrac{3}{8}x = \dfrac{1}{10}$

$$\frac{1}{5}\left(-\frac{4}{7}\right) - \frac{3}{8}\left(-\frac{4}{7}\right) \overset{?}{=} \frac{1}{10}$$ *Substitute $-\dfrac{4}{7}$ for each x.*

$$-\frac{4}{35} + \frac{3}{14} \overset{?}{=} \frac{1}{10}$$ *Divide out common factors, then multiply fractions.*

$$-\frac{8}{70} + \frac{15}{70} \overset{?}{=} \frac{7}{70}$$ *Write each fraction with the LCD, 70.*

$$\frac{7}{70} = \frac{7}{70}$$ *True*

▸ **Now Try Exercise 99**

Helpful Hint

In Example 9, we multiplied both sides of the equation by the LCD, 40. When solving equations containing fractions, multiplying both sides of the equation by *any* common denominator will eventually lead to the correct answer (if you don't make a mistake), but you may have to work with larger numbers. In Example 9, if you multiplied both sides of the equation by 80, 120, or 160, for example, you would eventually obtain the answer $-\dfrac{4}{7}$. When solving equations containing fractions, you should multiply both sides of the equation by the LCD. But if you mistakenly multiply both sides of the equation by a different common denominator to clear fractions, you will still obtain the correct answer. To show that other common denominators may be used, solve Example 9 now by multiplying both sides of the equation by the common denominator 80 instead of the LCD, 40.

When checking solutions to equations that contain fractions, you may sometimes want to perform the check using a calculator. When checking a solution using a calculator, work with each side of the equation separately. Below we show the steps used to evaluate the left side of the equation in Example 9 for $x = -\dfrac{4}{7}$ using a scientific calculator.*

$$\frac{1}{5}x - \frac{3}{8}x = \frac{1}{10}$$

$$\frac{1}{5}\left(-\frac{4}{7}\right) - \frac{3}{8}\left(-\frac{4}{7}\right) = \frac{1}{10}$$

Evaluate the left side of the equation.

$$\boxed{1}\;\boxed{\div}\;\boxed{5}\;\boxed{\times}\;\boxed{4}\;\boxed{+/-}\;\boxed{\div}\;\boxed{7}\;\boxed{-}\;\boxed{3}\;\boxed{\div}\;\boxed{8}\;\boxed{\times}\;\boxed{4}\;\boxed{+/-}\;\boxed{\div}\;\boxed{7}\;\boxed{=}\;0.1.$$

Since the right side of the equation, $\dfrac{1}{10} = 0.1$, the answer checks.

————————————

*Keystrokes may differ on some scientific calculators. Read the instruction manual for your calculator.

Helpful Hint

Some of the most commonly used terms in algebra are "evaluate," "simplify," "solve," and "check." Make sure you understand what each term means and when each term is used.

Evaluate: To *evaluate an expression* means to find its numerical value.

Evaluate

$$16 \div 2^2 + 36 \div 4$$
$$= 16 \div 4 + 36 \div 4$$
$$= 4 + 36 \div 4$$
$$= 4 + 9$$
$$= 13$$

Evaluate

$$-x^2 + 3x - 2 \text{ when } x = 4$$
$$= -4^2 + 3(4) - 2$$
$$= -16 + 3(4) - 2$$
$$= -16 + 12 - 2$$
$$= -4 - 2$$
$$= -6$$

Simplify: To *simplify an expression* means to perform the operations and combine like terms.

Simplify $3(x - 2) - 4(2x + 3)$

$$3(x - 2) - 4(2x + 3) = 3x - 6 - 8x - 12$$
$$= -5x - 18$$

Note that when you simplify an expression containing variables you do not generally end up with just a numerical value unless all the variable terms happen to add to zero.

Solve: To *solve an equation* means to find the value or the values of the variable that make the equation a true statement.

Solve

$$2x + 3(x + 1) = 18$$
$$2x + 3x + 3 = 18$$
$$5x + 3 = 18$$
$$5x = 15$$
$$x = 3$$

Check: To *check the proposed solution to an equation*, substitute the value in the original equation. If this substitution results in a true statement, then the answer checks. For example, to check the solution to the equation just solved, we substitute 3 for x in the original equation.

Check

$$2x + 3(x + 1) = 18$$
$$2(3) + 3(3 + 1) \stackrel{?}{=} 18$$
$$2(3) + 3(4) \stackrel{?}{=} 18$$
$$6 + 12 \stackrel{?}{=} 18$$
$$18 = 18 \qquad \textit{True}$$

Since we obtained a true statement, the 3 checks.

It is important to realize that expressions may be evaluated or simplified (depending on the type of problem) and equations are solved and then checked.

EXERCISE SET 2.4

MathXL®   *MyMathLab*

Concept/Writing Exercises

1. Does the equation $x + 3 = 2x + 5$ contain a variable on only one side of the equation? Explain.

2. Does the equation $2x - 4 = 3$ contain a variable on only one side of the equation? Explain.

3. If $1x = \dfrac{1}{3}$, what does x equal?

4. If $1x = -\dfrac{3}{5}$, what does x equal?

5. If $-x = \dfrac{1}{2}$ what does x equal?

6. If $-x = \dfrac{7}{8}$, what does x equal?

7. If $-x = -\dfrac{4}{9}$, what does x equal?

8. If $-x = -\dfrac{3}{5}$, what does x equal?

9. Do you evaluate or solve an expression? Explain.

10. Do you evaluate or solve an equation? Explain.

11. a) Write the general procedure for solving an equation where the variable appears on only one side of the equal sign.

b) Refer to page 120 to see whether you omitted any steps.

12. When solving equations that contain fractions, what is the first step in the process of solving the equation?

13. a) Explain, in a step-by-step manner, how to solve the equation $2(3x + 4) = -4$.

b) Solve the equation by following the steps you listed in part **a)**.

14. a) Explain, step-by-step, how to solve the equation $4x - 2(x + 3) = 4$.

b) Solve the equation by following the steps you listed in part **a)**.

Practice the Skills

Solve each equation. You may wish to use a calculator to solve equations containing decimal numbers.

15. $5x - 6 = 19$

16. $2x - 4 = 8$

17. $-4w - 5 = 11$

18. $-4x + 6 = 20$

19. $3x + 6 = 12$

20. $6 - 3x = 18$

21. $5x - 2 = 10$

22. $-2t + 9 = 21$

23. $-5k - 4 = -19$

24. $-4x - 7 = -6$

25. $12 - x = 9$

26. $-3x - 3 = -12$

27. $8 + 3x = 19$

28. $-2x + 7 = -10$

29. $16x + 5 = -14$

30. $19 = 25 + 4x$

31. $-4.2 = 3x + 25.8$

32. $-24 + 16x = -24$

33. $7r - 16 = -2$

34. $-2w + 4 = -8$

35. $60 = -5s + 9$

36. $15 = 7x + 1$

37. $14 = 5x + 8 - 3x$

38. $15 = 6x - 3 + 3x$

39. $2.3x - 9.34 = 6.3$

40. $x + 0.05x = 21$

41. $0.91y + 2.25 - 0.01y = 5.85$

42. $0.15 = 0.05x - 1.35 - 0.20x$

43. $28.8 = x + 1.40x$

44. $8.40 = 2.45x - 1.05x$

45. $\dfrac{1}{7}(x + 6) = 4$

46. $\dfrac{m - 6}{5} = 2$

47. $\dfrac{d + 3}{7} = 9$

48. $\dfrac{1}{5}(x + 2) = -3$

49. $\dfrac{1}{3}(t - 5) = -6$

50. $\dfrac{2}{3}(n - 3) = 8$

51. $\dfrac{3}{4}(x - 5) = -12$

52. $\dfrac{1}{4} = \dfrac{z + 1}{4}$

53. $\dfrac{x + 4}{7} = \dfrac{3}{7}$

54. $\dfrac{4x + 5}{6} = \dfrac{7}{2}$

55. $\dfrac{3}{4} = \dfrac{4m - 5}{6}$

56. $\dfrac{5}{6} = \dfrac{5t - 4}{2}$

57. $4(n + 2) = 8$

58. $3(x - 2) = 12$

59. $-2(x - 3) = 26$

60. $5(3 - x) = 15$

61. $-4 = -(x + 5)$

62. $-3(2 - 3x) = 9$

63. $12 = 4(x - 3)$

64. $-2(x + 8) - 5 = 1$

65. $2x - 3(x + 5) = 6$

66. $5(3x + 1) - 12x = -2$

67. $-3r - 4(r + 2) = 11$

68. $9 = -2(a - 3)$

69. $x - 3(2x + 3) = 11$

70. $3y - (y + 5) = 9$

71. $5x + 3x - 4x - 7 = 9$

72. $4(x + 2) = 13$

73. $0.7(x - 3) = 1.4$

74. $21 + (c - 9) = 24$

75. $2.5(4q - 3) = 0.5$

76. $0.1(2.4x + 5) = 1.7$

77. $3 - 2(x + 3) + 2 = 1$

78. $2(3x - 4) - 4x = 12$

79. $1 + (x + 3) + 6x = 6$

80. $5x - 2x + 7x = -81$

81. $4.85 - 6.4x + 1.11 = 22.6$

82. $5.76 - 4.24x - 1.9x = 27.864$

83. $7 = 8 - 5(m + 3)$

84. $4 = \dfrac{3t + 1}{7}$

85. $10 = \dfrac{2s + 4}{5}$

86. $12 = \dfrac{4d - 1}{3}$

87. $x + \dfrac{2}{3} = \dfrac{3}{5}$

88. $n - \dfrac{1}{4} = \dfrac{1}{2}$

89. $\dfrac{r}{3} + 2r = 7$

90. $\dfrac{x}{4} - 6x = 23$

91. $\dfrac{3}{7} = \dfrac{3t}{4} + 1$

92. $\dfrac{5}{8} = \dfrac{5t}{6} + 2$

93. $\dfrac{1}{2}r + \dfrac{1}{5}r = 7$

94. $\dfrac{x}{3} - \dfrac{3x}{4} = \dfrac{1}{12}$

95. $\dfrac{2}{8} + \dfrac{3}{4} = \dfrac{w}{5}$

96. $\dfrac{x}{4} - \dfrac{x}{6} = \dfrac{1}{4}$

97. $\dfrac{1}{2}x + 4 = \dfrac{1}{6}$

98. $\dfrac{4}{5} + n = \dfrac{1}{3}$

99. $\dfrac{4}{5}s - \dfrac{3}{4}s = \dfrac{1}{10}$

100. $\dfrac{1}{3}x - \dfrac{3}{4}x = \dfrac{1}{5}$

101. $\dfrac{4}{9} = \dfrac{1}{3}(n - 7)$

102. $-\dfrac{3}{8} = \dfrac{1}{8} - \dfrac{2x}{7}$

103. $-\dfrac{3}{5} = -\dfrac{1}{9} - \dfrac{3}{4}x$

104. $-\dfrac{3}{5} = -\dfrac{1}{6} - \dfrac{5}{4}m$

Problem Solving

105. a) Explain why it is easier to solve the equation $3x + 2 = 11$ by first subtracting 2 from both sides of the equation rather than by first dividing both sides of the equation by 3.

b) Solve the equation.

106. a) Explain why it is easier to solve the equation $5x - 3 = 12$ by first adding 3 to both sides of the equation rather than by first dividing both sides of the equation by 5.

b) Solve the equation.

Challenge Problems

For exercises 107–109, solve the equation.

107. $3(x - 2) - (x + 5) - 2(3 - 2x) = 18$

108. $-6 = -(x - 5) - 3(5 + 2x) - 4(2x - 4)$

109. $4[3 - 2(x + 4)] - (x + 3) = 13$

110. Solve the equation $\square \odot - \triangledown = @$ for $\odot$.

Group Activity

In Chapter 3, we will discuss procedures for writing application problems as equations. Let's look at an application now.

Birthday Party John Logan purchased 2 large chocolate bars and a birthday card. The birthday card cost $3. The total cost was $9. What was the price of a single chocolate bar?

This problem can be represented by the equation $2x + 3 = 9$, which can be used to solve the problem. Solving the equation we find that x, the price of a single chocolate bar, is $3.

*For Exercises 111 and 112, each group member should do parts **a)** and **b)**. Then do part **c)** as a group.*

a) *Obtain an equation that can be used to solve the problem.*

b) *Solve the equation and answer the question.*

c) *Compare and check each other's work.*

111. Stationery Eduardo Verner purchased three boxes of stationery. He also purchased wrapping paper and thank-you cards. If the wrapping paper and thank-you cards together cost $6, and the total he paid was $42, find the cost of a box of stationery.

112. Candies Mahandi Ison purchased three rolls of peppermint candies and the local newspaper. The newspaper cost 50 cents. He paid $2.75 in all. What did a roll of candies cost?

Cumulative Review Exercises

[1.4] **113.** True or false: Every real number is a rational number.

[1.9] **114.** Evaluate $[5(2 - 6) + 3(8 \div 4)^2]^2$.

[2.2] **115.** To solve an equation, what do you need to do to the variable?

[2.3] **116.** To solve the equation $7 = -4x$, would you add 4 to both sides of the equation or divide both sides of the equation by -4? Explain your answer.

Mid-Chapter Test: 2.1–2.4

To find out how well you understand the chapter material to this point, take this brief test. The answers, and the section where the material was initially discussed, are given in the back of the book. Review any questions that you answered incorrectly.

In Exercises 1 and 2, combine like terms.

1. $5x - 9y - 12 + 4y - 7x + 6$

2. $\dfrac{2}{5}x - 8 - \dfrac{3}{4}x + \dfrac{1}{2}$

In Exercises 3 and 4, use the distributive property to remove parentheses.

3. $-4(2a - 3b + 6)$

4. $1.6(2.1x - 3.4y - 5.2)$

5. Simplify $5(t - 3) - 3(t + 7) - 2$.

6. Is $x = 2$ a solution of $3(x - 4) = -2(x + 1)$?

7. Is $p = \dfrac{2}{5}$ a solution of $7p - 3 = 2p - 5$?

In Exercises 8–10, solve each equation and check your solution.

8. $x - 5 = -9$

9. $12 + x = -4$

10. $-16 = 7 + y$

11. When solving the equation $\dfrac{x}{4} = 5$, what would you do to isolate the variable? Explain.

In Exercises 12–15, solve each equation and check your solution.

12. $6 = 12y$

13. $\dfrac{x}{8} = 3$

14. $-\dfrac{x}{5} = -2$

15. $-x = \dfrac{3}{7}$

In Exercises 16–20, solve each equation.

16. $6x - 3 = 12$

17. $-4 = -2w - 7$

18. $\dfrac{3}{4} = \dfrac{4n - 1}{6}$

19. $-5(x + 4) - 7 = 3$

20. $8 - 9(y + 4) + 6 = -2$

2.5 Solving Linear Equations with the Variable on Both Sides of the Equation

1 Solve equations with the variable on both sides of the equal sign.

2 Solve equations containing decimal numbers or fractions.

3 Identify identities and contradictions.

1 Solve Equations with the Variable on Both Sides of the Equal Sign

The equation $4x + 6 = 2x + 4$ contains the variable, x, on both sides of the equal sign. To solve equations of this type, we must use the appropriate properties to rewrite the equation with all terms containing the variable on only one side of the equal sign and all terms not containing the variable on the other side of the equal sign. Following is a general procedure, similar to the one outlined in Section 2.4, that can be used to solve linear equations with the variable on both sides of the equal sign. The steps in the procedure are only guidelines to use. For example, there may be times when you may choose to use the distributive property, step 2, before multiplying both sides of the equation by the LCD, step 1. We will illustrate this in Examples 8 and 9.

To Solve Linear Equations with the Variable on Both Sides of the Equal Sign

1. If the equation contains fractions, multiply **both** sides of the equation by the least common denominator. This will eliminate fractions from the equation.
2. Use the distributive property to remove parentheses.
3. Combine like terms on the same side of the equal sign.
4. Use the addition property to rewrite the equation with all terms containing the variable on one side of the equal sign and all terms not containing the variable on the other side of the equal sign. It may be necessary to use the addition property twice to accomplish this goal. You will eventually get an equation of the form $ax = b$.
5. Use the multiplication property to isolate the variable. This will give a solution of the form $x =$ some number.
6. Check the solution in the original equation.

The steps listed on page 129 are basically the same as the steps listed in the boxed procedure on page 120, except that in step 4 you may need to use the addition property more than once to obtain an equation of the form $ax = b$.

Remember that our goal in solving an equation is to isolate the variable, that is, to get the variable alone on one side of the equation.

Consider the equation $3x + 4 = x + 12$ which contains no fractions or parentheses, and no like terms on the same side of the equal sign. Therefore, we start with step 4, the addition property. We will use the addition property twice in order to obtain an equation where the variable appears on only one side of the equal sign. We begin by subtracting x from both sides of the equation to get all the terms containing the variable on the left side of the equation. This will give the following:

Equation

$$3x + 4 = x + 12$$
$$3x - x + 4 = x - x + 12 \qquad \textit{Addition property}$$
$$\text{or} \qquad 2x + 4 = 12 \qquad \textit{Variable appears only on left side of equal sign.}$$

Notice that the variable, x, now appears on only one side of the equation. However, $+4$ still appears on the same side of the equal sign as the $2x$. We use the addition property a second time to get the term containing the variable by itself on one side of the equation. Subtracting 4 from both sides of the equation gives $2x = 8$, which is an equation of the form $ax = b$.

Equation

$$2x + 4 = 12$$
$$2x + 4 - 4 = 12 - 4 \qquad \textit{Addition property}$$
$$2x = 8 \qquad \textit{x-term is now isolated.}$$

The x-term, $2x$, is now by itself on one side of the equation. Therefore, we have isolated the x-term on the left side of the equation. We can now use the multiplication property, step 5, to isolate the variable and solve the equation for x. We divide both sides of the equation by 2 to isolate the variable and solve the equation.

$$2x = 8$$
$$\frac{\overset{1}{\cancel{2}}x}{\underset{1}{\cancel{2}}} = \frac{\overset{4}{\cancel{8}}}{\underset{1}{\cancel{2}}} \qquad \textit{Multiplication property}$$
$$x = 4 \qquad \textit{x is now isolated.}$$

The solution to the equation is 4.

EXAMPLE 1 ▶ Solve the equation $4x + 6 = 2x + 4$.

Solution   We start by getting all the terms with the variable on one side of the equal sign and all terms without the variable on the other side. The terms with the variable may be collected on either side of the equal sign. Many methods can be used to get the terms with the variable by themselves on one side of the equal sign. We will illustrate two. In method 1, we will collect all terms with the variable on the left side of the equation. In method 2, we will collect all terms with the variable on the right side of the equation. In both methods, we will follow the steps given in the box on page 129. Since this equation does not contain fractions or parentheses, and there are no like terms on the same side of the equal sign, we begin with step 4.

Method 1: Isolate the variable term on the left.

$$4x + 6 = 2x + 4$$

Step 4 $4x - 2x + 6 = 2x - 2x + 4 \qquad \textit{Subtract 2x from both sides.}$
$$2x + 6 = 4$$

Step 4 $2x + 6 \boxed{- 6} = 4 \boxed{- 6}$ *Subtract 6 from both sides.*

$$2x = -2$$

Step 5 $\dfrac{2x}{2} = \dfrac{-2}{2}$ *Divide both sides by 2.*

$$x = -1$$

Method 2: Isolate the variable term on the right.

$$4x + 6 = 2x + 4$$

Step 4 $4x \boxed{- 4x} + 6 = 2x \boxed{- 4x} + 4$ *Subtract 4x from both sides.*

$$6 = -2x + 4$$

Step 4 $6 \boxed{- 4} = -2x + 4 \boxed{- 4}$ *Subtract 4 from both sides.*

$$2 = -2x$$

Step 5 $\dfrac{2}{-2} = \dfrac{-2x}{-2}$ *Divide both sides by −2.*

$$-1 = x$$

The same answer is obtained whether we collect the terms with the variable on the left or right side. However, we need to divide both sides of the equation by a negative number in method 2.

Step 6   Check: $4x + 6 = 2x + 4$

$$4(-1) + 6 \stackrel{?}{=} 2(-1) + 4$$

$$-4 + 6 \stackrel{?}{=} -2 + 4$$

$$2 = 2 \qquad \textit{True}$$

Since the check is true, the solution is −1.

▶ **Now Try Exercise 19**

EXAMPLE 2 ▶ Solve the equation $2x - 3 - 5x = 13 + 4x - 2$.

Solution   We will choose to collect the terms containing the variable on the right side of the equation in order to create a positive coefficient of x. Since there are like terms *on the same side of the equal sign*, we will begin by combining these like terms.

$$2x - 3 - 5x = 13 + 4x - 2$$

Step 3 $-3x - 3 = 4x + 11$ *Like terms were combined.*

Step 4 $-3x \boxed{+ 3x} - 3 = 4x \boxed{+ 3x} + 11$ *Add 3x to both sides.*

$$-3 = 7x + 11$$

Step 4 $-3 \boxed{- 11} = 7x + 11 \boxed{- 11}$ *Subtract 11 from both sides.*

$$-14 = 7x$$

Step 5 $\dfrac{-14}{7} = \dfrac{7x}{7}$ *Divide both sides by 7.*

$$-2 = x$$

Step 6   Check: $2x - 3 - 5x = 13 + 4x - 2$

$$2(-2) - 3 - 5(-2) \stackrel{?}{=} 13 + 4(-2) - 2$$

$$-4 - 3 + 10 \stackrel{?}{=} 13 - 8 - 2$$

$$-7 + 10 \stackrel{?}{=} 5 - 2$$

$$3 = 3 \qquad \textit{True}$$

Since the check is true, the solution is −2.

▶ **Now Try Exercise 29**

The solution to Example 2 could be condensed as follows:

$$2x - 3 - 5x = 13 + 4x - 2$$

$-3x - 3 = 4x + 11$	*Like terms were combined.*
$-3 = 7x + 11$	*3x was added to both sides.*
$-14 = 7x$	*11 was subtracted from both sides.*
$-2 = x$	*Both sides were divided by 7.*

We solved Example 2 by moving the terms containing the variable to the right side of the equation. Now rework the problem by moving the terms containing the variable to the left side of the equation. You should obtain the same answer.

EXAMPLE 3 ▸ Solve the equation $2(p + 3) = -3p + 10$.

Solution

$$2(p + 3) = -3p + 10$$

Step 2 $2p + 6 = -3p + 10$ *Distributive property was used.*

Step 4 $2p + 3p + 6 = -3p + 3p + 10$ *Add 3p to both sides.*

$$5p + 6 = 10$$

Step 4 $5p + 6 - 6 = 10 - 6$ *Subtract 6 from both sides.*

$$5p = 4$$

Step 5 $\dfrac{5p}{5} = \dfrac{4}{5}$ *Divide both sides by 5.*

$$p = \dfrac{4}{5}$$

The solution is $\dfrac{4}{5}$.

▸ **Now Try Exercise 27**

The solution to Example 3 could be condensed as follows:

$$2(p + 3) = -3p + 10$$

$2p + 6 = -3p + 10$	*Distributive property was used.*
$5p + 6 = 10$	*3p was added to both sides.*
$5p = 4$	*6 was subtracted from both sides.*
$p = \dfrac{4}{5}$	*Both sides were divided by 5.*

Helpful Hint

After the distributive property was used in Example 3, we obtained the equation $2p + 6 = -3p + 10$. Then we had to decide whether to collect terms with the variable on the left or the right side of the equal sign. If we wish the sum of the terms containing a variable to be positive, we use the addition property to eliminate the variable term with the *smaller* numerical coefficient from one side of the equation. Since -3 is smaller than 2, we added $3p$ to both sides of the equation. This eliminated $-3p$ from the right side of the equation and resulted in the sum of the variable terms on the left side of the equation, $5p$, being positive.

EXAMPLE 4 ▸ Solve the equation $2(x - 5) + 3 = 3x + 9$.

Solution

$$2(x - 5) + 3 = 3x + 9$$

Step 2 $2x - 10 + 3 = 3x + 9$ *Distributive property was used.*

Step 3 $2x - 7 = 3x + 9$ *Like terms were combined.*

Step 4 $-7 = x + 9$ *2x was subtracted from both sides.*

Step 4 $-16 = x$ *9 was subtracted from both sides.*

The solution is -16.

▸ **Now Try Exercise 35**

EXAMPLE 5 ▶ Solve the equation $7 - 2x + 5x = -2(-3x + 4)$.

Solution $\quad\quad\quad 7 - 2x + 5x = -2(-3x + 4)$

Step 2	$7 - 2x + 5x = 6x - 8$	*Distributive property was used.*
Step 3	$7 + 3x = 6x - 8$	*Like terms were combined.*
Step 4	$7 = 3x - 8$	*3x was subtracted from both sides.*
Step 4	$15 = 3x$	*8 was added to both sides.*
Step 5	$5 = x$	*Both sides were divided by 3.*

The solution is 5.

▶ **Now Try Exercise 63**

2 Solve Equations Containing Decimal Numbers or Fractions

Now we will solve an equation that contains decimal numbers. As explained in the previous section, equations containing decimal numbers may be solved by a number of different procedures. We will illustrate two procedures for solving Example 6.

EXAMPLE 6 ▶ Solve the equation $5.74x + 5.42 = 2.24x - 9.28$.

Solution

Method 1 We first notice that there are no like terms on the same side of the equal sign that can be combined. We will elect to collect the terms containing the variable on the left side of the equation.

$$5.74x + 5.42 = 2.24x - 9.28$$

Step 4 $\quad 5.74x - 2.24x + 5.42 = 2.24x - 2.24x - 9.28$ *Subtract 2.24x from both sides.*

$$3.50x + 5.42 = -9.28$$

Step 4 $\quad 3.50x + 5.42 - 5.42 = -9.28 - 5.42$ *Subtract 5.42 from both sides.*

$$3.50x = -14.70$$

Step 5 $\quad \dfrac{3.50x}{3.50} = \dfrac{-14.70}{3.50}$ *Divide both sides by 3.50.*

$$x = -4.20$$

The solution is −4.20.

Method 2 In the previous section, we introduced a procedure to eliminate decimal numbers from equations. If the equation contains decimals given in tenths, multiply both sides of the equation by 10. If the equation contains decimals given in hundredths, multiply both sides of the equation by 100, and so on. Since the given equation has numbers given in hundredths, we will multiply both sides of the equation by 100.

$$5.74x + 5.42 = 2.24x - 9.28$$

$$100(5.74x + 5.42) = 100(2.24x - 9.28)$$ *Multiply both sides by 100.*

$$100(5.74x) + 100(5.42) = 100(2.24x) - 100(9.28)$$ *Distributive property*

$$574x + 542 = 224x - 928$$

Step 4 $\quad 574x + 542 - 542 = 224x - 928 - 542$ *Subtract 542 from both sides.*

$$574x = 224x - 1470$$

Step 4 $\quad 574x - 224x = 224x - 224x - 1470$ *Subtract 224x from both sides.*

$$350x = -1470$$

Step 5 $\quad \dfrac{350x}{350} = \dfrac{-1470}{350}$ *Divide both sides by 350.*

$$x = -4.20$$

Notice we obtain the same answer using either method. You may use either method to solve equations of this type.

▶ **Now Try Exercise 25**

Now let's solve some equations that contain fractions.

EXAMPLE 7 ▶ Solve the equation $\frac{1}{2}a = \frac{3}{4}a + \frac{1}{5}$.

Solution

Step 1   In this equation we are solving for a. The least common denominator is 20. Begin by multiplying both sides of the equation by the LCD.

$$\frac{1}{2}a = \frac{3}{4}a + \frac{1}{5}$$

Step 2 $20\left(\frac{1}{2}a\right) = 20\left(\frac{3}{4}a + \frac{1}{5}\right)$ *Multiply both sides by the LCD, 20.*

Step 3 $10a = \overset{5}{20}\left(\frac{3}{4}a\right) + \overset{4}{20}\left(\frac{1}{5}\right)$ *Distributive property*

$$10a = 15a + 4$$

Step 4 $-5a = 4$ *15a was subtracted from both sides.*

Step 5 $a = -\frac{4}{5}$ *Both sides were divided by −5.*

Step 6   Check: $\frac{1}{2}a = \frac{3}{4}a + \frac{1}{5}$

$$\frac{1}{2}\left(-\frac{4}{5}\right) \overset{?}{=} \frac{3}{4}\left(-\frac{4}{5}\right) + \frac{1}{5}$$

$$-\frac{2}{5} \overset{?}{=} -\frac{3}{5} + \frac{1}{5}$$

$$-\frac{2}{5} = -\frac{2}{5}$$ *True*

The solution is $-\frac{4}{5}$.

▶ **Now Try Exercise 43**

Helpful Hint

The equation in Example 7, $\frac{1}{2}a = \frac{3}{4}a + \frac{1}{5}$ could have been written as $\frac{a}{2} = \frac{3a}{4} + \frac{1}{5}$ because $\frac{1}{2}a$ is the same as $\frac{a}{2}$, and $\frac{3}{4}a$ is the same as $\frac{3a}{4}$. You would solve the equation $\frac{a}{2} = \frac{3a}{4} + \frac{1}{5}$ the same way you solved the equation in Example 7.

You would begin by multiplying both sides of the equation by the LCD, 20.

EXAMPLE 8 ▶ Solve the equation $\frac{x}{4} + 3 = 2(x - 2)$.

Solution   We will begin by multiplying both sides of the equation by the LCD, 4.

$$\frac{x}{4} + 3 = 2(x - 2)$$

$$4\left(\frac{x}{4} + 3\right) = 4[2(x - 2)]$$ *Multiply both sides by the LCD, 4.*

$$4\left(\frac{x}{4}\right) + 4(3) = 4[2(x - 2)] \quad \text{\textit{Distributive property (used on left)}}$$

$$x + 12 = 8(x - 2)$$

$$x + 12 = 8x - 16 \quad \text{\textit{Distributive property (used on right)}}$$

$$12 = 7x - 16 \quad \text{\textit{x was subtracted from both sides.}}$$

$$28 = 7x \quad \text{\textit{16 was added to both sides.}}$$

$$4 = x \quad \text{\textit{Both sides were divided by 7.}}$$

A check will show that 4 is the solution.

▶ **Now Try Exercise 61**

Helpful Hint

Notice the equation in Example 8 had *two terms* on the left side of the equal sign, $\frac{x}{4}$ and 3.

The equation had only *one term* on the right side of the equal sign, $2(x - 2)$. Therefore, after we multiplied both sides of the equation by 4, the next step was to use the distributive property on the left side of the equation.

In Example 8, we began the solution by multiplying both sides of the equation by the LCD. In Example 9, we will solve the same equation, but this time we will begin by using the distributive property.

EXAMPLE 9 ▶ Solve the equation in Example 8, $\frac{x}{4} + 3 = 2(x - 2)$, by first using the distributive property.

Solution   Begin by using the distributive property.

$$\frac{x}{4} + 3 = 2(x - 2)$$

$$\frac{x}{4} + 3 = 2x - 4 \quad \text{\textit{Distributive property}}$$

$$4\left(\frac{x}{4} + 3\right) = 4(2x - 4) \quad \text{\textit{Multiply both sides by the LCD, 4.}}$$

$$4\left(\frac{x}{4}\right) + 4(3) = 4(2x) - 4(4) \quad \text{\textit{Distributive property (used on left and right)}}$$

$$x + 12 = 8x - 16$$

$$12 = 7x - 16 \quad \text{\textit{x was subtracted from both sides.}}$$

$$28 = 7x \quad \text{\textit{16 was added to both sides.}}$$

$$4 = x \quad \text{\textit{Both sides were divided by 7.}}$$

The solution is 4.

▶ **Now Try Exercise 65**

Notice that we obtained the same answer in Examples 8 and 9. You may work problems of this type using either procedure unless your instructor asks you to work problems of this type using a specific method.

EXAMPLE 10 ▶ Solve the equation $\frac{1}{2}(2x + 3) = \frac{2}{3}(x - 6) + 4$.

Notice that this equation contains one term on the left side of the equal sign and two terms on the right side of the equal sign.

Solution   We will work this problem by first using the distributive property.

$$\frac{1}{2}(2x + 3) = \frac{2}{3}(x - 6) + 4$$

$$\frac{1}{2}(2x) + \frac{1}{2}(3) = \frac{2}{3}(x) - \frac{2}{3}(6) + 4 \qquad \textit{Distributive property (used on left and right)}$$

$$x + \frac{3}{2} = \frac{2}{3}x - 4 + 4$$

$$x + \frac{3}{2} = \frac{2}{3}x \qquad \textit{Like terms were combined.}$$

$$6\left(x + \frac{3}{2}\right) = 6\left(\frac{2}{3}x\right) \qquad \textit{Multiply both sides by the LCD, 6.}$$

$$6x + 6\left(\frac{3}{2}\right) = 6\left(\frac{2}{3}x\right) \qquad \textit{Using the distributive property}$$

$$6x + \overset{3}{6}\left(\frac{3}{2}\right) = \overset{2}{6}\left(\frac{2}{3}x\right)$$

$$6x + 9 = 4x$$

$$2x + 9 = 0 \qquad \textit{4x was subtracted from both sides.}$$

$$2x = -9 \qquad \textit{9 was subtracted from both sides.}$$

$$x = -\frac{9}{2} \qquad \textit{Both sides were divided by 2.}$$

Check: Substitute $-\frac{9}{2}$ for each x in the equation.

$$\frac{1}{2}(2x + 3) = \frac{2}{3}(x - 6) + 4$$

$$\frac{1}{2}\left[2\left(-\frac{9}{2}\right) + 3\right] \overset{?}{=} \frac{2}{3}\left(-\frac{9}{2} - 6\right) + 4$$

$$\frac{1}{2}(-9 + 3) \overset{?}{=} \frac{2}{3}\left(-\frac{9}{2} - \frac{12}{2}\right) + 4$$

$$\frac{1}{2}(-6) \overset{?}{=} \frac{2}{3}\left(-\frac{21}{2}\right) + 4$$

$$-3 \overset{?}{=} -7 + 4$$

$$-3 = -3 \qquad \textit{True}$$

The solution is $-\frac{9}{2}$.

▶ **Now Try Exercise 75**

In Example 10, we began by using the distributive property. We could have also begun by multiplying both sides of the equation by the LCD, 6, before using the distributive property. Work Example 10 again now by first multiplying both sides of the equation by the LCD, 6. You should obtain the same answer, $x = -\frac{9}{2}$.

Example 10 could have also been written as $\dfrac{2x + 3}{2} = \dfrac{2(x - 6)}{3} + 4$. If you were given the equation in this form, you could begin by using the distributive property on $2(x - 6)$ or you could begin by multiplying both sides of the equation by the LCD, 6. Because this is just another way of writing the equation in Example 10, the answer would be $-\dfrac{9}{2}$.

We will discuss solving equations containing fractions in more detail later in the book.

3 Identify Identities and Contradictions

Thus far all the equations we have solved have had a single value for a solution. Equations of this type are called **conditional equations**, for they are only true under specific conditions. Some equations, as in Example 11, are true for infinitely many values of x. Equations that are true for infinitely many values of x are called **identities**. A third type of equation, as in Example 12, has no solution and is called a **contradiction**.

EXAMPLE 11 ▶ Solve the equation $5x - 5 - 2x = 3(x - 2) + 1$.

Solution

$$5x - 5 - 2x = 3(x - 2) + 1$$
$$5x - 5 - 2x = 3x - 6 + 1 \qquad \text{\textit{Distributive property was used.}}$$
$$3x - 5 = 3x - 5 \qquad \text{\textit{Like terms were combined.}}$$

Since the same expression appears on both sides of the equal sign, the statement is true for infinitely many values of x. If we continue to solve this equation further, we might obtain

$$3x - 5 = 3x - 5$$
$$3x = 3x \qquad \text{\textit{5 was added to both sides.}}$$
$$0 = 0 \qquad \text{\textit{3x was subtracted from both sides.}}$$

NOTE: The solution process could have been stopped at $3x - 5 = 3x - 5$. Since one side is identical to the other side, the equation is true for infinitely many values of x. *The solution to this equation is all real numbers.* **When solving an equation like the equation in Example 11, that is always true, write your answer as "all real numbers."**

▶ **Now Try Exercise 47**

EXAMPLE 12 ▶ Solve the equation $-2x + 5 + 3x = 5x - 4x + 7$.

Solution

$$-2x + 5 + 3x = 5x - 4x + 7$$
$$x + 5 = x + 7 \qquad \text{\textit{Like terms were combined.}}$$
$$x - x + 5 = x - x + 7 \qquad \text{\textit{Subtract x from both sides.}}$$
$$5 = 7 \qquad \text{\textit{False}}$$

NOTE: When solving an equation, if you obtain an obviously false statement, as in this example, the equation has *no solution*. No value of x will make the equation a true statement. **When solving an equation like the equation in Example 12, that is never true, write your answer as "no solution."** An answer left blank may be marked wrong.

▶ **Now Try Exercise 31**

Helpful Hint

Some students start solving equations correctly but do not complete the solution. Sometimes they are not sure that what they are doing is correct and they give up for lack of confidence. You must have confidence in yourself. As long as you follow the procedure on page 129, you should obtain the correct solution even if it takes quite a few steps. Remember two important things: (1) your goal is to isolate the variable, and (2) whatever you do to one side of the equation you must also do to the other side. That is, you must treat both sides of the equation equally.

EXERCISE SET 2.5   Math XL   MyMathLab
MathXL®   MyMathLab

Concept/Writing Exercises

1. **a)** Write the general procedure for solving an equation that does not contain fractions where the variable appears on both sides of the equation.

   b) Refer to page 129 to see whether you omitted any steps.

2. What is a conditional equation?

3. **a)** What is an identity?

   b) What is the solution to the equation $3x + 5 = 3x + 5$?

4. When solving an equation, how will you know if the equation is an identity?

5. Explain why the equation $x + 5 = x + 5$ must be an identity.

6. **a)** What is a contradiction?

   b) What is the solution to a contradiction?

7. When solving an equation, how will you know if the equation has no solution?

8. Explain why the equation $x + 5 = x + 4$ must be a contradiction.

9. **a)** Explain, step-by-step, how to solve the equation $4x + 3(x + 2) = 5x - 10$.

   b) Solve the equation by following the steps you listed in part **a)**.

10. **a)** Explain, step-by-step, how to solve the equation $4(x + 3) = 6(x - 5)$.

    b) Solve the equation by following the steps you listed in part **a)**.

Practice the Skills

Solve each equation.

11. $3x = -2x + 15$

12. $x + 4 = 2x - 7$

13. $-4x + 10 = 6x$

14. $3a = 4a + 8$

15. $5x + 3 = 6$

16. $-6x = 2x + 16$

17. $21 - 6p = 3p - 2p$

18. $8 - 3x = 4x + 50$

19. $2x - 4 = 3x - 6$

20. $5x + 7 = 3x + 5$

21. $6 - 2y = 9 - 8y + 6y$

22. $-4 + 2y = 2y - 6 + y$

23. $124.8 - 9.4x = 4.8x + 32.5$

24. $9 - 0.5x = 4.5x + 8.5$

25. $0.62x - 0.65 = 9.75 - 2.63x$

26. $8.71 - 2.44x = 11.02 - 5.74x$

27. $5x + 3 = 2(x + 6)$

28. $x - 14 = 3(x + 2)$

29. $4y - 2 - 8y = 19 + 5y - 3$

30. $3x - 5 + 9x = 2 + 4x + 9$

31. $2(x - 2) = 4x - 6 - 2x$

32. $4r = 10 - 2(r - 4)$

33. $-(w + 2) = -6w + 32$

34. $7(-3m + 5) = 3(10 - 6m)$

35. $-3(2t - 5) + 5 = 3t + 13$

36. $4(x - 3) + 2 = 2x + 8$

37. $\dfrac{a}{5} = \dfrac{a - 3}{2}$

38. $\dfrac{b}{16} = \dfrac{b - 6}{4}$

39. $\dfrac{n}{10} = 9 - \dfrac{n}{5}$

40. $6 - \dfrac{x}{4} = \dfrac{x}{8}$

41. $\dfrac{5}{2} - \dfrac{x}{3} = 3x$

42. $\dfrac{x}{4} - 3 = -2x$

43. $\dfrac{5}{8} + \dfrac{1}{4}a = \dfrac{1}{2}a$

44. $\dfrac{3}{4}x + \dfrac{1}{2} = \dfrac{1}{2}x$

45. $0.1(x + 10) = 0.3x - 4$

46. $5(3.2x - 3) = 2(x - 4)$

47. $2(x + 4) = 4x + 3 - 2x + 5$

48. $3(y - 1) + 9 = 8y + 6 - 5y$

49. $5(3n + 3) = 2(5n - 4) + 6n$

50. $-4(-3z - 5) = -(10z + 8) - 2z$

51. $-(3 - p) = -(2p + 3)$

52. $12 - 2x - 3(x + 2) = 4x + 6 - x$

53. $-(x + 4) + 5 = 4x + 1 - 5x$

54. $18x + 3(4x - 9) = -6x + 81$

55. $35(2x - 1) = 7(x + 4) + 3x$

56. $10(x - 10) + 5 = 5(2x - 20)$

57. $0.4(x + 0.7) = 0.6(x - 4.2)$

58. $0.5(6x - 8) = 1.4(x - 5) - 0.2$

59. $\dfrac{3}{5}x - 2 = x + \dfrac{1}{3}$

60. $\dfrac{3}{5}x + 4 = \dfrac{1}{5}x + 5$

61. $\dfrac{y}{5} + 2 = 3(y - 4)$

62. $2(x - 4) = \dfrac{x}{5} + 10$

63. $12 - 3x + 7x = -2(-5x + 6)$

64. $-2x - 3 - x = -3(-2x + 7)$

65. $3(x - 6) - 4(3x + 1) = x - 22$

66. $-2(-3x + 5) + 6 = 4(x - 2)$

67. $5 + 2x = 6(x + 1) - 5(x - 3)$

68. $4 - (6x + 6) = -(-2x + 10)$

69. $7 - (-y - 5) = 2(y + 3) - 6(y + 1)$

70. $12 - 6x + 3(2x + 3) = 2x + 5$

71. $\dfrac{3}{5}(x - 6) = \dfrac{2}{3}(3x - 5)$

72. $\dfrac{1}{2}(2d + 4) = \dfrac{1}{3}(4d - 4)$

73. $\dfrac{3(2r - 5)}{5} = \dfrac{3r - 6}{4}$

74. $\dfrac{3(x - 4)}{4} = \dfrac{5(2x - 3)}{3}$

75. $\dfrac{2}{7}(5x + 4) = \dfrac{1}{2}(3x - 4) + 1$

76. $\dfrac{5}{12}(x + 2) = \dfrac{2}{3}(2x + 1) + \dfrac{1}{6}$

77. $\dfrac{a - 5}{2} = \dfrac{3a}{4} + \dfrac{a - 25}{6}$

78. $\dfrac{a - 7}{3} = \dfrac{a + 5}{2} - \dfrac{7a - 1}{6}$

Problem Solving

79. a) Construct a *conditional equation* containing three terms on the left side of the equal sign and two terms on the right side of the equal sign.

  b) Explain how you know your answer to part **a)** is a conditional equation.

  c) Solve the equation.

80. a) Construct a *conditional equation* containing two terms on the left side of the equal sign and three terms on the right side of the equal sign.

  b) Explain how you know your answer to part **a)** is a conditional equation.

  c) Solve the equation.

81. a) Construct an *identity* containing three terms on the left side of the equal sign and two terms on the right side of the equal sign.

  b) Explain how you know your answer to part **a)** is an identity.

  c) What is the solution to the equation?

82. a) Construct an *identity* containing two terms on the left side of the equal sign and three terms on the right side of the equal sign.

  b) Explain how you know your answer to part **a)** is an identity.

  c) What is the solution to the equation?

83. a) Construct a *contradiction* containing three terms on the left side of the equal sign and two terms on the right side of the equal sign.

  b) Explain how you know your answer to part **a)** is a contradiction.

  c) What is the solution to the equation?

84. a) Construct a *contradiction* containing three terms on the left side of the equal sign and four terms on the right side of the equal sign.

  b) Explain how you know your answer to part **a)** is a contradiction.

  c) What is the solution to the equation?

Challenge Problems

85. Solve the equation $5* - 1 = 4* + 5*$ for $*$.

86. Solve the equation $2\triangle - 4 = 3\triangle + 5 - \triangle$ for $\triangle$.

87. Solve the equation $3\odot - 5 = 2\odot - 5 + \odot$ for $\odot$.

88. Solve $-2(x + 3) + 5x = 3(4 - 2x) - (x + 2)$.

89. Solve $4 - [5 - 3(x + 2)] = x - 3$.

Group Activity

Discuss and answer Exercise 90 as a group. In the next chapter, we will be discussing procedures for writing application problems as equations. Let's get some practice now.

90. Chocolate Bars Consider the following word problem. Mary Kay purchased two large chocolate bars. The total cost of the two chocolate bars was equal to the cost of one chocolate bar plus $6. Find the cost of one chocolate bar.

  a) Each group member: Represent this problem as an equation with the variable x.

  b) Each group member: Solve the equation you determined in part **a)**.

  c) As a group, check your equation and your answer to make sure that it makes sense.

Cumulative Review Exercises

[1.5] **91.** Evaluate

  a) $|4|$ **b)** $|-7|$ **c)** $|0|$.

[1.9] **92.** Evaluate $\left(\dfrac{2}{3}\right)^5$ on your calculator.

[2.1] **93.** Explain the difference between factors and terms.

94. Simplify $2(x - 3) + 4x - (4 - x)$.

[2.4] **95.** Solve $2(x - 3) + 4x - (4 - x) = 0$.

96. Solve $(x + 4) - (4x - 3) = 16$.

2.6 Formulas

1 Use the simple interest formula and the distance formula.

2 Use geometric formulas.

3 Solve for a variable in a formula.

A **formula** is an equation commonly used to express a specific relationship mathematically. For example, the formula for the area of a rectangle is

$$\text{area} = \text{length} \cdot \text{width} \quad \text{or} \quad A = lw$$

To **evaluate a formula**, substitute the appropriate numerical values for the variables and perform the indicated operations.

1 Use the Simple Interest Formula and the Distance Formula

A formula commonly used in banking is the **simple interest formula**.

Simple Interest Formula

$$\text{interest} = \text{principal} \cdot \text{rate} \cdot \text{time} \quad \text{or} \quad i = prt$$

This formula is used to determine the simple interest, i, earned on some savings accounts, or the simple interest an individual must pay on certain loans. In the simple interest formula $i = prt$, p is the principal (the amount invested or borrowed), r is the interest rate in decimal form, and t is the amount of time of the investment or loan.

EXAMPLE 1 ▶ **Auto Loan** To buy a car, Mary Beth Orrange borrowed $10,000 from a bank for 3 years. The bank charged 5% simple annual interest for the loan. How much interest will Mary Beth owe the bank?

Solution Understand and Translate Since the bank charged simple interest, we use the simple interest formula to solve the problem. We are given that the rate, r, is 5%, or 0.05 in decimal form. The principal, p, is $10,000 and the time, t, is 3 years. We substitute these values in the simple interest formula and solve for the interest, i.

$$i = prt$$

Carry Out

$$i = 10,000(0.05)(3)$$

$$i = 1500$$

Check   There are various ways to check this problem. First ask yourself "Is the answer realistic?" $1500 is a realistic answer. The interest on $10,000 for 1 year at 5% is $500. Therefore for 3 years, an interest of $1500 is correct.

Answer   Mary Beth will pay $1500 interest. After 3 years, when she repays the loan, she will pay the principal, $10,000, plus the interest, $1500 for a total of $11,500.

▶ **Now Try Exercise 89**

EXAMPLE 2 ▶ **Savings Account** John Starmack invests $4000 in a savings account that earns simple interest for 2 years. If the interest earned from the account is $500, find the rate.

Solution Understand and Translate We use the simple interest formula, $i = prt$. We are given the principal, p, the time, t, and the interest, i. We are asked to find the rate, r. We substitute the given values in the simple interest formula and solve the resulting equation for r.

$$i = prt$$
$$500 = 4000(r)(2)$$

Carry Out
$$500 = 8000r$$
$$\frac{500}{8000} = \frac{8000r}{8000}$$
$$0.0625 = r$$

Check and Answer   The simple interest rate of 0.0625 or 6.25% per year is realistic. If we substitute $p = \$4000$, $r = 0.0625$ and $t = 2$, we obtain the interest, $i = \$500$. Thus, the answer checks. The simple interest rate is 6.25%.

▶ **Now Try Exercise 91**

We will use the simple interest formula again in Section 3.4.
Another important formula is the distance formula.

Distance Formula

distance = rate · time   or $d = r \cdot t$

Example 3 illustrates the use of the distance formula.

EXAMPLE 3 ▶ **Auto Race** At a NASCAR auto race, Dale Earnhart, Jr., completed the race in 3.2 hours at an average speed of 156.25 miles per hour. Determine the distance of the race.

Solution   Understand and Translate   We are given the rate, 156.25 miles per hour, and the time is 3.2 hours. We are asked to find the distance.

$$\text{distance} = \text{rate} \cdot \text{time}$$
Carry Out
$$= (156.25)(3.2) = 500$$

Answer   Thus, the distance of the race was 500 miles.

▶ **Now Try Exercise 95**

Let's look at the units in Example 3. The rate is given in miles per hour and the time is given in hours. If we analyze the units (a process called *dimensional analysis*), we see that the answer is given in miles.

$$\text{distance} = \text{rate} \cdot \text{time}$$
$$= \frac{\text{miles}}{\text{hour}} \cdot \text{hour}$$
$$= \text{miles}$$

We will use the distance formula again in Section 3.4.
Now we will discuss geometric formulas that will be used throughout the book.

2 Use Geometric Formulas

The **perimeter**, P, is the sum of the lengths of the sides of a figure. Perimeters are measured in the same common unit as the sides. For example, perimeter may be measured in centimeters, inches, or feet. The **area**, A, is the total surface within the figure's boundaries. Areas are measured in square units. For example, area may be measured in square centimeters, square inches, or square feet. **Table 2.1** on page 142 gives the formulas for finding the areas and perimeters of triangles and quadrilaterals. **Quadrilateral** is a general name for a four-sided figure.

In **Table 2.1**, the letter h is used to represent the *height* of the figure. In the figure of the trapezoid, the sides b and d are called the *bases* of the trapezoid. In the triangle, the side labeled b is called the *base* of the triangle.

TABLE 2.1   Formulas for Areas and Perimeters of Quadrilaterals and Triangles*

Figure	Sketch	Area	Perimeter
Square	s	$A = s^2$	$P = 4s$
Rectangle	w l	$A = lw$	$P = 2l + 2w$
Parallelogram	h w l	$A = lh$	$P = 2l + 2w$
Trapezoid	b a h c d	$A = \frac{1}{2}h(b + d)$	$P = a + b + c + d$
Triangle	a h c b	$A = \frac{1}{2}bh$	$P = a + b + c$

EXAMPLE 4 ▶ **Building an Exercise Area** Dr. Alex Taurke, a veterinarian, decides to fence in a large rectangular area in the yard behind his office for exercising dogs that are boarded overnight. The part of the yard to be fenced in will be 40 feet long and 23 feet wide (see **Fig. 2.2**).

a) How much fencing is needed?

b) How large, in square feet, will the fenced in area be?

Solution

a) Understand   To find the amount of fencing required, we need to find the perimeter of the rectangular area to be fenced in. To find the perimeter, P, substitute 40 for the length, l, and 23 for the width, w, in the perimeter formula, $P = 2l + 2w$.

$$P = 2l + 2w$$

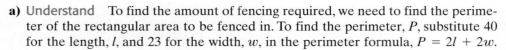

FIGURE 2.2

Carry Out $P = 2(40) + 2(23) = 80 + 46 = 126$

Check and Answer   By looking at **Figure 2.2**, we can see that a perimeter of 126 feet is a reasonable answer. Thus, 126 feet of fencing will be needed to fence in the area for the dogs to exercise.

b) To find the fenced in area, substitute 40 for the length and 23 for the width in the formula for the area of a rectangle. Both the length and width are measured in feet. Since we are multiplying an amount measured in feet by a second amount measured in feet, the answer will be in square feet (or ft^2).

$$A = lw$$
$$= 40(23) = 920 \text{ square feet (or } 920 \text{ ft}^2)$$

Based upon the data given, an area of 920 ft^2 is reasonable. The area to be fenced in will be 920 square feet.

▶ **Now Try Exercise 97**

*See Appendix C for additional information on geometry and geometric figures.

EXAMPLE 5 ▶ Panoramic Photo Heather Hunter enlarges rectangular panoramic photos, like the one shown below. One of her enlarged panoramic photos has a perimeter of 116 inches and a length of 40 inches. Find the width of the photo.

Solution Understand and Translate   The perimeter, P, is 116 inches and the length, l, is 40 inches. Substitute these values into the formula for the perimeter of a rectangle and solve for the width, w.

$$P = 2l + 2w$$
$$116 = 2(40) + 2w$$

Carry Out

$$116 = 80 + 2w$$
$$116 - 80 = 80 - 80 + 2w \qquad \text{Subtract 80 from both sides.}$$
$$36 = 2w$$
$$\frac{36}{2} = \frac{2w}{2} \qquad \text{Divide both sides by 2.}$$
$$18 = w$$

Check and Answer   By considering what you know about panoramic photos, and comparing their length and width, you should realize that the dimensions of a length of 40 inches and a width of 18 inches is reasonable. The answer is, the width of the photo is 18 inches.

▶ **Now Try Exercise 25**

EXAMPLE 6 ▶ Sailboat A small sailboat has a triangular sail that has an area of 30 square feet and a base of 5 feet (see **Fig. 2.3**). Determine the height of the sail.

Solution Understand and Translate   We use the formula for the area of a triangle given in **Table 2.1**.

$$A = \frac{1}{2}bh$$
$$30 = \frac{1}{2}(5)h$$

Carry Out

$$2 \cdot 30 = 2 \cdot \frac{1}{2}(5)h \qquad \text{Multiply both sides by 2.}$$
$$60 = 5h$$
$$\frac{60}{5} = \frac{5h}{5} \qquad \text{Divide both sides by 5.}$$
$$12 = h$$

Check and Answer   The height of the triangle is 12 feet. By looking at **Figure 2.3**, and by your knowledge of sailboat sails, you may realize that a sail 12 feet tall and 5 feet wide at the base is reasonable. Thus, the height of the sail is 12 feet.

▶ **Now Try Exercise 99**

5 ft

The Pythagoras

FIGURE 2.3

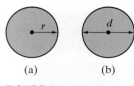

(a)   (b)

FIGURE 2.4

Another figure that we see and use daily is the circle. The **circumference**, C, is the length (or perimeter) of the curve that forms a circle. The **radius**, r, is the line segment from the center of the circle to any point on the circle (**Fig. 2.4a**). The **diameter** of a circle is a line segment through the center whose endpoints both lie on the circle (**Fig. 2.4b**). *Note that the length of the diameter is twice the length of the radius.*

The formulas for both the area and the circumference of a circle are given in **Table 2.2**.

TABLE 2.2	Formulas for Circles	
Circle	**Area**	**Circumference**
	$A = \pi r^2$	$C = 2\pi r$

The value of **pi**, symbolized by the Greek lowercase letter π, is an irrational number which cannot be exactly expressed as a decimal number or a numerical fraction. Pi is *approximately* 3.14.

USING YOUR CALCULATOR

Scientific and graphing calculators have a key for finding the value of π. If you press the $\boxed{\pi}$ key, your calculator may display 3.1415927. This is only an approximation of π. If you own a scientific or graphing calculator, use the $\boxed{\pi}$ key when evaluating expressions containing π. If your calculator does not have a $\boxed{\pi}$ key, use 3.14 to approximate it. *When evaluating an expression containing π, we will use the $\boxed{\pi}$ key on a calculator to obtain the answer.* The final answer displayed in the text or answer section may therefore be slightly different (and more accurate) than yours if you use 3.14 for π.

EXAMPLE 7 ▶ **Pizza** A Pizza Hut large pizza has a diameter of 14 inches. Determine the area and circumference of the pizza.

Solution   The radius is half its diameter, so $r = \dfrac{14}{2} = 7$ inches.

$$A = \pi r^2 \qquad\qquad\qquad C = 2\pi r$$
$$A = \pi (7)^2 \qquad\qquad\quad\; C = 2\pi (7)$$
$$A = \pi (49) \qquad\qquad\quad\;\; C \approx 43.98 \text{ inches}$$
$$A \approx 153.94 \text{ square inches}$$

To obtain our answers 153.94 and 43.98, we used the $\boxed{\pi}$ key on a calculator and rounded our final answer to the nearest hundredth. If you do not have a calculator with a $\boxed{\pi}$ key and use 3.14 for π, your answer for the area would be 153.86.

▶ **Now Try Exercise 101**

Table 2.3 on page 145 gives formulas for finding the volume of certain **three-dimensional figures**. **Volume** may be considered the space occupied by a figure, and it is measured in cubic units, such as cubic centimeters or cubic feet.

EXAMPLE 8 ▶ **Spaceship Earth** The inside of Spaceship Earth at Epcot Center in Disney World, Florida, is a sphere with a diameter of 165 feet (see photo). Determine the volume of Spaceship Earth.

Solution   Understand and Translate   **Table 2.3** gives the formula for the volume of a sphere. The formula involves the radius. Since the diameter is 165 feet, its radius is $\dfrac{165}{2} = 82.5$ feet.

$$V = \frac{4}{3}\pi r^3$$

Carry Out $V = \frac{4}{3}\pi (82.5)^3 = \frac{4}{3}\pi (561{,}515.625) \approx 2{,}352{,}071.15$

Check and Answer   The volume inside the sphere is very large, 2,352,071.15 cubic feet. Since a rollercoaster-type ride is inside the sphere, the volume must be very large, and so the answer is reasonable.

▶ **Now Try Exercise 109**

TABLE 2.3   Formulas for Volumes of Three-Dimensional Figures

Figure	Sketch	Volume
Rectangular solid		$V = lwh$
Right circular cylinder		$V = \pi r^2 h$
Right circular cone		$V = \frac{1}{3}\pi r^2 h$
Sphere		$V = \frac{4}{3}\pi r^3$

3 Solve for a Variable in a Formula

Often in this course and in other mathematics and science courses, you will be given an equation or formula solved for one variable and have to solve it for a different variable. We will now learn how to do this. This material will reinforce what you learned about solving equations earlier in this chapter. We will use the procedures learned here to solve problems in many other sections of the text.

To solve for a variable in a formula, treat each of the quantities, except the one for which you are solving, as if they were constants. Then solve for the desired variable by isolating it on one side of the equation.

EXAMPLE 9 ▶ **Distance Formula** Solve the distance formula $d = rt$ for t.

Solution   We must get t all by itself on one side of the equal sign. Since t is multiplied by r, we divide both sides of the equation by r to isolate the t.

$$d = rt$$
$$\frac{d}{r} = \frac{rt}{r} \qquad \text{Divide both sides by } r.$$
$$\frac{d}{r} = t$$

Therefore, $t = \dfrac{d}{r}$.

Now Try Exercise 45

EXAMPLE 10 ▶ Perimeter of Rectangle The formula for the perimeter of a rectangle is $P = 2l + 2w$. Solve this formula for the length, l.

Solution   We must get l all by itself on one side of the equation. We begin by removing the $2w$ from the right side of the equation to isolate the term containing the l.

$$P = 2l + 2w$$
$$P - 2w = 2l + 2w - 2w \qquad \text{Subtract 2w from both sides.}$$
$$P - 2w = 2l$$
$$\frac{P - 2w}{2} = \frac{2l}{2} \qquad \text{Divide both sides by 2.}$$
$$\frac{P - 2w}{2} = l \quad \left(\text{or} \quad l = \frac{P}{2} - w \right)$$

▶ **Now Try Exercise 53**

Some formulas contain fractions. When a formula contains a fraction, we can eliminate the fraction by multiplying both sides of the equation by the least common denominator, as illustrated in Example 11. We use the multiplication property of equality, as explained in Section 2.3.

EXAMPLE 11 ▶ The formula for the area of a triangle is $A = \frac{1}{2}bh$. Solve this formula for h.

Solution   We begin by multiplying both sides of the equation by the LCD, 2, to eliminate the fraction. We then isolate the variable h.

$$A = \frac{1}{2}bh$$
$$2 \cdot A = 2 \cdot \frac{1}{2}bh \qquad \text{Multiply both sides by 2.}$$
$$2A = bh$$
$$\frac{2A}{b} = \frac{bh}{b} \qquad \text{Divide both sides by b.}$$
$$\frac{2A}{b} = h$$

Thus, $h = \frac{2A}{b}$.

▶ **Now Try Exercise 51**

Write Equations in $y = mx + b$ Form

When discussing graphing later in this book, we will need to solve many equations for the variable y, and write the equation in the form $y = mx + b$, where m and b represent real numbers. Examples of equations in this form are $y = 2x + 4$, $y = -\frac{1}{2}x - 3$, and $y = \frac{4}{5}x + \frac{1}{3}$. The procedure to write equations in $y = mx + b$ form is illustrated in Examples 12 and 13.

EXAMPLE 12 ▶ Solve the equation $6x + 3y = 12$ for y. Write the answer in $y = mx + b$ form.

Solution   Begin by isolating the term containing the variable y.

$$6x + 3y = 12$$

$$6x \boxed{-\ 6x} + 3y = 12 \boxed{-\ 6x} \qquad \textit{Subtract 6x from both sides.}$$

$$3y = 12 - 6x$$

$$\frac{\cancel{3}y}{\cancel{3}} = \frac{12 - 6x}{3} \qquad \textit{Divide both sides by 3.}$$

$$y = \frac{12 - 6x}{3}$$

$$y = \frac{12}{3} - \frac{6x}{3} \qquad \textit{Write as two fractions.}$$

$$y = 4 - 2x$$

$$y = -2x + 4$$

▸ **Now Try Exercise 67**

Helpful Hint

Notice that in Example 12, when we obtained $y = \dfrac{12 - 6x}{3}$, we had solved the equation for y since the y was isolated on one side of the equation. When we wrote the answer as $y = -2x + 4$, we wrote the equation in $y = mx + b$ form.

EXAMPLE 13 ▸ Solve the equation $y - \dfrac{1}{3} = \dfrac{1}{4}(x - 6)$ for y. Write the answer in $y = mx + b$ form.

Solution Multiply both sides of the equation by the LCD, 12.

$$y - \frac{1}{3} = \frac{1}{4}(x - 6)$$

$$\boxed{12}\left(y - \frac{1}{3}\right) = \boxed{12} \cdot \frac{1}{4}(x - 6) \qquad \textit{Multiply both sides by 12.}$$

$$12y - 4 = 3(x - 6) \qquad \textit{Distributive property used on left}$$

$$12y - 4 = 3x - 18 \qquad \textit{Distributive property used on right}$$

$$12y = 3x - 14 \qquad \textit{Add 4 to both sides}$$

$$y = \frac{3x - 14}{12} \qquad \textit{Divided both sides by 12.}$$

$$y = \frac{3x}{12} - \frac{14}{12} \qquad \textit{Write as two fractions.}$$

$$y = \frac{1}{4}x - \frac{7}{6}$$

▸ **Now Try Exercise 79**

In Example 13, you may wish to use the distributive property on the right side of the equation before multiplying both sides of the equation by the LCD, 12. Try working the example using this method now to see which procedure you prefer.

EXERCISE SET 2.6   *Math* XL   *MyMathLab*

Concept/Writing Exercises

1. What is a formula?
2. What does it mean to *evaluate a formula*?
3. Write the simple interest formula, then indicate what each letter in the formula represents.

4. What is a quadrilateral?
5. Write the distance formula, then indicate what each letter in the formula represents.
6. What is the relationship between the radius and the diameter of a circle?

7. Is π equal to 3.14? Explain your answer.

8. a) What is the perimeter of a figure?

  b) What is the area of a figure?

9. By using any formula for area, explain why area is measured in square units.

10. By using any formula for volume, explain why volume is measured in cubic units.

Practice the Skills

Use the formula to find the value of the variable indicated. Use a calculator to save time and where necessary, round your answer to the nearest hundredth.

11. $d = rt$ (distance formula); find d when $r = 60$ and $t = 4$.

12. $P = 4s$ (perimeter of a square); find P when $s = 6$.

13. $A = lw$ (area of a rectangle); find A when $l = 12$ and $w = 8$.

14. $A = s^2$ (area of a square); find A when $s = 7$.

15. $i = prt$ (simple interest formula); find i when $p = 2000$, $r = 0.06$, and $t = 3$.

16. $c = 2.54i$ (to change inches to centimeters); find c when $i = 12$.

17. $P = 2l + 2w$ (perimeter of a rectangle); find P when $l = 8$ and $w = 5$.

18. $f = 1.47m$ (to change speed from mph to ft/sec); find f when $m = 60$.

19. $A = \pi r^2$ (area of a circle); find A when $r = 5$.

20. $A = \dfrac{m + n}{2}$ (mean of two values); find A when $m = 16$ and $n = 56$.

21. $A = \dfrac{a + b + c}{3}$ (mean of three values); find A when $a = 72$, $b = 81$, and $c = 93$.

22. $p = i^2 r$ (formula for finding electrical power); find r when $p = 2000$ and $i = 4$.

23. $z = \dfrac{x - m}{s}$ (statistics formula for finding the z-score); find z when $x = 100$, $m = 80$, and $s = 10$.

24. $A = \dfrac{1}{2}bh$ (area of a triangle); find b when $A = 30$ and $h = 10$

25. $P = 2l + 2w$ (perimeter of a rectangle); find l when $P = 28$ and $w = 6$.

26. $A = P(1 + rt)$ (banking formula to find the amount in an account); find r when $A = 1050$, $t = 1$, and $P = 1000$.

27. $V = \pi r^2 h$ (volume of a cylinder); find h when $V = 678.24$ and $r = 6$.

28. $V = \dfrac{4}{3}\pi r^3$ (volume of a sphere); find V when $r = 8$.

29. $B = \dfrac{703w}{h^2}$ (for finding body mass index); find w when $B = 24$ and $h = 61$.

30. $S = C + rC$ (for determining selling price when an item is marked up); find S when $C = 160$ and $r = 0.12$ (or 12%).

In Exercises 31–36, use **Tables 2.1, 2.2,** *and* **2.3** *to find the formula for the area or volume of the figure. Then determine either the area or volume.*

31.

8 ft

32.

4 in.

6 in.

33.
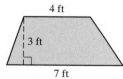
4 ft

3 ft

7 ft

34.
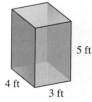
5 ft

4 ft   3 ft

35.

4 cm

9 cm

36.
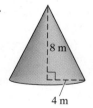
8 m

4 m

In Exercises 37 and 38, use the formula $C = \dfrac{5}{9}(F - 32)$ to find the Celsius temperature (C) equivalent to the given Fahrenheit temperature (F).

37. $F = 50°$

38. $F = 86°$

In Exercises 39 and 40, use the formula $F = \dfrac{9}{5}C + 32$, to find the Fahrenheit temperature (F) equivalent to the given Celsius temperature (C).

39. $C = 25°$

40. $C = 10°$

In Exercises 41–44, find the missing quantity. Use the ideal gas law, $P = KT/V$, where P is pressure, T is temperature, V is volume, and K is a constant.

41. $T = 20$, $K = 2$, $V = 1$

42. $P = 80$, $T = 100$, $V = 5$

43. $T = 30$, $P = 3$, $K = 0.5$

44. $P = 100$, $K = 2$, $V = 6$

In Exercises 45–66, solve for the indicated variable.

45. $A = lw$, for w

46. $P = 4s$, for s

47. $d = rt$, for t

48. $C = \pi d$, for d

49. $i = prt$, for t

50. $V = lwh$, for l

51. $A = \frac{1}{2}bh$, for b

52. $E = IR$, for I

53. $P = 2l + 2w$, for w

54. $PV = KT$, for T

55. $3 - 2r = n$, for r

56. $4m + 5n = 25$, for n

57. $y = mx + b$, for b

58. $y = mx + b$, for x

59. $d = a + b + c$, for b

60. $ax + by = c$, for y

61. $ax + by + c = 0$, for y

62. $V = \pi r^2 h$, for h

63. $V = \frac{1}{3}\pi r^2 h$, for h

64. $A = \frac{m + 2d}{3}$, for d

65. $A = \frac{m + d}{2}$, for m

66. $L = \frac{c + 2d}{4}$, for d

In Exercises 67–82, solve each equation for y. Write the answer in y = mx + b form. See Examples 12 and 13.

67. $2x + y = 5$

68. $6x + 2y = -12$

69. $-3x + 3y = -15$

70. $-2y + 4x = -8$

71. $4x = 6y - 8$

72. $15 = 3y - x$

73. $5y = -10 + 3x$

74. $-2y = -3x - 18$

75. $-6y = 15 - 3x$

76. $-12 = -2x - 3y$

77. $-8 = -x - 2y$

78. $4x + 3y = 20$

79. $y + 3 = -\frac{1}{3}(x - 4)$

80. $y - 3 = \frac{2}{3}(x + 4)$

81. $y - \frac{1}{5} = 2\left(x + \frac{1}{3}\right)$

82. $y + 5 = \frac{3}{4}\left(x + \frac{1}{2}\right)$

Problem Solving

83. When using the distance formula, what happens to the distance if the rate is doubled and the time is halved? Explain.

84. When using the simple interest formula, what happens to the simple interest if both the principal and rate are doubled but the time is halved? Explain.

85. Consider the formula for the area of a square, $A = s^2$. If the length of the side of a square, s, is doubled, what is the change in its area? Explain.

86. Consider the formula for the volume of a cube, $V = s^3$. If the length of the side of a cube, s, is doubled, what is the change in its volume? Explain.

87. Which would have the greater area, a square whose side has a length of s inches, or a circle whose diameter has a length of s inches? Explain, using a sketch.

88. Which would have the greater area, a square whose diagonal has a length of s inches, or a circle whose diameter has a length of s inches? Explain, using a sketch.

In Exercises 89–92, use the simple interest formula.

89. Auto Loan Thang Tran decided to borrow $6000 from Citibank to help pay for a car. His loan was for 3 years at a simple interest rate of 8%. How much interest will Thang pay?

90. Simple Interest Loan Holly Broesamle lent her brother $4000 for a period of 2 years. At the end of the 2 years, her brother repaid the $4000 plus $640 interest. What simple interest rate did her brother pay?

91. Savings Account Mary Seitz invested a certain amount of money in a savings account paying 3% simple interest per year. When she withdrew her money at the end of 3 years, she received $450 in interest. How much money did Mary place in the savings account?

92. Savings Account Peter Ostroushko put $6000 in a savings account earning $3\frac{1}{2}$% simple interest per year. When he withdrew his money, he received $840 in interest. How long had he left his money in the account?

In Exercises 93–96, use the distance formula.

93. Average Speed On her way from Omaha, Nebraska, to Kansas City, Kansas, Peg Hovde traveled 150 miles in 3 hours. What was her average speed?

94. Walk Lisa Feintech went for a walk where she walked at an average speed of 3.4 miles per hour for 2 hours. How far did she walk?

95. Fastest Car The fastest speed recorded on land was about 763.2 miles per hour by a jet powered car called ThrustSSC. If, during the speed trial, the car traveled for 0.01 hour, how far had the car traveled?

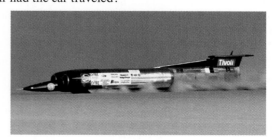

The ThrustSSC

96. Fastest Plane The fastest aircraft is the Lockheed SR-71 Blackbird. If, during the trial run, the plane covered a distance of 660 miles in 0.3 hours, determine the plane's average speed.

Use the formulas given in **Tables 2.1, 2.2,** *and* **2.3** *to work Exercises 97–110.*

97. **DVD Player** A portable DVD player has a screen with a length of 8 inches and a width (or height) of 6 inches. Determine the area of the screen.

98. **Television** A plasma television has a screen with a length of 34.9 inches and a width (or height) of 19.6 inches. Determine the perimeter of the screen.

99. **Yield Sign** A yield traffic sign is triangular with a base of 36 inches and a height of 31 inches. Find the area of the sign.

100. **Fencing** Milt McGowen has a rectangular lot that measures 100 feet by 60 feet. If Milt wants to fence in his lot, how much fencing will he need?

101. **Swimming Pool** A circular above-ground swimming pool has a diameter of 24 feet. Determine the circumference of the pool.

102. **Living Room Table** A round living room table top has a diameter of 3 feet. Find the area of the table top.

103. **Kite** Below we show a kite. Determine the area of the kite.

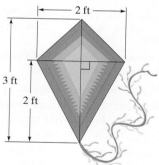

104. **Suitcase** A suitcase measures 26 inches long by 19.5 inches wide and 11 inches deep. Determine the volume of the suitcase.

105. **Trapezoidal Sign** Canter Martin made a sign to display at a baseball game. The sign was in the shape of a trapezoid. Its bases are 4 feet and 3 feet, and its height is 2 feet. Find the area of the sign.

106. **Jacuzzi** The inside of a circular jacuzzi is 8 feet in diameter. If the water inside the jacuzzi is 3 feet deep, determine, in cubic feet, the volume of water in the jacuzzi.

107. **Banyan Tree** The largest banyan tree in the continental United States is at the Edison House in Fort Myers, Florida. The circumference of the aerial roots of the tree is 390 feet. Find the *diameter* of the aerial roots to the nearest tenth of a foot.

108. **Amphitheater** The seats in an amphitheater are inside a trapezoidal area as shown in the figure.

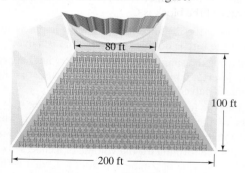

The bases of the trapezoidal area are 80 feet and 200 feet, and the height is 100 feet. Find the area of the floor occupied by seats.

109. **Basketball** Find the volume of a basketball if its diameter is 9 inches.

110. **Oil Drum** Roberto Sanchez has an empty oil drum that he uses for storage. The oil drum is 4 feet high and has a diameter of 24 inches. Find the volume of the drum in cubic feet.

111. **Body Mass Index** A person's body mass index (BMI) is found by multiplying a person's weight, w, in pounds by 703, then dividing this product by the square of the person's height, h, in inches.

  a) Write a formula to find the BMI.

  b) Brandy Belmont is 5 feet 3 inches tall and weighs 135 pounds. Find her BMI.

112. **Body Mass Index** Refer to Exercise 111. Mario Guzza's weight is 162 pounds, and he is 5 feet 7 inches tall. Find his BMI.

Challenge Problems

113. Cereal Box A cereal box is to be made by folding the cardboard along the dashed lines as shown in the figure on the right.

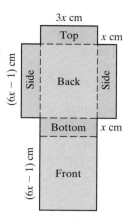

a) Using the formula

$$\text{volume} = \text{length} \cdot \text{width} \cdot \text{height}$$

write an equation for the volume of the box.

b) Find the volume of the box when $x = 7$ cm.

c) Write an equation for the surface area of the box.

d) Find the surface area when $x = 7$ cm.

Group Activity

114. Square Face on Cube Consider the following photo. The front of the figure is a square with a smaller black square painted on the center of the larger square. Suppose the length of one side of the larger square is A, and length of one side of the smaller (the black square) is B. Also the thickness of the block is C.

a) Group member one: Determine an expression for the surface area of the black square.

b) Group member two: Determine an expression for the surface area of the larger square (which includes the smaller square).

c) Group member three: Determine the surface area of the larger square minus the black square (the purple area shown).

d) As a group, write an expression for the volume of the entire solid block.

e) As a group, determine the volume of the entire solid block if its length is 1.5 feet and its width is 0.8 feet.

Cumulative Review Exercises

[1.7] **115.** Evaluate $-\dfrac{4}{15} + \dfrac{2}{5}$.

116. Evaluate $-6 + 7 - 4 - 3$.

[1.9] **117.** Evaluate. $[4(12 \div 2^2 - 3)^2]^2$.

[2.4] **118.** Solve the equation $\dfrac{r}{2} + 2r = 20$.

2.7 Ratios and Proportions

1 Understand ratios.

2 Solve proportions using cross-multiplication.

3 Solve applications.

4 Use proportions to change units.

5 Use proportions to solve problems involving similar figures.

The Incredibles

1 Understand Ratios

A **ratio** is a quotient of two quantities. Ratios provide a way to compare two numbers or quantities. The ratio of the number a to the number b may be written

$$a \text{ to } b, \quad a:b, \quad \text{or} \quad \frac{a}{b}$$

where a and b are called the **terms of the ratio**. Notice that the symbol : can be used to indicate a ratio.

EXAMPLE 1 ▶ **Favorite Movies** Several children in sixth grade were asked to name their favorite movie of 2004. The results are indicated on the graph in **Figure 2.5**.

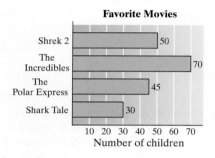

Favorite Movies

Shrek 2 — 50
The Incredibles — 70
The Polar Express — 45
Shark Tale — 30

10 20 30 40 50 60 70
Number of children

FIGURE 2.5

a) Find the ratio of the number of children who selected *The Incredibles* to those who selected *The Polar Express*. Then write the ratio in lowest terms.

b) Find the ratio of the number of children who selected *Shrek 2* to the total number in the survey.

Solution We will use our five-step problem-solving procedure.

a) Understand and Translate The ratio we are seeking is

Number who selected *The Incredibles* : Number who selected *The Polar Express*

Carry Out We substitute the appropriate values into the ratio. This gives

$$70:45$$

To write the ratio in *lowest terms*, we simplify by dividing each number in the ratio by 5, the greatest number that divides both terms in the ratio. This gives

$$14:9$$

Check and Answer Our division is correct. The ratio is $14:9$.

b) We use the same procedure as in part **a)**. Fifty children selected *Shrek 2*. There were $50 + 70 + 45 + 30$ or 195 children surveyed. Thus, the ratio is $50:195$, which simplifies to

$$10:39$$

▶ **Now Try Exercise 33**

The answer in Example 1, part **a)** could have also been written $\frac{14}{9}$ or 14 to 9.
The answer in part **b)** could have also been written $\frac{10}{39}$ or 10 to 39.

EXAMPLE 2 ▸ **Cholesterol Level** There are two types of cholesterol: low-density lipoprotein, (LDL—considered the harmful type of cholesterol) and high-density lipoprotein (HDL—considered the healthful type of cholesterol). Some doctors recommend that the ratio of low- to high-density cholesterol be less than or equal to 4:1. Mr. Suarez's cholesterol test showed that his low-density cholesterol measured 167 milligrams per deciliter, and his high-density cholesterol measured 40 milligrams per deciliter. Is Mr. Suarez's ratio of low- to high-density cholesterol less than or equal to the recommended 4:1 ratio?

Solution   Understand   We need to determine if Mr. Suarez's low- to high-density cholesterol is less than or equal to 4:1.

Translate   Mr. Suarez's low- to high-density cholesterol is 167:40. To make the second term equal to 1, we divide both terms in the ratio by the second term, 40.

Carry Out
$$\frac{167}{40} : \frac{40}{40}$$
or 4.175:1

Check and Answer   Our division is correct. Therefore, Mr. Suarez's ratio is not less than or equal to the desired 4:1 ratio.

▸ **Now Try Exercise 87**

EXAMPLE 3 ▸ **Gas–Oil Mixture** Some power equipment, such as chainsaws and blowers, use a gas–oil mixture to run the engine. The instructions on a particular chainsaw indicate that 5 gallons of gasoline should be mixed with 40 ounces of special oil to obtain the proper gas–oil mixture. Find the ratio of gasoline to oil in the proper mixture.

Solution   Understand   To express these quantities in a ratio, both quantities must be in the same units. We can either convert 5 gallons to ounces or 40 ounces to gallons.

Translate   Let's change 5 gallons to ounces. Since there are 128 ounces in 1 gallon, 5 gallons of gas equals 5(128) or 640 ounces. The ratio we are seeking is

ounces of gasoline : ounces of oil

Carry Out   640:40

or 16:1   *Divide both terms by 40 to simplify.*

Check and Answer   Our simplification is correct. The correct ratio of gas to oil for this chainsaw is 16:1.

▸ **Now Try Exercise 23**

EXAMPLE 4 ▸ **Gear Ratio** The *gear ratio* of two gears is defined as

$$\text{gear ratio} = \frac{\text{number of teeth on the driving gear}}{\text{number of teeth on the driven gear}}$$

Find the gear ratio of the gears shown in **Figure 2.6**.

Solution   Understand and Translate   To find the gear ratio we need to substitute the appropriate values.

Carry Out $\text{gear ratio} = \dfrac{\text{number of teeth on driving gear}}{\text{number of teeth on driven gear}} = \dfrac{60}{8} = \dfrac{15}{2}$

Thus, the gear ratio is 15:2. Gear ratios are generally given as some quantity to 1. If we divide both terms of the ratio by the second term, we will obtain a ratio of some number to 1. Dividing both 15 and 2 by 2 gives a gear ratio of 7.5:1.

Check and Answer   The gear ratio is 7.5:1. This means that as the driving gear goes around once the driven gear goes around 7.5 times. (A typical first gear ratio on a passenger car may be 3.545:1.)

▸ **Now Try Exercise 27**

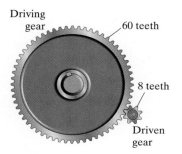

Driving gear   60 teeth
8 teeth
Driven gear

FIGURE 2.6

2 Solve Proportions Using Cross-Multiplication

A **proportion** is a special type of equation. It is a statement of equality between two ratios. One way of denoting a proportion is $a:b = c:d$, which is read "a is to b as c is to d." In this text we write proportions as

$$\frac{a}{b} = \frac{c}{d}$$

The a and d are referred to as the **extremes**, and the b and c are referred to as the **means** of the proportion. In Sections 2.4 and 2.5, we solved equations containing fractions by multiplying both sides of the equation by the LCD to eliminate fractions. For example, for the proportion

$$\frac{x}{3} = \frac{35}{15}$$

$$15\left(\frac{x}{3}\right) = 15\left(\frac{35}{15}\right) \quad \textit{Multiply both sides by the LCD, 15.}$$

$$5x = 35$$

$$x = 7$$

Another method that can be used to solve proportions is **cross-multiplication**. This process of cross-multiplication gives the same results as multiplying both sides of the equation by the LCD. However, many students prefer to use cross-multiplication because they do not have to determine the LCD of the fractions, and then multiply both sides of the equation by the LCD.

Cross-Multiplication

If $\frac{a}{b} = \frac{c}{d}$, then $ad = bc$.

Note that *the product of the extremes is equal to the product of the means.*

If any three of the four quantities of a proportion are known, the fourth quantity can easily be found.

EXAMPLE 5 ▶ Solve $\frac{x}{3} = \frac{35}{15}$ for x by cross-multiplying.

Solution

$$\frac{x}{3} = \frac{35}{15}$$

$$x \cdot 15 = 3 \cdot 35$$

$$15x = 105$$

$$x = \frac{105}{15} = 7$$

Check:
$$\frac{x}{3} = \frac{35}{15}$$
$$\frac{7}{3} \overset{?}{=} \frac{35}{15}$$
$$\frac{7}{3} = \frac{7}{3} \quad \textit{True}$$

▶ **Now Try Exercise 37**

Before we introduced cross-multiplication, we solved the proportion $\frac{x}{3} = \frac{35}{15}$ by multiplying both sides of the equation by 15. In Example 5, we solved the same proportion using cross-multiplication. Notice we obtained the same solution, 7, in each case. When you solve an equation using cross-multiplication, you are in effect multiplying both sides of the equation by the product of the two denominators, and then dividing out the common factors. However, this process is not shown.

EXAMPLE 6 ▸ Solve $\dfrac{-8}{3} = \dfrac{64}{x}$ for x by cross-multiplying.

Solution

$$\frac{-8}{3} = \frac{64}{x}$$

$$-8 \cdot x = 3 \cdot 64$$

$$-8x = 192$$

$$\frac{-8x}{-8} = \frac{192}{-8}$$

$$x = -24$$

Check:

$$\frac{-8}{3} = \frac{64}{x}$$

$$\frac{-8}{3} \stackrel{?}{=} \frac{64}{-24}$$

$$\frac{-8}{3} \stackrel{?}{=} \frac{8}{-3}$$

$$\frac{-8}{3} = \frac{-8}{3} \quad \textit{True}$$

▸ **Now Try Exercise 41**

3 Solve Applications

Often, practical problems can be solved using proportions. To solve such problems, use the five-step problem-solving procedure we have been using throughout the book. Below we give that procedure with more specific directions for translating problems into proportions.

To Solve Problems Using Proportions

1. Understand the problem.
2. Translate the problem into mathematical language.
   a) First, represent the unknown quantity by a variable (a letter).
   b) Second, set up the proportion by listing the given ratio on the left side of the equal sign, and the unknown and the other given quantity on the right side of the equal sign. When setting up the right side of the proportion, the same respective quantities should occupy the same respective positions on the left and the right. For example, an acceptable proportion might be

$$\textit{Given ratio} \left\{ \frac{\text{miles}}{\text{hour}} = \frac{\text{miles}}{\text{hour}} \right.$$

3. Carry out the mathematical calculations necessary to solve the problem.
   a. Once the proportion is correctly written, drop the units and cross-multiply.
   b. Solve the resulting equation.
4. Check the answer obtained in step 3.
5. Make sure you have answered the question.

   Note that the two ratios * **used in a proportion must have the same units.** For example, if one ratio is given in miles/hour and the second ratio is given in feet/hour, one of the ratios must be changed before setting up the proportion.

EXAMPLE 7 ▸ **Painting** A gallon of paint will cover an area of 575 square feet.

a) How many gallons of paint are needed to cover a house with a surface area of 6525 square feet?

b) If a gallon of paint costs \$24.99, what will it cost (before tax) to paint the house?

Solution

a) Understand   The given ratio is 1 gallon per 575 square feet. The unknown quantity is the number of gallons necessary to cover 6525 square feet.

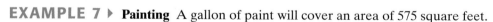

* Strictly speaking, a quotient of two quantities with different units, such as $\dfrac{6\ miles}{1\ hour}$, is called a *rate*. However, few books make the distinction between ratios and rates when discussing proportions.

Translate   Let x = number of gallons.

$$\text{Given ratio} \begin{cases} \dfrac{1 \text{ gallon}}{575 \text{ square feet}} = \dfrac{x \text{ gallons}}{6525 \text{ square feet}} \end{cases}$$

← Unknown

← Given quantity

Note how the amount, in gallons, and the area, in square feet, are given in the same relative positions.

Carry Out

$$\frac{1}{575} = \frac{x}{6525}$$

$$1(6525) = 575x \qquad \textit{Cross-multiply.}$$

$$6525 = 575x \qquad \textit{Solve.}$$

$$\frac{6525}{575} = x$$

$$11.3 \approx x$$

Check   Using a calculator, we determine that both ratios in the proportion, $\dfrac{1}{575}$ and $\dfrac{11.3}{6525}$, have approximately the same value of 0.00173. Thus, the answer of about 11.3 gallons checks.

Answer   The amount of paint needed to cover an area of 6525 square feet is about 11.3 gallons.

b) Assuming the painter only buys full gallons of paint, he will need to buy 12 gallons in order to paint the house. Since each gallon costs \$24.99, the cost (before tax) to paint the house is found by multiplication.

$$12 \times 24.99 = \$299.88$$

The cost (before tax) to paint the house is \$299.88.

▶ **Now Try Exercise 61**

EXAMPLE 8 ▶ **Charity Luncheon**   Each year in Tampa, Florida, the New York Yankees host a charity luncheon, with the proceeds going to support the Tampa Boys and Girls Clubs. At the luncheon, the guests meet and get autographs from members of the team. If a particular player signs, on the average, 33 autographs in 4 minutes, how much time must be allowed for him to sign 350 autographs?

Solution   The unknown quantity is the time needed for the player to sign 350 autographs. We are given that, on the average, he signs 33 autographs in 4 minutes. We will use this given ratio in setting up our proportion.

Translate   We will let x represent the time to sign 350 autographs.

$$\text{Given ratio} \begin{cases} \dfrac{33 \text{ autographs}}{4 \text{ minutes}} = \dfrac{350 \text{ autographs}}{x \text{ minutes}} \end{cases}$$

Carry Out

$$\frac{33}{4} = \frac{350}{x}$$

$$33x = 4(350)$$

$$33x = 1400$$

$$x = \frac{1400}{33} \approx 42.4$$

Check and Answer   Using a calculator, we can determine that both ratios in the proportion, $\dfrac{33}{4}$ and $\dfrac{350}{42.4}$, have approximately the same value of 8.25. Thus, about 42.4 minutes would be needed for the player to sign the 350 autographs.

▶ **Now Try Exercise 55**

EXAMPLE 9 ▶ **Drug Dosage** A doctor asks a nurse to give a patient 250 milligrams of the drug simethicone. The drug is available only in a solution whose concentration is 40 milligrams of simethicone per 0.6 milliliter of solution. How many milliliters of solution should the nurse give the patient?

Solution Understand and Translate We can set up the proportion using the medication on hand as the given ratio and the number of milliliters needed to be given as the unknown.

$$\text{Given ratio (medication on hand)} \begin{cases} \dfrac{40 \text{ milligrams}}{0.6 \text{ milliliter}} = \dfrac{250 \text{ milligrams}}{x \text{ milliliters}} \end{cases} \begin{matrix} \longleftarrow \text{ Desired medication} \\ \longleftarrow \text{ Unknown} \end{matrix}$$

Carry Out

$$\frac{40}{0.6} = \frac{250}{x}$$

$$40x = 0.6(250) \qquad \textit{Cross-multiply.}$$

$$40x = 150 \qquad \textit{Solve.}$$

$$x = \frac{150}{40} = 3.75$$

Check and Answer The nurse should administer 3.75 milliliters of the simethicone solution.

▶ **Now Try Exercise 69**

Helpful Hint

When you are setting up a proportion, it does not matter which unit in the given ratio is in the numerator and which is in the denominator as long as the units in the other ratio are *in the same relative position*. For example,

$$\frac{60 \text{ miles}}{1.5 \text{ hours}} = \frac{x \text{ miles}}{4.2 \text{ hours}} \quad \text{and} \quad \frac{1.5 \text{ hours}}{60 \text{ miles}} = \frac{4.2 \text{ hours}}{x \text{ miles}}$$

will both give the same answer of 168 (try it and see). When setting up the proportion, set it up so that it makes the most sense to you. Notice that when setting up a proportion containing different units, the same units should not be multiplied by themselves during cross multiplication.

$$\begin{matrix} \text{Correct} \\ \dfrac{\text{miles}}{\text{hour}} = \dfrac{\text{miles}}{\text{hour}} \end{matrix} \qquad \begin{matrix} \text{Incorrect} \\ \dfrac{\text{miles}}{\text{hour}} \diagdown \diagup \dfrac{\text{hour}}{\text{miles}} \end{matrix}$$

4 Use Proportions to Change Units

Proportions can also be used to convert from one quantity to another. For example, you can use a proportion to convert a measurement in feet to a measurement in meters, or to convert from pounds to kilograms. The following examples illustrate converting units.

EXAMPLE 10 ▶ **Kilometers to Miles** There are approximately 1.6 kilometers in 1 mile. What is the distance, in miles, of 78 kilometers?

Solution Understand and Translate We know that 1 mile ≈1.6 kilometers. We use this known fact in one ratio of our proportion. In the second ratio, we set the quantities with the same units in the same respective positions. The unknown quantity is the number of miles, which we will call x.

$$\text{Known ratio} \begin{cases} \dfrac{1 \text{ mile}}{1.6 \text{ kilometers}} = \dfrac{x \text{ miles}}{78 \text{ kilometers}} \end{cases}$$

Note that both numerators contain the same units, and both denominators contain the same units.

Carry Out   Now drop the units and solve for x by cross-multiplying.

$$\frac{1}{1.6} = \frac{x}{78}$$

$$1(78) = 1.6x \qquad \textit{Cross-multiply.}$$

$$78 = 1.6x \qquad \textit{Solve.}$$

$$\frac{78}{1.6} = \frac{1.6x}{1.6}$$

$$48.75 = x$$

Check and Answer   Thus, 78 kilometers equals about 48.75 miles.

▶ **Now Try Exercise 75**

EXAMPLE 11 ▶ **Exchanging Currency**   When people travel to a foreign country they often need to exchange currency. Donna Boccio visited Cancun, Mexico. She stopped by a local bank and was told that $1 U.S. could be exchanged for 10.96 pesos.

a) How many pesos would she get if she exchanged $150 U.S.?

b) Later that same day, Donna went to the city market where she purchased a ceramic figurine. The price she negotiated for the figurine was 245 pesos. Using the exchange rate given, determine the cost of the figurine in U.S. dollars.

Solution

a) Understand   We are told that $1 U.S. can be exchanged for 10.96 Mexican pesos. We use this known fact for one ratio in our proportion. In the second ratio, we set the quantities with same units in the same respective positions.

Translate   The unknown quantity is the number of pesos, which we shall call x.

$$\text{Given ratio} \left\{ \frac{\$1 \text{ U.S.}}{10.96 \text{ pesos}} = \frac{\$150 \text{ U.S.}}{x \text{ pesos}} \right.$$

Note that both numerators contain U.S. dollars and both denominators contain pesos.

Carry Out

$$\frac{1}{10.96} = \frac{150}{x}$$

$$1x = 10.96(150)$$

$$x = 1644$$

Check and Answer   Thus, $150 U.S. could be exchanged for 1644 Mexican pesos.

b) Understand and Translate   We use the same given ratio that we used in part **a)**. Now we must find the equivalent in U.S. dollars of 245 Mexican pesos. Let's call the equivalent U.S. dollars x.

$$\text{Given ratio} \left\{ \frac{\$1 \text{ U.S.}}{10.96 \text{ pesos}} = \frac{\$x \text{ U.S.}}{245 \text{ pesos}} \right.$$

Carry Out

$$\frac{1}{10.96} = \frac{x}{245}$$

$$1(245) = 10.96x$$

$$245 = 10.96x$$

$$22.35 \approx x$$

Check and Answer   The cost of the figurine in U.S. dollars is $22.35.

▶ **Now Try Exercise 85**

Helpful Hint

Some of the problems we have just worked using proportions could have been done without using proportions. However, when working problems of this type, students often have difficulty in deciding whether to multiply or divide to obtain the correct answer. By setting up a proportion, you may be better able to understand the problem and have more success in obtaining the correct answer.

5 Use Proportions to Solve Problems Involving Similar Figures

Proportions can also be used to solve problems in geometry and trigonometry. The following examples illustrate how proportions may be used to solve problems involving **similar figures**. Two figures are said to be similar when their corresponding angles are equal and their corresponding sides are in proportion. Two similar figures will have the same shape.

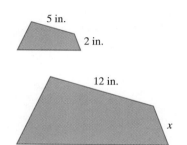

EXAMPLE 12 ▶ The figures to the left are similar. Find the length of the side indicated by the x.

Solution We set up a proportion of corresponding sides to find the length of side x.

Lengths from smaller figure   Lengths from larger figure

5 inches and 12 inches are corresponding sides of similar figures.   ⟶

2 inches and x are corresponding sides of similar figures.   ⟶

$$\frac{5}{2} = \frac{12}{x}$$

$$5x = 24$$

$$x = \frac{24}{5} = 4.8$$

Thus, the side indicated by x is 4.8 inches in length.

▶ **Now Try Exercise 51**

Note in Example 12 that the proportion could have also been set up as

$$\frac{5}{12} = \frac{2}{x}$$

because one pair of corresponding sides is in the numerators and another pair is in the denominators.

EXAMPLE 13 ▶ Triangles ABC and $AB'C'$ are similar triangles. Find the length of side AB'.

Solution We set up a proportion of corresponding sides to find the length of side AB'. We will let x represent the length of side AB'. One proportion we can use is

$$\frac{\text{length of } AB}{\text{length of } BC} = \frac{\text{length of } AB'}{\text{length of } B'C'}$$

Now we insert the proper values and solve for the variable, x.

$$\frac{15}{9} = \frac{x}{7.2}$$

$$(15)(7.2) = 9x$$

$$108 = 9x$$

$$12 = x$$

Thus, the length of side AB' is 12 inches.

▶ **Now Try Exercise 53**

EXERCISE SET 2.7

Math XL
MathXL®

MyMathLab
MyMathLab

Concept/Writing Exercises

1. What is a ratio?
2. In the ratio $a:b$, what are the a and b called?
3. List three ways to write the ratio of c to d.
4. What is a proportion?
5. As you have learned, proportions can be used to solve a wide variety of problems. What information is needed for a problem to be set up and solved using a proportion?
6. What are similar figures?
7. Must similar figures have the same shape? Explain.
8. Must similar figures be the same size? Explain.

In Exercises 9–12, is the proportion set up correctly? Explain.

9. $\dfrac{\text{gal}}{\text{min}} = \dfrac{\text{gal}}{\text{min}}$

10. $\dfrac{\text{mi}}{\text{hr}} = \dfrac{\text{mi}}{\text{hr}}$

11. $\dfrac{\text{ft}}{\text{sec}} = \dfrac{\text{sec}}{\text{ft}}$

12. $\dfrac{\text{tax}}{\text{cost}} = \dfrac{\text{cost}}{\text{tax}}$

Practice the Skills

The results of a mathematics examination are 6 A's, 4 B's, 9 C's, 3 D's, and 2 F's. Write the following ratios in lowest terms.

13. A's to C's
14. F's to total grades
15. D's to A's
16. Grades better than C to total grades
17. Total grades to D's
18. Grades better than C to grades less than C

Determine the following ratios. Write each ratio in lowest terms.

19. 7 gallons to 4 gallons
20. 50 dollars to 60 dollars
21. 5 ounces to 15 ounces
22. 18 liters to 24 liters
23. 3 hours to 30 minutes
24. 6 feet to 4 yards
25. 7 dimes to 12 nickels
26. 26 ounces to 4 pounds

In Exercises 27 and 28, find the gear ratio. Write the ratio as some quantity to 1. (See Example 4.)

27. Driving gear, 40 teeth; driven gear, 5 teeth
28. Driving gear, 30 teeth; driven gear, 8 teeth

In Exercises 29–32, **a)** *Determine the indicated ratio, and* **b)** *write the ratio as some quantity to 1.*

29. **American Consumers** According to a report issued by the U.S. Department of Agriculture's Economic Research Service, each year the average American consumer drinks approximately 50 gallons of soft drinks compared to 26 gallons of coffee, 23 gallons of milk, and less than 10 gallons of fruit juices. What is the ratio of the number of gallons of soft drinks consumed to the number of gallons of milk consumed?

30. **Mail Letter** In January 2006, the cost to mail a one-ounce letter was 39 cents and the cost to mail a two-ounce letter was 63 cents. What is the ratio of the cost to mail a one-ounce letter to the cost to mail a two-ounce letter?

31. **Minimum Wage** The United States minimum wage in 1985 was \$3.35 per hour, and the United States minimum wage in 2005 was \$5.15 per hour. What is the ratio of the U.S. minimum wage in 2005 to the U.S. minimum wage in 1985?

32. **Population** The United States population in 1990 was about 249 million, and the population in 2006 was about 299 million. What is the ratio of the U.S. population in 2006 to the U.S. population in 1990?

Exercises 33–36 show graphs. For each exercise, find the indicated ratio.

33. **Traveling the Toll Roads**
    a) Determine the ratio of the toll rate on the Delaware Turnpike to the toll rate on the Garden State Parkway.
    b) Determine the ratio of the toll rate on the Massachusetts Turnpike (Boston ext.) to the toll rate on the New York State Thruway (Current).

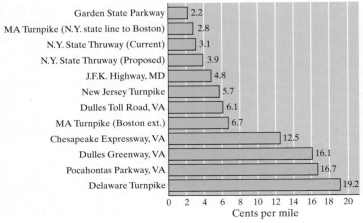

How Toll Rates Compare

	Cents per mile
Garden State Parkway	2.2
MA Turnpike (N.Y. state line to Boston)	2.8
N.Y. State Thruway (Current)	3.1
N.Y. State Thruway (Proposed)	3.9
J.F.K. Highway, MD	4.8
New Jersey Turnpike	5.7
Dulles Toll Road, VA	6.1
MA Turnpike (Boston ext.)	6.7
Chesapeake Expressway, VA	12.5
Dulles Greenway, VA	16.1
Pocahontas Parkway, VA	16.7
Delaware Turnpike	19.2

Source: Rochester Democrat & Chronicle, April, 2005

34. Number of Passports Processed

a) Estimate the ratio of passports processed in the United States in 1996 to those processed in the United States in 2004.

b) Estimate the ratio of passports processed in the United States in 2002 to those processed in the United States in 1997.

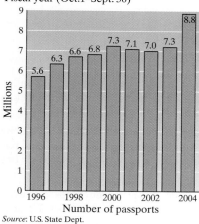

Passports Processed in the United States
Fiscal year (Oct.1–Sept. 30)

Number of passports

Source: U.S. State Dept.

35. Favorite Doughnut

See figure at the top of the next column.

a) Determine the ratio of people whose favorite doughnut is glazed to people whose favorite doughnut is filled.

b) Determine the ratio of people whose favorite doughnut is frosted to people whose favorite doughnut is plain.

Favorite Doughnut Flavors

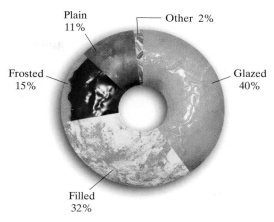

Plain 11%   Other 2%
Frosted 15%   Glazed 40%
Filled 32%

Source: The Heller Research Group

36. Commuting to Work

a) Determine the ratio of workers who drove alone to work to those who carpooled to work.

b) Determine the ratio of workers who walked to work to those who worked at home.

Commuting to Work, 2003

Drove alone 78%
Other means 1%
Worked at home 4%
Walked 2%
Public transportation (including taxis) 5%
Carpooled 10%

Source: www.factfinder.census.gov

Solve each proportion for the variable by cross-multiplying.

37. $\dfrac{x}{3} = \dfrac{20}{5}$

38. $\dfrac{x}{8} = \dfrac{24}{48}$

39. $\dfrac{5}{3} = \dfrac{75}{a}$

40. $\dfrac{x}{3} = \dfrac{90}{30}$

41. $\dfrac{-7}{3} = \dfrac{21}{p}$

42. $\dfrac{-12}{13} = \dfrac{36}{x}$

43. $\dfrac{15}{45} = \dfrac{x}{-6}$

44. $\dfrac{y}{6} = \dfrac{7}{42}$

45. $\dfrac{3}{z} = \dfrac{-1.5}{27}$

46. $\dfrac{3}{12} = \dfrac{-1.4}{z}$

47. $\dfrac{9}{12} = \dfrac{x}{8}$

48. $\dfrac{2}{20} = \dfrac{x}{200}$

The following figures are similar. For each pair, find the length of the side indicated by x.

49.

3 in.   8 in.   12 in.   x

50.

2 ft   1.8 ft   0.8 ft   x

51.

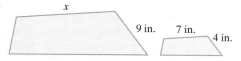

x   9 in.   7 in.   4 in.

52.

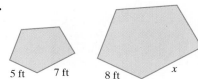

5 ft   7 ft   8 ft   x

53.

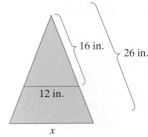

54.

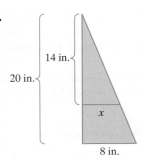

Problem Solving

In Exercises 55–74, write a proportion that can be used to solve the problem. Then solve the equation to obtain the answer.

55. Washing Clothes A bottle of liquid Tide contains 100 fluid ounces. If one wash load requires 4 ounces of the detergent, how many washes can be done with one bottle of Tide?

56. Laying Cable A telephone cable crew is laying cable at a rate of 42 feet an hour. How long will it take them to lay 252 feet of cable?

57. Truck Mileage A 2004 Chevy S-10 pickup truck with a 4.3-liter engine is rated to get 19 miles per gallon (highway driving). How far can it travel on 14.2 gallons of gas?

58. Purchasing Stock If 2 shares of stock can be purchased for $38.25, how many shares can be purchased for $344.25?

59. Model Train A model train set is in a ratio of 1 : 20. That is, one foot of the model represents 20 feet of the original train. If a caboose is 30 feet long, how long should the model be?

60. Property Tax The property tax in the city of Hendersonville, North Carolina, is $9.475 per $1000 of assessed value. If the Estever's house is assessed at $145,000, how much property tax will they owe?

 61. Insecticide Application The instructions on a bottle of liquid insecticide say "use 3 teaspoons of insecticide per gallon of water." If your sprayer has an 8-gallon capacity, how much insecticide should be used to fill the sprayer?

62. Spreading Fertilizer If a 40-pound bag of fertilizer covers 5000 square feet, how many pounds of fertilizer are needed to cover an area of 26,000 square feet?

63. Blue Heron The photograph shows a blue heron. If the blue heron, that measures 3.5 inches in the photo is actually 3.75 feet tall, approximately how long is its beak if it measures 0.4 inch in the photo?

64. Maps On a map, 0.5 inch represents 22 miles. What will be the length on a map that corresponds to a distance of 55 miles?

65. Onion Soup A recipe for 6 servings of French onion soup requires $1\frac{1}{2}$ cups of thinly sliced onions. If the recipe were to be made for 15 servings, how many cups of onions would be needed?

66. John Grisham Novel Karen Estes is currently reading a John Grisham novel. If she reads 72 pages in 1.3 hours, how long will it take her to read the entire 656-page novel?

67. Wall Street Bull Suppose the famous bull by the New York Stock Exchange (see photo below) is a replica of a real bull in a ratio of 2.95 to 1. That is, the metal bull is 2.95 times greater than the regular bull. If the length of the Wall Street bull is 28 feet long, approximately how long is the bull that served as its model?

68. Flood When they returned home from vacation, the Duncans had a foot of water in their basement. They contacted their fire department, which sent equipment to pump out the water. After the pump had been on for 30 minutes, 3 inches of water had been removed. How long, from the time they started pumping, will it take to remove all the water from the basement?

69. Drug Dosage A nurse must administer 220 micrograms of atropine sulfate. The drug is available in solution form. The concentration of the atropine sulfate solution is 400 micrograms per milliliter. How many milliliters should be given?

70. Dosage by Body Surface A doctor asks a nurse to administer 0.7 gram of meprobamate per square meter of body surface. The patient's body surface is 0.6 square meter. How much meprobamate should be given?

71. Reading a Novel Mary read 40 pages of a novel in 30 minutes. If she continues reading at the same rate, how long will it take her to read the entire 760-page book?

72. Swimming Laps Jason Abbott swims 3 laps in 2.3 minutes. Approximately how long will it take him to swim 30 laps if he continues to swim at the same rate?

73. Prader-Willi Syndrome It is estimated that each year in the United States about 1 in every 12,000 (1 : 12,000) people is born with a genetic disorder called Prader-Willi syndrome. If there were approximately 4,063,000 births in the United States in 2005, approximately how may children were born with Prader-Willi syndrome?

74. Scrapbooking Penelope Penna completed 4 scrapbook pages in 20.5 minutes. Approximately how long will it take her to complete 36 scrapbook pages if she continues to complete the scrapbook at the same rate?

In Exercises 75–86, use a proportion to make the conversion. Round your answers to two decimal places.

75. Convert 78 inches to feet.

76. Convert 22,704 feet to miles (5280 feet = 1 mile).

77. Convert 26.1 square feet to square yards (9 square feet = 1 square yard).

78. Convert 146.4 ounces to pounds.

79. Newborn One inch equals 2.54 centimeters. Find the length of a newborn, in inches, if it measures 50.8 centimeters.

80. Distance One mile equals approximately 1.6 kilometers. Find the distance, in kilometers, from San Diego, California, to San Francisco, California—a distance of 520 miles.

San Francisco, California

81. Home Run Record Barry Bonds, who plays for the San Francisco Giant's baseball team, holds the record for the most home runs, 73, in a 162 game season. In the first 50 games of a season, how many home runs would a player need to hit to be on schedule to break Bonds' record?

82. Topsoil A 40 pound bag of topsoil covers 12 square feet (one inch deep). How many pounds of the top soil are needed to cover 350 square feet (one inch deep)?

83. Interest on Savings Jim Chao invests a certain amount of money in a savings account. If he earned $110.52 in 180 days, how much interest would he earn in 500 days assuming the interest rate stays the same?

84. Gold If gold is selling for $408 per 480 grains (a troy ounce), what is the cost per grain?

85. Mexican Pesos Suppose that the exchange rate from U.S. dollars to Mexican pesos is $1 per 10.567 pesos. How many pesos would Elizabeth Averbeck receive if she exchanged $200 U.S.?

86. Currency Exchange When Mike Weatherbee visited the United States from Canada, he exchanged $13.50 Canadian for $10 U.S.. If he exchanges his remaining $600 Canadian for U.S. dollars, how much more in dollars will he receive?

The Peace Bridge Connecting the United States and Canada

87. Cholesterol Mrs. Ruff's low-density cholesterol level is 127 milligrams per deciliter (mg/dL). Her high-density cholesterol level is 60 mg/dL. Is Mrs. Ruff's ratio of low-to high-density cholesterol level less than or equal to the 4 : 1 recommended level? (See Example 2.)

88. Cholesterol

   a) Another ratio used by some doctors when measuring cholesterol level is the ratio of total cholesterol to high-density cholesterol.* Is this ratio increased or decreased if the total cholesterol remains the same but the high-density level is increased? Explain.

   b) Doctors recommend that the ratio of total cholesterol to high-density cholesterol be less than or equal to 4.5 : 1. If Mike's total cholesterol is 220 mg/dL and his high-density cholesterol is 50 mg/dL, is his ratio less than or equal to 4.5 : 1? Explain.

89. For the proportion $\frac{a}{b} = \frac{c}{d}$, if a increases while b and d stay the same, what must happen to c? Explain.

90. For the proportion $\frac{a}{b} = \frac{c}{d}$, if a and c remain the same while d decreases, what must happen to b? Explain.

---

*Total cholesterol includes both low- and high-density cholesterol, plus other types of cholesterol.

Challenge Problems

91. Wear on Tires A new Goodyear tire has a tread of about 0.34 inch. After 5000 miles the tread is about 0.31 inch. If the legal minimum amount of tread for a tire is 0.06 inch, how many more miles will the tires last? (Assume no problems with the car or tires and that the tires wear at an even rate.)

92. Apple Pie The recipe for the filling for an apple pie calls for

12 cups sliced apples $\frac{1}{4}$ teaspoon salt

$\frac{1}{2}$ cup flour $1\frac{1}{2}$ cups sugar

1 teaspoon nutmeg   2 tablespoons butter or margarine

1 teaspoon cinnamon

Determine the amount of each of the other ingredients that should be used if only 8 cups of apples are available.

93. Insulin Insulin comes in 10-cubic-centimeter (cc) vials labeled in the number of units of insulin per cubic centimeter. Thus, a vial labeled U40 means there are 40 units of insulin per cubic centimeter of fluid. If a patient needs 25 units of insulin, how many cubic centimeters of fluid should be drawn up into a syringe from the U40 vial?

Group Activity

Discuss and answer Exercises 94 and 95 as a group.

94. a) Each group member: Find the ratio of your height to your arm span (finger tips to finger tips) when your arms are extended horizontally outward. You will need help from your group in getting these measurements.

b) If a box were to be drawn about your body with your arms extended, would the box be a square or a rectangle? If a rectangle, would the longer length be your arm span or your height measurement? Explain.

c) Compare these results with other members of your group.

d) What one ratio would you use to report the height to arm span for your group as a whole? Explain.

95. A special ratio in mathematics is called the *golden ratio*. Do research in a history of mathematics book or on the Internet, and as a group write a paper that explains what the golden ratio is and why it is important.

Cumulative Review Exercises

[1.10] *Name each illustrated property.*

96. $x + 3 = 3 + x$

97. $3(xy) = (3x)y$

98. $2(x - 3) = 2x - 6$

[2.5] **99.** Solve $3(4x - 3) = 6(2x + 1) - 15$

[2.6] **100.** Solve $y = mx + b$ for m.

Chapter 2 Summary

IMPORTANT FACTS AND CONCEPTS	EXAMPLES
Section 2.1	
The **terms** of an expression are the parts that are added.	$2x^2 - 3xy + 5$ has 3 terms: $2x^2$, $-3xy$, and 5.
The numerical part of a term is called its **numerical coefficient**.	The numerical coefficient of $\frac{3x}{4}$ is $\frac{3}{4}$.
A **constant** is a term that is a number without a variable.	In $3x^2 + 2x - 7$, the -7 is a constant.
Like terms have the same variables with the same exponents.	$7x$ and x; $6y^2$ and $2y^2$; $3(x + 4)$ and $-8(x + 4)$
To Combine Like Terms 1. Determine which terms are like terms. 2. Add or subtract the coefficients of the like terms. 3. Multiply the number found in step 2 by the common variable(s).	$-3x^2 + 4y - 7x^2 + 6 - y - 9$ $= -3x^2 - 7x^2 + 4y - y + 6 - 9$ $= -10x^2 + 3y - 3$
Distributive Property For any real numbers a, b, and c, $a(b + c) = ab + ac$	$-5(3r - 6) = -15r + 30$

IMPORTANT FACTS AND CONCEPTS	EXAMPLES

Section 2.1 (continued)

To Simplify an Expression

1. Use the distributive property to remove any parentheses.
2. Combine like terms.

$$2(3c - 1) - 5(c + 4) - 6$$
$$= 6c - 2 - 5c - 20 - 6$$
$$= c - 28$$

When two or more expressions are multiplied, each expression is a **factor** of the product.

Since $7 \cdot 8 = 56$, the 7 and the 8 are factors of 56.

Section 2.2

A **linear equation** in one variable is an equation that can be written in the form

$$ax + b = c$$

where a, b, and c are real numbers and $a \neq 0$.

$$9x - 2 = 16$$

The **solution to an equation** is the number or numbers that when substituted for the variable or variables make the equation a true statement.

The solution to $2x + 3 = 9$ is 3.

The solution to an equation may be **checked** by substituting the value that is believed to be the solution for the variable in the original equation.

To check whether -2 is the solution to $-7x + 1 = 15$:

$$-7x + 1 = 15$$
$$-7(-2) + 1 \stackrel{?}{=} 15$$
$$14 + 1 \stackrel{?}{=} 15$$
$$15 = 15 \quad \textit{True}$$

Thus, -2 is the solution.

Two or more equations with the same solution are called **equivalent equations**.

$-4x = 12$, $2x - 3 = -9$, and $x = -3$ are equivalent equations

---

Addition Property of Equality

If $a = b$, then $a + c = b + c$ for any real numbers a, b, and c.

Solve the equation $x - 9 = -2$.

$$x - 9 = -2$$
$$x - 9 + 9 = -2 + 9$$
$$x = 7$$

Section 2.3

Two numbers are **reciprocals** of each other when their product is 1.

3 and $\frac{1}{3}$ are reciprocals since $3 \cdot \frac{1}{3} = 1$.

Multiplication Property of Equality

If $a = b$, then $a \cdot c = b \cdot c$ for any real numbers a, b, and c.

Solve the equation $\frac{3}{7}x = 6$.

$$\frac{3}{7}x = 6$$
$$\frac{7}{3} \cdot \frac{3}{7}x = \frac{7}{3} \cdot 6$$
$$x = 14$$

IMPORTANT FACTS AND CONCEPTS	EXAMPLES

Section 2.4

To Solve Linear Equations with a Variable on Only One Side of the Equal Sign

1. If the equation contains fractions, multiply **both** sides of the equation by the least common denominator (LCD).
2. Use the distributive property to remove parentheses.
3. Combine like terms on the same side of the equal sign.
4. Use the addition property to obtain an equation with the term containing the variable on one side of the equal sign and a constant on the other side.
5. Use the multiplication property to isolate the variable.
6. Check the solution in the original equation.

Solve the equation $3(x - 5) - 6x = -2$.

$$3(x - 5) - 6x = -2$$
$$3x - 15 - 6x = -2$$
$$-3x - 15 = -2$$
$$-3x - 15 + 15 = -2 + 15$$
$$-3x = 13$$
$$\frac{-3x}{-3} = \frac{13}{-3}$$
$$x = -\frac{13}{3}$$

A check will show that $-\dfrac{13}{3}$ is the solution.

Section 2.5

To Solve Linear Equations with the Variable on Both Sides of the Equal Sign

1. If the equation contains fractions, multiply **both** sides of the equation by the LCD.
2. Use the distributive property to remove parentheses.
3. Combine like terms on the same side of the equal sign.
4. Use the addition property to rewrite the equation with all terms containing the variable on one side of the equal sign and all terms not containing the variable on the other side of the equal sign.
5. Use the multiplication property to isolate the variable.
6. Check the solution in the original equation.

Solve the equation $9 - 3x - 2(x + 5) = 4x + 7 - x$.

$$9 - 3x - 2(x + 5) = 4x + 7 - x$$
$$9 - 3x - 2x - 10 = 4x + 7 - x$$
$$-5x - 1 = 3x + 7$$
$$-5x + 5x - 1 = 3x + 5x + 7$$
$$-1 = 8x + 7$$
$$-1 - 7 = 8x + 7 - 7$$
$$-8 = 8x$$
$$\frac{-8}{8} = \frac{8x}{8}$$
$$-1 = x$$

A check will show that -1 is the solution.

A **conditional equation** is an equation that has a single value for a solution.

An **identity** is an equation that is true for infinitely many values of the variable.

A **contradiction** is an equation that has no solution.

$3x - 2 = 8$ is a conditional equation since its solution is $\dfrac{10}{3}$.

$-4(x + 3) = -5x - 12 + x$ is an identity because the equation is true for any real number.

$-9x + 7 + 6x = -5x + 1 + 2x$ is a contradiction because the equation is never true and has no solution.

Section 2.6

Simple Interest Formula

$$\text{interest} = \text{principal} \cdot \text{rate} \cdot \text{time} \quad \text{or} \quad i = prt$$

Determine the interest earned on a $5000 investment at 3% simple interest for 2 years.

$$i = prt$$
$$i = 5000(0.03)(2)$$
$$i = \$300$$

Distance Formula

$$\text{distance} = \text{rate} \cdot \text{time} \quad \text{or} \quad d = r \cdot t$$

Timothy John completed a snowmobile race in 2.4 hours at an average speed of 75 miles per hour. Determine the distance of the race.

$$d = rt$$
$$d = (75)(2.4)$$
$$d = 180 \text{ miles}$$

IMPORTANT FACTS AND CONCEPTS	EXAMPLES

Section 2.6 (continued)

Area is the total surface within the figure's boundaries. Areas are measured in square units.

Perimeter is the sum of the lengths of the sides of a figure. Perimeters are measured in the same common unit as the sides.

Perimeter and area formulas can be found in **Table 2.1** on page 142.

Determine the perimeter and area of the following trapezoid.

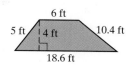

$$P = a + b + c + d$$
$$P = 5 + 6 + 10.4 + 18.6$$
$$P = 40 \text{ ft}$$

$$A = \frac{1}{2}h(b + d)$$

$$A = \frac{1}{2}(4)(6 + 18.6)$$

$$A = 49.2 \text{ ft}^2$$

The **circumference** of a circle is the length (or perimeter) of the curve that forms a circle.

Circle formulas can be found in **Table 2.2** on page 144.

Determine the area and circumference of the following circle.

$$A = \pi r^2$$
$$A = \pi(4)^2$$
$$A = \pi(16)$$
$$A \approx 50.27 \text{ cm}^2$$
$$C = 2\pi r$$
$$C = 2\pi(4)$$
$$C = 8\pi$$
$$C \approx 25.13 \text{ cm}$$

Volume may be considered the space occupied by a figure. Volume is measured is cubic units.

Volume formulas can be found in **Table 2.3** on page 145.

Determine the volume of the following figure.

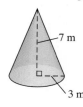

$$V = \frac{1}{3}\pi r^2 h$$

$$V = \frac{1}{3}\pi(3)^2(7)$$

$$V = \frac{1}{3}\pi(9)(7)$$

$$V = 21\pi$$

$$V \approx 65.97 \text{ m}^3$$

To **solve for a variable in a formula**, treat each of the quantities, except the one for which you are solving, as if they were constants. Then solve for the desired variable by isolating it on one side of the equation.

Solve $V = lwh$, for h.

$$V = lwh$$

$$\frac{V}{lw} = \frac{lwh}{lw}$$

$$\frac{V}{lw} = h$$

IMPORTANT FACTS AND CONCEPTS	EXAMPLES
Section 2.7	

IMPORTANT FACTS AND CONCEPTS	EXAMPLES
A **ratio** is a quotient of two quantities.	3 to 5, 3:5, $\dfrac{3}{5}$
A **proportion** is a statement of equality between two ratios. In the proportion $\dfrac{a}{b} = \dfrac{c}{d}$, the a and d are called the **extremes**, and the b and c are called the **means** of the proportion.	$\dfrac{7}{10} = \dfrac{21}{30}$ 7 and 30 are the extremes. 10 and 21 are the means.

Cross-Multiplication

$$\text{If } \frac{a}{b} = \frac{c}{d}, \text{ then } ad = bc.$$

Solve $\dfrac{-9}{2} = \dfrac{126}{x}$ for x by cross-multiplying.

$$\frac{-9}{2} = \frac{126}{x}$$
$$-9 \cdot x = 2 \cdot 126$$
$$-9x = 252$$
$$\frac{-9x}{-9} = \frac{252}{-9}$$
$$x = -28$$

To Solve Problems Using Proportions

1. Understand the problem.
2. Translate the problem into mathematical language.
3. Carry out the mathematical calculations necessary to solve the problem.
4. Check the answer obtained in step 3.
5. Make sure you have answered the question.

See page 155 for more details on proportions.

Melanie Jo can type 40 words per minute. If she types for 20.5 minutes, how many words will she type?

$$\frac{40 \text{ words}}{1 \text{ minute}} = \frac{x \text{ words}}{20.5 \text{ minutes}}$$
$$40(20.5) = 1(x)$$
$$820 = x$$

Melanie Jo will type 820 words.

Similar figures are figures whose corresponding angles are equal and whose corresponding sides are in proportion.

These two figures are similar.

12 in.

21.6 in.

5 in.

9 in.

Chapter 2 Review Exercises

[2.1] *Use the distributive property to simplify.*

1. $3(x + 4)$
2. $5(x - 2)$
3. $-2(x + 4)$
4. $-(x + 2)$
5. $-(m + 3)$
6. $-4(4 - x)$
7. $5(5 - p)$
8. $6(4x - 5)$
9. $-5(5x - 5)$
10. $4(-x + 3)$
11. $\dfrac{1}{2}(2x + 4)$
12. $-\dfrac{1}{3}(3 + 6y)$
13. $-(x + 2y - z)$
14. $-3(2a - 5b + 7)$

[2.1] *Simplify.*

15. $7x - 3x$

16. $5 - 3y + 3$

17. $1 + 3x + 2x$

18. $-2x - x + 3y$

19. $4m + 2n + 4m + 6n$

20. $9x + 3y + 2$

21. $6x - 2x + 3y + 6$

22. $x + 8x - 9x + 3$

23. $-4x^2 - 8x^2 + 3$

24. $-2(3a^2 - 4) + 6a^2 - 8$

25. $2x + 3(x + 4) - 5$

26. $-4 + 2(3 - 2b) + b$

27. $6 - (-7x + 6) - 7x$

28. $2(2x + 5) - 10 - 4$

29. $-6(4 - 3x) - 18 + 4x$

30. $4y - 3(x + y) + 6x^2$

31. $\frac{1}{4}d + 2 - \frac{3}{5}d + 5$

32. $3 - (x - y) + (x - y)$

33. $\frac{5}{6}x - \frac{1}{3}(2x - 6)$

34. $\frac{2}{3} - \frac{1}{4}n - \frac{1}{3}(n + 2)$

[2.2–2.5] *Solve.*

35. $-3x = -3$

36. $x + 6 = -7$

37. $x - 4 = 7$

38. $\frac{x}{3} = -9$

39. $5x + 1 = 12$

40. $14 = 3 + 2x$

41. $4c + 3 = -21$

42. $9 - 2a = 15$

43. $-x = -12$

44. $3(x - 2) = 6$

45. $-12 = 3(2x - 8)$

46. $4(6 + 2x) = 0$

47. $-6n + 2n + 6 = 0$

48. $-3 = 3w - (4w + 6)$

49. $6 - (2n + 3) - 4n = 6$

50. $4x + 6 - 7x + 9 = 18$

51. $5 + 3(x - 1) = 3(x + 1) - 1$

52. $8.4r - 6.3 = 6.3 + 2.1r$

53. $19.6 - 21.3t = 80.1 - 9.2t$

54. $0.35(c - 5) = 0.45(c + 4)$

55. $0.2(x + 6) = -0.3(2x - 1)$

56. $-2.3(x - 8) = 3.7(x + 4)$

57. $\frac{p}{3} + 2 = \frac{1}{4}$

58. $\frac{d}{6} + \frac{1}{7} = 2$

59. $\frac{3}{5}(r - 6) = 3r$

60. $\frac{2}{3}w = \frac{1}{7}(w - 2)$

61. $8x - 5 = -4x + 19$

62. $-(w + 2) = 2(3w - 6)$

63. $2x + 6 = 3x + 9 - 3$

64. $-5a + 3 = 2a + 10$

65. $5p - 2 = -2(-3p + 6)$

66. $3x - 12x = 24 - 9x$

67. $4(2x - 3) + 4 = 8x - 8$

68. $4 - c - 2(4 - 3c) = 3(c - 4)$

69. $2(x + 7) = 6x + 9 - 4x$

70. $-5(3 - 4x) = -6 + 20x - 9$

71. $4(x - 3) - (x + 5) = 0$

72. $-2(4 - x) = 6(x + 2) + 3x$

73. $\frac{x + 3}{2} = \frac{x}{2}$

74. $\frac{x}{6} = \frac{x - 4}{2}$

75. $\frac{1}{5}(3s + 4) = \frac{1}{3}(2s - 8)$

76. $\frac{2(2t - 4)}{5} = \frac{3t + 6}{4} - \frac{3}{2}$

77. $\frac{2}{5}(2 - x) = \frac{1}{6}(-2x + 2)$

78. $\frac{x}{4} + \frac{x}{6} = \frac{1}{2}(x + 3)$

[2.6] *Use the formula to find the value of the variable indicated.*

79. $y = mx + b$ (slope-intercept form of a line); find m when $y = 7$, $x = 2$, and $b = 1$.

80. $A = \frac{1}{2}h(b + d)$ (area of a trapezoid); find A when $h = 12$, $b = 3$, and $d = 5$.

Determine the area or volume of the figure.

81.

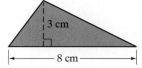

3 cm

8 cm

82.

2 in.

For Exercises 83–85, solve for the indicated variable.

83. $P = 2l + 2w$, for l

84. $y - y_1 = m(x - x_1)$, for m

85. $-x + 3y = 2$, for y

86. Spring Break Yong Wolfer traveled to Florida for spring break at an average speed of 61.7 miles per hour for 5 hours. How for did he travel?

87. Flower Garden Chrishawn Miller has a rectangular flower garden that measures 20 feet by 12 feet. What is the area of Chrishawn's flower garden?

88. Tuna Fish Find the volume of a tuna fish can if its diameter is 4 inches and its height is 2 inches.

[2.7] *Determine the following ratios. Write each ratio in lowest terms.*

89. 12 feet to 20 feet

90. 80 ounces to 12 pounds

91. 4 minutes : 40 seconds

Solve each proportion.

92. $\dfrac{x}{4} = \dfrac{8}{16}$

93. $\dfrac{5}{20} = \dfrac{x}{80}$

94. $\dfrac{3}{x} = \dfrac{15}{45}$

95. $\dfrac{20}{45} = \dfrac{15}{x}$

96. $\dfrac{6}{5} = \dfrac{-12}{x}$

97. $\dfrac{b}{6} = \dfrac{8}{-3}$

98. $\dfrac{-7}{9} = \dfrac{-12}{y}$

99. $\dfrac{x}{-15} = \dfrac{30}{-5}$

The following pairs of figures are similar. For each pair, find the length of the side indicated by x.

100.

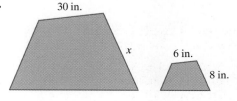

30 in.

x

6 in.

8 in.

101.

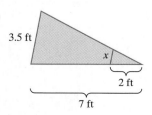

3.5 ft

x

2 ft

7 ft

Set up a proportion and solve each problem.

102. Boat Trip A boat travels 40 miles in 1.8 hours. If it travels at the same rate, how long will it take for it to travel 140 miles?

103. Washing Dishes If Adam Kloza can wash 12 dishes in 3.5 minutes, how many dishes can he wash in 21 minutes?

104. Copy Machine If a copy machine can copy 20 pages per minute, how many pages can be copied in 22 minutes?

105. Map Scale If the scale of a map is 1 inch to 60 miles, what distance on the map represents 380 miles?

106. Model Car Bryce Winston builds a model car to a scale of 1 inch to 1.5 feet. If the completed model is 10.5 inches, what is the size of the actual car?

107. Money Exchange Suppose that one U.S. dollar can be exchanged for 9.165 Mexican pesos, find the value of 1 peso in terms of U.S. dollars.

108. Ketchup If a machine can fill and cap 80 bottles of ketchup in 50 seconds, how many bottles of ketchup can it fill and cap in 2 minutes?

Chapter 2 Practice Test

To find out how well you understand the chapter material, take this practice test. The answers, and the section where the material was initially discussed, are given in the back of the book. Each problem is also fully worked out on the **Chapter Test Prep Video CD**. *Review any questions that you answered incorrectly.*

Use the distributive property to simplify.

1. $-3(4 - 2x)$

2. $-(x + 3y - 4)$

Simplify.

3. $5x - 8x + 4$

4. $4 + 2x - 3x + 6$

5. $-y - x - 4x - 6$

6. $a - 2b + 6a - 6b - 3$

7. $2x^2 + 3 + 2(3x - 2)$

Solve exercises 8–16.

8. $2.4x - 3.9 = 3.3$

9. $\dfrac{5}{6}(x - 2) = x - 3$

10. $2x - 3(-2x + 4) = -13 + x$

11. $3x - 4 - x = 2(x + 5)$

12. $-3(2x + 3) = -2(3x + 1) - 7$

13. $ax + by + c = 0$, for x

14. $-6x + 5y = -2$, for y

15. $\dfrac{1}{7}(2x - 5) = \dfrac{3}{8}x - \dfrac{5}{7}$

16. $\dfrac{9}{x} = \dfrac{3}{-15}$

17. What do we call an equation that has
  a) exactly one solution,
  b) no solution,
  c) all real numbers as its solution?

18. The following figures are similar. Find the length of side x.

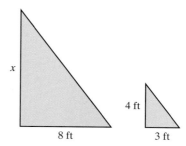

19. Simple Interest Loan Laura Hoye lent her sister $2000 for a period of 1 year. At the end of 1 year, her sister repaid the $2000 plus $80 interest. What simple interest rate did her sister pay?

20. Peanut Butter Pie A peanut butter pie has a diameter of 9 inches. Determine the circumference of the pie.

21. Travel Time While traveling, you notice that you traveled 25 miles in 35 minutes. If your speed does not change, how long will it take you to travel 125 miles?

Cumulative Review Test

Take the following test and check your answers with those given in the back of the book. Review any questions that you answered incorrectly. The section where the material was covered is indicated after the answer.

1. Multiply $\dfrac{52}{15} \cdot \dfrac{10}{13}$

2. Divide $\dfrac{5}{24} \div \dfrac{2}{9}$

3. Insert $<$, $>$, or $=$ in the shaded area to make a true statement: $|-2|$ ▨ 1.

4. Evaluate $-5 - (-4) + 12 - 8$.

5. Subtract -6 from -7.

6. Evaluate $20 - 6 \div 3 \cdot 2$.

7. Evaluate $3[6 - (4 - 3^2)] - 30$.

8. Evaluate $-2x^2 - 6x + 8$ when $x = -2$.

9. Name the illustrated property.
$$-5(x - 3y - 4z) = -5x + 15y + 20z$$

Simplify.

10. $8x + 2y + 4x - y$

11. $9 - \dfrac{2}{3}x + 16 + \dfrac{3}{4}x$

12. $3x^2 + 5 + 4(2x - 7)$

Solve.

13. $7x + 3 = -4$

14. $\dfrac{1}{4}x = -11$

15. $4(x - 2) = 5(x - 1) + 3x + 2$

16. $\dfrac{3}{4}n - \dfrac{1}{5} = \dfrac{2}{3}n$

17. $A = \dfrac{a + b + c}{3}$, for b

18. $\dfrac{40}{30} = \dfrac{3}{x}$

19. **Trampoline** A circular trampoline has a diameter of 22 feet. Determine the area of the trampoline.

20. **Earnings** If Samuel earns \$10.50 after working for 2 hours mowing a lawn, how much does he earn after 8 hours?

3 Applications of Algebra

WE OFTEN PURCHASE items that need to be delivered to our homes or apartments. When this occurs there may be a delivery charge, and certain other charges. In Example 5 on page 192 we set up and solve an equation dealing with a delivery charge and the purchase of sod for the lawn.

3.1 Changing Application Problems into Equations

1 Translate phrases into mathematical expressions.

2 Express the relationship between two related quantities.

3 Write expressions involving multiplication.

4 Translate applications into equations.

1 Translate Phrases into Mathematical Expressions

Helpful Hint   *Study Tip*

It is important that you prepare for this chapter carefully. Make sure you read the book and work the examples carefully. *Attend class every day, and most of all, work all the exercises assigned to you.*

As you read through the examples in the rest of the chapter, think about how they can be expanded to other, similar problems. For example, in Example 1a) we will state that the distance, d, increased by 10 miles, can be represented by $d + 10$. You can generalize this to other, similar problems. For example, a weight, w, increased by 15 pounds, can be represented as $w + 15$.

One practical advantage of knowing algebra is that you can use it to solve everyday problems involving mathematics. For algebra to be useful in solving everyday problems, you must first be able to *translate application problems into mathematical language.* One purpose of this section is to help you take an application problem, also referred to as a *word* or *verbal problem*, and write it as a mathematical equation.

Often the most difficult part of solving an application problem is translating it into an equation. Before you can translate a problem into an equation, you must understand the meaning of certain words and phrases and how they are expressed mathematically. **Table 3.1** is a list of selected words and phrases and the operations they imply. We used the variable x. However, any variable could have been used.

TABLE 3.1

Word or Phrase	Operation	Statement	Algebraic Form
Added to		8 *added to* a number	$x + 8$
More than	Addition	6 *more than* a number	$x + 6$
Increased by		A number *increased by* 3	$x + 3$
The sum of		*The sum of* a number and 4	$x + 4$
Subtracted from		6 *subtracted from* a number	$x - 6$
Less than	Subtraction	2 *less than* a number	$x - 2$
Decreased by		A number *decreased by* 5	$x - 5$
The difference between		The *difference between* a number and 9	$x - 9$
Multiplied by		A number *multiplied by* 6	$6x$
The product of	Multiplication	*The product of* 4 and a number	$4x$
Twice a number, 3 times a number, etc.		*Twice a number*	$2x$
Of, when used with a percent or fraction		20% *of* a number	$0.20x$
Divided by	Division	A number *divided by* 8	$\dfrac{x}{8}$
The quotient of		*The quotient of* a number and 6	$\dfrac{x}{6}$

Often a statement contains more than one operation. The following chart provides some examples of this.

Statement	Algebraic Form
Four more than twice a number	$\underbrace{2x}_{\text{Twice a number}} + 4$
Five less than 3 times a number	$\underbrace{3x}_{\text{Three times a number}} - 5$
Three times the sum of a number and 8	$3\underbrace{(x + 8)}_{\text{The sum of a number and 8}}$
Twice the difference between a number and 4	$2\underbrace{(x - 4)}_{\text{The difference between a number and 4}}$

Avoiding Common Errors

Subtraction is not commutative. That is, $a - b \neq b - a$. Therefore, you must be very careful when writing expressions involving subtraction. Study the following examples.

5 less than 3 times a number

CORRECT	INCORRECT
$3x - 5$	~~$5 - 3x$~~

5 subtracted from 3 times a number

CORRECT	INCORRECT
$3x - 5$	~~$5 - 3x$~~

the difference between $3x$ and 5

CORRECT	INCORRECT
$3x - 5$	~~$5 - 3x$~~

To give you more practice with the mathematical terms, we will also convert some algebraic expressions into statements. Often an algebraic expression can be written in several different ways. Following is a list of some of the possible statements that can be used to represent the given algebraic expression.

Algebraic	Statements
$2x + 3$	Three more than twice a number The sum of twice a number and 3 Twice a number, increased by 3 Three added to twice a number
$3x - 4$	Four less than 3 times a number Three times a number, decreased by 4 The difference between 3 times a number and 4 Four subtracted from 3 times a number

EXAMPLE 1 ▶ Express each statement as an algebraic expression.

a) The distance, d, increased by 10 miles
b) Six times the height, h
c) Eight less than twice the area, a
d) Four pounds more than 5 times the weight, w

Solution

a) $d + 10$ b) $6h$ c) $2a - 8$ d) $5w + 4$

▶ Now Try Exercise 13

2 Express the Relationship between Two Related Quantities

Sometimes in a problem, two numbers are related to each other in a certain way. We often represent the simplest, or most basic, number that needs to be expressed as a variable, and the other as an expression containing that variable. Some examples follow.

Statement	One Number	Second Number
Two numbers differ by 5	x	$x + 5$
Mike's age now and Mike's age in 8 years	x	$x + 8$
One number is 6 times the other number	x	$6x$
One number is 12% less than the other	x	$x - 0.12x$

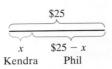

FIGURE 3.1

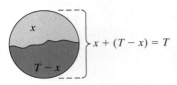

FIGURE 3.2

Note that often more than one pair of expressions can be used to represent the two numbers. For example, "two numbers differ by 5" can also be expressed as x and $x - 5$.

Now let's consider when one quantity is divided into two parts. For example, suppose $25 is divided between Kendra and Phil. If Kendra gets $10, how much does Phil get? Phil will get $25 − $10 or $15. If Kendra gets $12, then Phil will get $25 − $12 or $13. In general, if we let x represent the amount Kendra gets, then Phil will get $25 − x. Note that the sum of x and $25 − x$ is 25 (see **Fig. 3.1**).

In general, if T represents the total amount to be divided into two parts, then if one part is called x, the other part will be $T - x$. In **Figure 3.2**, if T represents the total area of the circle and x is the area of one part of the circle, then the area of the remaining part must be $T - x$.

There are many sections in this book where you will need to read a word problem and write an expression or equation to represent the word problem. The first thing you will need to do to write an expression or equation is to determine what quantity to let the variable represent. For example, if x is the variable, you will need to write "let $x =$." When doing so, in general, if one quantity is given in terms of another quantity, we let the variable represent the basic (or the original quantity) on which the second quantity is based. For example, suppose we are given that "Paul is 6 years older than George." Since Paul's age is given in terms of George's age, we let the variable represent George's age. Thus, we would write "let $x =$ George's age."

In Example 2, we give you some practice in identifying the item to call x when reading a word problem.

EXAMPLE 2 ▶ For each part, determine what to let $x =$.

a) Mary weighs 15 pounds more than Sue.

b) The length of a rectangle is 4 inches more than twice its width.

c) Joe earns $56.20 less than Larry.

d) The theater sold 525 tickets, some were adult and some were children's tickets.

Solution   In general, when writing expressions to represent word problems, if a quantity A is expressed in terms of a quantity B, then we let $x =$ quantity B.

a) Since Mary's weight is expressed in terms of Sue's weight, we let $x =$ Sue's weight.

b) Since the length of a rectangle is expressed in terms of its width, we let $x =$ width of the rectangle.

c) Since the amount Joe earns is expressed in terms of what Larry earns, we let $x =$ amount Larry makes.

d) Here, since neither quantity of adult tickets nor children's tickets is expressed in terms of the other's quantity, we can let $x =$ number of adult tickets sold or let $x =$ number of children's tickets sold. If we let $x =$ number of adult tickets sold, then $525 - x$ will equal the number of children's tickets sold. If we let $x =$ number of children's tickets sold, then $525 - x$ will equal the number of adult tickets sold.

▶ **Now Try Exercise 37**

In Example 2 we let x represent the variable. Although x is often used as the variable, other letters may be used. For example, consider the statement "Paul is 6 years older than George." If we choose to use the letter g as the variable, then we would write "let $g =$ George's age." The variable selected is just a matter of choice. If for example, the question discusses time, we may let the variable be t. If the question discusses the cost of an item, we may let the variable be c, and so on.

In Example 3, you will select the variable to use to represent quantities.

EXAMPLE 3 ▶ For each relationship, select a variable to represent one quantity and state what that variable represents. Then express the second quantity in terms of the variable selected.

a) The Dukes scored 12 points more than the Chiefs.

b) An adult robin is 4.3 times the weight of a baby robin.

c) Bill and Mary share $75.

d) Kim has 7 more than 5 times the amount Sylvia has.

e) The length of a rectangle is 3 feet less than 4 times its width.

Solution   To express the relationships, we must first decide which quantity we will let the variable represent. To give you practice with variables other than x, we will select different letters to represent the variable. We will select variables that relate to the expression given.

a) Since the number of points scored by the Dukes is expressed in terms of the number of points scored by the Chiefs, we will select to use the variable c.

$$\text{Let } c = \text{number of points scored by the Chiefs.}$$
$$\text{Then } c + 12 = \text{number of points scored by the Dukes.}$$

It is important that you realize that any other letter could have been selected for the variable. Suppose you used p since we are discussing points scored. Then you would have

$$\text{Let } p = \text{number of points scored by the Chiefs.}$$
$$\text{Then } p + 12 = \text{number of points scored by the Dukes.}$$

There is no "correct" variable to use; it is just a matter of preference. In the answers in the back of the book, if we listed two related quantities as c and $c + 12$, and you wrote the two related quantities as p and $p + 12$, you would still be correct.

b) The weight of an adult robin is given in terms of the weight of a baby robin.

$$\text{Let } w = \text{weight of a baby robin.}$$
$$\text{Then } 4.3w = \text{weight of an adult robin.}$$

c) We are not told how much of the $75 each person receives. In this case we can let the variable represent the amount either person receives. We will let a represent the amount Bill receives.

$$\text{Let } a = \text{amount Bill receives.}$$
$$\text{Then } 75 - a = \text{amount Mary receives.}$$

d) The amount Kim has is given in terms of the amount Sylvia has.

$$\text{Let } s = \text{amount Sylvia has.}$$
$$\text{Then } 5s + 7 = \text{amount Kim has.}$$

e) The length of the rectangle is given in terms of the width of the rectangle.

$$\text{Let } w = \text{width of the rectangle.}$$
$$\text{Then } 4w - 3 = \text{length of the rectangle.}$$

▶ **Now Try Exercise 47**

3 Write Expressions Involving Multiplication

Consider the statement "the cost of 3 items at $5 each." How would you represent this quantity using mathematical symbols? You would probably reason that the cost would be 3 times $5 and write $3 \cdot 5$ or $3(5)$.

Now consider the statement "the cost of x items at $5 each." How would you represent this statement using mathematical symbols? If you use the same reasoning, you might write $x \cdot 5$ or $x(5)$. Another way to write this product is $5x$. Thus, the cost of x items at $5 each could be represented as $5x$.

Finally, consider the statement "the cost of x items at y dollars each." Following the reasoning used in the previous two illustrations, you might write $x \cdot y$ or $x(y)$. Since these products can be written as xy, the cost of x items at y dollars each can be represented as xy.

EXAMPLE 4 ▶ Write each statement as an algebraic expression.

a) The cost of purchasing x pens at $2 each
b) A 5% commission on x dollars in sales
c) The dollar amount earned in h hours if a person earns $6.50 per hour
d) The number of cents in q quarters
e) The number of ounces in x pounds

Solution

a) We can reason like this: One pen would cost 1(2) dollars, two pens would cost 2(2) dollars, three pens 3(2) dollars, four pens 4(2) dollars, and so on. Continuing this reasoning process, we can see that x pens would cost $x(2)$ or $2x$ dollars.

b) A 5% commission on $1 sales would be 0.05(1), on $2 sales 0.05(2), on $3 sales 0.05(3), on $4 sales 0.05(4), and so on. Therefore, the commission on sales of x dollars would be $0.05(x)$ or $0.05x$.
    Note: If you need a review on changing a percent to a decimal number, review Appendix A.

c) In one hour the person would earn 1($6.50). In two hours the person would earn 2($6.50), and in h hours the person would earn $h(\$6.50)$ or $\$6.50h$.

d) We know that each quarter is worth 25 cents. Thus, one quarter is 1(25) cents. Two quarters is 2(25) cents, and so on. Therefore, q quarters is $q(25)$ cents or $25q$ cents.

e) Each pound is equal to 16 ounces. Using the same reasoning as in part **d)**, we see that x pounds is $16x$ ounces.

▶ **Now Try Exercise 71**

EXAMPLE 5 ▶ Truck Rental Maria Mears rented a truck for 1 day. She paid a daily fee of $38 and a mileage fee of 25 cents per mile. Write an expression that represents her total cost when she drives x miles.

Solution   Maria's total cost consists of two parts, the daily fee and the mileage fee. Notice the daily fee is given in terms of dollars, and the mileage fee is given in cents. When writing an expression to represent the total cost, we want the units to be the same. Therefore, we will use a mileage fee of $0.25 per mile, which is equal to 25 cents per mile.

$$\text{Let } x = \text{number of miles driven.}$$

$$\text{Then } 0.25x = \text{cost of driving } x \text{ miles}$$

$$\overbrace{\text{daily fee} + \text{mileage fee}}^{\text{total cost}}$$
$$38 \quad + \quad 0.25x$$

Thus, the expression that represents Maria's total cost is $38 + 0.25x$.

▶ **Now Try Exercise 75**

EXAMPLE 6 ▶ Write a Sum or Difference In a bus the number of males was 3 more than twice the number of females. Write an expression for

a) the sum of the number of males and females
b) the difference between the number of males and females
c) the difference between the number of females and males

Solution   Since the number of males is expressed in terms of the number of females, we let the variable represent the number of females. We will choose x to represent the variable.

Let x = number of females.

Then $2x + 3$ = number of males

a) The expression for the sum of the number of males and females is

$$\underbrace{(2x + 3)}_{\substack{\text{number} \\ \text{of males}}} \quad + \quad \underbrace{x}_{\substack{\text{number} \\ \text{of females}}}$$

b) The expression for the difference between the number of males and females is

$$\underbrace{(2x + 3)}_{\substack{\text{number} \\ \text{of males}}} \quad - \quad \underbrace{x}_{\substack{\text{number} \\ \text{of females}}}$$

c) The expression for the difference between the number of females and males is

$$\underbrace{x}_{\substack{\text{number} \\ \text{of females}}} \quad - \quad \underbrace{(2x + 3)}_{\substack{\text{number} \\ \text{of males}}}$$

Notice that parentheses are needed around the $2x + 3$ since both terms $2x$ and 3 are being subtracted.

▶ **Now Try Exercise 87**

Helpful Hint

This Helpful Hint will discuss the use of parentheses when writing the sum or difference of quantities.

Sum: When writing the sum of two quantities, parentheses may be used to help in the understanding of the problem, but they are not necessary.

Examples	Answers
Find the sum of x and $2x - 3$.	$x + (2x - 3)$ or $x + 2x - 3$
Find the sum of $3c - 4$ and $c + 5$.	$(3c - 4) + (c + 5)$ or $3c - 4 + c + 5$

Difference: When writing the difference of two quantities, when only *a single term* is being subtracted, parentheses may be used in the understanding of the problem, but they are not necessary.

Examples	Answers
Subtract r from $3r - 2$.	$(3r - 2) - r$ or $3r - 2 - r$
Find the difference between $2s + 6$ and s.	$(2s + 6) - s$ or $2s + 6 - s$

When writing the difference of two quantities, *when two or more terms are being subtracted, parentheses **must** be placed around all the terms being subtracted*, since all the terms are being subtracted and not just the first term.

Examples	Answers
Subtract $x + 2$ from $3x$.	$3x - (x + 2)$
Subtract $3t - 4$ from $5t$.	$5t - (3t - 4)$
Subtract $r - 5$ from $2r + 3$.	$(2r + 3) - (r - 5)$ or $2r + 3 - (r - 5)$
Find the difference between 6 and $m + 3$.	$6 - (m + 3)$
Find the difference between $4n - 9$ and $2n - 3$.	$(4n - 9) - (2n - 3)$ or $4n - 9 - (2n - 3)$

Expressions Involving Percent

Example 7 involves percent. Before we leave this section, let's discuss expressions that involve percent further. Since percents are used so often, you must have a clear understanding of how to write expressions involving percent. Whenever we perform a calculation involving percent, we change the percent to a decimal number or a fraction first.

When shopping we may see a "25% off" sign. We assume that this means 25% of the *original cost*, even though this is not stated. If we let c represent the original cost, then 25% of the original cost would be represented as $0.25c$. Twenty-five percent off the original cost means the original cost, c, decreased by 25% of the original cost. Twenty five percent off the original cost would be represented as $c - 0.25c$.

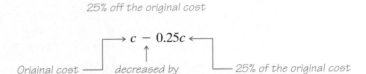

25% off the original cost

$$c - 0.25c$$

Original cost ——— decreased by ——— 25% of the original cost

Now let's work an example involving percent.

EXAMPLE 7 ▶ Write each statement as an algebraic expression.

a) The cost of a pair of boots c, increased by 6%

b) The population in the town of Brooksville, p, decreased by 12%

Solution

a) The question asks for the cost increased by 6%. We assume that this means the cost increased by 6% of the original cost. Therefore, the answer is $c + 0.06c$.

b) Using the same reasoning as in part **a)**, the answer is $p - 0.12p$.

▶ **Now Try Exercise 79**

Avoiding Common Errors

In Example 7**a)** we asked you to represent a cost, c, increased by 6%. Note, the answer is $c + 0.06c$. Often, students write the answer to this question as $c + 0.06$. It is important to realize that a percent of a quantity must always be a percent multiplied by some number or letter. Some phrases involving the word percent and the correct and incorrect interpretations follow.

PHRASE	CORRECT	INCORRECT
A $7\frac{1}{2}$% sales tax on c dollars	$0.075c$	~~0.075~~
The cost, c, increased by a $7\frac{1}{2}$% sales tax	$c + 0.075c$	~~$c + 0.075$~~
The cost, c, reduced by 25%	$c - 0.25c$	~~$c - 0.25$~~

---

4 ▸ Translate Applications into Equations

Now we will explain how to write application problems as equations. The word *is* in an application problem often means *is equal to* and is represented by an equal sign. Some examples of statements written as equations follow.

Statement	Equation
Six times a number *is* 42.	$6x = 42$
Five more than twice a number *is* 4.	$2x + 5 = 4$
A number decreased by 4 *is* 3 more than twice the number.	$x - 4 = 2x + 3$
The sum of a number and the number increased by 4 *is* 60.	$x + (x + 4) = 60$
Twice the difference of a number and 3 *is* the sum of the number and 20.	$2(x - 3) = x + 20$
A number increased by 15% *is* 120.	$x + 0.15x = 120$
Six less than three times a number *is* one-fourth the number.	$3x - 6 = \dfrac{1}{4}x$

Now let's translate some equations into statements. Some examples of equations written as statements follow. We will write only two statements for each equation, but remember there are other ways these equations can be written.

Equation	Statements
$3x - 4 = 4x + 3$	Four less than 3 times a number *is* 3 more than 4 times the number.
	Three times a number, decreased by 4 *is* 4 times the number, increased by 3.
$3(x - 2) = 6x - 4$	Three times the difference between a number and 2 *is* 4 less than 6 times the number.
	The product of 3 and the difference between a number and 2 *is* 6 times the number, decreased by 4.

Now let's work some examples where we write equations.

EXAMPLE 8 ▶ **Translate Words into Equations** Write each problem as an equation.

a) The population in the town of Rush is increasing by 500 people per year. The increase in population in *t* years *is* 2500.

b) The number of cents in *d* dimes *is* 120.

c) The cost of *x* gallons of gasoline at $3.20 per gallon *is* $35.20.

Solution

a) In one year the population increases by 1(500). In two years the population increases by 2(500), and in *t* years the increase in population is $t(500)$ or $500t$. Since the increase in population in *t* years *is* 2500, the equation is $500t = 2500$.

b) The number of cents in 1 dime is 1(10). The number of cents in 2 dimes is 2(10), and the number of cents in *d* dimes is $d(10)$ or $10d$. Since the number of cents in *d* dimes *is* 120, the equation is $10d = 120$.

c) Using similar reasoning as in parts **a)** and **b)**, the equation is $3.20x = 35.20$.

▶ **Now Try Exercise 109**

Helpful Hint

In a written expression certain other words may be used in place of *is* to represent the equal sign. Some of these are *will be*, *was*, *yields*, and *gives*. For example,

"When 4 is added to a number, the sum *will be* 20" can be expressed as $x + 4 = 20$.

"Six subtracted from a number *was* $\dfrac{1}{2}$ the number" can be expressed as $x - 6 = \dfrac{1}{2}x$.

"A rental car cost $25 per day. The cost for renting the car for *x* days *was* $150" can be expressed as $25x = 150$.

EXAMPLE 9 ▶ **Translate Words into an Equation** Write the problem as an equation.

One number is 4 less than twice the other. Their sum is 14.

Solution   First, we express the two numbers in terms of the variable. We will use the variable x to represent one of the numbers.

$$\text{Let } x = \text{one number.}$$
$$\text{Then } 2x - 4 = \text{second number.}$$

Now we write the equation using the information given.

$$\text{first number} + \text{second number} = 14$$
$$x + (2x - 4) = 14$$

▶ **Now Try Exercise 99**

Helpful Hint

If you examine Example 9, the word *is* is used twice, once in each sentence. However, only one equal sign appears in the equation. When the word *is* appears more than once, generally one *is* is being used to express the relationship between the numbers, and the other *is* will represent the equal sign in the equation. When you come across this situation, read the question carefully to determine which *is* will represent the equal sign in the equation.

In Example 10 we will use the term *consecutive even integers*. **Consecutive integers** are integers that differ by 1 unit. For example, the integers 6 and 7 are consecutive integers. Two consecutive integers may be represented as x and $x + 1$. **Consecutive even integers** are even integers that differ by 2 units. For example, 6 and 8 are consecutive even integers. **Consecutive odd integers** also differ by 2 units. For example, 7 and 9 are consecutive odd integers. Two consecutive even integers or two consecutive odd integers may be represented as x and $x + 2$, where x is always the smaller of the integers and $x + 2$ is always the larger of the integers.

EXAMPLE 10 ▶ **Consecutive Even Integers** Write the problem as an equation.

For two consecutive even integers, the sum of the smaller and 3 times the larger is 22.

Solution   First, we express the two consecutive even integers in terms of the variable.

$$\text{Let } x = \text{smaller consecutive even integer.}$$
$$\text{Then } x + 2 = \text{larger consecutive even integer.}$$

Now we write the equation using the information given.

$$\text{smaller} + 3 \text{ times the larger} = 22$$
$$x + 3(x + 2) = 22$$

▶ **Now Try Exercise 107**

EXAMPLE 11 ▶ **Translate Words into an Equation** Write the problem as an equation.

One train travels 3 miles more than twice the distance another train travels. The total distance traveled by both trains is 800 miles.

Solution   First express the distance traveled by each train in terms of the variable.

$$\text{Let } x = \text{distance traveled by one train.}$$
$$\text{Then } 2x + 3 = \text{distance traveled by second train.}$$

Now write the equation using the information given.

$$\text{distance of train 1} + \text{distance of train 2} = \text{total distance}$$
$$x + (2x + 3) = 800$$

▶ **Now Try Exercise 121**

EXAMPLE 12 ▸ **Translate Words into an Equation** Write the problem as an equation.

Lori Soushon is 4 years older than 3 times the age of her son Ron. The difference in Lori's age and Ron's age is 26 years.

Solution   Since Lori's age is given in terms of Ron's age, we will let the variable represent Ron's age.

$$\text{Let } x = \text{Ron's age.}$$

$$\text{Then } 3x + 4 = \text{Lori's age.}$$

We are told that the difference in Lori's age and Ron's age is 26 years. The word *difference* indicates subtraction. Since Lori is older than Ron, we must subtract Ron's age from Lori's age to get a positive number.

$$\text{Lori's age } - \text{ Ron's age } = 26$$

$$(3x + 4) - x = 26$$

▸ **Now Try Exercise 117**

Example 13 will involve percent. Make sure you understand the example.

EXAMPLE 13 ▸ **Translate Words into an Equation.** Write the problem as an equation.

The 2007 property tax for Danielle's house was 3.9% greater than her property tax in 2006. Her property tax in 2007 was $4008.

Solution   In this example, we will choose to use the variable t, for tax. Since the 2007 property tax is based upon the 2006 property tax, we will let the variable represent the 2006 property tax.

$$\text{Let } t = \text{2006 property tax.}$$

$$\text{Then } t \underbrace{+ 0.039t}_{\substack{this\ represents \\ the\ 3.9\%\ increase}} = \text{2007 property tax.}$$

Since Danielle's 2007 property tax *was* $4008, the equation we write is $t + 0.039t = 4008$.

▸ **Now Try Exercise 123**

Helpful Hint

It is important that you understand this section and work all your assigned homework problems. You will use the material learned in this section in the next three sections, and throughout the book.

In the examples in this section, we used different letters to represent the variable. Often the letter x is used to represent the variable, but other letters might be used. For example, when discussing an expression or equation involving distance, you may use x to represent the distance, or you may choose to use d, or another variable to represent the distance. Thus to represent the expression "the distance increased by 20 miles," you may write $x + 20$ or $d + 20$. Both are correct.

If your teacher, or the exercise, does not indicate which letter to use to represent the variable, you may select the letter you wish to use. If the answer appendix has an answer as $d + 20$ and your answer is $x + 20$, your answer would be correct, if x and d represent the same quantity.

EXERCISE SET 3.1

Math XL
MathXL®

MyMathLab
MyMathLab

Concept/Writing Exercises

1. Give four phrases that indicate the operation of addition.

2. Give four phrases that indicate the operation of subtraction.

3. Give four phrases that indicate the operation of multiplication.

4. Give four phrases that indicate the operation of division.

5. Explain why $c + 0.25$ *does not* represent the cost of an item increased by 25 percent.

6. Explain why $c - 0.10$ *does not* represent the cost of an item decreased by 10 percent.

7. A 25-foot length of rope is cut into two pieces. If one of the pieces has length x, what is the length of the other piece?

8. A 42-foot length of board is cut into two pieces. If one of the pieces has length y, what is the length of the other piece?

9. Give at least four words or phrases that may be used to indicate the use of an equal sign.

10. Consider the statement "Juan is 3 years older than twice Paul's age." Is Juan's age expressed in terms of Paul's age or is Paul's age expressed in terms of Juan's age. Explain.

Practice the Skills

In Exercises 11–30, express the statement as an algebraic expression. See Example 1.

11. The height, h, increased by 4 inches

12. The weight, w, increased by 20 pounds

13. The age, a, decreased by 5 years

14. The time, t, decreased by 3 hours

15. Five times the height, h

16. Seven times the length, l

17. Twice the distance, d

18. Three times the rate, r

19. One-half the age, a

20. One-third the weight, w

21. Five subtracted from r

22. Nine subtracted from p

23. m subtracted from 8

24. n subtracted from 4

25. Eight pounds more than twice the weight, w

26. Six inches more than 3 times the height, h

27. Four years less than 5 times the age, a

28. One mile more than $\frac{1}{2}$ the distance, d

29. One-third the weight, w, decreased by 7 pounds

30. One-fifth the height, h, increased by 2 feet

In Exercises 31–44, determine what x =. See Example 2.

31. Paul is 4 inches taller than Sonya.

32. The length of a rectangle is 5 inches greater than its width.

33. The length of Tortuga Beach is 60 feet shorter than the length of Jones Beach.

34. Wilma ran 4 miles per hour faster than Natasha.

35. The United States won 3 times the number of medals that Finland won.

36. The distance to Georgia is $\frac{1}{2}$ the distance to Tennessee.

37. The Cadillac costs $200 more than twice the cost of the Chevy.

38. Noah received 25 more votes than 3 times the number of votes that Tawnya received.

39. June's grade was 2 points less than twice Teri's grade.

40. Alberto's salary was $2000 greater than 4 times Nick's salary.

41. $60 divided between Kristen and Yvonne.

42. Drawka has 25 marbles. They are either red or blue marbles.

43. Together Don and Angela weigh 270 pounds.

44. Together Oliver and Dalane have 1053 clients.

In Exercises 45–56, select a variable to represent one quantity and state what that variable represents. Express the second quantity in terms of the variable selected. See Example 3. Note that the variable you select may be different than the variable used in the answers in the back of the book.

45. The table costs 3 times as much as the chair.

46. Joan's house is 810 square feet larger than Alfredo's house.

47. The area of the living room is 20 square feet greater than twice the area of the kitchen.

48. The amount in Darla's savings account is $250 less than 4 times the amount in Carmen's savings account.

49. The length of the rectangle is 2 inches less than 5 times the width.

50. A total of $600 is to be divided between Evita and Brian.

51. A total of 38 medals were won by Sweden and Brazil.

52. A movie theatre sold a total of 220 more adult tickets than children tickets.

53. Mike's age is 2 years more than $\frac{1}{2}$ George's age.

54. The book *Golden Angels* sold 4 copies less than 5 times the amount the book *Flycatcher* sold.

55. Jan and Edward used two different treadmills for exercise. The total distance walked between them was 6.4 miles.

56. On a 540-mile trip Cheng and Elsie shared the driving.

In Exercises 57–84, write the indicated expression. See Examples 3–7.

57. Age Dan Graber is n years old now. Write an expression that represents his age in 8 years.

58. Speed-Reading John Debruzzi used to read p words per minute. After taking a speed-reading course, his speed increased by 60 words per minute. Write an expression that represents his new reading speed.

59. Motorcycle Melissa Blum is selling her motorcycle. She was asking x dollars for the motorcycle but has cut the price in half. Write an expression that represents the new price.

60. Age Cathy Bennett's son is one-third as old as Cathy, c. Write an expression for her son's age.

61. Age Gayle Krzemien's age is one less than twice Mary Lou Baker's age, a. Write an expression for Gayle's age.

62. Calories The calories in a serving of mixed nuts is 280 calories less than twice the number of calories in a serving of cashew nuts, c. Write an expression for the number of calories in a serving of mixed nuts.

63. Temperature The average daily temperature in Jacksonville, Florida in July is 30° less than twice its average daily temperature in January, t. Write an expression for the average daily temperature in July.

64. Population In 2005, the population of China was 20 million more than 1.2 times the population of India, p. Write an expression for the population of China.

65. Weight Anika Angel weighed p pounds at birth. At age 6 months her weight was 2.3 pounds less than twice her birth weight. Write an expression for Anika's weight at age 6 months.

66. Population Increase The city of Clarkville has a population of 4000. If the population increases by 300 people per year, write an expression that represents the population after n years.

67. Profits Monica and Julia share in the profits of a toy store. If the total profit is $80,000 and m is the amount Monica receives, write an expression for the amount Julia receives.

68. Charity Event A total of 83 men and women attended a charity event. If the number of men who attended is m, write an expression for the number of women who attended.

69. Home Runs In professional baseball, as of this writing the all-time home run-leader is Hank Aaron, and Babe Ruth is second. Hank Aaron had 673 less than twice the number of home runs Babe Ruth had, r. Write an expression for the number of home runs Hank Aaron had.

70. Life Expectancy In 2005, the country with the highest average life expectancy was Japan, and the country with the lowest average life expectancy was Botswana. The average life expectancy in Japan was 3.2 years greater than twice that in Botswana, b. Write an expression for the average life expectancy in Japan. (*Source:* Population Connection).

71. Money Carolyn Curley found that she had x dimes in her handbag. Write an expression that represents this quantity of money in cents.

72. Weight Jason Mahar's weight is w pounds. Write an expression that represents his weight in ounces.

73. Money Susan Grady has d dollars in her purse. Write an expression that represents this quantity of money in cents.

74. Soil A total of six hundred pounds of soil is put onto two trucks. If a pounds of soil is placed onto one truck, write an expression for the amount of soil placed onto the other truck.

75. Truck Rental Bob Melina rented a truck for a trip. He paid a daily fee of $45 and a mileage fee of 40 cents a mile. Write an expression that represents his total cost when he travels x miles in one day.

76. Soil Delivery Mary Vachon had top soil delivered to her house. The total cost included a delivery charge of $48 plus $60 per cubic yard of soil. Write an expression for the total cost if Mary has x cubic yards of soil delivered.

77. Sales Increase Barry Cogan is a sales representative for a medical supply company. His 2006 sales increased by 20% over his 2005 sales, s. Write an expression for his 2006 sales.

78. Salary Increase Charles Idion, an engineer, had a salary increase of 15% over last year's salary, s. Write an expression for this year's salary.

79. Electricity Use Jean Olson's electricity use in 2006 decreased by 12% from her 2005 electricity use, e. Write an expression for her 2006 electricity use.

80. Shirt Sale At a 25% off everything sale, Bill Winchief purchased a new shirt. If c represents the original cost, write an expression for the sale price of the shirt.

81. Car Cost The cost of a new car purchased in Collier County included a 7% sales tax. If c represents the cost of the car before tax, write an expression for the total cost, including the sales tax.

82. Film The number of rolls of film sold in 2005 was 43.9% fewer than the number of rolls sold in 2000. Write an expression for the number of rolls sold in 2005 if r represents the number of rolls sold in 2000. (*Source:* Photo Marketing Association)

83. Median Age In 2005, the state with the greatest median age was Maine and the state with the lowest was Utah. The median age in Utah was 31.3% less than that of Maine. If m represents the median age in Maine, write an expression for the median age in Utah. (*Source:* U.S. Census Bureau)

84. Growing Las Vegas Las Vegas, Nevada, according to the U.S. Census Bureau, had the greatest population growth from 2003 to 2004 of any major city in the United States. Its population in 2004 was 4.1% greater than its population in 2003. If p represents Las Vegas's population in 2003, write an expression for its 2004 population.

Las Vegas at Night

In Exercises 85–98, write the indicated expression. You will need to select the variable to use. Note that the variable you use may be different than the variable used in the answer section. See Example 6.

85. Weight Jennifer's weight is 15 pounds more than Frieda's weight. Write an expression for the sum of their weights.

86. Length The Chestnut-mandibled toucan is 3 inches longer than the keel-billed toucan (shown). Write an expression for the sum of their lengths.

87. Height Armando is taller than his son Luis. Armando is 1 inch less than twice Luis's height. Write an expression for the *difference* in Armando's and Luis's height.

88. Profits A company's profits in 2005 were $100 less than the company's profits in 2006. Write an expression for the *difference* in the 2006 and 2005 profits.

89. Numbers The smaller of two numbers is 40 less than 3 times the larger number. Write an expression for the smaller number *subtracted from* the larger number.

90. Numbers The smaller of two numbers is 16 less than twice the larger number. Write an expression for the smaller number *subtracted* from the larger number.

91. Weight Jill's four-year-old child weighs 3 pounds less than twice the weight of her two-year-old child. Write an expression for the sum of their weights.

92. Assessed Value The assessed value of Glen Sandifer's house is $200 more than twice that of Olga Thompson's house. Write an expression for the sum of the assessed values.

93. Land Area The largest state in area is Alaska and the smallest is Rhode Island. The area of Alaska is 462 square miles more than 479 times the area of Rhode Island. Write an expression for the sum of the areas of the two states.

94. Black Bear The number of black bear sightings in 2006 in Ocala National Forest was 6 less than twice the number of sightings in 2005. Write an expression for the number of 2005 sightings subtracted from the number of 2006 sightings.

95. Stocks The price of Apple Computer stock is 6% greater than the cost of National Bank stock. Write an expression for the sum of the prices of National Bank and Apple Computer stocks.

96. Car Cost The cost of a 2006 Chevrolet Corvette increased by 1.2% over the cost of the 2005 Corvette. Write an expression for the sum of the costs of a 2005 and a 2006 Corvette.

97. Bank Assets The 2006 assets in Third National Bank were 2.3% lower then their 2005 assets. Write an expression for the 2006 assets subtracted from the 2005 assets.

98. Spam The number of pieces of spam (or junk mail) that Laura Hoye received in 2006 was 12% less than the number she received in 2005. Write an expression for the difference between the number of pieces of spam she received in 2005 and 2006.

In Exercises 99–130, write an equation to represent the problem. See Examples 8–13.

99. Two Numbers One number is 4 times another. The sum of the two numbers is 20.

100. Age Marie is 6 years older than Denise. The sum of their ages is 48.

101. Consecutive Integers The sum of two consecutive integers is 41.

102. Even Integers The product of two consecutive even integers is 74.

103. Numbers Twice a number, decreased by 8 is 12.

104. Consecutive Integers For two consecutive integers, the sum of the smaller and twice the larger is 29.

105. Numbers One-fifth of the sum of a number and 10 is 150.

106. Numbers One-third of the sum of a number and 12 is 5.

107. Even Integers For two consecutive even integers, the sum of the smaller and twice the larger is 22.

108. Odd Integers For two consecutive odd integers, the sum of 3 times the smaller and the larger is 14.

109. Earnings John Jones earns $12.50 per hour. If he works h hours his earnings will be $150.

110. Employees The T. W. Wilson company plans to increase its number of employees by 20 per year. The increase in the number of employees in t years will be 120.

111. Top Soil Abe Mantell purchased x bags of top soil at a cost of $2.99 a bag. The total he paid was $17.94.

112. Plants Mark Ernsthausen purchased p plants at a nursery for $5.99 each. The total he paid was $65.89.

113. Quarters The number of cents in q quarters is 175.

114. Seconds The number of seconds in m minutes is 480.

115. Age Darta Aguilar is 1 year older than twice Julie Chesser's age. The sum of their ages is 52.

116. Horses Marc Campbell owns more horses than Selina Jones owns. The number of horses that Marc owns is 3 less than five times the number that Selina owns. The difference in the number of horses owned by Marc and Selina is 5.

117. Baseball Cards Jakob Meyer owns 300 more than twice the number of baseball cards that Saul Gonzales owns. The difference in the number of cards that Jakob and Saul own is 420.

118. Jogging David Ostrow jogs 5 times as far as Jennifer Freer. The total distance traveled by both people is 8 miles.

119. Amtrak An Amtrak train travels 4 miles less than twice the distance traveled by a Southern Pacific train. The total distance traveled by both trains is 890 miles.

120. Wagon Ride On a wagon ride the number of girls was 6 less than twice the number of boys. The total number of boys and girls on the wagon was 18.

121. Distance Walked Donna Douglas walked 2 miles less than 3 times as far as Malik Oamar walked. Together they walked a total of 12.6 miles.

122. Distance Lilia Orlova rollerbladed 4 miles less than twice the distance she ran. The total distance she traveled was 15 miles.

123. Viper The cost of a new Dodge Viper increased by 2.3% over last year's price. The new price is $89,600.

124. Income Dan Tadeo's 2006 income was 4.6% greater than his 2005 income. His income in 2006 was $56,900.

125. Population The population of the town of Tom's Valley decreased by 1.9%. The population after the decrease was 12,087.

126. Video Cassette At the Better Buy Warehouse, Anne Long purchased a video cassette recorder that was reduced by 10% for $208.

127. New Car Carlotta Diaz bought a new car. The cost of the car plus a 7% sales tax was $32,600.

128. Sport Coat David Gillespie purchased a sport coat at a 25% off sale. He paid $195 for the sport coat.

129. Cost of Meal Beth Rechsteiner ate at a steakhouse. The cost of the meal plus a 15% tip was $42.50.

130. Railroad In a narrow gauge railway, the distance between the tracks is about 64% of the distance between the tracks in a standard gauge railroad. The difference in the distances between the tracks in the two types of railroads is about 1.67 feet.

Challenge Problem

131. Time

a) Write an algebraic expression for the number of seconds in d days, h hours, m minutes, and s seconds.

b) Use the expression found in part **a)** to determine the number of seconds in 4 days, 6 hours, 15 minutes, and 25 seconds.

Group Activity

Exercises 132 and 133 will help prepare you for the next section, where we set up and solve application problems. Discuss and work each exercise as a group. For each exercise, write down the quantity you are being asked to find and represent this quantity with a variable. Then write an equation containing your variable that can be used to solve the problem. Do not solve the equation.

132. Water Usage An average bath uses 30 gallons of water and an average shower uses 6 gallons of water per minute. How long a shower would result in the same water usage as a bath?

133. Salary Plans An employee has a choice of two salary plans. Plan A provides a weekly salary of $200 plus a 5% commission on the employee's sales. Plan B provides a weekly salary of $100 plus an 8% commission on the employee's sales. What must be the weekly sales for the two plans to give the same weekly salary?

Cumulative Review Exercises

[1.9] **134.** Evaluate $3[(4 - 16) \div 2] + 5^2 - 3$.

[2.6] **135.** $P = 2l + 2w$; find l when $P = 40$ and $w = 5$.

136. Solve $3x - 2y = 6$ for y.

[2.7] **137.** Solve the proportion $\dfrac{3.6}{x} = \dfrac{10}{7}$.

138. Write the ratio 26 ounces to 4 pounds in lowest terms.

3.2 Solving Application Problems

1 Use the problem-solving procedure.

2 Set up and solve number application problems.

3 Set up and solve application problems involving money.

4 Set up and solve applications concerning percent.

1 Use the Problem-Solving Procedure

Many types of application problems can be solved using algebra. In this section we introduce several types. In Sections 3.3 and 3.4 we introduce additional types of applications. They are also presented in many other sections and exercise sets throughout the book. Your instructor may not have time to cover all the applications given in this book. If not, you may still wish to spend some time on your own reading those problems just to get a feel for the types of applications presented.

To be prepared for this section, you must understand the material presented in Section 3.1. The best way to learn how to set up an application or word problem is to practice. The more problems you study and attempt, the easier it will become to solve them.

The general problem-solving procedure given in Section 1.2 can be used to solve all types of verbal problems. Below, we present the **five-step problem-solving procedure** again so you can easily refer to it. We have included some additional information under steps 1 and 2, since in this section we are going to emphasize translating application problems into equations.

Problem-Solving Procedure for Solving Applications

1. **Understand the problem.**   Identify the quantity or quantities you are being asked to find.

2. **Translate the problem into mathematical language (express the problem as an equation).**

   a) Choose a variable to represent one quantity, *and write down exactly what it represents*. Represent any other quantity to be found in terms of this variable.

   b) Using the information from step a), write an equation that represents the application.

3. **Carry out the mathematical calculations (solve the equation).**

4. **Check the answer (using the** *original* **application).**

5. **Answer the question asked.**

Sometimes we will combine two steps in the problem-solving procedure when it helps to clarify the explanation. We may not show the check of a problem to save space. Even if we do not show a check, you should check the problem yourself and make sure your answer is reasonable and makes sense.

Let's now set up and solve some application problems using this procedure.

2 Set Up and Solve Number Application Problems

The examples presented under this objective involve information and data but do not contain percents. When we work the examples we will follow the five-step problem-solving procedure.

EXAMPLE 1 ▸ **An Unknown Number** Two subtracted from 4 times a number is 10. Find the number.

Solution   Understand   To solve this problem, we need to express the statement given as an equation. We are asked to find the unknown number. We use the information learned in the previous section to write the equation.

Translate   Let $x =$ the unknown number. Now write the equation.

$$\underbrace{\text{2 subtracted from 4 times a number}}_{4x - 2} \overset{\text{is}}{=} \overset{10}{10}$$

Carry Out
$$4x = 12$$
$$x = 3$$

Check   Substitute 3 for the number in the original problem, two subtracted from 4 times a number is 10.

$$4(3) - 2 \stackrel{?}{=} 10$$
$$10 = 10 \quad \textit{True}$$

Answer   Since the solution checks, the unknown number is 3.

▸ **Now Try Exercise 3**

EXAMPLE 2 ▸ **Number Problem** The sum of two numbers is 26. Find the two numbers if the larger number is 2 less than three times the smaller number.

Solution   Understand   This problem involves finding two numbers. When finding two numbers, if a second number is expressed in terms of a first number, we generally let the variable represent the first number. Then we represent the second number as an expression containing the variable used for the first number. In this example, we are given that "the larger number is 2 less than three times the smaller number." Notice that the larger number is expressed in terms of the smaller number. Therefore, we will let the variable represent the smaller number.

Translate   Let x = smaller number.
                          Then $3x - 2$ = larger number.

The sum of the two numbers is 26. Therefore, we write the equation

$$\text{smaller number} + \text{larger number} = 26$$
$$x + (3x - 2) = 26$$

Carry Out   Now we solve the equation.

$$4x - 2 = 26$$
$$4x = 28$$
$$x = 7$$

The smaller number is 7. Now we find the larger number.

$$\text{larger number} = 3x - 2$$
$$= 3(7) - 2 \quad \textit{Substitute 7 for x.}$$
$$= 19$$

The larger number is 19.

Check   The sum of the two numbers is 26.

$$7 + 19 \stackrel{?}{=} 26$$
$$26 = 26 \quad \textit{True}$$

Answer   The two numbers are 7 and 19.

▸ **Now Try Exercise 9**

Helpful Hint

When reading a word problem, ask yourself, "How many answers are required?" In Example 2, the question asked for the two numbers. The answer is 7 and 19. It is important that you read the question and identify what you are being asked to find. If the question had asked "Find the *smaller* of the two numbers if the larger number is 2 less than three times the smaller number," then the answer would have been only the 7. If the question had asked to find the *larger* of the two numbers, then the answer would have been only 19. *Make sure you answer the question asked in the problem.*

EXAMPLE 3 ▶ **2006 Winter Olympics** In the 2006 Winter Olympics in Torino, Italy, Germany won the most medals and the United States won the second greatest number of medals. Germany won 21 less than twice the number of medals won by the United States. If the difference between the number of medals won by Germany and the United States was 4, determine the number of medals won by Germany.

Solution Understand The word *difference* in the problem indicates that this problem will involve subtraction. We are asked to find the number of medals won by Germany. Since the number of medals won by Germany is given in terms of the number of medals won by the United States, we will let the variable represent the number of medals won by the United States. We will use the variable w.

Translate   Let w = medals won by United States.

Then $2w - 21$ = medals won by Germany.

Now we write the equation using the given information. In real-life problems, in order to get a positive answer, the smaller amount will always be subtracted from the larger amount. Since the difference in medals between Germany and the United States is 4, we write the following equation.

$$\underbrace{\text{number of medals won by Germany}}_{2w - 21} - \underbrace{\text{number of medals won by United States.}}_{w} = 4$$

Carry Out $2w - 21 - w = 4$

$w - 21 = 4$

$w = 25$

Check and Answer The answer is not 25. Remember w represents the number of medals won by the United States. We are asked to find the number of medals won by Germany. The number of medals won by Germany is $2w - 21$ or $2(25) - 21 = 50 - 21$ or 29. Notice the difference in the number of medals won by Germany and the United States is $29 - 25$ or 4, so the answer checks.

▶ **Now Try Exercise 23**

EXAMPLE 4 ▶ **Bicycles** The Chain Wheel Drive Bicycle Company presently manufactures 800 bicycles a month. Each month after this month the company plans to increase production by 150 bicycles a month until its monthly production reaches 1700 bicycles. How long will it take the company to reach its production goal?

Solution Understand We are asked to find the *number of months* that it will take for the company's production to reach 1700 bicycles a month. Next month its production will increase by 150 bicycles. In two months, its production will increase by 2(150) over the present month's production. In n months, its production will increase by $n(150)$ or $150n$. We will use this information when we write the equation to solve the problem.

Translate   Let n = number of months.

Then $150n$ = increase in production over n months.

$$(\text{present production}) + \left(\begin{array}{c}\text{increased production}\\ \text{over } n \text{ months}\end{array}\right) = \text{future production}$$

$$800 + 150n = 1700$$

Carry Out $150n = 900$

$$n = \frac{900}{150}$$

$$n = 6 \text{ months}$$

Check and Answer   As a check, let's list the number of bicycles produced this month and for the next 6 months.

Presently	Next month	Month 2	Month 3	Month 4	Month 5	Month 6
↓	↓	↓	↓	↓	↓	↓
800	950	1100	1250	1400	1550	1700

Thus, in 6 months the company will produce 1700 bicycles per month.

▶ **Now Try Exercise 19**

3 Set Up and Solve Application Problems Involving Money

When setting up an equation that involves money, you must make sure that all the monetary units entered into the equation are the same, either all dollars or all cents. When given problems to convert to equations, some information might be given in dollars and other information might be given in cents. When this happens, we generally convert the amount given in cents to an equivalent amount of dollars. For example, when renting a truck the cost may be $30 a day plus 20 cents a mile. When writing the equation, we would write the 20 cents a mile as $0.20 a mile. The cost of traveling x miles at 20 cents a mile would be written $0.20x$.

Now let's work some examples.

EXAMPLE 5 ▶ **Grub Problem** Part of Kim Martello's lawn was destroyed by grubs. She decided to purchase new sod (or grass) to lay down. The cost of the sod is 45 cents per square foot plus a delivery charge of $59. If the total cost of delivery plus the sod was $284, how many square feet of sod was delivered?

Solution   Understand   The total cost consists of two parts, a variable cost of 45 cents per square foot of sod, plus a fixed delivery charge of $59. We need to determine the number of square feet of sod that will result in a total cost of $284. Since the fixed cost is given in dollars, we write the variable cost, or the cost of the sod, in dollars also.

Translate   Let x = number of square feet of sod.

Then $0.45x$ = cost of x square feet of sod.

sod cost + delivery cost = total cost

$$0.45x + 59 = 284$$

Carry Out $0.45x = 225$ *59 was subtracted from both sides.*

$$\frac{0.45x}{0.45} = \frac{225}{0.45}$$

$$x = 500$$

Check   The cost of 500 square feet of sod at 45 cents a square foot is $500(0.45) = \$225$. Adding the $225 to the delivery cost of $59 gives $284, so the answer checks.

Answer   Five hundred square feet of sod was delivered.

▶ **Now Try Exercise 31**

EXAMPLE 6 ▶ **Photo Printer** Elsie Newman is going to purchase a photo printer to print pictures from her digital camera. She is considering a Hewlett-Packard (HP) printer and a Lexmark™ printer. The HP printer costs $419 and the cost for the ink and paper is 14 cents per photo printed. The Lexmark printer costs $299 and the cost for the ink and paper is 18 cents per photo. How many photos would need to be printed for the total cost of the printers, ink, and paper to be the same?

Solution Understand   The Hewlett-Packard printer has a greater initial cost ($419 versus $299); however, its cost per photo printed is less (14 cents versus 18 cents). We are asked to find the number of photos printed so that the total cost of the two printers will be the same.

Translate

Let n = number of photos.

Then $0.14n$ = cost for printing n photos with the HP printer

and $0.18n$ = cost for printing n photos with the Lexmark printer.

total cost of HP printer = total cost of Lexmark printer

$$\binom{\text{initial}}{\text{cost}} + \binom{\text{cost}}{\text{for } n \text{ photos}} = \binom{\text{initial}}{\text{cost}} + \binom{\text{cost}}{\text{for } n \text{ photos}}$$

$$419 + 0.14n = 299 + 0.18n$$

Carry Out

$$120 + 0.14n = 0.18n \qquad \text{299 was subtracted from both sides.}$$

$$120 = 0.04n \qquad \text{0.14n was subtracted from both sides.}$$

$$\frac{120}{0.04} = \frac{0.04n}{0.04}$$

$$3000 = n$$

Check and Answer   The total cost would be the same when 3000 photos were printed. We will leave the check of this answer for you.

▶ **Now Try Exercise 35**

4 Set Up and Solve Applications Concerning Percent

Now we'll look at some application problems that involve percent. Remember that a percent is always a percent of something. Thus if the cost of an item, c, is increased by 8%, we would represent the new cost as $c + 0.08c$, and not $c + 0.08$. See the Avoiding Common Errors box on page 175.

EXAMPLE 7 ▶ **Water Bike Rental**   At a beachfront hotel, the cost for a water bike rental is $30 per half hour, which includes a $7\frac{1}{2}$% sales tax. Find the cost of the rental before tax.

Solution Understand   We are asked to find the cost of the water bike rental before tax. The cost of the rental before tax plus the tax on the water bike must equal $30.

Translate

Let x = cost of the rental before tax.

Then $0.075x$ = tax on the rental.

(cost of the water bike rental before tax) + (tax on the rental) = 30

$$x + 0.075x = 30$$

Carry Out

$$1.075x = 30$$

$$x = \frac{30}{1.075}$$

$$x \approx 27.91$$

Check and Answer   A check will show that if the cost of the rental is $27.91, the cost of the rental including a $7\frac{1}{2}$% tax is about $30.

▶ **Now Try Exercise 43**

EXAMPLE 8 ▸ **Caloric Intake** According to the Centers for Disease Control and Prevention, from 1971 to 2005 the average caloric intake for men in the United States increased by 8% to 2646 calories. Determine the average caloric intake for men in 1971.

Solution   Understand   We can represent the 2005 caloric intake in terms of the 1971 caloric intake. Therefore, we will select a variable to represent the 1971 caloric intake and represent the 2005 caloric intake in terms of the variable selected. The 2005 caloric intake must be 8% greater than the 1971 caloric intake.

Translate   Let c = the 1971 caloric intake.

Then $c + 0.08c$ = the 2005 caloric intake.

Since the 2005 caloric intake is 2646, we set up the following equation.

2005 caloric intake is 2646

$$c + 0.08c = 2646$$

Carry Out $$1.08c = 2646$$

$$c = \frac{2646}{1.08}$$

$$c = 2450$$

Check and Answer   Since c represents the 1971 caloric intake, and it is less than the 2005 caloric intake, our answer is reasonable. The 1971 caloric intake for men was 2450 calories.

▸ **Now Try Exercise 45**

EXAMPLE 9 ▸ **Salary Plans** Jeanne Pirie recently graduated from college and has accepted a position selling medical supplies and equipment. During her first year, she is given a choice of salary plans. Plan 1 is a $450 weekly base salary plus a 3% commission of weekly sales. Plan 2 is a straight 10% commission of weekly sales. What weekly sales, in dollars, would result in Jeanne receiving the same salary from both plans?

Solution   Understand   We are asked to find the *dollar sales* that will result in Jeanne receiving the same total salary from both plans. To solve this problem, we write expressions to represent the salary from each of the plans. We then obtain the desired equation by setting the salaries from the two plans equal to one another.

Translate   Let x = dollar sales.

Then $0.03x$ = commission from plan 1 sales

and $0.10x$ = commission from plan 2 sales.

salary from plan 1 = salary from plan 2
base salary + 3% commission = 10% commission

$$450 + 0.03x = 0.10x$$

Carry Out $$450 = 0.07x$$

or $$0.07x = 450$$

$$\frac{0.07x}{0.07} = \frac{450}{0.07}$$

$$x \approx 6428.57$$

Check   We will leave it up to you to show that sales of $6428.57 result in Jeanne receiving the same weekly salary from both plans.

Answer   Jeanne's weekly salary will be the same from both plans if she sells $6428.57 worth of medical supplies and equipment.

▸ **Now Try Exercise 55**

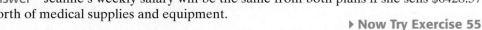

Helpful Hint

Here are some suggestions if you find you are having some difficulty with application problems.

1. **Instructor**—Make an appointment to see your instructor. Make sure you have read the material in the book and attempted all the homework problems. Go with specific questions for your instructor.

2. **Video CDs**—Find out if the video CDs that accompany this book are available at your college. If so, view the videos that go with this chapter. Using the pause control, you can watch the videos at your own pace.

3. **Tutoring**—If your college learning center offers free tutoring, as many colleges do, you may wish to take advantage of tutoring.

4. **Study Group**—Form a study group with classmates. Exchange phone numbers and e-mail addresses. You may be able to help one another.

5. **Student's Solutions Manual**—If you get stuck on an exercise you may want to use the Student's Solutions Manual to help you understand a problem. Do not use the Solutions Manual in place of working the exercises. In general, the Solutions Manual should be used only to check your work.

6. **MyMathLab**—MyMathLab provides free-response exercises correlated to the text that are algorithmically generated for unlimited practice and mastery. In addition, online tools such as video lectures, animations, and a multimedia textbook are available to help you understand the material. Check with your instructor to determine if MyMathLab is available.

7. **Math XL®**—MathXL is a powerful online homework, tutorial, and assessment system correlated specifically to this text. You can take chapter tests in MathXL and receive a personalized study plan based on your test results. The study plan links directly to tutorial exercises for the objectives you need to study or retest. Check with your instructor to determine if MathXL is available.

8. **Prentice Hall Mathematics Tutor**—Once the program has been initiated by your instructor, you can get individual tutoring by phone, fax, or e-mail.

It is important that you keep trying! Remember, the more you practice, the better you will become at solving application problems.

EXERCISE SET 3.2   Math XL   MyMathLab

Concept/Writing Exercises

1. Outline the five-step problem-solving procedure we use.

2. If you are having difficulty with some of the material in this section, list some of the things you may be able to do to help you. See the Helpful Hint above.

Practice the Skills/Problem Solving

Exercises 3–28 involve finding a number or numbers. Review Examples 1–4, then set up an equation that can be used to solve the problem. Solve the equation and **answer the question asked.** *Use a calculator where you feel it is appropriate.*

3. Unknown Number Three subtracted from 4 times a number is 17. Find the number.

4. Unknown Number Five subtracted from 6 times a number is 13. Find the number.

5. Consecutive Integers The sum of two consecutive integers is 87. Find the numbers.

6. Consecutive Integers The sum of two consecutive integers is 113. Find the numbers.

7. Odd Integers The sum of two consecutive odd integers is 96. Find the numbers.

8. Even Integers The sum of two consecutive even integers is 146. Find the numbers.

9. Sum of Numbers One number is 3 more than twice a second number. Their sum is 27. Find the numbers.

10. Sum of Numbers One number is 5 less than 3 times a second number. Their sum is 43. Find the numbers.

11. Difference of Numbers The larger of two numbers is 4 less than five times the smaller. When the smaller number is subtracted from the larger, the difference is 4. Find the two numbers.

12. Sum of Numbers One number is 2 less than 3 times a second number. Their sum is 26. Find the numbers.

 13. Difference of Numbers The larger of two integers is 8 less than twice the smaller. When the smaller number is subtracted from the larger, the difference is 17. Find the two numbers.

14. Facing Pages The sum of the two facing page numbers in an open book is 145. What are the page numbers?

15. Grandma's Gifts Grandma gave some baseball cards to Richey and some to Erin. She gave 3 times the amount to Erin as she did to Richey. If the total amount she gave to both of them was 260 cards, how many cards did she give to Richey?

16. Ski Shop The Alpine Valley Ski Shop sold 6 times as many downhill skis as cross-country skis. Determine the number of pairs of cross-country skis sold if the difference in the number of pairs of downhill and cross-country skis sold is 1800.

17. Animal Art Joseph Murray built a horse for display at a baseball stadium. It took him 1.4 hours more than twice the number of hours to attach the baseball gloves to the horse than to build the horse. If the total time it took him to build the horse and attach the gloves was 32.6 hours, how long did it take to attach the gloves?

The author, Allen R. Angel, is shown in this photo.

18. Candle Shop A candle shop makes 60 candles per week. It plans to increase the number of candles it makes by 8 per week until it reaches a production of 132 candles per week. How many weeks will it take for the shop to reach its production schedule?

19. Collecting Frogs Mary Shapiro collects ceramic and stuffed frogs. She presently has 422 frogs. She wishes to add 6 a week to her collection until her collection reaches a total of 500 frogs. How long will it take Mary's frog collection to reach 500 frogs?

20. Population The town of Dover currently has a population of 6500. If its population is increasing at a rate of 1200 people per year, how long will it take for the population to reach 20,600?

21. Circuit Boards The FGN Company produces circuit boards. It now has 4600 employees nationwide. It wishes to reduce the number of employees by 250 per year through retirements, until its total employment is 2200. How long will this take?

22. Computers The CTN Corporation has a supply of 3600 computers. It wishes to ship 120 computers each week until its supply drops to 2000. How long will this take?

23. Tornados According to The Weather Channel, the greatest number of tornados in the United States occurs in June and the fewest number occurs in December. The average number of tornados in June is 16 less than 11 times the average number of tornados in December. If the difference between the average number of tornados in June and December is 204, determine the average number of tornados in December and June.

24. Watching TV According to the Kaiser Family Foundation, in the United States, the amount of time per day spent by 8–18-year-olds watching television is 16 minutes more than 5 times the number of minutes they spend reading. If the total amount of time per day reading and watching television is 274 minutes, determine the number of minutes spent watching television.

25. Housekeepers The average hourly wage paid to hotel housekeepers in New York City is $1.46 more than twice the average wage paid to hotel housekeepers in New Orleans. Determine the average hourly wage paid to housekeepers in New York City if the difference in their average hourly wages is $8.10.

26. Albums According to *Wikipedia*, the best-selling album in 2005 was Mariah Carey's *The Emancipation of Mimi*. The best-selling album in 2004 was the album *Confession* by Usher. *Confession* sold 0.94 million copies less than twice the amount *The Emancipation of Mimi* sold. The sum of these two albums was 13.97 million copies. Determine the sales of both albums.

27. Cost to Produce Shirt According to the World Trade Organization, in 2005 the cost to produce a shirt in Northern China was less than in any other area. The cost (in U.S. dollars) to produce a shirt in Mexico is 16 cents less than 3 times the amount to produce a shirt in Northern China. The cost to produce one shirt in each country totals $3.28. Determine the cost to produce a shirt in Northern China and in Mexico.

28. Oil Use According to the International Energy Agency, the demand for oil in the United States is far greater than in any other country. In 2004, the United States used an average of 1.6 million gallons of oil per day more than 3 times that used by China, the second greatest user. The total used by both countries in one day was 26.8 million gallons. Determine the number of gallons of oil used in one day in the United States in 2004.

Exercises 29–42 involve money. Read Examples 5–6, then set up an equation that can be used to solve the problem. Solve the equation and **answer the question asked.**

29. Gasoline Luvia Rivera has only $48 to purchase gasoline. If gasoline costs $3.20 per gallon, determine how many gallons of gasoline Luvia can purchase.

30. Truck Rental Carol Battle rents a truck for one day and pays $50 per day plus 30 cents a mile. How far can Carol drive in one day if she has only $92?

31. Copy Machine Yamil Bernz purchased a copy machine for $2100 and a one-year maintenance protection plan that costs 2 cents per copy made. If he spends a total of $2462 in a year, which includes the cost of the machine and the copies made, determine the number of copies he made.

32. Gym Membership At Goldies Gym there is a one-time membership fee of $300 plus dues of $40 per month. If Carlos Manieri has spent a total of $700 for Goldies Gym, how long has he been a member?

33. Television Miles Potier's Time Warner cable bill costs $72.68 per month plus $3.95 for each On Demand movie he watches that month. If his cable bill for December was $96.38, determine the number of On Demand movies he watched in December.

34. Hardwood Floors Ruth Zasada is having hardwood floors installed in her living room. The cost for the material is $2840 plus an installation charge of $1.90 per square foot. If the total cost for the material plus installation is $5120, determine the area of her living room.

35. Truck Rental Howard Sporn is considering two companies from which to rent a truck. American Truck Rental charges $20 per day and 25 cents a mile. SavMor Truck Rental charges $35 a day and 15 cents a mile. How far would Howard need to drive in one day for the both companies to have the same total cost?

36. Washing Machines Scott Montgomery is considering two washing machines, a Kenmore® and a Neptune®. The Neptune costs $454 while the Kenmore costs $362. The energy guides indicate that the Kenmore will cost an estimated $84 per year to operate and the Neptune will cost an estimated $38 per year to operate. How long will it be before the total cost is the same for both washing machines?

37. Salaries Brooke Mills is being recruited by a number of high-tech companies. Data Technology Corporation has offered her an annual salary of $40,000 per year plus a $2400 increase per year. Nuteck has offered her an annual salary of $49,600 per year plus a $800 increase per year. In how many years will the salaries from the companies be the same?

38. Racquet Club The Coastline Racquet Club has two payment plans for its members. Plan 1 has a monthly fee of $20 plus $8 per hour for court time. Plan 2 has no monthly fee, but court time is $16.25 per hour. If court time is rented in 1-hour intervals, how many hours would you have to play per month so that plan 1 becomes a better buy?

39. Printers Hector Hanna will purchase one of two laser printers, a Hewlett-Packard (HP) or a Lexmark®. The HP costs $499 and the Lexmark costs $419. Suppose, because of the price of the ink cartridges, the cost of printing a page on the HP is $0.06 per page and the cost of printing a page on the Lexmark is $0.08 per page. How many pages would need to be printed for the two printers to have the same total cost?

40. Satellite or Cable Sean Stewart is deciding whether to select a satellite receiver or cable for his television programming. The satellite receiver costs $298.90 and the monthly charge is $68.70. With cable there is no initial cost to purchase equipment, but the monthly charge for comparable channels is $74.80. After how many months will the the total cost of the two systems be equal?

41. Newsletter Neil Simpson had a professional organization newsletter printed and sent out to all the members. The total cost included a $600 printing cost plus a 39 cents mailing cost for each envelope. If the total cost was $1380, determine how many newsletters were mailed.

42. Patio Resurfacing Elizabeth Chu is having her patio resurfaced using cement pavers. She is considering two companies for the job. A & E Pavers charge $1500 for the pavers plus $40 per hour for labor to install the pavers. The Jerilyn Fairman Company charges $1800 for the pavers plus $25 per hour to install the pavers. How many hours of labor would result in the same total cost with both companies?

*Exercises 43–64 involve percents. Read Examples 7–9, then set up an equation that can be used to solve the problem. Solve the equation and **answer the questions asked.***

43. Airfare The airfare for a flight from Amarillo, Texas, to New Orleans cost $280, which includes a 7% sales tax. What is the cost of the flight before tax?

44. New Car Yoliette Fournier purchased a new car. The cost of the car, including a 7.5% sales tax was $24,600. What was the cost of the car before tax?

45. Salary Increase Zhen Tong just received a job offer that will pay him 30% more than his present job does. If the salary at his new job will be $30,200, determine his present salary.

46. New Headquarters Tarrach and Associates plan on increasing the size of its headquarters by 20%. If its new headquarters is to be 14,200 square feet, determine the size of its present headquarters.

47. Oysters The number of bushels of oysters harvested in Chesapeake Bay has been consistently decreasing. There has been a 93% decrease in the number of bushels of oysters harvested in 2004 from the number harvested in 2001. If in 2004, about 26 thousand bushels of oysters were harvested, determine the number of bushels of oysters harvested in 2001. (*Source:* Maryland Department of Natural Resources)

48. Retirement Income Ray and Mary Burnham have decided to retire. They estimate their annual income after retirement will be reduced by 15% from their pre-retirement income. If they estimate their retirement income to be $42,000, determine their pre-retirement income.

49. Autographs A tennis star was hired to sign autographs at a convention. She was paid $3000 plus 3% of all admission fees collected at the door. The total amount she received for the day was $3750. Find the total amount collected at the door.

50. Sale At a 1-day 20% off sale, Jane Demsky purchased a hat for $25.99. What is the regular price of the hat?

51. Wage Cut A manufacturing plant is running at a deficit. To avoid layoffs, the workers agree on a temporary wage cut of 2%. If the average salary in the plant after the wage cut is $38,600, what was the average salary before the wage cut?

52. Teachers During the 2006 contract negotiations, the city school board approved a 5% pay increase for its teachers effective in 2007. If Dana Frick, a first-grade teacher, projects his 2007 annual salary to be $46,400, what is his present salary?

53. Earnings and Education The U.S. Census Bureau reported that in 2005, graduates with an associate's degree earned an average of 24.6% less than graduates with a bachelor's degree. If, in 2005, the average graduate with an associate's degree earned $37,600, determine the average salary of a graduate with a bachelor's degree.

54. Sales Volume Mona Fabricant receives a weekly salary of $350. She also receives a 6% commission on the total sales she makes. What must her sales be in a week, if she is to make a total of $710?

55. Salary Plans Vince McAdams, a salesman, is given a choice of two salary plans. Plan 1 is a weekly salary of $600 plus 2% commission of sales. Plan 2 is a straight commission of 10% of sales. How much in sales must Vince make in a week for both plans to result in the same salary?

56. Area The Johnson Performing Arts Center has increased in size. The area of the new building is 42% larger than the area of the original building. If the area of the new building is 56,000 square feet, determine the area of the original building.

57. Book The number of pages in the third edition of a book was 4% less than the number of pages in the second edition. If the number of pages in the third edition is 480, determine the number of pages in the second edition.

58. Financial Planning Belen Poltorade, a financial planner, is offering her customers two financial plans for managing their assets. With plan 1 she charges a planning fee of $1000 plus 1% of the assets she will manage for the customers. With plan 2 she charges a planning fee of $500 plus 2% of the assets she will manage. How much in customer assets would result in both plans having the same total fees?

59. Salary Plans Becky Schwartz, a saleswoman, is offered two salary plans. Plan 1 is $400 per week salary plus a 2% commission of sales. Plan 2 is a $250 per week salary plus a 16% commission of sales. How much would Becky need to make in sales for the salary to be the same from both plans?

60. Art Show Bill Rush wants to rent a building for a week to show his artwork and has been offered two rental plans. Plan 1 is a rental fee of $500 plus 3% of the dollar sales he makes. Plan 2 is $100 plus 15% of the dollar sales he makes. What dollar sales would result in both plans having the same total cost?

61. Eating Out After Linda Kodama is seated in a restaurant, she realizes that she has only $30. From this $30 she must pay a 7% tax and she wishes to leave a 15% tip on the price of the meal before tax. What is the maximum price for a meal that she can afford to pay?

62. Membership Fees The Holiday Health Club has reduced its annual membership fee by 10%. In addition, if you sign up on a Monday, the Club will take an additional $20 off the already reduced price. If Jorge Sanchez purchases a year's membership on a Monday and pays $250, what is the regular membership fee?

63. Estate Phil Dodge left an estate valued at $140,000. In his will, he specified that his wife will get 25% more of his estate than his daughter. How much will his wife receive?

64. Charitable Giving Charles Ford made a $200,000 cash contribution to two charities, the American Red Cross and the United Way. The amount received by the American Red Cross was 30% greater than the amount received by the United Way. How much did the United Way receive?

Challenge Problems

65. Average Value To find the *average* of a set of values, you find the sum of the values and divide the sum by the number of values.

a) If Paul Lavenski's first three test grades are 74, 88, and 76, write an equation that can be used to find the grade that Paul must get on his fourth exam to have an 80 average.

b) Solve the equation from part **a)** and determine the grade Paul must receive.

66. Driver Education A driver education course costs $45 but saves those under age twenty-five 10% of their annual insurance premiums until they reach age twenty-five. Scott Day has just turned 18, and his insurance costs $600 per year.

a) How long will it take for the amount saved from insurance to equal the price of the course?

b) Including the cost of the course, when Scott turns 25, how much will he have saved?

Cumulative Review Exercises

[1.9] **67.** Evaluate $4[(4 - 6) \div 2] + 3^2 - 1$.

[1.10] **68.** Name the following property: $3x + 4 = 4 + 3x$.

[2.6] **69.** Solve the formula $A = \frac{1}{2}bh$ for h.

[2.7] **70.** Solve the proportion $\frac{4.5}{6} = \frac{9}{x}$.

Mid-Chapter Test: 3.1–3.2

To find out how well you understand the chapter material to this point, take this brief test. The answers and the section where the material was initially discussed are given in the back of the book. Review any questions you answered incorrectly.

In Exercises 1–6, express each statement as an algebraic expression.

1. Six times the weight, w

2. Five inches more than 3 times the height, h

3. Represent the cost, c, increased by 20%, as a mathematical expression.

4. Dennis Donahue rents a truck for $40 per day plus 25 cents per mile, m. Write an expression for the total cost of the rental for one day.

5. Write an expression for the number of cents in n half-dollars.

6. Twenty-five dollars is divided between Amy Keyser and Sherry Norris. If Amy gets x dollars, how much will Sherry get?

7. Explain why the cost, c, of an item at a 25% off sale is not $c - 25$. Write the correct algebraic expression for the cost of an item at a 25% off sale.

8. In the statement, determine what $x =$.

    A Gaudy Leaf Frog is 2 centimeters longer than 3 times the length of a Poison Dart Frog (see photos).

Gaudy Leaf Frog

Poison Dart Frog

9. Select a variable to represent one quantity and state what that variable represents. Express the second quantity in terms of the variable selected.

    The distance Mary traveled is 6 miles more than 4 times the distance Pedro traveled.

10. The value of a car in 2005 was 18% less than its value in 2006. Write an expression for the difference in the value of the car from 2006 to 2005.

In Exercises 11 and 12, write the problem as an equation. Do not solve.

11. The population of Cedar Oaks increased by 12%. The population after the increase was 38,619.

12. For two consecutive odd integers, the sum of the smaller and 3 times the larger is 26.

In Exercises 13–20, write an equation that can be used to solve the problem. Solve the equation and answer the question asked.

13. **Consecutive Integers** The sum of two consecutive integers is 93. Find the numbers.

14. **Numbers** The larger of two integers is one less than 3 times the smaller. When the smaller number is subtracted from the larger, the difference is 7. Find the numbers.

15. **Candy** A candy manufacturer presently produces 240 boxes of candy a day and wants to increase production by 20 boxes per day until it produces 600 boxes of candy per day. How many days will it take for production to reach 600 boxes per day?

16. **Tennis** Kristina Schmid is considering joining one of two tennis clubs. At Dale's Tennis Club the monthly fee is $90, and court time is $4 per hour. At Abel's Tennis Club the monthly fee is $30, but court time is $8 per hour. How many hours in a month would Kristina need to play for the total cost to be the same with both clubs?

17. **Television** The cost of a television plus a 7% sales tax is $749. Find the cost of the television before tax.

18. **Clients** Anita and Betty together have a total of 600 clients. If Anita has 12 more than twice the number of clients Betty has, determine the number of clients each person has.

19. **Truck Rental** A truck cost $36 a day plus 18 cents a mile to rent. If the total cost for a one-day rental is $45.36, how many miles were driven?

20. **Salary Plans** A salesman is offered two salary plans. Plan 1 is $200 per week plus 8% commission of the dollar sales he makes. Plan 2 is $300 per week plus 6% of the dollar sales he makes. How much in sales must the salesman make in a week for the two plans to have the same total salary?

3.3 Geometric Problems

1 Solve geometric problems.

1 Solve Geometric Problems

This section serves two purposes. One is to reinforce the geometric formulas introduced in Section 2.6. The second is to reinforce procedures for setting up and solving verbal problems discussed in Sections 3.1 and 3.2. The more practice you have at setting up and solving the verbal problems, the better you will become at solving them.

FIGURE 3.3

EXAMPLE 1 ▸ **Sandbox** Mrs. Christine O'Connor is planning to build a sandbox for her daughter. She has 30 feet of lumber with which to build the perimeter: What should be the dimensions of the rectangular sandbox if the length is to be 3 feet longer than the width (**Fig. 3.3**)?

Solution Understand We are asked to find the dimensions of the sandbox that Christine plans to build. Since the amount of lumber that will be used to make the frame is 30 feet, the perimeter is 30 feet. Since the length is given in terms of the width, we will let the variable represent the width. Then we can express the length in terms of the variable selected for the width. To solve this problem, we use the formula for the perimeter of a rectangle, $P = 2l + 2w$, where $P = 30$ feet.

Translate   Let w = width of the sandbox.

Then $w + 3$ = length of the sandbox.

$$P = 2l + 2w$$
$$30 = 2(w + 3) + 2w$$

Carry Out

$$30 = 2w + 6 + 2w$$
$$30 = 4w + 6$$
$$24 = 4w$$
$$6 = w$$

The width is 6 feet. Since the length is 3 feet longer than the width, the length is $6 + 3 = 9$ feet.

Check We will check the solution by substituting the appropriate values in the perimeter formula.

$$P = 2l + 2w$$
$$30 \stackrel{?}{=} 2(9) + 2(6)$$
$$30 = 30 \qquad \textit{True}$$

Answer The width of the sandbox will be 6 feet and the length will be 9 feet.

▸ **Now Try Exercise 23**

EXAMPLE 2 ▸ **Corner Lot** A triangle that contains two sides of equal length is called an **isosceles triangle**. In isosceles triangles, the angles opposite the two sides of equal length have equal measures. Mr. and Mrs. Harmon Katz have a corner lot that is in the shape of an isosceles triangle. Two angles of their triangular lot are the same and the third angle is 30° greater than the other two. Find the measure of all three angles (see **Fig. 3.4**).

Solution Understand To solve this problem, you must know that the sum of the angles of any triangle measures 180°. We are asked to find the measure of each of the three angles, where the two smaller angles have the same measure. We will let the variable represent the measure of the smaller angles, and then we will express the larger angle in terms of the variable selected for the smaller angles.

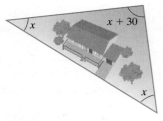

FIGURE 3.4

Translate   Let x = the measure of each smaller angle.

Then $x + 30$ = the measure of the larger angle.

$$\text{sum of the 3 angles} = 180$$
$$x + x + (x + 30) = 180$$

Carry Out

$$3x + 30 = 180$$
$$3x = 150$$
$$x = \frac{150}{3} = 50$$

The two smaller angles are each 50°. The larger angle is $x + 30°$ or $50° + 30° = 80°$.

Check and Answer   Since $50° + 50° + 80° = 180°$, the answer checks. The two smaller angles are each 50° and the larger angle is 80°.

▸ **Now Try Exercise 11**

Recall from Section 2.6 that a quadrilateral is a four-sided figure. Quadrilaterals include squares, rectangles, parallelograms, and trapezoids. The sum of the measures of the angles of any quadrilateral is 360°. We will use this information in Example 3.

EXAMPLE 3 ▸ **Water Trough** Sarah Fuqua owns horses and uses a water trough whose ends are trapezoids. The measure of the two bottom angles of the trapezoid are the same, and the measure of the two top angles are the same. The bottom angles measure 15° less than twice the measure of the top angles. Find the measure of each angle.

Solution   Understand   To help visualize the problem, we draw a picture of the trapezoid, as in **Figure 3.5**. We use the fact that the sum of the measures of the four angles of a quadrilateral is 360°.

Translate   Let $x =$ the measure of each of the two smaller angles.

Then $2x - 15 =$ the measure of each of the two larger angles.

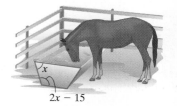

FIGURE 3.5

$$\left(\begin{array}{c}\text{measure of the}\\ \text{two smaller angles}\end{array}\right) + \left(\begin{array}{c}\text{measure of the}\\ \text{two larger angles}\end{array}\right) = 360$$

$$x + x + (2x - 15) + (2x - 15) = 360$$

Carry Out

$$x + x + 2x - 15 + 2x - 15 = 360$$
$$6x - 30 = 360$$
$$6x = 390$$
$$x = 65$$

Each smaller angle is 65°. Each larger angle is $2x - 15 = 2(65) - 15 = 115°$.

Check and Answer   Since $65° + 65° + 115° + 115° = 360°$, the answer checks. Each smaller angle is 65° and each larger angle is 115°.

▸ **Now Try Exercise 27**

EXAMPLE 4 ▸ **Fenced-In Area** Ronald Yates recently started an ostrich farm. He is separating the ostriches by fencing in three equal rectangular areas, as shown in **Figure 3.6**. The length of the fenced-in area, l, is to be 30 feet greater than the width and the total amount of fencing available is 660 feet. Find the length and width of the fenced-in area.

Solution   Understand   The fencing consists of four pieces of fence of length w, and two pieces of fence of length l.

Translate   Let $w =$ width of fenced-in area.

Then $w + 30 =$ length of fenced-in area.

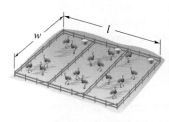

FIGURE 3.6

$$\left(\begin{array}{c}\text{4 pieces of fence}\\ \text{of length } w\end{array}\right) + \left(\begin{array}{c}\text{2 pieces of fence}\\ \text{of length } w + 30\end{array}\right) = 660$$

$$4w + 2(w + 30) = 660$$

Carry Out

$$4w + 2w + 60 = 660$$
$$6w + 60 = 660$$
$$6w = 600$$
$$w = 100$$

Since the width is 100 feet, the length is $w + 30$ or $100 + 30$ or 130 feet.

Check and Answer Since 4(100) + 2(130) = 660, the answer checks. The width of the ostrich farm is 100 feet and the length is 130 feet.

▶ **Now Try Exercise 37**

EXERCISE SET 3.3

Math XL
MathXL®

MyMathLab
MyMathLab

Concept/Writing Exercises

1. In the equation $A = l \cdot w$, what happens to the area if the length is doubled and the width is halved? Explain your answer.

2. In the equation $A = s^2$, what happens to the area if the length of a side, s, is tripled? Explain your answer.

3. In the equation $V = l \cdot w \cdot h$, what happens to the volume if the length, width, and height are all doubled? Explain your answer.

4. In the equation $C = 2\pi r$, what happens to the circumference if the radius is tripled? Explain your answer.

5. In the equation $A = \pi r^2$, what happens to the area if the radius is tripled? Explain your answer.

6. In the equation $V = \frac{4}{3}\pi r^3$, what happens to the volume if the radius is tripled? Explain your answer.

7. What is an isosceles triangle?

8. What is a quadrilateral?

9. What is the sum of the measures of the angles of a triangle?

10. What is the sum of the measures of the angles of a quadrilateral?

Practice the Skills/Problem Solving

*Solve the following geometric problems.**

11. **Isosceles Triangle** In an isosceles triangle, one angle is 42° greater than the other two equal angles. Find the measure of all three angles. See Example 2.

12. **Triangular Building** This building in New York City, referred to as the Flatiron Building, is in the shape of an isosceles triangle. If the shortest side of the building is 50 feet shorter than the two longer sides, and the perimeter around the building is 196 feet, determine the length of the three sides of the building.

13. **A Special Triangle** An **equilateral triangle** is a triangle that has three sides of the same length. The perimeter of an equilateral triangle is 34.5 inches. Find the length of each side.

14. **Equilateral Triangle** The perimeter of an equilateral triangle is 48.6 centimeters. Find the length of each side. See Exercise 13.

15. **Complementary Angles** Two angles are **complementary angles** if the sum of their measures is 90°. Angle A and angle B are complementary angles, and angle A is 21° more than twice angle B. Find the measures of angle A and angle B.

Complementary Angles

16. **Complementary Angles** Angles A and B are complementary angles, and angle B is 14° less than angle A. Find the measures of angle A and angle B. See Exercise 15.

17. **Supplementary Angles** Two angles are **supplementary angles** if the sum of their measures is 180°. Angle A and angle B are supplementary angles, and angle B is 8° less than three times angle A. Find the measures of angle A and angle B.

Supplementary Angles

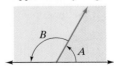

18. **Supplementary Angles** Angles A and B are supplementary angles and angle A is 2° more than 4 times angle B. Find the measures of angle A and angle B. See Exercise 17.

*See Appendix C for more material on geometry.

19. Vertical Angles When two lines cross, the opposite angles are called **vertical angles**. Vertical angles have equal measures. Determine the measures of the vertical angles indicated in the following figure.

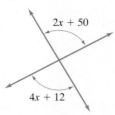

$2x + 50$

$4x + 12$

20. Vertical Angles A pair of vertical angles is indicated in the following figure. Determine the measure of the vertical angles indicated. See Exercise 19.

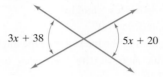

$3x + 38$ $5x + 20$

21. Unknown Angles One angle of a triangle is 10° greater than the smallest angle, and the third angle is 30° less than twice the smallest angle. Find the measures of the three angles.

22. Unknown Angles One angle of a triangle is 20° larger than the smallest angle, and the third angle is 6 times as large as the smallest angle. Find the measures of the three angles.

23. Dimensions of Rectangle The length of a rectangle is 6 feet more than its width. What are the dimensions of the rectangle if the perimeter is 44 feet?

24. Dimensions of Rectangle The perimeter of a rectangle is 120 feet. Find the length and width of the rectangle if the length is twice the width.

25. Tennis Court The length of a regulation tennis court is 6 feet greater than twice its width. The perimeter of the court is 228 feet. Find the length and width of the court.

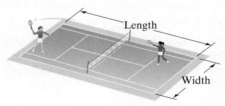

Length

Width

26. Patio Rikki Blair is building a rectangular patio. The perimeter of the patio is to be 96 feet. Determine the dimensions of the patio if the length is to be 6 feet less than twice the width.

27. Parallelogram In a parallelogram the opposite angles have the same measures. Each of the two larger angles in a parallelogram is 20° less than 3 times the smaller angles. Find the measure of each angle.

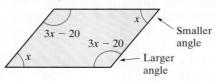

$3x - 20$ x
$3x - 20$ Smaller
                      angle
x Larger
                      angle

28. Parallelogram The two smaller angles of a parallelogram have equal measures, and the two larger angles each measure 27° less than twice each smaller angle. Find the measure of each angle.

29. Rhombus A rhombus is a parallelogram with four equal sides. Each of the two larger angles of a rhombus is 5 times as large as the two smaller angles. Find the measure of each of the four angles.

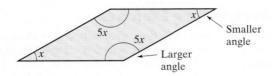

$5x$ x Smaller
      $5x$ angle
x Larger
                 angle

30. Rhombus Each of the two larger angles of a rhombus are 20° less than four times the two smaller angles. Find the measure of each of the four angles.

31. Quadrilateral The measure of one angle of a quadrilateral is 10° greater than the smallest angle; the third angle is 14° greater than twice the smallest angle; and the fourth angle is 21° greater than the smallest angle. Find the measures of the four angles of the quadrilateral.

32. Quadrilateral The measure of one angle of a quadrilateral is twice the smallest angle; the third angle is 20° greater than the smallest angle; and the fourth angle is 20° less than twice the smallest angle. Find the measures of the four angles of the quadrilateral.

33. Building a Bookcase A bookcase is to have four shelves, including the top, as shown. The height of the bookcase is to be 3 feet more than the width. Find the width and height of the bookcase if only 30 feet of lumber is available.

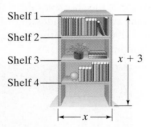

Shelf 1
Shelf 2
Shelf 3 $x + 3$
Shelf 4

x

34. Bookcase A bookcase is to have four shelves as shown. The height of the bookcase is to be 2 feet more than the width, and only 20 feet of lumber is available. What should be the width and height of the bookcase?

1
2
3
4

35. Bookcase What should be the width and height of the bookcase in Exercise 34 if the height is to be twice the width?

36. Storage Shelves Carlotta plans to build storage shelves as shown. She has only 45 feet of lumber for the entire unit and wishes the width to be 3 times the height. Find the width and height of the unit.

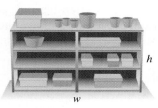

37. Fenced-In Area A rectangular area is to be fenced in along a straight river bank as illustrated. The length of the fenced-in area is to be 5 feet greater than the width, and the total amount of fencing to be used is 71 feet. Find the width and length of the fenced-in area.

38. Gardening Trina Zimmerman is placing a border around and within a garden where she intends to plant flowers (see the figure). She has 60 feet of bordering, and the length of the garden is to be 2 feet greater than the width. Find the length and width of the garden. The red shows the location of all the bordering in the figure.

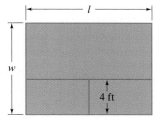

Challenge Problems

39. One way to express the area of the figure on the right is $(a + b)(c + d)$. Can you determine another expression, using the area of the four rectangles, to represent the area of the figure?

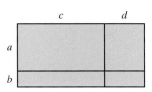

Group Activity

Discuss and answer Exercise 40 as a group.

40. Consider the four pieces shown. Two are squares and two are rectangles.

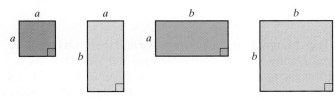

a) Individually, rearrange and place the four pieces together to form one square.

b) The area of the square you constructed is $(a + b)^2$. Write another expression for the area of the square by adding the four individual areas.

c) Compare your answers. If each member of the group did not get the same answers to parts **a)** and **b)**, work together to determine the correct answer.

d) Answer the following question as a group. If b is twice the length of a, and the perimeter of the square you created is 54 inches, find the length of a and b.

e) Use the values of a and b found in part **d)** to find the area of the square you created.

f) Use the values of a and b found in part **d)** to find the areas of the four individual pieces that make up the large square.

g) Does the sum of the areas of the four pieces found in part **f)** equal the area of the large square found in part **e)**? Is this what you expected? Explain.

Cumulative Review Exercises

Insert either $>$, $<$, or $=$ in each shaded area to make the statement true.

[1.5] **41.** $-|-6|$ ▢ $|-4|$

42. $|-3|$ ▢ $-|3|$

[1.7] **43.** Evaluate $-8 - (-2) + (-4)$.

[2.1] **44.** Simplify $-7y + x - 3(x - 2) + 2y$.

[2.6] **45.** Solve $6x + 3y = 9$ for y.

3.4 Motion, Money, and Mixture Problems

1 Solve motion problems involving two rates.

2 Solve money problems.

3 Solve mixture problems.

We now discuss three additional types of applications: motion, money, and mixture problems. These problems are grouped in the same section because, as you will learn shortly, you use the same general multiplication procedure to solve them. We begin by discussing motion problems.

1 Solve Motion Problems Involving Two Rates

A **motion problem** is one in which an object is moving at a specific rate for a specific period of time. A car traveling at a constant speed or a person walking at a constant speed may be considered motion problems. Motion problems are often solved using the distance formula, distance = rate × time, that was given in Section 2.6. In this section, we will discuss motion problems that involve *two rates*, such as two trains traveling at different speeds. In these problems, we generally begin by letting the variable represent one of the unknown quantities, and then we represent the second unknown quantity in terms of the first unknown quantity. For example, suppose that one train travels 20 miles per hour faster than another train. We might let r represent the rate of the slower train and $r + 20$ represent the rate of the faster train.

To solve problems of this type that use the distance formula and involve two different rates, we can construct a table, like the following one, to organize the information. The formula at the top of the table shows how the distance in the last column is calculated.

Rate × Time = Distance

Item	Rate	Time	Distance
Item 1			distance 1
Item 2			distance 2

Depending upon the information given in the problem, we generally add the two distances, or subtract the smaller distance from the larger, or set the two distances equal to each other. We can set up one of two types of equations, as indicated below, to solve the problem.

$$\text{distance 1} + \text{distance 2} = \text{total distance}$$
$$\text{distance 1} - \text{distance 2} = \text{difference in distance}$$
$$(\text{or distance 2} - \text{distance 1} = \text{difference in distance})$$

Examples 1 and 2 illustrate the procedure used.

EXAMPLE 1 ▸ **Camping Trip** Maryanne and Paul Justinger and their son Danny are on a canoe trip on the Erie Canal. Danny is in one canoe and Paul and Maryanne are in a second canoe. Both canoes start at the same time from the same point and travel in the same direction. The parents paddle their canoe at 2 miles per hour and their son paddles his canoe at 4 miles per hour. In how many hours will the two canoes be 5 miles apart?

Solution Understand and Translate   We are asked to find the time it takes for the canoes to become separated by 5 miles. We will construct a table to aid us in setting up the problem.

Let t = time when canoes are 5 miles apart.

We draw a sketch to help visualize the problem (**Fig. 3.7**). When the two canoes are 5 miles apart, each has traveled for the same number of hours, t.

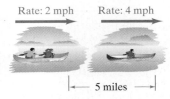

Rate: 2 mph   Rate: 4 mph

|← 5 miles →|

FIGURE 3.7

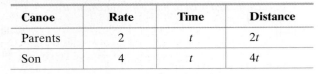

Canoe	Rate	Time	Distance
Parents	2	t	$2t$
Son	4	t	$4t$

Since the canoes are traveling in the same direction, the distance between them is found by subtracting the distance traveled by the slower canoe from the distance traveled by the faster canoe.

$$\left(\begin{array}{c} \text{distance traveled} \\ \text{by faster canoe} \end{array} \right) - \left(\begin{array}{c} \text{distance traveled} \\ \text{by slower canoe} \end{array} \right) = 5 \text{ miles}$$

Carry Out

$$4t - 2t = 5$$
$$2t = 5$$
$$t = 2.5$$

Answer   After 2.5 hours the two canoes will be 5 miles apart.

▶ **Now Try Exercise 1**

EXAMPLE 2 ▶ **Paving Roads** Two highway paving crews are 20 miles apart working toward each other. One crew paves 0.4 mile of road per day more than the other crew, and the two crews meet after 10 days. Find the rate at which each crew paves the road.

Solution   Understand and Translate   We are asked to find the two rates. We are told that both crews work for 10 days.

$$\text{Let } r = \text{rate of slower crew.}$$
$$\text{Then } r + 0.4 = \text{rate of faster crew.}$$

We make a sketch (**Fig. 3.8**) and set up a table of values.

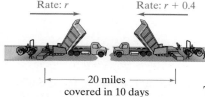

Rate: r Rate: $r + 0.4$

├── 20 miles ──┤
covered in 10 days

FIGURE 3.8

Crew	Rate	Time	Distance
Slower	r	10	$10r$
Faster	$r + 0.4$	10	$10(r + 0.4)$

The total distance covered by both crews is 20 miles. Since the crews are moving in opposite directions, the distance between them is found by adding the two distances.

$$\left(\begin{array}{c} \text{distance covered} \\ \text{by slower crew} \end{array} \right) + \left(\begin{array}{c} \text{distance covered} \\ \text{by faster crew} \end{array} \right) = 20 \text{ miles}$$

Carry Out

$$10r + 10(r + 0.4) = 20$$
$$10r + 10r + 4 = 20$$
$$20r + 4 = 20$$
$$20r = 16$$
$$\frac{20r}{20} = \frac{16}{20}$$
$$r = 0.8$$

Answer   The slower crew paves 0.8 mile of road per day and the faster crew paves $r + 0.4$ or $0.8 + 0.4 = 1.2$ miles of road per day.

▶ **Now Try Exercise 17**

Helpful Hint

Notice in Example 1 that the canoes are traveling in the *same direction* and the solution involves *subtracting* the smaller distance from the larger distance. In Example 2, the crews are traveling in the *opposite directions* and the solution involves *addition*. In general, when working with two different moving items, if the items are moving in the same direction, the solution will involve subtracting the smaller distance from the larger distance. If the items are moving in opposite directions, the solution will involve adding the distances together.

2 Solve Money Problems

Now we will work some examples that involve *two items*, where each is assigned a specific amount of money. We place these examples here because they are solved using a procedure very similar to the procedure used to solve motion problems with two rates. One type of money problem involves simple interest. The simple interest formula, interest = principal · rate · time, was discussed in Section 2.6. In Example 3, we use the simple interest formula, where two different principals are involved.

When working with interest problems involving two amounts, we often let the variable represent one amount of money, then we represent the second amount in terms of the variable. For example, if we know the total amount invested in two simple interest accounts is $20,000, we might let x represent the amount in one account, then $20,000 - x$ would be the amount invested in the other account. When solving an interest problem involving two amounts, we can use a table, as illustrated below, just as we did when working with motion problems.

Principal × Rate × Time = Interest

Account	Principal	Rate	Time	Interest
Account 1				interest 1
Account 2				interest 2

After determining the interest columns on the right, we generally use one of the following formulas, depending on the question, to determine the answer.

$$\text{interest 1} + \text{interest 2} = \text{total interest}$$
$$\text{interest 1} - \text{interest 2} = \text{difference in interest}$$
$$(\text{or interest 2} - \text{interest 1} = \text{difference in interest})$$
$$\text{interest 1} = \text{interest 2}$$

Do you see the similarities with the motion problems? Now let's work an example.

EXAMPLE 3 ▶ **Investments** Carmine DeSanto has $15,000 to invest. He is considering two investments. One is a loan he can make to another party that pays him 8% simple interest for a year. A second investment is a 1-year certificate of deposit that pays 5%. Carmine decides that he wants to place some money in each investment, and he wants to earn a total of $1125 interest in 1 year from the two investments. How much money should Carmine put in each investment?

Solution Understand and Translate We use the simple interest formula, interest = principal · rate · time, to solve this problem.

$$\text{Let } x = \text{amount to be invested at 5\%.}$$
$$\text{Then } 15,000 - x = \text{amount to be invested at 8\%.}$$

Account	Principal	Rate	Time	Interest
CD	x	0.05	1	$0.05x$
Loan	$15,000 - x$	0.08	1	$0.08(15,000 - x)$

Since the sum of the interest from the two investments is $1125, we write the equation

$$\left(\begin{array}{c}\text{interest from} \\ \text{5\% CD}\end{array}\right) + \left(\begin{array}{c}\text{interest from} \\ \text{8\% investment}\end{array}\right) = \text{total interest}$$

$$0.05x + 0.08(15,000 - x) = 1125$$

Carry Out
$$0.05x + 0.08(15,000) - 0.08(x) = 1125$$
$$0.05x + 1200 - 0.08x = 1125$$
$$-0.03x + 1200 = 1125$$
$$-0.03x = -75$$
$$x = \frac{-75}{-0.03} = 2500$$

Check and Answer   Thus, $2500 should be invested at 5% interest. The amount to be invested at 8% is

$$15{,}000 - x = 15{,}000 - 2500 = 12{,}500$$

The total amount invested is $2500 + $12,500 = $15,000, which checks with the information given.

▶ Now Try Exercise 25

In Example 3, we let x represent the amount invested at 5%. If we had let x represent the amount invested at 8%, the answer would not have changed. Rework Example 3 now, letting x represent the amount invested at 8%.

In other types of problems involving two amounts of money, we generally set up similar tables, as illustrated in the next example.

EXAMPLE 4 ▶ **Rocking Chairs** Johnson's Patio Furniture Store sells two types of rocking chairs. The single-person rocking chair sells for $130 each and the two-person rocking chair sells for $240 each. On a given day 10 rocking chairs were sold for a total of $1740. Determine the number of single-person and the number of two-person rocking chairs that were sold.

Solution   Understand and Translate   We are asked to find the number of each type of rocking chair sold.

Let x = number of single-person rocking chairs sold.

Then $10 - x$ = number of two-person rocking chairs sold.

The income received from the sale of the single-person rocking chairs is found by multiplying the number of single-person rocking chairs sold by the cost of a single-person rocking chair. The income received from the sale of the two-person rocking chairs is found by multiplying the number of two-person rocking chairs sold by the cost of a two-person rocking chair.

$$\left(\begin{array}{c}\text{Number of}\\\text{Rocking chairs}\end{array}\right) \times \left(\begin{array}{c}\text{Cost of}\\\text{Rocking chairs}\end{array}\right) = \left(\begin{array}{c}\text{Income from}\\\text{Rocking chairs}\end{array}\right)$$

Rocking Chair	Number of Rocking Chairs	Cost	Income from Rocking Chairs
Single	x	130	$130x$
Double	$10 - x$	240	$240(10 - x)$

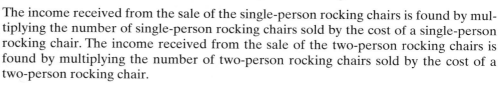

$$\left(\begin{array}{c}\text{income from}\\\text{single-person}\\\text{rocking chairs}\end{array}\right) + \left(\begin{array}{c}\text{income from}\\\text{two-person}\\\text{rocking chairs}\end{array}\right) = \text{total income}$$

Carry Out
$$130x + 240(10 - x) = 1740$$
$$130x + 2400 - 240x = 1740$$
$$-110x + 2400 = 1740$$
$$-110x = -660$$
$$x = \frac{-660}{-110} = 6$$

Check and Answer   Six single-person rocking chairs and $10 - 6$ or 4 two-person rocking chairs were sold.

Check

income from 6 single-person rocking chairs =   780
income from 4 two-person rocking chairs =   960
total = 1740   *True*

▶ Now Try Exercise 35

3 Solve Mixture Problems

Now we will work some mixture problems. Any problem in which two or more quantities are combined to produce a different quantity or a single quantity is separated into two or more different quantities may be considered a **mixture problem**. Mixture problems are familiar to everyone, as we can see in the everyday examples that follow.

Mixture problems in this section will generally be one of two types. In one type, we will mix two solids, as illustrated in **Figure 3.9a**, and be concerned about the value or cost of the mixture. In the second type, we will mix two liquids or solutions, as illustrated in **Figure 3.9b**, and be concerned about the content or strength of the mixture.

<table>
<tr><td align="center">Type 1
Mixing solids together</td><td align="center">Type 2
Mixing liquids, or solutions, together</td></tr>
</table>

<table>
<tr><td align="center">(a) Concerned about the value or
the cost of the mixture</td><td align="center">(b) Concerned about the content or
the strength of the mixture</td></tr>
</table>

FIGURE 3.9

As we did with motion problems involving two rates and money problems, we will use a table to help analyze mixture problems.

When we construct a table for mixture problems, our table will generally have three rows instead of two as with motion and money problems. One row will be for each of the two individual items being mixed, and the third row will be for the mixture of the two items.

Type 1—Mixing Solids

When working with mixture problems involving solids, we often let the variable represent one unknown quantity, and then we represent a second unknown quantity in terms of the first unknown quantity. For example, if we know the total weight of a mixture of two items is 10 pounds, and one of the items in the mixture weighs x pounds, then the weight of the other item in the mixture will be $10 - x$ pounds.

When working with mixture problems involving solids, we generally use the fact that the value (or cost) of one part of the mixture plus the value (or cost) of the second part of the mixture is equal to the total value (or total cost) of the mixture.

When we are combining two solid items and are interested in the *value* of the mixture, the following table, or a variation of it, is often used.

Quantity × Price (per unit) = Value of Item

Item	Quantity	Price	Value of Item
item 1			value of item 1
item 2			value of item 2
mixture			value of mixture

When we use this table, we generally use the following formula to solve the problem.

value of item 1 + value of item 2 = value of mixture

Now let's look at a mixture problem where we discuss the value or cost of the mixture.

EXAMPLE 5 ▶ Grass Seed Scott's® Family grass seed sells for $2.65 per pound, and Scott's Spot Filler grass seed sells for $2.30 per pound. How many pounds of each should be mixed to get a 10-pound mixture that sells for $2.40 per pound?

Solution **a) Understand and Translate** We are asked to find the number of pounds of each type of grass seed.

Let x = number of pounds of Family grass seed.

Then $10 - x$ = number of pounds of Spot Filler grass seed.

We make a sketch of the situation (**Fig. 3.10**), then construct a table.

FIGURE 3.10

The cost or value of the seeds is found by multiplying the number of pounds by the price per pound.

Type of Seeds	Number of Pounds	Cost per Pound	Cost of Seeds
Family	x	2.65	$2.65x$
Spot Filler	$10 - x$	2.30	$2.30(10 - x)$
Mixture	10	2.40	$2.40(10)$

$$\left(\begin{array}{c}\text{cost of}\\\text{Family Seed}\end{array}\right) + \left(\begin{array}{c}\text{cost of Spot}\\\text{Filler Seed}\end{array}\right) = \text{cost of mixture}$$

$$2.65x + 2.30(10 - x) = 2.40(10)$$

Carry Out

$$2.65x + 23.0 - 2.30x = 24.0$$
$$0.35x + 23.0 = 24.0$$
$$0.35x = 1.00$$
$$x \approx 2.86$$

Answer Thus, about 2.86 pounds of the Family grass seed must be mixed with $10 - x$ or $10 - 2.86 = 7.14$ pounds of the Spot Filler grass seeds to make a mixture that sells for $2.40 a pound.

▶ **Now Try Exercise 39**

Now we will first discuss the second type of mixture problem, involving mixing solutions.

Type 2—Mixing Solutions

When solving mixture problems involving solutions, we often let the variable represent one unknown quantity, and then we represent a second unknown quantity in terms of the first unknown quantity. For example, if we know that when two solutions are mixed they make a total of 8 liters, we may represent the number of liters of one of the solutions as x and the number of liters of the second solution as $8 - x$. Note that when we add x and $8 - x$ we get 8, the total amount.

We generally solve mixture problems involving solutions by using the fact that the amount of one part of the mixture plus the amount of the second part of the mixture is equal to the total amount of the mixture.

When working with solutions, we use the formula, *amount of substance in the solution = quantity of solution × strength of solution (in percent)*. When we are mixing two quantities and are interested in the *composition* of the mixture, we generally use the following table or a variation of the table.

Quantity × Strength (in percent) = Amount of Substance

Solution	Quantity	Strength	Amount of Substance
Solution 1			amount of substance in solution 1
Solution 2			amount of substance in solution 2
Mixture			amount of substance in mixture

When using this table, we generally use the following formula to solve the problem.

$$\left(\begin{array}{c}\text{amount of substance}\\\text{in solution 1}\end{array}\right) + \left(\begin{array}{c}\text{amount of substance}\\\text{in solution 2}\end{array}\right) = \left(\begin{array}{c}\text{amount of substance}\\\text{in mixture}\end{array}\right)$$

Let us now look at an example of a mixture problem where two solutions are combined.

EXAMPLE 6 ▶ Mixing Acid Solutions Mr. Dave Lumsford needs a 10% acetic acid solution for a chemistry experiment. After checking the store room, he finds that there are only 5% and 20% acetic acid solutions available. Mr. Lumsford decides to make the 10% solution by combining the 5% and 20% solutions. How many liters of the 5% solution must he add to 8 liters of the 20% solution to get a solution that is 10% acetic acid?

Solution Understand and Translate   We are asked to find the number of liters of the 5% acetic acid solution to mix with 8 liters of the 20% acetic acid solution.

Let x = number of liters of 5% acetic acid solution.

Let's draw a sketch of the problem (**Fig. 3.11**).

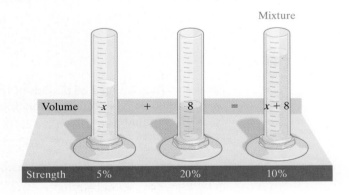

FIGURE 3.11

The amount of acid in a given solution is found by multiplying the number of liters by the percent strength.

Solution	Liters	Strength	Amount of Acetic Acid
5%	x	0.05	$0.05x$
20%	8	0.20	$0.20(8)$
Mixture	$x + 8$	0.10	$0.10(x + 8)$

$$\left(\begin{array}{c}\text{amount of acid}\\\text{in 5\% solution}\end{array}\right) + \left(\begin{array}{c}\text{amount of acid}\\\text{in 20\% solution}\end{array}\right) = \left(\begin{array}{c}\text{amount of acid}\\\text{in 10\% mixture}\end{array}\right)$$

Carry Out

$$0.05x + 0.20(8) = 0.10(x + 8)$$
$$0.05x + 1.6 = 0.10x + 0.8$$
$$0.05x + 0.8 = 0.10x$$
$$0.8 = 0.05x$$
$$\frac{0.8}{0.05} = x$$
$$16 = x$$

Answer   Sixteen liters of 5% acetic acid solution must be added to the 8 liters of 20% acetic acid solution to get a 10% acetic acid solution. The total number of liters that will be obtained is $16 + 8$ or 24.

▶ **Now Try Exercise 47**

EXAMPLE 7 ▶ Nicole Pappas, a medical researcher, has 40% and 5% solutions of phenobarbital. How much of each solution must she mix to get 0.6 liter of a 20% phenobarbital solution?

Solution   Understand and Translate   We are asked to find how much of the 40% and 5% phenobarbital solutions must be mixed to get 0.6 liter of a 20% solution. We can choose to let x be the amount of either the 40% or the 5% solution. We will choose as follows:

$$\text{Let } x = \text{number of liters of the 40\% solution.}$$
$$\text{Then } 0.6 - x = \text{number of liters of the 5\% solution.}$$

Remember from Section 3.1 that if a total of 0.6 liter is divided in two, if one part is x, the other part is $0.6 - x$.

Let's draw a sketch of the problem (**Fig. 3.12**).

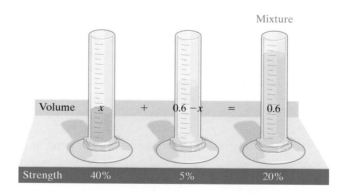

FIGURE 3.12

The amount of phenobarbital in a given solution is found by multiplying the number of liters by the percent strength.

Solution	Liters	Strength	Amount of Phenobarbital
40%	x	0.40	$0.40x$
5%	$0.6 - x$	0.05	$0.05(0.6 - x)$
Mixture	0.6	0.20	$0.6(0.20)$

$$\left(\begin{array}{c}\text{amount of phenobarbital}\\\text{in 40\% solution}\end{array}\right) + \left(\begin{array}{c}\text{amount of phenobarbital}\\\text{in 5\% solution}\end{array}\right) = \left(\begin{array}{c}\text{amount of phenobarbital}\\\text{in mixture}\end{array}\right)$$

$$0.40x + 0.05(0.6 - x) = (0.6)(0.20)$$

Carry Out

$$0.40x + 0.03 - 0.05x = 0.12$$
$$0.35x + 0.03 = 0.12$$
$$0.35x = 0.09$$
$$x \approx 0.26$$

Answer   Since the answer was less than 0.6 liter, the answer is reasonable. About 0.26 liter of the 40% solution must be mixed with about $0.6 - x = 0.60 - 0.26 = 0.34$ liter of the 5% solution to get 0.6 liter of the 20% mixture.

▶ **Now Try Exercise 55**

In Example 7, we chose to let $x =$ number of liters of the 40% solution. We could have selected to let $x =$ number of liters of the 5% solution. Then $0.6 - x$ would be the number of liters of the 40% solution. Had you worked the problem out like this, you would have found that x was approximately 0.34 liter. Try reworking Example 7 now letting $x =$ number of liters of the 5% solution.

EXAMPLE 8 ▶ An orange punch contains 4% orange juice. If 5 ounces of water is added to 8 ounces of the punch, determine the percent of orange juice in the mixture.

Solution   **Understand and Translate**   We are asked to find the percent of orange juice in the mixture.

Let $x =$ percent of orange juice in the mixture.

We will again set up a table.

Solution	Ounces	Percent of Juice	Amount of Juice
Punch	8	0.04	8(0.04)
Water	5	0.00	5(0.00)
Mixture	13	x	13x

$$\left(\begin{array}{c} \text{amount of juice} \\ \text{in punch} \end{array} \right) + \left(\begin{array}{c} \text{amount of juice} \\ \text{in water} \end{array} \right) = \left(\begin{array}{c} \text{amount of juice} \\ \text{in mixture} \end{array} \right)$$

$$8(0.04) + 5(0.00) = 13x$$

Carry Out

$$0.32 + 0.00 = 13x$$
$$0.32 = 13x$$
$$0.025 \approx x$$

Answer   Therefore, the percent of juice in the mixture is about 2.5%.

▶ **Now Try Exercise 51**

EXERCISE SET 3.4

Practice the Skills/Problem Solving

In Exercises 1–58, set up an equation that can be used to solve each problem. Solve the equation, and answer the question. Use a calculator when you feel it is appropriate.

In Exercises 1–24, solve the motion problem. See Examples 1 and 2.

1. **Ferries** Two high-speed ferries leave at the same time from Ft. Myers, Florida, going to Key West, Florida. The first ferry, the *Cat*, travels at 34 miles per hour. The second ferry, the *Bird*, travels at 28 miles per hour. In how many hours will the two ferries be 6 miles apart?

2. **Trains** Two trains in New York City start at the same station going in the same direction on sets of parallel tracks. The local train stops often and averages 18.4 miles per hour. The express train stops less frequently and averages 30.2 miles per hour. In how many hours will the two trains be 5.9 miles apart?

3. **Horseback Riding** Two friends, Jodi Cotton and Abe Mantell, go horseback riding on the same trail in the same direction. Jodi's horse travels at 8 miles per hour while Abe's horse travels at a slower pace. After 2 hours they are 4 miles apart. Find the speed at which Abe's horse is traveling.

4. **Camel Riding** In the Outback in Australia, Betty Sue Adams and Carl Minieri go camel riding in the same direction along the same path. Betty Sue's camel travels at 6 miles per hour while Carl's camel travels at a slower pace. After 3 hours they are 2.4 miles apart. Find the speed of Carl's camel.

5. **Airplanes** A Jet Blue airplane leaves Chicago for New York at the same time a Southwest airplane leaves New York for Chicago. The distance from Chicago to New York is 821 miles. If the Jet Blue plane travels at 560 miles per hour and the Southwest plane travels at 580 miles per hour, how long into their flights will the two planes pass each other?

6. **Walking** Barb Dansky and Sandy Spears are at opposite ends of a shopping mall 4780.4 feet apart walking toward each other. If Barb walks 1.5 feet per second and Sandy walks 2.2 feet per second, how long will it be before they meet?

7. **Walkie-Talkies** Willie and Shanna Johnston have walkie-talkies that have a range of 16.8 miles. Willie and Shanna start at the same point and walk in opposite directions. If Willie walks 3 miles per hour and Shanna walks 4 miles per hour, how long will it take before they are out of range?

8. **Blue Angels** At a Navy Blue Angel air show two F/A-18 Hornet jets travel toward each other, both at a speed of 1000 miles per hour. After they pass each other, if they were to keep flying at the same speed in opposite directions, how long would it take for them to be 500 miles apart?

9. **Product Testing** The Goodyear Tire Company is testing new tires by placing them on a machine that can simulate the tires riding on a road. First, the machine runs the tires for 7.2 hours at 60 miles per hour. Then the tires are run for 6.8 hours at a different speed. After this 14-hour period, the machine indicates that the tires have traveled the equivalent of 908 miles. Find the second speed to which the machine was set.

10. **Ski Lifts** To get to the top of Whistler Mountain, people must use two different ski lifts. The first lift travels 4 miles per hour for 0.2 hours. The second lift travels for 0.3 hours to reach the top of the mountain. If the total distance traveled up the mountain is 1.2 miles, find the average speed of the second ski lift.

11. **Earthquakes** Earthquakes generate circular *p*-waves and *s*-waves, which travel outward (see the figure). Suppose the *p*-waves have a velocity of 3.6 miles per second and the *s*-waves have a velocity of 1.8 miles per second. How long after the earthquake will *p*-waves and *s*-waves be 80 miles apart?

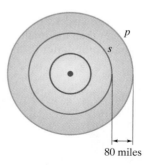

80 miles

12. **Navigating O'Hare Airport** Sadie Bragg and Jose Cruz are in Chicago's O'Hare airport walking between terminals. Sadie walks on the moving walkway (like a flat escalator moving along the floor). Her speed (relative to the ground) is 220 feet per minute. Jose starts walking at the same time and walks alongside the walkway at a speed of 100 feet per minute. How long have they been walking when Sadie is 600 feet ahead of Jose?

13. Disabled Boat Two Coast Guard cutters are 225 miles apart traveling toward each other, one from the east and the other from the west, searching for a disabled boat. The eastbound cutter travels 5 miles per hour faster than the westbound cutter. If the two cutters pass each other after 3 hours, find the average speed of each cutter.

14. Snowplowing Two snowplowing crews are clearing snow from a 2.56-mile runway at an airport. Crew A starts at one end of the runway and crew B starts at the other end. They start cleaning at the same time and they work toward each other. Crew A clears the runway at a speed of 0.4 miles per hour faster than crew B. If they meet 0.5 hours after they start, find the average speed of each snowplow.

15. Round Trip Samia Metwali walks for a time at 4 miles per hour, then slows down and walks at 3.2 miles per hour. The total distance she walked was 6 miles. If she walked for 0.5 hour more at 3.2 miles per hour than she did at 4 miles per hour, determine the time Samia walked at 4 miles per hour.

16. Visit to Grandchild Chuck Neumann drove from his house in Auburn Hills, Michigan, to visit his grandchild in Pasadena, Texas, a distance of 1343 miles. Part of the way he drove at 60 miles per hour and part of the way he drove at 70 miles per hour. If he drove for 0.5 hour more at 60 miles per hour than he did at 70 miles per hour, determine the time Chuck drove at 60 miles per hour.

17. Paving Road Two crews are laying blacktop on a road. They start at the same time at opposite ends of a 12-mile road and work toward one another. One crew lays blacktop at an average rate of 0.75 mile a day faster than the other crew. If the two crews meet after 3.2 days, find the rate of each crew.

18. Beach Clean-Up On Earth Day, two groups of people clean a 7-mile stretch of Myrtle Beach. Auturo's group and Jane's group start at the same time at opposite ends of the beach and walk toward each other. Auturo's group is traveling at a rate of 0.5 mile per hour faster than Jane's group, and they meet in 2 hours. Find the speed of each group.

19. Sailing Two sailboats are 9.8 miles apart and sailing toward each other. The larger boat, the *Pythagoras*, sails 4 miles per hour faster than the smaller boat, the *Apollo*. The two boats pass each other after 0.7 hour. Find the speed of each boat.

20. Exercising Dien and Phuong Vu belong to a health club and exercise together regularly. They start running on two treadmills at the same time. Dien's machine is set for 6 miles per hour and Phuong's machine is set for 4 miles per hour. When they finish, they compare the distances and find that together they have run a total of 11 miles. How long had they run?

21. Traffic Jam Betty Truitt drives for a number of hours at 70 miles per hour. When traffic slows, she drives at 50 miles per hour. She travels at 50 miles per hour for 0.5 hour longer than she traveled at 70 miles per hour. The difference in the distance traveled at 50 mph and 70 mph is 5 miles. Determine how long Betty traveled at 50 miles per hour.

22. Salt Mine The ore at a mine must travel on two different conveyer belts to be loaded onto a train. The second conveyer belt travels at a rate of 0.6 foot per second faster than the first conveyer belt. The ore travels 180 seconds on the first belt and 160 seconds on the second belt. If the total distance traveled by the ore is 1116 feet, determine the speed of the second belt.

23. Ironman Triathlon A triathlon consists of three parts: swimming, cycling, and running. Participants in the Ironman Triathlon in Hawaii must swim, cycle, and run certain distances. The 2005 woman's winner was Switzerland's Natascha Badmann. She swam at an average of 2.31 miles per hour for 1.04 hours then cycled at an average of 23.00 miles per hour for 4.87 hours. Finally, she ran at an average of 8.45 miles per hour for about 3.10 hours.

   a) Estimate the distance that Natascha swam.

   b) Estimate the distance that Natascha cycled.

   c) Estimate the distance that Natascha ran.

   d) Estimate the total distance covered during the triathlon.

   e) Estimate the winning time of the triathlon.*

24. Ironman Triathlon Refer to Exercise 23, about the Ironman Triathlon. The 2005 men's winner of the Ironman was Faris Al-Sultan of Germany. He swam at an average speed of 2.89 miles per hour for 0.83 hour, then cycled at an average speed of 25.4 miles per hour for 4.40 hours, then ran at an average speed of 8.98 miles per hour for 2.92 hours. Answer parts a)–e) of the question in Exercise 23.

---

*The actual distances are: swim, 2.4 mi; cycle, 112 mi; run 26 mi, 385 yd.

In Exercises 25–38, solve the money problems. See Examples 3 and 4.

25. Simple Interest Paul and Donna Petrie invested $12,000, part at 5% simple interest and the rest at 7% simple interest for a period of 1 year. How much did they invest at each rate if their total annual interest from both investments was $800? (Use interest = principal·rate·time.)

26. Simple Interest Jerry Correa invested $7000, part at 8% simple interest and the rest at 5% simple interest for a period of 1 year. If he received a total annual interest of $476 from both investments, how much did he invest at each rate?

27. Simple Interest Aleksandra Tomich invested $6000, part at 6% simple interest and part at 4% simple interest for a period of 1 year. How much did she invest at each rate if each account earned the same interest?

28. Simple Interest Susan Foreman invested $12,500, part at 7% simple interest and part at 6% simple interest for a period of 1 year. How much was invested at each rate if each account earned the same interest?

29. Simple Interest Míng Wang invested $10,000, part at 4% and part at 5% simple interest for a period of 1 year. How much was invested in each account if the interest earned in the 5% account was $320 greater than the interest earned in the 4% account?

30. Simple Interest Sharon Sledge invested $20,000, part at 5% and part at 7% simple interest for a period of 1 year. How much was invested in each account if the interest earned in the 7% account was $440 greater than the interest earned in the 5% account?

31. Rate Increase Patricia Burgess knows that at some point during the calendar year her basic monthly telephone rate increased from $17.10 to $18.40. If she paid a total of $207.80 for basic telephone service for the calendar year, in what month did the rate increase take effect?

32. Cable TV Violet Kokola knows that her subscription rate for the basic tier of cable television increased from $18.20 to $19.50 at some point during the calendar year. She paid a total of $230.10 for the year to the cable company. Determine the month of the rate increase.

33. Wages Mihály Sarett holds two part-time jobs. One job, at Home Depot, pays $6.50 an hour and the second job, at a veterinary clinic, pays $7.00 per hour. Last week Mihály worked a total of 18 hours and earned $122.00. How many hours did Mihály work at each job?

34. Rock and Roll Hall of Fame At the Rock and Roll Hall of Fame in Cleveland, Ohio, adult admission is $20 and children admission is $11. During one day, a total of 1551 adult and children admissions were collected, and $24,441 in admission fees were collected. How many children admissions were collected?

35. Baseball Hall of Fame At the Baseball Hall of Fame in Cooperstown, New York, adult admission is $14.50 and children admission is $5.00. During one day, a total of 2100 adult and children admissions were collected, and $21,900 in admission fees was collected. How many adult admissions were collected?

36. Ticket Sales At a movie theater an evening show cost $7.50 and a matinee cost $4.75. On one day there was one matinee and one evening showing of *Superman Returns*. On that day a total of 310 adult tickets were sold, which resulted in ticket sales of $2022.50. How many adults went to the matinee and how many went to the evening show?

37. Stock Purchase Suppose Nike stock is selling at $78 a share and Kellogg stock is selling at $33 a share. Mike Moussa has a maximum of $10,000 to invest. He wishes to purchase five times as many shares of Kellogg as of Nike. Only whole shares of stock can be purchased.
a) How many shares of each will he purchase?
b) How much money will be left over?

38. Stock Purchase Suppose Wal-Mart stock is selling at $59 a share and Mattel stock is selling at $28 a share. Amy Waller has a maximum of $6000 to invest. She wishes to purchase four times as many shares of Wal-Mart as of Mattel. Only whole shares of stock can be purchased.
a) How many shares of each will she purchase?
b) How much money will be left over?

In Exercises 39–58 solve the mixture problem. See Examples 5–8.

39. Nut Shop Jean Valjean owns a nut shop where walnuts cost $6.80 per pound and almonds cost $6.40 per pound. Jean gets an order that specifically requests a 30-pound mixture of walnuts and almonds that will cost $6.65 per pound. How many pounds of each type of nut should Jean mix to get the desired mixture?

40. Grass Seed Scott's® Family grass seed sells for $2.45 per pound and Scott's Spot Filler grass seed sells for $2.10 per pound. How many pounds of each should be mixed to get a 10-pound mixture that sells for $2.20 per pound?

41. Top Soil Strained top soil sells for $160 per cubic yard and unstrained top soil sells for $120 per cubic yard. How many cubic yards of each should be mixed to make 8 cubic yards of a mixture that sells for $150 per cubic yard?

42. Bird Food At Agway Gardens, bird food is sold in bulk. In one barrel are sunflower seeds that sell for $1.80 per pound. In a second barrel is cracked corn that sells for $1.40 per pound. If a mixture is made by taking 2.5 pounds of the sunflower seeds and 1 pound of the cracked corn, what should the mixture cost per pound?

43. **Bulk Candies** A grocery store sells certain candies in bulk. The Good and Plenty cost $2.49 per pound and Sweet Treats cost $2.89 per pound. If Jane Strange takes 3 scoops of Good and Plenty and mixes it with 5 scoops of Sweet Treats, how much per pound should the mixture sell for? Assume each scoop contained the same weight of candy.

44. **Starbucks** Ruth Cordeff runs a coffee house where chocolate almond coffee beans sell for $7.00 per pound and hazelnut coffee beans sell for $6.10 per pound. A customer asks Ruth to make a 6-pound mixture using the two kinds of beans. How many pounds of each should be used if the mixture is to cost $6.40 per pound?

45. **Beef Wellington** To make beef Wellington, Chef Ramon uses a marinade that is a blend of two red wines. He mixes 5 liters of a wine that is 12% alcohol by volume with 2 liters of a wine that is 9% alcohol by volume. Determine the alcohol content of the mixture.

46. **Pharmacy** Susan Staples, a pharmacist, has a 60% solution of sodium iodite. She also has a 25% solution of the same drug. She gets a prescription calling for a 40% solution of the drug. How much of each solution should she mix to make 0.5 liter of the 40% solution?

47. **Sulfuric Acid** In chemistry class, Todd Corbin has 1 liter of a 20% sulfuric acid solution. How much of a 12% sulfuric acid solution must he mix with the 1 liter of 20% solution to make a 15% sulfuric acid solution?

48. **Paint** Nick Pappas has two cans of white paint, one with a 2% yellow pigment and the other with a 5% yellow pigment. Nick wants to mix the two paints to get paint with a 4% yellow pigment. How much of the 5% yellow pigment paint should be mixed with 0.4 gallon of the 2% yellow pigment paint to get the desired paint?

49. **Mouthwash** The label on the Listerine® Cool Mint Antiseptic mouthwash says that it is 21.6% alcohol by volume. The label on the Scope Original Mint mouthwash says that it is 15.0% alcohol by volume. If Hans mixes 6 ounces of the Listerine with 4 ounces of the Scope, what is the percent alcohol content of the mixture?

50. **Clorox** Clorox® bleach is 5.25% sodium hypochlorite. The instructions on the Clorox bottle say to add 1 cup of Clorox to 4 cups (a quart) of water. Find the percent of sodium hypochlorite in the mixture.

51. **Orange Juice** Mary Ann Terwilliger has made 6 quarts of an orange juice punch for a party. The punch contains 12% orange juice. She feels that she may need more punch, but she has no more orange juice so she adds $\frac{1}{2}$ quart of water to the punch. Find the percent of orange juice in the new mixture.

52. **Insecticide** The active ingredient in one type of Ortho® insect spray is 50% malathion. The instructions on the bottle says to add 1 fluid ounce to a gallon (128 fluid ounces) of water. What percent of malathion will be in the mixture?

53. **Hawaiian Punch** The label on a 12-ounce can of frozen concentrate Hawaiian Punch® indicates that when the can of concentrate is mixed with 3 cans (36 ounces) of cold water, the resulting mixture is 10% juice. Find the percent of pure juice in the concentrate.

54. **Salt Concentration** Suppose the dolphins at Sea World must be kept in salt water with an 0.8% salt content. After a week of warm weather, the salt content has increased to 0.9% due to water evaporation. How much water with 0% salt content must be added to 50,000 gallons of the 0.9% salt water to lower the salt concentration to 0.8%?

55. **Bleach** Clorox® bleach is 5.25% sodium hypochlorite and swimming pool shock treatment is 10.5% sodium hypochlorite. How much of each item must be mixed to get 6 cups of a mixture that is 7.2% sodium hypochlorite?

56. **Antifreeze** Prestone® antifreeze contains 12% ethylene glycose, and Xeres® antifreeze contains 9% ethylene glycose. How much of each type antifreeze should be mixed to get 10 quarts of a mixture that is 10% ethylene glycose?

57. **Alcohol Solution** How many pints of a 12% isotrophic alcohol solution should Jenny Crawford mix with 15 pints of a 5% isotrophic alcohol solution to get a mixture that is 8% isotrophic alcohol?

58. **Plant Food** Miracle-Gro® All Purpose liquid plant food has 12% nitrogen. Miracle-Gro Quick Start liquid plant food has 4% nitrogen. If 2 cups of the All Purpose plant food are mixed with 3 cups of the Quick Start plant food, determine the percent of nitrogen in the mixture.

Challenge Problems

59. Fat Albert The home base of the Navy's Blue Angels is in Pensacola, Florida. The Angels spend winters in El Centro, California. Assume they fly at about 900 miles per hour when they fly from Pensacola to El Centro in their F/A-18 Hornets. On every trip, their C-130 transport (affectionately called Fat Albert) leaves before them, carrying supplies and support personnel. The C-130 generally travels at about 370 miles per hour. On a trip from Pensacola to El Centro, how long before the Hornets leave should Fat Albert leave if it is to arrive 3 hours before the Hornets? The flying distance between Pensacola and El Centro is 1720 miles.

Group Activity

Discuss and answer Exercises 60 and 61 as a group.

60. Race Horse According to the *Guinness Book of World Records*, the fastest race horse speed recorded was by a horse called Big Racket on February 5, 1945, in Mexico City, Mexico. Big Racket ran a $\frac{3}{4}$-mile race in 62.41 seconds. Find Big Racket's speed in miles per hour. Round your answer to the nearest hundredth.

61. Garage Door Opener An automatic garage door opener is designed to begin to open when a car is 100 feet from the garage. At what rate will the garage door have to open if it is to raise 6 feet by the time a car traveling at 4 miles per hour reaches it? (1 mile per hour ≈ 1.47 feet per second.)

Cumulative Review Exercises

[1.3] **62. a)** Divide $2\frac{3}{4} \div 1\frac{5}{8}$.

　　　　b) Add $2\frac{3}{4} + 1\frac{5}{8}$.

[2.5] **63.** Solve the equation $6(x - 3) = 4x - 18 + 2x$.

[2.7] **64.** Solve the proportion $\frac{6}{x} = \frac{72}{9}$.

[3.2] **65. Consecutive Integers** The sum of two consecutive integers is 77. Find the numbers.

Chapter 3 Summary

IMPORTANT FACTS AND CONCEPTS	EXAMPLES
Section 3.1	
a subtracted from b means $b - a$.	3 subtracted from $5x$ is $5x - 3$.
The difference between a and b means $a - b$.	The difference between 3 and $5x$ is $3 - 5x$.
If T represents the total amount to be divided in two parts, and if x represents one of the parts, then $T - x$ represents the other part.	If $800 is divided between Jay and Rose and if Rose gets x dollars, then Jay gets $800 - x$ dollars.
A percent is always a percent of some number.	An 8% sales tax on r dollars is $0.08r$. The cost of an item c increased by 5% is $c + 0.05c$.
When writing equations, the words *is, was, will be, yields, gives*, often mean =.	"Two less than 5 times a number *is* the number increased by 3" can be expressed as $5x - 2 = x + 3$.
Consecutive integers, such as 23 and 24, differ by 1 unit.	x and $x + 1$ represent consecutive integers.
Consecutive even integers, such as 24 and 26, differ by 2 units.	x and $x + 2$ represent consecutive even integers.
Consecutive odd integers, such as 23 and 25, differ by 2 units.	x and $x + 2$ represent consecutive odd integers.

IMPORTANT FACTS AND CONCEPTS	EXAMPLES

Section 3.2

Problem-Solving Procedure

1. Understand the problem.
2. Translate the problem into mathematical language.
3. Carry out the mathematical calculation.
4. Check the answer.
5. Answer the question asked.

(See page 189 for more detailed information.)

Three subtracted from 4 times a number is 17. Find the number.

Solution Understand We need to express the information given as an equation. We are asked to find the unknown number.

Translate Let x = the unknown number, then we can write the equation

$$4x - 3 = 17$$

Carry out
$$4x = 20$$
$$x = 5$$

Check   Substitute 5 for the unknown number
$$4(5) - 3 = 17$$
$$17 = 17, \ \text{true}$$

Answer   The unknown number is 5.

If you are having difficulty with this section, seek help.

See Helpful Hint on page 201 for possible sources for help.

Section 3.3

An **isosceles triangle** has two sides of the same length. The angles opposite the sides of equal length have equal measures.

Two angles of an isosceles triangle are each 30° greater than the smallest angle. Find the measures of the three angles.

Solution   Let x = the smallest angle.

Then $x + 30$ = the measure of each larger angle.
$$x + (x + 30) + (x + 30) = 180$$
$$3x + 60 = 180$$
$$3x = 120$$
$$x = 40$$

The smallest angle is 40°, and the two larger angles are 40° + 30° or 70°. Note that 40° + 70° + 70° = 180°, so the answer checks.

Section 3.4

A **motion problem** can involve two rates.

Peter and Paul go walking. They start at the same point at the same time and walk in the same direction. Peter walks at 4 mph and Paul walks at 3.5 mph. In how many hours will they be 1 mile apart?

Solution   Let t = time when they are 1 mile apart.
$$4t - 3.5t = 1$$
$$0.5t = 1$$
$$t = 2$$
In 2 hours they will be 1 mile apart.

A **money problem** can involve two rates of interest or two different costs.

John sold 12 paintings in one day. Some sold at $80 and the others sold at $125. If he collected a total of $1275, how many of each type did he sell?

Solution   Let x = number of $80 paintings.

Then $12 - x$ = number of $125 paintings.
$$80x + 125(12 - x) = 1275$$
$$80x + 1500 - 125x = 1275$$
$$-45x + 1500 = 1275$$
$$-45x = -225$$
$$x = 5$$

Five $80 paintings and $12 - 5$ or 7 $125 paintings were sold.

IMPORTANT FACTS AND CONCEPTS	EXAMPLES

Section 3.4 (continued)

A **mixture problem** may involve mixing different strengths or different types of solutions, or mixing solid items, such as nuts.

How many liters of an 8% acetic acid solution must be mixed with 6 liters of a 15% acetic acid solution to get a 10% acetic acid solution?

Solution Let x = number of liters of the 8% solution.

$$0.08x + 0.15(6) = 0.10(x + 6)$$
$$0.08x + 0.9 = 0.10x + 0.6$$
$$0.9 = 0.02x + 0.6$$
$$0.3 = 0.02x$$
$$15 = x$$

Fifteen liters of the 8% solution must be mixed with the 6 liters of the 15% solution.

Chapter 3 Review Exercises

[3.1]

1. **Age** Charles's age is 7 more than 3 times Norman's age, n. Write an expression for Charles's age.

2. **Gasoline** The cost of a gallon of gasoline in California is 1.2 times the cost of a gallon of gasoline in Georgia, g. Write an expression for the cost of a gallon of gasoline in California.

3. **Dress** Write an expression for the cost of a dress, d, reduced by 25%.

4. **Pounds** Write an expression for the number of ounces in y pounds.

5. **Money** Two hundred dollars is divided between Jishing Wang and Norma Agras. If Jishing gets x dollars, write an expression for the amount Norma gets.

6. **Age** Mario is 6 years older than seven times Dino's age. Select a variable to represent one quantity and state what the variable represents. Express the second quantity in terms of the variable selected.

7. **Robberies** The number of robberies in 2006 was 12% less than the number of robberies in 2005. Write an expression for the difference in the number of robberies between 2005 and 2006.

8. **Numbers** The smaller of two numbers is 24 less than 3 times the larger. When the smaller is subtracted from the larger, the difference is 8. Write an equation to represent this information. Do not solve the equation.

[3.2] *In Exercises 9–18, set up an equation that can be used to solve the problem. Solve the equation and answer the question.*

9. **Numbers** One number is 8 more than the other. Find the two numbers if their sum is 74.

10. **Consecutive Integers** The sum of two consecutive integers is 237. Find the two integers.

11. **Numbers** The larger of two integers is 3 more than 5 times the smaller integer. Find the two numbers if the smaller subtracted from the larger is 31.

12. **New Car** Shaana recently purchased a new car. What was the cost of the car before tax if the total cost including a 7% tax was $23,260?

13. **Bagels** A bakery currently ships 520 bagels per month to various outlets. They wish to increase the shipment of bagels by 20 per month until reaching a shipment level of 900 bagels. How long will this take?

14. **Salary Comparison** Irene Doo, a salesperson, receives a salary of $600 per week plus a 3% commission on all sales she makes. Her company is planning on changing her salary to $500 per week, plus an 8% commission on all sales she makes. What weekly dollar sales would she have to make for the total salaries from each plan to be the same?

15. **Sales Price** During a going-out-of-business sale, all prices were reduced by 20%. If, during the sale Kathy Golladay purchased a camcorder for $495, what was the original price of the camcorder?

16. **Landscaping** Two Brothers Nursery charges $400 for a tree and $45 per hour to plant the tree. ABC Nursery charges $200 for the same size tree and $65 per hour to plant the tree. How many hours for planting would result in the total price for both landscapers being the same?

17. House Price According to the National Association of Realtors, the median sale price for single-family homes increased 11.7% from June 2003 to January 2005. If the median price was $191,000 in 2005, find the median price in 2003.

18. Tax Refund According to the Internal Revenue Service, the average refund in 2004 was $282 less than 4 times the 1980 annual refund. If the average annual refund in 2004 was $2454, find the average annual refund in 1980.

[3.3] *Solve each problem.*

19. Unknown Angles One angle of a triangle measures 10° greater than the smallest angle, and the third angle measures 10° less than twice the smallest angle. Find the measures of the three angles.

20. Unknown Angles One angle of a trapezoid measures 10° greater than the smallest angle; a third angle measures five times the smallest angle; and the fourth angle measures 20° greater than four times the smallest angle. Find the measure of the four angles.

21. Garden Steve Rodi has a rectangular garden whose length is 4 feet longer than its width. The perimeter of the garden is 70 feet. Find the width and length of the garden.

22. Designing a House The figure shows plans for a rectangular basement of a house. The dots represents poles in the ground and the lines represent string that has been attached to the poles. The length of the basement is to be 30 feet greater than the width. If a total of 310 feet of string was used to mark off the rooms, find the width and length of the basement.

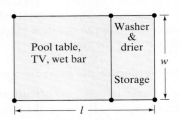

23. Rhombus The two larger angles of a rhombus are each 3 times the measure of the two smaller angles. Find the measure of each angle.

24. Bookcase Wade Ellis is building a bookcase with 4 shelves. The length of the bookcase is to be twice the height, as illustrated. If only 20 feet of lumber is available, what should be the length and height of the bookcase?

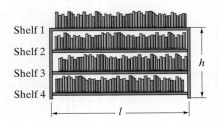

[3.4] *Solve each problem.*

25. Jogging Two joggers follow the same route. Harold Lowe jogs at 8 kilometers per hour and Susan Karney Fackert at 6 kilometers per hour. If they leave at the same time, how long will it take for them to be 4 kilometers apart?

26. Trains Leaving Two trains going in opposite directions leave from the same station on parallel tracks. One train travels at 50 miles per hour and the other at 60 miles per hour. How long will it take for the trains to be 440 miles apart?

27. Pittsburgh Incline The Duquesne Incline in Pittsburgh, Pennsylvania, is shown below. The two cars start at the same time at opposite ends of the incline and travel toward each other at the same speed. The length of the incline is 400 feet and the time it takes for the cars to be at the halfway point is about 22.73 seconds. Determine, in feet per second, the speed the cars travel.

28. Savings Accounts Tatiana wishes to place part of $12,000 into a savings account earning 8% simple interest and part into a savings account earning $7\frac{1}{4}$% simple interest. How much should she invest in each if she wishes to earn $900 in interest for the year?

29. Savings Accounts Aimee Tait invests $4000 into two savings accounts. One account pays 3% simple interest and the other account pays 3.5% simple interest. If the interest earned in the account paying 3.5% simple interest is $94.50 more than the interest earned in the 3% account, how much was invested in each account?

30. Holiday Punch Marcie Waderman is having a holiday party at her house. She made 2 gallons of a punch solution that contains 2% alcohol. How much pure punch must Marcie add to the punch to reduce the alcohol level to 1.5%?

31. Wind Chimes Alan Carmell makes and then sells wind chimes. He makes two types, a smaller one that sells for $8 and a larger one that sells for $20. At an arts and crafts show he sells a total of 30 units, and his total receipts were $492. How many of each type of chime did he sell?

32. Acid Solution Bruce Kennan, a chemist, wishes to make 2 liters of an 8% acid solution by mixing a 10% acid solution and a 5% acid solution. How many liters of each should he use?

[3.2–3.4] *Solve each problem.*

33. **Numbers** The sum of two consecutive odd integers is 208. Find the two integers.

34. **Television** What is the cost of a television before tax if the total cost, including a 6% tax, is $477?

35. **Medical Supplies** Mr. Chang sells medical supplies. He receives a weekly salary of $300 plus a 5% commission on the sales he makes. If Mr. Chang earned $900 last week, what were his sales in dollars?

36. **Triangle** One angle of a triangle is 8° greater than the smallest angle. The third angle is 4° greater than twice the smallest angle. Find the measure of the three angles of the triangle.

37. **Increase Staff** The Darchelle Leggett Company plans to increase its number of employees by 25 per year. If the company now has 427 employees, how long will it take before they have 627 employees?

38. **Parallelogram** The two larger angles of a parallelogram each measure 40° greater than the two smaller angles. Find the measure of the four angles.

39. **Copy Centers** Copy King charges a monthly fee of $20 plus 4 cents per copy. King Kopie charges a monthly fee of $25 plus 3 cents a copy. How many copies made in a month would result in both companies charging the same amount?

40. **Swimming** Rita Gonzales and Jim Ham are going swimming in Putnam Lake. They start swimming in the same direction at the same time. Rita swims at 1 mph and Jim swims at a slower pace. After 0.5 hour they are 0.2 mile apart. Find the speed that Jim is swimming.

41. **Butcher** A butcher combined ground beef that cost $3.50 per pound with ground beef that cost $4.10 per pound. How many pounds of each were used to make 80 pounds of a mixture that sells for $3.65 per pound?

42. **Speed Traveled** Two brothers who are 230 miles apart start driving toward each other at the same time. The younger brother travels 5 miles per hour faster than the older brother, and the brothers meet after 2 hours. Find the speed traveled by each brother.

43. **Acid Solution** How many liters of a 30% acid solution must be mixed with 2 liters of a 12% acid solution to obtain a 15% acid solution?

44. **Fencing** Kathy Tomaino is partitioning a rectangular yard using fencing, as shown below. If the length is 1.5 times the width and if only 96 feet of fencing is available, find the length and width of the partitioned area.

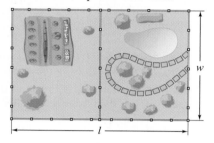

45. **Acid Solution** Six liters of a 3% sulfuric acid solution is mixed with an 8% sulfuric acid solution. If the mixture is a 4% sulfuric acid solution, determine how many liters of the 8% sulfuric acid solution were used in the mixture.

Chapter 3 Practice Test

To find out how well you understand the chapter material, take this practice test. The answers, and the section where the material was initially discussed, are given in the back of the book. Each problem is also fully worked out on the **Chapter Test Prep Video CD**. *Review any questions that you answered incorrectly.*

1. **Money** Five hundred dollars was divided between Boris and Monique. If Monique received *n* dollars, write an expression for the amount Boris received.

2. **Restaurants** The money Sally Sestini earned was $6000 more than twice what William Rowley earned. Write an expression for what Sally earned.

3. Write an expression for the number of seconds in *t* minutes.

4. **Season Tickets** The cost of a season ticket for the Dallas Cowboys football games increased by 6% from the previous years cost, *c*. Write an expression for the cost of a season ticket for this year.

In Exercises 5–6, select a variable to represent one quantity and state what it represents. Express the second quantity in terms of the variable selected.

5. **Tic Tacs** The number of packages of peppermint flavored Tic Tacs® sold was 105 packages less than 7 times the number of packages of orange flavored Tic Tacs sold

6. **Men and Women** At a play there were 600 men and women.

7. **Numbers** The larger of two numbers is 1 less than twice the smaller number. Write an expression for the smaller number subtracted from the larger number

8. **Liquid Tide** The number of fluid ounces in a large bottle of Tide® is 18 fluid ounces more than the amount in a smaller bottle. Write an expression for the sum of the amounts in a small and large bottle.

9. **Nuts** The cost of a can of Planters® Deluxe Nuts is 84% greater than the cost of a can of Planters Peanuts. Write an expression for the difference in cost between the Deluxe Nuts and the Peanuts.

In Exercises 10–25, set up an equation that can be used to solve the problem. Solve the problem and answer the question asked.

10. **Integers** The sum of two integers is 158. Find the two integers if the larger is 10 less than twice the smaller.

11. **Consecutive Odd Integers** For two consecutive odd integers, the sum of the smaller and 4 times the larger is 33. Find the integers.

12. **Numbers** One number is 12 less than 5 times the other. Find the two numbers if their sum is 42.

13. **Lawn Furniture** Dona Bishop purchased a set of lawn furniture. The cost of the furniture, including a 6% tax, was $2650. Find the cost of the furniture before tax.

14. **Eating Out** Mark Sullivan has only $40. He wishes to leave a 15% tip. Find the price of the most expensive meal that he can order.

15. **Business Venture** Julie Burgmeier receives twice the profit in a business venture than Peter Ancona does. If the profit for the year was $120,000, how much will each receive?

16. **Snowplowing** William Echols is going to hire a snow plowing service. Elizabeth Suco charges an annual fee of $80, plus $5 each time she plows. Jon Wilkins charges an annual fee of $50, plus $10 each time he plows. How many times would the snow need to be plowed for the cost of both plans to be the same?

17. **Laser Printers** A Delta laser printer cost $499. The cost of printing each page of text is 1 cent. A TexMar laser printer cost $350. The cost of printing each page is 3 cents. How many pages would need to be printed for the total cost of both printers to be the same?

18. **Triangle** A triangle has a perimeter of 75 inches. Find the three sides if one side is 15 inches larger than the smallest side, and the third side is twice the smallest side.

19. **American Flag** Carlos Mendoze's American flag has a perimeter of 28 feet. Find the dimensions of the flag if the length is 4 feet less than twice its width.

20. **Trapezoid** The two larger angles of a trapezoid measure 3° more than twice the two smaller angles. Determine the four angles of the trapezoid.

21. **Laying Cable** Ellis and Harlene Matza are digging a shallow 67.2-foot-long trench to lay electrical cable to a new outdoor light fixture they just installed. They start digging at the same time at opposite ends of where the trench is to go, and dig toward each other. Ellis digs at a rate of 0.2 foot per minute faster than Harlene, and they meet after 84 minutes. Find the speed that each digs.

22. **Running** Alice and Bonnie start running at the same time from the same point and run in the same direction. Alice runs at 8 mph while Bonnie runs at a slower pace. After 2 hours they are 4 miles apart. Determine the speed at which Bonnie is running.

23. **Bulk Candy** A candy shop sells candy in bulk. In one bin is Jelly Belly candy, which sells for $2.20 per pound, and in a second bin is Kits, which sells for $2.75 per pound. How much of each type should be mixed to obtain a 3 pound mixture, which sells for $2.40 per pound?

24. **Salt Solution** How many liters of 20% salt solution must be added to 60 liters of 40% salt solution to get a solution that is 35% salt?

25. **Acid Solution** A chemist wishes to make 3 liters of a 6% acid solution by mixing an 8% acid solution and a 5% acid solution. How many liters of each should she mix?

Cumulative Review Test

Take the following test and check your answers with those given in the back of the book. Review any questions that you answered incorrectly. The section where the material was covered is indicated after the answer.

1. Social Security The following circle graph shows what the typical retiree receives in social security, as a percent of their total income.

Where Social Security Recipients Get Their Income

Source: Newsweek

If Emily receives $40,000 per year, and her income is typical of all social security recipients, how much is she receiving in social security?

2. Engineers The following graph appeared in the Febuary 2, 2006 issue of *USA Today*. The article said that by 2010 about 90% of the world's scientists and engineers will live in Asia.

Percentage of Undergraduate Degrees Awarded in Engineering

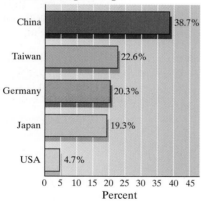

Source: National Science Foundation
(2000, latest data available)

a) What is the percentage difference in the Engineering degrees awarded in China and the United States?

b) If 1.3 million undergraduate degrees were awarded in the United States in 2006, how many were Engineering degrees?

3. Carbon Dioxide Levels David Warner, an environmentalist, was checking the level of carbon dioxide in the air. On five readings, he got the results indicated in the table above on the right.

Test	Carbon Dioxide (parts per million)
1	5
2	6
3	8
4	12
5	5

a) Find the mean level of carbon dioxide detected.

b) Find the median level of carbon dioxide detected.

4. Evaluate $\dfrac{5}{12} \div \dfrac{3}{4}$.

5. How much larger is $\dfrac{2}{3}$ inch than $\dfrac{1}{8}$ inch?

6. a) List the set of natural numbers.

b) List the set of whole numbers.

c) What is a rational number?

7. a) Evaluate $|-4|$.

b) Which is greater, $|-5|$ or $|-3|$? Explain.

8. Evaluate $2 - 6^2 \div 2 \cdot 2$.

9. Simplify $4(2x - 3) - 2(3x + 5) - 6$.

In Exercises 10–13, solve the equation.

10. $5x - 6 = x + 14$

11. $6r = 2(r + 3) - (r + 5)$

12. $2(x + 5) = 3(2x - 4) - 4x$

13. $\dfrac{4.8}{x} = -\dfrac{3}{5}$

14. If $A = \pi r^2$, find A when $r = 6$.

15. Consider the equation $4x + 8y = 16$.

a) Solve the equation for y and write the equation in $y = mx + b$ form.

b) Find y when $x = -4$.

16. Solve the formula $P = 2l + 2w$ for w.

17. Gas Needed If Lisa Shough's car can travel 50 miles on 2 gallons of gasoline, how many gallons of gas will it need to travel 225 miles?

18. Calling Plan Lori Sypher is considering two cellular telephone plans. Plan A has a monthly charge of $19.95 plus 35 cents per minute. Plan B has a monthly charge of $29.95 plus 10 cents per minute. How long would Lori need to talk in a month for the two plans to have the same total cost?

19. Sum of Numbers The sum of two numbers is 29. Find the two numbers if the larger is 11 greater than twice the smaller.

20. Quadrilateral One angle of a quadrilateral measures 5° larger than the smallest angle; the third angle measures 50° larger than the smallest angle; and the fourth angle measures 25° greater than 4 times the smallest angle. Find the measure of each angle of the quadrilateral.

Graphing Linear Equations

WE SEE GRAPHS daily. They are very important in both mathematics and in everyday living. Graphs are used to display information. For example, in Exercise 77 on page 244, we use a graph to illustrate the total cost of renting a truck under various conditions.

4.1 The Cartesian Coordinate System and Linear Equations in Two Variables

1 Plot points in the Cartesian coordinate system.

2 Determine whether an ordered pair is a solution to a linear equation.

René Descartes

1 Plot Points in the Cartesian Coordinate System

Many algebraic relationships are easier to understand if we can see a picture of them. A **graph** shows the relationship between two variables in an equation. In this chapter we discuss several procedures that can be used to draw graphs using the **Cartesian (or rectangular) coordinate system**. The Cartesian coordinate system is named for its developer, the French mathematician and philosopher René Descartes (1596–1650).

The Cartesian coordinate system provides a means of locating and identifying points just as the coordinates on a map help us find cities and other locations. Consider the map of the Great Smoky Mountains (see **Fig. 4.1**). Can you find Cades Cove on the map? If we tell you that it is in grid A3, you can probably find it much more quickly and easily.

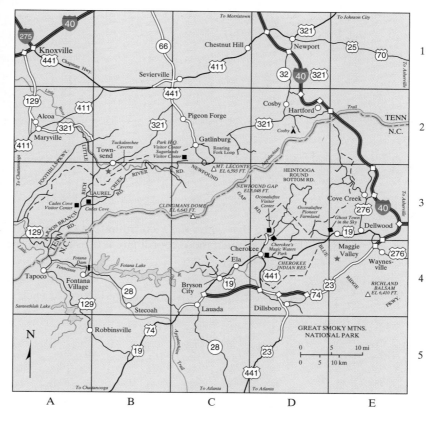

FIGURE 4.1

The Cartesian coordinate system is a grid system, like that of a map, except that it is formed by two axes (or number lines) drawn perpendicular to each other. The two intersecting axes form four **quadrants**, numbered I through IV in **Figure 4.2**.

The horizontal axis is called the **x-axis**. The vertical axis is called the **y-axis**. The point of intersection of the two axes is called the **origin**. At the origin the value of x is 0 and the value of y is 0. Starting from the origin and moving to the right along the x-axis, the numbers increase (**Fig. 4.3**). Starting from the origin and moving to the left, the numbers decrease. Starting from the origin and moving up the y-axis, the numbers increase. Starting from the origin and moving down, the numbers decrease.

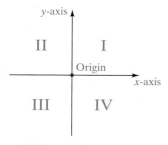

FIGURE 4.2

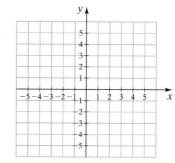

FIGURE 4.3

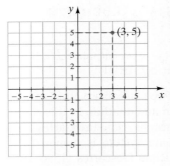

FIGURE 4.4

To locate a point, it is necessary to know both the value of x and the value of y, or the **coordinates**, of the point. When the x- and y-coordinates of a point are placed in parentheses, *with the x-coordinate listed first*, we have an **ordered pair**. In the ordered pair $(3, 5)$ the x-coordinate is 3 and the y-coordinate is 5. The point corresponding to the ordered pair $(3, 5)$ is plotted in **Figure 4.4**. The phrase "the point corresponding to the ordered pair $(3, 5)$" is often abbreviated "the point $(3, 5)$." For example, if we write "the point $(-1, 2)$," it means "the point corresponding to the ordered pair $(-1, 2)$."

EXAMPLE 1 ▶ Plot (or mark) each point on the same axes.

a) $A(5, 3)$ b) $B(2, 4)$ c) $C(-3, 1)$

d) $D(4, 0)$ e) $E(-2, -5)$ f) $F(0, -3)$

g) $G(0, 2)$ h) $H\left(6, -\dfrac{9}{2}\right)$ i) $I\left(-\dfrac{3}{2}, -\dfrac{5}{2}\right)$

Solution The first number in each ordered pair is the x-coordinate and the second number is the y-coordinate. The points are plotted in **Figure 4.5**.

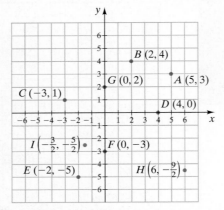

FIGURE 4.5

Note that when the x-coordinate is 0, as in Example 1 **f)** and 1 **g)**, the point is on the y-axis. When the y-coordinate is 0, as in Example 1 **d)**, the point is on the x-axis.

▶ **Now Try Exercise 29**

EXAMPLE 2 ▶ List the ordered pairs for each point shown in **Figure 4.6**.

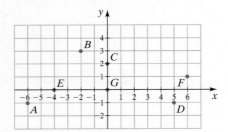

FIGURE 4.6

Solution Remember to give the x-value first in the ordered pair.

Point	Ordered Pair
A	$(-6, -1)$
B	$(-2, 3)$
C	$(0, 2)$
D	$(5, -1)$
E	$(-4, 0)$
F	$(6, 1)$
G	$(0, 0)$

▶ **Now Try Exercise 27**

2 Determine Whether an Ordered Pair Is a Solution to a Linear Equation

In Section 4.2 we will learn to graph linear equations in two variables. Below we explain how to identify a linear equation in two variables.

Linear Equations in Two Variables

A **linear equation in two variables** is an equation that can be put in the form

$$ax + by = c$$

where a, b, and c are real numbers.

The graphs of equations of the form $ax + by = c$ are straight lines. For this reason such equations are called linear. Linear equations may be written in various forms, as we will show later. A linear equation in the form $ax + by = c$ is said to be in **standard form**.

Examples of Linear Equations
$$4x - 3y = 12$$
$$y = 5x + 3$$
$$x - 3y + 4 = 0$$

Note in the examples that only the equation $4x - 3y = 12$ is in standard form. However, the bottom two equations can be written in standard form, as follows:

$$y = 5x + 3 \qquad\qquad x - 3y + 4 = 0$$
$$-5x + y = 3 \qquad\qquad x - 3y = -4$$

Most of the equations we have discussed thus far have contained only one variable. Exceptions to this include formulas used in application sections. Consider the linear equation in *one* variable, $2x + 3 = 5$. What is its solution?

$$2x + 3 = 5$$
$$2x = 2$$
$$x = 1$$

This equation has only one solution, 1.

Check
$$2x + 3 = 5$$
$$2(1) + 3 \stackrel{?}{=} 5$$
$$5 = 5 \quad \text{\textit{True}}$$

Now consider the linear equation in *two* variables, $y = x + 1$. What is the solution? Since the equation contains two variables, its solutions must contain two numbers, one for each variable. One pair of numbers that satisfies this equation is $x = 1$ and $y = 2$. To see that this is true, we substitute both values into the equation and see that the equation checks.

Check
$$y = x + 1$$
$$2 \stackrel{?}{=} 1 + 1$$
$$2 = 2 \quad \text{\textit{True}}$$

We write this answer as an ordered pair by writing the x- and y-values within parentheses separated by a comma. Remember that the x-value is always listed first because the form of an ordered pair is (x, y). Therefore, one possible solution to this equation is the ordered pair $(1, 2)$. The equation $y = x + 1$ has other possible solutions. Below we show three other solutions and their checks.

Solution	Solution	Solution
$x = 2, y = 3$	$x = -3, y = -2$	$x = -\dfrac{1}{3}, y = \dfrac{2}{3}$

Check
$$y = x + 1$$
$$3 \stackrel{?}{=} 2 + 1$$
$$3 = 3 \quad \textit{True}$$

$$y = x + 1$$
$$-2 \stackrel{?}{=} -3 + 1$$
$$-2 = -2 \quad \textit{True}$$

$$y = x + 1$$
$$\frac{2}{3} \stackrel{?}{=} -\frac{1}{3} + 1$$
$$\frac{2}{3} = \frac{2}{3} \quad \textit{True}$$

Solution Written as an Ordered Pair

$$(2, 3)$$ $$(-3, -2)$$ $$\left(-\frac{1}{3}, \frac{2}{3}\right)$$

How many possible solutions does the equation $y = x + 1$ have? The equation $y = x + 1$ has an unlimited or *infinite number* of possible solutions. Since it is not possible to list all the specific solutions, the solutions are illustrated with a graph.

Graph of an Equation

A **graph** of an equation in two variables is an illustration of a set of points whose coordinates satisfy the equation.

Figure 4.7a shows the points $(2, 3)$, $(-3, -2)$, and $\left(-\dfrac{1}{3}, \dfrac{2}{3}\right)$ plotted in the Cartesian coordinate system. **Figure 4.7b** shows a straight line drawn through the three points. Arrowheads are placed at the ends of the line to show that the line continues in both directions. Every point on this line will satisfy the equation $y = x + 1$, so this graph illustrates all the solutions of $y = x + 1$. The ordered pair $(1, 2)$, which is on the line, also satisfies the equation.

In **Figure 4.7b**, what do you notice about the points $(2, 3)$, $(1, 2)$, $\left(-\dfrac{1}{3}, \dfrac{2}{3}\right)$, and $(-3, -2)$? You probably noticed that they are in a straight line. A set of points that are in a straight line are said to be **collinear**. *In Section 4.2 when you graph linear equations by plotting points, the points you plot should all be collinear.*

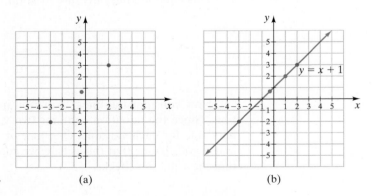

FIGURE 4.7   (a)   (b)

EXAMPLE 3 ▸ Determine whether the three points appear to be collinear.

a) $(2, 7)$, $(0, 3)$, and $(-2, -1)$

b) $(0, 5)$, $\left(\dfrac{5}{2}, 0\right)$, and $(5, -5)$

c) $(-2, -5)$, $(0, 1)$, and $(6, 8)$

Solution We plot the points to determine whether they appear to be collinear. The solution is shown in **Figure 4.8**.

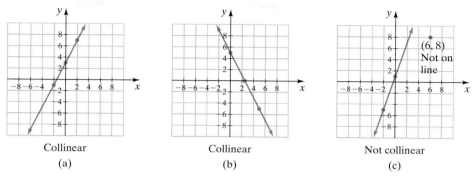

Collinear	Collinear	Not collinear
(a)	(b)	(c)

FIGURE 4.8

▶ **Now Try Exercise 33**

To graph an equation, you will need to determine ordered pairs that satisfy the equation and then plot the points.

How many points do you need to graph a linear equation? As mentioned earlier, *the graph of every linear equation of the form ax + by = c will be a straight line*. Since only two points are needed to draw a straight line, only two points are needed to graph a linear equation. However, it is always a good idea to plot at least three points. See the Helpful Hint that follows.

Helpful Hint

Only two points are needed to graph a linear equation because the graph of every linear equation is a straight line. However, if you graph a linear equation using only two points and you have made an error in determining or plotting one of those points, your graph will be wrong and you will not know it. In **Figures 4.9a** and **b** we plot only two points to show that if only one of the two points plotted is incorrect, the graph will be wrong. In both **Figures 4.9a** and **b** we use the ordered pair $(-2, -2)$. However, in **Figure 4.9a** the second point is $(1, 2)$, while in **Figure 4.9b** the second point is $(2, 1)$. Notice how the two graphs differ.

If you use at least three points to plot your graph, as in **Figure 4.7b** on page 230, and they appear to be collinear, you probably have not made a mistake.

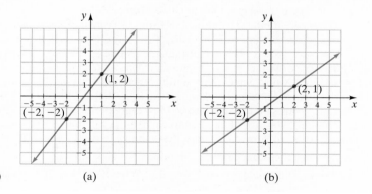

FIGURE 4.9   (a)   (b)

EXAMPLE 4 ▶

a) Determine which of the following ordered pairs satisfy the equation $2x + y = 4$.
$$(2, 0), (0, 4), (3, 1), (-1, 6)$$

b) Plot all the points that satisfy the equation on the same axes and draw a straight line through the points.

c) What does this straight line represent?

Solution

a) We substitute values for x and y into the equation $2x + y = 4$ and determine whether they check.

Check

$$(2, 0)$$
$$2x + y = 4$$
$$2(2) + 0 \overset{?}{=} 4$$
$$4 = 4 \quad \textit{True}$$

$$(0, 4)$$
$$2x + y = 4$$
$$2(0) + 4 \overset{?}{=} 4$$
$$4 = 4 \quad \textit{True}$$

$$(3, 1)$$
$$2x + y = 4$$
$$2(3) + 1 \overset{?}{=} 4$$
$$7 = 4 \quad \textit{False}$$

$$(-1, 6)$$
$$2x + y = 4$$
$$2(-1) + 6 \overset{?}{=} 4$$
$$4 = 4 \quad \textit{True}$$

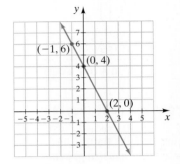

FIGURE 4.10

The ordered pairs $(2, 0)$, $(0, 4)$, and $(-1, 6)$ satisfy the equation. The ordered pair $(3, 1)$ does not satisfy the equation.

b) **Figure 4.10** shows the three points that satisfy the equation. A straight line drawn through the three points shows that they appear to be collinear.

c) The straight line represents all solutions of $2x + y = 4$. The coordinates of every point on this line satisfy the equation $2x + y = 4$.

▶ **Now Try Exercise 37**

USING YOUR GRAPHING CALCULATOR

Some of you may have graphing calculators. In this chapter we will give a number of Using Your Graphing Calculator boxes with some information on using your calculator. Because the instructions will be general, you may need to refer to your calculator manual for more specific instructions. The keystrokes you will use will depend on the brand and model of your calculator. In this book all graphing calculator windows displayed will be from TI-83 Plus or TI-84 Plus graphing calculator. Both calculators display the same window. In using Using Your Graphing Calculator boxes throughout the book, we will sometimes indicate that the keystrokes and windows shown are for a TI-84 Plus graphing calculator. The same keystrokes and windows will also apply to the TI-83 Plus graphing calculator even though we may not indicate this in the boxes.

A primary use of a graphing calculator is to graph equations. A graphing calculator *window* is the rectangular screen in which a graph is displayed. **Figure 4.11** shows a calculator window with labels added. **Figure 4.12** shows the meaning of the information given in **Figure 4.11**. These are the *standard window settings* for a graphing calculator screen.

The x-axis on the standard window goes from -10 (the minimum value of x, Xmin) to 10 (the maximum value of x, Xmax) with a scale of 1. Therefore, each tick mark represents 1 unit (Xscl = 1). The y-axis goes from -10 (the minimum value of y, Ymin) to 10 (the maximum value of y, Ymax) with a scale of 1 (Yscl = 1). The numbers below the graph in **Figure 4.11** indicate, in order, the window settings: Xmin, Xmax, Xscl, Ymin, Ymax, Yscl. *When no settings are shown below a graph, always assume the standard window setting is used.* Since the window is rectangular, the distance between tick marks on the standard window is greater on the x-axis than on the y-axis.

When graphing, you will often need to change the window settings. Read your graphing calculator manual to learn how to change the window settings. On a TI-84 Plus you press the WINDOW key and then change the settings.

Now, turn on your calculator and press the WINDOW key. If necessary, adjust the window so that it looks like the window in **Figure 4.12**. Use the (−) key, if necessary, to make negative numbers. (You can also obtain the standard window settings on a TI-84 Plus by pressing the ZOOM key and then pressing option 6, ZStandard.) Next press the GRAPH key. Your screen should resemble the screen in **Figure 4.11** (without the labels that were added). Now press the WINDOW key again. Then use the appropriate keys to change the window setting so it is the same as that shown in **Figure 4.13** on page 233.

Now press the GRAPH key again. You should get the screen shown in **Figure 4.14** on page 233. In **Figure 4.14**, the x-axis starts at 0 and goes to 50, and each tick mark represents 5 units (represented by the first 3 numbers under the window). The y-axis starts at 0 and goes to 100, and each tick mark represents 10 units (represented by the last 3 numbers under the window).

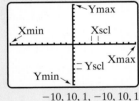

$-10, 10, 1, -10, 10, 1$

FIGURE 4.11

```
WINDOW
 Xmin=-10
 Xmax=10
 Xscl=1
 Ymin=-10
 Ymax=10
 Yscl=1
 Xres=1
```

FIGURE 4.12

(continued on the next page)

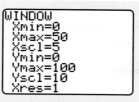

FIGURE 4.13

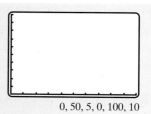

0, 50, 5, 0, 100, 10

FIGURE 4.14

EXERCISES

For Exercises 1 and 2, set your window to the values shown. Then use the GRAPH *key to shown the axes formed.*

1. Xmin = −20, Xmax = 40, Xscl = 5,
   Ymin = −10, Ymax = 60, Yscl = 10

2. Xmin = −200, Xmax = 400, Xscl = 100,
   Ymin = −500, Ymax = 1000, Yscl = 200

3. Consider the screen in **Figure 4.15**. The numbers under the window have been omitted. If Xmin = −300 and Xmax = 400, find Xscl. Explain how you determined your answer.

4. In **Figure 4.15**, if Ymin = −200 and Ymax = 1000, find Yscl. Explain how you determined your answer.

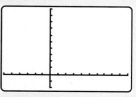

FIGURE 4.15

EXERCISE SET 4.1

Concept/Writing Exercises

1. In an ordered pair, which coordinate is always listed first?

2. What is another name for the Cartesian coordinate system?

3. **a)** Is the *horizontal axis* the *x*- or *y*-axis in the Cartesian coordinate system?
   b) Is the *vertical axis* the *x*- or *y*-axis?

4. What is the *origin* in the Cartesian coordinate system?

5. We can refer to the *x-axis* and we can refer to the *y-axis*. We can also refer to the *x*- and *y-axes*. Explain when we use the word *axis* and when we use the word *axes*.

6. Explain how to plot the point (−2, 4) in the Cartesian coordinate system.

7. What does the graph of a linear equation illustrate?

8. Why are arrowheads added to the ends of graphs of linear equations?

9. **a)** How many points are needed to graph a linear equation?
   b) Why is it always a good idea to use three or more points when graphing a linear equation?

10. What will the graph of a linear equation look like?

11. What is the standard form of a linear equation?

12. When graphing linear equations, the points that are plotted should all be *collinear*. Explain what this means.

13. In the Cartesian coordinate system there are four quadrants. Draw the *x*- and *y*-axes and mark the four quadrants, I through IV, on your axes.

14. How many solutions does a linear equation in two variables have?

Practice the Skills

Indicate the quadrant in which each of the points belongs.

15. (−4, 3)   16. (−3, 1)   17. (5, −6)   18. (2, −3)

19. (8, 5)   20. (4, 30)   21. (−17, −87)   22. (63, 47)

23. (−124, −132)   24. (75, 200)   25. (−8, 42)   26. (76, −92)

27. List the ordered pairs corresponding to each point.

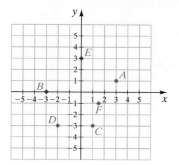

28. List the ordered pairs corresponding to each point.

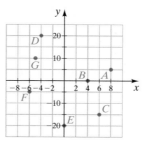

Plot each point on the same axes.

29. $A(3, 2), B(-4, 1), C(0, -3), D(-2, 0), E(-3, -4), F\left(-4, -\dfrac{5}{2}\right)$

30. $A(-3, -1), B(2, 0), C(3, 2), D\left(\dfrac{1}{2}, -4\right), E(-4, 2), F(0, 5)$

31. $A(4, 0), B(-1, 3), C(2, 4), D(0, -2), E(-3, -3), F(2, -3)$

32. $A(-3, 4), B(2, 3), C(0, 3), D(-1, 0), E(-2, -2), F(2, -4)$

Plot the following points. Then determine whether they appear to be collinear.

33. $A(1, -1), B(5, 3), C(-3, -5), D(0, -2), E(2, 0)$

34. $A(1, -2), B(0, -5), C(4, 1), D(-1, -8), E\left(\dfrac{1}{2}, -\dfrac{7}{2}\right)$

35. $A(1, 5), B\left(-\dfrac{1}{2}, \dfrac{1}{2}\right), C(0, 2), D(-5, -3), E(-2, -4)$

36. $A(1, -1), B(3, 5), C(0, -3), D(-2, -7), E(2, 1)$

In Exercises 37–42, **a)** *determine which of the four ordered pairs does not satisfy the given equation.* **b)** *Plot all the points that satisfy the equation on the same axes and draw a straight line through the points.*

37. $y = x + 2$,   **a)** $(2, 4)$ **b)** $(-2, 0)$ **c)** $(-1, 5)$ **d)** $(0, 2)$

38. $2x + y = -4$,   **a)** $(-2, 0)$ **b)** $(2, 3)$ **c)** $(0, -4)$ **d)** $(-1, -2)$

39. $3x - 2y = 6$,   **a)** $(4, 0)$ **b)** $(2, 0)$ **c)** $\left(\dfrac{2}{3}, -2\right)$ **d)** $\left(\dfrac{4}{3}, -1\right)$

40. $4x - 3y = 0$,   **a)** $(3, 4)$ **b)** $(-3, -4)$ **c)** $(0, 0)$ **d)** $(2, 5)$

41. $\dfrac{1}{2}x + 4y = 4$,   **a)** $(-2, 3)$ **b)** $\left(2, \dfrac{3}{4}\right)$ **c)** $(0, 1)$ **d)** $\left(-4, \dfrac{3}{2}\right)$

42. $y = \dfrac{1}{2}x + 2$,   **a)** $(0, 2)$ **b)** $(-4, 3)$ **c)** $(-2, 1)$ **d)** $(4, 4)$

Problem Solving

Consider the linear equation $y = 3x - 4$. In Exercises 43–46, find the value of y that makes the given ordered pair a solution to the equation.

43. $(2, y)$ 44. $(-1, y)$ 45. $(0, y)$ 46. $(3, y)$

Consider the linear equation $2x + 3y = 12$. In Exercises 47–50, find the value of x that makes the given ordered pair a solution to the equation.

47. $(x, 2)$ 48. $(x, 4)$ 49. $\left(x, \dfrac{11}{3}\right)$ 50. $\left(x, \dfrac{22}{3}\right)$

51. What is the value of y at the point where a straight line crosses the x-axis? Explain.

52. What is the value of x at the point where a straight line crosses the y-axis? Explain.

53. **Longitude and Latitude** Another type of coordinate system that is used to identify a location or position on earth's surface involves *latitude* and *longitude*. On a globe, the longitudinal lines are lines that go from top to bottom; on a world map they go up and down. The latitudinal lines go around the globe, or left to right on a world map. The locations of Hurricane Georges and Tropical Storm Hermine are indicated on the map on the right.

   a) Estimate the latitude and longitude of Hurricane Georges.

   b) Estimate the latitude and longitude of Tropical Storm Hermine.

   c) Estimate the latitude and longitude of the city of Miami.

   d) Use either a map or a globe to estimate the latitude and longitude of your college.

Source: National Weather Service

Group Activity

*In Section 4.2 we discuss how to find ordered pairs to plot when graphing linear equations. Let's see if you can draw some graphs now. Individually work parts **a)** through **c)** in Exercises 54–56.*

a) *Select any three values for x and find the corresponding values of y.*

b) *Plot the points (they should appear to be collinear).*

c) *Draw the graph.*

d) *As a group, compare your answers. You should all have the same lines.*

54. $y = x$ **55.** $y = 2x$ **56.** $y = -2x$

Cumulative Review Exercises

[2.2] **57.** Give the general form of a linear equation in one variable.

[2.5] **58.** What is a conditional equation in one variable?

  59. Solve $-2(-3x + 5) + 6 = 4(x - 2)$.

[2.6] **60.** Find the circumference and area of the circle shown below.

  61. Solve the equation $2x - 5y = 6$ for y.

4.2 Graphing Linear Equations

1 Graph linear equations by plotting points.

2 Graph linear equations of the form $ax + by = 0$.

3 Graph linear equations using the x- and y-intercepts.

4 Graph horizontal and vertical lines.

5 Study applications of graphs.

In Section 4.1 we explained the Cartesian coordinate system, how to plot points, and how to recognize linear equations in two variables. Now we are ready to graph linear equations. *In this section we discuss two methods that can be used to graph linear equations: (1) graphing by plotting points and (2) graphing using the x- and y-intercepts.*

1 Graph Linear Equations by Plotting Points

Graphing by plotting points is the most versatile method of graphing because we can also use it to graph second- and higher-degree equations. In Chapter 10, we will graph quadratic equations, which are second-degree equations, by plotting points.

To Graph Linear Equations by Plotting Points

1. Solve the linear equation for the variable y. That is, get the variable y by itself on the left side of the equal sign.

2. Select a value for the variable x. Substitute this value in the equation for x and find the corresponding value of y. Record the ordered pair (x, y).

3. Repeat step 2 with two different values of x. This will give you two additional ordered pairs.

4. Plot the three ordered pairs. The three points should appear to be collinear. If they do not, recheck your work for mistakes.

5. With a straightedge, draw a straight line through the three points. Draw arrowheads on each end of the line to show that the line continues indefinitely in both directions.

In step 1 you are asked to solve the equation for y. Although it is not necessary to do this to graph the equation, it can provide insight as to which values to select for the variable in step 2. If you have forgotten how to solve the equation for y, review Section 2.6. When selecting values in step 2, you should select integer values of x that result in integer values of y if possible. Also, you should select values of x that are small enough so that the ordered pairs obtained can be plotted on the axes. Since y is often easy to find when $x = 0$, 0 is always a good value to select for x.

EXAMPLE 1 ▶ Graph $y = 3x + 6$.

Solution   First we determine that this is a linear equation. Its graph must therefore be a straight line. The equation is already solved for y. We select three values for x, substitute them in the equation, and find the corresponding values for y. We will arbitrarily select the values $-2, 0$, and 1 for x. The calculations that follow show that when $x = -2$, $y = 0$, when $x = 0$, $y = 6$, and when $x = 1$, $y = 9$.

x	$y = 3x + 6$	Ordered Pair
-2	$y = 3(-2) + 6 = 0$	$(-2, 0)$
0	$y = 3(0) + 6 = 6$	$(0, 6)$
1	$y = 3(1) + 6 = 9$	$(1, 9)$

x	y
-2	0
0	6
1	9

It is convenient to list the x- and y-values in a table. Then we plot the three ordered pairs on the same axes (**Fig. 4.16**).

Since the three points appear to be collinear, the graph appears correct. Connect the three points with a straight line and place arrowheads at the ends of the line to show that the line continues infinitely in both directions.

▶ **Now Try Exercise 25**

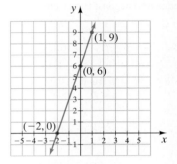

FIGURE 4.16

To graph the equation $y = 3x + 6$, we arbitrarily used the three values $x = -2$, $x = 0$, and $x = 1$. We could have selected three entirely different values and obtained exactly the same graph. When selecting values to substitute for x, use values that make the equation easy to evaluate.

The graph drawn in Example 1 represents the set of *all* ordered pairs that satisfy the equation $y = 3x + 6$. If we select any point on this line, the ordered pair represented by that point will be a solution to the equation $y = 3x + 6$. Similarly, any solution to the equation will be represented by a point on the line. Let's select some points on the line, say, $(-1, 3)$ and $(-3, -3)$, and verify that they are solutions to the equation (**Fig. 4.17**).

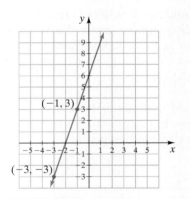

FIGURE 4.17

Check $(-1, 3)$	Check $(-3, -3)$
$y = 3x + 6$	$y = 3x + 6$
$3 \stackrel{?}{=} 3(-1) + 6$	$-3 \stackrel{?}{=} 3(-3) + 6$
$3 \stackrel{?}{=} -3 + 6$	$-3 \stackrel{?}{=} -9 + 6$
$3 = 3$ *True*	$-3 = -3$ *True*

Remember, a graph of an equation is an illustration of the set of points whose coordinates satisfy the equation.

EXAMPLE 2 ▶ Graph $3y = 5x - 6$.

Solution   We begin by solving the equation for y. This will help us in selecting values to use for x. To solve the equation for y, we divide both sides of the equation by 3.

$$3y = 5x - 6$$

$$y = \frac{5x - 6}{3}$$

$$y = \frac{5x}{3} - \frac{6}{3}$$

$$y = \frac{5}{3}x - 2$$

Now we can see that if we select values for x that are multiples of the denominator, 3, the values we obtain for y will be integers. Let's select the values $-3, 0$, and 3 for x.

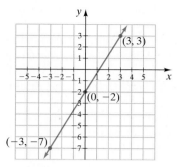

FIGURE 4.18

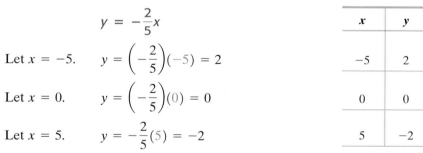

$$y = \frac{5}{3}x - 2$$

			x	y
Let $x = -3$.	$y = \frac{5}{3}(-3) - 2 = -5 - 2 = -7$		-3	-7
Let $x = 0$.	$y = \frac{5}{3}(0) - 2 = -2$		0	-2
Let $x = 3$.	$y = \frac{5}{3}(3) - 2 = 5 - 2 = 3$		3	3

Finally, we plot the points and draw the straight line (**Figure 4.18**).

▸ **Now Try Exercise 31**

2 Graph Linear Equations of the Form $ax + by = 0$

In Example 3 we graph an equation of the form $ax + by = 0$, which is a linear equation whose constant is 0.

EXAMPLE 3 ▸ Graph $2x + 5y = 0$.

Solution   We begin by solving the equation for y.

$$2x + 5y = 0$$
$$5y = -2x$$
$$y = -\frac{2x}{5} \quad \text{or} \quad y = -\frac{2}{5}x$$

Now we select values for x and find the corresponding values of y. Which values shall we select for x? Notice that the coefficient of the x-term is a fraction, with the denominator 5. If we select values for x that are multiples of the denominator, such as $\ldots, -15, -10, -5, 0, 5, 10, 15, \ldots$, the 5 in the denominator will divide out. This will give us integer values for y. We will arbitrarily select the values $x = -5$, $x = 0$, and $x = 5$.

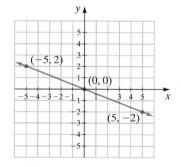

FIGURE 4.19

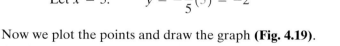

		x	y
	$y = -\frac{2}{5}x$		
Let $x = -5$.	$y = \left(-\frac{2}{5}\right)(-5) = 2$	-5	2
Let $x = 0$.	$y = \left(-\frac{2}{5}\right)(0) = 0$	0	0
Let $x = 5$.	$y = -\frac{2}{5}(5) = -2$	5	-2

Now we plot the points and draw the graph (**Fig. 4.19**).

▸ **Now Try Exercise 37**

The graph in Example 3 passes through the origin. The graph of every linear equation with a constant of 0 (equations of the form $ax + by = 0$) will pass through the origin.

3 Graph Linear Equations Using the x- and y-Intercepts

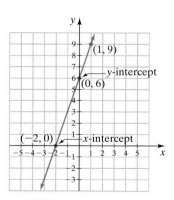

FIGURE 4.20

Now we discuss graphing linear equations using the x- and y-intercepts. The **x-intercept** is the point at which a graph crosses the x-axis and the **y-intercept** is the point at which the graph crosses the y-axis. Consider the graph in **Figure 4.20**, which is the graph we drew in Example 1. Note that the graph crosses the x-axis at -2. Therefore, $(-2, 0)$ is the x-intercept. Since the graph crosses the x-axis at -2, we might say the x-intercept is *at* −2 (on the *x*-axis). In general, the *x*-intercept is $(x, 0)$, and the *x*-intercept is *at x* (on the *x*-axis).

Note that the graph in **Figure 4.20** crosses the *y*-axis at 6. Therefore, $(0, 6)$ is the *y*-intercept. Since the graph crosses the *y*-axis at 6, we might say the *y*-intercept is *at* 6 (on the *y*-axis). In general, the *y*-intercept is $(0, y)$, and the *y*-intercept is *at y* (on the *y*-axis).

Note that the graph in **Figure 4.19** crosses both the *x*- and *y*-axes at the origin. Thus, both the *x*- and *y*-intercepts of this graph are $(0, 0)$.

It is often convenient to graph linear equations by finding their *x*- and *y*-intercepts. To graph an equation using the *x*- and *y*-intercepts, use the following procedure.

To Graph Linear Equations Using the *x*- and *y*-Intercepts

1. Find the *y*-intercept by setting *x* in the given equation equal to 0 and finding the corresponding value of *y*.
2. Find the *x*-intercept by setting *y* in the given equation equal to 0 and finding the corresponding value of *x*.
3. Determine a check point by selecting a nonzero value for *x* and finding the corresponding value of *y*.
4. Plot the *y*-intercept (where the graph crosses the *y*-axis), the *x*-intercept (where the graph crosses the *x*-axis), and the check point. The three points should appear to be collinear. If not, recheck your work.
5. Using a straightedge, draw a straight line through the three points. Draw an arrowhead at both ends of the line to show that the line continues indefinitely in both directions.

Helpful Hint

Since only two points are needed to determine a straight line, it is not absolutely necessary to determine and plot the check point in step 3. However, if you use only the *x*- and *y*-intercepts to draw your graph and one of those points is wrong, your graph will be incorrect and you will not know it. It is always a good idea to use three points when graphing a linear equation.

EXAMPLE 4 ▶ Graph $3y = 6x + 12$ using the *x*- and *y*-intercepts.

Solution   To find the *y*-intercept (where the graph crosses the *y*-axis), set $x = 0$ and find the corresponding value of *y*.

$$3y = 6x + 12$$
$$3y = 6(0) + 12$$
$$3y = 0 + 12$$
$$3y = 12$$
$$y = \frac{12}{3} = 4$$

The graph crosses the *y*-axis at 4. The ordered pair representing the *y*-intercept is $(0, 4)$. To find the *x*-intercept (where the graph crosses the *x*-axis), set $y = 0$ and find the corresponding value of *x*.

$$3y = 6x + 12$$
$$3(0) = 6x + 12$$
$$0 = 6x + 12$$
$$-12 = 6x$$
$$\frac{-12}{6} = x$$
$$-2 = x$$

The graph crosses the *x*-axis at -2. The ordered pair representing the *x*-intercept is $(-2, 0)$. Now plot the intercepts (**Fig. 4.21**).

Before graphing the equation, select a nonzero value for *x*, find the corresponding value of *y*, and make sure that it is collinear with the *x*- and *y*-intercepts. This third point is the check point.

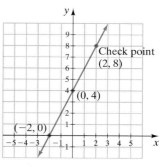

FIGURE 4.21

$$\text{Let } x = 2.$$
$$3y = 6x + 12$$
$$3y = 6(2) + 12$$
$$3y = 12 + 12$$
$$3y = 24$$
$$y = \frac{24}{3} = 8$$

Plot the check point $(2, 8)$. Since the three points appear to be collinear, draw the straight line through all three points.

▶ **Now Try Exercise 55**

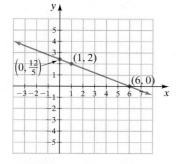

FIGURE 4.22

EXAMPLE 5 ▶ Graph $2x + 5y = 12$ using the *x*- and *y*-intercepts.

Solution

Find *y*-intercept	Find *x*-intercept	Check Point
Let $x = 0$.	Let $y = 0$.	Let $x = 1$.
$2x + 5y = 12$	$2x + 5y = 12$	$2x + 5y = 12$
$2(0) + 5y = 12$	$2x + 5(0) = 12$	$2(1) + 5y = 12$
$0 + 5y = 12$	$2x + 0 = 12$	$2 + 5y = 12$
$5y = 12$	$2x = 12$	$5y = 10$
$y = \dfrac{12}{5}$	$x = 6$	$y = 2$

The three ordered pairs are $\left(0, \dfrac{12}{5}\right)$, $(6, 0)$, and $(1, 2)$.

The three points appear to be collinear. Draw a straight line through all three points (**Fig. 4.22**).

▶ **Now Try Exercise 45**

EXAMPLE 6 ▶ Graph $y = 20x + 60$.

Solution

Find *y*-Intercept	Find *x*-Intercept	Check Point
Let $x = 0$.	Let $y = 0$.	Let $x = 3$.
$y = 20x + 60$	$y = 20x + 60$	$y = 20x + 60$
$y = 20(0) + 60$	$0 = 20x + 60$	$y = 20(3) + 60$
$y = 60$	$-60 = 20x$	$y = 60 + 60$
	$-3 = x$	$y = 120$

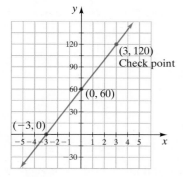

FIGURE 4.23

The three ordered pairs are $(0, 60)$, $(-3, 0)$, and $(3, 120)$. Since the values of *y* are large, we let each interval on the *y*-axis be 15 units rather than 1 (**Fig. 4.23**). Sometimes you will have to use different scales on the *x*- and *y*-axes, as illustrated, to accommodate the graph. Now we plot the points and draw the graph.

▶ **Now Try Exercise 59**

When selecting the scales for your axes, you should realize that different scales will result in the same equation having a different appearance. Consider the graphs shown in **Figure 4.24**. Both graphs represent the same equation, $y = x$. In **Figure 4.24a** both the x- and y-axes have the same scale. In **Figure 4.24b**, the x- and y-axes do not have the same scale. Both graphs are correct in that each represents the graph of $y = x$. The difference in appearance is due to the difference in scales on the x-axis. When possible, keep the scales on the x- and y-axes the same, as in **Figure 4.24a**.

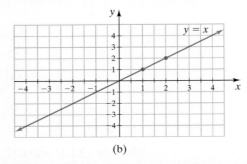

FIGURE 4.24   (a)   (b)

USING YOUR GRAPHING CALCULATOR

To graph an equation on a graphing calculator, use the following steps.

1. Solve the equation for y, if necessary.
2. Press the $\boxed{Y =}$ key and enter the equation.
3. Press the $\boxed{\text{GRAPH}}$ key (to see the graph). You may need to adjust the window, as explained in the Using Your Graphing Calculator box on page 232.

When we solve the equation $2y = 4x - 12$ for y we obtain $y = 2x - 6$. If you press $\boxed{Y =}$ and enter $2x - 6$ as Y_1, and then press $\boxed{\text{GRAPH}}$ you should get the graph shown in **Figure 4.25**.

If you do not get the graph in **Figure 4.25**, press $\boxed{\text{WINDOW}}$ and determine whether you have the standard window $-10, 10, 1, -10, 10, 1$. If not, change to the standard window and press the $\boxed{\text{GRAPH}}$ key again.

It is possible to graph two or more equations on your graphing calculator. If, for example, you wanted to graph both $y = 2x - 6$ and $y = -3x + 4$ on the same screen, you would begin by pressing the $\boxed{Y =}$ key. Then you would let $Y_1 = 2x - 6$ and $Y_2 = -3x + 4$. After you enter both equations and press the $\boxed{\text{GRAPH}}$ key, both equations will be graphed. Try graphing both equations on your graphing calculator now.

EXERCISES

Graph each equation on your graphing calculator.

1. $y = 3x - 5$
2. $y = -2x + 6$
3. $2x - 3y = 6$
4. $5x + 10y = 20$

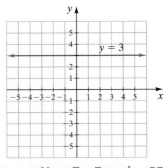

FIGURE 4.25

4 Graph Horizontal and Vertical Lines

When a linear equation contains only one variable, its graph will be either a horizontal or a vertical line, as is explained in Examples 7 and 8.

EXAMPLE 7 ▶ Graph $y = 3$.

Solution This equation can be written as $y = 3 + 0x$. Thus, for any value of x selected, y will be 3. The graph of $y = 3$ is illustrated in **Figure 4.26**.

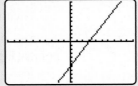

FIGURE 4.26

▶ Now Try Exercise 23

The graph of an equation of the form $y = b$ is a **horizontal line** whose y-intercept is $(0, b)$.

EXAMPLE 8 ▸ Graph $x = -2$.

Solution This equation can be written as $x = -2 + 0y$. Thus, for any value of y selected, x will have a value of -2. The graph of $x = -2$ is illustrated in **Figure 4.27**.

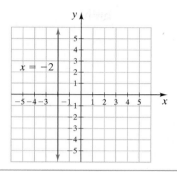

FIGURE 4.27

▸ **Now Try Exercise 21**

The graph of an equation of the form $x = a$ is a **vertical line** *whose x-intercept is $(a, 0)$.*

5 Study Applications of Graphs

Before we leave this section, let's look at an application of graphing. We will see additional applications of graphing linear equations in Section 4.4.

EXAMPLE 9 ▸ **Weekly Salary** Carol Bradley recently graduated from college. She accepted a position as a sales manager trainee at a furniture store where she is paid a weekly salary plus a commission on her sales. She will receive a salary of $300 per week plus a 7% commission on all her sales, s.

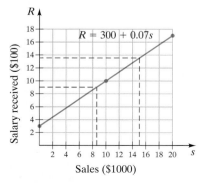

a) Write an equation for the salary Carol will receive, R, in terms of the sales, s.

b) Graph the salary for sales of $0 up to and including $20,000.

c) From the graph, estimate Carol's salary if her weekly sales are $15,000.

d) From the graph, estimate the sales needed for Carol to earn a weekly salary of $900.

Solution

a) Since s is the amount of sales, a 7% commission on s dollars in sales is $0.07s$.

$$\text{salary received} = \$300 + \text{commission}$$
$$R = 300 + 0.07s$$

b) We select three values for s and find the corresponding values of R.

$R = 300 + 0.07s$

Let $s = 0$.	$R = 300 + 0.07(0) = 300$
Let $s = 10{,}000$.	$R = 300 + 0.07(10{,}000) = 1000$
Let $s = 20{,}000$.	$R = 300 + 0.07(20{,}000) = 1700$

s	R
0	300
10,000	1000
20,000	1700

The graph is illustrated in **Figure 4.28**. Notice that since we only graph the equation for values of s from $0 to $20,000, we do not place arrowheads on the ends of the graph.

c) To determine Carol's weekly salary on sales of $15,000, locate $25,000 on the sales axis. Then draw a vertical line up to where it intersects the graph, the *red* line in **Figure 4.28**. Now draw a horizontal line across to the salary axis. Since the horizontal line crosses the salary axis at about $1350, weekly sales of $15,000 would result in a weekly salary of about $1350. We can

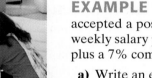

FIGURE 4.28

find the exact salary by substituting 15,000 for s in the equation $R = 300 + 0.07s$ and finding the value of R. Do this now.

d) To find the sales needed for Carol to earn a weekly salary of $900, we find $900 on the salary axis. We then draw a horizontal line from that point to the graph, as shown with the *green* line in **Figure 4.28** on page 241. We then draw a vertical line from the point of intersection of the graph to the sales axis. This value on the sales axis represents the sales needed for Carol to earn $900. Thus, sales of about $8600 per week would result in a salary of $900. We can find an exact answer by substituting 900 for R in the equation $R = 300 + 0.07s$ and solving the equation for s. Do this now.

▶ **Now Try Exercise 75**

USING YOUR GRAPHING CALCULATOR

Before we leave this section, let's spend more time discussing the graphing calculator. In the previous Using Your Graphing Calculator, we graphed $y = 2x - 6$. The graph of $y = 2x - 6$ is shown again using the standard window in **Figure 4.29**. After obtaining the graph, if you press the $\boxed{\text{TRACE}}$ key, you may (depending on your calculator) obtain the graph in **Figure 4.30**.

Notice that the cursor, the blinking box in **Figure 4.30**, is on the y-intercept, that is, where $x = 0$ and $y = -6$. By pressing the left or the right arrow keys, you can move the cursor. The corresponding values of x and y change with the position of the cursor. You can magnify the part of the graph by the cursor by using the $\boxed{\text{ZOOM}}$ key. Read your calculator manual to learn how to use the ZOOM feature and the various ZOOM options available.

Another important key on many graphing calculators is the TABLE feature. On the TI-84 Plus when you press $\boxed{2^{\text{nd}}}$ $\boxed{\text{GRAPH}}$ to get the TABLE feature, you may get the display shown in **Figure 4.31**.

If your display does not show the screen in **Figure 4.31**, use your up and down arrows until you get the screen shown. The table gives x-values and corresponding y-values for the graph. Notice from the table that when $x = 3$, $y = 0$ (the x-intercept) and when $x = 0$, $y = -6$ (the y-intercept).

You can change the table features by pressing $\boxed{\text{TBLSET}}$ (press $\boxed{2^{\text{nd}}}$ $\boxed{\text{WINDOW}}$ on the TI-84 Plus). For example, if you want the table to give values of x in tenths, you could change ΔTbl to 0.1 instead of the standard 1 unit.

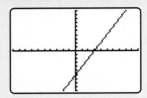

FIGURE 4.29

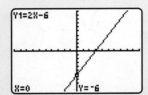

FIGURE 4.30

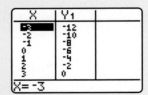

FIGURE 4.31

EXERCISES

Use the TABLE feature of your calculator to find the x-and y-intercepts of the graphs of the following equations. In Exercises 3 and 4, you will need to set the table to tenths.

1. $y = 4x - 8$
2. $y = -2x - 6$
3. $4x - 5y = 10$
4. $2y = 6x - 9$

EXERCISE SET 4.2

Concept/Writing Exercises

1. Explain how to find the x- and y-intercepts of a line.

2. How many points are needed to graph a straight line? How many points should be used? Why?

3. What will the graph of $y = b$ look like for any real number b?

4. What will the graph of $x = a$ look like for any real number a?

5. In Example 9c and 9d, we made an estimate. Why is it sometimes not possible to obtain an exact answer from a graph?

6. In Example 9 does the salary, R, depend on the sales, s, or do the sales depend on the salary? Explain.

7. Will the equation $2x - 4y = 0$ go through the origin? Explain.

8. Write an equation, other than the ones given in this section, whose graph will go through the origin. Explain how you determined your answer.

Practice the Skills

Find the missing coordinate if the ordered pair is to be a solution to the equation $3x + y = 9$.

9. $(3, ?)$ **10.** $(-4, ?)$ **11.** $(?, -6)$

12. $(?, -9)$ **13.** $(?, 0)$ **14.** $\left(\dfrac{3}{2}, ?\right)$

Find the missing coordinate in the given solutions for $3x - 2y = 8$.

15. $(4, ?)$ **16.** $(0, ?)$ **17.** $(?, 0)$

18. $\left(?, -\dfrac{5}{2}\right)$ **19.** $(-2, ?)$ **20.** $(?, 5)$

Graph each equation.

21. $x = -3$ **22.** $x = \dfrac{3}{2}$ **23.** $y = 4$ **24.** $y = -\dfrac{5}{3}$

Graph by plotting points. Plot at least three points for each graph.

25. $y = 3x - 1$ **26.** $y = -x + 3$ **27.** $y = 4x - 2$ **28.** $y = x - 4$

29. $x + 2y = 6$ **30.** $-3x + 3y = 6$ **31.** $3x - 2y = 4$ **32.** $3x - 2y = 6$

33. $4x + 3y = -9$ **34.** $6y - 12x = 18$ **35.** $6x + 5y = 30$ **36.** $2x + 3y = -6$

37. $-4x + 5y = 0$ **38.** $3x + 2y = 0$ **39.** $y = -20x + 60$ **40.** $2y - 100x = 50$

41. $y = \dfrac{4}{3}x$ **42.** $y = -\dfrac{3}{5}x$ **43.** $y = \dfrac{1}{2}x + 4$ **44.** $y = -\dfrac{2}{5}x + 2$

Graph using the x- and y-intercepts.

45. $y = 3x + 3$ **46.** $y = -3x + 6$ **47.** $y = -4x + 2$ **48.** $y = -2x + 5$

49. $y = 4x + 16$ **50.** $y = -5x + 4$ **51.** $4y + 6x = 24$ **52.** $4x = 3y - 9$

53. $\dfrac{1}{2}x + 2y = 4$ **54.** $x + \dfrac{1}{2}y = 2$ **55.** $12x - 24y = 48$ **56.** $25x + 50y = 100$

57. $8y = 6x - 12$ **58.** $6y = -4x + 12$ **59.** $y = 15x + 45$ **60.** $y = -10x + 30$

61. $\dfrac{1}{3}x + \dfrac{1}{4}y = 12$ **62.** $\dfrac{1}{4}x - \dfrac{2}{3}y = 60$ **63.** $\dfrac{1}{2}x = \dfrac{2}{5}y - 80$ **64.** $\dfrac{2}{3}y = \dfrac{5}{4}x + 120$

Write the equation represented by the given graph.

65. 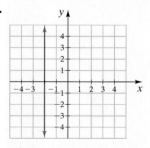 **66.** 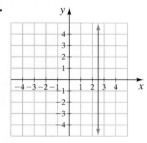 **67.** 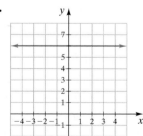 **68.**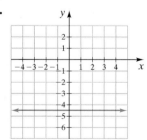

Problem Solving

69. What is the value of a if the graph of $ax + 3y = 10$ is to have an x-intercept of $(2, 0)$?

70. What is the value of a if the graph of $ax + 7y = 6$ is to have an x-intercept of $(3, 0)$?

71. What is the value of b if the graph of $3x + by = 14$ is to have a y-intercept of $(0, 7)$?

72. What is the value of b if the graph of $4x + by = 15$ is to have a y-intercept of $(0, -3)$?

The bar graphs in Exercises 73 and 74 display information. State whether the graph displays a linear relationship. Explain your answer.

73.

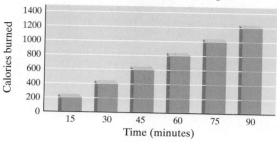

Calories Burned by Average 150–Pound Person Walking at 4.5 mph

74.

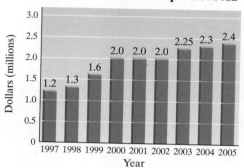

Price of a 30-Second Super Bowl Ad

Source: NFL Research, St. Petersburg Times Research

Review Example 9 before working Exercises 75–80.

75. Telephone Calls Ken Judd's telephone plan consists of a monthly fee of $15 plus 10 cents per minute for long-distance calls made.

   a) Write an equation for the total monthly cost, *C*, when *n* minutes are used for long-distance calls.

   b) Graph the equation for up to and including 100 minutes of long-distance calls made.

   c) Estimate the total monthly cost if 40 minutes of long-distance calls are made.

   d) If the totally monthly bill is $25, estimate the number of minutes used for long-distance calls.

76. Distance Traveled Distance traveled is calculated using the formula,

$$\text{distance} = \text{rate} \cdot \text{time or } d = rt.$$

   Assume the rate of a car is a constant 30 miles per hour.

   a) Write an equation for the distance, *d*, in terms of time, *t*.

   b) Graph the equation for times of 0 to 20 hours inclusive.

   c) Estimate the distance traveled in 12 hours.

   d) If the distance traveled is 150 miles, estimate the time traveled.

77. Truck Rental Lynn Brown needs a large truck to move some furniture. She found that the cost *C*, of renting a truck is $40 per day plus $1 per mile, *m*.

   a) Write an equation for the cost in terms of the miles driven.

   b) Graph the equation for values up to and including 100 miles.

   c) Estimate the cost of driving 60 miles in one day.

   d) Estimate the miles driven if the cost for one day is $70.

78. Simple Interest Simple interest is calculated by the simple interest formula,

$$\text{interest} = \text{principal} \cdot \text{rate} \cdot \text{time or } I = prt.$$

   Suppose the principal is $10,000 and the rate is 5%.

   a) Write an equation for simple interest in terms of time.

   b) Graph the equation for times of 0 to 20 years inclusive.

   c) What is the simple interest for 10 years?

   d) If the simple interest is $500, find the length of time.

79. Video Store Profit The weekly profit, *P* of a video rental store can be approximated by the formula $P = 1.5n - 200$, where *n* is the number of videos rented weekly.

   a) Draw a graph of profit in terms of video rentals for up to and including 1000 tapes.

   b) Estimate the weekly profit if 500 videos are rented.

   c) Estimate the number of videos rented if the week's profit is $1000.

80. Playing Tennis The cost, *C*, of playing tennis in the Downtown Tennis Club includes an annual $200 membership fee plus $10 per hour, *h*, of court time.

   a) Write an equation for the annual cost of playing tennis at the Downtown Tennis Club in terms of hours played.

   b) Graph the equation for up to and including 300 hours.

   c) Estimate the cost for playing 50 hours in a year.

   d) If the annual cost for playing tennis was $1700, estimate how many hours of tennis were played.

Determine the coefficients to be placed in the shaded areas so that the graph of the equation will be a line with the x- and y-intercepts specified. Explain how you determined your answer.

81. ▨*x* + ▨*y* = 6; *x*-intercept at 2; *y*-intercept at 3

82. ▨*x* + ▨*y* = 18; *x*-intercept at −3, *y*-intercept at 6

83. ▨*x* − ▨*y* = −12; *x*-intercept at −2, *y*-intercept at 3

84. ▨*x* − ▨*y* = 30; *x*-intercept at −10, *y*-intercept at −15

Challenge Problems

85. Consider the following equations: $y = 2x - 1$, $y = -x + 5$.

   a) Carefully graph both equations on the same axes.

   b) Determine the point of intersection of the two graphs.

   c) Substitute the values for x and y at the point of intersection into each of the two equations and determine whether the point of intersection satisfies each equation.

   d) Do you believe there are any other ordered pairs that satisfy both equations? Explain your answer. (We will study equations like these, called systems of equations, in Chapter 9.)

86. In Chapter 12 we will be graphing quadratic equations. The graphs of quadratic equations are *not* straight lines. Graph the quadratic equation $y = x^2 - 4$ by selecting values for x and find the corresponding values of y, then plot the points. Make sure you plot a sufficient number of points to get an accurate graph.

Group Activity

Discuss and answer Exercise 87 as a group.

87. Let's study the graphs of the equations $y = 2x + 4$, $y = 2x + 2$, and $y = 2x - 2$ to see how they are similar and how they differ. Each group member should start with the same axes.

   a) Group member 1: Graph $y = 2x + 4$.

   b) Group member 2: Graph $y = 2x + 2$.

   c) Group member 3: Graph $y = 2x - 2$.

   d) Now transfer all three graphs onto the same axes. (You can use one of the group members' graphs or you can construct new axes.)

   e) Explain what you notice about the three graphs.

   f) Explain what you notice about the y-intercepts.

Cumulative Review Exercises

[1.9] **88.** Evaluate $2[6 - (4 - 5)] \div 2 - 8^2$.

   89. Evaluate $\dfrac{-3^2 \cdot 4 \div 2}{\sqrt{9} - 2^2}$.

[2.7] **90.** **House Cleaning** According to the instructions on a bottle of concentrated household cleaner, 8 ounces of the cleaner should be mixed with 3 gallons of water. If your bucket holds only 2.5 gallons of water, how much cleaner should you use?

[3.2] **91.** **Integers** The larger of two integers is 1 more than 3 times the smaller. If the sum of the two integers is 37, find the two integers.

4.3 Slope of a Line

1   Find the slope of a line.

2   Recognize positive and negative slopes.

3   Examine the slopes of horizontal and vertical lines.

4   Examine the slopes of parallel and perpendicular lines.

1   Find the Slope of a Line

In this section we discuss the *slope* of a line. In the following Helpful Hint we discuss similarities between slope as commonly used and the slope of a line.

Helpful Hint *Slope*

We often come across slopes in everyday life. A highway (or a ramp) may have a grade (or slope) of 8%. A roof may have a pitch (or slope) of $\dfrac{6}{15}$. The slope is a measure of steepness which can be determined by dividing the vertical change, called the *rise*, by the horizontal change, called the *run*.

(continued on the next page)

Suppose a road has an 8% grade. Since $8\% = \dfrac{8}{100}$, this means the road drops (or rises) 8 feet for each 100 feet of horizontal length. A roof pitch of $\dfrac{6}{15}$ means the roof drops 6 feet for each 15 feet of horizontal length.

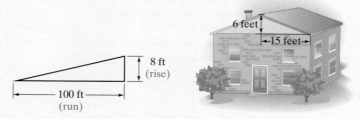

When we find the slope of a line we are also finding a ratio of the vertical change to the horizontal change. The major difference is that when we find the slope of a non-horizontal, non-vertical line, the slope can be a positive number or a negative number, as will be explained shortly.

The *slope of a line* is a measure of the *steepness* of the line. The slope of a line is an important concept in many areas of mathematics. A knowledge of slope is helpful in understanding linear equations. We now define the slope of a line.

Slope of a Line

The **slope of a line** is a ratio of the vertical change to the horizontal change between any two selected points on the line.

As an example, consider the line that goes through the two points $(3, 6)$ and $(1, 2)$. (see **Fig. 4.32a**).

If we draw a line parallel to the x-axis through the point $(1, 2)$ and a line parallel to the y-axis through the point $(3, 6)$, the two lines intersect at $(3, 2)$, see **Figure 4.32b**. From the figure, we can determine the slope of the line. The vertical change (along the y-axis) is $6 - 2$, or 4 units. The horizontal change (along the x-axis) is $3 - 1$, or 2 units.

$$\text{slope} = \frac{\text{vertical change}}{\text{horizontal change}} = \frac{4}{2} = 2$$

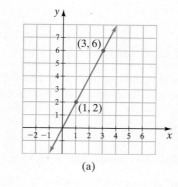

(a)

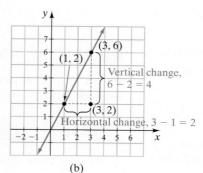

(b)

FIGURE 4.32

FIGURE 4.33

Thus, the slope of the line through these two points is 2. By examining the line connecting these two points, we can see that as the graph moves up 2 units on the y-axis it moves to the right 1 unit on the x-axis (**Fig. 4.33**).

Now we present the procedure to find the slope of a line between any two points (x_1, y_1) and (x_2, y_2). Consider **Figure 4.34**.

The vertical change can be found by subtracting y_1 from y_2. The horizontal change can be found by subtracting x_1 from x_2.

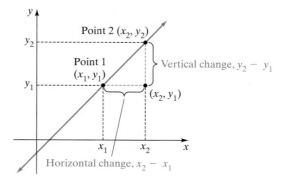

FIGURE 4.34

Slope of a Line through the Points (x_1, y_1) and (x_2, y_2)

$$\text{slope} = \frac{\text{change in } y \text{ (vertical change)}}{\text{change in } x \text{ (horizontal change)}} = \frac{y_2 - y_1}{x_2 - x_1}$$

It makes no difference which two points are selected when finding the slope of a line. It also makes no difference which point you label (x_1, y_1) or (x_2, y_2). The Greek capital letter delta, Δ, is often used to represent the words *the change in*. So, Δy is read, the change in y, and Δx is read, the change in x. Thus, the slope, which is symbolized by the letter m, is indicated as

$$m = \frac{\Delta y}{\Delta x} = \frac{y_2 - y_1}{x_2 - x_1}$$

EXAMPLE 1 ▸ Find the slope of the line through the points $(-6, 1)$ and $(3, 5)$.

Solution   We will designate $(-6, 1)$ as (x_1, y_1) and $(3, 5)$ as (x_2, y_2).

$$m = \frac{y_2 - y_1}{x_2 - x_1}$$

$$= \frac{5 - 1}{3 - (-6)}$$

$$= \frac{5 - 1}{3 + 6} = \frac{4}{9}$$

Thus, the slope is $\dfrac{4}{9}$.

If we had designated $(3, 5)$ as (x_1, y_1) and $(-6, 1)$ as (x_2, y_2), we would have obtained the same results.

$$m = \frac{y_2 - y_1}{x_2 - x_1}$$

$$= \frac{1 - 5}{-6 - 3} = \frac{-4}{-9} = \frac{4}{9}$$

▸ **Now Try Exercise 13**

Avoiding Common Errors

Students sometimes subtract the x's and y's in the slope formula in the wrong order. For instance, using the problem in Example 1:

$$m = \frac{\cancel{y_2 - y_1}}{\cancel{x_1 - x_2}} = \frac{5 - 1}{-6 - 3} = \frac{5 - 1}{-6 - 3} = \frac{4}{-9} = -\frac{4}{9}$$

Notice that subtracting in this incorrect order results in a negative slope, when the actual slope of the line is positive. The same sign error will occur each time subtraction is done incorrectly in this manner.

2 Recognize Positive and Negative Slopes

A straight line for which the value of y increases as x increases has a **positive slope**; see **Figure 4.35a**. A line with a positive slope rises as it moves from left to right. A straight line for which the value of y decreases as x increases has a **negative slope**; see **Figure 4.35b**. A line with a negative slope falls as it moves from left to right.

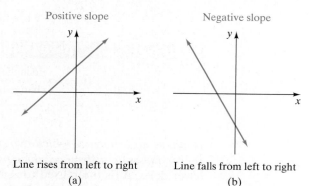

Positive slope

Negative slope

Line rises from left to right
(a)

Line falls from left to right
(b)

FIGURE 4.35

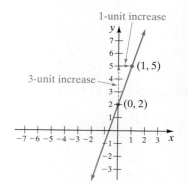

FIGURE 4.36

EXAMPLE 2 ▸ Consider the line in **Figure 4.36**.

a) Determine the slope of the line by observing the vertical change and horizontal change between the points $(1, 5)$ and $(0, 2)$ on the graph.

b) Calculate the slope of the line using the two given points.

Solution

a) The first thing you should notice is that the slope is positive since the line rises from left to right. Now determine the vertical change between the two points. The vertical change is $+3$ units. Next determine the horizontal change between the two points. The horizontal change is $+1$ unit. Since the slope is the ratio of the vertical change to the horizontal change between any two points, and since the slope is positive, the slope of the line is $\frac{3}{1}$ or 3.

b) We can use any two points on the line to determine its slope. Since we are given the ordered pairs $(1, 5)$ and $(0, 2)$, we will use them.

Let (x_2, y_2) be $(1, 5)$. Let (x_1, y_1) be $(0, 2)$.

$$m = \frac{y_2 - y_1}{x_2 - x_1} = \frac{5 - 2}{1 - 0} = \frac{3}{1} = 3$$

Note that the slope obtained in part **b)** agrees with the slope obtained in part **a)**. If we had designated $(1, 5)$ as (x_1, y_1) and $(0, 2)$ as (x_2, y_2), the slope would not have changed. Try it and see that you will still obtain a slope of 3.

▸ **Now Try Exercise 25**

EXAMPLE 3 ▶ Find the slope of the line in **Figure 4.37** using the vertical change and horizontal change between the two points shown.

Solution Since the graph falls from left to right, you should realize that the line has a negative slope. The vertical change between the two given points is −3 units since it is decreasing. The horizontal change between the two given points is 4 units since it is increasing. Since the ratio of the vertical change to the horizontal change is −3 units to 4 units, the slope of this line is $\dfrac{-3}{4}$ or $-\dfrac{3}{4}$.

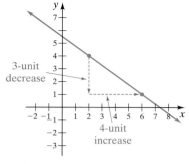

FIGURE 4.37

▶ **Now Try Exercise 29**

Using the two points shown in **Figure 4.37** and the definition of slope, calculate the slope of the line in Example 3. You should obtain the same answer.

EXAMPLE 4 ▶ **Poverty Rate in the United States** The poverty rate in the United States fell in the years from 1993 to 2000. In the years from 2000 to 2003, the poverty rate increased. The poverty rate fell almost linearly from 1993 to 1996 and from 1996 to 2000. It increased almost linearly from 2000 to 2003.

The graph in **Figure 4.38** closely approximates the poverty rates over these 11 years.

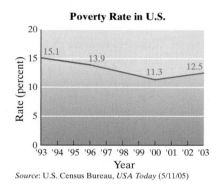

Poverty Rate in U.S.

FIGURE 4.38

Source: U.S. Census Bureau, *USA Today* (5/11/05)

a) Use the points (1996, 13.9) and (2000, 11.3), to determine the slope of the red line segment.

b) Use the points (2000, 11.3) and (2003, 12.5), to determine the slope of the blue line segment.

Solution:

a) To determine the slope of the red line segment, divide the change in poverty rates, on the vertical axis, by the change in years, on the horizontal axis. We will use (2000, 11.3) as (x_2, y_2) and (1996, 13.9) as (x_1, y_1).

$$m = \frac{y_2 - y_1}{x_2 - x_1} = \frac{\text{poverty rate in 2000} - \text{poverty rate in 1996}}{2000 - 1996}$$

$$= \frac{11.3 - 13.9}{4} = \frac{-2.6}{4} \times \frac{10}{10} = \frac{-26}{40} = -\frac{13}{20}$$

Thus the slope of the red line segment is $-\dfrac{13}{20}$.

b) To find the slope of the blue line segment, we will use (2003, 12.5) as (x_2, y_2) and (2000, 11.3) as (x_1, y_1).

$$m = \frac{y_2 - y_1}{x_2 - x_1} = \frac{\text{poverty rate in 2003} - \text{poverty rate in 2000}}{2003 - 2000}$$

$$= \frac{12.5 - 11.3}{3} = \frac{1.2}{3} \times \frac{10}{10} = \frac{12}{30} = \frac{2}{5}$$

Thus the slope of the blue line segment is $\frac{2}{5}$.

▶ **Now Try Exercise 71**

In Example 4, observe that the slope of the red line segment is a negative number since the poverty rates were decreasing for the years from 1996 to 2000. The slope of the blue line segment is a positive number since the poverty rates were increasing for the years from 2000 to 2003.

3 Examine the Slopes of Horizontal and Vertical Lines

Now we consider the slope of horizontal and vertical lines.

Consider the graph of $y = 5$ (**Fig. 4.39**). What is its slope?

The graph is parallel to the x-axis and goes through the points (2, 5) and (6, 5). Arbitrarily select (6, 5) as (x_2, y_2) and (2, 5) as (x_1, y_1). Then the slope of the line is

$$m = \frac{y_2 - y_1}{x_2 - x_1} = \frac{5 - 5}{6 - 2} = \frac{0}{4} = 0$$

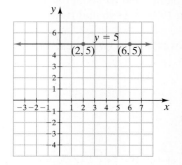

FIGURE 4.39

Since there is no change in y, this line has a slope of 0. Note that *any* two points on the line would yield the same slope, 0.

> **Slope of a Horizontal Line**
> Every horizontal line has a slope of 0.

Now we discuss vertical lines. Consider the graph of $x = 2$ (**Fig. 4.40**). What is its slope?

The graph is parallel to the y-axis and goes through the points (2, 1) and (2, 4). Arbitrarily select (2, 4) as (x_2, y_2) and (2, 1) as (x_1, y_1). Then the slope of the line is

$$m = \frac{y_2 - y_1}{x_2 - x_1} = \frac{4 - 1}{2 - 2} = \frac{3}{0}$$

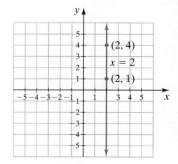

FIGURE 4.40

We learned in Section 1.8 that $\frac{3}{0}$ is undefined. Thus, we say that the slope of this line is undefined.

> **Slope of a Vertical Line**
> The slope of any vertical line is undefined.

4 Examine the Slopes of Parallel and Perpendicular Lines

Two lines are **parallel** when they do not intersect, no matter how far they are extended. **Figure 4.41** on page 251 illustrates two parallel lines.

If we compute the slope of line 1 using the given points, we obtain a slope of 3. If we compute the slope of line 2, we obtain a slope of 3. (You should compute the slopes of both lines now to verify this.) Notice both lines have the same slopes. Any two non-vertical lines that have the same slope are parallel lines.

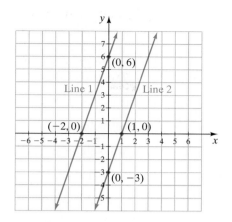

FIGURE 4.41

Parallel Lines

Two nonvertical lines with the same slope and different *y*-intercepts are parallel lines. Any two vertical lines are parallel to each other.

EXAMPLE 5 ▶

a) Draw a line with a slope of $\frac{1}{2}$ through the point $(2, 3)$.

b) On the same set of axes, draw a line with a slope of $\frac{1}{2}$ through the point $(-1, -3)$.

c) Are the two lines in part a) and b) parallel? Explain.

Solution

a) Place a dot at $(2, 3)$. Because the slope is a positive $\frac{1}{2}$, from the point $(2, 3)$ move *up* 1 unit and to the *right* 2 units to get a second point. Draw a line through the two points; see the blue line in **Figure 4.42**.

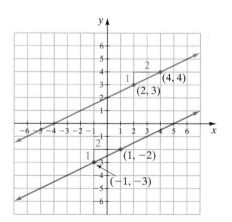

FIGURE 4.42

b) Place a dot at $(-1, -3)$. From the point $(-1, -3)$ move up 1 unit and to the right 2 units to get a second point. Draw a line through the two points; see the red line in **Figure 4.42**.

c) The lines appear to be parallel on the graph. Since both lines have the same slope, $\frac{1}{2}$, they are parallel lines.

▶ **Now Try Exercise 73**

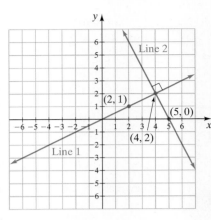

FIGURE 4.43

Now let's consider perpendicular lines. Two lines are **perpendicular** when they meet and form a right (90°) angle. **Figure 4.43** illustrates two perpendicular lines.

If we compute the slope of line 1 using the given points, we obtain a slope of $\frac{1}{2}$. If we compute the slope of line 2 using the given points, we obtain a slope of -2.

(You should compute the slopes of both lines now to verify this.) Notice the product of their slopes, $\frac{1}{2}(-2)$, is -1. Any two numbers whose product is -1 are said to be **negative reciprocals** of each other. In general, if m represents a number, its negative reciprocal will be $-\frac{1}{m}$ because $m\left(-\frac{1}{m}\right) = -1$. Any two lines with slopes that are negative reciprocals of each other are perpendicular lines.

> **Perpendicular Lines**
>
> Two lines whose slopes are negative reciprocals of each other are perpendicular lines. Any vertical line is perpendicular to any horizontal line.

EXAMPLE 6 ▶

a) Draw a line with a slope of -3 through the point $(2, 3)$.

b) On the same set of axes, draw a line with a slope of $\frac{1}{3}$ through the point $(-1, -3)$.

c) Are the two lines in parts **a)** and **b)** perpendicular? Explain.

Solution

a) Place a dot at $(2, 3)$. A slope of -3 means $\frac{-3}{1}$. Because the slope is *negative*, from the point $(2, 3)$ move *down* 3 units and to the *right* 1 unit to get a second point. Draw a line through the two points; see the blue line in **Figure 4.44**.

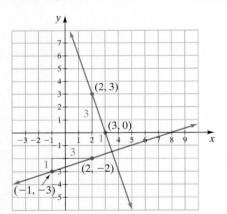

FIGURE 4.44

b) Place a dot at $(-1, -3)$. Because the slope is a *positive* $\frac{1}{3}$, from this point move *up* 1 unit and to the *right* 3 units. Draw a line through the two points; see the red line in **Figure 4.44**.

c) The lines appear to be perpendicular on the graph. To determine if they are perpendicular, multiply the slopes of the two lines together. If their product is -1, then the slopes are negative reciprocals, and the lines are perpendicular.

$$\text{Slope of line 1 (or } m_1) = -3, \text{ slope of line 2 (or } m_2) = \frac{1}{3}$$

$$m_1 \cdot m_2 = (-3)\left(\frac{1}{3}\right) = -1$$

Since the slopes are negative reciprocals, the two lines are perpendicular.

▶ **Now Try Exercise 75**

EXAMPLE 7 ▶ If m_1 represents the slope of line 1 and m_2 represents the slope of line 2, determine if line 1 and line 2 are parallel, perpendicular or neither.

a) $m_1 = \dfrac{5}{6}, m_2 = \dfrac{5}{6}$ **b)** $m_1 = \dfrac{2}{5}, m_2 = 4$ **c)** $m_1 = \dfrac{3}{5}, m_2 = -\dfrac{5}{3}$

Solution

a) Since the slopes are the same, both $\dfrac{5}{6}$, the lines are parallel.

b) Since the slopes are not the same, the lines are not parallel. Since $m_1 \cdot m_2 = \left(\dfrac{2}{5}\right)(4) \neq -1$, the slopes are not negative reciprocals and the lines are not perpendicular. Thus the answer is neither.

c) Since the slopes are not the same, the lines are not parallel. Since $m_1 \cdot m_2 = \dfrac{3}{5}\left(-\dfrac{5}{3}\right) = -1$, the slopes are negative reciprocals and the lines are perpendicular.

▶ **Now Try Exercise 53**

EXERCISE SET 4.3

Math XL®   MathXL®   MyMathLab   MyMathLab

Concept/Writing Exercises

1. Explain what is meant by the slope of a line.

2. Explain how to find the slope of a line.

3. Describe the appearance of a line that has a positive slope.

4. Describe the appearance of a line that has a negative slope.

5. Explain how to tell by observation whether a line has a positive slope or negative slope.

6. What is the slope of any horizontal line? Explain your answer.

7. Do vertical lines have a slope? Explain.

8. What letter is used to represent the slope?

9. If two non-vertical lines are parallel, what do we know about the slopes of the two lines?

10. If two non-vertical lines are perpendicular, what do we know about the slopes of the two lines?

Practice the Skills

Using the slope formula, find the slope of the line through the given points.

11. $(4, 1)$ and $(6, 5)$

12. $(9, -2)$ and $(7, -4)$

13. $(8, 0)$ and $(4, -2)$

14. $(-4, 2)$ and $(6, 5)$

15. $\left(9, \dfrac{1}{2}\right)$ and $\left(-3, \dfrac{1}{2}\right)$

16. $(-8, 6)$ and $(-2, 6)$

17. $(5, -6)$ and $(8, -3)$

18. $(9, 3)$ and $(5, -6)$

19. $(6, 4)$ and $(6, 2)$

20. $(-7, 8)$ and $(3, -1)$

21. $(6, 0)$ and $(-2, 3)$

22. $(-2, 3)$ and $(-2, -5)$

23. $\left(0, \dfrac{5}{2}\right)$ and $\left(-\dfrac{3}{4}, 2\right)$

24. $(-1, 8)$ and $\left(\dfrac{1}{3}, -1\right)$

By observing the vertical and horizontal change of the line between the two points indicated, determine the slope of each line.

25.

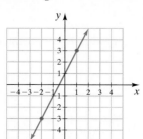

26.

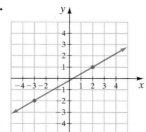

27.

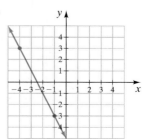

28.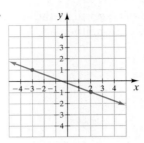

29.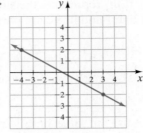

30.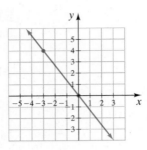

31.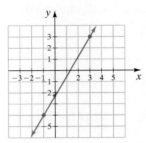

32.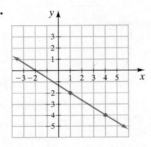

33.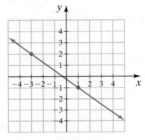

34.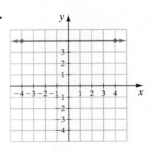

35.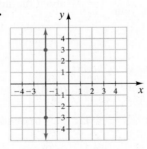

36.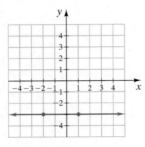

37.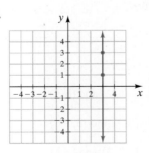

38.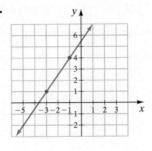

In Exercises 39–48, graph the line with the given slope that goes through the given point.

39. Through $(3, -1)$ with $m = 2$.

40. Through $(-1, -2)$ with $m = -2$

41. Through $(0, -2)$ with $m = \dfrac{1}{2}$

42. Through $(0, 2)$ with $m = -\dfrac{1}{2}$

43. Through $(0, 0)$ with $m = -\dfrac{1}{3}$

44. Through $(-3, 4)$ with $m = -\dfrac{2}{3}$

45. Through $(-3, 2)$ with $m = 0$

46. Through $(-1, 3)$ with $m = 0$

47. Through $(2, -2)$ with slope undefined

48. Through $(-1, 5)$ with slope undefined

In Exercises 49–64, m_1 represents the slope of line 1, and m_2 represents the slope of the distinct line, line 2. Indicate whether line 1 and line 2 are parallel, perpendicular, or neither.

49. $m_1 = 3, m_2 = 3$

50. $m_1 = \dfrac{1}{2}, m_2 = -6$

51. $m_1 = \dfrac{1}{4}, m_2 = -4$

52. $m_1 = -1, m_2 = -1$

53. $m_1 = \dfrac{2}{3}, m_2 = -\dfrac{3}{2}$

54. $m_1 = 7, m_2 = -7$

55. $m_1 = 6, m_2 = \dfrac{2}{3}$

56. $m_1 = -\dfrac{1}{3}, m_2 = -3$

57. $m_1 = \dfrac{1}{4}, m_2 = 4$

58. $m_1 = 6, m_2 = -\dfrac{1}{6}$

59. $m_1 = 0, m_2 = 0$

60. $m_1 = 0, m_2 = -\dfrac{2}{5}$

61. m_1 is undefined, m_2 is undefined

62. $m_1 = 0, m_2$ is undefined

63. m_1 is undefined, $m_2 = 0$

64. $m_1 = 5, m_2 = -5$

65. The slope of a given line is 3. If a line is to be drawn parallel to the given line, what will be its slope?

66. The slope of a given line is -2. If a line is to be drawn parallel to the given line, what will be its slope?

67. The slope of a given line is -4. If a line is to be drawn perpendicular to the given line, what will be its slope?

68. The slope of a given line is 5. If a line is to be drawn perpendicular to the given line, what will be its slope?

Problem Solving

In Exercises 69 and 70, determine which line (the first or second) has the greater slope. Explain your answer. Notice that the scales on the x- and y-axes are different.

69.
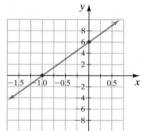 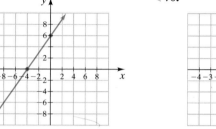

70.
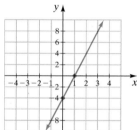 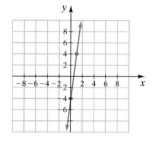

In Exercises 71 and 72, find the slope of the line segments indicated in **a)** *red and* **b)** *blue.*

71.
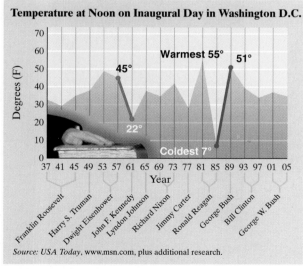

72.
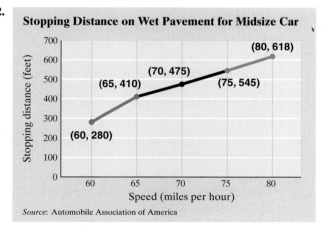

73. A given line goes through the points $(2, 6)$ and $(4, -2)$. If a line is to be drawn parallel to the given line, what will be its slope?

74. A given line goes through the points $(-2, 5)$ and $(4, 7)$. If a line is to be drawn parallel to the given line, what will be its slope?

75. A given line goes through the points $(1, -7)$ and $(2, 1)$. If a line is to be drawn perpendicular to the given line, what will be its slope?

76. A given line goes through the points $(-3, 0)$ and $(-2, 3)$. If a line is to be drawn perpendicular to the given line, what will be its slope?

Challenge Problems

77. Find the slope of the line through the points $\left(\frac{1}{2}, -\frac{3}{8}\right)$ and $\left(-\frac{4}{9}, -\frac{7}{2}\right)$.

78. If one point on a line is $(6, -4)$ and the slope of the line is $-\frac{5}{3}$, identify another point on the line.

79. A quadrilateral (a four-sided figure) has four vertices (the points where the sides meet). Vertex A is at $(0, 1)$, vertex B is at $(6, 2)$, vertex C is at $(5, 4)$, and vertex D is at $(1, -1)$.
a) Graph the quadrilateral in the Cartesian coordinate system.
b) Find the slopes of sides AC, CB, DB, and AD.
c) Do you think this figure is a parallelogram? Explain.

80. Population The following graph shows the world's population estimated to the year 2016.

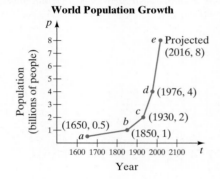

World Population Growth

a) Find the slope of the line segment between each pair of points, that is, ab, bc, and so on. Remember, the second coordinate is in billions. Thus, for example, 0.5 billion is actually 500,000,000.

b) Would you say that this graph represents a linear equation? Explain.

81. Consider the graph below.

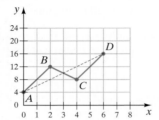

a) Determine the slope of each of the three solid blue lines.

b) Determine the average of the three slopes found in part a).

c) Determine the slope of the red dashed line from A to D.

d) Determine whether the slope of the red dashed line from A to D is the same as the (mean) average of the slopes of the three solid blue lines.

e) Explain what this example illustrates.

Group Activity

Discuss and answer Exercise 82 as a group, according to the instructions.

82. The slope of a hill and the slope of a line both measure steepness. However, there are several important differences.

a) As a group, explain how you think the slope of a hill is determined.

b) Is the slope of a line, graphed in the Cartesian coordinate system, measured in any specific unit?

c) Is the slope of a hill measured in any specific unit?

Cumulative Review Exercises

[1.9] **83.** Evaluate $4x^2 + 9x + \frac{x}{3}$ when $x = 0$.

[2.3] **84. a)** If $-x = -\frac{5}{2}$, what is the value of x?

   b) If $8x = 0$, what is the value of x?

[2.4] **85.** Solve $\frac{2}{3}x + \frac{1}{7}x = 4$.

[2.6] **86.** Solve the equation $d = a + b + c$ for c.

[4.2] **87.** Find the x- and y-intercepts for the line whose equation is $5x - 3y = 30$.

Mid-Chapter Test: 4.1–4.3

To find out how well you understand the chapter material to this point, take this brief test. The answers, and the section where the material was initially discussed, are given in the back of the book. Review any questions that you answered incorrectly.

1. In which quadrant does the point $(3, -4)$ belong?

2. Plot the points $A(2, 6)$, $B(-3, 1)$, $C(-5, -2)$, $D(0, -4)$, $E(4, -7)$ on the same axes.

3. Determine which of the three ordered pairs does not satisfy the equation $\frac{1}{3}x + y = -2$.

    a) $(3, -3)$ **b)** $(0, 2)$ **c)** $(-6, 0)$

4. Find the value of y that makes the ordered pair $(-1, y)$ a solution to the linear equation $y = 5x + 1$.

5. Find the value of x that makes the ordered pair $(x, 2)$ a solution to the equation $3x - 4y = 1$.

6. What does the graph of an equation illustrate?

Graph each equation.

7. $x = \frac{5}{2}$

8. $y = -2$

Graph by plotting points.

9. $y = 3x + 1$

10. $y = -\frac{1}{2}x + 4$

Graph using the x- and y-intercepts.

11. $3x - 4y = 12$

12. $\frac{1}{2}x + \frac{1}{5}y = 10$

Find the slope of the line through each pair of points.

13. $(-1, 5)$ and $(6, 3)$ **14.** $(4, 2)$ and $(7, 2)$

15. $(-3, 0)$ and $(-3, 5)$

Graph the line with the given slope that goes through the given point.

16. Through $(-2, 3)$ with $m = -\frac{1}{2}$

17. Through $(4, 1)$ with $m = \frac{3}{5}$.

Indicate whether the lines with the following slopes are parallel, perpendicular, or neither.

18. $m_1 = 5$ and $m_2 = \frac{1}{5}$

19. $m_1 = \frac{6}{7}$ and $m_2 = -\frac{7}{6}$

20. **Interest** The following graph illustrates the interest obtaind when \$1000 is invested for 1 year of various interest rates from 0% to 10%. Determine the slope of the line in the graph.

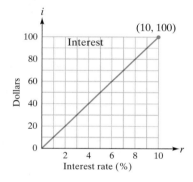

4.4 Slope-Intercept and Point-Slope Forms of a Linear Equation

1 Write a linear equation in slope-intercept form.

2 Graph a linear equation using the slope and y-intercept.

3 Use the slope-intercept form to determine the equation of a line.

4 Use the point-slope form to determine the equation of a line.

5 Compare the three methods of graphing linear equations.

In Section 4.1 we introduced the *standard form* of a linear equation, $ax + by = c$. In this section we introduce two more forms, the slope-intercept form and the point-slope form. We begin our discussion with the slope-intercept form.

1 Write a Linear Equation in Slope-Intercept Form

A very important form of a linear equation is the **slope-intercept form, $y = mx + b$** (recall that we briefly discussed the $y = mx + b$ form in Section 2.6). The graph of an equation of the form $y = mx + b$ will always be a straight line with a **slope of m** and a **y-intercept $(0, b)$**. For example, the graph of the equation $y = 3x - 4$ will be a straight line with a slope of 3 and a y-intercept $(0, -4)$. The graph of $y = -2x + 5$ will be a straight line with a slope of -2 and a y-intercept $(0, 5)$.

Slope-Intercept Form of a Linear Equation

$$y = mx + b$$

where m is the slope, and $(0, b)$ is the y-intercept of the line.

slope   where the graph crosses the y-axis

$$y = mx + b$$

Equations in Slope-Intercept Form	Slope	y-Intercept
$y = 4x - 6$	4	$(0, -6)$
$y = \dfrac{1}{2}x + \dfrac{3}{2}$	$\dfrac{1}{2}$	$\left(0, \dfrac{3}{2}\right)$
$y = -5x + 3$	-5	$(0, 3)$

Writing an Equation in Slope-Intercept Form

To write a linear equation in slope-intercept form, solve the equation for y.

Once the equation is solved for y, the numerical coefficient of the x-term will be the slope, and the constant term will give the y-intercept.

EXAMPLE 1 ▶ Write the equation $-3x + 4y = 8$ in slope-intercept form. State the slope and y-intercept.

Solution   To write this equation in slope-intercept form, we solve the equation for y.

$$-3x + 4y = 8$$
$$4y = 3x + 8$$
$$y = \frac{3x + 8}{4}$$
$$y = \frac{3}{4}x + \frac{8}{4}$$
$$y = \frac{3}{4}x + 2$$

The slope is $\dfrac{3}{4}$, and the y-intercept is $(0, 2)$.

▶ **Now Try Exercise 11**

EXAMPLE 2 ▶ Determine whether the two equations represent lines that are parallel, perpendicular, or neither.

a. $2x + y = 9$
  $2y = -4x + 5$

b. $3x - 2y = 7$
  $6y + 4x = -6$

Solution   We learned in Section 4.3 that two lines that have the same slope are parallel lines, and two lines whose slopes are negative reciprocals are perpendicular lines. We can determine the slope of each line by solving each equation for y. The coefficient of the x term will be the slope.

a) $2x + y = 9$

$$y = -2x + 9$$

$$2y = -4x + 5$$
$$y = \frac{-4x + 5}{2}$$
$$y = -2x + \frac{5}{2}$$

Since both equations have the same slope, -2, the equations represent lines that are parallel. Notice the equations represent two different lines because their y-intercepts are different.

b)

$$3x - 2y = 7$$
$$-2y = -3x + 7$$
$$y = \frac{-3x + 7}{-2}$$
$$y = \frac{3}{2}x - \frac{7}{2}$$

$$6y + 4x = -6$$
$$6y = -4x - 6$$
$$y = \frac{-4x - 6}{6}$$
$$y = -\frac{2}{3}x - 1$$

The slope of one line is $\frac{3}{2}$ and the slope of the other line is $-\frac{2}{3}$. Multiplying the slopes we obtain $\left(\frac{3}{2}\right)\left(-\frac{2}{3}\right) = -1$. Since the product is -1, the slopes are negative reciprocals. Therefore the equations represent lines that are perpendicular.

▸ **Now Try Exercise 39**

2 Graph a Linear Equation Using the Slope and *y*-Intercept

In Section 4.2 we discussed two methods of graphing a linear equation. They were (1) by plotting points and (2) using the *x*- and *y*-intercepts. Now we present a third method. This method makes use of the slope and the *y*-intercept. Remember that when we solve an equation for *y* we put the equation in slope-intercept form. Once it is in this form, we can determine the slope and *y*-intercept of the graph from the equation.

We graph equations using the slope and *y*-intercept in a manner very similar to the way we worked Examples 5 and 6 in Section 4.3. However, when graphing using the slope-intercept form, our starting point is always the *y*-intercept. After you determine the *y*-intercept, a second point can be obtained by moving up and to the right if the slope is positive, or down and to the right if the slope is negative.

EXAMPLE 3 ▸ Write the equation $-3x + 4y = 8$ in slope-intercept form; then use the slope and *y*-intercept to graph $-3x + 4y = 8$.

Solution   In Example 1 we solved $-3x + 4y = 8$ for *y*. We found that

$$y = \frac{3}{4}x + 2$$

The slope of the line is $\frac{3}{4}$ and the *y*-intercept is $(0, 2)$. We mark the first point, the *y*-intercept at 2 on the *y*-axis (**Fig. 4.45**). Now we use the slope $\frac{3}{4}$, to find a second point. Since the slope is positive, we move 3 units up and 4 units to the right to find the second point. A second point will be at $(4, 5)$. We can continue this process to obtain a third point at $(8, 8)$. Now we draw a straight line through the three points. Notice that the line has a positive slope, which is what we expected.

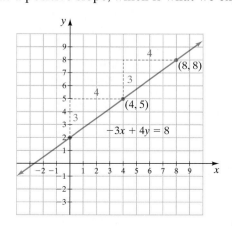

FIGURE 4.45

▸ **Now Try Exercise 19**

EXAMPLE 4 ▶ Graph $5x + 3y = 12$ by using the slope and y-intercept.

Solution Solve the equation for y.

$$5x + 3y = 12$$
$$3y = -5x + 12$$
$$y = \frac{-5x + 12}{3}$$
$$= -\frac{5}{3}x + 4$$

Thus, the slope is $-\dfrac{5}{3}$ and the y-intercept is $(0, 4)$. Begin by marking a point at 4 on the y-axis (**Fig. 4.46**). Then move 5 units down and 3 units to the right to determine the next point. Move down and to the right because the slope is negative and a line with a negative slope must fall as it goes from left to right. You can follow this procedure again to obtain a third point. Finally, draw the straight line between the plotted points.

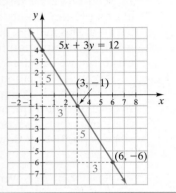

FIGURE 4.46

▶ Now Try Exercise 21

3 Use the Slope-Intercept Form to Determine the Equation of a Line

Now that we know how to use the slope-intercept form of a line, we can use it to write the equation of a given line. To do so, we need to determine the slope, m, and y-intercept of the line. Once we determine these values we can write the equation in slope-intercept form, $y = mx + b$. For example, if we determine the slope of a line is -4 and the y-intercept is at 6, the equation of the line is $y = -4x + 6$.

EXAMPLE 5 ▶ Determine the equation of the line shown in **Figure 4.47**.

Solution The graph shows that the y-intercept is at -5. Now we need to determine the slope of the line. Since the graph falls from left to right, it has a negative slope. We can see that the vertical change is 3 units for each horizontal change of 1 unit. Thus, the slope of the line is -3. The slope can also be determined by selecting any two points on the line and calculating the slope. Let's use the point $(-2, 1)$ to represent (x_2, y_2) and the point $(0, -5)$ to represent (x_1, y_1).

$$m = \frac{\Delta y}{\Delta x} = \frac{y_2 - y_1}{x_2 - x_1}$$
$$= \frac{1 - (-5)}{-2 - 0}$$
$$= \frac{1 + 5}{-2} = \frac{6}{-2} = -3$$

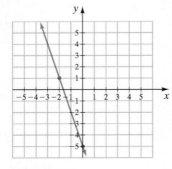

FIGURE 4.47

Again we obtain a slope of -3, Substituting -3 for m and -5 for b into the slope-intercept form of a line gives us the equation of the line in **Figure 4.47** on page 260, which is $y = -3x - 5$.

▶ **Now Try Exercise 29**

Now let's look at an application of graphing.

EXAMPLE 6 ▶ **Artistic Vases** Kris, a pottery artist, makes ceramic vases that he sells at art shows. His business has a fixed monthly cost (booth rental, advertising, cell phone, etc.) and a variable cost per vase made (cost of materials, cost of labor, cost for kiln use, etc.). The total monthly cost for making x vases is illustrated in the graph in **Figure 4.48**.

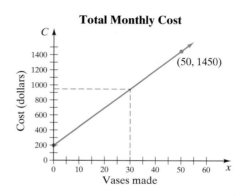

FIGURE 4.48

a) Find the equation of the total monthly cost when x vases are made.

b) Use the equation found in part a) to find the total monthly cost if 30 vases are made.

c) Use the graph in **Figure 4.48** to see whether your answer in part b) appears correct.

Solution a) **Understand and Translate** Notice that the vertical axis is cost, C, and not y. The letters or names used on the axes do not change the way we solve the problem. We will use slope-intercept form to write the equation of the line. However, since y is replaced by C, we will use $C = mx + b$, in which b is where the graph crosses the vertical or C-axis. We first note that the graph crosses the vertical axis at 200. Thus, b is 200. Now we need to find the slope of the line. Let's use the point $(0, 200)$ as (x_1, y_1) and $(50, 1450)$ as (x_2, y_2).

Carry Out

$$m = \frac{y_2 - y_1}{x_2 - x_1}$$

$$= \frac{1450 - 200}{50 - 0} = \frac{1250}{50} = 25$$

Answer The slope is 25. The equation in slope-intercept form is

$$C = mx + b$$
$$= 25x + 200$$

b) To find the monthly cost when 30 vases are sold, we substitute 30 for x.

$$C = 25x + 200$$
$$= 25(30) + 200$$
$$= 750 + 200 = 950$$

The monthly cost when 30 vases are made is $950.

c) If we draw a vertical line up from 30 on the x-axis (the red line), we see that the corresponding cost is about $950. Thus, our answer in part b) appears correct.

▶ **Now Try Exercise 65**

4 **Use the Point-Slope Form to Determine the Equation of a Line**

Thus far, we have discussed the standard form of a linear equation, $ax + by = c$, and the slope-intercept form of a linear equation, $y = mx + b$. Now we will discuss another form, called the *point-slope form*.

When the slope of a line and a point on the line are known, we can use the point-slope form to determine the equation of the line. The **point-slope form** can be obtained by beginning with the slope between any selected point (x, y) and a fixed point (x_1, y_1) on a line.

$$m = \frac{y - y_1}{x - x_1} \quad \text{or} \quad \frac{m}{1} = \frac{y - y_1}{x - x_1}$$

Now cross-multiply to obtain

$$m(x - x_1) = y - y_1 \quad \text{or} \quad y - y_1 = m(x - x_1)$$

Point-Slope Form of a Linear Equation

$$y - y_1 = m(x - x_1)$$

where m is the slope of the line and (x_1, y_1) is a point on the line.

EXAMPLE 7 ▶ Write an equation, in slope-intercept form, of the line that goes through the point $(4, 2)$ and has a slope of 5.

Solution   Since we are given a point on the line and the slope of the line, we begin by writing the equation in point-slope form. The slope m is 5. The point on the line is $(4, 2)$; we will use this point for (x_1, y_1) in the formula. We substitute 5 for m, 4 for x_1, and 2 for y_1 in the point-slope form of a linear equation.

$$y - y_1 = m(x - x_1)$$
$$y - 2 = 5(x - 4) \qquad \textit{Equation in point-slope form}$$
$$y - 2 = 5x - 20 \qquad \textit{Distributive property}$$
$$y = 5x - 18 \qquad \textit{Equation in slope-intercept form}$$

The graph of $y = 5x - 18$ has a slope of 5 and passes through the point $(4, 2)$.

▶ **Now Try Exercise 51**

The answer to Example 7 was given in slope-intercept form. If we were asked to give the answer in standard form, two acceptable answers would be $-5x + y = -18$ or $5x - y = 18$. Your instructor may specify the form in which the equation is to be given.

In Example 8, we will work an example very similar to Example 7. However, in the solution to Example 8, the equation will contain a fraction. We solved some equations that contained fractions in Chapter 2. Recall that to simplify an equation that contains a fraction, we multiply both sides of the equation by the denominator of the fraction.

EXAMPLE 8 ▶ Write an equation, in slope-intercept form, of the line that goes through the point $(6, -2)$ and has a slope of $\frac{2}{3}$.

Solution   We will begin with the point-slope form of a line, where m is $\frac{2}{3}$, 6 is x_1, and -2 is y_1.

$$y - y_1 = m(x - x_1)$$

$$y - (-2) = \frac{2}{3}(x - 6) \qquad \text{Equation in point-slope form}$$

$$y + 2 = \frac{2}{3}(x - 6)$$

$$3(y + 2) = 3 \cdot \frac{2}{3}(x - 6) \qquad \text{Multiply both sides by 3.}$$

$$3y + 6 = 2(x - 6) \qquad \text{Distributive property}$$

$$3y + 6 = 2x - 12 \qquad \text{Distributive property}$$

$$3y = 2x - 18 \qquad \text{Subtract 6 from both sides.}$$

$$y = \frac{2x - 18}{3} \qquad \text{Divide both sides by 3.}$$

$$y = \frac{2}{3}x - 6 \qquad \text{Equation in slope-intercept form}$$

▶ **Now Try Exercise 53**

Helpful Hint

We have discussed three forms of a linear equation. We summarize the three forms below. It is important that you memorize these forms.

Standard Form	Examples
$ax + by = c$	$2x - 3y = 8$
	$-5x + y = -2$

Slope-Intercept Form	Examples
$y = mx + b$	$y = 2x - 5$
m is the slope, $(0, b)$ is the y-intercept	$y = -\frac{3}{2}x + 2$

Point-Slope Form	Examples
$y - y_1 = m(x - x_1)$	$y - 3 = 2(x + 4)$
m is the slope, (x_1, y_1) is a point on the line	$y + 5 = -4(x - 1)$

We now discuss how to use the point-slope form to determine the equation of a line when two points on the line are known.

EXAMPLE 9 ▶ Find an equation of the line through the points $(-1, 3)$ and $(-5, 1)$. Write the equation in slope-intercept form.

Solution To use the point-slope form, we must first find the slope of the line through the two points. To determine the slope, let's designate $(-1, 3)$ as (x_1, y_1) and $(-5, 1)$ as (x_2, y_2).

$$m = \frac{y_2 - y_1}{x_2 - x_1} = \frac{1 - 3}{-5 - (-1)} = \frac{1 - 3}{-5 + 1} = \frac{-2}{-4} = \frac{1}{2}$$

The slope is $\frac{1}{2}$. We can use either point (one at a time) in determining the equation of the line. This example will be worked out using both points to show that the solutions obtained are identical.

Using the point $(-1, 3)$ as (x_1, y_1),

$$y - y_1 = m(x - x_1)$$

$$y - 3 = \frac{1}{2}[x - (-1)]$$

$$y - 3 = \frac{1}{2}(x + 1)$$

$$2 \cdot (y - 3) = 2 \cdot \frac{1}{2}(x + 1) \qquad \textit{Multiply both sides by the LCD, 2.}$$

$$2y - 6 = x + 1$$

$$2y = x + 7$$

$$y = \frac{x + 7}{2} \quad \text{or} \quad y = \frac{1}{2}x + \frac{7}{2}$$

Using the point $(-5, 1)$ as (x_1, y_1),

$$y - y_1 = m(x - x)$$

$$y - 1 = \frac{1}{2}[x - (-5)]$$

$$y - 1 = \frac{1}{2}(x + 5)$$

$$2 \cdot (y - 1) = 2 \cdot \frac{1}{2}(x + 5) \qquad \textit{Multiply both sides by the LCD, 2.}$$

$$2y - 2 = x + 5$$

$$2y = x + 7$$

$$y = \frac{x + 7}{2} \quad \text{or} \quad y = \frac{1}{2}x + \frac{7}{2}$$

Note that the equations for the line are identical.

▸ **Now Try Exercise 57**

Helpful Hint

In the exercise set at the end of this section, you will be asked to write a linear equation in slope-intercept form. Even though you will eventually write the equation in slope-intercept form, you may need to start your work with the point-slope form. Below we indicate the initial form to use to solve the problem.

Begin with the **slope-intercept form** if you know

   The slope of the line and the y-intercept

Begin with the **point-slope form** if you know

a)   The slope of the line and a point on the line, or

b)   Two points on the line (first find the slope, then use the point-slope form)

---

5 Compare the Three Methods of Graphing Linear Equations

We have discussed three methods to graph a linear equation: (1) plotting points, (2) using the x- and y-intercepts, and (3) using the slope and y-intercept. In Example 10 we graph an equation using all three methods. No single method is always the easiest to use. If the equation is given in slope-intercept form, $y = mx + b$, then graphing by plotting points or by using the slope and y-intercept might be easier. If the equation is given in standard form, $ax + by = c$, then graphing using the intercepts might be easier. Unless your teacher specifies that you should graph by a specific method, you may use the method with which you feel most comfortable. Graphing by plotting points is the most versatile method since it can also be used to graph equations that are not straight lines.

EXAMPLE 10 ▶ Graph $3x - 2y = 8$

a) by plotting points;

b) using the x- and y-intercepts;

c) using the slope and y-intercept.

Solution   For parts **a)** and **c)** we will write the equation in slope-intercept form.

$$3x - 2y = 8$$
$$-2y = -3x + 8$$
$$y = \frac{-3x + 8}{-2} = \frac{3}{2}x - 4$$

a) Plotting Points To find ordered pairs to plot, we substitute values for x and find the corresponding values of y. Three ordered pairs are indicated in the following table. Next we plot the ordered pairs and draw the graph (**Fig. 4.49**).

$$y = \frac{3}{2}x - 4$$

x	y
0	-4
2	-1
4	2

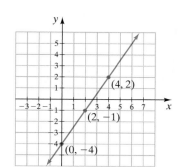

FIGURE 4.49

b) Intercepts We find the x- and y-intercepts and a check point. Then we plot the points and draw the graph (**Fig. 4.50**).

$$3x - 2y = 8$$

x-Intercept	y-Intercept	Check Point
Let $y = 0$.	Let $x = 0$.	Let $x = 2$.
$3x - 2y = 8$	$3x - 2y = 8$	$3x - 2y = 8$
$3x - 2(0) = 8$	$3(0) - 2y = 8$	$3(2) - 2y = 8$
$3x = 8$	$-2y = 8$	$6 - 2y = 8$
$x = \dfrac{8}{3}$	$y = -4$	$-2y = 2$
		$y = -1$

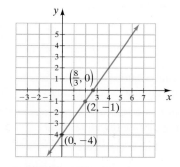

FIGURE 4.50

The three ordered pairs are $\left(\dfrac{8}{3}, 0\right)$, $(0, -4)$, and $(2, -1)$.

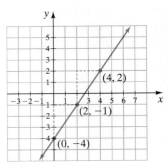

FIGURE 4.51

c) **Slope and y-intercept** The y-intercept is $(0, -4)$; therefore, we place a point at -4 on the y-axis. Since the slope is $\frac{3}{2}$, we obtain a second point by moving 3 units up and 2 units to the right. The graph is illustrated in **Figure 4.51**.
Notice that we get the same line by all three methods.

▶ **Now Try Exercise 27**

USING YOUR GRAPHING CALCULATOR

In Example 10, we graphed the equation $y = \frac{3}{2}x - 4$. Let's see what the graph of $y = \frac{3}{2}x - 4$ looks like on a graphing calculator using three different window settings. **Figure 4.52** shows the graph of $y = \frac{3}{2}x - 4$ using the *standard window setting*.

To obtain the standard window, press the $\boxed{\text{ZOOM}}$ key and then choose option 6, ZStandard. Notice that in the standard window setting, the units on the y-axis are not as long as the units on the x-axis.

In **Figure 4.53**, we illustrate the graph of $y = \frac{3}{2}x - 4$ using the *square window setting*. To obtain the square window setting, press the $\boxed{\text{ZOOM}}$ key and then choose option 5, ZSquare. This setting makes the units on both axes the same size. This allows the line to be displayed at the correct orientation to the axes.

In **Figure 4.54**, we illustrate the graph using the *decimal window setting*. To obtain the decimal window setting, press the $\boxed{\text{ZOOM}}$ key and then choose option 4, ZDecimal. The decimal window setting also makes the units the same on both axes, but sets the increment from one pixel (dot) to the next at 0.1 on both axes.

STANDARD WINDOW SETTING ($\boxed{\text{ZOOM}}$: OPTION 6)	SQUARE WINDOW SETTING ($\boxed{\text{ZOOM}}$: OPTION 5)	DECIMAL WINDOW SETTING ($\boxed{\text{ZOOM}}$: OPTION 4)
	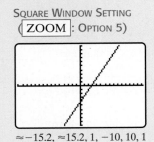	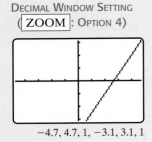
$-10, 10, 1, -10, 10, 1$	$\approx-15.2, \approx15.2, 1, -10, 10, 1$	$-4.7, 4.7, 1, -3.1, 3.1, 1$
FIGURE 4.52	**FIGURE 4.53**	**FIGURE 4.54**

EXERCISES

*Graph each equation using the **a)** standard window, **b)** square window, and **c)** decimal window settings.*

1. $y = 4x - 6$

2. $y = -\frac{1}{5}x + 4$

EXERCISE SET 4.4   MathXL   MyMathLab

MathXL®   MyMathLab

Concept/Writing Exercises

1. Give the slope-intercept form of a linear equation.

2. When you are given an equation in a form other than slope-intercept form, how can you change it to slope-intercept form?

3. What is the equation of a line, in slope-intercept form, if the slope is 3 and the y-intercept is at -5?

4. What is the equation of a line, in slope-intercept form, if the slope is -3 and the y-intercept is at 5?

5. Explain how you can determine whether two equations represent parallel lines without graphing the equations.

6. Explain how you can determine whether two equations represent the same line without graphing the equations.

7. Give the point-slope form of a linear equation.

8. Assume the slope of a line is 2 and the line goes through the origin. Write the equation of the line in point-slope form.

Practice the Skills

Determine the slope and y-intercept of the line represented by the given equation.

9. $y = 2x - 6$

10. $y = -3x + 17$

11. $4x - 3y = 21$

12. $7x = 5y + 25$

Determine the slope and y-intercept of the line represented by each equation. Graph the line using the slope and y-intercept.

13. $y = x - 3$

14. $y = -x + 5$

 15. $y = 3x + 2$

16. $3x + y = 4$

17. $y = 2x$

18. $y = -4x$

19. $-2x + y = -3$

20. $3x + 3y = 9$

21. $5x - 2y = 10$

22. $-x + 2y = 8$

23. $6x + 12y = 18$

24. $16y = 8x + 32$

25. $-6x + 2y - 8 = 0$

26. $4x = 6y + 9$

27. $3x = 2y - 4$

28. $20x = 80y + 40$

Determine the equation of each line.

29.
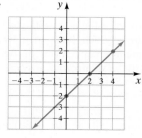

30.
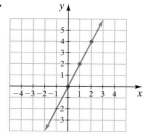

31.
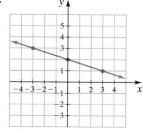

32.
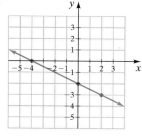

33.
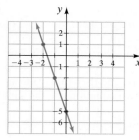

34.
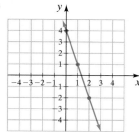

35.
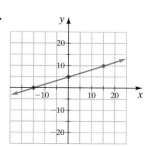

36.
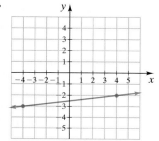

Determine whether each pair of lines are parallel, perpendicular, or neither.

37. $y = 4x + 6$
$y = 4x - 1$

38. $2x + 3y = 1$
$y = -\dfrac{2}{3}x + 3$

39. $4x + 2y = 7$
$4x = 8y + 12$

40. $3x - 5y = 7$
$5y + 3x = 4$

41. $3x + 5y = 9$
$6x = -10y + 9$

42. $8x + 2y = 10$
$x - 7 = 4y$

43. $y = \dfrac{1}{2}x - 2$
$2y = 6x + 9$

44. $3y - 4 = -5x$
$y = -\dfrac{5}{3}x - 3$

45. $5y = 2x + 9$
$-10x = 4y + 11$

46. $3x - 9y = 21$
$-3x + 9y = 27$

47. $3x + 7y = 21$
$7x + 3y = 21$

48. $5x - 6y = 18$
$-6x + 5y = 10$

Problem Solving

Write the equation of each line, with the given properties, in slope-intercept form.

49. Slope $= 3$, through $(0, 2)$

50. Slope $= 2$, through $(4, 3)$

51. Slope $= -3$, through $(-4, 5)$

52. Slope $= -3$, through $(2, 0)$

53. Slope $= \frac{1}{2}$, through $(-1, -3)$

54. Slope $= -\frac{2}{3}$, through $(4, -5)$

55. Slope $= \frac{2}{3}$, y-intercept is $(0, 6)$

56. Slope $= \frac{1}{9}$, y-intercept is $\left(0, -\frac{2}{5}\right)$

57. Through $(-4, -2)$ and $(-2, 4)$

58. Through $(7, 4)$ and $(6, 3)$

59. Through $(-6, 9)$ and $(8, -12)$

60. Through $(3, 0)$ and $(-3, 5)$

61. Through $(10, 3)$ and $(0, -2)$

62. Through $(-6, -2)$ and $(5, -3)$

63. Slope $= 7.4$, y-intercept is $(0, -4.5)$

64. Slope $= -\frac{7}{8}$, y-intercept is $\left(0, -\frac{3}{10}\right)$

65. Weight Loss Clinic Stacy Best owns a weight loss clinic. She charges her clients a one-time membership fee. She also charges per pound of weight lost. Therefore, the more successful she is at helping clients lose weight, the more income she will receive. The following graph shows a client's cost for losing weight.

Cost of Losing Weight

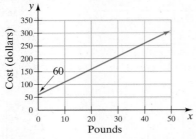

a) Find the equation that represents the cost for a client who loses x pounds.

b) Use the equation found in part **a)** to determine the cost for a client who loses 30 pounds.

66. Submarine Submerges A submarine is submerged below sea level. Tom Johnson, the captain, orders the ship to dive slowly. The following graph illustrates the submarine's depth at a time t minutes after the submarine begins to dive.

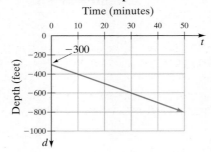

Submarine's Depth

a) Find the equation that represents the depth at time t.

b) Use the equation found in part **a)** to find the submarine's depth after 30 minutes.

67. Suppose that you were asked to write the equation of a line with the properties given below. Which form of a linear equation—standard form, slope-intercept form, or point-slope form—would you start with? Explain your answer.

a) The slope of the line and the y-intercept of the line

b) The slope and a point on the line

c) Two points on the line

68. Consider the two equations $20x - 30y = 50$ and $-20x + 30y = 40$.

a) When these equations are graphed, will the two lines have the same slope? Explain how you determined your answer.

b) When these two equations are graphed, will they be parallel lines?

69. Assume the slope of a line is 2 and two points on the line are $(-5, -4)$ and $(3, 12)$.

a) If you use $(-5, -4)$ as (x_1, y_1) and then $(3, 12)$ as (x_1, y_1) will the appearance of the two equations be the same in point-slope form? Explain.

b) Find the equation, in point-slope form, using $(-5, -4)$ as (x_1, y_1).

c) Find the equation, in point-slope form, using $(3, 12)$ as (x_1, y_1).

d) Write the equation obtained in part **b)** in slope-intercept form.

e) Write the equation obtained in part **c)** in slope-intercept form.

f) Are the equations obtained in parts **d)** and **e)** the same? If not, explain why.

70. Assume the slope of a line is -3 and two points on the line are $(-1, 8)$ and $(2, -1)$.

a) If you use $(-1, 8)$ as (x_1, y_1) and then $(2, -1)$ as (x_1, y_1) will the appearance of the two equations be the same in point-slope form? Explain.

b) Find the equation, in point-slope form, using $(-1, 8)$ as (x_1, y_1).

c) Find the equation, in point-slope form, using $(2, -1)$ as (x_1, y_1).

d) Write the equation obtained in part **b)** in slope-intercept form.

e) Write the equation obtained in part **c)** in slope-intercept form.

f) Are the equations obtained in parts **d)** and **e)** the same? If not, explain why.

Challenge Problems

71. Unit Conversions The following graph shows the approximate relationship between speed in miles per hour and feet per second.

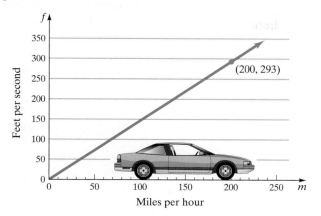

a) Determine the slope of the line.

b) Determine the equation of the line.

c) At the 2006 Daytona 500, Jimmie Johnson, the winner, had an average speed of 142.7 miles per hour. (See photo). Use the equation you obtained in part **b)** to determine the speed in feet per second.

d) Use the graph to estimate a speed of 100 miles per hour in feet per second.

e) Use the graph to estimate a speed of 80 feet per second in miles per hour.

72. Temperature The following graph shows the relationship between Fahrenheit temperature and Celsius temperature.

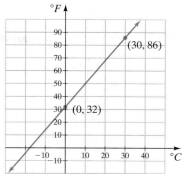

a) Determine the slope of the line.

b) Determine the equation of the line in slope-intercept form.

c) Use the equation (or formula) you obtained in part **b)** to find the Fahrenheit temperature when the Celsius temperature is $20°$.

d) Use the graph to estimate the Celsius temperature when the Fahrenheit temperature is $100°$.

e) Estimate the Celsius temperature that corresponds to a Fahrenheit temperature of $0°$.

73. Determine the equation of the line with y-intercept at 5 that is parallel to the line whose equation is $2x + y = 6$. Explain how you determined your answer.

74. Will a line through the points $(60, 30)$ and $(20, 90)$ be parallel to the line with x-intercept at 2 and y-intercept at 3? Explain how you determined your answer.

75. Write an equation of the line parallel to the graph of $3x - 4y = 6$ that passes through the point $(-8, -1)$.

76. Determine the equation of the straight line that intersects the greatest number of shaded points on the following graph.

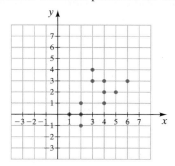

Group Activity

Discuss and answer Exercise 77 as a group, according to the instructions.

77. Consider the equation $-3x + 2y = 4$.

a) Group member 1: Explain how to graph this equation by plotting points. Then graph the equation by plotting points.

b) Group member 2: Explain how to graph this equation using the intercepts. Then graph the equation using the intercepts.

c) Group member 3: Explain how to graph this equation using the slope and y-intercept. Then graph the equation using the slope and y-intercept.

d) As a group, compare your graphs. Did you all obtain the same graph? If not, determine why.

Cumulative Review Exercises

[1.5] **78.** Insert either $>$, $<$, or $=$ in the shaded area to make the statement true: $|-4|$ ▓ $|-9|$.

[1.8] *Indicate whether each statement is true or false.*

79. The product of two negative numbers is always a positive number.

80. The sum of two negative numbers is always a negative number.

81. The difference of two negative numbers is always a negative number.

82. The quotient of two negative numbers is always a negative number.

[1.9] **83.** Evaluate 4^3.

[2.6] **84.** Solve $i = prt$ for r.

Chapter 4 Summary

IMPORTANT FACTS AND CONCEPTS	EXAMPLES

Section 4.1

The **Cartesian coordinate system** is formed by two axes drawn perpendicular to each other. The point of intersection is called the **origin**. The horizontal axis is called the **x-axis**. The vertical axis is called the **y-axis**. **Ordered pairs** are of the form (x, y).

Cartesian coordinate system

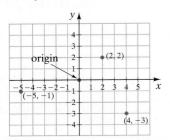

A **linear equation in two variables** is an equation that can be put in the form

$$ax + by = c$$

where a, b, and c are real numbers. This form is also called the **standard form** for a linear equation.

$$3x + 7y = 2, \quad -2x - y = 9$$

A **graph** of an equation in two variables is an illustration of a set of points whose coordinates satisfy the equation.
Points that lie in a straight line are **collinear**.

Every point on the graph satisfies the equation $y = 2x - 1$.

The points on the graph are collinear and the graph is a straight line.

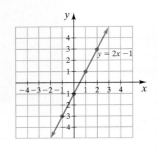

Section 4.2

To Graph Linear Equations by Plotting Points

1. Solve the linear equation for the variable y.
2. Select a value for the variable x. Substitute this value in the equation for x and find the corresponding value of y. Record the ordered pair (x, y).
3. Repeat step 2 with two different values of x.
4. Plot the three ordered pairs.
5. Draw a straight line through the three points. Draw arrowheads on each end of the line to show that the line continues indefinitely in both directions.

Table

x	y
-1	3
0	2
2	0

Graph $y = -x + 2$.

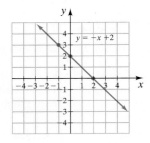

IMPORTANT FACTS AND CONCEPTS	EXAMPLES

Section 4.2 (continued)

The **x-intercept** is the point where the graph crosses the *x*-axis. The **y-intercept** is the point where the graph crosses the *y*-axis.	On the previous graph, the *x*-intercept is $(2, 0)$ and the *y*-intercept is $(0, 2)$.

To Graph Linear Equations using the x- and y-Intercepts

1. Find the *y*-intercept by setting *x* in the given equation equal to 0 and finding the corresponding value of *y*.
2. Find the *x*-intercept by setting *y* in the given equation equal to 0 and finding the corresponding value of *x*.
3. Determine a checkpoint by selecting a nonzero value for *x* and finding the corresponding value of *y*.
4. Plot the *y*-intercept, the *x*-intercept, and the checkpoint.
5. Draw a straight line through the three points. Draw an arrowhead at both ends of the line.

Graph $4x + 2y = 8$ using the *x*- and *y*-intercepts.

Let $x = 0$:
$$4(0) + 2y = 8$$
$$2y = 8$$
$$y = 4.$$
y-intercept: $(0, 4)$

Let $y = 0$:
$$4x + 2(0) = 8$$
$$4x = 8$$
$$x = 2$$
x-intercept: $(2, 0)$ Checkpoint: $(1, 2)$

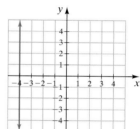

Horizontal Line
The graph of an equation of the form $y = b$ is a horizontal line whose *y*-intercept is $(0, b)$.

Vertical Line
The graph of an equation of the form $x = a$ is a vertical line whose *x*-intercept is $(a, 0)$.

Graph $y = 2$.   Graph $x = -4$.

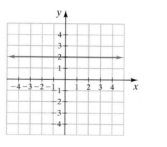

Section 4.3

The **slope of a line** is a ratio of the vertical change to the horizontal change between any two selected points on the line. The slope of a line through the points (x_1, y_1) and (x_2, y_2) is $$\text{slope} = \frac{\text{change in } y \text{ (vertical change)}}{\text{change in } x \text{ (horizontal change)}} = \frac{y_2 - y_1}{x_2 - x_1}$$	The slope of the line through $(-1, 3)$ and $(5, 7)$ is $$m = \frac{7 - 3}{5 - (-1)} = \frac{4}{6} = \frac{2}{3}$$

A straight line where the value of *y* increases as *x* increases has a **positive slope**.

A straight line where the value of *y* decreases as *x* increases has a **negative slope**.

A horizontal line has a **slope of 0**.

The slope of a vertical line is **undefined**.

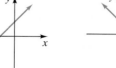

Positive slope   Negative slope
(rises to right)   (falls to right)

Slope is 0.   Slope is undefined.
(horizontal line)   (vertical line)

Two nonvertical lines with the same slope and different *y*-intercepts are **parallel lines**. Any two vertical lines are parallel to each other.	The graphs of the equations $y = 2x + 3$ and $y = 2x + 4$ are parallel lines since the graphs have the same slope, 2, and different *y*-intercepts.

IMPORTANT FACTS AND CONCEPTS	EXAMPLES

Section 4.3 (continued)

Two lines whose slopes are negative reciprocals of each other are **perpendicular lines**. Any vertical line is perpendicular to any horizontal line.	The graphs of the equations $y = 2x + 4$ and $y = -\frac{1}{2}x + 3$ are perpendicular lines since the slopes of the graphs are negative reciprocals of each other.

Section 4.4

Slope-Intercept Form of a Linear Equation $$y = mx + b$$ where m is the slope, and $(0, b)$ is the y-intercept of the line.	The graph of $y = 3x - 4$ has a slope of 3 and a y-intercept of $(0, -4)$. The equation of a line with a slope of $-\frac{1}{2}$ and a y-intercept of $(0, 6)$ is $y = -\frac{1}{2}x + 6$.
To graph $ax + by = c$, write the equation in slope-intercept form by solving the equation for y. Then use the slope and y-intercept to sketch the graph.	$2x + 4y = 8$ written in slope-intercept form is $y = -\frac{1}{2}x + 2$. The slope is $-\frac{1}{2}$ and the y-intercept is $(0, 2)$. 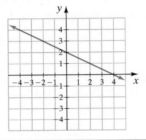
Point-Slope Form of a Linear Equation $$y - y_1 = m(x - x_1)$$ where m is the slope of the line and (x_1, y_1) is a point on the line.	The graph of $y - 7 = \frac{5}{6}(x - 3)$ has a slope of $\frac{5}{6}$ and contains the point $(3, 7)$.

Chapter 4 Review Exercises

[4.1]

1. Plot each ordered pair on the same axes.

    a) $A(5, 3)$ **b)** $B(0, 6)$ **c)** $C\left(5, \frac{1}{2}\right)$

    d) $D(-4, 3)$ **e)** $E(-6, -1)$ **f)** $F(-2, 0)$

2. Determine whether the following points are collinear.

         $(0, -4), (6, 8), (-2, 0), (4, 5)$

3. Which of the following ordered pairs satisfy the equation $2x + 3y = 9$?

    a) $\left(5, -\frac{1}{3}\right)$ **b)** $(3, 1)$

    c) $(-2, 4)$ **d)** $\left(2, \frac{5}{3}\right)$

[4.2]

4. Find the missing coordinate in the following solutions to $3x - 2y = 8$.

    a) $(2, ?)$ **b)** $(0, ?)$ **c)** $(?, 5)$ **d)** $(?, 0)$

Graph each equation using the method of your choice.

    5. $y = 4$ **6.** $x = 2$ **7.** $y = 3x$ **8.** $y = 2x - 1$

    9. $y = -2x + 5$ **10.** $2y + x = 8$ **11.** $-2x + 3y = 6$ **12.** $5x + 2y + 10 = 0$

    13. $5x + 10y = 20$ **14.** $\frac{2}{3}x = \frac{1}{4}y + 20$

[4.3] *Find the slope of the line through the given points.*

    15. $(6, -4)$ and $(1, 5)$ **16.** $(-4, -6)$ and $(8, -7)$ **17.** $(-2, -3)$ and $(-4, 1)$

18. What is the slope of a horizontal line?

19. What is the slope of a vertical line?

20. Define the slope of a straight line.

Find the slope of each line.

21.

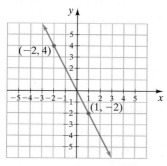

22.

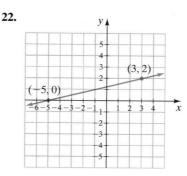

Assume that line 1 and line 2 are distinct lines. If m_1 represents the slope of line 1 and m_2 represents the slope of line 2, determine if line 1 and line 2 are parallel, perpendicular, or neither.

23. $m_1 = \dfrac{7}{8}, m_2 = -\dfrac{7}{8}$

24. $m_1 = -3, m_2 = \dfrac{1}{3}$

25. The following graph shows the number of manatee deaths in Florida. Find the slope of the line segment in

a) red,

b) blue.

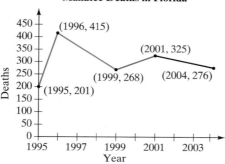

Manatee Deaths in Florida

Source: Florida Fish and Wildlife Conservation Commission, www.floridamarine.com

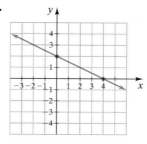

[4.4] *Determine the slope and y-intercept of the graph of each equation.*

26. $6x + 7y = 21$

27. $2x + 7 = 0$

28. $4y + 12 = 0$

Write the equation of each line.

29.

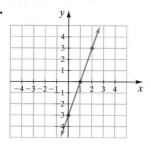

30.

Determine whether each pair of lines is parallel, perpendicular, or neither.

31. $y = 2x - 7$

$6y = 12x + 18$

32. $2x - 3y = 15$

$3x + 2y = 12$

Find the equation of each line with the given properties.

33. Slope $= 3$ through $(2, 7)$

34. Slope $= -\dfrac{2}{3}$, through $(3, 2)$

35. Slope $= 0$, through $(6, 2)$

36. Slope is undefined, through $(4, 1)$

37. Through $(-2, 4)$ and $(0, -3)$

38. Through $(-5, -2)$ and $(-5, 3)$

Chapter 4 Practice Test

To find out how well you understand the chapter material, take this practice test. The answers, and the section where the material was initially discussed, are given in the back of the book. Each problem is also fully worked out on the **Chapter Test Prep Video CD.** *Review any questions that you answered incorrectly.*

1. What is a graph?

2. In which quadrants do the following points lie?

   a) $(3, -5)$

   b) $\left(-2, \dfrac{1}{2}\right)$

3. a) What is the standard form of a linear equation?

   b) What is the slope-intercept form of a linear equation?

   c) What is the point-slope form of a linear equation?

4. Which of the following ordered pairs satisfy the equation $3y = 5x - 9$?

   a) $(4, 2)$

   b) $\left(\dfrac{9}{5}, 0\right)$

   c) $(-1, -10)$

   d) $(0, -3)$

5. Find the slope of the line through the points $(-2, 5)$ and $(4, -3)$.

6. Find the slope and y-intercept of $4x - 9y = 15$.

7. Write an equation of the graph in the accompanying figure.

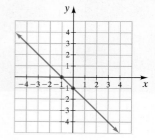

8. Graph $x = -4$.

9. Graph $y = 2$.

10. Graph $y = 3x - 2$ by plotting points.

11. a) Solve the equation $3x - 6y = 12$ for y.

    b) Graph the equation by plotting points.

12. Graph $3x + 5y = 15$ using the intercepts.

13. Write, in slope-intercept form, an equation of the line with a slope of 4 passing through the point $(2, -5)$

14. Write, in slope-intercept form, an equation of the line passing through the points $(3, -1)$ and $(-4, 2)$.

15. Determine whether the following equations represent parallel lines. Explain how you determined your answer.

$$2y = 3x - 6 \quad \text{and} \quad y - \frac{3}{2}x = -5$$

16. Graph $y = 3x - 4$ using the slope and y-intercept.

17. Graph $4x - 2y = 6$ using the slope and y-intercept.

18. **Weekly Income** Kate Moore, a salesperson, has a weekly income, i, that can be determined by the function $i = 200 + 0.05s$, where s is her weekly sales.

    a) Draw a graph of her weekly income for sales from $0 to $10,000.

    b) Estimate her weekly income if her sales are $5000.

Cumulative Review Test

Take the following test and check your answers with those given in the back of the book. Review any questions that you answered incorrectly. The section where the material was covered is indicated after the answer.

1. Write the set of
   a) natural numbers.
   b) whole numbers.

2. Evaluate $4 - 9 - (-10) + 13$.

3. Evaluate $(8 \div 4)^3 + 9^2 \div 3$.

4. Evaluate $(10 \div 5 \cdot 5 + 5 - 5)^2$.

5. Name each indicated property.
   a) $3(x + 2) = 3x + 3 \cdot 2$
   b) $a + b = b + a$

6. Solve $2x + 5 = 3(x - 5)$.

7. Solve $3(x - 1) - (x + 4) = 2x - 7$

8. Solve the proportion $\dfrac{2}{20} = \dfrac{x}{200}$.

9. **Chicken Soup** At Tsong Hsu's Grocery Store, 3 cans of chicken soup sell for $1.50. Find the cost of 8 cans.

10. Solve $v = lwh$ for w.

11. **Numbers** Eleven increased by twice a number is 19. Find the number.

12. **Rectangle** The length of a rectangle is 3 more than twice the width. Find the length and width of the rectangle if its perimeter is 36 feet.

13. **Running** Two runners start at the same point and run in opposite directions. One runs at 6 mph and the other runs at 8 mph. In how many hours will they be 28 miles apart?

14. Give three ordered pairs that satisfy the equation $2x + 4y = 8$.

15. Graph $y = 3x - 5$ by plotting points.

16. Graph $6x - 3y = -12$ using the intercepts.

17. Find the slope of the line through the points $(5, -6)$ and $(6, -5)$.

18. The slope of a given line is -7. If a line is to be drawn parallel to the given line, what will be its slope?

19. Graph $y = \dfrac{2}{3}x - 3$ using the slope and y-intercept.

20. Write the equation, in point-slope form, of the line with a slope of 3 passing through the point $(5, 2)$.

5 Exponents and Polynomials

GOALS OF THIS CHAPTER

The major emphasis of this chapter is to teach you how to work with exponents, polynomials, and scientific notation. You must understand the rules of exponents presented in the first two sections in order to be successful with the remaining material in the chapter.

You will use the rules learned in this chapter throughout the book, especially in Chapter 6, when factoring is introduced.

MOST AREAS OF science and technology deal with very small and very large numbers. Computers, for example, perform calculations in microseconds and their hard drives hold gigabytes of data. Scientific notation is a convenient way to work with small and large quantities. In Exercise 84 on page 302 we use scientific notation to determine how long it takes light from the sun to reach the earth.

5.1 Exponents

1 Review exponents.

2 Learn the rules of exponents.

3 Simplify an expression before using the expanded power rule.

1 Review Exponents

To be able to work the examples and exercises in this chapter, we need to expand our knowledge of exponents. Exponents were introduced in Section 1.9. Let's review the fundamental concepts. In the expression x^n, x is referred to as the **base** and n is called the **exponent**. x^n is read "x to the nth power."

$$x^2 = \underbrace{x \cdot x}_{2 \text{ factors of } x}$$

$$x^4 = \underbrace{x \cdot x \cdot x \cdot x}_{4 \text{ factors of } x}$$

$$x^m = \underbrace{x \cdot x \cdot x \cdot \cdots \cdot x}_{m \text{ factors of } x}$$

EXAMPLE 1 ▶ Write $xxxxyyy$ using exponents.

Solution

$$\underbrace{xxxx}_{\substack{4 \text{ factors} \\ \text{of } x}} \quad \underbrace{yyy}_{\substack{3 \text{ factors} \\ \text{of } y}} = x^4 y^3$$

▶ **Now Try Exercise 7**

Remember, when a term containing a variable is given without a numerical coefficient, the numerical coefficient of the term is assumed to be 1. For example, $x = 1x$ and $x^2 y = 1x^2 y$.

Also recall that when a variable or numerical value is given without an exponent, the exponent of that variable or numerical value is assumed to be 1. For example, $x = x^1$, $xy = x^1 y^1$, $x^2 y = x^2 y^1$, and $2xy^2 = 2^1 x^1 y^2$.

2 Learn the Rules of Exponents

Now we will learn the rules of exponents.

EXAMPLE 2 ▶ Multiply $x^4 \cdot x^3$.

Solution

$$\overbrace{x \cdot x \cdot x \cdot x}^{x^4} \cdot \overbrace{x \cdot x \cdot x}^{x^3} = x^7$$

▶ **Now Try Exercise 11**

Example 2 illustrates that when multiplying expressions with the same base we keep the base and *add* the exponents. This is the **product rule for exponents**.

Product Rule for Exponents
$x^m \cdot x^n = x^{m+n}$

In Example 2, we showed that $x^4 \cdot x^3 = x^7$. This problem could also be done using the product rule: $x^4 \cdot x^3 = x^{4+3} = x^7$.

EXAMPLE 3 ▶ Multiply each expression using the product rule.

a) $3^2 \cdot 3$ **b)** $2^4 \cdot 2^2$ **c)** $x \cdot x^4$ **d)** $x^3 \cdot x^6$ **e)** $y^4 \cdot y^7$

Solution

a) $3^2 \cdot 3 = 3^2 \cdot 3^1 = 3^{2+1} = 3^3$ or 27   **b)** $2^4 \cdot 2^2 = 2^{4+2} = 2^6$ or 64

c) $x \cdot x^4 = x^1 \cdot x^4 = x^{1+4} = x^5$ **d)** $x^3 \cdot x^6 = x^{3+6} = x^9$

e) $y^4 \cdot y^7 = y^{4+7} = y^{11}$

▶ **Now Try Exercise 19**

Avoiding Common Errors

Note in Example 3a) that $3^2 \cdot 3^1$ is 3^3 and not 9^3. When multiplying powers of the same base, *do not multiply the bases.*

CORRECT	INCORRECT
$3^2 \cdot 3^1 = 3^3$	$3^2 \cdot 3^1 = 9^3$

Example 4 will help you understand the **quotient rule for exponents**.

EXAMPLE 4 ▶ Divide $x^5 \div x^3$.

Solution

$$\frac{x^5}{x^3} = \frac{\overset{1}{\cancel{x}} \cdot \overset{1}{\cancel{x}} \cdot \overset{1}{\cancel{x}} \cdot x \cdot x}{\underset{1}{\cancel{x}} \cdot \underset{1}{\cancel{x}} \cdot \underset{1}{\cancel{x}}} = \frac{1x^2}{1} = x^2$$

▶ **Now Try Exercise 23**

When dividing expressions with the same base, keep the base and *subtract* the exponent in the denominator from the exponent in the numerator.

Quotient Rule for Exponents

$$\frac{x^m}{x^n} = x^{m-n}, \qquad x \neq 0$$

In Example 4, we showed that $\frac{x^5}{x^3} = x^2$. This problem could also be done using the quotient rule: $\frac{x^5}{x^3} = x^{5-3} = x^2$.

EXAMPLE 5 ▶ Divide each expression using the quotient rule.

a) $\dfrac{3^5}{3^2}$ **b)** $\dfrac{6^4}{6}$ **c)** $\dfrac{x^{12}}{x^5}$ **d)** $\dfrac{y^{10}}{y^8}$ **e)** $\dfrac{z^8}{z}$

Solution

a) $\dfrac{3^5}{3^2} = 3^{5-2} = 3^3$ or 27

b) $\dfrac{6^4}{6} = \dfrac{6^4}{6^1} = 6^{4-1} = 6^3$ or 216

c) $\dfrac{x^{12}}{x^5} = x^{12-5} = x^7$

d) $\dfrac{y^{10}}{y^8} = y^{10-8} = y^2$

e) $\dfrac{z^8}{z} = \dfrac{z^8}{z^1} = z^{8-1} = z^7$

▶ **Now Try Exercise 25**

Avoiding Common Errors

Note in Example 5a) that $\dfrac{3^5}{3^2}$ is 3^3 and not 1^3. When dividing powers of the same base, *do not divide out the bases.*

CORRECT	INCORRECT
$\dfrac{3^3}{3^1} = 3^2$ or 9	

The answer to Example 5c), $\dfrac{x^{12}}{x^5}$, is x^7. We obtained this answer using the quotient rule. This answer could also be obtained by dividing out the common factors in both the numerator and denominator as follows.

$$\frac{x^{12}}{x^5} = \frac{(\cancel{x} \cdot \cancel{x} \cdot \cancel{x} \cdot \cancel{x} \cdot \cancel{x}) \cdot x \cdot x \cdot x \cdot x \cdot x \cdot x \cdot x}{(\cancel{x} \cdot \cancel{x} \cdot \cancel{x} \cdot \cancel{x} \cdot \cancel{x})} = x^7$$

We divided out the product of five x's, which is x^5. We can indicate this process in shortened form as follows.

$$\frac{x^{12}}{x^5} = \frac{\overset{1}{\cancel{x^5}} \cdot x^7}{\cancel{x^5}} = x^7$$

In this section, to simplify an expression when the numerator and denominator have the same base and the exponent in the denominator is greater than the exponent in the numerator, we divide out common factors. For example, $\dfrac{x^5}{x^{12}}$ can be simplified by dividing out the common factor, x^5, as follows.

$$\frac{x^5}{x^{12}} = \frac{\overset{1}{\cancel{x^5}}}{\underset{1}{\cancel{x^5}} \cdot x^7} = \frac{1}{x^7}$$

We will now simplify some expressions by dividing out common factors.

EXAMPLE 6 ▶ Simplify each expression by dividing out a common factor in both the numerator and denominator.

a) $\dfrac{x^9}{x^{12}}$ **b)** $\dfrac{y^4}{y^9}$

Solution

a) Since the numerator is x^9, we write the denominator with a factor of x^9. Since $x^9 \cdot x^3 = x^{12}$, we rewrite x^{12} as $x^9 \cdot x^3$.

$$\frac{x^9}{x^{12}} = \frac{\overset{1}{\cancel{x^9}}}{\underset{1}{\cancel{x^9}} \cdot x^3} = \frac{1}{x^3}$$

b) $\dfrac{y^4}{y^9} = \dfrac{\overset{1}{\cancel{y^4}}}{\underset{1}{\cancel{y^4}} \cdot y^5} = \dfrac{1}{y^5}$

▶ **Now Try Exercise 27**

In the next section, we will show another way to simplify expressions like $\dfrac{x^9}{x^{12}}$ by using the negative exponent rule.

Example 7 leads us to our next rule, the **zero exponent rule**.

EXAMPLE 7 ▶ Divide $\dfrac{x^3}{x^3}$.

Solution By the quotient rule,

$$\frac{x^3}{x^3} = x^{3-3} = x^0$$

However,

$$\frac{x^3}{x^3} = \frac{1x^3}{1x^3} = \frac{1 \cdot \cancel{x} \cdot \cancel{x} \cdot \cancel{x}}{1 \cdot \cancel{x} \cdot \cancel{x} \cdot \cancel{x}} = \frac{1}{1} = 1$$

Since $\dfrac{x^3}{x^3} = x^0$ and $\dfrac{x^3}{x^3} = 1$, then x^0 must equal 1.

▶ **Now Try Exercise 29**

Zero Exponent Rule

$$x^0 = 1, \qquad x \neq 0$$

By the zero exponent rule, any real number, except 0, raised to the zero power equals 1. Note that 0^0 is undefined.

EXAMPLE 8 ▶ Simplify each expression. Assume $x \neq 0$.

a) 3^0 b) x^0 c) $3x^0$ d) $(3x)^0$ e) $4x^2y^3z^0$

Solution

a) $3^0 = 1$

b) $x^0 = 1$

c) $3x^0 = 3(x^0)$ *Remember, the exponent refers only to the immediately preceding symbol*
   $= 3 \cdot 1 = 3$ *unless parentheses are used.*

d) $(3x)^0 = 1$

e) $4x^2y^3z^0 = 4x^2y^3 \cdot 1 = 4x^2y^3$

▶ **Now Try Exercise 37**

Avoiding Common Errors

An expression raised to the zero power is not equal to 0; it is equal to 1.

CORRECT
$x^0 = 1$
$5^0 = 1$

INCORRECT
$x^0 = 0$
$5^0 = 0$

The **power rule** will be explained with the aid of Example 9.

EXAMPLE 9 ▶ Simplify $(x^3)^2$.

Solution

$$(x^3)^2 = \underbrace{x^3 \cdot x^3}_{2 \text{ factors of } x^3} = x^{3+3} = x^6$$

▶ **Now Try Exercise 45**

Power Rule for Exponents

$$(x^m)^n = x^{m \cdot n}$$

The power rule indicates that when we raise an exponential expression to a power, we keep the base and *multiply* the exponents. Example 9 could also be simplified using the power rule: $(x^3)^2 = x^{3 \cdot 2} = x^6$.

EXAMPLE 10 ▶ Simplify each expression.

a) $(x^3)^5$ b) $(3^4)^2$ c) $(y^5)^7$

Solution

a) $(x^3)^5 = x^{3 \cdot 5} = x^{15}$ b) $(3^4)^2 = 3^{4 \cdot 2} = 3^8$ c) $(y^5)^7 = y^{5 \cdot 7} = y^{35}$

▶ **Now Try Exercise 51**

Helpful Hint

Students often confuse the product and power rules. Note the difference carefully.

Product Rule	Power Rule
$x^m \cdot x^n = x^{m+n}$	$(x^m)^n = x^{m \cdot n}$
$2^3 \cdot 2^5 = 2^{3+5} = 2^8$	$(2^3)^5 = 2^{3 \cdot 5} = 2^{15}$

Example 11 will help us in explaining the **expanded power rule**. As the name suggests, this rule is an expansion of the power rule.

EXAMPLE 11 ▶ Simplify $\left(\dfrac{ax}{by}\right)^4$.

Solution

$$\left(\frac{ax}{by}\right)^4 = \frac{ax}{by} \cdot \frac{ax}{by} \cdot \frac{ax}{by} \cdot \frac{ax}{by}$$

$$= \frac{a \cdot a \cdot a \cdot a \cdot x \cdot x \cdot x \cdot x}{b \cdot b \cdot b \cdot b \cdot y \cdot y \cdot y \cdot y} = \frac{a^4 \cdot x^4}{b^4 \cdot y^4} = \frac{a^4 x^4}{b^4 y^4}$$

▶ **Now Try Exercise 67**

Expanded Power Rule for Exponents

$$\left(\frac{ax}{by}\right)^m = \frac{a^m x^m}{b^m y^m}, \qquad b \neq 0, \; y \neq 0$$

The expanded power rule illustrates that every factor within parentheses is raised to the power outside the parentheses when the expression is simplified.

EXAMPLE 12 ▶ Simplify each expression.

a) $(5x)^2$ **b)** $(-y)^3$ **c)** $(4xy)^4$ **d)** $\left(\dfrac{-2y}{3z}\right)^2$

Solution

a) $(5x)^2 = 5^2 x^2 = 25x^2$ **b)** $(-y)^3 = (-1y)^3 = (-1)^3 y^3 = -1y^3 = -y^3$

c) $(4xy)^4 = 4^4 x^4 y^4 = 256 x^4 y^4$ **d)** $\left(\dfrac{-2y}{3z}\right)^2 = \dfrac{(-2)^2 y^2}{3^2 z^2} = \dfrac{4y^2}{9z^2}$

▶ **Now Try Exercise 69**

3 Simplify an Expression Before Using the Expanded Power Rule

Whenever we have an expression raised to a power, it helps to simplify the expression in parentheses before using the expanded power rule. This procedure is illustrated in Examples 13 and 14.

EXAMPLE 13 ▶ Simplify $\left(\dfrac{9x^3 y^2}{3xy^2}\right)^3$.

Solution We first simplify the expression within parentheses by dividing out common factors.

$$\left(\frac{9x^3 y^2}{3xy^2}\right)^3 = \left(\frac{9}{3} \cdot \frac{x^3}{x} \cdot \frac{y^2}{y^2}\right)^3 = (3x^2)^3$$

Now we use the expanded power rule to simplify further.

$$(3x^2)^3 = 3^3(x^2)^3 = 27x^6$$

Thus, $\left(\dfrac{9x^3y^2}{3xy^2}\right)^3 = 27x^6$.

▶ Now Try Exercise 93

Helpful Hint *Study Tip*

Be very careful when writing exponents. Since exponents are generally smaller than regular text, take your time and write them clearly, and position them properly. If exponents are not written clearly it is very easy to confuse exponents such as 2 and 3, or 1 and 4, or 0 and 6. If you write down or carry an exponent from step to step incorrectly, you will obtain an incorrect answer.

EXAMPLE 14 ▶ Simplify $\left(\dfrac{25x^4y^3}{5x^2y^7}\right)^4$.

Solution Begin by simplifying the expression within parentheses.

$$\left(\frac{25x^4y^3}{5x^2y^7}\right)^4 = \left(\frac{25}{5}\cdot\frac{x^4}{x^2}\cdot\frac{y^3}{y^7}\right)^4 = \left(\frac{5x^2}{y^4}\right)^4$$

Now use the expanded power rule to simplify further.

$$\left(\frac{5x^2}{y^4}\right)^4 = \frac{5^4(x^2)^4}{(y^4)^4} = \frac{625x^8}{y^{16}}$$

Thus, $\left(\dfrac{25x^4y^3}{5x^2y^7}\right)^4 = \dfrac{625x^8}{y^{16}}$.

▶ Now Try Exercise 95

Avoiding Common Errors

Students sometimes make errors in simplifying expressions containing exponents. One of the most common errors follows. Study this error carefully to make sure you do not make the same mistake.

CORRECT

$$\frac{4}{2x} = \frac{\overset{2}{\cancel{4}}}{\underset{1}{\cancel{2}}x} = \frac{2}{x}$$

$$\frac{x}{xy} = \frac{\overset{1}{\cancel{x}}}{\cancel{x}y} = \frac{1}{y}$$

$$\frac{5x^3y^2}{y^2} = \frac{5x^3\overset{1}{\cancel{y^2}}}{\underset{1}{\cancel{y^2}}} = 5x^3$$

INCORRECT

$$\frac{4}{x+2} = \frac{\overset{2}{\cancel{4}}}{x+\underset{1}{\cancel{2}}} = \frac{2}{x+1}$$

$$\frac{x}{x+y} = \frac{\overset{1}{\cancel{x}}}{\cancel{x}+y} = \frac{1}{1+y}$$

$$\frac{5x^3+y^2}{y^2} = \frac{5x^3+\overset{1}{\cancel{y^2}}}{\underset{1}{\cancel{y^2}}} = 5x^3+1$$

The simplifications on the right side are not correct because only common *factors* can be divided out (remember, factors are multiplied together). In the first denominator on the right, $x + 2$, the x and 2 are terms, not factors, since they are being added. Similarly, in the second denominator, $x + y$, the x and the y are terms, not factors, since they are being added. Also, in the numerator $5x^3 + y^2$, the $5x^3$ and y^2 are terms, not factors, since they are being added. No common factors can be divided out in the fractions on the right.

EXAMPLE 15 ▶ Simplify $(2a^5b^3)^5(3a^2b)$.

Solution   First simplify $(2a^5b^3)^5$ by using the expanded power rule.

$$(2a^5b^3)^5 = 2^5a^{5\cdot5}b^{3\cdot5} = 32a^{25}b^{15}$$

Now use the product rule to simplify further.

$$(2a^5b^3)^5(3a^2b) = (32a^{25}b^{15})(3a^2b^1)$$
$$= 32 \cdot 3 \cdot a^{25} \cdot a^2 \cdot b^{15} \cdot b^1$$
$$= 96a^{25+2}b^{15+1}$$
$$= 96a^{27}b^{16}$$

Thus, $(2a^5b^3)^5(3a^2b) = 96a^{27}b^{16}$.

▶ **Now Try Exercise 125**

Summary of the Rules of Exponents Presented in This Section

1. $x^m \cdot x^n = x^{m+n}$ **product rule**

2. $\dfrac{x^m}{x^n} = x^{m-n}, \quad x \neq 0$ **quotient rule**

3. $x^0 = 1, \quad x \neq 0$ **zero exponent rule**

4. $(x^m)^n = x^{m \cdot n}$ **power rule**

5. $\left(\dfrac{ax}{by}\right)^m = \dfrac{a^m x^m}{b^m y^m}, \quad b \neq 0, \quad y \neq 0$ **expanded power rule**

EXERCISE SET 5.1

MathXL   MathXL®   MyMathLab   MyMathLab

Concept/Writing Exercises

1. In the exponential expression t^p, what is the t called? What is the p called?

2. **a)** Write the product rule for exponents.
   b) Explain the product rule.

3. **a)** Write the quotient rule for exponents.
   b) Explain the quotient rule.

4. **a)** Write the zero exponent rule.
   b) Explain the zero exponent rule.

5. **a)** Write the power rule for exponents.
   b) Explain the power rule.

6. **a)** Write the expanded power rule for exponents.
   b) Explain the expanded power rule.

7. Write *aabbbbb* using exponents.

8. Write *pppqqqqrrrrrr* using exponents.

9. Explain the difference between the product rule and the power rule. Give an example of each.

10. For what value of x is $x^0 \neq 1$?

Practice the Skills

Multiply.

11. $x^5 \cdot x^4$

12. $x^6 \cdot x$

13. $-z^4 \cdot z$

14. $x^7 \cdot x^2$

 15. $y^3 \cdot y^2$

16. $4^2 \cdot 4^3$

17. $3^2 \cdot 3^3$

18. $-x^3 \cdot x^4$

19. $z^3 \cdot z^5$

20. $2^4 \cdot 2^2$

Divide.

21. $\dfrac{6^2}{6}$

22. $\dfrac{x^4}{x^3}$

23. $\dfrac{x^{10}}{x^3}$

24. $\dfrac{y^9}{y}$

25. $\dfrac{3^6}{3^2}$

26. $\dfrac{4^5}{4^3}$

27. $\dfrac{y^4}{y^6}$

28. $\dfrac{a^7}{a^9}$

29. $\dfrac{c^4}{c^4}$

30. $\dfrac{5^4}{5^4}$

31. $\dfrac{a^3}{a^9}$

32. $\dfrac{x^9}{x^{13}}$

Simplify.

33. x^0

34. 5^0

35. $3x^0$

36. $-7x^0$

37. $4(5d)^0$

38. $-2(8x)^0$

39. $-9(-4y)^0$

40. $-(-x)^0$

41. $6x^3y^2z^0$

42. $-5xy^2z^0$

43. $-8r(st)^0$

44. $-3(a^2b^5c^3)^0$

Simplify.

45. $(x^4)^2$

46. $(a^5)^3$

47. $(x^5)^5$

48. $(y^5)^2$

49. $(x^3)^1$

50. $(x^6)^2$

51. $(x^4)^3$

52. $(x^5)^4$

53. $(n^6)^3$

54. $(1.3x)^2$

55. $(-2w^2)^3$

56. $(-3x)^2$

57. $(-3x^3)^3$

58. $(-xy)^4$

59. $(4x^3y^2)^3$

60. $(3a^2b^4)^3$

Simplify.

61. $\left(\dfrac{x}{3}\right)^2$

62. $\left(\dfrac{-2}{x}\right)^3$

63. $\left(\dfrac{y}{x}\right)^4$

64. $\left(\dfrac{2}{y}\right)^4$

65. $\left(\dfrac{-6}{x}\right)^3$

66. $\left(\dfrac{4m}{n}\right)^3$

67. $\left(\dfrac{2x}{y}\right)^3$

68. $\left(\dfrac{3s}{t^2}\right)^2$

69. $\left(\dfrac{4p}{5}\right)^2$

70. $\left(\dfrac{2y^3}{x}\right)^4$

71. $\left(\dfrac{3x^4}{y}\right)^3$

72. $\left(\dfrac{-4x^2}{5}\right)^2$

Simplify.

73. $\dfrac{a^8b}{ab^4}$

74. $\dfrac{x^3y^5}{x^7y}$

75. $\dfrac{5x^{12}y^2}{10xy^9}$

76. $\dfrac{10x^3y^8}{2xy^{10}}$

77. $\dfrac{30y^5z^3}{5yz^6}$

78. $\dfrac{3ab}{27a^3b^4}$

79. $\dfrac{35x^4y^9}{15x^9y^{12}}$

80. $\dfrac{6m^3n^9}{9m^7n^{12}}$

81. $-\dfrac{36xy^7z}{12x^4y^5z}$

82. $\dfrac{4x^4y^7z^3}{32x^5y^4z^9}$

83. $-\dfrac{6x^2y^7z}{3x^5y^9z^6}$

84. $-\dfrac{25x^4y^{10}}{30x^3y^7z}$

Simplify.

85. $\left(\dfrac{10x^4}{5x^6}\right)^3$

86. $\left(\dfrac{4x^4}{8x^8}\right)^3$

87. $\left(\dfrac{6y^6}{2y^3}\right)^3$

88. $\left(\dfrac{4xy^5}{y}\right)^3$

89. $\left(\dfrac{6a^2b^4}{3a^7b^9}\right)^0$

90. $\left(\dfrac{16y^6}{24y^{10}}\right)^3$

91. $\left(\dfrac{x^4y^3}{x^2y^5}\right)^2$

92. $\left(\dfrac{2x^7y^2}{4xy}\right)^3$

93. $\left(\dfrac{9y^2z^7}{18y^9z}\right)^4$

94. $\left(\dfrac{y^7z^5}{y^8z^4}\right)^{10}$

95. $\left(\dfrac{25s^4t}{5s^6t^4}\right)^3$

96. $\left(\dfrac{-64xy^6}{32xy^9}\right)^4$

Simplify.

97. $(3xy^4)^2$

98. $(4ab^3)^3$

99. $(5ab^3)(b)$

100. $(6xy^5)(3x^2y^4)$

101. $(-2xy)(3xy)$

102. $(-3x^4y^2)(5x^2y)$

103. $(5x^2y)(3xy^5)$

104. $(-5xy)(-2xy^6)$

105. $(-3p^2q)^2(-p^2q)$

106. $(2c^3d^2)^2(3cd)^0$

107. $(7r^3s^2)^2(9r^3s^4)^0$

108. $(3x^2)^4(2xy^5)$

Simplify.

109. $(-x)^2$

110. $(2xy^4)^3$

111. $\left(\dfrac{x^5y^5}{xy^5}\right)^3$

112. $(2x^2y^5)(3x^5y^4)^3$

113. $(2.5x^3)^2$

114. $(-3a^2b^3c^4)^3$

115. $\dfrac{x^9y^3}{x^2y^7}$

116. $(xy^4)(xy^4)^3$

117. $\left(-\dfrac{m^4}{n^3}\right)^3$

118. $\left(-\dfrac{12x}{16x^7y^2}\right)^2$

119. $(-6x^3y^2)^3$

120. $(3x^6y)^2(4xy^8)$

121. $(-2x^4y^2z)^3$

122. $\left(\dfrac{z}{4}\right)^3$

123. $(9r^4s^5)^3$

124. $(5x^4z^{10})^2(2x^2z^8)$

125. $(4x^2y)(3xy^2)^3$

126. $\dfrac{x^2y^6}{x^4y}$

127. $(7.3x^2y^4)^2$

128. $\left(\dfrac{-3x^3}{4}\right)^3$

129. $(x^7y^5)(xy^2)^4$

130. $(4c^3d^2)(2c^5d^3)^2$

131. $\left(\dfrac{-x^4z^7}{x^2z^5}\right)^4$

132. $(x^4y^6)^3(3x^2y^5)$

Study the Avoiding Common Errors box on page 281. Simplify the following expressions by dividing out common factors. If the expression cannot be simplified by dividing out common factors, so state.

133. $\dfrac{a+b}{b}$

134. $\dfrac{xy}{x}$

135. $\dfrac{y^2+3}{y}$

136. $\dfrac{a+9}{3}$

137. $\dfrac{6yz^4}{yz^2}$

138. $\dfrac{x}{x+1}$

139. $\dfrac{a^2+b^2}{a^2}$

140. $\dfrac{x^4}{x^2y}$

Problem Solving

141. What is the value of a^3b if $a=2$ and $b=5$?

142. What is the value of xy^2 if $x=-3$ and $y=-4$?

143. What is the value of $(xy)^0$ if $x=-5$ and $y=3$?

144. What is the value of $(xy)^0$ if $x=2$ and $y=4$?

145. Consider the expression $(-9x^4y^6)^8$. When the expanded power rule is used to simplify the expression, will the *sign* of the simplified expression be positive or negative? Explain how you determined your answer.

146. Consider the expression $(-x^5y^7)^9$. When the expanded power rule is used to simplify the expression, will the *sign* of the simplified expression be positive or negative? Explain how you determined your answer.

Write an expression for the total area of the figure or figures shown.

147.

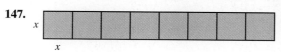

148.

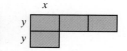

149.

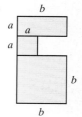

150.

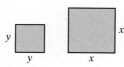

Challenge Problems

Simplify.

151. $(3yz^2)^2\left(\dfrac{2y^3z^5}{10y^6z^4}\right)^0(4y^2z^3)^3$

152. $\left(\dfrac{3x^4y^5}{6x^6y^8}\right)^3\left(\dfrac{9x^7y^8}{3x^3y^5}\right)^2$

Group Activity

Discuss and answer Exercise 153 as a group, according to the instructions.

153. In the next section we will be working with negative exponents. To prepare for that work, use the expression $\dfrac{3^2}{3^3}$ to work parts **a)** through **c)**. Work parts **a)** through **d)** individually, then part **e)** as a group.

   a) Divide out common factors in the numerator and denominator and determine the value of the expression.

   b) Use the quotient rule on the given expression and write down your results.

   c) Write a statement of equality using the results of part **a)** and **b)** above.

   d) Repeat parts **a)** through **c)** for the expression $\dfrac{2^3}{2^4}$.

   e) As a group, compare your answers to parts **a)** through **d)**, then write an exponential expression for $\dfrac{1}{x^m}$.

Cumulative Review Exercises

[1.9] **154.** Evaluate $3^4 \div 3^3 - (5-8) + 7$.

[2.1] **155.** Simplify $-4(x-3) + 5x - 2$.

[2.5] **156.** Solve the equation
$$2(x+4) - 3 = 5x + 4 - 3x + 1.$$

[2.6] **157.** **a)** Use the formula $P = 2l + 2w$ to find the length of the sides of the rectangle shown if the perimeter of the rectangle is 26 inches.

   b) Solve the formula $P = 2l + 2w$ for w.

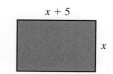

5.2 Negative Exponents

1 Understand the negative exponent rule.

2 Simplify expressions containing negative exponents.

1 ▶ Understand the Negative Exponent Rule

One additional rule that involves exponents is the negative exponent rule. You will need to understand negative exponents to be successful with scientific notation in the next section.

The negative exponent rule will be developed using the quotient rule illustrated in Example 1.

EXAMPLE 1 ▶ Simplify $\dfrac{x^3}{x^5}$ by **a)** using the quotient rule and **b)** dividing out common factors.

Solution

a) By the quotient rule,

$$\frac{x^3}{x^5} = x^{3-5} = x^{-2}$$

b) By dividing out common factors,

$$\frac{x^3}{x^5} = \frac{\cancel{x} \cdot \cancel{x} \cdot \cancel{x}}{\cancel{x} \cdot \cancel{x} \cdot \cancel{x} \cdot x \cdot x} = \frac{1}{x^2}$$

▶ **Now Try Exercise 35**

In Example 1, we see that $\dfrac{x^3}{x^5}$ is equal to both x^{-2} and $\dfrac{1}{x^2}$. Therefore, x^{-2} must equal $\dfrac{1}{x^2}$. That is, $x^{-2} = \dfrac{1}{x^2}$. This is an example of the **negative exponent rule**.

Negative Exponent Rule

$$x^{-m} = \frac{1}{x^m}, \qquad x \neq 0$$

When a variable or number is raised to a negative exponent, the expression may be rewritten as 1 divided by the variable or number raised to that positive exponent.

Examples

$$x^{-6} = \frac{1}{x^6} \qquad 4^{-2} = \frac{1}{4^2} = \frac{1}{16}$$

$$y^{-7} = \frac{1}{y^7} \qquad 5^{-3} = \frac{1}{5^3} = \frac{1}{125}$$

Avoiding Common Errors

Students sometimes believe that a negative exponent automatically makes the value of the expression negative. This is not true.

EXPRESSION	CORRECT	INCORRECT	ALSO INCORRECT
3^{-2}	$\dfrac{1}{3^2} = \dfrac{1}{9}$	$\cancel{-3^2}$	$\cancel{-\dfrac{1}{3^2}}$
x^{-3}	$\dfrac{1}{x^3}$	$\cancel{-x^3}$	$\cancel{-\dfrac{1}{x^3}}$

To help you see that the negative exponent rule makes sense, consider the following sequence of exponential expressions and their corresponding values.

$$2^3 = 8, \quad 2^2 = 4, \quad 2^1 = 2, \quad 2^0 = 1, \quad 2^{-1} = \frac{1}{2^1} \text{ or } \frac{1}{2}, \quad 2^{-2} = \frac{1}{2^2} \text{ or } \frac{1}{4}, \quad 2^{-3} = \frac{1}{2^3} \text{ or } \frac{1}{8}$$

Note that each time the exponent decreases by 1, the value of the expression is halved. For example, when we go from 2^3 to 2^2, the value of the expression goes from 8 to 4. If we continue decreasing the exponents beyond $2^0 = 1$, the next exponent in the pattern is -1. If we take half of 1 we get $\dfrac{1}{2}$. This pattern illustrates that $x^{-m} = \dfrac{1}{x^m}$.

2 Simplify Expressions Containing Negative Exponents

Generally, when you are asked to simplify an exponential expression **your final answer should contain no negative exponents**. You may simplify exponential expressions using the negative exponent rule and the rules of exponents presented in the previous section. The following examples indicate how exponential expressions containing negative exponents may be simplified.

EXAMPLE 2 ▶ Use the negative exponent rule to write each expression with a positive exponent. Simplify the expressions further when possible.

   a) y^{-5} **b)** x^{-4} **c)** 2^{-3} **d)** 6^{-1} **e)** -5^{-3} **f)** $(-5)^{-3}$

Solution

a) $y^{-5} = \dfrac{1}{y^5}$ **b)** $x^{-4} = \dfrac{1}{x^4}$

c) $2^{-3} = \dfrac{1}{2^3} = \dfrac{1}{8}$ **d)** $6^{-1} = \dfrac{1}{6}$

e) $-5^{-3} = -\dfrac{1}{5^3} = -\dfrac{1}{125}$ **f)** $(-5)^{-3} = \dfrac{1}{(-5)^3} = \dfrac{1}{-125} = -\dfrac{1}{125}$

                                                                ▶ **Now Try Exercise 11**

EXAMPLE 3 ▶ Use the negative exponent rule to write each expression with a positive exponent.

   a) $\dfrac{1}{x^{-2}}$ **b)** $\dfrac{1}{4^{-1}}$

Solution First use the negative exponent rule on the denominator. Then simplify further.

a) $\dfrac{1}{x^{-2}} = \dfrac{1}{\dfrac{1}{x^2}} = 1 \div \dfrac{1}{x^2} = \dfrac{1}{1} \cdot \dfrac{x^2}{1} = x^2$ **b)** $\dfrac{1}{4^{-1}} = \dfrac{1}{\dfrac{1}{4}} = 1 \div \dfrac{1}{4} = \dfrac{1}{1} \cdot \dfrac{4}{1} = 4$

                                                                ▶ **Now Try Exercise 15**

Helpful Hint

From Examples 2 and 3, we can see that when a factor is moved from the denominator to the numerator or from the numerator to the denominator, the sign of the *exponent* changes.

$$x^{-4} = \dfrac{1}{x^4} \qquad\qquad \dfrac{1}{x^{-4}} = x^4$$

$$3^{-5} = \dfrac{1}{3^5} \qquad\qquad \dfrac{1}{3^{-5}} = 3^5$$

Now let's look at additional examples that combine two or more of the rules presented so far.

EXAMPLE 4 ▶ Simplify.   **a)** $(z^{-5})^4$ **b)** $(4^2)^{-3}$

Solution

a) $(z^{-5})^4 = z^{(-5)(4)}$ *By the power rule*

$\qquad\qquad = z^{-20}$

$\qquad\qquad = \dfrac{1}{z^{20}}$ *By the negative exponent rule*

b) $(4^2)^{-3} = 4^{(2)(-3)}$ *By the power rule*

$\qquad\qquad = 4^{-6}$

$\qquad\qquad = \dfrac{1}{4^6}$ *By the negative exponent rule*

▶ **Now Try Exercise 25**

EXAMPLE 5 ▶ Simplify.   **a)** $x^3 \cdot x^{-5}$ **b)** $3^{-4} \cdot 3^{-7}$

Solution

a) $x^3 \cdot x^{-5} = x^{3+(-5)}$ *By the product rule*

$\qquad\qquad = x^{-2}$

$\qquad\qquad = \dfrac{1}{x^2}$ *By the negative exponent rule*

b) $3^{-4} \cdot 3^{-7} = 3^{-4+(-7)}$ *By the product rule*

$\qquad\qquad = 3^{-11}$

$\qquad\qquad = \dfrac{1}{3^{11}}$ *By the negative exponent rule*

▶ **Now Try Exercise 51**

Avoiding Common Errors

What is the sum of $3^2 + 3^{-2}$? Look carefully at the correct solution.

CORRECT	INCORRECT
$3^2 + 3^{-2} = 9 + \dfrac{1}{9}$	$3^2 + 3^{-2} = 0$
$\qquad\qquad = 9\dfrac{1}{9}$	

Note that $3^2 \cdot 3^{-2} = 3^{2+(-2)} = 3^0 = 1$.

EXAMPLE 6 ▶ Simplify.   **a)** $\dfrac{b^{-5}}{b^{13}}$ **b)** $\dfrac{6^{-8}}{6^{-5}}$

Solution

a) $\dfrac{b^{-5}}{b^{13}} = b^{-5-13}$ *By the quotient rule*

$\qquad\quad = b^{-18}$

$\qquad\quad = \dfrac{1}{b^{18}}$ *By the negative exponent rule*

b) $\dfrac{6^{-8}}{6^{-5}} = 6^{-8-(-5)}$ *By the quotient rule*

$\quad = 6^{-8+5}$

$\quad = 6^{-3}$

$\quad = \dfrac{1}{6^3}$ or $\dfrac{1}{216}$ *By the negative exponent rule*

▶ **Now Try Exercise 81**

The following Helpful Hint should be read carefully.

Helpful Hint

Consider a division problem where a variable has a negative exponent in either its numerator or its denominator, such as in Example 6a). Another way to simplify such an expression is to move the variable with the negative exponent from the numerator to the denominator, or from the denominator to the numerator, and change the sign of the exponent. For example,

$$\frac{x^{-4}}{x^5} = \frac{1}{x^5 \cdot x^4} = \frac{1}{x^{5+4}} = \frac{1}{x^9}$$

$$\frac{y^3}{y^{-7}} = y^3 \cdot y^7 = y^{3+7} = y^{10}$$

Now consider a division problem where a number or variable has a negative exponent in both its numerator and its denominator, such as in Example 6b). Another way to simplify such an expression is to move the variable with the more negative exponent from the numerator to the denominator, or from the denominator to the numerator, and change the sign of the exponent from negative to positive. For example,

$$\frac{x^{-8}}{x^{-3}} = \frac{1}{x^8 \cdot x^{-3}} = \frac{1}{x^{8-3}} = \frac{1}{x^5}$$ *Note that $-8 < -3$.*

$$\frac{y^{-4}}{y^{-7}} = y^7 \cdot y^{-4} = y^{7-4} = y^3$$ *Note that $-7 < -4$.*

EXAMPLE 7 ▶ Simplify.

a) $7x^4(6x^{-9})$ **b)** $\dfrac{16r^3 s^{-3}}{8rs^2}$ **c)** $\dfrac{2x^2 y^5}{8x^7 y^{-3}}$

Solution

a) $7x^4(6x^{-9}) = 7 \cdot 6 \cdot x^4 \cdot x^{-9} = 42x^{-5} = \dfrac{42}{x^5}$

b) $\dfrac{16r^3 s^{-3}}{8rs^2} = \dfrac{16}{8} \cdot \dfrac{r^3}{r} \cdot \dfrac{s^{-3}}{s^2}$

$\quad = 2 \cdot r^2 \cdot \dfrac{1}{s^5} = \dfrac{2r^2}{s^5}$

c) $\dfrac{2x^2 y^5}{8x^7 y^{-3}} = \dfrac{2}{8} \cdot \dfrac{x^2}{x^7} \cdot \dfrac{y^5}{y^{-3}}$

$\quad = \dfrac{1}{4} \cdot \dfrac{1}{x^5} \cdot y^8 = \dfrac{y^8}{4x^5}$

▶ **Now Try Exercise 121**

In Example 7**b)**, the variable with the negative exponent, s^{-3}, was moved from the numerator to the denominator. In Example 7**c)**, the variable with the negative exponent, y^{-3}, was moved from the denominator to the numerator. In each case, the sign of the exponent was changed from negative to positive when the variable factor was moved.

EXAMPLE 8 ▶ Simplify.

a) $(5x^{-3})^{-2}$ **b)** $(-5x^{-3})^{-2}$ **c)** $(-5x^{-3})^{-3}$

Solution   Begin by using the expanded power rule.

a) $(5x^{-3})^{-2} = 5^{-2}x^{(-3)(-2)}$

$$= 5^{-2}x^6$$

$$= \frac{1}{5^2}x^6$$

$$= \frac{x^6}{25}$$

b) $(-5x^{-3})^{-2} = (-5)^{-2}x^{(-3)(-2)}$

$$= \frac{1}{(-5)^2}x^6$$

$$= \frac{x^6}{25}$$

c) $(-5x^{-3})^{-3} = (-5)^{-3}x^{(-3)(-3)}$

$$= \frac{1}{(-5)^3}x^9$$

$$= \frac{1}{-125}x^9$$

$$= -\frac{x^9}{125}$$

▶ **Now Try Exercise 105**

Avoiding Common Errors

Can you explain why the simplification on the right is incorrect?

CORRECT

$$\frac{x^3 y^{-2}}{w} = \frac{x^3}{wy^2}$$

INCORRECT

$$\frac{x^3 + y^{-2}}{w} = \frac{x^3}{w + y^2}$$

The simplification on the right is incorrect because in the numerator $x^3 + y^{-2}$, the y^{-2} *is not a factor*; it is a term. We will learn how to simplify expressions like this when we study complex fractions in Section 7.5.

EXAMPLE 9 ▶ Simplify $\left(\dfrac{2}{3}\right)^{-2}$.

Solution   By the expanded power rule, we may write

$$\left(\frac{2}{3}\right)^{-2} = \frac{2^{-2}}{3^{-2}} = \frac{\dfrac{1}{2^2}}{\dfrac{1}{3^2}} = \frac{1}{2^2} \cdot \frac{3^2}{1} = \frac{3^2}{2^2} = \frac{9}{4}$$

▶ **Now Try Exercise 95**

If we examine the results of Example 9, we see that

$$\left(\frac{2}{3}\right)^{-2} = \frac{3^2}{2^2} = \left(\frac{3}{2}\right)^2.$$

This example illustrates that $\left(\dfrac{a}{b}\right)^{-m} = \left(\dfrac{b}{a}\right)^{m}$ when $a \neq 0$ and $b \neq 0$. Thus, for example, $\left(\dfrac{3}{4}\right)^{-5} = \left(\dfrac{4}{3}\right)^{5}$ and $\left(\dfrac{5}{9}\right)^{-3} = \left(\dfrac{9}{5}\right)^{3}$. We can summarize this information as follows.

A Fraction Raised to a Negative Exponent Rule

For a fraction of the form $\dfrac{a}{b}$, $a \neq 0$ and $b \neq 0$, $\left(\dfrac{a}{b}\right)^{-m} = \left(\dfrac{b}{a}\right)^{m}$.

EXAMPLE 10 ▸ Simplify.   **a)** $\left(\dfrac{4}{5}\right)^{-3}$ **b)** $\left(\dfrac{x^5}{y^7}\right)^{-4}$

Solution   We use the above rule to simplify.

a) $\left(\dfrac{4}{5}\right)^{-3} = \left(\dfrac{5}{4}\right)^{3} = \dfrac{5^3}{4^3} = \dfrac{125}{64}$ **b)** $\left(\dfrac{x^5}{y^7}\right)^{-4} = \left(\dfrac{y^7}{x^5}\right)^{4} = \dfrac{y^{7\cdot4}}{x^{5\cdot4}} = \dfrac{y^{28}}{x^{20}}$

▸ **Now Try Exercise 97**

EXAMPLE 11 ▸ Simplify.   **a)** $\left(\dfrac{x^2 y^{-3}}{z^4}\right)^{-5}$ **b)** $\left(\dfrac{2x^{-3} y^2 z}{x^2}\right)^{2}$

Solution

a) We will work part **a)** using two different methods. In method 1, we begin by using the expanded power rule. In method 2, we use a fraction raised to a negative exponent rule before we use the expanded power rule. You may use either method.

Method 1
$$\left(\frac{x^2 y^{-3}}{z^4}\right)^{-5} = \frac{x^{2(-5)} y^{(-3)(-5)}}{z^{4(-5)}} \qquad \textit{Expanded power rule}$$

$$= \frac{x^{-10} y^{15}}{z^{-20}} \qquad \textit{Multiply exponents}$$

$$= \frac{y^{15} z^{20}}{x^{10}} \qquad \textit{Negative exponent rule}$$

Method 2
$$\left(\frac{x^2 y^{-3}}{z^4}\right)^{-5} = \left(\frac{z^4}{x^2 y^{-3}}\right)^{5} \qquad \left(\frac{a}{b}\right)^{-m} = \left(\frac{b}{a}\right)^{m}$$

$$= \left(\frac{y^3 z^4}{x^2}\right)^{5} \qquad \textit{Simplify expression within parentheses}$$

$$= \frac{y^{3\cdot5} z^{4\cdot5}}{x^{2\cdot5}} \qquad \textit{Expanded power rule}$$

$$= \frac{y^{15} z^{20}}{x^{10}} \qquad \textit{Multiply exponents}$$

b) First simplify the expression within parentheses, then square the results. To simplify, we note that $\dfrac{x^{-3}}{x^2}$ becomes $\dfrac{1}{x^5}$.

$$\left(\frac{2x^{-3} y^2 z}{x^2}\right)^{2} = \left(\frac{2y^2 z}{x^5}\right)^{2} = \frac{2^2 y^{2\cdot2} z^{1\cdot2}}{x^{5\cdot2}} = \frac{4y^4 z^2}{x^{10}}$$

▸ **Now Try Exercise 125**

Summary of Rules of Exponents

1. $x^m \cdot x^n = x^{m+n}$ **product rule**

2. $\dfrac{x^m}{x^n} = x^{m-n}, \quad x \neq 0$ **quotient rule**

3. $x^0 = 1, \quad x \neq 0$ **zero exponent rule**

4. $(x^m)^n = x^{m \cdot n}$ **power rule**

5. $\left(\dfrac{ax}{by}\right)^m = \dfrac{a^m x^m}{b^m y^m}, \quad b \neq 0, y \neq 0$ **expanded power rule**

6. $x^{-m} = \dfrac{1}{x^m}, \quad x \neq 0$ **negative exponent rule**

7. $\left(\dfrac{a}{b}\right)^{-m} = \left(\dfrac{b}{a}\right)^m, \quad a \neq 0, b \neq 0$ **a fraction raised to a negative exponent rule**

EXERCISE SET 5.2

MathXL®   MyMathLab

Concept/Writing Exercises

1. Describe the negative exponent rule.

2. Is the expression x^{-2} simplified? Explain. If not simplified, then simplify.

3. Is the expression $x^5 y^{-3}$ simplified? Explain. If not simplified, then simplify.

4. Can the expression $a^6 b^{-2}$ be simplified to $\dfrac{1}{a^6 b^{-2}}$? If not, what is the correct simplification? Explain.

5. Can the expression 5^{-2} be simplified to -25? If not, what is the correct simplification? Explain.

6. Are the following expressions simplified? If an expression is not simplified, explain why and then simplify.

   a) $\dfrac{5}{x^3}$ b) n^{-5}

   c) $\dfrac{a^{-4}}{2}$ d) $\dfrac{x^{-4}}{x^4}$

7. a) Identify the term or terms in the numerator of the expression $\dfrac{x^5 y^2}{z^3}$.

   b) Identify the factors in the numerator of the expression.

8. a) Identify the term or terms in the numerator of the expression $\dfrac{x^{-4} y^3}{z^5}$.

   b) Identify the factors in the numerator of the expression.

9. Describe what happens to the exponent of a factor when the factor is moved from the numerator to the denominator of a fraction.

10. Describe what happens to the exponent of a factor when the factor is moved from the denominator to the numerator of a fraction.

Practice the Skills

Simplify.

11. x^{-6} 12. y^{-5} 13. 5^{-1} 14. 7^{-2}

15. $\dfrac{1}{x^{-3}}$ 16. $\dfrac{1}{b^{-4}}$ 17. $\dfrac{1}{a^{-1}}$ 18. $\dfrac{1}{y^{-4}}$

19. $\dfrac{1}{6^{-2}}$ 20. $\dfrac{1}{4^{-3}}$ 21. $(x^{-2})^3$ 22. $(m^{-5})^{-2}$

23. $(y^{-5})^4$ 24. $(a^5)^{-4}$ 25. $(x^4)^{-2}$ 26. $(x^{-9})^{-2}$

27. $(3^{-2})^{-1}$ 28. $(2^{-3})^2$ 29. $y^4 \cdot y^{-2}$ 30. $x^{-3} \cdot x^1$

31. $x^7 \cdot x^{-5}$ 32. $d^{-3} \cdot d^{-4}$ 33. $3^{-2} \cdot 3^4$ 34. $6^{-3} \cdot 6^6$

35. $\dfrac{r^5}{r^6}$ 36. $\dfrac{x^2}{x^{-1}}$ 37. $\dfrac{p^0}{p^{-3}}$ 38. $\dfrac{x^{-2}}{x^5}$

39. $\dfrac{x^{-7}}{x^{-3}}$

40. $\dfrac{z^{-11}}{z^{-12}}$

41. $\dfrac{3^2}{3^{-1}}$

42. $\dfrac{4^2}{4^{-1}}$

43. 5^{-3}

44. x^{-7}

45. $\dfrac{1}{z^{-9}}$

46. $\dfrac{1}{3^{-3}}$

47. $(p^{-4})^{-6}$

48. $(x^{-3})^{-4}$

49. $(y^{-2})^{-3}$

50. $z^9 \cdot z^{-12}$

51. $x^3 \cdot x^{-7}$

52. $x^{-3} \cdot x^{-5}$

53. $x^{-8} \cdot x^{-7}$

54. $8^{-3} \cdot 8^3$

55. -4^{-2}

56. $(-4)^{-2}$

57. $-(-4)^{-2}$

58. -2^{-3}

59. $(-2)^{-3}$

60. $-(-2)^{-3}$

61. $(-6)^{-2}$

62. -6^{-2}

63. $\dfrac{x^{-5}}{x^5}$

64. $\dfrac{y^6}{y^{-8}}$

65. $\dfrac{n^{-5}}{n^{-7}}$

66. $\dfrac{3^{-4}}{3}$

67. $\dfrac{9^{-3}}{9^{-3}}$

68. $(7q^5r^2)^0$

69. $(2^{-1} + 3^{-1})^0$

70. $(3^{-1} + 4^2)^0$

71. $\dfrac{2}{2^{-5}}$

72. $(z^{-5})^{-9}$

73. $(x^{-4})^{-2}$

74. $(x^{-7})^0$

75. $(x^0)^{-2}$

76. $(3^{-2})^{-1}$

77. $2^{-3} \cdot 2$

78. $7^5 \cdot 7^{-3}$

79. $7^{-5} \cdot 7^3$

80. $\dfrac{z^{-3}}{z^{-7}}$

81. $\dfrac{x^{-1}}{x^{-4}}$

82. $\dfrac{r^6}{r}$

83. $(4^2)^{-1}$

84. $(2^{-2})^{-2}$

85. $\dfrac{5}{5^{-2}}$

86. $\dfrac{x^6}{x^7}$

87. $\dfrac{3^{-4}}{3^{-2}}$

88. $x^{-10} \cdot x^8$

89. $\dfrac{8^{-1}}{8^{-1}}$

90. $2x^{-1}y$

91. $(-6x^2)^{-2}$

92. $(-3z^3)^{-2}$

93. $3x^{-2}y^2$

94. $-5x^4y^{-1}$

95. $\left(\dfrac{1}{2}\right)^{-2}$

96. $\left(\dfrac{3}{5}\right)^{-2}$

97. $\left(\dfrac{5}{4}\right)^{-3}$

98. $\left(\dfrac{3}{5}\right)^{-3}$

99. $\left(\dfrac{c^4}{d^2}\right)^{-2}$

100. $\left(\dfrac{x^2}{y}\right)^{-2}$

101. $-\left(\dfrac{r^4}{s}\right)^{-4}$

102. $-\left(\dfrac{m^3}{n^4}\right)^{-5}$

103. $-7a^{-3}b^{-4}$

104. $(3x^2y^3)^{-2}$

105. $(4x^5y^{-3})^{-3}$

106. $2w(3w^{-5})$

107. $(3z^{-4})(6z^{-5})$

108. $2x^5(3x^{-6})$

109. $4x^4(-2x^{-4})$

110. $(9x^5)(-3x^{-7})$

111. $(4x^2y)(3x^3y^{-1})$

112. $(7a^{-6}b^{-1})(a^9b^0)$

113. $(-5y^2)(4y^{-3}z^5)$

114. $(-3y^{-2})(5x^{-1}y^3)$

115. $\dfrac{24d^{12}}{3d^8}$

116. $\dfrac{8z^{-4}}{32z^{-2}}$

117. $\dfrac{36x^{-4}}{9x^{-2}}$

118. $\dfrac{18m^{-3}n^0}{6m^5n^9}$

119. $\dfrac{3x^4y^{-2}}{6y^3}$

120. $\dfrac{16x^{-7}y^{-2}}{4x^5y^2}$

121. $\dfrac{32x^4y^{-2}}{4x^{-2}y^0}$

122. $\dfrac{21x^{-3}z^2}{7xz^{-3}}$

123. $\left(\dfrac{5x^4y^{-7}}{z^3}\right)^{-2}$

124. $\left(\dfrac{b^4c^{-2}}{2d^{-3}}\right)^{-1}$

125. $\left(\dfrac{2r^{-5}s^9}{t^{12}}\right)^{-4}$

126. $\left(\dfrac{5m^{-1}n^{-3}}{p^2}\right)^{-3}$

127. $\left(\dfrac{x^3y^{-4}z}{y^{-2}}\right)^{-6}$

128. $\left(\dfrac{3p^{-1}q^{-2}r^3}{p^2}\right)^3$

129. $\left(\dfrac{p^6q^{-3}}{4p^8}\right)^2$

130. $\left(\dfrac{x^{12}y^5}{y^{-3}z}\right)^{-4}$

Problem Solving

131. a) Does $p^{-1}q^{-1} = \dfrac{1}{pq}$? Explain your answer.

 b) Does $p^{-1} + q^{-1} = \dfrac{1}{p+q}$? Explain your answer.

132. a) Does $\dfrac{x^{-1}y^2}{z} = \dfrac{y^2}{xz}$? Explain your answer.

 b) Does $\dfrac{x^{-1} + y^2}{z} = \dfrac{y^2}{x+z}$? Explain your answer.

Evaluate.

133. $4^2 + 4^{-2}$

134. $3^2 + 3^{-2}$

135. $5^3 + 5^{-3}$

136. $6^{-3} + 6^3$

Evaluate.

137. $5^0 - 3^{-1}$

138. $4^{-1} - 3^{-1}$

139. $2^{-3} - 2^3 \cdot 2^{-3}$

140. $2 \cdot 4^{-1} + 4 \cdot 3^{-1}$

141. $2 \cdot 4^{-1} - 4 \cdot 3^{-1}$

142. $2 \cdot 4^{-1} - 3^{-1}$

143. $3 \cdot 5^0 - 5 \cdot 3^{-2}$

144. $7 \cdot 2^{-3} - 2 \cdot 4^{-1}$

Determine the number that when placed in the shaded area makes the statement true.

145. $3^{\blacksquare} = \dfrac{1}{9}$

146. $\dfrac{1}{2^{\blacksquare}} = 64$

147. $\dfrac{1}{6^{\blacksquare}} = 216$

148. $4^{\blacksquare} = \dfrac{1}{256}$

Challenge Problems

In Exercises 149–151, determine the number (or numbers) that when placed in the shaded area (or areas) make the statement true.

149. $(x^{\blacksquare} y^3)^{-2} = \dfrac{x^4}{y^6}$

150. $(x^4 y^{-3})^{\blacksquare} = \dfrac{y^9}{x^{12}}$

151. $(\blacksquare x^{\blacksquare} y^{-2})^3 = \dfrac{8}{x^9 y^6}$

152. For any nonzero real number a, if $a^{-1} = x$, describe the following in terms of x.

　　a) $-a^{-1}$ 　　　　**b)** $\dfrac{1}{a^{-1}}$

153. Consider $(3^{-1} + 2^{-1})^0$. We know this is equal to 1 by the zero exponent rule. Determine the error in the following calculation. Explain your answer.

$$(3^{-1} + 2^{-1})^0 = (3^{-1})^0 + (2^{-1})^0$$
$$= 3^{-1(0)} + 2^{-1(0)}$$
$$= 3^0 + 2^0$$
$$= 1 + 1 = 2$$

Group Activity

Discuss and answer Exercise 154 as a group.

154. Often problems involving exponents can be done in more than one way. Consider

$$\left(\dfrac{3x^2 y^3}{x}\right)^{-2}$$

　　a) Group member 1: Simplify this expression by first simplifying the expression within parentheses.

b) Group member 2: Simplify this expression by first using the expanded power rule.

c) Group member 3: Simplify this expression by first using the negative exponent rule.

d) Compare your answers. If you did not all get the same answers, determine why.

e) As a group, decide which method—**a), b),** or **c)**—was the easiest way to simplify this expression.

Cumulative Review Exercises

[2.7] **155. Racing** If a race car travels 104 miles in 52 minutes, how far will it travel in 93 minutes (assuming all conditions stay the same)?

[3.2] **156. Even Integers** The sum of two consecutive even integers is 190. Find the numbers.

[3.3] **157. Shed** Michael Beattie is building a rectangular shed. The perimeter of the shed is to be 56 feet. Determine the dimensions of the shed if the length is to be 8 feet less than twice the width.

[3.4] **158. Simple Interest** Mia Kattee invested $9000, part at 3% and part at 4% simple interest for a period of one year. How much was invested in each account if the interest earned in the 3% account was $32 greater than the interest earned in the 4% account?

[5.1] **159.** Simplify $(6xy^5)(3x^2 y^4)$.

5.3 Scientific Notation

1 Convert numbers to and from scientific notation.

2 Recognize numbers in scientific notation with a coefficient of 1.

3 Do calculations using scientific notation.

1 Convert Numbers to and from Scientific Notation

We often see, and sometimes use, very large or very small numbers. For example, in September 2006, the world population was about 6,539,000,000 people. You may have read that an influenza virus is about 0.0000001 meters in diameter. Because it is difficult to work with many zeros, we can express such numbers using exponents. For example, the number 6,539,000,000 could be written 6.539×10^9 and the number 0.0000001 could be written 1.0×10^{-7}.

Numbers such as 6.539×10^9 and 1.0×10^{-7} are in a form called **scientific notation**. Each number written in scientific notation is written as a number greater than or equal to 1 and less than 10 ($1 \leq a < 10$) multiplied by some power of 10. The exponent on the 10 must be an integer.

Examples of Numbers in Scientific Notation

$$1.2 \times 10^6$$
$$3.762 \times 10^3$$
$$8.07 \times 10^{-2}$$
$$1.0 \times 10^{-5}$$

Below we change the number 68,400 to scientific notation.

$$68,400 = 6.84 \times 10,000$$
$$= 6.84 \times 10^4 \qquad \textit{Note that } 10,000 = 10 \cdot 10 \cdot 10 \cdot 10 = 10^4.$$

Therefore, $68,400 = 6.84 \times 10^4$. To go from 68,400 to 6.84 the decimal point was moved four places to the left. Note that the exponent on the 10, the 4, is the same as the number of places the decimal point was moved to the left.

Following is a simplified procedure for writing a number in scientific notation.

To Write a Number in Scientific Notation

1. Move the decimal point in the original number to the right of the first nonzero digit. This will give a number greater than or equal to 1 and less than 10.

2. Count the number of places you moved the decimal point to obtain the number in step 1. If the original number was 10 or greater, the count is considered positive. If the original number was less than 1, the count is considered negative.

3. Multiply the number obtained in step 1 by 10 raised to the count (power) found in step 2.

EXAMPLE 1 ▶ Write the following numbers in scientific notation.

a) 18,500   **b)** 0.0000416   **c)** 3,721,000   **d)** 0.0093

Solution

a) The original number is greater than 10; therefore, the exponent is positive. The decimal point in 18,500 belongs after the last zero.

$$18,500. = 1.85 \times 10^4$$
$$\text{4 places}$$

b) The original number is less than 1; therefore, the exponent is negative.

$$0.0000416 = 4.16 \times 10^{-5}$$
$$\text{5 places}$$

c) $3{,}721{,}000 = 3.721 \times 10^6$

⤹ *6 places*

d) $0.0093 = 9.3 \times 10^{-3}$

⤸ *3 places*

▸ **Now Try Exercise 13**

When we write a number in scientific notation, we are allowed to leave our answer with a negative exponent, as in Example 1**b)** and 1**d)**.

Now we explain how to write a number in scientific notation as a number without exponents, or in decimal form.

To Convert a Number from Scientific Notation to Decimal Form

1. Observe the exponent of the power of 10.

2. **a)** If the exponent is positive, move the decimal point in the number (greater than or equal to 1 and less than 10) to the right the same number of places as the exponent. It may be necessary to add zeros to the number. This will result in a number greater than or equal to 10.

   b) If the exponent is 0, do not move the decimal point. Drop the factor 10^0 since it equals 1. This will result in a number greater than or equal to 1 but less than 10.

   c) If the exponent is negative, move the decimal point in the number to the left the same number of places as the exponent (dropping the negative sign). It may be necessary to add zeros to the number. This will result in a number less than 1.

EXAMPLE 2 ▸ Write each number without exponents.

a) 2.9×10^4 **b)** 6.28×10^{-3} **c)** 7.95×10^8

Solution

a) Move the decimal point four places to the right.

$$2.9 \times \boxed{10^4} = 2.9 \times \boxed{10{,}000} = 29{,}000$$

b) Move the decimal point three places to the left.

$$6.28 \times 10^{-3} = 0.00628$$

c) Move the decimal point eight places to the right.

$$7.95 \times 10^8 = 795{,}000{,}000$$

▸ **Now Try Exercise 29**

---

2 Recognize Numbers in Scientific Notation with a Coefficient of 1

We often hear terms like kilograms, milligrams, and gigabytes. For example, an aspirin tablet bottle may indicate that each aspirin contains 325 milligrams of aspirin. Your hard drive on your computer may hold 40 gigabytes of memory. The prefixes kilo, milli, and giga are some of the prefixes used in the *metric system*. The metric system is used in every westernized nation except the United States as the main system of measurement. The prefixes are always used with some type of base unit. The base unit may be measures like meter, m (a unit of length); gram, g (a unit of mass); liter, ℓ (a unit of volume); bits, b (a unit of computer memory); or hertz, Hz (a measure of frequency).

For example, a *milli*meter is $\dfrac{1}{1000}$ meter. A *mega*gram is 1,000,000 grams, and so on. The following table illustrates the meaning of some prefixes.*

Prefix	Meaning	Symbol	Meaning as a Decimal Number
nano	10^{-9}	n	$\dfrac{1}{1,000,000,000}$ or 0.000000001
micro	10^{-6}	μ	$\dfrac{1}{1,000,000}$ or 0.000001
milli	10^{-3}	m	$\dfrac{1}{1000}$ or 0.001
base unit**	10^{0}		1
kilo	10^{3}	k	1000
mega	10^{6}	M	1,000,000
giga	10^{9}	G	1,000,000,000

**The base unit is not a prefix. We included this row to include 10^0 in the chart.

You will sometimes see numbers written as powers of 10, but without a numerical coefficient, as in the table above. If no numerical coefficient is indicated, the numerical coefficient is always assumed to be 1. Thus, for example, $10^{-3} = 1.0 \times 10^{-3}$ and $10^9 = 1.0 \times 10^9$. A computer hard drive that contains 40 gigabytes (40 Gb) contains about $40(1.0 \times 10^9) = 40 \times 10^9 = 40,000,000,000$ bytes. Fifty micrometers (50 μm) is $50(1.0 \times 10^{-6}) = 50 \times 10^{-6} = 0.00005$ meter. Three hundred twenty-five milligrams (325 mg) is $325(1.0 \times 10^{-3}) = 325 \times 10^{-3} = 0.325$ gram. Notice that in the table, each prefix represents a value that is 10^3 or 1000 times greater than the prefix above it. For example, a micrometer is 10^3 or 1000 times greater than a nanometer. A gigameter is 1000 times larger than a megameter, and so on.

In **Figure 5.1**, we see that the frequency of FM radio and VHF TV is about 10^8 hertz (or cycles per second). Thus, the frequency of FM radio is $10^8 = 1.0 \times 10^8 = 100,000,000$ hertz. This number, one hundred million hertz, can also be expressed as 100×10^6 or 100 megahertz, 100 MHz.

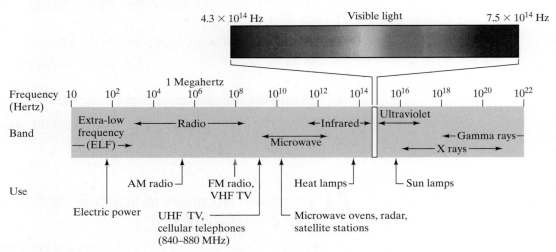

FIGURE 5.1

Now that we know how to interpret powers of 10 that are given without numerical coefficients, we will work some problems using scientific notation.

*There are other prefixes not listed. For example, centi is 10^{-2} or 0.01 times the base unit.

EXAMPLE 3 ▸ Write each quantity without the metric prefix.

  a) 52 kilograms   **b)** 183 nanoseconds

Solution

  a) 52 kilograms (52 kg) $= 52 \times 10^3$ grams $= 52{,}000$ grams

  b) 183 nanoseconds (183 ns) $= 183 \times 10^{-9}$ seconds $= 0.000000183$ second

▸ **Now Try Exercise 45**

Helpful Hint *Study Tip*

Think about how often in your daily life you come across large and small quantities that may be expressed using scientific notation. This may give you more of an appreciation for scientific notation.

3 Do Calculations Using Scientific Notation

We can use the rules of exponents presented in Sections 5.1 and 5.2 when working with numbers written in scientific notation.

EXAMPLE 4 ▸ Multiply $(4.2 \times 10^6)(2.0 \times 10^{-4})$. Write the answer in decimal form.

Solution By the commutative and associative properties of multiplication we can rearrange the expression as follows.

$$(4.2 \times 10^6)(2.0 \times 10^{-4}) = (4.2 \times 2.0)(10^6 \times 10^{-4})$$

$$= 8.4 \times 10^{6+(-4)} \qquad \textit{By the product rule}$$

$$= 8.4 \times 10^2 \qquad \textit{Scientific notation}$$

$$= 840 \qquad \textit{Decimal form}$$

▸ **Now Try Exercise 55**

EXAMPLE 5 ▸ Divide $\dfrac{3.2 \times 10^{-6}}{5.0 \times 10^{-3}}$. Write the answer in scientific notation.

Solution
$$\frac{3.2 \times 10^{-6}}{5.0 \times 10^{-3}} = \left(\frac{3.2}{5.0}\right)\left(\frac{10^{-6}}{10^{-3}}\right)$$

$$= 0.64 \times 10^{-6-(-3)} \qquad \textit{By the quotient rule}$$

$$= 0.64 \times 10^{-6+3}$$

$$= 0.64 \times 10^{-3}$$

$$= 6.4 \times 10^{-4} \qquad \textit{Scientific notation}$$

The answer to Example 5 in decimal form would be 0.00064.

▸ **Now Try Exercise 59**

USING YOUR CALCULATOR

What will your calculator show when you multiply very large or very small numbers? The answer depends on whether your calculator has the ability to display an answer in scientific notation. On calculators without the ability to express numbers in scientific notation, you will probably get an error message because the answer will be too large or too small for the display. For example, on a calculator without scientific notation:

$$8000000 \boxed{\times} 600000 \boxed{=} \boxed{\text{Error}}$$

On scientific calculators and graphing calculators, the answer to this example might be displayed in the following ways.

Possible displays

$$8000000 \boxed{\times} 600000 \boxed{=} \boxed{4.8 \qquad 12}$$

$$8000000 \boxed{\times} 600000 \boxed{=} \boxed{4.8 \qquad ^{12}}$$

$$8000000 \boxed{\times} 600000 \boxed{=} \boxed{4.8E12}$$

Each answer means 4.8×10^{12}. Let's look at one more example.

Possible displays

$$0.0000003 \boxed{\times} 0.004 \boxed{=} \boxed{1.2 \qquad -9}$$

$$0.0000003 \boxed{\times} 0.004 \boxed{=} \boxed{1.2 \qquad ^{-9}}$$

$$0.0000003 \boxed{\times} 0.004 \boxed{=} \boxed{1.2E-9}$$

Each display means 1.2×10^{-9}. On some calculators you will press the $\boxed{\text{ENTER}}$ key instead of the $\boxed{=}$ key. The TI-84 Plus graphing calculator displays answers using E, such as 4.8E12.

EXAMPLE 6 ▶ **Comparing Big Ships** The *Disney Magic* cruise ship gross tonnage is about 8.3×10^4 tons. The Carnival line's *Destiny* cruise ship gross tonnage is about 1.02×10^5 tons.

a) How much greater is the gross tonnage of the *Destiny* than the *Disney Magic*?

b) How many times greater is the gross tonnage of the *Destiny* than the *Disney Magic*?

Disney Characters, © Disney Enterprises, Inc.
Used by permission from Disney Enterprises, Inc.

Solution

a) Understand   We need to subtract 8.3×10^4 from 1.02×10^5. To add or subtract numbers in scientific notation, we generally make the exponents on the 10's the same. This will allow us to add or subtract the numerical values preceding the base while maintaining the common base and exponent on the base.

Translate   We can write 1.02×10^5 as 10.2×10^4. Now subtract as follows.

Carry Out

$$
\begin{array}{r}
10.2 \times 10^4 \\
-8.3 \times 10^4 \\
\hline
1.9 \times 10^4
\end{array}
$$

Notice that in subtraction, we did not subtract the 10^4's. This subtraction could also be done as $(10.2 \times 10^4) - (8.3 \times 10^4) = (10.2 - 8.3) \times 10^4 = 1.9 \times 10^4$.

Check   We can check by writing the numbers out in decimal form.

$$
\begin{array}{r}
102{,}000 \\
-83{,}000 \\
\hline
19{,}000 \text{ or } 1.9 \times 10^4
\end{array}
$$

Answer   Since we obtain the same results, the difference is 1.9×10^4 (or 19,000) tons.

b) Understand   Part **b)** may seem similar to part **a)**, but it is a different question because we are asked to find the *number of times* greater rather than *how much greater*. To find the number of times greater, we perform division.

Translate   Divide the gross tonnage of the *Destiny* by the gross tonnage of the *Disney Magic*.

Carry Out

$$
\frac{1.02 \times 10^5}{8.3 \times 10^4} = \frac{1.02}{8.3} \times \frac{10^5}{10^4}
$$

$$
\approx 0.12 \times 10^{5-4} \qquad \textit{Quotient rule of exponents}
$$

$$
\approx 0.12 \times 10^1
$$

$$
\approx 1.2
$$

Check   We can check by writing the numbers out in decimal form.

$$
\frac{102{,}000}{83{,}000} \approx 1.2
$$

Answer   Since we obtain the same result, the tonnage of the *Destiny* is about 1.2 times that of the *Disney Magic*.

▶ **Now Try Exercise 75**

EXAMPLE 7 ▶ **Fastest Computer**   As of June 2005, the fastest computer in the world, called the Blue Gene/L System, located at the Lawrence Livermore National Laboratory in California, could perform a single calculation in about 0.0000000000000073 second. How long would it take this computer to perform 7 billion (7,000,000,000) calculations? *Source:* www.top500.org/news/articles/article_68.php

Solution   Understand   The computer could perform 1 calculation in 1(0.0000000000000073) second, 2 calculations in 2(0.0000000000000073) second, 3 calculations in 3(0.0000000000000073) second, and 7 billion operations in 7,000,000,000(0.0000000000000073) second.

The Blue Gene/L System

Translate   We will multiply by converting each number to scientific notation.

$$7{,}000{,}000{,}000(0.0000000000000073) = (7.0 \times 10^9)(7.3 \times 10^{-15})$$

Carry Out
$$= (7.0 \times 7.3)(10^9 \times 10^{-15})$$
$$= 51.1 \times 10^{-6}$$
$$= 5.11 \times 10^{-5}$$
$$= 0.0000511$$

Answer   The Blue Gene/L System would take about 0.0000511 of a second to perform 7 billion calculations.

▶ **Now Try Exercise 79**

EXERCISE SET 5.3

Concept/Writing Exercises

1. Describe the form of a number given in scientific notation.

2. **a)** Describe how to write a number 10 or greater in scientific notation.
   b) Using the procedure described in part **a)**, write 315,200 in scientific notation.

3. **a)** Describe how to write a number less than 1 in scientific notation.
   b) Using the procedure described in part **a)**, write 0.0000723 in scientific notation.

4. How many places, and in what direction, will you move the decimal point when you convert a number from scientific notation to decimal form when the exponent on the base 10 is -9?

5. How many places, and in what direction, will you move the decimal point when you convert a number from scientific notation to decimal form when the exponent on the base 10 is 6?

6. When changing a number to scientific notation, under what conditions will the exponent on the base 10 be negative?

7. When changing a number to scientific notation, under what conditions will the exponent on the base 10 be positive?

8. In writing the number 112,546 in scientific notation, will the exponent on the base 10 be positive or negative? Explain.

9. In writing the number 0.000937 in scientific notation, will the exponent on the base 10 be positive or negative? Explain.

10. Write the number 1,000,000 in scientific notation.

11. Write the number 0.000001 in scientific notation.

12. **a)** Is 82.39×10^4 written in scientific notation? If not, how should it be written?
    b) Is 0.083×10^{-5} written in scientific notation? If not, how should it be written?

Practice the Skills

Express each number in scientific notation.

13. 350,000
14. 3,610,000
15. 7950
16. 0.000089
17. 0.053
18. 19,000
19. 0.000726
20. 0.00000186
21. 5,260,000,000
22. 0.0075
23. 0.00000914
24. 74,100
25. 220,300
26. 0.08
27. 0.005104
28. 416,000

Express each number in decimal form (without exponents).

29. 4.3×10^4
30. 1.63×10^{-4}
31. 9.32×10^{-6}
32. 6.15×10^5
33. 2.13×10^{-5}
34. 7.26×10^{-6}
35. 6.25×10^5
36. 4.6×10^1
37. 9.0×10^6
38. 6.475×10^1
39. 5.35×10^2
40. 3.14×10^{-2}
41. 7.73×10^{-7}
42. 6.201×10^{-4}
43. 1.0×10^4
44. 7.13×10^{-4}

In Exercises 45–52, write the quantity without metric prefixes. See Example 3.

45. 8 micrometers
46. 29 micrograms
47. 125 gigawatts
48. 8.7 nanoseconds
49. 15.3 kilometers
50. 80.2 megahertz
51. 48.2 millimeters
52. 3.12 milligrams

Perform each indicated operation and express each number in decimal form (without exponents).

53. $(2.0 \times 10^2)(3.0 \times 10^5)$ **54.** $(2.0 \times 10^{-3})(3.0 \times 10^2)$ **55.** $(2.7 \times 10^{-6})(9.0 \times 10^4)$

56. $(1.3 \times 10^{-8})(1.74 \times 10^6)$ **57.** $(1.6 \times 10^{-2})(4.0 \times 10^{-3})$ **58.** $(4.0 \times 10^5)(1.2 \times 10^{-4})$

59. $\dfrac{3.9 \times 10^{-5}}{3.0 \times 10^{-2}}$ **60.** $\dfrac{6.0 \times 10^{-3}}{3.0 \times 10^{1}}$ **61.** $\dfrac{7.5 \times 10^{6}}{3.0 \times 10^{3}}$

62. $\dfrac{1.4 \times 10^{7}}{4.0 \times 10^{8}}$ **63.** $\dfrac{2.0 \times 10^{4}}{8.0 \times 10^{-2}}$ **64.** $\dfrac{1.6 \times 10^{4}}{8.0 \times 10^{-3}}$

Perform each indicated operation by first converting each number to scientific notation. Write the answer in scientific notation.

65. $(700{,}000)(6{,}000{,}000)$ **66.** $(0.003)(0.00015)$ **67.** $(0.0004)(320)$

68. $(67{,}000)(200{,}000)$ **69.** $\dfrac{5{,}600{,}000}{8000}$ **70.** $\dfrac{0.00004}{200}$

71. $\dfrac{0.00035}{0.000002}$ **72.** $\dfrac{150{,}000}{0.0005}$

73. List the following numbers from smallest to largest: 7.3×10^2, 3.3×10^{-4}, 1.75×10^6, 5.3.

74. List the following numbers from smallest to largest: 4.8×10^5, 3.2×10^{-1}, 4.6, 8.3×10^{-4}.

Problem Solving

In Exercises 75–92, write the answer in decimal form (without exponents) unless asked to do otherwise.

75. Population In 2006, the U.S. population was about 2.99×10^8 people and the world population was about 6.55×10^9 people.

 a) How many people lived outside the United States in 2006?

 b) How many times greater is the world population than the U.S. population?

76. Diapers Laid end to end the 18 billion disposable diapers thrown away in the United States each year would reach the moon and back seven times (seven round trips).

 a) Write 18 billion in scientific notation.

 b) If the distance from earth to the moon is 2.38×10^5 miles, what is the length of all these diapers placed end to end? Write your answer in scientific notation and as a number without exponents.

77. Niagara Falls A treaty between the United States and Canada requires that during the tourist season a minimum of 100,000 cubic feet of water per second flows over Niagara Falls (another 130,000 to 160,000 cubic feet/sec is diverted for power generation). Find the minimum volume of water that will flow over the falls in a 24-hour period during the tourist season.

78. PGA Tour The top five money leaders from the 2004 PGA Tour are listed below.

Golfer	Approximate Money
1. Vijay Singh	$10,910,000
2. Ernie Els	$5,790,000
3. Phil Michelson	$5,780,000
4. Tiger Woods	$5,370,000
5. Stewart Cink	$4,450,000

Source: USA Today

 a) Write, in scientific notation, Tiger Woods's approximate money total.

 b) Determine, in scientific notation, the average daily amount that Tiger Woods earned in 2004. Recall that 2004 was a leap year.

79. Computer Speed If a computer can do a calculation in 0.000002 second, how long, in seconds, would it take the computer to do 8 trillion (8,000,000,000,000) calculations? Write your answer in scientific notation.

80. Fortune Cookies Who writes the fortunes in those fortune cookies? Fortune cookies are an American invention. Steven Yang, who runs M & Y Trading Company, a San Francisco Company, prints about 90% of all the 1.02×10^9 fortunes in fortune cookies in America in any given year.

 a) How many fortunes does Yang print in a year?

 b) How many fortunes does Yang print in a day?

81. Movies The gross ticket sales of the top five movies in the United States as of June 29, 2005, are listed below.

Movie	Year Released	Approximate U.S. Gross Ticket Sales
1. *Titanic*	1997	$601,000,000
2. *Star Wars*	1977	$461,000,000
3. *Shrek 2*	2004	$437,000,000
4. *E.T.*	1982	$433,000,000
5. *Star Wars—The Phantom Menace*	1999	$431,000,000

a) How much greater was the gross ticket sales of *Titanic* than *E.T.*? Write your answer in scientific notation.

b) How many times greater was the gross ticket sales of *Titanic* than *E.T.*?

82. Melanoma The number of new cases of melanoma has increased from 4.03×10^4 in 1997 to about 5.5×10^4 in 2003.

a) How much greater was the number of cases of melanoma in 2003 than in 1997?

b) How many times greater was the number of cases of melanoma in 2003 than in 1997?

Source: Rochester Democrat & Chronicle

83. World's Richest The worth of Bill Gates (the world's richest person in 2005) was 4.65×10^{10}. The worth of Pierre Omidyar (the world's thirty-fifth richest person in 2005) was 9.9×10^9. *Source:* www.forbes.com

a) How much greater was the worth of Bill Gates than the worth of Pierre Omidyar in 2005?

b) How many times greater was the worth of Bill Gates than the worth of Pierre Omidyar in 2005?

84. Light from the Sun The sun is 9.3×10^7 miles from Earth. Light travels at a speed of 1.86×10^5 miles per second. How long, in both seconds and in minutes, does it take light from the Sun to reach Earth?

85. Astronomy The mass of Earth, Earth's Moon, and the planet Jupiter are listed below.

Earth: 5,794,000,000,000,000,000,000 metric tons

Moon: 73,400,000,000,000,000,000 metric tons

Jupiter: 1,899,000,000,000,000,000,000,000 metric tons

a) Write the mass of Earth, the Moon, and Jupiter in scientific notation.

b) How many times greater is the mass of Earth than the mass of the Moon?

c) How many times greater is the mass of Jupiter than the mass of Earth?

86. Missing Persons Of the nearly 47,600 active adult missing-person FBI cases as of May 1, 2005, about 53% were men. Determine, in scientific notation, the approximate number of active adult missing-person FBI cases that were men. *Source: Rochester Democrat & Chronicle*

87. Camera Products and Services The circle graph below shows the percentage of revenue from digital products and services and from traditional products and services during the first quarter of 2005 for Kodak. The company's total revenue for the first quarter of 2005 was $2,800,000,000. How much revenue was from digital products and services? Write your answer in scientific notation.

Kodak First Quarter 2005 Revenue

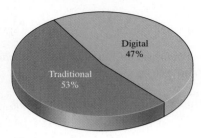

Source: Rochester Democrat & Chronicle

88. Chesapeake Bay Oysters The following bar graph shows the approximate number of bushels of oysters harvested in the Chesapeake Bay.

Chesapeake Oyster Harvests Declining

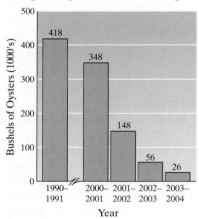

Source: USA Today

a) Give, in scientific notation, the approximate number of bushels of oysters harvested in 1990–1991.

b) Give, in scientific notation, the approximate number of bushels of oysters harvested in 2003–2004.

c) Determine, in scientific notation, the approximate difference between the number of bushels harvested in 2003–2004 and 1990–1991.

89. Federal Government Employees The following bar graph shows the number of federal government employees, by age.

Federal Government Employees by Age (2004)

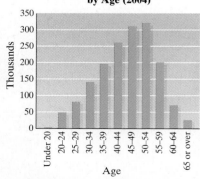

Source: Office of Personnel Management

a) Write, in scientific notation, the approximate number of federal government employees in 2004 who were age 55–59 and the approximate number of federal government employees in 2004 who were age 20–24.

b) Determine, in scientific notation, the approximate total number of federal government employees in 2004 who were age 55–59 or age 20–24.

c) Determine, in scientific notation, the approximate difference between the number of federal government employees in 2004 who were age 55–59 and age 20–24.

90. The Spice Girls The Spice Girls hold the Guinness World Record as having the highest ever annual earnings by a girl band. The four group members had a total income of $49,000,000 in 1998. If each member of the band received an equal amount, determine, in scientific notation, the amount earned by each group member in 1998.

91. Structure of Matter An article in *Scientific American* states that physicists have created a *standard model* that describes the structure of matter down to 10^{-18} meters. If this number is written without exponents, how many zeroes would there be to the right of the decimal point?

92. A Large Number Avogadro's number, named after the nineteenth century Italian chemist Amedeo Avogadro, is roughly 6.02×10^{23}. It represents the number of atoms in 12 grams of pure carbon.

a) If this number were written out in decimal form, how many digits would it contain?

b) What is the number of atoms in 1 gram of pure carbon? Write your answer in scientific notation.

Challenge Problems

93. The Movie Contact In the movie *Contact*, Jodie Foster plays an astronomer who makes the statement "There are 400 billion stars out there just in our universe alone. If only one out of a million of those had planets, and if just one out of a million of those had life, and if just one out of a million of those had intelligent life, there would be literally millions of civilizations out there." Do you believe this statement is correct? Explain your answer.

94. How many times smaller is 1 nanosecond than one millisecond?

95. How many times, either greater or smaller, is 10^{-12} meters than 10^{-18} meters?

96. Light Year Light travels at a speed of 1.86×10^5 miles per second. A *light year* is the distance light travels in one year. Determine the number of miles in a light year.

Group Activity

Discuss and answer Exercise 97 as a group.

97. A Million versus a Billion Do you have any idea of the difference in size between a million (1,000,000), a billion (1,000,000,000), and a trillion (1,000,000,000,000)?

a) Write a million, a billion, and a trillion in scientific notation.

b) Group member 1: Determine how long it would take to spend a million dollars if you spent $1000 a day.

c) Group member 2: Repeat part **b)** for a billion dollars.

d) Group member 3: Repeat part **b)** for a trillion dollars.

e) As a group, determine how many times greater a billion dollars is than a million dollars.

Cumulative Review Exercises

[1.9] **98.** Evaluate $4x^2 + 3x + \dfrac{x}{2}$ when $x = 0$.

[2.3] **99. a)** If $-x = -\dfrac{3}{2}$, what is the value of x?

    b) If $5x = 0$, what is the value of x?

[2.5] **100.** Solve the equation $2x - 3(x - 2) = x + 2$.

[5.1] **101.** Simplify $\left(\dfrac{-2x^5 y^7}{8x^8 y^3} \right)^3$.

Mid-Chapter Test: 5.1–5.3

To find out how well you understand the chapter material to this point, take this brief test. The answers, and the section where the material was initially discussed, are given in the back of the book. Review any questions that you answered incorrectly.

Simplify.

1. $y^{11} \cdot y^2$

2. $\dfrac{x^{13}}{x^{10}}$

3. $(-3x^5y^7)(-2xy^6)$

4. $\dfrac{6a^{12}b^8}{9a^7b^2}$

5. $(-4x^2y^4)^3$

6. $\left(\dfrac{-5s^4t^6}{10s^6t^3}\right)^2$

7. $(7x^9y^5)(-3xy^4)^2$

8. $\dfrac{p^{-3}}{p^5}$

9. $x^{-4} \cdot x^{-6}$

10. $(3^{-1} + 5^2)^0$

11. $\left(\dfrac{3}{7}\right)^{-2}$

12. $(8x^{-2}y^5)(4x^3y^{-6})$

13. $\dfrac{6m^{-4}n^{-7}}{2m^0n^{-1}}$

14. $\left(\dfrac{2x^{-3}y^{-4}}{x^3yz^{-2}}\right)^{-2}$

15. **a)** Describe how to write a number 10 or greater in scientific notation.

    b) Describe how to write a number less than 1 in scientific notation.

16. Express 6,540,000,000 in scientific notation.

17. Express 3.27×10^{-5} in decimal form (without exponents).

18. Write 18.9 kilometers without metric prefixes.

19. Multiply $(3.4 \times 10^{-6})(7.0 \times 10^3)$ and express the answer in decimal form (without exponents).

20. Divide $\dfrac{0.00006}{200}$ by first converting to scientific notation. Write the answer in scientific notation.

5.4 Addition and Subtraction of Polynomials

1. Identify polynomials.

2. Add polynomials.

3. Subtract polynomials.

4. Subtract polynomials in columns.

1 Identify Polynomials

A **polynomial in x** is an expression containing the sum of a finite number of terms of the form ax^n, for any real number a and any *whole number n*.

Examples of Polynomials	Not Polynomials	
$8x$	$4x^{1/2}$	*(Fractional exponent)*
$\dfrac{1}{3}x - 4$	$3x^2 + 4x^{-1} + 5$	*(Negative exponent)*
$x^2 - 2x + 1$	$4 + \dfrac{1}{x}$	$\left(\dfrac{1}{x} = x^{-1},\ negative\ exponent\right)$

A polynomial is written in **descending order** (or **descending powers**) **of the variable** when the exponents on the variable decrease from left to right.

Example of Polynomial in Descending Order
$$2x^4 + 4x^2 - 6x + 3$$

Note in the example that the constant term 3 is last because it can be written as $3x^0$. Remember that $x^0 = 1$.

A polynomial can be in more than one variable. For example, $3xy + 2$ is a polynomial in two variables, x and y.

A polynomial with one term is called a **monomial**. A **binomial** is a two-termed polynomial. A **trinomial** is a three-termed polynomial. Polynomials containing more than three terms are not given special names. The prefix "poly" means "many." The chart that follows summarizes this information.

Type of Polynomial	Number of Terms	Examples
Monomial	One	$8,\ 4x,\ -6x^2$
Binomial	Two	$x + 5,\ x^2 - 6,\ 4y^2 - 5y$
Trinomial	Three	$x^2 - 2x + 3,\ 3z^2 - 6z + 7$

The **degree of a term** of a polynomial in one variable is the exponent on the variable in that term.

Term	Degree of Term	
$4x^2$	Second	
$2y^5$	Fifth	
$-5x$	First	*($-5x$ can be written $-5x^1$.)*
3	Zero	*(3 can be written $3x^0$.)*

For a polynomial in two or more variables, the degree of a term is the sum of the exponents on the variables. For example, the degree of the term $4x^2y^3$ is 5 because $2 + 3 = 5$. The degree of the term $5a^4bc^3$ is 8 because $4 + 1 + 3 = 8$.

The **degree of a polynomial** is the same as that of its highest-degree term.

Polynomial	Degree of Polynomial	
$8x^3 + 2x^2 - 3x + 4$	Third	*($8x^3$ is highest-degree term.)*
$x^2 - 4$	Second	*(x^2 is highest-degree term.)*
$6x - 5$	First	*($6x$ or $6x^1$ is highest-degree term.)*
4	Zero	*(4 or $4x^0$ is highest-degree term.)*
$x^2y^4 + 2x + 3$	Sixth	*(x^2y^4 is highest-degree term.)*

2 Add Polynomials

In Section 2.1, we stated that like terms are terms having the same variables and the same exponents. That is, like terms may differ only in their numerical coefficients.

Examples of Like Terms

$3,\ -5$

$2x,\ x$

$-2x^2,\ 4x^2$

$3y^2,\ 5y^2$

$3xy^2,\ 5xy^2$

To Add Polynomials

To add polynomials, combine the like terms of the polynomials.

EXAMPLE 1 ▶ Add $(4x^2 + 6x + 3) + (2x^2 + 5x - 1)$.

Solution   Remember that $(4x^2 + 6x + 3) = 1(4x^2 + 6x + 3)$ and $(2x^2 + 5x - 1) = 1(2x^2 + 5x - 1)$. We can use the distributive property to remove the parentheses, as shown below.

$$(4x^2 + 6x + 3) + (2x^2 + 5x - 1)$$
$$= 1(4x^2 + 6x + 3) + 1(2x^2 + 5x - 1)$$
$$= 4x^2 + 6x\ \ + 3\ \ + 2x^2 + 5x - 1 \qquad \textit{Use the distributive property.}$$
$$= \underline{4x^2 + 2x^2}\ \ \underline{+ 6x + 5x}\ \ \underline{+3\ \ - 1} \qquad \textit{Rearrange terms.}$$
$$= \qquad 6x^2 \quad + \quad 11x \quad + \quad 2 \qquad \textit{Combine like terms.}$$

▶ **Now Try Exercise 67**

In the following examples, we will not show the multiplication by 1 as was shown in Example 1.

EXAMPLE 2 ▶ Add $(5a^2 + 3a + b) + (a^2 - 7a + 3)$.

Solution
$$(5a^2 + 3a + b) + (a^2 - 7a + 3)$$
$$= 5a^2 + 3a + b + a^2 - 7a + 3 \qquad \textit{Remove parentheses.}$$
$$= \underline{5a^2 + a^2} + \underline{3a - 7a} + b + 3 \qquad \textit{Rearrange terms.}$$
$$= \quad 6a^2 \quad - \quad 4a \quad + b + 3 \qquad \textit{Combine like terms.}$$

▶ **Now Try Exercise 75**

EXAMPLE 3 ▶ Add $(3x^2y - 4xy + y) + (x^2y + 2xy + 3y)$.

Solution
$$(3x^2y - 4xy + y) + (x^2y + 2xy + 3y)$$
$$= 3x^2y - 4xy + y + x^2y + 2xy + 3y \qquad \textit{Remove parentheses.}$$
$$= \underline{3x^2y + x^2y} \; \underline{- 4xy + 2xy} + \underline{y + 3y} \qquad \textit{Rearrange terms.}$$
$$= \quad 4x^2y \quad - \quad 2xy \quad + \quad 4y \qquad \textit{Combine like terms.}$$

▶ **Now Try Exercise 77**

Usually, when we add polynomials, we will do so as in Examples 1 through 3. That is, we will list the polynomials horizontally. However, in Section 5.6, when we divide polynomials, there will be steps where we add polynomials in columns.

> **To Add Polynomials in Columns**
> 1. Arrange polynomials in descending order, one under the other with like terms in the same columns.
> 2. Add the terms in each column.

EXAMPLE 4 ▶ Add $5x^2 - 9x - 3$ and $-3x^2 - 4x + 6$ using columns.

Solution
$$\begin{array}{r} 5x^2 - 9x - 3 \\ -3x^2 - 4x + 6 \\ \hline 2x^2 - 13x + 3 \end{array}$$

▶ **Now Try Exercise 83**

EXAMPLE 5 ▶ Add $5w^3 + 2w - 4$ and $2w^2 - 6w - 3$ using columns.

Solution Since the polynomial $5w^3 + 2w - 4$ does not have a w^2 term, we will add the term $0w^2$ to the polynomial. This procedure sometimes helps in aligning like terms.

$$\begin{array}{r} 5w^3 + 0w^2 + 2w - 4 \\ 2w^2 - 6w - 3 \\ \hline 5w^3 + 2w^2 - 4w - 7 \end{array}$$

▶ **Now Try Exercise 85**

3 Subtract Polynomials

Now let's subtract polynomials.

> **To Subtract Polynomials**
> 1. Use the distributive property to remove parentheses. (This will have the effect of changing the sign of *every* term within the parentheses of the polynomial being subtracted.)
> 2. Combine like terms.

EXAMPLE 6 ▶ Subtract $(3x^2 - 2x + 5) - (x^2 - 3x + 4)$.

Solution $(3x^2 - 2x + 5)$ means $1(3x^2 - 2x + 5)$ and $(x^2 - 3x + 4)$ means $1(x^2 - 3x + 4)$. We use this information in the solution, as shown below.

$$(3x^2 - 2x + 5) - (x^2 - 3x + 4) = 1(3x^2 - 2x + 5) - 1(x^2 - 3x + 4)$$

$$= 3x^2 - 2x + 5 - x^2 + 3x - 4 \qquad \text{\textit{Remove parentheses.}}$$

$$= \underbrace{3x^2 - x^2}\ \underbrace{- 2x + 3x}\ \underbrace{+ 5 - 4} \qquad \text{\textit{Rearrange terms.}}$$

$$= \quad 2x^2 \quad + \quad x \quad + \quad 1 \qquad \text{\textit{Combine like terms.}}$$

▶ **Now Try Exercise 97**

Remember from Section 2.1 that when a negative sign precedes the parentheses, the sign of every term within the parentheses is changed when the parentheses are removed. This was shown in Example 6. In Example 7, we will not show the multiplication by -1, as was done in Example 6.

EXAMPLE 7 ▶ Subtract $(-3x^2 - 5x + 3)$ from $(x^3 + 2x + 6)$.

Solution $(x^3 + 2x + 6) - (-3x^2 - 5x + 3)$

$$= x^3 + 2x + 6 + 3x^2 + 5x - 3 \qquad \text{\textit{Remove parentheses.}}$$

$$= x^3 + 3x^2 \underbrace{+ 2x + 5x}\ \underbrace{+ 6 - 3} \qquad \text{\textit{Rearrange terms.}}$$

$$= x^3 + 3x^2 + \quad 7x \quad + \quad 3 \qquad \text{\textit{Combine like terms.}}$$

▶ **Now Try Exercise 107**

Avoiding Common Errors

One of the most common mistakes occurs when subtracting polynomials. When subtracting one polynomial from another, **the sign of each term in the polynomial being subtracted must be changed, not just the sign of the first term.**

CORRECT	INCORRECT
$6x^2 - 4x + 3 - (2x^2 - 3x + 4)$	$6x^2 - 4x + 3 - (2x^2 - 3x + 4)$
$= 6x^2 - 4x + 3\ -\ 2x^2\ +\ 3x\ -\ 4$	$= 6x^2 - 4x + 3 - 2x^2 - 3x + 4$
$= 4x^2 - x - 1$	$= 4x^2 - 7x + 7$

Do not make this mistake!

4 Subtract Polynomials in Columns

Polynomials can be subtracted as well as added using columns.

To Subtract Polynomials in Columns

1. Write *the polynomial being subtracted* below the polynomial from which it is being subtracted. List like terms in the same column.

2. *Change the sign of each term* in the polynomial being subtracted. (This step can be done mentally, if you like.)

3. Add the terms in each column.

EXAMPLE 8 ▶ Subtract $(2x^2 - 4x + 6)$ from $(4x^2 + 5x + 8)$ using columns.

Solution   Align like terms in columns (step 1).

$$\begin{aligned} 4x^2 + 5x + 8 \\ -(2x^2 - 4x + 6) \end{aligned} \qquad \text{\textit{Align like terms.}}$$

Change *all* signs in the second row (step 2); then add (step 3).

$$4x^2 + 5x + 8$$
$$\underline{-2x^2 + 4x - 6}\qquad \textit{Change all signs.}$$
$$2x^2 + 9x + 2\qquad \textit{Add.}$$

▶ **Now Try Exercise 113**

EXAMPLE 9 ▶ Subtract $(2x^2 - 6)$ from $(-3x^3 + 4x - 3)$ using columns.

Solution   To help align like terms, write each expression in descending order. If any power of x is missing, write that term with a numerical coefficient of 0.

$$-3x^3 + 4x - 3 = -3x^3 + 0x^2 + 4x - 3$$

$$2x^2 - 6 = 2x^2 + 0x - 6$$

Align like terms.

$$-3x^3 + 0x^2 + 4x - 3$$
$$\underline{-(2x^2 + 0x - 6)}$$

Change all signs in the second row; then add the terms in each column.

$$-3x^3 + 0x^2 + 4x - 3$$
$$\underline{- 2x^2 - 0x + 6}$$
$$-3x^3 - 2x^2 + 4x + 3$$

▶ **Now Try Exercise 115**

NOTE: Many of you will find that you can change the signs mentally and can therefore align and change the signs in one step.

EXERCISE SET 5.4   Math XL   MyMathLab
MathXL®   MyMathLab

Concept/Writing Exercises

1. What is a polynomial?

2. Is $6m^3 - 5m^{1/2}$ a polynomial? Explain.

3. Is $4x^{-3} + 9$ a polynomial? Explain.

4. Is $5x + \dfrac{2}{x}$ a polynomial? Explain.

5. a) What is a monomial? Make up three examples.
   b) What is a binomial? Make up three examples.
   c) What is a trinomial? Make up three examples.

6. a) Explain how to find the degree of a term in one variable.
   b) Explain how to find the degree of a polynomial in one variable.

7. Make up your own sixth-degree polynomial with four terms. Explain why it is a sixth-degree polynomial with four terms.

8. Explain how to find the degree of a term in a polynomial in more than one variable.

9. Which of the following are fourth-degree terms? Explain your answer.
   a) $3xy^2$ b) $6r^2s^2$ c) $-2mn^3$

10. Explain why $(3x + 2) - (4x - 6) \neq 3x + 2 - 4x - 6$.

11. Explain how to write a polynomial in one variable in descending order of the variable.

12. Why is the constant term always written last when writing a polynomial in descending order?

13. Explain how to add polynomials.

14. a) Describe how to add polynomials in columns.
    b) How will you rewrite $4x^3 + 5x - 7$ in order to add it to $3x^3 + x^2 - 4x + 8$ using columns? Explain.

15. Explain how to subtract polynomials.

16. a) Describe how to subtract polynomials in columns.
    b) How will you rewrite $(3x^2 - 4x + 5) - (x^2 + 7x - 4)$ in order to subtract?

Practice the Skills

Indicate the degree of each term.

17. x^5

18. z^{11}

19. $-3b^8$

20. $5a^4$

21. x^2y

22. a^4b^3

23. $3r^2s^8$

24. $6m^5n^8$

25. $-8x^3y^5z$

26. $-12p^4q^7r$

Indicate which expressions are polynomials. If the polynomial has a specific name—monomial, binomial, or trinomial—give that name.

27. $2x^2 - 6x + 7$

28. $x^2 + 3$

29. -6

30. $4x^{-2}$

31. $9a^4 - 5$

32. $7x + 8$

33. $8x^9$

34. $3x^{1/2} + 2x$

35. $a^{-1} + 4$

36. $x^3 - 8x^2 + 8$

37. $6n^3 - 5n^2 + 4n - 3$

38. $10x^2$

39. $4 - 2b^2 - 5b$

40. $5p^{-3}$

41. $\dfrac{2}{3}x^2 - \dfrac{1}{x}$

42. $0.6r^4 - \dfrac{1}{2}r^3 - 0.4r^2 - \dfrac{1}{3}$

Express each polynomial in descending order. If the polynomial is already in descending order, so state. Give the degree of each polynomial.

43. 8

44. $4 + 5x$

45. $-4 + x^2 - 2x$

46. $6x - 5$

47. $x + 3x^2 - 8$

48. $4 - 3p^3$

49. $-a - 3$

50. $2x^2 + 5x - 8$

51. $6w^2 - 5w + 9$

52. 15

53. $-4 + x - 3x^2 + 4x^3$

54. $1 - x^3 + 3x$

55. $5x + 3x^2 - 6 - 2x^4$

56. $-3r - 5r^2 + 2r^4 - 6$

Add.

57. $(9x - 2) + (x - 7)$

58. $(5x - 6) + (2x - 3)$

59. $(-3x + 8) + (2x + 3)$

60. $(-7x - 9) + (-2x + 9)$

61. $(t + 7) + (-3t - 8)$

62. $(4x - 3) + (3x - 3)$

63. $(x^2 + 2.6x - 3) + (4x + 3.8)$

64. $(-4p^2 - 3p - 2) + (-p^2 - 4)$

65. $(4m - 3) + (5m^2 - 4m + 7)$

66. $(-x^2 - 2x - 4) + (4x^2 + 3)$

67. $(2x^2 - 3x + 5) + (-x^2 + 6x - 8)$

68. $(x^2 - 6x + 7) + (-x^2 + 3x + 5)$

69. $(-x^2 - 4x + 8) + \left(5x - 2x^2 + \dfrac{1}{2}\right)$

70. $(8x^2 + 3x - 5) + \left(x^2 + \dfrac{1}{2}x + 2\right)$

71. $(5.2n^2 - 6n + 1.7) + (3n^2 + 1.2n - 2.3)$

72. $(8x^2 + 4) + (-2.6x^2 - 5x - 2.3)$

73. $(-7x^3 - 3x^2 + 4) + (4x + 5x^3 - 7)$

74. $(6x^3 - 4x^2 - 7) + (3x^2 + 3x - 3)$

75. $(8x^2 + 2x - y) + (3x^2 - 9x + 5)$

76. $(-7a^2 + 3a - b) + (4a^2 - 2a - 8)$

77. $(2x^2y + 2x - 3) + (3x^2y - 5x + 5)$

78. $(x^2y + x - y) + (2x^2y + 2x - 6y + 3)$

Add using columns.

79. Add $8x - 7$ and $3x + 4$.

80. Add $-x + 5$ and $-4x - 5$.

81. Add $4y^2 - 2y + 4$ and $3y^2 + 1$.

82. Add $6m^2 - 2m + 1$ and $-10m^2 + 8$.

83. Add $-x^2 - 3x + 3$ and $5x^2 + 5x - 7$.

84. Add $-2s^2 - s + 5$ and $3s^2 - 6s$.

85. Add $2x^3 + 3x^2 + 6x - 9$ and $7 - 4x^2$.

86. Add $-3x^3 + 3x + 9$ and $2x^2 - 4$.

87. Add $4n^3 - 5n^2 + n - 6$ and $-n^3 - 6n^2 - 2n + 8$.

88. Add $7x^3 + 5x - 6$ and $3x^3 - 4x^2 - x + 8$.

Subtract.

89. $(4x - 4) - (2x + 2)$

90. $(6x - 5) - (2x - 3)$

91. $(-2x - 3) - (-5x - 7)$

92. $(10x - 3) - (-2x + 7)$

93. $(-r + 5) - (2r + 5)$

94. $(4x + 8) - (3x + 9)$

95. $(-y^2 + 4y - 5.2) - (5y^2 + 2.1y + 7.5)$

96. $(9x^2 + 7x - 5) - (3x^2 + 3.5)$

97. $(5x^2 - x - 1) - (-3x^2 - 2x - 5)$

98. $(-a^2 + 3a + 12) - (-4a^2 - 3)$

99. $(-4.1n^2 - 3n) - (2.3n^2 - 9n + 7.6)$

100. $(7x - 0.6) - (-2x^2 + 4x - 8)$

101. $(8x^3 - 2x^2 - 4x + 5) - (5x^2 + 8)$

102. $\left(9x^3 - \dfrac{1}{5}\right) - (x^2 + 5x)$

103. $(2x^3 - 4x^2 + 5x - 7) - \left(3x + \dfrac{3}{5}x^2 - 5\right)$

104. $(-3x^2 + 4x - 7) - \left(x^3 + 4x^2 - \dfrac{3}{4}x\right)$

105. Subtract $(7x + 4)$ from $(8x + 2)$.

106. Subtract $(-4x + 7)$ from $(-3x - 9)$.

107. Subtract $(5x - 6)$ from $(2x^2 - 4x + 8)$.

108. Subtract $(3x^2 - 5x - 3)$ from $(-x^2 + 3x + 10)$.

109. Subtract $(-2c^2 + 7c - 7)$ from $(-5c^3 - 6c^2 + 7)$.

110. Subtract $(4x^3 - 6x^2)$ from $(3x^3 + 5x^2 + 9x - 7)$.

Subtract using columns.

111. Subtract $(3x - 3)$ from $(6x + 5)$.

112. Subtract $(6x + 8)$ from $(2x - 5)$.

113. Subtract $(2a^2 + 3a - 9)$ from $(5a^2 - 13a + 19)$.

114. Subtract $(4x^2 - 7x + 3)$ from $(8x^2 + 5x - 9)$.

115. Subtract $(6x^2 - 1)$ from $(7x^2 - 3x - 4)$.

116. Subtract $(5n^3 + 7n - 9)$ from $(2n^3 - 6n + 3)$.

117. Subtract $(5x^2 + 4)$ from $(x^2 + 4)$.

118. Subtract $(-5m^2 + 6m)$ from $(m - 6)$.

119. Subtract $(x^2 + 6x - 7)$ from $(4x^3 - 6x^2 + 7x - 9)$.

120. Subtract $(2x^3 + 4x^2 - 9x)$ from $(-5x^3 + 4x - 12)$.

Problem Solving

121. Make up your own addition problem where the sum of two binomials is $-5x - 1$.

122. Make up your own addition problem where the sum of two trinomials is $3x^3 - 2x - 4$.

123. Make up your own subtraction problem where the difference of two trinomials is $7x - 3$.

124. Make up your own subtraction problem where the difference of two trinomials is $-x^2 + 4x - 5$.

125. When two binomials are added, will the sum always, sometimes, or never be a binomial? Explain your answer and give examples to support your answer.

126. When one binomial is subtracted from another, will the difference always, sometimes, or never be a binomial? Explain your answer and give examples to support your answer.

127. When two trinomials are added, will the sum always, sometimes, or never be a trinomial? Explain your answer and give examples to support your answer.

128. When one trinomial is subtracted from another, will the difference always, sometimes, or never be a trinomial? Explain your answer and give examples to support your answer.

129. Write a fourth-degree trinomial in the variable x that has neither a second-degree term nor a constant term.

130. Write a sixth-degree trinomial in the variable x that has no fifth-, fourth-, first-, or zero-degree terms.

131. Is it possible to have a fourth-degree trinomial in x that has no third-, second-, or zero-degree terms and contains no like terms? Explain.

132. Is it possible to have a fifth-degree trinomial in x that has no fourth-, third-, second-, or first-degree terms and contains no like terms? Explain.

Write a polynomial that represents the area of each figure shown.

133.

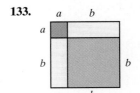

134.

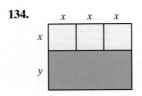

135.

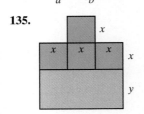

136.

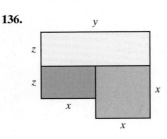

Challenge Problems

Simplify.

137. $(3x^2 - 6x + 3) - (2x^2 - x - 6) - (x^2 + 7x - 9)$

138. $3x^2y - 6xy - 2xy + 9xy^2 - 5xy + 3x$

139. $4(x^2 + 2x - 3) - 6(2 - 4x - x^2) - 2x(x + 2)$

Group Activity

Discuss and answer Exercise 140 as a group.

140. Make up a trinomial, a binomial, and a different trinomial such that (first trinomial) + (binomial) − (second trinomial) = 0.

Cumulative Review Exercises

[2.7] **141.** Solve $\dfrac{5}{9} = \dfrac{2.5}{x}$ by cross-multiplying.

[3.1] **142.** Linda Rudolph is n years old now. Write an expression that represents her age 5 years ago.

[4.2] **143.** Graph the equation $3y = 9$.

[4.3] **144.** Graph $y = -2x - 4$.

[5.1] **145.** Simplify $\left(\dfrac{4x^3 y^5}{12x^7 y^4}\right)^3$.

[5.3] **146.** Write 3.5×10^{-2} without using exponents.

5.5 Multiplication of Polynomials

1 Multiply a monomial by a monomial.

2 Multiply a polynomial by a monomial.

3 Multiply binomials using the distributive property.

4 Multiply binomials using the FOIL method.

5 Multiply binomials using formulas for special products.

6 Multiply any two polynomials.

1 Multiply a Monomial by a Monomial

We begin our discussion of multiplication of polynomials by multiplying a monomial by a monomial. To multiply two monomials, multiply their coefficients and use the product rule of exponents to determine the exponents on the variables. Problems of this type were done in Section 5.1.

EXAMPLE 1 ▸ Multiply.

a) $(7x^3)(6x^5)$ **b)** $(4b^2)(-9b^7)$

Solution

a) $(7x^3)(6x^5) = 7 \cdot 6 \cdot x^3 \cdot x^5 = 42x^{3+5} = 42x^8$

b) $(4b^2)(-9b^7) = (4)(-9) \cdot b^2 \cdot b^7 = -36b^{2+7} = -36b^9$

▸ **Now Try Exercise 15**

EXAMPLE 2 ▸ Multiply $(5x^2 y)(8x^5 y^4)$.

Solution   Remember that when a variable is given without an exponent we assume that the exponent on the variable is 1.

$$(5x^2 y)(8x^5 y^4) = 40x^{2+5} y^{1+4} = 40x^7 y^5$$

▸ **Now Try Exercise 19**

EXAMPLE 3 ▸ Multiply.

a) $6xy^2 z^5 (-3x^4 y^7 z)$ **b)** $(-4x^4 z^9)(-3xy^7 z^3)$

Solution

a) $6xy^2 z^5 (-3x^4 y^7 z) = -18x^5 y^9 z^6$

b) $(-4x^4 z^9)(-3xy^7 z^3) = 12x^5 y^7 z^{12}$

▸ **Now Try Exercise 21**

2 Multiply a Polynomial by a Monomial

To multiply a polynomial by a monomial, we use the distributive property presented earlier.

$$a(b + c) = ab + ac$$

The distributive property can be expanded to

$$a(b + c + d + \cdots + n) = ab + ac + ad + \cdots + an$$

EXAMPLE 4 ▸ Multiply $3x(2x^2 + 4)$.

Solution $3x(2x^2 + 4) = (3x)(2x^2) + (3x)(4)$

$$= 6x^3 + 12x$$

▸ **Now Try Exercise 29**

Notice that the use of the distributive property results in monomials being multiplied by monomials. If we study Example 4, we see that the $3x$ and $2x^2$ are both monomials, as are the $3x$ and 4.

EXAMPLE 5 ▸ Multiply $-3n(4n^2 - 2n - 1)$.

Solution $-3n(4n^2 - 2n - 1) = (-3n)(4n^2) + (-3n)(-2n) + (-3n)(-1)$

$$= -12n^3 + 6n^2 + 3n$$

▸ **Now Try Exercise 33**

EXAMPLE 6 ▸ Multiply $5x^2(4x^3 - 2x + 7)$.

Solution $5x^2(4x^3 - 2x + 7) = (5x^2)(4x^3) + (5x^2)(-2x) + (5x^2)(7)$

$$= 20x^5 - 10x^3 + 35x^2$$

▸ **Now Try Exercise 37**

EXAMPLE 7 ▸ Multiply $2x(3x^2y - 6xy + 5)$.

Solution $2x(3x^2y - 6xy + 5) = (2x)(3x^2y) + (2x)(-6xy) + (2x)(5)$

$$= 6x^3y - 12x^2y + 10x$$

▸ **Now Try Exercise 39**

In Example 8, we perform a multiplication where the monomial is placed to the right of the polynomial. Each term of the polynomial is multiplied by the monomial, as illustrated in the example.

EXAMPLE 8 ▸ Multiply $(3x^3 - 2xy + 3)4x$.

Solution $(3x^3 - 2xy + 3)4x = (3x^3)(4x) + (-2xy)(4x) + (3)(4x)$

$$= 12x^4 - 8x^2y + 12x$$

▸ **Now Try Exercise 41**

The problem in Example 8 could be written as $4x(3x^3 - 2xy + 3)$ by the commutative property of multiplication, and then simplified as in Examples 4 through 7.

3 Multiply Binomials Using the Distributive Property

Now we will discuss multiplying a binomial by a binomial. Before we explain how to do this, consider the multiplication problem $43 \cdot 12$.

$$43 \longleftarrow \text{Multiplicand}$$
$$\underline{12} \longleftarrow \text{Multiplier}$$
$$2(4) \longrightarrow 86 \longleftarrow 2(3)$$
$$1(4) \longrightarrow \underline{43} \longleftarrow 1(3)$$
$$516 \longleftarrow \text{Product}$$

Note how the 2 multiplies both the 3 and the 4, and the 1 also multiplies both the 3 and the 4. That is, every digit in the multiplier multiplies every digit in the multiplicand. We can also illustrate the multiplication process as follows.

$$\begin{aligned}
(43)(12) &= (40 + 3)(10 + 2) \\
&= (40 + 3)(10) + (40 + 3)(2) \\
&= (40)(10) + (3)(10) + (40)(2) + (3)(2) \\
&= 400 + 30 + 80 + 6 \\
&= 516
\end{aligned}$$

Whenever any two polynomials are multiplied, the same process must be followed. That is, **every term in one polynomial must multiply every term in the other polynomial**.

Consider multiplying $(a + b)(c + d)$. Treating $(a + b)$ as a single term and using the distributive property, we get

$$(a + b)(c + d) = (a + b)c + (a + b)d$$

Using the distributive property a second time gives

$$= ac + bc + ad + bd$$

Notice how each term of the first polynomial was multiplied by each term of the second polynomial, and all the products were added to obtain the answer.

EXAMPLE 9 ▸ Multiply $(3x + 2)(x - 5)$.

Solution
$$\begin{aligned}
(3x + 2)(x - 5) &= (3x + 2)x + (3x + 2)(-5) \\
&= 3x(x) + 2(x) + 3x(-5) + 2(-5) \\
&= 3x^2 + 2x - 15x - 10 \\
&= 3x^2 - 13x - 10
\end{aligned}$$

▸ **Now Try Exercise 43**

Note that after performing the multiplication like terms must be combined.

EXAMPLE 10 ▸ Multiply $(x - 4)(y + 3)$.

Solution
$$\begin{aligned}
(x - 4)(y + 3) &= (x - 4)y + (x - 4)3 \\
&= xy - 4y + 3x - 12
\end{aligned}$$

▸ **Now Try Exercise 65**

4 Multiply Binomials Using the FOIL Method

A common method used to multiply two binomials is the **FOIL method**. This procedure also results in each term of one binomial being multiplied by each term in the other binomial. Students often prefer to use this method when multiplying two binomials.

The FOIL method is not actually a different method used to multiply binomials but rather an acronym to help students remember to correctly apply the distributive property. We could have used IFOL or any other arrangement of the four letters. However, FOIL is easier to remember than the other arrangements.

The FOIL Method

Consider $(a + b)(c + d)$.

F   stands for **first**—multiply the first terms of each binomial together:

$$\overset{F}{(a + b)(c + d)} \qquad \text{product } ac$$

O   stands for **outer**—multiply the two outer terms together:

$$\overset{O}{(a + b)(c + d)} \qquad \text{product } ad$$

I   stands for **inner**—multiply the two inner terms together:

$$\overset{I}{(a + b)(c + d)} \qquad \text{product } bc$$

L   stands for **last**—multiply the last terms together:

$$\overset{L}{(a + b)(c + d)} \qquad \text{product } bd$$

The product of the two binomials is the sum of these four products.

$$(a + b)(c + d) = ac + ad + bc + bd$$

EXAMPLE 11 ▶ Using the FOIL method, multiply $(2x - 3)(x + 4)$.

Solution

$$(2x - 3)(x + 4)$$

$$= \overset{F}{(2x)(x)} + \overset{O}{(2x)(4)} + \overset{I}{(-3)(x)} + \overset{L}{(-3)(4)}$$

$$= 2x^2 \quad + \quad 8x \quad - \quad 3x \quad - \quad 12$$

$$= 2x^2 + 5x - 12$$

Thus, $(2x - 3)(x + 4) = 2x^2 + 5x - 12$.

▶ **Now Try Exercise 45**

EXAMPLE 12 ▶ Multiply $(4 - 2x)(6 - 5x)$.

Solution

$$(4 - 2x)(6 - 5x)$$

$$= \overset{F}{4(6)} + \overset{O}{4(-5x)} + \overset{I}{(-2x)(6)} + \overset{L}{(-2x)(-5x)}$$

$$= 24 \quad - \quad 20x \quad - \quad 12x \quad + \quad 10x^2$$

$$= 10x^2 - 32x + 24$$

Thus, $(4 - 2x)(6 - 5x) = 10x^2 - 32x + 24$.

▶ **Now Try Exercise 63**

EXAMPLE 13 ▶ Multiply $(4p + 5)(4p - 5)$.

Solution

$$\begin{array}{ccccccc} & F & & O & & I & & L \\ (4p + 5)(4p - 5) = & (4p)(4p) & + & (4p)(-5) & + & (5)(4p) & + & (5)(-5) \\ = & 16p^2 & - & 20p & + & 20p & - & 25 \\ = & 16p^2 - 25 \end{array}$$

Thus, $(4p + 5)(4p - 5) = 16p^2 - 25$.

▶ **Now Try Exercise 57**

Helpful Hint *Study Tip*

Make sure you have a thorough understanding of multiplication of polynomials. In the next chapter, we will be studying factoring, which is the reverse process of multiplication of polynomials. To understand factoring, you must first understand multiplication of polynomials.

5 Multiply Binomials Using Formulas for Special Products

Example 13 illustrates a special product, the product of the sum and difference of the same two terms.

Product of the Sum and Difference of the Same Two Terms

$$(a + b)(a - b) = a^2 - b^2$$

In this special product, a represents one term and b the other term. Then $(a + b)$ is the sum of the terms and $(a - b)$ is the difference of the terms. This special product is also called the **difference of two squares formula** because the expression on the right side of the equal sign is the difference of two squares. Since multiplication is commutative $(a + b)(a - b) = a^2 - b^2$ can also be written $(a - b)(a + b) = a^2 - b^2$.

EXAMPLE 14 ▶ Use the rule for finding the product of the sum and difference of two quantities to multiply each expression.

a) $(x + 5)(x - 5)$ **b)** $(2x + 4)(2x - 4)$ **c)** $(3x - 2y)(3x + 2y)$

Solution

a) If we let $x = a$ and $5 = b$, then

$$(a + b)(a - b) = a^2 - b^2$$
$$\downarrow \quad \downarrow \downarrow \quad \downarrow \qquad \downarrow \qquad \downarrow$$
$$(x + 5)(x - 5) = (x)^2 - (5)^2$$
$$= x^2 - 25$$

b)
$$(a + b)(a - b) = a^2 - b^2$$
$$\downarrow \quad \downarrow \downarrow \quad \downarrow \qquad \downarrow \qquad \downarrow$$
$$(2x + 4)(2x - 4) = (2x)^2 - (4)^2$$
$$= 4x^2 - 16$$

c)
$$(a - b) \quad (a + b) = a^2 - b^2$$
$$\downarrow \quad \downarrow \quad \downarrow \quad \downarrow \qquad \downarrow \qquad \downarrow$$
$$(3x - 2y)(3x + 2y) = (3x)^2 - (2y)^2$$
$$= 9x^2 - 4y^2$$

▶ **Now Try Exercise 77**

Example 14 could also have been done using the FOIL method.

Chapter 5   Exponents and Polynomials

EXAMPLE 15 ▸ Using the FOIL method, multiply $(x + 3)^2$.

Solution $(x + 3)^2 = (x + 3)(x + 3)$

$$
\begin{array}{cccc}
F & O & I & L
\end{array}
$$

$$= x(x) + x(3) + 3(x) + (3)(3)$$

$$= x^2 + 3x + 3x + 9$$

$$= x^2 + 6x + 9$$

▸ **Now Try Exercise 59**

Example 15 illustrates the **square of a binomial**, another special product.

Square of Binomial Formulas

$$(a + b)^2 = (a + b)(a + b) = a^2 + 2ab + b^2$$
$$(a - b)^2 = (a - b)(a - b) = a^2 - 2ab + b^2$$

To square a binomial, add the square of the first term, twice the product of the terms, and the square of the second term.

EXAMPLE 16 ▸ Use the square of a binomial formula to multiply each expression.

a) $(x + 5)^2$ **b)** $(2x - 4)^2$ **c)** $(3r + 2s)^2$ **d)** $(x - 3)(x - 3)$

Solution

a) If we let $x = a$ and $5 = b$, then

$$
(a + b)(a + b) = a^2 + 2\ a\ b + b^2
$$
$$\downarrow \quad \downarrow \ \downarrow \ \downarrow \qquad \downarrow \qquad \downarrow \ \downarrow \qquad \downarrow$$
$$(x + 5)^2 = (x + 5)(x + 5) = (x)^2 + 2(x)(5) + (5)^2$$

$$= x^2 + 10x + 25$$

b)
$$
(a - b)\ (a - b) = a^2 - 2\ a\ b + b^2
$$
$$\downarrow \ \downarrow \ \downarrow \ \downarrow \qquad \downarrow \qquad \downarrow \ \downarrow \qquad \downarrow$$
$$(2x - 4)^2 = (2x - 4)(2x - 4) = (2x)^2 - 2(2x)(4) + (4)^2$$

$$= 4x^2 - 16x + 16$$

c)
$$
(a + b)\ (a + b) = a^2 + 2\ a\ b + b^2
$$
$$\downarrow \ \searrow \ \downarrow \ \searrow \qquad \downarrow \qquad \downarrow \ \downarrow \qquad \downarrow$$
$$(3r + 2s)^2 = (3r + 2s)(3r + 2s) = (3r)^2 + 2(3r)(2s) + (2s)^2$$

$$= 9r^2 + 12rs + 4s^2$$

d)
$$
(a - b)(a - b) = a^2 - 2\ a\ b + b^2
$$
$$\downarrow \ \downarrow \ \downarrow \ \downarrow \qquad \downarrow \qquad \downarrow \ \downarrow \qquad \downarrow$$
$$(x - 3)(x - 3) = (x - 3)(x - 3) = (x)^2 - 2(x)(3) + (3)^2$$

$$= x^2 - 6x + 9$$

▸ **Now Try Exercise 83**

Example 16 could also have been done using the FOIL method.

Avoiding Common Errors

CORRECT	INCORRECT
$(a + b)^2 = a^2 + 2ab + b^2$	$\cancel{(a + b)^2 = a^2 + b^2}$
$(a - b)^2 = a^2 - 2ab + b^2$	$\cancel{(a - b)^2 = a^2 - b^2}$

Do not forget the middle term when you square a binomial.

$$(x + 2)^2 \neq x^2 + 4$$
$$(x + 2)^2 = (x + 2)(x + 2)$$
$$= x^2 + 4x + 4$$

6 Multiply Any Two Polynomials

When multiplying a binomial by a binomial, we saw that every term in the first binomial was multiplied by every term in the second binomial. When multiplying any two polynomials, each term of one polynomial must be multiplied by each term of the other polynomial. In the multiplication $(3x + 2)(4x^2 - 5x - 3)$, we use the distributive property as follows:

$$(3x + 2)(4x^2 - 5x - 3)$$
$$= 3x(4x^2 - 5x - 3) + 2(4x^2 - 5x - 3)$$
$$= 12x^3 - 15x^2 - 9x + 8x^2 - 10x - 6$$
$$= 12x^3 - 7x^2 - 19x - 6$$

Thus, $(3x + 2)(4x^2 - 5x - 3) = 12x^3 - 7x^2 - 19x - 6$.

Multiplication problems can be performed by using the distributive property, as we just illustrated. However, many students prefer to multiply a polynomial by a polynomial using a vertical procedure. On page 313, we showed that when multiplying the number 43 by the number 12, we multipy each digit in the number 43 by each digit in the number 12. Review that example now. We can follow a similar procedure when multiplying a polynomial by a polynomial, as illustrated in the following examples. We must be careful, however, to align like terms in the same columns when performing the individual multiplications.

EXAMPLE 17 ▶ Multiply $(3x + 4)(2x + 5)$.

Solution   First write the polynomials one beneath the other.

$$\begin{array}{r} 3x + 4 \\ \underline{2x + 5} \end{array}$$

Next, multiply each term in $(3x + 4)$ by 5.

$$\begin{array}{r} 3x + 4 \\ \underline{2x + 5} \end{array}$$
$$5(3x + 4) \longrightarrow \begin{array}{r} 15x + 20 \end{array}$$

Next, multiply each term in $(3x + 4)$ by $2x$ and align like terms.

$$\begin{array}{r} 3x + 4 \\ \underline{2x + 5} \\ 15x + 20 \end{array}$$
$$2x(3x + 4) \longrightarrow \begin{array}{r} 6x^2 + 8x \\ \hline 6x^2 + 23x + 20 \end{array} \quad \textit{Add like terms in columns.}$$

▶ **Now Try Exercise 55**

The same answer for Example 17 would be obtained using the FOIL method.

EXAMPLE 18 ▶ Multiply $(4y + 3)(2y^2 - 7y - 5)$.

Solution   For convenience, we place the shorter expression on the bottom, as illustrated.

$$
\begin{array}{r}
2y^2 - 7y - 5 \\
4y + 3 \\
\hline
6y^2 - 21y - 15 \\
8y^3 - 28y^2 - 20y \\
\hline
8y^3 - 22y^2 - 41y - 15
\end{array}
$$

Multiply the top polynomial by 3.
Multiply the top polynomial by 4y; align like terms.
Add like terms in columns.

▶ **Now Try Exercise 95**

EXAMPLE 19 ▶ Multiply $(x^2 - 3x + 2)(2x^2 - 3)$.

Solution

$$
\begin{array}{r}
x^2 - 3x + 2 \\
2x^2 - 3 \\
\hline
-3x^2 + 9x - 6 \\
2x^4 - 6x^3 + 4x^2 \\
\hline
2x^4 - 6x^3 + x^2 + 9x - 6
\end{array}
$$

Multiply the top polynomial by −3.
Multiply the top polynomial by 2x²; align like terms.
Add like terms in columns.

▶ **Now Try Exercise 103**

EXAMPLE 20 ▶ Multiply $(3x^3 - 2x^2 + 4x + 6)(x^2 - 5x)$.

Solution

$$
\begin{array}{r}
3x^3 - 2x^2 + 4x + 6 \\
x^2 - 5x \\
\hline
-15x^4 + 10x^3 - 20x^2 - 30x \\
3x^5 - 2x^4 + 4x^3 + 6x^2 \\
\hline
3x^5 - 17x^4 + 14x^3 - 14x^2 - 30x
\end{array}
$$

Multiply the top polynomial by −5x.
Multiply the top polynomial by x²; align like terms.
Add like terms in columns.

▶ **Now Try Exercise 105**

EXERCISE SET 5.5

Concept/Writing Exercises

1. Explain how to multiply a monomial by a monomial.

2. What is the name of the property used when multiplying a polynomial by a monomial?

3. What do the letters in the acronym FOIL represent?

4. How does the FOIL method work when multiplying two binomials?

5. When multiplying two binomials, will you get the same answer if you multiply using the order LOIF instead of FOIL? Explain your answer.

6. Why is the special product $(a + b)(a - b) = a^2 - b^2$ also called the difference of two squares formula?

7. Write the square of binomial formulas.

8. Describe how to square a binomial.

9. Does $(x - 2)^2 = x^2 - 2^2$? Explain. If not, what is the correct result?

10. Does $(x + 5)^2 = x^2 + 5^2$? Explain. If not, what is the correct result?

11. Make up a multiplication problem where a monomial in x is multiplied by a binomial in x. Determine the product.

12. Make up a multiplication problem where two binomials in x are multiplied. Determine the product.

13. Make up a multiplication problem where a monomial in y is multiplied by a trinomial in y. Determine the product.

14. When multiplying two polynomials, is it necessary for each term in one polynomial to multiply each term in the other polynomial?

Practice the Skills

Multiply.

15. $(3x^4)(-8x^2)$

16. $(-7p^5)(-2p^3)$

17. $5x^3y^5(4x^2y)$

18. $-5x^2y^4(2x^3y^2)$

19. $(4xy^6)(-7x^2y^9)$

20. $(4a^3b^7)(6a^2b)$

21. $9xy^6(6x^5y^8)$

22. $(6m^3n^4)(3n^5)$

23. $(6x^2y)\left(\frac{1}{2}x^4\right)$

24. $\frac{3}{4}x(8x^2y^3)$

25. $(3.3x^4)(1.8x^4y^3)$

26. $(2.3x^5)(4.1x^2y^4)$

Multiply.

27. $9(x-5)$

28. $4(x+3)$

29. $-3x(2x-2)$

30. $-4p(-3p+6)$

31. $-2(8y+5)$

32. $2x(x^2+3x-1)$

33. $-2x(x^2-2x+5)$

34. $-6c(-3c^2+5c-6)$

35. $5x(-4x^2+6x-4)$

36. $(3x^2+x-6)x$

37. $0.5x^2(x^3-6x^2-1)$

38. $2.3b^2(2b^2-b+3)$

39. $0.3x(2xy+5x-6y)$

40. $-\frac{1}{2}x^3(2x^2+4x-6y^2)$

41. $(x^2-4y^3-3)y^4$

42. $\frac{1}{4}y^4(y^2-12y+4x)$

Multiply.

43. $(5x-2)(x+4)$

44. $(2x-3)(x+5)$

45. $(2x+5)(3x-6)$

46. $(4a-1)(a+4)$

47. $(2x-4)(2x+4)$

48. $(4+5w)(3+w)$

49. $(8-5x)(6+x)$

50. $(-x+3)(2x+5)$

51. $(6x-1)(-2x+5)$

52. $(7n-3)(2n+1)$

53. $(x-2)(4x-2)$

54. $(2x+3)(x+5)$

55. $(3k-6)(4k-2)$

56. $(3d-5)(4d-1)$

57. $(x-2)(x+2)$

58. $(3x-8)(2x+3)$

59. $(2x-3)(2x-3)$

60. $(7x+3)(2x+4)$

61. $(6z-4)(7-z)$

62. $(6-2m)(5m-3)$

63. $(9-2x)(7-4x)$

64. $(2-5x)(7-2x)$

65. $(x+7)(y-3)$

66. $(z+2y)(4z-3)$

67. $(2x-3y)(3x+2y)$

68. $(2x+3)(2y-5)$

69. $(9x+y)(4-3x)$

70. $(2x-0.1)(x+2.4)$

71. $(x+0.6)(x+0.3)$

72. $(3x-6)\left(x+\frac{1}{3}\right)$

73. $(x+4)\left(x-\frac{1}{2}\right)$

74. $(2y-4)\left(\frac{1}{2}x-1\right)$

Multiply using a special product formula.

75. $(x+6)(x-6)$

76. $(x+3)^2$

77. $(3x-8)(3x+8)$

78. $(r-4)(r-4)$

79. $(x+y)^2$

80. $(2x-7)(2x+7)$

81. $(x-0.2)^2$

82. $(a+3b)(a-3b)$

83. $(4x+5)(4x+5)$

84. $(5x+4)(5x-4)$

85. $(0.4x+y)^2$

86. $\left(x-\frac{1}{2}y\right)^2$

87. $(4c-5d)(4c+5d)$

88. $(4+3w)(4-3w)$

89. $(-2x+6)(-2x-6)$

90. $(-3m+2n)(-3m-2n)$

91. $(7s-3t)^2$

92. $(7a+2)^2$

Multiply.

93. $(4m+3)(4m^2-5m+6)$

94. $(x+4)(3x^2+4x-1)$

95. $(3x+2)(4x^2-x+5)$

96. $(x-1)(3x^2+3x+2)$

97. $(-2x^2-4x+1)(7x-3)$

98. $(4x^2+9x-2)(x-2)$

99. $(a+b)(a^2-ab+b^2)$

100. $(a-b)(a^2+ab+b^2)$

101. $(3x^2 - 2x + 4)(2x^2 + 3x + 1)$

103. $(x^2 - x + 3)(x^2 - 2x)$

105. $(2x^3 - 6x^2 + x - 3)(x^2 + 4x)$

102. $(x^2 - 2x + 3)(x^2 - 4)$

104. $(6x + 4)(2x^2 + 2x - 4)$

106. $(3y^3 + 4y^2 - y + 7)(y^2 - 5y)$

Determine the cube of each expression by writing the expression as the square of an expression multiplied by another expression. For example, $(x + 3y)^3 = (x + 3y)^2(x + 3y)$

107. $(b - 1)^3$

109. $(3a - 5)^3$

108. $(x + 2)^3$

110. $(2z + 3)^3$

Problem Solving

111. Will the product of a monomial and a monomial always be a monomial? Explain your answer.

112. Will the product of a monomial and a binomial ever be a trinomial? Explain your answer.

113. Will the product of two binomials after like terms are combined always be a trinomial? Explain your answer.

114. Will the product of any polynomial and a binomial always be a polynomial? Explain.

Consider the multiplications in Exercises 115 and 116. Determine the exponents to be placed in the shaded areas.

115. $3x^2(2x^{\blacksquare} - 5x^{\blacksquare} + 3x^{\blacksquare}) = 6x^8 - 15x^5 + 9x^3$.

116. $4x^3(x^{\blacksquare} + 2x^{\blacksquare} - 5x^{\blacksquare}) = 4x^7 + 8x^5 - 20x^4$.

117. Suppose that one side of a rectangle is represented as $x + 2$ and a second side is represented as $2x + 1$.

  a) Express the area of the rectangle in terms of x.

  b) Find the area if $x = 4$ feet.

  c) What value of x, in feet, would result in the rectangle being a square? Explain how you determined your answer.

118. Consider the figure below.

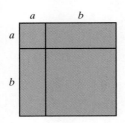

  a) Write an expression for the length of the top.

  b) Write an expression for the length of the left side.

  c) Is this figure a square? Explain.

  d) Express the area of this square as the square of a binomial.

  e) Determine the area of the square by summing the areas of the four individual pieces.

  f) Using the figure and your answer to part **e)**, complete the following.
$$(a + b)^2 = \text{?}$$

119. Suppose that a rectangular solid has length $x + 5$, width $3x + 4$, and height $2x - 2$ (see the figure).

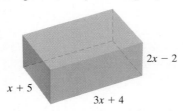

  a) Write a polynomial that represents the area of the base by multiplying the length by the width.

  b) The volume of the figure can be found by multiplying the area of the base by the height. Write a polynomial that represents the volume of the figure.

  c) Using the polynomial in part **b)**, find the volume of the figure if x is 4 feet.

  d) Using the binomials given for the length, width, and height, find the volume if x is 4 feet.

  e) Are your answers to parts **c)** and **d)** the same? If not, explain why.

Challenge Problems

Multiply.

120. $(2x^3 - 6x^2 + 5x - 3)(3x^3 - 6x + 4)$

121. $\left(\dfrac{1}{2}x + \dfrac{2}{3}\right)\left(\dfrac{2}{3}x - \dfrac{2}{5}\right)$

Group Activity

122. Consider the trinomial $2x^2 + 7x + 3$.

  a) As a group, determine whether there is a maximum number of pairs of binomials whose product is $2x^2 + 7x + 3$. That is, how many different pairs of binomials can go in the shaded areas?
$$2x^2 + 7x + 3 = (\blacksquare)(\blacksquare)$$

  b) Individually, find a pair of binomials whose product is $2x^2 + 7x + 3$.

  c) Compare your answer to part **b)** with the other members of your group. If you did not all arrive at the same answer, explain why.

Cumulative Review Exercises

[2.5] **123.** Solve the equation $3(x + 7) - 5 = 3x - 17$.

[3.3] **124. Complementary Angles** Angles C and D are complementary angles, and angle D is $16°$ less than angle C. Find the measures of angle C and angle D.

[5.1] **125.** Simplify $\left(\dfrac{3xy^4}{6y^6}\right)^4$.

[5.1–5.2] **126.** Evaluate the following.
   a) -6^3
   b) 6^{-3}

[5.4] **127.** Subtract $4x^2 - 4x - 9$ from $-x^2 - 6x + 5$.

5.6 Division of Polynomials

1 Divide a polynomial by a monomial.

2 Divide a polynomial by a binomial.

3 Check division of polynomial problems.

4 Write polynomials in descending order when dividing.

1 Divide a Polynomial by a Monomial

Now let's see how to divide polynomials. We begin by dividing a polynomial by a monomial.

To Divide a Polynomial by a Monomial

To divide a polynomial by a monomial, divide each term of the polynomial by the monomial.

EXAMPLE 1 ▶ Divide.

a) $\dfrac{4x + 20}{4}$ **b)** $\dfrac{9x^2 - 6x}{3x}$

Solution

a) $\dfrac{4x + 20}{4} = \dfrac{4x}{4} + \dfrac{20}{4} = x + 5$

b) $\dfrac{9x^2 - 6x}{3x} = \dfrac{9x^2}{3x} - \dfrac{6x}{3x} = 3x - 2$

▶ **Now Try Exercise 17**

Avoiding Common Errors

CORRECT

$$\dfrac{x + 2}{2} = \dfrac{x}{2} + \dfrac{2}{2} = \dfrac{x}{2} + 1$$

$$\dfrac{x + 2}{x} = \dfrac{x}{x} + \dfrac{2}{x} = 1 + \dfrac{2}{x}$$

INCORRECT

$$\dfrac{x + \overset{1}{\cancel{2}}}{\underset{1}{\cancel{2}}} = \dfrac{x + 1}{1} = x + 1$$

$$\dfrac{\overset{1}{\cancel{x}} + 2}{\underset{1}{\cancel{x}}} = \dfrac{1 + 2}{1} = 3$$

Can you explain why the procedures on the right are not correct?

EXAMPLE 2 ▶ Divide $\dfrac{4t^5 - 6t^4 + 8t - 3}{2t^2}$.

Solution $\dfrac{4t^5 - 6t^4 + 8t - 3}{2t^2} = \dfrac{4t^5}{2t^2} - \dfrac{6t^4}{2t^2} + \dfrac{8t}{2t^2} - \dfrac{3}{2t^2}$

$$= 2t^3 - 3t^2 + \dfrac{4}{t} - \dfrac{3}{2t^2}$$

▶ **Now Try Exercise 37**

EXAMPLE 3 ▶ Divide $\dfrac{3x^3 - 6x^2 + 4x - 1}{-3x}$.

Solution   A negative sign appears in the denominator. Usually, it is easier to divide if the divisor is positive. We can multiply both numerator and denominator by -1 to get a positive denominator.

$$\frac{(-1)(3x^3 - 6x^2 + 4x - 1)}{(-1)(-3x)} = \frac{-3x^3 + 6x^2 - 4x + 1}{3x}$$

$$= \frac{-3x^3}{3x} + \frac{6x^2}{3x} - \frac{4x}{3x} + \frac{1}{3x}$$

$$= -x^2 + 2x - \frac{4}{3} + \frac{1}{3x}$$

▶ **Now Try Exercise 41**

---

2 Divide a Polynomial by a Binomial

We divide a polynomial by a binomial in much the same way as we perform long division. This procedure will be explained in Example 4.

EXAMPLE 4 ▶ Divide $\dfrac{x^2 + 6x + 8}{x + 2}$.   ← dividend
   ← divisor

Solution   Rewrite the division problem in the following form:

$$x + 2 \overline{)\, x^2 + 6x + 8}$$

Divide x^2 (the first term in the dividend) by x (the first term in the divisor).

$$\frac{x^2}{x} = x$$

Place the quotient, x, above the like term containing x in the dividend.

$$\begin{array}{r} x \\ x + 2 \overline{)\, x^2 + 6x + 8} \end{array}$$

Next, multiply the x by $x + 2$ as you would do in long division and place the terms of the product under their like terms.

$$\begin{array}{r} \textit{Times} \qquad\quad x \\ x + 2 \overline{)\, x^2 + 6x + 8} \\ \textit{Equals} \searrow x^2 + 2x \quad \leftarrow\; x(x + 2) \end{array}$$

Now subtract $x^2 + 2x$ from $x^2 + 6x$. When subtracting, remember to change the sign of the terms being subtracted and then add the like terms.

$$\begin{array}{r} x \\ x + 2 \overline{)\, x^2 + 6x + 8} \\ \underline{x^2 + 2x} \\ 4x \end{array}$$

Next, bring down the 8, the next term in the dividend.

$$\begin{array}{r} x \\ x + 2 \overline{)\, x^2 + 6x + 8} \\ \underline{x^2 + 2x} \\ 4x + 8 \end{array}$$

Now divide $4x$, the first term at the bottom, by x, the first term in the divisor.

$$\frac{4x}{x} = +4$$

Write the $+4$ in the quotient above the constant in the dividend.

$$
\begin{array}{r}
x + 4 \\
x + 2 \overline{)\, x^2 + 6x + 8} \\
\underline{x^2 + 2x} \\
4x + 8
\end{array}
$$

Multiply the $x + 2$ by 4 and place the terms of the product under their like terms.

$$
\begin{array}{r}
\textit{Times} \\
x + 4 \\
x + 2 \overline{)\, x^2 + 6x + 8} \\
\underline{x^2 + 2x} \\
\textit{Equals} \quad 4x + 8 \\
4x + 8 \longleftarrow 4(x + 2)
\end{array}
$$

Now subtract.

$$
\begin{array}{r}
x + 4 \quad \longleftarrow \textit{Quotient} \\
x + 2 \overline{)\, x^2 + 6x + 8} \\
\underline{x^2 + 2x} \\
4x + 8 \\
\underline{4x + 8} \\
0 \quad \longleftarrow \textit{Remainder}
\end{array}
$$

Thus,

$$\frac{x^2 + 6x + 8}{x + 2} = x + 4$$

There is no remainder.

▶ **Now Try Exercise 43**

EXAMPLE 5 ▶ Divide $\dfrac{3x^2 + x - 12}{x + 2}$.

Solution First write the problem in the following form:
$$x + 2 \overline{)\, 3x^2 + x - 12}$$

Since $3x^2$ divided by x is $3x$, place $3x$ above the x-term in the dividend.

$$
\begin{array}{r}
3x \\
x + 2 \overline{)\, 3x^2 + x - 12}
\end{array}
$$

Then multiply $3x(x + 2)$ and write the product $3x^2 + 6x$ as shown below. Then subtract to get a difference of $-5x$.

$$
\begin{array}{r}
3x \\
x + 2 \overline{)\, 3x^2 + \ x - 12} \\
\underline{3x^2 + 6x} \\
-5x
\end{array}
$$

Next bring down the -12. Then divide $-5x$ by x, which gives -5. Place -5 over -12 in the dividend, as shown below. Then multiply $-5(x + 2)$. Write the product, $-5x - 10$ below the $-5x - 12$. Then subtract to get a remainder of -2.

$$
\begin{array}{r}
3x - 5 \\
x + 2 \overline{)\, 3x^2 + \ x - 12} \\
\underline{3x^2 + 6x} \\
-5x - 12 \\
\underline{-5x - 10} \\
-2
\end{array}
$$

When there is a remainder, as in this example, list the quotient plus the remainder above the divisor. Thus,

$$\frac{3x^2 + x - 12}{x + 2} = 3x - 5 - \frac{2}{x + 2}.$$

▶ Now Try Exercise 45

EXAMPLE 6 ▶ Divide $\dfrac{6x^2 - 5x + 5}{2x + 3}$.

Solution

$$\frac{6x^2}{2x} \qquad \frac{-14x}{2x}$$

$$
\begin{array}{r}
3x - 7 \\
2x + 3 \overline{)\ 6x^2 - 5x + 5} \\
\underline{6x^2 + 9x} \quad \longleftarrow \ 3x(2x+3) \\
-14x + 5 \\
\underline{+14x +21} \quad \longleftarrow \ -7(2x+3) \\
26 \quad \longleftarrow \ \text{Remainder}
\end{array}
$$

Thus, $\dfrac{6x^2 - 5x + 5}{2x + 3} = 3x - 7 + \dfrac{26}{2x + 3}.$

▶ Now Try Exercise 57

3 Check Division of Polynomial Problems

The answer to a division problem can be checked. Consider the division problem $13 \div 5$.

$$
\begin{array}{r}
2 \\
5 \overline{)13} \\
\underline{10} \\
3
\end{array}
$$

Note that the divisor times the quotient, plus the remainder, equals the dividend:

$$(\text{divisor} \times \text{quotient}) + \text{remainder} = \text{dividend}$$
$$(5 \cdot 2) + 3 \stackrel{?}{=} 13$$
$$10 + 3 \stackrel{?}{=} 13$$
$$13 = 13 \qquad \textit{True}$$

This same procedure can be used to check all division problems.

To Check Division of Polynomials
$(\text{divisor} \times \text{quotient}) + \text{remainder} = \text{dividend}$

Let's check the answer to Example 6. The divisor is $2x + 3$, the quotient is $3x - 7$, the remainder is 26, and the dividend is $6x^2 - 5x + 5$.

Check

$$(\text{divisor} \times \text{quotient}) + \text{remainder} = \text{dividend}$$
$$(2x + 3)(3x - 7) + 26 \stackrel{?}{=} 6x^2 - 5x + 5$$
$$(6x^2 - 5x - 21) + 26 \stackrel{?}{=} 6x^2 - 5x + 5$$
$$6x^2 - 5x + 5 = 6x^2 - 5x + 5 \qquad \textit{True}$$

4 Write Polynomials in Descending Order When Dividing

When dividing a polynomial by a binomial, both the polynomial and binomial should be listed in descending order. If a given power term is missing, it is often helpful to include that term with a numerical coefficient of 0 as a place-holder. This will help keep like terms aligned. For example, to divide $(6x^2 + x^3 - 4)/(x - 2)$, we begin by writing $(x^3 + 6x^2 + 0x - 4)/(x - 2)$.

EXAMPLE 7 ▶ Divide $(-x + 9x^3 - 28)$ by $(3x - 4)$.

Solution First we rewrite the dividend in descending order to get $(9x^3 - x - 28) \div (3x - 4)$. Since there is no x^2 term in the dividend, we will add $0x^2$ to help align like terms.

$$\frac{9x^3}{3x} \quad \frac{12x^2}{3x} \quad \frac{15x}{3x}$$
$$\downarrow \qquad \downarrow \qquad \downarrow$$

$$
\begin{array}{r}
3x^2 + 4x + 5 \\
3x - 4 \overline{)9x^3 + 0x^2 - x - 28} \\
\underline{9x^3 - 12x^2} \qquad\qquad 3x^2(3x - 4) \\
12x^2 - x \\
\underline{12x^2 - 16x} \qquad\qquad 4x(3x - 4) \\
15x - 28 \\
\underline{15x - 20} \qquad\qquad 5(3x - 4) \\
-8 \qquad\qquad Remainder
\end{array}
$$

Thus, $\dfrac{-x + 9x^3 - 28}{3x - 4} = 3x^2 + 4x + 5 - \dfrac{8}{3x - 4}$. Check this division yourself using the procedure just discussed.

▶ **Now Try Exercise 55**

EXERCISE SET 5.6   *Math XL* (MathXL®) *MyMathLab* (MyMathLab)

Concept/Writing Exercises

1. Explain how to divide a polynomial by a monomial.

2. Explain how to check a division problem.

3. Explain why $\dfrac{2x + 8}{2} \ne \dfrac{x + 8}{1}$. Then, correctly divide the binomial by the monomial.

4. Explain why $\dfrac{y + 5}{y} \ne \dfrac{1 + 5}{1}$. Then, correctly divide the binomial by the monomial.

5. How should the terms of a polynomial and binomial be listed when dividing a polynomial by a binomial?

6. How would you rewrite $\dfrac{x^3 - 14x + 15}{x - 3}$ so that it is easier to complete the division?

7. How would you rewrite $\dfrac{x^2 - 7}{x - 2}$ so that it is easier to complete the division?

8. Show that $\dfrac{x^2 - 3x + 7}{x + 2} = x - 5 + \dfrac{17}{x + 2}$ by checking the division.

9. Show that $\dfrac{x^2 + 2x - 17}{x - 3} = x + 5 - \dfrac{2}{x - 3}$ by checking the division.

10. Show that $\dfrac{x^3 + 2x - 3}{x + 1} = x^2 - x + 3 - \dfrac{6}{x + 1}$ by checking the division.

Rewrite each multiplication problem as a division problem. There is more than one correct answer.

11. $(x - 7)(x + 6) = x^2 - x - 42$

12. $(x + 3)(3x - 1) = 3x^2 + 8x - 3$

⊙ **13.** $(2x + 3)(x + 1) = 2x^2 + 5x + 3$ **14.** $(2x - 5)(x + 1) = 2x^2 - 3x - 5$

15. $(2x + 3)(2x - 3) = 4x^2 - 9$ **16.** $(3n + 4)(n - 5) = 3n^2 - 11n - 20$

Practice the Skills

Divide.

17. $\dfrac{3x + 6}{3}$ **18.** $\dfrac{4x - 6}{2}$ **19.** $\dfrac{4n + 10}{2}$

20. $(-3x - 8) \div 4$ **21.** $\dfrac{7x + 6}{3}$ **22.** $\dfrac{5x - 10}{5}$

23. $\dfrac{-6x + 4}{2}$ **24.** $\dfrac{-5a + 4}{-3}$ **25.** $\dfrac{-9x - 3}{-3}$

26. $\dfrac{8x - 3}{-8}$ **27.** $\dfrac{2x + 16}{4}$ **28.** $\dfrac{2p - 3}{2p}$

29. $\dfrac{4 - 10w}{-4}$ **30.** $\dfrac{6 - 5x}{-3x}$ **31.** $(4x^2 + 8x - 12) \div 4x^2$

32. $\dfrac{12x^2 - 6x + 3}{3}$ ⊙ **33.** $\dfrac{-4x^5 + 6x + 8}{2x^2}$ **34.** $\dfrac{6t^2 + 3t + 8}{2}$

35. $(x^5 + 3x^4 - 3) \div x^3$ **36.** $(6x^2 - 7x + 9) \div 3x$ **37.** $\dfrac{6x^5 - 4x^4 + 12x^3 - 5x^2}{2x^3}$

38. $\dfrac{9x^2 + 18x - 7}{-9}$ **39.** $\dfrac{8k^3 + 6k^2 - 8}{-4k}$ **40.** $\dfrac{-12x^4 + 6x^2 - 15x + 4}{-3x}$

41. $\dfrac{12x^5 + 3x^4 - 10x^2 - 9}{-3x^2}$ **42.** $\dfrac{-15m^3 - 6m^2 + 15}{-5m^3}$

Divide.

⊙ **43.** $\dfrac{x^2 + 4x + 3}{x + 1}$ **44.** $(2x^2 + 3x - 35) \div (x + 5)$ **45.** $\dfrac{5y^2 - 34y - 7}{y - 7}$

46. $\dfrac{2p^2 - 7p - 15}{p - 5}$ **47.** $\dfrac{6x^2 + 16x + 8}{3x + 2}$ **48.** $\dfrac{3r^2 + 5r - 8}{r - 1}$

49. $\dfrac{x^2 - 16}{-4 + x}$ **50.** $\dfrac{6t^2 - 7t - 20}{3t + 4}$ **51.** $(2x^2 + 7x - 18) \div (2x - 3)$

52. $(4a^2 - 25) \div (2a - 5)$ **53.** $\dfrac{x^2 - 36}{x - 6}$ **54.** $\dfrac{9x^2 - 16}{3x - 4}$

55. $\dfrac{-x + 9x^3 - 16}{3x - 4}$ **56.** $\dfrac{10x + 3x^2 + 6}{x + 2}$ **57.** $\dfrac{6x + 8x^2 - 12}{2x + 3}$

58. $\dfrac{x^3 + 5x^2 + 2x - 8}{x + 2}$ **59.** $\dfrac{7x^3 + 28x^2 - 5x - 20}{x + 4}$ **60.** $\dfrac{2x^3 - 3x^2 - 3x + 6}{x - 1}$

61. $\dfrac{2x^3 - 4x^2 + 12}{x - 2}$ **62.** $\dfrac{2x^3 + 6x - 4}{x + 4}$ **63.** $(w^3 - 8) \div (w - 3)$

64. $\dfrac{x^3 + 8}{x + 2}$ **65.** $\dfrac{x^3 - 27}{x - 3}$ **66.** $\dfrac{x^3 + 64}{x + 4}$

⊙ **67.** $\dfrac{4x^3 - 5x}{2x - 1}$ **68.** $\dfrac{9x^3 - x + 3}{3x - 2}$ **69.** $\dfrac{-m^3 - 6m^2 + 2m - 3}{m - 1}$

70. $\dfrac{-x^3 + 3x^2 + 14x + 16}{x + 3}$ **71.** $\dfrac{4t^3 - t + 4}{t + 2}$ **72.** $\dfrac{9n^3 - 6n + 4}{3n - 3}$

Problem Solving

73. When dividing a binomial by a monomial, must the quotient be a binomial? Explain and give an example to support your answer.

74. When dividing a trinomial by a monomial, must the quotient be a trinomial? Explain and give an example to support your answer.

75. If the divisor is $x + 4$, the quotient is $2x + 3$, and the remainder is 4, find the dividend (or the polynomial being divided).

76. If the divisor is $2x - 3$, the quotient is $3x - 1$, and the remainder is -2, find the dividend.

77. If a fourth-degree polynomial in x is divided by a first-degree polynomial in x, what will be the degree of the quotient? Explain.

78. If a second-degree polynomial in x is divided by a first-degree polynomial in x, what will be the degree of the quotient? Explain.

Determine the expression to be placed in the shaded area to make a true statement. Explain how you determined your answer.

79. $\dfrac{16x^4 + 20x^3 - 4x^2 + 12x}{\boxed{}} = 4x^3 + 5x^2 - x + 3$

80. $\dfrac{9x^5 - 6x^4 + 3x^2 + 12}{\boxed{}} = 3x^3 - 2x^2 + 1 + \dfrac{4}{x^2}$

Determine the exponents to be placed in the shaded areas to make a true statement. Explain how you determined your answer.

81. $\dfrac{8x^{\square} + 4x^{\square} - 20x^{\square} - 5x^{\square}}{2x^2} = 4x^3 + 2x - 10 - \dfrac{5}{2x}$

82. $\dfrac{15x^{\square} + 25x^{\square} + 5x^{\square} + 10x^{\square}}{5x^2} = 3x^5 + 5x^4 + x^2 + 2$

Challenge Problems

Divide. The quotients in Exercises 83 and 84 will contain fractions.

83. $\dfrac{3x^3 - 5}{3x - 2}$

84. $\dfrac{4x^3 - 4x + 6}{2x + 3}$

85. $\dfrac{3x^2 + 6x - 10}{-x - 3}$

Group Activity

Discuss and answer Exercises 86 and 87 as a group. Determine the polynomial that when substituted in the shaded area results in a true statement. Explain how you determined your answer.

86. $\dfrac{\boxed{}}{x + 4} = x + 2 + \dfrac{2}{x + 4}$

87. $\dfrac{\boxed{}}{x + 3} = x + 1 - \dfrac{1}{x + 3}$

Cumulative Review Exercises

[1.4] **88.** Consider the set of numbers

$$\left\{ 2, -5, 0, \sqrt{7}, \frac{2}{5}, -6.3, \sqrt{3}, -\frac{23}{34} \right\}.$$

List those that are

a) natural numbers;

b) whole numbers;

c) rational numbers;

d) irrational numbers;

e) real numbers.

[1.8] **89. a)** To what is $\dfrac{0}{1}$ equal?

b) How do we refer to an expression like $\dfrac{1}{0}$?

[1.9] **90.** Give the order of operations to be followed when evaluating a mathematical expression.

[2.5] **91.** Solve the equation $2(x + 3) + 2x = x + 4$.

[3.2] **92. Sale** At a 30% off sale Jennifer Lucking purchased a sweater for $27.65. What is the original price of the sweater?

[5.2] **93.** Simplify $\dfrac{x^9}{x^{-4}}$.

Chapter 5 Summary

IMPORTANT FACTS AND CONCEPTS	EXAMPLES

Section 5.1

In the expression x^n, x is called the **base** and n is called the **exponent**.

base $\rightarrow 3^4 \nwarrow$ exponent

Rules of Exponents

Simplify.

1. $x^m \cdot x^n = x^{m+n}$ product rule

1. $x^5 \cdot x^4 = x^{5+4} = x^9$

2. $\dfrac{x^m}{x^n} = x^{m-n}, \quad x \neq 0$ quotient rule

2. $\dfrac{x^{12}}{x^7} = x^{12-7} = x^5$

3. $x^0 = 1, \quad x \neq 0$ zero exponent rule

3. $(-3ab^4)^0 = 1$

4. $(x^m)^n = x^{m \cdot n}$ power rule

4. $(x^6)^3 = x^{6 \cdot 3} = x^{18}$

5. $\left(\dfrac{ax}{by}\right)^m = \dfrac{a^m x^m}{b^m y^m}, \quad b \neq 0, \quad y \neq 0$ expanded power rule

5. $\left(\dfrac{4x}{5y}\right)^2 = \dfrac{4^2 x^2}{5^2 y^2} = \dfrac{16x^2}{25y^2}$

Section 5.2

Negative Exponent Rule

$$x^{-m} = \frac{1}{x^m}, \quad x \neq 0$$

$$x^{-2} = \frac{1}{x^2}$$

$$\frac{1}{y^{-6}} = y^6$$

A Fraction Raised to a Negative Exponent Rule

For a fraction of the form $\dfrac{a}{b}$, $a \neq 0$ and $b \neq 0$, $\left(\dfrac{a}{b}\right)^{-m} = \left(\dfrac{b}{a}\right)^m$

$$\left(\frac{7}{8}\right)^{-2} = \left(\frac{8}{7}\right)^2 = \frac{8^2}{7^2} = \frac{64}{49}$$

Section 5.3

Each number written in **scientific notation** is written as a number greater than or equal to 1 and less than 10 multiplied by some power of 10.

1.3×10^7

4.76×10^{-2}

To Write a Number in Scientific Notation

1. Move the decimal point in the original number to the right of the first nonzero digit.
2. Count the number of places you moved the decimal point in step 1. If the original number was 10 or greater, the count is positive. If the original number was less than 1, the count is negative.
3. Multiply the number obtained in step 1 by 10 raised to the count (power) found in step 2.

$25{,}700 = 2.57 \times 10^4$

$0.0000346 = 3.46 \times 10^{-5}$

To Convert a Number from Scientific Notation to Decimal Form

1. Observe the exponent of the power of 10.
2. **a)** If the exponent is positive, move the decimal point in the number to the right the same number of places as the exponent.
   b) If the exponent is 0, do not move the decimal point.
   c) If the exponent is negative, move the decimal point in the number to the left the same number of places as the exponent (dropping the negative sign).

$9.8 \times 10^6 = 9{,}800{,}000$

$5.17 \times 10^{-3} = 0.00517$

IMPORTANT FACTS AND CONCEPTS	EXAMPLES

Section 5.4

A **polynomial in x** is an expression containing the sum of a finite number of terms of the form ax^n, for any real number a and any whole number n.	$\frac{1}{5}x - 2$ and $x^2 - 4x + 7$ are both polynomials in x.
A polynomial is written in **descending order of the variable** when the exponents on the variable decrease from left to right.	$5x^4 - 3x^3 + 7x^2 - 6x + 9$ is written in descending order of the variable.
A **monomial** is a polynomial with one term.	$-7y^2$ is a monomial.
A **binomial** is a two-termed polynomial.	$x^2 - 8$ is a binomial.
A **trinomial** is a three-termed polynomial.	$4z^2 - 9z + 1$ is a trinomial.

The **degree of a term** of a polynomial in **one variable** is the exponent on the variable in that term.	$2y^6$ is a sixth-degree term.
The **degree of a term** of a polynomial in **two or more variables** is the sum of the exponents on those variables.	$3x^2y^5$ is a seventh-degree term
The **degree of a polynomial** is the same as that of its highest-degree term.	$9x^3 + 2x^2 - 5x + 4$ is a third-degree polynomial.

To Add Polynomials
To add polynomials, combine the like terms of the polynomials.

$$(3x^2 - 9x + 4) + (2x^2 - 3x - 5) = \underbrace{3x^2 + 2x^2}\ \underbrace{-9x - 3x}\ \underbrace{+4 - 5}$$
$$= 5x^2\quad -12x\quad -1$$

To Subtract Polynomials

1. Use the distributive property to remove parentheses.
2. Combine like terms.

$$(9a^2 - 6a + 1) - (a^2 - 5a - 3)$$
$$= 9a^2 - 6a + 1 - a^2 + 5a + 3$$
$$= \underbrace{9a^2 - a^2}\ \underbrace{-6a + 5a}\ \underbrace{+1 + 3}$$
$$= 8a^2\quad -a\quad +4$$

Section 5.5

FOIL Method to Multiply Two Binomials (First, Outer, Inner, Last)

$$(a + b)(c + d)$$

$$(3x - 5)(x + 4) = \overset{F}{(3x)(x)} + \overset{O}{(3x)(4)} + \overset{I}{(-5)(x)} + \overset{L}{(-5)(4)}$$
$$= 3x^2\ +\ 12x\ -\ 5x\ -\ 20$$
$$= 3x^2\ +\ 7x\ -\ 20$$

Product of Sum and Difference of the Same Two Terms (also called the difference of two squares):

$$(a + b)(a - b) = a^2 - b^2$$

$$(y + 6)(y - 6) = (y)^2 - (6)^2$$
$$= y^2 - 36$$

Square of a Binomial

$$(a + b)^2 = a^2 + 2ab + b^2$$
$$(a - b)^2 = a^2 - 2ab + b^2$$

1. $(x + 7)^2 = (x)^2 + 2(x)(7) + (7)^2$
$$= x^2 + 14x + 49$$
2. $(z - 3)^2 = (z)^2 - 2(z)(3) + (3)^2$
$$= z^2 - 6z + 9$$

IMPORTANT FACTS AND CONCEPTS	EXAMPLES

Section 5.5 (continued)

To Multiply Any Two Polynomials

To multiply any two polynomials, each term of one polynomial must multiply each term of the second polynomial.

$(x^2 + 3x + 5)(x - 2)$ or

$$\begin{array}{r} x^2 + 3x + 5 \\ x - 2 \\ \hline -2x^2 - 6x - 10 \\ x^3 + 3x^2 + 5x \\ \hline x^3 + x^2 - x - 10 \end{array}$$

Section 5.6

To Divide a Polynomial by a Monomial

To divide a polynomial by a monomial, divide each term of the polynomial by the monomial.

$$\frac{6x + 24}{6} = \frac{6x}{6} + \frac{24}{6} = x + 4$$

To Divide a Polynomial by a Binomial

To divide a polynomial by a binomial we perform division in much the same way as we perform long division.

$$\frac{x^2 - 4x + 3}{x + 2}$$

$$\begin{array}{r} x - 6 \\ x + 2 \overline{) x^2 - 4x + 3} \\ \underline{x^2 + 2x} \\ -6x + 3 \\ \underline{-6x - 12} \\ 15 \end{array}$$

$$\frac{x^2 - 4x + 3}{x + 2} = x - 6 + \frac{15}{x + 2}$$

Chapter 5 Review Exercises

[5.1] *Simplify.*

1. $x^5 \cdot x^2$

2. $x^2 \cdot x^4$

3. $3^2 \cdot 3^3$

4. $2^4 \cdot 2$

5. $\dfrac{x^4}{x}$

6. $\dfrac{a^5}{a^5}$

7. $\dfrac{5^5}{5^3}$

8. $\dfrac{4^4}{4}$

9. $\dfrac{x^6}{x^8}$

10. $\dfrac{y^4}{y}$

11. x^0

12. $7y^0$

13. $(-6z)^0$

14. 6^0

15. $(5x)^2$

16. $(3a)^3$

17. $(-3x)^3$

18. $(6s)^3$

19. $(2x^2)^4$

20. $(-x^4)^6$

21. $(-p^8)^4$

22. $\left(-\dfrac{2x^3}{y}\right)^2$

23. $\left(-\dfrac{5y^2}{2b}\right)^2$

24. $6x^2 \cdot 4x^3$

25. $\dfrac{16x^2y}{4xy^2}$

26. $2x(3xy^3)^3$

27. $\left(\dfrac{9x^2y}{3xy}\right)^2$

28. $(2x^2y)^3(3xy^4)$

29. $4x^2y^3(2x^3y^4)^2$

30. $3c^2(2c^4d^3)$

31. $\left(\dfrac{9a^3b^2}{3ab^7}\right)^3$

32. $\left(\dfrac{21x^4y^3}{7y^2}\right)^3$

[5.2] *Simplify.*

33. b^{-9}

34. 3^{-3}

35. 5^{-2}

36. $\dfrac{1}{z^{-2}}$

37. $\dfrac{1}{x^{-7}}$

38. $\dfrac{1}{4^{-2}}$

39. $y^5 \cdot y^{-8}$

40. $x^{-2} \cdot x^{-3}$

41. $p^{-6} \cdot p^4$

42. $a^{-2} \cdot a^{-3}$

43. $\dfrac{m^5}{m^{-5}}$

44. $\dfrac{x^5}{x^{-2}}$

45. $\dfrac{x^{-3}}{x^3}$

46. $(3x^4)^{-2}$

47. $(4x^{-3}y)^{-3}$

48. $(-2m^{-3}n)^2$

49. $6y^{-2} \cdot 2y^4$

50. $(-5y^{-3}z)^3$

51. $(-4x^{-2}y^3)^{-2}$

52. $2x(3x^{-2})$

53. $(5x^{-2}y)(2x^4y)$

54. $4y^{-2}(3x^2y)$

55. $4x^5(6x^{-7}y^2)$

56. $\dfrac{6xy^4}{2xy^{-1}}$

57. $\dfrac{12x^{-2}y^3}{3xy^2}$

58. $\dfrac{49x^2y^{-3}}{7x^{-3}y}$

59. $\dfrac{4x^8y^{-2}}{8x^7y^3}$

60. $\dfrac{36x^4y^7}{9x^5y^{-3}}$

[5.3] *Express each number in scientific notation.*

61. 1,720,000

62. 0.153

63. 0.00763

64. 47,000

65. 5760

66. 0.000314

Express each number without exponents.

67. 7.5×10^{-3}

68. 6.52×10^{-4}

69. 8.9×10^6

70. 5.12×10^4

71. 3.14×10^{-5}

72. 1.103×10^7

Write each of the following as a base unit without metric prefixes.

73. 92 milliliters

74. 6 gigameters

75. 12.8 micrograms

76. 19.2 kilograms

Perform each indicated operation and write your answer without exponents.

77. $(2.5 \times 10^2)(3.4 \times 10^{-4})$

78. $(4.2 \times 10^{-3})(3.0 \times 10^5)$

79. $(3.5 \times 10^{-2})(7.0 \times 10^3)$

80. $\dfrac{7.94 \times 10^6}{2.0 \times 10^{-2}}$

81. $\dfrac{1.5 \times 10^{-2}}{5.0 \times 10^2}$

82. $\dfrac{6.5 \times 10^4}{2.0 \times 10^6}$

Convert each number to scientific notation. Then calculate. Express your answer in scientific notation.

83. $(14,000)(260,000)$

84. $(0.00053)(40,000)$

85. $(12,500)(400,000)$

86. $\dfrac{250}{500,000}$

87. $\dfrac{0.000068}{0.02}$

88. $\dfrac{850,000}{0.025}$

89. Milk Tank A milk tank holds 6.4×10^6 fluid ounces of milk. If one gallon is 1.28×10^2 fluid ounces, determine the number of gallons of milk the tank holds.

90. Social Security In 2004 there was about $1.5 trillion in the social security trust fund.

   a) Write this amount without scientific notation.

   b) Using scientific notation, determine the annual amount of interest obtained in a year if the interest rate obtained is 7.5% per year. Write the answer in scientific notation.

[5.4] *Indicate whether each expression is a polynomial. If the polynomial is not written in descending order, rewrite it in descending order. If the polynomial has a specific name, give that name. State the degree of each polynomial.*

91. $x^{-4} - 8$

92. 7

93. $x^2 - 4 + 3x$

94. $-3 - x + 4x^2$

95. $4x^{1/2} - 6$

96. $13x^3 - 4$

97. $x - 4x^2$

98. $y^5 + y^{-3} - 9$

99. $2x^3 - 7 + 4x^2 - 3x$

[5.4–5.6] *Perform each indicated operation.*

100. $(x + 8) + (4x - 11)$

101. $(2d - 3) + (5d + 7)$

102. $(-x - 10) + (-2x + 5)$

103. $(-3x^2 + 9x + 5) + (-x^2 + 2x - 12)$

104. $(-m^2 + 5m - 8) + (6m^2 - 5m - 2)$

105. $(6.2p - 4.3) + (1.9p + 7.1)$

106. $(-6y - 7) - (-3y + 8)$

107. $(4x^2 - 9x) - (3x + 15)$

108. $(5a^2 - 6a - 9) - (2a^2 - a + 12)$

109. $(x^2 + 7x - 3) - (x^2 + 3x - 5)$

110. $(-2x^2 + 8x - 7) - (3x^2 + 12)$

111. $\dfrac{1}{7}x(21x + 21)$

112. $-3x(5x + 4)$

113. $3x(2x^2 - 4x + 7)$

114. $-c(2c^2 - 3c + 5)$

115. $-7b(-4b^2 - 3b - 5)$

116. $(x + 4)(x + 5)$

117. $(3x + 6)(-4x + 1)$

118. $(-5x + 3)^2$

119. $(6 - 2x)(2 + 3x)$

120. $(r + 5)(r - 5)$

121. $(x - 1)(3x^2 + 4x - 6)$

122. $(3x + 1)(x^2 + 2x + 4)$

123. $(-4x + 2)(3x^2 - x + 7)$

124. $\dfrac{2x + 4}{2}$

125. $\dfrac{12y + 18}{3}$

126. $\dfrac{8x^2 + 4x}{x}$

127. $\dfrac{6x^2 + 9x - 4}{3}$

128. $\dfrac{6w^2 - 5w + 3}{3w}$

129. $\dfrac{16x^6 - 8x^5 - 3x^3 + 1}{4x}$

130. $\dfrac{8m - 4}{-2}$

131. $\dfrac{5x^3 + 10x + 2}{2x^2}$

132. $\dfrac{5x^2 - 6x + 15}{3x}$

133. $\dfrac{x^2 + x - 12}{x - 3}$

134. $\dfrac{5x^2 + 28x - 10}{x + 6}$

135. $\dfrac{6n^2 + 19n + 3}{6n + 1}$

136. $\dfrac{4x^3 + 12x^2 + x - 12}{2x + 3}$

137. $\dfrac{4x^2 - 12x + 9}{2x - 3}$

Chapter 5 Practice Test

 To find out how well you understand the chapter material, take this practice test. The answers, and the section where the material was initially discussed, are given in the back of the book. Each problem is also fully worked out on the **Chapter Test Prep Video CD***. Review any questions that you answered incorrectly.*

Simplify each expression.

1. $5x^4 \cdot 3x^2$

2. $(3xy^2)^3$

3. $\dfrac{24p^7}{3p^2}$

4. $\left(\dfrac{3x^2y}{6xy^3}\right)^3$

5. $(2x^3y^{-2})^{-2}$

6. $(4x^0)(3x^2)^0$

7. $\dfrac{30x^6y^2}{45x^{-1}y}$

Convert each number to scientific notation and then determine the answer. Express your answer in scientific notation.

8. $(285{,}000)(50{,}000)$

9. $\dfrac{0.0008}{4000}$

Determine whether each expression is a polynomial. If the polynomial has a specific name, give that name.

10. $4x$

11. $-8c + 5$

12. $x^{-2} + 4$

13. Write the polynomial $-5 + 6x^3 - 2x^2 + 5x$ in descending order, and give its degree.

In Exercises 14–24, perform each indicated operation.

14. $(6x - 4) + (2x^2 - 5x - 3)$

15. $(y^2 - 7y + 3) - (4y^2 - 5y - 2)$

16. $(4x^2 - 5) - (x^2 + x - 8)$

17. $-5d(-3d + 8)$

18. $(5x + 8)(3x - 4)$

19. $(9 - 4c)(5 + 3c)$

20. $(3x - 5)(2x^2 + 4x - 5)$

21. $\dfrac{16x^2 + 8x - 4}{4}$

22. $\dfrac{-12x^2 - 6x + 5}{-3x}$

23. $\dfrac{8x^2 - 2x - 15}{2x - 3}$

24. $\dfrac{12x^2 + 7x - 12}{4x + 5}$

25. Half-Life The half-life of an element is the time it takes one half the amount of a radioactive element to decay. The half-life of carbon 14 (C^{14}) is 5730 years. The half-life of uranium 238 (U^{238}) is 4.46×10^9 years.

a) Write the half-life of C^{14} in scientific notation.

b) How many times longer is the half-life of U^{238} than C^{14}?

Cumulative Review Test

Take the following test and check your answers with those given in the back of the book. Review any questions that you answered incorrectly. The section where the material was covered is indicated after the answer.

1. Evaluate $12 + 8 \div 2^2 + 3$.

2. Simplify $7 - (2x - 3) + 2x - 8(1 - x)$.

3. Evaluate $-4x^2 + x - 7$ when $x = -2$.

4. Solve $\dfrac{5}{8} = \dfrac{5t}{6} + 2$

5. Solve $3x + 5 = 4(x - 2)$.

6. Solve the equation $3(x + 2) + 3x - 5 = 4x + 1$.

7. Solve the equation $3x - 2 = y - 7$ for y.

8. Find the slope of the line through the points $(1, 3)$ and $(5, 1)$.

9. Determine whether the following pair of lines is parallel.

$$3x - 5y = 7$$
$$5y + 3x = 2$$

10. Simplify $\left(\dfrac{5xy^{-3}}{x^{-2}y^5} \right)^2$.

11. Write the polynomial $-5x + 2 - 7x^2$ in descending order and give the degree.

Perform each indicated operation.

12. $(x^2 + 4x - 3) + (2x^2 + 5x + 1)$

13. $(6a^2 + 3a + 2) - (a^2 - 3a - 3)$

14. $(5t - 3)(2t - 1)$

15. $(2x - 1)(3x^2 - 5x + 2)$

16. $\dfrac{10d^2 + 12d - 8}{4d}$

17. $\dfrac{6x^2 + 11x - 10}{3x - 2}$

18. Chicken Soup At Art's Grocery Store, three cans of chicken soup sell for $1.25. Find the cost of eight cans.

19. Average Speed Bob Dolan drives from Jackson, Mississippi, to Tallulah, Louisiana, a distance of 60 miles. At the same time, Nick Reide starts driving from Tallulah to Jackson along the same route. If Bob and Nick meet after 0.5 hour and Nick's average speed was 7 miles per hour greater than Bob's, find the average speed of each car.

20. Rectangle The length of a rectangle is 2 less than 3 times the width. Find the dimensions of the rectangle if its perimeter is 28 feet.

Factoring

WE SEE RECTANGLES in our daily lives. Often quadratic equations are used in helping design rectangular structures. For example, in Exercise 21 on page 388, we use quadratic equations to determine the dimensions of a rectangular garden that is to have a specific area.

The major emphasis of this chapter is to teach you how to factor polynomials. Factoring polynomials is the reverse process of multiplying polynomials.

In the first five sections of this chapter, you will learn how to factor a monomial from a polynomial, factor by grouping, factor trinomials of the form $ax^2 + bx + c$ when $a = 1$ and $a \neq 1$, and factor by using special factoring formulas. In the last two sections of this chapter, you will learn how to solve quadratic equations using factoring and how to solve applications of quadratic equations.

It is essential that you have a thorough understanding of factoring, especially Sections 6.3 through 6.5, to complete Chapter 7 successfully.

6.1 Factoring a Monomial from a Polynomial

1 Identify factors.

2 Determine the greatest common factor of two or more numbers.

3 Determine the greatest common factor of two or more terms.

4 Factor a monomial from a polynomial.

1 Identify Factors

In Chapter 5, you learned how to multiply polynomials. In this chapter, we focus on factoring, the reverse process of multiplication. In Section 5.5, we showed that $3x(2x^2 + 4) = 6x^3 + 12x$. In this chapter, we start with an expression like $6x^3 + 12x$ and determine that its factors are $3x$ and $2x^2 + 4$, and write $6x^3 + 12x = 3x(2x^2 + 4)$. To **factor an expression** means to write the expression as a product of its factors. Factoring is important because it can be used to solve equations and perform operations on fractions.

If $a \cdot b = c$, then a and b are said to be *factors* of c.

$3 \cdot 5 = 15$; so 3 and 5 are factors of 15.

$x^3 \cdot x^4 = x^7$; so x^3 and x^4 are factors of x^7.

$x(x + 2) = x^2 + 2x$; so x and $x + 2$ are factors of $x^2 + 2x$.

$(x - 1)(x + 3) = x^2 + 2x - 3$; so $x - 1$ and $x + 3$ are factors of $x^2 + 2x - 3$.

A given number or expression may have many factors. Consider the number 30.

$$1 \cdot 30 = 30, \quad 2 \cdot 15 = 30, \quad 3 \cdot 10 = 30, \quad 5 \cdot 6 = 30$$

So, the positive factors of 30 are 1, 2, 3, 5, 6, 10, 15, and 30. Factors can also be negative. Since $(-1)(-30) = 30$, -1 and -30 are also factors of 30. In fact, for each factor a of an expression, $-a$ must also be a factor. Other factors of 30 are therefore $-1, -2, -3, -5, -6, -10, -15$, and -30. When asked to list the factors of an expression that contains a positive numerical coefficient with a variable, we generally list only positive factors.

EXAMPLE 1 ▶ List the factors of $6x^3$.

Solution

factors	factors
$1 \cdot 6x^3 = 6x^3$	$x \cdot 6x^2 = 6x^3$
$2 \cdot 3x^3 = 6x^3$	$2x \cdot 3x^2 = 6x^3$
$3 \cdot 2x^3 = 6x^3$	$3x \cdot 2x^2 = 6x^3$
$6 \cdot x^3 = 6x^3$	$6x \cdot x^2 = 6x^3$

The factors of $6x^3$ are $1, 2, 3, 6, x, 2x, 3x, 6x, x^2, 2x^2, 3x^2, 6x^2, x^3, 2x^3, 3x^3$, and $6x^3$. The opposite (or negative) of each of these factors is also a factor, but these opposites are generally not listed unless specifically asked for.

▶ **Now Try Exercise 7**

Here are examples of multiplying and factoring. Notice again that factoring is the reverse process of multiplying.

Multiplying	Factoring
$3(2x + 5) = 6x + 15$	$6x + 15 = 3(2x + 5)$
$4y(y - 7) = 4y^2 - 28y$	$4y^2 - 28y = 4y(y - 7)$
$(x + 1)(x + 3) = x^2 + 4x + 3$	$x^2 + 4x + 3 = (x + 1)(x + 3)$

2 Determine the Greatest Common Factor of Two or More Numbers

To factor a monomial from a polynomial, we make use of the *greatest common factor (GCF)*. If after studying the following material you wish to see additional material on obtaining the GCF, you may read Appendix B, where one of the topics discussed is finding the GCF.

Recall from Section 1.3 that the **greatest common factor** of two or more numbers is the greatest number that divides into all the numbers. The greatest common factor of the numbers 6 and 8 is 2. Two is the greatest number that divides into both 6 and 8. What is the GCF of 48 and 60? When the GCF of two or more numbers is not easily found, we can find it by writing each number as a product of prime numbers. A **prime number** is an integer greater than 1 that has exactly two factors, itself and one. The first 15 prime numbers are

$$2, 3, 5, 7, 11, 13, 17, 19, 23, 29, 31, 37, 41, 43, 47$$

A positive integer (other than 1) that is not prime is called **composite**. The number 1 is neither prime nor composite, it is called a **unit**. The first 15 composite numbers are

$$4, 6, 8, 9, 10, 12, 14, 15, 16, 18, 20, 21, 22, 24, 25$$

Every even number greater than 2 is a composite number since it has more than two factors, itself, 1, and 2.

To write a number as a product of prime numbers, follow the procedure illustrated in Examples 2 and 3.

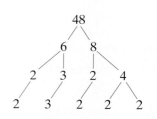

FIGURE 6.1

EXAMPLE 2 ▶ Write 48 as a product of prime numbers.

Solution   Select any two numbers whose product is 48. Two possibilities are $6 \cdot 8$ and $4 \cdot 12$, but there are other choices. Continue breaking down the factors until all the factors are prime, as illustrated in **Figure 6.1**

Note that no matter how you select your initial factors,

$$48 = 2 \cdot 2 \cdot 2 \cdot 2 \cdot 3 = 2^4 \cdot 3$$

▶ **Now Try Exercise 9**

In Example 2, we found that $48 = 2 \cdot 2 \cdot 2 \cdot 2 \cdot 3 = 2^4 \cdot 3$. The $2 \cdot 2 \cdot 2 \cdot 2 \cdot 3$ or $2^4 \cdot 3$ may also be referred to as **prime factorizations** of 48.

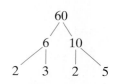

FIGURE 6.2

EXAMPLE 3 ▶ Write 60 as a product of its prime factors.

Solution   One way to find the prime factors is shown in **Figure 6.2**. Therefore, $60 = 2 \cdot 2 \cdot 3 \cdot 5 = 2^2 \cdot 3 \cdot 5$.

▶ **Now Try Exercise 11**

In the following box, we give the procedure to determine the greatest common factor of two or more numbers.

To Determine the GCF of Two or More Numbers

1. Write each number as a product of prime factors.
2. Determine the prime factors common to all the numbers.
3. Multiply the common factors found in step 2. The product of these factors is the GCF.

EXAMPLE 4 ▶ Determine the greatest common factor of 48 and 60.

Solution   From Examples 2 and 3, we know that

Step 1

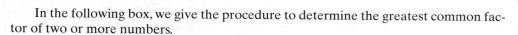

$$48 = 2 \cdot 2 \cdot 2 \cdot 2 \cdot 3 = 2^4 \cdot 3$$
$$60 = 2 \cdot 2 \cdot 3 \cdot 5 = 2^2 \cdot 3 \cdot 5$$

Step 2   The common factors are circled. Two factors of 2 and one factor of 3 are common to both numbers. The product of these factors is the GCF of 48 and 60:

Step 3 $$\text{GCF} = 2 \cdot 2 \cdot 3 = 12$$

The GCF of 48 and 60 is 12. Twelve is the greatest number that divides into both 48 and 60.

▶ **Now Try Exercise 15**

EXAMPLE 5 ▶ Determine the GCF of 18 and 24.

Solution

$$18 = 2 \cdot 3 \cdot 3 = 2 \cdot 3^2$$
$$24 = 2 \cdot 2 \cdot 2 \cdot 3 = 2^3 \cdot 3$$

One factor of 2 and one factor of 3 are common to both 18 and 24.

$$\text{GCF} = 2 \cdot 3 = 6$$

▶ **Now Try Exercise 19**

---

3   Determine the Greatest Common Factor of Two or More Terms

The GCF of several terms containing variables is easily found. Consider the terms $x^3, x^4, x^5,$ and x^6. The GCF of these terms is x^3, since x^3 is the largest number of x's common to all four terms. We can illustrate this by writing the terms in factored form, with x^3 as one factor.

$$x^3 = x^3 \cdot 1$$
$$x^4 = x^3 \cdot x$$
$$x^5 = x^3 \cdot x^2$$
$$x^6 = x^3 \cdot x^3$$

GCF of all four terms is x^3.

Notice that x^3 divides all four terms,

$$\frac{x^3}{x^3} = 1, \quad \text{and} \quad \frac{x^4}{x^3} = x, \quad \text{and} \quad \frac{x^5}{x^3} = x^2, \quad \text{and} \quad \frac{x^6}{x^3} = x^3.$$

EXAMPLE 6 ▶ Determine the GCF of the terms $m^9, m^5, m^7,$ and m^3.

Solution   The GCF is m^3 because m^3 is the largest number of m's common to all the terms.

▶ **Now Try Exercise 21**

EXAMPLE 7 ▶ Determine the GCF of the terms x^2y^3, x^3y^2 and xy^4.

Solution   The largest number of x's common to all three terms is x^1 or x. The largest number of y's common to all three terms is y^2. So the GCF of the three terms is xy^2.

▶ **Now Try Exercise 29**

To Determine the Greatest Common Factor of Two or More Terms

To determine the GCF of two or more terms, take each factor the *largest* number of times that it appears in all of the terms.

EXAMPLE 8 ▶ Determine the GCF of the terms $xy, x^2y^2,$ and x^3.

Solution   The GCF is x. The largest power of x that is common to all three terms is x^1, or x. Since the term x^3 does not contain a power of y, the GCF does not contain y.

▶ **Now Try Exercise 25**

EXAMPLE 9 ▶ Determine the GCF of each group of terms.

a) $18y^2, 15y^3, 27y^5$ **b)** $-20x^2, 12x, 40x^3$ **c)** $5s^4, s^7, s^3$

Solution

a) The GCF of 18, 15, and 27 is 3. The GCF of $y^2, y^3,$ and y^5 is y^2. Therefore, the GCF of the three terms is $3y^2$.

b) The GCF of -20, 12, and 40 is 4. The GCF of x^2, x, and x^3 is x. Therefore, the GCF of the three terms is $4x$.

c) The GCF of 5, 1, and 1 is 1. The GCF of s^4, s^7, and s^3 is s^3. Therefore, the GCF of the three terms is $1s^3$, which we write as s^3.

▶ **Now Try Exercise 33**

EXAMPLE 10 ▶ Determine the GCF of each pair of terms.

a) $a(a - 6)$ and $3(a - 6)$ **b)** $t(t + 4)$ and $t + 4$
c) $3(p + q)$ and $4p(p + q)$

Solution

a) The GCF is $(a - 6)$.

b) $t + 4$ can be written as $1(t + 4)$. Therefore, the GCF of $t(t + 4)$ and $1(t + 4)$ is $t + 4$.

c) The GCF is $(p + q)$.

▶ **Now Try Exercise 39**

4   Factor a Monomial from a Polynomial

In Section 5.5 we multiplied factors. Factoring is the reverse process of multiplying factors. As mentioned earlier, to *factor an expression* means to write the expression as a product of its factors.

> **To Factor a Monomial from a Polynomial**
>
> **1.** Determine the greatest common factor of all terms in the polynomial.
> **2.** Write each term as the product of the GCF and its other factor.
> **3.** Use the distributive property to factor out the GCF.

In step 3 of the process, we indicate that we use the distributive property. The distributive property is actually used in reverse. For example, if we have $4 \cdot x + 4 \cdot 2$, we use the distributive property in reverse to write $4(x + 2)$.

EXAMPLE 11 ▶ Factor $6x + 18$.

Solution   The GCF is 6.

$$6x + 18 = \boxed{6} \cdot x + \boxed{6} \cdot 3 \qquad \text{\textit{Write each term as a product of the GCF and its other factor.}}$$

$$= \boxed{6}(x + 3) \qquad \text{\textit{Distributive property}}$$

▶ **Now Try Exercise 49**

To check the factoring process, multiply the factors using the distributive property. If the factoring is correct, the product will be the polynomial you started with. Following is a check of the factoring in Example 11.

Check $6(x + 3) = 6x + 18$

EXAMPLE 12 ▶ Factor $15x - 20$.

Solution   The GCF is 5.

$$15x - 20 = \boxed{5} \cdot 3x - \boxed{5} \cdot 4$$
$$= \boxed{5}(3x - 4)$$

Check that the factoring is correct by multiplying.

▶ **Now Try Exercise 51**

EXAMPLE 13 ▶ Factor $6y^2 + 9y^5$.

Solution The GCF is $3y^2$.

$$6y^2 + 9y^5 = \boxed{3y^2} \cdot 2 + \boxed{3y^2} \cdot 3y^3$$
$$= \boxed{3y^2}\,(2 + 3y^3)$$

Check that the factoring is correct by multiplying.

▶ **Now Try Exercise 59**

EXAMPLE 14 ▶ Factor $8q^3 - 20q^2 - 12q$.

Solution The GCF is $4q$.

$$8q^3 - 20q^2 - 12q = \boxed{4q} \cdot 2q^2 - \boxed{4q} \cdot 5q - \boxed{4q} \cdot 3$$
$$= \boxed{4q}\,(2q^2 - 5q - 3)$$

Check $4q(2q^2 - 5q - 3) = 8q^3 - 20q^2 - 12q$

▶ **Now Try Exercise 81**

EXAMPLE 15 ▶ Factor $35x^2 - 25x + 5$.

Solution The GCF is 5.

$$35x^2 - 25x + 5 = \boxed{5} \cdot 7x^2 - \boxed{5} \cdot 5x + \boxed{5} \cdot 1$$
$$= \boxed{5}\,(7x^2 - 5x + 1)$$

Check that the factoring is correct by multiplying.

▶ **Now Try Exercise 85**

EXAMPLE 16 ▶ Factor $4x^3 + x^2 + 8x^2y$.

Solution The GCF is x^2.

$$4x^3 + x^2 + 8x^2y = \boxed{x^2} \cdot 4x + \boxed{x^2} \cdot 1 + \boxed{x^2} \cdot 8y$$
$$= \boxed{x^2}\,(4x + 1 + 8y)$$

Check that the factoring is correct by multiplying.

▶ **Now Try Exercise 89**

Notice in Examples 15 and 16 that when one of the terms is itself the GCF, we express it in factored form as the product of the term itself and 1.

EXAMPLE 17 ▶ Factor $x(5x - 2) + 7(5x - 2)$.

Solution The GCF of $x(5x - 2)$ and $7(5x - 2)$ is $(5x - 2)$. Factoring out the GCF gives

$$x(5x - 2) + 7(5x - 2) = (5x - 2)(x + 7)$$

Check that the factoring is correct by multiplying.

▶ **Now Try Exercise 95**

EXAMPLE 18 ▶ Factor $4x(3x - 5) - 7(3x - 5)$.

Solution The GCF of $4x(3x - 5)$ and $-7(3x - 5)$ is $(3x - 5)$. Factoring out the GCF gives

$$4x(3x - 5) - 7(3x - 5) = (3x - 5)(4x - 7)$$

Check that the factoring is correct by multiplying.

Recall from Section 1.10 that the commutative property of multiplication states that the order in which any two real numbers are multiplied does not matter. Therefore, $(3x - 5)(4x - 7)$ can also be written $(4x - 7)(3x - 5)$. In the book, we will place the common factor on the left.

▸ **Now Try Exercise 97**

EXAMPLE 19 ▸ Factor $2x(x + 3) - 5(x + 3)$.

Solution   The GCF of $2x(x + 3)$ and $-5(x + 3)$ is $(x + 3)$. Factoring out the GCF gives

$$2x(x + 3) - 5(x + 3) = (x + 3)(2x - 5)$$

Check that the factoring is correct by multiplying.

▸ **Now Try Exercise 93**

Important: Whenever you are factoring a polynomial by any of the methods presented in this chapter, the first step will always be to see if there is a common factor (other than 1) to all the terms in the polynomial. If so, factor the greatest common factor from each term using the distributive property.

Helpful Hint

Checking a Factoring Problem

Every factoring problem may be checked by multiplying the factors. The product of the factors should be identical to the expression that was originally factored. You should check all factoring problems.

EXERCISE SET 6.1

Math XL MathXL®   *MyMathLab* MyMathLab

Concept/Writing Exercises

1. What does it mean to factor an expression?

2. What is a prime number?

3. What is a composite number?

4. Is the number 1 a prime number? If not, what is it called?

5. What is the greatest common factor of two or more numbers?

6. Explain how to factor a monomial from a polynomial.

7. List the factors of $4x^2$.

8. How may any factoring problem be checked?

Practice the Skills

Write each number as a product of prime numbers.

9. 56   **10.** 120   **11.** 90   **12.** 540   **13.** 248   **14.** 144

Determine the greatest common factor for each pair of numbers.

15. 20, 24   **16.** 45, 27   **17.** 70, 98   **18.** 120, 96   **19.** 80, 126   **20.** 88, 160

Determine the greatest common factor for each group of terms.

21. x^5, x, x^2

22. y^3, y^5, y^2

23. $3x, 6x^2, 9x^3$

24. $6p, 4p^2, 8p^3$

25. a, ab, ab^2

26. x, y, z

27. q^3r, q^2r^2, qr^4

28. $4x^2y^2, 3xy^4, 2xy^2$

29. $x^3y^7, x^7y^{12}, x^5y^5$

30. $6x, 12y, 18x^2$

31. $-3, 20x, 30x^2$

32. $24s^5, 6r^3s^3, 15rs^2$

33. $9x^3y^4, 8x^2y^4, 12x^4y^2$

34. $16x^9y^{12}, 8x^5y^3, 20x^4y^2$

35. $40x^3, 27x, 30x^4y^2$

36. $6p^4q^3, 9p^2q^5, 9p^4q^2$

37. $8(x - 4), 7(x - 4)$

38. $4(x - 5), 3x(x - 5)$

39. $x^2(2x - 3), 5(2x - 3)$

40. $x(9x - 3), 9x - 3$

41. $3w + 5, 6(3w + 5)$

42. $b(b + 3), b + 3$

43. $x - 4, y(x - 4)$

44. $3y(x + 2), 3(x + 2)$

45. $3(x - 1), 5(x - 1)^2$

46. $5(n + 2), 7(n + 2)^2$

47. $(x - 9)(x + 6), (x - 9)(x + 3)$

48. $(a + 4)(a - 3), 5(a - 3)$

Factor the GCF from each term in the expression.

49. $4x - 8$

50. $4x + 2$

51. $15x - 5$

52. $12x + 15$

53. $7q + 28$

54. $3t^2 - 10t$

55. $9x^2 - 12x$

56. $24y - 6y^2$

57. $7x^5 - 9x^4$

58. $9x + 27x^3$

59. $3x^5 - 12x^2$

60. $26p^2 - 8p$

61. $36x^{12} + 24x^8$

62. $45y^{12} + 30y^{10}$

63. $27y^{15} - 9y^3$

64. $30w^5 + 25w^3$

65. $y + 6x^3y$

66. $4x^2y - 6x$

67. $7a^4 + 3a^2$

68. $3x^2y + 6x^2y^2$

69. $16xy^2z + 4x^3y$

70. $48m^4n^2 - 16mn^2$

71. $80x^5y^3z^4 - 36x^2yz^3$

72. $56xy^5z^{13} - 24y^4z^2$

73. $25x^2yz^3 + 25x^3yz$

74. $13y^5z^3 - 11xy^2z^5$

75. $19x^4y^{12}z^{13} - 8x^5y^3z^9$

76. $16r^4s^5t^3 - 20r^5s^4t$

77. $8c^2 - 4c - 32$

78. $x^3 - 4x^2 - 3x$

79. $9x^2 + 18x + 3$

80. $4x^2 + 8x + 24$

81. $4x^3 - 8x^2 + 12x$

82. $12a^3 - 16a^2 - 4a$

83. $40b^2 - 48c + 24$

84. $5x^3 - xy^2 + x$

85. $15p^2 - 6p + 9$

86. $45y^3 - 63y^2 + 27y$

87. $9a^4 - 6a^3 + 3ab$

88. $45v^4w^2 + 10v^2x - 20vw^5$

89. $8x^2y + 12xy^2 + 5xy$

90. $52x^2y^2 + 16xy^3 + 26z$

91. $x(x - 7) + 6(x - 7)$

92. $9x(3x - 4) - 4(3x - 4)$

93. $3b(a - 2) - 4(a - 2)$

94. $3x(7x + 1) - 2(7x + 1)$

95. $4x(2x + 1) + 1(2x + 1)$

96. $4m(5m - 1) - 3(5m - 1)$

97. $5x(2x + 1) + 2x + 1$

98. $3x(4x - 5) + 4x - 5$

99. $3c(6c + 7) - 2(6c + 7)$

100. $5t(t - 2) - 3(t - 2)$

Problem Solving

Factor each expression, if possible. Treat the unknown symbol as if it were a variable.

101. $12\nabla - 6\nabla^2$

102. $3\star + 6$

103. $12\square^3 - 4\square^2 + 4\square$

104. $\copyright + 11\Delta$

Challenge Problems

105. Factor $6x^5(2x + 7) + 4x^3(2x + 7) - 2x^2(2x + 7)$.

106. Factor $4x^2(x - 3)^3 - 6x(x - 3)^2 + 4(x - 3)$.

107. Factor $x^2 + 2x + 3x + 6$. (*Hint:* Factor the first two terms, then factor the last two terms, then factor the resulting two terms. We will discuss factoring problems of this type in Section 6.2.)

Cumulative Review Exercises

[2.1] **108.** Simplify $2x - (x - 5) + 4(3 - x)$.

[2.5] **109.** Solve the equation $4 + 3(x - 8) = x - 4(x + 2)$.

[2.6] **110.** Solve the equation $4x - 5y = 20$ for y.

111. Find the volume of the cone shown below.

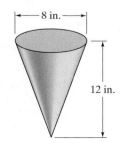

[3.2] **112.** The sum of two numbers is 41. Find the two numbers if the larger number is one less than twice the smaller number.

[5.1] **113.** Simplify $\left(\dfrac{3x^2y^3}{2x^5y^2}\right)^2$.

6.2 Factoring by Grouping

1 Factor a polynomial
containing four terms by
grouping.

1 Factor a Polynomial Containing Four Terms by Grouping

It may be possible to factor a polynomial containing four or more terms by removing common factors from groups of terms. This process is called **factoring by grouping**. In Sections 6.3 and 6.4, we discuss factoring trinomials. One of the methods we will use in Section 6.4 requires a knowledge of factoring by grouping. Example 1 illustrates the procedure for factoring by grouping.

EXAMPLE 1 ▸ Factor $ax + ay + bx + by$ by grouping.

Solution   There is no factor (other than 1) common to all four terms. However, a is common to the first two terms and b is common to the last two terms. Factor a from the first two terms and b from the last two terms.

$$ax + ay + bx + by = a(x + y) + b(x + y)$$

This factoring gives two terms, and $(x + y)$ is common to both terms. Proceed to factor $(x + y)$ from each term, as shown below.

$$a(x + y) + b(x + y) = (x + y)(a + b)$$

Notice that when $(x + y)$ is factored out we are left with $a + b$, which becomes the other factor. Thus, $ax + ay + bx + by = (x + y)(a + b)$.
▸ **Now Try Exercise 7**

To Factor a Four-Term Polynomial Using Grouping

1. Determine whether there are any factors common to all four terms. If so, factor the greatest common factor from each of the four terms.

2. If necessary, arrange the four terms so that the first two terms have a common factor and the last two have a common factor.

3. Use the distributive property to factor each group of two terms.

4. Factor the greatest common factor from the results of step 3.

EXAMPLE 2 ▸ Factor $x^2 + 3x + 4x + 12$ by grouping.

Solution   No factor is common to all four terms. However, you can factor x from the first two terms and 4 from the last two terms.

$$x^2 + 3x + 4x + 12 = x(x + 3) + 4(x + 3)$$

Notice that the expression on the right of the equal sign has two *terms* and that the *factor* $(x + 3)$ is common to both terms. Factor out the $(x + 3)$ using the distributive property.

$$x(x + 3) + 4(x + 3) = (x + 3)(x + 4)$$

Thus, $x^2 + 3x + 4x + 12 = (x + 3)(x + 4)$
▸ **Now Try Exercise 11**

In Example 2, the $3x$ and $4x$ are like terms and may be combined. However, if we were to combine them we would not be able to factor the four terms by grouping. Some four-term polynomials, such as in Example 9, have no like terms that can be combined.

EXAMPLE 3 ▶ Factor $15x^2 + 10x + 12x + 8$ by grouping.

Solution

$15x^2 + 10x + 12x + 8 = 5x(3x + 2) + 4(3x + 2)$ *Factor 5x from the first two terms and 4 from the last two terms.*

$$= (3x + 2)(5x + 4)$$

▶ **Now Try Exercise 17**

A factoring by grouping problem can be checked by multiplying the factors using the FOIL method. If you have not made a mistake, your result will be the polynomial you began with. Here is a check of Example 3.

Check

$$\begin{array}{cccc} F & O & I & L \end{array}$$

$$(3x + 2)(5x + 4) = (3x)(5x) + (3x)(4) + (2)(5x) + (2)(4)$$
$$= 15x^2 + 12x + 10x + 8$$
$$= 15x^2 + 10x + 12x + 8$$

We are able to write $12x + 10x$ as $10x + 12x$ because of the commutative property of addition. Since this is the polynomial we started with, the factoring is correct.

Helpful Hint

In Example 3, when we factored $15x^2 + 10x + 12x + 8$ we obtained

$$5x(3x + 2) + 4(3x + 2)$$

When we factored out the $(3x + 2)$, to be consistent with the way we factored out common factors in Section 6.1, we placed the common factor $(3x + 2)$ on the left. That gave $(3x + 2)(5x + 4)$. We could have just as well placed the common factor on the right to obtain $(5x + 4)(3x + 2)$.

Both answers are correct since $(3x + 2)(5x + 4) = (5x + 4)(3x + 2)$ by the commutative property of multiplication.

EXAMPLE 4 ▶ Factor $15x^2 + 12x + 10x + 8$ by grouping.

Solution $15x^2 + 12x + 10x + 8 = 3x(5x + 4) + 2(5x + 4)$
$$= (5x + 4)(3x + 2)$$

▶ **Now Try Exercise 19**

Notice that Example 4 is the same as Example 3 with the two middle terms interchanged. The answers to Examples 3 and 4 are equivalent since only the order of the factors are changed. When factoring by grouping, if the two middle terms are like terms, the two like terms may be interchanged and the answer will remain the same.

EXAMPLE 5 ▶ Factor $x^2 - 3x + x - 3$ by grouping.

Solution   In the first two terms, x is the common factor. Is there a common factor in the last two terms? Yes; remember that 1 is a factor of every term. Factor 1 from the last two terms.

$$x^2 - 3x + x - 3 = x^2 - 3x + 1 \cdot x - 1 \cdot 3$$
$$= x(x - 3) + 1(x - 3)$$
$$= (x - 3)(x + 1)$$

Note that $x - 3$ was expressed as $1 \cdot x - 1 \cdot 3 = 1(x - 3)$.

▶ **Now Try Exercise 21**

EXAMPLE 6 ▶ Factor $6x^2 - 3x - 2x + 1$ by grouping.

Solution   When $3x$ is factored from the first two terms, we get

$$6x^2 - 3x - 2x + 1 = 3x(2x - 1) - 2x + 1$$

What should we factor from the last two terms? We wish to factor $-2x + 1$ in such a manner that we end up with an expression that is a multiple of $(2x - 1)$.

Whenever we wish to change the sign *of each term of an expression, we can factor out a negative number from each term.* In this case, we factor out -1.

$$-2x + 1 = -1(2x - 1)$$

Now, we rewrite $-2x + 1$ as $-1(2x - 1)$.

$$3x(2x - 1)\ -2x + 1 = 3x(2x - 1)\ -1(2x - 1)$$

Now we factor out the common factor $(2x - 1)$.

$$3x(2x - 1) - 1(2x - 1) = (2x - 1)(3x - 1)$$

▶ **Now Try Exercise 23**

EXAMPLE 7 ▶ Factor $q^2 + 3q - q - 3$ by grouping.

Solution

$$\begin{aligned} q^2 + 3q - q - 3 &= q(q + 3) - q - 3 && \text{\textit{Factor out q.}} \\ &= q(q + 3) - 1(q + 3) && \text{\textit{Factor out} -1.} \\ &= (q + 3)(q - 1) && \text{\textit{Factor out} $(q + 3)$.} \end{aligned}$$

Note that we factored -1 from $-q - 3$ to get $-1(q + 3)$.

▶ **Now Try Exercise 25**

EXAMPLE 8 ▶ Factor $3x^2 - 6x - 4x + 8$ by grouping.

Solution

$$\begin{aligned} 3x^2 - 6x - 4x + 8 &= 3x(x - 2) - 4(x - 2) \\ &= (x - 2)(3x - 4) \end{aligned}$$

Note: $-4x + 8 = -4(x - 2)$.

▶ **Now Try Exercise 27**

Helpful Hint

When factoring four terms by grouping, if the coefficient of the third term is positive, as in Examples 2 through 5, you will generally factor out a positive coefficient from the last two terms. *If the coefficient of the third term is negative*, as in Examples 6 through 8, *you will generally factor out a negative coefficient from the last two terms.* The sign of the coefficient of the third term in the expression *must be included* so that the factoring results in two terms. For example,

$$2x^2 + 8x\ +\ 3x + 12 = 2x(x + 4)\ +\ 3(x + 4) = (x + 4)(2x + 3)$$
$$3x^2 - 15x\ -\ 2x + 10 = 3x(x - 5)\ -\ 2(x - 5) = (x - 5)(3x - 2)$$

In the examples illustrated so far, the two middle terms have been like terms. This need not be the case, as illustrated in Example 9.

EXAMPLE 9 ▶ Factor $xy + 3x - 2y - 6$ by grouping.

Solution   This problem contains two variables, x and y. The procedure to factor here is basically the same as before. Factor x from the first two terms and -2 from the last two terms.

$$\begin{aligned} xy + 3x - 2y - 6 &= x(y + 3) - 2(y + 3) \\ &= (y + 3)(x - 2) && \text{\textit{Factor out} $(y + 3)$.} \end{aligned}$$

▶ **Now Try Exercise 41**

EXAMPLE 10 ▶ Factor $2x^2 + 4xy + 3xy + 6y^2$.

Solution   We will factor out $2x$ from the first two terms and $3y$ from the last two terms.

$$2x^2 + 4xy + 3xy + 6y^2 = 2x(x + 2y) + 3y(x + 2y)$$

Now we factor out the common factor $(x + 2y)$ from each term on the right.

$$2x(x + 2y) + 3y(x + 2y) = (x + 2y)(2x + 3y)$$

Check   F   O   I   L

$$\begin{aligned}
(x + 2y)(2x + 3y) &= (x)(2x) + (x)(3y) + (2y)(2x) + (2y)(3y) \\
&= 2x^2 + 3xy + 4xy + 6y^2 \\
&= 2x^2 + 4xy + 3xy + 6y^2
\end{aligned}$$

▶ **Now Try Exercise 29**

If Example 10 were given as $2x^2 + 3xy + 4xy + 6y^2$, would the results be the same? Try it and see.

EXAMPLE 11 ▶ Factor $15a^2 - 10ab + 12ab - 8b^2$.

Solution   Factor $5a$ from the first two terms and $4b$ from the last two terms.

$$\begin{aligned}
15a^2 - 10ab + 12ab - 8b^2 &= 5a(3a - 2b) + 4b(3a - 2b) \\
&= (3a - 2b)(5a + 4b)
\end{aligned}$$

▶ **Now Try Exercise 31**

EXAMPLE 12 ▶ Factor $3x^2 - 15x + 6x - 30$.

Solution   *The first step in any factoring problem is to determine whether all the terms have a common factor. If so, we factor out that common factor.* In this polynomial, 3 is common to every term. Therefore, we begin by factoring out the 3.

$$3x^2 - 15x + 6x - 30 = 3(x^2 - 5x + 2x - 10)$$

Now we factor the expression in parentheses by grouping. We factor out x from the first two terms and 2 from the last two terms.

$$\begin{aligned}
3(x^2 - 5x + 2x - 10) &= 3[x(x - 5) + 2(x - 5)] \\
&= 3[(x - 5)(x + 2)] \\
&= 3(x - 5)(x + 2)
\end{aligned}$$

Thus, $3x^2 - 15x + 6x - 30 = 3(x - 5)(x + 2)$.

▶ **Now Try Exercise 49**

EXERCISE SET 6.2   *Math* XP   *MyMathLab*
                      MathXL®   MyMathLab

Concept/Writing Exercises

✎ **1.** What is the first step in any factoring by grouping problem?

✎ **2.** How can you check the answer to a factoring by grouping problem?

✎ **3.** A polynomial of four terms is factored by grouping and the result is $(x - 2y)(x - 3)$. Find a polynomial that was factored, and explain how you determined the answer.

✎ **4.** A polynomial of four terms is factored by grouping and the result is $(x - 2)(x + 4)$. Find a polynomial that was factored, and explain how you determined the answer.

5. What number when factored from each term in an expression changes the sign of each term in the original expression?

✎ **6.** Describe the steps you take to factor a polynomial of four terms by grouping.

Practice the Skills

Factor by grouping.

7. $x^2 + 3x + 2x + 6$

8. $x^2 + 7x + 3x + 21$

9. $x^2 + 5x + 4x + 20$

10. $x^2 - x + 3x - 3$

11. $x^2 + 2x + 5x + 10$

12. $x^2 - 6x + 5x - 30$

13. $c^2 - 4c + 7c - 28$

14. $r^2 - 4r + 6r - 24$

15. $4x^2 - 6x + 6x - 9$

16. $4b^2 - 10b + 10b - 25$

17. $3x^2 + 9x + x + 3$

18. $a^2 + a + 3a + 3$

19. $6x^2 + 3x - 2x - 1$

20. $5x^2 + 30x - 3x - 18$

21. $8x^2 + 32x + x + 4$

22. $9w^2 - 6w - 6w + 4$

23. $12t^2 - 8t - 3t + 2$

24. $12x^2 + 42x - 10x - 35$

25. $x^2 + 9x - x - 9$

26. $35x^2 - 40x + 21x - 24$

27. $6p^2 + 15p - 4p - 10$

28. $10c^2 + 25c - 6c - 15$

29. $x^2 + 2xy - 3xy - 6y^2$

30. $x^2 - 3xy + 4xy - 12y^2$

31. $3x^2 + 2xy - 9xy - 6y^2$

32. $3x^2 - 18xy + 4xy - 24y^2$

33. $10x^2 - 12xy - 25xy + 30y^2$

34. $6a^2 - 3ab + 4ab - 2b^2$

35. $x^2 - bx - ax + ab$

36. $x^2 + bx + ax + ab$

37. $xy + 9x - 5y - 45$

38. $x^2 - 2x + ax - 2a$

39. $a^2 + 3a + ab + 3b$

40. $3x^2 - 15x - 2xy + 10y$

41. $xy - x + 5y - 5$

42. $y^2 - yb + ya - ab$

43. $12 + 8y - 3x - 2xy$

44. $7y - 49 - xy + 7x$

45. $z^3 + 5z^2 + z + 5$

46. $x^3 - 3x^2 + 2x - 6$

47. $x^3 - 5x^2 + 8x - 40$

48. $y^3 - 3y + 2y^2 - 6$

49. $2x^2 - 12x + 8x - 48$

50. $3x^2 - 3x - 3x + 3$

51. $4x^2 + 8x + 8x + 16$

52. $3z^4 - 3z^3 - 7z^3 + 7z^2$

53. $6x^3 + 9x^2 - 2x^2 - 3x$

54. $9x^3 + 6x^2 - 45x^2 - 30x$

55. $p^3 - 6p^2q + 2p^2q - 12pq^2$

56. $18x^2 + 27xy + 12xy + 18y^2$

Rearrange the terms so that the first two terms have a common factor and the last two terms have a common factor (other than 1). Then factor by grouping. There may be more than one way to arrange the factors. However, the answer should be equivalent regardless of the arrangement selected.

57. $5x + 3y + xy + 15$

58. $5m + 2w + mw + 10$

59. $6x + 5y + xy + 30$

60. $ax - 10 - 5x + 2a$

61. $ax + by + ay + bx$

62. $ax - 21 - 3a + 7x$

63. $rs - 42 + 6s - 7r$

64. $ca - 2b + 2a - cb$

65. $dc + 3c - ad - 3a$

66. $ac - bd - ad + bc$

Problem Solving

67. If you know that a polynomial with four terms is factorable by a specific arrangement of the terms, then will *any* arrangement of the terms be factorable by grouping? Explain, and support your answer with an example.

Factor each expression, if possible. Treat the unknown symbol as if it were a variable.

68. $\heartsuit^2 + 3\heartsuit + 4\heartsuit + 12$

69. $\odot^2 + 3\odot - 5\odot - 15$

70. $\Delta^2 + 2\Delta - \Delta + 6$

Challenge Problems

*In Section 6.4, we will factor trinomials of the form $ax^2 + bx + c, a \neq 1$ using grouping. To do this we rewrite the middle term of the trinomial, bx, as a sum or difference of two terms. Then we factor the resulting polynomial of four terms by grouping. For Exercises 71–76, **a)** rewrite the trinomial as a polynomial of four terms by replacing the bx-term with the sum or difference given. **b)** Factor the polynomial of four terms. Note that the factors obtained are the factors of the trinomial.*

71. $2x^2 - 11x + 15, -11x = -5x - 6x$

72. $3x^2 + 10x + 8, 10x = 4x + 6x$

73. $2x^2 - 11x + 15, -11x = -6x - 5x$

74. $3x^2 + 10x + 8, 10x = 6x + 4x$

75. $4x^2 - 17x - 15, -17x = 3x - 20x$

76. $4x^2 - 17x - 15, -17x = -20x + 3x$

Factor each expression, if possible. Treat the unknown symbols as if they were variables.

77. $\star\odot + 3\star + 2\odot + 6$

78. $2\Delta^2 - 4\Delta\star - 8\Delta\star + 16\star^2$

Cumulative Review Exercises

[2.5] **79.** Solve $5 - 3(2x - 7) = 4(x + 5) - 6$.

[3.4] **80. Special Mixture** Ed and Beatrice Petrie own a small grocery store near Leesport, Pennsylvania. The store carries a variety of bulk candy. To celebrate the tenth anniversary of the store's opening, the Petries decide to create a special candy mixture containing jelly beans and gumdrops. The jelly beans sell for $6.25 per pound and the gumdrops sell for $2.50 per pound. How many pounds of each type of candy will be needed to make a 50-pound mixture that will sell for $4.75 per pound?

[5.6] **81.** Divide $\dfrac{15x^3 - 6x^2 - 9x + 5}{3x}$.

82. Divide $\dfrac{a^2 - 16}{a + 4}$.

See Exercise 80.

6.3 Factoring Trinomials of the Form $ax^2 + bx + c, a = 1$

1 Factor trinomials of the form $ax^2 + bx + c$, where $a = 1$.

2 Remove a common factor from a trinomial.

An Important Note Regarding Factoring Trinomials

Factoring trinomials is important in algebra, higher-level mathematics, physics, and other science courses. Because it is important, and also to be successful in Chapter 7, you should study and learn Sections 6.3 and 6.4 well.

In this section, we learn to factor trinomials of the form $ax^2 + bx + c$, where a, the numerical coefficient of the squared term, is 1. That is, we will be factoring trinomials of the form $x^2 + bx + c$. One example of this type of trinomial is $x^2 + 5x + 6$. Recall that x^2 means $1x^2$.

In Section 6.4, we will learn to factor trinomials of the form $ax^2 + bx + c$, where $a \neq 1$. One example of this type of trinomial is $2x^2 + 7x + 3$.

1 Factor Trinomials of the Form $ax^2 + bx + c$, where $a = 1$

Now we discuss how to factor trinomials of the form $ax^2 + bx + c$, where a, the numerical coefficient of the squared term, is 1. Examples of such trinomials are

$$x^2 + 7x + 12 \qquad\qquad x^2 - 2x - 24$$
$$a = 1, b = 7, c = 12 \qquad\qquad a = 1, b = -2, c = -24$$

Recall that factoring is the reverse process of multiplication. We can show with the FOIL method of multiplying binomials that

$$(x + 3)(x + 4) = x^2 + 7x + 12 \quad \text{and} \quad (x - 6)(x + 4) = x^2 - 2x - 24$$

Therefore, $x^2 + 7x + 12$ and $x^2 - 2x - 24$ factor as follows:

$$x^2 + 7x + 12 = (x + 3)(x + 4) \quad \text{and} \quad x^2 - 2x - 24 = (x - 6)(x + 4)$$

Notice that each of these trinomials when factored results in the product of two binomials in which the first term of each binomial is x and the second term is a number (including its sign). In general, when we factor a trinomial of the form $x^2 + bx + c$ we will get a pair of binomial factors as follows:

$$x^2 + bx + c = (x + \blacksquare)(x + \blacksquare)$$

Numbers go here.

If, for example, we find that the numbers that go in the shaded areas of the factors are 4 and -6, the factors are written $(x + 4)$ and $(x - 6)$. Notice that instead of listing the second factor as $(x + (-6))$, we list it as $(x - 6)$.

To determine the numbers to place in the shaded areas when factoring a trinomial of the form $x^2 + bx + c$, write down factors of the form $(x + \blacksquare)(x + \blacksquare)$ and then try different sets of factors of the constant, c, in the shaded areas of the parentheses. We multiply each pair of factors using the FOIL method, and continue until we find the pair whose sum of the products of the outer and inner terms is the same as the x-term in the trinomial. For example, to factor the trinomial $x^2 + 7x + 12$ we determine the possible factors of 12. Then we try each pair of factors until we obtain a pair whose product from the FOIL method contains $7x$, the same x-term as in the trinomial. This method for factoring is called **trial and error**. In Example 1, we factor $x^2 + 7x + 12$ by trial and error.

EXAMPLE 1 ▶ Factor $x^2 + 7x + 12$ by trial and error.

Solution Begin by listing the factors of 12 (see the left-hand column of the chart below.) Then list the possible factors of the trinomial, and the products of these factors. Finally, determine which, if any, of these products gives the correct middle term, $7x$.

Factors of 12	Possible Factors of Trinomial	Product of Factors
$(1)(12)$	$(x + 1)(x + 12)$	$x^2 + 13x + 12$
$(2)(6)$	$(x + 2)(x + 6)$	$x^2 + 8x + 12$
$(3)(4)$	$(x + 3)(x + 4)$	$x^2 + 7x + 12$
$(-1)(-12)$	$(x - 1)(x - 12)$	$x^2 - 13x + 12$
$(-2)(-6)$	$(x - 2)(x - 6)$	$x^2 - 8x + 12$
$(-3)(-4)$	$(x - 3)(x - 4)$	$x^2 - 7x + 12$

In the last column, we find the trinomial we are seeking in the third line. Thus,

$$x^2 + 7x + 12 = (x + 3)(x + 4)$$

▶ **Now Try Exercise 17**

Now let's consider how we may more easily determine the correct factors of 12 to place in the shaded areas when factoring the trinomial in Example 1. In Section 5.5, we illustrated how the FOIL method is used to multiply two binomials. Let's multiply $(x + 3)(x + 4)$ using the FOIL method.

$$(x + 3)(x + 4) = x^2 + 4x + 3x + 12$$
$$= x^2 + 7x + 12$$

We see that $(x + 3)(x + 4) = x^2 + 7x + 12$.

Note that the *sum of the outer and inner terms is 7x and the product of the last terms is 12.* To factor $x^2 + 7x + 12$, we look for two numbers whose product is 12 and whose sum is 7. We list the factors of 12 first and then list the sum of the factors.

Factors of 12	Sum of Factors
$(1)(12) = 12$	$1 + 12 = 13$
$(2)(6) = 12$	$2 + 6 = 8$
$(3)(4) = 12$	$3 + 4 = 7$
$(-1)(-12) = 12$	$-1 + (-12) = -13$
$(-2)(-6) = 12$	$-2 + (-6) = -8$
$(-3)(-4) = 12$	$-3 + (-4) = -7$

The only factors of 12 whose sum is a positive 7 are 3 and 4. The factors of $x^2 + 7x + 12$ will therefore be $(x + 3)$ and $(x + 4)$.

$$x^2 + 7x + 12 = (x + 3)(x + 4)$$

In the previous illustration, all the possible factors of 12 were listed so that you could see them. However, when working a problem, once you find the specific factors you are seeking you need go no further.

To Factor Trinomials of the Form $ax^2 + bx + c$, where $a = 1$

1. Find two numbers whose product equals the constant, c, and whose sum equals the coefficient of the x-term, b.

2. Use the two numbers found in step 1, including their signs, to write the trinomial in factored form. The trinomial in factored form will be

$$(x + \text{one number})(x + \text{second number})$$

How do we find the two numbers mentioned in steps 1 and 2? The sign of the constant, c, is a key in finding the two numbers. *The Helpful Hint that follows is very important and useful. Study it carefully.*

Helpful Hint

When asked to factor a trinomial of the form $x^2 + bx + c$, first observe the sign of the constant.

a) If the constant, c, is positive, both numbers in the factors will have the same sign, either both positive or both negative. Furthermore, that common sign will be the same as the sign of the coefficient of the x-term of the trinomial being factored. That is, if b is positive, both factors will contain positive numbers, and if b is negative, both factors will contain negative numbers.
   Example:

$$x^2 + 7x + 12 = (x + 3)(x + 4)$$

Both factors have positive numbers.

The coefficient, b, is positive   *The constant, c, is positive*   *positive*   *positive*

(continued on the next page)

Example:

$$x^2 - 5x + 6 = (x - 2)(x - 3)$$

Both factors have negative numbers.

The coefficient, b, is negative   The constant, c, is positive   negative   negative

b) If the constant is negative, the two numbers in the factors will have opposite signs. That is, one number will be positive and the other number will be negative.

Example:

$$x^2 + x - 6 = (x + 3)(x - 2)$$

One factor has a positive number and the other factor has a negative number.

The coefficient, b, is positive   The constant, c, is negative   positive   negative

Example:

$$x^2 - 3x - 10 = (x + 2)(x - 5)$$

One factor has a positive number and the other factor has a negative number.

The coefficient, b, is negative   The constant, c, is negative   positive   negative

We will use this information as a starting point when factoring trinomials.

EXAMPLE 2 ▶ Consider a trinomial of the form $x^2 + bx + c$. Use the signs of b and c given below to determine the signs of the numbers in the factors.

a) b is negative and c is positive   **b)** b is negative and c is negative

c) b is positive and c is negative   **d)** b is positive and c is positive

Solution   In each case we look at the sign of the constant, c, first.

a) Since the constant, c, is positive, both numbers must have the same sign. Since the coefficient of the x-term, b, is negative, both factors will contain negative numbers.

b) Since the constant, c, is negative, one factor will contain a positive number and the other will contain a negative number.

c) Since the constant, c, is negative, one factor will contain a positive number and the other will contain a negative number.

d) Since the constant, c, is positive, both numbers must have the same sign. Since the coefficient of the x-term, b, is positive, both factors will contain positive numbers.

▶ **Now Try Exercise 5**

EXAMPLE 3 ▶ Factor $x^2 + x - 6$.

Solution   We must find two numbers whose product is the constant, -6, and whose sum is the coefficient of the x-term, 1. Remember that x means $1x$. Since the constant is negative, one number must be positive and the other negative. Recall that the product of two numbers with unlike signs is a negative number. We now list the factors of -6 and look for the two factors whose sum is 1.

Factors of -6	Sum of Factors
$1(-6) = -6$	$1 + (-6) = -5$
$2(-3) = -6$	$2 + (-3) = -1$
$3(-2) = -6$	$3 + (-2) = 1$
$6(-1) = -6$	$6 + (-1) = 5$

Note that the factors 1 and -6 in the top row are different from the factors -1 and 6 in the bottom row, and their sums are different.

    The numbers 3 and -2 have a product of -6 and a sum of 1. Thus, the factors are $(x + 3)$ and $(x - 2)$.

$$x^2 + x - 6 = (x + 3)(x - 2)$$

The order of the factors is not crucial. Therefore, $x^2 + x - 6 = (x - 2)(x + 3)$ is also an acceptable answer.

▶ **Now Try Exercise 19**

    As mentioned earlier, **trinomial factoring problems can be checked by multiplying the factors using the FOIL method**. If the factoring is correct, the product obtained using the FOIL method will be identical to the original trinomial. Let's check the factors obtained in Example 3.

Check $(x + 3)(x - 2) = x^2 - 2x + 3x - 6 = x^2 + x - 6$

Since the product of the factors is identical to the original trinomial, the factoring is correct.

EXAMPLE 4 ▶ Factor $x^2 - x - 6$.

Solution   The factors of -6 are illustrated in Example 3. The factors whose product is -6 and whose sum is -1 are 2 and -3.

Factors of -6	Sum of Factors
$2(-3) = -6$	$2 + (-3) = -1$

Therefore, $x^2 - x - 6 = (x + 2)(x - 3)$

▶ **Now Try Exercise 25**

EXAMPLE 5 ▶ Factor $x^2 - 5x + 6$.

Solution   We must find two numbers whose product is 6 and whose sum is -5. Since the constant, 6, is positive, both numbers must have the same sign. Since the coefficient of the x-term, -5, is negative, both numbers must be negative. Recall that the product of a negative number and a negative number is positive. We now list the negative factors of 6 and look for the pair whose sum is -5.

Factors of 6	Sum of Factors
$(-1)(-6)$	$-1 + (-6) = -7$
$(-2)(-3)$	$-2 + (-3) = -5$

The factors of 6 whose sum is -5 are -2 and -3.

$$x^2 - 5x + 6 = (x - 2)(x - 3)$$

▶ **Now Try Exercise 29**

    In Example 5, suppose we were asked to factor $-5x + x^2 + 6$. In order to factor a trinomial, we always write the expression in the form $x^2 + bx + c$ if it is not given in that form. Therefore, the first step in the factoring process would be to rewrite $-5x + x^2 + 6$ as $x^2 - 5x + 6$. Then we would proceed as in Example 5.

EXAMPLE 6 ▶ Factor $r^2 + 2r - 24$.

Solution   In this example, the variable is r, but the factoring procedure is the same. We must find the two factors of -24 whose sum is 2. Since the constant is negative, one factor will be positive and the other factor will be negative.

Factors of -24	Sum of Factors
$(1)(-24)$	$1 + (-24) = -23$
$(2)(-12)$	$2 + (-12) = -10$
$(3)(-8)$	$3 + (-8) = -5$
$(4)(-6)$	$4 + (-6) = -2$
$(6)(-4)$	$6 + (-4) = \boxed{2}$

Since we have found the two numbers, 6 and -4, whose product is -24 and whose sum is 2, we need go no further.

$$r^2 + 2r - 24 = (r + 6)(r - 4)$$

▶ **Now Try Exercise 33**

EXAMPLE 7 ▶ Factor $x^2 - 8x + 16$.

Solution   We must find the factors of 16 whose sum is -8. Both factors must be negative. (Can you explain why?) The two factors whose product is 16 and whose sum is -8 are -4 and -4.

$$x^2 - 8x + 16 = (x - 4)(x - 4)$$
$$= (x - 4)^2$$

▶ **Now Try Exercise 41**

EXAMPLE 8 ▶ Factor $x^2 - 11x - 60$.

Solution   We must find two numbers whose product is -60 and whose sum is -11. Since the constant is negative, one number must be positive and the other negative. The desired numbers are -15 and 4 because $(-15)(4) = -60$ and $-15 + 4 = -11$.

$$x^2 - 11x - 60 = (x - 15)(x + 4)$$

▶ **Now Try Exercise 47**

EXAMPLE 9 ▶ Factor $x^2 + 5x + 12$.

Solution   Let's first find the two numbers whose product is 12 and whose sum is 5. Since both the constant and the coefficient of the x-term are positive, the two numbers must also be positive.

Factors of 12	Sum of Factors
$(1)(12)$	$1 + 12 = 13$
$(2)(6)$	$2 + 6 = 8$
$(3)(4)$	$3 + 4 = 7$

Note that there are no two integers whose product is 12 and whose sum is 5. When two integers cannot be found to satisfy the given conditions, the trinomial cannot be factored using only integer factors. *A polynomial that cannot be factored using only integer coefficients is called a* **prime polynomial**. If you come across a polynomial that cannot be factored using only integer coefficients, as in Example 9, do not leave the answer blank. Instead, write *prime*. However, before you write the answer *prime*, recheck your work and make sure you have tried every possible combination.

▶ **Now Try Exercise 31**

When factoring a trinomial of the form $x^2 + bx + c$, there is at most one pair of numbers whose product is c and whose sum is b. For example, when factoring

$x^2 - 12x + 32$, the two numbers whose product is 32 and whose sum is -12 are -4 and -8. No other pair of numbers will satisfy these specific conditions. Thus, the only factors of $x^2 - 12x + 32$ are $(x - 4)$ and $(x - 8)$.

In Examples 10 and 11, we will factor trinomials of the form $x^2 + bxy + cy^2$, where b and c are real numbers. An example of a trinomial in this form is $x^2 - 2xy - 15y^2$. When factoring trinomials of this form, the last terms in both factors must contain a y-term in order for the product of the last terms of the factors to give a y^2-term. That is, the factors must be of the form as follows.

$$x^2 + bxy + cy^2 = (x + \blacksquare y)(x + \blacksquare y)$$

Numbers go here.

When finding the factors of $x^2 + bxy + cy^2$, we look for two numbers whose product is c and whose sum is b, just as we did when we factored trinomials of the form $x^2 + bx + c$. When we determine the numbers, we place the numbers in the shaded areas of the factors.

EXAMPLE 10 ▶ Factor $x^2 + 2xy + y^2$.

Solution In this problem, the second term contains two variables, x and y, and the last term is not a constant. The procedure used to factor this trinomial is similar to that outlined previously. You should realize, however, that the product of the first terms of the factors we are looking for must be x^2, and the product of the last terms of the factors must be y^2.

We must find two numbers whose product is 1 (from $1y^2$) and whose sum is 2 (from $2xy$). The two numbers are 1 and 1. Thus

$$x^2 + 2xy + y^2 = (x + 1y)(x + 1y) = (x + y)(x + y) = (x + y)^2$$

▶ **Now Try Exercise 65**

EXAMPLE 11 ▶ Factor $x^2 - 2xy - 15y^2$.

Solution Find two numbers whose product is -15 and whose sum is -2. The numbers are -5 and 3. The last terms must be $-5y$ and $3y$ to obtain $-15y^2$.

$$x^2 - 2xy - 15y^2 = (x - 5y)(x + 3y)$$

▶ **Now Try Exercise 69**

2 Remove a Common Factor from a Trinomial

Sometimes each term of a trinomial has a common factor. When this occurs, factor out the common factor first, as explained in Section 6.1. **The first step in any factoring problem is to factor out any factors common to all the terms in the polynomial. Whenever the numerical coefficient of the highest-degree term is not 1, you should check for a common factor.** After factoring out any common factor, you should factor the remaining trinomial further, if possible.

EXAMPLE 12 ▶ Factor $2x^2 + 2x - 12$.

Solution Since the numerical coefficient of the squared term is not 1, we check for a common factor. Because 2 is common to each term of the polynomial, we factor it out.

$$2x^2 + 2x - 12 = 2(x^2 + x - 6) \qquad \textit{Factor out the common factor.}$$

Now we factor the remaining trinomial $x^2 + x - 6$ into $(x + 3)(x - 2)$. Thus,

$$2x^2 + 2x - 12 = 2(x + 3)(x - 2).$$

Note that the trinomial $2x^2 + 2x - 12$ is now completely factored into *three* factors: two binomial factors, $x + 3$ and $x - 2$, and a monomial factor, 2. After 2 has been factored out, it plays no part in the factoring of the remaining trinomial.

▶ **Now Try Exercise 71**

EXAMPLE 13 ▶ Factor $3n^3 + 24n^2 - 60n$.

Solution   We see that $3n$ divides into each term of the polynomial and therefore is a common factor. After factoring out the $3n$, we factor the remaining trinomial.

$$3n^3 + 24n^2 - 60n = 3n(n^2 + 8n - 20) \quad \text{\textit{Factor out the common factor.}}$$
$$= 3n(n + 10)(n - 2) \quad \text{\textit{Factor the remaining trinomial.}}$$

▶ **Now Try Exercise 79**

EXERCISE SET 6.3

Concept/Writing Exercises

For each trinomial, determine the signs that will appear in the binomial factors. Explain how you determined your answer.

1. $x^2 + 92x + 960$
2. $x^2 - 500x + 4000$
3. $b^2 - 20b - 1500$
4. $q^2 - 10q - 7200$
5. $x^2 - 240x + 8000$
6. $x^2 + 50x + 600$

Write the trinomial whose factors are listed. Explain how you determined your answer.

7. $(x - 3)(x - 8)$
8. $(x - 2y)(x + 6y)$
9. $2(x - 5y)(x + y)$
10. $5(c + d)(c - d)$

11. On an exam, a student factored $2x^2 - 6x + 4$ as $(2x - 4)(x - 1)$. Even though $(2x - 4)(x - 1)$ does multiply out to $2x^2 - 6x + 4$, why did his or her professor deduct points?

12. On an exam, a student factored $3x^2 + 3x - 18$ as $(3x - 6)(x + 3)$. Even though $(3x - 6)(x + 3)$ does multiply out to $3x^2 + 3x - 18$, why did his or her professor deduct points?

13. How can a trinomial factoring problem be checked?

14. Explain how to determine the factors when factoring a trinomial of the form $x^2 + bx + c$.

Practice the Skills

Factor each polynomial. If the polynomial is prime, so state.

15. $x^2 - 7x + 10$
16. $x^2 + 8x + 15$
17. $x^2 + 6x + 8$
18. $x^2 - 3x + 2$
19. $x^2 + 5x - 24$
20. $x^2 - x - 12$
21. $x^2 + 4x - 6$
22. $y^2 - 6y + 8$
23. $y^2 - 13y + 12$
24. $x^2 + 3x - 54$
25. $a^2 - 2a - 8$
26. $p^2 + 3p - 10$
27. $r^2 - 2r - 15$
28. $x^2 - 6x + 8$
29. $b^2 - 11b + 18$
30. $x^2 + 11x - 30$
31. $x^2 - 8x - 15$
32. $x^2 - 8x + 7$
33. $q^2 + 4q - 45$
34. $x^2 + 10x + 25$
35. $x^2 - 7x - 30$
36. $b^2 - 9b - 36$
37. $x^2 + 4x + 4$
38. $x^2 - 4x + 4$
39. $s^2 - 8s + 16$
40. $u^2 + 2u + 1$
41. $p^2 - 12p + 36$
42. $x^2 - 10x - 25$
43. $-18w + w^2 + 45$
44. $-11x + x^2 + 10$
45. $10x - 39 + x^2$
46. $-3x + 8 + x^2$
47. $x^2 - x - 20$
48. $t^2 - 28t - 60$
49. $y^2 + 13y + 40$
50. $r^2 + 14r + 48$
51. $x^2 + 12x - 64$
52. $x^2 - 18x + 80$
53. $s^2 + 14s - 24$
54. $x^2 - 13x + 36$
55. $x^2 - 20x + 64$
56. $x^2 + 19x + 48$
57. $a^2 - 20a + 99$
58. $x^2 + 5x - 24$
59. $x^2 + 2 + 3x$

60. $m^2 - 11 - 10m$

61. $7w - 18 + w^2$

62. $30 + y^2 - 13y$

63. $x^2 - 8xy + 15y^2$

64. $x^2 - 2xy + y^2$

65. $m^2 - 6mn + 9n^2$

66. $b^2 - 2bc - 3c^2$

67. $x^2 + 8xy + 12y^2$

68. $x^2 + 16xy - 17y^2$

69. $m^2 - 5mn - 24n^2$

70. $c^2 + 2cd - 24d^2$

Factor completely.

71. $6x^2 - 30x + 24$

72. $2a^2 - 12a - 32$

73. $5x^2 + 20x + 15$

74. $4x^2 + 12x - 16$

75. $2x^2 - 18x + 40$

76. $3y^2 - 33y + 54$

77. $b^3 - 7b^2 + 10b$

78. $c^3 + 8c^2 - 48c$

79. $3z^3 - 21z^2 - 54z$

80. $3x^3 - 36x^2 + 33x$

81. $x^3 + 8x^2 + 16x$

82. $2x^3y - 12x^2y + 10xy$

83. $7a^2 - 35ab + 42b^2$

84. $3x^3 + 3x^2y - 18xy^2$

85. $3r^3 + 6r^2t - 24rt^2$

86. $r^2s + 7rs^2 + 12s^3$

87. $x^4 - 4x^3 - 21x^2$

88. $2z^5 + 14z^4 + 12z^3$

Problem Solving

89. The first two columns in the following table describe the signs of the coefficient of the x-term and constant term of a trinomial of the form $x^2 + bx + c$. Determine whether the third column should contain "both positive," "both negative," or "one positive and one negative." Explain how you determined your answer.

Sign of Coefficient of x-term	Sign of Constant of Trinomial	Sign of Constant Terms in the Binomial Factors
−	+	
−	−	
+	−	
+	+	

90. Assume that a trinomial of the form $x^2 + bx + c$ is factorable. Determine whether the constant terms in the factors

are "both positive," "both negative," or "one positive and one negative" for the given signs of b and c. Explain your answer.

a) $b < 0, c > 0$

b) $b > 0, c > 0$

c) $b > 0, c < 0$

d) $b < 0, c < 0$

91. Write a trinomial whose binomial factors contain constant terms that sum to −12 and have a product of 32. Show the factoring of the trinomial.

92. Write a trinomial whose binomial factors contain constant terms that sum to 5 and have a product of 4. Show the factoring of the trinomial.

93. Write a trinomial whose binomial factors contain constant terms that sum to −2 and have a product of −35. Show the factoring of the trinomial.

94. Write a trinomial whose binomial factors contain constant terms that sum to 5 and have a product of −14. Show the factoring of the trinomial.

Challenge Problems

Factor.

95. $x^2 + 0.6x + 0.08$

96. $x^2 - 0.5x - 0.06$

97. $x^2 + \frac{2}{5}x + \frac{1}{25}$

98. $x^2 - \frac{2}{7}x + \frac{1}{49}$

99. $x^2 - 24x - 256$

100. $x^2 + 5x - 300$

Cumulative Review Exercises

[2.5] **101.** Solve the equation $4(2x - 4) = 5x + 11$.

[3.4] **102. Mixing Solutions** Karen Moreau, a chemist, mixes 4 liters of an 18% acid solution with 1 liter of a 26% acid solution. Find the strength of the mixture.

[4.2] **103.** Graph $y = 3x - 2$.

[5.5] **104.** Multiply $(2x^2 + 5x - 6)(x - 2)$.

[5.6] **105.** Divide $3x^2 - 10x - 10$ by $x - 4$.

[6.2] **106.** Factor $20x^2 + 8x - 15x - 6$ by grouping.

See Exercise 102.

6.4 Factoring Trinomials of the Form $ax^2 + bx + c, a \neq 1$

1 Factor trinomials of the form $ax^2 + bx + c, a \neq 1$, by trial and error.

2 Factor trinomials of the form $ax^2 + bx + c, a \neq 1$, by grouping.

An Important Note

In this section, we discuss two methods of factoring trinomials of the form $ax^2 + bx + c, a \neq 1$. That is, we will be factoring trinomials whose squared term has a numerical coefficient not equal to 1, after removing any common factors. Examples of trinomials with $a \neq 1$ are

$$2x^2 + 11x + 12 \ (a = 2) \qquad 4x^2 - 3x + 1 \ (a = 4)$$

The methods we discuss are (1) **factoring by trial and error** and (2) **factoring by grouping.** We present two different methods for factoring these trinomials because some students, and some instructors, prefer one method, while others prefer the second method. You may use either method unless your instructor asks you to use a specific method. We will use the same examples to illustrate both methods so that you can make a comparison. Each method is treated independently of the other. If your teacher asks you to use a specific method, either factoring by trial and error or factoring by grouping, you need only read the material related to that specific method. Factoring by trial and error was introduced in Section 6.3 and factoring by grouping was introduced in Section 6.2.

1 Factor Trinomials of the Form $ax^2 + bx + c, a \neq 1$, by Trial and Error

Let's now discuss factoring trinomials of the form $ax^2 + bx + c, a \neq 1$, by the trial and error method, introduced in Section 6.3. It may be helpful for you to reread that material before going any further.

Recall that factoring is the reverse of multiplying. Consider the product of the following two binomials:

$$
\begin{aligned}
(2x + 3)(x + 5) &= \overset{F}{2x(x)} + \overset{O}{(2x)(5)} + \overset{I}{3(x)} + \overset{L}{3(5)} \\
&= 2x^2 + 10x + 3x + 15 \\
&= 2x^2 + 13x + 15
\end{aligned}
$$

Notice that the product of the first terms of the binomials gives the x-squared term of the trinomial, $2x^2$. Also notice that the product of the last terms of the binomials gives the last term, or constant, of the trinomial, $+15$. Finally, notice that the sum of the products of the outer terms and inner terms of the binomials gives the middle term of the trinomial, $+13x$. When we factor a trinomial using trial and error, we make use of these important facts. Note that $2x^2 + 13x + 15$ in factored form is $(2x + 3)(x + 5)$.

$$2x^2 + 13x + 15 = (2x + 3)(x + 5)$$

When factoring a trinomial of the form $ax^2 + bx + c$ by trial and error, the product of the x-terms in the binomial factors must equal the first term of the trinomial, ax^2. Also, the product of the constants in the binomial factors, including their signs, must equal the constant, c, of the trinomial.

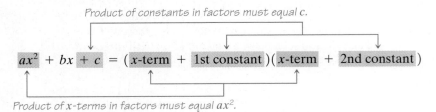

For example, when factoring the trinomial $2x^2 + 7x + 6$, each of the following pairs of factors has a product of the first terms equal to $2x^2$ and a product of the last terms equal to 6.

Trinomial	Possible Factors	Product of First Terms	Product of Last Terms
$2x^2 + 7x + 6$	$(2x + 1)(x + 6)$	$2x(x) = 2x^2$	$1(6) = 6$
	$(2x + 2)(x + 3)$	$2x(x) = 2x^2$	$2(3) = 6$
	$(2x + 3)(x + 2)$	$2x(x) = 2x^2$	$3(2) = 6$
	$(2x + 6)(x + 1)$	$2x(x) = 2x^2$	$6(1) = 6$

Each of these pairs of factors is a possible answer, but only one has the correct factors. How do we determine which is the correct factoring of the trinomial $2x^2 + 7x + 6$? The key lies in the x-term. We know that when we multiply two binomials using the FOIL method the sum of the products of the outer and inner terms gives us the x-term of the trinomial. We use this concept in reverse to determine the correct pair of factors. We need to find the pair of factors whose sum of the products of the outer and inner terms is equal to the x-term of the trinomial.

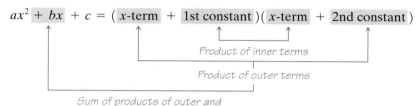

Now look at the possible pairs of factors we obtained for $2x^2 + 7x + 6$ to see if any yield the correct x-term, $7x$.

Trinomial	Possible Factors	Product of the First Terms	Product of the Last Terms	Sum of the Products of Outer and Inner Terms
$2x^2 + 7x + 6$	$(2x + 1)(x + 6)$	$2x^2$	6	$2x(6) + 1(x) = 13x$
	$(2x + 2)(x + 3)$	$2x^2$	6	$2x(3) + 2(x) = 8x$
	$(2x + 3)(x + 2)$	$2x^2$	6	$2x(2) + 3(x) = 7x$
	$(2x + 6)(x + 1)$	$2x^2$	6	$2x(1) + 6(x) = 8x$

Since $(2x + 3)(x + 2)$ yields the correct x-term, $7x$, the factors of the trinomial $2x^2 + 7x + 6$ are $(2x + 3)$ and $(x + 2)$.

$$2x^2 + 7x + 6 = (2x + 3)(x + 2)$$

We can check this factoring using the FOIL method.

Check

$$\begin{aligned} (2x + 3)(x + 2) &= \overset{F}{2x(x)} + \overset{O}{2x(2)} + \overset{I}{3(x)} + \overset{L}{3(2)} \\ &= 2x^2 + 4x + 3x + 6 \\ &= 2x^2 + 7x + 6 \end{aligned}$$

Since we obtained the original trinomial, our factoring is correct.

Note in the preceding illustration that $(2x + 1)(x + 6)$ are different factors than $(2x + 6)(x + 1)$, because in one case 1 is paired with $2x$ and in the second case 1 is paired with x. The factors $(2x + 1)(x + 6)$ and $(x + 6)(2x + 1)$ are, however, the same set of factors with their order reversed.

Helpful Hint

When factoring a trinomial of the form $ax^2 + bx + c$, remember that the sign of the constant, c, and the sign of the x-term, bx, offer valuable information. When factoring a trinomial by trial and error, first check the sign of the constant. If it is positive, the signs in both factors will be the same as the sign of the x-term. If the constant is negative, one factor will contain a plus sign and the other a negative sign.

Now we outline the procedure to factor trinomials of the form $ax^2 + bx + c, a \neq 1$, by trial and error. Keep in mind that the more you practice, the better you will become at factoring.

> **To Factor Trinomials of the Form $ax^2 + bx + c, a \neq 1$, by Trial and Error**
>
> 1. Determine whether there is a factor common to all three terms. If so, factor it out.
> 2. Write all pairs of factors of the coefficient of the squared term, a.
> 3. Write all pairs of factors of the constant term, c.
> 4. Try various combinations of these factors until the correct middle term, bx, is found.

When factoring using this procedure, if there is more than one pair of numbers whose product is a, we generally begin with the middle-size pair. We will illustrate the procedure in Examples 1 through 8.

EXAMPLE 1 ▶ Factor $3x^2 + 20x + 12$.

Solution　We first determine that all three terms have no common factors other than 1. Since the first term is $3x^2$, one factor must contain a $3x$ and the other an x. Therefore, the factors will be of the form $(3x + \blacksquare)(x + \blacksquare)$. Now we must find the numbers to place in the shaded areas. The product of the last terms in the factors must be 12. Since the constant and the coefficient of the x-term are both positive, only the positive factors of 12 need be considered. We will list the positive factors of 12, the possible factors of the trinomial, and the sum of the products of the outer and inner terms. Once we find the factors of 12 that yield the proper sum of the products of the outer and inner terms, $20x$, we can write the answer.

Factors of 12	Possible Factors of Trinomial	Sum of the Products of the Outer and Inner Terms
1(12)	$(3x + 1)(x + 12)$	$37x$
2(6)	$(3x + 2)(x + 6)$	$20x$
3(4)	$(3x + 3)(x + 4)$	$15x$
4(3)	$(3x + 4)(x + 3)$	$13x$
6(2)	$(3x + 6)(x + 2)$	$12x$
12(1)	$(3x + 12)(x + 1)$	$15x$

Since the product of $(3x + 2)$ and $(x + 6)$ yields the correct x-term, $20x$, they are the correct factors.

$$3x^2 + 20x + 12 = (3x + 2)(x + 6)$$

▶ **Now Try Exercise 5**

In Example 1, our first factor could have been written with an x and the second with a $3x$. Had we done this, we still would have obtained the correct answer: $(x + 6)(3x + 2)$. We also could have stopped once we found the pair of factors that yielded the $20x$. Instead, we listed all the factors so that you could study them.

EXAMPLE 2 ▶ Factor $5x^2 - 7x - 6$.

Solution   There are no factors common to all three terms. Since the first term is $5x^2$, one factor must contain a $5x$ and the other an x. We now list the factors of -6 and look for the pair of factors that yields $-7x$.

Factors of -6	Possible Factors	Sum of the Products of the Outer and Inner Terms
$-1(6)$	$(5x - 1)(x + 6)$	$29x$
$-2(3)$	$(5x - 2)(x + 3)$	$13x$
$-3(2)$	$(5x - 3)(x + 2)$	$7x$
$-6(1)$	$(5x - 6)(x + 1)$	$-x$

Since we did not obtain the desired quantity, $-7x$, by writing the negative factor with the $5x$, we will now try listing the negative factor with the x.

Factors of -6	Possible Factors	Sum of the Products of the Outer and Inner Terms
$1(-6)$	$(5x + 1)(x - 6)$	$-29x$
$2(-3)$	$(5x + 2)(x - 3)$	$-13x$
$3(-2)$	$(5x + 3)(x - 2)$	$-7x$
$6(-1)$	$(5x + 6)(x - 1)$	x

We see that $(5x + 3)(x - 2)$ gives the $-7x$ we are looking for. Thus,

$$5x^2 - 7x - 6 = (5x + 3)(x - 2)$$

Again we listed all the possible combinations for you to study.

▶ **Now Try Exercise 9**

Helpful Hint

In Example 2, we were asked to factor $5x^2 - 7x - 6$. When we considered the product of $-3(2)$ in the first set of possible factors, we obtained

Factors of -6	Possible Factors	Sum of the Products of the Outer and Inner Terms
$-3(2)$	$(5x - 3)(x + 2)$	$7x$

Later in the solution we tried the factors $3(-2)$ and obtained the correct answer.

$3(-2)$	$(5x + 3)(x - 2)$	$-7x$

When factoring a trinomial with a *negative constant*, if you obtain the x-term whose sign is the opposite of the one you are seeking, *reverse the signs on the constants* in the factors. This should give you the set of factors you are seeking.

EXAMPLE 3 ▶ Factor $8x^2 + 33x + 4$.

Solution   There are no factors common to all three terms. Since the first term is $8x^2$, there are a number of possible combinations for the first terms in the factors. Since $8 = 8 \cdot 1$ and $8 = 4 \cdot 2$, the possible factors may be of the form $(8x\quad)(x\quad)$ or $(4x\quad)(2x\quad)$. When this situation occurs, we will generally start with the middle-size pair of factors. Thus, we begin with $(4x\quad)(2x\quad)$. If this pair does not lead to the solution, we will then try $(8x\quad)(x\quad)$. We now list the factors of the constant, 4. Since all signs are positive, we list only the positive factors of 4.

Factors of 4	Possible Factors	Sum of the Products of the Outer and Inner Terms
1(4)	$(4x + 1)(2x + 4)$	$18x$
2(2)	$(4x + 2)(2x + 2)$	$12x$
4(1)	$(4x + 4)(2x + 1)$	$12x$

Since we did not obtain the correct factors with $(4x\quad)(2x\quad)$, we now try $(8x\quad)(x\quad)$.

Factors of 4	Possible Factors	Sum of the Products of the Outer and Inner Terms
1(4)	$(8x + 1)(x + 4)$	$33x$
2(2)	$(8x + 2)(x + 2)$	$18x$
4(1)	$(8x + 4)(x + 1)$	$12x$

Since the product of $(8x + 1)$ and $(x + 4)$ yields the correct x-term, $33x$, they are the correct factors.

$$8x^2 + 33x + 4 = (8x + 1)(x + 4)$$

▶ Now Try Exercise 19

EXAMPLE 4 ▶ Factor $25t^2 - 10t + 1$.

Solution   The factors must be of the form $(25t\quad)(t\quad)$ or $(5t\quad)(5t\quad)$. We will start with the middle-size factors $(5t\quad)(5t\quad)$. Since the constant is positive and the coefficient of the x-term is negative, both factors must be negative.

Factors of 1	Possible Factors	Sum of the Products of the Outer and Inner Terms
$(-1)(-1)$	$(5t - 1)(5t - 1)$	$-10t$

Since we found the correct factors, we can stop.

$$25t^2 - 10t + 1 = (5t - 1)(5t - 1) = (5t - 1)^2$$

▶ Now Try Exercise 13

EXAMPLE 5 ▶ Factor $2x^2 + 3x + 7$.

Solution   The factors will be of the form $(2x\quad)(x\quad)$. We need only consider the positive factors of 7. Can you explain why?

Factors of 7	Possible Factors	Sum of the Products of the Outer and Inner Terms
1(7)	$(2x + 1)(x + 7)$	$15x$
7(1)	$(2x + 7)(x + 1)$	$9x$

Since we have tried all possible combinations and we have not obtained the x-term, $3x$, this trinomial *cannot be factored using only integer factors*. As explained in Section 6.3, the trinomial $2x^2 + 3x + 7$ is a *prime polynomial*.

▶ Now Try Exercise 17

EXAMPLE 6 ▶ Factor $6a^2 + 19ab + 3b^2$.

Solution   This trinomial is different from the other trinomials in that the last term is not a constant but contains b^2. Don't let this scare you. The factoring process is the same, except that the second term of both factors will contain b. We begin by considering factors of the form $(3a \quad)(2a \quad)$. If we cannot find the factors, then we try factors of the form $(6a \quad)(a \quad)$.

Factors of 3	Possible Factors	Sum of the Products of the Outer and Inner Terms
1(3)	$(3a + b)(2a + 3b)$	$11ab$
3(1)	$(3a + 3b)(2a + b)$	$9ab$
1(3)	$(6a + b)(a + 3b)$	$19ab$
3(1)	$(6a + 3b)(a + b)$	$9ab$

$$6a^2 + 19ab + 3b^2 = (6a + b)(a + 3b)$$

Check $(6a + b)(a + 3b) = 6a^2 + 18ab + ab + 3b^2 = 6a^2 + 19ab + 3b^2$

▶ **Now Try Exercise 55**

EXAMPLE 7 ▶ Factor $6x^2 - 13xy - 8y^2$.

Solution   We begin with factors of the form $(3x \quad)(2x \quad)$. If we cannot find the solution from these, we will try $(6x \quad)(x \quad)$. Since the last term, $-8y^2$, is negative, one factor will contain a plus sign and the other will contain a minus sign.

Factors of −8	Possible Factors	Sum of the Products of the Outer and Inner Terms
1(−8)	$(3x + y)(2x - 8y)$	$-22xy$
2(−4)	$(3x + 2y)(2x - 4y)$	$-8xy$
4(−2)	$(3x + 4y)(2x - 2y)$	$2xy$
8(−1)	$(3x + 8y)(2x - y)$	$13xy$

We are looking for $-13xy$. When we considered 8(−1), we obtained $13xy$. As explained in the Helpful Hint on page 359, if we reverse the signs of the numbers in the factors, we will obtain the factors we are seeking.

$$(3x + 8y)(2x - y) \quad \text{Gives } 13xy$$
$$(3x - 8y)(2x + y) \quad \text{Gives } -13xy$$

Therefore, $6x^2 - 13xy - 8y^2 = (3x - 8y)(2x + y)$.

▶ **Now Try Exercise 57**

Now we will look at an example in which all the terms of the trinomial have a common factor.

EXAMPLE 8 ▶ Factor $6x^3 + 15x^2 - 36x$.

Solution   *The first step in any factoring problem is to determine whether all the terms contain a common factor. If so, factor out that common factor first.* In this example, $3x$ is common to all three terms. We begin by factoring out the $3x$. Then we continue factoring by trial and error.

$$6x^3 + 15x^2 - 36x = 3x(2x^2 + 5x - 12)$$
$$= 3x(2x - 3)(x + 4)$$

▶ **Now Try Exercise 47**

2 Factor Trinomials of the Form $ax^2 + bx + c, a \neq 1$, by Grouping

We will now discuss the use of grouping. The steps in the box that follow give the procedure for factoring trinomials by grouping.

> ### To Factor Trinomials of the Form $ax^2 + bx + c, a \neq 1$, by Grouping
>
> 1. Determine whether there is a factor common to all three terms. If so, factor it out.
> 2. Find two numbers whose product is equal to the product of a times c, and whose sum is equal to b.
> 3. Rewrite the middle term, bx, as the sum or difference of two terms using the numbers found in step 2.
> 4. Factor by grouping as explained in Section 6.2.

This process will be made clear in Example 9. We will rework Examples 1 through 8 here using factoring by grouping. Example 9, which follows, is the same trinomial given in Example 1. After you study this method and try some exercises, you will gain a feel for which method you prefer using.

EXAMPLE 9 ▶ Factor $3x^2 + 20x + 12$.

Solution   First determine whether there is a factor common to all the terms of the polynomial. There are no common factors (other than 1) to the three terms.

$$a = 3 \quad b = 20 \quad c = 12$$

1. We must find two numbers whose product is $a \cdot c$ and whose sum is b. We must therefore find two numbers whose product equals $3 \cdot 12 = 36$ and whose sum equals 20. Only the positive factors of 36 need be considered since all signs of the trinomial are positive.

Factors of 36	Sum of Factors
(1)(36)	$1 + 36 = 37$
(2)(18)	$2 + 18 = 20$
(3)(12)	$3 + 12 = 15$
(4)(9)	$4 + 9 = 13$
(6)(6)	$6 + 6 = 12$

   The desired factors are 2 and 18.

2. Rewrite $20x$ as the sum or difference of two terms using the values found in step 1. Therefore, we rewrite $20x$ as $2x + 18x$.

$$3x^2 + 20x + 12$$
$$= 3x^2 + 2x + 18x + 12$$

3. Now factor by grouping. Start by factoring out a common factor from the first two terms and a common factor from the last two terms. This procedure was discussed in Section 6.2.

   x is common factor   *6 is common factor*

$$= 3x^2 + 2x + 18x + 12$$
$$= x(3x + 2) + 6(3x + 2)$$
$$= (3x + 2)(x + 6)$$

▶ Now Try Exercise 7

Note that in step 2 of Example 9 we rewrote $20x$ as $2x + 18x$. Would it have made a difference if we had written $20x$ as $18x + 2x$? Let's work it out and see.

$$3x^2 + 20x + 12$$
$$= 3x^2 \overbrace{+18x +2x} + 12$$

3x is common factor	2 is common factor

$$= 3x^2 + 18x + 2x + 12$$
$$= 3x(x + 6) + 2(x + 6)$$
$$= (x + 6)(3x + 2)$$

Since $(x + 6)(3x + 2) = (3x + 2)(x + 6)$, the factors are the same. We obtained the same answer by writing the $20x$ as either $2x + 18x$ or $18x + 2x$. *In general, when rewriting the middle term of the trinomial using the specific factors found, the terms may be listed in either order.* You should, however, check after you list the two terms to make sure that the sum of the terms you listed equals the middle term.

EXAMPLE 10 ▶ Factor $5x^2 - 7x - 6$.

Solution   There are no common factors other than 1.

$$a = 5, \quad b = -7, \quad c = -6$$

The product of a times c is $5(-6) = -30$. We must find two numbers whose product is -30 and whose sum is -7.

Factors of -30	Sum of Factors
$(-1)(30)$	$-1 + 30 = 29$
$(-2)(15)$	$-2 + 15 = 13$
$(-3)(10)$	$-3 + 10 = 7$
$(-5)(6)$	$-5 + 6 = 1$
$(-6)(5)$	$-6 + 5 = -1$
$(-10)(3)$	$-10 + 3 = -7$
$(-15)(2)$	$-15 + 2 = -13$
$(-30)(1)$	$-30 + 1 = -29$

Rewrite the middle term of the trinomial, $-7x$, as $-10x + 3x$.

$$5x^2 - 7x - 6$$
$$= 5x^2 \overbrace{-10x +3x} - 6 \qquad \textit{Now factor by grouping.}$$
$$= 5x(x - 2) + 3(x - 2)$$
$$= (x - 2)(5x + 3)$$

▶ **Now Try Exercise 31**

In Example 10, we could have expressed the $-7x$ as $3x - 10x$ and obtained the same answer. Try working Example 10 by rewriting $-7x$ as $3x - 10x$.

Helpful Hint

Notice in Example 10 that we were looking for two factors of -30 whose sum was -7. When we considered the factors -3 and 10, we obtained a sum of 7, which is the opposite of -7. The factors we eventually obtained that gave a sum of -7 were 3 and -10. Note that when the *constant of the trinomial is negative*, if we switch the signs of the constants in the factors, the sign of the sum of the factors changes. Thus, when trying pairs of factors to obtain the middle term, if you obtain the opposite of the coefficient you are seeking, reverse the signs in the factors. This should give you the coefficient you are seeking.

USING YOUR GRAPHING CALCULATOR

In Example 10, we began factoring $5x^2 - 7x - 6$ by finding two numbers whose product equals -30 and whose sum equals -7. The graphing calculator can be used to help find those numbers.

We can represent two numbers whose product is -30 by x and $-\dfrac{30}{x}$. Notice $x\left(-\dfrac{30}{x}\right) = -30$. Let $Y_1 = -\dfrac{30}{x}$, the second factor in $x\left(-\dfrac{30}{x}\right)$. Let $Y_2 = -\dfrac{30}{x} + x$, the sum of the factors. Use the TABLE feature to create pairs of numbers whose product is -30 (the product of columns X and Y_1 is -30) and whose sum is in column Y_2 (see **Figure 6.3**). Look down the Y_2 column until you find the sum you are looking for, -7.

Use the value for X, 3, and the value for Y_1, -10, to rewrite the middle term of the trinomial, $-7x$, as $3x - 10x$. Continue factoring by grouping.

EXERCISES

Use your graphing calculator as an aid in factoring each trinomial.

1. $6x^2 - 13x - 28$
2. $10x^2 - x - 24$
3. $12x^2 + 16x - 35$
4. $27x^2 + 57x + 20$

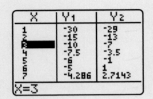

FIGURE 6.3

EXAMPLE 11 ▶ Factor $8x^2 + 33x + 4$.

Solution   There are no common factors other than 1. We must find two numbers whose product is $8 \cdot 4$ or 32 and whose sum is 33. The numbers are 1 and 32.

Factors of 32	Sum of Factors
$(1)(32)$	$1 + 32 = 33$

Rewrite $33x$ as $32x + x$. Then factor by grouping.

$$8x^2 + 33x + 4$$
$$= 8x^2 + 32x + x + 4$$
$$= 8x(x + 4) + 1(x + 4)$$
$$= (x + 4)(8x + 1)$$

▶ Now Try Exercise 25

Notice in Example 11 that we rewrote $33x$ as $32x + x$ rather than $x + 32x$. We did this to reinforce factoring out 1 from the last two terms of an expression. You should obtain the same answer if you rewrite $33x$ as $x + 32x$. Try this now.

EXAMPLE 12 ▶ Factor $25t^2 - 10t + 1$.

Solution   There are no common factors other than 1. We must find two numbers whose product is $25 \cdot 1$ or 25 and whose sum is -10. Since the product of a times c is positive and the coefficient of the t-term is negative, both numerical factors must be negative.

Factors of 25	Sum of Factors
$(-1)(-25)$	$-1 + (-25) = -26$
$(-5)(-5)$	$-5 + (-5) = -10$

The desired factors are -5 and -5.

$$25t^2 - 10t + 1$$

$$= 25t^2 \underbrace{-5t - 5t} + 1 \qquad \textit{Rewrite } -10t \textit{ as } -5t - 5t.$$

$$= 5t(5t - 1) - 5t + 1$$

$$= 5t(5t - 1) - 1(5t - 1) \qquad \textit{Rewrite } -5t + 1 \textit{ as } -1(5t - 1).$$

$$= (5t - 1)(5t - 1) \text{ or } (5t - 1)^2$$

▶ **Now Try Exercise 29**

Helpful Hint

When attempting to factor a trinomial, if there are no two integers whose product equals $a \cdot c$ and whose sum equals b, the trinomial cannot be factored.

EXAMPLE 13 ▶ Factor $2x^2 + 3x + 7$.

Solution   There are no common factors other than 1. We must find two numbers whose product is 14 and whose sum is 3. We need consider only positive factors of 14. Why?

Factors of 14	Sum of Factors
(1)(14)	$1 + 14 = 15$
(2)(7)	$2 + 7 = 9$

Since there are no factors of 14 whose sum is 3, we conclude that this trinomial cannot be factored. This is an example of a *prime polynomial*.

▶ **Now Try Exercise 21**

EXAMPLE 14 ▶ Factor $6a^2 + 19ab + 3b^2$.

Solution   There are no common factors other than 1. This trinomial contains two variables. It is factored in basically the same manner as the previous examples. Find two numbers whose product is $6 \cdot 3$ or 18 and whose sum is 19. The two numbers are 18 and 1.

$$6a^2 + 19ab + 3b^2$$

$$= 6a^2 \underbrace{+ 18ab + ab} + 3b^2$$

$$= 6a(a + 3b) + b(a + 3b)$$

$$= (a + 3b)(6a + b)$$

▶ **Now Try Exercise 61**

EXAMPLE 15 ▶ Factor $6x^2 - 13xy - 8y^2$.

Solution   There are no common factors other than 1. Find two numbers whose product is $6(-8)$ or -48 and whose sum is -13. Since the product is negative, one factor must be positive and the other negative. Some factors are given below.

Product of Factors	Sum of Factors
(1)(−48)	$1 + (-48) = -47$
(2)(−24)	$2 + (-24) = -22$
(3)(−16)	$3 + (-16) = -13$

There are many other factors, but we have found the pair we were looking for. The two numbers whose product is -48 and whose sum is -13 are 3 and -16.

$$6x^2 - 13xy - 8y^2$$

$$= 6x^2 \underbrace{+ 3xy - 16xy} - 8y^2$$

$$= 3x(2x + y) - 8y(2x + y)$$

$$= (2x + y)(3x - 8y)$$

Check $(2x + y)(3x - 8y)$

$$
\begin{array}{cccccc}
& F & O & I & L \\
\end{array}
$$

$$= (2x)(3x) + (2x)(-8y) + (y)(3x) + (y)(-8y)$$

$$= \quad 6x^2 \quad - \quad 16xy \quad + \quad 3xy \quad - \quad 8y^2$$

$$= 6x^2 - 13xy - 8y^2$$

▶ **Now Try Exercise 63**

If you rework Example 15 by writing $-13xy$ as $-16xy + 3xy$, what answer would you obtain? Try it now and see.

Remember that in any factoring problem our first step is to determine whether all terms in the polynomial have a common factor other than 1. If so, we use the distributive property to factor the GCF from each term. We then continue to factor the trinomial, if possible.

EXAMPLE 16 ▶ Factor $6x^3 + 15x^2 - 36x$.

Solution   The factor $3x$ is common to all three terms. Factor the $3x$ from each term of the polynomial.

$$6x^3 + 15x^2 - 36x = 3x(2x^2 + 5x - 12)$$

Now continue by factoring $2x^2 + 5x - 12$. The two numbers whose product is $2(-12)$ or -24 and whose sum is 5 are 8 and -3.

$$2x(2x^2 + 5x - 12)$$

$$= 2x(2x^2 + 8x - 3x - 12)$$

$$= 2x[2x(x + 4) - 3(x + 4)]$$

$$= 2x(x + 4)(2x - 3)$$

▶ **Now Try Exercise 49**

Helpful Hint

Which Method Should You Use to Factor a Trinomial?

If your instructor asks you to use a specific method, you should use that method. If your instructor does not require a specific method, you should use the method you feel most comfortable with. You may wish to start with the trial-and-error method if there are only a few possible factors to try. If you cannot find the factors by trial and error or if there are many possible factors to consider, you may wish to use the grouping procedure. With time and practice you will learn which method you feel most comfortable with and which method gives you greater success.

EXERCISE SET 6.4

Concept/Writing Exercises

1. What is the relationship between factoring trinomials and multiplying binomials?

2. When factoring a trinomial of the form $ax^2 + bx + c$, what must the product of the constants in the binomial factors equal?

3. When factoring a trinomial of the form $ax^2 + bx + c$, what must the product of the first terms of the binomial factors equal?

4. Explain the procedure used to factor a trinomial of the form $ax^2 + bx + c, a \neq 1$.

Practice the Skills

Factor completely. If the polynomial is prime, so state.

5. $2x^2 + 11x + 5$

6. $2x^2 + 9x + 4$

7. $3x^2 + 14x + 8$

8. $7x^2 + 37x + 10$

9. $5x^2 - 9x - 2$

10. $3y^2 + 17y + 10$

11. $3r^2 + 13r - 10$

12. $3x^2 - 2x - 8$

13. $4z^2 - 12z + 9$

14. $4n^2 - 9n + 5$

15. $6z^2 + z - 12$

16. $5m^2 - 17m + 6$

17. $5a^2 - 12a + 6$

18. $2x^2 - x - 6$

19. $8x^2 + 19x + 6$

20. $6y^2 - 11y + 4$

21. $3x^2 + 11x + 4$

22. $3a^2 + 7a - 20$

23. $5y^2 - 16y + 3$

24. $5x^2 + 2x + 9$

25. $7x^2 + 43x + 6$

26. $7x^2 - 8x + 1$

27. $4x^2 + 4x - 15$

28. $15x^2 - 19x + 6$

29. $49t^2 - 14t + 1$

30. $16z^2 - 8z + 1$

31. $5z^2 - 6z - 8$

32. $3z^2 - 11z - 6$

33. $4y^2 + 5y - 6$

34. $5y^2 - 3y - 1$

35. $10x^2 - 27x + 5$

36. $6a^2 + 7a - 10$

37. $10d^2 - 7d - 12$

38. $6x^2 + 13x + 3$

39. $8x^2 - 46x - 12$

40. $12x^2 - 13x - 35$

41. $10t + 3 + 7t^2$

42. $n - 30 + n^2$

43. $6x^2 + 16x + 10$

44. $12z^2 + 32z + 20$

45. $6x^3 - 5x^2 - 4x$

46. $8x^3 + 8x^2 - 6x$

47. $12x^3 + 28x^2 + 8x$

48. $18x^3 - 21x^2 - 9x$

49. $4x^3 - 2x^2 - 12x$

50. $300x^2 - 400x - 400$

51. $48c^2 + 8c - 16$

52. $28x^2 - 28x + 7$

53. $4p - 12 + 8p^2$

54. $72 + 3r^2 - 30r$

55. $8c^2 + 41cd + 5d^2$

56. $8x^2 - 8xy - 6y^2$

57. $15x^2 - xy - 6y^2$

58. $2x^2 - 7xy + 3y^2$

59. $12x^2 + 10xy - 8y^2$

60. $12a^2 - 34ab + 24b^2$

61. $7p^2 + 13pq + 6q^2$

62. $24x^2 - 92x + 80$

63. $6m^2 - mn - 2n^2$

64. $8m^2 + 4mn - 4n^2$

65. $8x^3 + 10x^2y + 3xy^2$

66. $8a^2b + 10ab^2 + 3b^3$

67. $4x^4 + 8x^3y + 3x^2y^2$

68. $26u^2v + 6uv^2 + 24u^3$

Problem Solving

For Exercises 69–74, write the polynomial whose factors are listed. Explain how you determined your answer.

69. $3x + 1, x - 7$

70. $6y - 5, 4y - 3$

71. $5, x + 3, 2x + 1$

72. $3, 2x + 3, x - 4$

73. $t^2, t + 4, 3t - 1$

74. $5x^2, 3x - 7, 2x + 3$

75. a) If you know one binomial factor of a trinomial, explain how you can use division to find the second binomial factor of the trinomial (see Section 5.6).

   b) One factor of $18x^2 + 93x + 110$ is $3x + 10$. Use division to find the second factor.

76. One factor of $30x^2 - 17x - 247$ is $6x - 19$. Find the other factor.

Challenge Problems

Factor each trinomial.

77. $18x^2 + 9x - 20$

78. $9p^2 - 104p + 55$

79. $15x^2 - 124x + 160$

80. $16x^2 - 62x - 45$

81. $105a^2 - 220a - 160$

82. $72x^2 + 417x - 420$

83. Two factors of $6x^3 + 235x^2 + 2250x$ are x and $3x + 50$, determine the other factor. Explain how you determined your answer.

84. Two factors of the polynomial $2x^3 + 11x^2 + 3x - 36$ are $x + 3$ and $2x - 3$. Determine the third factor. Explain how you determined your answer.

Cumulative Review Exercises

[1.9]   **85.** Evaluate $-x^2 - 4(y + 3) + 2y^2$ when $x = -3$ and $y = -5$.

[2.6]   **86. Daytona 500** Jimmie Johnson won the 2006 Daytona 500 in a time of about 3.56 hours. If the race covered 507.5 miles, find the average speed of Johnson's car.

[6.1]   **87.** Factor $36x^4y^3 - 12xy^2 + 24x^5y^6$.

[6.3]   **88.** Factor $b^2 + 4b - 96$.

See Exercise 86.

Mid-Chapter Test: 6.1–6.4

To find out how well you understand the chapter material to this point, take this brief test. The answers, and the section where the material was initially discussed, are given in the back of the book. Review any questions that you answered incorrectly.

1. How may any factoring problem be checked?

2. Determine the greatest common factor of $18xy^2$, $27x^3y^4$, and $12x^2y^3$.

In Exercises 3–5, factor the GCF from each term in the expression.

3. $4a^2b^3 - 24a^3b$

4. $5c(d - 6) - 3(d - 6)$

5. $7x(2x + 9) + 2x + 9$

In Exercises 6–10, factor by grouping.

6. $x^2 + 4x + 7x + 28$

7. $x^2 + 5x - 3x - 15$

8. $6a^2 + 15ab - 2ab - 5b^2$

9. $5x^2 - 2xy - 45x + 18y$

10. $8x^3 + 4x^2 - 48x^2 - 24x$

In Exercises 11–20, factor each polynomial completely. If the polynomial is prime, so state.

11. $x^2 - 10x + 21$

12. $t^2 + 9t + 20$

13. $p^2 - 3p - 8$

14. $x^2 + 16x + 64$

15. $m^2 - 4mn - 45n^2$

16. $3x^2 + 17x + 10$

17. $4z^2 - 11z + 6$

18. $3y^2 + 13y + 6$

19. $9x^2 - 6x + 1$

20. $6a^2 + 3ab - 3b^2$

6.5 Special Factoring Formulas and a General Review of Factoring

1 Factor the difference of two squares.

2 Factor the sum and difference of two cubes.

3 Learn the general procedure for factoring a polynomial.

There are special formulas for certain types of factoring problems that are often used. The special formulas we focus on in this section are the *difference of two squares, the sum of two cubes, and the difference of two cubes*. There is no special formula for the sum of two squares; this is because the sum of two squares cannot be factored using the set of real numbers. *You will need to memorize the three highlighted formulas in this section* so that you can use them whenever you need them.

1 Factor the Difference of Two Squares

Let's begin with the difference of two squares. Consider the binomial $x^2 - 9$. Note that each term of the binomial can be expressed as the square of some expression.

$$x^2 - 9 = x^2 - 3^2$$

This is an example of a **difference of two squares**. To factor the difference of two squares, it is convenient to use the difference of two squares formula (which was introduced in Section 5.5).

Difference of Two Squares

$$a^2 - b^2 = (a + b)(a - b)$$

EXAMPLE 1 ▶ Factor $x^2 - 9$.

Solution If we write $x^2 - 9$ as a difference of two squares, we have $(x)^2 - (3)^2$. Using the difference of two squares formula, where a is replaced by x and b is replaced by 3, we obtain the following:

$$\underset{\downarrow}{a}^2 - \underset{\downarrow}{b}^2 = (\underset{\downarrow}{a} + \underset{\downarrow}{b})(\underset{\downarrow}{a} - \underset{\downarrow}{b})$$

$$x^2 - 9 = (x)^2 - (3)^2 = (x + 3)(x - 3)$$

Thus, $x^2 - 9 = (x + 3)(x - 3)$.

▶ **Now Try Exercise 13**

EXAMPLE 2 ▶ Factor using the difference of two squares formula.

a) $x^2 - 16$ **b)** $25x^2 - 4$ **c)** $36x^2 - 49y^2$

Solution

a) $x^2 - 16 = (x)^2 - (4)^2$
$$= (x + 4)(x - 4)$$

b) $25x^2 - 4 = (5x)^2 - (2)^2$
$$= (5x + 2)(5x - 2)$$

c) $36x^2 - 49y^2 = (6x)^2 - (7y)^2$
$$= (6x + 7y)(6x - 7y)$$

▶ **Now Try Exercise 21**

EXAMPLE 3 ▶ Factor each difference of two squares.

a) $16x^4 - 9y^4$ **b)** $x^6 - y^4$

Solution

a) Rewrite $16x^4$ as $(4x^2)^2$ and $9y^4$ as $(3y^2)^2$, then use the difference of two squares formula.

$$16x^4 - 9y^4 = (4x^2)^2 - (3y^2)^2$$
$$= (4x^2 + 3y^2)(4x^2 - 3y^2)$$

b) Rewrite x^6 as $(x^3)^2$ and y^4 as $(y^2)^2$, then use the difference of two squares formula.

$$x^6 - y^4 = (x^3)^2 - (y^2)^2$$
$$= (x^3 + y^2)(x^3 - y^2)$$

▶ **Now Try Exercise 29**

EXAMPLE 4 ▶ Factor $9x^2 - 36y^2$ using the difference of two squares formula.

Solution First factor out the common factor, 9.

$$9x^2 - 36y^2 = 9(x^2 - 4y^2)$$

Now use the formula for the difference of two squares.

$$9(x^2 - 4y^2) = 9[(x)^2 - (2y)^2]$$
$$= 9(x + 2y)(x - 2y)$$

▶ **Now Try Exercise 23**

Notice in Example 4 that $9x^2 - 36y^2$ is the difference of two squares, $(3x)^2 - (6y)^2$. If you factor this difference of squares without first factoring out the common factor 9, the factoring may be more difficult. After you factor this difference of squares you will need to factor out the common factor 3 from each binomial factor, as illustrated below.

$$9x^2 - 36y^2 = (3x)^2 - (6y)^2$$
$$= (3x + 6y)(3x - 6y)$$
$$= 3(x + 2y)3(x - 2y)$$
$$= 9(x + 2y)(x - 2y)$$

We obtain the same answer as we did in Example 4. However, since we did not factor out the common factor 9 first, we had to work a little harder to obtain the answer.

EXAMPLE 5 ▶ Factor $z^4 - 16$ using the difference of two squares formula.

Solution We rewrite z^4 as $(z^2)^2$ and 16 as $(4)^2$, then use the difference of two squares formula.

$$z^4 - 16 = (z^2)^2 - (4)^2$$
$$= (z^2 + 4)(z^2 - 4)$$

Notice that the second factor, $z^2 - 4$, is also the difference of two squares. To complete the factoring, we use the difference of two squares formula again to factor $z^2 - 4$.

$$= (z^2 + 4)(z^2 - 4)$$
$$= (z^2 + 4)(z + 2)(z - 2)$$

▶ **Now Try Exercise 35**

Avoiding Common Errors

The difference of two squares can be factored. However, a sum of two squares, where there is no common factor to the two terms, cannot be factored using real numbers.

CORRECT	INCORRECT
$a^2 - b^2 = (a + b)(a - b)$	$a^2 + b^2 = (a + b)(a + b)$

---

2 Factor the Sum and Difference of Two Cubes

We begin our discussion of the sum and difference of two cubes with a multiplication of polynomials problem. Consider the product of $(a + b)(a^2 - ab + b^2)$.

$$
\begin{array}{r}
a^2 - ab + b^2 \\
a + b \\
\hline
a^2b - ab^2 + b^3 \qquad \leftarrow b(a^2 - ab + b^2) \\
a^3 - a^2b + ab^2 \qquad\qquad \leftarrow a(a^2 - ab + b^2) \\
\hline
a^3 \qquad\qquad\quad + b^3 \quad \leftarrow \text{Sum of terms}
\end{array}
$$

Thus, $(a + b)(a^2 - ab + b^2) = a^3 + b^3$. Since factoring is the opposite of multiplying, we may factor $a^3 + b^3$ as follows:

$$a^3 + b^3 = (a + b)(a^2 - ab + b^2)$$

We see, using the same procedure, that $a^3 - b^3 = (a - b)(a^2 + ab + b^2)$. The expression $a^3 + b^3$ is a sum of two cubes and the expression $a^3 - b^3$ is a difference of two cubes. The formulas for factoring **the sum and the difference of two cubes** follow.

Sum of Two Cubes

$$a^3 + b^3 = (a + b)(a^2 - ab + b^2)$$

Difference of Two Cubes

$$a^3 - b^3 = (a - b)(a^2 + ab + b^2)$$

Note that the trinomials $a^2 - ab + b^2$ and $a^2 + ab + b^2$ cannot be factored further. Now let's solve some factoring problems using the sum and the difference of two cubes.

EXAMPLE 6 ▶ Factor $x^3 + 8$.

Solution   We rewrite $x^3 + 8$ as a sum of two cubes: $x^3 + 8 = (x)^3 + (2)^3$. Using the sum of two cubes formula, if we let a correspond to x and b correspond to 2, we get

$$a^3 + b^3 = (a + b)(a^2 - a \cdot b + b^2)$$

$$x^3 + 8 = (x)^3 + (2)^3 = (x + 2)[x^2 - x \cdot 2 + 2^2]$$
$$= (x + 2)(x^2 - 2x + 4)$$

You can check the factoring by multiplying $(x + 2)(x^2 - 2x + 4)$. If factored correctly, the product of the factors will equal the original expression, $x^3 + 8$. Try it and see.

▶ **Now Try Exercise 41**

Helpful Hint

When factoring the sum or difference of two cubes remember that the sign between the terms in the *binomial factor* will be the same as the sign between the terms of the expression you are factoring. Furthermore, the sign of the *ab* term will be the opposite of the sign between the terms of the binomial factor. The last term in the trinomial factor will always be positive. Consider

$$a^3 + b^3 = (a + b)(a^2 - ab + b^2)$$

same sign
opposite sign
always positive

$$a^3 - b^3 = (a - b)(a^2 + ab + b^2)$$

same sign
opposite sign
always positive

EXAMPLE 7 ▶ Factor $y^3 - 125$.

Solution   We rewrite $y^3 - 125$ as a difference of two cubes: $(y)^3 - (5)^3$. Using the difference of two cubes formula, if we let a correspond to y and b correspond to 5, we get

$$a^3 - b^3 = (a - b)(a^2 + a \cdot b + b^2)$$

$$y^3 - 125 = (y)^3 - (5)^3 = (y - 5)[y^2 + y \cdot 5 + 5^2]$$
$$= (y - 5)(y^2 + 5y + 25)$$

▶ **Now Try Exercise 43**

EXAMPLE 8 ▶ Factor $64a^3 - b^3$.

Solution   We rewrite $64a^3 - b^3$ as a difference of two cubes. Since $(4a)^3 = 64a^3$, we write

$$64a^3 - b^3 = (4a)^3 - (b)^3$$

$$= (4a - b)[(4a)^2 + (4a)(b) + b^2]$$

$$= (4a - b)(16a^2 + 4ab + b^2)$$

▶ **Now Try Exercise 49**

EXAMPLE 9 ▶ Factor $8r^3 + 27s^3$.

Solution   We rewrite $8r^3 + 27s^3$ as a sum of two cubes. Since $8r^3 = (2r)^3$ and $27s^3 = (3s)^3$, we write

$$8r^3 + 27s^3 = (2r)^3 + (3s)^3$$

$$= (2r + 3s)[(2r)^2 - (2r)(3s) + (3s)^2]$$

$$= (2r + 3s)(4r^2 - 6rs + 9s^2)$$

▶ **Now Try Exercise 53**

Avoiding Common Errors

Recall that $a^2 + b^2 \neq (a + b)^2$ and $a^2 - b^2 \neq (a - b)^2$. The same principle applies to the sum and difference of two cubes.

CORRECT	INCORRECT
$a^3 + b^3 = (a + b)(a^2 - ab + b^2)$	$a^3 + b^3 = (a + b)^3$
$a^3 - b^3 = (a - b)(a^2 + ab + b^2)$	$a^3 - b^3 = (a - b)^3$

Since $(a + b)^3 = (a + b)(a + b)(a + b)$, it cannot possibly equal $a^3 + b^3$. Also, since $(a - b)^3 = (a - b)(a - b)(a - b)$, it cannot possibly equal $a^3 - b^3$. At this point, we suggest you determine the products of $(a + b)(a + b)(a + b)$ and $(a - b)(a - b)(a - b)$.

It may be easier to see that, for example, $a^3 + b^3 = (a + b)(a^2 - ab + b^2)$ and not $(a + b)^3$ by substituting numbers for a and b. Suppose $a = 3$ and $b = 4$, then

$$3^3 + 4^3 = (3 + 4)[3^2 - 3(4) + 4^2]$$

$$27 + 64 = 7(13)$$

$$91 = 91$$

but $3^3 + 4^3 \neq (3 + 4)^3$

$$91 \neq 343$$

---

3 Learn the General Procedure for Factoring a Polynomial

In this chapter, we have presented several methods of factoring. We now combine techniques from this and previous sections to give you an overview of a general factoring procedure.

Here is a general procedure for factoring any polynomial:

General Procedure for Factoring a Polynomial

1. If all the terms of the polynomial have a greatest common factor other than 1, factor it out.

2. If the polynomial has two terms (or is a binomial), determine whether it is a difference of two squares or a sum or a difference of two cubes. If so, factor using the appropriate formula.

3. If the polynomial has three terms, factor the trinomial using the methods discussed in Sections 6.3 and 6.4.

4. If the polynomial has more than three terms, try factoring by grouping.

5. As a final step, examine your factored polynomial to determine whether the terms in any factors have a common factor. If you find a common factor, factor it out at this point.

EXAMPLE 10 ▶ Factor $3x^4 - 27x^2$.

Solution First determine whether the terms have a greatest common factor other than 1. Since $3x^2$ is common to both terms, factor it out.

$$3x^4 - 27x^2 = 3x^2(x^2 - 9)$$
$$= 3x^2(x + 3)(x - 3)$$

Note that $x^2 - 9$ is a difference of two squares.

▶ Now Try Exercise 69

EXAMPLE 11 ▶ Factor $2m^2n^2 + 6m^2n - 36m^2$.

Solution Begin by factoring the GCF, $2m^2$, from each term. Then factor the remaining trinomial.

$$2m^2n^2 + 6m^2n - 36m^2 = 2m^2(n^2 + 3n - 18)$$
$$= 2m^2(n + 6)(n - 3)$$

▶ Now Try Exercise 81

EXAMPLE 12 ▶ Factor $15c^2d - 10cd + 20d$.

Solution

$$15c^2d - 10cd + 20d = 5d(3c^2 - 2c + 4)$$

Since $3c^2 - 2c + 4$ cannot be factored, we stop here.

▶ Now Try Exercise 77

EXAMPLE 13 ▶ Factor $3xy + 6x + 3y + 6$.

Solution Always begin by determining whether all the terms in the polynomial have a common factor. In this example, 3 is the GCF. Factor 3 from each term.

$$3xy + 6x + 3y + 6 = 3(xy + 2x + y + 2)$$

Now factor by grouping.

$$= 3[x(y + 2) + 1(y + 2)]$$
$$= 3(y + 2)(x + 1)$$

▶ Now Try Exercise 79

In Example 13, what would happen if we forgot to factor out the common factor 3? Let's rework the problem without first factoring out the 3, and see what happens. Factor $3x$ from the first two terms, and 3 from the last two terms.

$$3xy + 6x + 3y + 6 = 3x(y + 2) + 3(y + 2)$$
$$= (y + 2)(3x + 3)$$

In step 5 of the general factoring procedure on page 373, we are reminded to examine the factored polynomial to see whether the terms in any factor have a common factor. If we study the factors, we see that the factor $3x + 3$ has a common factor of 3. If we factor out the 3 from $3x + 3$ we will obtain the same answer obtained in Example 13.

$$(y + 2)(3x + 3) = 3(y + 2)(x + 1)$$

EXAMPLE 14 ▶ Factor $12x^2 + 12x - 9$.

Solution   First factor out the common factor, 3. Then factor the remaining trinomial by one of the methods discussed in Section 5.4 (either by grouping or trial and error).

$$12x^2 + 12x - 9 = 3(4x^2 + 4x - 3)$$
$$= 3(2x + 3)(2x - 1)$$

▶ Now Try Exercise 59

EXAMPLE 15 ▶ Factor $2x^4y + 54xy$.

Solution   First factor out the common factor, $2xy$.

$$2x^4y + 54xy = 2xy(x^3 + 27)$$
$$= 2xy(x + 3)(x^2 - 3x + 9)$$

Note that $x^3 + 27$ is a sum of two cubes.

▶ Now Try Exercise 95

EXERCISE SET 6.5   *Math* XL   **MyMathLab**
MathXL®   MyMathLab

Concept/Writing Exercises

1. a) Write the formula for factoring the difference of two squares.
  b) Explain how to factor the difference of two squares.

2. a) Write the formula for factoring the difference of two cubes.
  b) Explain how to factor the difference of two cubes.

3. a) Write the formula for factoring the sum of two cubes.

  b) Explain how to factor the sum of two cubes.

4. Why is it important to memorize the special factoring formulas?

5. Is there a special formula for factoring the sum of two squares?

6. Describe the general procedure for factoring a polynomial.

In Exercises 7–12, the binomial is a sum of squares. There is no formula for factoring the sum of squares. However, sometimes a common factor can be factored out from a sum of squares. Factor those polynomials that are factorable. If the polynomial is not factorable, write the word prime.

7. $x^2 + 9$

8. $4y^2 + 1$

9. $3b^2 + 48$

10. $16s^2 + 64t^2$

11. $16m^2 + 36n^2$

12. $9y^2 + 16z^2$

Practice the Skills

Factor each difference of two squares.

13. $y^2 - 25$

14. $x^2 - 4$

15. $81 - z^2$

16. $64 - z^2$

17. $x^2 - 49$

18. $c^2 - d^2$

19. $x^2 - y^2$

20. $16x^2 - 9$

21. $9y^2 - 25z^2$

22. $64z^2 - 9$

23. $64a^2 - 36b^2$

24. $100x^2 - 81y^2$

25. $36 - 49x^2$

26. $100 - y^4$

27. $z^4 - 81x^2$

28. $9x^4 - 81y^2$

29. $25x^4 - 49y^4$

30. $4x^4 - 25y^4$

31. $36m^4 - 49n^2$

32. $10x^2 - 160$

33. $2x^4 - 50y^2$

34. $4x^3 - xy^2$

35. $x^4 - 81$

36. $36x^4 - 4y^2$

Factor each sum or difference of two cubes.

37. $x^3 + y^3$

38. $a^3 - b^3$

39. $x^3 - y^3$

40. $a^3 + b^3$

41. $x^3 + 64$

42. $x^3 - 8$

43. $x^3 - 27$

44. $a^3 + 27$

45. $a^3 + 1$

46. $t^3 - 1$

47. $27x^3 - 1$

48. $64y^3 + 125$

49. $27a^3 - 125$

50. $125 + x^3$

51. $27 - 8y^3$

52. $8 + 27y^3$

53. $64m^3 + 27n^3$

54. $64x^3 - 125y^3$

55. $8a^3 - 27b^3$

56. $27c^3 + 125d^3$

Factor completely.

57. $4x^2 - 24x + 36$

58. $3x^2 - 9x - 12$

59. $50x^2 - 10x - 12$

60. $3x^2 - 48$

61. $2d^2 + 16d + 32$

62. $3x^2 + 9x + 12x + 36$

63. $5x^2 - 10x - 15$

64. $3xy - 6x + 9y - 18$

65. $5x^2 - 20$

66. $x^2y + 2xy - 6xy - 12y$

67. $2x^2 - 50$

68. $4a^2y - 64y^3$

69. $2x^2y - 18y$

70. $3x^3 - 147x$

71. $3x^3y^2 + 3y^2$

72. $x^4 - 125x$

73. $2x^3 - 16$

74. $x^3 - 27y^3$

75. $18x^2 - 50$

76. $54a^3 - 16$

77. $6t^2r - 15tr + 21r$

78. $12n^2 + 4n - 16$

79. $6x^2 - 4x + 24x - 16$

80. $4ab^2 + 4ab - 24a$

81. $2rs^2 - 10rs - 48r$

82. $4x^4 - 26x^3 + 30x^2$

83. $4x^2 + 5x - 6$

84. $12a^2 + 36a + 27$

85. $25b^2 - 100$

86. $3b^2 - 75c^2$

87. $a^5b^2 - 4a^3b^4$

88. $12x^2 + 36x - 3x - 9$

89. $5x^4 + 10x^3 + 5x^2$

90. $3c^6 + 12c^4d^2$

91. $x^3 + 25x$

92. $8y^2 - 23y - 3$

93. $y^4 - 16$

94. $36a^2 - 15ab - 6b^2$

95. $16m^3 + 250$

96. $2ab - 3b + 4a - 6$

97. $ac + 2a + bc + 2b$

98. $x^3 - 100x$

99. $9 - 9y^4$

Problem Solving

100. Explain why the sum of two squares, $a^2 + b^2$, cannot be factored using real numbers.

101. Have you ever seen the proof that 1 is equal to 2? Here it is.

Let $a = b$, then square both sides of the equation:

$$a^2 = b^2$$

$$a^2 = b \cdot b$$

$$a^2 = ab \qquad \text{Substitute } a = b.$$

$$a^2 - b^2 = ab - b^2 \qquad \text{Subtract } b^2 \text{ from both sides of the equation.}$$

$$(a + b)(a - b) = b(a - b) \qquad \text{Factor both sides of the equation.}$$

$$\frac{(a + b)\cancel{(a - b)}}{\cancel{(a - b)}} = \frac{b\cancel{(a - b)}}{\cancel{(a - b)}} \qquad \text{Divide both sides of the equation by } (a - b) \text{ and divide out common factors.}$$

$$a + b = b$$

$$b + b = b \qquad \text{Substitute } a = b.$$

$$2b = b$$

$$\frac{\overset{1}{\cancel{2b}}}{\underset{1}{\cancel{b}}} = \frac{\overset{1}{\cancel{b}}}{\underset{1}{\cancel{b}}} \qquad \text{Divide both sides of the equation by } b.$$

$$2 = 1$$

Obviously, $2 \neq 1$. Therefore, we must have made an error somewhere. Can you find it?

Factor each expression. Treat the unknown symbols as if they were variables.

102. $\blacklozenge\maltese + 2\blacklozenge + \odot\maltese + 2\odot$

103. $2\blacklozenge^6 + 4\blacklozenge^4\maltese^2$

104. $4\blacklozenge^2\maltese - 6\blacklozenge\maltese - 20\maltese\blacklozenge + 30\maltese$

Challenge Problems

105. Factor $x^6 - 27y^9$.

106. Factor $x^6 + 1$.

107. Factor $x^2 - 6x + 9 - 4y^2$. (*Hint:* Write the first three terms as the square of a binomial.)

108. Factor $x^6 - y^6$. (*Hint:* Factor initially as the difference of two squares.)

109. Factor $x^2 + 10x + 25 - y^2 + 4y - 4$. (*Hint:* Group the first three terms and the last three terms.)

Cumulative Review Exercises

[2.5] **110.** Solve $7x - 2(x + 6) = 2x - 5$.

[2.6] **111.** Use the formula $A = \dfrac{1}{2}h(b + d)$ to find h in the following trapezoid if the area of the trapezoid is 36 square inches.

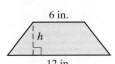

[5.1] **112.** Simplify $-9\,(a^3\, b^2\, c^6)^0$.

[5.1] **113.** Simplify $\left(\dfrac{4x^4 y}{6x y^5}\right)^3$.

[5.2] **114.** Simplify $a^{-4} a^{-7}$.

6.6 Solving Quadratic Equations Using Factoring

1 Recognize quadratic equations.

2 Solve quadratic equations using factoring.

1 Recognize Quadratic Equations

In this section, we introduce **quadratic equations**, which are equations that contain a second-degree term and no term of a higher degree.

> **Quadratic Equation**
>
> Quadratic equations have the form
>
> $$ax^2 + bx + c = 0$$
>
> where a, b, and c are real numbers, $a \neq 0$.

Examples of Quadratic Equations

$$x^2 + 4x - 12 = 0$$
$$2x^2 - 5x = 0$$
$$3x^2 - 2 = 0$$

Quadratic equations like these, in which one side of the equation is written in descending order of the variable and the other side of the equation is 0, are said to be in **standard form**.

Some quadratic equations can be solved by factoring. Two methods for solving quadratic equations that cannot be solved by factoring are given in Chapter 12. To solve a quadratic equation by factoring, we use the **zero-factor property**.

You know that if you multiply by 0, the product is 0. That is, if $a = 0$ or $b = 0$, then $ab = 0$. The reverse is also true. If a product equals 0, at least one of its factors must be 0.

> **Zero-Factor Property**
>
> If $ab = 0$, then $a = 0$ or $b = 0$.

We now illustrate how the zero-factor property is used in solving equations.

EXAMPLE 1 ▶ Solve the equation $(x + 3)(x + 4) = 0$.

Solution   Since the product of the factors equals 0, according to the zero-factor property, one or both factors must equal 0. Set each factor equal to 0, and solve each resulting equation.

$$x + 3 = 0 \qquad \text{or} \qquad x + 4 = 0$$
$$x + 3 - 3 = 0 - 3 \qquad \qquad x + 4 - 4 = 0 - 4$$
$$x = -3 \qquad \qquad \qquad x = -4$$

Thus, if x is either -3 or -4, the product of the factors is 0. The solutions to the equation are -3 and -4.

Check $x = -3$ $x = -4$

$$(x + 3)(x + 4) = 0 \qquad \qquad (x + 3)(x + 4) = 0$$
$$(-3 + 3)(-3 + 4) \overset{?}{=} 0 \qquad \qquad (-4 + 3)(-4 + 4) \overset{?}{=} 0$$
$$0(1) \overset{?}{=} 0 \qquad \qquad \qquad -1(0) \overset{?}{=} 0$$
$$0 = 0 \quad \textit{True} \qquad \qquad \qquad 0 = 0 \quad \textit{True}$$

▶ **Now Try Exercise 7**

EXAMPLE 2 ▶ Solve the equation $(3x - 2)(4x + 1) = 0$.

Solution   Set each factor equal to 0 and solve for x.

$$3x - 2 = 0 \qquad \text{or} \qquad 4x + 1 = 0$$
$$3x = 2 \qquad \qquad \qquad 4x = -1$$
$$x = \frac{2}{3} \qquad \qquad \qquad x = -\frac{1}{4}$$

The solutions to the equation are $\frac{2}{3}$ and $-\frac{1}{4}$.

▶ **Now Try Exercise 11**

2 Solve Quadratic Equations Using Factoring

Now we give a general procedure for solving quadratic equations using factoring.

To Solve a Quadratic Equation Using Factoring

1. Write the equation in standard form with the squared term having a positive coefficient. This will result in one side of the equation being 0.
2. Factor the side of the equation that is not 0.
3. Set each factor *containing a variable* equal to 0 and solve each equation.
4. Check each solution found in step 3 in the *original* equation.

EXAMPLE 3 ▶ Solve the equation $3x^2 = 12x$.

Solution   To make the right side of the equation equal to 0, we subtract $12x$ from both sides of the equation. Then we factor out $3x$ from both terms. Why did we make the right side of the equation equal to 0 instead of the left side?

$$3x^2 = 12x$$
$$3x^2 - 12x = 12x - 12x$$
$$3x^2 - 12x = 0$$
$$3x(x - 4) = 0$$

Now set each factor equal to 0.

$$3x = 0 \qquad \text{or} \qquad x - 4 = 0$$

$$x = \frac{0}{3} \qquad\qquad\qquad x = 4$$

$$x = 0$$

The solutions to the quadratic equation are 0 and 4. Check by substituting $x = 0$, then $x = 4$ in $3x^2 = 12x$.

▶ **Now Try Exercise 47**

EXAMPLE 4 ▶ Solve the equation $x^2 + 10x + 28 = 4$.

Solution   To make the right side of the equation equal to 0, we subtract 4 from both sides of the equation. Then we factor and solve.

$$x^2 + 10x + 24 = 0$$

$$(x + 4)(x + 6) = 0$$

$$x + 4 = 0 \qquad \text{or} \qquad x + 6 = 0$$

$$x = -4 \qquad\qquad x = -6$$

The solutions are -4 and -6. We will check these values in the original equation.

Check
$$x = -4 \qquad\qquad\qquad\qquad x = -6$$
$$x^2 + 10x + 28 = 4 \qquad\qquad\qquad x^2 + 10x + 28 = 4$$
$$(-4)^2 + 10(-4) + 28 \stackrel{?}{=} 4 \qquad\qquad (-6)^2 + 10(-6) + 28 \stackrel{?}{=} 4$$
$$16 - 40 + 28 \stackrel{?}{=} 4 \qquad\qquad\qquad 36 - 60 + 28 \stackrel{?}{=} 4$$
$$-24 + 28 \stackrel{?}{=} 4 \qquad\qquad\qquad\qquad -24 + 28 \stackrel{?}{=} 4$$
$$4 = 4 \quad \textit{True} \qquad\qquad\qquad\qquad 4 = 4 \quad \textit{True}$$

▶ **Now Try Exercise 23**

EXAMPLE 5 ▶ Solve the equation $4y^2 + 5y - 20 = -11y$.

Solution   Since all terms are not on the same side of the equation, add $11y$ to both sides of the equation.

$$4y^2 + 16y - 20 = 0$$

Factor out the common factor.

$$4(y^2 + 4y - 5) = 0$$

Factor the remaining trinomial.

$$4(y + 5)(y - 1) = 0$$

Now solve for y.

$$y + 5 = 0 \qquad \text{or} \qquad y - 1 = 0$$
$$y = -5 \qquad\qquad y = 1$$

Since 4 is a factor that does not contain a variable, we do not set it equal to 0. The solutions to the quadratic equation are -5 and 1.

▶ **Now Try Exercise 25**

EXAMPLE 6 ▶ Solve the equation $-x^2 + 5x + 6 = 0$.

Solution When the squared term is negative, we generally make it positive by multiplying both sides of the equation by -1.

$$-1(-x^2 + 5x + 6) = -1 \cdot 0$$
$$x^2 - 5x - 6 = 0$$

Note that the sign of each term on the left side of the equation changed and that the right side of the equation remained 0. Why? Now proceed as before.

$$x^2 - 5x - 6 = 0$$
$$(x - 6)(x + 1) = 0$$
$$x - 6 = 0 \quad \text{or} \quad x + 1 = 0$$
$$x = 6 \qquad\qquad x = -1$$

A check using the original equation will show that the solutions are 6 and -1.

▶ **Now Try Exercise 33**

Avoiding Common Errors

Be careful not to confuse factoring a polynomial with using factoring as a method to solve an equation.

CORRECT   INCORRECT

Factor: $x^2 + 3x + 2$ Factor: $x^2 + 3x + 2$
$(x + 2)(x + 1)$ $(x + 2)(x + 1)$
                                        ~~$x + 2 = 0$ or $x + 1 = 0$~~
                                        ~~$x = -2$ $x = -1$~~

Do you know what is wrong with the example on the right? It goes too far. The expression $x^2 + 3x + 2$ is a polynomial (a trinomial), not an equation. Since it is not an equation, it cannot be solved. When you are given a polynomial, you cannot just include "= 0" to change it to an equation.

CORRECT

Solve: $x^2 + 3x + 2 = 0$
$(x + 2)(x + 1) = 0$
$x + 2 = 0 \quad \text{or} \quad x + 1 = 0$
$x = -2 \qquad\qquad x = -1$

EXAMPLE 7 ▶ Solve the equation $x^2 = 49$.

Solution Subtract 49 from both sides of the equation; then factor using the difference of two squares formula.

$$x^2 - 49 = 0$$
$$(x + 7)(x - 7) = 0$$
$$x + 7 = 0 \quad \text{or} \quad x - 7 = 0$$
$$x = -7 \qquad\qquad x = 7$$

The solutions are -7 and 7.

▶ **Now Try Exercise 45**

EXAMPLE 8 ▶ Solve the equation $(x - 3)(x + 1) = 5$.

Solution   Begin by multiplying the factors, then write the quadratic equation in standard form.

$$(x - 3)(x + 1) = 5$$
$$x^2 - 2x - 3 = 5 \qquad \textit{Multiply the factors.}$$
$$x^2 - 2x - 8 = 0 \qquad \textit{Write the equation in standard form.}$$
$$(x - 4)(x + 2) = 0 \qquad \textit{Factor.}$$
$$x - 4 = 0 \quad \text{or} \quad x + 2 = 0 \qquad \textit{Zero-factor property}$$
$$x = 4 \qquad\qquad\qquad x = -2$$

The solutions are 4 and −2. We will check these values in the original equation.

Check $x = 4$ $x = -2$

$$(x - 3)(x + 1) = 5 \qquad\qquad (x - 3)(x + 1) = 5$$
$$(4 - 3)(4 + 1) \stackrel{?}{=} 5 \qquad\qquad (-2 - 3)(-2 + 1) \stackrel{?}{=} 5$$
$$1(5) \stackrel{?}{=} 5 \qquad\qquad\qquad (-5)(-1) \stackrel{?}{=} 5$$
$$5 = 5 \quad \textit{True} \qquad\qquad\qquad 5 = 5 \quad \textit{True}$$

▶ **Now Try Exercise 51**

Helpful Hint

In Example 8, you might have been tempted to start the problem by writing

$$x - 3 = 5 \quad \text{or} \quad x + 1 = 5.$$

This would lead to an incorrect solution. Remember, the zero-factor property only holds when one side of the equation is equal to 0. In Example 8, once we obtained $(x - 4)(x + 2) = 0$, we were able to use the zero-factor property.

USING YOUR GRAPHING CALCULATOR

If you solve the equation $x^2 - x - 6 = 0$ by factoring, you should obtain the solutions −2 and 3. Solve the equation now. The solutions to this equation can be found on a graphing calculator in a number of ways. If we replace the 0 with y we obtain $y = x^2 - x - 6$. **Figure 6.4** shows the graph of $Y_1 = x^2 - x - 6$ using the standard window settings.

Some graphing calculators have a TABLE feature which shows a table of values. On the TI-84 Plus press $\boxed{2^{\text{nd}}}$ $\boxed{\text{GRAPH}}$ to use the TABLE feature. In the Using Your Graphing Calculator box on page 242, we explained how to use the TABLE feature. Please re-read that material now and notice how to change the table settings. From the graph, observe that both x-intercepts occur to the right of $x = -3$. We will therefore set TblStart $= -3$. We will let ΔTbl $= 1$. After you get the table, you can scroll up and down the table using the up and down arrows. **Figure 6.5** shows a table of values for $Y_1 = x^2 - x - 6$. Note that $Y_1 = 0$ when $X = -2$ and $X = 3$. These values are the solution to the equation $x^2 - x - 6 = 0$. The values −2 and 3 are also called **zeros**, for when they are substituted for x in $y = x^2 - x - 6$, y has a value of zero. One drawback to using the TABLE feature to find the zeros is that if you do not use the appropriate table settings, you may not find the zeroes. Observe the TABLE when we use ΔTbl $= 2$, (see **Figure 6.6**.) Note that the value of Y_1 is zero only when $X = 3$. The other zero is not displayed because only odd values of x are shown in the table. If an x-intercept is not at an integer value, then it will not show up in a table in ΔTbl $= 1$. Can you explain why? If this happens you can change your table setting, or use the CALC menu, as explained on the next page.

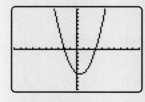

FIGURE 6.4   FIGURE 6.5   FIGURE 6.6

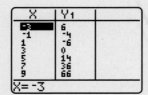

(continued on the next page)

Another method that can be used to find the solution to $x^2 - x - 6 = 0$ is to graph $y = x^2 - x - 6$ and use the CALC (which stands for calculate) menu to find the x-intercepts. On the TI-84 Plus you press $\boxed{2^{nd}}$ $\boxed{\text{TRACE}}$ to get to the CALC menu. The CALC menu is shown in **Figure 6.7**.

Scroll down to option 2: zero and press $\boxed{\text{ENTER}}$ which then gives the screen in **Figure 6.8**. Under the words *Left Bound*? you see $x = 0$. Use the arrow key to move the cursor to the right until it is just to the left of the positive x-intercept (this puts the cursor slightly below the x-axis). Note that the value of Y is negative. Press $\boxed{\text{ENTER}}$. The screen now shows *Right Bound*? Move the cursor so it is slightly to the right of the x-intercept (this puts the cursor slightly above the x-axis), then press $\boxed{\text{ENTER}}$.

Note that the sign of the value of Y has changed, and is positive. The screen now shows *Guess*? Press $\boxed{\text{ENTER}}$ again and the screen shows that the zero is 3 (see **Figure 6.9**.) That is, when $x = 3$, $y = 0$. If you make a mistake, clear your calculator, re-enter $Y_1 = x^2 - x - 6$, and start the procedure again.

FIGURE 6.7

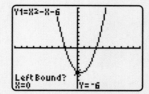

FIGURE 6.8

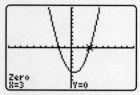

FIGURE 6.9

You can also use this procedure to determine the other zero, or the other solution to $x^2 - x - 6 = 0$. To obtain the other zero, press $\boxed{2^{nd}}$ $\boxed{\text{TRACE}}$ to get to the CALC menu again. Then using the left arrow, move the cursor from its current position to just left of the other x-intercept. In this case, it will be slightly *above* the x-axis, and the value of Y will be *positive*. Press $\boxed{\text{ENTER}}$ to get *Left Bound*? Finish the procedure by moving the cursor right until it is slightly *below* the x-axis. Press $\boxed{\text{ENTER}}$ to get the *Right Bound*? and once more to get *Guess*? The screen shows that the zero is -2.

EXERCISES

Use your graphing calculator to determine the solutions to each of the following quadratic equations. You may have to adjust the window settings to see the entire graph.

1. $x^2 - 2x - 8 = 0$

2. $x^2 - 7x + 10 = 0$

3. $2x^2 - 6x + 4 = 0$

4. $2x^2 - 8x - 10 = 0$

EXERCISE SET 6.6

 MathXL MathXL®   **MyMathLab** MyMathLab

Concept/Writing Exercises

1. Explain the zero-factor property.

2. What is a quadratic equation?

3. What is the standard form of a quadratic equation?

4. Explain the procedure to use to solve a quadratic equation.

5. a) When solving the equation $(x + 1)(x - 2) = 4$, explain why we **cannot** solve the equation by first writing

$x + 1 = 4$ or $x - 2 = 4$ and then solving each equation for x.

b) Solve the equation $(x + 1)(x - 2) = 4$.

6. When solving an equation such as $3(x - 4)(x + 5) = 0$, we set the factors $x - 4$ and $x + 5$ equal to 0, but we do not set the 3 equal to 0. Can you explain why?

Practice the Skills

Solve.

7. $(x + 6)(x - 7) = 0$

8. $-2x(x + 9) = 0$

9. $7x(x - 8) = 0$

10. $(x + 3)(x + 5) = 0$

11. $(3x + 7)(2x - 11) = 0$

12. $(3x - 2)(x - 5) = 0$

13. $x^2 - 16 = 0$

14. $x^2 - 9 = 0$

15. $x^2 - 12x = 0$

16. $9x^2 + 27x = 0$

17. $x^2 + 7x = 0$

18. $a^2 - 4a - 12 = 0$

19. $x^2 - 8x + 16 = 0$

20. $x^2 + 12x + 36 = 0$

21. $x^2 + 12x = -20$

22. $3y^2 - 4 = -4y$

23. $x^2 + 12x + 22 = 2$

24. $3x^2 = -21x - 18$

25. $2x^2 - 5x - 24 = -3x$

26. $x^2 = 4x + 21$

27. $23p - 24 = -p^2$

28. $3x^2 - 9x - 30 = 0$

29. $33w + 90 = -3w^2$

30. $t^2 + 44 + 15t = 0$

31. $-2x - 15 = -x^2$

32. $-9x + 20 = -x^2$

33. $-x^2 + 29x + 30 = 0$

34. $12y - 11 = y^2$

35. $-15 = 4m^2 + 17m$

36. $z^2 + 8z = -16$

37. $9p^2 = -21p - 6$

38. $2x^2 - 5 = 3x$

39. $3r^2 + 13r = 10$

40. $3x^2 = 7x + 20$

41. $4x^2 + 4x - 48 = 0$

42. $6x^2 - 7x - 5 = 0$

43. $8x^2 + 2x = 3$

44. $2x^2 + 4x - 6 = 0$

45. $c^2 = 64$

46. $2n^2 + 36 = -18n$

47. $2x^2 = 50x$

48. $4x^2 - 25 = 0$

49. $x^2 = 100$

50. $3x^2 - 48 = 0$

51. $(x - 2)(x - 1) = 12$

52. $(x + 2)(x + 5) = -2$

53. $(3x + 2)(x + 1) = 4$

54. $(x - 1)(2x - 5) = 9$

55. $2(a^2 + 9) = 15a$

56. $x(x + 5) = 6$

Problem Solving

In Exercises 57–60, create a quadratic equation with the given solutions. Explain how you determined your answers.

57. $6, -4$ **58.** $-3, -5$ **59.** $6, 0$ **60.** $0, -9$

61. The solutions to a quadratic equation are $\dfrac{1}{2}$ and $-\dfrac{1}{3}$.

     a) What factors with integer coefficients were set equal to 0 to obtain these solutions?

     b) Write a quadratic equation whose solutions are $\dfrac{1}{2}$ and $-\dfrac{1}{3}$.

62. The solutions to a quadratic equation are $\dfrac{2}{3}$ and $-\dfrac{3}{4}$.

     a) What factors with integer coefficients were set equal to 0 to obtain these solutions?

     b) Write a quadratic equation whose solutions are $\dfrac{2}{3}$ and $-\dfrac{3}{4}$.

Challenge Problems

63. Solve the equation
$(2x - 3)(x - 4) = (x - 5)(x + 3) + 7$.

64. Solve the equation
$(x - 3)(x - 2) = (x + 5)(2x - 3) + 21$.

65. Solve the equation $x(x - 3)(x + 2) = 0$.

66. Solve the equation $x^3 - 10x^2 + 24x = 0$.

Cumulative Review Exercises

[1.7]   **67.** Subtract $\dfrac{3}{5} - \dfrac{2}{9}$.

[2.5]   **68. a)** What is the name given to an equation that has an infinite number of solutions?

          b) What is the name given to an equation that has no solution?

[2.7]   **69. Cyprus Gardens** At Cyprus Gardens, there is a long line of people waiting to go through the entrance. If 160 people are admitted in 13 minutes, how many people will be admitted in 60 minutes? Assume the rate stays the same.

[5.1]   **70.** Simplify $\left(\dfrac{3p^5q^7}{p^9q^8}\right)^2$.

Cyprus Gardens; see Exercise 69

[5.4]   *Identify the following as a monomial, binomial, trinomial or not a polynomial. If an expression is not a polynomial, explain why.*

71. $2x$ **72.** $x - 3$ **73.** $\dfrac{1}{x}$ **74.** $x^2 - 6x + 9$

6.7 Applications of Quadratic Equations

1 Solve applications by factoring quadratic equations.

2 Learn the Pythagorean Theorem.

1 Solve Applications by Factoring Quadratic Equations

In Section 6.6, we learned how to solve quadratic equations by factoring. In this section, we will discuss and solve application problems that require solving quadratic equations to obtain the answer. In Example 1, we will solve a problem involving a relationship between two numbers.

EXAMPLE 1 ▶ **Number Problem** The product of two numbers is 78. Find the two numbers if one number is 7 more than the other.

Solution Understand and Translate   Our goal is to find the two numbers.

$$\text{Let } x = \text{smaller number.}$$

$$x + 7 = \text{larger number.}$$

$$x(x + 7) = 78$$

Carry Out
$$x^2 + 7x = 78$$

$$x^2 + 7x - 78 = 0$$

$$(x - 6)(x + 13) = 0$$

$$x - 6 = 0 \quad \text{or} \quad x + 13 = 0$$

$$x = 6 \qquad\qquad x = -13$$

Remember that x represents the smaller of the two numbers. This problem has two possible solutions.

	Solution 1	Solution 2
Smaller number	6	-13
Larger number	$x + 7 = 6 + 7 = 13$	$x + 7 = -13 + 7 = -6$

Thus, the two possible solutions are 6 and 13, and -13 and -6.

Check	6 and 13	-13 and -6
Product of the two numbers is 78.	$6 \cdot 13 = 78$	$(-13)(-6) = 78$
One number is 7 more than the other number.	13 is 7 more than 6.	-6 is 7 more than -13.

Answer   One solution is: smaller number 6, larger number 13. A second solution is: smaller number -13, larger number -6. You must give both solutions. If the question had stated "the product of two *positive* numbers is 78," the only solution would be 6 and 13.

▶ **Now Try Exercise 13**

Now let us work an application problem involving geometry.

EXAMPLE 2 ▶ **Advertising** The marketing department of a large publishing company is planning to make a large rectangular sign to advertise a new book at a convention. They want the length of the sign to be 3 feet longer than the width (**Fig. 6.10** on page 384). Signs at the convention may have a maximum area of 54 square feet. Find the length and width of the sign if the area is to be 54 squarate feet.

FIGURE 6.10

Solution   Understand and Translate   We need to find the length and width of the sign. We will use the formula for the area of a rectangle.

$$\text{Let } x = \text{width.}$$
$$x + 3 = \text{length.}$$
$$\text{area} = \text{length} \cdot \text{width}$$

Carry Out
$$54 = (x + 3)x$$
$$54 = x^2 + 3x$$
$$0 = x^2 + 3x - 54$$
$$\text{or} \quad x^2 + 3x - 54 = 0$$
$$(x - 6)(x + 9) = 0$$
$$x - 6 = 0 \quad \text{or} \quad x + 9 = 0$$
$$x = 6 \qquad\qquad x = -9$$

Check and Answer   Since the width of the sign cannot be a negative number, the only solution is

$$\text{width} = x = 6 \text{ feet}, \quad \text{length} = x + 3 = 6 + 3 = 9 \text{ feet}$$

The area, length · width, is 54 square feet, and the length is 3 feet more than the width, so the answer checks.

▶ **Now Try Exercise 21**

EXAMPLE 3 ▶ **Earth's Gravitational Field**   In earth's gravitational field, the distance, d, in feet, that an object falls t seconds after it has been released is given by the formula $d = 16t^2$. While at the top of a roller coaster, a rider's eyeglasses slide off his head and fall out of the cart. How long does it take the eyeglasses to reach the ground 64 feet below?

Solution   Understand and Translate   Substitute 64 for d in the formula and then solve for t.

$$d = 16t^2$$
$$64 = 16t^2$$

Carry Out
$$\frac{64}{16} = t^2$$
$$4 = t^2$$

Now subtract 4 from both sides of the equation and write the equation with 0 on the right side to put the quadratic equation in standard form.

$$4 - 4 = t^2 - 4$$
$$0 = t^2 - 4$$
$$\text{or} \quad t^2 - 4 = 0$$
$$(t + 2)(t - 2) = 0$$
$$t + 2 = 0 \quad \text{or} \quad t - 2 = 0$$
$$t = -2 \qquad\qquad t = 2$$

Check and Answer   Since t represents the number of seconds, it must be a positive number. Thus, the only possible answer is 2 seconds. It takes 2 seconds for the eyeglasses (or any other object falling under the influence of gravity) to fall 64 feet.

▶ **Now Try Exercise 25**

2 Learn The Pythagorean Theorem

Now we will introduce the Pythagorean Theorem, which describes an important relationship between the length of the sides of a right triangle. The Pythagorean Theorem is named after Pythagoras of Samos ($\approx$569 B.C.–475 B.C.) who was born in Samos, Ionia. Pythagoras is often described as the first pure mathematician. Unlike many later

Pythagoras of Samos

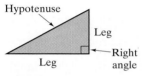

FIGURE 6.11

Greek mathematicians, relatively little is known about his life. The society he led, the Pythagorians, was half religious and half scientific. They followed a code of secrecy and did not publish any of their writings. There is fairly good agreement on the main events of Pythagoras's life, but many of the dates are disputed by scholars. Now let's discuss the Pythagorean Theorem.

A **right triangle** is a triangle that contains a right, or 90°, angle (**Fig. 6.11**). The two shorter sides of a right triangle are called the **legs** and the largest side, which is always opposite the right angle is called the **hypotenuse**. The **Pythagorean Theorem** expresses the relationship between the lengths of the legs of a right triangle and its hypotenuse.

Pythagorean Theorem

The square of the hypotenuse of a right triangle is equal to the sum of the squares of the two legs.

$$(\text{leg})^2 + (\text{leg})^2 = (\text{hypotenuse})^2$$

If a and b represent the legs, and c represents the hypotenuse, then

$$a^2 + b^2 = c^2$$

When you use the Pythagorean Theorem, it makes no difference which leg you designate as a and which leg you designate as b, but the hypotenuse is always designated as c.

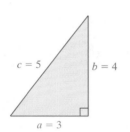

FIGURE 6.12

EXAMPLE 4 ▶ Verifying Right Triangles Determine if a right triangle can have the following sides.

a) 3 inches, 4 inches, 5 inches   **b)** 2 inches, 5 inches, 7 inches

Solution

a) Understand   To determine if a right triangle can have the sides given, we will use the Pythagorean Theorem. If the results show the Pythagorean Theorem holds true, the triangle can have the given sides. If the results are false, then the sides given cannot be those of a right triangle.

Translate   We must always select the largest size to represent the hypotenuse, c. We will designate the length of leg a to be 3 inches and the length of leg b to be 4 inches. The length of the hypotenuse, c, will be 5 inches. See **Figure 6.12**.

$$a^2 + b^2 = c^2$$
$$3^2 + 4^2 \overset{?}{=} 5^2$$

Carry Out
$$9 + 16 \overset{?}{=} 25$$
$$25 = 25 \qquad \textit{True}$$

Check and Answer   Since using the Pythagorean Theorem results in a true statement, a right triangle can have the given sides.

b) We will work this part using the Pythagorean Theorem, as we did in part **a)**. We will let leg a have a length of 2 inches, leg b have a length of 5 inches, and the hypotenuse, c, have a length of 7 inches.

$$a^2 + b^2 = c^2$$
$$2^2 + 5^2 \overset{?}{=} 7^2$$
$$4 + 25 \overset{?}{=} 49$$
$$29 = 49 \qquad \textit{False}$$

Since 29 is not equal to 49, the Pythagorean Theorem does not hold for these lengths. Therefore, no right triangle can have sides with lengths of 2 inches, 5 inches and 7 inches.

▶ **Now Try Exercise 27**

Helpful Hint

When drawing a right triangle, the hypotenuse, c, is always the side opposite the right angle. See Figures **6.13 (a)–(d)**.

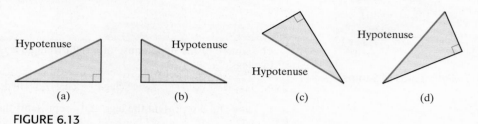

FIGURE 6.13

Notice that the hypotenuse is always the longest side of a right triangle.

EXAMPLE 5 ▶ **Using the Pythagorean Theorem** One leg of a right triangle is 7 feet longer than the other leg. The hypotenuse is 13 feet. Find the dimensions of the right triangle.

Solution   Understand and Translate   We will first draw a diagram of the situation. See **Figure 6.14**.

Now we will use the Pythagorean Theorem to determine the dimensions of the right triangle.

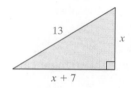

FIGURE 6.14

Carry Out

$$a^2 + b^2 = c^2$$
$$x^2 + (x + 7)^2 = 13^2$$
$$x^2 + (x^2 + 14x + 49) = 169$$
$$2x^2 + 14x - 120 = 0$$
$$2(x^2 + 7x - 60) = 0$$
$$2(x + 12)(x - 5) = 0$$
$$x + 12 = 0 \quad \text{or} \quad x - 5 = 0$$
$$x = -12 \qquad x = 5$$

Check and Answer   Since a length cannot be a negative number, the only answer is 5. The dimensions of the right triangle are 5 feet, $x + 7$ or 12 feet, and 13 feet. One leg is 5 feet, the other leg is 12 feet, and the hypotenuse is 13 feet.

▶ **Now Try Exercise 35**

EXAMPLE 6 ▶ **Sand and Water Table** Clayton Jackson is building a rectangular table for his son to hold sand in one area and water in a different area (**Fig. 6.15**). The length of the table will be 2 feet less than twice the width. He is placing a divider that is 5 feet in length along the diagonal of the table to separate the sand from the water. Find the dimensions of the table.

Solution   Understand and Translate   Our goal is to find the dimensions of the table. After looking at **Figure 6.15**, you should realize that we are working with a right triangle. Therefore, we will use the Pythagorean Theorem to answer the question. To do so, we must express the dimensions of the table in mathematical terms.

Let w = width of the table.

Then $2w - 2$ = length of the table.

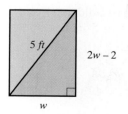

FIGURE 6.15

Figure 6.15 illustrates the relationship. Now we use the Pythagorean Theorem. We will let w represent leg a, $2w - 2$ represent leg b, and 5 represent the hypotenuse, c.

Carry Out

$$a^2 + b^2 = c^2$$
$$w^2 + (2w - 2)^2 = 5^2$$
$$w^2 + 4w^2 - 8w + 4 = 25$$
$$5w^2 - 8w + 4 = 25$$
$$5w^2 - 8w - 21 = 0$$
$$(5w + 7)(w - 3) = 0$$
$$5w + 7 = 0 \quad \text{or} \quad w - 3 = 0$$
$$5w = -7 \qquad\qquad w = 3$$
$$w = -\frac{7}{5}$$

Check and Answer   Since the width of the table cannot be a negative number, the only solution is 3. Therefore, the width of the sand and water table is 3 feet. The length of the sand and water table is $2w - 2 = 2(3) - 2 = 6 - 2 = 4$ feet.

▶ **Now Try Exercise 37**

In the Exercise Set, we use the terms consecutive integers, consecutive even integers, and consecutive odd integers. Recall from Section 3.1 that **consecutive integers** may be represented as x and $x + 1$. **Consecutive even** or **consecutive odd integers** may be represented as x and $x + 2$.

EXERCISE SET 6.7

Concept/Writing Exercises

1. What is a right triangle?

2. **a)** What are the smaller sides of a right triangle called?
   b) What is the longest side of a right triangle called?

3. State the Pythagorean Theorem formula.

4. When designating the legs of a right triangle as a and b, does it make any difference which leg you call a and which leg you call b? Explain your answer.

Practice the Skills

In Exercises 5–8, determine the value of the question mark.

5.

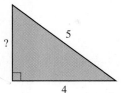

6.

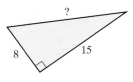

7.

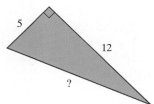

8.

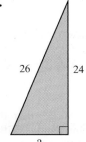

In Exercises 9–12, a and b represent two legs of a right triangle and c represents the hypotenuse. Determine the value of the question mark (?).

9. $a = 24, c = 30, b = ?$

10. $a = 12, b = 16, c = ?$

11. $a = 15, b = 36, c = ?$

12. $b = 20, c = 25, a = ?$

Problem Solving

Express each problem as an equation, then solve.

13. **Product of Numbers** The product of two positive numbers is 117. Determine the two numbers if one is 4 more than the other.

14. **Positive Numbers** The product of two positive numbers is 245. Determine the two numbers if one number is 5 times the other.

15. **Positive Numbers** The product of two positive numbers is 84. Find the two numbers if one number is 2 more than twice the other.

16. **Consecutive Integers** The product of two consecutive positive integers is 56. Find the two integers.

17. **Consecutive Even Integers** The product of two consecutive positive even integers is 288. Find the two integers.

18. **Consecutive Odd Integers** The product of two consecutive positive odd integers is 143. Determine the two integers.

19. **Area of Rectangle** The area of a rectangle is 36 square feet. Determine the length and width if the length is 4 times the width.

20. **Rectangular Scrapbook** A scrapbook page has an area of 180 square inches. Find the length and width if the width is 3 inches less than the length.

21. **Rectangular Garden** Maureen Woolhouse has a rectangular garden whose width is 2/3 its length. If its area is 150 square feet, determine the length and width of the garden.

22. **Buying Wallpaper** Alejandro Ibanez wishes to buy a wallpaper border to go along the top of one wall in his living room. The length of the wall is 7 feet greater than its height.

   a) Find the length and height of the wall if the area of the wall is 120 square feet.

   b) What is the length of the border he will need?

   c) If the border costs $4 per linear foot, how much will the border cost?

23. **Square** If each side of a square is increased by 5 meters, the area becomes 81 square meters. Determine the length of a side of the original square.

24. **Sign** If the length of the sign in Example 2 is to be 2 feet longer than the width and the area is to be 35 square feet, determine the dimensions of the sign.

25. **Dropped Egg** How long would it take for an egg dropped from a helicopter to fall 256 feet to the ground? See Example 3.

26. **Falling Rock** How long would it take a rock that falls from a cliff 400 feet above the sea to hit the sea?

In Exercises 27–30, determine if a right triangle can have the following sides where a and b represent the legs and c represents the hypotenuse. Explain your answer.

27. $a = 7, b = 24, c = 25$

28. $a = 16, c = 20, b = 22$

29. $a = 9, b = 40, c = 41$

30. $a = 13, b = 18, c = 28$

In Exercises 31–34, find the value of the question mark.

31.

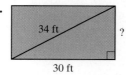

32.

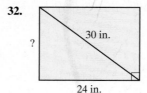

33.

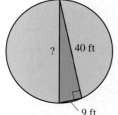

34.

35. **Triangle** One leg of a right triangle is 2 feet longer than the other leg. The hypotenuse is 10 feet. Find the lengths of the three sides of the triangle.

36. **Triangle** One leg of a right triangle is two inches more than twice the other leg. The hypotenuse is 13 inches. Find the lengths of the three sides of the triangle.

37. **Artwork** Rachel bought a framed piece of artwork as a souvenir from her trip to Disney World. The diagonal of the frame is 15 inches. If the length of the frame is 3 inches greater than its width, find the dimensions of the frame.

38. **Laptop** The top of a new experimental rectangular laptop computer has a diagonal of 17 inches. If the length of the computer is 1 inch less than twice its width, find the dimensions of the computer.

39. **Rectangular Garden** Mary Ann Tuerk has constructed a rectangular garden. The length of the garden is 3 feet more than three times its width. The diagonal of the garden is 4 feet more than three times the width. Find the length and width of the garden.

40. **Height of Tree** A tree is supported by ropes. One rope goes from the top of the tree to a point on the ground. The height of the tree is 4 feet more than twice the distance between the base of the tree and the rope anchored in the ground. The length of the rope is 6 feet more than twice the distance between the base of the tree and the rope anchored in the ground. Find the height of the tree.

41. **Book Store** A book store owner finds that her daily profit, P, is approximated by the formula $P = x^2 - 15x - 50$, where x is the number of books she sells. How many books must she sell in a day for her profit to be $400?

42. **Water Sprinklers** The cost, C, for manufacturing x water sprinklers is given by the formula $C = x^2 - 27x - 20$. Determine the number of water sprinklers manufactured at a cost of $70.

43. **Sum of Numbers** The sum, s, of the first n even numbers is given by the formula $s = n^2 + n$. Determine n for the given sums:

a) $s = 20$ b) $s = 90$

44. **Telephone Lines** For a switchboard that handles n telephone lines, the maximum number of telephone connections, C, that it can make simultaneously is given by the formula

$$C = \frac{n(n - 1)}{2}.$$

a) How many telephone connections can a switchboard make simultaneously if it handles 15 lines?

b) How many lines does a switchboard have if it can make 55 telephone connections simultaneously?

Challenge Problems

45. **Area** Determine the area of the rectangle.

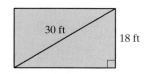

46. **Area** Determine the area of the circle.

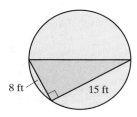

47. Solve the equation $x^3 - 4x^2 - 32x = 0$.

48. Solve the equation $x^3 + 2x^2 - 15x = 0$.

49. Create an equation whose solutions are $-2, 0$, and 3. Explain how you determined your answer.

50. **Numbers** The product of two numbers is -63. Determine the numbers if their sum is -2.

51. **Numbers** The sum of two numbers is 9. The sum of the squares of the two numbers is 45. Determine the two numbers.

52. **Wizard of Oz** Near the end of the movie *The Wizard of Oz*, when the scarecrow receives his brain, he starts talking rapidly and seems to impress us with his new found intelligence. One of the things that the scarecrow *attempts* to say is the Pythagorean Theorem. The only problem is that he says it incorrectly! Rent or borrow a videotape or DVD of the movie and write down exactly what the scarecrow says about the Pythagorean Theorem. Then see if you can rewrite the script to make it mathematically correct.

Group Activity

Discuss and solve Exercises 53 and 54 in groups.

53. Cost and Revenue The break-even point for a manufacturer occurs when its cost of production, C, is equal to its revenue, R. The cost equation for a company is $C = 2x^2 - 20x + 600$ and its revenue equation is $R = x^2 + 50x - 400$, where x is the number of units produced and sold. How many units must be produced and sold for the manufacturer to break even? There are two values.

54. Cannonball When a certain cannon is fired, the height, in feet, of the cannonball at time t can be found by using the formula $h = -16t^2 + 128t$.

   a) Determine the height of the cannonball 3 seconds after being fired.

   b) Determine the time it takes for the cannonball to hit the ground. (*Hint:* What is the value of h at impact?)

Cumulative Review Exercises

[3.1] **55.** Express the statement "seven less than three times a number" as a mathematical expression.

[5.4] **56.** Subtract $x^2 - 4x + 6$ from $3x + 2$.

[5.5] **57.** Multiply $(3x^2 + 2x - 4)(2x - 1)$.

[5.6] **58.** Divide $\dfrac{6x^2 - 19x + 15}{3x - 5}$ by dividing the numerator by the denominator.

[6.4] **59.** Divide $\dfrac{6x^2 - 19x + 15}{3x - 5}$ by factoring the numerator and dividing out common factors.

Chapter 6 Summary

IMPORTANT FACTS AND CONCEPTS	EXAMPLES
Section 6.1	
To **factor an expression** means to write the expression as a product of its factors.	$x^2 + 2x - 35 = (x + 7)(x - 5)$
If $a \cdot b = c$, then a and b are **factors** of c.	6 and 4 are factors of 24.
The **greatest common factor (GCF)** of two or more numbers is the greatest number that divides into all the numbers.	The GCF of 36 and 48 is 12.
A **prime number** is an integer greater than 1 that has exactly two factors, itself and 1. A positive integer (other than 1) that is not prime is called **composite**. The number 1 is neither prime nor composite; it is called a **unit**.	23 is a prime number. 72 is a composite number.
To Determine the GCF of Two or More Numbers 1. Write each number as a product of prime factors. 2. Determine the prime factors common to all the numbers. 3. Multiply the common factors found in step 2. The product is the GCF.	$40 = 2^3 \cdot 5$ $140 = 2^2 \cdot 5 \cdot 7$ The GCF of 40 and 140 is $2^2 \cdot 5 = 4 \cdot 5 = 20$.
To Determine the Greatest Common Factor of Two or More Terms To determine the GCF of two or more terms, take each factor the *largest* number of times that it appears in all of the terms.	The GCF of xy^2, x^3y^4, and x^2y^3 is xy^2.
To Factor a Monomial from a Polynomial 1. Determine the greatest common factor of all terms in the polynomial. 2. Write each term as the product of the GCF and its other factor. 3. Use the distributive property to factor out the GCF.	$6a^4 + 27a^3 - 18a^2 = 3a^2(2a^2 + 9a - 6)$

IMPORTANT FACTS AND CONCEPTS	EXAMPLES

Section 6.2

To Factor a Four-Term Polynomial Using Grouping

1. Determine whether there are any factors common to all four terms. If so, factor the GCF from each of the four terms.
2. If necessary, arrange the four terms so that the first two terms have a common factor and the last two have a common factor.
3. Use the distributive property to factor each group of two terms.
4. Factor the GCF from the results of step 3.

$$xy + 5x - 3y - 15 = x(y + 5) - 3(y + 5)$$
$$= (y + 5)(x - 3)$$

Section 6.3

To Factor Trinomials of the Form $ax^2 + bx + c$, where $a = 1$

1. Find two numbers whose product equals the constant, c, and whose sum equals the coefficient of the x-term, b.
2. Use the two numbers found in step 1, including their signs, to write the trinomial in factored form. The trinomial in factored form will be

$$(x + \text{one number})(x + \text{second number}).$$

Factor $x^2 + 5x - 36$.

The two numbers whose product is -36 and whose sum is 5 are 9 and -4.

Therefore,

$$x^2 + 5x - 36 = (x + 9)(x - 4).$$

A **prime polynomial** is a polynomial that cannot be factored using only integer coefficients

$x^2 - 7x + 11$ is a prime polynomial.

Section 6.4

To Factor Trinomials of the Form $ax^2 + bx + c, a \neq 1$, by Trial and Error

1. Factor out any factor common to all three terms.
2. Write all pairs of factors of the coefficient of the squared term, a.
3. Write all pairs of factors of the constant term, c.
4. Try various combinations of these factors until the correct middle term, bx, is found.

Factor $3x^2 - 2x - 8$.

Factors of -8	Possible Factors	Sum of the Products of the Outer and Inner Terms
$-1(8)$	$(3x - 1)(x + 8)$	$23x$
$-2(4)$	$(3x - 2)(x + 4)$	$10x$
$-4(2)$	$(3x - 4)(x + 2)$	$2x$
$-8(1)$	$(3x - 8)(x + 1)$	$-5x$
$4(-2)$	$(3x + 4)(x - 2)$	$-2x$

Therefore, $3x^2 - 2x - 8 = (3x + 4)(x - 2)$.

To Factor Trinomials of the Form $ax^2 + bx + c, a \neq 1$, by Grouping

1. Factor out any factor common to all three terms.
2. Find two numbers whose product is equal to the product of a times c, and whose sum is equal to b.
3. Rewrite the middle term, bx, as the sum or difference of two terms using the numbers found in step 2.
4. Factor by grouping as explained in Section 6.2.

Factor $4x^2 + 19x - 30$.

$$ac = 4(-30) = -120$$

Two numbers whose product is -120 and whose sum is 19 are 24 and -5.

$$4x^2 + 19x - 30 = 4x^2 + \overbrace{24x - 5x}^{19x} - 30$$
$$= 4x(x + 6) - 5(x + 6)$$
$$= (x + 6)(4x - 5)$$

IMPORTANT FACTS AND CONCEPTS	EXAMPLES

Section 6.5

Difference of Two Squares

$$a^2 - b^2 = (a + b)(a - b)$$

$$y^2 - 49 = (y + 7)(y - 7)$$

Sum of Two Cubes

$$a^3 + b^3 = (a + b)(a^2 - ab + b^2)$$

$$\begin{aligned} 8p^3 + q^3 &= (2p)^3 + (q)^3 \\ &= (2p + q)[(2p)^2 - (2p)(q) + (q)^2] \\ &= (2p + q)(4p^2 - 2pq + q^2) \end{aligned}$$

Difference of Two Cubes

$$a^3 - b^3 = (a - b)(a^2 + ab + b^2)$$

$$\begin{aligned} 8p^3 - q^3 &= (2p)^3 - (q)^3 \\ &= (2p - q)[(2p)^2 + (2p)(q) + (q)^2] \\ &= (2p - q)(4p^2 + 2pq + q^2) \end{aligned}$$

General Procedure for Factoring a Polynomial

1. If all the terms of the polynomial have a GCF other than 1, factor it out.
2. If the polynomial has two terms, determine whether it is a difference of two squares or a sum or a difference of two cubes. If so, factor using the appropriate formula.
3. If the polynomial has three terms, factor the trinomial using the methods discussed in Sections 6.3 and 6.4.
4. If the polynomial has more than three terms, try factoring by grouping.
5. Examine your factored polynomial to determine whether the terms in any factors have a common factor. If you find a common factor, factor it out.

Factor $3x^2 - 48$.

$$\begin{aligned} 3x^2 - 48 &= 3(x^2 - 16) \\ &= 3(x + 4)(x - 4) \end{aligned}$$

Section 6.6

Quadratic Equation

Quadratic equations have the form

$$ax^2 + bx + c = 0$$

where a, b, and c are real numbers, $a \neq 0$.

$6x^2 - 7x + 3 = 0$ is a quadratic equation in **standard form**.

Zero-Factor Property

If $ab = 0$, then $a = 0$ or $b = 0$.

If $(x + 3)(x - 1) = 0$, then $x + 3 = 0$ or $x - 1 = 0$.

To Solve a Quadratic Equation Using Factoring

1. Write the equation in standard form with the squared term having a positive coefficient.
2. Factor the side of the equation that is not 0.
3. Set each factor *containing a variable* equal to 0 and solve each equation.
4. Check each solution found in step 3 in the *original* equation.

Solve the equation $x^2 - 3x - 52 = -12$.

$$\begin{aligned} x^2 - 3x - 40 &= 0 \\ (x - 8)(x + 5) &= 0 \end{aligned}$$

$$x - 8 = 0 \quad \text{or} \quad x + 5 = 0$$
$$x = 8 \qquad\qquad x = -5$$

IMPORTANT FACTS AND CONCEPTS	EXAMPLES

Section 6.7

A **right triangle** is a triangle that contains a right, or 90°, angle.

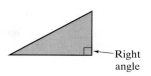
← Right angle

Pythagorean Theorem

If a and b represent the legs of a right triangle, and c represents the hypotenuse, then

$$a^2 + b^2 = c^2$$

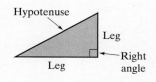
Hypotenuse
Leg
→ Right angle
Leg

c
b
a

Determine the value of x.

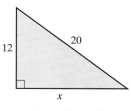
12   20
x

$$12^2 + x^2 = 20^2$$
$$144 + x^2 = 400$$
$$x^2 = 256$$
$$x = 16$$

Chapter 6 Review Exercises

[6.1] *Find the greatest common factor for each set of terms.*

1. $3y^5, y^4, y^3$

2. $3p, 6p^2, 9p^3$

3. $18c^4, 12c^2, 30c^5$

4. $20x^2y^3, 25x^3y^4, 10x^5y^2z$

5. $9xyz, 12xz, 36, x^2y$

6. $9st, 16s^2t, 24s, s^3t^3$

7. $8(x - 3), x - 3$

8. $x(x + 5), x + 5$

Factor each expression. If an expression is prime, so state.

9. $7x - 35$

10. $35x - 5$

11. $24y^2 - 4y$

12. $55p^3 - 20p^2$

13. $60a^2b - 36ab^2$

14. $9xy - 36x^3y^2$

15. $20x^3y^2 + 8x^9y^3 - 16x^5y^2$

16. $24x^2 - 13y^2 + 6xy$

17. $14a^2b - 7b - a^3$

18. $x(5x + 3) - 2(5x + 3)$

19. $3x(x - 1) + 4(x - 1)$

20. $2x(4x - 3) + 4x - 3$

[6.2] *Factor by grouping.*

21. $x^2 + 6x + 2x + 12$

22. $x^2 - 5x + 4x - 20$

23. $y^2 - 6y - 6y + 36$

24. $3xy + 3x + 2y + 2$

25. $4a^2 - 4ab - a + b$

26. $2x^2 + 12x - x - 6$

27. $x^2 + 3x - 2xy - 6y$

28. $5x^2 - xy + 20xy - 4y^2$

29. $4x^2 + 12xy - 5xy - 15y^2$

30. $6a^2 - 10ab - 3ab + 5b^2$

31. $pq - 3q + 4p - 12$

32. $3x^2 - 9xy + 2xy - 6y^2$

33. $7a^2 + 14ab - ab - 2b^2$

34. $8x^2 - 4x + 6x - 3$

[6.3] *Factor completely. If an expression is prime, so state.*

35. $x^2 - x - 6$

36. $x^2 + 4x - 15$

37. $x^2 + 11x + 18$

38. $n^2 + 3n - 40$

39. $b^2 + b - 20$

40. $x^2 - 15x + 56$

41. $c^2 - 10c - 20$

42. $y^2 - 10y - 22$

43. $x^3 - 17x^2 + 72x$

44. $t^3 - 5t^2 - 36t$

45. $x^2 - 2xy - 15y^2$

46. $4x^3 + 32x^2y + 60xy^2$

[6.4] *Factor completely. If an expression is prime, so state.*

47. $2x^2 - x - 15$

48. $6x^2 - 29x - 5$

49. $4x^2 - 9x + 5$

50. $5m^2 - 14m + 8$

51. $16y^2 + 8y - 3$

52. $5x^2 - 32x + 12$

53. $2t^2 + 14t + 9$

54. $5x^2 + 37x - 24$

55. $6s^2 + 13s + 5$

56. $6x^2 + 11x - 10$

57. $12x^2 + 2x - 4$

58. $25x^2 - 30x + 9$

59. $9x^3 - 12x^2 + 4x$

60. $18x^3 + 12x^2 - 16x$

61. $4a^2 - 16ab + 15b^2$

62. $16a^2 - 22ab - 3b^2$

[6.5] *Factor completely.*

63. $x^2 - 100$

64. $x^2 - 36$

65. $3x^2 - 48$

66. $81x^2 - 9y^2$

67. $81 - a^2$

68. $64 - x^2$

69. $16x^4 - 49y^2$

70. $64x^6 - 49y^6$

71. $a^3 + b^3$

72. $x^3 - y^3$

73. $x^3 - 1$

74. $x^3 + 8$

75. $a^3 + 27$

76. $b^3 - 64$

77. $125a^3 + b^3$

78. $27 - 8y^3$

79. $3x^3 - 192y^3$

80. $27x^4 - 75y^2$

[6.1–6.5] *Factor completely.*

81. $x^2 - 14x + 48$

82. $3x^2 - 18x + 27$

83. $5q^2 - 5$

84. $8x^2 + 16x - 24$

85. $4y^2 - 36$

86. $x^2 - 6x - 27$

87. $9x^2 - 6x + 1$

88. $7x^2 + 25x - 12$

89. $6b^3 - 6$

90. $x^3y - 27y$

91. $a^2b - 2ab - 15b$

92. $6x^3 + 30x^2 + 9x^2 + 45x$

93. $x^2 - 4xy + 3y^2$

94. $3m^2 + 2mn - 8n^2$

95. $4x^2 + 12xy + 9y^2$

96. $25a^2 - 49b^2$

97. $xy - 7x + 2y - 14$

98. $16y^5 - 25y^7$

99. $6x^2 + 5xy - 21y^2$

100. $4x^3 + 18x^2y + 20xy^2$

101. $16x^4 - 8x^3 - 3x^2$

102. $d^4 - 16$

[6.6] *Solve.*

103. $x(x + 9) = 0$

104. $(a - 2)(a + 6) = 0$

105. $(x + 5)(4x - 3) = 0$

106. $x^2 + 7x = 0$

107. $6x^2 + 30x = 0$

108. $6x^2 + 18x = 0$

109. $r^2 + 9r + 18 = 0$

110. $x^2 - 3x = -2$

111. $x^2 - 12 = -x$

112. $15x + 12 = -3x^2$

113. $x^2 - 6x + 8 = 0$

114. $3p^2 + 6p = 45$

115. $8x^2 - 3 = -10x$

116. $3p^2 - 11p = 4$

117. $4x^2 - 16 = 0$

118. $49x^2 - 100 = 0$

119. $8x^2 - 14x + 3 = 0$

120. $-48x = -12x^2 - 45$

[6.7]

121. State the Pythagorean Theorem.

122. What is the longest side of a right triangle called?

In Exercises 123 and 124, determine the value of the question mark (?).

123.
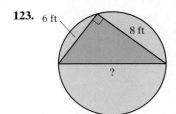

124.

Express each problem as an equation, then solve.

125. Product of Integers The product of two consecutive positive odd integers is 99. Determine the two integers.

126. Product of Integers The product of two positive integers is 56. Determine the integers if the larger is 6 more than twice the smaller.

127. Area of Rectangle The area of a rectangle is 180 square feet. Determine the length and width of the rectangle if the length is 3 feet greater than the width.

128. Right Triangle One leg of a right triangle is 7 feet longer than the other leg. The hypotenuse is 9 feet longer than the shortest leg. Find the lengths of the three sides of the triangle.

129. Square The length of each side of a square is made smaller by 4 inches. If the area of the resulting square is 25 square inches, determine the length of a side of the original square.

130. Table Brian has a rectangular table. The length of the table is 2 feet greater than the width of the table. A diagonal across the table is 4 feet greater than the width of the table. Find the length of the diagonal across the table.

131. Falling Pear How long would it take a pear that falls off a 16-foot tree to hit the ground?

132. Baking Brownies The Pine Hills Neighborhood Association has determined that the cost, C, to make x dozen brownies can be estimated by the formula $C = x^2 - 79x + 20$. If they have \$100 to be used to make the brownies, how many dozen brownies can the association make to sell at a fund-raiser?

Chapter 6 Practice Test

*To find how well you understand the chapter material, take this practice test. The answers, and the section where the material was initially discussed, are given in the back of the book. Each problem is also fully worked out on the **Chapter Test Prep Video CD**. Review any questions that you answered incorrectly..*

1. Determine the greatest common factor of $9y^5, 15y^3$, and $27y^4$.

2. Determine the greatest common factor of $8p^3q^2, 32p^2q^5$, and $24p^4q^3$.

Factor completely.

3. $5x^2y^3 - 15x^5y^2$

4. $8a^3b - 12a^2b^2 + 28a^2b$

5. $4x^2 - 20x + x - 5$

6. $a^2 - 4ab - 5ab + 20b^2$

7. $r^2 + 5r - 24$

8. $25a^2 - 5ab - 6b^2$

9. $4x^2 - 16x - 48$

10. $2y^3 - y^2 - 3y$

11. $12x^2 - xy - 6y^2$

12. $x^2 - 9y^2$

13. $x^3 - 64$

Solve.

14. $(6x - 5)(x + 3) = 0$ **15.** $x^2 - 6x = 0$

16. $x^2 = 64$ **17.** $x^2 + 18x + 81 = 0$

18. $x^2 - 7x + 12 = 0$ **19.** $x^2 + 6 = -5x$

20. Right Triangle Find the length of the side indicated with a question mark.

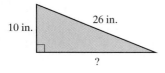

10 in.   26 in.

?

21. Right Triangle In a right triangle, one leg is 2 feet less than twice the length of the smaller leg. The hypotenuse is 2 feet more than twice the length of the smaller leg. Determine the hypotenuse of the triangle.

22. Product of Integers The product of two positive integers is 36. Determine the two integers if the larger is 1 more than twice the smaller.

23. Consecutive Even Integers The product of two positive consecutive even integers is 168. Determine the integers.

24. Rectangle The area of a rectangle is 24 square meters. Determine the length and width of the rectangle if its length is 2 meters greater than its width.

25. Fallen Object How long would it take for an object dropped from a hot air balloon to fall 1600 feet to the ground?

Cumulative Review Test

Take the following test and check your answers with those given in the back of the book. Review any questions that you answered incorrectly. The section where the material was covered is indicated after the answer.

1. Evaluate $4 - 5(2x + 4x^2 - 21)$ when $x = -4$.

2. Evaluate $5x^2 - 3y + 7(2 + y^2 - 4x)$ when $x = 3$ and $y = -2$.

3. Motel Room The cost of a motel room including a 12% state tax and 3% county tax is $103.50. Determine the cost of the room before tax.

4. Consider the set of numbers

$$\left\{-6, -0.2, \frac{3}{5}, \sqrt{7}, -\sqrt{2}, 7, 0, -\frac{5}{9}, 1.34\right\}$$

List the elements that are

a) natural numbers.

b) rational numbers.

c) irrational numbers.

d) real numbers.

5. Which is greater, $|-8|$ or $-|8|$? Explain your answer.

6. Solve the equation $4x - 2 = 4(x - 7) + 2x$ for x.

7. Solve the proportion $\dfrac{5}{12} = \dfrac{8}{x}$ for x by cross-multiplying.

8. Painting A gallon of paint covers 825 square feet. How much paint is needed to cover a house with a surface area of 5775 square feet?

9. Solve the equation $4x + 3y = 7$ for y. Write your answer in $y = mx + b$ form.

10 **Acid Solution** How many liters of a 10% acid solution must be mixed with three liters of a 4% acid solution to get an 8% acid solution?

11. Cross-Country Skiing Two cross-country skiers follow the same trail in a local park. Brooke Stoner skis at a rate of 8 kilometers per hour and Bob Thoresen skis at a rate of 4 kilometers per hour. How long will it take Brooke to catch Bob if she leaves 15 minutes after he does?

12. Graph $y = -\dfrac{3}{5}x + 1$ using the slope and y-intercept.

13. Find an equation of a line through the points $(3, 7)$ and $(-2, 4)$. Write the equation in slope-intercept form.

14. Simplify $(2x^{-3})^{-2}(4x^{-3}y^2)^3$.

15. Subtract $(4x^3 - 3x^2 + 7)$ from $(x^3 - x^2 + 6x - 5)$.

Perform the operations indicated.

16. $(3x - 2)(x^2 + 5x - 6)$

17. $\dfrac{x^2 - 2x + 6}{x + 3}$

18. Factor $qr + 2q - 8r - 16$ by grouping.

19. Factor $5x^2 - 7x - 6$.

20. Factor $7y^3 - 63y$.

7 Rational Expressions and Equations

WHEN TWO PEOPLE perform a task, it takes less time than for each person to do the task separately. In Exercise 29 on page 452, we will determine how long it takes two people working together to load baskets of peaches onto a truck when we know how long it takes for each person to do the task separately.

7.1 Simplifying Rational Expressions

1 Determine the values for which a rational expression is defined.

2 Understand the three signs of a fraction.

3 Simplify rational expressions.

4 Factor a negative 1 from a polynomial.

1 Determine the Values for Which a Rational Expression Is Defined

We begin this chapter by defining a *rational expression*.

> **Rational Expression**
>
> A **rational expression** is an expression of the form p/q, where p and q are polynomials and $q \neq 0$.

Examples of Rational Expressions

$$\frac{4}{5}, \quad \frac{x-6}{x}, \quad \frac{x^2+8x}{x-3}, \quad \frac{a}{a^2-4}$$

The denominator of a rational expression cannot equal 0 since division by 0 is not defined. In the expression $\frac{x+3}{x}$, the value of x cannot be 0 because the denominator would then equal 0. We say that the expression $\frac{x+3}{x}$ is *defined* for all real numbers except 0. It is *undefined* when x is 0. In $\frac{x^2+8x}{x-3}$, the value of x cannot be 3 because the denominator would then be 0. What values of x cannot be used in the expression $\frac{x}{x^2-4}$? If you answered 2 and -2, you answered correctly. **Whenever we have a rational expression containing a variable in the denominator, we always assume that the value or values of the variable that make the denominator 0 are excluded.**

One method that can be used to determine the value or values of the variable that are excluded is to set the denominator equal to 0 and then solve the resulting equation for the variable.

EXAMPLE 1 ▶ Determine the value or values of the variable for which the rational expression is defined.

a) $\dfrac{7}{x-4}$ **b)** $\dfrac{3x+4}{2x-7}$ **c)** $\dfrac{x+5}{x^2+6x-7}$

Solution

a) We need to determine the value or values of x that make $x-4$ equal to 0 and exclude these values. By looking at the denominator we can see that when $x=4$, the denominator is $4-4$ or 0. Thus, we do not consider $x=4$ when we consider the rational expression $\dfrac{7}{x-4}$. This expression is defined for all real numbers except 4. We will sometimes shorten our answer and write $x \neq 4$.

b) We need to determine the value or values of x that make $2x-7$ equal to 0 and exclude these. We can do this by setting $2x-7$ equal to 0 and solving the equation for x.

$$2x-7=0$$
$$2x=7$$
$$x=\frac{7}{2}$$

Thus, we do not consider $x=\dfrac{7}{2}$ when we consider the rational expression $\dfrac{3x+4}{2x-7}$. This expression is defined for all real numbers except $x=\dfrac{7}{2}$.

We will sometimes shorten our answer and write $x \neq \dfrac{7}{2}$.

c) To determine the value or values that are excluded, we set the denominator equal to zero and solve the equation for the variable.

$$x^2 + 6x - 7 = 0$$
$$(x + 7)(x - 1) = 0$$
$$x + 7 = 0 \quad \text{or} \quad x - 1 = 0$$
$$x = -7 \quad\quad\quad\quad x = 1$$

Therefore, we do not consider the values $x = -7$ or $x = 1$ when we consider the rational expression $\dfrac{x + 5}{x^2 + 6x + 7}$. Both $x = -7$ and $x = 1$ make the denominator zero. This expression is defined for all real numbers except $x = -7$ and $x = 1$. Thus, $x \neq -7$ and $x \neq 1$.

▶ **Now Try Exercise 15**

2   Understand the Three Signs of a Fraction

Three **signs** are associated with any fraction: the sign of the numerator, the sign of the denominator, and the sign of the fraction.

$$\text{Sign of fraction} \longrightarrow + \dfrac{-a}{+b}$$

Sign of numerator (pointing to top); *Sign of denominator* (pointing to bottom)

Whenever any of the three signs is omitted, we assume it to be positive. For example,

$$\dfrac{a}{b} \quad \text{means} \quad + \dfrac{+a}{+b}$$

$$\dfrac{-a}{b} \quad \text{means} \quad + \dfrac{-a}{+b}$$

$$-\dfrac{a}{b} \quad \text{means} \quad - \dfrac{+a}{+b}$$

Negative Fractions

Changing any two of the three signs of a fraction does not change the value of a fraction. Thus,

$$\dfrac{-a}{b} = -\dfrac{a}{b} = \dfrac{a}{-b}$$

Generally, we do not write a fraction with a negative denominator. For example, the expression $\dfrac{2}{-5}$ would be written as either $\dfrac{-2}{5}$ or $-\dfrac{2}{5}$. The expression $\dfrac{x}{-(4 - x)}$ can be written $\dfrac{x}{x - 4}$ since $-(4 - x) = -4 + x$ or $x - 4$.

3   Simplify Rational Expressions

A rational expression is **simplified** or **reduced to its lowest terms** when the numerator and denominator have no common factors other than 1. The fraction $\dfrac{9}{12}$ is not simplified because 9 and 12 both contain the common factor 3. When the 3 is factored out, the simplified fraction is $\dfrac{3}{4}$.

$$\dfrac{9}{12} = \dfrac{\overset{1}{\cancel{3}} \cdot 3}{\underset{1}{\cancel{3}} \cdot 4} = \dfrac{3}{4}$$

The rational expression $\dfrac{ab - b^2}{2b}$ is not simplified because both the numerator and denominator have a common factor, b. To simplify this expression, factor b from each term in the numerator, then divide it out.

$$\frac{ab - b^2}{2b} = \frac{\overset{1}{\cancel{b}}(a - b)}{2\underset{1}{\cancel{b}}} = \frac{a - b}{2}$$

Thus, $\dfrac{ab - b^2}{2b}$ becomes $\dfrac{a - b}{2}$ when simplified.

> ### To Simplify Rational Expressions
>
> 1. Factor both the numerator and denominator as completely as possible.
> 2. Divide out any factors common to both the numerator and denominator.

EXAMPLE 2 ▶ Simplify $\dfrac{5x^3 + 10x^2 - 35x}{10x^2}$.

Solution Factor the greatest common factor, $5x$, from each term in the numerator. Since $5x$ is a factor common to both the numerator and denominator, divide it out.

$$\frac{5x^3 + 10x^2 - 35x}{10x^2} = \frac{\overset{1}{\cancel{5x}}(x^2 + 2x - 7)}{\underset{1}{\cancel{5x}} \cdot 2x} = \frac{x^2 + 2x - 7}{2x}$$

▶ **Now Try Exercise 33**

Helpful Hint

Study the expression on the left. The explanation on the right indicates whether a common factor may be divided out of both the numerator and denominator. If an expression can be simplified, the simplified expression is shown on the right.

Expression	Simplified Expression
$\dfrac{x}{x + 3}$	This expression cannot be simplified further because the x in the denominator is not a factor of the denominator.
$\dfrac{x}{3x}$	Since the x is a factor of both the numerator and the denominator, we can factor out an x and we can simplify as follows: $$\frac{\overset{1}{\cancel{x}}}{3\underset{1}{\cancel{x}}} = \frac{1}{3}$$
$\dfrac{(x + 3)(x - 4)}{x + 3}$	Since $x + 3$ can be written as $1(x + 3)$, the $x + 3$ can be considered a factor of the denominator. Since $(x + 3)$ is also a factor of the numerator, the $x + 3$ can be factored out of both the numerator and denominator, as shown below. $$\frac{\overset{1}{\cancel{(x + 3)}}(x - 4)}{1\underset{1}{\cancel{(x + 3)}}} = \frac{x - 4}{1} = x - 4$$

Helpful Hint

In Example 2 we *simplified* a polynomial divided by a monomial using factoring. In Section 5.6 we *divided* polynomials by monomials by writing each term in the numerator over the expression in the denominator. For example,

$$\frac{5x^3 + 10x^2 - 35x}{10x^2} = \frac{5x^3}{10x^2} + \frac{10x^2}{10x^2} - \frac{35x}{10x^2}$$

$$= \frac{x}{2} + 1 - \frac{7}{2x}$$

The answer above, $\frac{x}{2} + 1 - \frac{7}{2x}$, is equivalent to the answer $\frac{x^2 + 2x - 7}{2x}$, which was obtained by factoring in Example 2. We show this below.

$$\frac{x}{2} + 1 - \frac{7}{2x}$$

$$= \frac{x}{x} \cdot \frac{x}{2} + \frac{2x}{2x} \cdot 1 - \frac{7}{2x} \qquad \textit{Write each term with LCD 2x.}$$

$$= \frac{x^2}{2x} + \frac{2x}{2x} - \frac{7}{2x}$$

$$= \frac{x^2 + 2x - 7}{2x}$$

When asked to *simplify* an expression we will factor numerators and denominators, when possible, then divide out common factors. This process was illustrated in Example 2 and will be further illustrated in Examples 3 through 5.

EXAMPLE 3 ▸ Simplify $\dfrac{x^2 - x - 12}{x + 3}$.

Solution   Factor the numerator; then divide out the common factor.

$$\frac{x^2 - x - 12}{x + 3} = \frac{\overset{1}{\cancel{(x + 3)}}(x - 4)}{\underset{1}{\cancel{x + 3}}} = x - 4$$

▸ **Now Try Exercise 35**

EXAMPLE 4 ▸ Simplify $\dfrac{r^2 - 25}{r - 5}$.

Solution   Factor the numerator; then divide out common factors.

$$\frac{r^2 - 25}{r - 5} = \frac{(r + 5)\overset{1}{\cancel{(r - 5)}}}{\underset{1}{\cancel{r - 5}}} = r + 5$$

▸ **Now Try Exercise 61**

EXAMPLE 5 ▸ Simplify $\dfrac{3x^2 - 10x - 8}{x^2 + 3x - 28}$.

Solution   Factor both the numerator and denominator, then divide out common factors.

$$\frac{3x^2 - 10x - 8}{x^2 + 3x - 28} = \frac{(3x + 2)\overset{1}{\cancel{(x - 4)}}}{(x + 7)\underset{1}{\cancel{(x - 4)}}} = \frac{3x + 2}{x + 7}.$$

Note that $\dfrac{3x + 2}{x + 7}$ cannot be simplified any further.

▸ **Now Try Exercise 41**

Avoiding Common Errors

Remember: Only common *factors* can be divided out from expressions.

CORRECT

$$\frac{\overset{5}{\cancel{20}}\,\overset{x}{\cancel{x^2}}}{\underset{1}{\cancel{4}}\,\underset{1}{\cancel{x}}} = 5x$$

INCORRECT

$$\cancel{\frac{\overset{x}{x^2} - \overset{5}{20}}{\underset{1}{x} - \underset{1}{4}}}$$

In the denominator of the example on the left, $4x$, the 4 and x are factors since they are *multiplied* together. The 4 and the x are also both factors of the numerator $20x^2$, since $20x^2$ can be written $4 \cdot x \cdot 5x$.

Some students incorrectly divide out *terms*. In the expression $\dfrac{x^2 - 20}{x - 4}$, the x and -4 are *terms* of the denominator, not factors, and therefore cannot be divided out.

USING YOUR GRAPHING CALCULATOR

The graphing calculator can be used to determine whether a rational expression has been simplified correctly. To check the simplification below,

$$\frac{3x + 6}{6x^2 - 24} = \frac{1}{2(x - 2)}$$

we let $Y_1 = \dfrac{3x + 6}{6x^2 - 24}$ and $Y_2 = \dfrac{1}{2(x - 2)}$. Then we use the TABLE feature to compare results, as shown in **Figure 7.1**.

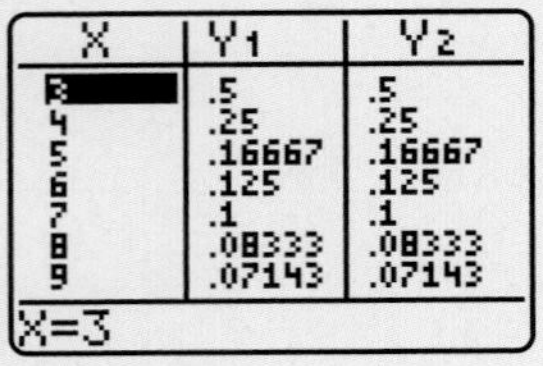

X	Y₁	Y₂
3	.5	.5
4	.25	.25
5	.16667	.16667
6	.125	.125
7	.1	.1
8	.08333	.08333
9	.07143	.07143

X=3

FIGURE 7.1

Since Y_1 and Y_2 have the same values for each value of x, we have not made a mistake. This procedure can only tell you if a mistake has been made. It cannot tell you if the fraction has been simplified completely. For example, $\dfrac{3x + 6}{6x^2 - 24}$ and $\dfrac{3}{6x - 12}$ will give you the same set of values, but $\dfrac{3}{6x - 12}$ is not simplified completely. Try this example again, using TblMin $= -3$. What do you notice about the values for Y_1 and Y_2? Do you know why the table is slightly different?

EXERCISES

Use the TABLE feature of your graphing calculator to determine whether the following rational expressions are equal for all values of x where the expression on the left is defined.

1. $\dfrac{x^2 - 4x + 4}{x^2 - 2x}$; $\dfrac{x - 2}{x}$

2. $\dfrac{x + 8}{x^2 - 64}$; $\dfrac{1}{x - 8}$

3. $\dfrac{x^2 - 4x + 4}{x^2 - 3x - 10}$; $\dfrac{x - 2}{x - 5}$

4. $\dfrac{5x^2 + 30x}{x + 6}$; $5x$

4 Factor a Negative 1 from a Polynomial

Recall from Section 6.2 that when -1 is factored from a polynomial, the sign of each term in the polynomial changes.

Examples

$$-3x + 10 = -1(3x - 10) = -(3x - 10)$$
$$5 - 2x = -1(-5 + 2x) = -(2x - 5)$$
$$-2x^2 + 3x - 4 = -1(2x^2 - 3x + 4) = -(2x^2 - 3x + 4)$$

Whenever the terms in a numerator and denominator differ only in their signs (one is the opposite or additive inverse of the other), we can factor out -1 from either the numerator or denominator and then divide out the common factor. This procedure is illustrated in Example 6.

EXAMPLE 6 ▶ Simplify $\dfrac{3x - 8}{8 - 3x}$.

Solution   Since each term in the numerator differs only in sign from its like term in the denominator, we will factor -1 from each term in the denominator.

$$\frac{3x - 8}{8 - 3x} = \frac{3x - 8}{-1(-8 + 3x)}$$

$$= \frac{\overset{1}{\cancel{3x - 8}}}{-\underset{1}{\cancel{(3x - 8)}}}$$

$$= -1$$

▶ **Now Try Exercise 43**

Helpful Hint

In Example 6 we found that $\dfrac{3x - 8}{8 - 3x} = -1$. Note that the numerator, $3x - 8$, and the denominator, $8 - 3x$, are opposites since they differ only in sign. That is,

$$\frac{3x - 8}{8 - 3x} = \frac{3x - 8}{-(3x - 8)} = -1$$

Whenever we have the quotient of two expressions that are opposites, such as $\dfrac{a - b}{b - a}, a \neq b$, the quotient can be replaced by -1.

We will use this Helpful Hint in Example 7.

EXAMPLE 7 ▶ Simplify $\dfrac{4n^2 - 23n - 6}{6 - n}$.

Solution $\dfrac{4n^2 - 23n - 6}{6 - n} = \dfrac{(4n + 1)(n - 6)}{6 - n}$ *The terms in $n - 6$ differ only in sign from the terms in $6 - n$.*

$$= (4n + 1)(-1)$$ *Replace $\dfrac{n - 6}{6 - n}$ with -1.*

$$= -(4n + 1)$$

Note that $-4n - 1$ is also an acceptable answer.

▶ **Now Try Exercise 49**

EXERCISE SET 7.1   Math XL   MyMathLab
MathXL®   MyMathLab

Concept/Writing Exercises

1. a) Define a rational expression.
  b) Give three examples of rational expressions.

2. Explain how to determine the value or values of the variable that makes a rational expression undefined.

3. In any rational expression with a variable in the denominator, what do we always assume about the variable?

4. Explain how to simplify a rational expression.

Explain why the following expressions cannot be simplified.

5. $\dfrac{4 + 3x}{9}$

6. $\dfrac{5x + 7y}{12xy}$

Explain why x can represent any real number in the following expressions.

7. $\dfrac{x - 3}{x^2 + 2}$

8. $\dfrac{x + 6}{x^2 + 4}$

In Exercises 9 and 10, determine what values, if any, x cannot represent. Explain.

9. $\dfrac{x + 5}{x - 2}$

10. $\dfrac{x}{(x + 4)^2}$

11. Is $-\dfrac{x + 8}{8 - x}$ equal to -1? Explain.

12. Is $-\dfrac{3x + 2}{-3x - 2}$ equal to 1? Explain.

Practice the Skills

Determine the value or values of the variable where each expression is defined.

13. $\dfrac{x - 2}{x}$

14. $\dfrac{8}{r + 3}$

15. $\dfrac{7}{4n - 16}$

16. $\dfrac{7}{2x - 3}$

17. $\dfrac{x + 4}{x^2 - 4}$

18. $\dfrac{9}{x^2 + 4x - 5}$

19. $\dfrac{2x - 3}{2x^2 - 9x + 9}$

20. $\dfrac{x^2 + 2}{2x^2 - 13x + 15}$

21. $\dfrac{x}{x^2 + 36}$

22. $\dfrac{4}{9 + x^2}$

23. $\dfrac{p + 8}{4p^2 - 25}$

24. $\dfrac{10}{9r^2 - 16}$

Simplify the following rational expressions which contain a monomial divided by a monomial. This material was covered in Sections 5.1 and 5.2, and it will help prepare you for the next section.

25. $\dfrac{8x^3y}{24x^2y^5}$

26. $\dfrac{18x^3y^2}{30x^4y^5}$

27. $\dfrac{(2a^4b^5)^3}{2a^{12}b^{20}}$

28. $\dfrac{(5r^2s^3)^2}{(3r^4s)^3}$

Simplify.

29. $\dfrac{2x}{x + xy}$

30. $\dfrac{12x}{3x + 9}$

31. $\dfrac{5x + 15}{x + 3}$

32. $\dfrac{3x^2 + 6x}{3x^2 + 9x}$

33. $\dfrac{x^3 + 6x^2 + 7x}{2x}$

34. $\dfrac{x^2y^2 - 2xy + 5y}{y}$

35. $\dfrac{r^2 - r - 2}{r - 2}$

36. $\dfrac{b - 6}{b^2 - 8b + 12}$

37. $\dfrac{x^2 + 2x}{x^2 + 4x + 4}$

38. $\dfrac{x^2 + 3x - 18}{4x - 12}$

39. $\dfrac{z^2 - 10z + 25}{z^2 - 25}$

40. $\dfrac{k^2 - 6k + 9}{k^2 - 9}$

41. $\dfrac{x^2 - 2x - 3}{x^2 - x - 6}$

42. $\dfrac{4x^2 - 12x - 40}{2x^2 - 16x + 30}$

43. $\dfrac{4x - 3}{3 - 4x}$

44. $\dfrac{10a - 6}{3 - 5a}$

45. $\dfrac{x^2 - 2x - 8}{4 - x}$

46. $\dfrac{7 - s}{s^2 - 12s + 35}$

47. $\dfrac{x^2 + 3x - 18}{-2x^2 + 6x}$

48. $\dfrac{3p^2 - 13p - 10}{3p + 2}$

49. $\dfrac{2x^2 + 5x - 3}{1 - 2x}$

50. $\dfrac{x^2 - 25}{x^2 - 2x - 15}$

51. $\dfrac{m - 2}{4m^2 - 13m + 10}$

52. $\dfrac{2x^2 - 13x + 21}{(x - 3)^2}$

53. $\dfrac{x^2 - 25}{(x + 5)^2}$

54. $\dfrac{16x^2 + 24x + 9}{4x + 3}$

55. $\dfrac{6x^2 - 13x + 6}{3x - 2}$

56. $\dfrac{6t^2 - 7t - 5}{3t - 5}$

57. $\dfrac{x^2 - 3x + 4x - 12}{x + 4}$

58. $\dfrac{x^2 - 2x + 4x - 8}{2x^2 + 3x + 8x + 12}$

59. $\dfrac{2x^2 - 8x + 3x - 12}{2x^2 + 8x + 3x + 12}$

60. $\dfrac{x^3 - 125}{x^2 - 25}$

61. $\dfrac{a^3 - 8}{a - 2}$

62. $\dfrac{x^3 + 1}{x + 1}$

63. $\dfrac{9s^2 - 16t^2}{3s - 4t}$

64. $\dfrac{a + 6b}{a^2 - 36b^2}$

65. $\dfrac{6x + 9y}{2x^2 + xy - 3y^2}$

66. $\dfrac{3k^2 + 6kr - 9r^2}{k^2 + 5kr + 6r^2}$

Problem Solving

Simplify the following expressions, if possible. Treat the unknown symbol as if it were a variable.

67. $\dfrac{3\text{☺}}{15}$

68. $\dfrac{\text{☺}}{\text{☺} + 8\text{☺}^2}$

69. $\dfrac{7\Delta}{14\Delta + 63}$

70. $\dfrac{\Delta^2 + 2\Delta}{\Delta^2 + 4\Delta + 4}$

71. $\dfrac{3\Delta - 4}{4 - 3\Delta}$

72. $\dfrac{(\Delta - 3)^2}{\Delta^2 - 6\Delta + 9}$

Determine the denominator that will make each statement true. Explain how you obtained your answer.

73. $\dfrac{x^2 - x - 6}{\Box} = x - 3$

74. $\dfrac{2x^2 + 11x + 12}{\Box} = 2x + 3$

Determine the numerator that will make each statement true. Explain how you obtained your answer.

75. $\dfrac{\Box}{x + 4} = x + 5$

76. $\dfrac{\Box}{x - 5} = 2x - 1$

Challenge Problems

*In Exercises 77–79, **a)** determine the value or values that x cannot represent. **b)** Simplify the expression.*

77. $\dfrac{x - 2}{x^2 - 2x + 3x - 6}$

78. $\dfrac{x - 4}{2x^2 - 5x - 8x + 20}$

79. $\dfrac{x + 5}{2x^3 + 7x^2 - 15x}$

Simplify. Explain how you determined your answer.

80. $\dfrac{\frac{1}{6}x^5 - \frac{2}{3}x^4}{x^4}$

81. $\dfrac{\frac{1}{5}x^5 - \frac{2}{3}x^4}{\frac{1}{5}x^5 - \frac{2}{3}x^4}$

82. $\dfrac{\frac{1}{5}x^5 - \frac{2}{3}x^4}{\frac{2}{3}x^4 - \frac{1}{5}x^5}$

Group Activity

Discuss and answer Exercise 83 as a group.

83. a) As a group, determine the values of the variable where the expression $\dfrac{x^2 - 25}{x^3 + 2x^2 - 15x}$ is undefined.

b) As a group, simplify the rational expression.

c) Group member 1: Substitute 6 in the *original expression* and evaluate.

d) Group member 2: Substitute 6 in the *simplified expression* from part **b)** and compare your result to that of Group member 1.

e) Group member 3: Substitute −2 in the original expression and in the simplified expression in part **b)**. Compare your answers.

f) As a group, discuss the results of your work in parts **c)**–**e)**.

g) Now, as a group, substitute −5 in the original expression and in the simplified expression. Discuss your results.

h) Is $\dfrac{x^2 - 25}{x^3 + 2x^2 - 15x}$ always equal to its simplified form for *any* value of x? Explain your answer.

Cumulative Review Exercises

[2.6] **84.** Solve the formula $z = \dfrac{x - y}{4}$ for y.

[3.3] **85. Triangle** Find the measures of the three angles of a triangle if one angle is 30° greater than the smallest angle, and the third angle is 10° greater than 3 times the smallest angle.

[5.1] **86.** Simplify $\left(\dfrac{5x^2y^2}{9x^4y^3}\right)^2$.

[5.4] **87.** Subtract $3x^2 - 4x - 8 - (-5x^2 + 6x + 11)$.

[6.3] **88.** Factor $3a^2 - 6a - 72$ completely.

[6.7] **89.** Find the length of the hypotenuse of the right triangle.

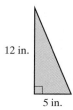

12 in.

5 in.

7.2 Multiplication and Division of Rational Expressions

1 Multiply rational expressions.

2 Divide rational expressions.

1 Multiply Rational Expressions

In Section 1.3 we reviewed multiplication of numerical fractions. Recall that to multiply two fractions we multiply their numerators together and multiply their denominators together.

> **To Multiply Two Fractions**
>
> $$\frac{a}{b} \cdot \frac{c}{d} = \frac{a \cdot c}{b \cdot d}, \quad b \neq 0 \quad \text{and} \quad d \neq 0$$

EXAMPLE 1 ▸ Multiply $\left(\frac{3}{5}\right)\left(\frac{-2}{9}\right)$.

Solution First divide out common factors; then multiply.

$$\frac{\overset{1}{\cancel{3}}}{5} \cdot \frac{-2}{\underset{3}{\cancel{9}}} = \frac{1 \cdot (-2)}{5 \cdot 3} = -\frac{2}{15}$$

▸ **Now Try Exercise 7**

The same principles apply when multiplying rational expressions containing variables. Before multiplying, you should first divide out any factors common to both a numerator and a denominator.

> **To Multiply Rational Expressions**
>
> 1. Factor all numerators and denominators completely.
> 2. Divide out common factors.
> 3. Multiply numerators together and multiply denominators together.

EXAMPLE 2 ▸ Multiply $\dfrac{3x^2}{2y} \cdot \dfrac{4y^3}{3x}$.

Solution This problem can be represented as

$$\frac{3x^2}{2y} \cdot \frac{4y^3}{3x} = \frac{3xx}{2y} \cdot \frac{4yyy}{3x}$$

$$= \frac{\overset{1}{\cancel{3}}\overset{1}{\cancel{x}}x}{2y} \cdot \frac{4yyy}{\underset{1}{\cancel{3}}\underset{1}{\cancel{x}}} \qquad \textit{Divide out the 3's and x's.}$$

$$= \frac{\overset{1}{\cancel{3}}\overset{1}{\cancel{x}}x}{\underset{1}{\cancel{2}}\underset{1}{\cancel{y}}} \cdot \frac{\overset{2}{\cancel{4}}\overset{1}{\cancel{y}}yy}{\underset{1}{\cancel{3}}\underset{1}{\cancel{x}}} \qquad \textit{Divide both the 4 and the 2 by 2, and divide out the y's.}$$

Now we multiply the remaining numerators together and the remaining denominators together.

$$= \frac{2xy^2}{1} \quad \text{or} \quad 2xy^2$$

▸ **Now Try Exercise 19**

Rather than illustrating this entire process when multiplying rational expressions, we will often proceed as follows:

$$\frac{3x^2}{2y} \cdot \frac{4y^3}{3x}$$

$$= \frac{\overset{1}{\cancel{3}}\overset{x}{\cancel{x^2}}}{\underset{1}{\cancel{2}}\underset{1}{\cancel{y}}} \cdot \frac{\overset{2}{\cancel{4}}\overset{y^2}{\cancel{y^3}}}{\underset{1}{\cancel{3}}\underset{1}{\cancel{x}}} = 2xy^2$$

EXAMPLE 3 ▸ Multiply $-\dfrac{5y^2}{2x^3} \cdot \dfrac{3x^2}{7y^2}$.

Solution

$$-\frac{\overset{1}{\cancel{5}}\overset{}{\cancel{y^2}}}{\underset{x}{\cancel{2x^3}}} \cdot \frac{\overset{1}{\cancel{3}}\overset{}{\cancel{x^2}}}{\underset{1}{\cancel{7}}\underset{}{\cancel{y^2}}} = -\frac{15}{14x}$$

▸ **Now Try Exercise 53**

From this point on when a common factor is factored out from a numerator and a denominator, we will generally not show the 1s above the expression factored out in the numerator and below the expression factored out in the denominator. After the common factor is factored out, if the remaining expression is any number or expression other than 1, we will indicate that by using small numbers above the numerator or below the denominator.

EXAMPLE 4 ▸ Multiply $(x - 6) \cdot \dfrac{4}{x^3 - 6x^2}$.

Solution $(x - 6) \cdot \dfrac{4}{x^3 - 6x^2} = \dfrac{\cancel{x-6}}{1} \cdot \dfrac{4}{x^2\cancel{(x-6)}} = \dfrac{4}{x^2}$

▸ **Now Try Exercise 27**

EXAMPLE 5 ▸ Multiply $\dfrac{(x + 2)^2}{6x^2} \cdot \dfrac{3x}{x^2 - 4}$.

Solution

$$\frac{(x + 2)^2}{6x^2} \cdot \frac{3x}{x^2 - 4} = \frac{(x + 2)(x + 2)}{6x^2} \cdot \frac{3x}{(x + 2)(x - 2)}$$

$$= \frac{\cancel{(x+2)}(x + 2)}{\underset{2}{\cancel{6}}\underset{x}{x^2}} \cdot \frac{\overset{1}{\cancel{3}}\overset{1}{\cancel{x}}}{\cancel{(x+2)}(x - 2)} = \frac{x + 2}{2x(x - 2)}$$

▸ **Now Try Exercise 61**

In Example 5 we could have multiplied the factors in the denominator to get $\dfrac{x + 2}{2x^2 - 4x}$. This is also a correct answer. In this section we will leave rational answers with the numerator as a polynomial (in unfactored form) and the denominators in factored form, as was given in Example 5. This is consistent with how we will leave rational answers when we add and subtract rational expressions in later sections.

EXAMPLE 6 ▶ Multiply $\dfrac{b-4}{5b} \cdot \dfrac{10b}{4-b}$.

Solution

$$\dfrac{b-4}{\overset{}{\underset{1}{5\!\!\!/b}}} \cdot \dfrac{\overset{2}{10\!\!\!/b}}{4-b} = \dfrac{2(b-4)}{4-b}.$$

This solution is still not complete. In Section 7.1 we showed that $4 - b$ is $-1(-4 + b)$ or $-1(b - 4)$. Thus,

$$\dfrac{2(b-4)}{4-b} = \dfrac{2(b-4)}{-1(b-4)} = -2.$$

▶ **Now Try Exercise 23**

Helpful Hint

When only the signs differ in a numerator and denominator in a multiplication problem, factor out -1 *from either the numerator or denominator*; then divide out the common factor.

$$\dfrac{a-b}{x} \cdot \dfrac{y}{b-a} = \dfrac{a-b}{x} \cdot \dfrac{y}{-1(a-b)} = -\dfrac{y}{x}$$

EXAMPLE 7 ▶ Multiply $\dfrac{3x+2}{2x-1} \cdot \dfrac{4-8x}{3x+2}$.

Solution

$$\dfrac{3x+2}{2x-1} \cdot \dfrac{4-8x}{3x+2} = \dfrac{3x+2}{2x-1} \cdot \dfrac{4(1-2x)}{3x+2} \qquad \textit{Factor.}$$

$$= \dfrac{3x+2}{2x-1} \cdot \dfrac{4(1-2x)}{3x+2}. \qquad \textit{Divide out common factors.}$$

Note that the factor $(1 - 2x)$ in the numerator of the second fraction differs only in sign from $2x - 1$, the denominator of the first fraction. We will therefore factor -1 from each term of the $(1 - 2x)$ in the numerator of the second fraction.

$$= \dfrac{3x+2}{2x-1} \cdot \dfrac{4(-1)(2x-1)}{3x+2} \qquad \textit{Factor } -1 \textit{ from the second numerator.}$$

$$= \dfrac{3x+2}{2x-1} \cdot \dfrac{-4(2x-1)}{3x+2} \qquad \textit{Divide out common factors.}$$

$$= \dfrac{-4}{1} = -4$$

▶ **Now Try Exercise 25**

EXAMPLE 8 ▶ Multiply $\dfrac{2x^2+5x-12}{6x^2-11x+3} \cdot \dfrac{3x^2+2x-1}{x^2+5x+4}$.

Solution Factor all numerators and denominators, and then divide out common factors.

$$\dfrac{2x^2+5x-12}{6x^2-11x+3} \cdot \dfrac{3x^2+2x-1}{x^2+5x+4} = \dfrac{(2x-3)(x+4)}{(2x-3)(3x-1)} \cdot \dfrac{(3x-1)(x+1)}{(x+1)(x+4)}$$

$$= \dfrac{(2x-3)(x+4)}{(2x-3)(3x-1)} \cdot \dfrac{(3x-1)(x+1)}{(x+1)(x+4)} = 1$$

▶ **Now Try Exercise 67**

EXAMPLE 9 ▶ Multiply $\dfrac{2x^3 - 18x^2 + 16x}{6y^2} \cdot \dfrac{-2y}{3x^2 - 3x}$.

Solution

$$\dfrac{2x^3 - 18x^2 + 16x}{6y^2} \cdot \dfrac{-2y}{3x^2 - 3x} = \dfrac{2x(x^2 - 9x + 8)}{6y^2} \cdot \dfrac{-2y}{3x(x - 1)}$$

$$= \dfrac{2x(x - 8)(x - 1)}{6y^2} \cdot \dfrac{-2y}{3x(x - 1)}$$

$$= \dfrac{2x(x - 8)(x - 1)}{\underset{3\ y}{6y^2}} \cdot \dfrac{-2y}{3x(x - 1)}$$

$$= \dfrac{-2(x - 8)}{9y} = \dfrac{-2x + 16}{9y}$$

▶ **Now Try Exercise 29**

EXAMPLE 10 ▶ Multiply $\dfrac{x^2 - y^2}{x + y} \cdot \dfrac{x + 2y}{2x^2 - 3xy + y^2}$.

Solution

$$\dfrac{x^2 - y^2}{x + y} \cdot \dfrac{x + 2y}{2x^2 - 3xy + y^2} = \dfrac{(x + y)(x - y)}{x + y} \cdot \dfrac{x + 2y}{(2x - y)(x - y)}$$

$$= \dfrac{(x + y)(x - y)}{x + y} \cdot \dfrac{x + 2y}{(2x - y)(x - y)}$$

$$= \dfrac{x + 2y}{2x - y}$$

▶ **Now Try Exercise 33**

2 Divide Rational Expressions

In Chapter 1 we learned that to divide one fraction by a second fraction, we multiply the first fraction by the reciprocal of the second fraction (or by the reciprocal of the divisor).

> **To Divide Two Fractions**
>
> $$\dfrac{a}{b} \div \dfrac{c}{d} = \dfrac{a}{b} \cdot \dfrac{d}{c} = \dfrac{ad}{bc}, \quad b \neq 0, \quad d \neq 0, \quad \text{and} \quad c \neq 0$$

EXAMPLE 11 ▶ Divide.

a) $\dfrac{2}{7} \div \dfrac{9}{7}$ **b)** $\dfrac{3}{4} \div \dfrac{10}{6}$

Solution

a) $\dfrac{2}{7} \div \dfrac{9}{7} = \dfrac{2}{\underset{1}{7}} \cdot \dfrac{\overset{1}{7}}{9} = \dfrac{2 \cdot 1}{1 \cdot 9} = \dfrac{2}{9}$

b) $\dfrac{3}{4} \div \dfrac{10}{6} = \dfrac{3}{\underset{2}{4}} \cdot \dfrac{\overset{3}{6}}{10} = \dfrac{3 \cdot 3}{2 \cdot 10} = \dfrac{9}{20}$

▶ **Now Try Exercise 13**

The same principles are used to **divide rational expressions**.

> **To Divide Rational Expressions**
>
> Multiply the first fraction by the reciprocal of the second fraction.

EXAMPLE 12 ▶ Divide $\dfrac{8x^3}{z} \div \dfrac{5z^3}{3}$.

Solution   Multiply the first fraction by the reciprocal of the second fraction.

$$\frac{8x^3}{z} \div \frac{5z^3}{3} = \frac{8x^3}{z} \cdot \frac{3}{5z^3} = \frac{24x^3}{5z^4}$$

▶ **Now Try Exercise 35**

EXAMPLE 13 ▶ Divide $\dfrac{x^2 - 9}{x + 4} \div \dfrac{x - 3}{x + 4}$.

Solution

$$\frac{x^2 - 9}{x + 4} \div \frac{x - 3}{x + 4} = \frac{x^2 - 9}{x + 4} \cdot \frac{x + 4}{x - 3}$$

Multiply the first fraction by the reciprocal of the second fraction.

$$= \frac{(x + 3)\cancel{(x - 3)}}{\cancel{x + 4}} \cdot \frac{\cancel{x + 4}}{\cancel{x - 3}} = x + 3$$

Factor, and divide out common factors.

▶ **Now Try Exercise 49**

EXAMPLE 14 ▶ Divide $\dfrac{-1}{2x - 3} \div \dfrac{8}{3 - 2x}$.

Solution

$$\frac{-1}{2x - 3} \div \frac{8}{3 - 2x} = \frac{-1}{2x - 3} \cdot \frac{3 - 2x}{8}$$

Multiply the first fraction by the reciprocal of the second fraction.

$$= \frac{-1}{\cancel{2x - 3}} \cdot \frac{-1\cancel{(2x - 3)}}{8}$$

Factor out -1, then divide out common factors.

$$= \frac{(-1)(-1)}{(1)(8)} = \frac{1}{8}$$

▶ **Now Try Exercise 41**

EXAMPLE 15 ▶ Divide $\dfrac{w^2 - 11w + 30}{w^2} \div (w - 5)^2$.

Solution $(w - 5)^2$ means $\dfrac{(w - 5)^2}{1}$. Factor the numerator of the first fraction and multiply by the reciprocal of the second fraction.

$$\frac{w^2 - 11w + 30}{w^2} \div (w - 5)^2 = \frac{w^2 - 11w + 30}{w^2} \cdot \frac{1}{(w - 5)^2}$$

$$= \frac{(w - 6)\cancel{(w - 5)}}{w^2} \cdot \frac{1}{\cancel{(w - 5)}(w - 5)}$$

$$= \frac{w - 6}{w^2(w - 5)}$$

▶ **Now Try Exercise 43**

EXAMPLE 16 ▸ Divide $\dfrac{12x^2 - 22x + 8}{7x} \div \dfrac{3x^2 + 2x - 8}{2x^2 + 4x}$.

Solution

$$\frac{12x^2 - 22x + 8}{7x} \div \frac{3x^2 + 2x - 8}{2x^2 + 4x} = \frac{12x^2 - 22x + 8}{7x} \cdot \frac{2x^2 + 4x}{3x^2 + 2x - 8}$$

$$= \frac{2(6x^2 - 11x + 4)}{7x} \cdot \frac{2x(x + 2)}{(3x - 4)(x + 2)}$$

$$= \frac{2(3x - 4)(2x - 1)}{7x} \cdot \frac{2x(x + 2)}{(3x - 4)(x + 2)}$$

$$= \frac{4(2x - 1)}{7} = \frac{8x - 4}{7}$$

▸ **Now Try Exercise 45**

EXERCISE SET 7.2

Math XL
MathXL®

MyMathLab
MyMathLab

Concept/Writing Exercises

1. Explain how to multiply rational expressions.

2. Explain how to divide rational expressions.

What polynomial should be in the shaded area of the second fraction to make each statement true? Explain how you determined your answer.

3. $\dfrac{x + 3}{x - 4} \cdot \dfrac{\blacksquare}{x + 3} = x + 5$

4. $\dfrac{x - 5}{x + 2} \cdot \dfrac{\blacksquare}{x - 5} = 2x - 3$

5. $\dfrac{x - 5}{x + 5} \cdot \dfrac{x + 5}{\blacksquare} = \dfrac{1}{x + 7}$

6. $\dfrac{2x - 1}{x - 3} \cdot \dfrac{x - 3}{\blacksquare} = \dfrac{1}{x - 6}$

Practice the Skills

Multiply or divide as indicated.

7. $\left(\dfrac{2}{5}\right)\left(\dfrac{15}{19}\right)$

8. $\left(\dfrac{3}{8}\right)\left(-\dfrac{9}{33}\right)$

9. $\left(\dfrac{6}{8}\right)\left(-\dfrac{10}{14}\right)$

10. $\left(\dfrac{7}{9}\right)\left(\dfrac{81}{98}\right)$

11. $\left(-\dfrac{4}{11}\right)\left(-\dfrac{55}{64}\right)$

12. $\left(-\dfrac{12}{13}\right)\left(-\dfrac{65}{42}\right)$

13. $\dfrac{3}{7} \div \dfrac{5}{7}$

14. $\dfrac{3}{8} \div \dfrac{15}{44}$

15. $-\dfrac{2}{9} \div \dfrac{32}{39}$

16. $\left(-\dfrac{3}{4}\right) \div \left(-\dfrac{15}{16}\right)$

Multiply.

17. $\dfrac{6x}{4y} \cdot \dfrac{y^2}{12}$

18. $\dfrac{15x^3y^2}{2z} \cdot \dfrac{z}{5xy^3}$

19. $\dfrac{14x^2}{y^4} \cdot \dfrac{5x^2}{y^2}$

20. $\dfrac{7n^3}{16m} \cdot \dfrac{-4}{21m^2n^3}$

21. $\dfrac{6x^5y^3}{5z^3} \cdot \dfrac{6x^4}{5yz^4}$

22. $\dfrac{x^2 - 4}{x^2 - 16} \cdot \dfrac{x - 4}{x - 2}$

23. $\dfrac{3x - 2}{3x + 2} \cdot \dfrac{x - 1}{1 - x}$

24. $\dfrac{m - 5}{2m + 5} \cdot \dfrac{8m}{-m + 5}$

25. $\dfrac{x^2 + 7x + 6}{x + 6} \cdot \dfrac{1}{x + 1}$

26. $\dfrac{b^2 + 7b + 12}{6b} \cdot \dfrac{b^2 - 4b}{b^2 - b - 12}$

27. $\dfrac{a}{a^2 - b^2} \cdot \dfrac{a + b}{a^2 + ab}$

28. $\dfrac{t^2 - 36}{t^2 + t - 30} \cdot \dfrac{t - 5}{4t}$

29. $\dfrac{6x^2 - 14x - 12}{6x + 4} \cdot \dfrac{2x + 4}{2x^2 - 2x - 12}$

30. $\dfrac{2x^2 - 9x + 9}{8x - 12} \cdot \dfrac{4x}{x^2 - 3x}$

31. $\dfrac{3x^2 - 13x - 10}{x^2 - 2x - 15} \cdot \dfrac{x^2 + x - 2}{3x^2 - x - 2}$

32. $\dfrac{2t^2 - t - 6}{2t^2 - 3t - 2} \cdot \dfrac{2t^2 - 5t - 3}{2t^2 + 11t + 12}$

33. $\dfrac{x + 9}{x - 3} \cdot \dfrac{x^3 - 27}{x^2 + 3x + 9}$

34. $\dfrac{x^3 + 8}{x^2 - x - 6} \cdot \dfrac{x + 5}{x^2 - 2x + 4}$

Divide.

35. $\dfrac{12x^3}{y^2} \div \dfrac{3x}{y^3}$

36. $\dfrac{9x^3}{5} \div \dfrac{1}{20y^2}$

37. $\dfrac{15xy^2}{4z} \div \dfrac{5x^2y^2}{12z^2}$

38. $\dfrac{36y}{5z^2} \div \dfrac{3xy}{2z}$

39. $\dfrac{11xy}{7ab^2} \div \dfrac{6xy}{7}$

40. $3xz \div \dfrac{6xy}{z}$

41. $\dfrac{12r + 6}{r} \div \dfrac{2r + 1}{r^3}$

42. $\dfrac{x - 3}{10y^2} \div \dfrac{x^2 - 9}{5xy}$

43. $\dfrac{x^2 + 11x + 18}{x} \div \dfrac{x + 2}{x}$

44. $\dfrac{1}{x^2 + 7x - 18} \div \dfrac{1}{x^2 - 17x + 30}$

45. $\dfrac{x^2 - 12x + 32}{x^2 - 6x - 16} \div \dfrac{x^2 - x - 12}{x^2 - 5x - 24}$

46. $\dfrac{a - b}{9a + 9b} \div \dfrac{a^2 - b^2}{a^2 + 2a + 7}$

47. $\dfrac{2x^2 + 9x + 4}{x^2 + 7x + 12} \div \dfrac{2x^2 - x - 1}{(x + 3)^2}$

48. $\dfrac{a^2 - b^2}{9} \div \dfrac{3a - 3b}{27x^2}$

49. $\dfrac{x^2 - y^2}{x^2 - 2xy + y^2} \div \dfrac{x + y}{y - x}$

50. $\dfrac{9x^2 - 9y^2}{18x^2y^2} \div \dfrac{3x + 3y}{12x^2y^5}$

51. $\dfrac{5x^2 - 4x - 1}{5x^2 + 6x + 1} \div \dfrac{x^2 - 5x + 4}{x^2 + 2x + 1}$

52. $\dfrac{7n^2 - 15n + 2}{n^2 + n - 6} \div \dfrac{n^2 - 3n - 10}{n^2 - 2n - 15}$

Perform each indicated operation.

53. $\dfrac{11z}{6y^2} \cdot \dfrac{24x^2y^4}{11z}$

54. $\dfrac{5z^3}{7} \cdot \dfrac{9x^2}{15z}$

55. $\dfrac{63a^2b^3}{20c^3} \cdot \dfrac{4c^4}{9a^3b^5}$

56. $\dfrac{-2xw}{y^5} \div \dfrac{8x^2}{y^6}$

57. $\dfrac{-xy}{a} \div \dfrac{-2ax}{6y}$

58. $\dfrac{27x}{8y^2} \div 3x^2y^2$

59. $\dfrac{64m^6}{21x^5y^7} \cdot \dfrac{14x^{12}y^5}{16m^5}$

60. $\dfrac{-18x^2y}{11z^2} \cdot \dfrac{22z^3}{x^2y^5}$

61. $\dfrac{(x + 3)^2}{5x^2} \cdot \dfrac{10x}{x^2 - 9}$

62. $\dfrac{1}{4x - 3} \cdot (24x - 18)$

63. $\dfrac{1}{5x^2y^2} \div \dfrac{1}{35x^3y}$

64. $\dfrac{x^2y^5}{3z} \div \dfrac{7z}{2x}$

65. $\dfrac{(4m)^2}{8n^3} \div \dfrac{m^6n^8}{2}$

66. $\dfrac{11r^5s^2}{(r^2s^3)^3} \cdot \dfrac{6r^4}{4s}$

67. $\dfrac{r^2 + 5r + 6}{r^2 + 9r + 18} \cdot \dfrac{r^2 + 4r - 12}{r^2 - 5r + 6}$

68. $\dfrac{z^2 - z - 20}{z^2 - 3z - 10} \cdot \dfrac{(z + 2)^2}{(z + 4)^2}$

69. $\dfrac{x^2 - 12x + 36}{x^2 - 8x + 12} \div \dfrac{x^2 - 7x + 12}{x^2 - 6x + 8}$

70. $\dfrac{p^2 - 5p + 6}{p^2 - 10p + 16} \div \dfrac{p^2 + 2p}{p^2 - 6p - 16}$

71. $\dfrac{2w^2 + 3w - 35}{w^2 - 7w - 8} \cdot \dfrac{w^2 - 5w - 24}{w^2 + 8w + 15}$

72. $\dfrac{3z^2 - 4z - 4}{z^2 - 4} \cdot \dfrac{2z^2 + 5z + 2}{2z^2 - 3z - 2}$

73. $\dfrac{q^2 - 11q + 30}{2q^2 - 7q - 15} \div \dfrac{q^2 - 2q - 24}{q^2 - q - 20}$

74. $\dfrac{2x^2 - 19x + 24}{x^2 - 12x + 32} \div \dfrac{2x^2 + x - 6}{x^2 + 7x + 10}$

75. $\dfrac{4n^2 - 9}{9n^2 - 1} \cdot \dfrac{3n^2 - 2n - 1}{2n^2 - 5n + 3}$

76. $\dfrac{2z^2 + 9z + 9}{4z^2 - 9} \div \dfrac{(z + 3)^2}{(2z - 3)^2}$

Problem Solving

Perform each indicated operation. Treat Δ and ☺ as if they were variables.

77. $\dfrac{6\Delta^2}{13} \cdot \dfrac{13}{36\Delta^5}$

78. $\dfrac{\Delta - 7}{2\Delta + 5} \cdot \dfrac{3\Delta}{-\Delta + 7}$

79. $\dfrac{\Delta - ☺}{9\Delta - 9☺} \div \dfrac{\Delta^2 - ☺^2}{\Delta^2 + 2\Delta☺ + ☺^2}$

80. $\dfrac{\Delta^2 - ☺^2}{\Delta^2 - 2\Delta☺ + ☺^2} \div \dfrac{\Delta + ☺}{☺ - \Delta}$

For each equation, fill in the shaded area with a binomial or trinomial to make the statement true. Explain how you determined your answer.

81. $\dfrac{\boxed{}}{x + 2} = x + 3$

82. $\dfrac{x + 5}{\boxed{}} = \dfrac{1}{x - 5}$

83. $\dfrac{\boxed{}}{x - 6} = x + 2$

84. $\dfrac{\boxed{}}{x^2 - 7x + 10} = \dfrac{1}{x - 2}$

85. $\dfrac{\boxed{}}{x^2 - 4} \cdot \dfrac{x + 2}{x - 1} = 1$

86. $\dfrac{x + 4}{x^2 + 9x + 20} \cdot \dfrac{\boxed{}}{x - 2} = 1$

Challenge Problems

Simplify.

87. $\left(\dfrac{x + 2}{x^2 - 4x - 12} \cdot \dfrac{x^2 - 9x + 18}{x - 2} \right) \div \dfrac{x^2 + 5x + 6}{x^2 - 4}$

88. $\left(\dfrac{x^2 + 4x + 3}{x^2 - 6x - 16} \right) \div \left(\dfrac{x^2 + 5x + 6}{x^2 - 9x + 8} \cdot \dfrac{x^2 - 1}{x^2 + 4x + 4} \right)$

89. $\left(\dfrac{x^2 - x - 6}{2x^2 - 9x + 9} \div \dfrac{x^2 + x - 12}{x^2 + 3x - 4} \right) \cdot \dfrac{2x^2 - 5x + 3}{x^2 + x - 2}$

90. $\left(\dfrac{x^2 + 4x + 3}{x^2 - 6x - 16} \div \dfrac{x^2 + 5x + 6}{x^2 - 9x + 8} \right) \cdot \left(\dfrac{x^2 - 1}{x^2 + 4x + 4} \right)$

For Exercises 91 and 92, determine the polynomials that when placed in the shaded areas make the statement true. Explain how you determined your answer.

91. $\dfrac{\boxed{}}{\boxed{}} \cdot \dfrac{x^2 + 3x - 4}{x^2 - 4x + 3} = \dfrac{x - 2}{x - 5}$

92. $\dfrac{\boxed{}}{x^2 + x - 2} \cdot \dfrac{x^2 + 6x + 8}{\boxed{}} = \dfrac{x + 3}{x + 5}$

Group Activity

93. Consider the three problems that follow:

1. $\left(\dfrac{x + 2}{x - 3} \right) \div \left(\dfrac{x^2 - 5x + 6}{x - 2} \cdot \dfrac{x + 2}{x - 3} \right)$

2. $\left(\dfrac{x + 2}{x - 3} \div \dfrac{x^2 - 5x + 6}{x - 2} \right) \cdot \left(\dfrac{x + 2}{x - 3} \right)$

3. $\left(\dfrac{x + 2}{x - 3} \right) \div \left(\dfrac{x^2 - 5x + 6}{x - 2} \right) \cdot \left(\dfrac{x + 2}{x - 3} \right)$

a) Without working the problem, decide as a group which of the problems will have the same answer. Explain.

b) Individually, simplify each of the three problems.

c) Compare your answers to part **b)** with the other members of your group. If you did not get the same answers, determine why.

Cumulative Review Exercises

[3.4] **94. Tug Boat** A tug boat leaves its dock traveling at an average of 15 miles per hour towards a barge it is to pull back to the dock. On the return trip, pulling the barge, the tug averages 5 miles per hour. If the trip back to the dock took 2 hours longer than the trip out, find the time it took the tug boat to reach the barge.

[5.5] **95.** Multiply $(4x^3 y^2 z^4)(3xy^3 z^7)$.

[5.6] **96.** Divide $\dfrac{4x^3 - 5x}{2x - 1}$.

[6.4] **97.** Factor $6x^2 - 18x - 60$.

[6.6] **98.** Solve $3x^2 - 9x - 30 = 0$.

7.3 Addition and Subtraction of Rational Expressions with a Common Denominator and Finding the Least Common Denominator

1 Add and subtract rational expressions with a common denominator.

2 Find the least common denominator.

1 Add and Subtract Rational Expressions with a Common Denominator

Recall that when adding (or subtracting) two arithmetic fractions with a common denominator we add (or subtract) the numerators while keeping the common denominator.

To Add or Subtract Two Fractions

$$\frac{a}{c} + \frac{b}{c} = \frac{a+b}{c}, c \neq 0 \qquad \frac{a}{c} - \frac{b}{c} = \frac{a-b}{c}, c \neq 0$$

EXAMPLE 1 ▶ **a)** Add $\dfrac{7}{16} + \dfrac{8}{16}$.   **b)** Subtract $\dfrac{4}{9} - \dfrac{1}{9}$.

Solution

a) $\dfrac{7}{16} + \dfrac{8}{16} = \dfrac{7+8}{16} = \dfrac{15}{16}$ **b)** $\dfrac{4}{9} - \dfrac{1}{9} = \dfrac{4-1}{9} = \dfrac{3}{9} = \dfrac{1}{3}$

▶ **Now Try Exercise 13**

Note in Example **1a)** that we did not simplify $\dfrac{8}{16}$ to $\dfrac{1}{2}$. The fractions are given with a common denominator, 16. If $\dfrac{8}{16}$ was simplified to $\dfrac{1}{2}$, you would lose the common denominator that is needed to add or subtract fractions.

The same principles apply when **adding or subtracting rational expressions** containing variables.

To Add or Subtract Rational Expressions with a Common Denominator

1. Add or subtract the numerators.
2. Place the sum or difference of the numerators found in step 1 over the common denominator.
3. Simplify the fraction if possible.

EXAMPLE 2 ▶ Add $\dfrac{3}{x-4} + \dfrac{x+8}{x-4}$.

Solution $\dfrac{3}{x-4} + \dfrac{x+8}{x-4} = \dfrac{3+(x+8)}{x-4} = \dfrac{x+11}{x-4}$

▶ **Now Try Exercise 19**

EXAMPLE 3 ▶ Add $\dfrac{2x^2+7}{x+3} + \dfrac{6x-7}{x+3}$.

Solution $\dfrac{2x^2+7}{x+3} + \dfrac{6x-7}{x+3} = \dfrac{(2x^2+7)+(6x-7)}{x+3}$

$$= \dfrac{2x^2+7+6x-7}{x+3}$$

$$= \dfrac{2x^2+6x}{x+3}$$

Now factor $2x$ from each term in the numerator and simplify.

$$= \frac{2x\cancel{(x+3)}}{\cancel{x+3}} = 2x$$

▶ Now Try Exercise 21

EXAMPLE 4 ▶ Add $\dfrac{x^2 + 2x - 2}{(x + 5)(x - 2)} + \dfrac{5x + 12}{(x + 5)(x - 2)}$.

Solution

$$\frac{x^2 + 2x - 2}{(x + 5)(x - 2)} + \frac{5x + 12}{(x + 5)(x - 2)} = \frac{(x^2 + 2x - 2) + (5x + 12)}{(x + 5)(x - 2)} \quad \text{Write as a single fraction.}$$

$$= \frac{x^2 + 2x - 2 + 5x + 12}{(x + 5)(x - 2)} \quad \text{Remove parentheses in the numerator.}$$

$$= \frac{x^2 + 7x + 10}{(x + 5)(x - 2)} \quad \text{Combine like terms.}$$

$$= \frac{\cancel{(x + 5)}(x + 2)}{\cancel{(x + 5)}(x - 2)} \quad \text{Factor, divide out common factor.}$$

$$= \frac{x + 2}{x - 2}$$

▶ Now Try Exercise 27

When subtracting rational expressions, be sure to subtract the entire numerator of the fraction being subtracted. Study the following Avoiding Common Errors box very carefully.

Avoiding Common Errors

Consider the subtraction

$$\frac{4x}{x - 2} - \frac{2x + 1}{x - 2}$$

Many people begin problems of this type incorrectly. Here are the correct and incorrect ways of working this problem.

CORRECT

$$\frac{4x}{x - 2} - \frac{2x + 1}{x - 2} = \frac{4x - (2x + 1)}{x - 2}$$

$$= \frac{4x - 2x - 1}{x - 2}$$

$$= \frac{2x - 1}{x - 2}$$

INCORRECT

$$\cancel{\frac{4x}{x - 2} - \frac{2x + 1}{x - 2}} \quad \cancel{\frac{4x - 2x + 1}{x - 2}}$$

Note that the entire numerator of the second fraction (not just the first term) **must be subtracted.** Also note that the sign of *each* term of the numerator being subtracted will change when the parentheses are removed.

EXAMPLE 5 ▶ Subtract $\dfrac{x^2 - 6x + 3}{x^2 + 7x + 12} - \dfrac{x^2 - 8x - 5}{x^2 + 7x + 12}$.

Solution

$$\dfrac{x^2 - 6x + 3}{x^2 + 7x + 12} - \dfrac{x^2 - 8x - 5}{x^2 + 7x + 12} = \dfrac{(x^2 - 6x + 3) - (x^2 - 8x - 5)}{x^2 + 7x + 12}$$ *Write as a single fraction.*

$$= \dfrac{x^2 - 6x + 3 - x^2 + 8x + 5}{x^2 + 7x + 12}$$ *Remove parentheses.*

$$= \dfrac{2x + 8}{x^2 + 7x + 12}$$ *Combine like terms.*

$$= \dfrac{2\cancel{(x + 4)}}{(x + 3)\cancel{(x + 4)}}$$ *Factor, divide out common factor.*

$$= \dfrac{2}{x + 3}$$

▶ **Now Try Exercise 43**

The variable used when working with rational expressions is irrelevant. In Example 6 we work with rational expressions in variable *r*.

EXAMPLE 6 ▶ Subtract $\dfrac{6r}{r - 5} - \dfrac{4r^2 - 17r + 15}{r - 5}$.

Solution

$$\dfrac{6r}{r - 5} - \dfrac{4r^2 - 17r + 15}{r - 5} = \dfrac{6r - (4r^2 - 17r + 15)}{r - 5}$$ *Write as a single fraction.*

$$= \dfrac{6r - 4r^2 + 17r - 15}{r - 5}$$ *Remove parentheses.*

$$= \dfrac{-4r^2 + 23r - 15}{r - 5}$$ *Combine like terms.*

$$= \dfrac{-(4r^2 - 23r + 15)}{r - 5}$$ *Factor out −1.*

$$= \dfrac{-(4r - 3)\cancel{(r - 5)}}{\cancel{r - 5}}$$ *Factor, divide out common factor.*

$$= -(4r - 3) \quad \text{or} \quad -4r + 3$$

▶ **Now Try Exercise 31**

2 Find the Least Common Denominator

To add two fractions with unlike denominators, we must first obtain a common denominator. Now we explain how to find the **least common denominator** for rational expressions. We will use this information in Section 7.4 when we add and subtract rational expressions.

EXAMPLE 7 ▶ Add $\dfrac{4}{7} + \dfrac{2}{3}$.

Solution The least common denominator (LCD) of the fractions $\dfrac{4}{7}$ and $\dfrac{2}{3}$ is 21.

Twenty-one is the smallest number that is divisible by both denominators, 7 and 3. Rewrite each fraction so that its denominator is 21.

$$\dfrac{4}{7} + \dfrac{2}{3} = \dfrac{3}{3} \cdot \dfrac{4}{7} + \dfrac{2}{3} \cdot \dfrac{7}{7}$$

$$= \dfrac{12}{21} + \dfrac{14}{21} = \dfrac{26}{21} \quad \text{or} \quad 1\dfrac{5}{21}$$

▶ **Now Try Exercise 95**

To add or subtract rational expressions, we must write each expression with a common denominator.

> ## To Find the Least Common Denominator of Rational Expressions
>
> **1.** Factor each denominator completely. Any factors that occur more than once should be expressed as powers. For example, $(x - 3)(x - 3)$ should be expressed as $(x - 3)^2$.
>
> **2.** List all different factors (other than 1) that appear in any of the denominators. When the same factor appears in more than one denominator, write that factor with the highest power that appears.
>
> **3.** The least common denominator is the product of all the factors listed in step 2.

EXAMPLE 8 ▸ Find the least common denominator.

$$\frac{1}{7} + \frac{1}{y}$$

Solution   The only factor (other than 1) of the first denominator is 7. The only factor (other than 1) of the second denominator is y. The LCD is therefore $7 \cdot y = 7y$.

▸ **Now Try Exercise 53**

EXAMPLE 9 ▸ Find the LCD.

$$\frac{8}{x^2} - \frac{3}{5x}$$

Solution   The factors that appear in the denominators are 5 and x. List each factor with its highest power. The LCD is the product of these factors.

$$\text{LCD} = 5 \cdot \overset{\text{Highest power of } x}{x^2} = 5x^2$$

▸ **Now Try Exercise 59**

EXAMPLE 10 ▸ Find the LCD.

$$\frac{11}{18x^3y} + \frac{5}{27x^2y^3}$$

Solution   Write both 18 and 27 as products of prime factors: $18 = 2 \cdot 3^2$ and $27 = 3^3$. *If you have forgotten how to write a number as a product of prime factors, read Section 6.1 or Appendix B now.*

$$\frac{11}{18x^3y} + \frac{5}{27x^2y^3} = \frac{11}{2 \cdot 3^2 x^3 y} + \frac{5}{3^3 x^2 y^3}$$

The factors that appear are 2, 3, x, and y. List the highest powers of each of these factors.

$$\text{LCD} = 2 \cdot 3^3 \cdot x^3 \cdot y^3 = 54x^3y^3$$

▸ **Now Try Exercise 61**

EXAMPLE 11 ▸ Find the LCD.

$$\frac{9}{x} - \frac{2z}{x + 3}$$

Solution   The factors in the denominators are x and $x + 3$. *Note that the x in the second denominator, $x + 3$, is a term, not a factor.*

$$\text{LCD} = x(x + 3)$$

▸ **Now Try Exercise 65**

EXAMPLE 12 ▶ Find the LCD.

$$\frac{7}{3x^2 - 6x} + \frac{8x^2}{x^2 - 4x + 4}$$

Solution   Factor both denominators.

$$\frac{7}{3x^2 - 6x} + \frac{8x^2}{x^2 - 4x + 4} = \frac{7}{3x(x - 2)} + \frac{8x^2}{(x - 2)(x - 2)}$$

$$= \frac{7}{3x(x - 2)} + \frac{8x^2}{(x - 2)^2}$$

The factors in the denominators are 3, x, and $x - 2$. List the highest power of each of these factors.

$$\text{LCD} = 3 \cdot x \cdot (x - 2)^2 = 3x(x - 2)^2.$$

▶ **Now Try Exercise 85**

EXAMPLE 13 ▶ Find the LCD.

$$\frac{11x}{x^2 - x - 12} - \frac{6x^2}{x^2 - 7x + 12}$$

Solution   Factor both denominators.

$$\frac{11x}{x^2 - x - 12} - \frac{6x^2}{x^2 - 7x + 12} = \frac{11x}{(x + 3)(x - 4)} - \frac{6x^2}{(x - 3)(x - 4)}$$

The factors in the denominators are $x + 3$, $x - 4$, and $x - 3$.

$$\text{LCD} = (x + 3)(x - 4)(x - 3)$$

Although $x - 4$ is a common factor of each denominator, the highest power of that factor that appears in each denominator is 1.

▶ **Now Try Exercise 81**

EXAMPLE 14 ▶ Find the LCD.

$$\frac{6w}{w^2 - 14w + 45} + w + 8$$

Solution   Factor the denominator of the first term.

$$\frac{6w}{w^2 - 14w + 45} + w + 8 = \frac{6w^2}{(w - 5)(w - 9)} + w + 8$$

Since the denominator of $w + 8$ is 1, the expression can be rewritten as

$$\frac{6w}{(w - 5)(w - 9)} + \frac{w + 8}{1}$$

The LCD is therefore $1(w - 5)(w - 9)$ or simply $(w - 5)(w - 9)$.

▶ **Now Try Exercise 89**

EXERCISE SET 7.3   *Math XL*   **MyMathLab**
MathXL®   MyMathLab

Concept/Writing Exercises

1. Explain how to add or subtract rational expressions with a common denominator.

2. When subtracting rational expressions, what must happen to the sign of each term of the numerator being subtracted?

3. Explain how to find the least common denominator of two rational expressions.

4. In the addition $\dfrac{1}{x} + \dfrac{1}{x + 1}$, is the least common denominator x, $x + 1$, or $x(x + 1)$? Explain.

Determine the LCD to be used to perform each indicated operation. Explain how you determined the LCD. Do not perform the operations.

5. $\dfrac{9}{x+6} - \dfrac{2}{x}$

6. $\dfrac{10}{x-2} + \dfrac{3}{7}$

7. $\dfrac{2}{x+3} + \dfrac{1}{x} + \dfrac{1}{4}$

8. $\dfrac{6}{x-3} + \dfrac{1}{x} - \dfrac{1}{8}$

In Exercises 9–12, **a)** *Explain why the expression on the left side of the equal sign is not equal to the expression on the right side of the equal sign.* **b)** *Show what the expression on the right side should be for it to be equal to the one on the left.*

9. $\dfrac{4x-3}{5x+4} - \dfrac{2x-9}{5x+4} \neq \dfrac{4x-3-2x-9}{5x+4}$

10. $\dfrac{5x}{2x-3} - \dfrac{-3x-7}{2x-3} \neq \dfrac{5x+3x-7}{2x-3}$

11. $\dfrac{8x-2}{x^2-4x+3} - \dfrac{3x^2-4x+5}{x^2-4x+3} \neq \dfrac{8x-2-3x^2-4x+5}{x^2-4x+3}$

12. $\dfrac{2x+5}{x^2-6x} - \dfrac{-x^2+3x+6}{x^2-6x} \neq \dfrac{2x+5+x^2+3x+6}{x^2-6x}$

Practice the Skills

Add or subtract.

13. $\dfrac{4}{7} + \dfrac{2}{7}$

14. $\dfrac{8}{5} - \dfrac{6}{5}$

15. $\dfrac{5r+2}{4} - \dfrac{3}{4}$

16. $\dfrac{3x+6}{2} - \dfrac{x}{2}$

17. $\dfrac{2}{x} + \dfrac{x+4}{x}$

18. $\dfrac{3x+1}{x+1} + \dfrac{6x+8}{x+1}$

19. $\dfrac{6}{n+1} + \dfrac{n+2}{n+1}$

20. $\dfrac{7}{x-2} - \dfrac{x+4}{x-2}$

21. $\dfrac{x}{x-3} + \dfrac{4x+9}{x-3}$

22. $\dfrac{4x-3}{x-7} - \dfrac{2x+8}{x-7}$

23. $\dfrac{4t+7}{5t^2} - \dfrac{3t+4}{5t^2}$

24. $\dfrac{3w+6}{w^2+2w+1} + \dfrac{-2w-5}{w^2+2w+1}$

25. $\dfrac{5x+4}{x^2-x-12} + \dfrac{-4x-1}{x^2-x-12}$

26. $\dfrac{-x-6}{x^2-16} + \dfrac{2(x+5)}{x^2-16}$

27. $\dfrac{2m+5}{(m+4)(m-3)} - \dfrac{m+1}{(m+4)(m-3)}$

28. $\dfrac{x^2+3x}{(x+6)(x-3)} - \dfrac{x+15}{(x+6)(x-3)}$

29. $\dfrac{2p-6}{p-5} - \dfrac{p+6}{p-5}$

30. $\dfrac{x^2-6}{3x} - \dfrac{x^2+4x-11}{3x}$

31. $\dfrac{x^2+4x+1}{x+2} - \dfrac{5x+7}{x+2}$

32. $\dfrac{-4x+2}{3x+6} + \dfrac{4(x-1)}{3x+6}$

33. $\dfrac{3x+13}{2x+10} - \dfrac{2(x+4)}{2x+10}$

34. $\dfrac{x^2}{x+4} - \dfrac{16}{x+4}$

35. $\dfrac{b^2-2b-2}{b^2-b-6} + \dfrac{b-4}{b^2-b-6}$

36. $\dfrac{4x+17}{3-x} - \dfrac{3x+20}{3-x}$

37. $\dfrac{t-3}{t+3} - \dfrac{-3t-15}{t+3}$

38. $\dfrac{x+8}{3x+2} - \dfrac{x+8}{3x+2}$

39. $\dfrac{3x^2+15x}{x^3+2x^2-8x} + \dfrac{2x^2+5x}{x^3+2x^2-8x}$

40. $\dfrac{x^2-12}{x+5} - \dfrac{13}{x+5}$

41. $\dfrac{3x^2-9x}{4x^2-8x} + \dfrac{3x}{4x^2-8x}$

42. $\dfrac{x^3-10x^2+35x}{x(x-6)} - \dfrac{x^2+5x}{x(x-6)}$

43. $\dfrac{3x^2-4x+6}{3x^2+7x+2} - \dfrac{10x+11}{3x^2+7x+2}$

44. $\dfrac{x^2-2}{x^2+6x-7} - \dfrac{-4x+19}{x^2+6x-7}$

45. $\dfrac{x^2+3x-6}{x^2-5x+4} - \dfrac{-2x^2+4x-4}{x^2-5x+4}$

46. $\dfrac{4x^2+15}{9x^2-64} - \dfrac{x^2-x+39}{9x^2-64}$

47. $\dfrac{5x^2+30x+8}{x^2-64} + \dfrac{x^2+19x}{x^2-64}$

48. $\dfrac{20x^2+8x+1}{6x^2+x-2} - \dfrac{8x^2-9x-5}{6x^2+x-2}$

Find the least common denominator for each expression.

49. $\dfrac{x}{5} + \dfrac{x+4}{5}$

50. $\dfrac{2+r}{7} - \dfrac{12}{7}$

51. $\dfrac{3}{n} + \dfrac{1}{9n}$

52. $\dfrac{6}{x+1} - \dfrac{4}{7}$

53. $\dfrac{3}{5x} + \dfrac{7}{3}$

54. $\dfrac{1}{8} + \dfrac{1}{z}$

55. $\dfrac{6}{p} + \dfrac{9}{p^3}$

56. $\dfrac{2x}{x+3} + \dfrac{6}{x-9}$

57. $\dfrac{m+3}{3m-4} + m$

58. $\dfrac{x+4}{2x} + \dfrac{8}{7x}$

59. $\dfrac{x}{6x} + \dfrac{4}{x^2}$

60. $\dfrac{x}{5x^2} + \dfrac{9}{7x^3}$

61. $\dfrac{x+1}{12x^2y} - \dfrac{7}{9x^3}$

62. $\dfrac{-3}{8x^2y^2} + \dfrac{5}{12x^4y^5}$

63. $\dfrac{4}{2r^4s^5} - \dfrac{5}{9r^3s^7}$

64. $\dfrac{5}{4w^5z^4} + \dfrac{4}{9wz^2}$

65. $\dfrac{3}{m} - \dfrac{17m}{m+2}$

66. $\dfrac{x-3}{17} - \dfrac{6}{x-5}$

67. $\dfrac{5x-2}{x^2+x} - \dfrac{13}{x}$

68. $\dfrac{3t}{t-5} + \dfrac{2}{5-t}$

69. $\dfrac{n}{4n-1} + \dfrac{n-8}{1-4n}$

70. $\dfrac{3}{-2a+3b} - \dfrac{10}{2a-3b}$

71. $\dfrac{3}{4k-5r} - \dfrac{10}{-4k+5r}$

72. $\dfrac{p}{4p^2+2p} - \dfrac{7}{2p+1}$

73. $\dfrac{4}{2q^2 + 2q} - \dfrac{5}{9q}$

74. $\dfrac{10}{(x + 4)(x + 2)} - \dfrac{6 + x}{x + 2}$

75. $\dfrac{21}{24x^2 y} + \dfrac{x + 4}{15xy^3}$

76. $\dfrac{p^2 + 4}{p^2 - 25} + \dfrac{9}{p - 5}$

77. $\dfrac{11}{3x + 12} + \dfrac{3x + 1}{2x + 4}$

78. $6x^2 + \dfrac{8x}{x - 7}$

79. $\dfrac{9x + 4}{x + 1} - \dfrac{2x - 6}{x + 8}$

80. $\dfrac{x + 3}{x^2 + 11x + 18} - \dfrac{x^2 - 11}{x^2 - 3x - 10}$

81. $\dfrac{x - 2}{x^2 - 5x - 24} + \dfrac{3}{x^2 + 11x + 24}$

82. $\dfrac{6n}{n^2 - 4} - \dfrac{n - 3}{n^2 - 5n - 14}$

83. $\dfrac{5}{(a - 4)^2} - \dfrac{a + 2}{a^2 - 7a + 12}$

84. $\dfrac{3x + 5}{x^2 - 1} + \dfrac{x^2 - 18}{(x + 1)^2}$

85. $\dfrac{9x}{x^2 + 6x + 5} - \dfrac{5x^2}{x^2 + 4x + 3}$

86. $\dfrac{6x + 5}{x + 2} + \dfrac{3x}{(x + 2)^2}$

87. $\dfrac{3x - 5}{x^2 - 6x + 9} + \dfrac{3}{x - 3}$

88. $\dfrac{2n + 11}{(n + 5)(n + 2)} - \dfrac{3n - 5}{(n - 3)(n + 5)}$

89. $\dfrac{8x^2}{x^2 - 7x + 6} + x - 9$

90. $\dfrac{2x - 1}{x^2 - 25} + x - 10$

91. $\dfrac{t - 1}{3t^2 + 10t - 8} - \dfrac{11}{3t^2 + 11t - 4}$

92. $\dfrac{-4x + 9}{2x^2 + 5x + 2} + \dfrac{x^2}{3x^2 + 4x - 4}$

93. $\dfrac{3x - 1}{4x^2 + 4x + 1} + \dfrac{x^2 + x - 9}{8x^2 + 10x + 3}$

94. $\dfrac{3x + 7}{6x^2 + 11x - 10} + \dfrac{x^2 - 8}{9x^2 - 12x + 4}$

Add or subtract using the technique from Example 7.

95. $\dfrac{1}{7} + \dfrac{2}{5}$

96. $\dfrac{3}{8} + \dfrac{1}{4}$

97. $\dfrac{2}{9} + \dfrac{3}{4}$

98. $\dfrac{5}{6} - \dfrac{1}{3}$

99. $\dfrac{5}{9} - \dfrac{1}{2}$

100. $\dfrac{6}{5} - \dfrac{3}{10}$

Problem Solving

List the polynomial to be placed in each shaded area to make a true statement. Explain how you determined your answer.

101. $\dfrac{x^2 - 6x + 3}{x + 3} + \dfrac{\boxed{}}{x + 3} = \dfrac{2x^2 - 5x - 6}{x + 3}$

102. $\dfrac{4x^2 - 6x - 7}{x^2 - 4} - \dfrac{\boxed{}}{x^2 - 4} = \dfrac{2x^2 + x - 3}{x^2 - 4}$

103. $\dfrac{-x^2 - 4x + 3}{2x + 5} + \dfrac{\boxed{}}{2x + 5} = \dfrac{5x - 7}{2x + 5}$

104. $\dfrac{-3x^2 - 9}{(x + 4)(x - 2)} - \dfrac{\boxed{}}{(x + 4)(x - 2)} = \dfrac{x^2 + 3x}{(x + 4)(x - 2)}$

Find the least common denominator of each expression.

105. $\dfrac{3}{\text{☺}} + \dfrac{4}{5\text{☺}}$

106. $\dfrac{5}{8\Delta^2 \text{☺}^2} + \dfrac{6}{5\Delta^4 \text{☺}^5}$

107. $\dfrac{8}{\Delta^2 - 9} - \dfrac{2}{\Delta + 3}$

108. $\dfrac{6}{\Delta + 3} - \dfrac{\Delta + 5}{\Delta^2 - 4\Delta + 3}$

Challenge Problems

Perform each indicated operation.

109. $\dfrac{4x - 1}{x^2 - 25} - \dfrac{3x^2 - 8}{x^2 - 25} + \dfrac{8x - 7}{x^2 - 25}$

110. $\dfrac{x^2 - 8x + 2}{x + 7} + \dfrac{2x^2 - 5x}{x + 7} - \dfrac{3x^2 + 7x + 10}{x + 7}$

Find the least common denominator for each expression.

111. $\dfrac{17}{6x^5 y^9} - \dfrac{9}{2x^3 y} + \dfrac{6}{5x^{12} y^2}$

112. $\dfrac{2x}{x - 3} - \dfrac{3}{x^2 - 9} + \dfrac{5}{x + 3}$

113. $\dfrac{3x}{x^2 - x - 12} + \dfrac{2}{x^2 - 6x + 8} + \dfrac{3}{x^2 + x - 6}$

114. $\dfrac{9}{x^2 - 4} - \dfrac{8}{3x^2 + 5x - 2} + \dfrac{7}{3x^2 - 7x + 2}$

Cumulative Review Exercises

[1.3] **115.** Subtract $4\dfrac{3}{5} - 2\dfrac{5}{9}$.

[2.5] **116.** Solve $6x + 4 = -(x + 2) - 3x + 4$.

[2.7] **117. Hummingbird Food** The instructions on a bottle of concentrated hummingbird food indicate that 6 ounces of the concentrate should be mixed with 1 gallon (128 ounces) of water. If you wish to mix the concentrate with only 48 ounces of water, how much concentrate should you use?

[3.2] **118. Tennis Club** A Tennis Club has two payment plans. Plan 1 is a yearly membership fee of $250 plus $5.00 per hour for use of the tennis court. Plan 2 is an annual membership fee of $600 with no charge for court time. How many hours would Malcolm Wu have to play in a year to make the cost of Plan 1 equal to the cost of Plan 2?

[4.3] **119.** What is the slope of the line in the graph below?

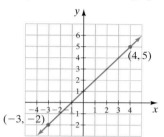

[5.3] **120.** Use scientific notation to evaluate $\dfrac{840,000,000}{0.0021}$. Leave your answer in scientific notation.

[6.6] **121.** Solve $2x^2 - 3 = x$.

7.4 Addition and Subtraction of Rational Expressions

1 Add and subtract rational expressions.

In Section 7.3 we discussed how to add and subtract rational expressions with a common denominator. Now we discuss adding and subtracting rational expressions that are not given with a common denominator.

1 Add and Subtract Rational Expressions

The method used to add and subtract rational expressions with unlike denominators is outlined in Example 1.

EXAMPLE 1 ▸ Add $\dfrac{7}{x} + \dfrac{6}{y}$.

Solution   First we determine the LCD as outlined in Section 7.3.

$$\text{LCD} = xy$$

We write each fraction with the LCD. We do this by multiplying **both** the numerator and denominator of each fraction by any factors needed to obtain the LCD.

In this problem, the fraction on the left must be multiplied by y/y and the fraction on the right must be multiplied by x/x.

$$\frac{7}{x} + \frac{6}{y} = \frac{y}{y} \cdot \frac{7}{x} + \frac{6}{y} \cdot \frac{x}{x} = \frac{7y}{xy} + \frac{6x}{xy}$$

By multiplying both the numerator and denominator by the same factor, we are in effect multiplying by 1, which does not change the value of the fraction, only its appearance. Thus, the new fraction is equivalent to the original fraction.

Now we add the numerators, while leaving the LCD alone.

$$\frac{7y}{xy} + \frac{6x}{xy} = \frac{7y + 6x}{xy} \quad \text{or} \quad \frac{6x + 7y}{xy}$$

▸ **Now Try Exercise 7**

To Add or Subtract Two Rational Expressions with Unlike Denominators

1. Determine the LCD.
2. Rewrite each fraction as an equivalent fraction with the LCD. This is done by multiplying both the numerator and denominator of each fraction by any factors needed to obtain the LCD.
3. Add or subtract the numerators while maintaining the LCD.
4. When possible, factor the remaining numerator and simplify the fraction.

EXAMPLE 2 ▶ Add $\dfrac{1}{4x^2y} + \dfrac{3}{14xy^3}$.

Solution   The LCD is $28x^2y^3$. We must write each fraction with the denominator $28x^2y^3$. To do this, we multiply the fraction on the left by $\dfrac{7y^2}{7y^2}$ and the fraction on the right by $\dfrac{2x}{2x}$.

$$\frac{1}{4x^2y} + \frac{3}{14xy^3} = \boxed{\frac{7y^2}{7y^2}}\cdot\frac{1}{4x^2y} + \frac{3}{14xy^3}\cdot\boxed{\frac{2x}{2x}}$$

$$= \frac{7y^2}{28x^2y^3} + \frac{6x}{28x^2y^3}$$

$$= \frac{7y^2 + 6x}{28x^2y^3} \quad\text{or}\quad \frac{6x + 7y^2}{28x^2y^3}$$

▶ **Now Try Exercise 15**

Helpful Hint

In Example 2 we multiplied the first fraction by $\dfrac{7y^2}{7y^2}$ and the second fraction by $\dfrac{2x}{2x}$ to get two fractions with a common denominator. How did we know what to multiply each fraction by? Many of you can determine this by observing the LCD and then determining what each denominator needs to be multiplied by to get the LCD. If this is not obvious, you can divide the LCD by the given denominator to determine what the numerator and denominator of each fraction should be multiplied by. In Example 2, the LCD is $28x^2y^3$. If we divide $28x^2y^3$ by each given denominator, $4x^2y$ and $14xy^3$, we can determine what the numerator and denominator of each respective fraction should be multiplied by

$$\frac{28x^2y^3}{4x^2y} = \boxed{7y^2} \qquad \frac{28x^2y^3}{14xy^3} = \boxed{2x}$$

Thus, $\dfrac{1}{4x^2y}$ should be multiplied by $\dfrac{7y^2}{7y^2}$ and $\dfrac{3}{14xy^3}$ should be multiplied by $\dfrac{2x}{2x}$ to obtain the LCD $28x^2y^3$.

EXAMPLE 3 ▶ Add $\dfrac{3}{x+2} + \dfrac{5}{x}$.

Solution   We must write each fraction with the LCD, which is $x(x+2)$. To do this, we multiply the fraction on the left by x/x and the fraction on the right by $(x+2)/(x+2)$.

$$\frac{3}{x+2} + \frac{5}{x} = \boxed{\frac{x}{x}}\cdot\frac{3}{x+2} + \frac{5}{x}\cdot\boxed{\frac{x+2}{x+2}}$$

$$= \frac{3x}{x(x+2)} + \frac{5(x+2)}{x(x+2)} \qquad \textit{Rewrite each fraction as an equivalent fraction with the LCD.}$$

$$= \frac{3x}{x(x+2)} + \frac{5x+10}{x(x+2)} \qquad \textit{Distributive property.}$$

$$= \frac{3x+(5x+10)}{x(x+2)} \qquad \textit{Write as a single fraction.}$$

$$= \frac{3x+5x+10}{x(x+2)} \qquad \textit{Remove parentheses in the numerator.}$$

$$= \frac{8x+10}{x(x+2)} \qquad \textit{Combine like terms in the numerator.}$$

▶ **Now Try Exercise 25**

Helpful Hint

Look at the answer to Example 3, $\dfrac{8x + 10}{x(x + 2)}$. Notice that the numerator could have been factored to obtain $\dfrac{2(4x + 5)}{x(x + 2)}$. Also notice that the denominator could have been multiplied to get $\dfrac{8x + 10}{x^2 + 2x}$. All three of these answers are equivalent and each is correct. In this section, when writing answers, unless there is a common factor in the numerator and denominator we will leave the numerator in unfactored form and the denominator in factored form. If both the numerator and denominator have a common factor, we will factor the numerator and simplify the fraction.

EXAMPLE 4 ▶ Subtract $\dfrac{w}{w - 7} - \dfrac{6}{w - 4}$.

Solution   The LCD is $(w - 7)(w - 4)$. The fraction on the left must be multiplied by $(w - 4)/(w - 4)$ to obtain the LCD. The fraction on the right must be multiplied by $(w - 7)/(w - 7)$ to obtain the LCD.

$$\frac{w}{w - 7} - \frac{6}{w - 4} = \frac{w - 4}{w - 4} \cdot \frac{w}{w - 7} - \frac{6}{w - 4} \cdot \frac{w - 7}{w - 7}$$

$$= \frac{w(w - 4)}{(w - 4)(w - 7)} - \frac{6(w - 7)}{(w - 4)(w - 7)} \qquad \text{\textit{Rewrite each fraction as an equivalent fraction with the LCD.}}$$

$$= \frac{w^2 - 4w}{(w - 4)(w - 7)} - \frac{6w - 42}{(w - 4)(w - 7)} \qquad \text{\textit{Distributive property.}}$$

$$= \frac{(w^2 - 4w) - (6w - 42)}{(w - 4)(w - 7)} \qquad \text{\textit{Write as a single fraction.}}$$

$$= \frac{w^2 - 4w - 6w + 42}{(w - 4)(w - 7)} \qquad \text{\textit{Remove parentheses in the numerator.}}$$

$$= \frac{w^2 - 10w + 42}{(w - 4)(w - 7)} \qquad \text{\textit{Combine like terms in the numerator.}}$$

▶ **Now Try Exercise 29**

EXAMPLE 5 ▶ Subtract $\dfrac{x + 2}{x - 4} - \dfrac{x + 3}{x + 4}$.

Solution   The LCD is $(x - 4)(x + 4)$.

$$\frac{x + 2}{x - 4} - \frac{x + 3}{x + 4} = \frac{x + 4}{x + 4} \cdot \frac{x + 2}{x - 4} - \frac{x + 3}{x + 4} \cdot \frac{x - 4}{x - 4}$$

$$= \frac{(x + 4)(x + 2)}{(x + 4)(x - 4)} - \frac{(x + 3)(x - 4)}{(x + 4)(x - 4)} \qquad \text{\textit{Rewrite each fraction as an equivalent fraction with the LCD.}}$$

Use the FOIL method to multiply each numerator.

$$= \frac{x^2 + 6x + 8}{(x + 4)(x - 4)} - \frac{x^2 - x - 12}{(x + 4)(x - 4)}$$

$$= \frac{(x^2 + 6x + 8) - (x^2 - x - 12)}{(x + 4)(x - 4)} \qquad \text{\textit{Write as a single fraction.}}$$

$$= \frac{x^2 + 6x + 8 - x^2 + x + 12}{(x + 4)(x - 4)} \qquad \text{\textit{Remove parentheses in the numerator.}}$$

$$= \frac{7x + 20}{(x + 4)(x - 4)} \qquad \text{\textit{Combine like terms in the numerator.}}$$

▶ **Now Try Exercise 37**

Consider the problem

$$\frac{13}{x-2} + \frac{x+8}{2-x}$$

How do we add these rational expressions? We could write each fraction with the denominator $(x-2)(2-x)$. However, there is an easier way. Study the following Helpful Hint.

Helpful Hint

When adding or subtracting fractions whose denominators are opposites (and therefore differ only in signs), multiply both the numerator *and* the denominator of *either* fraction by -1. Then both fractions will have the same denominator.

$$\frac{x}{a-b} + \frac{y}{b-a} = \frac{x}{a-b} + \frac{y}{b-a} \cdot \frac{-1}{-1}$$

$$= \frac{x}{a-b} + \frac{-y}{a-b}$$

$$= \frac{x-y}{a-b}$$

EXAMPLE 6 ▸ Add $\dfrac{4}{x-2} + \dfrac{x+3}{2-x}$.

Solution   Since the denominators differ only in sign, we may multiply both the numerator and the denominator of either fraction by -1. Here we will multiply the numerator and denominator of the second fraction by -1 to obtain the common denominator $x-2$.

$$\frac{4}{x-2} + \frac{x+3}{2-x} = \frac{4}{x-2} + \frac{x+3}{2-x} \cdot \frac{-1}{-1} \qquad \textit{Multiply numerator and denominator by } -1.$$

$$= \frac{4}{x-2} + \frac{(-x-3)}{x-2}$$

$$= \frac{4 + (-x-3)}{x-2} \qquad \textit{Write as a single fraction.}$$

$$= \frac{4-x-3}{x-2} \qquad \textit{Remove parentheses in the numerator.}$$

$$= \frac{-x+1}{x-2} \qquad \textit{Combine like terms in the numerator.}$$

▸ **Now Try Exercise 31**

Let's work another example where the denominators differ only in sign.

EXAMPLE 7 ▸ Subtract $\dfrac{a-9}{3a-4} - \dfrac{2a-5}{4-3a}$.

Solution   The denominators of the two fractions differ only in sign. We will work this problem in a similar manner to how we worked Example 6. We will multiply both the numerator and denominator of the second fraction by -1 to obtain the common denominator $3a-4$.

$$\frac{a-9}{3a-4} - \frac{2a-5}{4-3a} = \frac{a-9}{3a-4} - \frac{2a-5}{4-3a} \cdot \frac{-1}{-1} \qquad \textit{Multiply numerator and denominator by } -1.$$

$$= \frac{a-9}{3a-4} - \frac{(-2a+5)}{3a-4}$$

$$= \frac{(a - 9) - (-2a + 5)}{3a - 4} \qquad \textit{Write as a single fraction.}$$

$$= \frac{a - 9 + 2a - 5}{3a - 4} \qquad \textit{Remove parentheses in the numerator.}$$

$$= \frac{3a - 14}{3a - 4} \qquad \textit{Combine like terms in the numerator.}$$

▶ **Now Try Exercise 33**

EXAMPLE 8 ▶ Add $\dfrac{3}{x^2 + 5x + 6} + \dfrac{1}{3x^2 + 8x - 3}$.

Solution

$$\frac{3}{x^2 + 5x + 6} + \frac{1}{3x^2 + 8x - 3} = \frac{3}{(x + 2)(x + 3)} + \frac{1}{(3x - 1)(x + 3)}$$

The LCD is $(x + 2)(x + 3)(3x - 1)$.

$$= \frac{3x - 1}{3x - 1} \cdot \frac{3}{(x + 2)(x + 3)} + \frac{1}{(3x - 1)(x + 3)} \cdot \frac{x + 2}{x + 2}$$

$$= \frac{9x - 3}{(3x - 1)(x + 2)(x + 3)} + \frac{x + 2}{(3x - 1)(x + 2)(x + 3)}$$

$$= \frac{(9x - 3) + (x + 2)}{(3x - 1)(x + 2)(x + 3)}$$

$$= \frac{9x - 3 + x + 2}{(3x - 1)(x + 2)(x + 3)}$$

$$= \frac{10x - 1}{(3x - 1)(x + 2)(x + 3)}$$

▶ **Now Try Exercise 55**

EXAMPLE 9 ▶ Subtract $\dfrac{5}{x^2 - 5x} - \dfrac{x}{5x - 25}$.

Solution $\qquad \dfrac{5}{x^2 - 5x} - \dfrac{x}{5x - 25} = \dfrac{5}{x(x - 5)} - \dfrac{x}{5(x - 5)}$

The LCD is $5x(x - 5)$.

$$= \frac{5}{5} \cdot \frac{5}{x(x - 5)} - \frac{x}{5(x - 5)} \cdot \frac{x}{x}$$

$$= \frac{25}{5x(x - 5)} - \frac{x^2}{5x(x - 5)}$$

$$= \frac{25 - x^2}{5x(x - 5)}$$

$$= \frac{(5 - x)(5 + x)}{5x(x - 5)} \qquad \textit{Factor the numerator.}$$

$$= \frac{-1(x - 5)(x + 5)}{5x(x - 5)} \qquad 5 - x = -1(x - 5)$$

$$= \frac{-1\cancel{(x - 5)}(x + 5)}{5x\cancel{(x - 5)}} \qquad \textit{Simplify.}$$

$$= \frac{-1(x + 5)}{5x} \quad \text{or} \quad -\frac{x + 5}{5x}$$

▶ **Now Try Exercise 65**

Avoiding Common Errors

A common error in an addition or subtraction problem is to add or subtract the numerators and the denominators. Here is one such example.

CORRECT

$$\frac{1}{x} + \frac{x}{1} = \frac{1}{x} + \frac{x}{1} \cdot \boxed{\frac{x}{x}}$$

$$= \frac{1}{x} + \frac{x^2}{x}$$

$$= \frac{1 + x^2}{x} \text{ or } \frac{x^2 + 1}{x}$$

INCORRECT

$$\frac{1}{x} + \frac{x}{1} \ne \frac{1 + x}{x + 1}$$

$$\frac{1}{x} - \frac{x}{1} \ne \frac{1 - x}{x - 1}$$

Remember that to add or subtract fractions you must first have a common denominator. Then you add or subtract the numerators while maintaining the common denominator.

Another common mistake is to treat an addition or subtraction problem as a multiplication problem. You can divide out common factors only when *multiplying* expressions, not when adding or subtracting them.

CORRECT

$$\frac{1}{x} \cdot \frac{x}{1} = \frac{1}{\cancel{x}} \cdot \frac{\cancel{x}^{1}}{1}$$

$$= 1 \cdot 1 = 1$$

INCORRECT

$$\frac{1}{x} + \frac{x}{1} \ne \frac{1}{\cancel{x}} + \frac{\cancel{x}^{1}}{1}$$

$$= 1 + 1 = 2$$

EXERCISE SET 7.4

Concept/Writing Exercises

1. When adding or subtracting fractions with unlike denominators, how can you determine what each denominator should be multiplied by to get the LCD?

2. When you multiply both the numerator and denominator of a fraction by the factors needed to obtain the LCD, why are you not changing the value of the fraction?

3. a) Give a step-by-step procedure to add or subtract two rational expressions that have unlike denominators.

b) Using the procedure outlined in part **a)**, add $\dfrac{x}{x^2 - x - 6} + \dfrac{3}{x^2 - 4}$.

4. Explain how to add or subtract fractions whose denominators are opposites. Give an example.

5. Consider $\dfrac{y}{4z} + \dfrac{5}{6z^2}$

a) What is the LCD?

b) Perform the indicated operation.

c) If you mistakenly used $24z^2$ for the LCD when adding, would you eventually obtain the correct answer? Explain.

6. Would you use the LCD to perform the following indicated operations? Explain.

a) $\dfrac{3}{x + 3} - \dfrac{4}{x} + \dfrac{5}{3}$ **b)** $\dfrac{1}{x + 2} \cdot \dfrac{5}{x}$

c) $x + \dfrac{2}{9}$ **d)** $\dfrac{5}{x^2 - 9} \div \dfrac{2}{x - 3}$

Practice the Skills

Add or subtract.

7. $\dfrac{2}{x} + \dfrac{3}{y}$

8. $\dfrac{5}{x} - \dfrac{1}{y}$

9. $\dfrac{5}{x^2} + \dfrac{1}{2x}$

10. $7 - \dfrac{1}{x^2}$

11. $3 + \dfrac{8}{x}$

12. $\dfrac{5}{6y} + \dfrac{3}{5y^2}$

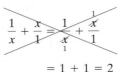 **13.** $\dfrac{2}{x^2} + \dfrac{3}{5x}$

14. $\dfrac{6}{x} - \dfrac{5}{x^2}$

15. $\dfrac{9}{4x^2y} + \dfrac{3}{5xy^2}$

16. $\dfrac{7}{12x^4y} - \dfrac{1}{5x^2y^3}$

17. $4y + \dfrac{x}{y}$

18. $x + \dfrac{2x}{y}$

19. $\dfrac{3a - 1}{2a} + \dfrac{2}{3a}$

20. $\dfrac{11}{n} + 5$

21. $\dfrac{6x}{y} + \dfrac{2y}{xy}$

22. $\dfrac{4}{5p} - \dfrac{5}{2p^2}$

23. $\dfrac{9}{b} - \dfrac{4}{5a^2}$

24. $\dfrac{x-3}{x} - \dfrac{1}{6x}$

25. $\dfrac{4}{x} + \dfrac{9}{x-3}$

26. $10 - \dfrac{3}{x-3}$

27. $\dfrac{9}{p+3} + \dfrac{2}{p}$

28. $\dfrac{2a}{a+b} + \dfrac{a-b}{a}$

29. $\dfrac{5}{d+1} - \dfrac{d}{3d+5}$

30. $\dfrac{2}{x-3} - \dfrac{5}{x-1}$

31. $\dfrac{8}{p-3} + \dfrac{2}{3-p}$

32. $\dfrac{3}{n-5} - \dfrac{7}{5-n}$

33. $\dfrac{9}{x+7} - \dfrac{5}{-x-7}$

34. $\dfrac{6}{7x-1} - \dfrac{3}{1-7x}$

35. $\dfrac{8}{a-2} + \dfrac{a}{2a-4}$

36. $\dfrac{4}{y-1} + \dfrac{3}{y+1}$

37. $\dfrac{x+5}{x-5} - \dfrac{x-5}{x+5}$

38. $\dfrac{x+7}{x+3} - \dfrac{x+3}{x+7}$

39. $\dfrac{5}{6n+3} - \dfrac{2}{n}$

40. $\dfrac{x}{4x-4} - \dfrac{1}{3x}$

41. $\dfrac{3}{2w+10} + \dfrac{6}{w+2}$

42. $\dfrac{5k}{4k-8} - \dfrac{k}{k+2}$

43. $\dfrac{z}{z^2-16} + \dfrac{4}{z+4}$

44. $\dfrac{5}{(x+4)^2} + \dfrac{3}{x+4}$

45. $\dfrac{x+2}{x^2-4} - \dfrac{2}{x+2}$

46. $\dfrac{3}{(x-2)(x+3)} + \dfrac{5}{(x+2)(x+3)}$

47. $\dfrac{3r+4}{r^2-10r+24} - \dfrac{2}{r-6}$

48. $\dfrac{x+9}{x^2-3x-10} - \dfrac{2}{x-5}$

49. $\dfrac{x^2-3}{x^2+2x-8} - \dfrac{x-4}{x+4}$

50. $\dfrac{x+8}{x^2-4x+4} - \dfrac{x+1}{x-2}$

51. $\dfrac{x-6}{x^2+10x+25} + \dfrac{x-3}{x+5}$

52. $\dfrac{x}{x^2-xy} - \dfrac{y}{xy-x^2}$

53. $\dfrac{5}{a^2-9a+8} - \dfrac{6}{a^2-6a-16}$

54. $\dfrac{4}{a^2+2a-15} - \dfrac{1}{a^2-9}$

55. $\dfrac{2}{x^2+6x+9} + \dfrac{7}{x^2+x-6}$

56. $\dfrac{x}{2x^2+7x-4} + \dfrac{4}{x^2-x-20}$

57. $\dfrac{x}{2x^2+7x+3} - \dfrac{5}{3x^2+7x-6}$

58. $\dfrac{x}{6x^2+7x+2} + \dfrac{5}{2x^2-3x-2}$

59. $\dfrac{x}{4x^2+11x+6} - \dfrac{2}{8x^2+2x-3}$

60. $\dfrac{x}{5x^2-9x-2} - \dfrac{1}{3x^2-7x+2}$

61. $\dfrac{3w+12}{w^2+w-12} - \dfrac{2}{w-3}$

62. $\dfrac{5x+10}{x^2-5x-14} - \dfrac{4}{x-7}$

63. $\dfrac{4r}{2r^2-10r+12} + \dfrac{4}{r-2}$

64. $\dfrac{6m}{3m^2-24m+48} - \dfrac{2}{m-4}$

65. $\dfrac{4}{x^2-4x} - \dfrac{x}{4x-16}$

66. $\dfrac{6}{x^2-6x} - \dfrac{x}{6x-36}$

Problem Solving

For what value(s) of x is each expression defined?

67. $\dfrac{8}{x} + 6$

68. $\dfrac{6}{x-1} - \dfrac{5}{x}$

69. $\dfrac{3}{x-4} + \dfrac{7}{x+6}$

70. $\dfrac{4}{x^2-9} - \dfrac{9}{x+3}$

Add or subtract. Treat the unknown symbols as if they were variables.

71. $\dfrac{3}{\Delta-2} - \dfrac{4}{2-\Delta}$

72. $\dfrac{\Delta}{2\Delta^2+7\Delta-4} + \dfrac{2}{\Delta^2-\Delta-20}$

Challenge Problems

Under what conditions is each expression defined? Explain your answers.

73. $\dfrac{5}{a+b} + \dfrac{4}{a}$

74. $\dfrac{x+2}{x+5y} - \dfrac{y-2}{3x}$

Perform each indicated operation.

75. $\dfrac{x}{x^2-9} + \dfrac{2x}{x+3} + \dfrac{2x^2-5x}{9-x^2}$

76. $\dfrac{8x+9}{x^2+x-6} + \dfrac{x}{x+3} - \dfrac{5}{x-2}$

77. $\dfrac{x+6}{4-x^2} - \dfrac{x+3}{x+2} + \dfrac{x-3}{2-x}$

78. $\dfrac{3x-1}{x+2} + \dfrac{x}{x-3} - \dfrac{4}{2x+3}$

79. $\dfrac{2}{x^2-x-6} + \dfrac{3}{x^2-2x-3} + \dfrac{1}{x^2+3x+2}$

80. $\dfrac{3x}{x^2-4} + \dfrac{4}{x^3+8}$

Group Activity

Discuss and answer Exercise 81 as a group.

81. a) As a group, find the LCD of

$$\frac{x + 3y}{x^2 + 3xy + 2y^2} + \frac{y - x}{2x^2 + 3xy + y^2}$$

b) As a group, perform the indicated operation, but do not simplify your answer.

c) As a group, simplify your answer.

d) Group member 1: Substitute 2 for x and 1 for y in the fraction on the left in part **a)** and evaluate.

e) Group member 2: Substitute 2 for x and 1 for y in the fraction on the right in part **a)** and evaluate.

f) Group member 3: Add the numerical fractions found in parts **d)** and **e)**.

g) Individually, substitute 2 for x and 1 for y in the expression obtained in part **b)** and evaluate.

h) Individually, substitute 2 for x and 1 for y in the expression obtained in part **c)**, evaluate, and compare your answers.

i) As a group, discuss what you discovered from this activity.

j) Do you think your results would have been similar for any numbers substituted for x and y (for which the denominator is not 0)? Why?

Cumulative Review Exercises

[2.7] **82. White Pass Railroad** The White Pass Railroad is a narrow gauge railroad that travels slowly through the mountains of Alaska. If the train travels 22 miles in 0.8 hours, how long will it take to travel 42 miles? Assume the train travels at the same rate throughout the trip.

[4.3] **83.** A line passes through the points $(-4, 6)$ and $(3, -2)$. Find the slope of the line.

[5.6] **84.** Divide $(8x^2 + 6x - 15) \div (2x + 3)$.

[7.2] **85.** Multiply $\dfrac{x^2 + xy - 6y^2}{x^2 - xy - 2y^2} \cdot \dfrac{y^2 - x^2}{x^2 + 2xy - 3y^2}$.

Mid-Chapter Test: 7.1–7.4

To find out how well you understand the chapter material to this point, take this brief test. The answers, and the section where the material was initially discussed, are given in the back of the book. Review any questions that you answered incorrectly.

Determine the value or values of the variable where each expression is defined.

1. $\dfrac{9}{3x - 2}$

2. $\dfrac{2x + 1}{x^2 - 5x - 14}$

Simplify each rational expression.

3. $\dfrac{9x + 18}{x + 2}$

4. $\dfrac{2x^2 + 13x + 15}{3x^2 + 14x - 5}$

5. $\dfrac{25r^2 - 36t^2}{5r - 6t}$

Multiply or divide as indicated.

6. $\dfrac{15x^2}{2y} \cdot \dfrac{4y^4}{5x^5}$

7. $\dfrac{m - 3}{m + 4} \cdot \dfrac{m^2 + 8m + 16}{3 - m}$

8. $\dfrac{x^3 + 27}{x^2 - 2x - 15} \cdot \dfrac{x^2 - 7x + 10}{x^2 - 3x + 9}$

9. $\dfrac{5x - 1}{x^2 + 11x + 10} \div \dfrac{10x - 2}{x^2 + 17x + 70}$

10. $\dfrac{5x^2 + 7x + 2}{x^2 + 6x + 5} \div \dfrac{7x^2 - 39x - 18}{x^2 - x - 30}$

Add or subtract as indicated.

11. $\dfrac{x^2}{x+6} - \dfrac{36}{x+6}$

12. $\dfrac{2x^2 - 2x}{2x+5} + \dfrac{x-15}{2x+5}$

13. $\dfrac{3x^2 - x}{4x^2 - 9x + 2} - \dfrac{3x+4}{4x^2 - 9x + 2}$

Find the least common denominator.

14. $\dfrac{2m}{6m^2 + 3m} + \dfrac{m+7}{2m+1}$

15. $\dfrac{9x+8}{2x^2 - 5x - 12} + \dfrac{2x+3}{x^2 - 9x + 20}$

For Exercises 16–19, add or subtract as indicated.

16. $\dfrac{x+1}{2x} + \dfrac{4x-3}{5x}$

17. $\dfrac{2a+5}{a+3} - \dfrac{3a+1}{a-4}$

18. $\dfrac{x^2 + 5}{2x^2 + 13x + 6} + \dfrac{3x-1}{2x+1}$

19. $\dfrac{x}{x^2 + 3x + 2} - \dfrac{4}{x^2 - x - 6}$

20. To add the rational expressions $\dfrac{7}{x+1} + \dfrac{8}{x}$, Samuel Ditsi decided to add both numerators and then add both denominators to get $\dfrac{7+8}{(x+1)+x}$, which simplified to $\dfrac{15}{2x+1}$. This procedure is wrong. Why is it wrong? Explain your answer. Then add the rational expressions $\dfrac{7}{x+1} + \dfrac{8}{x}$ correctly.

7.5 Complex Fractions

1 Simplify complex fractions by combining terms.

2 Simplify complex fractions using multiplication first to clear fractions.

1 Simplify Complex Fractions by Combining Terms

A **complex fraction** is one that has a fraction in its numerator or its denominator or in both its numerator and denominator.

Examples of Complex Fractions

$$\dfrac{\dfrac{3}{5}}{7} \qquad \dfrac{\dfrac{x+9}{x}}{4x} \qquad \dfrac{\dfrac{x}{y}}{x+1} \qquad \dfrac{\dfrac{a+b}{a}}{\dfrac{a-b}{b}}$$

The expression above the main fraction line is the numerator of the complex fraction, and the expression below the main fraction line is the denominator of the complex fraction.

Numerator of complex fraction $\left\{ \dfrac{a+b}{a} \right.$

$\longleftarrow$ Main fraction line

Denominator of complex fraction $\left\{ \dfrac{a+b}{a} \right.$

There are two methods to simplify complex fractions. The first reinforces many of the concepts used in this chapter because we may need to add, subtract, multiply, and divide simpler fractions as we simplify the complex fraction. Many students prefer to use the second method because the answer may be obtained more quickly. We will give two examples using the first method and then work three examples using the second method.

Method 1—To Simplify a Complex Fraction by Combining Terms

1. Add or subtract the fractions in both the numerator and denominator of the complex fraction to obtain single fractions in both the numerator and the denominator.

2. Multiply the fraction in the numerator by the reciprocal of the divisor (that is, by the reciprocal of the fraction in the denominator).

3. Simplify further if possible.

EXAMPLE 1 ▶ Simplify $\dfrac{\dfrac{ab^2}{c^3}}{\dfrac{a}{bc^2}}$.

Solution Since both numerator and denominator are already single fractions, we omit step 1 and begin with step 2. When we multiply the fraction in the numerator by the reciprocal of the fraction in the denominator we obtain the following.

$$\dfrac{\dfrac{ab^2}{c^3}}{\dfrac{a}{bc^2}} = \dfrac{\cancel{a}b^2}{\underset{c}{\cancel{c^3}}} \cdot \dfrac{b\overset{1}{\cancel{c^2}}}{\cancel{a}} = \dfrac{b^3}{c}$$

Thus the expression simplifies to $\dfrac{b^3}{c}$.

▶ **Now Try Exercise 11**

EXAMPLE 2 ▶ Simplify $\dfrac{a + \dfrac{1}{x}}{x + \dfrac{1}{a}}$.

Solution Express both the numerator and denominator of the complex fraction as single fractions. The LCD of the numerator is x and the LCD of the denominator is a.

$$\dfrac{a + \dfrac{1}{x}}{x + \dfrac{1}{a}} = \dfrac{\boxed{\dfrac{x}{x}} \cdot a + \dfrac{1}{x}}{\boxed{\dfrac{a}{a}} \cdot x + \dfrac{1}{a}} = \dfrac{\dfrac{ax}{x} + \dfrac{1}{x}}{\dfrac{ax}{a} + \dfrac{1}{a}} = \dfrac{\dfrac{ax + 1}{x}}{\dfrac{ax + 1}{a}}$$

Now multiply the numerator by the reciprocal of the denominator.

$$= \dfrac{\cancel{ax + 1}}{x} \cdot \dfrac{a}{\cancel{ax + 1}} = \dfrac{a}{x}$$

▶ **Now Try Exercise 25**

In Example 4 on page 431 we will rework Example 2. However, at that time we will work the example using method 2. Most students will agree that method 2 is simpler to use for problems of this type, where the numerator or denominator consists of a sum or difference of terms. We illustrated Example 2 here to show you that method 1 works for problems of this type, and to give you more practice with method 1.

2 Simplify Complex Fractions Using Multiplication First to Clear Fractions

Here is the second method for simplifying complex fractions.

Method 2—To Simplify a Complex Fraction Using Multiplication First

1. Find the least common denominator of *all* the denominators appearing in the complex fraction.

2. Multiply both the numerator and denominator of the complex fraction by the LCD found in step 1.

3. Simplify when possible.

EXAMPLE 3 ▸ Simplify $\dfrac{\frac{2}{3} + \frac{1}{5}}{\frac{4}{5} - \frac{1}{3}}$.

Solution The denominators in the complex fraction are 3 and 5. The LCD of 3 and 5 is 15. Thus 15 is the LCD of the complex fraction. Multiply both the numerator and denominator of the complex fraction by 15.

$$\frac{\frac{2}{3} + \frac{1}{5}}{\frac{4}{5} - \frac{1}{3}} = \frac{15}{15} \cdot \frac{\left(\frac{2}{3} + \frac{1}{5}\right)}{\left(\frac{4}{5} - \frac{1}{3}\right)} = \frac{15\left(\frac{2}{3}\right) + 15\left(\frac{1}{5}\right)}{15\left(\frac{4}{5}\right) - 15\left(\frac{1}{3}\right)}$$

Now simplify.

$$= \frac{10 + 3}{12 - 5} = \frac{13}{7} \text{ or } 1\frac{6}{7}$$

▸ **Now Try Exercise 9**

Now we will rework Example 2 using method 2.

EXAMPLE 4 ▸ Simplify $\dfrac{a + \frac{1}{x}}{x + \frac{1}{a}}$.

Solution The denominators in the complex fraction are x and a. Therefore, the LCD of the complex fraction is ax. Multiply both the numerator and denominator of the complex fraction by ax.

$$\frac{a + \frac{1}{x}}{x + \frac{1}{a}} = \frac{ax}{ax} \cdot \frac{\left(a + \frac{1}{x}\right)}{\left(x + \frac{1}{a}\right)} = \frac{a^2x + a}{ax^2 + x}$$

$$= \frac{a\,\cancel{(ax + 1)}}{x\,\cancel{(ax + 1)}} = \frac{a}{x}$$

▸ **Now Try Exercise 25**

Note that the answers to Examples 2 and 4 are the same.

EXAMPLE 5 ▸ Simplify $\dfrac{y^2}{\frac{1}{x} + \frac{1}{y}}$.

Solution The denominators in the complex fraction are x and y. Therefore, the LCD of the complex fraction is xy. Multiply both the numerator and denominator of the complex fraction by xy.

$$\frac{y^2}{\frac{1}{x} + \frac{1}{y}} = \frac{xy}{xy} \cdot \frac{y^2}{\left(\frac{1}{x} + \frac{1}{y}\right)}$$

$$= \frac{xy^3}{xy\left(\frac{1}{x}\right) + xy\left(\frac{1}{y}\right)}$$

$$= \frac{xy^3}{y + x}$$

▸ **Now Try Exercise 33**

When asked to simplify a complex fraction, you may use either method unless you are told by your instructor to use a specific method. Read the Helpful Hint that follows.

Helpful Hint

We have presented two methods for simplifying complex fractions. Which method should you use? Although either method can be used to simplify complex fractions, most students prefer to use method 1 when both the numerator and denominator consist of a single term, as in Example 1. When the complex fraction has a sum or difference of expressions in either the numerator or denominator, as in Examples 2, 3, 4, or 5, most students prefer to use method 2.

EXERCISE SET 7.5

Concept/Writing Exercises

1. What is a complex fraction?

2. What is the numerator and denominator of each complex fraction?

a) $\dfrac{\dfrac{5}{3}}{\dfrac{3}{x^2 + 5x + 6}}$ **b)** $\dfrac{\dfrac{5}{3}}{x^2 + 5x + 6}$

3. What is the numerator and denominator of each complex fraction?

a) $\dfrac{\dfrac{x + 9}{4}}{\dfrac{7}{x^2 + 5x + 6}}$ **b)** $\dfrac{\dfrac{1}{2y} + x}{\dfrac{3}{y} + x^2}$

4. a) Select the method you prefer to use to simplify complex fractions. Then write a step-by-step procedure for simplifying complex fractions using that method.

b) Using the procedure you wrote in part **a)**, simplify the following complex fraction.

$$\dfrac{\dfrac{4}{x} - \dfrac{3}{y}}{x + \dfrac{1}{y}}$$

Practice the Skills

Simplify.

5. $\dfrac{4 + \dfrac{2}{3}}{5 + \dfrac{1}{3}}$

6. $\dfrac{2 + \dfrac{4}{5}}{1 - \dfrac{9}{16}}$

7. $\dfrac{2 + \dfrac{3}{8}}{1 + \dfrac{1}{3}}$

8. $\dfrac{\dfrac{1}{4} + \dfrac{5}{6}}{\dfrac{2}{3} + \dfrac{3}{5}}$

9. $\dfrac{\dfrac{2}{3} + \dfrac{1}{4}}{\dfrac{5}{6} - \dfrac{1}{3}}$

10. $\dfrac{\dfrac{5}{8} + \dfrac{1}{3}}{\dfrac{11}{12} - \dfrac{1}{6}}$

11. $\dfrac{\dfrac{xy^2}{7}}{\dfrac{3}{x^2}}$

12. $\dfrac{\dfrac{11a}{b^3}}{\dfrac{b^2}{4}}$

13. $\dfrac{\dfrac{6a^2b}{7}}{\dfrac{9ac^2}{b^2}}$

14. $\dfrac{\dfrac{18x^4}{5y^4z^5}}{\dfrac{9xy^2}{15z^5}}$

15. $\dfrac{a - \dfrac{a}{b}}{\dfrac{3 + a}{b}}$

16. $\dfrac{a + \dfrac{2}{b}}{\dfrac{a}{b}}$

17. $\dfrac{\dfrac{9}{x} + \dfrac{3}{x^2}}{3 + \dfrac{1}{x}}$

18. $\dfrac{\dfrac{4}{a} + \dfrac{1}{2a}}{a + \dfrac{a}{2}}$

19. $\dfrac{5 - \dfrac{1}{x}}{4 - \dfrac{1}{x}}$

20. $\dfrac{\dfrac{2x}{x - y}}{\dfrac{x^2}{y}}$

21. $\dfrac{\dfrac{m}{n} - \dfrac{n}{m}}{\dfrac{m + n}{n}}$

22. $\dfrac{5}{\dfrac{1}{x} + y}$

23. $\dfrac{\dfrac{a^2}{b} - b}{\dfrac{b^2}{a} - a}$

24. $\dfrac{\dfrac{1}{x^2} - \dfrac{8}{x}}{3 + \dfrac{1}{x^2}}$

25. $\dfrac{2 - \dfrac{a}{b}}{\dfrac{a}{b} - 2}$

26. $\dfrac{\dfrac{x}{y} - 9}{\dfrac{-x}{y} + 9}$

27. $\dfrac{\dfrac{4}{x^2} + \dfrac{4}{x}}{\dfrac{4}{x} + \dfrac{4}{x^2}}$

28. $\dfrac{\dfrac{a^2 - b^2}{a}}{\dfrac{a + b}{a^4}}$

29. $\dfrac{\dfrac{1}{a} - \dfrac{1}{b}}{\dfrac{1}{ab}}$

30. $\dfrac{\dfrac{1}{a} - \dfrac{1}{b}}{\dfrac{1}{a} + \dfrac{1}{b}}$

31. $\dfrac{\dfrac{a}{b} + \dfrac{1}{a}}{\dfrac{b}{a} + \dfrac{1}{a}}$

32. $\dfrac{\dfrac{2}{a} + \dfrac{3}{b}}{\dfrac{1}{a}}$

33. $\dfrac{x}{\dfrac{1}{x} - \dfrac{1}{y}}$

34. $\dfrac{\dfrac{1}{a} + \dfrac{1}{b}}{ab}$

35. $\dfrac{\dfrac{5}{a} + \dfrac{5}{a^2}}{\dfrac{5}{b} + \dfrac{5}{b^2}}$

36. $\dfrac{\dfrac{x}{y} - \dfrac{2}{x}}{\dfrac{y}{x} + \dfrac{1}{y}}$

Problem Solving

For the complex fractions in Exercises 37–40,

  a) *Determine which of the two methods discussed in this section you would use to simplify the fraction. Explain why.*

  b) *Simplify by the method you selected in part **a**).*

  c) *Simplify by the method you did not select in part **a**). If your answers to parts **b**) and **c**) are not the same, explain why.*

37. $\dfrac{5 + \dfrac{3}{5}}{\dfrac{1}{8} - 4}$

38. $\dfrac{\dfrac{x + y}{x^3} - \dfrac{1}{x}}{\dfrac{x - y}{x^5} + 5}$

39. $\dfrac{\dfrac{x - y}{x + y} + \dfrac{6}{x + y}}{2 - \dfrac{7}{x + y}}$

40. $\dfrac{\dfrac{25}{x - y} + \dfrac{2}{x + y}}{\dfrac{5}{x - y} - \dfrac{3}{x + y}}$

*In Exercises 41 and 42, **a)** write the complex fraction, and **b)** simplify the complex fraction.*

41. The numerator of the complex fraction consists of one term: 5 is divided by $12x$. The denominator of the complex fraction consists of two terms: 4 divided by $3x$ is subtracted from 8 divided by x^2.

42. The numerator of the complex fraction consists of two terms: 3 divided by $2x$ is subtracted from 6 divided by x. The denominator of the complex fraction consists of two terms: the sum of x and the quantity 1 divided by x.

Challenge Problems

Simplify. (Hint: Refer to Section 5.2, which discusses negative exponents.)

43. $\dfrac{x^{-1} + y^{-1}}{3}$

44. $\dfrac{x^{-1} + y^{-1}}{y^{-1}}$

45. $\dfrac{x^{-1} + y^{-1}}{x^{-1}y^{-1}}$

46. $\dfrac{x^{-2} - y^{-2}}{y^{-1} - x^{-1}}$

47. Jack The efficiency of a jack, E, is expressed by the formula $E = \dfrac{\dfrac{1}{2}h}{h + \dfrac{1}{2}}$, where h is determined by the pitch of the jack's thread. Determine the efficiency of a jack if h is

  a) $\dfrac{2}{3}$ **b)** $\dfrac{4}{5}$

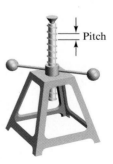
Pitch

Simplify.

48. $\dfrac{\dfrac{x}{y} + \dfrac{y}{x} + \dfrac{2}{x}}{\dfrac{x}{y} + y}$

49. $\dfrac{\dfrac{a}{b} + b - \dfrac{1}{a}}{\dfrac{a}{b^2} - \dfrac{b}{a} + \dfrac{3}{a^2}}$

50. $\dfrac{x}{4 + \dfrac{x}{1 + x}}$

Cumulative Review Exercises

[2.5] **51.** Solve the equation

$$2x - 8(5 - x) = 9x - 3(x + 2).$$

[5.4] **52.** What is a polynomial?

[6.3] **53.** Factor $x^2 - 13x + 40$.

[7.4] **54.** Subtract $\dfrac{x}{3x^2 + 17x - 6} - \dfrac{2}{x^2 + 3x - 18}$.

7.6 Solving Rational Equations

1 Solve rational equations with integer denominators.

2 Solve rational equations where a variable appears in a denominator.

1 Solve Rational Equations with Integer Denominators

In Sections 7.1 through 7.5, we focused on how to add, subtract, multiply, and divide rational expressions. Now we are ready to solve rational equations. A **rational equation** is one that contains one or more rational (or fractional) expressions. A rational equation may be one that contains rational coefficients, such as $\dfrac{1}{2}x + \dfrac{3}{5}x = 8$ or $\dfrac{x}{2} + \dfrac{3x}{5} = 8$, or one that contains rational terms, with a variable in a denominator, such as $\dfrac{4}{x - 2} = 5$. We solved linear equations with rational coefficients in Sections 5.4 and 5.5.

The emphasis of this section will be on solving rational equations where a variable appears in a denominator. The following procedure that we will use to solve rational equations in this section is very similar to the procedure we used in Chapter 2.

> ### To Solve Rational Equations
>
> 1. Determine the least common denominator (LCD) of all fractions in the equation.
> 2. Multiply **both** sides of the equation by the LCD. **This will result in every term in the equation being multiplied by the LCD**.
> 3. Remove any parentheses and combine like terms on each side of the equation.
> 4. Solve the equation using the properties discussed in earlier chapters.
> 5. Check your solution in the *original* equation.

The purpose of multiplying both sides of the equation by the LCD (step 2) is to eliminate all fractions from the equation. After both sides of the equation are multiplied by the LCD, the resulting equation should contain no fractions. We will omit some of the checks to save space.

Before we solve rational equations where a variable appears in a denominator, let's review how to solve equations with rational coefficients. Examples 1 and 2 illustrate the procedure to follow.

EXAMPLE 1 ▸ Solve $\dfrac{t}{4} - \dfrac{t}{5} = 1$ for t.

Solution The LCD of 4 and 5 is 20. Multiply both sides of the equation by 20.

$$\frac{t}{4} - \frac{t}{5} = 1$$

$$20\left(\frac{t}{4} - \frac{t}{5}\right) = 20 \cdot 1 \qquad \textit{Multiply both sides by the LCD, 20.}$$

$$20\left(\frac{t}{4}\right) - 20\left(\frac{t}{5}\right) = 20 \qquad \textit{Distributive property}$$

$$5t - 4t = 20$$

$$t = 20$$

Check $\dfrac{t}{4} - \dfrac{t}{5} = 1$

$\dfrac{20}{4} - \dfrac{20}{5} \overset{?}{=} 1$

$5 - 4 \overset{?}{=} 1$

$1 = 1$ *True*

The solution is 20.

▸ **Now Try Exercise 13**

EXAMPLE 2 ▸ Solve $\dfrac{x-5}{30} = \dfrac{4}{5} - \dfrac{x-1}{10}$.

Solution   The least common denominator is 30. Multiply both sides of the equation by 30.

$$\frac{x-5}{30} = \frac{4}{5} - \frac{x-1}{10}$$

$$30\left(\frac{x-5}{30}\right) = 30\left(\frac{4}{5} - \frac{x-1}{10}\right) \qquad \text{\textit{Multiply both sides by the LCD, 30.}}$$

$$x - 5 = 30\left(\frac{4}{5}\right) - 30\left(\frac{x-1}{10}\right) \qquad \text{\textit{Distributive property}}$$

$$x - 5 = 24 - 3(x-1)$$

$$x - 5 = 24 - 3x + 3 \qquad \text{\textit{Distributive property}}$$

$$x - 5 = -3x + 27 \qquad \text{\textit{Combine like terms.}}$$

$$4x - 5 = 27 \qquad \text{\textit{3x was added to both sides.}}$$

$$4x = 32 \qquad \text{\textit{5 was added to both sides.}}$$

$$x = 8 \qquad \text{\textit{Both sides were divided by 4.}}$$

A check will show that the answer is 8. We suggest you check this answer now to get practice at checking answers.

▸ **Now Try Exercise 27**

In Example 2, the equation could also have been written as

$$\frac{1}{30}(x - 5) = \frac{4}{5} - \frac{1}{10}(x - 1).$$

For additional examples of solving rational equations with integers in the denominators, review Sections 2.4 and 2.5.

2 Solve Rational Equations Where a Variable Appears in a Denominator

Now we are ready to solve rational equations where a variable appears in a denominator. When solving a rational equation where a variable appears in any denominator, you *must* check your answer. See the following warning!

Warning   **Whenever a variable appears in any denominator of a rational equation, it is necessary to check your answer in the original equation. If the answer obtained makes any denominator equal to zero, that value is not a solution to the equation.** Such values are called **extraneous roots** or **extraneous solutions**.

EXAMPLE 3 ▶ Solve $4 - \dfrac{5}{x} = \dfrac{3}{2}$.

Solution   Multiply both sides of the equation by the LCD, $2x$.

$$2x\left(4 - \frac{5}{x}\right) = \left(\frac{3}{2}\right) \cdot 2x \qquad \textit{Multiply both sides by the LCD, 2x.}$$

$$2x(4) - 2x\left(\frac{5}{x}\right) = \left(\frac{3}{2}\right) \cdot 2x \qquad \textit{Distributive property}$$

$$8x - 10 = 3x$$

$$5x - 10 = 0 \qquad \textit{3x was subtracted from both sides.}$$

$$5x = 10 \qquad \textit{10 was added to both sides.}$$

$$x = 2$$

Check $4 - \dfrac{5}{x} = \dfrac{3}{2}$

$$4 - \frac{5}{2} \stackrel{?}{=} \frac{3}{2}$$

$$\frac{8}{2} - \frac{5}{2} \stackrel{?}{=} \frac{3}{2}$$

$$\frac{3}{2} = \frac{3}{2} \qquad \textit{True}$$

Since 2 does check, it is the solution to the equation.

▶ **Now Try Exercise 17**

EXAMPLE 4 ▶ Solve $\dfrac{p - 5}{p + 3} = \dfrac{1}{5}$.

Solution   The LCD is $5(p + 3)$. Multiply both sides of the equation by the LCD.

$$5(p + 3) \cdot \frac{(p - 5)}{p + 3} = \frac{1}{5} \cdot 5(p + 3)$$

$$5(p - 5) = 1(p + 3)$$

$$5p - 25 = p + 3$$

$$4p - 25 = 3$$

$$4p = 28$$

$$p = 7$$

A check will show that 7 is the solution.

▶ **Now Try Exercise 43**

In Section 2.7 we illustrated that proportions of the form

$$\frac{a}{b} = \frac{c}{d}$$

can be cross-multiplied to obtain $a \cdot d = b \cdot c$. Example 4 is a proportion and can also be solved by cross-multiplying, as done in Example 5.

EXAMPLE 5 ▸ Use cross-multiplication to solve $\dfrac{9}{x+1} = \dfrac{5}{x-3}$.

Solution

$$\frac{9}{x+1} = \frac{5}{x-3}$$

$$9(x-3) = 5(x+1) \qquad \textit{Cross-multiply.}$$

$$9x - 27 = 5x + 5 \qquad \textit{Distributive property}$$

$$4x - 27 = 5$$

$$4x = 32$$

$$x = 8$$

A check will show that 8 is the solution to the equation.

▸ **Now Try Exercise 41**

Now let's examine some examples that involve quadratic equations. Recall from Section 6.6 that quadratic equations have the form $ax^2 + bx + c = 0$, where $a \neq 0$.

EXAMPLE 6 ▸ Solve $x + \dfrac{12}{x} = -7$.

Solution

$$x + \frac{12}{x} = -7$$

$$\boxed{x} \cdot \left(x + \frac{12}{x} \right) = -7 \cdot \boxed{x} \qquad \textit{Multiply both sides by x.}$$

$$x(x) + x\left(\frac{12}{x}\right) = -7x \qquad \textit{Distributive property}$$

$$x^2 + 12 = -7x$$

$$x^2 + 7x + 12 = 0 \qquad \textit{7x was added to both sides.}$$

$$(x+3)(x+4) = 0 \qquad \textit{Factor.}$$

$$x + 3 = 0 \quad \text{or} \quad x + 4 = 0 \qquad \textit{Zero-factor property}$$

$$x = -3 \qquad\qquad x = -4$$

Check

$$x = -3 \qquad\qquad\qquad x = -4$$

$$x + \frac{12}{x} = -7 \qquad\qquad x + \frac{12}{x} = -7$$

$$-3 + \frac{12}{-3} \overset{?}{=} -7 \qquad\qquad -4 + \frac{12}{-4} \overset{?}{=} -7$$

$$-3 + (-4) \overset{?}{=} -7 \qquad\qquad -4 + (-3) \overset{?}{=} -7$$

$$-7 = -7 \quad \textit{True} \qquad\qquad -7 = -7 \quad \textit{True}$$

The solutions are -3 and -4.

▸ **Now Try Exercise 57**

EXAMPLE 7 ▶ Solve $\dfrac{x^2 - 2x}{x - 6} = \dfrac{24}{x - 6}$.

Solution   If we try to solve this equation using cross-multiplication we will get a cubic equation. We will solve this equation by multiplying both sides of the equation by the LCD, $x - 6$.

$$\frac{x^2 - 2x}{x - 6} = \frac{24}{x - 6}$$

$$\cancel{x - 6} \cdot \frac{x^2 - 2x}{\cancel{x - 6}} = \frac{24}{\cancel{x - 6}} \cdot \cancel{x - 6} \qquad \textit{Multiply both sides by the LCD, } x - 6.$$

$$x^2 - 2x = 24$$

$$x^2 - 2x - 24 = 0 \qquad \textit{24 was subctracted from both sides.}$$

$$(x + 4)(x - 6) = 0 \qquad \textit{Factor}$$

$$x + 4 = 0 \quad \text{or} \quad x - 6 = 0 \qquad \textit{Zero-factor property}$$

$$x = -4 \qquad\qquad x = 6$$

Check

$$x = -4 \qquad\qquad\qquad\qquad x = 6$$

$$\frac{x^2 - 2x}{x - 6} = \frac{24}{x - 6} \qquad\qquad \frac{x^2 - 2x}{x - 6} = \frac{24}{x - 6}$$

$$\frac{(-4)^2 - 2(-4)}{-4 - 6} \overset{?}{=} \frac{24}{-4 - 6} \qquad\qquad \frac{6^2 - 2(6)}{6 - 6} \overset{?}{=} \frac{24}{6 - 6}$$

$$\frac{16 + 8}{-10} \overset{?}{=} \frac{24}{-10} \qquad\qquad\qquad \frac{24}{0} = \frac{24}{0}$$

$$\frac{24}{-10} = \frac{24}{-10} \quad \textit{True} \qquad\qquad \uparrow \qquad \uparrow$$

$$\qquad\qquad\qquad\qquad\qquad\quad \textit{Since the denominator is 0, and we}$$
$$\qquad\qquad\qquad\qquad\qquad\quad \textit{cannot divide by 0, 6 is not a solution.}$$

Since $\dfrac{24}{0}$ is not a real number, 6 is an extraneous solution. Thus, this equation has only one solution, -4.

▶ **Now Try Exercise 47**

Helpful Hint

Remember, when solving a rational equation in which a variable appears in a denominator, you must check *all* your answers to make sure that none is an extraneous root. Extraneous roots can usually be spotted quickly. If any of your answers make any denominator 0, that answer is an extraneous root and not a true solution.

Generally speaking, if none of your answers makes any denominator 0, and you have not made a mistake when solving the equation, the answers you obtained when solving the equation will be solutions to the equation.

EXAMPLE 8 ▶ Solve $\dfrac{5w}{w^2 - 4} + \dfrac{1}{w - 2} = \dfrac{4}{w + 2}$.

Solution   First factor $w^2 - 4$.

$$\frac{5w}{(w + 2)(w - 2)} + \frac{1}{w - 2} = \frac{4}{w + 2}$$

Multiply both sides of the equation by the LCD, $(w + 2)(w - 2)$.

$$(w + 2)(w - 2)\left[\frac{5w}{(w + 2)(w - 2)} + \frac{1}{w - 2}\right] = \frac{4}{w + 2} \cdot (w + 2)(w - 2)$$

$$(w + 2)(w - 2) \cdot \frac{5w}{(w + 2)(w - 2)} + (w + 2)(w - 2) \cdot \frac{1}{w - 2} = \frac{4}{w + 2} \cdot (w + 2)(w - 2)$$

$$\cancel{(w + 2)}\cancel{(w - 2)} \cdot \frac{5w}{\cancel{(w + 2)}\cancel{(w - 2)}} + (w + 2)\cancel{(w - 2)} \cdot \frac{1}{\cancel{w - 2}} = \frac{4}{\cancel{w + 2}} \cdot \cancel{(w + 2)}(w - 2)$$

$$5w + (w + 2) = 4(w - 2)$$

$$6w + 2 = 4w - 8$$

$$2w + 2 = -8$$

$$2w = -10$$

$$w = -5$$

A check will show that -5 is the solution to the equation.

▸ **Now Try Exercise 65**

Helpful Hint

Some students confuse adding and subtracting rational expressions with solving rational equations. When adding or subtracting rational expressions, we must rewrite each expression with a common denominator. When solving a rational equation, we multiply both sides of the equation by the LCD to eliminate fractions from the equation. Consider the following two problems. Note that the one on the right is an equation because it contains an equal sign. We will work both problems. The LCD for both problems is $x(x + 4)$.

Adding Rational Expressions

$$\frac{x + 2}{x + 4} + \frac{3}{x}$$

We rewrite each fraction with the LCD, $x(x + 4)$.

$$= \frac{x}{x} \cdot \frac{x + 2}{x + 4} + \frac{3}{x} \cdot \frac{x + 4}{x + 4}$$

$$= \frac{x(x + 2)}{x(x + 4)} + \frac{3(x + 4)}{x(x + 4)}$$

$$= \frac{x^2 + 2x}{x(x + 4)} + \frac{3x + 12}{x(x + 4)}$$

$$= \frac{x^2 + 2x + 3x + 12}{x(x + 4)}$$

$$= \frac{x^2 + 5x + 12}{x(x + 4)}$$

Solving Rational Equations

$$\frac{x + 2}{x + 4} = \frac{3}{x}$$

We eliminate fractions by multiplying both sides of the equation by the LCD, $x(x + 4)$.

$$(x)(x + 4)\left(\frac{x + 2}{x + 4}\right) = \frac{3}{x}(x)(x + 4)$$

$$x(x + 2) = 3(x + 4)$$

$$x^2 + 2x = 3x + 12$$

$$x^2 - x - 12 = 0$$

$$(x - 4)(x + 3) = 0$$

$$x - 4 = 0 \quad \text{or} \quad x + 3 = 0$$

$$x = 4 \qquad\qquad x = -3$$

The numbers 4 and -3 on the right will both check and are thus solutions to the equation.

Note that when adding and subtracting rational expressions we usually end up with an algebraic expression. When solving rational equations, the solution will be a numerical value or values. The equation on the right could also be solved using cross-multiplication.

EXERCISE SET 7.6   Math XP   MyMathLab
MathXL®   MyMathLab

Concept/Writing Exercises

1. a) Explain the steps to use to solve rational equations.

   b) Using the procedure you wrote in part **a)**, solve the equation $\dfrac{1}{x-1} - \dfrac{1}{x+1} = \dfrac{3x}{x^2-1}$.

2. a) Without solving the equations, determine whether the answer to the two equations that follow will be the same or different. Explain your answer.

$$\dfrac{3}{x-2} + \dfrac{2}{x+2} = \dfrac{4x}{x^2-4}, \quad \dfrac{4x}{x^2-4} - \dfrac{2}{x+2} = \dfrac{3}{x-2}$$

   b) Determine the answer to both questions.

3. Consider the following problems.

   Simplify: Solve:

   $\dfrac{x}{3} - \dfrac{x}{4} + \dfrac{1}{x-1}$ $\dfrac{x}{3} - \dfrac{x}{4} = \dfrac{1}{x-1}$

   a) Explain the difference between the two types of problems.

   b) Explain how you would work each problem to obtain the correct answer.

   c) Find the correct answer to each problem.

4. Consider the following problems.

   Simplify: Solve:

   $\dfrac{x}{2} - \dfrac{x}{3} + \dfrac{5}{2x+7}$ $\dfrac{x}{2} - \dfrac{x}{3} = \dfrac{5}{2x+7}$

   a) Explain the difference between the two types of problems.

   b) Explain how you would work each problem to obtain the correct answer.

   c) Find the correct answer to each problem.

5. Under what conditions must you check rational equations for extraneous solutions?

6. Which of the following numbers, 3, 1, or, 0 cannot be a solution to the equation $3 - \dfrac{1}{x} = 4$? Explain.

7. Which of the following numbers, 0, 3, or 2 cannot be a solution to the equation $\dfrac{3}{x-2} + 5x = 6$? Explain.

8. Which of the following numbers, 0, −5 or 2 cannot be a solution to the equation $4x - \dfrac{2}{x+5} = 3$? Explain.

In Exercises 9–12, indicate whether it is necessary to check the solution obtained to see if it is extraneous. Explain your answer.

9. $\dfrac{4}{7} + \dfrac{x-2}{6} = 5$

10. $\dfrac{5}{x-2} + \dfrac{1}{x} = 6$

11. $\dfrac{2}{x+4} + 1 = 6$

12. $\dfrac{z}{3} + \dfrac{7}{z} = 8$

Practice the Skills

Solve each equation and check your solution. See Examples 1 and 2.

13. $\dfrac{x}{3} - \dfrac{x}{4} = 1$

14. $\dfrac{t}{5} - \dfrac{t}{6} = 2$

15. $\dfrac{r}{6} = \dfrac{r}{4} + \dfrac{1}{3}$

16. $\dfrac{n}{5} = \dfrac{n}{6} + \dfrac{2}{3}$

17. $\dfrac{z}{2} + 6 = \dfrac{z}{5}$

18. $\dfrac{3w}{5} - 6 = w$

19. $\dfrac{z}{6} + \dfrac{2}{3} = \dfrac{z}{5} - \dfrac{1}{3}$

20. $\dfrac{m-2}{6} = \dfrac{2}{3} + \dfrac{m}{12}$

21. $d + 7 = \dfrac{3}{2}d + 5$

22. $\dfrac{q}{5} + \dfrac{q}{2} = \dfrac{21}{10}$

23. $3k + \dfrac{1}{6} = 4k - 4$

24. $\dfrac{p}{4} + \dfrac{1}{4} = \dfrac{p}{3} - \dfrac{1}{2}$

25. $\dfrac{n+6}{3} = \dfrac{5(n-8)}{10}$

26. $\dfrac{3(x-6)}{5} = \dfrac{4(x+2)}{8}$

27. $\dfrac{x-5}{15} = \dfrac{3}{5} - \dfrac{x-4}{10}$

28. $\dfrac{z+4}{6} = \dfrac{3}{2} - \dfrac{2z+2}{12}$

29. $\dfrac{-p+1}{4} + \dfrac{13}{20} = \dfrac{p}{5} - \dfrac{p-1}{2}$

30. $\dfrac{1}{10} - \dfrac{n+1}{6} = \dfrac{1}{5} - \dfrac{n+10}{15}$

31. $\dfrac{d-3}{4} + \dfrac{1}{15} = \dfrac{2d+1}{3} - \dfrac{34}{15}$

32. $\dfrac{t+4}{5} = \dfrac{5}{8} + \dfrac{t+7}{40}$

Solve each equation and check your solution. See Examples 3–8.

33. $2 + \dfrac{3}{x} = \dfrac{11}{4}$

34. $3 - \dfrac{1}{x} = \dfrac{14}{5}$

35. $7 - \dfrac{5}{x} = \dfrac{9}{2}$

36. $4 + \dfrac{3}{z} = \dfrac{9}{2}$

37. $\dfrac{4}{n} - \dfrac{3}{2n} = \dfrac{1}{2}$

38. $\dfrac{5}{3x} + \dfrac{2}{x} = 1$

39. $\dfrac{x-1}{x-5} = \dfrac{4}{x-5}$

40. $\dfrac{2x+3}{x+2} = \dfrac{3}{2}$

41. $\dfrac{5}{a + 3} = \dfrac{4}{a + 1}$

42. $\dfrac{5}{x + 2} = \dfrac{1}{x - 4}$

43. $\dfrac{y + 3}{y - 3} = \dfrac{6}{4}$

44. $\dfrac{x}{x + 6} = \dfrac{2}{5}$

45. $\dfrac{2x - 3}{x - 4} = \dfrac{5}{x - 4}$

46. $\dfrac{3}{x} + 9 = \dfrac{3}{x}$

47. $\dfrac{x^2}{x - 3} = \dfrac{9}{x - 3}$

48. $\dfrac{x^2}{x + 5} = \dfrac{25}{x + 5}$

49. $\dfrac{n - 3}{n + 2} = \dfrac{n + 4}{n + 10}$

50. $\dfrac{x + 5}{x + 1} = \dfrac{x - 6}{x - 3}$

51. $\dfrac{1}{r} = \dfrac{3r}{8r + 3}$

52. $\dfrac{1}{r} = \dfrac{2r}{r + 15}$

53. $\dfrac{k}{k + 2} = \dfrac{3}{k - 2}$

54. $\dfrac{3a - 2}{2a + 2} = \dfrac{3}{a - 1}$

55. $\dfrac{4}{r} + r = \dfrac{20}{r}$

56. $a + \dfrac{5}{a} = \dfrac{14}{a}$

57. $x + \dfrac{20}{x} = -9$

58. $x - \dfrac{32}{x} = 4$

59. $\dfrac{3y - 2}{y + 1} = 4 - \dfrac{y + 2}{y - 1}$

60. $\dfrac{2b}{b + 1} = 2 - \dfrac{5}{2b}$

61. $\dfrac{1}{x + 3} + \dfrac{1}{x - 3} = \dfrac{-5}{x^2 - 9}$

62. $\dfrac{t + 2}{t - 5} - \dfrac{3}{4} = \dfrac{6}{t - 5}$

63. $\dfrac{x}{x - 3} + \dfrac{3}{2} = \dfrac{3}{x - 3}$

64. $\dfrac{y}{2y + 2} + \dfrac{2y - 16}{4y + 4} = \dfrac{y - 3}{y + 1}$

65. $\dfrac{3}{x - 5} - \dfrac{4}{x + 5} = \dfrac{11}{x^2 - 25}$

66. $\dfrac{2n^2 - 15}{n^2 + n - 6} = \dfrac{n + 1}{n + 3} + \dfrac{n - 3}{n - 2}$

67. $\dfrac{3x}{x^2 - 9} + \dfrac{1}{x - 3} = \dfrac{3}{x + 3}$

68. $\dfrac{3}{x + 3} + \dfrac{5}{x + 4} = \dfrac{12x + 7}{x^2 + 7x + 12}$

69. $\dfrac{1}{y - 1} + \dfrac{1}{2} = \dfrac{2}{y^2 - 1}$

70. $\dfrac{2y}{y + 2} = \dfrac{y}{y + 3} - \dfrac{3}{y^2 + 5y + 6}$

71. $\dfrac{3t}{6t + 6} + \dfrac{t}{2t + 2} = \dfrac{2t - 3}{t + 1}$

72. $\dfrac{2}{x - 2} - \dfrac{1}{x + 1} = \dfrac{2}{x^2 - x - 2}$

Problem Solving

In Exercises 73–78, determine the solution by observation. Explain how you determined your answer.

73. $\dfrac{3}{x - 2} = \dfrac{x - 2}{x - 2}$

74. $\dfrac{1}{2} + \dfrac{x}{2} = \dfrac{5}{2}$

75. $\dfrac{x}{x - 6} + \dfrac{x}{x - 6} = 0$

76. $\dfrac{x}{4} + \dfrac{3x}{4} = x$

77. $\dfrac{x - 2}{3} + \dfrac{x - 2}{3} = \dfrac{2x - 4}{3}$

78. $\dfrac{3}{x} - \dfrac{1}{x} = \dfrac{2}{x}$

79. Optics A formula frequently used in optics is

$$\frac{1}{p} + \frac{1}{q} = \frac{1}{f}$$

where p represents the distance of the object from a mirror (or lens), q represents the distance of the image from the mirror (or lens), and f represents the focal length of the mirror (or lens). If a mirror has a focal length of 10 centimeters, how far from the mirror will the image appear when the object is 30 centimeters from the mirror?

Challenge Problems

80. a) Explain why the equation $\dfrac{x^2}{x - 3} = \dfrac{9}{x - 3}$ cannot be solved by cross-multiplying using the material presented in the book.

b) Solve the equation given in part a).

81. Solve the equation $\dfrac{x - 4}{x^2 - 2x} = \dfrac{-4}{x^2 - 4}$

82. Electrical Resistance In electronics the total resistance R_T, of resistors wired in a parallel circuit is determined by the formula

$$\frac{1}{R_T} = \frac{1}{R_1} + \frac{1}{R_2} + \frac{1}{R_3} + \cdots + \frac{1}{R_n}$$

where $R_1, R_2, R_3, \ldots, R_n$ are the resistances of the individual resistors (measured in ohms) in the circuit.

a) Find the total resistance if two resistors, one of 200 ohms and the other of 300 ohms, are wired in a parallel circuit.

b) If three identical resistors are to be wired in parallel, what should be the resistance of each resistor if the total resistance of the circuit is to be 300 ohms?

83. Can an equation of the form $\dfrac{a}{x} + 1 = \dfrac{a}{x}$ have a real number solution for any real number a? Explain your answer.

Group Activity

Discuss and answer Exercise 84 as a group.

84. a) As a group, discuss two different methods you can use to solve the equation $\dfrac{x+3}{5} = \dfrac{x}{4}$.

b) Group member 1: Solve the equation by obtaining a common denominator.
Group member 2: Solve the equation by cross-multiplying.
Group member 3: Check the results of group member 1 and group member 2.

c) Individually, create another equation by taking the reciprocal of each term in the equation in part **a)**. Compare your results. Do you think that the reciprocal of the answer you found in part **b)** will be the solution to this equation? Explain.

d) Individually, solve the equation you found in part **c)** and check your answer. Compare your work with the other group members. Was the conclusion you came to in part **c)** correct? Explain.

e) As a group, solve the equation $\dfrac{1}{x} + \dfrac{1}{3} = \dfrac{2}{x}$. Check your result.

f) As a group, create another equation by taking the reciprocal of each term of the equation in part **e)**. Do you think that the reciprocal of the answer you found in part **e)** will be the solution to this equation? Explain.

g) Individually, solve the equation you found in part **f)** and check your answer. Compare your work with the other group members. Did your group make the correct conclusion in part **f)**? Explain.

h) As a group, discuss the relationship between the solution to the equation $\dfrac{7}{x-9} = \dfrac{3}{x}$ and the solution to the equation $\dfrac{x-9}{7} = \dfrac{x}{3}$. Explain your answer.

Cumulative Review Exercises

[3.2] **85. Internet Plans** An Internet service offers two plans for its customers. One plan includes 5 hours of use and costs $7.95 per month. Each additional minute after the 5 hours costs $0.15. The second plan costs $19.95 per month and provides unlimited Internet access. How many hours would Jake LaRue have to use the Internet monthly to make the second plan the less expensive?

86. Filling a Jacuzzi How long will it take to fill a 600-gallon Jacuzzi if water is flowing into the Jacuzzi at a rate of 4 gallons a minute?

[3.3] **87. Supplementary Angles** Two angles are supplementary angles if the sum of their measures is 180°. Find the two supplementary angles if the smaller angle is 30° less than half the larger angle.

[5.6] **88.** Multiply $(3.4 \times 10^{-5})(2 \times 10^{13})$.

[6.6] **89.** Explain the difference between a linear equation and a quadratic equation, and give an example of each.

7.7 Rational Equations: Applications and Problem Solving

1 Set up and solve applications containing rational expressions.

2 Set up and solve motion problems.

3 Set up and solve work problems.

1 Set Up and Solve Applications Containing Rational Expressions

Many applications of algebra involve rational equations. After we represent the application as an equation, we solve the rational equation as we did in Section 7.6.

The first type of application we will consider is a *geometry problem*.

EXAMPLE 1 ▶ A New Rug Mary and Larry Armstrong are interested in purchasing a carpet whose area is 60 square feet. Determine the length and width if the width is 5 feet less than $\dfrac{3}{5}$ of the length, see **Figure 7.2**.

FIGURE 7.2

Solution

Understand and Translate   Let x = length.

$$\text{Then } \frac{3}{5}x - 5 = \text{width.}$$

$$\text{area} = \text{length} \cdot \text{width}$$

$$60 = x\left(\frac{3}{5}x - 5\right)$$

Carry Out

$$60 = \frac{3}{5}x^2 - 5x$$

$$5\,(60) = 5\left(\frac{3}{5}x^2 - 5x\right) \qquad \textit{Multiply both sides by 5.}$$

$$300 = 3x^2 - 25x \qquad \textit{Distributive property}$$

$$0 = 3x^2 - 25x - 300 \qquad \textit{Subtract 300 from both sides.}$$

$$\text{or} \quad 3x^2 - 25x - 300 = 0$$

$$(3x + 20)(x - 15) = 0 \qquad \textit{Factor.}$$

$$3x + 20 = 0 \quad \text{or} \quad x - 15 = 0 \qquad \textit{Zero-factor property}$$

$$3x = -20 \qquad\qquad x = 15$$

$$x = -\frac{20}{3}$$

Check and Answer   Since the length of a rectangle cannot be negative, we can eliminate $-\dfrac{20}{3}$ as an answer to our problem.

$$\text{length} = x = 15 \text{ feet}$$

$$\text{width} = \frac{3}{5}(15) - 5 = 4 \text{ feet}$$

Check

$$a = lw$$

$$60 \overset{?}{=} 15(4)$$

$$60 = 60 \qquad \textit{True}$$

Therefore, the length is 15 feet and the width is 4 feet.

▶ **Now Try Exercise 5**

Now we will work with a problem that expresses the relationship between two numbers. Problems like this are sometimes referred to as *number problems*.

EXAMPLE 2 ▶ Reciprocals One number is 4 times another number. The sum of their reciprocals is $\dfrac{5}{2}$. Determine the numbers.

Solution   Understand and Translate

$$\text{Let } x = \text{first number.}$$

$$\text{Then } 4x = \text{second number.}$$

The reciprocal of the first number is $\dfrac{1}{x}$ and the reciprocal of the second number is $\dfrac{1}{4x}$.

The sum of their reciprocals is $\dfrac{5}{2}$, thus:

$$\frac{1}{x} + \frac{1}{4x} = \frac{5}{2}$$

Carry Out

$$4x\left(\frac{1}{x} + \frac{1}{4x}\right) = 4x\left(\frac{5}{2}\right)$$ *Multiply both sides by the LCD, 4x.*

$$4x\left(\frac{1}{x}\right) + 4x\left(\frac{1}{4x}\right) = 10x$$ *Distributive property*

$$4 + 1 = 10x$$

$$5 = 10x$$

$$\frac{5}{10} = x$$

$$\frac{1}{2} = x$$

Check The first number is $\frac{1}{2}$. The second number is therefore $4x = 4\left(\frac{1}{2}\right) = 2$. Let's now check if the sum of the reciprocals is $\frac{5}{2}$. The reciprocal of $\frac{1}{2}$ is 2. The reciprocal of 2 is $\frac{1}{2}$. The sum of the reciprocals is

$$2 + \frac{1}{2} = \frac{4}{2} + \frac{1}{2} = \frac{5}{2}$$

Answer Since the sum of the reciprocals is $\frac{5}{2}$, the two numbers are 2 and $\frac{1}{2}$.

▶ **Now Try Exercise 11**

2 Set Up and Solve Motion Problems

In Chapter 3 we discussed *motion problems.* Recall that

$$\text{distance} = \text{rate} \cdot \text{time}$$

If we solve this equation for time, we obtain

$$\text{time} = \frac{\text{distance}}{\text{rate}} \quad \text{or} \quad t = \frac{d}{r}$$

This equation is useful in solving motion problems when the total time of travel for two objects or the time of travel between two points is known.

EXAMPLE 3 ▶ **Canoeing** Cindy Kilborn went canoeing in the Colorado River. The current in the river was 2 miles per hour. If it took Cindy the same amount of time to travel 10 miles downstream as 2 miles upstream, determine the speed at which Cindy's canoe would travel in still water.

Solution Understand and Translate

Let r = the canoe's speed in still water.

Then $r + 2$ = the canoe's speed traveling downstream (with current)

and $r - 2$ = the canoe's speed traveling upstream (against current.)

Direction	Distance	Rate	Time
Downstream	10	$r + 2$	$\dfrac{10}{r + 2}$
Upstream	2	$r - 2$	$\dfrac{2}{r - 2}$

Since the time it takes to travel 10 miles downstream is the same as the time to travel 2 miles upstream, we set the times equal to each other and then solve the resulting equation.

$$\text{time downstream} = \text{time upstream}$$

$$\frac{10}{r+2} = \frac{2}{r-2}$$

Carry Out

$$10(r-2) = 2(r+2) \qquad \textit{Cross-multiply.}$$

$$10r - 20 = 2r + 4$$

$$8r = 24$$

$$r = 3$$

Check and Answer   Since this rational equation contains a variable in a denominator, the solution must be checked. A check will show that 3 satisfies the equation. Thus, the canoe would travel at 3 miles per hour in still water.

▸ **Now Try Exercise 15**

EXAMPLE 4 ▸ **Scenic Route** Mazie Akana drives along Route 72 along the shore-line in Oahu, Hawaii. Because of the beautiful scenery she drives an average of 20 miles per hour. Then she drives inland and averages 65 miles per hour. If the total distance she drove was 100 miles and the total time she drove was 3.5 hours, how long did she drive at each speed?

Solution   Understand and Translate

Let d = distance traveled at 20 miles per hour.

Then $100 - d$ = distance traveled at 65 miles per hour.

Direction	Distance	Rate	Time
Shoreline	d	20	$\dfrac{d}{20}$
Inland	$100 - d$	65	$\dfrac{100 - d}{65}$

Since the total time spent driving is 3.5 hours, we write

$$\text{time along shoreline} + \text{time inland} = 3.5 \text{ hours}$$

$$\frac{d}{20} + \frac{100 - d}{65} = 3.5$$

Carry Out

$$260\left(\frac{d}{20} + \frac{100 - d}{65}\right) = (260)(3.5) \qquad \textit{Multiply both sides by the LCD, 260.}$$

$$\overset{13}{\cancel{260}}\left(\frac{d}{\cancel{20}}\right) + \overset{4}{\cancel{260}}\left(\frac{100-d}{\cancel{65}}\right) = 910 \qquad \textit{Distributive property}$$

$$13d + 4(100 - d) = 910$$

$$13d + 400 - 4d = 910$$

$$9d + 400 = 910$$

$$9d = 510$$

$$d = \frac{510}{9}$$

$$d \approx 57$$

Answer The answer to the problem is not 57. Remember that the question asked us to *find the time spent* traveling at each speed. The variable d does not represent time but represents the distance traveled at 20 miles per hour. To find the time traveled and to answer the question asked, we need to evaluate $\dfrac{d}{20}$ and $\dfrac{100-d}{65}$ for $d=57$.

Time at 20 mph

$$\frac{d}{20}=\frac{57}{20}\approx 2.9$$

Time at 65 mph

$$\frac{100-d}{65}=\frac{100-57}{65}=\frac{43}{65}\approx 0.6$$

Thus, Mazie drove about 2.9 hours along the shoreline and about 0.6 hours inland. The total time was $2.9+0.6$ or 3.5 hours.

▶ **Now Try Exercise 17**

EXAMPLE 5 ▶ **Distance of a Marathon** At a fund-raising marathon participants can either bike, walk, or run. Kim Clark, who rode a bike, completed the entire distance of the marathon with an average speed of 16 kilometers per hour (kph). Steve Schwartz, who jogged, completed the entire distance with an average speed of 5 kph. If Kim completed the race in 2.75 hours less time than Steve did, determine the distance the marathon covered.

Solution   **Understand and Translate** Let d = the distance from the start to the finish of the marathon. Then we can construct the following table. To determine the time, we divide the distance by the rate.

Person	Distance	Rate	Time
Kim	d	16	$\dfrac{d}{16}$
Steve	d	5	$\dfrac{d}{5}$

We are given that Kim completed the marathon in 2.75 hours less time than Steve did. Therefore, to make Kim's and Steve's times equal, we need to subtract 2.75 hours from Steve's time (or add 2.75 hours to Kim's time; see the paragraph following this example). We will subtract 2.75 hours from Steve's time and use the following equation to solve the problem.

$$\text{Time for Kim} = \text{Time for Steve} - 2.75 \text{ hours}$$

$$\frac{d}{16}=\frac{d}{5}-2.75$$

Carry Out

$$80\left(\frac{d}{16}\right)=80\left(\frac{d}{5}-2.75\right) \qquad \textit{Multiply both sides by the LCD, 80.}$$

$$5d=80\left(\frac{d}{5}\right)-80(2.75) \qquad \textit{Distributive property}$$

$$5d=16d-220$$

$$-11d=-220$$

$$d=20$$

Check and Answer The distance from the starting point to the ending point of the marathon appears to be 20 kilometers. To check this answer we will determine the times it took Kim and Steve to complete the marathon and see if the difference between the times is 2.75 hours. To determine the times, divide the distance, 20 kilometers, by the rate.

$$\text{Kim's time} = \frac{20}{16}=1.25 \text{ hours}$$

$$\text{Steve's time} = \frac{20}{5}=4 \text{ hours}$$

Since $4-1.25=2.75$ hours, the answer checks. Therefore the distance the marathon covers is 20 kilometers.

▶ **Now Try Exercise 25**

In Example 5 we subtracted 2.75 hours from Steve's time to obtain an equation. We could have added 2.75 hours to Kim's time to obtain an equivalent equation. Rework Example 5 now by adding 2.75 hours to Kim's time.

3 Set Up and Solve Work Problems

Problems in which two or more machines or people work together to complete a specific task are sometimes referred to as *work problems*. Work problems often involve equations containing fractions. Generally, work problems are based on the fact that the fractional part of the work done by person 1 (or machine 1) plus the fractional part of the work done by person 2 (or machine 2) is equal to the total amount of work done by both people (or both machines). *We represent the total amount of work done by the number 1, which represents one whole job completed.*

$$\begin{pmatrix}\text{part of task done} \\ \text{by first person} \\ \text{or machine}\end{pmatrix} + \begin{pmatrix}\text{part of task done} \\ \text{by second person} \\ \text{or machine}\end{pmatrix} = \begin{pmatrix}1 \\ \text{(one whole task} \\ \text{completed)}\end{pmatrix}$$

To determine the part of the task completed by each person or machine, we use the formula

part of task completed = rate · time

This formula is very similar to the formula

amount = rate · time

that was discussed in Section 3.4. To determine the part of the task completed, we need to determine the rate. Suppose that Paul can do a particular task in 6 hours. Then he would complete 1/6 of the task per hour. Thus, his rate is 1/6 of the task per hour. If Audrey can do a particular task in 5 minutes, her rate is 1/5 of the task per minute. In general, if a person or machine can complete a task in t units of time, the rate is $1/t$.

EXAMPLE 6 ▶ **Plowing a Field** Bob Smith can plow a field by himself in 20 hours. His wife, Mary, by herself can plow the same field in 30 hours. How long will it take them to plow the field if they work together?

Solution   Understand and Translate   Let t = the time, in hours, for Mr. and Mrs. Smith working together, to plow the field. We will construct a table to help us in finding the part of the task completed by Mr. Smith and Mrs. Smith in t hours.

Landscaper	Rate of Work (part of the task completed per hour)	Time Worked	Part of Task
Mr. Smith	$\dfrac{1}{20}$	t	$\dfrac{t}{20}$
Mrs. Smith	$\dfrac{1}{30}$	t	$\dfrac{t}{30}$

$$\begin{pmatrix}\text{part of the field plowed} \\ \text{by Mr. Smith in } t \text{ hours}\end{pmatrix} + \begin{pmatrix}\text{part of the field plowed} \\ \text{by Mrs. Smith in } t \text{ hours}\end{pmatrix} = 1(\text{entire field plowed})$$

$$\frac{t}{20} \qquad + \qquad \frac{t}{30} \qquad = \qquad 1$$

Carry Out Now multiply both sides of the equation by the LCD, 60.

$$60\left(\frac{t}{20} + \frac{t}{30}\right) = 60 \cdot 1$$

$$\overset{3}{60}\left(\frac{t}{20}\right) + \overset{2}{60}\left(\frac{t}{30}\right) = 60 \qquad \textit{Distributive property}$$

$$3t + 2t = 60$$

$$5t = 60$$

$$t = 12$$

Answer Thus, Mr. and Mrs. Smith working together can plow the field in 12 hours. We leave the check for you.

▶ **Now Try Exercise 27**

Helpful Hint

In Example 6, Mr. Smith could plow the field by himself in 20 hours, and Mrs. Smith could plow the field by herself in 30 hours. We determined that together they could plow the field in 12 hours. Does this answer make sense? Since you would expect the time to plow the field together to be less than the time either of them could plow it alone, the answer makes sense. When solving a work problem, always examine your answer to see if it makes sense. If not, rework the problem and find your error.

EXAMPLE 7 ▶ **Storing Wine** At a winery in Napa Valley, California, one pipe can fill a tank with wine in 3 hours and another pipe can empty the tank in 5 hours. If the valves to both pipes are open, how long will it take to fill the empty tank?

Solution **Understand and Translate** Let t = amount of time to fill the tank with both valves open.

Pipe	Rate of Work	Time	Part of Task
Pipe filling tank	$\frac{1}{3}$	t	$\frac{t}{3}$
Pipe emptying tank	$\frac{1}{5}$	t	$\frac{t}{5}$

As one pipe is filling, the other is emptying the tank. Thus, the pipes are working against each other. Therefore, instead of adding the parts of the task, as was done in Example 6 where the people worked together, we will subtract the parts of the task.

$$\left(\begin{array}{c}\text{part of tank}\\\text{filled in } t \text{ hours}\end{array}\right) - \left(\begin{array}{c}\text{part of tank}\\\text{emptied in } t \text{ hours}\end{array}\right) = 1(\text{total tank filled})$$

$$\frac{t}{3} - \frac{t}{5} = 1$$

Carry Out

$$15\left(\frac{t}{3} - \frac{t}{5}\right) = 15 \cdot 1 \qquad \textit{Multiply both sides by the LCD, 15.}$$

$$\overset{5}{15}\left(\frac{t}{3}\right) - \overset{3}{15}\left(\frac{t}{5}\right) = 15 \qquad \textit{Distributive property}$$

$$5t - 3t = 15$$

$$2t = 15$$

$$t = 7\frac{1}{2}$$

Check and Answer The tank will be filled in $7\frac{1}{2}$ hours. This answer is reasonable because we expect it to take longer than 3 hours when the tank is being drained at the same time.

▸ **Now Try Exercise 35**

EXAMPLE 8 ▸ **Cleaning Service** Linda and John Franco own a house cleaning service. When Linda cleans Damon's house by herself, it takes 7 hours. When Linda and John work together, they can clean the house in 4 hours. How long will it take John to clean the house by himself?

Solution   Let t = time for John to clean the house by himself. Then John's rate is $\frac{1}{t}$. Let's make a table to help analyze the problem. Since Linda can clean the house by herself in 7 hours, her rate is $\frac{1}{7}$ of the job per hour. In the table, we use the fact that together they can clean the house in 4 hours.

Worker	Rate of Work	Time	Part of Task
Linda	$\frac{1}{7}$	4	$\frac{4}{7}$
John	$\frac{1}{t}$	4	$\frac{4}{t}$

$$\left(\begin{array}{c}\text{part of house}\\\text{cleaned by Linda}\end{array}\right) + \left(\begin{array}{c}\text{part of house}\\\text{cleaned by John}\end{array}\right) = 1$$

$$\frac{4}{7} \quad + \quad \frac{4}{t} \quad = 1$$

Carry Out

$$7t\left(\frac{4}{7} + \frac{4}{t}\right) = 7t \cdot 1 \qquad \textit{Multiply both sides by the LCD, 7t.}$$

$$7t\left(\frac{4}{7}\right) + 7t\left(\frac{4}{t}\right) = 7t \qquad \textit{Distributive property}$$

$$4t + 28 = 7t$$

$$28 = 3t$$

$$\frac{28}{3} = t$$

$$9\frac{1}{3} = t$$

Check and Answer Thus, it takes John, $9\frac{1}{3}$ hours, or 9 hours 20 minutes, to clean the house by himself. This answer is reasonable because we expect it to take longer for John to clean the house by himself than it would for Linda and John working together.

▸ **Now Try Exercise 39**

EXAMPLE 9 ▸ **Thank-You Notes** Peter and Kaitlyn Kewin are handwriting thank-you notes to guests who attended their 20th wedding anniversary party. Kaitlyn by herself could write all the notes in 6 hours and Peter could write all the notes by himself in 10 hours. After Kaitlyn has been writing thank-you notes for 4 hours by herself, she must leave town on business. Peter then continues the task of writing the thank-you notes. How long will it take Peter to finish writing the remaining notes?

Solution   **Understand and Translate**   Let t = time it will take Peter to finish writing the notes.

Person	Rate of Work	Time	Part of Task
Kaitlyn	$\dfrac{1}{6}$	4	$\dfrac{4}{6} = \dfrac{2}{3}$
Peter	$\dfrac{1}{10}$	t	$\dfrac{t}{10}$

$$\left(\begin{array}{c}\text{part of notes written}\\ \text{by Kaitlyn}\end{array}\right) + \left(\begin{array}{c}\text{part of notes written}\\ \text{by Peter}\end{array}\right) = 1$$

$$\frac{2}{3} \qquad + \qquad \frac{t}{10} \qquad = 1$$

Carry Out

$$30\left(\frac{2}{3} + \frac{t}{10}\right) = 30 \cdot 1 \qquad \textit{Multiply both sides by the LCD, 30.}$$

$$\overset{10}{\cancel{30}}\left(\frac{2}{\cancel{3}}\right) + \overset{3}{\cancel{30}}\left(\frac{t}{\cancel{10}}\right) = 30 \qquad \textit{Distributive property}$$

$$20 + 3t = 30$$

$$3t = 10$$

$$t = \frac{10}{3} \quad \text{or} \quad 3\frac{1}{3}$$

Answer   Thus, it will take Peter $3\frac{1}{3}$ hours to complete the cards.

▸ **Now Try Exercise 37**

EXERCISE SET 7.7

   MyMathLab

Concept/Writing Exercises

1. Geometric formulas, which were discussed in Section 2.6, are often rational equations. Give three formulas that are rational equations.

2. Suppose that car 1 and car 2 are traveling to the same location 60 miles away and that car 1 travels 10 miles per hour faster than car 2.

   a) Let r represent the speed of car 2. Write an expression, using time = $\dfrac{\text{distance}}{\text{rate}}$, for the time it takes car 1 to reach its destination.

   b) Now let r represent the speed of car 1. Write an expression, using time = $\dfrac{\text{distance}}{\text{rate}}$, for the time it takes car 2 to reach its destination.

3. In an equation for a work problem, one side of the equation is set equal to 1. What does the 1 represent in the problem?

4. Suppose Tracy Augustine can complete a particular task in 3 hours, while the same task can be completed by John Bailey in 7 hours. How would you represent the part of the task completed by Tracy in 1 hour? By John in 1 hour?

Practice the Skills/Problem Solving

In Exercises 5–36, solve the problem and answer the questions.

Geometry Problems; see Example 1.

5. Packaging Computers The Phillips Paper Company makes rectangular pieces of cardboard for packing computers. The sheets of cardboard are to have an area of 99 square inches, and the length of a sheet is to be 5 inches more than $\frac{2}{3}$ its width. Determine the length and width of the cardboard to be manufactured.

6. Carry-on Luggage On most airlines carry-on luggage can have a maximum width of 10 inches with a maximum volume of 3840 cubic inches, see the figure. If the height of the luggage is $\frac{2}{3}$ the length, determine the dimensions of the largest piece of carry-on luggage. Use $V = lwh$.

7. Triangles of Dough Pillsbury Crescent Rolls are packaged in tubes that contain perforated triangles of dough. The base of the triangular piece of dough is about 5 centimeters more than its height. Determine the base and height of a piece of dough if the area is about 42 square centimeters.

8. Yield Sign Yield right of way signs used in the United States are triangles. The area of the sign is about 558 square inches. The height of the sign is about 5 inches less than its base. Determine the length of the base of a yield right of way sign.

9. Triangular Garden A triangular area is 20 square feet. Find the base of the triangular area if the height is 1 foot more than $\frac{1}{2}$ the base.

10. Billboard A billboard sign is in the shape of a trapezoid. The area of the sign is 10 square feet. If the height of the sign is $\frac{1}{5}$ the length of the sum of the 2 bases, determine the height of the sign.

Number Problems; see Example 2.

11. Difference of Numbers One number is 9 times larger than another. The difference of their reciprocals is 1. Determine the two numbers.

12. Sum of Numbers One number is 3 times larger than another. The sum of their reciprocals is $\frac{4}{3}$. Determine the two numbers.

13. Increased Numerator The numerator of the fraction $\frac{3}{4}$ is increased by an amount so that the value of the resulting fraction is $\frac{5}{2}$. Determine the amount by which the numerator was increased.

14. Decreased Denominator The denominator of the fraction $\frac{8}{21}$ is decreased by an amount so that the value of the resulting fraction is $\frac{1}{2}$. Determine the amount by which the denominator was decreased.

Motion Problems; see Examples 3–5.

15. Paddleboat Ride In the Mississippi River near New Orleans, the Creole Queen paddleboat travels 6 miles upstream (against the current) in the same amount of time it travels 12 miles downstream (with the current). If the current of the river is 3 miles per hour, determine the speed of the Creole Queen in still water.

16. Kayak Ride Kathy Boothby-Sestak can paddle her kayak 6 miles per hour in still water. It takes her as long to paddle 5 miles upstream as 10 miles downstream in the Wabash River near Lafayette, Indiana. Determine the river's current.

17. Trolley Ride A trolley travels in one direction at an average of 12 miles per hour, then turns around and travels on the same track in the opposite direction at 12 miles per hour. If the total time traveling on the trolley is $2\frac{1}{2}$ hours, how far did the trolley travel in one direction?

18. Motorcycle Trip Brandy Dawson and Jason Dodge start a motorcycle trip at the same point a little north of Fort Worth, Texas. Both are traveling to San Antonio, Texas, a distance of about 400 kilometers. Brandy rides 30 kilometers per hour faster than Jason does. When Brandy reaches her destination, Jason has only traveled to Austin, Texas, a distance of about 250 kilometers. Determine the approximate speed of each motorcycle.

19. **Jet Flight** Elenore Morales traveled 1600 miles by commercial jet from Kansas City, Missouri, to Spokane, Washington. She then traveled an additional 500 miles on a private propeller plane from Spokane to Billings, Montana. If the speed of the jet was 4 times the speed of the propeller plane and the total time in the air was 6 hours, determine the speed of each plane.

20. **Exercise Regimen** As part of his exercise regimen, Chris Barker walks a distance of 2 miles on an indoor track and then jogs at twice his walking speed for another 2 miles. If the total time spent on the track was one hour, determine the speeds at which he walks and jogs.

21. **No Wake Zone** Alisha is traveling by motorboat from her dock to Paradise Island. While she is in a no wake zone, her average speed is 4 miles per hour. Once she leaves the no wake zone, her average speed is 28 miles per hour. If the total distance traveled from her dock to the island is 36.6 miles and the total time of the trip is 1.7 hours, determine the distance from her dock to the end of the no wake zone and the distance from the end of the no wake zone to Paradise Island.

22. **Jogging and Walking** Kristen Taylor jogs a certain distance and then walks a certain distance. When she jogs she averages 5 miles per hour and when she walks she averages 3 miles per hour. If she walks and jogs a total of 5 miles in a total of 1.4 hours, how far does she jog and how far does she walk?

23. **Headwind and Tailwind** A Boeing 747 flew from San Francisco to Honolulu, a distance of 2900 miles. Flying with the wind, it averaged 600 miles per hour. When the wind changed from a tailwind to a headwind, the plane's speed dropped to 550 miles per hour. If the total time of the trip was 5 hours, determine the length of time it flew at each speed.

24. **Thalys Train** The Thalys train in Europe has been known to travel an average 240 kilometers per hour (kph). Prior to using the Thalys (bullet) trains in Europe, trains traveled an average speed of 120 kph. If a Thalys train traveling from Brussels to Amsterdam can complete its trip in 0.88 hour less

time than an older train, determine the distance from Brussels to Amsterdam.

25. **Water Skiers** At a water show a boat pulls a water skier at a speed of 30 feet per second. When it reaches the end of the lake, more skiers are added to be pulled by the boat, so the boat's speed drops to 25 feet per second. If the boat traveled the same distance with additional skiers as it did with the single skier, and the trip back with the additional skiers took 8 seconds longer than the trip with the single skier, how far, in feet, in one direction, had the boat traveled?

26. **Cross-Country Skiing** Alana Bradley and her father Tim begin skiing the same cross-country ski trail in Elmwood Park in Sioux Falls, South Dakota, at the same time. If Alana, who averages 9 miles per hour, finishes the trail 0.25 hours sooner than her father, who averages 6 miles per hour, determine the length of the trail.

Work Problems; see Examples 6–9.

27. **Wallpaper** Reynaldo and Felicia Fernandez decide to wallpaper their family room. Felicia, who has wallpapering experience, can wallpaper the room in 6 hours. Reynaldo can wallpaper the same room in 8 hours. How long will it take them to wallpaper the family room if they work together?

28. **Conveyor Belt** At a salt mine, one conveyor belt requires 20 minutes to fill a large truck with ore. A second conveyor belt requires 30 minutes to fill the same truck with ore. How long would it take if both conveyor belts were working together to fill the truck with ore?

29. **Picking Peaches** In a peach orchard in Williamson, New York, Gary Rominger can load his truck with peaches in 6 hours. His friend, Alex Taurke, takes twice as long to load Gary's truck with peaches. How long will it take them working together to load the truck with peaches?

30. **Watering Plants** In a small nursery, Becky Hailey can water all the plants in 30 minutes. Her co-worker, Karen Grizzaffi, can water all the plants in 20 minutes. How long will it take them working together to water the plants?

31. Painting a Room Eric Kweeder can paint a room in 60 minutes. His brother, Jessup, can paint the same room in 40 minutes. How long will it take them working together to paint the room?

32. Laying a Floor Robert Struckland can lay a new hardwood floor in 6.5 hours. His neighbor, Andrea Vorwark, can lay the same floor in 7 hours. How long will it take them working together to lay the hardwood floor?

33. Hot Tub Pam and Loren Fornieri know that their hot tub can be filled in 40 minutes and drained completely in 60 minutes. If the water is turned on and the drain is left open, how long would it take the tub to fill completely?

34. Filling a Tank During a rainstorm, the rain is flowing into a large holding tank. At the rate the rain is falling, the empty tank would fill in 8 hours. At the bottom of the tank is a spigot to dispense water. Typically, it takes about 12 hours with the spigot wide open to empty the water in a full tank. If the tank is empty and the spigot has been accidentally left open, and the rain falls at the constant rate, how long would it take for the tank to fill completely?

35. Payroll Checks At the Community Savings Bank, it takes a computer 40 minutes to process and print payroll checks. When a second computer is used and the two computers work together, the checks can be processed and printed in 24 minutes. How long would it take the second computer by itself to process and print the payroll checks?

36. Flowing Water When the water is turned on and passes through a one-half inch diameter hose, a small above-ground pool can be filled in 6 hours. When the water is turned on at two spigots and passes through both the one-half inch diameter hose and a three-quarter inch diameter hose, the pool can be filled in 2 hours. How long would it take to fill the pool using only the three-quarter inch diameter hose?

37. Digging a Trench A construction company with two backhoes has contracted to dig a long trench for drainage pipes. The larger backhoe can dig the entire trench by itself in 12 days. The smaller backhoe can dig the entire trench by itself in 15 days. The large backhoe begins working on the trench by itself, but after 5 days it is transferred to a different job and the smaller backhoe begins working on the trench. How long will it take for the smaller backhoe to complete the job?

38. Delivery of Food Ian and Nicole Murphy deliver food to various restaurants. If Ian drove the entire trip, the trip would take about 10 hours. If Nicole drove the entire trip, the trip would take about 8 hours. After Nicole had been driving for 4 hours, Ian takes over the driving. About how much longer will Ian drive before they reach their final destination?

39. Snowstorm Following a severe snowstorm, Ken and Bettina Reeves must clear their driveway and sidewalk. Ken can clear the snow by himself in 4 hours, and Bettina can clear the snow by herself in 6 hours. After Bettina has been working for 3 hours, Ken is able to join her. How much longer will it take them working together to remove the rest of the snow?

40. Farm Cooperative A large farming cooperative near Hutchinson, Kansas, owns three hay balers. The oldest baler can pick up and bale an acre of hay in 3 hours and each of the two new balers can work an acre in 2 hours.
   a) How long would it take the three balers working together to pick up and bale 1 acre?
   b) How long would it take for the three balers working together to pick up and bale the farm's 375 acres of hay?

41. Skimming Oil A boat designed to skim oil off the surface of the water has two skimmers. One skimmer can fill the boat's holding tank in 60 hours while the second skimmer can fill the boat's holding tank in 50 hours. There is also a valve in the holding tank that is used to transfer the oil to a larger vessel. If no new oil is coming into the holding tank, a full holding tank of skimmed oil can be transferred to a larger tank in 30 hours. If both skimmers begin skimming and the valve on the holding tank is opened, how long will it take for the empty holding tank on the boat to fill?

42. Flower Garden Bob can plant a flower garden by himself in 8 hours. Mary can plant the same garden by herself in 10 hours, and Gloria can plant the same garden by herself in 12 hours. How long would it take them working together to plant the garden?

Challenge Problems

43. **Reciprocal of a Number** If 2 times a number is added to 3 times the reciprocal of the number, the answer is 7. Determine the number(s).

44. **Determine a Number** The reciprocal of the difference of a certain number and 5 is twice the reciprocal of the difference of twice the number and 10. Determine the number(s).

45. **Picking Blueberries** Ed and Samantha Weisman, whose parents own a fruit farm, must each pick the same number of pints of blueberries each day during the season. Ed picks an

average of 8 pints per hour, while Samantha picks an average of 4 pints per hour. If Ed and Samantha begin picking blueberries at the same time, and Samantha finishes 1 hour after Ed, how many pints of blueberries must each pick?

46. **Sorting Mail** A mail processing machine can sort a large bin of mail in 1 hour. A newer model can sort the same quantity of mail in 30 minutes. If they operate together, how long will it take them to sort the bin of mail?

Cumulative Review Exercises

[2.1] **47.** Simplify $\frac{1}{2}(x + 3) - (2x + 5)$.

[4.2] **48.** Graph $x = -2$.

[7.2] **49.** Divide $\dfrac{x^2 - 14x + 48}{x^2 - 5x - 24} \div \dfrac{2x^2 - 13x + 6}{2x^2 + 5x - 3}$.

[7.4] **50.** Subtract $\dfrac{x}{6x^2 - x - 15} - \dfrac{5}{9x^2 - 12x - 5}$.

7.8 Variation

1 Solve direct variation problems.

2 Solve inverse variation problems.

3 Solve joint variation problems.

4 Solve combined variation problems.

In Sections 7.6 and 7.7, we saw many applications of equations containing rational expressions. In this section, we see still more.

1 Solve Direct Variation Problems

Many scientific formulas are expressed as variations. A **variation** is an equation that relates one variable to one or more other variables using the operations of multiplication or division (or both operations). There are essentially three types of variation problems: direct, inverse, and joint variation.

In **direct variation**, the two related variables will both increase together or both decrease together; that is, as one increases so does the other, and as one decreases so does the other.

Consider a car traveling at 30 miles an hour. The car travels 30 miles in 1 hour, 60 miles in 2 hours, and 90 miles in 3 hours. Notice that as the time increases, the distance traveled increases.

The formula used to calculate distance traveled is

$$\text{distance} = \text{rate} \cdot \text{time}$$

Since the rate is a constant, 30 miles per hour, the formula can be written

$$d = 30t$$

We say that distance varies directly as time or that distance is directly proportional to time. This is an example of a direct variation.

> ### Direct Variation
>
> If a variable y varies directly as a variable x, then
>
> $$y = kx$$
>
> where k is the **constant of proportionality** (or the variation constant).

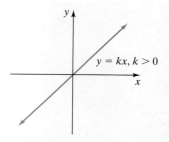

FIGURE 7.3

The graph of $y = kx$, $k > 0$, is always a straight line that goes through the origin (see **Fig. 7.3**). The slope of the line depends on the value of k. The greater the value of k, the greater the slope.

EXAMPLE 1 ▸ **Circle** The circumference of a circle, C, is directly proportional to (or varies directly as) its radius, r. Write the equation for the circumference of a circle if the constant of proportionality, k, is 2π.

Solution $C = kr$ (C varies directly as r)

$C = 2\pi r$ (constant of proportionality is 2π)

▸ **Now Try Exercise 11**

EXAMPLE 2 ▸ **Administering a Drug** The amount, a, of the drug theophylline given to patients is directly proportional to the patient's mass, m, in kilograms.

a) Write this variation as an equation.

b) If 150 mg is given to a boy whose mass is 30 kg, find the constant of proportionality.

c) How much of the drug should be given to a patient whose mass is 62 kg?

Solution   **a)** We are told this is a direct variation. That is, the greater a person's mass the more of the drug that will need to be given. We therefore set up a direct variation.

$$a = km$$

b) Understand and Translate   To determine the value of the constant of proportionality, we substitute the given values for the amount and mass. We then solve for k.

$$a = km$$

$$150 = k(30) \qquad \text{Substitute the given values.}$$

Carry Out $\qquad\qquad\qquad 5 = k$

Answer   Thus, $k = 5$ mg. Five milligrams of the drug should be given for each kilogram of a person's mass.

c) Understand and Translate   Now that we know the constant of proportionality we can use it to determine the amount of the drug to use for a person's mass. We set up the variation and substitute the values of k and m.

$$a = km$$

$$a = 5(62) \qquad \text{Substitute the given values.}$$

Carry Out $\qquad\qquad a = 310$

Answer   Thus, 310 mg of theophylline should be given to a person whose mass is 62 kg.

▸ **Now Try Exercise 57**

EXAMPLE 3 ▸ y varies directly as the square of z. If y is 80 when z is 20, find y when z is 45.

Solution   Since y varies directly as the *square of z*, we begin with the formula $y = kz^2$. Since the constant of proportionality is not given, we must first find k using the given information.

$$y = kz^2$$

$$80 = k(20)^2 \qquad \text{Substitute the given values.}$$

$$80 = 400k \qquad \text{Solve for } k.$$

$$\frac{80}{400} = \frac{400k}{400}$$

$$0.2 = k$$

We now use $k = 0.2$ to find y when z is 45.

$$y = kz^2$$
$$y = 0.2(45)^2 \quad \text{\textit{Substitute the given values.}}$$
$$y = 405$$

Thus, when z equals 45, y equals 405.

▶ **Now Try Exercise 35**

2 Solve Inverse Variation Problems

A second type of variation is **inverse variation**. When two quantities vary inversely, it means that as one quantity increases, the other quantity decreases, and vice versa.

To explain inverse variation, we again use the formula, distance = rate · time. If we solve for time, we get time $= \dfrac{\text{distance}}{\text{rate}}$. Assume that the distance is fixed at 120 miles; then

$$\text{time} = \frac{120}{\text{rate}}$$

At 120 miles per hour, it would take 1 hour to cover this distance. At 60 miles an hour, it would take 2 hours. At 30 miles an hour, it would take 4 hours. Note that as the rate (or speed) decreases the time increases, and vice versa.

The equation above can be written

$$t = \frac{120}{r}$$

This equation is an example of an inverse variation. The time and rate are inversely proportional. The constant of proportionality is 120.

> ### Inverse Variation
>
> If a variable y varies inversely as a variable x, then
>
> $$y = \frac{k}{x} \quad \text{(or } xy = k\text{)}$$
>
> where k is the constant of proportionality.

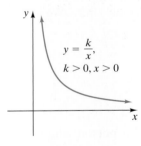

$y = \dfrac{k}{x}$,
$k > 0, x > 0$

FIGURE 7.4

Two quantities vary inversely, or are inversely proportional, when as one quantity increases the other quantity decreases. The graph of $y = \dfrac{k}{x}$, for $k > 0$ and $x > 0$, will have the shape illustrated in **Figure 7.4**. The graph of an inverse variation is not defined at $x = 0$ because 0 is not in the domain of the function $y = \dfrac{k}{x}$.

EXAMPLE 4 ▶ **Melting Ice** The amount of time, t, it takes a block of ice to melt in water is inversely proportional to the water's temperature, T.

a) Write this variation as an equation.

b) If a block of ice takes 15 minutes to melt in 60°F water, determine the constant of proportionality.

c) Determine how long it will take a block of ice of the same size to melt in 50°F water.

Solution   a) The hotter the water temperature, the shorter the time for the block of ice to melt. The inverse variation is

$$t = \frac{k}{T}$$

b) Understand and Translate To determine the constant of proportionality, we substitute the values for the temperature and time and solve for k.

$$t = \frac{k}{T}$$

$$15 = \frac{k}{60} \qquad \textit{Substitute the given values.}$$

Carry Out

$$900 = k$$

Answer The constant of proportionality is 900.

c) Understand and Translate Now that we know the constant of proportionality, we can use it to determine how long it will take for the same size block of ice to melt in 50°F water. We set up the proportion, substitute the values for k and T, and solve for t.

$$t = \frac{k}{T}$$

$$t = \frac{900}{50} \qquad \textit{Substitute the given values.}$$

Carry Out

$$t = 18$$

Answer It will take 18 minutes for the block of ice to melt in the 50°F water.

▶ **Now Try Exercise 61**

EXAMPLE 5 ▶ **Lighting** The illuminance, I, of a light source varies inversely as the square of the distance, d, from the source. Assuming that the illuminance is 75 units at a distance of 4 meters, find the formula that expresses the relationship between the illuminance and the distance.

Solution **Understand and Translate** Since the illuminance varies inversely as the *square* of the distance, the general form of the equation is

$$I = \frac{k}{d^2} \qquad (\text{or } Id^2 = k)$$

To find k, we substitute the given values for I and d.

$$75 = \frac{k}{4^2} \qquad \textit{Substitute the given values.}$$

Carry Out

$$75 = \frac{k}{16} \qquad \textit{Solve for k.}$$

$$(75)(16) = k$$

$$1200 = k$$

Answer The formula is $I = \dfrac{1200}{d^2}$.

▶ **Now Try Exercise 65**

---

3 Solve Joint Variation Problems

One quantity may vary as a product of two or more other quantities. This type of variation is called **joint variation.**

Joint Variation

If y varies jointly as x and z, then

$$y = kxz$$

where k is the constant of proportionality.

EXAMPLE 6 ▸ **Area of a Triangle** The area, A, of a triangle varies jointly as its base, b, and height, h. If the area of a triangle is 48 square inches when its base is 12 inches and its height is 8 inches, find the area of a triangle whose base is 15 inches and height is 40 inches.

Solution   Understand and Translate   First write the joint variation; then substitute the known values and solve for k.

$$A = kbh$$
$$48 = k(12)(8) \qquad \text{\textit{Substitute the given values.}}$$

Carry Out
$$48 = k(96) \qquad \text{\textit{Solve for k.}}$$
$$\frac{48}{96} = k$$
$$k = \frac{1}{2}$$

Now solve for the area of the given triangle.

$$A = kbh$$
$$= \frac{1}{2}(15)(40) \qquad \text{\textit{Substitute the given values.}}$$
$$= 300$$

Answer   The area of the triangle is 300 square inches.

▸ **Now Try Exercise 69**

Summary of Variations		
DIRECT	INVERSE	JOINT
$y = kx$	$y = \dfrac{k}{x}$	$y = kxz$

4 Solve Combined Variation Problems

Often in real-life situations one variable varies as a combination of variables. The following examples illustrate the use of **combined variations.**

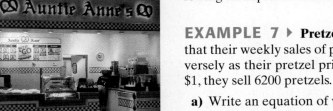

EXAMPLE 7 ▸ **Pretzel Shop** The owners of a Auntie Anne's Pretzel Shop find that their weekly sales of pretzels, S, vary directly as their advertising budget, A, and inversely as their pretzel price, P. When their advertising budget is $400 and the price is $1, they sell 6200 pretzels.

a) Write an equation of variation expressing S in terms of A and P. Include the value of the constant.

b) Find the expected sales if the advertising budget is $600 and the price is $1.20.

Solution   **a)** Understand and Translate   We begin with the equation

$$S = \frac{kA}{P}$$

$$6200 = \frac{k(400)}{1} \qquad \text{\textit{Substitute the given values.}}$$

Carry Out
$$6200 = 400k \qquad \text{\textit{Solve for k.}}$$

$$15.5 = k$$

Answer   Therefore, the equation for the sales of pretzels is $S = \dfrac{15.5A}{P}$.

b) Understand and Translate   Now that we know the combined variation equation, we can use it to determine the expected sales for the given values.

$$S = \frac{15.5A}{P}$$

$$= \frac{15.5(600)}{1.20} \quad \textit{Substitute the given values.}$$

Carry Out $\qquad\qquad = 7750$

Answer   They can expect to sell 7750 pretzels.

▸ **Now Try Exercise 71**

EXAMPLE 8 ▸ **Electrostatic Force**   The electrostatic force, F, of repulsion between two positive electrical charges is jointly proportional to the two charges, q_1 and q_2, and inversely proportional to the square of the distance, d, between the two charges. Express F in terms of q_1, q_2, and d.

Solution

$$F = \frac{kq_1q_2}{d^2}$$

▸ **Now Try Exercise 75**

EXERCISE SET 7.8   Math XL   MyMathLab
MathXL®   MyMathLab

Concept/Writing Exercises

1. a) Explain what it means when two items vary directly.

b) Give your own example of two quantities that vary directly.

c) Write the direct variation for your example in part **b)**.

2. a) Explain what it means when two items vary inversely.

b) Give your own example of two quantities that vary inversely.

c) Write the inverse variation for your example in part **b)**.

3. What is meant by joint variation?

4. What is meant by combined variation?

5. a) In the equation $y = \dfrac{17}{x}$, as x increases, does the value for y increase or decrease?

b) Is this an example of direct or inverse variation? Explain.

6. a) In the equation $z = 0.8x^3$, as x increases, does the value for z increase or decrease?

b) Is this an example of direct or inverse variation? Explain.

Variation   *Use your intuition to determine whether the variation between the indicated quantities is direct or inverse.*

7. The speed and the distance covered by a person riding a bike on the Mount Vernon bike path in Alexandria, Virginia

8. The number of pages Tom can read in a 2-hour period and his reading speed

9. The speed of an athlete and the time it takes him to run a 10-kilometer race

10. Barbara's weekly salary and the amount of money withheld for state income taxes

11. The radius of a circle and its area

12. The side of a cube and its volume

13. The radius of a balloon and its volume

14. The diameter of a circle and its circumference

15. The diameter of a hose and the volume of water coming out of the hose

16. The weight of a rocket (due to Earth's gravity) and its distance from Earth

17. The time it takes an ice cube to melt in water and the temperature of the water

18. The distance between two cities on a map and the actual distance between the two cities

19. The shutter opening of a camera and the amount of sunlight that reaches the film

20. The cubic-inch displacement in liters and the horsepower of an engine

21. The length of a board and the force needed to break the board at the center

22. The number of calories eaten and the amount of exercise required to burn off those calories

23. The light illuminating an object and the distance the light is from the object

24. The number of calories in a cheeseburger and the size of the cheeseburger

Practice the Skills

For Exercises 25–32, **a)** *write the variation and* **b)** *find the quantity indicated.*

25. x varies directly as y. Find x when $y = 12$ and $k = 6$.

26. C varies directly as the square of Z. Find C when $Z = 9$ and $k = \dfrac{3}{4}$.

27. y varies directly as R. Find y when $R = 180$ and $k = 1.7$.

28. x varies inversely as y. Find x when $y = 25$ and $k = 5$.

29. R varies inversely as W. Find R when $W = 160$ and $k = 8$.

30. L varies inversely as the square of P. Find L when $P = 4$ and $k = 100$.

31. A varies directly as B and inversely as C. Find A when $B = 12$, $C = 4$, and $k = 3$.

32. A varies jointly as R_1 and R_2 and inversely as the square of L. Find A when $R_1 = 120$, $R_2 = 8$, $L = 5$, and $k = \dfrac{3}{2}$.

For Exercises 33–42, **a)** *write the variation and* **b)** *find the quantity indicated.*

33. x varies directly as y. If x is 12 when y is 3, find x when y is 5.

34. Z varies directly as W. If Z is 7 when W is 28, find Z when W is 140.

35. y varies directly as the square of R. If y is 5 when R is 5, find y when R is 10.

36. P varies directly as the square of Q. If P is 32 when Q is 4, find P when Q is 7.

37. S varies inversely as G. If S is 12 when G is 0.4, find S when G is 5.

38. C varies inversely as J. If C is 7 when J is 0.7, find C when J is 12.

39. x varies inversely as the square of P. If x is 4 when P is 5, find x when P is 2.

40. R varies inversely as the square of T. If R is 3 when T is 6, find R when T is 2.

41. F varies jointly as M_1 and M_2 and inversely as d. If F is 20 when $M_1 = 5$, $M_2 = 10$, and $d = 0.2$, find F when $M_1 = 10$, $M_2 = 20$, and $d = 0.4$.

42. F varies jointly as q_1 and q_2 and inversely as the square of d. If F is 8 when $q_1 = 2$, $q_2 = 8$, and $d = 4$, find F when $q_1 = 28$, $q_2 = 12$, and $d = 2$.

Problem Solving

43. Assume a varies directly as b. If b is doubled, how will it affect a? Explain.

44. Assume a varies directly as b^2. If b is doubled, how will it affect a? Explain.

45. Assume y varies inversely as x. If x is doubled, how will it affect y? Explain.

46. Assume y varies inversely as a^2. If a is doubled, how will it affect y? Explain.

In Exercises 47–52, use the formula $F = \dfrac{km_1m_2}{d^2}$.

47. If m_1 is doubled, how will it affect F?

48. If m_1 is quadrupled and d is doubled, how will it affect F?

49. If m_1 is doubled and m_2 is halved, how will it affect F?

50. If d is halved, how will it affect F?

51. If m_1 is halved and m_2 is quadrupled, how will it affect F?

52. If m_1 is doubled, m_2 is quadrupled, and d is quadrupled, how will it affect F?

In Exercises 53 and 54, determine if the variation is of the form $y = kx$ *or* $y = \dfrac{k}{x}$, *and find* k.

53.

x	y
2	$\dfrac{5}{2}$
5	1
10	$\dfrac{1}{2}$
20	$\dfrac{1}{4}$

54.

x	y
6	2
9	3
15	5
27	9

55. **Profit** The profit from selling lamps is directly proportional to the number of lamps sold. When 150 lamps are sold, the profit is $2542.50. Find the profit when 520 lamps are sold.

56. **Profit** The profit from selling stereos is directly proportional to the number of stereos sold. When 65 stereos are sold, the profit is $4056. Find the profit when 80 stereos are sold.

57. Antibiotic The recommended dosage, d, of the antibiotic drug vancomycin is directly proportional to a person's weight. If Phuong Kim, who is 132 pounds, is given 2376 milligrams, find the recommended dosage for Nathan Brown, who weighs 172 pounds.

58. Dollars and Pesos Converting American dollars to Mexican pesos is a direct variation. The more dollars you convert the more pesos you receive. Last week, Carlos Manuel converted $275 into 2433.75 pesos. Today, he received $400 from his aunt. If the conversion rate is unchanged when he converts the $400 into pesos, how many pesos will be receive?

59. Hooke's Law Hooke's law states that the length a spring will stretch, S, varies directly with the force (or weight), F, attached to the spring. If a spring stretches 1.4 inches when 20 pounds is attached, how far will it stretch when 15 pounds is attached?

60. Distance When a car travels at a constant speed, the distance traveled, d, is directly proportional to the time, t. If a car travels 150 miles in 2.5 hours, how far will the same car travel in 4 hours?

61. Pressure and Volume The volume of a gas, V, varies inversely as its pressure, P. If the volume, V, is 800 cubic centimeters when the pressure is 200 millimeters (mm) of mercury, find the volume when the pressure is 25 mm of mercury.

62. Building a Brick Wall The time, t, required to build a brick wall varies inversely as the number of people, n, working on it. If it takes 8 hours for five bricklayers to build a wall, how long will it take four bricklayers to build a wall?

63. Running a Race The time, t, it takes a runner to cover a specified distance is inversely proportional to the runner's speed. If Jann Avery runs at an average of 6 miles per hour, she will finish a race in 2.6 hours. How long will it take Jackie Donofrio who runs at 5 miles per hour, to finish the same race?

64. Pitching a Ball When a ball is pitched in a professional baseball game, the time, t, it takes for the ball to reach home plate varies inversely with the speed, s, of the pitch.* A ball pitched at 90 miles per hour takes 0.459 second to reach the plate. How long will it take a ball pitched at 75 miles per hour to reach the plate?

65. Intensity of Light The intensity, I, of light received at a source varies inversely as the square of the distance, d, from the source. If the light intensity is 20 foot-candles at 15 feet, find the light intensity at 10 feet.

66. Tennis Ball When a tennis player serves the ball, the time it takes for the ball to hit the ground in the service box is inversely proportional to the speed the ball is traveling. If Andy Roddick serves at 122 miles per hour, it takes 0.21 second for the ball to hit the ground after striking his racquet. How long will it take the ball to hit the ground if he serves at 80 miles per hour?

Andy Roddick

67. Stopping Distance Assume that the stopping distance of a van varies directly with the square of the speed. A van traveling 40 miles per hour can stop in 60 feet. If the van is traveling 56 miles per hour, what is its stopping distance?

68. Falling Rock A rock is dropped from the top of a cliff. The distance it falls in feet is directly proportional to the square of the time in seconds. If the rock falls 4 feet in $\frac{1}{2}$ second, how far will it fall in 3 seconds?

69. Volume of a Pyramid The volume, V, of a pyramid varies jointly as the area of its base, B, and its height, h (see the figure). If the volume of the pyramid is 160 cubic meters when the area of its base is 48 square meters and its height is 10 meters, find the volume of a pyramid when the area of its base is 42 square meters and its height is 9 meters.

70. Mortgage Payment The monthly mortgage payment, P, you pay on a mortgage varies jointly as the interest rate, r, and the amount of the mortgage, m. If the monthly mortgage payment on a $50,000 mortgage at a 7% interest rate is $332.50, find the monthly payment on a $66,000 mortgage at 7%.

71. DVD Rental The weekly DVD rentals, R, at Busterblock Video vary directly with their advertising budget, A, and inversely with the daily rental price, P. When their advertising budget is $400 and the rental price is $2 per day, they rent 4600 DVDs per week. How many DVDs would they rent per week if they increased their advertising budget to $500 and raised their rental price to $2.50?

*A ball slows down on its way to the plate due to wind resistance. For a 95-mph pitch, the ball is about 8 mph faster when it leaves the pitcher's hand than when it crosses the plate.

72. Electrical Resistance The electrical resistance of a wire, R, varies directly as its length, L, and inversely as its cross-sectional area, A. If the resistance of a wire is 0.2 ohm when the length is 200 feet and its cross-sectional area is 0.05 square inch, find the resistance of a wire whose length is 5000 feet with a cross-sectional area of 0.01 square inch.

73. Weight of an Object The weight, w, of an object in Earth's atmosphere varies inversely with the square of the distance, d, between the object and the center of Earth. A 140-pound person standing on Earth is approximately 4000 miles from Earth's center. Find the weight (or gravitational force of attraction) of this person at a distance 100 miles from Earth's surface.

74. Wattage Rating The wattage rating of an appliance, W, varies jointly as the square of the current, I, and the resistance, R. If the wattage is 3 watts when the current is 0.1 ampere and the resistance is 100 ohms, find the wattage when the current is 0.4 ampere and the resistance is 250 ohms.

75. Phone Calls The number of phone calls between two cities during a given time period, N, varies directly as the populations p_1 and p_2 of the two cities and inversely as the distance, d, between them. If 100,000 calls are made between two cities 300 miles apart and the populations of the cities are 60,000 and 200,000, how many calls are made between two cities with populations of 125,000 and 175,000 that are 450 miles apart?

76. Water Bill In a specific region of the country, the amount of a customer's water bill, W, is directly proportional to the average daily temperature for the month, T, the lawn area, A, and the square root of F, where F is the family size, and inversely proportional to the number of inches of rain, R.

In one month, the average daily temperature is 78°F and the number of inches of rain is 5.6. If the average family of four, who has 1000 square feet of lawn, pays $68 for water, estimate the water bill in the same month for the average family of six, who has 1500 square feet of lawn.

77. Intensity of Illumination An article in the magazine *Outdoor and Travel Photography* states, "If a surface is illuminated by a point-source of light (a flash), the intensity of illumination produced is inversely proportional to the square of the distance separating them."

If the subject you are photographing is 4 feet from the flash, and the illumination on this subject is $\frac{1}{16}$ of the light of the flash, what is the intensity of illumination on an object that is 7 feet from the flash?

78. Force of Attraction One of Newton's laws states that the force of attraction, F, between two masses is directly proportional to the masses of the two objects, m_1 and m_2, and inversely proportional to the square of the distance, d, between the two masses.

a) Write the formula that represents Newton's law.

b) What happens to the force of attraction if one mass is doubled, the other mass is tripled, and the distance between the objects is halved?

79. Pressure on an Object The pressure, P, in pounds per square inch (psi) on an object x feet below the sea is 14.70 psi plus the product of a constant of proportionality, k, and the number of feet, x, the object is below sea level (see the figure). The 14.70 represents the weight, in pounds, of the column of air (from sea level to the top of the atmosphere) standing over a 1-inch-by-1-inch square of ocean. The kx represents the weight, in pounds, of a column of water 1 inch by 1 inch by x feet.

a) Write a formula for the pressure on an object x feet below sea level.

b) If the pressure gauge in a submarine 60 feet deep registers 40.5 psi, find the constant k.

c) A submarine is built to withstand a pressure of 160 psi. How deep can the submarine go?

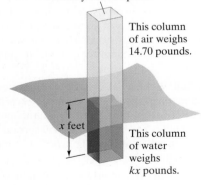

1 inch by 1 inch square

This column of air weighs 14.70 pounds.

x feet

This column of water weighs kx pounds.

Cumulative Review Exercises

[5.6] **80.** Divide $\dfrac{8x^2 + 6x - 21}{4x + 9}$.

[6.1] **81.** Factor $y(z - 2) + 8(z - 2)$.

[6.6] **82.** Solve $3x^2 - 24 = -6x$.

[7.2] **83.** Multiply $\dfrac{x + 8}{x - 3} \cdot \dfrac{x^3 - 27}{x^2 + 3x + 9}$.

Chapter 7 Summary

IMPORTANT FACTS AND CONCEPTS	EXAMPLES

Section 7.1

A **rational expression** is an expression of the form p/q, where p and q are polynomials and $q \neq 0$.

$\dfrac{x + 2}{x}$ and $\dfrac{x^2 + x}{x - 1}$ are rational expressions.

Whenever we have a rational expression containing a variable in the denominator, we always assume that the value or values of the variable that make the denominator 0 are excluded.

The rational expression $\dfrac{x - 5}{x - 3}$ is defined for all real numbers except 3.

A rational expression is simplified or reduced to its lowest term when the numerator and denominator have no common factors other than 1.

$\dfrac{1}{8}$ and $\dfrac{x^2 + x + 1}{2x^2 + 3x + 7}$ are expressions reduced to lowest terms.

To Simplify Rational Expressions

1. Factor both the numerator and denominator completely.
2. Divide out common factors.

$$\dfrac{12x^2 - 11x - 5}{9x^2 + 6x + 1} = \dfrac{\cancel{(3x + 1)}(4x - 5)}{\cancel{(3x + 1)}(3x + 1)} = \dfrac{4x - 5}{3x + 1}$$

Section 7.2

To Multiply Two Fractions

$$\dfrac{a}{b} \cdot \dfrac{c}{d} = \dfrac{a \cdot c}{b \cdot d}, \quad b \neq 0 \quad \text{and} \quad d \neq 0$$

$$\dfrac{1}{3} \cdot \dfrac{4}{5} = \dfrac{1 \cdot 4}{3 \cdot 5} = \dfrac{4}{15}$$

To Multiply Rational Expressions

1. Factor all numerators and denominators completely.
2. Divide out common factors.
3. Multiply numerators together and multiply denominators together.

$$\dfrac{4x^3}{7y^2} \cdot \dfrac{14y^3}{6x} = \dfrac{\overset{2}{\cancel{4}}\,\overset{x^2}{\cancel{x^3}} \cdot \overset{2}{\cancel{14}}\,\overset{y}{\cancel{y^3}}}{\cancel{7}\,y^2 \cdot \underset{3}{\cancel{6}}\,\cancel{x}} = \dfrac{4x^2 y^2}{3}$$

To Divide Two Fractions

$$\dfrac{a}{b} \div \dfrac{c}{d} = \dfrac{a}{b} \cdot \dfrac{d}{c} = \dfrac{ad}{bc}, \quad b \neq 0, \quad d \neq 0, \quad \text{and} \quad c \neq 0$$

$$\dfrac{3}{5} \div \dfrac{6}{7} = \dfrac{3}{5} \cdot \dfrac{7}{\underset{2}{\cancel{6}}} = \dfrac{7}{10}$$

To Divide Rational Expressions

Multiply the first fraction by the reciprocal of the second fraction.

$$\dfrac{x + 4}{x + 3} \div \dfrac{3x + 12}{x + 3} = \dfrac{\cancel{x + 4}}{\cancel{x + 3}} \cdot \dfrac{\cancel{x + 3}}{3\cancel{(x + 4)}} = \dfrac{1}{3}$$

Section 7.3

To Add or Subtract Two Fractions

$$\dfrac{a}{c} + \dfrac{b}{c} = \dfrac{a + b}{c}, c \neq 0 \quad \dfrac{a}{c} - \dfrac{b}{c} = \dfrac{a - b}{c}, c \neq 0$$

$$\dfrac{3}{11} + \dfrac{4}{11} = \dfrac{7}{11}, \quad \dfrac{18}{19} - \dfrac{5}{19} = \dfrac{13}{19}$$

To Add or Subtract Rational Expressions with a Common Denominator

1. Add or subtract the numerators.
2. Place the sum or difference of the numerators over the common denominator.
3. Simplify the fraction if possible.

$$\dfrac{2}{x - 3} + \dfrac{x + 5}{x - 3} = \dfrac{2 + x + 5}{x - 3} = \dfrac{x + 7}{x - 3}$$

IMPORTANT FACTS AND CONCEPTS	EXAMPLES

Section 7.3 (continued)

To Find the Least Common Denominator of Rational Expressions

1. Factor each denominator completely.
2. List all different factors of each denominator. When the same factor appears in more than one denominator, write that factor with the highest power that appears.
3. The least common denominator is the product of all the factors listed in step 2.

Find the least common denominator.

$$\frac{1}{9x^3y^4} + \frac{1}{6x^5y^3}$$

$$9x^3y^4 = 3 \cdot 3x^3y^4$$
$$6x^5y^3 = 3 \cdot 2x^5y^3$$

LCD is $2 \cdot 3^2 x^5 y^4 = 18x^5 y^4$.

Section 7.4

To Add or Subtract Two Rational Expressions with Unlike Denominators

1. Determine the LCD.
2. Rewrite each fraction as an equivalent fraction with the LCD.
3. Add or subtract the numerators while maintaining the LCD.
4. When possible, factor the remaining numerator and simplify the fraction.

$$\frac{9}{m} + \frac{5}{m-1} = \frac{m-1}{m-1} \cdot \frac{9}{m} + \frac{5}{m-1} \cdot \frac{m}{m}$$

$$= \frac{9(m-1)}{m(m-1)} + \frac{5m}{m(m-1)}$$

$$= \frac{9m - 9 + 5m}{m(m-1)}$$

$$= \frac{14m - 9}{m(m-1)}$$

Section 7.5

A **complex fraction** is one that has a fraction in its numerator or its denominator or in both its numerator and denominator.

$$\frac{\frac{2}{3}}{\frac{4}{7}}, \quad \frac{\frac{1}{x} + \frac{1}{y}}{\frac{1}{a} + \frac{1}{b}}$$

Method 1—To Simplify a Complex Fraction by Combining Terms

1. Add or subtract the fractions in both the numerator and denominator of the complex fraction to obtain single fractions in both.
2. Multiply the fraction in the numerator by the reciprocal of the fraction in the denominator.
3. Simplify further if possible.

$$\frac{1 + \frac{1}{x}}{x} = \frac{\frac{x}{x} + \frac{1}{x}}{x} = \frac{\frac{x+1}{x}}{x}$$

$$= \frac{x+1}{x} \cdot \frac{1}{x} = \frac{x+1}{x^2}$$

Method 2—To Simplify a Complex Fraction Using Multiplication First

1. Find the LCD of *all* the denominators appearing in the complex fraction.
2. Multiply both the numerator and denominator of the complex fraction by the LCD found in step 1.
3. Simplify when possible.

$$\frac{1 + \frac{1}{x}}{x} = \frac{x}{x} \cdot \frac{1 + \frac{1}{x}}{x} = \frac{x(1) + x\left(\frac{1}{x}\right)}{x(x)} = \frac{x+1}{x^2}$$

Note: LCD = x.

Section 7.6

A **rational equation** is an equation that contains one or more rational expressions.

$$\frac{1}{3}x - \frac{1}{7}x = 10, \qquad x + \frac{9}{x} = \frac{1}{3}$$

IMPORTANT FACTS AND CONCEPTS	EXAMPLES

Section 7.6 (continued)

To Solve Rational Equations

1. Determine the LCD of all fractions in the equation.
2. Multiply **both** sides of the equation by the LCD.
3. Remove any parentheses and combine like terms on each side of the equation.
4. Solve the equation using the properties discussed in earlier chapters.
5. Check your solution in the *original* equation.

$$\frac{x}{5} - \frac{x}{8} = 1$$

$$40\left(\frac{x}{5} - \frac{x}{8}\right) = 40(1)$$

$$8x - 5x = 40$$

$$3x = 40$$

$$x = \frac{40}{3}$$

A check shows that $\frac{40}{3}$ is the solution.

Section 7.7

Applications

A **geometric problem** involves geometric figures and formulas.

A rectangle has an area of 70 square meters. Find the dimensions if the width is 3 meters shorter than the length.

The answer is 7 meters by 10 meters.

A **motion problem** involves distance, rate, and time and uses the formula

$$\text{distance} = \text{rate} \cdot \text{time}$$

or

$$\text{time} = \frac{\text{distance}}{\text{rate}}$$

or

$$\text{rate} = \frac{\text{distance}}{\text{time}}$$

A cyclist can travel 20 miles with the wind to his back in the same time he can travel 12 miles going into the wind. If the wind is blowing at 2 miles per hour, find the speed of the cyclist without any wind.

The answer is 8 miles per hour.

A **work problem** involves two or more machines or people working together to complete a specific task.

Tom can paint a room in 6 hours and Bill can paint the same room in 4 hours. How long will it take them working together to paint this room?

The answer is 2.4 hours.

Section 7.8

A **variation** is an equation that relates one variable to one or more other variables using the operations of multiplication or division.

Direct Variation

If a variable y varies directly as a variable x, then $y = kx$, where k is the constant of proportionality.

$$y = 3x$$

Inverse Variation

If a variable y varies inversely as a variable x, then

$$y = \frac{k}{x} \quad (\text{or } xy = k)$$

where k is the constant of proportionality.

$$y = \frac{3}{x}$$

Joint Variation

If y varies jointly as x and z, then

$$y = kxz$$

where k is the constant of proportionality.

$$y = 3xz$$

Chapter 7 Review Exercises

[7.1] *Determine the values of the variable for which the following expressions are defined.*

1. $\dfrac{5}{2x - 18}$

2. $\dfrac{2x + 1}{x^2 - 8x + 15}$

3. $\dfrac{7x - 1}{5x^2 + 4x - 1}$

Simplify.

4. $\dfrac{y}{xy - 3y}$

5. $\dfrac{x^3 + 5x^2 + 12x}{x}$

6. $\dfrac{9x^2 + 3xy}{3x}$

7. $\dfrac{x^2 + 2x - 8}{x - 2}$

8. $\dfrac{a^2 - 81}{a - 9}$

9. $\dfrac{-2x^2 + 7x + 4}{x - 4}$

10. $\dfrac{b^2 - 7b + 10}{b^2 - 3b - 10}$

11. $\dfrac{4x^2 - 11x - 3}{4x^2 - 7x - 2}$

12. $\dfrac{2x^2 - 21x + 40}{4x^2 - 4x - 15}$

[7.2] *Multiply.*

13. $\dfrac{5a^2}{6b} \cdot \dfrac{2}{4a^2b}$

14. $\dfrac{30x^2y^3}{3z} \cdot \dfrac{6z^3}{5xy^3}$

15. $\dfrac{20a^3b^4}{7c^3} \cdot \dfrac{14c^7}{5a^5b}$

16. $\dfrac{1}{x - 4} \cdot \dfrac{4 - x}{9}$

17. $\dfrac{-m + 4}{15m} \cdot \dfrac{10m}{m - 4}$

18. $\dfrac{a - 2}{a + 3} \cdot \dfrac{a^2 + 4a + 3}{a^2 - a - 2}$

Divide.

19. $\dfrac{9x^6}{y^2} \div \dfrac{x^4}{4y}$

20. $\dfrac{5xy^2}{z} \div \dfrac{x^4y^2}{4z^2}$

21. $\dfrac{6a + 6b}{a^2} \div \dfrac{a^2 - b^2}{a^2}$

22. $\dfrac{1}{a^2 + 8a + 15} \div \dfrac{8}{a + 5}$

23. $(t + 8) \div \dfrac{t^2 + 5t - 24}{t - 3}$

24. $\dfrac{x^2 + xy - 2y^2}{2y} \div \dfrac{x + 2y}{12y^2}$

[7.3] *Add or subtract.*

25. $\dfrac{n}{n + 5} - \dfrac{2}{n + 5}$

26. $\dfrac{4x}{x + 7} + \dfrac{28}{x + 7}$

27. $\dfrac{5x - 4}{x + 8} + \dfrac{44}{x + 8}$

28. $\dfrac{7x - 3}{x^2 + 7x - 30} - \dfrac{3x + 9}{x^2 + 7x - 30}$

29. $\dfrac{5h^2 + 12h - 1}{h + 5} - \dfrac{h^2 - 5h + 14}{h + 5}$

30. $\dfrac{6x^2 - 4x}{2x - 3} - \dfrac{-3x + 12}{2x - 3}$

Find the least common denominator for each expression.

31. $\dfrac{a}{8} + \dfrac{5a}{3}$

32. $\dfrac{10}{x + 3} + \dfrac{2x}{x + 3}$

33. $\dfrac{10}{4xy^3} - \dfrac{11}{10x^2y}$

34. $\dfrac{6}{x - 3} - \dfrac{2}{x}$

35. $\dfrac{8}{n + 5} + \dfrac{2n - 3}{n - 4}$

36. $\dfrac{5x - 12}{x^2 + 2x} - \dfrac{4}{x + 2}$

37. $\dfrac{2r + 1}{r - s} - \dfrac{6}{r^2 - s^2}$

38. $\dfrac{3x^2}{x - 9} + 10x^3$

39. $\dfrac{19x - 5}{x^2 + 2x - 35} + \dfrac{-10x + 1}{x^2 + 9x + 14}$

[7.4] *Add or subtract.*

40. $\dfrac{5}{3y^2} + \dfrac{y}{2y}$

41. $\dfrac{3x}{xy} + \dfrac{1}{4x}$

42. $\dfrac{5x}{3xy} - \dfrac{6}{x^2}$

43. $7 - \dfrac{2}{x + 2}$

44. $\dfrac{x - y}{y} - \dfrac{x + y}{x}$

45. $\dfrac{7}{x + 4} + \dfrac{2}{x}$

46. $\dfrac{2}{3x} - \dfrac{3}{3x - 6}$

47. $\dfrac{1}{(z + 5)} + \dfrac{9}{(z + 5)^2}$

48. $\dfrac{x + 2}{x^2 - x - 6} + \dfrac{x - 3}{x^2 - 8x + 15}$

[7.2–7.4] *Perform each indicated operation.*

49. $\dfrac{x + 4}{x + 6} - \dfrac{x - 5}{x + 2}$

50. $2 + \dfrac{x}{x - 4}$

51. $\dfrac{a + 2}{b} \div \dfrac{a - 2}{5b^2}$

52. $\dfrac{x + 5}{x^2 - 9} + \dfrac{2}{x + 3}$

53. $\dfrac{6p + 12q}{p^2q} \cdot \dfrac{p^4}{p + 2q}$

54. $\dfrac{8}{(x + 2)(x - 3)} - \dfrac{6}{(x - 2)(x + 2)}$

55. $\dfrac{x + 7}{x^2 + 9x + 14} - \dfrac{x - 10}{x^2 - 49}$

56. $\dfrac{x - y}{x + y} \cdot \dfrac{xy + x^2}{x^2 - y^2}$

57. $\dfrac{3x^2 - 27y^2}{30} \div \dfrac{(x - 3y)^2}{6}$

58. $\dfrac{a^2 - 11a + 30}{a - 6} \cdot \dfrac{a^2 - 8a + 15}{a^2 - 10a + 25}$

59. $\dfrac{a}{a^2 - 1} - \dfrac{3}{3a^2 - 2a - 5}$

60. $\dfrac{2x^2 + 6x - 20}{x^2 - 2x} \div \dfrac{x^2 + 7x + 10}{2x^2 - 8}$

[7.5] *Simplify each complex fraction.*

61. $\dfrac{5 + \frac{2}{3}}{\frac{3}{4}}$

62. $\dfrac{1 + \frac{5}{8}}{3 - \frac{9}{16}}$

63. $\dfrac{\frac{12ab}{9c}}{\frac{4a}{c^2}}$

64. $\dfrac{18x^4y^2}{\frac{9xy^5}{4z^2}}$

65. $\dfrac{a - \frac{a}{b}}{\frac{1 + a}{b}}$

66. $\dfrac{r^2 + \frac{7}{s}}{s^2}$

67. $\dfrac{\frac{3}{x} + \frac{2}{x^2}}{5 - \frac{1}{x}}$

68. $\dfrac{\frac{x}{x + y}}{\frac{x^2}{4x + 4y}}$

69. $\dfrac{\frac{9}{x}}{\frac{9}{x^2}}$

70. $\dfrac{\frac{1}{a} + 2}{\frac{1}{a} + \frac{3}{a}}$

71. $\dfrac{\frac{1}{x^2} - \frac{1}{x}}{\frac{1}{x^2} + \frac{1}{x}}$

72. $\dfrac{\frac{8x}{y} - x}{\frac{y}{x} - 1}$

[7.6] *Solve.*

73. $\dfrac{5}{9} = \dfrac{10}{x + 3}$

74. $\dfrac{x}{4} = \dfrac{x - 3}{2}$

75. $\dfrac{12}{n} + 2 = \dfrac{n}{4}$

76. $\dfrac{10}{m} + \dfrac{3}{2} = \dfrac{m}{10}$

77. $\dfrac{-4}{d} = \dfrac{3}{2} + \dfrac{4 - d}{d}$

78. $\dfrac{1}{x - 7} + \dfrac{1}{x + 7} = \dfrac{1}{x^2 - 49}$

79. $\dfrac{x - 3}{x - 2} + \dfrac{x + 1}{x + 3} = \dfrac{2x^2 + x + 1}{x^2 + x - 6}$

80. $\dfrac{a}{a^2 - 64} + \dfrac{4}{a + 8} = \dfrac{3}{a - 8}$

81. $\dfrac{d}{d - 4} - 4 = \dfrac{4}{d - 4}$

[7.7] *Solve.*

82. **Sandcastles** It takes John and Amy Brogan 6 hours to build a sandcastle. It takes Paul and Cindy Carter 4 hours to make the same sandcastle. How long will it take all four people together to build the sandcastle?

83. **Filling a Pool** A $\frac{3}{4}$-inch-diameter hose can fill a swimming pool in 7 hours. A $\frac{5}{16}$-inch-diameter hose can siphon all the water out of a full pool in 12 hours. How long will it take to fill the pool if while one hose is filling the pool the other hose is siphoning water from the pool?

84. **Sum of Numbers** One number is six times as large as another. The sum of their reciprocals is 7. Determine the numbers.

85. **Rollerblading and Motorcycling** Robert Johnston can travel 3 miles on his rollerblades in the same time Tran Lee can travel 8 miles on his mountain bike. If Tran's speed on his bike is 3.5 miles per hour faster than that of Robert on his rollerblades, determine Robert's and Tran's speeds.

[7.8]

86. W is directly proportional to the square of L and inversely proportional to A. If $W = 4$ when $L = 2$ and $A = 10$, find W when $L = 5$ and $A = 20$.

87. z is jointly proportional to x and y and inversely proportional to the square of r. If $z = 12$ when $x = 20$, $y = 8$, and $r = 8$, find z when $x = 10$, $y = 80$, and $r = 3$.

88. **Drug Dosage** The recommended dosage, d, of the antibiotic drug vancomycin is directly proportional to a person's weight, w. If Carmen Brown, who is 132 pounds is given 182 milligrams, find the recommended dosage for Bill Glenn, who is 198 pounds.

89. **Runners' Speed** The time, t, it takes a runner to cover a specified distance is inversely proportional to the runner's speed. If Nhat Chung runs at an average of 6 miles per hour, he will finish a race in 1.4 hours. How long will it take Leif Lundgren, who runs at 5 miles per hour, to finish the same race?

Chapter 7 Practice Test

 To find out how well you understand the chapter material, take this practice test. The answers, and the section where the material was initially discussed, are given in the back of the book. Each problem is also fully worked out on the **Chapter Test Prep Video CD***. Review any questions that you answered incorrectly.*

Simplify.

1. $\dfrac{-8 + x}{x - 8}$

2. $\dfrac{x^3 - 1}{x^2 - 1}$

Perform each indicated operation.

3. $\dfrac{20x^2y^3}{4z^2} \cdot \dfrac{8xz^3}{5xy^4}$

4. $\dfrac{a^2 - 9a + 14}{a - 2} \cdot \dfrac{a^2 - 4a - 21}{(a - 7)^2}$

5. $\dfrac{x^2 - x - 6}{x^2 - 9} \cdot \dfrac{x^2 - 6x + 9}{x^2 + 4x + 4}$

6. $\dfrac{x^2 - 1}{x + 2} \cdot \dfrac{x + 2}{1 - x^2}$

7. $\dfrac{x^2 - 4y^2}{5x + 20y} \div \dfrac{x + 2y}{x + 4y}$

8. $\dfrac{15}{y^2 + 2y - 15} \div \dfrac{5}{y - 3}$

9. $\dfrac{m^2 + 3m - 18}{m - 3} \div \dfrac{m^2 - 8m + 15}{3 - m}$

10. $\dfrac{4x + 3}{8y} + \dfrac{2x - 5}{8y}$

11. $\dfrac{7x^2 - 4}{x + 3} - \dfrac{6x + 9}{x + 3}$

12. $\dfrac{2}{xy} - \dfrac{8}{xy^3}$

13. $3 - \dfrac{5z}{z - 5}$

14. $\dfrac{x - 5}{x^2 - 16} - \dfrac{x - 2}{x^2 + 2x - 8}$

Simplify.

15. $\dfrac{2 + \dfrac{1}{2}}{3 - \dfrac{1}{5}}$

16. $\dfrac{x + \dfrac{x}{y}}{\dfrac{7}{x}}$

17. $\dfrac{4 + \dfrac{3}{x}}{\dfrac{9}{x} - 5}$

Solve.

18. $2 + \dfrac{8}{x} = 6$

19. $\dfrac{2x}{3} - \dfrac{x}{4} = x + 1$

20. $\dfrac{x}{x - 8} + \dfrac{6}{x - 2} = \dfrac{x^2}{x^2 - 10x + 16}$

Solve.

21. **Working Together** Mr. Jackson, on his tractor, can clear a 1-acre field in 10 hours. Mr. Hackett, on his tractor, can clear a 1-acre field in 15 hours. If they work together, how long will it take them to clear a 1-acre field?

22. **Determine a Number** The sum of a positive number and its reciprocal is 2. Determine the number.

23. **Area of Triangle** The area of a triangle is 30 square inches. If the height is 2 inches less than 2 times the base, determine the height and base of the triangle.

24. **Exercising** LaConya Bertrell exercises for $1\frac{1}{2}$ hours each day. During the first part of her routine, she rides a bicycle and averages 10 miles per hour. For the remainder of the time, she rollerblades and averages 4 miles per hour. If the total distance she travels is 12 miles, how far did she travel on the rollerblades?

25. **Making Music** The wavelength of sound waves, w, is inversely proportional to the frequency, f (or pitch). If a frequency of 263 cycles per second (middle C on a piano) produces a wavelength of about 4.3 feet, determine the length of a wavelength of a frequency of 1000 cycles per second.

Cumulative Review Test

Take the following test and check your answers with those given in the back of the book. Review any questions that you answered incorrectly. The section where the material was covered is indicated after the answer.

1. Evaluate $3x^2 - 5xy^2 - 7$, when $x = -4$ and $y = -2$.

2. Evaluate $[6 - [3(8 \div 4)]^2 + 9 \cdot 4]^2$.

3. Solve $5z + 4 = -3(z - 7)$.

4. The cost of a 2006 Chevrolet Corvette increased by 1.2% over the cost of the 2005 Corvette. Write an expression for the sum of the costs of a 2006 and a 2005 Corvette.

5. Find the slope of the line through the points $(-7, 8)$ and $(3, 8)$.

6. Write the equation $3x - 4y = 12$ in slope-intercept form.

7. Simplify $(6x^2 - 3x - 5) - (-2x^2 - 8x - 19)$.

8. Multiply $(3n^2 - 4n + 3)(2n - 5)$.

9. Divide $\dfrac{4x - 38}{8}$.

10. Factor $8a^2 - 8a - 5a + 5$.

11. Factor $13x^2 + 26x - 39$.

12. Solve $2x^2 = 11x - 12$.

13. Multiply $\dfrac{x^2 + x - 12}{x^2 - x - 6} \cdot \dfrac{x^2 - 2x - 8}{2x^2 - 7x - 4}$.

14. Subtract $\dfrac{r}{r + 2} - \dfrac{3}{r - 5}$.

15. Add $\dfrac{4}{x^2 - 3x - 10} + \dfrac{6}{x^2 + 5x + 6}$.

16. Solve $\dfrac{x}{9} - \dfrac{x}{6} = \dfrac{1}{12}$.

17. Solve $\dfrac{7}{x + 3} + \dfrac{5}{x + 2} = \dfrac{5}{x^2 + 5x + 6}$.

18. **Medical Plans** A school district allows its employees to choose from two medical plans. With plan 1, the employee pays 10% of all medical bills (the school district pays the balance). With plan 2, the employee pays the school district a one-time payment of $150, then the employee pays 5% of all medical bills. What total medical bills would result in the employee paying the same amount with the two plans?

19. **Bird Seed** A feed store owner wishes to make his own store-brand mixture of bird seed by mixing sunflower seed that costs $0.50 per pound with a premixed assorted seed that costs $0.20 per pound. How many pounds of each will he have to use to make a 50-pound mixture that will cost $16.00?

20. **Sailing** During the first leg of a race, the sailboat *Thumper* sailed at an average speed of 6.5 miles per hour. During the second leg of the race, the winds increased and *Thumper* sailed at an average speed of 9.5 miles per hour. If the total distance sailed by *Thumper* was 12.75 miles, and the total time spent racing was 1.5 hours, determine the distance traveled by *Thumper* on each leg of the race.

Functions and Their Graphs

8

WE SEE GRAPHS DAILY in newspapers and magazines. You will see many such graphs in this chapter. For example, in Exercise 108 on page 497, a graph is used to show the growth in the shipment of LCD monitors.

8.1 More on Graphs

1 Graph equations by plotting points.

2 Interpret graphs.

In this section we will expand upon some of the concepts we introduced in Chapter 4. Although we will briefly review the concepts before discussing them, you may find it helpful to review Chapter 4 before studying Chapter 8.

1 Graph Equations by Plotting Points

In Section 4.2, we graphed linear equations by plotting points. Linear equations are also called *first degree equations* since the greatest exponent that appears on any variable is one. Recall that to graph a linear equation by plotting points we select values for x and find the corresponding values of y. Example 1 will refresh your memory on how we graph linear equations by plotting points.

EXAMPLE 1 ▶ Graph $y = -\frac{1}{3}x + 1$.

Solution We will select some values for x, find the corresponding values of y, and then draw the graph. When we select values for x, we will select some positive values, some negative values, and 0. The graph is illustrated in **Figure 8.1**. (To conserve space, we will not always list a column in the table for ordered pairs.)

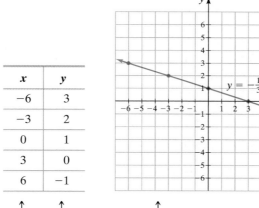

x	y
-6	3
-3	2
0	1
3	0
6	-1

FIGURE 8.1

1. Select values for x.
2. Compute y.
3. Plot the points and draw the graph.

▶ **Now Try Exercise 13**

In Example 1, notice that we selected values of x that were multiples of 3 so we would not have to work with fractions.

If we are asked to graph an equation not solved for y, such as $x + 3y = 3$, our first step will be to solve the equation for y. For example, if we solve $x + 3y = 3$ for y using the procedure discussed in Section 2.6, we obtain

$$x + 3y = 3$$
$$3y = -x + 3 \qquad \textit{Subtract x from both sides.}$$
$$y = \frac{-x + 3}{3} \qquad \textit{Divide both sides by 3.}$$
$$y = \frac{-x}{3} + \frac{3}{3} = -\frac{1}{3}x + 1$$

The resulting equation, $y = -\frac{1}{3}x + 1$, is the same equation we graphed in Example 1.

Therefore, the graph of $x + 3y = 3$ is also illustrated in **Figure 8.1**.

Up until this point we have only graphed equations whose graphs are straight lines. There are many equations whose graphs are not straight lines. Such equations are called **nonlinear equations**. To graph nonlinear equations by plotting points, we follow the same procedure used to graph linear equations. However, since the graphs are not straight lines, we may need to plot more points to draw the graphs.

EXAMPLE 2 ▶ Graph $y = x^2 - 4$.

Solution   We select some values for x and find the corresponding values of y. Then we plot the points and connect them with a smooth curve. When we substitute values for x and evaluate the right side of the equation, we follow the order of operations discussed in Section 1.9. For example, if $x = -3$, then $y = (-3)^2 - 4 = 9 - 4 = 5$. The graph is shown in **Figure 8.2**.

x	y
-3	5
-2	0
-1	-3
0	-4
1	-3
2	0
3	5

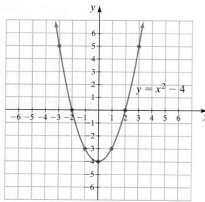

FIGURE 8.2

If we substituted 4 for x, y would equal 12. When $x = 5$, $y = 21$. Notice that this graph rises steeply as x moves away from the origin.

▶ **Now Try Exercise 19**

EXAMPLE 3 ▶ Graph $y = \dfrac{1}{x}$.

Solution   We begin by selecting values for x and finding the corresponding values of y. We then plot the points and draw the graph. Notice that if we substitute 0 for x, we obtain $y = \dfrac{1}{0}$. Since $\dfrac{1}{0}$ is undefined, we cannot use 0 as a first coordinate. There will be no part of the graph at $x = 0$. We will plot points to the left of $x = 0$, and points to the right of $x = 0$ separately. Select points close to 0 to see what happens to the graph as x gets close to $x = 0$. Note, for example, that when $x = -\dfrac{1}{2}$, $y = \dfrac{1}{-\dfrac{1}{2}} = -2$. This graph has two branches, one to the left of the y-axis and one to the right of the y-axis, as shown in **Figure 8.3**.

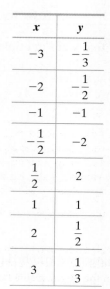

x	y
-3	$-\dfrac{1}{3}$
-2	$-\dfrac{1}{2}$
-1	-1
$-\dfrac{1}{2}$	-2
$\dfrac{1}{2}$	2
1	1
2	$\dfrac{1}{2}$
3	$\dfrac{1}{3}$

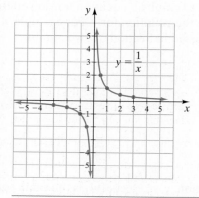

FIGURE 8.3

▶ **Now Try Exercise 29**

In the graph for Example 3, notice that for values of x far to the right of 0, or far to the left of 0, the graph approaches the x-axis but does not touch it. For example when

$x = 1000$, $y = 0.001$ and when $x = -1000$, $y = -0.001$. Can you explain why y can never have a value of 0?

EXAMPLE 4 ▶ Graph $y = |x|$.

Solution Recall that $|x|$ is read "the absolute value of x." Absolute values were discussed in Section 1.5. To graph this absolute value equation, we select some values for x and find the corresponding values of y. For example, if $x = -4$, then $y = |-4| = 4$. Then we plot the points and draw the graph.

Notice that this graph is V-shaped, as shown in **Figure 8.4**.

x	y
-4	4
-3	3
-2	2
-1	1
0	0
1	1
2	2
3	3
4	4

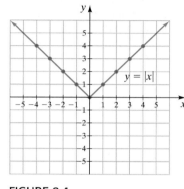

FIGURE 8.4   ▶ **Now Try Exercise 23**

Avoiding Common Errors

When graphing nonlinear equations, many students do not plot enough points to get a true picture of the graph. For example, when graphing $y = \dfrac{1}{x}$ many students consider only integer values of x. Following is a table of values for the equation and two graphs that contain the points indicated in the table.

x	-3	-2	-1	1	2	3
y	$-\dfrac{1}{3}$	$-\dfrac{1}{2}$	-1	1	$\dfrac{1}{2}$	$\dfrac{1}{3}$

CORRECT   INCORRECT

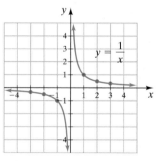   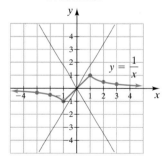

FIGURE 8.5   FIGURE 8.6

If you select and plot fractional values of x near 0, as was done in Example 3, you get the graph in **Figure 8.5**. The graph in **Figure 8.6** cannot be correct because the equation is not defined when x is 0 and therefore the graph cannot cross the y-axis. Whenever you plot a graph that contains a variable in the denominator, select values for the variable that are very close to the value that makes the denominator 0 and observe what happens. For example, when graphing $y = \dfrac{1}{x - 3}$ you should use values of x close to 3, such as 2.9 and 3.1 or 2.99 and 3.01, and see what values you obtain for y.

Also, when graphing nonlinear equations, it is a good idea to consider both positive and negative values. For example, if you used only positive values of x when graphing $y = |x|$, the graph would appear to be a straight line going through the origin, instead of the V-shaped graph shown in **Figure 8.4** above.

2 Interpret Graphs

We see many different types of graphs daily in newspapers, in magazines, on television, and so on. Throughout this book, we present a variety of graphs. Since being able to draw and interpret graphs is very important, we will study this further in Section 8.2. In Example 5 you must understand and interpret graphs to answer the question.

EXAMPLE 5 ▶ When Jim Herring went to see his mother in Cincinnati, he boarded a Southwest Airlines plane. The plane sat on the runway for 20 minutes and then took off. The plane flew at about 600 miles per hour for about 2 hours. It then reduced its speed to about 300 miles per hour and circled the Cincinnati Airport for about 15 minutes before it came in for a landing. After landing, the plane taxied to the gate and stopped. Which graph in **Figures 8.7a–8.7d** best illustrates this situation?

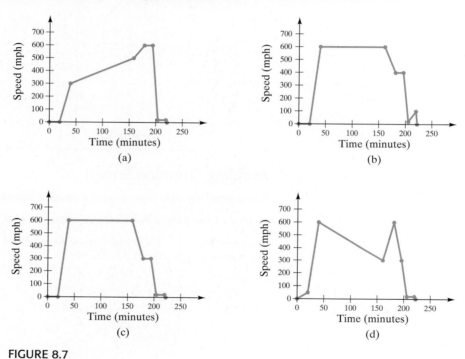

FIGURE 8.7

Solution   The graph that depicts the situation described is (c), reproduced with annotations in **Figure 8.8**. The graph shows speed versus time, with time on the horizontal axis. While the plane sat on the runway for 20 minutes its speed was 0 miles per hour (the horizontal line at 0 from 0 to 20 minutes). After 20 minutes the plane took off, and its speed increased to 600 miles per hour (the near-vertical line going from 0 to 600

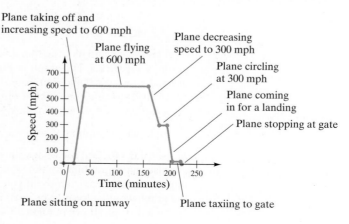

FIGURE 8.8

mph). The plane then flew at about 600 miles per hour for 2 hours (the horizontal line at about 600 mph). It then slowed down to about 300 miles per hour (the near-vertical line from 600 mph to 300 mph). Next the plane circled at about 300 miles per hour for about 15 minutes (the horizontal line at about 300 mph). The plane then came in for a landing (the near-vertical line from about 300 mph to about 20 mph). It then taxied to the gate (the horizontal line at about 20 mph). Finally, it stopped at the gate (the near-vertical line when the speed dropped to 0 mph).

▶ **Now Try Exercise 59**

EXERCISE SET 8.1

Concept/Writing Exercises

1. When graphing a nonlinear equation, will the graph always have a y-intercept? Explain.

2. When graphing a nonlinear equation, will you always get an accurate graph if you plot enough integer values? Explain.

3. When graphing the equation $y = \dfrac{1}{x}$, what value cannot be substituted for x? Explain.

4. In a graph that illustrates the speed of an airplane during the time of a flight, what would a horizontal line represent?

Practice the Skills

Graph each equation.

5. $y = x + 1$

6. $y = 3x$

7. $y = -3x - 5$

8. $y = -2x + 2$

9. $y = 2x + 4$

10. $y = x + 2$

11. $y = \dfrac{1}{2}x$

12. $y = -\dfrac{1}{3}x$

13. $y = \dfrac{1}{2}x - 1$

14. $y = -\dfrac{1}{2}x - 3$

15. $y = -\dfrac{1}{3}x + 2$

16. $y = -\dfrac{1}{3}x + 4$

17. $y = x^2$

18 $y = x^2 - 2$

19. $y = -x^2$

20. $y = -x^2 + 4$

21. $y = |x| + 1$

22 $y = |x| + 2$

23. $y = -|x|$

24. $y = -|x| - 3$

25. $y = x^3$

26. $y = -x^3$

27. $y = x^3 + 1$

28. $y = \dfrac{1}{x}$

29. $y = -\dfrac{1}{x}$

30. $x^2 = 1 + y$

31. $x = |y|$

32. $x = y^2$

In Exercises 33–40, use a calculator to obtain at least eight points that are solutions to the equation. Then graph the equation by plotting the points.

33. $y = x^3 - x^2 - x + 1$

34. $y = -x^3 + x^2 + x - 1$

35. $y = \dfrac{1}{x + 1}$

36. $y = \dfrac{1}{x} + 1$

37. $y = \sqrt{x}$

38. $y = \sqrt{x + 4}$

39. $y = \dfrac{1}{x^2}$

40. $y = \dfrac{|x^2|}{2}$

41. Is the point represented by the ordered pair $\left(\dfrac{1}{3}, \dfrac{1}{12}\right)$ on the graph of the equation $y = \dfrac{x^2}{x + 1}$? Explain.

42. Is the point represented by the ordered pair $\left(-\dfrac{1}{2}, -\dfrac{3}{5}\right)$ on the graph of the equation $y = \dfrac{x^2 + 1}{x^2 - 1}$? Explain.

43. a) Plot the points $A(2, 7), B(2, 3),$ and $C(6, 3)$, and then draw $\overline{AB}, \overline{AC},$ and $\overline{BC}$. ($\overline{AB}$ represents the line segment from A to B.)

   b) Find the area of the figure.

44. a) Plot the points $A(-4, 5), B(2, 5), C(2, -3),$ and $D(-4, -3)$, and then draw $\overline{AB}, \overline{BC}, \overline{CD},$ and $\overline{DA}$.

   b) Find the area of the figure.

45. Golf Courses The following graph shows that the average golf course length at the majors has been on the increase in recent years.

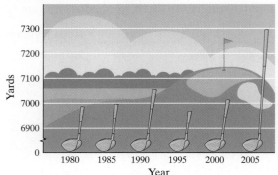

Average Golf Course Length at Majors

Source: Rees Jones Inc., PGA Tour, *USA TODAY* research

a) Estimate the average length of the golf course at the majors in 1980.

b) Estimate the average length of the golf course at the majors in 2005.

c) In which years was the average length of the golf courses at the majors greater than 7000 yards?

d) Does the increase in the average length of golf courses in the majors from 1995 to 2005 appear to be linear? Explain.

46. E-Commerce The following graph shows that E-commerce (sales on the Internet) has been constantly rising. The graph shows the sales, in the first quarter of each year, for the years from 2000 to 2005.

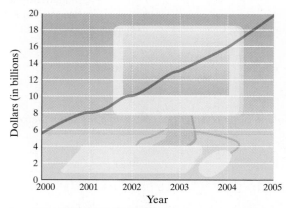

E-Commerce Soars

Source: Census Bureau, *USA TODAY* (8/9/05)

a) Estimate Internet sales in the first quarter of 2000.

b) Estimate Internet sales in the first quarter of 2005.

c) In what years was Internet sales in the first quarter greater than $12 billion?

d) Does the increase in Internet sales in the first quarter from 2000 to 2005 appear to be approximately linear? Explain.

47. Graph $y = x + 1$, $y = x + 3$, and $y = x - 1$ on the same axes.

a) What do you notice about the graphs of the equations and the values where the graphs intersect the y-axis?

b) Do all the graphs seem to have the same slant (or slope)?

48. Graph $y = \frac{1}{2}x$, $y = \frac{1}{2}x + 3$, and $y = \frac{1}{2}x - 4$ on the same axes.

a) What do you notice about the graphs of equations and the values where the graphs intersect the y-axis?

b) Do all of these graphs seem to have the same slant (or slope)?

49. Graph $y = 2x$. Determine the *rate of change* of y with respect to x. That is, by how many units does y change compared to each unit change in x?

50. Graph $y = 4x$. Determine the rate of change of y with respect to x.

51. Graph $y = 3x + 2$. Determine the rate of change of y with respect to x.

52. Graph $y = \frac{1}{2}x$. Determine the rate of change of y with respect to x.

53. The ordered pair $(3, -7)$ represents one point on the graph of a linear equation. If y increases 4 units for each unit increase in x on the graph, find two other solutions to the equation.

54. The ordered pair $(1, -4)$ represents one point on the graph of a linear equation. If y increases 3 units for each unit increase in x on the graph, find two other solutions to the equation.

Match Exercises 55–58 with the corresponding graph of elevation above sea level versus time, labeled a–d, on the next page.

55. Mary Leeseberg walked for 5 minutes on level ground. Then for 5 minutes she climbed a slight hill. Then she walked on level ground for 5 minutes. Then for the next 5 minutes she climbed a steep hill. During the next 10 minutes she descended uniformly until she reached the height at which she had started.

56. Don Gordon walked on level ground for 5 minutes. Then he walked down a steep hill for 10 minutes. For the next 5 minutes he walked on level ground. For the next 5 minutes he walked back up to his starting height. For the next 5 minutes he walked on level ground.

57. Nancy Johnson started out by walking up a steep hill for 5 minutes. For the next 5 minutes she walked down a steep hill to an elevation lower than her starting point. For the next 10 minutes she walked on level ground. For the next 10 minutes she walked up a slight hill, at which time she reached her starting elevation.

58. James Condor started out by walking up a hill for 5 minutes. For the next 10 minutes he walked down a hill to an elevation equal to his starting elevation. For the next 10 minutes he walked on level ground. For the next 5 minutes he walked downhill.

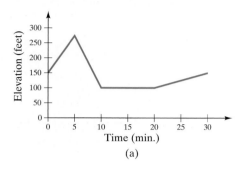

(a)

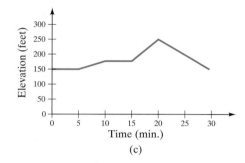

(c)

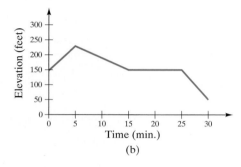

(b)

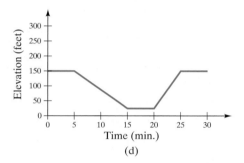

(d)

Match Exercises 59–62 with the corresponding graph of speed versus time, labeled a–d, below.

59. To go to work, Cletidus Hunt walked for 3 minutes, waited for the train for 5 minutes, rode the train for 15 minutes, then walked for 7 minutes.

60. To go to work, Tyrone Williams drove in stop-and-go traffic for 5 minutes, then drove on the expressway for 20 minutes, then drove in stop-and-go traffic for 5 minutes.

61. To go to work, Sheila Washington drove on a country road for 10 minutes, then drove on a highway for 12 minutes, then drove in stop-and-go traffic for 8 minutes.

62. To go to work, Brenda Pinkney rode her bike uphill for 10 minutes, then rode downhill for 15 minutes, then rode on a level street for 5 minutes.

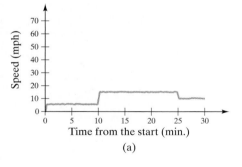

(a)

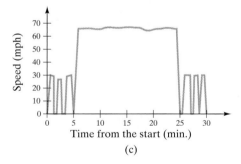

(c)

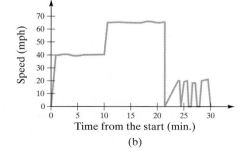

(b)

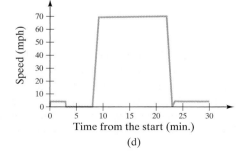

(d)

Match Exercises 63–66 with the corresponding graph of speed versus time, labeled a–d below.

63. Christina Dwyer walked for 5 minutes to warm up, jogged for 20 minutes, and then walked for 5 minutes to cool down.

64. Annie Droullard went for a leisurely bike ride at a constant speed for 30 minutes.

65. Michael Odu took a 30-minute walk through his neighborhood. He stopped very briefly on 7 occasions to pick up trash.

66. Richard Dai walked through his neighborhood and stopped 3 times to chat with his neighbors. He was gone from his house a total of 30 minutes.

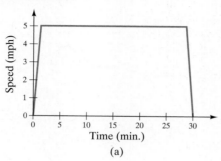

(a)

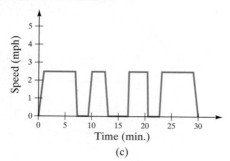

(c)

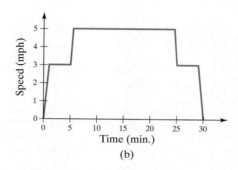

(b)

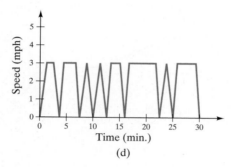

(d)

Match Exercises 67–70 with the corresponding graph of distance traveled versus time, labeled a–d. Recall from Chapter 2 that distance = rate × time. Selected distances are indicated on the graphs.

67. Train A traveled at a speed of 40 mph for 1 hour, then 80 mph for 2 hours, and then 60 mph for 3 hours.

68. Train C traveled at a speed of 80 mph for 2 hours, then stayed in a station for 1 hour, and then traveled 40 mph for 3 hours.

69. Train B traveled at a speed of 20 mph for 2 hours, then 60 mph for 3 hours, and then 80 mph for 1 hour.

70. Train D traveled at 30 mph for 1 hour, then 65 mph for 2 hours, and then 30 mph for 3 hours.

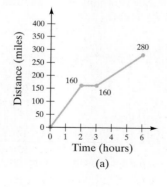

(a)

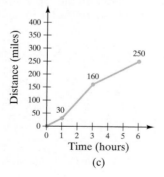

(c)

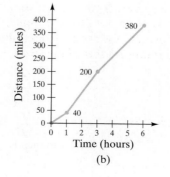

(b)

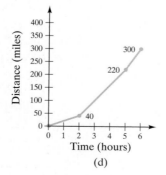

(d)

Use a graphing calculator to graph each function. Make sure you select values for the window that will show the curvature of the graph. Then, if your calculator can display tables, display a table of values in which the x-values extend by units, from 0 to 6.

71. $y = 2x - 3$

72. $y = \frac{1}{3}x + 2$

73. $y = x^2 - 2x - 8$

74. $y = -x^2 + 16$

75. $y = x^3 - 2x + 4$

76. $y = 2x^3 - 6x^2 - 1$

Challenge Problems

Graph each equation.

77. $y = |x - 2|$

78. $x = y^2 + 2$

Group Activity

Discuss and work Exercises 79–80 as a group.

79. a) Group member 1: Plot the points $(-2, 4)$ and $(6, 8)$. Determine the *midpoint* of the line segment connecting these points.

Group member 2: Follow the above instructions for the points $(-3, -2)$ and $(5, 6)$.

Group member 3: Follow the above instructions for the points $(4, 1)$ and $(-2, 4)$.

b) As a group, determine a formula for the midpoint of the line segment connecting the points (x_1, y_1) and (x_2, y_2). (*Note*: We will discuss the midpoint formula further in Chapter 14.)

80. Three points on a parallelogram are $A(3, 5)$, $B(8, 5)$, and $C(-1, -3)$.

a) Individually determine a fourth point D that completes the parallelogram.

b) Individually compute the area of your parallelogram.

c) Compare your answers. Did you all get the same answers? If not, why not?

d) Is there more than one point that can be used to complete the parallelogram? If so, give the points and find the corresponding areas of each parallelogram.

Cumulative Review Exercises

[1.9] **81.** Evaluate $x^2 - 4x + 5$ when $x = -3$.

[4.3] **82.** Determine the slope of the line through the points $(-5, 3)$ and $(6, 7)$.

[7.6] Solve.

83. $\dfrac{2x}{x^2 - 4} + \dfrac{1}{x - 2} = \dfrac{2}{x + 2}$

84. $\dfrac{4x}{x^2 + 6x + 9} - \dfrac{2x}{x + 3} = \dfrac{x + 1}{x + 3}$

8.2 Functions

1 Understand set builder notation and interval notation.

2 Understand relations.

3 Recognize functions.

4 Use the vertical line test.

5 Understand function notation.

6 Study applications of functions in daily life.

1 Understand Set Builder Notation and Interval Notation

We will begin discussing functions shortly. However, before we do, we need to introduce two methods used to indicate sets of numbers: *set builder notation* and *interval notation*. We begin with set builder notation.

When we list a set of elements, where the elements are separated by commas, the set is said to be in *roster form*. We gave sets of numbers in roster form in Section 1.4. For example, the set of natural numbers in roster form is $\{1, 2, 3, 4, \ldots\}$. As you will see shortly, sometimes it is not possible to list a set of elements in roster form. A set of numbers that cannot be listed in roster form generally can be written using *set builder notation*. In set builder notation, symbols are often used to represent sets of numbers. Recall from Section 1.4 that $\mathbb{R}$ is used to represent the set of real numbers. The following chart gives some other symbols used to represent sets of numbers.

Sets of Numbers

SET OF NUMBERS	SYMBOL
Set of Natural or Counting Numbers	N
Set of Whole Numbers	W
Set of Integers	I

Another symbol used in set builder notation is $\in$, which is read "is an element of." For example, $x \in N$ is read "x is an element of the set of natural numbers." Set builder notation also generally uses inequality symbols. In Section 1.5, we introduced the "is less than symbol," $<$, and "is greater than symbol," $>$. The following chart reviews these inequality symbols and introduces some new inequality symbols.

Inequality Symbols

$>$ is read "is greater than."

$\geq$ is read "is greater than or equal to."

$<$ is read "is less than."

$\leq$ is read "is less than or equal to."

$\neq$ is read "is not equal to."

You may recall from Section 1.5 that $<$ and $>$ can be explained using a real number line (**Fig. 8.9**).

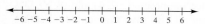

FIGURE 8.9

The number a is greater than the number b, $a > b$, when a is to the right of b on the number line (**Fig. 8.10**). We can also state that the number b is less than a, $b < a$, when b is to the left of a on the number line. The inequality $a \neq b$ means either $a < b$ or $a > b$.

FIGURE 8.10

We use the notation $x > 2$, read "x is greater than 2," to represent *all* real numbers greater than 2. We use the notation $x \leq -3$, read "x is less than or equal to -3," to represent all real numbers that are less than or equal to -3. The notation $-4 \leq x < 3$ means all real numbers that are greater than or equal to -4 and also less than 3. In the inequalities $x > 2$ and $x \leq -3$, the 2 and the -3 are called **endpoints**. In the inequality $-4 \leq x < 3$, the -4 and 3 are the endpoints. The solutions to inequalities that use either $<$ or $>$ do not include the endpoints, but the solutions to inequalities that use either $\leq$ or $\geq$ do include the endpoints. When inequalities are illustrated on the number line, a solid circle is used to show that the endpoint is included in the answer, and an open circle is used to show that the endpoint is not included. Following are some illustrations of how certain inequalities are indicated on the number line.

Inequality	Inequality Indicated on the Number Line
$x > 2$	
$x \leq -1$	
$-4 \leq x < 3$	

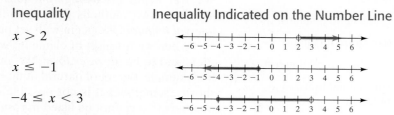

Some students misunderstand the word *between*. The word *between* indicates that the endpoints are not included in the answer. For example, the set of natural numbers between 2 and 6 is $\{3, 4, 5\}$. If we wish to include the endpoints, we can use the word *inclusive*. For example, the set of natural numbers between 2 and 6 inclusive is $\{2, 3, 4, 5, 6\}$.

The last inequality symbol, $\neq$, can be represented as an inequality involving both $<$ and $>$. For instance $2 \neq 3$ can be represented by $2 < 3$ or $2 > 3$. In words we have, 2 is not equal to 3 means that either 2 is less than 3 or that 2 is greater than 3. We will discuss inequalities of this type, called *compound inequalities*, in Section 10.1.

Now that we have discussed the inequality symbols we will show an example of **set builder notation**.

$$E = \{x | x \text{ is a natural number greater than 7}\}$$

This is read "Set E is the set of all elements x, such that x is a natural number greater than 7." In roster form, this set is written

$$E = \{8, 9, 10, 11, 12, \ldots\}$$

The general form of set builder notation is

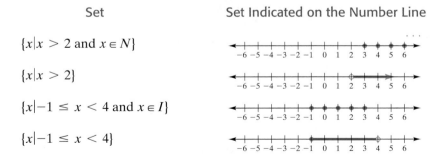

We often will use the variable x when using set builder notation, although any variable can be used.

Two condensed ways of writing set $E = \{x | x \text{ is a natural number greater than 7}\}$ in set builder notation follow.

$$E = \{x | x > 7 \text{ and } x \in N\} \quad \text{or} \quad E = \{x | x \geq 8 \text{ and } x \in N\}$$

The set $A = \{x | -3 < x \leq 4 \text{ and } x \in I\}$ is the set of integers greater than -3 and less than or equal to 4. The set written in roster form is $\{-2, -1, 0, 1, 2, 3, 4\}$. Notice that the endpoint -3 is not included in the set but the endpoint 4 is included.

How do the sets $B = \{x | x > 2 \text{ and } x \in N\}$ and $C = \{x | x > 2\}$ differ? Can you write each set in roster form? Can you illustrate both sets on the number line? Set B contains only the natural numbers greater than 2, that is, $\{3, 4, 5, 6, \ldots\}$. Set C contains not only the natural numbers greater than 2 but also fractions and decimal numbers greater than 2. If you attempted to write set C in roster form, where would you begin? What is the smallest number greater than 2? Is it 2.1 or 2.01 or 2.001? Since there is no smallest number greater than 2, this set cannot be written in roster form. Below we illustrate these two sets on the number line. We have also illustrated two other sets.

Set	Set Indicated on the Number Line
$\{x | x > 2 \text{ and } x \in N\}$	
$\{x | x > 2\}$	
$\{x | -1 \leq x < 4 \text{ and } x \in I\}$	
$\{x | -1 \leq x < 4\}$	

Interval notation is another method that is used to describe a set of numbers. In interval notation, a *parenthesis* is used to indicate that an endpoint is **not included** in the interval and a *bracket* is used to symbolize that the endpoint **is included** in the interval. The symbol ∞ is read "infinity"; it indicates that the set continues indefinitely. Whenever ∞ is used in interval notation, a parenthesis must be used on the corresponding side of the interval notation. On the next page are some examples of inequalities and how the sets they represent would be described using set builder notation and interval notation.

Inequality	Set Represented in Set Builder Notation	Set Represented in Interval Notation
$x > a$	$\{x \mid x > a\}$	(a, ∞)
$x \geq a$	$\{x \mid x \geq a\}$	$[a, \infty)$
$x < a$	$\{x \mid x < a\}$	$(-\infty, a)$
$x \leq a$	$\{x \mid x \leq a\}$	$(-\infty, a]$
$a < x < b$	$\{x \mid a < x < b\}$	(a, b)
$a \leq x \leq b$	$\{x \mid a \leq x \leq b\}$	$[a, b]$
$a < x \leq b$	$\{x \mid a < x \leq b\}$	$(a, b]$
$a \leq x < b$	$\{x \mid a \leq x < b\}$	$[a, b)$
$x \geq 5$	$\{x \mid x \geq 5\}$	$[5, \infty)$
$x < 3$	$\{x \mid x < 3\}$	$(-\infty, 3)$
$2 < x \leq 6$	$\{x \mid 2 < x \leq 6\}$	$(2, 6]$
$-6 \leq x \leq -1$	$\{x \mid -6 \leq x \leq -1\}$	$[-6, -1]$

The set of real numbers, $\mathbb{R}$, is represented in interval notation as $(-\infty, \infty)$. It is also possible to represent a single value in interval notation. For example, the value $x = 2$ would be represented as $[2, 2]$.

2 Understand Relations

In real life we often find that one quantity is related to a second quantity. For example, the amount you spend for oranges is related to the number of oranges you purchase. The speed of a sailboat is related to the speed of the wind. And the income tax you pay is related to the income you earn.

Suppose oranges cost 30 cents apiece. Then one orange costs 30 cents, two oranges cost 60 cents, three oranges cost 90 cents, and so on. We can list this information, or relationship, as a set of ordered pairs by listing the number of oranges first and the cost, in cents, second. The ordered pairs that represent this situation are $(1, 30)$, $(2, 60)$, $(3, 90)$, and so on. An equation that represents this situation is $c = 30n$, where c is the cost, in cents, and n is the number of oranges. Since the cost depends on the number of oranges, we say that the cost is the *dependent variable* and the number of oranges is the *independent variable*.

Now consider the equation $y = 2x + 3$. In this equation, the value obtained for y depends on the value selected for x. Therefore, x is the *independent variable* and y is the *dependent variable*. Note that in this example, unlike with the oranges, there is no physical connection between x and y. The variable x is the independent variable and y is the dependent variable simply because of their placement in the equation.

For an equation in variables x and y, if the value of y depends on the value of x, then y is the **dependent variable** and x is the **independent variable**. Since related quantities can be represented as ordered pairs, the concept of a **relation** can be defined as follows.

> **Relation**
>
> A **relation** is any set of ordered pairs.

3 Recognize Functions

We now develop the idea of a **function**—one of the most important concepts in mathematics. A function is a special type of relation in which each element in one set (called the domain) corresponds to *exactly one* element in a second set (called the range).

Consider the oranges that cost 30 cents apiece that we just discussed. We can illustrate the number of oranges and the cost of the oranges using **Figure 8.11**.

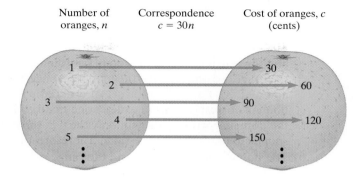

FIGURE 8.11

Notice that each number in the set of numbers of oranges, n, corresponds to (or is mapped to) exactly one number in the set of cost of oranges, c. Therefore, this correspondence is a function. The set consisting of the number of oranges, $\{1, 2, 3, 4, 5, \dots\}$, is called the **domain**. The set consisting of the costs in cents, $\{30, 60, 90, 120, 150, \dots\}$, is called the **range**. In general, the set of values for the independent variable is called the **domain**. The set of values for the dependent variable is called the **range**; see **Figure 8.12**.

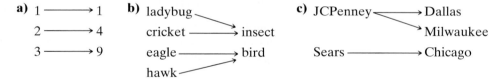

FIGURE 8.12

EXAMPLE 1 ▶ Determine whether each correspondence is a function.

Solution

a) For a correspondence to be a function, each element in the domain must correspond with exactly one element in the range. Here the domain is $\{1, 2, 3\}$ and the range is $\{1, 4, 9\}$. Since each element in the domain corresponds to exactly one element in the range, this correspondence is a function.

b) Here the domain is {ladybug, cricket, eagle, hawk} and the range is {insect, bird}. Even though the domain has four elements and the range has two elements, each element in the domain corresponds with exactly one element in the range. Thus, this correspondence is a function.

c) Here the domain is {JCPenney, Sears} and the range is {Dallas, Milwaukee, Chicago}. Notice that JCPenney corresponds to both Dallas and Milwaukee. Therefore each element in the domain *does not* correspond to exactly one element in the range. Thus, this correspondence is a relation but *not* a function.

▶ **Now Try Exercise 19**

Now we will formally define function.

Function

A **function** is a correspondence between a first set of elements, the domain, and a second set of elements, the range, such that each element of the domain corresponds to *exactly one* element in the range.

EXAMPLE 2 ▶ Which of the following relations are functions?

    a) $\{(1, 4), (2, 3), (3, 5), (-1, 3), (0, 6)\}$

    b) $\{(-1, 3), (4, 2), (3, 1), (2, 6), (3, 5)\}$

Solution

    a) The domain is the set of first coordinates in the set of ordered pairs, $\{1, 2, 3, -1, 0\}$, and the range is the set of second coordinates, $\{4, 3, 5, 6\}$. Notice that when listing the range, we only include the number 3 once, even though it appears in both $(2, 3)$ and $(-1, 3)$. Examining the set of ordered pairs, we see that each number in the domain corresponds with exactly one number in the range. For example, the 1 in the domain corresponds with only the 4 in the range, and so on. No *x*-value corresponds to more than one *y*-value. Therefore, this relation *is a function*.

    b) The domain is $\{-1, 4, 3, 2\}$ and the range is $\{3, 2, 1, 6, 5\}$. Notice that 3 appears as the first coordinate in two ordered pairs even though it is listed only once in the set of elements that represent the domain. Since the ordered pairs $(3, 1)$ and $(3, 5)$ have *the same first coordinate* and a different second coordinate, each value in the domain does not correspond to exactly one value in the range. Therefore, this relation is *not a function*.

▶ **Now Try Exercise 25**

Example 2 leads to an alternate definition of function.

> **Function**
>
> A **function** is a set of ordered pairs in which no *first* coordinate is repeated.

If the second coordinate in a set of ordered pairs repeats, the set of ordered pairs may still be a function, as in Example 2 **a)**. However, if two or more ordered pairs contain the same first coordinate, as in Example 2 **b)**, the set of ordered pairs is not a function.

4 Use the Vertical Line Test

The **graph of a function or relation** is the graph of its set of ordered pairs. The two sets of ordered pairs in Example 2 **a)** and **b)** are graphed in **Figures 8.13a** and **8.13b** respectively. Notice that in the function in **Figure 8.13a** it is not possible to draw a vertical line that intersects two points. We should expect this because, in a function, each *x*-value must correspond to exactly one *y*-value. In **Figure 8.13b** we *can* draw a vertical line through the points $(3, 1)$ and $(3, 5)$. This shows that each *x*-value does not correspond to exactly one *y*-value, and the graph does not represent a function.

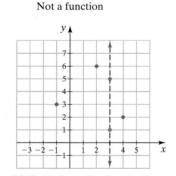

FIGURE 8.13   (a) First set of ordered pairs   (b) Second set of ordered pairs

This method of determining whether a graph represents a function is called the **vertical line test.**

> **Vertical Line Test**
>
> If a vertical line can be drawn through any part of the graph and the line intersects another part of the graph, the graph does not represent a function. If a vertical line cannot be drawn to intersect the graph at more than one point, the graph represents a function.

We use the vertical line test to show that **Figure 8.14b** represents a function and **Figures 8.14a** and **8.14c** do not represent functions.

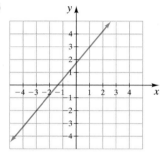

FIGURE 8.14   (a)   (b)   (c)

EXAMPLE 3 ▶ Use the vertical line test to determine whether the following graphs represent functions. Also determine the domain and range of each function or relation.

a)   **b)**

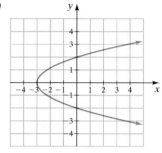

FIGURE 8.15   FIGURE 8.16

Solution

a) A vertical line cannot be drawn to intersect the graph in **Figure 8.15** at more than one point. Thus this is the graph of a function. Since the line extends indefinitely in both directions, every value of x will be included in the domain. The domain is the set of real numbers.

$$\text{Domain:} \quad \mathbb{R} \quad \text{or} \quad (-\infty, \infty)$$

The range is also the set of real numbers since all values of y are included on the graph.

$$\text{Range:} \quad \mathbb{R} \quad \text{or} \quad (-\infty, \infty)$$

b) Since a vertical line can be drawn to intersect the graph in **Figure 8.16** at more than one point, this is *not* the graph of a function. The domain of this relation is the set of values greater than or equal to -3.

$$\text{Domain:} \quad \{x | x \geq -3\} \quad \text{or} \quad [-3, \infty)$$

The range is the set of y-values, which can be any real number.

$$\text{Range:} \quad \mathbb{R} \quad \text{or} \quad (-\infty, \infty)$$

▶ **Now Try Exercise 67**

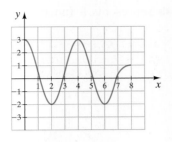

FIGURE 8.17

EXAMPLE 4 ▶ Consider the graph shown in **Figure 8.17**.

a) What member of the range is paired with 4 in the domain?

b) What members of the domain are paired with −2 in the range?

c) What is the domain of the function?

d) What is the range of the function?

Solution

a) The range is the set of *y*-values. The *y*-value paired with the *x*-value of 4 is 3.

b) The domain is the set of *x*-values. The *x*-values paired with the *y*-value of −2 are 2 and 6.

c) The domain is the set of *x*-values, 0 through 8. Thus the domain is

$$\{x|0 \le x \le 8\} \quad \text{or} \quad [0, 8]$$

d) The range is the set of *y*-values, −2 through 3. Thus, the range is

$$\{y|-2 \le y \le 3\} \quad \text{or} \quad [-2, 3]$$

▶ **Now Try Exercise 73**

EXAMPLE 5 ▶ **Figure 8.18** illustrates a graph of speed versus time of a man out for a walk and run. Write a story about the man's outing that corresponds to this function.

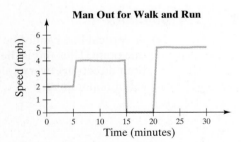

FIGURE 8.18

Solution   Understand   The horizontal axis is time and the vertical axis is speed. When the graph is horizontal it means the person is traveling at the constant speed indicated on the vertical axis. The near-vertical lines that increase with time (or have a positive slope, as will be discussed later) indicate an increase in speed, whereas the near-vertical lines that decrease with time (or have a negative slope) indicate a decrease in speed.

Answer   Here is one possible interpretation of the graph. The man walks for about 5 minutes at a speed of about 2 miles per hour. Then the man speeds up to about 4 miles per hour and walks fast or runs at about this speed for about 10 minutes. Then the man slows down and stops, and then rests for about 5 minutes. Finally, the man speeds up to about 5 miles per hour and runs at this speed for about 10 minutes.

▶ **Now Try Exercise 99**

5 Understand Function Notation

In Section 8.1 we graphed a number of equations, as summarized in **Table 8.1**. If you examine each equation in the table, you will see that they are all functions, since their graphs pass the vertical line test.

TABLE 8.1	Example				
Section 8.1 example	**Equation graphed**	**Graph**	**Does the graph represent a function?**	**Domain**	**Range**
1	$y = -\dfrac{1}{3}x + 1$	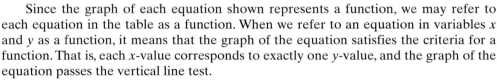	Yes	$(-\infty, \infty)$	$(-\infty, \infty)$
2	$y = x^2 - 4$		Yes	$(-\infty, \infty)$	$[-4, \infty)$
3	$y = \dfrac{1}{x}$		Yes	$(-\infty, 0) \cup (0, \infty)$	$(-\infty, 0) \cup (0, \infty)$
4	$y = |x|$		Yes	$(-\infty, \infty)$	$[0, \infty)$

Since the graph of each equation shown represents a function, we may refer to each equation in the table as a function. When we refer to an equation in variables x and y as a function, it means that the graph of the equation satisfies the criteria for a function. That is, each x-value corresponds to exactly one y-value, and the graph of the equation passes the vertical line test.

Not all equations are functions, as you will see in a later chapter in this book, Conic Sections, where we discuss equations of circles and ellipses. However, until we get to this chapter, all equations that we discuss will be functions.

Consider the equation $y = 3x + 2$. By applying the vertical line test to its graph (**Fig. 8.19**), we can see that the graph represents a function. When an equation in variables x and y is a function, we often write the equation using **function notation**, $f(x)$, read "f of x." Since the equation $y = 3x + 2$ is a function, and the value of y depends on the value of x, we say that **y is a function of x**. When we are given a linear equation in variables x and y, *that is solved for y*, we can write the equation in function notation by substituting $f(x)$ for y. In this case, we can write the equation in function notation as $f(x) = 3x + 2$. The notation $f(x)$ represents the dependent variable *and does not mean f times x*. Other letters may be used to indicate functions. For example, $g(x)$ and $h(x)$ also represent functions of x.

Functions written in function notation are also equations since they contain an equal sign. We may refer to $y = 3x + 2$ as either an equation or a function. Similarly, we may refer to $f(x) = 3x + 2$ as either a function or an equation.

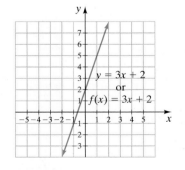

FIGURE 8.19

If y is a function of x, the notation $f(5)$, read "f of 5," means the value of y when x is 5. To evaluate a function for a specific value of x, substitute that value for x in the function. For example, if $f(x) = 3x + 2$, then $f(5)$ is found as follows:

$$f(x) = 3x + 2$$
$$f(5) = 3(5) + 2 = 17$$

Therefore, when x is 5, y is 17. The ordered pair $(5, 17)$ would appear on the graph of $y = 3x + 2$.

Helpful Hint

Linear equations that are not solved for y can be written using function notation by solving the equation for y, then replacing y with $f(x)$. For example, the equation $-9x + 3y = 6$ becomes $y = 3x + 2$ when solved for y. We can therefore write $f(x) = 3x + 2$.

EXAMPLE 6 ▶ If $f(x) = -4x^2 + 3x - 2$, find

a) $f(2)$ **b)** $f(-1)$ **c)** $f(a)$

Solution

a) $f(x) = -4x^2 + 3x - 2$

$f(2) = -4(2)^2 + 3(2) - 2 = -4(4) + 6 - 2 = -16 + 6 - 2 = -12$

b) $f(-1) = -4(-1)^2 + 3(-1) - 2 = -4(1) - 3 - 2 = -4 - 3 - 2 = -9$

c) To evaluate the function at a, we replace each x in the function with an a.

$$f(x) = -4x^2 + 3x - 2$$
$$f(a) = -4a^2 + 3a - 2$$

▶ **Now Try Exercise 79**

EXAMPLE 7 ▶ Determine each indicated function value.

a) $g(-2)$ for $g(t) = \dfrac{1}{t + 8}$

b) $h(5)$ for $h(s) = 2|s - 6|$

c) $j(-3)$ for $j(r) = \sqrt{22 - r}$

Solution   In each part, substitute the indicated value into the function and evaluate.

a) $g(-2) = \dfrac{1}{-2 + 8} = \dfrac{1}{6}$

b) $h(5) = 2|5 - 6| = 2|-1| = 2(1) = 2$

c) $j(-3) = \sqrt{22 - (-3)} = \sqrt{22 + 3} = \sqrt{25} = 5$

▶ **Now Try Exercise 83**

6 Study Applications of Functions in Daily Life

Many of the applications that we discussed in Chapter 3 were functions. However, we had not defined a function at that time. Now we examine additional applications of functions.

EXAMPLE 8 ▶ **Business Jets** The graph in **Figure 8.20** is taken from the November 11, 2004, issue of *USA Today*. The graph shows the number of business jets manufactured for the years from 1994 through 2004, projected through 2013.

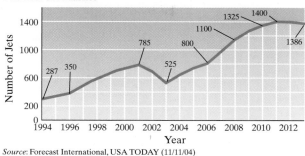

Business-Jet Market

FIGURE 8.20   *Source*: Forecast International, USA TODAY (11/11/04)

a) Explain why the graph in **Figure 8.20** represents a function.

b) Determine the number of business jets projected to be manufactured in 2010.

c) Determine the projected percent increase in the number of business jets to be manufactured from 2003 to 2011.

d) Determine the percent decrease in the number of business jets manufactured from 2001 to 2003.

Solution

a) The graph represents a function since each year corresponds to a specific number of business jets manufactured. Notice that the graph passes the vertical line test.

b) In 2010, the graph shows that 1325 business jets are projected to be manufactured. If we let the function be represented by J, then $J(2010) = 1325$.

c) We will follow the problem-solving procedure to solve this problem.

Understand and Translate   We need to determine the percent increase in the number of business jets to be manufactured from 2003 to 2011. To do this, use the formula

$$\text{percent change (increase or decrease)} = \frac{\left(\begin{array}{c}\text{value in}\\\text{latest period}\end{array}\right) - \left(\begin{array}{c}\text{value in}\\\text{previous period}\end{array}\right)}{\text{value in previous period}}$$

The latest period is 2011 and the previous period is 2003. Substituting the values, we get

$$\text{percent change} = \frac{1400 - 525}{525}$$

Carry Out

$$= \frac{875}{525} \approx 1.667 = 166.7\%$$

Check and Answer   Our calculations appear correct. There is projected to be about a 166.7% increase in the number of business jets manufactured from 2003 to 2011.

d) To find the percent decrease from 2001 to 2003, we follow the same procedure as in part **c)**. The latest period is 2003 and the previous period is 2001.

$$\text{percent change (increase or decrease)} = \frac{\left(\begin{array}{c}\text{value in}\\\text{latest period}\end{array}\right) - \left(\begin{array}{c}\text{value in}\\\text{previous period}\end{array}\right)}{\text{value in previous period}}$$

$$= \frac{525 - 785}{785} = \frac{-260}{785} \approx -0.331 = -33.1\%$$

The negative sign preceding the 33.1% indicates a percent decrease. Thus, there was about a 33.1% decrease in the number of business jets manufactured from 2001 to 2003.

▶ **Now Try Exercise 105**

EXAMPLE 9 ▸ **Immigration** The size of the U.S. foreign-born population is at an all-time high. The graph in **Figure 8.21** shows the U.S. foreign-born population, in millions, from 1890 to 2004 and projected to 2010.

a) Using the graph in **Figure 8.21**, explain why this set of points represents a function.

b) Using the graph in **Figure 8.22**, estimate the foreign-born population in 2008.

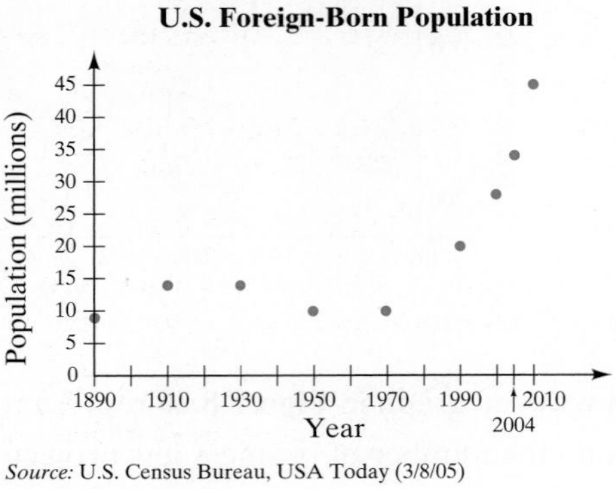

U.S. Foreign-Born Population

Source: U.S. Census Bureau, USA Today (3/8/05)

FIGURE 8.21

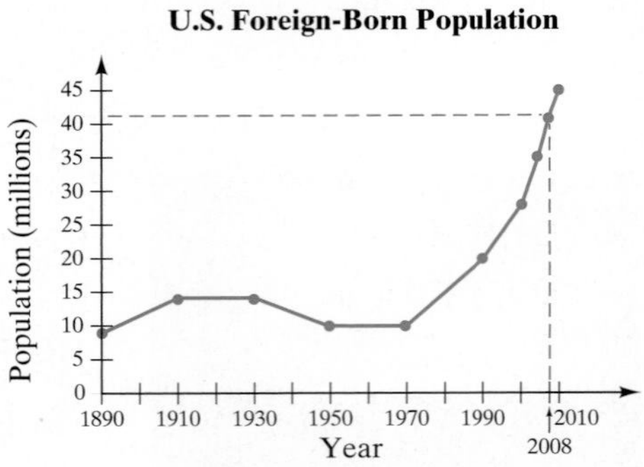

U.S. Foreign-Born Population

FIGURE 8.22

Solution

a) Since each year corresponds with exactly one population, this set of points represents a function. Notice that this graph passes the vertical line test.

b) We can connect the points with straight line segments as in **Figure 8.22**. Then we can estimate from the graph that there were about 41 million foreign-born Americans in 2008. If we call the function f, then $f(2008) = 41$.

▸ **Now Try Exercise 109**

In Section 2.6 we learned to use formulas. Consider the formula for the area of a circle, $A = \pi r^2$. In the formula, π is a constant that is approximately 3.14. For each specific value of the radius, r, there corresponds exactly one area, A. Thus the area of a circle is a function of its radius. We may therefore write

$$A(r) = \pi r^2$$

Often formulas are written using function notation like this.

EXAMPLE 10 ▸ The Celsius temperature, C, is a function of the Fahrenheit temperature, F.

$$C(F) = \frac{5}{9}(F - 32)$$

Determine the Celsius temperature that corresponds to 50°F.

Solution   We need to find $C(50)$. We do so by substitution.

$$C(F) = \frac{5}{9}(F - 32)$$

$$C(50) = \frac{5}{9}(50 - 32)$$

$$= \frac{5}{9}(18) = 10$$

Therefore, 50°F = 10°C.

▸ **Now Try Exercise 89**

In Example 10, F is the independent variable and C is the dependent variable. If we solved the function for F, we would obtain $F(C) = \frac{9}{5}C + 32$. In this formula, C is the independent variable and F is the dependent variable.

EXERCISE SET 8.2

Math XL MathXL®   *MyMathLab* MyMathLab

Concept/Writing Exercises

1. What is a function?

2. What is a relation?

3. Are all functions also relations? Explain.

4. Are all relations also functions? Explain.

5. List the set of integers *between* 4 and 9.

6. List the set of integers *between* 4 and 9 *inclusive*.

7. Explain how to use the vertical line test to determine if a relation is a function.

8. What is the domain of a function?

9. What is the range of a function?

10. What are the domain and range of the function $f(x) = 3x - 2$? Explain your answer.

11. What are the domain and range of a function of the form $f(x) = ax + b, a \neq 0$? Explain your answer.

12. Consider the absolute value function $y = |x|$. What is its domain and range? Explain.

13. What is a dependent variable?

14. What is an independent variable?

15. How is "$f(x)$" read?

16. Are all functions that are given in function notation also equations? Explain.

Practice the Skills

In Exercises 17–22, **a)** *determine if the relation illustrated is a function.* **b)** *Give the domain and range of each function or relation.*

17. twice a number

    $3 \longrightarrow 6$

    $5 \longrightarrow 10$

    $11 \longrightarrow 22$

18. nicknames

    Robert $\longrightarrow$ Bobby

    $\longrightarrow$ Rob

    Margaret $\longrightarrow$ Peggy

    $\longrightarrow$ Maggie

19. number of siblings

    Cameron $\longrightarrow 3$

    Tyrone $\longrightarrow 6$

    Vishnu $\longrightarrow$

20. a number squared

    $4 \longrightarrow 16$

    $5 \longrightarrow 25$

    $7 \longrightarrow 49$

21. cost of a stamp

    $1990 \longrightarrow 20$

    $2001 \longrightarrow 34$

    $2002 \longrightarrow 37$

22. absolute value

    $|-8| \longrightarrow 8$

    $|8| \longrightarrow$

    $|0| \longrightarrow 0$

In Exercises 23–30, **a)** *determine which of the following relations are also functions.* **b)** *Give the domain and range of each relation or function.*

23. $\{(1, 4), (2, 2), (3, 5), (4, 3), (5, 1)\}$

24. $\{(1, 0), (4, 2), (9, 3), (1, -1), (4, -2), (9, -3)\}$

25. $\{(3, -1), (5, 0), (1, 2), (4, 4), (2, 2), (7, 9)\}$

26. $\{(-1, 1), (0, -3), (3, 4), (4, 5), (-2, -2)\}$

27. $\{(1, 4), (2, 5), (3, 6), (2, 2), (1, 1)\}$

28. $\{(6, 3), (-3, 4), (0, 3), (5, 2), (3, 5), (2, 8)\}$

29. $\{(0, 3), (1, 3), (2, 2), (1, -1), (2, -7)\}$

30. $\{(3, 5), (2, 5), (1, 5), (0, 5), (-1, 5)\}$

In Exercises 31–40, list each set in roster form.

31. $A = \{x | -1 < x < 1 \text{ and } x \in I\}$

32. $B = \{y | y \text{ is an odd natural number less than } 6\}$

33. $C = \{z | z \text{ is an even integer greater than } 16 \text{ and less than or equal to } 20\}$

34. $D = \{x | x \geq -3 \text{ and } x \in I\}$

35. $E = \{x | x < 3 \text{ and } x \in W\}$

36. $F = \left\{ x \left| -\dfrac{6}{5} \leq x < \dfrac{15}{4} \text{ and } x \in N \right. \right\}$

37. $H = \{x | x \text{ is a whole number multiple of } 7\}$

38. $L = \{x | x \text{ is an integer greater than } -5\}$

39. $J = \{x | x > 0 \text{ and } x \in I\}$

40. $K = \{x | x \text{ is a whole number between } 9 \text{ and } 10\}$

In Exercises 41 and 42, **a)** *write out how you would read each set;* **b)** *write the set in roster form.*

41. $A = \{x | x < 7 \text{ and } x \in N\}$

42. $B = \{x | x \text{ is one of the last five capital letters in the English alphabet}\}$

Illustrate each set on a number line.

43. $\{x | x \geq 0\}$

44. $\{w | w > -5\}$

45. $\{z | z \leq 2\}$

46. $\{y | y < 4\}$

47. $\{p | -6 \leq p < 3\}$

48. $\{x | -1.67 \leq x < 5.02\}$

49. $\{q | q > -3 \text{ and } q \in N\}$

50. $\{x | -1.93 \leq x \leq 2 \text{ and } x \in I\}$

51. $\{r | r \leq \pi \text{ and } r \in W\}$

52. $\left\{ x \left| \dfrac{5}{12} < x \leq \dfrac{7}{12} \text{ and } x \in N \right. \right\}$

Express in set builder notation each set of numbers that are indicated on the number line.

53.

54.

55.

56.

57.

58.

59.

60.

61.

62.

In Exercises 63–74, **a)** *determine whether the graph illustrated represents a function.* **b)** *Give the domain and range of each function or relation.* **c)** *Approximate the value or values of x where y = 2.*

63.

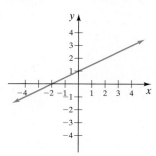

64.

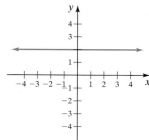

65.

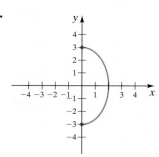

66.

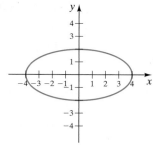

67.

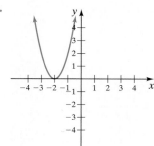

68.

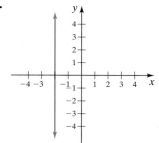

69.

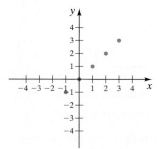

70.

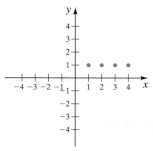

71.

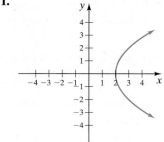

72.

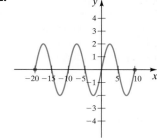

73.

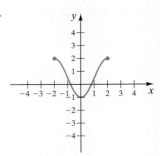

74.

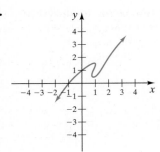

Evaluate each function at the indicated values.

75. $f(x) = -2x + 7$; find
  a) $f(2)$.
  b) $f(-3)$.

76. $f(a) = \dfrac{1}{3}a + 4$; find
  a) $f(0)$.
  b) $f(-12)$.

77. $h(x) = x^2 - x - 6$; find
  a) $h(0)$.
  b) $h(-1)$.

78. $g(x) = -2x^2 + 7x - 11$; find
  a) $g(2)$.
  b) $g\left(\dfrac{1}{2}\right)$.

79. $r(t) = -t^3 - 2t^2 + t + 4$; find
  a) $r(1)$.
  b) $r(-2)$.

80. $g(t) = 4 - 3t + 16t^2 - 2t^3$; find
  a) $g(0)$.
  b) $g(3)$.

81. $h(z) = |5 - 2z|$; find
  a) $h(6)$.
  b) $h\left(\dfrac{5}{2}\right)$.

82. $q(x) = -2|x + 8| + 13$; find
  a) $q(0)$.
  b) $q(-4)$.

83. $s(t) = \sqrt{t + 3}$; find
  a) $s(-3)$.
  b) $s(6)$.

84. $f(t) = \sqrt{5 - 2t}$; find
  a) $f(-2)$.
  b) $f(2)$.

85. $g(x) = \dfrac{x^3 - 2}{x - 2}$; find
  a) $g(0)$.
  b) $g(2)$.

86. $h(x) = \dfrac{x^2 + 4x}{x + 6}$; find
  a) $h(-3)$.
  b) $h\left(\dfrac{2}{5}\right)$.

Problem Solving

87. Area of a Rectangle The formula for the area of a rectangle is $A = lw$. If the length of a rectangle is 6 feet, then the area is a function of its width, $A(w) = 6w$. Find the area when the width is
  a) 4 feet.
  b) 6.5 feet.

88. Simple Interest The formula for the simple interest earned for a period of 1 year is $i = pr$, where p is the principal invested and r is the simple interest rate. If $1000 is invested, the simple interest earned in 1 year is a function of the simple interest rate, $i(r) = 1000r$. Determine the simple interest earned in 1 year if the interest rate is
  a) 2.5%.
  b) 4.25%.

89. Area of a Circle The formula for the area of a circle is $A = \pi r^2$. The area is a function of the radius.

  a) Write this function using function notation.
  b) Determine the area when the radius is 12 yards.

90. Perimeter of a Square The formula for the perimeter of a square is $P = 4s$ where s represents the length of any one of the sides of the square.

  a) Write this function using function notation.
  b) Determine the perimeter of a square with sides of length 7 meters.

91. Temperature The formula for changing Fahrenheit temperature into Celsius temperature is $C = \dfrac{5}{9}(F - 32)$. The Celsius temperature is a function of Fahrenheit temperature.

  a) Write this function using function notation.
  b) Find the Celsius temperature that corresponds to $-31°F$.

92. Volume of a Cylinder The formula for the volume of a right circular cylinder is $V = \pi r^2 h$. If the height, h, is 3 feet, then the volume is a function of the radius, r.

a) Write this formula in function notation, where the height is 3 feet.

b) Find the volume if the radius is 2 feet.

93. Sauna Temperature The temperature, T, in degrees Celsius, in a sauna n minutes after being turned on is given by the function $T(n) = -0.03n^2 + 1.5n + 14$. Find the sauna's temperature after

a) 3 minutes.   b) 12 minutes.

94. Stopping Distance The stopping distance, d, in meters for a car traveling v kilometers per hour is given by the function $d(v) = 0.18v + 0.01v^2$. Find the stopping distance for the following speeds:

a) 60 km/hr   b) 25 km/hr

95. Air Conditioning When an air conditioner is turned on maximum in a bedroom at $80°$, the temperature, T, in the room after A minutes can be approximated by the function $T(A) = -0.02A^2 - 0.34A + 80, 0 \le A \le 15$.

a) Estimate the room temperature 4 minutes after the air conditioner is turned on.

b) Estimate the room temperature 12 minutes after the air conditioner is turned on.

96. Accidents The number of accidents, n, in 1 month involving drivers x years of age can be approximated by the function $n(x) = 2x^2 - 150x + 4000$. Find the approximate number of accidents in 1 month that involved

a) 18-year-olds.

b) 25-year-olds.

97. Oranges The total number of oranges, T, in a square pyramid whose base is n by n oranges is given by the function

$$T(n) = \frac{1}{3}n^3 + \frac{1}{2}n^2 + \frac{1}{6}n$$

Find the number of oranges if the base is

a) 6 by 6 oranges.

b) 8 by 8 oranges.

98. Rock Concert If the cost of a ticket to a rock concert is increased by x dollars, the estimated increase in revenue, R, in thousands of dollars is given by the function $R(x) = 24 + 5x - x^2, x < 8$. Find the increase in revenue if the cost of the ticket is increased by

a) $1.

b) $4.

Review Example 5 before working Exercises 99–104.

99. Heart Rate The following graph shows a person's heart rate while doing exercise. Write a story that this graph may represent.

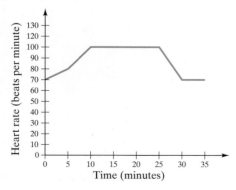

100. Water Level The following graph shows the water level at a certain point during a flood. Write a story that this graph may represent.

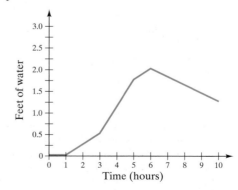

101. Height above Sea Level The following graph shows height above sea level versus time when a man leaves his house and goes for a walk. Write a story that this graph may represent.

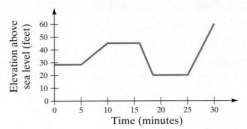

102. Water Level in a Bathtub The following graph shows the level of water in a bathtub versus time. Write a story that this graph may represent.

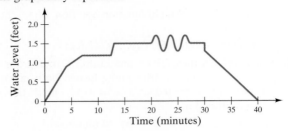

103. Speed of a Car The following graph shows the speed of a car versus time. Write a story that this graph may represent.

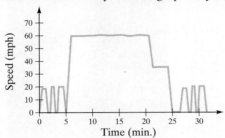

104. Distance Traveled The following graph shows the distance traveled by a person in a car versus time. Write a story that this graph may represent.

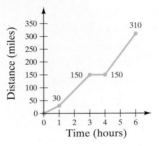

105. Home Prices The following graph compares the median sales price of homes in the United States and in California's zip code 95129.

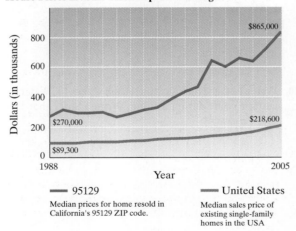

House Prices in 95129 Area Top U.S. Average

- 95129
  Median prices for home resold in California's 95129 ZIP code.
- United States
  Median sales price of existing single-family homes in the USA

Source: DataQuick Information Systems, San Diego: National Association of Realtors, USA Today (8/2/05)

a) Do both lines shown represent functions? Explain.

b) In this graph, what is the independent variable?

c) If f represents the average sales price of the homes in the United States, determine $f(2005)$.

d) If g represents the average sales price in the 95129 zip code, determine $g(2005)$.

e) Determine the percent increase in sales price of a single family home in the United States from 1988 to 2005.

106. College Savings Plans The 529 college saving plans have increased in number in the United States from 2002 to 2005, as illustrated in the following graph.

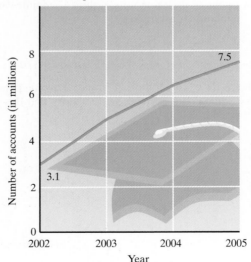

529 Plans are Popular

Source: College Savings Plans Network, *USA Today* (8/9/05)

a) Does this graph represent a function? Explain.

b) In this graph, what is the dependent variable?

c) If n represents the number of 529 plans, determine $n(2005)$.

d) Determine the percent increase in the number of 529 plans from 2002 to 2005.

107. Morning Shows The following graph shows the number of viewers of *The Today Show* (NBC) and *Good Morning America* (ABC) from the 1992–1993 season to the 2004–2005 season.

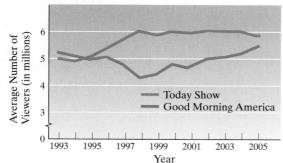

Morning Show Viewers

Source: Nielsen Media Reasearch, *New York Times* (8/9/05)

a) Do both lines represent functions? Explain.

b) If *f* represents the number of viewers of *The Today Show*, estimate $f(1998)$.

c) If *g* represents the number of viewers of *Good Morning America*, estimate $g(1998)$.

d) Do both lines appear to be approximately linear from 1998 to 2005? Explain.

e) If this trend continues, estimate when the two shows will have the same number of viewers.

108. Shipments of LCD monitors Shipments of LCD monitors are expected to grow in the years to come. The following graph shows the shipments of LCD monitors, in millions of units, for the years from 2002 to 2008.

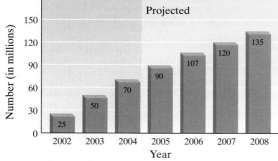

Shipments of LCD Monitors

Source: DisplaySearch, Market Intelligence Center, *Wall Street Journal* (3/24/05)

a) Draw a line graph that displays this information.

b) Does the graph you drew in part **a)** appear to be approximately linear? Explain.

c) Assuming this trend continues, from the line graph you drew, estimate the number of LCD monitors to be shipped in 2009.

d) Does the bar graph represent a function?

e) Does the line graph you drew in part **a)** represent a function?

109. Super Bowl Commercials The average price of the cost of a 30-second commercial during the Super Bowl has been increasing over the years. The following chart gives the approximate cost of a 30-second commercial for selected years from 1981 through 2005.

Year	Cost ($1000s)
1981	280
1985	500
1989	740
1993	970
1997	1200
2001	2000
2005	2400

a) Draw a line graph that displays this information.

b) Does the graph appear to be approximately linear? Explain.

c) From the graph, estimate the cost of a 30-second commercial in 2004.

110. Household Expenditures The average annual household expenditure is a function of the average annual household income. The average expenditure can be estimated by the function

$$f(i) = 0.6i + 5000 \quad \$3500 \le i \le \$50,000$$

where $f(i)$ is the average household expenditure and *i* is the average household income.

a) Draw a graph showing the relationship between average household income and the average household expenditure.

b) Estimate the average household expenditure for a family whose average household income is $30,000.

111. Supply and Demand The price of commodities, like soybeans, is determined by **supply and demand**. If too many soybeans are produced, the supply will be greater than the demand, and the price will drop. If not enough soybeans are produced, the demand will be greater than the supply, and the price of soybeans will rise. Thus the price of soybeans is a function of the number of bushels of soybeans produced. The price of a bushel of soybeans can be estimated by the function

$$f(Q) = -0.00004Q + 4.25, \quad 10,000 \le Q \le 60,000$$

where $f(Q)$ is the price of a bushel of soybeans and *Q* is the annual number of bushels of soybeans produced.

a) Construct a graph showing the relationship between the number of bushels of soybeans produced and the price of a bushel of soybeans.

b) Estimate the cost of a bushel of soybeans if 40,000 bushels of soybeans are produced in a given year.

Group Activity

*In many real-life situations, more than one function may be needed to represent a problem. This often occurs where two or more different rates are involved. For example, when discussing federal income taxes, there are different tax rates. When two or more functions are used to represent a problem, the function is called a **piecewise function**. Following are two examples of piecewise functions and their graphs.*

$$f(x) = \begin{cases} -x + 2, & 0 \le x < 4 \\ 2x - 10, & 4 \le x < 8 \end{cases}$$

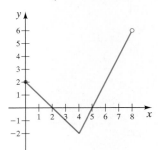

$$f(x) = \begin{cases} 2x - 1, & -2 \le x < 2 \\ x - 2, & 2 \le x < 4 \end{cases}$$

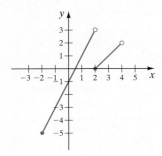

As a group, graph the following piecewise functions.

112. $f(x) = \begin{cases} x + 3, & -1 \le x < 2 \\ 7 - x, & 2 \le x < 4 \end{cases}$

113. $g(x) = \begin{cases} 2x + 3, & -3 < x < 0 \\ -3x + 1, & 0 \le x < 2 \end{cases}$

Cumulative Review Exercises

[2.5] **114.** Solve $3x - 2 = \dfrac{1}{3}(3x - 3)$.

[2.6] **115.** Solve the following formula for p_2.

$$E = a_1 p_1 + a_2 p_2 + a_3 p_3$$

[4.4] **116.** Write the equation $3x + 6y = 9$ in slope-intercept form and indicate the slope and the y-intercept.

[7.2] **117.** Divide $\dfrac{3x^2 - 16x - 12}{3x^2 - 10x - 8} \div \dfrac{x^2 - 7x + 6}{3x^2 - 11x - 4}$.

8.3 Linear Functions

1 Graph linear functions.

2 Graph linear functions using intercepts.

3 Study applications of functions.

4 Solve linear equations in one variable graphically.

1 Graph Linear Functions

Now that we have defined functions, we can discuss linear functions. Consider the equation $y = 2x + 4$ whose graph is illustrated in **Figure 8.23**. Notice that this graph passes the vertical line test and therefore represents a function. Since the graph is linear and is a function, we refer to it as a **linear function**.

A **linear function** is a function of the form $f(x) = ax + b$. The graph of any linear function is a straight line. The domain of any function is the set of real numbers for which the function is a real number. The domain of any linear function is the set of all real numbers, $\mathbb{R}$. Any real number, x, substituted in a linear function will result in $f(x)$ being a real number. We will discuss domains of functions further in Section 8.5.

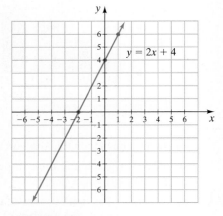

FIGURE 8.23

To graph a linear function, we treat $f(x)$ as y and follow the same procedure used to graph linear equations.

EXAMPLE 1 ▸ Graph $f(x) = \dfrac{1}{2}x - 1$.

Solution   We construct a table of values by substituting values for x and finding corresponding values of $f(x)$ or y. Then we plot the points and draw the graph, as illustrated in **Figure 8.24**.

x	$f(x)$
-2	-2
0	-1
2	0

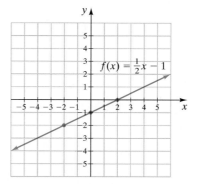

FIGURE 8.24

▸ **Now Try Exercise 9**

Note that the vertical axis in **Figure 8.24** may also be labeled as $f(x)$ instead of y. In this book we will continue to label it y.

2 Graph Linear Functions Using Intercepts

In Section 4.2 we discussed graphing linear equations using their intercepts. In the box below we refresh your memory on how to find the intercepts.

To Find the x- and y-Intercepts

To find the y-intercept, set $x = 0$ and solve for y.

To find the x-intercept, set $y = 0$ and solve for x.

To graph linear functions using intercepts just remember that $f(x)$ is the same as y. Example 2 explains how to graph a linear functions using its intercepts.

EXAMPLE 2 ▸ Graph $f(x) = -\dfrac{1}{3}x - 1$ using the x- and y-intercepts.

Solution   Treat $f(x)$ the same as y. To find the y-intercept, set $x = 0$ and solve for $f(x)$.

$$f(x) = -\frac{1}{3}x - 1$$

$$f(x) = -\frac{1}{3}(0) - 1 = -1$$

The y-intercept is $(0, -1)$.

To find the x-intercept, set $f(x) = 0$ and solve for x.

$$f(x) = -\frac{1}{3}x - 1$$

$$0 = -\frac{1}{3}x - 1$$

$$3(0) = 3\left(-\frac{1}{3}x - 1\right) \qquad \textit{Multiply both sides by 3.}$$

$$0 = -x - 3 \qquad \textit{Distributive property}$$

$$x = -3 \qquad \textit{Add x to both sides.}$$

The x-intercept is $(-3, 0)$. The graph is shown in **Figure 8.25**.

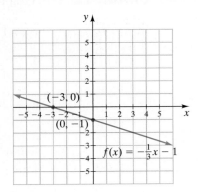

FIGURE 8.25

▸ **Now Try Exercise 15**

USING YOUR GRAPHING CALCULATOR

Sometimes it may be difficult to estimate the intercepts of a graph accurately. When this occurs, you might want to use a graphing calculator. We demonstrate how in the following example.

EXAMPLE   Determine the x- and y-intercepts of the graph of $y = 1.3(x - 3.2)$.

Solution   Press the Y= key, and then assign $1.3(x - 3.2)$ to Y_1. Then press the GRAPH key to graph the function $y = 1.3(x - 3.2)$, as shown in **Figure 8.26**.

From the graph it may be difficult to determine the intercepts. One way to find the y-intercept is to use the TRACE feature. **Figure 8.27** shows a TI-84 Plus screen after the TRACE key is pressed. Notice the y-intercept is at -4.16.

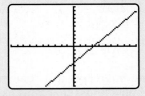

FIGURE 8.26

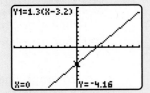

FIGURE 8.27

Some graphing calculators have the ability to find the x-intercepts of a function by pressing just a few keys. A **zero** (or **root**) of a function is a value of x such that $f(x) = 0$. A zero (or root) of a function is the x-coordinate of the x-intercept of the graph of the function. Read your calculator manual to learn how to find the zeros or roots of a function. On a TI-84 Plus you press the keys 2nd TRACE to get to the CALC menu (which stands for calculate). Then you choose option 2, *zero*. Once the zero feature has been selected, the calculator will display

Left bound?

At this time, move the cursor along the curve until it is to the *left* of the zero. Then press ENTER . The calculator now displays

Right bound?

Move the cursor along the curve until it is to the *right* of the zero. Then press ENTER . The calculator now displays

Guess?

Now press ENTER for the third time and the zero is displayed at the bottom of the screen, as in **Figure 8.28**. Thus the x-intercept of the function is at 3.2. For practice at finding the intercepts on your calculator, work Exercises 49–52.

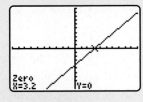

FIGURE 8.28

3 Study Applications of Functions

Graphs are often used to show the relationship between variables. The axes of a graph do not have to be labeled x and y. They can be any designated variables. Consider the following example.

EXAMPLE 3 ▸ **Tire Store Profit** The yearly profit, p, of a tire store can be estimated by the function $p(n) = 20n - 30,000$, where n is the number of tires sold per year.

a) Draw a graph of profit versus tires sold for up to and including 6000 tires.

b) Estimate the number of tires that must be sold for the company to break even.

c) Estimate the number of tires sold if the company has a $70,000 profit.

Solution **a)** Understand The profit, p, is a function of the number of tires sold, n. The horizontal axis will therefore be labeled Number of tires sold (the independent variable) and the vertical axis will be labeled Profit (the dependent variable). Since the minimum number of tires that can be sold is 0, negative values do not have to be listed on the horizontal axis. The horizontal axis will therefore go from 0 to 6000 tires. We will graph this equation by determining and plotting the intercepts.

Translate and Carry Out To find the p-intercept, we set $n = 0$ and solve for $p(n)$.

$$p(n) = 20n - 30,000$$
$$p(n) = 20(0) - 30,000 = -30,000$$

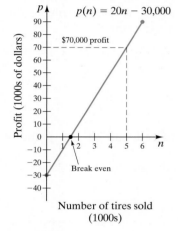

FIGURE 8.29

Thus, the p-intercept is $(0, -30,000)$.

To find the n-intercept, we set $p(n) = 0$ and solve for n.

$$p(n) = 20n - 30,000$$
$$0 = 20n - 30,000$$
$$30,000 = 20n$$
$$1500 = n$$

Thus the n-intercept is $(1500, 0)$.

Answer Now we use the p- and n-intercepts to draw the graph (see **Fig. 8.29**).
b) The break-even point is the number of tires that must be sold for the company to have neither a profit nor a loss. The break-even point is where the graph intersects the n-axis, for this is where the profit, p, is 0. To break even, approximately 1500 tires must be sold.
c) To make $70,000, approximately 5000 tires must be sold (shown by the dashed red line in **Fig. 8.29**).

▸ **Now Try Exercise 31**

Sometimes it is difficult to read an exact answer from a graph. To determine the exact number of tires needed to break even in Example 3, substitute 0 for $p(n)$ in the function $p(n) = 20n - 30,000$ and solve for n. To determine the exact number of tires needed to obtain a $70,000 profit, substitute 70,000 for $p(n)$ and solve the equation for n.

EXAMPLE 4 ▸ **Toy Store Sales** Andrew Gestrich is the owner of a toy store. His monthly salary consists of $200 plus 10% of the store's sales for that month.

a) Write a function expressing his monthly salary, m, in terms of the store's sales, s.

b) Draw a graph of his monthly salary for sales up to and including $20,000.

c) If the store's sales for the month of April are $15,000, what will Andrew's salary be for April?

s	m
0	200
10,000	1200
20,000	2200

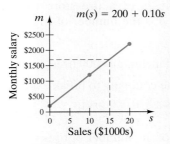

FIGURE 8.30

Solution

a) Andrew's monthly salary is a function of sales. His monthly salary, m, consists of $200 plus 10% of the sales, s. Ten percent of s is $0.10s$. Thus the function for finding his salary is

$$m(s) = 200 + 0.10s$$

b) Since monthly salary is a function of sales, sales will be represented on the horizontal axis and monthly salary will be represented on the vertical axis. Since sales can never be negative, the monthly salary can never be negative. Thus both axes will be drawn with only positive numbers. We will draw this graph by plotting points. We select values for s, find the corresponding values of m, and then draw the graph. We can select values of s that are between $0 and $20,000 (**Fig. 8.30**).

c) By reading our graph carefully, we can estimate that when the store's sales are $15,000, Andrew's monthly salary is about $1700.

▸ **Now Try Exercise 33**

4 Solve Linear Equations in One Variable Graphically

Earlier we discussed the graph of $f(x) = 2x + 4$. In **Figure 8.31** below we illustrate the graph of $f(x)$ along with the graph of $g(x) = 0$. Notice that the two graphs intersect at $(-2, 0)$. We can obtain the x-coordinate of the ordered pair by solving the equation $f(x) = g(x)$. Remember $f(x)$ and $g(x)$ both represent y, and by solving this equation for x we are obtaining the value of x where the y's are equal.

$$f(x) = g(x)$$
$$\overbrace{2x + 4} = \overbrace{0}$$
$$2x = -4$$
$$x = -2$$

Note that we obtain -2, the x-coordinate in the ordered pair at the point of intersection.

Now let's find the x-coordinate of the point at which the graphs of $f(x) = 2x + 4$ and $g(x) = 2$ intersect. We solve the equation $f(x) = g(x)$.

$$f(x) = g(x)$$
$$\overbrace{2x + 4} = \overbrace{2}$$
$$2x = -2$$
$$x = -1$$

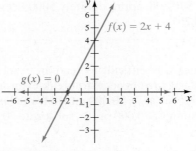

FIGURE 8.31

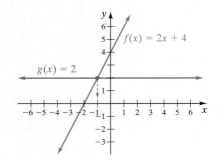

FIGURE 8.32

The x-coordinate of the point of intersection of the two graphs is -1, as shown in **Figure 8.32**. Notice that $f(-1) = 2(-1) + 4 = 2$.

In general, if we are given an equation in one variable, we can regard each side of the equation as a separate function. To obtain the solution to the equation, we can graph the two functions. The x-coordinate of the point of intersection will be the solution to the equation.

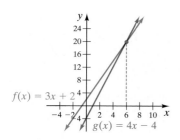

FIGURE 8.33

EXAMPLE 5 ▶ Solve the equation $3x + 2 = 4x - 4$ graphically.

Solution Let $f(x) = 3x + 2$ and $g(x) = 4x - 4$. The graph of these functions is illustrated in **Figure 8.33**. The x-coordinate of the point of intersection is 6. Thus, the solution to the equation is 6. Check the solution now.

▶ **Now Try Exercise 45**

USING YOUR GRAPHING CALCULATOR

In Example 5, we solved an equation in one variable by graphing two functions. In the following example, we explain how to find the point of intersection of two functions on a graphing calculator.

EXAMPLE Use a graphing calculator to find the solution to $2(x + 3) = \frac{1}{2}x + 4$.

Solution Assign $2(x + 3)$ to Y_1 and assign $\frac{1}{2}x + 4$ to Y_2 to get

$$Y_1 = 2(x + 3)$$
$$Y_2 = \frac{1}{2}x + 4$$

Now press the $\boxed{\text{GRAPH}}$ key to graph the functions. The graph of the functions is shown in **Figure 8.34**.

By examining the graph can you determine the x-coordinate of the point of intersection? Is it -1, or -1.5, or some other value? We can determine the point of intersection in a number of different ways. One method involves using the TRACE and ZOOM features. **Figure 8.35** shows the window of a TI-84 Plus after the TRACE feature has been used and the cursor has been moved close to the point of intersection. (Note that pressing the up and down arrows switches the cursor from one function to the other.)

At the bottom of the screen in **Figure 8.35**, you see the x- and y-coordinates at the cursor. To get a closer view around the area of the cursor, you can *zoom in* using the $\boxed{\text{ZOOM}}$ key. After you zoom in, you can move the cursor closer to the point of intersection and get a better reading (**Fig. 8.36**). You can do this over and over until you get as accurate an answer as you need. It appears from **Figure 8.36** that the x-coordinate of the intersection is about -1.33.

Graphing calculators can also display the intersection of two graphs with the use of certain keys. The keys to press depend on your calculator. Read your calculator manual to determine how to do this. This procedure is generally quicker and easier to use to find the point of intersection of two graphs.

On the TI-84 Plus, select option 5: INTERSECT from the CALC menu to find the intersection. Once the INTERSECT feature has been selected, the calculator will display

First curve?

At this time, move the cursor along the first curve until it is close to the point of intersection. Then press $\boxed{\text{ENTER}}$. The calculator will next display

Second curve?

The cursor will then appear on the second curve. If the cursor is not close to the point of intersection, move it along this curve until it is close to the intersection. Then press $\boxed{\text{ENTER}}$. Next the calculator will display

Guess?

Now press $\boxed{\text{ENTER}}$ again, and the point of intersection will be displayed.

Figure 8.37 shows the window after this procedure has been done. We see that the x-coordinate of the point of intersection is $-1.333\ldots$ or $-1\frac{1}{3}$ and the y-coordinate of the point of intersection is $3.333\ldots$ or $3\frac{1}{3}$.

For practice in using a graphing calculator to solve an equation in one variable, work Exercises 45–48.

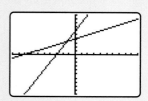

FIGURE 8.34

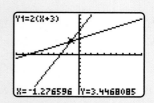

FIGURE 8.35

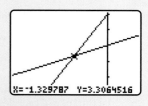

FIGURE 8.36

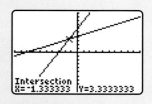

FIGURE 8.37

EXERCISE SET 8.3
Math XL MathXL®   *MyMathLab* MyMathLab

Concept/Writing Exercises

1. What does the graph of a linear function look like?

2. If you are given a linear equation in standard form, and wish to write the equation using function notation, how would you do it?

3. Explain how to find the x- and y-intercepts of the graph of a linear function.

4. What terms do graphing calculators use to indicate the x-intercepts?

5. Explain how to solve an equation in one variable graphically.

6. Explain how to solve the equation $2(x - 1) = 3x - 5$ graphically.

Practice the Skills

Graph each function by plotting points.

7. $f(x) = 3x - 2$

8. $g(x) = -2x + 3$

9. $g(x) = -\frac{2}{3}x + 4$

10. $f(x) = \frac{1}{2}x - 6$

11. $f(x) = \frac{3}{4}x + 1$

12. $g(x) = 6x - 6$

Graph each function using the x- and y-intercepts.

13. $f(x) = 3x - 6$

14. $g(x) = -2x + 6$

15. $f(x) = 2x + 3$

16. $f(x) = -6x + 5$

17. $g(x) = 4x - 8$

18. $p(x) = -\frac{3}{4}x + 3$

19. $s(x) = \frac{4}{3}x + 3$

20. $g(x) = -\frac{1}{4}x + 2$

21. $h(x) = -\frac{6}{5}x + 2$

22. $f(x) = \frac{1}{2}x - 2$

23. $g(x) = -\frac{1}{2}x + 2$

24. $p(x) = -\frac{1}{4}x - 6$

25. $w(x) = \frac{1}{3}x - 2$

26. $h(x) = -\frac{1}{3}x + 4$

27. $s(x) = -\frac{4}{3}x + 48$

28. $d(x) = -\frac{1}{3}x - 2$

Problem Solving

29. **Distance** Using the distance formula

$$\text{distance} = \text{rate} \cdot \text{time, or } d = rt$$

draw a graph of distance versus time for a constant rate of 30 miles per hour.

30. **Simple Interest** Using the simple interest formula

$$\text{interest} = \text{principal} \cdot \text{rate} \cdot \text{time, or } i = prt$$

draw a graph of interest versus time for a principal of $1000 and a rate of 3%.

31. **Bicycle Profit** The profit of a bicycle manufacturer can be approximated by the function $p(x) = 60x - 80,000$, where x is the number of bicycles produced and sold.

a) Draw a graph of profit versus the number of bicycles sold (for up to and including 5000 bicycles).

b) Estimate the number of bicycles that must be sold for the company to break even.

c) Estimate the number of bicycles that must be sold for the company to make $150,000 profit.

32. **Taxi Operating Costs** Raul Lopez's weekly cost of operating a taxi is $75 plus 15¢ per mile.

a) Write a function expressing Raul's weekly cost, c, in terms of the number of miles, m.

b) Draw a graph illustrating weekly cost versus the number of miles, for up to and including 200, driven per week.

c) If during 1 week, Raul drove the taxi 150 miles, what would be the cost?

d) How many miles would Raul have to drive for the weekly cost to be $135?

33. **Salary Plus Commission** Jayne Haydack's weekly salary at Charter Network is $500 plus 15% commission on her weekly sales.

a) Write a function expressing Jayne's weekly salary, s, in terms of her weekly sales, x.

b) Draw a graph of Jayne's weekly salary versus her weekly sales, for up to and including $5000 in sales.

c) What is Jayne's weekly salary if her sales were $3000?

d) If Jayne's weekly salary for the week was $1100, what were her weekly sales?

34. Salary Plus Commission Lynn Hicks, a real estate agent, makes $100 per week plus a 3% sales commission on each property she sells.

a) Write a function expressing her weekly salary, s, in terms of sales, x.

b) Draw a graph of her salary versus her weekly sales, for sales up to $100,000.

c) If she sells one house per week for $75,000, what will her weekly salary be?

35. Weight of Girls The following graph shows weight, in kilograms, for girls (up to 36 months of age) versus length (or height), in centimeters. The red line is the average weight for all girls of the given length, and the green lines represent the upper and lower limits of the normal range.

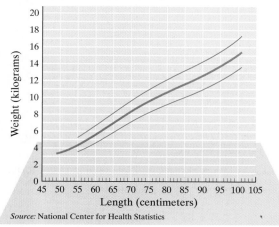

Girls: Birth to 36 Months Physical Growth

Source: National Center for Health Statistics

a) Explain why the red line represents a function.

b) What is the independent variable? What is the dependent variable?

c) Is the graph of weight versus length approximately linear?

d) What is the weight in kilograms of the average girl who is 85 centimeters long?

e) What is the average length in centimeters of the average girl with a weight of 7 kilograms?

f) What weights are considered normal for a girl 95 centimeters long?

g) What is happening to the normal range as the lengths increase? Is this what you would expect to happen? Explain.

36. Compound Interest The following graph shows the effect of compound interest.

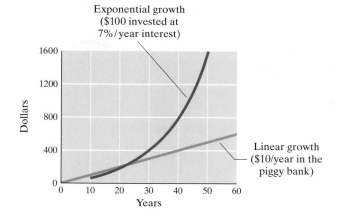

If a child puts $10 each year in a piggy bank, the savings will grow linearly, as shown by the lower curve. If, at age 10, the child invests $100 at 7% interest compounded annually, that $100 will grow exponentially.

a) Explain why both graphs represent functions.

b) What is the independent variable? What is the dependent variable?

c) Using the linear growth curve, determine how long it would take to save $600.

d) Using the exponential growth curve, which begins at year 10, determine how long after the account is opened would the amount reach $600?

e) Starting at year 20, how long would it take for the money growing at a linear rate to double?

f) Starting at year 20, how long would it take for the money growing exponentially to double? (Exponential growth will be discussed at length in Chapter 9.)

37. When, if ever, will the x- and y-intercepts of a graph be the same? Explain.

38. Write two linear functions whose x- and y-intercepts are both $(0, 0)$.

39. Write a function whose graph will have no x-intercept but will have a y-intercept at $(0, 4)$.

40. Write an equation whose graph will have no y-intercept but will have an x-intercept at -5.

41. If the x- and y-intercepts of a linear function are at 1 and -3, respectively, what will be the new x- and y-intercepts if the graph is moved (or translated) up 3 units?

42. If the x- and y-intercepts of a linear function are -1 and 3, respectively, what will be the new x- and y-intercepts if the graph is moved (or translated) down 4 units?

*In Exercises 43 and 44, we give two ordered pairs, which are on a graph. **a)** Plot the points and draw the line through the points. **b)** Find the change in y, or the vertical change, between the points. **c)** Find the change in x, or the horizontal change, between the points. **d)** Find the ratio of the vertical change to the horizontal change between these two points. Do you know what this ratio represents? (We will discuss this further in Section 8.4.)*

43. $(0, 2)$ and $(-4, 0)$

44. $(3, 5)$ and $(-1, -1)$

Solve each equation for x as done in Example 5. Use a graphing calculator if one is available. If not, draw the graphs yourself.

45. $2x + 5 = 8x - 1$

46. $3(x + 2) + 1 = 2(x - 1) + 7$

47. $0.3(x + 5) = -0.6(x + 2)$

48. $2x + \dfrac{1}{4} = 5x - \dfrac{1}{2}$

Find the x- and y-intercepts of the graph of each equation using your graphing calculator.

49. $y = 2(x + 3.2)$

50. $5x - 2y = 7$

51. $-4x - 3.2y = 8$

52. $y = \dfrac{3}{5}x - \dfrac{1}{2}$

Cumulative Review Exercises

[5.4]   **53.** Add $7x^2 - 3x - 100$ and $-4x^2 - 9x + 12$.

[6.4]   **54.** Factor $3x^2 - 12x - 96$.

[7.1]   **55.** Simplify $\dfrac{x - 2}{x^2 - 4}$.

[7.6]   **56.** Solve $x + \dfrac{24}{x} = 10$.

Mid-Chapter Test: 8.1–8.3

To find out how well you understand the chapter material to this point, take this brief test. The answers, and the section where the material was initially discussed, are given in the back of the book. Review any questions you answered incorrectly.

Graph each equation.

1. $y = 3x + 2$

2. $y = -x^2 + 3$

3. $y = |x| - 4$

4. $y = \sqrt{x - 4}$

5. a) What is a relation?

    b) What is a function?

    c) Is every relation a function? Explain.

    d) Is every function a relation? Explain.

In Exercises 6–8, determine which of the following relations are also functions. Give the domain and range of each relation or function.

6. $\{(1, 5), (2, -3), (7, -1), (-5, 6)\}$

7.

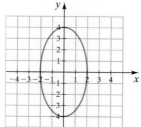

8.

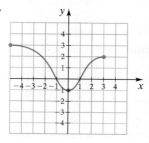

9. List $E = \{x \mid x$ is a whole multiple of 9$\}$ in roster form.

10. Illustrate $\{z \mid z \le -5\}$ on a number line.

11. If $g(x) = 2x^2 + 8x - 13$, find $g(-2)$.

12. The height, h, in feet of an apple thrown from the top of a building is

$$h(t) = -6t^2 + 3t + 150$$

where t is time in seconds. Find the height of the apple 3 seconds after it is thrown.

Graph each function.

13. $f(x) = 2x + 3$

14. $g(x) = \dfrac{1}{2}x - 4$

15. Graph $m(x) = -\dfrac{1}{3}x + 2$ using the x- and y- intercepts.

16. Profit The daily profit, in dollars, for a shoe company is $p(x) = 30x - 660$, where x is the number of pairs of shoes manufactured and sold.

    a) Draw a graph of profit versus the number of pairs of shoes sold (for up to 40 pairs).

    b) Determine the number of pairs of shoes that must be sold for the company to break even.

    c) Determine the number of pairs of shoes that must be sold for the company to make a daily profit of $360.

8.4 Slope, Modeling, and Linear Relationships

1 Recognize slope as a rate of change.

2 Use the slope-intercept form to construct models from graphs.

3 Use the point-slope form to construct models from graphs.

4 Recognize vertical translations.

5 Use slope to identify and construct perpendicular lines.

We introduced slope, the slope-intercept form of a linear equation, and the point-slope form of a linear equation in Sections 4.3 and 4.4. In this section, we briefly review these topics and then we will expand upon what was discussed in Chapter 4.

1 Recognize Slope as a Rate of Change

Recall from Section 4.3 that the slope of a line is the ratio of the vertical change to the horizontal change between any two selected points on the line.

Slope

The **slope** of the line through the distinct points (x_1, y_1) and (x_2, y_2) is

$$\text{slope} = \frac{\text{change in } y \text{ (vertical change)}}{\text{change in } x \text{ (horizontal change)}} = \frac{y_2 - y_1}{x_2 - x_1}$$

provided that $x_1 \neq x_2$.

It makes no difference which two points on the line are selected when finding the slope of a line. It also makes no difference which point you label (x_1, y_1) or (x_2, y_2). As mentioned before, the letter m is used to represent the slope of a line. The Greek capital letter delta, Δ, is used to represent the words *the change in*. Thus, the slope is sometimes indicated as

$$m = \frac{\Delta y}{\Delta x} = \frac{y_2 - y_1}{x_2 - x_1}$$

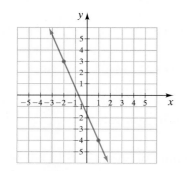

FIGURE 8.38

EXAMPLE 1 ▶ Find the slope of the line in **Figure 8.38**.

Solution   Two points on the line are $(-2, 3)$ and $(1, -4)$. Let $(x_2, y_2) = (-2, 3)$ and $(x_1, y_1) = (1, -4)$. Then

$$m = \frac{y_2 - y_1}{x_2 - x_1} = \frac{3 - (-4)}{-2 - 1} = \frac{3 + 4}{-3} = -\frac{7}{3}$$

The slope of the line is $-\dfrac{7}{3}$. Note that if we had let $(x_1, y_1) = (-2, 3)$ and $(x_2, y_2) = (1, -4)$, the slope would still be $-\dfrac{7}{3}$. Try it and see.

▶ **Now Try Exercise 11**

Sometimes it is helpful to describe slope as a *rate of change*. Consider a slope of $\dfrac{5}{3}$. This means that the y-value increases 5 units for each 3-unit increase in x. Equivalently, we can say that the y-value increases $\dfrac{5}{3}$ units, or $1.\overline{6}$ units, for each 1-unit increase in x.

When we give the change in y per unit change in x we are giving the slope as a **rate of change**. When discussing real-life situations or when creating mathematical models, it is often useful to discuss slope as a rate of change.

EXAMPLE 2 ▸ Public Debt The following table of values and the corresponding graph (**Fig. 8.39**) illustrate the U.S. public debt in billions of dollars from 1910 through 2005.

Year	U.S. Public Debt (billions of dollars)
1910	1.1
1930	16.1
1950	256.1
1970	370.1
1990	3323.3
2002	5957.2
2005	7832.6

Source: U.S. Dept. of the Treasury, Bureau of Public Debt.

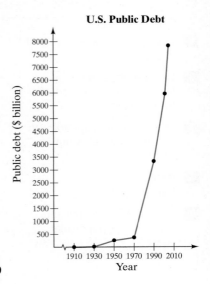

U.S. Public Debt

FIGURE 8.39

a) Determine the slope of the line segments between 1910 and 1930 and between 2002 and 2005.

b) Compare the two slopes found in part **a)** and explain what this means in terms of the U.S. public debt.

Solution Understand **a)** To find the slope between any 2 years, find the ratio of the change in debt to the change in years.

Slope from 1910 to 1930

$$m = \frac{16.1 - 1.1}{1930 - 1910} = \frac{15}{20} = 0.75$$

The U.S. public debt from 1910 to 1930 increased at a rate of $0.75 billion per year.

Slope from 2002 to 2005

$$m = \frac{7832.6 - 5957.2}{2005 - 2002} = \frac{1875.4}{3} \approx 625.13$$

The U.S. public debt from 2002 to 2005 increased at a rate of about $625.13 billion per year.

b) Slope measures a rate of change. Comparing the slopes for the two periods shows that there was a much greater increase in the average rate of change in the public debt from 2002 to 2005 than from 1910 to 1930. The slope of the line segment from 2002 to 2005 is greater than the slope of any other line segment on the graph. This indicates that the public debt from 2002 to 2005 grew at a faster rate than at any other time period illustrated.

▸ **Now Try Exercise 51**

Since the slope of a line can be a positive number, a negative number, or zero, the rate of change can be increasing, decreasing, or zero. We will now consider a function that has a zero rate of change.

Many amusement parks charge a flat fee to enter the park for the day. The cost for a customer is the same regardless of the number of rides the customer takes. If the daily admission is $34.00, the cost, c, as a function of the number of rides can be represented as $c(x) = 34$. The graph of this function is illustrated in **Figure 8.40** on the next page. Since this graph is a horizontal line, and the slope of any horizontal line is 0, the slope of the graph of the function is 0. Let us select two points on the graph $(10, 34)$

and $(30, 34)$. If we determine the slope by the formula, we find the slope is indeed 0. Let $(x_2, y_2) = (30, 34)$ and $(x_1, y_1) = (10, 34)$.

$$m = \frac{y_2 - y_1}{x_2 - x_1} = \frac{34 - 34}{30 - 10} = \frac{0}{20} = 0$$

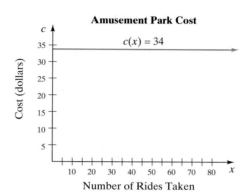

FIGURE 8.40

Notice that this function, $c(x) = 34$, is defined using only a constant. It is called a **constant function**. Any constant function will have the form $f(x) = b$ and its graph will be a horizontal line whose slope is 0.

2 Use the Slope-Intercept Form to Construct Models from Graphs

Earlier we introduced the slope-intercept form of a linear equation. We repeat it again for your convenience.

Slope-Intercept Form

The **slope-intercept form of a linear equation** is

$$y = mx + b$$

where **m is the slope** of the line and **$(0, b)$ is the y-intercept** of the line.

Examples of Equations in Slope-Intercept Form

$$y = 3x - 6 \qquad y = \frac{1}{2}x + \frac{3}{2}$$

Slope ⟶ ↓   ↓ ⟵ y-intercept is $(0, b)$

$$y = mx + b$$

Equation	Slope	y-Intercept
$y = 3x - 6$	3	$(0, -6)$
$y = \frac{1}{2}x + \frac{3}{2}$	$\frac{1}{2}$	$\left(0, \frac{3}{2}\right)$

Writing an Equation in Slope-Intercept Form

To write an equation in slope-intercept form, solve the equation for y.

Often we can use the slope-intercept form of a linear equation to determine a function that models a real-life situation. Example 3 shows how this may be done.

EXAMPLE 3 ▶ **Newspapers** Consider the purple graph in **Figure 8.41**, which shows the declining number of adults who read the daily newspaper. Notice that the graph is somewhat linear. The dashed red line is a linear function which was drawn to approximate the purple graph.

a) Write a linear function to represent the dashed red line.

b) Assuming this trend continues, use the function determined in part **a)** to estimate the percent of adults who will read a newspaper in 2012.

Percentage of U.S. Adults Who Read a Newspaper

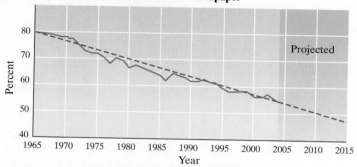

FIGURE 8.41   *Source*: NAA Market & Business Analysis; *Newsweek* Projection, *The Washington Post* (2/20/05)

Solution

a) To make the numbers easier to work with, we will select 1965 as a *reference year*. Then we can replace 1965 with 0, 1966 with 1, 1967 with 2, and so on. Then 2004 would be 39 and 2005 would be 40 (see **Fig. 8.42**).

Percentage of U.S. Adults Who Read a Newspaper

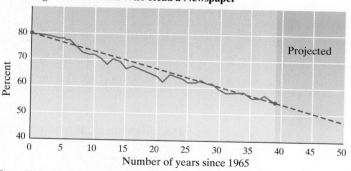

FIGURE 8.42   *Source*: NAA Market & Business Analysis; *Newsweek* Projection, *The Washington Post* (2/20/05)

We will select two points an the graph that will allow us to find the slope of the graph. If we call the vertical axis y and the horizontal axis x, then the y-intercept is 80. Thus, one point on the graph is $(0, 80)$. In 2004, or year 39 in **Figure 8.42**, it appears that about 55% of the adult population read a daily newspaper. Let's select $(39, 55)$ as a second point on the graph of the straight line in **Figure 8.42**. We designate $(39, 55)$ as (x_2, y_2) and $(0, 80)$ as (x_1, y_1).

$$\text{slope} = \frac{\text{change in percent}}{\text{change in year}} = \frac{y_2 - y_1}{x_2 - x_1} = \frac{55 - 80}{39 - 0} = \frac{-25}{39} \approx -0.641$$

Since the slope is approximately -0.641 and the y-intercept is $(0, 80)$, the equation of the straight line is $y = -0.641x + 80$. This equation in function notation is

$f(x) = -0.641x + 80$. To use this function remember that $x = 0$ represents 1965, $x = 1$ represents 1966, and so on. Note that $f(x)$, the percent, is a function of x, the number of years since 1965.

b) To determine the approximate percent of readers in 2012, and since $2012 - 1965 = 47$, we substitute 47 for x in the function.

$$f(x) = -0.641x + 80$$
$$f(47) = -0.641(47) + 80$$
$$= -30.127 + 80$$
$$= 49.873$$

Thus, if the current trend continues, about 49.9% of adults will read a daily newspaper in 2012.

▸ **Now Try Exercise 55**

3 Use the Point-Slope Form to Construct Models from Graphs

Earlier we introduced the point-slope form of a linear equation. We repeat it here for your convenience.

Point-Slope Form

The **point-slope form of a linear equation** is
$$y - y_1 = m(x - x_1)$$
where m **is the slope** of the line and (x_1, y_1) is a point on the line.

Now let's look at an application where we use the point-slope form to determine a function that models a given situation.

EXAMPLE 4 ▸ **Burning Calories** The number of calories burned in 1 hour riding a bicycle is a linear function of the speed of the bicycle. The average person riding at 12 mph will burn about 564 calories in 1 hour and while riding at 18 mph will burn about 846 calories in 1 hour. This information is shown in **Figure 8.43**.

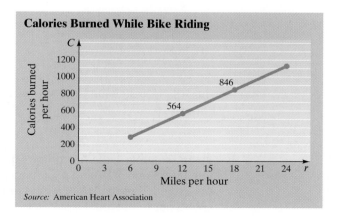

Calories Burned While Bike Riding

Source: American Heart Association

FIGURE 8.43

a) Determine a linear function that can be used to estimate the number of calories, C, burned in 1 hour when a bicycle is ridden at r mph, for $6 \le r \le 24$.

b) Use the function determined in part **a)** to estimate the number of calories burned in 1 hour when a bicycle is ridden at 20 mph.

c) Use the function determined in part **a)** to estimate the speed at which a bicycle should be ridden to burn 800 calories in 1 hour.

Solution   **a) Understand and Translate**   In this example, instead of using the variables x and y as we used in Examples 1 and 3, we use the variables r (for rate or speed) and C (for calories). Regardless of the variables used, the procedure used to determine the equation of the line remains the same. To find the necessary function, we will use the points $(12, 564)$ and $(18, 846)$. We will first calculate the slope and then use the point–slope form to determine the equation of the line.

Carry Out

$$m = \frac{C_2 - C_1}{r_2 - r_1}$$

$$= \frac{846 - 564}{18 - 12} = \frac{282}{6} = 47$$

Now we write the equation using the point-slope form. We will choose the point $(12, 564)$ for (r_1, C_1).

$$C - C_1 = m(r - r_1)$$
$$C - 564 = 47(r - 12) \qquad \textit{Point-slope form}$$
$$C - 564 = 47r - 564$$
$$C = 47r \qquad \textit{Slope-intercept form}$$

Answer   Since the number of calories burned, C, is a function of the rate, r, the function we are seeking is

$$C(r) = 47r$$

b) To estimate the number of calories burned in 1 hour while riding at 20 mph, we substitute 20 for r in the function.

$$C(r) = 47r$$
$$C(20) = 47(20) = 940$$

Therefore, 940 calories are burned while riding at 20 mph for 1 hour.

c) To estimate the speed at which a bicycle should be ridden to burn 800 calories in 1 hour, we substitute 800 for $C(r)$ in the function.

$$C(r) = 47r$$
$$800 = 47r$$
$$\frac{800}{47} = r$$
$$r \approx 17.02$$

Thus the bicycle would need to be ridden at about 17.02 mph to burn 800 calories in 1 hour.

▶ **Now Try Exercise 71**

In Example 4, the function we determined was $C(r) = 47r$. The graph of this function has a slope of 47 and a y-intercept at $(0, 0)$. If the graph in **Figure 8.43** on page 511 was extended to the left, it would intersect the origin. This makes sense since a rate of 0 miles per hour would result in 0 calories being burned by riding in 1 hour.

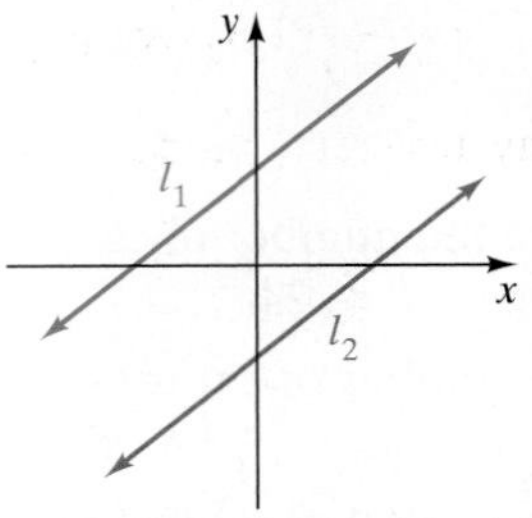

4 Recognize Vertical Translations

We also use slope to identify the relationship between two lines. If the slopes of two lines are the same, we say that the lines are parallel. **Figure 8.44** illustrates two parallel lines. In Chapter 4, we defined parallel lines as lines that never intersect. We found that lines that do not intersect have the same slope.

Parallel lines

FIGURE 8.44

Consider the three linear functions

$$f(x) = 2x + 3$$
$$g(x) = 2x$$
$$h(x) = 2x - 3$$

Each function is graphed in **Figure 8.45**.

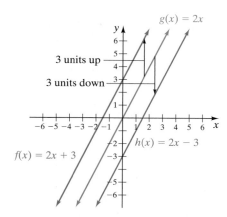

FIGURE 8.45

Since the lines have the same slope, 2, we know that they are parallel and will never meet. What are the y-intercepts of $f(x) = 2x + 3$, $g(x) = 2x$ (or $g(x) = 2x + 0$), and $h(x) = 2x - 3$? The y-intercepts are $(0, 3)$, $(0, 0)$, and $(0, -3)$, respectively. Notice that the graph of $f(x) = 2x + 3$ is the graph of $g(x) = 2x$ shifted, or *translated,* up 3 units, and $h(x) = 2x - 3$ is the graph of $g(x) = 2x$ translated down 3 units. Since the graphs are moved, or translated vertically, we refer to these shifts as **vertical translations**.

When a graph of a linear function is translated up or down by a constant amount, the function has increased or decreased by that constant amount. The graph of the new function is parallel to the graph of the original function.

5 Use Slope to Identify and Construct Perpendicular Lines

Perpendicular lines

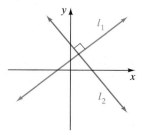

FIGURE 8.46

If two lines are not parallel, they will intersect. When two lines intersect to form a right (or 90°) angle, we say the lines are **perpendicular lines**. **Figure 8.46** illustrates perpendicular lines. The slopes of perpendicular lines have a special relationship.

> **Perpendicular Lines**
>
> Two lines are **perpendicular** when their slopes are *negative reciprocals*.

For any nonzero number a, its **negative reciprocal** is $\dfrac{-1}{a}$ or $-\dfrac{1}{a}$. For example, the negative reciprocal of 2 is $\dfrac{-1}{2}$ or $-\dfrac{1}{2}$. The product of any nonzero number and its negative reciprocal is -1.

$$a\left(-\frac{1}{a}\right) = -1$$

Note that any vertical line is perpendicular to any horizontal line even though the negative reciprocal cannot be applied. (Why not?)

EXAMPLE 5 ▶ Two points on l_1 are $(8, 5)$ and $(4, -1)$. Two points on l_2 are $(0, 2)$ and $(6, -2)$. Determine whether l_1 and l_2 are parallel lines, perpendicular lines, or neither.

Solution   Determine the slopes of l_1 and l_2.

$$m_1 = \frac{5 - (-1)}{8 - 4} = \frac{6}{4} = \frac{3}{2} \qquad m_2 = \frac{2 - (-2)}{0 - 6} = \frac{4}{-6} = -\frac{2}{3}$$

Since their slopes are different, l_1 and l_2 are not parallel. To see whether the lines are perpendicular, we need to determine whether the slopes are negative reciprocals. If $m_1 m_2 = -1$, the slopes are negative reciprocals and the lines are perpendicular.

$$m_1 m_2 = \frac{3}{2}\left(-\frac{2}{3}\right) = -1$$

Since the product of the slopes equals -1, the lines are perpendicular.

▶ **Now Try Exercise 19**

EXAMPLE 6 ▶ Consider the equation $2x + 4y = 8$. Determine the equation of the line that has a y-intercept of 5 and is **a)** parallel to the given line and **b)** perpendicular to the given line.

Solution

a) If we know the slope of a line and its y-intercept, we can use the slope-intercept form, $y = mx + b$, to write the equation. We begin by solving the given equation for y.

$$2x + 4y = 8$$
$$4y = -2x + 8$$
$$y = \frac{-2x + 8}{4}$$
$$y = -\frac{1}{2}x + 2$$

Two lines are parallel when they have the same slope. Therefore, the slope of the line parallel to the given line must be $-\frac{1}{2}$. Since its slope is $-\frac{1}{2}$ and its y-intercept is 5, its equation must be

$$y = -\frac{1}{2}x + 5$$

The graphs of $2x + 4y = 8$ and $y = -\frac{1}{2}x + 5$ are shown in **Figure 8.47**.

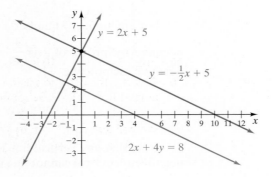

FIGURE 8.47

b) Two lines are perpendicular when their slopes are negative reciprocals. We know that the slope of the given line is $-\dfrac{1}{2}$. Therefore, the slope of the perpendicular line must be $-1 \Big/ \left(-\dfrac{1}{2}\right)$ or 2. The line perpendicular to the given line has a y-intercept of 5. Thus the equation is

$$y = 2x + 5$$

Figure 8.47 also shows the graph of $y = 2x + 5$.

▶ **Now Try Exercise 39**

EXAMPLE 7 ▶ Consider the equation $5y = -10x + 7$.

a) Determine the equation of a line that passes through $\left(4, \dfrac{1}{3}\right)$ that is perpendicular to the graph of the given equation. Write the equation in standard form.

b) Write the equation determined in part **a)** using function notation.

Solution

a) Determine the slope of the given line by solving the equation for y.

$$5y = -10x + 7$$
$$y = \frac{-10x + 7}{5}$$
$$y = -2x + \frac{7}{5}$$

Since the slope of the given line is -2, the slope of a line perpendicular to it must be the negative reciprocal of -2, which is $\dfrac{1}{2}$. The line we are seeking must pass through the point $\left(4, \dfrac{1}{3}\right)$. Using the point-slope form, we obtain

$$y - y_1 = m(x - x_1)$$
$$y - \frac{1}{3} = \frac{1}{2}(x - 4) \qquad \textit{Point slope form}$$

Now multiply both sides of the equation by the least common denominator, 6, to eliminate fractions.

$$6\left(y - \frac{1}{3}\right) = 6\left[\frac{1}{2}(x - 4)\right]$$
$$6y - 2 = 3(x - 4)$$
$$6y - 2 = 3x - 12$$

Now write the equation in standard form.

$$-3x + 6y - 2 = -12$$
$$-3x + 6y = -10 \qquad \textit{Standard form}$$

Note that $3x - 6y = 10$ is also an acceptable answer (see **Fig. 8.48**).

b) To write the equation using function notation, we solve the equation determined in part **a)** for y, and then replace y with $f(x)$.

We will leave it to you to show that the function is $f(x) = \dfrac{1}{2}x - \dfrac{5}{3}$.

▶ **Now Try Exercise 43**

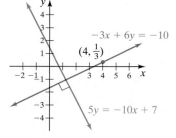

FIGURE 8.48

EXERCISE SET 8.4

Math XL   MyMathLab
MathXL®   MyMathLab

Concept/Writing Exercises

1. What does it mean when slope is given as a rate of change?

2. What is the slope of a constant function?

3. a) What does it mean when a graph is translated up 3 units?

   b) If the y-intercept of a graph is $(0, -4)$ and the graph is translated up 5 units, what will be its new y-intercept?

4. a) What does it mean when a graph is translated down 4 units?

   b) If the y-intercept of a graph is $(0, -3)$ and the graph is translated down 4 units, what will be its new y-intercept?

5. How can we determine whether two lines are perpendicular?

6. Why can't the negative reciprocal test be used to determine whether a vertical line is perpendicular to a horizontal line?

Practice the Skills

Find the slope of the line in each of the figures. If the slope of the line is undefined, so state. Then write an equation of the given line using function notation when possible.

 7.
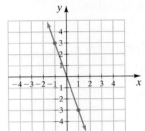

8.
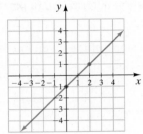

9.
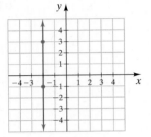

10.
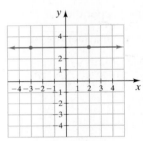

11.
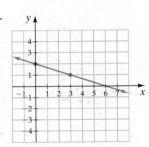

12.
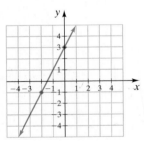

13.
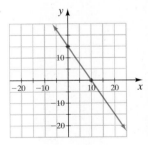

14.
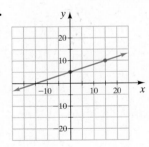

Graph each constant function.

15. $f(x) = -5$ 16. $g(x) = 2$ 17. $g(x) = -15$ 18. $h(x) = 75$

Two points on l_1 and two points on l_2 are given. Determine whether l_1 is parallel to l_2, l_1 is perpendicular to l_2, or neither.

19. l_1: $(2, 0)$ and $(0, 2)$; l_2: $(3, 0)$ and $(0, 3)$

20. l_1: $(7, 6)$ and $(3, 9)$; l_2: $(5, -1)$ and $(9, -4)$

21. l_1: $(4, 6)$ and $(5, 7)$; l_2: $(-1, -1)$ and $(1, 4)$

22. l_1: $(-3, 4)$ and $(4, -3)$; l_2: $(-5, -6)$ and $(6, -5)$

23. l_1: $(3, 2)$ and $(-1, -2)$; l_2: $(2, 0)$ and $(3, -1)$

24. l_1: $(3, 5)$ and $(9, 1)$; l_2: $(4, 0)$ and $(6, 3)$

Determine whether the two equations represent lines that are parallel, perpendicular, or neither.

25. $y = \dfrac{1}{5}x + 9$

$y = -5x + 2$

26. $2x + 3y = 11$

$y = -\dfrac{2}{3}x + 4$

27. $4x + 2y = 8$

$8x = 4 - 4y$

28. $2x - y = 4$

$3x + 6y = 18$

29. $2x - y = 4$

$-x + 4y = 4$

30. $6x + 2y = 8$

$4x - 5 = -y$

31. $y = \dfrac{1}{2}x - 6$

$-4y = 8x + 15$

32. $2y - 8 = -5x$

$y = -\dfrac{5}{2}x - 2$

33. $y = \dfrac{1}{2}x + 6$

$-2x + 4y = 8$

34. $-4x + 6y = 11$

$2x - 3y = 5$

35. $x - 2y = -9$

$y = x + 6$

36. $\dfrac{1}{2}x - \dfrac{3}{4}y = 1$

$\dfrac{3}{5}x + \dfrac{2}{5}y = -1$

Find the equation of a line with the properties given. Write the equation in the form indicated.

37. Through $(2, 5)$ and parallel to the graph of $y = 2x + 4$ (slope-intercept form)

38. Through $(-1, 6)$ and parallel to the graph of $4x - 2y = 6$ (slope-intercept form)

39. Through $(-3, -5)$ and parallel to the graph of $2x - 5y = 7$ (standard form)

40. Through $(-1, 4)$ and perpendicular to the graph of $y = -2x - 1$ (standard form)

41. With x-intercept $(3, 0)$ and y-intercept $(0, 5)$ (slope-intercept form)

42. Through $(-2, -1)$ and perpendicular to the graph of $f(x) = -\dfrac{1}{5}x + 1$ (function notation)

43. Through $(5, -2)$ and perpendicular to the graph of $y = \dfrac{1}{3}x + 1$ (function notation)

44. Through $(-3, 5)$ and perpendicular to the line with x-intercept $(2, 0)$ and y-intercept $(0, 2)$ (standard form)

45. Through $(6, 2)$ and perpendicular to the line with x-intercept $(2, 0)$ and y-intercept $(0, -3)$ (slope-intercept form)

46. Through the point $(1, 2)$ and parallel to the line through the points $(3, 5)$ and $(-2, 3)$ (function notation)

Problem Solving

47. If a line passes through the points $(6, 4)$ and $(-4, 2)$, find the change of y with respect to a 1-unit change in x.

48. If a line passes through the points $(-3, -4)$ and $(5, 2)$, find the change of y with respect to a 1-unit change in x.

TV Sales *For Exercises 49 and 50, use the graphs below. The graph on the left shows the projected digital TV sales (in millions) and the graph on the right shows the projected analog TV sales (in millions) for the years from 2004 to 2008.*

TV Sales

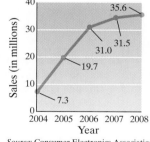

Projected digital TV Sales

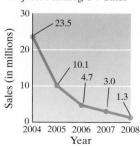

Projected analog TV Sales

Source: Consumer Electronics Association, *USA Today* (1/5/05)

49. a) For the graph of digital TV sales, determine the slope of the line segment from 2005 to 2006.

b) Is the slope of the line segment positive or negative?

c) Find the average rate of change from 2004 to 2008.

50. a) For the graph of analog TV sales, determine the slope of the line segment from 2005 to 2006.

b) Is the slope of the line segment positive or negative?

c) Find the average rate of change from 2004 to 2008.

51. Amtrak Expenses The National Railroad and Passenger Corporation, better known as Amtrak, continues to face economic struggles. The following table gives the expenses, in millions of dollars, of Amtrak for selected years.

Year	Amtrak Expenses (in millions of dollars)
1995	$ 2257
2000	$ 2876
2004	$ 3133
*2008	$ 3260

Source: Amtrak Fiscal Year 2004 Annual Report

*Projected

a) Plot these points on a graph.

b) Connect these points using line segments.

c) Determine the slopes of each of the three line segments.

d) During which period was there the greatest average rate of change? Explain.

52. Demand for Steel The world demand for steel has been on the rise in recent years. The following table gives the world demand for steel, in millions of metric tons, for the years from 2001 to 2004.

World Demand for Steel

Year	Demand (in millions of metric tons)
2001	740
2002	810
2003	880
2004	950

Source: "World Steel Dynamics," *Wall Street Journal* (12/8/04)

a) Plot these points on a graph.

b) Determine the slope of each line segment.

c) Is this graph an example of a linear function? Explain.

d) Determine a linear function that can be used to estimate the world demand for steel, d, from 2001 to 2004. Let t represent the number of years since 2001. (That is, 2001 corresponds to $t = 0$, 2002 corresponds to $t = 1$, and so on.)

e) Assuming this trend continues for the next 20 years, find the world demand for steel in 2016.

f) Assuming this trend continues, in which year will the demand reach 1230 metric tons?

53. Heart Rate The following bar graph shows the maximum recommended heart rate, in beats per minute, under stress for men of different ages. The bars are connected by a straight line.

a) Use the straight line to determine a function that can be used to estimate the maximum recommended heart rate, h, for $0 \le x \le 50$, where x is the number of years after age 20.

b) Using the function from part **a)**, determine the maximum recommended heart rate for a 34-year-old man.

Heart Rate vs. Age

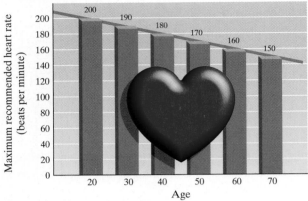

Source: The American Geriatric Society

54. Poverty Threshold The federal government defines the poverty threshold as an estimate of the annual family income necessary to have what society defines as a minimally acceptable standard of living. The following bar graph shows the poverty threshold for a family of four for the years 2000 through 2004.

U.S. Poverty Threshold for a Family of Four

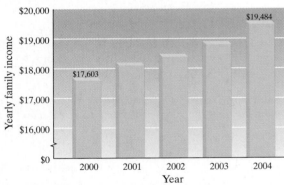

Source: U.S. Bureau of the Census: www.census.gov/hhes/poverty

a) Determine a linear function that can be used to estimate the poverty threshold for a family of four, P, from 2000 through 2004. Let t represent the number of years since 2000. (In other words, 2000 corresponds to $t = 0$, 2001 corresponds to $t = 1$, and so on.)

b) Using the function from part **a)**, determine the poverty threshold in 2003. Compare your answer with the graph to see whether the graph supports your answer.

c) Assuming this trend continues, determine the poverty threshold for a family of four in the year 2010.

d) Assuming this trend continues, in which year will the poverty threshold for a family of four reach $20,424.50?

55. Medicaid Spending The following graph shows the amount of money spent on Medicaid for the years from 1997 to 2004.

Medicaid Spending

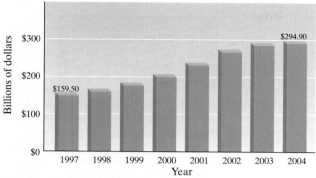

Source: Centers for Medicaid and Medicare Services, USA Today (8/2/05)

a) Using 1997 as a reference year, determine a linear function that can be used to estimate Medicaid spending (in billions of dollars), M, for the years from 1997 to 2004. In this function, let t represent the number of years since 1997.

b) Using the function from part **a)**, estimate Medicaid spending for the year 2003. Compare your answer with the graph to see whether the graph supports your answer.

c) Assuming this trend continues, what will be the Medicaid spending in 2010?

d) Assuming this trend continues, during what year will Medicaid spending reach $340 billion?

56. Purchasing Power of the Dollar The purchasing power of the dollar is measured by comparing the current price of items to the price of those same items in 1982. From the chart below you will see that the purchasing power of the dollar has steadily declined for the years 1990 through 2003. This means that $1 buys less each year.

Purchasing Power of the Dollar

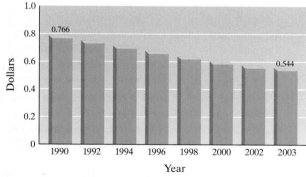

Source: U.S. Bureau of Economic Analysis

a) Using 1990 as the reference year, determine a linear function that can be used to estimate the purchasing power, P, for the years 1990 through 2003. In the function, let t represent the number of years since 1990.

b) Using the function from part **a)**, estimate the purchasing power of the dollar in 1994. Compare your answer with the graph to see whether the graph supports your answer.

c) Assuming this trend continues, what would be the purchasing power of the dollar in 2006?

d) Assuming this trend continues, when would the purchasing power of the dollar reach $0.426?

57. Teenagers Using Illicit Drugs The percent of teenagers who claim to have used illicit drugs (in the last 30 days) has been on the decline in the years from 2001 to 2004. From the graph below, the decline appears to be approximately linear.

Percent of Teenagers Using Illicit Drugs

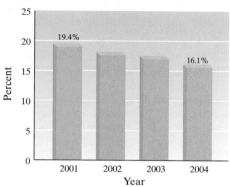

Source: University of Michigan, 2004 Monitoring the Future Study, *The Washington Post* (12/22/04)

a) Using 2001 as a reference year, determine a linear function that can be used to estimate the percent of teenagers using illicit drugs, P, for the years 2001 through 2004. In this function, let t represent the number of years since 2001.

b) Is the slope of the linear function positive or negative? Explain.

c) Using the function from part **a)**, estimate the percent of teenagers using illicit drugs in 2003. Compare your answer with the graph to see whether the graph supports your answer.

d) Assuming this trend continues, what would be the percent of teenagers using illicit drugs in 2010?

58. Personal Income Personal income was on the rise every month from June 2003 to November 2004. From the graph below, the increase appears to be approximately linear.

Personal Income

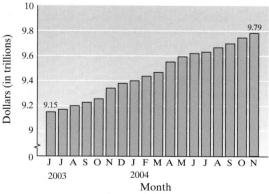

Source: Commerce Department, *The New York Times* (12/24/04)

a) Using June 2003 as a reference point, determine a linear function that can be used to estimate the personal income (in trillions of dollars), I, for the months June 2003 through November 2004. In this function, let t represent the number of months since June 2003 (that is, $t = 0$ corresponds to June 2003, $t = 1$ corresponds to July 2003, $t = 6$ corresponds to December 2003, $t = 17$ corresponds to November 2004, and so on).

b) Is the slope of this linear function positive or negative? Explain.

c) Using the function from part **a)**, estimate the personal income (in trillions of dollars) for February 2004 ($t = 8$). Compare your answer with the graph to see whether the graph supports your answer.

d) Assuming this trend continues, what would be the personal income in December 2005 ($t = 30$)?

59. Median Home Sale Price The median home sale price in the United States has been rising approximately linearly since 1995. The median home sale price in 1995 was $110,500. The median home sale price in 2004 was $185,200. Let P be the median home sale price and let t be the number of years since 1995.   *Source:* National Association of Realtors

a) Determine a function $P(t)$ that fits this data.

b) Use the function from part **a)** to estimate the median home sale price in 2000.

c) If this trend continues, estimate the median home sale price in 2010.

d) If this trend continues, in which year will the median home sale price reach $200,000?

60. Social Security The number of workers per social security beneficiary has been declining approximately linearly since 1970. In 1970 there were 3.7 workers per beneficiary. In 2050 it is projected there will be 2.0 workers per beneficiary. Let W be the workers per social security beneficiary and t be the number of years since 1970.

a) Find a function $W(t)$ that fits the data.

b) Estimate the number of workers per beneficiary in 2020.

61. Treadmill The number of calories burned in 1 hour on a treadmill is a function of the speed of the treadmill. The average person walking on a treadmill (at 0° incline) at a speed of 2.5 miles per hour will burn about 210 calories. At 6 miles per hour the average person will burn about 370 calories. Let C be the calories burned in 1 hour and s be the speed of the treadmill.

a) Determine a linear function $C(s)$ that fits the data.

b) Estimate the calories burned by the average person on a treadmill in 1 hour at a speed of 5 miles per hour.

62. Inclined Treadmill The number of calories burned for 1 hour on a treadmill going at a constant speed is a function of the incline of the treadmill. At 4 miles per hour an average person on a 5° incline will burn 525 calories. At 4 mph on a 15° incline the average person will burn 880 calories. Let C be the calories burned and d be the degrees of incline of the treadmill.

a) Determine a linear function $C(d)$ that fits the data.

b) Determine the number of calories burned by the average person in 1 hour on a treadmill going 4 miles per hour and at a 9° incline.

63. Demand for DVD Players The *demand* for a product is the number of items the public is willing to buy at a given price. Suppose the demand, d, for DVD players sold in 1 month is a linear function of the price, p, for $150 \le p \le 400$. If the price is $200, then 50 DVD players will be sold each month. If the price is $300, only 30 DVD players will be sold.

a) Using ordered pairs of the form (p, d), write an equation for the demand, d, as a function of price, p.

b) Using the function from part **a)**, determine the demand when the price of the DVD players is $260.

c) Using the function from part **a)**, determine the price charged if the demand for DVD players is 45.

64. Demand for New Sandwiches The marketing manager of Arby's restaurants determines that the demand, d, for a new chicken sandwich is a linear function of the price, p, for $0.80 \le p \le 4.00$. If the price is $1.00, then 530 chicken sandwiches will be sold each month. If the price is $2.00, only 400 chicken sandwiches will be sold each month.

a) Using ordered pairs of the form (p, d), write an equation for the demand, d, as a function of price, p.

b) Using the function from part **a)**, determine the demand when the price of the chicken sandwich is $2.60.

c) Using the function from part **a)**, determine the price charged if the demand for chicken sandwiches is 244 chicken sandwiches.

65. Supply of Kites The *supply* of a product is the number of items a seller is willing to sell at a given price. The maker of a new kite for children determines that the number of kites she is willing to supply, s, is a linear function of the selling price p for $2.00 \le p \le 4.00$. If a kite sells for $2.00, then 130 per month will be supplied. If a kite sells for $4.00, then 320 per month will be supplied.

a) Using ordered pairs of the form (p, s), write an equation for the supply, s, as a function of price, p.

b) Using the function from part **a)**, determine the supply when the price of a kite is $2.80.

c) Using the function from part **a)**, determine the price paid if the supply is 225 kites.

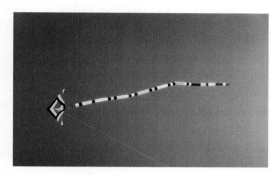

66. **Supply of Baby Strollers** The manufacturer of baby strollers determines that the supply, s, is a linear function of the selling price, p, for $200 \leq p \leq 300$. If a stroller sells for $210.00, then 20 strollers will be supplied per month. If a stroller sells for $230.00, then 30 strollers will be supplied per month.

a) Using ordered pairs of the form (p, s), write an equation for the supply, s, as a function of price, p.

b) Using the function from part **a)**, determine the supply when the price of a stroller is $220.00.

c) Using the function from part **a)**, determine the selling price if the supply is 35 strollers.

67. **High School Play** The income, i, from a high school play is a linear function of the number of tickets sold, t. When 80 tickets are sold, the income is $1000. When 200 tickets are sold, the income is $2500.

a) Use these data to write the income, i, as a function of the number of tickets sold, t.

b) Using the function from part **a)**, determine the income if 120 tickets are sold.

c) If the income is $2200, how many tickets were sold?

68. **Gas Mileage of a Car** The gas mileage, m, of a specific car is a linear function of the speed, s, at which the car is driven, for $30 \leq s \leq 60$. If the car is driven at a rate of 30 mph, the car's gas mileage is 35 miles per gallon. If the car is driven at 60 mph, the car's gas mileage is 20 miles per gallon.

a) Use this data to write the gas mileage, m, as a function of speed, s.

b) Using the function from part **a)**, determine the gas mileage if the car is driven at a speed of 48 mph.

c) Using the function from part **a)**, determine the speed at which the car must be driven to get gas mileage of 40 miles per gallon.

69. **Auto Registration** The registration fee, r, for a vehicle in a certain region is a linear function of the weight of the vehicle, w, for $1000 \leq w \leq 6000$ pounds. When the weight is 2000 pounds, the registration fee is $30. When the weight is 4000 pounds, the registration fee is $50.

a) Use these data to write the registration fee, r, as a function of the weight of the vehicle, w.

b) Using the function from part **a)**, determine the registration fee for a 2006 Ford Mustang if the weight of the vehicle is 3613 pounds.

c) If the cost of registering a vehicle is $60, determine the weight of the vehicle.

70. **Lecturer Salary** Suppose the annual salary of a lecturer at Chaumont University is a linear function of the number of years of teaching experience. A lecturer with 9 years of teaching experience is paid $41,350. A lecturer with 15 years of teaching experience is paid $46,687.

a) Use this data to write the annual salary of a lecturer, s, as a function of the number of years of teaching experience, n.

b) Using the function from part **a)**, determine the annual salary of a lecturer with 10 years of teaching experience.

c) Using the function from part **a)**, estimate the number of years of teaching experience a lecturer must have to obtain an annual salary of $44,908.

71. **Life Expectancy** As seen in the following graph, the expected number of remaining years of life of a person, y, *approximates* a linear function. The expected number of remaining years is a function of the person's current age, a, for $30 \leq a \leq 80$. For example, from the graph we see that a person who is currently 50 years old has a life expectancy of 36.0 *more* years.

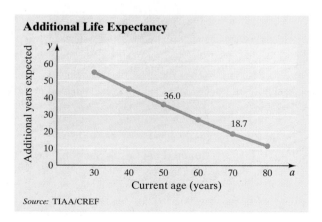

Source: TIAA/CREF

a) Using the two points on the graph, determine the function $y(a)$ that can be used to approximate the graph.

b) Using the function from part **a)**, estimate the life expectancy of a person who is currently 37 years old.

c) Using the function from part **a)**, estimate the current age of a person who has a life expectancy of 25 years.

72. Guarneri del Gesù Violin Handcrafted around 1735, Guarneri del Gesù violins are extremely rare and extremely valuable. The graph below shows that the projected value, v, of a Guarneri del Gesù violin is a linear function of the age, a, in years, of the violin, for $261 \le a \le 290$.

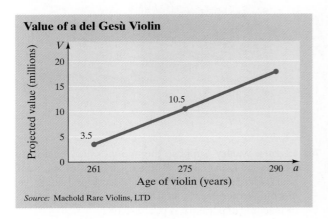

Value of a del Gesù Violin

Source: Machold Rare Violins, LTD

a) Determine the function $v(a)$ represented by this line.

b) Using the function from part **a)**, determine the projected value of a 265-year-old Guarneri del Gesù violin.

c) Using the function from part **a)**, determine the age of a Guarneri del Gesù violin with a projected value of $15 million.

Guarneri del Gesù, "Sainton," 1741

73. Boys' Weights Parents may recognize the following diagram from visits to the pediatrician's office. The diagram shows percentiles for boys' heights and weights from birth to age 36 months. Overall, the graphs shown are not linear functions. However, certain portions of the graphs can be approximated with a linear function. For example, the graph representing the 95th percentile of boys' weights (the top red line) from age 18 months to age 36 months is approximately linear.

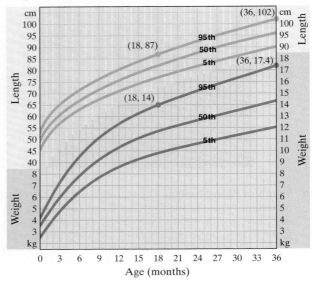

Boys: Birth to 36 months
Length-for-Age and Weight-for-Age Percentiles

Source: National Center for Health Statistics

a) Use the points shown on the graph of the 95th percentile to write weight, w, as a linear function of age, a, for boys between 18 and 36 months old.

b) Using the function from part **a)**, estimate the weight of a 22-month-old boy who is in the 95th percentile for weight. Compare your answer with the graph to see whether the graph supports your answer.

74. Boys' Lengths The diagram in Exercise 73 shows that the graph representing the 95th percentile of boys' lengths (the top yellow line) from age 18 months to age 36 months is approximately linear.

a) Use the points shown on the graph of the 95th percentile to write length, l, as a linear function of age, a, for boys between age 18 and 36 months.

b) Using the function from part **a)**, estimate the length of a 21-month-old boy who is in the 95th percentile. Compare your answer with the graph to see whether the graph supports your answer.

75. If the y-intercept on a graph is $(0, -13)$ and the line is translated up 17 units, find the y-intercept of the translated graph.

76. If the y-intercept on a graph is $(0, 17)$ and the line is translated up 38 units, find the y-intercept of the translated graph.

77. If the y-intercept on a graph is $(0, 6)$ and the line is translated down 9 units, find the y-intercept of the translated graph.

78. If the y-intercept on a graph is $(0, -9)$ and the line is translated down 8 units, find the y-intercept of the translated graph.

79. In the following graph, the green line is a vertical translation of the blue line.

    a) Determine the equation of the blue line.

    b) Use the equation of the blue line to determine the equation of the green line.

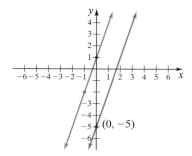

80. In the following graph, the red line is a vertical translation of the blue line.

    a) Determine the equation of the blue line.

    b) Use the equation of the blue line to determine the equation of the red line.

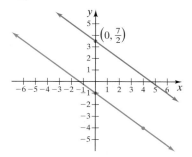

81. The graph of $y = x - 1$ is translated up 3 units. Determine

    a) the slope of the translated graph.

    b) the y-intercept of the translated graph.

    c) the equation of the translated graph.

82. The graph of $y = -\dfrac{3}{2}x + 3$ is translated down 4 units. Determine

    a) the slope of the translated graph.

    b) the y-intercept of the translated graph.

    c) the equation of the translated graph.

 83. The graph of $3x - 2y = 6$ is translated down 4 units. Find the equation of the translated graph.

84. The graph of $-3x - 5y = 15$ is translated up 2 units. Find the equation of the translated graph.

Suppose you are attempting to graph the equations shown and you get the screens shown. Explain how you know that you have made a mistake in entering each equation. The standard window setting is used on each graph.

85. $y = 3x + 6$

86. $y = -2x - 4$

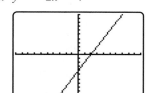

87. $y = \dfrac{1}{2}x + 4$

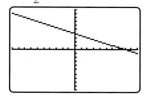

88. $y = -4x - 1$

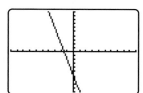

Challenge Problems

89. Castle The photo below is the Castle at Chichén Itzá, Mexico. Each side of the castle has a stairway consisting of 91 steps. The steps of the castle are quite narrow and steep, which makes them hard to climb. The total vertical distance of the 91 steps is 1292.2 inches. If a straight line were to be drawn connecting the tips of the steps, the absolute value of the slope of this line would be 2.21875. Find the average height and width of a step.

90. A **tangent line** is a straight line that touches a curve at a single point (the tangent line may cross the curve at a different point if extended). **Figure 8.49** shows three tangent lines to the curve at points a, b, and c. Note that the tangent line at point a has a positive slope, the tangent line at point b has a slope of 0, and the tangent line at point c has a negative slope. Now consider the curve in **Figure 8.50**. Imagine that tangent lines are drawn at all points on the curve except at endpoints a and e. Where on the curve in **Figure 8.50** would the tangent lines have a positive slope, a slope of 0, and a negative slope?

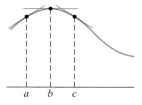

FIGURE 8.49

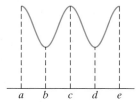

FIGURE 8.50

Group Activity

91. The following graph from *Consumer Reports* shows the depreciation on a typical car. The initial purchase price is represented as 100%.

a) Group member 1: Determine the 1-year period in which a car depreciates most. Estimate from the graph the percent a car depreciates during this period.

b) Group member 2: Determine between which years the depreciation appears linear or nearly linear.

c) Group member 3: Determine between which 2 years the depreciation is the lowest.

d) As a group, estimate the slope of the line segment from year 0 to year 1. Explain what this means in terms of rate of change.

Typical Depreciation Curve

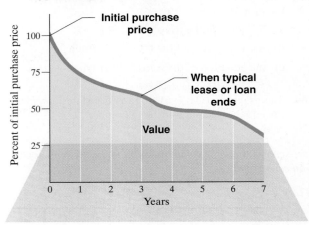

92. The graph on the right shows the growth of the circumference of a girl's head. The orange line is the average head circumference of all girls for the given age while the green lines represent the upper and lower limits of the normal range. Discuss and answer the following questions as a group.

a) Explain why the graph of the average head circumference represents a function.

b) What is the independent variable? What is the dependent variable?

c) What is the domain of the graph of the average head circumference? What is the range of the average head circumference graph?

d) What interval is considered normal for girls of age 18?

e) For this graph, is head circumference a function of age or is age a function of head circumference? Explain your answer.

f) Estimate the average girl's head circumference at age 10 and at age 14.

g) This graph appears to be nearly linear. Determine an equation or function that can be used to estimate the orange line between $(2, 48)$ and $(18, 55)$.

Head Circumference

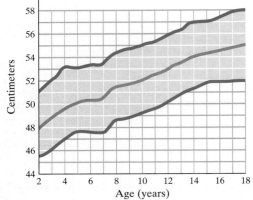

Source: National Center for Health Statistics

Cumulative Review Exercises

[1.9] **93.** Evaluate $\dfrac{-6^2 - 16 \div 2 \div |-4|}{5 - 3 \cdot 2 - 4 \div 2^2}$.

Solve each equation.

[2.4] **94.** $2.6x - (-1.4x + 3.4) = 6.2$

[2.5] **95.** $\dfrac{3}{4}x + \dfrac{1}{5} = \dfrac{2}{3}(x - 2)$.

[8.2] **96. a)** What is a relation?

   b) What is a function?

   c) Draw a graph that is a relation but not a function.

97. Find the domain and range of the function $\{(4, 3), (5, -2), (3, 2), (6, -1)\}$.

8.5 The Algebra of Functions

1 Find the sum, difference, product, and quotient of functions.

2 Graph the sum of functions.

1 Find the Sum, Difference, Product, and Quotient of Functions

Let's discuss some ways that functions can be combined. If we let $f(x) = x - 3$ and $g(x) = x^2 + 2x$, we can find $f(5)$ and $g(5)$ as follows.

$$f(x) = x - 3 \qquad\qquad g(x) = x^2 + 2x$$
$$f(5) = 5 - 3 = 2 \qquad\qquad g(5) = 5^2 + 2(5) = 35$$

If we add $f(x) + g(x)$, we get

$$f(x) + g(x) = (x - 3) + (x^2 + 2x)$$
$$= x^2 + 3x - 3$$

This new function formed by the sum of $f(x)$ and $g(x)$ is designated as $(f + g)(x)$. Therefore, we may write

$$(f + g)(x) = x^2 + 3x - 3$$

We find $(f + g)(5)$ as follows.

$$(f + g)(5) = 5^2 + 3(5) - 3$$
$$= 25 + 15 - 3 = 37$$

Notice that

$$f(5) + g(5) = (f + g)(5)$$
$$2 + 35 = 37 \qquad \textit{True}$$

In fact, for any real number substituted for x you will find that

$$f(x) + g(x) = (f + g)(x)$$

Similar notation exists for subtraction, multiplication, and division of functions.

Operations on Functions

If $f(x)$ represents one function, $g(x)$ represents a second function, and x is in the domain of both functions, then the following operations on functions may be performed:

Sum of functions: $(f + g)(x) = f(x) + g(x)$
Difference of functions: $(f - g)(x) = f(x) - g(x)$
Product of functions: $(f \cdot g)(x) = f(x) \cdot g(x)$
Quotient of functions: $(f/g)(x) = \dfrac{f(x)}{g(x)}$, provided that $g(x) \neq 0$

EXAMPLE 1 ▶ If $f(x) = x^2 + x - 6$ and $g(x) = x - 3$, find

a) $(f + g)(x)$ **b)** $(f - g)(x)$
c) $(g - f)(x)$ **d)** Does $(f - g)(x) = (g - f)(x)$?

Solution To answer parts **a)**–**c)**, we perform the indicated operation.

a) $(f + g)(x) = f(x) + g(x)$
$$= (x^2 + x - 6) + (x - 3)$$
$$= x^2 + x - 6 + x - 3$$
$$= x^2 + 2x - 9$$

b) $(f - g)(x) = f(x) - g(x)$
$$= (x^2 + x - 6) - (x - 3)$$
$$= x^2 + x - 6 - x + 3$$
$$= x^2 - 3$$

c) $(g - f)(x) = g(x) - f(x)$
$$= (x - 3) - (x^2 + x - 6)$$
$$= x - 3 - x^2 - x + 6$$
$$= -x^2 + 3$$

d) By comparing the answers to parts **b)** and **c)**, we see that
$$(f - g)(x) \neq (g - f)(x)$$

▶ Now Try Exercise 11

EXAMPLE 2 ▸ If $f(x) = x^2 - 4$ and $g(x) = x - 2$, find

a) $(f - g)(6)$ **b)** $(f \cdot g)(5)$ **c)** $(f/g)(8)$

Solution

a) $(f - g)(x) = f(x) - g(x)$
$$= (x^2 - 4) - (x - 2)$$
$$= x^2 - x - 2$$
$$(f - g)(6) = 6^2 - 6 - 2$$
$$= 36 - 6 - 2$$
$$= 28$$

We could have also found the solution as follows:

$$f(x) = x^2 - 4 \qquad\qquad g(x) = x - 2$$
$$f(6) = 6^2 - 4 = 32 \qquad\qquad g(6) = 6 - 2 = 4$$
$$(f - g)(6) = f(6) - g(6)$$
$$= 32 - 4 = 28$$

b) We will find $(f \cdot g)(5)$ using the fact that

$$(f \cdot g)(5) = f(5) \cdot g(5)$$
$$f(x) = x^2 - 4 \qquad\qquad g(x) = x - 2$$
$$f(5) = 5^2 - 4 = 21 \qquad\qquad g(5) = 5 - 2 = 3$$

Thus $f(5) \cdot g(5) = 21 \cdot 3 = 63$. Therefore, $(f \cdot g)(5) = 63$. We could have also found $(f \cdot g)(5)$ by multiplying $f(x) \cdot g(x)$ and then substituting 5 into the product.

c) We will find $(f/g)(8)$ by using the fact that

$$(f/g)(8) = f(8)/g(8)$$
$$f(x) = x^2 - 4 \qquad\qquad g(x) = x - 2$$
$$f(8) = 8^2 - 4 = 60 \qquad\qquad g(8) = 8 - 2 = 6$$

Then $f(8)/g(8) = 60/6 = 10$. Therefore, $(f/g)(8) = 10$. We could have also found $(f/g)(8)$ by dividing $f(x)/g(x)$ and then substituting 8 into the quotient.

▸ **Now Try Exercise 31**

Notice that we included the phrase "and x is in the domain of both functions" in the Operations on Functions box on page 525 . As we stated earlier, the domain of a function is the set of values that can be used for the independent variable. For example, the domain of the function $f(x) = 2x^2 - 6x + 5$ is all real numbers, because when x is any real number $f(x)$ will also be a real number. The domain of $g(x) = \dfrac{1}{x - 8}$ is all real numbers except 8, because when x is any real number except 8, the function $g(x)$ is a real number. When x is 8, the function is not a real number because $\dfrac{1}{0}$ is undefined.

2 Graph the Sum of Functions

Now we will explain how we can graph the sum, difference, product, or quotient of two functions. **Figure 8.51** shows two functions, $f(x)$ and $g(x)$.

To graph the sum of $f(x)$ and $g(x)$, or $(f + g)(x)$, we use $(f + g)(x) = f(x) + g(x)$. The table on the next page gives the integer values of x from -2 to 4, the values of $f(-2)$ through $f(4)$, and the values of $g(-2)$ through $g(4)$. These values are taken directly from **Figure 8.51** on the next page. The values of $(f + g)(-2)$ through $(f + g)(4)$ are determined by adding the values of $f(x)$ and $g(x)$. The graph of $(f + g)(x) = f(x) + g(x)$ is illustrated in green in **Figure 8.52** on the next page.

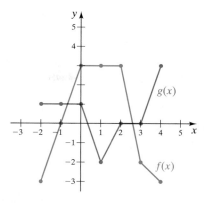

FIGURE 8.51

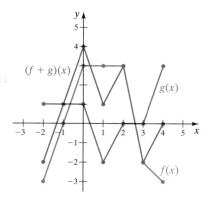

FIGURE 8.52

x	$f(x)$	$g(x)$	$(f + g)(x)$
-2	-3	1	$-3 + 1 = -2$
-1	0	1	$0 + 1 = 1$
0	3	1	$3 + 1 = 4$
1	3	-2	$3 + (-2) = 1$
2	3	0	$3 + 0 = 3$
3	-2	0	$-2 + 0 = -2$
4	-3	3	$-3 + 3 = 0$

We could graph the difference, product, or quotient of the two functions using a similar technique. For example, to graph the product function $(f \cdot g)(x)$, we would evaluate $(f \cdot g)(-2)$ as follows:

$$(f \cdot g)(-2) = f(-2) \cdot g(-2)$$
$$= (-3)(1) = -3$$

Thus, the graph of $(f \cdot g)(x)$ would have an ordered pair at $(-2, -3)$. Other ordered pairs would be determined by the same procedure.

In newspapers, magazines, and on the Internet we often find graphs that show the sum of two or more functions. Graphs that show the sum of functions are generally indicated in one of three ways: line graphs, bar graphs, or stacked (or cumulative) line graphs. Examples 3 through 5 show the three general methods. Each of these examples will use the same data pertaining to cholesterol.

EXAMPLE 3 ▶ **Line Graph** Jim Silverstone has kept a record of his bad cholesterol (low-density lipoprotein, or LDL) and his good cholesterol (high-density lipoprotein or HDL) from 2002 through 2006. **Table 8.2** shows his LDL and his HDL for these years.

TABLE 8.2	Cholesterol				
	2002	**2003**	**2004**	**2005**	**2006**
LDL	220	240	140	235	130
HDL	30	40	70	35	40

a) Explain why the data consisting of the years and the LDL values are a function, and the data consisting of the years and the HDL values are also a function.

b) Draw a line graph that shows the LDL, the HDL, and the total cholesterol from 2002 through 2006. The total cholesterol is the sum of the LDL and the HDL.

c) If L represents the amount of LDL and H represents the amount of HDL, show that $(L + H)(2006) = 170$.

d) By looking at the graph drawn in part **b)**, determine the years in which the LDL was less than 180.

Solution

a) The data consisting of the years and the LDL values are a function because for each year there is exactly one LDL value. Note that the year is the independent variable, and the LDL value is the dependent variable. The data consisting of the years and the HDL values are a function for the same reason.

b) For any given year, the total cholesterol is the sum of the LDL and HDL for that year. For example, for 2005, to find the cholesterol, we add $235 + 35 = 270$. The graph in **Figure 8.53** shows LDL, HDL, and total cholesterol for the years 2002 through 2006.

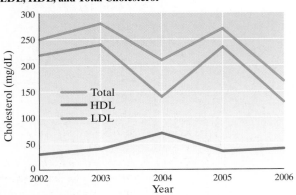

FIGURE 8.53

c) To find LDL + HDL, or the total cholesterol, we add the two values for 2006.

$$(L + H)(2006) = L(2006) + H(2006)$$
$$= 130 + 40 = 170$$

d) By looking at the graph drawn in part **b)**, we see that the years in which the LDL was less than 180 are 2004 and 2006.

▶ **Now Try Exercise 63a**

EXAMPLE 4 ▶ **Bar Graph**

a) Using the data given in **Table 8.2** on page 527, draw a bar graph that shows the LDL, HDL, and total cholesterol for the years 2002 through 2006.

b) If L represents the amount of LDL and H represents the amount of HDL, use the graph drawn in part **a)** to determine $(L + H)(2003)$.

c) By observing the graph drawn in part **a)**, determine in which years the total cholesterol was less than 220.

d) By observing the graph drawn in part **a)**, estimate the HDL in 2004.

Solution

a) To obtain a bar graph showing the total cholesterol, we add the HDL to the LDL for each given year. For example, for 2002, we start by drawing a bar up to 220 to represent the LDL. Directly on top of that bar we add a second bar of 30 units to represent the HDL. This brings the total bar to 220 + 30 or 250 units. We use the same procedure for each year from 2002 to 2006. The bar graph is shown in **Figure 8.54**.

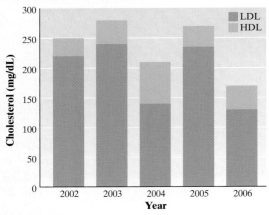

FIGURE 8.54

b) By observing the graph in **Figure 8.54**, we see that the $(L + H)(2003)$, or the total cholesterol for 2003, is about 280.

c) By observing the graph, we see that the total cholesterol was less than 220 in 2004 and 2006.

d) For 2004, the HDL bar begins at about 140 and ends at about 210. The difference in these amounts, $210 - 140 = 70$, represents the amount of HDL in 2004. Therefore, the HDL in 2004 was about 70.

▸ **Now Try Exercise 63b**

EXAMPLE 5 ▸ Stacked Line Graph

a) Using the data from **Table 8.2** on page 527, draw a stacked (or cumulative) line graph that shows the LDL, HDL, and total cholesterol for the years 2002–2006.

b) Using the graph drawn in part **a)**, determine which years the total cholesterol was greater than or equal to 200.

c) Using the graph drawn in part **a)**, estimate the amount of HDL in 2006.

d) Using the graph from part **a)**, determine the years in which the LDL was greater than or equal to 180 and the total cholesterol was less than or equal to 250.

Solution

a) To obtain a stacked line graph, draw the line to represent the LDL. This will be the same line that was drawn to represent the LDL in **Figure 8.53** on page 528. On top of this line draw a line to represent the HDL. One way to obtain the line for the HDL is to work year by year, and then connect the points for each year with straight-line segments. For example, in 2002, the HDL would start at 220, the LDL amount, and be increased by 30, the HDL amount, to get a total of 250. Use this procedure for each year. The graph is shown in **Figure 8.55**.

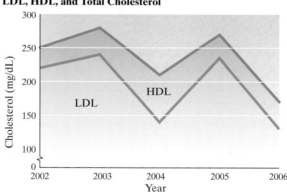

LDL, HDL, and Total Cholesterol

FIGURE 8.55

b) By looking at the graph, we see that the total cholesterol indicated by the dark green line, was greater than or equal to 200 in 2002, 2003, 2004, and 2005.

c) By looking at the HDL area of the graph, we can see that in 2006 the HDL starts at around 130 and ends at about 170. If we subtract, we obtain $170 - 130 = 40$. Therefore, the HDL in 2006 is about 40. If we let H represent the amount of HDL, then $H(2006) \approx 40$.

d) By looking at the graph, we can determine that the only year in which the LDL was greater than or equal to 180 and the total cholesterol was less than or equal to 250 was 2002.

▸ **Now Try Exercise 63c**

USING YOUR GRAPHING CALCULATOR

Graphing calculators can graph the sums, differences, products, and quotients of functions. One way to do this is to enter the individual functions. Then, following the instructions that come with your calculator, you can add, subtract, multiply, or divide the functions. For example, the screen in **Figure 8.56** shows a TI-84 Plus ready to graph $Y_1 = x - 3$, $Y_2 = 2x + 4$, and the sum of the functions, $Y_3 = Y_1 + Y_2$. On the TI-84 Plus, to get $Y_3 = Y_1 + Y_2$, you press the $\boxed{\text{VARS}}$ key. Then you move the cursor to Y-VARS, and the you select 1: Function. Next you press $\boxed{1}$ to enter Y_1. Next you press $\boxed{+}$. Then press $\boxed{\text{VARS}}$ and go to Y-VARS, and choose 1: Function. Finally, press $\boxed{2}$ to enter Y_2. **Figure 8.57** shows the graphs of the two functions, and the graph of the sum of the functions.

FIGURE 8.56

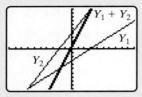

FIGURE 8.57

EXERCISE SET 8.5

 MathXL_®   MyMathLab

Concept/Writing Exercises

1. Does $f(x) + g(x) = (f + g)(x)$ for all values of x?

2. Does $f(x) - g(x) = (f - g)(x)$ for all values of x?

3. What restriction is placed on the property $f(x)/g(x) = (f/g)(x)$? Explain.

4. Does $(f + g)(x) = (g + f)(x)$ for all values of x? Explain and give an example to support your answer.

5. Does $(f - g)(x) = (g - f)(x)$ for all values of x? Explain and give an example to support your answer.

6. If $f(2) = 9$ and $g(2) = -3$, determine
   a) $(f + g)(2)$ b) $(f - g)(2)$
   c) $(f \cdot g)(2)$ d) $(f/g)(2)$

7. If $f(-2) = -3$ and $g(-2) = 5$, find
   a) $(f + g)(-2)$ b) $(f - g)(-2)$
   c) $(f \cdot g)(-2)$ d) $(f/g)(-2)$

8. If $f(7) = 10$ and $g(7) = 0$, determine
   a) $(f + g)(7)$ b) $(f - g)(7)$
   c) $(f \cdot g)(7)$ d) $(f/g)(7)$

Practice the Skills

*For each pair of functions, find **a)** $(f + g)(x)$, **b)** $(f + g)(a)$, and **c)** $(f + g)(2)$.*

9. $f(x) = x + 5, g(x) = x^2 + x$

10. $f(x) = x^2 - x - 8, g(x) = x^2 + 1$

11. $f(x) = -3x^2 + x - 4, g(x) = x^3 + 3x^2$

12. $f(x) = 4x^3 + 2x^2 - x - 1, g(x) = x^3 - x^2 + 2x + 6$

13. $f(x) = 4x^3 - 3x^2 - x, g(x) = 3x^2 + 4$

14. $f(x) = 3x^2 - x + 2, g(x) = 6 - 4x^2$

Let $f(x) = x^2 - 4$ and $g(x) = -5x + 3$. Find the following.

15. $f(2) + g(2)$

16. $f(5) + g(5)$

17. $f(4) - g(4)$

18. $f\left(\dfrac{1}{4}\right) - g\left(\dfrac{1}{4}\right)$

19. $f(3) \cdot g(3)$

20. $f(-1) \cdot g(-1)$

21. $\dfrac{f\left(\dfrac{3}{5}\right)}{g\left(\dfrac{3}{5}\right)}$

22. $f(-1)/g(-1)$

23. $g(-3) - f(-3)$

24. $g(6) \cdot f(6)$

25. $g(0)/f(0)$

26. $f(2)/g(2)$

Let $f(x) = 2x^2 - x$ *and* $g(x) = x - 6$. *Find the following.*

27. $(f + g)(x)$

28. $(f + g)(a)$

29. $(f + g)(2)$

30. $(f + g)(-3)$

31. $(f - g)(-2)$

32. $(f - g)(1)$

33. $(f \cdot g)(0)$

34. $(f \cdot g)(3)$

35. $(f/g)(-1)$

36. $(f/g)(6)$

37. $(g/f)(5)$

38. $(g - f)(4)$

39. $(g - f)(x)$

40. $(g - f)(r)$

Problem Solving

Using the graph on the right, find the value of the following.

41. $(f + g)(0)$

42. $(f - g)(0)$

43. $(f \cdot g)(2)$

44. $(f/g)(1)$

45. $(g - f)(-1)$

46. $(g + f)(-3)$

47. $(g/f)(4)$

48. $(g \cdot f)(-1)$

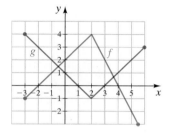

Using the graph on the right, find the value of the following.

49. $(f + g)(-2)$

50. $(f - g)(-1)$

51. $(f \cdot g)(1)$

52. $(g - f)(3)$

53. $(f/g)(4)$

54. $(g/f)(5)$

55. $(g/f)(2)$

56. $(g \cdot f)(0)$

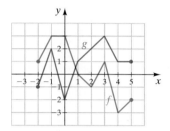

57. Retirement Account The following graph shows the amount of money Sharon and Frank Dangman have contributed to a joint retirement account for the years 2002 to 2006.

Retirement Account

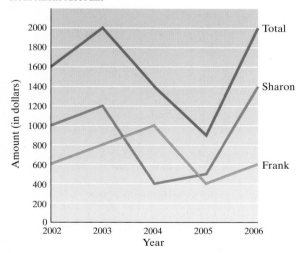

a) In which year did Frank contribute $1000?

b) In 2006, estimate how much more Sharon contributed to the retirement account than Frank contributed.

c) For this five-year period, estimate the total amount Sharon and Frank contributed to the joint retirement account.

d) Estimate $(F + S)(2005)$.

58. Genetically Modified Crops Worldwide production of genetically modified (or transgenic) crops—in both developing nations and industrial nations—is rapidly increasing. The following graph shows the land area devoted to genetically modified crops for developing nations, industrial nations, and total worldwide from 1995 to 2003. The total is determined by adding the amounts of both the developing and industrial nations. Land area is given in millions of hectares. A hectare is a metric system unit that is approximately equal to 2.471 acres.

Global Area of Genetically Modified (Transgenic) Crops

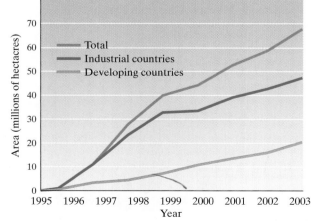

Source: Crop Biotech Net and www.isaaa.org/ko/global

a) Estimate the area in developing nations devoted to genetically modified crops in 2002.

b) Estimate the area in industrial nations devoted to genetically modified crops in 2002.

c) In what years from 1995 to 2003 was the total area devoted to genetically modified crops less than 23 million hectares?

d) In what years from 1995 to 2003 was the total area devoted to genetically modified crops greater than 50 million hectares?

59. Oil Consumption in China China's thirst for crude oil has been on the rise in recent years. The following bar graph shows China's total consumption, C, in millions of barrels, of crude oil per day. The red bars at the bottom represent China's import, I, of crude oil per day. The pink bars represent the crude oil produced in China per day for the years from 1995 to 2003.

Oil Consumption Per Day in China

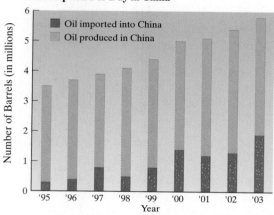

Sources: Bloomberg Financial Markets: BP Statistical Review: Bloomberg News: Customs General Administration of China, *New York Times* (12/23/04)

a) In what year was the import of crude oil to China the greatest? What was the amount imported each day?

b) In what years did the import of crude oil to China decrease from the year before?

c) Estimate I(2002).

d) Estimate the amount of crude oil produced in China per day in 2003.

60. Global Population The following graph shows the projected total global population and the projected population of children 0–14 years of age from 2002 to 2050.

Global Population

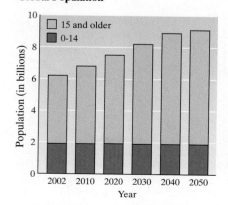

Source: U.S. Census Bureau, International Programs Center, International Data Base

a) Estimate the projected global population in 2050.

b) Estimate the projected number of children 0–14 years of age in 2050.

c) Estimate the projected number of people 15 years of age and older in 2050.

d) Estimate the projected difference in the total global population between 2002 and 2050.

61. House Sales In many regions of the country, houses sell better in the summer than at other times of the year. The graph below shows the total sales of houses in the town of Fuller from 2002 to 2006. The graph also shows the sale of houses in the summer, S, and in other times of the year, Y.

Houses Sold

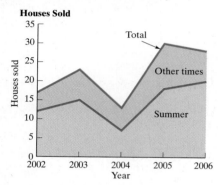

a) Estimate the number of houses sold in the summer of 2006.

b) Estimate the number of houses sold at other times in 2006.

c) Estimate Y (2005).

d) Estimate (S + Y) (2003).

62. Income Mark Whitaker owns a business where he does landscaping in the summer and snow removal in the winter. The graph below shows the total income, T, for the years 2002–2006 broken down into his landscaping income, L, and his snow removal income, S.

Income

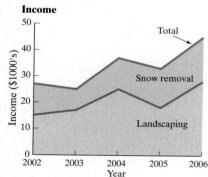

a) Estimate the total income for 2006.

b) Estimate L(2002).

c) Estimate S(2005).

d) Estimate (L + S)(2003).

63. Income The chart below shows Mr. and Mrs. Abrams's income for the years from 2002 to 2006.

	2002	2003	2004	2005	2006
Mr. Abrams	$15,500	$17,000	$8,000	$25,000	$20,000
Mrs. Abrams	$4,500	$18,000	$28,000	$7,000	$22,500

a) Draw a line graph illustrating Mr. Abram's income, Mrs. Abram's income, and their total income for the years 2002–2006. See Example 3.

b) Draw a bar graph illustrating the given information. See Example 4.

c) Draw a stacked line graph illustrating the given information. See Example 5.

64. Telephone Bills The chart below shows Kelly Lopez's home telephone bills and cellular telephone bills (rounded to the nearest $10) for the years from 2002 to 2006.

	2002	2003	2004	2005	2006
Home	$40	$50	$60	$50	$0
Cellular	$80	$50	$20	$50	$60

a) Draw a line graph illustrating the home telephone bills, the cellular phone bills, and the total phone bills for the years 2002–2006.

b) Draw a bar graph illustrating the given information.

c) Draw a stacked line graph illustrating the given information.

65. Taxes Maria Cisneros pays both federal and state income taxes. The following chart shows the amount of income taxes she paid to the federal government and to her state government (rounded to the nearest $100) for the years from 2002 to 2006.

	2002	2003	2004	2005	2006
Federal	$4000	$5000	$3000	$6000	$6500
State	$1600	$2000	$0	$1700	$1200

a) Draw a line graph illustrating the amount spent on federal taxes, the amount spent on state taxes, and the total amount spent on these two taxes for the years 2002–2006.

b) Draw a bar graph illustrating the given information.

c) Draw a stacked line graph illustrating the given information.

66. College Tuition The Olmert family has twin children, Justin and Kelly, who are attending different colleges. The tuition for Justin's and Kelly's colleges are given (to the nearest $1000) in the chart below for the years from 2004 to 2007.

	2004	2005	2006	2007
Justin	$12,000	$6000	$8000	$9000
Kelly	$2000	$8000	$8000	$5000

a) Draw a line graph illustrating the given information, including the total tuition spent on college for both Justin and Kelly for the years 2004–2007.

b) Draw a bar graph illustrating the given information.

c) Draw a stacked line graph illustrating the given information.

For Exercises 67–72, let f and g represent two functions that are graphed on the same axes.

67. If, at a, $(f + g)(a) = 0$, what must be true about $f(a)$ and $g(a)$?

68. If, at a, $(f \cdot g)(a) = 0$, what must be true about $f(a)$ and $g(a)$?

69. If, at a, $(f - g)(a) = 0$, what must be true about $f(a)$ and $g(a)$?

70. If, at a, $(f - g)(a) < 0$, what must be true about $f(a)$ and $g(a)$?

71. If, at a, $(f/g)(a) < 0$, what must be true about $f(a)$ and $g(a)$?

72. If, at a, $(f \cdot g)(a) < 0$, what must be true about $f(a)$ and $g(a)$?

Graph the following functions on your graphing calculator.

73. $y_1 = 2x + 3$
$y_2 = -x + 4$
$y_3 = y_1 + y_2$

74. $y_1 = x - 3$
$y_2 = 2x$
$y_3 = y_1 - y_2$

75. $y_1 = x$
$y_2 = x + 5$
$y_3 = y_1 \cdot y_2$

76. $y_1 = 2x^2 - 4$
$y_2 = x$
$y_3 = y_1/y_2$

Group Activity

77. SAT Scores The following graph shows the average math and verbal scores of entering college classes on the SAT college entrance exam for the years 1992 through 2004. Let f represent the math scores, and g represent the verbal scores, and let t represent the year. As a group, draw a graph that represents $(f + g)(t)$.

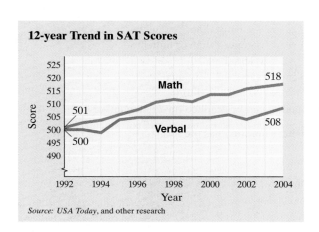

12-year Trend in SAT Scores

Source: *USA Today*, and other research

Cumulative Review Exercises

[1.9]   **78.** Evaluate $(-4)^{-3}$.

[2.6]   **79.** Solve the formula $A = \dfrac{1}{2}bh$ for h.

[3.2]   **80. Washing Machine** The cost of a washing machine, including a 6% sales tax, is $477. Determine the pre-tax cost of the washing machine.

[4.2]   **81.** Graph $3x - 4y = 12$.

[5.3]   **82.** Express 2,960,000 in scientific notation.

[8.1]   **83.** Graph $y = |x| - 2$.

Chapter 8 Summary

IMPORTANT FACTS AND CONCEPTS	EXAMPLES

Section 8.1

A **nonlinear equation** is an equation whose graph is not a straight line.	$y = x^2 + 2$ is a nonlinear equation whose graph is illustrated below. 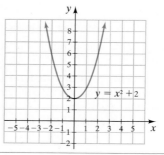
When graphing an equation where a variable appears in a denominator, the equation is not defined at the value that makes the denominator 0.	The graph of $y = \dfrac{1}{x}$ is not defined at $x = 0$. 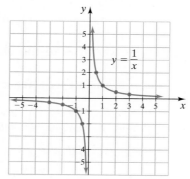

Section 8.2

Inequality Symbols

$>$ is read "is greater than."
$\geq$ is read "is greater than or equal to."
$<$ is read "is less than."
$\leq$ is read "is less than or equal to."
$\neq$ is read "is not equal to."

Inequalities can be graphed on a real number line.

$6 > 2$ is read 6 is greater than 2
$5 \geq 5$ is read 5 is greater than or equal to 5
$-4 < 3$ is read -4 is less than 3
$-10 \leq -1$ is read -10 is less than or equal to -1
$-5 \neq 17$ is read -5 is not equal to 17

$x > 3$

Set builder notation has the form

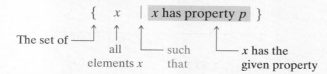

$\{x | -1 \leq x < 2\}$

$\{x | x > 4 \text{ and } x \in N\}$

IMPORTANT FACTS AND CONCEPTS	EXAMPLES

Section 8.2 (continued)

Interval Notation

When writing sets of numbers in interval notation, a parenthesis is used to indicate that an endpoint is not included in the interval and a bracket is used to indicate that an endpoint is included.

Set Builder Notation	Interval Notation
$\{x \mid -1 < x < 3\}$	$(-1, 3)$
$\{x \mid -1 \le x < 3\}$	$[-1, 3)$
$\{x \mid -1 < x \le 3\}$	$(-1, 3]$
$\{x \mid -1 \le x \le 3\}$	$[-1, 3]$
$\{x \mid -4 < x\}$	$(-4, \infty)$
$\{x \mid x \le 5\}$	$(-\infty, 5]$

For an equation in variables x and y, if the value of y depends on the value of x, then y is the **dependent variable** and x is the **independent variable**.

In the equation $y = 2x^2 + 3x - 4$, x is the independent variable and y is the dependent variable.

A **relation** is any set of ordered pairs.

A **function** is a correspondence between a first set of elements, the **domain**, and a second set of elements, the **range**, such that each element of the domain corresponds to exactly one element of the range.

Alternate Definition:

A **function** is a set of ordered pairs in which no first coordinate is repeated.

$\{(1, 2), (2, 3), (1, 4)\}$ is a relation, but not a function.

$\{(1, 6), (2, 7), (3, 10)\}$ is a relation. It is also a function since each element in the domain corresponds to exactly one element in the range.

domain: $\{1, 2, 3\}$, range $= \{6, 7, 10\}$

The **vertical line test** can be used to determine if a graph represents a function.

If a vertical line can be drawn through any part of the graph and the line intersects another part of the graph, the graph does not represent a function. If a vertical line cannot be drawn to intersect the graph at more than one point, the graph represents a function.

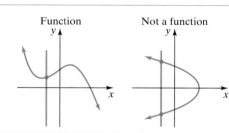

Function notation can be used to write an equation when y is a function of x. For function notation, replace y with $f(x)$, $g(x)$, $h(x)$, and so on.

$y = 7x - 9$ can be written as $f(x) = 7x - 9$

Given $y = f(x)$, to find $f(a)$, replace each x with a.

Let
$$f(x) = x^2 + 2x - 8.$$
Then
$$f(1) = 1^2 + 2(1) - 8 = -5$$
$$f(a) = a^2 + 2a - 8.$$

Section 8.3

A **linear function** is a function of the form $f(x) = ax + b$. The graph of a linear function is a straight line.

Graph $f(x) = \dfrac{1}{3}x - 2$.

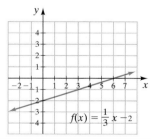

Many real-life applications can be represented as linear functions.

John earns a weekly salary of \$300 plus 10% of all dollar sales he makes. Represent his weekly salary, w, as a function of his weekly dollar sales, s.
$$w(s) = 300 + 0.10s$$

IMPORTANT FACTS AND CONCEPTS	EXAMPLES

Section 8.4

Slope can be represented as a **rate of change**. When we give the change in y per unit change in x, we are giving slope as a rate of change.

In 2005 the U.S. public debt was $7832.6 billion. In 2002 the U.S. public debt was $5957.2 billion. Determine the rate of change in the U.S. public debt from 2002 to 2005.

$$m = \frac{7832.6 - 5957.2}{2005 - 2002}$$

$$= \frac{1875.4}{3} \approx 625.13$$

The rate of change is $625.13 per year.

Vertical Translations

When a graph is shifted or translated vertically, we refer do it as a vertical translation.

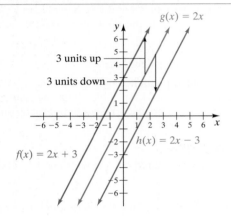

$g(x) = 2x$

3 units up

3 units down

$h(x) = 2x - 3$

$f(x) = 2x + 3$

Negative Reciprocals

The product of a number and its negative reciprocal is -1.

$$a\left(-\frac{1}{a}\right) = -1$$

The negative reciprocal of 9 is $-\frac{1}{9}$ since

$$9\left(-\frac{1}{9}\right) = -1.$$

Two lines are **parallel** if they have the same slope.

The graphs of $y = 2x + 4$ and $y = 2x + 7$ are parallel since both graphs have the same slope of 2, but different y-intercepts.

Two lines are **perpendicular** if their slopes are negative reciprocals. For any real number $a \neq 0$, its negative reciprocal is $-\frac{1}{a}$.

The graphs of $y = 3x - 5$ and $y = -\frac{1}{3}x + 8$ are perpendicular since one graph has a slope of 3 and the other graph has a slope of $-\frac{1}{3}$. The number $-\frac{1}{3}$ is the negative reciprocal of 3.

Section 8.5

Operations on Functions

   Sum of functions: $(f + g)(x) = f(x) + g(x)$

If $f(x) = x^2 + 2x - 5$ and $g(x) = x - 3$, then

$$(f + g)(x) = f(x) + g(x) = (x^2 + 2x - 5) + (x - 3)$$
$$= x^2 + 3x - 8$$

   Difference of functions: $(f - g)(x) = f(x) - g(x)$

$$(f - g)(x) = f(x) - g(x) = (x^2 + 2x - 5) - (x - 3)$$
$$= x^2 + x - 2$$

   Product of functions: $(f \cdot g)(x) = f(x) \cdot g(x)$

$$(f \cdot g)(x) = f(x) \cdot g(x)$$
$$= (x^2 + 2x - 5)(x - 3)$$
$$= x^3 - x^2 - 11x + 15$$

   Quotient of functions: $(f/g)(x) = \frac{f(x)}{g(x)}$, $g(x) \neq 0$

$$(f/g)(x) = \frac{f(x)}{g(x)} = \frac{x^2 + 2x - 5}{x - 3}, \quad x \neq 3$$

Chapter 8 Review Exercises

[8.1] *Graph each equation.*

1. $y = \dfrac{1}{2}x$

2. $y = -2x - 1$

3. $y = \dfrac{1}{2}x + 3$

4. $y = -\dfrac{3}{2}x + 1$

5. $y = x^2$

6. $y = x^2 - 1$

7. $y = |x|$

8. $y = |x| - 1$

9. $y = x^3$

10. $y = x^3 + 4$

[8.2]

11. Define function.

12. Is every relation a function? Is every function a relation? Explain.

Determine whether the following relations are functions. Explain your answers.

13.
$$
\begin{array}{l}
a \;\;\;\;\; 6 \\
b \;\;\;\;\; 7 \\
c \longrightarrow 8
\end{array}
$$

14. $\{(2, 5), (3, -4), (5, -9), (6, -1), (2, -2)\}$

List each set in roster form.

15. $A = \{x \mid 2 < x < 7 \text{ and } x \in N\}$

16. $B = \{x \mid x \ge 6 \text{ and } x \in I\}$

Represent each set using interval notation.

17. $\{x \mid x > 4\}$

18. $\{x \mid x \le -2\}$

19. $\{x \mid -1.5 < x \le 2.7\}$

For Exercises 20–23, **a)** *determine whether the following graphs represent functions;* **b)** *determine the domain and range of each relation or function.*

20.

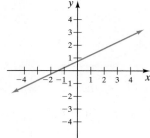

21.

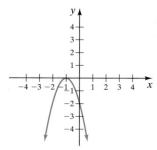

22.

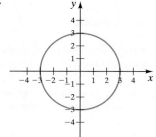

23.

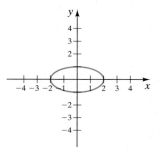

24. If $f(x) = -x^2 + 3x - 4$, find

    a) $f(2)$ and

    b) $f(h)$.

25. If $g(t) = 2t^3 - 3t^2 + 6$, find

    a) $g(-1)$ and

    b) $g(a)$.

26. Speed of Car Jane Covillion goes for a ride in a car. The following graph shows the car's speed as a function of time. Make up a story that corresponds to this graph.

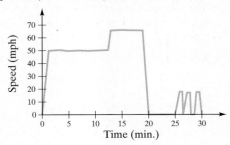

27. Apple Orchard The number of baskets of apples, N, that are produced by x trees in a small orchard ($x \leq 100$) is given by the function $N(x) = 40x - 0.2x^2$. How many baskets of apples are produced by

  a) 30 trees?

  b) 50 trees?

28. Falling Ball If a ball is dropped from the top of a 196-foot building, its height above the ground, h, at any time, t, can be found by the function $h(t) = -16t^2 + 196, 0 \leq t \leq 3.5$. Find the height of the ball at

  a) 1 second.

  b) 3 seconds.

[8.3] *Graph each function.*

29. $f(x) = \dfrac{1}{2}x - 4$

30. $f(x) = \dfrac{8}{3}x - 80$

31. $f(x) = 4$

32. Bagel Company The yearly profit, p, of a bagel company can be estimated by the function $p(x) = 0.1x - 5000$, where x is the number of bagels sold per year.

  a) Draw a graph of profits versus bagels sold for up to and including 250,000 bagels.

  b) Estimate the number of bagels that must be sold for the company to break even.

  c) Estimate the number of bagels sold if the company has $22,000 profit.

33. Interest Draw a graph illustrating the interest on a $12,000 loan for a 1-year period for various interest rates up to and including 20%. Use interest = principal · rate · time.

[8.4] *In Exercises 34–36,* **a)** *indicate if the graph represents a function.* **b)** *If so, express the equation of the graph using function notation.*

34.

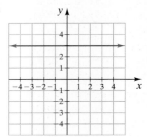

35.

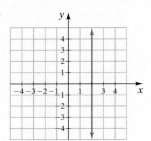

36.

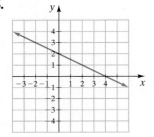

37. If the graph of $y = -2x + 5$ is translated down 4 units, determine

  a) the slope of the translated graph.

  b) the y-intercept of the translated graph.

  c) the equation of the translated graph.

38. If one point on a graph is $(-6, -4)$ and the slope is $\dfrac{2}{3}$, find the y-intercept of the graph.

39. Typhoid Fever The following chart shows the number of reported cases of typhoid fever in the United States for select years from 1970 through 2000.

  a) Plot each point and draw line segments from point to point.

  b) Compute the slope of the line segments.

  c) During which 10-year period did the number of reported cases of typhoid fever increase the most?

Year	Number of reported typhoid fever cases
1970	346
1980	510
1990	552
2000	317

Source: U.S. Dept. of Health and Human Services

40. Social Security The following graph shows the number of social security beneficiaries from 1980 projected through 2070. Use the slope-intercept form to find the function $n(t)$ (represented by the straight line) that can be used to represent this data.

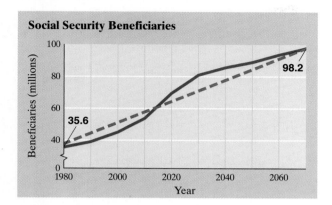

Social Security Beneficiaries

Determine whether the two given lines are parallel, perpendicular, or neither.

41. $2x - 3y = 10$

$y = \dfrac{2}{3}x - 5$

42. $2x - 3y = 7$

$-3x - 2y = 8$

43. $4x - 2y = 13$

$-2x + 4y = -9$

Find the equation of the line with the properties given. Write each answer in slope–intercept form.

44. Slope $= \dfrac{1}{2}$, through $(4, 9)$

45. Through $(-3, 1)$ and $(4, -6)$

46. Through $(0, 6)$ and parallel to the graph of $y = -\dfrac{2}{3}x + 1$

47. Through $(2, 8)$ and parallel to the graph whose equation is $5x - 2y = 7$

48. Through $(-3, 1)$ and perpendicular to the graph whose equation is $y = \dfrac{3}{5}x + 5$

49. Through $(4, 5)$ and perpendicular to the graph whose equation is $4x - 2y = 8$

Two points on l_1 and two points on l_2 are given. Determine whether l_1 is parallel to l_2, l_1 is perpendicular to l_2, or neither.

50. l_1: $(5, 3)$ and $(0, -3)$; l_2: $(1, -1)$ and $(2, -2)$

51. l_1: $(3, 2)$ and $(2, 3)$; l_2: $(4, 1)$ and $(1, 4)$

52. l_1: $(7, 3)$ and $(4, 6)$; l_2: $(5, 2)$ and $(6, 3)$

53. l_1: $(-3, 5)$ and $(2, 3)$; l_2: $(-4, -2)$ and $(-1, 2)$

54. Insurance Rates The monthly rates for $100,000 of life insurance from the General Financial Group for men increases approximately linearly from age 35 through age 50. The rate for a 35-year-old man is $10.76 per month and the rate for a 50-year-old man is $19.91 per month. Let r be the rate and let a be the age of a man between 35 and 50 years of age.

a) Determine a linear function $r(a)$ that fits these data.

b) Using the function in part **a)**, estimate the monthly rate for a 40 year-old man.

55. Burning Calories The number of calories burned in 1 hour of swimming, when swimming between 20 and 50 yards per minute, is a linear function of the speed of the swimmer. A person swimming at 30 yards per minute will burn about 489 calories in 1 hour. While swimming at 50 yards per minute a person will burn about 525 calories in 1 hour. This information is shown in the following graph.

Calories Burned while Swimming

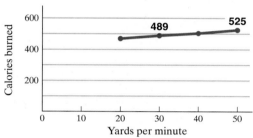

Source: Health Magazine Web Site, www.health.com

a) Determine a linear function that can be used to estimate the number of calories, C, burned in 1 hour when a person swims at r yards per minute.

b) Use the function determined in part **a)** to determine the number of calories burned in 1 hour when a person swims at 40 yards per minute.

c) Use the function determined in part **a)** to estimate the speed at which a person needs to swim to burn 600 calories in 1 hour.

[8.5] *Given $f(x) = x^2 - 3x + 4$ and $g(x) = 2x - 5$, find the following.*

56. $(f + g)(x)$

57. $(f + g)(4)$

58. $(g - f)(x)$

59. $(g - f)(-1)$

60. $(f \cdot g)(-1)$

61. $(f \cdot g)(3)$

62. $(f/g)(1)$

63. $(f/g)(2)$

64. Female Population According to the U.S. Census, the female population is expected to grow worldwide. The following graph shows the female population worldwide for selected years from 2002 to 2050.

Global Female Population

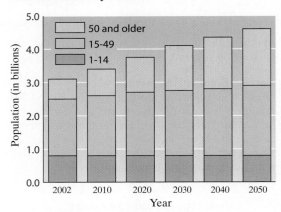

Source: U.S. Census Bureau, International Programs Center, International Data Base.

a) Estimate the projected female population worldwide in 2050.

b) Estimate the projected number of women 15–49 years of age in 2050.

c) Estimate the number of women who are projected to be in the 50 years and older age group in 2010.

d) Estimate the projected percent increase in the number of women 50 years and older from 2002 to 2010.

65. Retirement Income Ginny Jennings recently retired from her full-time job. The following graph shows her retirement income for the years 2003–2006.

Ginny's Retirement Income

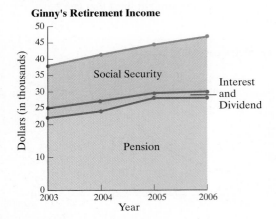

a) Estimate Ginny's total retirement income in 2006.

b) Estimate Ginny's pension income in 2005.

c) Estimate Ginny's interest and dividend income in 2003.

Chapter 8 Practice Test

To find out how well you understand the chapter material, take this practice test. The answers, and the section where the material was initially discussed, are given in the back of the book. Each problem is also fully worked out on the **Chapter Test Prep Video CD**. *Review any questions that you answered incorrectly.*

1. Graph $y = -2x + 1$.

2. Graph $y = \sqrt{x}$.

3. Graph $y = x^2 - 4$.

4. Graph $y = |x|$.

5. Define *function*.

6. Is the following set of ordered pairs a function? Explain your answer.

$$\{(3, 1), (-2, 6), (4, 6), (5, 2), (7, 3)\}$$

In Exercises 7 and 8, determine whether the graphs represent functions. Give the domain and range of the relation or function.

7.

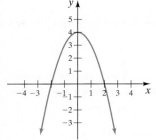

8.

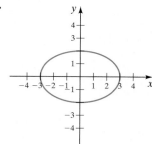

9. If $f(x) = 3x^2 - 6x + 5$, find $f(-2)$.

10. Graph $f(x) = -3$

11. Profit Graph The yearly profit, p, for Zico Publishing Company on the sales of a particular book can be estimated by the function $p(x) = 10.2x - 50,000$, where x is the number of books produced and sold.

   a) Draw a graph of profit versus books sold for up to and including 30,000 books.

   b) Use function $p(x)$ to estimate the number of books that must be sold for the company to break even.

   c) Use function $p(x)$ to estimate the number of books that the company must sell to make a \$100,000 profit.

12. U.S. Population Determine the function represented by the red line on the graph that can be used to estimate the projected U.S. population, p, from 2000 to 2050. Let 2000 be the reference year so that 2000 corresponds to $t = 0$.

U.S. Population Projections 2000–2050

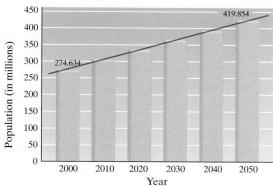

Source: U.S. Bureau of the Census, Statistical Abstract of the United States: 2004-2005

13. Determine whether the graphs of the two equations are parallel, perpendicular, or neither. Explain your answer.

$$2x - 3y = 12$$
$$4x + 10 = 6y$$

14. Determine the equation, in slope-intercept form, of the line that goes through the point $(6, -5)$ and is perpendicular to the graph of $y = \frac{1}{2}x + 1$.

15. Heart Disease Deaths due to heart disease has been declining approximately linearly since the year 2000. The bar graph below shows the number of deaths, per 100,000 deaths, due to heart disease in selected years, projected for 2006–2010.

Heart Disease Death Rate

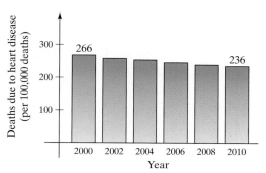

Source: U.S. Dept. of Health and Human Services

   a) Let r be the number of deaths due to heart disease per 100,000 deaths, and let t represent the years since 2000. Write the linear function $r(t)$ that can be used to approximate the data.

   b) Use the function from part **a)** to estimate the death rate due to heart disease in 2006.

   c) Assuming this trend continues until the year 2020, estimate the death rate due to heart disease in 2020.

In Exercises 16–18, if $f(x) = 2x^2 - x$ and $g(x) = x - 6$, find

16. $(f + g)(3)$ **17.** $(f/g)(-1)$

18. $f(a)$

19. Paper Use The following graph shows paper use in 1995 and projected paper use from 1995 through 2015.

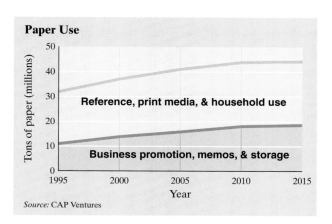

Source: CAP Ventures

   a) Estimate the total number of tons of paper to be used in 2010.

   b) Estimate the number of tons of paper to be used by businesses in 2010.

   c) Estimate the number of tons of paper to be used for reference, print media, and household use in 2010.

Cumulative Review Test

Take the following test and check your answers with those given in the back of the book. Review any questions that you answered incorrectly. The section where the material was covered is indicated after the answer.

1. Evaluate $4x^2 - 7xy^2 + 6$ when $x = -3$ and $y = -2$.

2. Evaluate $2 - \{3[6 - 4(6^2 \div 4)]\}$.

3. Solve $2(x + 4) - 5 = -3[x - (2x + 1)]$.

4. Solve the formula $P = 2E + 3R$ for R.

5. Write the equation $5x - 2y = 12$ in slope-intercept form.

6. Simplify $\left(\dfrac{6x^2y^3}{2x^5y}\right)^3$.

7. Simplify $(6x^2 - 3x - 2) - (-2x^2 - 8x - 1)$.

8. Multiply $(4x^2 - 6x + 3)(3x - 5)$.

9. Factor $6a^2 - 6a - 5a + 5$.

10. Solve $2x^2 = x + 15$.

11. Multiply $\dfrac{x^2 - 9}{x^2 - x - 6} \cdot \dfrac{x^2 - 2x - 8}{2x^2 - 7x - 4}$.

12. Subtract $\dfrac{x}{x + 4} - \dfrac{3}{x - 5}$.

13. Add $\dfrac{4}{x^2 - 3x - 10} + \dfrac{2}{x^2 + 5x + 6}$.

14. Solve $\dfrac{x}{9} - \dfrac{x}{6} = \dfrac{1}{12}$.

15. Solve $\dfrac{7}{x + 3} + \dfrac{5}{x + 2} = \dfrac{5}{x^2 + 5x + 6}$.

16. a) Determine whether the following graph represents a function.
    b) Find the domain and range of the graph.

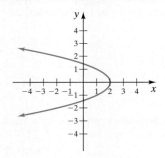

17. Determine whether the graphs of the two given equations are parallel, perpendicular, or neither.

$$2x - 5y = 6$$
$$5x - 2y = 9$$

In Exercises 18 and 19, if $f(x) = x^2 + 3x - 2$ and $g(x) = 4x - 6$, find

18. $(f + g)(x)$.

19. $(f \cdot g)(4)$.

20. **Natural Gas Consumption** The total consumption of natural gas in 2003 was 21.8 trillion cubic feet (2.18×10^{13}). The following pie chart shows the breakdown of consumption by sector.

Natural Gas Consumption by Sector
(2.18×10^{13} cubic feet)

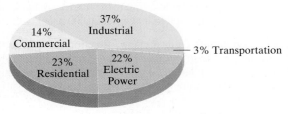

(Total is 99% due to rounding.)

Source: Energy Information Administration

Answer the following questions using scientific notation.

a) What was the amount of natural gas consumption by the commercial sector in 2003?

b) How much more natural gas was consumed by the industrial sector than by the transportation sector in 2003?

c) If the consumption of natural gas is expected to increase by a total of 10% from 2003 to 2006, what will be the consumption of natural gas in 2006?

9 Systems of Linear Equations

SYSTEMS OF EQUATIONS ARE frequently used to solve real-life problems. For example, in Example 6 on pages 580 and 581 we use a system of equations to determine how much of two given solutions a chemist must mix to get a third solution with the desired chemical composition.

9.1 Solving Systems of Equations Graphically

1 Determine if an ordered pair is a solution to a system of equations.

2 Determine if a system of equations is consistent, inconsistent, or dependent.

3 Solve a system of equations graphically.

1 Determine If an Ordered Pair Is a Solution to a System of Equations.

When we seek a common solution to two or more linear equations, the equations are called a **system of linear equations**. An example of a system of linear equations follows:

$$\left.\begin{array}{l}(1)\ y = x + 5 \\ (2)\ y = 2x + 4\end{array}\right\} \quad \textit{System of linear equations}$$

The **solution to a system of equations** is the ordered pair or pairs that satisfy all equations in the system. The solution to the system above is $(1, 6)$.

Check   In Equation (1)   In Equation (2)

$(1, 6)$ $(1, 6)$

$y = x + 5$ $y = 2x + 4$

$6 \overset{?}{=} 1 + 5$ $6 \overset{?}{=} 2(1) + 4$

$6 = 6$ *True* $6 = 6$ *True*

Because the ordered pair $(1, 6)$ satisfies *both* equations, it is a solution to the system of equations. Notice that the ordered pair $(3, 8)$ satisfies the first equation but does not satisfy the second equation.

Check   In Equation (1)   In Equation (2)

$(3, 8)$ $(3, 8)$

$y = x + 5$ $y = 2x + 4$

$8 \overset{?}{=} 3 + 5$ $8 \overset{?}{=} 2(3) + 4$

$8 = 8$ *True* $8 = 10$ *False*

Since the ordered pair $(3, 8)$ does not satisfy *both* equations, it is *not* a solution to the system of equations.

EXAMPLE 1 ▶ Determine which of the following ordered pairs satisfy the system of equations.

$$y = 2x - 8$$
$$2x + y = 4$$

a) $(1, -6)$ **b)** $(3, -2)$

Solution

a) Substitute 1 for x and -6 for y in each equation.

$y = 2x - 8$ $2x + y = 4$

$-6 \overset{?}{=} 2(1) - 8$ $2(1) + (-6) \overset{?}{=} 4$

$-6 \overset{?}{=} 2 - 8$ $2 - 6 \overset{?}{=} 4$

$-6 \overset{?}{=} -6$ *True* $-4 = 4$ *False*

Since $(1, -6)$ does not satisfy both equations, it is not a solution to the system of equations.

b) Substitute 3 for x and -2 for y in each equation.

$$y = 2x - 8 \qquad\qquad\qquad 2x + y = 4$$
$$-2 \overset{?}{=} 2(3) - 8 \qquad\qquad 2(3) + (-2) \overset{?}{=} 4$$
$$-2 \overset{?}{=} 6 - 8 \qquad\qquad\qquad 6 - 2 \overset{?}{=} 4$$
$$-2 = -2 \quad \textit{True} \qquad\qquad\qquad 4 = 4 \quad \textit{True}$$

Since $(3, -2)$ satisfies both equations, it is a solution to the system of linear equations.

▶ **Now Try Exercise 13**

In this chapter we discuss several methods for finding the solution to a system of equations including: the *graphical method*, the *substitution method*, and the *addition method*. In this section we discuss the graphical method.

2 Determine If a System of Equations Is Consistent, Inconsistent, or Dependent

The **solution to a system of linear equations** is the ordered pair (or pairs) common to all lines in the system when the lines are graphed. When two lines are graphed, three situations are possible, as illustrated in **Figure 9.1**.

In **Figure 9.1a**, lines 1 and 2 are not parallel lines. They intersect at exactly one point. This system of equations has *exactly one solution*. This is an example of a **consistent system of equations**. A consistent system of equations is a system of equations that has a solution.

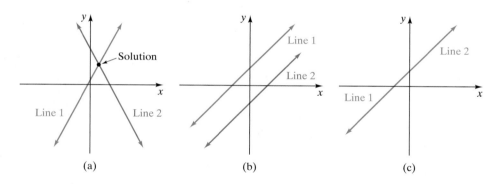

FIGURE 9.1   (a)   (b)   (c)

In **Figure 9.1b**, lines 1 and 2 are two different parallel lines. The lines do not intersect, and this system of equations has *no solution*. This is an example of an **inconsistent system of equations**. An inconsistent system of equations is a system of equations that has no solution.

In **Figure 9.1c**, lines 1 and 2 are actually the same line. In this case, every point on the line satisfies both equations and is a solution to the system of equations. This system has *an infinite number of solutions*. This is an example of a **dependent system of equations**. A dependent system of linear equations is a system of equations that has an infinite number of solutions. If a system of two linear equations is dependent, then both equations represent the same line. *Note that a dependent system is also a consistent system since it has a solution.*

We can determine if a system of linear equations is consistent, inconsistent, or dependent by writing each equation in slope-intercept form and comparing the slopes and y-intercepts. If the slopes of the lines are different (**Fig. 9.1a**), the system is consistent. If the slopes are the same but the y-intercepts are different (**Fig. 9.1b**), the system is inconsistent. If both the slopes and the y-intercepts are the same (**Fig. 9.1c**), the system is dependent.

EXAMPLE 2 ▶ Determine whether the following system has exactly one solution, no solution, or an infinite number of solutions.

$$2x + 4y = -15$$
$$-8y = 4x - 12$$

Solution   Write each equation in slope-intercept form and then compare the slopes and the *y*-intercepts.

$$2x + 4y = -15 \qquad\qquad -8y = 4x - 12$$

$$4y = -2x - 15 \qquad\qquad y = \frac{4x - 12}{-8}$$

$$y = \frac{-2x - 15}{4} \qquad\qquad y = -\frac{4}{8}x + \frac{12}{8}$$

$$y = -\frac{2}{4}x - \frac{15}{4} \qquad\qquad y = -\frac{1}{2}x + \frac{3}{2}$$

$$y = -\frac{1}{2}x - \frac{15}{4}$$

Since the lines have the same slope, $-\frac{1}{2}$, and different *y*-intercepts, the lines are parallel. This system of equations is therefore inconsistent and has no solution.

▶ **Now Try Exercise 29**

3   Solve a System of Equations Graphically

Now we will see how to solve systems of equations graphically.

To Obtain the Solution to a System of Equations Graphically

Graph each equation and determine the point or points of intersection.

EXAMPLE 3 ▶ Solve the following system of equations graphically.

$$2x + y = 11$$
$$x + 3y = 18$$

Solution   Find the *x*- and *y*-intercepts of each graph; then draw the graphs.

$2x + y = 11$	Ordered Pair	$x + 3y = 18$	Ordered Pair
Let $x = 0$; then $y = 11$.	$(0, 11)$	Let $x = 0$; then $y = 6$.	$(0, 6)$
Let $y = 0$; then $x = \frac{11}{2}$.	$\left(\frac{11}{2}, 0\right)$	Let $y = 0$; then $x = 18$.	$(18, 0)$

The two graphs (**Fig. 9.2**) appear to intersect at the point $(3, 5)$. The point $(3, 5)$ may be the solution to the system of equations. To be sure, however, we must check to see that $(3, 5)$ satisfies *both* equations.

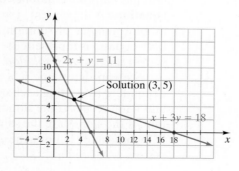

FIGURE 9.2

Check $2x + y = 11$ $x + 3y = 18$

$2(3) + 5 \overset{?}{=} 11$ $3 + 3(5) \overset{?}{=} 18$

$11 = 11$ *True* $18 = 18$ *True*

Since the ordered pair $(3, 5)$ checks in both equations, it is the solution to the system of equations. This system of equations is consistent.

▶ **Now Try Exercise 43**

EXAMPLE 4 ▶ Solve the following system of equations graphically.

$$2x + y = 3$$
$$4x + 2y = 12$$

Solution   Find the x- and y-intercepts of each graph; then draw the graphs.

$2x + y = 3$	Ordered Pair	$4x + 2y = 12$	Ordered Pair
Let $x = 0$; then $y = 3$.	$(0, 3)$	Let $x = 0$; then $y = 6$.	$(0, 6)$
Let $y = 0$; then $x = \dfrac{3}{2}$.	$\left(\dfrac{3}{2}, 0\right)$	Let $y = 0$; then $x = 3$.	$(3, 0)$

The two lines (**Fig. 9.3**) appear to be parallel.

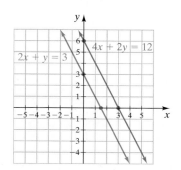

FIGURE 9.3

To show that the two lines are indeed parallel, write each equation in slope-intercept form.

$$2x + y = 3 \qquad\qquad 4x + 2y = 12$$
$$y = -2x + 3 \qquad\qquad 2y = -4x + 12$$
$$y = -2x + 6$$

Both equations have the same slope, -2, and different y-intercepts; thus the lines must be parallel. Since parallel lines do not intersect, this system of equations has no solution. This system of equations is inconsistent.

▶ **Now Try Exercise 55**

EXAMPLE 5 ▶ Solve the following system of equations graphically.

$$x - \frac{1}{2}y = 2$$
$$y = 2x - 4$$

Solution   Find the x- and y-intercepts of each graph; then draw the graphs.

$x - \dfrac{1}{2}y = 2$	Ordered Pair	$y = 2x - 4$	Ordered Pair
Let $x = 0$; then $y = -4$.	$(0, -4)$	Let $x = 0$; then $y = -4$.	$(0, -4)$
Let $y = 0$; then $x = 2$.	$(2, 0)$	Let $y = 0$; then $x = 2$.	$(2, 0)$

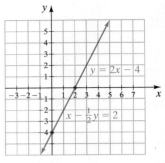

FIGURE 9.4

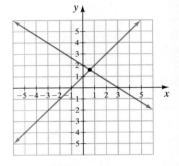

FIGURE 9.5

Because the two lines have the same x- and y-intercepts, both equations represent the same line (**Fig. 9.4**). When the equations are written in slope-intercept form, it becomes clear that the equations are identical and the system is dependent.

$$x - \frac{1}{2}y = 2 \qquad\qquad y = 2x - 4$$

$$2\left(x - \frac{1}{2}y\right) = 2(2)$$

$$2x - y = 4$$

$$-y = -2x + 4$$

$$y = 2x - 4$$

The solution to this system of equations is all the points on the line.

▶ **Now Try Exercise 53**

When graphing a system of equations, the intersection of the lines is not always easy to read on the graph. For example, can you determine the solution to the system of equations shown in **Figure 9.5**? You may estimate the solution to be $\left(\frac{7}{10}, \frac{3}{2}\right)$ when it may actually be $\left(\frac{4}{5}, \frac{8}{5}\right)$. The accuracy of your answer will depend on how carefully you draw the graphs and on the scale of the graph paper used. In Section 9.2 we present an algebraic method that gives exact solutions to systems of equations.

USING YOUR GRAPHING CALCULATOR

A graphing calculator can be used to solve or check systems of equations. To solve a system of equations graphically, graph both equations as was explained in Chapter 4. The point or points of intersection of the graphs is the solution to the system of equations.

The first step in solving the system on a graphing calculator is to solve each equation for y. For example, consider the system of equations given in Example 3.

System of Equations	Equations Solved for y
$2x + y = 11$	$y = -2x + 11$
$x + 3y = 18$	$y = -\frac{1}{3}x + 6$

Let $Y_1 = -2x + 11$

$$Y_2 = -\frac{1}{3}x + 6.$$

Figure 9.6 shows the screen of a TI-84 Plus with these two equations graphed.

To find the intersection of the two graphs, you can use the $\boxed{\text{TRACE}}$ and $\boxed{\text{ZOOM}}$ keys or the TABLE feature as was explained earlier. **Figure 9.7** shows a table of values for equations Y_1 and Y_2. Notice when $x = 3$, Y_1 and Y_2 both have the same value, 5. Therefore both graphs intersect at $(3, 5)$, and $(3, 5)$ is the solution.

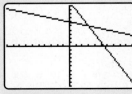

FIGURE 9.6

X	Y₁	Y₂
0	11	6
1	9	5.6667
2	7	5.3333
3	5	5
4	3	4.6667
5	1	4.3333
6	-1	4

X=3

FIGURE 9.7

(continued on the next page)

Some graphing calculators have a feature that displays the intersection of graphs by pressing a sequence of keys. For example, on the TI-84 Plus if you go to CALC (which stands for calculate) by pressing $\boxed{2^{nd}}$ $\boxed{\text{TRACE}}$, you get the screen shown in **Figure 9.8**. Now press $\boxed{5}$, intersect. Once the *intersect* feature has been selected, the calculator will display the graph and the question

FIRST CURVE?

At this time, move the cursor along the first curve until it is close to the point of intersection. Then press $\boxed{\text{ENTER}}$ The calculator now shows

SECOND CURVE?

and has the cursor on the second curve. If the cursor is not close to the point of intersection, move it along this curve until it is close to the intersection. Then press $\boxed{\text{ENTER}}$. The calculator now displays

GUESS?

Now press $\boxed{\text{ENTER}}$ and the point of intersection is displayed; see **Figure 9.9**. The point of intersection is $(3, 5)$.

FIGURE 9.8

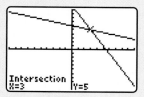

FIGURE 9.9

EXERCISES

Use your graphing calculator to find the solution to each system of equations. Round noninteger answers to the nearest tenth.

1. $x + 2y = -11$
  $2x - y = -2$

2. $x - 3y = -13$
  $-2x - 2y = 2$

3. $2x - y = 7.7$
  $-x - 3y = 1.4$

4. $3x + 2y = 7.8$
  $-x + 3y = 15.0$

In the example that follows, we will work an application problem using two variables and illustrate the solution in the form of a graph. Although an answer may sometimes be easier to obtain using only one variable, a graph of the situation may help you to better visualize the total picture.

EXAMPLE 6 ▸ **Security Systems** Meghan O'Donnell plans to install a security system in her house. She has narrowed her choices to two security dealers: Moneywell and Doile. Moneywell's system costs $3580 to install and the monitoring fee is $20 per month. Doile's equivalent system costs only $2620 to install, but the monitoring fee is $32 per month.

a) Assuming that the monthly monitoring fees do not change, in how many months would the total cost of Moneywell's and Doile's systems be the same?

b) If both dealers guarantee not to raise monthly fees for 10 years, and if Meghan plans to use the system for 10 years, which system would be the least expensive?

Solution **a)** Understand and Translate We need to determine the number of months for which both systems will have the same total cost.

Let n = number of months

c = total cost of the security system over n months.

Now we can write an equation to represent the cost of each system using the two variables c and n.

Moneywell	Doile
$\text{Total cost} = \begin{pmatrix} \text{initial} \\ \text{cost} \end{pmatrix} + \begin{pmatrix} \text{fees over} \\ n \text{ months} \end{pmatrix}$	$\text{Total cost} = \begin{pmatrix} \text{initial} \\ \text{cost} \end{pmatrix} + \begin{pmatrix} \text{fees over} \\ n \text{ months} \end{pmatrix}$
$c = 3580 + 20n$	$c = 2620 + 32n$

Thus, our system of equations is

$$c = 3580 + 20n$$
$$c = 2620 + 32n$$

Carry Out   Now let's graph each equation. Following are tables of values.

$$c = 3580 + 20n$$

Let $n = 0$. $c = 3580 + 20(0) = 3580$
Let $n = 100$. $c = 3580 + 20(100) = 5580$
Let $n = 160$. $c = 3580 + 20(160) = 6780$

n	c
0	3580
100	5580
160	6780

$$c = 2620 + 32n$$

Let $n = 0$. $c = 2620 + 32(0) = 2620$
Let $n = 100$. $c = 2620 + 32(100) = 5820$
Let $n = 160$. $c = 2620 + 32(160) = 7740$

n	c
0	2620
100	5820
160	7740

Figure 9.10 shows the graphs of the equations.

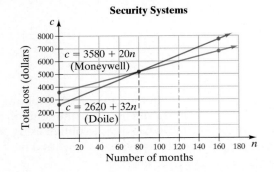

Security Systems

FIGURE 9.10

Check and Answer   The graph (**Fig. 9.10**) shows that the total cost of the two security systems would be the same in 80 months.
b) Since 10 years is 120 months, we draw a dashed vertical line at $n = 120$ months and see where it intersects the two lines. Since at 120 months the Doile line is higher than the Moneywell line, the cost for the Doile system for 120 months is more than the cost of the Moneywell system. Therefore, the cost of the Moneywell system would be less expensive for 10 years.

▸ **Now Try Exercise 71**

We will discuss applications of systems of equations further in Section 9.5.

Helpful Hint

An equation in one variable may be solved using a system of linear equations. Consider the equation $3x - 1 = x + 1$. Its solution is 1, as illustrated below.

$$3x - 1 = x + 1$$
$$2x - 1 = 1$$
$$2x = 2$$
$$x = 1$$

(continued on the next page)

Let's set each side of the equation $3x - 1 = x + 1$ equal to y to obtain the following system of equations:

$$y = 3x - 1$$
$$y = x + 1$$

The graphical solution of this system of equations is illustrated in **Figure 9.11**.

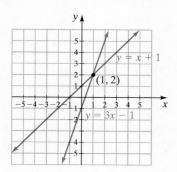

FIGURE 9.11

The solution to the system is $(1, 2)$. Notice that the x-coordinate of the solution of the system, 1, is the solution to the linear equation in one variable, $3x - 1 = x + 1$. If you have a difficult equation to solve and you have a graphing calculator, you can solve the equation on your calculator. The x-coordinate of the solution to the system will be the solution to the linear equation in one variable.

EXERCISE SET 9.1   Math XL   MyMathLab
MathXL®   MyMathLab

Concept/Writing Exercises

1. What does the solution to a system of equations represent?

2. a) What is a consistent system of equations?

  b) What is an inconsistent system of equations?

  c) What is a dependent system of equations?

3. Explain how to determine without graphing if a system of linear equations has exactly one solution, no solution, or an infinite number of solutions.

4. When a dependent system of two linear equations is graphed, what will be the result?

5. Explain why it may be difficult to obtain an exact answer to a system of equations graphically.

6. Is a dependent system of equations a consistent system or an inconsistent system? Explain.

Practice the Skills

Determine which, if any, of the following ordered pairs satisfy each system of linear equations.

7. $y = 3x - 6$
  $y = -3x$
  a) $(2, 0)$ **b)** $(0, 0)$ **c)** $(1, -3)$

8. $y = -4x$
  $y = -2x + 8$
  a) $(0, 8)$ **b)** $(-4, 16)$ **c)** $(3, -12)$

9. $y = 2x - 3$
  $y = x + 5$
  a) $(8, 13)$ **b)** $(4, 5)$ **c)** $(5, 7)$

10. $x + 2y = 4$
  $y = 3x - 5$
  a) $(0, 2)$ **b)** $(2, 1)$ **c)** $(4, 0)$

11. $4x + y = 15$
  $5x + y = 10$
  a) $(3, 3)$ **b)** $(2, 0)$ **c)** $(-1, 19)$

12. $y = 2x + 6$
  $y = 2x - 1$
  a) $(0, 6)$ **b)** $(3, 5)$ **c)** $(-2, 0)$

13. $4x - 6y = 12$

$y = \dfrac{2}{3}x - 2$

a) $(3, 0)$ **b)** $(9, 4)$ **c)** $(6, 1)$

14. $y = -x + 5$

$2y = -2x + 10$

a) $(6, -1)$ **b)** $(0, 5)$ **c)** $(-2, 3)$

15. $3x - 4y = 8$

$2y = \dfrac{2}{3}x - 4$

a) $(0, -2)$ **b)** $(1, -6)$ **c)** $\left(-\dfrac{1}{3}, -\dfrac{9}{4}\right)$

16. $2x + 3y = 6$

$-x + \dfrac{5}{2} = \dfrac{1}{2}y$

a) $\left(\dfrac{1}{2}, \dfrac{5}{3}\right)$ **b)** $\left(-2, \dfrac{10}{3}\right)$ **c)** $\left(\dfrac{9}{4}, \dfrac{1}{2}\right)$

Identify each system of linear equations (lines are labeled 1 and 2) as consistent, inconsistent, or dependent. State whether the system has exactly one solution, no solution, or an infinite number of solutions.

17.

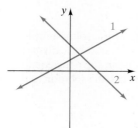

18.

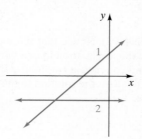

19.

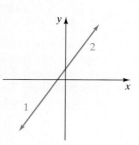

20.

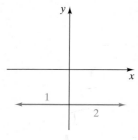

21.

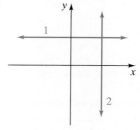

22.

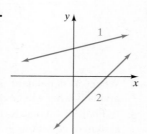

23.

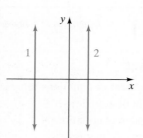

24.

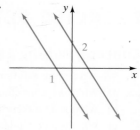

Express each equation in slope-intercept form. Without graphing the equations, state whether the system of equations has exactly one solution, no solution, or an infinite number of solutions.

25. $y = 2x - 1$

$3y = 5x - 6$

26. $x + y = 8$

$x - y = 8$

27. $2y = 3x + 3$

$y = \dfrac{3}{2}x - 2$

28. $y = \dfrac{1}{2}x + 3$

$2y = x + 6$

29. $2x = y - 6$

$3x = 3y + 5$

30. $x + 3y = 6$

$4x + y = 4$

31. $3x + 5y = -7$

$-3x - 5y = -10$

32. $x - y = 2$

$2x - 2y = -6$

33. $x = 3y + 5$

$2x - 6y = 10$

34. $x - y = 3$
$\frac{1}{2}x - 2y = -8$

35. $y = \frac{3}{2}x + \frac{1}{2}$
$3x - 2y = \frac{5}{2}$

36. $2y = \frac{7}{3}x - 9$
$4y = 8x + 9$

Determine the solution to each system of equations graphically. If the system is dependent or inconsistent, so state.

37. $y = x + 3$
$y = -x + 3$

38. $y = 2x + 4$
$y = -3x - 6$

39. $y = 3x - 6$
$y = -x + 6$

40. $y = 3x - 4$
$y = -x$

41. $4x = 8$
$y = -3$

42. $2x - y = 7$
$2y = 2x - 6$

43. $x + y = 5$
$-x + y = 1$

44. $-x + 2y = 7$
$2x - y = -2$

45. $y = -\frac{1}{2}x + 4$
$x + 2y = 6$

46. $-x + 2y = 0$
$2x - y = -3$

47. $x + 2y = 8$
$5x + 2y = 0$

48. $3x + y = -6$
$2x = 1 + y$

49. $2x + 3y = 6$
$4x = -6y + 12$

50. $2x + 5y - 6 = 0$
$2x - 3y = 6$

51. $y = 3$
$y = 2x - 3$

52. $x = 5$
$y = 2x - 8$

53. $x - 2y = 4$
$2x - 4y = 8$

54. $4x - y = 6$
$2y = 8x - 12$

55. $2x + y = -2$
$6x + 3y = 6$

56. $y = 2x - 1$
$2y = 4x + 5$

57. $4x - 3y = 6$
$2x + 4y = 14$

58. $2x + 6y = 12$
$y = -\frac{1}{3}x + 2$

59. $2x - 3y = 0$
$x + 2y = 0$

60. $4x = 4y - 12$
$-8x + 8y = 8$

Problem Solving

61. Given the system of equations $6x - 4y = 12$ and $12y = 18x - 24$, determine without graphing whether the graphs of the two equations will be parallel lines. Explain how you determined your answer.

62. Given the system of equations $4x - 8y = 12$ and $2x - 8 = 4y$, determine without graphing whether the graphs of the two equations will be parallel lines. Explain how you determined your answer.

63. If a system of linear equations has solutions $(4, 3)$ and $(6, 5)$, how many solutions does the system have? Explain.

64. If the slope of one line in a system of linear equations is 2 and the slope of the second line in the system is 3, how many solutions does the system have? Explain.

65. If two distinct lines are parallel, how many solutions does the system have? Explain.

66. If two different lines in a linear system of equations pass through the origin, must the solution to the system be $(0, 0)$? Explain.

67. Consider the system $x = 5$ and $y = 3$. How many solutions does the system have? What is the solution?

68. A system of linear equations has $(3, -1)$ as its solution. If one line in the system is vertical and the other line is horizontal, determine the equations in the system.

In Exercises 69–72, find each solution by graphing the system of equations.

69. Furnace Repair Edith Hall's furnace is 10 years old and has a problem. The furnace repair man indicates that it will cost Edith $600 to repair her furnace. She can purchase a new, more efficient furnace for $1800. Her present furnace averages about $650 per year for energy cost and the new furnace would average about $450 per year.

We can represent the total cost, *c*, of repair or replacement, plus energy cost over *n* years by the following system of equations.

$$
\begin{aligned}
(\text{repair}) \qquad & c = 600 + 650n \\
(\text{replacement}) \qquad & c = 1800 + 450n
\end{aligned}
$$

Find the number of years for which the total cost of repair would equal the total cost of replacement.

70. Security Systems Juan Varges is considering the two security systems discussed in Example 6. If Moneywell's system costs $4400 plus $15 per month and Doile's system costs $3400 plus $25 per month, after how many months would the total cost of the two systems be the same?

71. Boat Ride Rudy has visitors at his home and wants to take them out on a pontoon boat for a day. There are two pontoon boat rental agencies on the lake. Bob's Boat Rental charges $25 per hour for the boat rental, which includes all the gasoline used. Hopper's Rental charges $21 per hour plus a flat charge of $28 for the gasoline used. The equations that represent the total cost, *c*, follow. In the equations *h* represents the number of hours the boats are rented.

$$
\begin{aligned}
c &= 25h \\
c &= 21h + 28
\end{aligned}
$$

Determine the number of hours the boats must be rented for the total cost to be the same.

72. Landscaping The Evergreen Landscape Service charges a consultation fee of $200 plus $50 per hour for labor. The Out of Sight Landscape Service charges a consultation fee of $300 plus $40 per hour for labor. We can represent this situation with the system of equations

$$
\begin{aligned}
c &= 200 + 50h \\
c &= 300 + 40h
\end{aligned}
$$

where *c* is the total cost and *h* is the number of hours of labor. Find the number of hours of labor for the two services to have the same total cost.

Group Activity

Discuss and answer Exercises 73–78 as a group. Suppose that a system of three linear equations in two variables is graphed on the same axes. Find the maximum number of points where two or more of the lines can intersect if

73. the three lines have the same slope but different *y*-intercepts.

74. the three lines have the same slope and the same *y*-intercept.

75. two lines have the same slope but different *y*-intercepts and the third line has a different slope.

76. the three lines have different slopes but the same *y*-intercept.

77. the three lines have different slopes but two have the same *y*-intercept.

78. the three lines have different slopes and different *y*-intercepts.

Cumulative Review Exercises

[2.1] **79.** Simplify $3x - (x - 6) + 4(3 - x)$.

[2.5] **80.** Solve $2(x + 3) - x = 5x + 2$.

[2.6] **81.** If $A = p(1 + rt)$, find *r* when $A = 1000, t = 20$, and $p = 500$.

[4.2] **82. a)** Find the *x*- and *y*-intercepts of $2x + 3y = 12$.
     b) Use the intercepts to graph the equation.

[7.1] **83.** Simplify $\dfrac{x^2 - 9x + 14}{2 - x}$.

[7.6] **84.** Solve $\dfrac{4}{b} + 2b = \dfrac{38}{3}$.

9.2 Solving Systems of Equations by Substitution

1 Solve systems of equations by substitution.

As we stated in Section 9.1, a graphic solution to a system of equations may be inaccurate since you may need to estimate the coordinates of the point of intersection. When an exact solution is necessary, the system should be solved algebraically, either by substitution or by addition of equations.

1 Solve Systems of Equations by Substitution

The procedure for solving a system of equations by **substitution** is illustrated in Example 1. The procedure for solving by addition is presented in Section 9.3. Regardless of which of the two algebraic techniques is used to solve a system of equations, our immediate goal remains the same, that is, *to obtain one equation containing only one unknown.*

EXAMPLE 1 ▸ Solve the following system of equations by substitution.

$$x + 3y = 18$$
$$2x + y = 11$$

Solution Begin by solving for one of the variables in either of the equations. You may solve for any of the variables; however, if you solve for a variable with a numerical coefficient of 1 or -1, you may avoid working with fractions. In this system the x-term in $x + 3y = 18$ and the y-term in $2x + y = 11$; both have a numerical coefficient of 1.
   Let's solve for x in $x + 3y = 18$.

$$x + 3y = 18$$
$$x = -3y + 18$$

Next, substitute $-3y + 18$ for x in the *other equation*, $2x + y = 11$, and solve for the remaining variable, y.

$$2x + y = 11$$
$$2(\overbrace{-3y + 18}) + y = 11 \qquad \textit{Substitution; this is now an equation in only one variable, y.}$$
$$-6y + 36 + y = 11$$
$$-5y + 36 = 11$$
$$-5y = -25$$
$$y = 5$$

Finally, substitute $y = 5$ in the equation that is solved for x and find the value of x,

$$x = -3y + 18$$
$$x = -3(5) + 18$$
$$x = -15 + 18$$
$$x = 3$$

A check using $x = 3$ and $y = 5$ in both equations will show that the solution is the ordered pair $(3, 5)$.

▸ **Now Try Exercise 5**

   The system of equations in Example 1 could also be solved by substitution by solving the equation $2x + y = 11$ for y. Do this now. You should obtain the same answer.
   Note that the solution in Example 1 is identical to the graphical solution obtained in Example 3 of Section 9.1. Now we summarize the procedure for solving a system of equations by substitution.

To Solve a System of Equations by Substitution

1. Solve for a variable in either equation. (If possible, solve for a variable with a numerical coefficient of 1 or -1 to avoid working with fractions.)

2. Substitute the expression found for the variable in step 1 into the other equation.

3. Solve the equation determined in step 2 to find the value of one variable.

4. Substitute the value found in step 3 into the equation obtained in step 1 to find the value of the other variable.

5. Check by substituting both values in both original equations.

EXAMPLE 2 ▸ Solve the following system of equations by substitution.

$$2x + y = 3$$
$$4x + 2y = 12$$

Solution Solve for y in $2x + y = 3$.

$$2x + y = 3$$
$$y = -2x + 3$$

Now substitute the expression $-2x + 3$ for y in the *other equation*, $4x + 2y = 12$, and solve for x.

$$4x + 2y = 12$$

$$4x + 2(\overbrace{-2x + 3}) = 12 \qquad \textit{Substitution; this is now an equation in only one variable, x.}$$

$$4x - 4x + 6 = 12$$

$$6 = 12 \qquad \textit{False}$$

Since the statement 6 = 12 is false, the system has no solution. Therefore, the graphs of the equations will be parallel lines and the system is inconsistent because it has no solution.

▶ **Now Try Exercise 17**

Note that the solution in Example 2 is identical to the graphical solution obtained in Example 4 of Section 9.1. **Figure 9.3** on page 547 shows the parallel lines.

EXAMPLE 3 ▶ Solve the following system of equations by substitution.

$$x - \frac{1}{2}y = 2$$

$$y = 2x - 4$$

Solution   The equation $y = 2x - 4$ is already solved for y. Substitute $2x - 4$ for y in the other equation, $x - \frac{1}{2}y = 2$, and solve for x.

$$x - \frac{1}{2}y = 2$$

$$x - \frac{1}{2}(\overbrace{2x - 4}) = 2$$

$$x - x + 2 = 2$$

$$2 = 2 \qquad \textit{True}$$

Notice that the sum of the x terms is 0, and when simplified, x is no longer part of the equation. *Since the statement 2 = 2 is true, this system has an infinite number of solutions. Therefore, the graphs of the equations represent the same line and the system is dependent.*

▶ **Now Try Exercise 9**

Note that the solution in Example 3 is identical to the solution obtained graphically in Example 5 of Section 9.1. **Figure 9.4** on page 548 shows that the graphs of both equations are the same line.

EXAMPLE 4 ▶ Solve the following system of equations by substitution.

$$3x + 6y = 9$$

$$2x - 3y = 6$$

Solution   None of the variables in either equation has a numerical coefficient of 1. However, since the numbers 3, 6, and 9 are all divisible by 3, if you solve the first equation for x, you will avoid having to work with fractions.

$$3x + 6y = 9$$

$$3x = -6y + 9$$

$$\frac{3x}{3} = \frac{-6y + 9}{3}$$

$$x = -\frac{6}{3}y + \frac{9}{3}$$

$$x = -2y + 3$$

Now substitute $-2y + 3$ for x in the other equation, $2x - 3y = 6$, and solve for the remaining variable, y.

$$2x - 3y = 6$$
$$2(\overbrace{-2y + 3}) - 3y = 6$$
$$-4y + 6 - 3y = 6$$
$$-7y + 6 = 6$$
$$-7y = 0$$
$$y = 0$$

Finally, solve for x by substituting $y = 0$ in the equation previously solved for x.

$$x = -2y + 3$$
$$x = -2(0) + 3 = 0 + 3 = 3$$

The solution is $(3, 0)$.

▶ **Now Try Exercise 23**

Helpful Hint

Remember that a solution to a system of linear equations must contain both an x- and a y-value. Don't solve the system for one of the variables and forget to solve for the other. Write the solution as an ordered pair.

EXAMPLE 5 ▶ Solve the following system of equations by substitution.

$$6x + 8y = 3$$
$$3x = 3y + 5$$

Solution   We will elect to solve for x in the second equation.

$$3x = 3y + 5$$
$$x = \frac{3y + 5}{3}$$
$$x = y + \frac{5}{3}$$

Now substitute $y + \dfrac{5}{3}$ for x in the other equation.

$$6x + 8y = 3$$
$$6\left(\overbrace{y + \frac{5}{3}}\right) + 8y = 3$$
$$6y + 10 + 8y = 3$$
$$14y + 10 = 3$$
$$14y = -7$$
$$y = \frac{-7}{14} = -\frac{1}{2}$$

Finally, find the value of x.

$$x = y + \frac{5}{3}$$
$$x = -\frac{1}{2} + \frac{5}{3} = -\frac{3}{6} + \frac{10}{6} = \frac{7}{6}$$

The solution is the ordered pair $\left(\dfrac{7}{6}, -\dfrac{1}{2}\right)$.

▶ **Now Try Exercise 25**

EXAMPLE 6 ▶ Mule Trip In 2006 and 2007 a total of 2462 people took a mule trip to the bottom of the Grand Canyon. If the number who took the mule trip in 2007 was 372 more than the number who took the mule trip in 2006, determine the number of people who took the mule trip in 2006 and in 2007.

Solution Understand and Translate We are provided sufficient information to obtain two equations for our system of equations. Let x = number of people who took the mule trip in 2006 and let y = number of people who took the mule trip in 2007. Because the total for these two years was 2462, one equation is $x + y = 2462$. Since the number of people who took the trip in 2007 was 372 more than the number who took the trip in 2006, an equation that represents the number of people who took the trip in 2007 is $y = x + 372$. This is our second equation in the system of equations.

$$\text{System of equations} \begin{cases} x + y = 2462 \\ y = x + 372 \end{cases}$$

Carry Out Because the second equation, $y = x + 372$, is already solved for y, we will substitute $x + 372$ for y in the first equation.

$$x + y = 2462$$
$$x + \overbrace{x + 372} = 2462$$
$$2x + 372 = 2462$$
$$2x = 2090$$
$$x = 1045$$

Check and Answer The number of people who took the trip in 2006 was 1045. The number of people who took the trip in 2007 is therefore $x + 372 = 1045 + 372 = 1417$. Notice that $1045 + 1417 = 2462$.

▶ **Now Try Exercise 35**

EXERCISE SET 9.2

Math XL MathXL® *MyMathLab* MyMathLab

Concept/Writing Exercises

1. When solving the system of equations

$$3x + 6y = 12$$
$$4x + 3y = 8$$

by substitution, which variable, in which equation, would you choose to solve for to make the solution easier? Explain your answer.

2. When solving the system of equations

$$4x + 2y = 8$$
$$3x - 9y = 8$$

by substitution, which variable, in which equation, would you choose to solve for to make the solution easier? Explain your answer.

3. When solving a system of linear equations by substitution, how will you know if the system is inconsistent?

4. When solving a system of linear equations by substitution, how will you know if the system is dependent?

Practice the Skills

Find the solution to each system of equations by substitution.

5. $x + 2y = 6$
$2x - 3y = 5$

6. $y = 2x + 7$
$y = -x - 5$

7. $x + y = -2$
$x - y = 0$

8. $x + 2y = 6$
$4y = 12 - 2x$

9. $3x + y = 3$
$3x + y + 5 = 0$

10. $3x - y = 8$
$6x - 2y = 10$

11. $x = 3$

$x + y + 5 = 0$

12. $y = 2x + 4$

$y = -2$

13. $x = y + 1$

$4x + 2y = -14$

14. $2x + 3y = 7$

$6x - y = 1$

15. $2x + y = 11$

$y = 3x - 4$

16. $y = -2x + 7$

$x + 4y = 0$

17. $y = \frac{1}{3}x - 2$

$x - 3y = 6$

18. $x - \frac{1}{2}y = 7$

$y = 2x - 14$

19. $2x + 5y = 9$

$6x - 2y = 10$

20. $3x - 3y = 4$

$2x + 3y = 5$

21. $y = 2x - 13$

$-4x - 7 = 9y$

22. $5x - 2y = -7$

$8 = y - 4x$

23. $4x - 5y = -4$

$3x = 2y - 3$

24. $2x - 3y = 9$

$5x + 2y = -6$

25. $4x + 5y = -6$

$2x - \frac{10}{3}y = -4$

26. $4x + y = 1$

$10x - 5y = -2$

27. $3x - 4y = 15$

$-6x + 8y = -14$

28. $\frac{1}{2}x + 2y = 5$

$3x - \frac{1}{2}y = 5$

29. $4x - y = 1$

$10x + \frac{1}{2}y = 1$

30. $-\frac{1}{5}x + \frac{3}{7}y = 2$

$\frac{3}{5}x - \frac{2}{7}y = 1$

Problem Solving

31. Positive Integers The sum of two positive integers is 80. Find the integers if one number is 8 greater than the other.

32. Positive Integers The difference of two positive integers is 44. Find the integers if the larger number is twice the smaller number.

33. Rectangle The perimeter of a rectangle is 50 feet. Find the dimensions of the rectangle if the length is 9 feet greater than the width.

34. Rectangle The perimeter of a rectangle is 60 feet. Find the dimensions of the rectangle if its length is 5 times its width.

35. Rodeo Attendance At a rodeo the total paid attendance was 2500. If the number of people who received a discount on the admission fee was 622 less than the number who did not, determine the number of paid attendees that paid the full fee.

36. Wooden Horse To buy a statue of a wooden horse Billy and Jean combined their money. Together they had $530. If Jean had $130 more than Billy, how much money did each have?

37. Legal Settlement After a legal settlement, the client's portion of the award was three times as much money as the attorney's portion. If the total award was $40,000, how much did the client get?

38. Jelly Beans A candy store mixed green and red jelly beans in a barrel. The barrel contains 42 pounds of the mixture. If there are 3 times as many pounds of the green jelly beans as red jelly beans, find the number of pounds of green jelly beans and the number of pounds of red jelly beans in the barrel.

39. Refinancing Dona Boccio is considering refinancing her house. The cost of refinancing is a one-time charge of $1280. With her reduced mortgage rate, her monthly interest and principal payments would be $794 per month. Her total cost, c, for n months could be represented by $c = 1280 + 794n$. At her current rate her mortgage payments are $874 per month and the total cost for n months can be represented by $c = 874n$.

a) Determine the number of months for which both mortgage plans would have the same total cost.

b) If Dona plans to remain in her house for 12 years, should she refinance?

40. Temperatures In Seattle the temperature is 86°F, but it is decreasing by 2 degrees per hour. The temperature, T, at time, t, in hours, is represented by $T = 86 - 2t$. In Spokane the temperature is 59°F, but it is increasing by 2.5 degrees per hour. The temperature, T, can be represented by $T = 59 + 2.5t$.

a) If the temperature continues decreasing and increasing at the same rate in these cities, how long will it be before both cities have the same temperature?

b) When both cities have the same temperature, what will that temperature be?

Seattle, Washington

41. Traveling by Car Jean Woody's car is at the 80 mile marker on a highway. Roberta Kieronski's car is 15 miles behind Jean's car. Jean's car is traveling at 60 miles per hour. The mile marker that Jean's car will be at in t hours can be found by the equation $m = 80 + 60t$. Roberta's car is traveling at 72 miles per hour. The mile marker that Roberta's car will be at in t hours can be found by the equation $m = 65 + 72t$.

a) Determine the time it will take for Roberta's car to catch up with Jean's car.

b) At which mile marker will they be when they meet?

42. Computer Store Will Worthy's present salary consists of a fixed weekly salary of $300 plus a $20 bonus for each computer system he sells. His weekly salary can be represented by $s = 300 + 20n$, where n is the number of computer systems he sells. He is considering another position where his weekly salary would be $400 plus a $10 bonus for each computer system he sells. The other position's weekly salary can be represented by $s = 400 + 10n$. How many computer systems would Will need to sell in a week for his salary to be the same with both employers?

Challenge Problems

Answer parts **a)** *through* **d)** *on your own.*

43. Heat Transfer In a laboratory during an experiment on heat transfer, a large metal ball is heated to a temperature of 180°F. This metal ball is then placed in a gallon of oil at a temperature of 20°F. Assume that when the ball is placed in the oil it loses temperature at the rate of 10 degrees per minute while the oil's temperature rises at a rate of 6 degrees per minute.

a) Write an equation that can be used to determine the ball's temperature t minutes after being placed in the oil.

b) Write an equation that can be used to determine the oil's temperature t minutes after the ball is placed in it.

c) Determine how long it will take for the ball and oil to reach the same temperature.

d) When the ball and oil reach the same temperature, what will the temperature be?

Group Activity

Discuss and work Exercise 44 as a group.

44. In intermediate algebra you may solve systems containing three equations with three variables. As a group, solve the system of equations on the right. Your answer will be in the form of an **ordered triple** (x, y, z).

$$x = 4$$
$$2x - y = 6$$
$$-x + y + z = -3$$

Cumulative Review Exercises

[2.6] **45. Willow Tree** The diameter of a willow tree grows about 1.2 inches per year. What is the approximate age of a willow tree whose diameter is 27.6 inches?

[4.2] **46.** Find the missing coordinate in the given solution $(3, ?)$ for $4x + 5y = 22$.

47. Graph $4x - 8y = 16$ using the intercepts.

[4.4] **48.** Find the slope and y-intercept of the graph of the equation $3x - 5y = 25$.

49. Determine the equation of the following line.

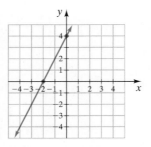

[5.5] **50.** Multiply $(6x + 7)(3x - 2)$.

9.3 Solving Systems of Equations by the Addition Method

1 Solve systems of equations by the addition method.

1 Solve Systems of Equations by the Addition Method

A third, and often the easiest, method of solving a system of equations is by the **addition (or elimination) method.** *The goal of this process is to obtain two equations whose sum will be an equation containing only one variable.* Always keep in mind that our immediate goal is to obtain one equation containing only one unknown.

In the addition method, we use the fact that if $a = b$ and $c = d$, then $a + c = b + d$. Suppose we have the following two equations.

$$x - 2y = 9$$
$$3x + 2y = 11$$

If we add the left sides of the two equations we obtain $4x$ because $-2y + 2y = 0$. If we add the right sides of the equations we get $9 + 11 = 20$. Thus, the sum of the equations is $4x = 20$, which is an equation in only one variable.

$$x - 2y = 9$$
$$\underline{3x + 2y = 11}$$
$$4x \qquad = 20$$

From the equation $4x = 20$, we determine that $x = 5$.

$$4x = 20$$
$$x = 5$$

We can now substitute 5 for x in either of the original equations to find y. We will substitute 5 for x in $x - 2y = 9$.

$$x - 2y = 9$$
$$5 - 2y = 9$$
$$-2y = 4$$
$$y = -2$$

The solution to the system is $(5, -2)$. This ordered pair will check in both equations. Check this solution now. Now let's work some examples.

EXAMPLE 1 ▶ Solve the following system of equations using the addition method.

$$x + y = 6$$
$$2x - y = 3$$

Solution   Note that one equation contains $+y$ and the other contains $-y$. By adding the equations, we can eliminate the variable y and obtain one equation containing only one variable, x. When added, $+y$ and $-y$ sum to 0, and the variable y is eliminated.

$$x + y = 6$$
$$\underline{2x - y = 3}$$
$$3x \qquad = 9$$

Now we solve for the remaining variable, x.

$$\frac{3x}{3} = \frac{9}{3}$$
$$x = 3$$

Finally, we solve for y by substituting $x = 3$ in either of the original equations.

$$x + y = 6$$
$$3 + y = 6$$
$$y = 3$$

The solution is $(3, 3)$.

We check the answer in *both* equations.

Check

$$x + y = 6 \qquad\qquad\qquad 2x - y = 3$$
$$3 + 3 \overset{?}{=} 6 \qquad\qquad\qquad 2(3) - 3 \overset{?}{=} 3$$
$$6 = 6 \quad \text{\textit{True}} \qquad\qquad\qquad 6 - 3 \overset{?}{=} 3$$
$$3 = 3 \quad \text{\textit{True}}$$

▶ **Now Try Exercise 7**

Now we summarize the procedure.

To Solve a System of Equations Using the Addition (or Elimination) Method

1. If necessary, rewrite each equation so that the terms containing variables appear on the left side of the equal sign and any constants appear on the right side of the equal sign.

2. If necessary, multiply one or both equations by a constant(s) so that when the equations are added the resulting sum will contain only one variable.

3. Add the equations. This will result in a single equation containing only one variable.

4. Solve for the variable in the equation from step 3.

5. **a)** Substitute the value found in step 4 into either of the original equations. Solve that equation to find the value of the remaining variable.

   or

   b) Repeat steps 2–4 to eliminate the other variable.

6. Check the values obtained in all original equations.

In step 5 we give two methods for finding the second variable after the first variable has been found. Method a) is generally used if the value obtained for the first variable is an integer. Method b) is often preferred if the value obtained for the first variable is a fraction.

In step 2 it may be necessary to multiply one or both equations by a constant. For example, suppose we have the system of equations, labeled (*eq.* 1) and (*eq.* 2) as shown below.

$$3x + 2y = 6 \quad (eq.\,1)$$
$$x - y = 8 \quad (eq.\,2)$$

If we wish to multiply equation 2 by -3, we will indicate this process as follows.

$$3x + 2y = 6 \qquad\qquad\qquad\qquad 3x + 2y = 6$$
$$-3(x - y) = -3(8) \quad (eq.\,2) \quad \text{\small Multiplied by -3} \quad \text{or} \quad -3x + 3y = -24$$

After performing the multiplication we will generally work with the revised system of equations, like the system given on the right.

EXAMPLE 2 ▶ Solve the following system of equations using the addition method.

$$x + 3y = 13 \quad (eq.\,1)$$
$$x + 4y = 18 \quad (eq.\,2)$$

Solution The goal of the addition process is to obtain two equations whose sum will be an equation containing only one variable. If we add these two equations, none of the variables will be eliminated. However, if we multiply either equation by -1 and then add, the terms containing x will sum to 0, and we will accomplish our goal. We will multiply (*eq.* 1) by -1.

$$-1(x + 3y) = -1 \cdot 13 \quad (eq.\,1) \quad \text{\small Multiplied by -1} \quad \text{or} \quad -x - 3y = -13$$
$$x + 4y = 18 \qquad\qquad (eq.\,2) \qquad\qquad\qquad\qquad x + 4y = 18$$

Remember that both sides of the equation must be multiplied by -1. This process changes the sign of each term in the equation being multiplied without changing the solution to the system of equations. Now add the two equations on the right on the bottom of the preceding page.

$$
\begin{array}{rcr}
-x - 3y &=& -13 \\
x + 4y &=& 18 \\
\hline
y &=& 5
\end{array}
$$

Now we solve for x in either of the original equations.

$$
\begin{aligned}
x + 3y &= 13 \\
x + 3(5) &= 13 \\
x + 15 &= 13 \\
x &= -2
\end{aligned}
$$

A check will show that the solution is $(-2, 5)$.

▶ **Now Try Exercise 13**

EXAMPLE 3 ▶ Solve the following system of equations using the addition method.

$$
\begin{aligned}
2x + y &= 11 && (eq.\,1) \\
x + 3y &= 18 && (eq.\,2)
\end{aligned}
$$

Solution   To eliminate the variable x, multiply $(eq.\,2)$ by -2 and add the two equations.

$$
\begin{aligned}
2x + y &= 11 && (eq.\,1) &&&& 2x + y = 11 \\
-2(x + 3y) &= -2 \cdot 18 && (eq.\,2) \quad \text{Multiplied by } -2 && \text{or} && -2x - 6y = -36
\end{aligned}
$$

Now add:

$$
\begin{array}{rcr}
2x + y &=& 11 \\
-2x - 6y &=& -36 \\
\hline
-5y &=& -25 \\
y &=& 5
\end{array}
$$

Solve for x.

$$
\begin{aligned}
2x + y &= 11 \\
2x + 5 &= 11 \\
2x &= 6 \\
x &= 3
\end{aligned}
$$

The solution is $(3, 5)$.

▶ **Now Try Exercise 23**

Note that the solution in Example 3 is the same as the solutions obtained graphically in Example 3 of Section 9.1 and by substitution in Example 1 of Section 9.2.

In Example 3, we could have multiplied the $(eq.\,1)$ by -3 to eliminate the variable y. At this time, rework Example 3 by eliminating the variable y to see that you get the same answer.

EXAMPLE 4 ▶ Solve the following system of equations using the addition method.

$$
\begin{aligned}
4x &= -2y - 18 \\
-5y &= 2x + 10
\end{aligned}
$$

Solution   Step 1 of the procedure indicates that we should rewrite each equation so that the terms containing variables appear on the left side of the equal sign and the constants appear on the right side of the equal sign. By adding $2y$ to both sides of the first equation and subtracting $2x$ from both sides of the second equation we obtain the following system of equations.

$$
\begin{aligned}
4x + 2y &= -18 && (eq.\,1) \\
-2x - 5y &= 10 && (eq.\,2)
\end{aligned}
$$

We now continue to solve the system. To eliminate the variable, x, we can multiply ($eq.\,2$) by 2 and then add.

$$4x + 2y = -18 \qquad (eq.\,1) \qquad\qquad\qquad 4x + 2y = -18$$
$$\boxed{2}(-2x - 5y) = \boxed{2} \cdot 10 \quad (eq.\,2) \quad \textit{Multiplied by 2} \quad \text{or} \quad -4x - 10y = 20$$

$$
\begin{array}{r}
4x + 2y = -18 \\
-4x - 10y = 20 \\
\hline
-8y = 2 \\
y = -\dfrac{1}{4}
\end{array}
$$

Solve for x.

$$4x + 2y = -18$$
$$4x + 2\left(-\frac{1}{4}\right) = -18$$
$$4x - \frac{1}{2} = -18$$
$$2\left(4x - \frac{1}{2}\right) = 2(-18) \qquad \textit{Multiply both sides by 2 to remove fractions.}$$
$$8x - 1 = -36$$
$$8x = -35$$
$$x = -\frac{35}{8}$$

The solution is $\left(-\dfrac{35}{8}, -\dfrac{1}{4}\right)$.

Check the solution $\left(-\dfrac{35}{8}, -\dfrac{1}{4}\right)$ in both equations.

Check

$$4x + 2y = -18 \qquad\qquad\qquad -2x - 5y = 10$$

$$4\left(-\frac{35}{8}\right) + 2\left(-\frac{1}{4}\right) \overset{?}{=} -18 \qquad -2\left(-\frac{35}{8}\right) - 5\left(-\frac{1}{4}\right) \overset{?}{=} 10$$

$$-\frac{35}{2} - \frac{1}{2} \overset{?}{=} -18 \qquad\qquad \frac{35}{4} + \frac{5}{4} \overset{?}{=} 10$$

$$-\frac{36}{2} \overset{?}{=} -18 \qquad\qquad\qquad \frac{40}{4} \overset{?}{=} 10$$

$$-18 = -18 \quad \textit{True} \qquad\qquad\qquad 10 = 10 \quad \textit{True}$$

▶ **Now Try Exercise 29**

Note that the solution to Example 4 contains fractions. You should not always expect to get integers as answers.

EXAMPLE 5 ▶ Solve the following system of equations using the addition method.

$$2x + 3y = 6 \qquad (eq.\,1)$$
$$5x - 4y = -8 \qquad (eq.\,2)$$

Solution The variable x can be eliminated by multiplying ($eq.\,1$) by -5 and ($eq.\,2$) by 2 and then adding the equations.

$$\boxed{-5}(2x + 3y) = \boxed{-5} \cdot 6 \qquad (eq.\,1) \quad \textit{Multiplied by} -5 \quad \text{or} \quad -10x - 15y = -30$$
$$\boxed{2}(5x - 4y) = \boxed{2} \cdot (-8) \quad (eq.\,2) \quad \textit{Multiplied by 2} \quad\;\; \text{or} \quad\;\; 10x - 8y = -16$$

$$
\begin{array}{r}
-10x - 15y = -30 \\
10x - 8y = -16 \\
\hline
-23y = -46 \\
y = 2
\end{array}
$$

Solve for x.

$$2x + 3y = 6$$
$$2x + 3(2) = 6$$
$$2x + 6 = 6$$
$$2x = 0$$
$$x = 0$$

The solution is $(0, 2)$.

▶ Now Try Exercise 19

In Example 5, the same value could have been obtained for y by multiplying the $(eq. 1)$ by 5 and $(eq. 2)$ by -2 and then adding. Try it now and see. We could have also begun by eliminating the variable y by multiplying $(eq. 1)$ by 4 and $(eq. 2)$ by 3.

EXAMPLE 6 ▶ Solve the following system of equations using the addition method.

$$2x + y = 3 \qquad (eq. 1)$$
$$4x + 2y = 12 \qquad (eq. 2)$$

Solution   The variable y can be eliminated by multiplying $(eq. 1)$ by -2 and then adding the two equations.

$$-2(2x + y) = -2 \cdot 3 \quad (eq. 1) \quad \textit{Multiplied by } -2 \quad \text{or} \quad -4x - 2y = -6$$
$$4x + 2y = 12 \qquad (eq. 2) \qquad\qquad\qquad\qquad 4x + 2y = 12$$

$$\begin{aligned} -4x - 2y &= -6 \\ 4x + 2y &= \;\;12 \\ \hline 0 &= \;\;\;6 \quad \textit{False} \end{aligned}$$

Don't panic when both variables drop out and you see an expression like $0 = 6$. Not all systems of equations have a solution. **Since $0 = 6$ is a false statement, this system has no solution**. The system is inconsistent. The graphs of the equations will be parallel lines.

▶ Now Try Exercise 25

Note that the solution in Example 6 is identical to the solutions obtained by graphing in Example 4 of Section 9.1 and by substitution in Example 2 of Section 9.2.

EXAMPLE 7 ▶ Solve the following system of equations using the addition method.

$$x - \frac{1}{2}y = 2$$
$$y = 2x - 4$$

Solution   First align the x- and y-terms on the left side of the equal sign by subtracting $2x$ from both sides of the second equation.

$$x - \frac{1}{2}y = 2 \qquad (eq. 1)$$
$$-2x + y = -4 \qquad (eq. 2)$$

Now proceed as in the previous examples. Begin by multiplying $(eq. 1)$ by 2 to remove fractions from the equation.

$$2\left(x - \frac{1}{2}y\right) = 2 \cdot 2 \quad (eq. 1) \quad \textit{Multiplied by 2} \quad \text{or} \quad 2x - y = 4$$
$$-2x + y = -4 \qquad (eq. 2) \qquad\qquad\qquad\qquad -2x + y = -4$$

$$\begin{aligned} 2x - y &= \;\;\;4 \\ -2x + y &= -4 \\ \hline 0 &= \;\;\;0 \quad \textit{True} \end{aligned}$$

Again both variables have dropped out. Here we are left with $0 = 0$. ***Since $0 = 0$ is a true statement, the system is dependent and has an infinite number of solutions.*** When graphed, both equations will be the same line.

▸ **Now Try Exercise 21**

The solution in Example 7 is the same as the solutions obtained by graphing in Example 5 of Section 9.1 and by substitution in Example 3 of Section 9.2.

EXAMPLE 8 ▸ Solve the following system of equations using the addition method.

$$2x + 3y = 7 \qquad (eq.\,1)$$
$$5x - 7y = -3 \qquad (eq.\,2)$$

Solution We can eliminate the variable x by multiplying $(eq.\,1)$ by -5 and $(eq.\,2)$ by 2.

$$-5(2x + 3y) = -5 \cdot 7 \quad (eq.\,1) \quad \text{Multiplied by } -5 \quad \text{or} \quad -10x - 15y = -35$$

$$2(5x - 7y) = 2(-3) \quad (eq.\,2) \quad \text{Multiplied by 2} \quad \text{or} \quad 10x - 14y = -6$$

$$\begin{array}{r} -10x - 15y = -35 \\ \underline{10x - 14y = -6} \\ -29y = -41 \end{array}$$

$$y = \frac{41}{29}$$

We can now find x by substituting $y = \dfrac{41}{29}$ into one of the original equations and solving for x. If you try this, you will see that although it can be done, the calculations are messy. An easier method of solving for x is to go back to the original equations and eliminate the variable y. We can do this by multiplying $(eq.\,1)$ by 7 and $(eq.\,2)$ by 3.

$$7(2x + 3y) = 7 \cdot 7 \quad (eq.\,1) \quad \text{Multiplied by 7} \quad \text{or} \quad 14x + 21y = 49$$

$$3(5x - 7y) = 3(-3) \quad (eq.\,2) \quad \text{Multiplied by 3} \quad \text{or} \quad 15x - 21y = -9$$

$$\begin{array}{r} 14x + 21y = 49 \\ \underline{15x - 21y = -9} \\ 29x \qquad\quad = 40 \end{array}$$

$$x = \frac{40}{29}$$

The solution is $\left(\dfrac{40}{29}, \dfrac{41}{29} \right)$.

▸ **Now Try Exercise 33**

Helpful Hint

We have illustrated three methods for solving a system of linear equations: graphing, substitution, and addition. When you are given a system of equations, which method should you use to solve the system? When you need an exact solution, graphing should not be used. Of the two algebraic methods, the addition method may be easier to use if there are no numerical coefficients of 1 in the system. If one or more of the variables have a coefficient of 1, you can use either substitution or addition.

EXERCISE SET 9.3

Concept/Writing Exercises

1. When solving the following system of equations by the addition method, what will your first step be in solving the system? Explain your answer. Do not solve the system.

$$-x + 3y = 4$$
$$2x + 5y = 2$$

2. When solving the following system of equations by the addition method, what will your first step be in solving the system? Explain your answer. Do not solve the system.

$$2x + 4y = -8$$
$$3x - 2y = 10$$

3. When solving a system of linear equations by the addition method, how will you know if the system is inconsistent?

4. When solving a system of linear equations by the addition method, how will you know if the system is dependent?

Practice the Skills

Solve each system of equations using the addition method.

5. $x + y = 6$
   $x - y = 4$

6. $x - y = 10$
   $x + y = 8$

7. $-x + y = 9$
   $-x + y = 1$

8. $5x + y = 14$
   $4x - y = 4$

9. $x + 2y = 21$
   $2x - 2y = -6$

10. $x - y = 3$
    $x + y = -3$

11. $4x + y = 6$
    $-8x - 2y = 20$

12. $x - y = 2$
    $3x - 3y = 1$

13. $-5x + y = 14$
    $-3x + y = -2$

14. $6x + 3y = 30$
    $2x + 3y = 18$

15. $2x + y = -6$
    $2x - 2y = 3$

16. $-4x + 3y = 0$
    $7x - 6y = 3$

17. $2y = 6x + 16$
    $y = -3x - 4$

18. $2x - 3y = 4$
    $2x + y = -4$

19. $5x + 3y = 12$
    $3x - 6y = 15$

20. $8x - 4y = 12$
    $2x - 8y = 3$

21. $-2y = -4x + 12$
    $y = 2x - 6$

22. $4x - 2y = 12$
    $4y = 8x - 24$

23. $5x - 4y = -3$
    $7y = 2x + 12$

24. $2x + 3y = -3$
    $-3x - 5y = 7$

25. $5x - 4y = 1$
    $-10x + 8y = -3$

26. $2x - 3y = 11$
    $5y = 3x - 17$

27. $5x - 6y = 0$
    $3x + 4y = 0$

28. $4x - 2y = -4$
    $5x - 6y = -26$

29. $-5x + 4y = -20$
    $3x - 2y = 15$

30. $4x - 3y = -4$
    $3x - 5y = 10$

31. $6x = 4y + 12$
    $3y - 5x = -6$

32. $5x = 2y - 4$
    $3x - 5y = 6$

33. $4x + 5y = 0$
    $3x = 6y + 4$

34. $4x - 3y = 8$
    $-3x + 4y = 9$

35. $x - \frac{1}{2}y = 4$
    $3x + y = 6$

36. $2x - \frac{1}{3}y = 6$
    $3x - \frac{5}{6}y = -13$

37. $3x - y = 4$
    $2x - \frac{2}{3}y = 8$

38. $-5x + 6y = -12$
    $\frac{5}{3}x - 4 = 2y$

Problem Solving

39. Sum of Numbers The sum of two numbers is 20. When the second number is subtracted from the first number, the difference is 8. Find the two numbers.

40. Sum of Numbers The sum of two numbers is 46. When the first number is subtracted from the second number, the difference is 6. Find the two numbers.

41. Sum of Numbers The sum of a number and twice a second number is 14. When the second number is subtracted from the first number the difference is 2. Find the two numbers.

42. Sum of Numbers The sum of two numbers is 9. Twice the first number subtracted from three times the second number is 7. Find the two numbers.

43. Rectangles When the length of a rectangle is x inches and the width is y inches, the perimeter is 18 inches. If the length is doubled and the width is tripled, the perimeter becomes 42 inches. Find the length and width of the original rectangle.

44. Perimeter of a Rectangle When the length of a rectangle is x inches and the width is y inches, the perimeter is 28 inches. If the length is doubled and the width is tripled, the perimeter becomes 66 inches. Find the length and width of the original rectangle.

45. Photograph A photograph has a perimeter of 36 inches. The difference between the photograph's length and width is 2 inches. Find the length and width of the photograph.

46. Rectangular Garden John has a large rectangular garden with a perimeter of 82 feet. The difference between the garden's length and width is 11 feet. Determine the length and width of his garden.

47. Construct a system of two equations that has no solution. Explain how you know the system has no solution.

48. Construct a system of two equations that has an infinite number of solutions. Explain how you know the system has an infinite number of solutions.

49. a) Solve the system of equations

$$4x + 2y = 1000$$
$$2x + 4y = 800$$

b) If we divide all the terms in the top equation by 2 we get the following system:

$$2x + y = 500$$
$$2x + 4y = 800$$

How will the solution to this system compare to the solution in part **a)**? Explain and then check your explanation by solving this system.

50. Suppose we divided all the terms in both equations given in Exercise 49 **a)** by 2, and then solved the system. How will the solution to this system compare to the solution in part **a)**? Explain and then check your explanation by solving each system.

Challenge Problems

In Exercises 51 and 52, solve each system of equations using the addition method. (Hint: First remove all fractions by multiplying both sides of the equation by the LCD.)

51. $\dfrac{x + 2}{2} - \dfrac{y + 4}{3} = 4$

$\dfrac{x + y}{2} = \dfrac{1}{2} + \dfrac{x - y}{3}$

52. $\dfrac{5}{2}x + 3y = \dfrac{9}{2} + y$

$\dfrac{1}{4}x - \dfrac{1}{2}y = 6x + 12$

In the next section we solve systems of three equations with three unknowns. Solve the following system.

53.
$$x + 2y - z = 2$$
$$2x - y + z = 3$$
$$3x + y + z = 8$$

Hint: Work with *one pair* of equations to get one equation in two unknowns. Then work with *a different pair* of the original equations to get another equation in the same two unknowns. Then solve the system of two equations in two unknowns. List your answer as an *ordered triple* of the form (x, y, z).

Group Activity

Work parts **a)** and **b)** of Exercise 54 on your own. Then discuss and work parts **c)** and **d)** as a group.

54. How difficult is it to construct a system of linear equations that has a specific solution? It is really not too difficult to do. Consider:

$$2(3) + 4(5) = 26$$
$$4(3) - 7(5) = -23$$

The system of equations

$$2x + 4y = 26$$
$$4x - 7y = -23$$

has solution $(3, 5)$.

a) Using the information provided, determine another system of equations that has $(3, 5)$ as a solution.

b) Determine a system of linear equations that has $(2, 3)$ as a solution.

c) Compare your answer with the answers of the other members of your group.

d) As a group, determine the number of systems of equations that have $(2, 3)$ as a solution.

Cumulative Review Exercises

[1.9] **55.** Evaluate 5^3.

[2.5] **56.** Solve the equation $2(2x - 3) = 2x + 8$.

[5.4] **57.** Simplify $(4x^2y - 3xy + y) - (2x^2y + 6xy - 3y)$.

[5.5] **58.** Multiply $(8a^4b^2c)(4a^2b^7c^4)$.

[6.2] **59.** Factor $xy + xc - ay - ac$ by grouping.

[8.2] **60.** If $f(x) = 2x^2 - 4$, find $f(-3)$.

9.4 Solving Systems of Linear Equations in Three Variables

1 Solve systems of linear equations in three variables.

2 Learn the geometric interpretation of a system of equations in three variables.

3 Recognize inconsistent and dependent systems.

1 Solve Systems of Linear Equations in Three Variables

The equation $2x - 3y + 4z = 8$ is an example of a linear equation in three variables. The solution to a linear equation in three variables is an *ordered triple* of the form (x, y, z). One solution to the equation given is $(1, 2, 3)$. Check now to verify that $(1, 2, 3)$ is a solution to the equation.

To solve systems of linear equations with three variables, we can use either substitution or the addition method, both of which were discussed in Sections 9.2 and 9.3, respectively.

EXAMPLE 1 ▸ Solve the following system by substitution.

$$x = -3$$
$$3x + 4y = 7$$
$$-2x - 3y + 5z = 19$$

Solution   Since we know that $x = -3$, we substitute -3 for x in the equation $3x + 4y = 7$ and solve for y.

$$3x + 4y = 7$$
$$3(-3) + 4y = 7$$
$$-9 + 4y = 7$$
$$4y = 16$$
$$y = 4$$

Now we substitute $x = -3$ and $y = 4$ into the last equation and solve for z.

$$-2x - 3y + 5z = 19$$
$$-2(-3) - 3(4) + 5z = 19$$
$$6 - 12 + 5z = 19$$
$$-6 + 5z = 19$$
$$5z = 25$$
$$z = 5$$

Check $x = -3$, $y = 4$, $z = 5$. The solution must be checked in *all three* original equations.

$$
\begin{array}{lll}
x = -3 & 3x + 4y = 7 & -2x - 3y + 5z = 19 \\
-3 = -3 \quad \textit{True} & 3(-3) + 4(4) \stackrel{?}{=} 7 & -2(-3) - 3(4) + 5(5) \stackrel{?}{=} 19 \\
& 7 = 7 \quad \textit{True} & 19 = 19 \quad \textit{True}
\end{array}
$$

The solution is the ordered triple $(-3, 4, 5)$. Remember that the ordered triple lists the x-value first, the y-value second, and the z-value third.

▸ **Now Try Exercise 3**

Not every system of linear equations in three variables can be solved by substitution. When such a system cannot be solved using substitution, we can find the solution by the addition method, as illustrated in Example 2.

EXAMPLE 2 ▸ Solve the following system of equations using the addition method.

$$3x + 2y + z = 4 \quad (eq.\,1)$$
$$2x - 3y + 2z = -7 \quad (eq.\,2)$$
$$x + 4y - z = 10 \quad (eq.\,3)$$

Solution To solve this system of equations, we must first obtain two equations containing the same two variables. We do so by selecting two equations and using the addition method to eliminate one of the variables. For example, by adding (*eq.* 1) and (*eq.* 3), the variable z will be eliminated. Then we use a different pair of equations [either (*eq.* 1) and (*eq.* 2) or (*eq.* 2) and (*eq.* 3)] and use the addition method to eliminate the *same* variable that was eliminated previously. If we multiply (*eq.* 1) by -2 and add it to (*eq.* 2), the variable z will again be eliminated. We will then have two equations containing only two unknowns. Let us begin by adding (*eq.* 1) and (*eq.* 3).

$$
\begin{array}{ll}
3x + 2y + z = \ \ 4 & (eq.\,1) \\
\underline{x + 4y - z = 10} & (eq.\,3) \\
4x + 6y \ \ \ \ \ \ = 14 & \text{Sum of equations, } (eq.\,4)
\end{array}
$$

Now let's use a different pair of equations and again eliminate the variable z.

$$
\begin{array}{lll}
-6x - 4y - 2z = \ \ -8 & (eq.\,1) & \text{Multiplied by } -2 \\
\underline{2x - 3y + 2z = \ \ -7} & (eq.\,2) & \\
-4x - 7y \ \ \ \ \ \ \ = -15 & \text{Sum of equations, } (eq.\,5)
\end{array}
$$

We now have a system consisting of two equations with two unknowns, (*eq.* 4) and (*eq.* 5). If we add these two equations, the variable x will be eliminated.

$$
\begin{array}{ll}
4x + 6y = \ \ 14 & (eq.\,4) \\
\underline{-4x - 7y = -15} & (eq.\,5) \\
-y = \ \ -1 & \text{Sum of equations} \\
y = \ \ \ \ 1 &
\end{array}
$$

Next we substitute $y = 1$ into either one of the two equations containing only two variables [(*eq.* 4) or (*eq.* 5)] and solve for x.

$$
\begin{array}{ll}
4x + 6y = 14 & (eq.\,4) \\
4x + 6(1) = 14 & \text{Substitute 1 for y in } (eq.\,4). \\
4x + 6 = 14 & \\
4x = 8 & \\
x = 2 &
\end{array}
$$

Finally, we substitute $x = 2$ and $y = 1$ into any of the original equations and solve for z.

$$
\begin{array}{ll}
3x + 2y + z = \ \ 4 & (eq.\,1) \\
3(2) + 2(1) + z = \ \ 4 & \text{Substitute 2 for x} \\
& \text{and 1 for y in } (eq.\,1). \\
6 + 2 + z = \ \ 4 & \\
8 + z = \ \ 4 & \\
z = -4 &
\end{array}
$$

The solution is the ordered triple $(2, 1, -4)$. Check this solution in *all three* original equations.

▶ **Now Try Exercise 15**

In Example 2 we chose first to eliminate the variable z by using (*eq.* 1) and (*eq.* 3) and then (*eq.* 1) and (*eq.* 2). We could have elected to eliminate either the variable x or the variable y first. For example, we could have eliminated variable x by multiplying (*eq.* 3) by -2 and then adding it to (*eq.* 2). We could also eliminate the variable x by multiplying (*eq.* 3) by -3 and then adding it to (*eq.* 1). Try solving the system in Example 2 by first eliminating the variable x.

EXAMPLE 3 ▶ Solve the following system of equations.

$$2x - 3y + 2z = -1 \quad (eq.\,1)$$
$$x + 2y \phantom{{}+2z} = 14 \quad (eq.\,2)$$
$$x \phantom{{}+2y} - 5z = -11 \quad (eq.\,3)$$

Solution   The third equation does not contain y. We will therefore work to obtain another equation that does not contain y. We will use $(eq.\,1)$ and $(eq.\,2)$ to do this.

$$4x - 6y + 4z = -2 \quad (eq.\,1) \quad \text{\textit{Multiplied by 2}}$$
$$\underline{3x + 6y \phantom{{}+4z} = 42} \quad (eq.\,2) \quad \text{\textit{Multiplied by 3}}$$
$$7x \phantom{{}+6y} + 4z = 40 \quad \text{\textit{Sum of equations, }}(eq.\,4)$$

We now have two equations containing only the variables x and z.

$$7x + 4z = 40 \quad (eq.\,4)$$
$$x - 5z = -11 \quad (eq.\,3)$$

Let's now eliminate the variable x.

$$7x + 4z = 40 \quad (eq.\,4)$$
$$\underline{-7x + 35z = 77} \quad (eq.\,3) \quad \text{\textit{Multiplied by }}-7$$
$$39z = 117 \quad \text{\textit{Sum of equations}}$$
$$z = 3$$

Now we solve for x by using one of the equations containing only the variables x and z. We substitute 3 for z in $(eq.\,3)$.

$$x - 5z = -11 \quad (eq.\,3)$$
$$x - 5(3) = -11 \quad \text{\textit{Substitute 3 for z in }}(eq.\,3).$$
$$x - 15 = -11$$
$$x = 4$$

Finally, we solve for y using any of the original equations that contains y.

$$x + 2y = 14 \quad (eq.\,2)$$
$$4 + 2y = 14 \quad \text{\textit{Substitute 4 for x in }}(eq.\,2).$$
$$2y = 10$$
$$y = 5$$

The solution is the ordered triple $(4, 5, 3)$.

Check   (eq. 1)   (eq. 2)   (eq. 3)

$$2x - 3y + 2z = -1 \qquad\quad x + 2y = 14 \qquad\qquad x - 5z = -11$$
$$2(4) - 3(5) + 2(3) \overset{?}{=} -1 \qquad 4 + 2(5) \overset{?}{=} 14 \qquad 4 - 5(3) \overset{?}{=} -11$$
$$8 - 15 + 6 \overset{?}{=} -1 \qquad\qquad 4 + 10 \overset{?}{=} 14 \qquad\quad 4 - 15 \overset{?}{=} -11$$
$$-1 = -1 \qquad\qquad\qquad\quad 14 = 14 \qquad\qquad -11 = -11$$
$$\text{\textit{True}} \qquad\qquad\qquad \text{\textit{True}} \qquad\qquad\qquad \text{\textit{True}}$$

▶ **Now Try Exercise 11**

Helpful Hint

If an equation in a system contains fractions, eliminate the fractions by multiplying each term in the equation by the least common denominator. Then continue to solve the system. If, for example, one equation in the system is $\dfrac{3}{4}x - \dfrac{5}{8}y + z = \dfrac{1}{2}$, multiply both sides of the equation by 8 to obtain the equivalent equation $6x - 5y + 8z = 4$.

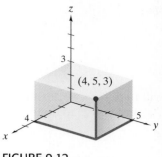

FIGURE 9.12

2 Learn the Geometric Interpretation of a System of Equations in Three Variables

When we have a system of linear equations in two variables, we can find its solution graphically using the Cartesian coordinate system. A linear equation in three variables, x, y, and z, can be graphed on a coordinate system with three axes drawn perpendicular to each other (see **Fig. 9.12**).

A point plotted in this three-dimensional system would appear to be a point in space. If we were to graph an equation such as $x + 2y + 3z = 4$, we would find that its graph would be a plane, not a line. In Example 3 we indicated the solution to be the ordered triple $(4, 5, 3)$. This means that the three planes, one from each of the three given equations, all intersect at the point $(4, 5, 3)$. In general, the ordered triple that is the solution to a system of equations in three variables is the point at which the three planes intersect. **Figure 9.12** shows the location of this point of intersection of the three planes. The drawing in Exercise 39 illustrates three planes intersecting at a point.

3 Recognize Inconsistent and Dependent Systems

We discussed inconsistent and dependent systems of equations in Section 9.1. Systems of linear equations in three variables may also be inconsistent or dependent. When solving a system of linear equations in three variables, if you obtain a false statement like $3 = 0$, the system is inconsistent and has no solution. This means that at least two of the planes are parallel, so the three planes cannot intersect. (See Exercises 37 and 38.)

When solving a system of linear equations in three variables, if you obtain the true statement $0 = 0$, it indicates that the system is dependent and has an infinite number of solutions. This may happen when all three equations represent the same plane or when the intersection of the planes is a line, as in the drawing in Exercise 40. Examples 4 and 5 illustrate an inconsistent system and a dependent system, respectively.

EXAMPLE 4 ▶ Solve the following system of equations.

$$-3x + 5y + z = -3 \quad (eq. 1)$$
$$6x - 10y - 2z = 1 \quad (eq. 2)$$
$$7x - 4y + 11z = -6 \quad (eq. 3)$$

Solution   We will begin by eliminating the variable x from ($eq. 1$) and ($eq. 2$).

$$-6x + 10y + 2z = -6 \quad (eq. 1) \quad \textit{Multiplied by 2}$$
$$\underline{6x - 10y - 2z = 1 \quad (eq. 2)}$$
$$0 = -5 \quad \textit{False}$$

Since we obtained the false statement $0 = -5$, this system is inconsistent and has no solution.

▶ **Now Try Exercise 31**

EXAMPLE 5 ▶ Solve the following system of equations.

$$x - y + z = 1 \quad (eq. 1)$$
$$x + 2y - z = 1 \quad (eq. 2)$$
$$x - 4y + 3z = 1 \quad (eq. 3)$$

Solution   We will begin by eliminating the variable x from ($eq. 1$) and ($eq. 2$) and then from ($eq. 1$) and ($eq. 3$).

$$-x + y - z = -1 \quad (eq. 1) \quad \textit{Multiplied by } -1$$
$$\underline{x + 2y - z = 1 \quad (eq. 2)}$$
$$3y - 2z = 0 \quad \textit{Sum of equations, } (eq. 4)$$

$$
\begin{array}{rll}
x - y + z = 1 & (eq.\,1) & \\
\underline{-x + 4y - 3z = -1} & (eq.\,3) & \textit{Multiplied by } -1 \\
3y - 2z = 0 & \textit{Sum of equations, } (eq.\,5) &
\end{array}
$$

Now we eliminate the variable y using $(eq.\,4)$ and $(eq.\,5)$.

$$
\begin{array}{rll}
-3y + 2z = 0 & (eq.\,4) & \textit{Multiplied by } -1 \\
\underline{3y - 2z = 0} & (eq.\,5) & \\
0 = 0 & \textit{True} &
\end{array}
$$

Since we obtained the true statement $0 = 0$, this system is dependent and has an infinite number of solutions.

Recall from Section 9.1 that systems of equations that are dependent are also consistent since they have a solution.

▶ **Now Try Exercise 33**

EXERCISE SET 9.4   *Math* **XL**   **MyMathLab**
MathXL®   MyMathLab

Concept/Writing Exercises

1. What will be the graph of an equation such as $3x - 4y + 2z = 1$?

2. Assume that the solution to a system of linear equations in three variables is $(1, 3, 5)$. What does this mean geometrically?

Practice the Skills

Solve by substitution.

3.
$x = 1$
$2x - y = 4$
$-3x + 2y - 2z = 1$

4. $-x + 3y - 5z = -7$
$2y - z = -1$
$z = 3$

5. $5x - 6z = -17$
$3x - 4y + 5z = -1$
$2z = -6$

6. $2x - 5y = 12$
$-3y = -9$
$2x - 3y + 4z = 8$

7. $x + 2y = 6$
$3y = 9$
$x + 2z = 12$

8. $x - y + 5z = -4$
$3x - 2z = 6$
$4z = 2$

Solve using the addition method.

9. $x - 2y = -3$
$3x + 2y = 7$
$2x - 4y + z = -6$

10. $x - y + 2z = 1$
$y - 4z = 2$
$-2x + 2y - 5z = 2$

11. $2y + 4z = 2$
$x + y + 2z = -2$
$2x + y + z = 2$

12. $2x + y - 8 = 0$
$3x - 4z = -3$
$2x - 3z = 1$

13. $3p + 2q = 11$
$4q - r = 6$
$6p + 7r = 4$

14. $3s + 5t = -12$
$2t - 2u = 2$
$-s + 6u = -2$

15. $p + q + r = 4$
$p - 2q - r = 1$
$2p - q - 2r = -1$

16. $x - 2y + 3z = -7$
$2x - y - z = 7$
$-4x + 3y + 2z = -14$

17. $2x - 2y + 3z = 5$
$2x + y - 2z = -1$
$4x - y - 3z = 0$

18. $2x - y - 2z = 3$
$x - 3y - 4z = 2$
$x + y + 2z = -1$

19. $r - 2s + t = 2$
$2r + 3s - t = -3$
$2r - s - 2t = 1$

20. $3a - 3b + 4c = -1$
$a - 2b + 2c = 2$
$2a - 2b - c = 3$

21. $2a + 2b - c = 2$
$3a + 4b + c = -4$
$5a - 2b - 3c = 5$

22. $x - 2y + 2z = 3$
$2x - 3y + 2z = 5$
$x + y + 6z = -2$

23. $-x + 3y + z = 0$
$-2x + 4y - z = 0$
$3x - y + 2z = 0$

24. $x + y + z = 0$
$-x - y + z = 0$
$-x + y + z = 0$

25. $-\frac{1}{4}x + \frac{1}{2}y - \frac{1}{2}z = -2$
$\frac{1}{2}x + \frac{1}{3}y - \frac{1}{4}z = 2$
$\frac{1}{2}x - \frac{1}{2}y + \frac{1}{4}z = 1$

26. $\frac{2}{3}x + y - \frac{1}{3}z = \frac{1}{3}$
$\frac{1}{2}x + y + z = \frac{5}{2}$
$\frac{1}{4}x - \frac{1}{4}y + \frac{1}{4}z = \frac{3}{2}$

27. $x - \frac{2}{3}y - \frac{2}{3}z = -2$
$\frac{2}{3}x + y - \frac{2}{3}z = \frac{1}{3}$
$-\frac{1}{4}x + y - \frac{1}{4}z = \frac{3}{4}$

28. $\frac{1}{8}x + \frac{1}{4}y + z = 2$
$\frac{1}{3}x + \frac{1}{4}y + z = \frac{17}{6}$
$-\frac{1}{4}x + \frac{1}{3}y - \frac{1}{2}z = -\frac{5}{6}$

29. $0.2x + 0.3y + 0.3z = 1.1$
$0.4x - 0.2y + 0.1z = 0.4$
$-0.1x - 0.1y + 0.3z = 0.4$

30. $0.6x - 0.4y + 0.2z = 2.2$
$-0.1x - 0.2y + 0.3z = 0.9$
$-0.2x - 0.1y - 0.3z = -1.2$

Determine whether the following systems are inconsistent, dependent, or neither.

31. $2x + y + 2z = 1$
$x - 2y - z = 0$
$3x - y + z = 2$

32. $2p - 4q + 6r = 8$
$-p + 2q - 3r = 6$
$3p + 4q + 5r = 8$

33. $x - 4y - 3z = -1$
$-3x + 12y + 9z = 3$
$2x - 10y - 7z = 5$

34. $5a - 4b + 2c = 5$
$-10a + 8b - 4c = -10$
$-7a - 4b + c = 7$

35. $x + 3y + 2z = 6$
$x - 2y - z = 8$
$-3x - 9y - 6z = -7$

36. $2x - 2y + 4z = 2$
$-3x + y = -9$
$2x - y + z = 5$

Problem Solving

An equation in three variables represents a plane. Consider a system of equations consisting of three equations in three variables. Answer the following questions.

37. If the three planes are parallel to one another as illustrated in the figure, how many points will be common to all three planes? Is the system consistent or inconsistent? Explain your answer.

38. If two of the planes are parallel to each other and the third plane intersects each of the other two planes, how many points will be common to all three planes? Is the system consistent or inconsistent? Explain your answer.

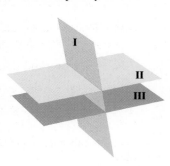

39. If the three planes are as illustrated in the figure, how many points will be common to all three planes? Is the system consistent or inconsistent? Explain your answer.

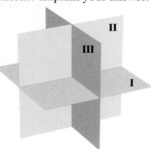

40. If the three planes are as illustrated in the figure, how many points will be common to all three planes? Is the system dependent? Explain your answer.

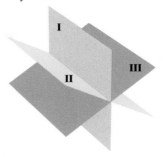

41. Is it possible for a system of linear equations in three variables to have exactly
   a) no solution,
   b) one solution,
   c) two solutions? Explain your answer.

42. In a system of linear equations in three variables, if the graphs of two equations are parallel planes, is it possible for the system to be
   a) consistent,
   b) dependent,
   c) inconsistent? Explain your answer.

43. Three solutions to the equation $Ax + By + Cz = 1$ are $(-1, 2, -1)$, $(-1, 1, 2)$, and $(1, -2, 2)$. Determine the values of A, B, and C and write the equation using the numerical values found.

44. Three solutions to the equation $Ax + By + Cz = 14$ are $(3, -1, 2)$, $(2, -2, 1)$, and $(-5, 3, -24)$. Find the values of A, B, and C and write the equation using the numerical values found.

In Exercises 45 and 46, write a system of linear equations in three variables that has the given solution. Explain how you determined your answer.

45. $(3, 1, 6)$

46. $(-2, 5, 3)$

47. a) Find the values of a, b, and c such that the points $(1, -1)$, $(-1, -5)$, and $(3, 11)$ lie on the graph of $y = ax^2 + bx + c$.
   b) Find the quadratic equation whose graph passes through the three points indicated. Explain how you determined your answer.

48. a) Find the values of a, b, and c such that the points $(1, 7)$, $(-2, -5)$, and $(3, 5)$ lie on the graph of $y = ax^2 + bx + c$.
   b) Find the quadratic equation whose graph passes through the three points indicated. Explain how you determined your answer.

Challenge Problems

Find the solution to the following systems of equations.

49. $3p + 4q = 11$
$2p + r + s = 9$
$q - s = -2$
$p + 2q - r = 2$

50. $3a + 2b - c = 0$
$2a + 2c + d = 5$
$a + 2b - d = -2$
$2a - b + c + d = 2$

Cumulative Review Exercises

[1.10] **51.** Name the properties illustrated.
   a) $x + 4 = 4 + x$
   b) $(3x)y = 3(xy)$
   c) $4(x + 2) = 4x + 8$

[2.5] **52.** Solve $3x + 4 = -(x - 6)$.

[3.2] **53. Cross-Country Skiing** Margie Steiner begins skiing along a trail at 3 miles per hour. Ten minutes $\left(\frac{1}{6}\text{hour}\right)$ later, her husband, David, begins skiing along the same trail at 5 miles per hour.
   a) How long after David leaves will he catch up to Margie?
   b) How far from the starting point will they be when they meet?

[3.3] **54. Perimeter** The perimeter of a rectangle is 22 feet. Find the dimensions of the rectangle if the length is two feet more than twice the width.

Mid-Chapter Test: 9.1–9.4

To find out how well you understand the chapter material to this point, take this brief test. The answers, and the section where the material was initially discussed, are given in the back of the book. Review any questions that you answered incorrectly.

Determine which of the following ordered pairs satisfy each system of equations.

1. $4x + 3y = -1$
$x - 2y = 8$
   a) $(-1, 1)$ **b)** $(2, -3)$

2. $6x - y = -2$
$7x + \dfrac{1}{2}y = 6$
   a) $\left(\dfrac{1}{2}, 5\right)$ **b)** $\left(\dfrac{1}{3}, 4\right)$

Without graphing, state whether the system of equations has exactly one solution, no solution, or an infinite number of solutions.

3. $2x + y = 8$
$3x - 4y = 1$

4. $\dfrac{1}{2}x - 3y = 5$
$-2x + 12y = -20$

5. $y = \dfrac{3}{2}x + \dfrac{5}{2}$
$3x - 2y = 7$

Determine the solution to each system of equations graphically. If the system is dependent or inconsistent, so state.

6. $y = 2x + 1$
$y = -x + 4$

7. $x = 5$
$y = -3$

Solve each system of equations by substitution.

8. $3x + y = -2$
$2x - 3y = -16$

9. $x - 3y = 2$
$4x + 9y = 1$

10. $3x - y = 5$
$x - \dfrac{1}{3}y = 2$

11. Rectangle The perimeter of a rectangle is 44 feet. Find the length and the width if the length is 8 feet greater than the width.

Solve each system of equations using the addition method.

12. $x + 3y = 1$
$2x - 3y = 11$

13. $4x + 3y = 4$
$-8x + 5y = 14$

14. $5x - 2y = 1$
$-10x + 4y = -2$

15. In solving the system of equations

$$3x - 5y = -16$$
$$2x + 3y = 21,$$

Hugo Platt stated that the solution was $x = 3$. This is incorrect. Why? Explain your answer. Give the correct solution to the system.

Solve each system of equations.

16. $x + y + z = 2$
$2x - y + 2z = -2$
$3x + 2y + 6z = 1$

17. $2x - y - z = 1$
$3x + 5y + 2z = 12$
$-6x - 4y + 5z = 3$

9.5 Systems of Linear Equations: Applications and Problem Solving

1 Use systems of equations to solve applications.

2 Use linear systems in three variables to solve applications.

1 Use Systems of Equations to Solve Applications

Many of the applications solved in earlier chapters using only one variable can now be solved using two variables. Following are some examples showing how applications can be represented by systems of equations.

Women and Men in the Workforce
(Percent of population in the civilian labor force)

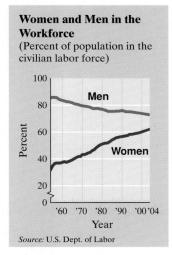

Source: U.S. Dept. of Labor

FIGURE 9.13

EXAMPLE 1 ▶ **A Changing Workforce** The graph in **Figure 9.13** indicates that the percent of males in the workforce is decreasing while the percent of females is increasing. The function $m(t) = -0.25t + 85.4$, where t = years since 1955, can be used to estimate the percent of males in the workforce, and the function $w(t) = 0.52t + 35.7$ can be used to estimate the percent of women in the workforce. If this trend continues, determine when the percent of women in the workforce will equal the percent of men.

Solution **Understand and Translate** Consider the two functions given above as the system of equations. To determine when the percent of women will equal the percent of men, we can set the two functions equal to one another and solve for time, t.

Carry Out
$$\text{percent of women} = \text{percent of men}$$
$$0.52t + 35.7 = -0.25t + 85.4$$
$$0.77t = 49.7$$
$$t \approx 64.5$$

Answer If this trend continues the percent of women in the workforce will equal the percent of men about 64.5 years after 1955. Since $1955 + 64.5 = 2019.5$, the percents will be equal in the year 2019.

▶ **Now Try Exercise 39**

EXAMPLE 2 ▶ **Land Area** The combined land area of Grenada and Guam is 890 square kilometers. The land area of Guam is 200 square kilometers more than the land area of Grenada. Find the land area of Guam and Grenada.

Solution **Understand** We need to determine the land area of Guam and Grenada.

Translate
Let a = land area of Guam

b = land area of Grenada.

Since the total land area of Grenada and Guam is 890 square kilometers, the first equation is
$$a + b = 890$$

Since the land area of Guam is 200 square kilometers greater than the land area of Grenada, the second equation is
$$a = b + 200$$

The system of equations is
$$a + b = 890 \qquad (eq. 1)$$
$$a = b + 200 \qquad (eq. 2)$$

Uruano Beach, Guam
Greg Vaughn/PacificStock.com

Carry Out We will use the substitution method, discussed in Section 4.1, to solve this system of equations.

Using (*eq. 2*), substitute $b + 200$ for a into the first equation to obtain

$$a + b = 890 \qquad \textit{First equation}$$
$$(b + 200) + b = 890 \qquad \textit{Substitute b + 200 for a.}$$
$$2b + 200 = 890 \qquad \textit{Simplify.}$$
$$2b = 690 \qquad \textit{Subtract 200 from both sides.}$$
$$b = 345 \qquad \textit{Divide both sides by 2.}$$

Thus, $b = 345$. To determine the value for a, substitute 345 for b into (*eq. 2*).

$$a = b + 200$$
$$a = 345 + 200$$
$$= 545$$

Answer The land area of Guam is 545 square kilometers and the land area of Grenada is 345 square kilometers.

▶ **Now Try Exercise 1**

EXAMPLE 3 ▶ **Canoe Speed** The Burnhams are canoeing on the Suwannee River. They travel at an average speed of 4.75 miles per hour when paddling with the current and 2.25 miles per hour when paddling against the current. Determine the speed of the canoe in still water and the speed of the current.

Solution **Understand** When they are traveling with the current, the canoe's speed is the canoe's speed in still water *plus* the current's speed. When traveling against the current, the canoe's speed is the canoe's speed in still water *minus* the current's speed.

Translate Let s = speed of the canoe in still water
c = speed of the current.

The system of equations is:

speed of the canoe traveling with the current: $s + c = 4.75$
speed of the canoe traveling against the current: $s - c = 2.25$

Carry Out We will use the addition method, as we discussed in Section 4.1, to solve this system of equations.

$$\begin{array}{r} s + c = 4.75 \\ \underline{s - c = 2.25} \\ 2s = 7.00 \\ s = 3.5 \end{array}$$

The speed of the canoe in still water is 3.5 miles per hour. We now determine the speed of the current.

$$s + c = 4.75$$
$$3.5 + c = 4.75$$
$$c = 1.25$$

Answer The speed of the current is 1.25 miles per hour, and the speed of the canoe in still water is 3.5 miles per hour.

▶ **Now Try Exercise 13**

EXAMPLE 4 ▶ **Salary** Yamil Bermudez, a salesman at Hancock Appliances, receives a weekly salary plus a commission, which is a percentage of his sales. One week, with sales of $3000, his total take-home pay was $850. The next week, with sales of $4000, his total take-home pay was $1000. Find his weekly salary and his commission rate.

Solution **Understand** Yamil's take-home pay consists of his weekly salary plus commission. We are given information about two specific weeks that we can use to find his weekly salary and his commission rate.

Translate   Let s = his weekly salary

r = his commission rate.

In week 1, his commission on \$3000 is 3000r, and in week 2 his commission on \$4000 is 4000r. Thus the system of equations is

salary + commission = take-home salary

1st week	$s + 3000r = 850$
2nd week	$s + 4000r = 1000$

System of equations

Carry Out

$$-s - 3000r = -850 \quad \text{\textit{1st week multiplied by} } -1$$
$$\underline{s + 4000r = 1000} \quad \text{\textit{2nd week}}$$
$$1000r = 150 \quad \text{\textit{Sum of equations}}$$
$$r = \frac{150}{1000}$$
$$r = 0.15$$

Yamil's commission rate is 15%. Now we find his weekly salary by substituting 0.15 for r in either equation.

$$s + 3000r = 850$$

$$s + 3000(0.15) = 850 \qquad \text{\textit{Substitute 0.15 for r in the}}$$
$$\text{\textit{1st-week equation.}}$$

$$s + 450 = 850$$

$$s = 400$$

Answer   Yamil's weekly salary is \$400 and his commission rate is 15%.

▶ **Now Try Exercise 15**

EXAMPLE 5 ▶ **Riding Horses** Ben Campbell leaves his ranch riding on his horse at 5 miles per hour. One-half hour later, Joe Campbell leaves the same ranch and heads along the same route on his horse at 8 miles per hour.

a) How long after Joe leaves the ranch will he catch up to Ben?

b) When Joe catches up to Ben, how far from the ranch will they be?

Solution   **a)** Understand   When Joe catches up to Ben, they both will have traveled the same distance. Joe will have traveled the distance in $\frac{1}{2}$ hour less time since he left $\frac{1}{2}$ hour after Ben. We will use the formula distance = rate $\cdot$ time to solve this problem.

Translate   Let b = time traveled by Ben

j = time traveled by Joe.

We will set up a table to organize the given information.

	Rate	Time	Distance
Ben	5	b	$5b$
Joe	8	j	$8j$

Since both Ben and Joe cover the same distance, we write

Ben's distance = Joe's distance

$$5b = 8j$$

Our second equation comes from the fact that Joe is traveling for $\frac{1}{2}$ hour less time than Ben. Therefore, $j = b - \frac{1}{2}$. Thus our system of equations is:

$$5b = 8j$$

$$j = b - \frac{1}{2}$$

Carry Out   We will solve this system of equations using substitution. Since $j = b - \frac{1}{2}$, substitute $b - \frac{1}{2}$ for j in the first equation and solve for b.

$$5b = 8j$$

$$5b = 8\left(b - \frac{1}{2}\right)$$

$$5b = 8b - 4$$

$$-3b = -4$$

$$b = \frac{-4}{-3} = 1\frac{1}{3}$$

Therefore, the time Ben has been traveling is $1\frac{1}{3}$ hours. To get the time Joe has been traveling, we will subtract $\frac{1}{2}$ hour from Ben's time.

$$j = b - \frac{1}{2}$$

$$j = 1\frac{1}{3} - \frac{1}{2}$$

$$j = \frac{4}{3} - \frac{1}{2} = \frac{8}{6} - \frac{3}{6} = \frac{5}{6}$$

Answer   Joe will catch up to Ben $\frac{5}{6}$ of an hour (or 50 minutes) after Joe leaves the ranch.

b) We can use either Ben's or Joe's distance to determine the distance traveled from the ranch. We will use Joe's distance.

$$d = 8j = 8\left(\frac{5}{6}\right) = \frac{\overset{4}{\cancel{8}}}{1} \cdot \frac{5}{\underset{3}{\cancel{6}}} = \frac{20}{3} = 6\frac{2}{3}$$

Thus, Joe will catch up to Ben when they are $6\frac{2}{3}$ miles from the ranch.

▶ **Now Try Exercise 33**

EXAMPLE 6 ▶ **Mixing Solutions** Chung Song, a chemist with Johnson and Johnson, wishes to create a new household cleaner containing 30% trisodium phosphate (TSP). Chung needs to mix a 16% TSP solution with a 72% TSP solution to get 6 liters of a 30% TSP solution. How many liters of the 16% solution and of the 72% solution will he need to mix?

Solution   Understand   To solve this problem we use the fact that the amount of TSP in a solution is found by multiplying the percent strength of the solution by the number of liters (the volume) of the solution. Chung needs to mix a 16% solution and a 72% solution to obtain 6 liters of a solution whose strength, 30%, is between the strengths of the two solutions being mixed.

Translate   Let x = number of liters of the 16% solution

y = number of liters of the 72% solution.

We will draw a sketch (**Fig. 9.14** on page 581) and then use a table to help analyze the problem.

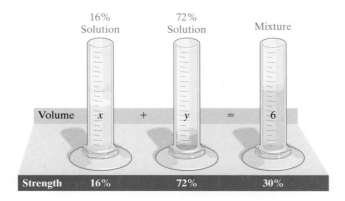

FIGURE 9.14

Solution	Strength of Solution	Number of Liters	Amount of TSP
16% solution	0.16	x	$0.16x$
72% solution	0.72	y	$0.72y$
Mixture	0.30	6	$0.30(6)$

Since the sum of the volumes of the 16% solution and the 72% solution is 6 liters, our first equation is

$$x + y = 6$$

The second equation comes from the fact that the solutions are mixed.

$$\left(\begin{array}{c}\text{amount of TSP}\\\text{in 16\% solution}\end{array}\right) + \left(\begin{array}{c}\text{amount of TSP}\\\text{in 72\% solution}\end{array}\right) = \left(\begin{array}{c}\text{amount of TSP}\\\text{in mixture}\end{array}\right)$$

$$0.16x \qquad + \qquad 0.72y \qquad = \qquad 0.30(6)$$

Therefore, the system of equations is

$$x + y = 6$$
$$0.16x + 0.72y = 0.30(6)$$

Carry Out Solving $x + y = 6$ for y, we get $y = -x + 6$. Substituting $-x + 6$ for y in the second equation gives us

$$0.16x + 0.72y = 0.30(6)$$
$$0.16x + 0.72(-x + 6) = 0.30(6)$$
$$0.16x - 0.72x + 4.32 = 1.8$$
$$-0.56x + 4.32 = 1.8$$
$$-0.56x = -2.52$$
$$x = \frac{-2.52}{-0.56} = 4.5$$

Therefore, Chung must use 4.5 liters of the 16% solution. Since the two solutions must total 6 liters, he must use $6 - 4.5$ or 1.5 liters of the 72% solution.

▶ **Now Try Exercise 17**

In Example 6, the equation $0.16x + 0.72y = 0.30(6)$ could have been simplified by multiplying both sides of the equation by 100. This would give the equation $16x + 72y = 30(6)$ or $16x + 72y = 180$. Then the system of equations would be $x + y = 6$ and $16x + 72y = 180$. If you solve this system, you should obtain the same solution. Try it and see.

2 Use Linear Systems in Three Variables to Solve Applications

Now let us look at some applications that involve three equations and three variables.

EXAMPLE 7 ▶ Bank Loans Tiny Tots Toys must borrow $25,000 to pay for an expansion. It is not able to obtain a loan for the total amount from a single bank, so it takes out loans from three different banks. It borrows some of the money at a bank that charges it 8% interest. At the second bank, it borrows $2000 more than one-half the amount borrowed from the first bank. The interest rate at the second bank is 10%. The balance of the $25,000 is borrowed from a third bank, where Tiny Tots pays 9% interest. The total annual interest Tiny Tots Toys pays for the three loans is $2220. How much does it borrow at each rate?

Solution   Understand   We are asked to determine how much is borrowed at each of the three different rates. Therefore, this problem will contain three variables, one for each amount borrowed. Since the problem will contain three variables, we will need to determine three equations to use in our system of equations.

Translate

Let x = amount borrowed at first bank

y = amount borrowed at second bank

z = amount borrowed at third bank.

Since the total amount borrowed is $25,000 we know that

$$x + y + z = 25,000 \qquad \text{\textit{Total amount borrowed is \$25,000.}}$$

At the second bank, Tiny Tots Toys borrows $2000 more than one-half the amount borrowed from the first bank. Therefore, our second equation is

$$y = \frac{1}{2}x + 2000 \qquad \text{\textit{Second, y, is \$2000 more than }} \tfrac{1}{2} \text{\textit{ of first, x.}}$$

Our last equation comes from the fact that the total annual interest charged by the three banks is $2220. The interest at each bank is found by multiplying the interest rate by the amount borrowed.

$$0.08x + 0.10y + 0.09z = 2220 \qquad \text{\textit{Total interest is \$2220.}}$$

Thus, our system of equations is

$$x + y + z = 25,000 \qquad (1)$$

$$y = \frac{1}{2}x + 2000 \qquad (2)$$

$$0.08x + 0.10y + 0.09z = 2220 \qquad (3)$$

Both sides of equation (2) can be multiplied by 2 to remove fractions.

$$2(y) = 2\left(\frac{1}{2}x + 2000\right)$$

$$2y = x + 4000 \qquad \text{\textit{Distributive property}}$$

$$-x + 2y = 4000 \qquad \text{\textit{Subtract x from both sides.}}$$

The decimals in equation (3) can be removed by multiplying both sides of the equation by 100. This gives

$$8x + 10y + 9z = 222,000$$

Our simplified system of equations is therefore

$$x + y + z = 25,000 \qquad (eq.\ 1)$$

$$-x + 2y \quad\;\;\; = 4000 \qquad (eq.\ 2)$$

$$8x + 10y + 9z = 222,000 \qquad (eq.\ 3)$$

Carry Out   There are various ways of solving this system. Let's use (*eq.* 1) and (*eq.* 3) to eliminate the variable z.

$$
\begin{array}{ll}
-9x - 9y - 9z = -225,000 & (eq.\ 1) \quad \text{Multiplied by } -9 \\
\underline{8x + 10y + 9z = 222,000} & (eq.\ 3) \\
-x + y = -3,000 & \text{Sum of equations, } (eq.\ 4)
\end{array}
$$

Now we use (*eq.* 2) and (*eq.* 4) to eliminate the variable x and solve for y.

$$
\begin{array}{ll}
x - 2y = -4000 & (eq.\ 2) \quad \text{Multiplied by } -1 \\
\underline{-x + y = -3000} & (eq.\ 4) \\
-y = -7000 & \text{Sum of equations} \\
y = 7000
\end{array}
$$

Now that we know the value of y we can solve for x.

$$
\begin{array}{ll}
-x + 2y = 4000 & (eq.\ 2) \\
-x + 2(7000) = 4000 & \text{Substitute 7000 for } y \text{ in } (eq.\ 2). \\
-x + 14,000 = 4000 & \\
-x = -10,000 & \\
x = 10,000 &
\end{array}
$$

Finally, we solve for z.

$$
\begin{array}{ll}
x + y + z = 25,000 & (eq.\ 1) \\
10,000 + 7000 + z = 25,000 & \\
17,000 + z = 25,000 & \\
z = 8000 &
\end{array}
$$

Answer   Tiny Tots Toys borrows \$10,000 at 8%, \$7000 at 10%, and \$8000 at 9% interest.

▶ **Now Try Exercise 55**

EXAMPLE 8 ▶ **Inflatable Boats** Hobson, Inc., has a small manufacturing plant that makes three types of inflatable boats: one-person, two-person, and four-person models. Each boat requires the service of three departments: cutting, assembly, and packaging. The cutting, assembly, and packaging departments are allowed to use a total of 380, 330, and 120 person-hours per week, respectively. The time requirements for each boat and department are specified in the following table. Determine how many of each type of boat Hobson must produce each week for its plant to operate at full capacity.

| | Time (person-hr) per Boat | | |
Department	One-Person Boat	Two-Person Boat	Four-Person Boat
Cutting	0.6	1.0	1.5
Assembly	0.6	0.9	1.2
Packaging	0.2	0.3	0.5

Solution   **Understand**   We are told that three different types of boats are produced and we are asked to determine the number of each type produced. Since this problem involves three amounts to be found, the system will contain three equations in three variables.

Translate   We will use the information given in the table.

$$
\begin{array}{l}
\text{Let } x = \text{number of one-person boats} \\
y = \text{number of two-person boats} \\
z = \text{number of four-person boats.}
\end{array}
$$

The total number of cutting hours for the three types of boats must equal 380 person-hours.

$$0.6x + 1.0y + 1.5z = 380$$

The total number of assembly hours must equal 330 person-hours.

$$0.6x + 0.9y + 1.2z = 330$$

The total number of packaging hours must equal 120 person-hours.

$$0.2x + 0.3y + 0.5z = 120$$

Therefore, the system of equations is

$$0.6x + 1.0y + 1.5z = 380 \quad (1)$$
$$0.6x + 0.9y + 1.2z = 330 \quad (2)$$
$$0.2x + 0.3y + 0.5z = 120 \quad (3)$$

Multiplying each equation in the system by 10 will eliminate the decimal numbers and give a simplified system of equations.

$$6x + 10y + 15z = 3800 \quad (eq. 1)$$
$$6x + 9y + 12z = 3300 \quad (eq. 2)$$
$$2x + 3y + 5z = 1200 \quad (eq. 3)$$

Carry Out Let's first eliminate the variable x using $(eq. 1)$ and $(eq. 2)$, and then $(eq. 1)$ and $(eq. 3)$.

$$
\begin{aligned}
6x + 10y + 15z &= 3800 \quad &(eq. 1) \\
-6x - 9y - 12z &= -3300 \quad &(eq. 2) \quad \text{Multiplied by } -1 \\
\hline
y + 3z &= 500 \quad &\text{Sum of equations, } (eq. 4)
\end{aligned}
$$

$$
\begin{aligned}
6x + 10y + 15z &= 3800 \quad &(eq. 1) \\
-6x - 9y - 15z &= -3600 \quad &(eq. 3) \quad \text{Multiplied by } -3 \\
\hline
y &= 200 \quad &\text{Sum of equations, } (eq. 5)
\end{aligned}
$$

Note that when we added the last two equations, both variables x and z were eliminated at the same time. Now we know the value of y and can solve for z.

$$y + 3z = 500 \quad (eq. 4)$$
$$200 + 3z = 500 \quad \text{Substitute 200 for } y.$$
$$3z = 300$$
$$z = 100$$

Finally, we find x.

$$6x + 10y + 15z = 3800 \quad (eq. 1)$$
$$6x + 10(200) + 15(100) = 3800$$
$$6x + 2000 + 1500 = 3800$$
$$6x + 3500 = 3800$$
$$6x = 300$$
$$x = 50$$

Answer Hobson should produce 50 one-person boats, 200 two-person boats, and 100 four-person boats per week.

▶ **Now Try Exercise 59**

EXERCISE SET 9.5

MathXL®   MyMathLab

Practice the Skills/Problem Solving

1. **Land Area** The combined land area of the countries of Georgia and Ireland is 139,973 square kilometers. Ireland is larger by 573 square kilometers. Determine the land area of each country.

Cliffs of Moher, Ireland

2. **Daytona 500 Wins** As of this writing, Richard Petty has won the Daytona 500 race the greatest number of times and Dale Yarborough has won the second greatest number of Daytona 500 races. Petty's number of wins is one less than twice Yarborough's number of wins. The total number of wins by the two drivers is 11. Determine the number of wins by Petty and by Yarborough.

3. **Fat Content** A nutritionist finds that a large order of fries at McDonald's has more fat than a McDonald's quarter-pound hamburger. The fries have 4 grams more than three times the amount of fat that the hamburger has. The difference in the fat content between the fries and the hamburger is 46 grams. Find the fat content of the hamburger and of the fries.

4. **Theme Parks** The two most visited theme parks in the United States in 2004 were Walt Disney's Magic Kingdom in Florida and Disneyland in California. The total number of visitors to these parks was 28.4 million people. The number of people who visited the Magic Kingdom was 1.8 million more than the number of people who visited Disneyland. How many people visited each of these parks in 2004? *Source:* www.coastergrotto.com

5. **Hot Dog Stand** At Big Al's hot dog stand, 2 hot dogs and 3 sodas cost $7. The cost of 4 hot dogs and 2 sodas is $10. Determine the cost of a hot dog and the cost of a soda.

6. **Water and Pretzel** At a professional football game, the cost of 2 bottles of water and 3 pretzels is $16.50. The cost of 4 bottles of water and 1 pretzel is $15.50. Determine the cost of a bottle of water and the cost of a pretzel.

7. **Digital Cameras** Ashley Dawn just bought a new digital camera, a 128-megabyte memory card, and a 512-megabyte memory card. The 512-MB memory card can store four times as many photos as the 128-MB memory card. Together the two memory cards can store 360 photos (of fine quality). Determine how many photos each memory card can store.

8. **Photo Printers** The July 2005 *Consumer Reports* magazine featured an article on photo printers that compared the costs to print photos with each printer. The most expensive of the 4 × 6 snapshot printers was the Olympus P-5100. The least expensive was the Epson Picture Mate. To print one photo on both printers would cost $0.80. The cost to print a photo on the Olympus is $0.20 more than twice the cost to print a photo on the Epson. Determine the cost to print a photo on each printer.

9. **Complementary Angles** Two angles are **complementary angles** if the sum of their measures is 90°. (See Section 2.3.) If the measure of the larger of two complementary angles is 15° more than two times the measure of the smaller angle, find the measures of the two angles.

10. **Complementary Angles** The difference between the measures of two complementary angles is 46°. Determine the measures of the two angles.

11. **Supplementary Angles** Two angles are **supplementary angles** if the sum of their measures is 180°. (See Section 2.3.) Find the measures of two supplementary angles if the measure of one angle is 28° less than three times the measure of the other.

12. **Supplementary Angles** Determine the measures of two supplementary angles if the measure of one angle is three and one half times larger than the measure of the other angle.

13. **Rowing Speed** The Heart O'Texas Rowing Team, while practicing in Austin, Texas rowed an average of 15.6 miles per hour with the current and 8.8 miles per hour against the current. Determine the team's rowing speed in still water and the speed of the current.

14. **Flying Speed** Jung Lee, in his Piper Cub airplane, flew an average of 121 miles per hour with the wind and 87 miles per hour against the wind. Determine the speed of the airplane in still air and the speed of the wind.

15. **Salary Plus Commission** Don Lavigne, an office equipment sales representative, earns a weekly salary plus a commission on his sales. One week his total compensation on sales of $4000 was $660. The next week his total compensation on sales of $6000 was $740. Find Don's weekly salary and his commission rate.

16. **Truck Rental** A truck rental agency charges a daily fee plus a mileage fee. Hugo was charged $85 for 2 days and 100 miles and Christina was charged $165 for 3 days and 400 miles. What is the agency's daily fee, and what is the mileage fee?

17. **Lavender Oil** Pola Sommers, a massage therapist, needs 3 ounces of a 20% lavender oil solution. She has only 5% and 30% lavender oil solutions available. How many ounces of each should Pola mix to obtain the desired solution?

18. **Fertilizer Solutions** Frank Ditlman needs to apply a 10% liquid nitrogen solution to his rose garden, but he only has a 4% liquid nitrogen solution and a 20% liquid nitrogen solution available. How much of the 4% solution and how much of the 20% solution should Frank mix together to get 10 gallons of the 10% solution?

19. **Weed Killer** Round-Up Concentrate Grass and Weed Killer consists of an 18% active ingredient glyphosate (and 82% inactive ingredients). The concentrate is to be mixed with water and the mixture applied to weeds. If the final mixture is to contain 0.9% active ingredient, how much concentrate and how much water should be mixed to make 200 gallons of the final mixture?

20. **Lawn Fertilizer** Scott's Winterizer Lawn Fertilizer is 22% nitrogen. Schultz's Lime with Lawn Fertilizer is 4% nitrogen. William Weaver, owner of Weaver's Nursery, wishes to mix these two fertilizers to make 400 pounds of a special 10% nitrogen mixture for midseason lawn feeding. How much of each fertilizer should he mix?

21. **Birdseed** Birdseed costs $0.59 a pound and sunflower seeds cost $0.89 a pound. Angela Leinenbachs' pet store wishes to make a 40-pound mixture of birdseed and sunflower seeds that sells for $0.76 per pound. How many pounds of each type of seed should she use?

22. **Coffee** Franco Manue runs a grocery store. He wishes to mix 30 pounds of coffee to sell for a total cost of $170. To obtain the mixture, he will mix coffee that sells for $5.20 per pound with coffee that sells for $6.30 per pound. How many pounds of each coffee should he use?

23. **Amtrak** Ann Marie Whittle has been pricing Amtrak fares for a group to visit New York. Three adults and four children would cost a total of $159. Two adults and three children would cost a total of $112. Determine the price of an adult ticket and a child's ticket.

24. **Buffalo Wings** The Wing House sells both regular size and jumbo size orders of Buffalo chicken wings. Three regular orders and five jumbo orders of wings cost $67. Four regular and four jumbo orders of wings cost $64. Determine the cost of a regular order of wings and a jumbo order of wings.

25. **Savings Accounts** Mr. and Mrs. Gamton invest a total of $10,000 in two savings accounts. One account pays 5% interest and the other 6%. Find the amount placed in each account if the accounts receive a total of $540 in interest after 1 year. Use interest = principal · rate · time.

26. **Investments** Louis Okonkwo invested $30,000, part at 9% and part at 5%. If he had invested the entire amount at 6.5%, his total annual interest would be the same as the sum of the annual interest received from the two other accounts. How much was invested at each interest rate?

27. **Milk** Becky Slaats is a plant engineer at Velda Farms Dairy Cooperative. She wishes to mix whole milk, which is 3.25% fat, and skim milk, which has no fat, to obtain 260 gallons of a mixture of milk that contains 2% fat. How many gallons of whole milk and how many gallons of skim milk should Becky mix to obtain the desired mixture?

28. **Quiche Lorraine** Lambert Marx's recipe for quiche lorraine calls for 2 cups (16 ounces) of light cream that is 20% butterfat. It is often difficult to find light cream with 20% butterfat at the supermarket. What is commonly found is heavy cream, which is 36% butterfat, and half-and-half, which is 10.5% butterfat. How much of the heavy cream and how much of the half-and-half should Lambert mix to obtain the mixture necessary for the recipe?

29. **Birdseed** By ordering directly through *www.birdseed.com*, the Carters can purchase Season's Choice birdseed for $1.79 per pound and Garden Mix birdseed for $1.19 per pound. If they wish to purchase 20 pounds and spend $28 on birdseed, how many pounds of each type should they buy?

30. **Juice** The Healthy Favorites Juice Company sells apple juice for 8.3¢ an ounce and raspberry juice for 9.3¢ an ounce. The company wishes to market and sell 8-ounce cans of apple-raspberry juice for 8.7¢ an ounce. How many ounces of each should be mixed?

31. **Car Travel** Two cars start at the same point in Alexandria, Virginia, and travel in opposite directions. One car travels 5 miles per hour faster than the other car. After 4 hours, the two cars are 420 miles apart. Find the speed of each car.

32. **Road Construction** Kip Ortiz drives from Atlanta to Louisville, a distance of 430 miles. Due to road construction and heavy traffic, during the first part of his trip, Kip drives at an average rate of 50 miles per hour. During the rest of his trip he drives at an average rate of 70 miles per hour. If his total trip takes 7 hours, how many hours does he drive at each speed?

33. **Avon Conference** Cabrina Wilson and Dabney Jefferson are Avon representatives who are attending a conference in Seattle. After the conference, Cabrina drives home to Boise at an average speed of 65 miles per hour and Dabney drives home to Portland at an average speed of 50 miles per hour. If the sum of their driving times is 11.4 hours and if the sum of the distances driven is 690 miles, determine the time each representative spent driving home.

34. **Exercise** For her exercise routine, Cynthia Harrison rides a bicycle for half an hour and then rollerblades for half an hour. Cynthia rides the bicycle at a speed that is twice the speed at which she rollerblades. If the total distance covered is 12 miles, determine the speed at which she bikes and rollerblades.

35. **Animal Diet** Animals in an experiment are on a strict diet. Each animal is to receive, among other nutrients, 20 grams of protein and 6 grams of carbohydrates. The scientist has only two food mixes available of the following compositions. How many grams of each mix should be used to obtain the right diet for a single animal?

Mix	Protein (%)	Carbohydrate (%)
Mix A	10	6
Mix B	20	2

36. **Chair Manufacturing** A company makes two models of chairs. Information about the construction of the chairs is given in the table shown. On a particular day the company allocated 46.4 person-hours for assembling and 8.8 person-hours for painting. How many of each chair can be made?

Model	Time to Assemble	Time to Paint
Model A	1 hr	0.5 hr
Model B	3.2 hr	0.4 hr

37. **Brass Alloy** By weight, one alloy of brass is 70% copper and 30% zinc. Another alloy of brass is 40% copper and 60% zinc. How many grams of each of these alloys need to be melted and combined to obtain 300 grams of a brass alloy that is 60% copper and 40% zinc?

38. **Silver Alloy** Sterling silver is 92.5% pure silver. How many grams of pure (100%) silver and how many grams of sterling silver must be mixed to obtain 250 g of a 94% silver alloy?

39. **Internal Revenue Service** The following graph shows the number of paper Form 1040 tax returns and the number of online Forms 1040, 1040A, 1040EZ tax returns filed with the IRS in the years from 2002 to 2005, and projected to 2010. If t represents the number of years since 2002, the number of paper form 1040 tax returns, in millions, filed with the IRS can be estimated by the function $P(t) = -2.73t + 58.37$ and the number of online Forms 1040, 1040A, 1040EZ tax returns, in millions, filed with the IRS can be estimated by the function $o(t) = 1.95t + 10.58$. Assuming this trend continues, solve this system of equations to determine the year that the number of paper Form 1040 tax returns will be the same as the number of online Forms 1040, 1040A, 1040EZ tax returns.

Federal Tax Return Method

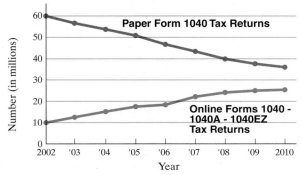

Source: www.irs.gov/pubs

40. Walking and Jogging Cuong Tham tries to exercise every day. He walks at 3 miles per hour and then jogs at 5 miles per hour. If it takes him 0.9 hours to travel a total of 3.5 miles, how long does he jog?

41. Texas Driving Tom Johnson and Melissa Acino started driving at the same time in different cars from Oklahoma City. They both traveled south on Route 35. When Melissa reached the Dallas/Ft. Worth area, a distance of 150 miles, Tom had only reached Denton, Texas, a distance of 120 miles. If Melissa averaged 15 miles per hour faster than Tom, find the average speed of each car.

42. Photocopy Costs At a local copy center two plans are available.

Plan 1: 10¢ per copy
Plan 2: an annual charge of $120 plus 4¢ per copy

a) Represent this information as a system of equations.
b) Graph the system of equations for up to 4000 copies made.
c) From the graph, estimate the number of copies a person would have to make in a year for the two plans to have the same total cost.
d) Solve the system algebraically. If your answer does not agree with your answer in part **c)**, explain why.

In Exercises 43–62, solve each problem using a system of three equations in three unknowns.

43. Mail Volume The average American household receives 24 pieces of mail each week. The number of bills and statements is two less than twice the number of pieces of personal mail. The number of advertisements is two more than five times the number of pieces of personal mail. How many pieces of personal mail, bills and statements, and advertisements does the average family get each week? *Source: Arthur D. Little, Inc.*

44. Submarine Personnel A 141-person crew is standard on a Los Angeles class submarine. The number of chief petty officers (enlisted) is four more than the number of commissioned officers. The number of other enlisted men is three less than eight times the number of commissioned officers. Determine the number of commissioned officers, chief petty officers, and other enlisted people on the submarine.

45. College Football Bowl Games Through 2004, the Universities of Alabama, Tennessee, and Texas have had the most appearances in college football bowl games. These three schools have had a total of 141 bowl appearances. Alabama has had 8 more appearances than Texas. Together, the number of appearances by Tennessee and Texas is 37 more than the number of appearances by Alabama. Determine the number of bowl appearances for each school and complete the following diagram.

Schools with the Most Appearances in a College Football Bowl Game (Includes 2004 Season):

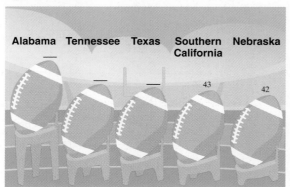

Source: NCAA, USA Today (12/22/04)

46. 2004 Summer Olympics In the 2004 Summer Olympics in Greece, the countries earning the most gold medals were the United States, China, and Russia. Together, these three countries earned a total of 94 gold medals. The United States earned 3 more gold medals than China. Together the number of gold medals earned by the United States and Russia is 2 less than twice the number of gold medals earned by China. Determine the number of gold medals each country earned and complete the following diagram. *Source:* www.athens2004.com

2004 Summer Olympics

Rank by Gold	Country	Number of Gold Medals
1	USA	
2	China	
3	Russia	
4	Australia	17
5	Japan	16
6	Germany	14

47. Top-10 Finishes on PGA Tour In the five years from 2000 through 2004, the three golfers with the most top-10 finishes in the Professional Golfers Association, or PGA, tour were Vijay Singh, Tiger Woods, and Phil Mickelson. Together, during these years, these three golfers had a total of 191 top-10 finishes. Tiger Woods had 8 more top-10 finishes than Phil Mickelson. Vijay Singh had 12 more top-10 finishes than Phil Mickelson. Determine the number of top-10 finishes for each golfer and complete the following diagram.

Swinging into the Top 10

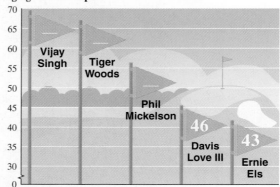

Source: PGA Tour, USA Today (1/12/05)

48. Nascar Racing The Nascar Nextel Cup Series consist of 36 races starting with the Daytona 500 in February and ending with the Ford 400 in Miami in November. In 2005, the top three drivers with the most points were, in order, Tony Stewart, Greg Biffle, and Carl Edwards. These three drivers had a total of 15 wins in the Nextel Cup series. Stewart had one more win than Edwards, and Biffle had two more wins than Edwards. Determine the number of wins each driver had.   Stewart: 5, Biffle: 6, Edwards: 4

49. New England Snowstorm During the last week of January 2005, New England had a record-breaking snowstorm that lasted several days. The coastal regions were hardest hit, with some cities receiving more than 3 feet of snow. In Massachusetts, the cities of Haverhill, Plymouth, and Salem had the most snow. The total snowfall for these three cities was 112.5 inches. Salem and Plymouth had the same amount of snow. Both had 1.5 inches more than Haverhill. Determine the snowfall amounts for each of these cities.   Haverhill: 36.5 in., Salem: 38 in., Plymouth: 38 in.

50. Football In the 2004 regular season of the National Football League (NFL), 19 players scored 100 or more points. The three players scoring the most points were Adam Vinatieri (NE), Jason Elam (DEN), and Jeff Reed (PIT). These three players scored a total of 394 points. Vinatieri scored 17 more points than Reed. Together, Vinatieri and Reed scored 7 more points than double the number of points scored by Elam. Determine the number of points Vinatieri Elam, and Reed scored. *Source:* www.nfl.com/stats/leaders
Vinatieri: 141, Elam: 129, Reed: 124

Jeff Reed of the Pittsburgh Steelers

51. Super Bowls Super Bowl XXXIX was held on February 6, 2005, in Jacksonville, Florida. Over the years, the states of Florida, California, and Louisiana, in this order, have hosted the most Super Bowls. These three states hosted a total of 32 Super Bowls. Florida hosted 3 more Super Bowls than Louisiana. Together, Florida and Louisiana hosted one less than twice the number California hosted. Determine the number of Super Bowls hosted by each of these three states. *Source:* NFL, *USA Today* (2/1/05)   Florida: 12, California: 11, Louisiana: 9

52. Concert Tickets Three kinds of tickets for a Soggy Bottom Boys concert are available: up-front, main-floor, and balcony. The most expensive tickets, the up-front tickets, are twice as expensive as balcony tickets. Balcony tickets are $10 less than main-floor tickets and $30 less than up-front tickets. Determine the price of each kind of ticket.

53. Triangle The sum of the measures of the angles of a triangle is 180°. The smallest angle of the triangle has a measure $\frac{2}{3}$ the measure of the second smallest angle. The largest angle has a measure that is 30° less than three times the measure of

the second smallest angle. Determine the measure of each angle.   30°, 45°, 105°

54. Triangle The largest angle of a triangle has a measure that is 10° less than three times the measure of the second smallest angle. The measure of the smallest angle is equal to the difference between the measure of the largest angle and twice the measure of the second smallest angle. Determine the measures of the three angles of the triangle.   30°, 40°, 110°

55. Investments Tam Phan received a check for $10,000. She decided to divide the money (not equally) into three different investments. She placed part of her money in a savings account paying 3% interest. The second amount, which was twice the first amount, was placed in a certificate of deposit paying 5% interest. She placed the balance in a money market fund paying 6% interest. If Tam's total interest over the period of 1 year was $525.00, how much was placed in each account?

56. Bonus Nick Pfaff, an attorney, divided his $15,000 holiday bonus check among three different investments. With some of the money, he purchased a municipal bond paying 5.5% simple interest. He invested twice the amount of money that he paid for the municipal bond in a certificate of deposit paying 4.5% simple interest. Nick placed the balance of the money in a money market account paying 3.75% simple interest. If Nick's total interest for 1 year was $692.50, how much was placed in each account? $4000 at 5.5%, $8000 at 4.5%, $3000 at 3.75%

57. Hydrogen Peroxide A 10% solution, a 12% solution, and a 20% solution of hydrogen peroxide are to be mixed to get 8 liters of a 13% solution. How many liters of each must be mixed if the volume of the 20% solution must be 2 liters less than the volume of the 10% solution?

58. Sulfuric Acid An 8% solution, a 10% solution, and a 20% solution of sulfuric acid are to be mixed to get 100 milliliters of a 12% solution. If the *volume of acid* from the 8% solution is to equal half the *volume of acid* from the other two solutions, how much of each solution is needed?

59. Furniture Manufacturing Donaldson Furniture Company produces three types of rocking chairs: the children's model, the standard model, and the executive model. Each chair is made in three stages: cutting, construction, and finishing. The time needed for each stage of each chair is given in the following chart. During a specific week the company has available a maximum of 154 hours for cutting, 94 hours for construction, and 76 hours for finishing. Determine how many of each type of chair the company should make to be operating at full capacity.   10 children's, 12 standard, 8 executive

Stage	Children's	Standard	Executive
Cutting	5 hr	4 hr	7 hr
Construction	3 hr	2 hr	5 hr
Finishing	2 hr	2 hr	4 hr

60. Bicycle Manufacturing The Jamis Bicycle Company produces three models of bicycles: Dakar, Komodo, and Aragon. Each bicycle is made in three stages: welding, painting, and assembling. The time needed for each stage of each bicycle is given in the chart on page 590. During a specific week, the company has available a maximum of 133 hours for welding, 78 hours for painting, and 96 hours for assembling. Determine how many of each type of bicycle the company should make to be operating at full capacity.

52. up-front: $60, main-floor: $40, balcony: $30   **55.** $1500 at 3%, $3000 at 5%, $5500 at 6%   **57.** 4 liters of 10%, 2 liters of 12%, 2 liters of 20%
58. 50 ml of 8% sol, 20 ml of 10% sol, 30 ml of 20% sol   **60.** 28 Dakars, 15 Komodos, 8 Aragons

Stage	Dakar	Komodo	Aragon
Welding	2	3	4
Painting	1	2	2.5
Assembling	1.5	2	3

61. Current Flow In electronics it is necessary to analyze current flow through paths of a circuit. In three paths ($A, B,$ and C) of a circuit, the relationships are the following:

$$I_A + I_B + I_C = 0$$
$$-8I_B + 10I_C = 0$$
$$4I_A - 8I_B = 6$$

where $I_A, I_B,$ and I_C represent the current in paths $A, B,$ and C, respectively. Determine the current in each path of the circuit.

62. Forces on a Beam In physics we often study the forces acting on an object. For three forces, $F_1, F_2,$ and F_3, acting on a beam, the following equations were obtained.

$$3F_1 + F_2 - F_3 = 2$$
$$F_1 - 2F_2 + F_3 = 0$$
$$4F_1 - F_2 + F_3 = 3$$

Find the three forces.

Group Activity

Discuss and answer Exercise 63 as a group.

63. Two Cars A *nonlinear system of equations* is a system of equations containing at least one equation that is not linear. (Nonlinear systems of equations will be discussed in Chapter 10.) The graph shows a nonlinear system of equations. The curves represent speed versus time for two cars.

a) Are the two curves functions? Explain.

b) Discuss the meaning of this graph.

c) At time $t = 0.5$ hr, which car is traveling at a greater speed? Explain your answer.

d) Assume the two cars start at the same position and are traveling in the same direction. Which car, A or B, traveled farther in 1 hour? Explain your answer.

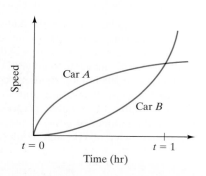

Cumulative Review Exercises

[1.9]   **64.** Evaluate $\frac{1}{2}x + \frac{2}{5}xy + \frac{1}{8}y$ when $x = -2, y = 5$.

[2.5]   **65.** Solve $4 - 2[(x - 5) + 2x] = -(x + 6)$.

[4.4]   **66.** Write an equation of the line that passes through points $(6, -4)$ and $(2, -8)$.

[8.2]   **67.** Explain how to determine whether a graph represents a function.

9.6 Solving Systems of Equations Using Matrices

1 Write an augmented matrix.

2 Solve systems of linear equations.

3 Solve systems of linear equations in three variables.

4 Recognize inconsistent and dependent systems.

1 Write an Augmented Matrix

A **matrix** is a rectangular array of numbers within brackets. The plural of *matrix* is **matrices**. Examples of matrices are

$$\begin{bmatrix} 4 & 6 \\ 9 & -2 \end{bmatrix} \quad \begin{bmatrix} 5 & 7 & 2 \\ -1 & 3 & 4 \end{bmatrix}$$

The numbers inside the brackets are referred to as **elements** of the matrix.

The matrix on the left contains 2 rows and 2 columns and is called a 2 by 2 (2×2) matrix. The matrix on the right contains 2 rows and 3 columns and is a 2 by 3 (2×3) matrix. The number of rows is the first dimension given, and the number of columns is the second dimension given when describing the dimensions of a matrix. A **square matrix** has the same number of rows as columns. Thus, the matrix on the left is a square matrix.

In this section, we will use matrices to solve systems of linear equations. The first step in solving a system of two linear equations using matrices is to write each equation in the form $ax + by = c$. The next step is to write the **augmented matrix**, which is made up of two smaller matrices separated by a vertical line. The numbers on the left of the vertical line are the coefficients of the variables in the system of equations, and the numbers on the right are the constants. For the system of equations

$$a_1 x + b_1 y = c_1$$
$$a_2 x + b_2 y = c_2$$

the augmented matrix is written

$$\left[\begin{array}{cc|c} a_1 & b_1 & c_1 \\ a_2 & b_2 & c_2 \end{array}\right]$$

Following is a system of equations and its augmented matrix.

System of Equations	Augmented Matrix

$$-x + \frac{1}{6}y = 4$$
$$-3x - 5y = -\frac{1}{2}$$

$$\left[\begin{array}{cc|c} -1 & \frac{1}{6} & 4 \\ -3 & -5 & -\frac{1}{2} \end{array}\right]$$

Notice that the bar in the augmented matrix separates the numerical coefficients from the constants. Since the matrix is just a shortened way of writing the system of equations, we can solve a linear system using matrices in a manner very similar to solving a system of equations using the addition method.

2 Solve Systems of Linear Equations

To solve a system of two linear equations using matrices, we rewrite the augmented matrix in **triangular form**,

$$\left[\begin{array}{cc|c} 1 & a & p \\ 0 & 1 & q \end{array}\right]$$

where the a, p, and q are constants. From this type of augmented matrix we can write an equivalent system of equations. This matrix represents the linear system

$$\begin{array}{c} 1x + ay = p \\ 0x + 1y = q \end{array} \quad \text{or} \quad \begin{array}{c} x + ay = p \\ y = q \end{array}$$

For example,

$$\left[\begin{array}{cc|c} 1 & 2 & 4 \\ 0 & 1 & 5 \end{array}\right] \quad \text{represents} \quad \begin{array}{c} x + 2y = 4 \\ y = 5 \end{array}$$

Note that the system above on the right can be easily solved by substitution. Its solution is $(-6, 5)$.

We use **row transformations** to rewrite the augmented matrix in triangular form. We will use three row transformation procedures.

Procedures for Row Transformations

1. All the numbers in a row may be multiplied (or divided) by any nonzero real number. (This is the same as multiplying both sides of an equation by a nonzero real number.)

2. All the numbers in a row may be multiplied by any nonzero real number. These products may then be added to the corresponding numbers in any other row. (This is equivalent to eliminating a variable from a system of equations using the addition method.)

3. The order of the rows may be switched. (This is equivalent to switching the order of the equations in a system of equations.)

Generally, when changing an element in the augmented matrix to 1 we use row transformation procedure 1, and when changing an element to 0 we use row transformation procedure 2. *Work by columns starting from the left.* Start with the first column, first row.

EXAMPLE 1 ▶ Solve the following system of equations using matrices.

$$2x - 3y = 10$$
$$4x + 5y = 9$$

Solution   First we write the augmented matrix.

$$\left[\begin{array}{cc|c} 2 & -3 & 10 \\ 4 & 5 & 9 \end{array}\right]$$

Our goal is to obtain a matrix of the form $\left[\begin{array}{cc|c} 1 & a & p \\ 0 & 1 & q \end{array}\right]$ We begin by using row transformation procedure 1 to change the 2 in the first column, first row, to 1. To do so, we multiply the first row of numbers by $\frac{1}{2}$. (We abbreviate this multiplication as $\frac{1}{2}R_1$ and place it to the right of the matrix in the same row where the operation was performed. This may help you follow the process more clearly.)

$$\left[\begin{array}{cc|c} 2\left(\frac{1}{2}\right) & -3\left(\frac{1}{2}\right) & 10\left(\frac{1}{2}\right) \\ 4 & 5 & 9 \end{array}\right]\begin{array}{c}\frac{1}{2}R_1\\ \\ \end{array}$$

This gives

$$\left[\begin{array}{cc|c} 1 & -\dfrac{3}{2} & 5 \\ 4 & 5 & 9 \end{array}\right]$$

The next step is to obtain 0 in the first column, second row. At present, 4 is in this position. We do this by multiplying the numbers in row 1 by −4, and adding the products to the numbers in row 2. (This is abbreviated $-4R_1 + R_2$.)

The numbers in the first row multiplied by −4 are

$$1(-4) \qquad -\frac{3}{2}(-4) \qquad 5(-4)$$

Now we add these products to their respective numbers in the second row. This gives

$$\left[\begin{array}{cc|c} 1 & -\dfrac{3}{2} & 5 \\ 4 + 1(-4) & 5 + \left(-\dfrac{3}{2}\right)(-4) & 9 + 5(-4) \end{array}\right]\begin{array}{c}\\ \\ -4R_1 + R_2\end{array}$$

Now we have

$$\left[\begin{array}{cc|c} 1 & -\dfrac{3}{2} & 5 \\ 0 & 11 & -11 \end{array}\right]$$

To obtain 1 in the second column, second row, we multiply the second row of numbers by $\frac{1}{11}$.

$$\left[\begin{array}{cc|c} 1 & -\dfrac{3}{2} & 5 \\ 0\left(\dfrac{1}{11}\right) & 11\left(\dfrac{1}{11}\right) & -11\left(\dfrac{1}{11}\right) \end{array}\right]\begin{array}{c}\\ \\ \frac{1}{11}R_2\end{array}$$

$$\left[\begin{array}{cc|c} 1 & -\dfrac{3}{2} & 5 \\ 0 & 1 & -1 \end{array}\right]$$

The matrix is now in the form we are seeking. The equivalent triangular system of equations is

$$x - \frac{3}{2}y = 5$$
$$y = -1$$

Now we can solve for x using substitution.

$$x - \frac{3}{2}y = 5$$
$$x - \frac{3}{2}(-1) = 5$$
$$x + \frac{3}{2} = 5$$
$$x = \frac{7}{2}$$

A check will show that the solution to the system is $\left(\frac{7}{2}, -1 \right)$.

▸ **Now Try Exercise 19**

3 Solve Systems of Linear Equations in Three Variables

Now we will use matrices to solve a system of three linear equations in three variables. We use the same row transformation procedures used when solving a system of two linear equations. Our goal is to obtain an augmented matrix in the triangular form

$$\begin{bmatrix} 1 & a & b & | & p \\ 0 & 1 & c & | & q \\ 0 & 0 & 1 & | & r \end{bmatrix}$$

where $a, b, c,$ and p, q and r are constants. This matrix represents the following system of equations.

$$
\begin{array}{ll}
1x + ay + bz = p & \qquad x + ay + bz = p \\
0x + 1y + cz = q \quad \text{or} & \qquad y + cz = q \\
0x + 0y + 1z = r & \qquad z = r
\end{array}
$$

When constructing the augmented matrix, *work by columns, from the left-hand column to the right-hand column. Always complete one column before moving to the next column. In each column, first obtain the 1 in the indicated position, and then obtain the zeros.* Example 2 illustrates this procedure.

Helpful Hint *Study Tip*

When using matrices, be careful to keep all the numbers lined up neatly in rows and columns. One slight mistake in copying numbers from one matrix to another will lead to an incorrect, and often frustrating, attempt at solving a system of equations.

$$
\begin{array}{r}
x - 3y + z = 3 \\
4x + 2y - 5z = 20 \\
-5x - y - 4z = 13
\end{array}
$$

For example, the system of equations, (above), when correctly represented

with the augmented matrix, $\begin{bmatrix} 1 & -3 & 1 & | & 3 \\ 4 & 2 & -5 & | & 20 \\ -5 & -1 & -4 & | & 13 \end{bmatrix}$, leads to the solution $(1, -2, -4)$.

However, a matrix that looks quite similar, $\begin{bmatrix} 1 & -3 & 1 & | & 3 \\ 4 & -1 & -5 & | & 20 \\ -5 & 2 & -4 & | & 13 \end{bmatrix}$, leads to the incorrect

ordered triple of $\left(-\frac{25}{53}, -\frac{130}{53}, -\frac{206}{53} \right)$.

EXAMPLE 2 ▶ Solve the following system of equations using matrices.

$$x - 2y + 3z = -7$$
$$2x - y - z = 7$$
$$-x + 3y + 2z = -8$$

Solution   First write the augmented matrix.

$$\begin{bmatrix} 1 & -2 & 3 & | & -7 \\ 2 & -1 & -1 & | & 7 \\ -1 & 3 & 2 & | & -8 \end{bmatrix}$$

Our next step is to use row transformations to change the first column to $\begin{smallmatrix}1\\0\\0\end{smallmatrix}$. Since the number in the first column, first row, is already a 1, we will work with the 2 in the first column, second row. Multiplying the numbers in the first row by -2 and adding those products to the respective numbers in the second row will result in the 2 changing to 0. The matrix is now

$$\begin{bmatrix} 1 & -2 & 3 & | & -7 \\ 0 & 3 & -7 & | & 21 \\ -1 & 3 & 2 & | & -8 \end{bmatrix} \quad -2R_1 + R_2$$

Continuing down the first column, we now change the -1 in the third row to 0. By multiplying the numbers in the first row by 1, and then adding the products to the third row, we get

$$\begin{bmatrix} 1 & -2 & 3 & | & -7 \\ 0 & 3 & -7 & | & 21 \\ 0 & 1 & 5 & | & -15 \end{bmatrix} 1R_1 + R_3$$

Now we work with the second column. We wish to change the numbers in the second column to the form $\begin{smallmatrix}a\\1\\0\end{smallmatrix}$ where a represents a number. Since there is presently a 1 in the third row, second column, and we want a 1 in the second row, second column, we switch the second and third rows of the matrix. This gives

$$\begin{bmatrix} 1 & -2 & 3 & | & -7 \\ 0 & 1 & 5 & | & -15 \\ 0 & 3 & -7 & | & 21 \end{bmatrix} \text{ Switch } R_2 \text{ and } R_3.$$

Continuing down the second column, we now change the 3 in the third row to 0 by multiplying the numbers in the second row by -3 and adding those products to the third row. This gives

$$\begin{bmatrix} 1 & -2 & 3 & | & -7 \\ 0 & 1 & 5 & | & -15 \\ 0 & 0 & -22 & | & 66 \end{bmatrix} -3R_2 + R_3$$

Now we work with the third column. We wish to change the numbers in the third column to the form $\begin{smallmatrix}b\\c\\1\end{smallmatrix}$ where b and c represent numbers. We must change the -22 in the third row to 1. We can do this by multiplying the numbers in the third row by $-\dfrac{1}{22}$. This results in the following.

$$\left[\begin{array}{ccc|c} 1 & -2 & 3 & -7 \\ 0 & 1 & 5 & -15 \\ 0 & 0 & 1 & -3 \end{array}\right] -\frac{1}{22}R_3$$

This matrix is now in the desired form. From this matrix we obtain the system of equations

$$x - 2y + 3z = -7$$
$$y + 5z = -15$$
$$z = -3$$

The third equation gives us the value of z in the solution. Now we can solve for y by substituting -3 for z in the second equation.

$$y + 5z = -15$$
$$y + 5(-3) = -15$$
$$y - 15 = -15$$
$$y = 0$$

Now we solve for x by substituting 0 for y and -3 for z in the first equation.

$$x - 2y + 3z = -7$$
$$x - 2(0) + 3(-3) = -7$$
$$x - 0 - 9 = -7$$
$$x - 9 = -7$$
$$x = 2$$

The solution is $(2, 0, -3)$. Check this now by substituting the appropriate values into each of the original equations.

▶ **Now Try Exercise 33**

4 Recognize Inconsistent and Dependent Systems

When solving a system of two equations, if you obtain an augmented matrix in which one row of numbers on the left side of the vertical line is all zeros but a zero does not appear in the same row on the right side of the vertical line, the system is inconsistent and has no solution. For example, a system of equations that yields the following augmented matrix is an inconsistent system.

$$\left[\begin{array}{cc|c} 1 & 2 & 5 \\ 0 & 0 & 3 \end{array}\right] \longleftarrow \textit{Inconsistent system}$$

The second row of the matrix represents the equation

$$0x + 0y = 3$$

which is never true.

If you obtain a matrix in which a 0 appears across an entire row, the system of equations is dependent. For example, a system of equations that yields the following augmented matrix is a dependent system.

$$\left[\begin{array}{cc|c} 1 & -3 & -4 \\ 0 & 0 & 0 \end{array}\right] \longleftarrow \textit{Dependent system}$$

The second row of the matrix represents the equation

$$0x + 0y = 0$$

which is always true.

Similar rules hold for systems with three equations.

$$\begin{bmatrix} 1 & 3 & 7 & 5 \\ 0 & 0 & 0 & -1 \\ 0 & 1 & -2 & 3 \end{bmatrix} \longleftarrow \textit{Inconsistent system}$$

$$\begin{bmatrix} 1 & 3 & -1 & 2 \\ 0 & 0 & 0 & 0 \\ 0 & 5 & 6 & -4 \end{bmatrix} \longleftarrow \textit{Dependent system}$$

USING YOUR GRAPHING CALCULATOR

Many graphing calculators have the ability to work with matrices. Such calculators have the ability to perform row operations on matrices. These graphing calculators can therefore be used to solve systems of equations using matrices.

Read the instruction manual that came with your graphing calculator to see if it can handle matrices. If so, learn how to use your graphing calculator to solve systems of equations using matrices.

EXERCISE SET 9.6

Concept/Writing Exercises

1. What is a square matrix?

2. Explain how to construct an augmented matrix.

3. If you obtain the following augmented matrix when solving a system of equations, what would be your next step in completing the process? Explain.

$$\begin{bmatrix} 1 & 3 & 6 \\ 0 & -2 & 14 \end{bmatrix}$$

4. If you obtained the following augmented matrix when solving a system of equations, what would be your next step in completing the process? Explain your answer.

$$\begin{bmatrix} 1 & -3 & 7 & -1 \\ 0 & -1 & 5 & 3 \\ 2 & 6 & 4 & -8 \end{bmatrix}$$

5. If you obtained the following augmented matrix when solving a system of linear equations, what would be your next step in completing the process? Explain your answer.

$$\begin{bmatrix} 1 & 4 & -7 & 7 \\ 0 & 5 & 2 & -1 \\ 0 & 1 & 6 & -2 \end{bmatrix}$$

6. If you obtained the following augmented matrix when solving a system of linear equations, what would be your next step in completing the process? Explain your answer.

$$\begin{bmatrix} 1 & 3 & -2 & 1 \\ 0 & 1 & 2 & -3 \\ 0 & 0 & -4 & -12 \end{bmatrix}$$

7. When solving a system of linear equations by matrices, if two rows are identical, will the system be consistent, dependent, or inconsistent?

8. When solving a system of equations using matrices, how will you know if the system is
   a) dependent,
   b) inconsistent?

Practice the Skills

Perform each row transformation indicated and write the new matrix.

9. $\begin{bmatrix} 5 & -10 & -25 \\ 3 & -7 & -4 \end{bmatrix}$ Multiply numbers in the first row by $\frac{1}{5}$.

10. $\begin{bmatrix} 1 & 8 & 3 \\ 0 & 4 & -3 \end{bmatrix}$ Multiply numbers in the second row by $\frac{1}{4}$.

11. $\begin{bmatrix} 4 & 7 & 2 & -1 \\ 3 & 2 & 1 & -5 \\ 1 & 1 & 3 & -8 \end{bmatrix}$ Switch row 1 and row 3.

12. $\begin{bmatrix} 1 & 5 & 7 & | & 2 \\ 0 & 8 & -1 & | & -6 \\ 0 & 1 & 3 & | & -4 \end{bmatrix}$ Switch row 2 and row 3.

13. $\begin{bmatrix} 1 & 3 & | & 12 \\ -4 & 11 & | & -6 \end{bmatrix}$ Multiply numbers in the first row by 4 and add the products to the second row.

14. $\begin{bmatrix} 1 & 5 & | & 6 \\ \frac{1}{2} & 10 & | & -4 \end{bmatrix}$ Multiply numbers in the first row by $-\frac{1}{2}$ and add the products to the second row.

15. $\begin{bmatrix} 1 & 0 & 8 & | & \frac{1}{4} \\ 5 & 2 & 2 & | & -2 \\ 6 & -3 & 1 & | & 0 \end{bmatrix}$ Multiply numbers in the first row by -5 and add the products to the second row.

16. $\begin{bmatrix} 1 & 2 & -1 & | & 6 \\ 0 & 1 & 5 & | & 0 \\ 0 & 0 & 3 & | & 12 \end{bmatrix}$ Multiply numbers in the third row by $\frac{1}{3}$.

Solve each system using matrices.

17. $x + 3y = 3$
$-x + y = -3$

18. $x + 2y = 10$
$3x - y = 9$

19. $x + 3y = -2$
$-2x - 7y = 3$

20. $3x + 6y = 0$
$2x - y = 10$

21. $5a - 10b = -10$
$2a + b = 1$

22. $3s - 2t = 1$
$-2s + 4t = -6$

23. $2x - 5y = -6$
$-4x + 10y = 12$

24. $-2m - 4n = 7$
$3m + 6n = -8$

25. $12x + 2y = 2$
$6x - 3y = -11$

26. $4r + 2s = -10$
$-2r + s = -7$

27. $-3x + 6y = 5$
$2x - 4y = 7$

28. $8x = 4y + 12$
$-2x + y = -3$

29. $12x - 8y = 6$
$-3x + 4y = -1$

30. $2x - 3y = 3$
$-5x + 9y = -7$

31. $10m = 8n + 15$
$16n = -15m - 2$

32. $8x = 9y + 4$
$16x - 27y = 11$

Solve each system using matrices.

33. $x - 3y + 2z = 5$
$2x + 5y - 4z = -3$
$-3x + y - 2z = -11$

34. $a - 3b + 4c = 7$
$4a + b + c = -2$
$-2a - 3b + 5c = 12$

35. $x + 2y = 5$
$y - z = -1$
$2x - 3z = 0$

36. $3a - 5c = 3$
$a + 2b = -6$
$7b - 4c = 5$

37. $x - 2y + 4z = 5$
$-3x + 4y - 2z = -8$
$4x + 5y - 4z = -3$

38. $3x + 5y + 2z = 3$
$-x - y - z = -2$
$2x - 2y + 5z = 11$

39. $2x - 5y + z = 1$
$3x - 5y + z = 3$
$-4x + 10y - 2z = -2$

40. $x + 2y + 3z = 1$
$4x + 5y + 6z = -3$
$7x + 8y + 9z = 0$

41. $4p - q + r = 4$
$-6p + 3q - 2r = -5$
$2p + 5q - r = 7$

42. $-4r + 3s - 6t = 14$
$4r + 2s - 2t = -3$
$2r - 5s - 8t = -23$

43. $2x - 4y + 3z = -12$
$3x - y + 2z = -3$
$-4x + 8y - 6z = 10$

44. $3x - 2y + 4z = -1$
$5x + 2y - 4z = 9$
$-6x + 4y - 8z = 2$

45. $5x - 3y + 4z = 22$
$-x - 15y + 10z = -15$
$-3x + 9y - 12z = -6$

46. $9x - 4y + 5z = -2$
$-9x + 5y - 10z = -1$
$9x + 3y + 10z = 1$

Problem Solving

47. When solving a system of linear equations using matrices, if two rows of matrices are switched, will the solution to the system change? Explain.

48. You can tell whether a system of two equations in two variables is consistent, dependent, or inconsistent by comparing the slopes and *y*-intercepts of the graphs of the equations. Can you tell, without solving, if a system of three equations in three variables is consistent, dependent, or inconsistent? Explain.

Solve using matrices.

49. Angles of a Roof In a triangular cross section of a roof, the largest angle is 55° greater than the smallest angle. The largest angle is 20° greater than the remaining angle. Find the measure of each angle.

50. Right Angle A right angle is divided into three smaller angles. The largest of the three angles is twice the smallest. The remaining angle is 10° greater than the smallest angle. Find the measure of each angle.

51. Bananas Sixty-five percent of the world's bananas are controlled by Chiquita, Dole, or Del Monte (all American companies). Chiquita, the largest, controls 12% more bananas than Del Monte. Dole, the second largest, controls 3% less than twice the percent that Del Monte controls. Determine the percents to be placed in each sector of the circle graph shown.

World's Bananas

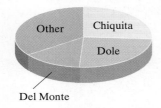

52. Impact Upon Businesses A sample of CEOs in a TEC International survey were asked to list the most important changes that could be made from 2004 to 2006 to strengthen their companies. The top three responses were, in order: reduce taxes, reform health care insurance, and strengthen the U.S. dollar. Seventy-seven percent of all the CEOs selected one of these three items as their top choice. Four percent more CEOs selected reducing taxes than reforming health care insurance. Reducing taxes was also two percent higher than three times the percent who selected strengthening the U.S. dollar. Determine the percent of CEOs that selected reducing taxes, reforming health care insurance, and strengthening the U.S. dollar. Then complete the graph below.

What Would Have the Largest Impact on Your Business?

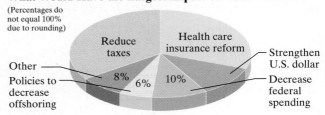

Source: TEC International survey of 2,300 CEOs

Cumulative Review Exercises

[5.6] **53.** Divide $\dfrac{3x^2 + 4x - 23}{x + 4}$.

[6.6] **54.** Solve $2x^2 - x - 36 = 0$.

[7.4] **55.** Subtract $\dfrac{1}{x^2 - 4} - \dfrac{2}{x - 2}$.

[7.7] **56. Stacking Wood** Sam Davidson can stack a cord of wood in 20 minutes. With his wife's help, they can stack the wood in 12 minutes. How long would in take his wife, Terry, to stack the wood by herself?

9.7 Solving Systems of Equations Using Determinants and Cramer's Rule

1 Evaluate a determinant of a 2 × 2 matrix.

2 Use Cramer's rule.

3 Evaluate a determinant of a 3 × 3 matrix.

4 Use Cramer's rule with systems in three variables.

1 Evaluate a Determinant of a 2 × 2 Matrix

We have discussed various ways of solving a system of linear equations, including: graphing, substitution, the addition (or elimination) method, and matrices. A system of linear equations may also be solved using determinants.

Associated with every square matrix is a number called its **determinant**. For a 2 × 2 matrix, its determinant is defined as follows.

> **Determinant**
>
> The **determinant** of a 2 × 2 matrix $\begin{bmatrix} a_1 & b_1 \\ a_2 & b_2 \end{bmatrix}$ is denoted $\begin{vmatrix} a_1 & b_1 \\ a_2 & b_2 \end{vmatrix}$ and is evaluated as
>
> $$\begin{vmatrix} a_1 & b_1 \\ a_2 & b_2 \end{vmatrix} = a_1 b_2 - a_2 b_1$$

EXAMPLE 1 ▶ Evaluate each determinant.

a) $\begin{vmatrix} 2 & -1 \\ 3 & -5 \end{vmatrix}$ **b)** $\begin{vmatrix} 2 & 3 \\ -1 & 4 \end{vmatrix}$

Solution

a) $a_1 = 2, a_2 = 3, b_1 = -1, b_2 = -5$

$$\begin{vmatrix} 2 & -1 \\ 3 & -5 \end{vmatrix} = 2(-5) - (3)(-1) = -10 + 3 = -7$$

b) $\begin{vmatrix} 2 & 3 \\ -1 & 4 \end{vmatrix} = (2)(4) - (-1)(3) = 8 + 3 = 11$

▶ **Now Try Exercise 7**

2 Use Cramer's Rule

If we begin with the equations

$$a_1 x + b_1 y = c_1$$
$$a_2 x + b_2 y = c_2$$

we can use the addition method to show that

$$x = \frac{c_1 b_2 - c_2 b_1}{a_1 b_2 - a_2 b_1} \quad \text{and} \quad y = \frac{a_1 c_2 - a_2 c_1}{a_1 b_2 - a_2 b_1}$$

(see Challenge Problem 65 on page 605). Notice that the *denominators* of x and y are both $a_1 b_2 - a_2 b_1$. Following is the determinant that yields this denominator. We have labeled this denominator D.

$$D = \begin{vmatrix} a_1 & b_1 \\ a_2 & b_2 \end{vmatrix} = a_1 b_2 - a_2 b_1$$

The *numerators* of x and y are different. Following are two determinants, labeled D_x and D_y, that yield the numerators of x and y.

$$D_x = \begin{vmatrix} c_1 & b_1 \\ c_2 & b_2 \end{vmatrix} = c_1 b_2 - c_2 b_1 \qquad D_y = \begin{vmatrix} a_1 & c_1 \\ a_2 & c_2 \end{vmatrix} = a_1 c_2 - a_2 c_1$$

We use determinants D, D_x, and D_y in Cramer's rule. **Cramer's rule** can be used to solve systems of linear equations.

Cramer's Rule for Systems of Linear Equations

For a system of linear equations of the form

$$a_1x + b_1y = c_1$$
$$a_2x + b_2y = c_2$$

$$x = \frac{\begin{vmatrix} c_1 & b_1 \\ c_2 & b_2 \end{vmatrix}}{\begin{vmatrix} a_1 & b_1 \\ a_2 & b_2 \end{vmatrix}} = \frac{D_x}{D} \quad \text{and} \quad y = \frac{\begin{vmatrix} a_1 & c_1 \\ a_2 & c_2 \end{vmatrix}}{\begin{vmatrix} a_1 & b_1 \\ a_2 & b_2 \end{vmatrix}} = \frac{D_y}{D}, \quad D \neq 0$$

Helpful Hint

The elements in determinant D are the numerical coefficients of the x and y terms in the two given equations, listed in the same order they are listed in the equations. To obtain the determinant D_x from determinant D, replace the coefficients of the x-terms (the values in the first column) with the constants of the two given equations. To obtain the determinant D_y from determinant D, replace the coefficients of the y-terms (the values in the second column) with the constants of the two given equations.

EXAMPLE 2 ▶ Solve the following system of equations using Cramer's rule.

$$3x + 5y = 7$$
$$4x - y = -6$$

Solution Both equations are given in the desired form, $ax + by = c$. When labeling a, b, and c, we will refer to $3x + 5y = 7$ as equation 1 and $4x - y = -6$ as equation 2 (in the subscripts).

$$
\begin{array}{ccc}
a_1 & b_1 & c_1 \\
\downarrow & \downarrow & \downarrow \\
3x + & 5y = & 7 \\
4x - & 1y = & -6 \\
\uparrow & \uparrow & \uparrow \\
a_2 & b_2 & c_2
\end{array}
$$

We now determine D, D_x, D_y.

$$D = \begin{vmatrix} a_1 & b_1 \\ a_2 & b_2 \end{vmatrix} = \begin{vmatrix} 3 & 5 \\ 4 & -1 \end{vmatrix} = 3(-1) - 4(5) = -3 - 20 = -23$$

$$D_x = \begin{vmatrix} c_1 & b_1 \\ c_2 & b_2 \end{vmatrix} = \begin{vmatrix} 7 & 5 \\ -6 & -1 \end{vmatrix} = 7(-1) - (-6)(5) = -7 + 30 = 23$$

$$D_y = \begin{vmatrix} a_1 & c_1 \\ a_2 & c_2 \end{vmatrix} = \begin{vmatrix} 3 & 7 \\ 4 & -6 \end{vmatrix} = 3(-6) - 4(7) = -18 - 28 = -46$$

Now we find the values of x and y.

$$x = \frac{D_x}{D} = \frac{23}{-23} = -1$$

$$y = \frac{D_y}{D} = \frac{-46}{-23} = 2$$

Thus the solution is $x = -1$, $y = 2$ or the ordered pair $(-1, 2)$. A check will show that this ordered pair satisfies both equations.

▶ **Now Try Exercise 15**

When the determinant D has a value of 0, Cramer's rule cannot be used since division by 0 is undefined. You must then use a different method to solve the system. Or you may evaluate D_x and D_y to determine whether the system is dependent or inconsistent.

> **When $D = 0$**
>
> If $D = 0$, $D_x = 0$, $D_y = 0$, then the system is dependent.
> If $D = 0$ and either $D_x \neq 0$ or $D_y \neq 0$, then the system is inconsistent.

3 Evaluate a Determinant of a 3 × 3 Matrix

For the determinant

$$\begin{vmatrix} a_1 & b_1 & c_1 \\ a_2 & b_2 & c_2 \\ a_3 & b_3 & c_3 \end{vmatrix}$$

the **minor determinant** of a_1 is found by crossing out the elements in the same row and same column in which the element a_1 appears. The remaining elements form the minor determinant of a_1. The minor determinants of other elements are found similarly.

$$\begin{vmatrix} a_1 & b_1 & c_1 \\ a_2 & b_2 & c_2 \\ a_3 & b_3 & c_3 \end{vmatrix} \qquad \begin{vmatrix} b_2 & c_2 \\ b_3 & c_3 \end{vmatrix} \qquad \textit{Minor determinant of } a_1$$

$$\begin{vmatrix} a_1 & b_1 & c_1 \\ a_2 & b_2 & c_2 \\ a_3 & b_3 & c_3 \end{vmatrix} \qquad \begin{vmatrix} b_1 & c_1 \\ b_3 & c_3 \end{vmatrix} \qquad \textit{Minor determinant of } a_2$$

$$\begin{vmatrix} a_1 & b_1 & c_1 \\ a_2 & b_2 & c_2 \\ a_3 & b_3 & c_3 \end{vmatrix} \qquad \begin{vmatrix} b_1 & c_1 \\ b_2 & c_2 \end{vmatrix} \qquad \textit{Minor determinant of } a_3$$

To evaluate determinants of a 3 × 3 matrix, we use minor determinants. The following box shows how such a determinant may be evaluated by **expansion by the minors of the first column**.

> **Expansion of the Determinant by the Minors of the First Column**
>
> $$\begin{vmatrix} a_1 & b_1 & c_1 \\ a_2 & b_2 & c_2 \\ a_3 & b_3 & c_3 \end{vmatrix} = a_1 \begin{vmatrix} b_2 & c_2 \\ b_3 & c_3 \end{vmatrix} - a_2 \begin{vmatrix} b_1 & c_1 \\ b_3 & c_3 \end{vmatrix} + a_3 \begin{vmatrix} b_1 & c_1 \\ b_2 & c_2 \end{vmatrix}$$
>
> where the terms are the Minor determinant of a_1, Minor determinant of a_2, Minor determinant of a_3.

EXAMPLE 3 ▶ Evaluate $\begin{vmatrix} 4 & -2 & 6 \\ 3 & 5 & 0 \\ 1 & -3 & -1 \end{vmatrix}$ using expansion by the minors of the first column.

Solution We will follow the procedure given in the box.

$$\begin{vmatrix} 4 & -2 & 6 \\ 3 & 5 & 0 \\ 1 & -3 & -1 \end{vmatrix} = 4 \begin{vmatrix} 5 & 0 \\ -3 & -1 \end{vmatrix} - 3 \begin{vmatrix} -2 & 6 \\ -3 & -1 \end{vmatrix} + 1 \begin{vmatrix} -2 & 6 \\ 5 & 0 \end{vmatrix}$$

$$= 4[5(-1) - (-3)0] - 3[(-2)(-1) - (-3)6] + 1[(-2)0 - 5(6)]$$
$$= 4(-5 + 0) - 3(2 + 18) + 1(0 - 30)$$
$$= 4(-5) - 3(20) + 1(-30)$$
$$= -20 - 60 - 30$$
$$= -110$$

The determinant has a value of -110.

▶ **Now Try Exercise 13**

4 Use Cramer's Rule with Systems in Three Variables

Cramer's rule can be extended to systems of equations in three variables as follows.

Cramer's Rule for a System of Equations in Three Variables

To solve the system

$$a_1 x + b_1 y + c_1 z = d_1$$
$$a_2 x + b_2 y + c_2 z = d_2$$
$$a_3 x + b_3 y + c_3 z = d_3$$

with

$$D = \begin{vmatrix} a_1 & b_1 & c_1 \\ a_2 & b_2 & c_2 \\ a_3 & b_3 & c_3 \end{vmatrix} \qquad D_x = \begin{vmatrix} d_1 & b_1 & c_1 \\ d_2 & b_2 & c_2 \\ d_3 & b_3 & c_3 \end{vmatrix}$$

$$D_y = \begin{vmatrix} a_1 & d_1 & c_1 \\ a_2 & d_2 & c_2 \\ a_3 & d_3 & c_3 \end{vmatrix} \qquad D_z = \begin{vmatrix} a_1 & b_1 & d_1 \\ a_2 & b_2 & d_2 \\ a_3 & b_3 & d_3 \end{vmatrix}$$

then

$$x = \frac{D_x}{D} \quad y = \frac{D_y}{D} \quad z = \frac{D_z}{D}, \quad D \neq 0$$

Note that the denominators of the expressions for x, y, and z are all the same determinant, D. Note that the d's replace the a's, the numerical coefficients of the x-terms, in D_x. The d's replace the b's, the numerical coefficients of the y-terms, in D_y. And the d's replace the c's, the numerical coefficients of the z-terms, in D_z.

EXAMPLE 4 ▶ Solve the following system of equations using determinants.

$$3x - 2y - z = -6$$
$$2x + 3y - 2z = 1$$
$$x - 4y + z = -3$$

Solution

$$\begin{array}{llll} a_1 = 3 & b_1 = -2 & c_1 = -1 & d_1 = -6 \\ a_2 = 2 & b_2 = 3 & c_2 = -2 & d_2 = 1 \\ a_3 = 1 & b_3 = -4 & c_3 = 1 & d_3 = -3 \end{array}$$

We will use expansion by the minor determinants of the first column to evaluate D, D_x, D_y, and D_z.

$$D = \begin{vmatrix} 3 & -2 & -1 \\ 2 & 3 & -2 \\ 1 & -4 & 1 \end{vmatrix} = 3 \begin{vmatrix} 3 & -2 \\ -4 & 1 \end{vmatrix} - 2 \begin{vmatrix} -2 & -1 \\ -4 & 1 \end{vmatrix} + 1 \begin{vmatrix} -2 & -1 \\ 3 & -2 \end{vmatrix}$$

$$= 3(-5) - 2(-6) + 1(7)$$
$$= -15 + 12 + 7 = 4$$

$$D_x = \begin{vmatrix} -6 & -2 & -1 \\ 1 & 3 & -2 \\ -3 & -4 & 1 \end{vmatrix} = -6 \begin{vmatrix} 3 & -2 \\ -4 & 1 \end{vmatrix} - 1 \begin{vmatrix} -2 & -1 \\ -4 & 1 \end{vmatrix} + (-3) \begin{vmatrix} -2 & -1 \\ 3 & -2 \end{vmatrix}$$

$$= -6(-5) - 1(-6) - 3(7)$$
$$= 30 + 6 - 21 = 15$$

$$D_y = \begin{vmatrix} 3 & -6 & -1 \\ 2 & 1 & -2 \\ 1 & -3 & 1 \end{vmatrix} = 3 \begin{vmatrix} 1 & -2 \\ -3 & 1 \end{vmatrix} - 2 \begin{vmatrix} -6 & -1 \\ -3 & 1 \end{vmatrix} + 1 \begin{vmatrix} -6 & -1 \\ 1 & -2 \end{vmatrix}$$

$$= 3(-5) - 2(-9) + 1(13)$$
$$= -15 + 18 + 13 = 16$$

$$D_z = \begin{vmatrix} 3 & -2 & -6 \\ 2 & 3 & 1 \\ 1 & -4 & -3 \end{vmatrix} = 3\begin{vmatrix} 3 & 1 \\ -4 & -3 \end{vmatrix} - 2\begin{vmatrix} -2 & -6 \\ -4 & -3 \end{vmatrix} + 1\begin{vmatrix} -2 & -6 \\ 3 & 1 \end{vmatrix}$$

$$= 3(-5) - 2(-18) + 1(16)$$

$$= -15 + 36 + 16 = 37$$

We found that $D = 4$, $D_x = 15$, $D_y = 16$, and $D_z = 37$. Therefore,

$$x = \frac{D_x}{D} = \frac{15}{4} \qquad y = \frac{D_y}{D} = \frac{16}{4} = 4 \qquad z = \frac{D_z}{D} = \frac{37}{4}$$

The solution to the system is $\left(\dfrac{15}{4}, 4, \dfrac{37}{4}\right)$. Note the ordered triple lists x, y, and z in this order.

▶ **Now Try Exercise 33**

When we have a system of equations in three variables in which one or more equations are missing a variable, we insert the variable with a coefficient of 0. Thus,

$$\begin{aligned} 2x - 3y + 2z &= -1 \\ x + 2y &= 14 \\ x - 3z &= -5 \end{aligned} \quad \text{is written} \quad \begin{aligned} 2x - 3y + 2z &= -1 \\ x + 2y + 0z &= 14 \\ x + 0y - 3z &= -5 \end{aligned}$$

Helpful Hint

When evaluating determinants, if any two rows (or columns) are identical, or identical except for opposite signs, the determinant has a value of 0. For example,

$$\begin{vmatrix} 5 & -2 \\ 5 & -2 \end{vmatrix} = 0 \quad \text{and} \quad \begin{vmatrix} 5 & -2 \\ -5 & 2 \end{vmatrix} = 0$$

$$\begin{vmatrix} 5 & -3 & 4 \\ 2 & 6 & 5 \\ 5 & -3 & 4 \end{vmatrix} = 0 \quad \text{and} \quad \begin{vmatrix} 5 & -3 & 4 \\ -5 & 3 & -4 \\ 6 & 8 & 2 \end{vmatrix} = 0$$

As with determinants of a 2×2 matrix, when the determinant D has a value of 0, Cramer's rule cannot be used since division by 0 is undefined. You must then use a different method to solve the system. Or you may evaluate D_x, D_y, and D_z to determine whether the system is dependent or inconsistent.

When $D = 0$

If $D = 0$, $D_x = 0$, $D_y = 0$, and $D_z = 0$, then the system is dependent.

If $D = 0$ and $D_x \neq 0$, $D_y \neq 0$, or $D_z \neq 0$, then the system is inconsistent.

USING YOUR GRAPHING CALCULATOR

In Section 9.6 we mentioned that some graphing calculators can handle matrices. Graphing calculators with matrix capabilities can also evaluate determinants of square matrices. Read your graphing calculator manual to learn if your calculator can find determinants. If so, learn how to do so on your calculator.

EXERCISE SET 9.7

Concept/Writing Exercises

1. Explain how to evaluate a 2×2 determinant.

2. Explain how to evaluate a 3×3 determinant by expansion by the minors of the first column.

3. Explain how you can determine whether a system of three linear equations is inconsistent using determinants.

4. Explain how you can determine whether a system of three linear equations is dependent using determinants.

5. While solving a system of two linear equations using Cramer's rule, you determine that $D = 4$, $D_x = 12$, and $D_y = -2$. What is the solution to this system?

6. While solving a system of three linear equations using Cramer's rule, you determine that $D = -2$, $D_x = 8$, $D_y = 14$, and $D_z = -2$. What is the solution to this system?

Practice the Skills

Evaluate each determinant.

7. $\begin{vmatrix} 2 & 4 \\ 1 & 5 \end{vmatrix}$

8. $\begin{vmatrix} 3 & 5 \\ -1 & -2 \end{vmatrix}$

9. $\begin{vmatrix} \frac{1}{2} & 3 \\ 2 & -4 \end{vmatrix}$

10. $\begin{vmatrix} 13 & -\frac{2}{3} \\ -1 & 0 \end{vmatrix}$

11. $\begin{vmatrix} 3 & 2 & 0 \\ 0 & 5 & 3 \\ -1 & 4 & 2 \end{vmatrix}$

12. $\begin{vmatrix} 4 & 1 & 1 \\ 0 & 0 & 3 \\ 2 & 2 & 9 \end{vmatrix}$

13. $\begin{vmatrix} 2 & 3 & 1 \\ 1 & -3 & -6 \\ -4 & 5 & 9 \end{vmatrix}$

14. $\begin{vmatrix} 5 & -8 & 6 \\ 3 & 0 & 4 \\ -5 & -2 & 1 \end{vmatrix}$

Solve each system of equations using determinants.

15. $x + 3y = 1$
    $-2x - 3y = 4$

16. $2x + 4y = -2$
    $-5x - 2y = 13$

17. $-x - 2y = 2$
    $x + 3y = -6$

18. $2r + 3s = -9$
    $3r + 5s = -16$

19. $6x = 4y + 7$
    $8x - 1 = -3y$

20. $6x + 3y = -4$
    $9x + 5y = -6$

21. $5p - 7q = -21$
    $-4p + 3q = 22$

22. $4x = -5y - 2$
    $-2x = y + 4$

23. $x + 5y = 3$
    $2x - 6 = -10y$

24. $9x + 6y = -3$
    $6x + 4y = -2$

25. $3r = -4s - 6$
    $3s = -5r + 1$

26. $x = y - 1$
    $3y = 2x + 9$

27. $5x - 5y = 3$
    $-x + y = -4$

28. $2x - 5y = -3$
    $-4x + 10y = 7$

29. $6.3x - 4.5y = -9.9$
    $-9.1x + 3.2y = -2.2$

30. $-1.1x + 8.3y = 36.5$
    $3.5x + 1.6y = -4.1$

Solve each system using determinants.

31. $x + y + z = 3$
    $-3y + 4z = 15$
    $-3x + 4y - 2z = -13$

32. $2x + 3y = 4$
    $3x + 7y - 4z = -3$
    $x - y + 2z = 9$

33. $3x - 5y - 4z = -4$
    $4x + 2y = 1$
    $6y - 4z = -11$

34. $2x + 5y + 3z = 2$
    $6x - 9y = 5$
    $3y + 2z = 1$

35. $x + 4y - 3z = -6$
    $2x - 8y + 5z = 12$
    $3x + 4y - 2z = -3$

36. $2x + y - 2z = 4$
    $2x + 2y - 4z = 1$
    $-6x + 8y - 4z = 1$

37. $a - b + 2c = 3$
    $a - b + c = 1$
    $2a + b + 2c = 2$

38. $-2x + y + 8 = -2$
    $3x + 2y + z = 3$
    $x - 3y - 5z = 5$

39. $a + 2b + c = 1$
    $a - b + 3c = 2$
    $2a + b + 4c = 3$

40. $4x - 2y + 6z = 2$
    $-6x + 3y - 9z = -3$
    $2x - 7y + 11z = -5$

41. $1.1x + 2.3y - 4.0z = -9.2$
    $-2.3x + 4.6z = 6.9$
    $-8.2y - 7.5z = -6.8$

42. $4.6y - 2.1z = 24.3$
    $-5.6x + 1.8y = -5.8$
    $2.8x - 4.7y - 3.1z = 7.0$

43. $-6x + 3y - 12z = -13$
$5x + 2y - 3z = 1$
$2x - y + 4z = -5$

44. $x - 2y + z = 2$
$4x - 6y + 2z = 3$
$2x - 3y + z = 0$

45. $2x + \dfrac{1}{2}y - 3z = 5$
$-3x + 2y + 2z = 1$
$4x - \dfrac{1}{4}y - 7z = 4$

46. $\dfrac{1}{4}x - \dfrac{1}{2}y + 3z = -3$
$2x - 3y + 2z = -1$
$\dfrac{1}{6}x + \dfrac{1}{3}y - \dfrac{1}{3}z = 1$

47. $0.3x - 0.1y - 0.3z = -0.2$
$0.2x - 0.1y + 0.1z = -0.9$
$0.1x + 0.2y - 0.4z = 1.7$

48. $0.6u - 0.4v + 0.5w = 3.1$
$0.5u + 0.2v + 0.2w = 1.3$
$0.1u + 0.1v + 0.1w = 0.2$

Problem Solving

49. Given a determinant of the form $\begin{vmatrix} a_1 & b_1 \\ a_2 & b_2 \end{vmatrix}$, how will the value of the determinant change if the a's are switched with each other and the b's are switched with each other, $\begin{vmatrix} a_2 & b_2 \\ a_1 & b_1 \end{vmatrix}$? Explain your answer.

50. Given a determinant of the form $\begin{vmatrix} a_1 & b_1 \\ a_2 & b_2 \end{vmatrix}$, how will the value of the determinant change if the a's are switched with the b's, $\begin{vmatrix} b_1 & a_1 \\ b_2 & a_2 \end{vmatrix}$? Explain your answer.

51. In a 2×2 determinant, if the rows are the same, what is the value of the determinant?

52. If all the numbers in one row or one column of a 2×2 determinant are 0, what is the value of the determinant?

53. If all the numbers in one row or one column of a 3×3 determinant are 0, what is the value of the determinant?

54. Given a 3×3 determinant, if all the numbers in one row are multiplied by -1, will the value of the new determinant change? Explain.

55. Given a 3×3 determinant, if the new first and second rows are switched, will the value of the determinant change? Explain.

56. In a 3×3 determinant, if any two rows are the same, can you make a generalization about the value of the determinant?

57. In a 3×3 determinant, if the numbers in the first row are multiplied by -1 and the numbers in the second row are multiplied by -1, will the value of the new determinant change? Explain.

58. In a 3×3 determinant, if the numbers in the second row are multiplied by -1 and the numbers in the third row are multiplied by -1, will the value of the new determinant change? Explain.

59. In a 3×3 determinant, if the numbers in the second row are multiplied by 2, will the value of the new determinant change? Explain.

60. In a 3×3 determinant, if the numbers in the first row are multiplied by 3 and the numbers in the third row are multiplied by 4, will the value of the new determinant change? Explain.

Solve for the given letter.

61. $\begin{vmatrix} 4 & 6 \\ -2 & y \end{vmatrix} = 32$

62. $\begin{vmatrix} b - 3 & -4 \\ b + 2 & -6 \end{vmatrix} = 14$

63. $\begin{vmatrix} 4 & 7 & y \\ 3 & -1 & 2 \\ 4 & 1 & 5 \end{vmatrix} = -35$

64. $\begin{vmatrix} 3 & 2 & -2 \\ 0 & 5 & -6 \\ -1 & x & -7 \end{vmatrix} = -31$

Challenge Problems

65. Use the addition method to solve the following system for **a)** x and **b)** y.

$$a_1 x + b_1 y = c_1$$
$$a_2 x + b_2 y = c_2$$

Cumulative Review Exercises

[2.6] **66.** Solve the equation $2x - 5y = 6$ for y.

Graph $3x + 4y = 8$ using the indicated method.

[4.2] **67.** By plotting points

68. Using the x- and y-intercepts

[4.4] **69.** Using the slope and y-intercept

Chapter 9 Summary

IMPORTANT FACTS AND CONCEPTS	EXAMPLES

Section 9.1

A **system of linear equations** is a system having two or more linear equations.

$$y = 2x - 5$$
$$y = -4x + 7$$

The **solution to a system of equations** is the ordered pair or pairs that satisfy all the equations in the system.

The solution to the above system is $(2, -1)$.

Possible Solutions to a System of Linear Equations.
A linear system of equations may have exactly one solution (intersecting lines), no solution (parallel lines), or an infinite number of solutions (same line).

A system of equations that has a solution is called a **consistent system**. A system of equations that has no solution is called an **inconsistent system**. A system of equations that has an infinite number of solutions is called a **dependent system**.

Consistent, exactly 1 solution

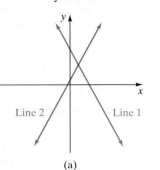

(a)

Inconsistent, no solution

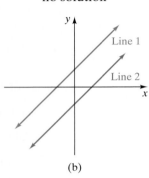

(b)

Dependent, infinite number of solutions

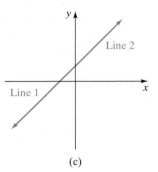

(c)

To **solve a system of equations graphically**, graph each equation and determine the point or points of intersection.

Solve the system of equations graphically.

$$y = -x + 5$$
$$y = x + 3$$

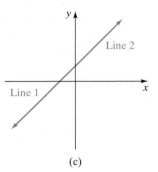

The solution is $(1, 4)$.

IMPORTANT FACTS AND CONCEPTS	EXAMPLES

Section 9.2

To Solve a System of Equations by Substitution

1. Solve for a variable in either equation.
2. Substitute the expression found for the variable in step 1 into the other equation.
3. Solve the equation determined in step 2 to find the value of one variable.
4. Substitute the value found in step 3 into the equation obtained in step 1 to find the value of the other variable.
5. Check by substituting both values in both original equations.

Solve the system of equations by substitution.

$$y = -x + 5$$
$$y = x + 3$$

Substitute $x + 3$ for y in the first equation.

$$x + 3 = -x + 5$$
$$2x + 3 = 5$$
$$2x = 2$$
$$x = 1$$

Now, solve for y.

$$y = -x + 5$$
$$y = -1 + 5$$
$$= 4$$

The solution is $(1, 4)$.

Section 9.3

To Solve a System of Equations by the Addition (or Elimination) Method

1. If necessary, rewrite each equation so that the terms containing variables appear on the left side of the equal sign and any constants appear on the right side of the equal sign.
2. If necessary, multiply one or both equations by a constant(s) so that when the equations are added the resulting sum will contain only one variable.
3. Add the equations.
4. Solve for the variable in the equation from step 3.
5. Substitute the value found in step 4 into either of the original equations. Solve that equation to find the value of the remaining variable.
6. Check the values obtained in all original equations.

Solve the system of equations by the addition method.

$$2x + 3y = -4 \quad (eq.\,1)$$
$$3x - y = -17. \quad (eq.\,2)$$

Multiply $(eq.\,2)$ by 3, then add:

$$2x + 3y = -4$$
$$\underline{9x - 3y = -51} \quad (eq.\,2) \; \textit{Multiplied by 3}$$
$$11x \qquad = -55$$
$$x = -5$$

Substitute -5 for x into $(eq.\,2)$.

$$3x - y = -17$$
$$3(-5) - y = -17$$
$$-15 - y = -17$$
$$-y = -2$$
$$y = 2$$

The solution is $(-5, 2)$.

Section 9.4

Three Linear Equations

To solve a system of three linear equations, use the substitution method or addition method.

Solve the system of equations.

$$x - y + 3z = -1 \quad (eq.\,1)$$
$$4y - 7z = 2 \quad (eq.\,2)$$
$$z = 2 \quad (eq.\,3)$$

Substitute 2 for z in $(eq.\,2)$ to obtain the value for y.

$$4y - 7z = 2$$
$$4y - 7(2) = 2$$
$$4y = 16$$
$$y = 4.$$

Substitute 4 for y and 2 for z in $(eq.\,1)$ to obtain the value for x.

$$x - y + 3z = -1$$
$$x - 4 + 3(2) = -1$$
$$x = -3.$$

A check shows that $(-3, 4, 2)$ is a solution to the system of equations.

IMPORTANT FACTS AND CONCEPTS	EXAMPLES

Section 9.5

Applications:
Systems of two linear equations in two unknowns.

The sum of the areas of two circles is 180 square meters. The difference of their areas is 20 square meters. Determine the area of each circle.

Solution

Let x be the area of the larger circle and y be the area of the smaller circle.

The two equations for this systems are

$$\begin{aligned} x + y &= 180 \quad \leftarrow \textit{Sum of areas} \\ x - y &= 20 \quad \leftarrow \textit{Difference of areas.} \\ \hline 2x \phantom{{}+y} &= 200 \\ x &= 100 \end{aligned}$$

Substitute 100 for x in the first equation to get

$$\begin{aligned} x + y &= 180 \\ 100 + y &= 180 \\ y &= 80 \end{aligned}$$

The area of the larger circle is 100 square meters and the area of the smaller circle is 80 square meters.

Section 9.6

A **matrix** is a rectangular array of numbers within brackets. The numbers inside the brackets are called **elements**.

$$\begin{bmatrix} 8 & 1 & 4 \\ -3 & 0 & 2 \end{bmatrix}, \begin{bmatrix} 5 & 0 \\ -2 & 8 \\ 6 & -11 \end{bmatrix} \text{ are matrices}$$

A **square matrix** has the same number of rows and columns.

$$\begin{bmatrix} 5 & -1 \\ 8 & 2 \end{bmatrix}, \begin{bmatrix} 3 & 0 & -6 \\ -1 & 5 & 2 \\ 9 & 10 & -7 \end{bmatrix} \text{ are square matrices}$$

An **augmented matrix** is a matrix separated by a vertical line. For a system of equations, the coefficients of the variables are placed on the left side of the vertical line and the constants are placed on the right side in the augmented matrix.

$$\begin{array}{cc} \text{SYSTEM} & \text{AUGMENTED MATRIX} \\ \begin{aligned} 2x - 3y &= 8 \\ 5x + 7y &= -4 \end{aligned} & \left[\begin{array}{cc|c} 2 & -3 & 8 \\ 5 & 7 & -4 \end{array}\right] \end{array}$$

The **triangular form** of an augmented matrix is

$$\left[\begin{array}{cc|c} 1 & a & p \\ 0 & 1 & q \end{array}\right]$$

where $a, p,$ and q are real numbers.

triangular form of an augmented matrix

$$\left[\begin{array}{cc|c} 1 & -6 & 2 \\ 0 & 1 & 9 \end{array}\right]$$

Row transformations can be used to rewrite a matrix into triangular form.

Procedures for Row Transformations

1. All the numbers in a row may be multiplied (or divided) by any nonzero real number.

2. All the numbers in a row may be multiplied by any nonzero real number. These products may then be added to the corresponding numbers in any other row.

3. The order of the rows may be switched.

Solve the system of equations

$$\begin{aligned} x + 4y &= -7 \\ 6x - 5y &= 16 \end{aligned}$$

The augmented matrix is

$$\left[\begin{array}{cc|c} 1 & 4 & -7 \\ 6 & -5 & 16 \end{array}\right] \text{ which becomes } \left[\begin{array}{cc|c} 1 & 4 & -7 \\ 0 & -29 & 58 \end{array}\right] \begin{array}{l} \\ -6R_1 + R_2 \end{array}$$

$$\text{or } \left[\begin{array}{cc|c} 1 & 4 & -7 \\ 0 & 1 & -2 \end{array}\right] -\frac{1}{29}R_2.$$

The equivalent system of equations is

$$\begin{aligned} x + 4y &= -7 \\ y &= -2 \end{aligned}$$

Substitute -2 for y into the first equation.

$$\begin{aligned} x + 4(-2) &= -7 \\ x - 8 &= -7 \\ x &= 1. \end{aligned}$$

The solution is $(1, -2)$.

IMPORTANT FACTS AND CONCEPTS	EXAMPLES

Section 9.6 (continued)

A system of equations is **inconsistent** and has **no solution** if you obtain an augmented matrix in which one row of numbers has zeros on the left side of the vertical line and a nonzero number on the right side of the vertical line.

$$\begin{bmatrix} 1 & 2 & -3 & | & 23 \\ 0 & 0 & 0 & | & 8 \\ -1 & 7 & 6 & | & 9 \end{bmatrix}$$

The second row shows this system is inconsistent and has no solution.

A system of equations is **dependent** and has an **infinite number of solutions** if you obtain an augmented matrix in which a 0 appears across an entire row.

$$\begin{bmatrix} 1 & 6 & -1 & | & 15 \\ 0 & 0 & 0 & | & 0 \\ 3 & 5 & 8 & | & -12 \end{bmatrix}$$

The second row shows this system is dependent and has an infinite number of solutions.

Section 9.7

The **determinant** of a 2×2 matrix $\begin{bmatrix} a_1 & b_1 \\ a_2 & b_2 \end{bmatrix}$ is denoted $\begin{vmatrix} a_1 & b_1 \\ a_2 & b_2 \end{vmatrix}$ and is evaluated as

$$\begin{vmatrix} a_1 & b_1 \\ a_2 & b_2 \end{vmatrix} = a_1 b_2 - a_2 b_1$$

$$\begin{vmatrix} 3 & -2 \\ 5 & 1 \end{vmatrix} = (3)(1) - (5)(-2) = 3 + 10 = 13$$

Cramer's Rule for Systems of Linear Equations

For a system of linear equations of the form

$$a_1 x + b_1 y = c_1$$
$$a_2 x + b_2 y = c_2$$

$$x = \frac{\begin{vmatrix} c_1 & b_1 \\ c_2 & b_2 \end{vmatrix}}{\begin{vmatrix} a_1 & b_1 \\ a_2 & b_2 \end{vmatrix}} = \frac{D_x}{D} \quad \text{and} \quad y = \frac{\begin{vmatrix} a_1 & c_1 \\ a_2 & c_2 \end{vmatrix}}{\begin{vmatrix} a_1 & b_1 \\ a_2 & b_2 \end{vmatrix}} = \frac{D_y}{D}, \quad D \neq 0$$

Solve the system of equations.

$$2x + y = 6$$
$$4x - 3y = -13$$

$$D = \begin{vmatrix} 2 & 1 \\ 4 & -3 \end{vmatrix} = -10$$

$$D_x = \begin{vmatrix} 6 & 1 \\ -13 & -3 \end{vmatrix} = -5 \qquad D_y = \begin{vmatrix} 2 & 6 \\ 4 & -13 \end{vmatrix} = -50$$

Then

$$x = \frac{D_x}{D} = \frac{-5}{-10} = \frac{1}{2}, \quad y = \frac{D_y}{D} = \frac{-50}{-10} = 5$$

The solution is $\left(\frac{1}{2}, 5 \right)$.

For the determinant

$$\begin{vmatrix} a_1 & b_1 & c_1 \\ a_2 & b_2 & c_2 \\ a_3 & b_3 & c_3 \end{vmatrix}$$

The **minor determinant of a_1** is found by crossing out the elements in the same row and column containing the element a_1.

Expansion of the Determinant by the Minors of the First Column

$$\begin{array}{ccc} \text{Minor} & \text{Minor} & \text{Minor} \\ \text{determinant} & \text{determinant} & \text{determinant} \\ \text{of } a_1 & \text{of } a_2 & \text{of } a_3 \\ \downarrow & \downarrow & \downarrow \end{array}$$

$$\begin{vmatrix} a_1 & b_1 & c_1 \\ a_2 & b_2 & c_2 \\ a_3 & b_3 & c_3 \end{vmatrix} = a_1 \begin{vmatrix} b_2 & c_2 \\ b_3 & c_3 \end{vmatrix} - a_2 \begin{vmatrix} b_1 & c_1 \\ b_3 & c_3 \end{vmatrix} + a_3 \begin{vmatrix} b_1 & c_1 \\ b_2 & c_2 \end{vmatrix}$$

For $\begin{vmatrix} 6 & 2 & -1 \\ 0 & 3 & 5 \\ 7 & 1 & 9 \end{vmatrix}$, the minor determinant of a_1 is $\begin{vmatrix} 3 & 5 \\ 1 & 9 \end{vmatrix}$.

Evaluate $\begin{vmatrix} 2 & 0 & 3 \\ -1 & -5 & 2 \\ 1 & 6 & -4 \end{vmatrix}$ using expansion by minors of the first column.

$$\begin{vmatrix} 2 & 0 & 3 \\ -1 & -5 & 2 \\ 3 & 6 & -4 \end{vmatrix} = 2 \begin{vmatrix} -5 & 2 \\ 6 & -4 \end{vmatrix} - (-1) \begin{vmatrix} 0 & 3 \\ 6 & -4 \end{vmatrix} + 3 \begin{vmatrix} 0 & 3 \\ -5 & 2 \end{vmatrix}$$

$$= 2(8) + 1(-18) + 3(15)$$
$$= 16 - 18 + 45$$
$$= 43$$

IMPORTANT FACTS AND CONCEPTS	EXAMPLES

Section 9.7 (continued)

Cramer's Rule for a System of Equations in Three Variables

To solve the system

$$a_1x + b_1y + c_1z = d_1$$
$$a_2x + b_2y + c_2z = d_2$$
$$a_3x + b_3y + c_3z = d_3$$

with

$$D = \begin{vmatrix} a_1 & b_1 & c_1 \\ a_2 & b_2 & c_2 \\ a_3 & b_3 & c_3 \end{vmatrix} \qquad D_x = \begin{vmatrix} d_1 & b_1 & c_1 \\ d_2 & b_2 & c_2 \\ d_3 & b_3 & c_3 \end{vmatrix}$$

$$D_y = \begin{vmatrix} a_1 & d_1 & c_1 \\ a_2 & d_2 & c_2 \\ a_3 & d_3 & c_3 \end{vmatrix} \qquad D_z = \begin{vmatrix} a_1 & b_1 & d_1 \\ a_2 & b_2 & d_2 \\ a_3 & b_3 & d_3 \end{vmatrix}$$

then

$$x = \frac{D_x}{D} \quad y = \frac{D_y}{D} \quad z = \frac{D_z}{D}, \quad D \neq 0$$

Solve the system of equations.

$$2x + y + z = 0$$
$$4x - y + 3z = -9$$
$$6x + 2y + 5z = -8$$

$$D = \begin{vmatrix} 2 & 1 & 1 \\ 4 & -1 & 3 \\ 6 & 2 & 5 \end{vmatrix} = -10 \qquad D_x = \begin{vmatrix} 0 & 1 & 1 \\ -9 & -1 & 3 \\ -8 & 2 & 5 \end{vmatrix} = -5$$

$$D_y = \begin{vmatrix} 2 & 0 & 1 \\ 4 & -9 & 3 \\ 6 & -8 & 5 \end{vmatrix} = -20 \qquad D_z = \begin{vmatrix} 2 & 1 & 0 \\ 4 & -1 & -9 \\ 6 & 2 & -8 \end{vmatrix} = 30$$

Then

$$x = \frac{D_x}{D} = \frac{-5}{-10} = \frac{1}{2}, \quad y = \frac{D_y}{D} = \frac{-20}{-10} = 2 \quad z = \frac{D_z}{D} = \frac{30}{-10} = -3$$

The solution is $\left(\frac{1}{2}, 2, -3\right)$.

Chapter 9 Review Exercises

[9.1] *Determine which, if any, of the ordered pairs satisfy each system of equations.*

1. $y = 4x - 2$
$2x + 3y = 8$

  a) $(2, 6)$ **b)** $(4, 0)$ **c)** $(1, 2)$

2. $y = -x + 4$
$3x + 5y = 15$

  a) $\left(\frac{5}{2}, \frac{3}{2}\right)$ **b)** $(-1, 5)$ **c)** $\left(\frac{1}{2}, \frac{3}{5}\right)$

Identify each system of linear equations as consistent, inconsistent, or dependent. State whether the system has exactly one solution, no solution, or an infinite number of solutions.

3.
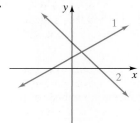

4.
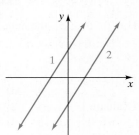

5.
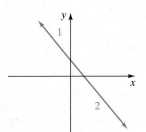

6.
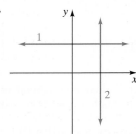

Write each equation in slope-intercept form. Without graphing or solving the system of equations, state whether the system of linear equations has exactly one solution, no solution, or an infinite number of solutions.

7. $x + 2y = 10$
$4x = -8y + 16$

8. $y = -2x + 5$
$2x + 5y = 10$

9. $y = \frac{1}{3}x + \frac{2}{3}$
$6y - 2x = 4$

10. $6x = 4y - 20$
$4x = 6y + 20$

Determine the solution to each system of equations graphically.

11. $y = x - 4$
$y = 2x - 7$

12. $x = -2$
$y = 5$

13. $x + 2y = 8$
$2x - y = -4$

14. $x + 4y = 8$
$y = 2$

15. $y = 3$
$y = -2x + 5$

16. $y = x - 3$
$3x - 3y = 9$

17. $3x + y = 0$
$3x - 3y = 12$

18. $x + 5y = 10$
$\dfrac{1}{5}x + y = -1$

[9.2] *Find the solution to each system of equations by substitution.*

19. $y = 4x - 18$
$2x - 5y = 0$

20. $x = 3y - 9$
$x + 2y = 1$

21. $2x - y = 7$
$x + 2y = 6$

22. $x = -3y$
$x + 4y = 5$

23. $4x - 2y = 7$
$y = 2x + 3$

24. $2x - 4y = 7$
$-4x + 8y = -14$

25. $2x - 3y = 8$
$6x - 5y = 20$

26. $3x - y = -5$
$x + 2y = 8$

[9.3] *Find the solution to each system of equations using the addition method.*

27. $x - y = -4$
$-x + 6y = -6$

28. $x + 2y = -3$
$5x - 2y = 9$

29. $x + y = 12$
$2x + y = 5$

30. $4x - 3y = 8$
$2x + 5y = 8$

31. $-2x + 3y = 15$
$7x + 3y = 6$

32. $2x + y = 3$
$-4x - 2y = 5$

33. $3x = -4y + 15$
$8y = -6x + 30$

34. $2x - 5y = 12$
$3x - 4y = -6$

[9.4] *Determine the solution to each system of equations using substitution or the addition method.*

35. $x - 2y - 4z = 13$
$3y + 2z = -2$
$5z = -20$

36. $2a + b - 2c = 5$
$3b + 4c = 1$
$3c = -6$

37. $x + 2y + 3z = 3$
$-2x - 3y - z = 5$
$3x + 3y + 7z = 2$

38. $-x - 4y + 2z = 1$
$2x + 2y + z = 0$
$-3x - 2y - 5z = 5$

39. $3y - 2z = -4$
$3x - 5z = -7$
$2x + y = 6$

40. $a + 2b - 5c = 19$
$2a - 3b + 3c = -15$
$5a - 4b - 2c = -2$

41. $x - y + 3z = 1$
$-x + 2y - 2z = 1$
$x - 3y + z = 2$

42. $-2x + 2y - 3z = 6$
$4x - y + 2z = -2$
$2x + y - z = 4$

[9.5] *Express each problem as a system of linear equations and use the method of your choice to find the solution to the problem.*

43. Ages Luan Baker is 10 years older than his niece, Jennifer Miesen. If the sum of their ages is 66, find Luan's age and Jennifer's age.

44. Air Speed An airplane can travel 560 miles per hour with the wind and 480 miles per hour against the wind. Determine the speed of the plane in still air and the speed of the wind.

45. Mixing Solutions Sally Dove has two acid solutions, as illustrated. How much of each must she mix to get 6 liters of a 40% acid solution?

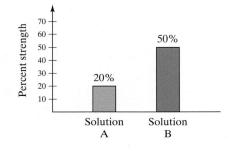

46. Ice Hockey The admission at an ice hockey game is $15 for adults and $11 for children. A total of 650 tickets were sold. Determine how many children's tickets and how many adult tickets were sold if a total of $8790 was collected.

47. Return to Space John Glenn was the first American astronaut to go into orbit around the Earth. Many years later he returned to space. The second time he returned to space he was 5 years younger than twice his age when he went into space for the first time. The sum of his ages for both times he was in space is 118. Find his age each time he was in space.

48. Savings Accounts Jorge Minez has a total of $40,000 invested in three different savings accounts. He has some money invested in one account that gives 7% interest. The second account has $5000 less than the first account and gives 5% interest. The third account gives 3% interest. If the total annual interest that Jorge receives in a year is $2300, find the amount in each account.

[9.6] *Solve each system of equations using matrices.*

49. $x + 5y = 1$
$-2x - 8y = -6$

50. $2x - 5y = 1$
$2x + 4y = 10$

51. $3y = 6x - 12$
$4x = 2y + 8$

52. $2x - y - z = 5$
$x + 2y + 3z = -2$
$3x - 2y + z = 2$

53. $3a - b + c = 2$
$2a - 3b + 4c = 4$
$a + 2b - 3c = -6$

54. $x + y + z = 3$
$3x + 4y = -1$
$y - 3z = -10$

[9.7] *Solve each system of equations using determinants.*

55. $7x - 8y = -10$
$-5x + 4y = 2$

56. $x + 4y = 5$
$5x + 3y = -9$

57. $9m + 4n = -1$
$7m - 2n = -11$

58. $p + q + r = 5$
$2p + q - r = -5$
$3p + 2q - 3r = -12$

59. $-2a + 3b - 4c = -7$
$2a + b + c = 5$
$-2a - 3b + 4c = 3$

60. $y + 3z = 4$
$-x - y + 2z = 0$
$x + 2y + z = 1$

Chapter 9 Practice Test

*To find out how well you understand the chapter material, take this practice test. The answers, and the section where the material was initially discussed, are given in the back of the book. Each problem is also fully worked out on the **Chapter Test Prep Video CD.** Review any questions that you answered incorrectly.*

1. Determine which, if any, of the ordered pairs satisfy the system of equations.

$$x + 2y = -6$$
$$3x + 2y = -12$$

   a) $(-6, 0)$ **b)** $\left(-3, -\dfrac{3}{2}\right)$ **c)** $(2, -4)$

Identify each system as consistent, inconsistent, or dependent. State whether the system has exactly one solution, no solution, or an infinite number of solutions.

2.

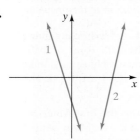

3.

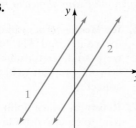

4.

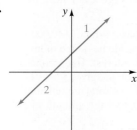

Determine, without solving the system, whether the system of equations is consistent, inconsistent, or dependent. State whether the system has exactly one solution, no solution, or an infinite number of solutions.

5. $5x + 2y = 4$
$6x = 3y - 7$

6. $5x + 3y = 9$
$2y = -\dfrac{10}{3}x + 6$

7. $5x - 4y = 6$
$-10x + 8y = -10$

Solve each system of equations by the method indicated.

8. $y = 3x - 2$
$y = -2x + 8$
graphically

9. $y = -x + 6$
$y = 2x + 3$
graphically

10. $y = 4x - 3$
$y = 5x - 4$
substitution

11. $4a + 7b = 2$
$5a + b = -13$
substitution

12. $0.3x = 0.2y + 0.4$
$-1.2x + 0.8y = -1.6$
addition

13. $\dfrac{3}{2}a + b = 6$
$a - \dfrac{5}{2}b = -4$
addition

14. $x + y + z = 2$
$-2x - y + z = 1$
$x - 2y - z = 1$
addition

15. Write the augmented matrix for the following system of equations.

$-2x + 3y + 7z = 5$
$3x - 2y + z = -2$
$x - 6y + 9z = -13$

16. Consider the following augmented matrix.

$$\begin{bmatrix} 6 & -2 & 4 & | & 4 \\ 4 & 3 & 5 & | & 6 \\ 2 & -1 & 4 & | & -3 \end{bmatrix}$$

Show the results obtained by multiplying the elements in the third row by -2 and adding the products to their corresponding elements in the second row.

Solve each system of equations using matrices.

17. $2x + 7y = 1$
$3x + 5y = 7$

18. $x - 2y + z = 7$
$-2x - y - z = -7$
$4x + 5y - 2z = 3$

Evaluate each determinant.

19. $\begin{vmatrix} 3 & -1 \\ 5 & -2 \end{vmatrix}$

20. $\begin{vmatrix} 8 & 2 & -1 \\ 3 & 0 & 5 \\ 6 & -3 & 4 \end{vmatrix}$

Solve each system of equations using determinants and Cramer's rule.

21. $4x + 3y = -6$
$-2x + 5y = 16$

22. $2r - 4s + 3t = -1$
$-3r + 5s - 4t = 0$
$-2r + s - 3t = -2$

Use the method of your choice to find the solution to each problem.

23. Sunflower Seed Mixture Agway Gardens has sunflower seeds, in a barrel, that sell for \$0.49 per pound and gourmet bird seed mix that sells for \$0.89 per pound. How much of each must be mixed to get a 20-pound mixture that sells for \$0.73 per pound?

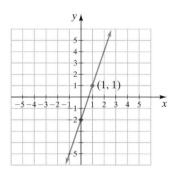

24. Mixing Solutions Tyesha Blackwell, a chemist, has 6% and 15% solutions of sulfuric acid. How much of each solution should she mix to get 10 liters of a 9% solution?

25. Sum of Numbers The sum of three numbers is 29. The greatest number is four times the smallest number. The remaining number is 1 more than twice the smallest number. Find the three numbers.

Cumulative Review Test

Take the following test and check your answers with those given in the back of the book. Review any questions that you answered incorrectly. The section where the material was covered is indicated after the answer.

1. Consider the following set of numbers.

$$\left\{ \frac{1}{2}, -4, 9, 0, \sqrt{3}, -4.63, 1 \right\}$$

List the elements of the set that are

a) natural numbers;

b) rational numbers;

c) real numbers.

2. Evaluate $16 \div \left\{ 4 \left[3 + \left(\dfrac{5 + 10}{5} \right)^2 \right] - 32 \right\}$.

3. Solve $-7(3 - x) = 4(x + 2) - 3x$.

4. Graph $4x - 8y = 16$.

5. Find the equation of the line that has a slope of $\dfrac{2}{5}$ and goes through the point $(-3, 1)$.

6. Write the equation of the graph in the accompanying figure.

7. Simplify $\dfrac{4a^3 b^{-5}}{28a^8 b}$.

8. Factor $3x^3 + 4x^2 + 6x + 8$.

9. Factor $x^2 - 16x + 28$.

10. Use factoring to solve $x^2 - 5x = 0$.

11. Simplify $\dfrac{y-5}{8} + \dfrac{2y+7}{8}$.

12. Solve $\dfrac{5}{y+2} = \dfrac{3}{2y-7}$.

13. Determine which of the following graphs represent functions. Explain.

a)

b)

c)

14. If $f(x) = \dfrac{x+3}{x^2-9}$, find

　a) $f(-4)$.　　**b)** $f(h)$.　　**c)** $f(3)$.

Solve each system of equations.

15. $3x + y = 6$
　　$y = 4x - 1$

16. $2p + 3q = 11$
　　$-3p - 5q = -16$

17. $x - 2y = 0$
　　$2x + z = 7$
　　$y - 2z = -5$

18. Angles of a Triangle If the largest angle of a triangle is nine times the measure of the smallest angle, and the middle-sized angle is 70° greater than the measure of the smallest angle, find the measure of the three angles.

19. Walking and Jogging Mark Simmons walks at 4 miles per hour and Judy Bolin jogs at 6 miles per hour. Mark begins walking $\dfrac{1}{2}$ hour before Judy starts jogging. If Judy jogs on the same path that Mark walks, how long after Judy begins jogging will she catch up to Mark?

20. Rock Concert There are two different prices of seats at a rock concert. The higher priced seats sell for $20 and the less expensive seats sell for $16. If a total of 1000 tickets are sold and the total ticket sales are $18,400, how many of each type of ticket are sold?

10 Inequalities in One and Two Variables

GOALS OF THIS CHAPTER

In this chapter we discuss inequalities. Inequalities are similar to equations in that they state a relationship between two expressions. We will do many of the same things to inequalities that we have done to equations. We will solve them, graph them, and work with systems involving them.

In Section 10.1, we solve linear inequalities in one variable. In Section 10.2 we expand on earlier material by solving equations and inequalities involving absolute value. In Section 10.3, we graph linear inequalities in two variables and then use this techinque to graph systems of linear inequalities.

WHEN MAILING A PACKAGE, the size of the package can come into play with respect to postage or even whether the package can be sent in the method you want. The restrictions on the size of a package can be expressed using inequalities. In Exercises 73 on page 629, you will work with package restrictions set by the United Parcel Service (UPS).

10.1 Solving Linear Inequalities in One Variable

1 Solve inequalities.

2 Graph solutions on a number line, in interval notation, and as a set.

3 Find the union and intersection of sets.

4 Solve compound inequalities involving *and*.

5 Solve compound inequalities involving *or*.

1 Solve Inequalities

Inequalities, interval notation, and set builder notation were discussed in Section 8.2. You may wish to review that section now. A review of the inequality symbols follows.*

Inequality Symbols	
$>$	is greater than
$\geq$	is greater than or equal to
$<$	is less than
$\leq$	is less than or equal to

A mathematical expression containing one or more of these symbols is called an **inequality**. The direction of the inequality symbol is sometimes called the **order** or **sense** of the inequality.

Examples of Inequalities in One Variable

$$2x + 3 \leq 5 \qquad 4x > 3x - 5 \qquad 1.5 \leq -2.3x + 4.5 \qquad \frac{1}{2}x + 3 \geq 0$$

To solve an inequality, we must isolate the variable on one side of the inequality symbol. To isolate the variable, we use the same basic techniques used in solving equations.

Properties Used to Solve Inequalities
1. If $a > b$, then $a + c > b + c$.
2. If $a > b$, then $a - c > b - c$.
3. If $a > b$, and $c > 0$, then $ac > bc$.
4. If $a > b$, and $c > 0$, then $\dfrac{a}{c} > \dfrac{b}{c}$.
5. If $a > b$, and $c < 0$, then $ac < bc$.
6. If $a > b$, and $c < 0$, then $\dfrac{a}{c} < \dfrac{b}{c}$.

The first two properties state that the same number can be added to or subtracted from both sides of an inequality. The third and fourth properties state that both sides of an inequality can be multiplied or divided by any positive real number. The last two properties indicate that **when both sides of an inequality are multiplied or divided by a negative number, the direction of the inequality symbol reverses**.

Example of Multiplication by a Negative Number

Multiply both sides of the inequality by −1 and reverse the direction of the inequality symbol.

$$4 > -2$$
$$-1(4) < -1(-2)$$
$$-4 < 2$$

Example of Division by a Negative Number

$$10 \geq -4$$
$$\frac{10}{-2} \leq \frac{-4}{-2}$$
$$-5 \leq 2$$

Divide both sides of the inequality by −2 and reverse the direction of the inequality symbol.

---

*$\neq$, is not equal to, is also an inequality. $\neq$ means $<$ or $>$. Thus, $2 \neq 3$ means $2 < 3$ or $2 > 3$.

Helpful Hint

Do not forget to reverse the direction of the inequality symbol when multiplying or dividing both sides of the inequality by a negative number.

Inequality	Direction of Inequality Symbol
$-3x < 6$	$\dfrac{-3x}{-3} > \dfrac{6}{-3}$
$-\dfrac{x}{2} > 5$	$(-2)\left(-\dfrac{x}{2}\right) < (-2)(5)$

EXAMPLE 1 ▶ Solve the inequalities. **a)** $5x - 7 \geq -17$ **b)** $-6x + 4 < -14$

Solution

a)
$$5x - 7 \geq -17$$
$$5x - 7 \boxed{+ 7} \geq -17 \boxed{+ 7} \qquad \textit{Add 7 to both sides.}$$
$$5x \geq -10$$
$$\frac{5x}{5} \geq \frac{-10}{5} \qquad \textit{Divide both sides by 5.}$$
$$x \geq -2$$

The solution set is $\{x | x \geq -2\}$. Any real number greater than or equal to -2 will satisfy the inequality.

b)
$$-6x + 4 < -14$$
$$-6x + 4 \boxed{- 4} < -14 \boxed{- 4} \qquad \textit{Subtract 4 from both sides.}$$
$$-6x < -18$$
$$\frac{-6x}{-6} > \frac{-18}{-6} \qquad \textit{Divide both sides by }-6\textit{ and reverse the direction of the inequality.}$$
$$x > 3$$

The solution set is $\{x | x > 3\}$. Any number greater than 3 will satisfy the inequality.

▶ **Now Try Exercise 17**

2 Graph Solutions on a Number Line, in Interval Notation, and as a Set

There are three ways to indicate the solution to an inequality. It can be shown on a number line, and it can be shown in the same way we described domains and ranges in Section 8.2 using interval notation and set builder notation. Most instructors have a preferred way to indicate the solution to an inequality.

Recall that *a solid circle on the number line indicates that the endpoint is part of the solution, and an open circle indicates that the endpoint is not part of the solution. In interval notation, brackets,* [ ], *are used to indicate that the endpoints are part of the solution and parentheses,* ( ), *indicate that the endpoints are not part of the solution.* The symbol ∞ is read "infinity"; it indicates that the solution set continues indefinitely. Whenever ∞ is used in interval notation, a *parenthesis* must be used on the corresponding side of the interval notation.

Solution of Inequality	Solution Set Indicated on Number Line	Solution Set Represented in Interval Notation
$x \geq 5$		$[5, \infty)$
$x < 3$		$(-\infty, 3)$
$2 < x \leq 6$		$(2, 6]$
$-6 \leq x \leq -1$		$[-6, -1]$
$x > a$		(a, ∞)
$x \geq a$		$[a, \infty)$
$x < a$		$(-\infty, a)$
$x \leq a$		$(-\infty, a]$
$a < x < b$		(a, b)
$a \leq x \leq b$		$[a, b]$
$a < x \leq b$		$(a, b]$
$a \leq x < b$		$[a, b)$

In the next example, we will solve an inequality which has fractions.

EXAMPLE 2 ▶ Solve the following inequality and give the solution both on a number line and in interval notation.

$$\frac{1}{4}z - \frac{1}{2} < \frac{2z}{3} + 2$$

Solution   We can eliminate fractions from an inequality by multiplying both sides of the inequality by the least common denominator, LCD, of the fractions. In this case we multiply both sides of the inequality by 12. We then solve the resulting inequality as we did in the previous example.

$$\frac{1}{4}z - \frac{1}{2} < \frac{2z}{3} + 2$$

$$12\left(\frac{1}{4}z - \frac{1}{2}\right) < 12\left(\frac{2z}{3} + 2\right) \qquad \textit{Multiply both sides by the LCD, 12.}$$

$$3z - 6 < 8z + 24 \qquad \textit{Distributive property}$$

$$3z - 8z - 6 < 8z - 8z + 24 \qquad \textit{Subtract 8z from both sides.}$$

$$-5z - 6 < 24$$

$$-5z - 6 + 6 < 24 + 6 \qquad \textit{Add 6 to both sides.}$$

$$-5z < 30$$

$$\frac{-5z}{-5} > \frac{30}{-5} \qquad \textit{Divide both sides by} -5 \textit{ and reverse the direction of inequality symbol.}$$

$$z > -6$$

<table>
<tr><td align="center">Number Line</td><td align="center">Interval Notation</td></tr>
</table>

$$(-6, \infty)$$

The solution set is $\{z|z > -6\}$.

▶ **Now Try Exercise 31**

In Example 2 we illustrated the solution on a number line, in interval notation, and as a solution set. Your instructor may indicate which form he or she prefers.

EXAMPLE 3 ▶ Solve the inequality $2(3p - 5) + 9 \le 8(p + 1) - 2(p - 3)$.

Solution

$$2(3p - 5) + 9 \le 8(p + 1) - 2(p - 3)$$
$$6p - 10 + 9 \le 8p + 8 - 2p + 6$$
$$6p - 1 \le 6p + 14$$
$$6p - 6p - 1 \le 6p - 6p + 14$$
$$-1 \le 14$$

Since -1 is always less than or equal to 14, the inequality is true for all real numbers. When an inequality is true for all real numbers, the solution set is *the set of all real numbers*, $\mathbb{R}$. The solution set to this example can also be indicated on a number line or given in interval notation.

$$\text{or} \quad (-\infty, \infty)$$

▶ **Now Try Exercise 23**

If Example 3 had resulted in the expression $-1 \ge 14$, the inequality would never have been true, since -1 is never greater than or equal to 14. When an inequality is never true, it has no solution. The solution set of an inequality that has no solution is the *empty or null set*, $\{\ \}$ or $\varnothing$. We will represent the empty set on the number line as follows, .

Helpful Hint

Generally, when writing a solution to an inequality, we write the variable on the left. For example, when solving an inequality, if we obtain $5 \ge y$ we would write the solution as $y \le 5$. For example,

$-6 < x$ means $x > -6$ (inequality symbol points to -6 in both cases)

$4 > x$ means $x < 4$ (inequality symbol points to x in both cases)

$a < x$ means $x > a$ (inequality symbol points to a in both cases)

$a > x$ means $x < a$ (inequality symbol points to x in both cases)

EXAMPLE 4 ▶ **Packages on a Boat** A small boat has a maximum weight load of 750 pounds. Millie Harrison has to transport packages weighing 42.5 pounds each.

a) Write an inequality that can be used to determine the maximum number of packages that Millie can safely place on the boat if she weighs 128 pounds.

b) Find the maximum number of packages that Millie can transport.

Solution **a)** Understand and Translate Let n = number of packages.

Millie's weight   +   weight of n packages ≤ 750

128   +   42.5n ≤ 750

b) Carry Out

$$128 + 42.5n \leq 750$$
$$42.5n \leq 622$$
$$n \leq 14.6$$

Answer   Therefore, Millie can transport up to 14 packages on the boat.

▶ Now Try Exercise 75

EXAMPLE 5 ▶ **Bowling Alley Rates** At the Corbin Bowl bowling alley in Tarzana, California, it costs $2.50 to rent bowling shoes and it costs $4.00 per game bowled.

a) Write an inequality that can be used to determine the maximum number of games that Ricky Olson can bowl if he has only $20.

b) Find the maximum number of games that Ricky can bowl.

Solution   **a)** Understand and Translate

Let g = number of games bowled.

Then $4.00g$ = cost of bowling g games.

cost of shoe rental + cost of bowling g games ≤ money Ricky has

    2.50   +   4.00g ≤   20

b) Carry Out

$$2.50 + 4.00g \leq 20$$
$$4.00g \leq 17.50$$
$$\frac{4.00g}{4.00} \leq \frac{17.50}{4.00}$$
$$g \leq 4.375$$

Answer and Check   Since he can't play a portion of a game, the maximum number of games that he can afford to bowl is 4. If Ricky were to bowl 5 games, he would owe $2.50 + 5($4.00) = $22.50, which is more money than the $20 that he has.

▶ Now Try Exercise 77

EXAMPLE 6 ▶ **Profit** For a business to realize a profit, its revenue (or income), R, must be greater than its cost, C. That is, a profit will be obtained when $R > C$ (the company breaks even when $R = C$). A company that produces playing cards has a weekly cost equation of $C = 1525 + 1.7x$ and a weekly revenue equation of $R = 4.2x$, where x is the number of decks of playing cards produced and sold in a week. How many decks of cards must be produced and sold in a week for the company to make a profit?

Solution   Understand and Translate   The company will make a profit when $R > C$, or

$$R > C$$
$$4.2x > 1525 + 1.7x$$

Carry Out

$$2.5x > 1525$$
$$x > \frac{1525}{2.5}$$
$$x > 610$$

Answer   The company will make a profit when more than 610 decks are produced and sold in a week.

▶ Now Try Exercise 79

EXAMPLE 7 ▸ **Tax Tables** The 2005 tax rate schedule for married couples in America who file a joint tax return is shown below.

Schedule Y-1 Use if your filing status is **Married filing jointly** or **Qualifying widow(er)**

If the Amount on Form 1040, Line 43 Is: Over—	But Not Over—	Enter on Form 1040, Line 44	of the Amount Over—
$0	$14,600	10%	$0
$14,600	$59,400	$1,460.00 + 15%	$14,600
$59,400	$119,950	$8,180.00 + 25%	$59,400
$119,950	$182,800	$23,317.50 + 28%	$119,950
$182,800	$326,450	$40,915.50 + 33%	$182,800
$326,450	∞	$88,320.00 + 35%	$326,450

a) Write, in interval notation, the amounts of taxable income (amount on Form 1040, line 43) that makes up each of the five listed tax brackets, that is, the 10%, 15%, 25%, 28%, 33%, and 35% tax brackets.

b) Determine the tax for a married couple filing jointly if their taxable income (line 43) is $13,500.

c) Determine the tax for a married couple filing jointly if their taxable income is $136,000.

Solution

a) The words *But Not Over* mean "less than or equal to." The taxable incomes that make up the six tax brackets are

   (0, 14,600] for the 10% tax bracket
   (14,600, 59,400] for the 15% tax bracket
   (59,400, 119,950] for the 25% tax bracket
   (119,950, 182,800] for the 28% tax bracket
   (182,800, 326,450] for the 33% tax bracket
   (326,450, ∞) for the 35% tax bracket

b) The tax for a married couple filing jointly with taxable income of $13,500 is 10% of $13,500. Therefore,

$$\text{tax} = 0.10(13{,}500) = \$1{,}350$$

The tax is $1350.

c) A taxable income of $136,000 places a married couple filing jointly in the 28% tax bracket. The tax is $23,317.50 + 28% of the taxable income over $119,950. The taxable income over $119,950 is $136,000 − $119,950 = $16,050. Therefore,

$$\text{tax} = 23{,}317.50 + 0.28(16{,}050) = 23{,}317.50 + 4494 = 27{,}811.50$$

The tax is $27,811.50.

▸ **Now Try Exercise 89**

3 Find the Union and Intersection of Sets

After finding the solution set to an inequality, it is sometimes necessary to combine the solution set with another solution set. Just as *operations,* such as addition and multiplication, are performed on numbers, operations can be performed on sets. Two set operations are *union* and *intersection*.

Union

The **union** of set A and set B, written $A \cup B$, is the set of elements that belong to either set A *or* set B.

The union is formed by combining, or joining together, the elements in set A with those in set B.

Examples of Union of Sets

$$A = \{1, 2, 3, 4, 5\}, \qquad B = \{3, 4, 5, 6, 7\}, \qquad A \cup B = \{1, 2, 3, 4, 5, 6, 7\}$$

$$A = \{a, b, c, d, e\}, \qquad B = \{x, y, z\}, \qquad A \cup B = \{a, b, c, d, e, x, y, z\}$$

In set builder notation we can express $A \cup B$ as

$$A \cup B = \{x | x \in A \text{ or } x \in B\}$$

Intersection

The **intersection** of set A and set B, written $A \cap B$, is the set of all elements that are common to both set A *and* set B.

Examples of Intersection of Sets

$$A = \{1, 2, 3, 4, 5\}, \qquad B = \{3, 4, 5, 6, 7\}, \qquad A \cap B = \{3, 4, 5\}$$

$$A = \{a, b, c, d, e\}, \qquad B = \{x, y, z\}, \qquad A \cap B = \{ \ \}$$

Note that in the last example, sets A and B have no elements in common. Therefore, their intersection is the empty set. In set builder notation we can express $A \cap B$ as

$$A \cap B = \{x | x \in A \text{ and } x \in B\}$$

4 Solve Compound Inequalities Involving *And*

A **compound inequality** is formed by joining two inequalities with the word *and* or *or*. Sometimes the word *and* is implied without being written.

Examples of Compound Inequalities

$$3 < x \quad \text{and} \quad x < 5$$

$$x + 4 > 2 \quad \text{or} \quad 2x - 3 < 6$$

$$4x - 6 \geq -3 \quad \text{and} \quad x - 6 < 17$$

In this objective, we discuss compound inequalities that use or imply the word *and*. The solution of a compound inequality using the word *and* is all the numbers that make *both* parts of the inequality true. Consider

$$3 < x \quad \text{and} \quad x < 5$$

What are the numbers that satisfy both inequalities? The numbers that satisfy both inequalities may be easier to see if we graph the solution to each inequality on a number line (see **Fig. 10.1**). Now we can see that the numbers that satisfy both inequalities are the numbers between 3 and 5. The solution set is $\{x|3 < x < 5\}$.

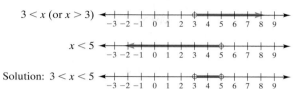

FIGURE 10.1

Recall that the intersection of two sets is the set of elements common to both sets. *To find the solution set of an inequality containing the word* **and**, *take the* **intersection** *of the solution sets of the two inequalities.*

EXAMPLE 8 ▶ Solve $x + 5 \leq 8$ and $2x - 9 > -7$

Solution   Begin by solving each inequality separately.

$$x + 5 \leq 8 \quad \text{and} \quad 2x - 9 > -7$$
$$x \leq 3 \qquad\qquad 2x > 2$$
$$\qquad\qquad x > 1$$

Now take the intersection of the sets $\{x|x \leq 3\}$ and $\{x|x > 1\}$. When we find $\{x|x \leq 3\} \cap \{x|x > 1\}$, we are finding the values of x common to both sets. **Figure 10.2** illustrates that the solution set is $\{x|1 < x \leq 3\}$. In interval notation, the solution is $(1, 3]$.

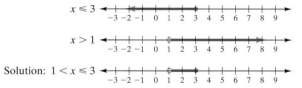

FIGURE 10.2

▶ **Now Try Exercise 67**

Sometimes a compound inequality using the word *and* can be written in a shorter form. For example, $3 < x$ and $x < 5$ can be written as $3 < x < 5$. The word *and* does not appear when the inequality is written in this form, but it is implied. The compound inequality $-1 < x + 3$ and $x + 3 \leq 5$ can be written $-1 < x + 3 \leq 5$.

EXAMPLE 9 ▶ Solve $-1 < x + 3 \leq 5$

Solution $-1 < x + 3 \leq 5$ means $-1 < x + 3$ and $x + 3 \leq 5$. Solve each inequality separately.

$$-1 < x + 3 \quad \text{and} \quad x + 3 \leq 5$$
$$-4 < x \qquad\qquad x \leq 2$$

Remember that $-4 < x$ means $x > -4$. **Figure 10.3** illustrates that the solution set is $\{x | -4 < x \leq 2\}$. In interval notation, the solution is $(-4, 2]$.

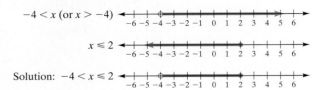

$-4 < x \ (\text{or } x > -4)$

$x \leq 2$

Solution: $-4 < x \leq 2$

FIGURE 10.3

▶ **Now Try Exercise 45**

The inequality in Example 9, $-1 < x + 3 \leq 5$, can be solved in another way. We can still use the properties discussed earlier to solve compound inequalities. However, when working with such inequalities, whatever we do to one part we must do to all three parts. In Example 9, we could have subtracted 3 from all three parts to isolate the variable in the middle and solve the inequality.

$$-1 < x + 3 \leq 5$$
$$-1 \boxed{-3} < x + \boxed{-3} \leq 5 \boxed{-3}$$
$$-4 < x \leq 2$$

Note that this is the same solution as obtained in Example 9.

EXAMPLE 10 ▶ Solve the inequality $-3 \leq 2t - 7 < 8$.

Solution   We wish to isolate the variable t. We begin by adding 7 to all three parts of the inequality.

$$-3 \leq 2t - 7 < 8$$
$$-3 \boxed{+7} \leq 2t - 7 \boxed{+7} < 8 \boxed{+7}$$
$$4 \leq 2t < 15$$

Now divide all three parts of the inequality by 2.

$$\frac{4}{2} \leq \frac{2t}{2} < \frac{15}{2}$$

$$2 \leq t < \frac{15}{2}$$

The solution may also be illustrated on a number line, written in interval notation, or written as a solution set. Below we show each form.

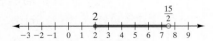

The answer in interval notation is $\left[2, \dfrac{15}{2}\right)$. The solution set is $\left\{t \,\middle|\, 2 \leq t < \dfrac{15}{2}\right\}$.

▶ **Now Try Exercise 51**

EXAMPLE 11 ▶ Solve the inequality $-2 < \dfrac{4 - 3x}{5} < 8$.

Solution   Multiply all three parts by 5 to eliminate the denominator.

$$-2 < \frac{4 - 3x}{5} < 8$$

$$-2(5) < 5\left(\frac{4 - 3x}{5}\right) < 8(5)$$

$$-10 < 4 - 3x < 40$$

$$-10 - 4 < 4 - 4 - 3x < 40 - 4$$

$$-14 < -3x < 36$$

Now divide all three parts of the inequality by -3. Remember that when we multiply or divide an inequality by a negative number, the direction of the inequality symbols reverse.

$$\frac{-14}{-3} > \frac{-3x}{-3} > \frac{36}{-3}$$

$$\frac{14}{3} > x > -12$$

Although $\dfrac{14}{3} > x > -12$ is correct, we generally write compound inequalities with the smaller value on the left. We will, therefore, rewrite the solution as

$$-12 < x < \frac{14}{3}$$

The solution may also be illustrated on a number line, written in interval notation, or written as a solution set.

The solution in interval notation is $\left(-12, \dfrac{14}{3}\right)$. The solution set is $\left\{x \,\middle|\, -12 < x < \dfrac{14}{3}\right\}$.

▶ **Now Try Exercise 53**

Helpful Hint

You must be very careful when writing the solution to a compound inequality. In Example 11 we can change the solution from

$$\frac{14}{3} > x > -12 \quad \text{to} \quad -12 < x < \frac{14}{3}$$

This is correct since both say that x is greater than -12 and less than $\dfrac{14}{3}$. Notice that the inequality symbol in both cases is pointing to the smaller number.

In Example 11, had we written the answer $\dfrac{14}{3} < x < -12$, we would have given the incorrect solution. Remember that the inequality $\dfrac{14}{3} < x < -12$ means that $\dfrac{14}{3} < x$ and $x < -12$. There is no number that is both greater than $\dfrac{14}{3}$ and less than -12. Also, by examining the inequality $\dfrac{14}{3} < x < -12$, it appears as if we are saying that -12 is a greater number than $\dfrac{14}{3}$, which is obviously incorrect.

It would also be incorrect to write the answer as

$$-12 > x > \frac{14}{3} \quad \text{or} \quad \frac{14}{3} < x < -12$$

EXAMPLE 12 ▶ **Calculating Grades** In an anatomy and physiology course, an average score greater than or equal to 80 and less than 90 will result in a final grade of B. Steve Reinquist received scores of 85, 90, 68, and 70 on his first four exams. For Steve to receive a final grade of B in the course, between which two scores must his fifth (and last) exam fall?

Solution Let x = Steve's last exam score.

$$80 \leq \text{average of five exams} < 90$$

$$80 \leq \frac{85 + 90 + 68 + 70 + x}{5} < 90$$

$$80 \leq \frac{313 + x}{5} < 90$$

$$400 \leq 313 + x < 450$$

$$400 - 313 \leq 313 - 313 + x < 450 - 313$$

$$87 \leq x < 137$$

Steve would need a minimum score of 87 on his last exam to obtain a final grade of B. If the highest score he could receive on the test is 100, is it possible for him to obtain a final grade of A (90 average or higher)? Explain.

▶ **Now Try Exercise 85**

5 Solve Compound Inequalities Involving *Or*

The solution to a compound inequality using the word *or* is all the numbers that make *either* of the inequalities a true statement. Consider the compound inequality

$$x > 3 \quad \text{or} \quad x < 5$$

What are the numbers that satisfy the compound inequality? Let's graph the solution to each inequality on the number line (see **Fig. 10.4**). Note that every real number satisfies at least one of the two inequalities. Therefore, the solution set to the compound inequality is the set of all real numbers, $\mathbb{R}$.

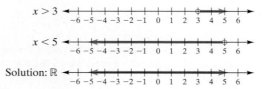

FIGURE 10.4

Recall that the *union* of two sets is the set of elements that belong to *either* of the sets. *To find the solution set of an inequality containing the word* **or***, take the* **union** *of the solution sets of the two inequalities that comprise the compound inequality.*

EXAMPLE 13 ▶ Solve $r - 2 \leq -6$ or $-4r + 3 < -5$.

Solution Solve each inequality separately.

$$r - 2 \leq -6 \quad \text{or} \quad -4r + 3 < -5$$
$$r \leq -4 \qquad\qquad -4r < -8$$
$$\qquad\qquad\qquad r > 2$$

Now graph each solution on number lines and then find the union (see **Fig. 10.5**). The union is $r \leq -4$ or $r > 2$.

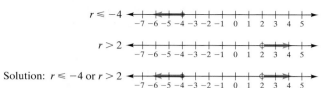

Solution: $r \leq -4$ or $r > 2$

FIGURE 10.5

The solution set is $\{r | r \leq -4\} \cup \{r | r > 2\}$, which can be written as $\{r | r \leq -4$ or $r > 2\}$. In interval notation, the solution is $(-\infty, -4] \cup (2, \infty)$.

▶ **Now Try Exercise 69**

We often encounter inequalities in our daily lives. For example, on a highway the minimum speed may be 45 miles per hour and the maximum speed may be 65 miles per hour. A restaurant may have a sign stating that maximum capacity is 300 people, and the minimum takeoff speed of an airplane may be 125 miles per hour.

Helpful Hint

There are various ways to write the solution to an inequality problem. Be sure to indicate the solution to an inequality problem in the form requested by your professor. Examples of various forms follow.

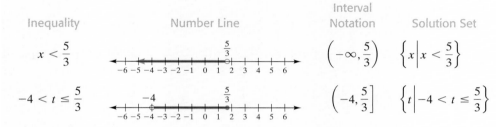

Inequality	Number Line	Interval Notation	Solution Set
$x < \dfrac{5}{3}$		$\left(-\infty, \dfrac{5}{3}\right)$	$\left\{ x \middle| x < \dfrac{5}{3} \right\}$
$-4 < t \leq \dfrac{5}{3}$		$\left(-4, \dfrac{5}{3}\right]$	$\left\{ t \middle| -4 < t \leq \dfrac{5}{3} \right\}$

EXERCISE SET 10.1   *Math XL*   **MyMathLab**
<small>MathXL®   MyMathLab</small>

Concept/Writing Exercises

1. When solving an inequality, when is it necessary to reverse the direction of the inequality symbol?

2. Explain the difference between $x < 7$ and $x \leq 7$.

3. a) When indicating a solution on a number line, when do you use open circles?

  b) When do you use closed circles?

  c) Give an example of an inequality whose solution on a number line would contain an open circle.

d) Give an example of an inequality whose solution on a number line would contain a closed circle.

4. What is a compound inequality? Give one example.

5. What does the inequality $a < x < b$ mean?

6. Explain why $\{x | 5 < x < 3\}$ is not an acceptable solution set for an inequality.

Practice the Skills

Express each inequality **a)** *using a number line,* **b)** *in interval notation, and* **c)** *as a solution set (use set builder notation).*

7. $x > -2$

8. $t > \dfrac{5}{3}$

9. $w \leq \pi$

10. $-4 < x < 3$

11. $-3 < q \leq \dfrac{4}{5}$

12. $x \geq -\dfrac{6}{5}$

13. $-7 < x \leq -4$

14. $-2\dfrac{7}{8} \leq k < -1\dfrac{2}{3}$

Solve each inequality and graph the solution on the number line.

15. $x - 9 > -6$

16. $2x + 3 > 4$

17. $3 - x < -4$

18. $12b - 5 \leq 8b + 7$

19. $4.7x - 5.48 \geq 11.44$

20. $1.4x + 2.2 < 2.6x - 0.2$

21. $4(x + 2) \leq 4x + 8$

22. $15.3 > 3(a - 1.4)$

23. $5b - 6 \geq 3(b + 3) + 2b$

24. $-6(d + 2) < -9d + 3(d - 1)$

25. $2y - 6y + 8 \leq 2(-2y + 9)$

26. $\dfrac{y}{2} + \dfrac{4}{5} \leq 3$

Solve each inequality and give the solution in interval notation.

27. $4 + \dfrac{4x}{3} < 6$

28. $4 - 3x < 5 + 2x + 17$

29. $\dfrac{v - 5}{3} - v \geq -3(v - 1)$

30. $\dfrac{h}{2} - \dfrac{5}{6} < \dfrac{7}{8} + h$

31. $\dfrac{t}{3} - t + 7 \leq -\dfrac{4t}{3} + 8$

32. $\dfrac{6(x - 2)}{5} > \dfrac{10(2 - x)}{3}$

33. $-3x + 1 < 3[(x + 2) - 2x] - 1$

34. $4[x - (3x - 2)] > 3(x + 5) - 15$

Find $A \cup B$ and $A \cap B$ for each set A and B.

35. $A = \{5, 6, 7\}, B = \{6, 7, 8\}$

36. $A = \{2, 4, 6, 8\}, B = \{1, 3, 5, 7\}$

37. $A = \{-2, -4, -5\}, B = \{-1, -2, -4, -6\}$

38. $A = \{-1, 0, 1\}, B = \{0, 2, 4, 6\}$

39. $A = \{\quad\}, B = \{0, 1, 2, 3\}$

40. $A = \{2, 4, 6\}, B = \{2, 4, 6, 8, \ldots\}$

41. $A = \{0, 2, 4, 6, 8\}, B = \{1, 3, 5, 7\}$

42. $A = \{1, 3, 5\}, B = \{1, 3, 5, 7, \ldots\}$

43. $A = \{0.1, 0.2, 0.3\}, B = \{0.2, 0.3, 0.4, 0.5, \ldots\}$

44. $A = \left\{1, \dfrac{1}{2}, \dfrac{1}{4}, \dfrac{1}{6}, \ldots\right\}, B = \left\{\dfrac{1}{4}, \dfrac{1}{6}, \dfrac{1}{8}\right\}$

Solve each inequality and give the solution in interval notation.

45. $-2 \leq t + 3 < 4$

46. $-7 < p - 6 \leq -5$

47. $-15 \leq -3z \leq 12$

48. $-16 < 5 - 3n \leq 13$

49. $4 \leq 2x - 4 < 7$

50. $-12 < 3x - 5 \leq -1$

51. $14 \leq 2 - 3g < 15$

52. $\dfrac{1}{2} < 3x + 4 < 13$

Solve each inequality and give the solution set.

53. $5 \leq \dfrac{3x + 1}{2} < 11$

54. $\dfrac{3}{5} < \dfrac{-x - 5}{3} < 2$

55. $-6 \leq -3(2x - 4) < 12$

56. $-6 < \dfrac{4 - 3x}{2} < \dfrac{2}{3}$

57. $0 \leq \dfrac{3(u - 4)}{7} \leq 1$

58. $-15 < \dfrac{3(x - 2)}{5} \leq 0$

Solve each inequality and indicate the solution set.

59. $c \leq 1$ and $c > -3$

60. $d > 0$ or $d \leq 8$

61. $x < 2$ and $x > 4$

62. $w \leq -1$ or $w > 6$

63. $x + 1 < 3$ and $x + 1 > -4$

64. $5x - 3 \leq 7$ or $-2x + 5 < -3$

Solve each inequality and give the solution in interval notation.

65. $2s + 3 < 7$ or $-3s + 4 \leq -17$

66. $4a + 7 \geq 9$ and $-3a + 4 \leq -17$

67. $4x + 5 \geq 5$ and $3x - 7 \leq -1$

68. $5 - 3x < -3$ and $5x - 3 > 10$

69. $4 - r < -2$ or $3r - 1 < -1$

70. $-x + 3 < 0$ or $2x - 5 \geq 3$

71. $2k + 5 > -1$ and $7 - 3k \leq 7$

72. $2q - 11 \leq -7$ or $2 - 3q < 11$

Problem Solving

73. UPS Packages The length plus the girth of a package to be shipped by United Parcel Service (UPS) can be no larger than 130 inches.

   a) Write an inequality that expresses this information, using l for the length and g for the girth.

   b) UPS has defined girth as twice the width plus twice the depth. Write an inequality using length, l, width, w, and depth, d, to indicate the maximum allowable dimensions of a package that may be shipped by UPS.

   c) If the length of a package is 40 inches and the width of a package is 20.5 inches, find the maximum allowable depth of the package.

74. Carry-On Luggage Many airlines have limited the size of luggage passengers may carry onboard on domestic flights. The length, l, plus the width, w, plus the depth, d, of the carry-on luggage must not exceed 45 inches.

   a) Write an inequality that describes this restriction, using l, w, and d as described above.

   b) If Ryan McHenry's luggage is 23 inches long and 12 inches wide, what is the maximum depth it can have and still be carried on the plane?

In Exercises 75–88, set up an inequality that can be used to solve the problem. Solve the problem and find the desired value.

75. Weight Limit Cal Worth, a janitor, must move a large shipment of books from the first floor to the fifth floor. The sign on the elevator reads "maximum weight 800 pounds." If each box of books weighs 70 pounds, find the maximum number of boxes that Cal should place in the elevator, if he does not ride.

76. Elevator Limit If the janitor in Exercise 75, weighing 195 pounds, must ride up with the boxes of books, find the maximum number of boxes that can be placed into the elevator.

77. Long Distance The telephone long-distance carrier Telecom-USA, which markets itself as 10-10-220, charges customers $0.99 for the first 20 minutes and then $0.07 for each minute (or any part thereof) beyond 20 minutes. If Patricia Lanz uses this carrier, how long can she talk for $5.00?

78. Parking Garage A downtown parking garage in Austin, Texas, charges $1.25 for the first hour and $0.75 for each additional hour or part thereof. What is the maximum length of time you can park in the garage if you wish to pay no more than $3.75?

79. Book Profit April Lemons is considering writing and publishing her own book. She estimates her revenue equation as $R = 6.42x$, and her cost equation as $C = 10{,}025 + 1.09x$, where x is the number of books she sells. Find the minimum number of books she must sell to make a profit. See Example 6.

80. Dry Cleaning Profit Peter Collinge is opening a dry cleaning store. He estimates his cost equation as $C = 8000 + 0.08x$ and his revenue equation as $R = 1.85x$, where x is the number of garments dry cleaned in a year. Find the minimum number of garments that must be dry cleaned in a year for Peter to make a profit.

81. First-Class Mail On January 8, 2006, the cost for mailing a package first class was $0.39 for the first ounce and $0.24 for each additional ounce. What is the maximum weight of a package that Richard Van Lommel can mail first class for $10.00?

82. Presorted First-Class Mail Companies can send pieces of mail weighing up to 1 ounce by using *presorted first-class* mail. The company must first purchase a bulk permit for $150 per year, and then pay $0.275 per piece sent. Without the permit, each piece would cost $0.37. Determine the minimum number of pieces of mail that would have to be mailed for it to be financially worthwhile for a company to use presorted first-class mail.

83. Comparing Payment Plans Melissa Pfistner recently accepted a sales position in Ohio. She can select between two payment plans. Plan 1 is a salary of $300 per week plus a 10% commission on sales. Plan 2 is a salary of $400 per week plus an 8% commission on sales. For what amount of weekly sales would Melissa earn more by plan 1?

84. College Employment To be eligible to continue her financial assistance for college, Katie Hanenberg can earn no more than $2000 during her 8-week summer employment. She already earns $90 per week as a day-care assistant. She is considering adding an evening job at a fast-food restaurant, where she will earn $6.25 per hour. What is the maximum number of hours she can work at the restaurant without jeopardizing her financial assistance?

85. A Passing Grade To pass a course, Corrina Schultz needs an average score of 60 or more. If Corrina's scores are 66, 72, 90, 49, and 59, find the minimum score that she can get on her sixth and last exam and pass the course.

86. Minimum Grade To receive an A in a course, Stephen Heasley must obtain an average score of 90 or higher on five exams. If Stephen's first four exam scores are 92, 87, 96, and 77, what is the minimum score that Stephen can receive on the fifth exam to get an A in the course?

87. Averaging Grades Calisha Mahoney's grades on her first four exams are 85, 92, 72, and 75. An average greater than or equal to 80 and less than 90 will result in a final grade of B. What range of grades on Calisha's fifth and last exam will result in a final grade of B? Assume a maximum grade of 100.

88. Clean Air For air to be considered "clean," the average of three pollutants must be less than 3.2 parts per million. If the first two pollutants are 2.7 and 3.42 ppm, what values of the third pollutant will result in clean air?

89. Income Taxes Refer to Example 7 on page 621. Su-hua and Ting-Fang Zheng file a joint tax return. Determine the 2005 income tax Su-hua and Ting-Fang will owe if their taxable income is **a)** $78,221.   **b)** $301,233.

90. Income Taxes Refer to Example 7 on page 621. Jose and Mildred Battiste file a joint tax return. Determine the 2005 income tax Jose and Mildred will owe if their taxable income is

a) $128,479.

b) $275,248.

Velocity

In a physics course, a positive velocity indicates that a projected object is traveling upward and a negative velocity indicates that the object has turned around and is traveling downward. Specifically, an object is traveling upward when velocity ≥ 0. The object has reached its maximum height when $v = 0$ and the object is traveling downward when velocity ≤ 0.

In Exercises 91–96, the velocity, $v(t)$, is given for an object that is projected upward. Using interval notation, determine the intervals when the object is traveling a) upward, b) downward.

91. $v(t) = -32t + 96$, $0 \leq t \leq 10$

92. $v(t) = -32t + 172.8$, $0 \leq t \leq 12$

93. $v(t) = -9.8t + 49$, $0 \leq t \leq 13$

94. $v(t) = -9.8t + 31.36$, $0 \leq t \leq 6$

95. $v(t) = -32t + 320$, $0 \leq t \leq 8$

96. $v(t) = -9.8t + 68.6$, $0 \leq t \leq 5$

97. Water Acidity Thomas Hayward is checking the water acidity in a swimming pool. The water acidity is considered normal when the average pH reading of three daily measurements is greater than 7.2 and less than 7.8. If the first two pH readings are 7.48 and 7.15, find the range of pH values for the third reading that will result in the acidity level being normal.

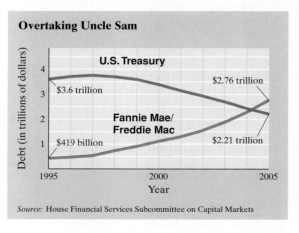

98. Comparing Debts Fannie Mae and Freddie Mac are government-sponsored corporations designed to lend money to people wishing to purchase homes. Since 1995, the debt of Fannie Mae and Freddie Mac has been sharply increasing while the debt of the U.S. Treasury has been sharply decreasing. The following graph displays the debts of Fannie Mae and Freddie Mac as well as the debts of the U.S. Treasury for the years from 1995 to 2005.

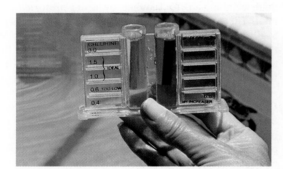

Source: House Financial Services Subcommittee on Capital Markets

a) During which years from 1995 to 2005 was the Fannie Mae/Freddie Mac debt below $1 trillion *and* the U.S. Treasury debt above $3 trillion? Explain how you determined your answer.

b) During which years from 1995 to 2005 was the Fannie Mae/Freddie Mac debt above $1 trillion *or* the U.S. Treasury debt below $3 trillion? Explain how you determined your answer.

99. Army Enlistment The graph below shows the enlistment goal of the U.S. Army and the actual number who enlisted from January through May of 2005.

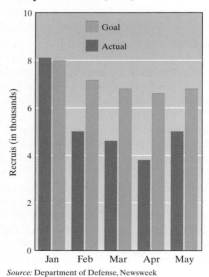

Source: Department of Defense, Newsweek

a) During which months has the goal been greater than 6000 *and* the number enlisted been greater than 4000?

b) During which months has the goal been less than 7000 *or* the number enlisted lower than 4000?

c) During which months has the goal been less than 7000 *and* the number enlisted lower than 4000?

100. If $a > b$, will a^2 always be greater than b^2? Explain and give an example to support your answer.

101. Insurance Policy A Blue Cross/Blue Shield insurance policy has a $100 deductible, after which it pays 80% of the total medical cost, c. The customer pays 20% until the customer has paid a total of $500, after which the policy pays 100% of the medical cost. We can describe this policy as shown on the right.

Blue Cross Pays

$$\begin{cases} 0, & \text{if } c \leq \$100 \\ 0.80(c - 100), & \text{if } \$100 < c \leq \$2100 \\ c - 500, & \text{if } c > \$2100 \end{cases}$$

Explain why this set of inequalities describes Blue Cross/Blue Shield's payment plan.

102. Explain why the inequality $a < bx + c < d$ cannot be solved for x unless additional information is given.

Growth Charts *In Exercises 103 and 104, we will consider growth charts for children from birth to age 36 months that were developed by the National Center for Health Statistics. In general, the nth percentile represents that value that n% of the items being measured are below and $(100 - n)$% of the items are above. For instance, suppose a score of 450 on a test represents the 70th percentile. This means that if a person had a score of 450, he or she surpassed about 70% of all others who took the same test and about $100 - 70 = 30$% surpassed that person's score.*

103. The following chart shows the weight-for-age percentiles for boys from birth to age 36 months. The red curve is the 50th percentile, which means that for any given age indicated, 50% of the weights are above the value indicated by the curve and 50% of the weights are below this value. The orange region is between the 10th percentile (blue curve) and the 90th percentile (green curve). That is, 80% of the weights are between the values represented by the blue curve and the green curve. Use this graph to determine, in interval notation, where 80% of the weights occur for boys of age

a) 9 months.

b) 21 months.

c) 36 months.

104. (See Exercise 103.) The following chart shows the weight-for-age percentiles for girls from birth to age 36 months. The orange region is between the 10th percentile (blue curve) and the 90th percentile (green curve) and 80% of the weights are in this region.

Use this graph to determine, in interval notation, where 80% of the weights occur for girls of age

a) 9 months.

b) 21 months.

c) 36 months.

Weight-for-age percentiles:
Boys, birth to 36 months

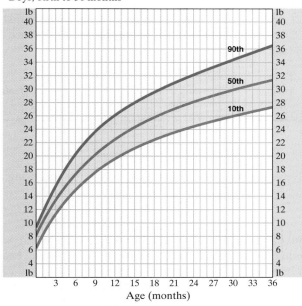

Source: National Center for Health Statistics

Weight-for-age percentiles:
Girls, birth to 36 months

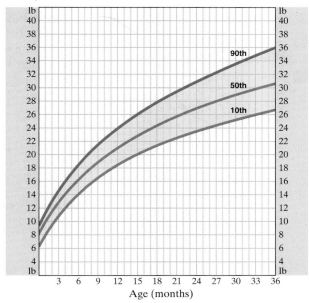

Source: National Center for Health Statistics

Challenge Problems

105. Calculating Grades Stephen Heasley's first five scores in European history were 82, 90, 74, 76, and 68. The final exam for the course is to count one-third in computing the final average. A final average greater than or equal to 80 and less than 90 will result in a final grade of B. What range of final exam scores will result in Stephen's receiving a final grade of B in the course? Assume that a maximum score of 100 is possible.

In Exercises 106–108, **a)** *explain how to solve the inequality, and* **b)** *solve the inequality and give the solution in interval notation.*

106. $x < 3x - 10 < 2x$

107. $x < 2x + 3 < 2x + 5$

108. $x + 5 < x + 3 < 2x + 2$

Cumulative Review Exercises

[1.4] **109.** For $A = \left\{ -3, 4, \dfrac{5}{2}, \sqrt{7}, 0, -\dfrac{29}{80} \right\}$, list the elements that are

    a) counting numbers.

    b) whole numbers.

    c) rational numbers.

    d) real numbers.

[1.10] *For Exercises 110 and 111, name each illustrated property.*

    110. $(3x + 6) + 4y = 3x + (6 + 4y)$

    111. $3x + y = y + 3x$

[2.6] **112.** Solve the formula $R = L + (V - D)r$ for V.

[10.1] **113.** For $A = \{1, 2, 6, 8, 9\}$ and $B = \{1, 3, 4, 5, 8\}$, find

    a) $A \cup B$.

    b) $A \cap B$.

10.2 Solving Equations and Inequalities Containing Absolute Values

1 Understand the geometric interpretation of absolute value.

2 Solve equations of the form $|x| = a, a > 0$.

3 Solve inequalities of the form $|x| < a, a > 0$.

4 Solve inequalities of the form $|x| > a, a > 0$.

5 Solve inequalities of the form $|x| > a$ or $|x| < a, a < 0$.

6 Solve inequalities of the form $|x| > 0$ or $|x| < 0$.

7 Solve equations of the form $|x| = |y|$.

1 Understand the Geometric Interpretation of Absolute Value

In Section 1.5 we introduced the concept of absolute value. We stated that the absolute value of a number may be considered the distance (without sign) from the number 0 on the number line. The absolute value of 3, written $|3|$, is 3 since it is 3 units from 0 on the number line. Similarly, the absolute value of -3, written $|-3|$, is also 3 since it is 3 units from 0 on the number line.

Consider the equation $|x| = 3$; what values of x make this equation true? We know that $|3| = 3$ and $|-3| = 3$. The solutions to $|x| = 3$ are 3 and -3. When solving the equation $|x| = 3$, we are finding the values whose distances are exactly 3 units from 0 on the number line (see **Fig. 10.6a**).

Now consider the inequality $|x| < 3$. To solve this inequality, we need to find the set of values whose distances are less than 3 units from 0 on the number line. These are the values of x between -3 and 3 (see **Fig. 10.6b**).

To solve the inequality $|x| > 3$, we need to find the set of values whose distances are greater than 3 units from 0 on the number line. These are the values that are either less than -3 or greater than 3 (see **Fig. 10.6c**).

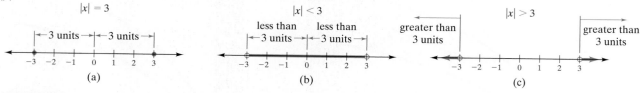

FIGURE 10.6

In this section we will solve equations and inequalities such as the following:

$$|2x - 1| = 5 \qquad |2x - 1| \le 5 \qquad |2x - 1| > 5$$

The geometric interpretation of $|2x - 1| = 5$ is similar to $|x| = 3$. When solving $|2x - 1| = 5$, we are determining the set of values that result in $2x - 1$ being exactly 5 units away from 0 on the number line.

The geometric interpretation of $|2x - 1| \le 5$ is similar to the geometric interpretation of $|x| \le 3$. When solving $|2x - 1| \le 5$, we are determining the set of values that result in $2x - 1$ being less than or equal to 5 units from 0 on the number line.

The geometric interpretation of $|2x - 1| > 5$ is similar to that of $|x| > 3$. When solving $|2x - 1| > 5$, we are determining the set of values that result in $2x - 1$ being greater than 5 units from 0 on the number line.

We will be solving absolute value equations and inequalities algebraically in the remainder of this section. We will first solve absolute value equations, then we will solve absolute value inequalities. We will end the section by solving absolute value equations where both sides of the equation contain an absolute value, for example, $|x + 3| = |2x - 5|$.

2 Solve Equations of the Form $|x| = a, a > 0$

When solving an equation of the form $|x| = a$, $a > 0$, we are finding the values that are exactly a units from 0 on the number line. The following procedure may be used to solve such problems.

> **To Solve Equations of the Form $|x| = a$**
>
> If $|x| = a$ and $a > 0$, then $x = a$ or $x = -a$.

EXAMPLE 1 ▶ Solve each equation.

a) $|x| = 7$ **b)** $|x| = 0$ **c)** $|x| = -7$

Solution

a) Using the procedure, we get $x = 7$ or $x = -7$. The solution set is $\{-7, 7\}$.

b) The only real number whose absolute value equals 0 is 0. Thus, the solution set for $|x| = 0$ is $\{0\}$.

c) The absolute value of a number is never negative, so there are no solutions to this equation. The solution set is $\varnothing$.

▶ **Now Try Exercise 15**

EXAMPLE 2 ▶ Solve the equation $|2w - 1| = 5$.

Solution   At first this might not appear to be of the form $|x| = a$. However, if we let $2w - 1$ be x and 5 be a, you will see that the equation is of this form. We are looking for the values of w such that $2w - 1$ is exactly 5 units from 0 on a number line. Thus, the quantity $2w - 1$ must be equal to 5 or -5.

$$2w - 1 = 5 \quad \text{or} \quad 2w - 1 = -5$$
$$2w = 6 \qquad\qquad\qquad 2w = -4$$
$$w = 3 \qquad\qquad\qquad\; w = -2$$

Check $w = 3$ $|2w - 1| = 5$ $w = -2$ $|2w - 1| = 5$

$$|2(3) - 1| \overset{?}{=} 5 \qquad\qquad |2(-2) - 1| \overset{?}{=} 5$$
$$|6 - 1| \overset{?}{=} 5 \qquad\qquad\quad |-4 - 1| \overset{?}{=} 5$$
$$|5| \overset{?}{=} 5 \qquad\qquad\qquad |-5| \overset{?}{=} 5$$
$$5 = 5 \quad \textit{True} \qquad\qquad\quad 5 = 5 \quad \textit{True}$$

The solutions 3 and -2 each result in $2w - 1$ being 5 units from 0 on the number line. The solution set is $\{-2, 3\}$.

▶ **Now Try Exercise 21**

Consider the equation $|2w - 1| - 3 = 2$. The first step in solving this equation is to isolate the absolute value term. We do this by adding 3 to both sides of the equation. This results in the equation we solved in Example 2.

3 Solve Inequalities of the Form $|x| < a, a > 0$

Now let's look at inequalities of the form $|x| < a$. Consider $|x| < 3$. This inequality represents the set of values that are less than 3 units from 0 on a number line (see **Fig. 10.6b** on page 632). The solution set is $\{x | -3 < x < 3\}$. The solution set to an inequality of the form $|x| < a$ is the set of values that are *less than a units from 0 on a number line*.

We can use the same reasoning process to solve more complicated problems, as shown in Example 3.

EXAMPLE 3 ▶ Solve the inequality $|2x - 3| < 5$.

Solution The solution to this inequality will be the set of values such that the distance between $2x - 3$ and 0 on a number line will be less than 5 units (see **Fig. 10.7**). Using **Figure 10.7**, we can see that $-5 < 2x - 3 < 5$.

FIGURE 10.7

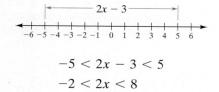

Solving, we get

$$-5 < 2x - 3 < 5$$
$$-2 < 2x < 8$$
$$-1 < x < 4$$

The solution set is $\{x | -1 < x < 4\}$. When x is any number between -1 and 4, the expression $2x - 3$ will represent a number that is less than 5 units from 0 on a number line (or a number between -5 and 5).

▶ **Now Try Exercise 33**

To solve inequalities of the form $|x| < a$, we can use the following procedure.

> **To Solve Inequalities of the Form $|x| < a$**
>
> If $|x| < a$ and $a > 0$, then $-a < x < a$.

EXAMPLE 4 ▶ Solve the inequality $|2x + 1| \leq 9$ and graph the solution on a number line.

Solution Since this inequality is of the form $|x| \leq a$, we write

$$-9 \leq 2x + 1 \leq 9$$
$$-10 \leq 2x \leq 8$$
$$-5 \leq x \leq 4$$

Any value of x greater than or equal to -5 and less than or equal to 4 would result in $2x + 1$ being less than or equal to 9 units from 0 on a number line.

▶ **Now Try Exercise 75**

EXAMPLE 5 ▶ Solve the inequality $|7.8 - 4x| - 5.3 < 14.1$ and graph the solution on a number line.

Solution First isolate the absolute value by adding 5.3 to both sides of the inequality. Then solve as in the previous examples.

$$|7.8 - 4x| - 5.3 < 14.1$$
$$|7.8 - 4x| < 19.4$$
$$-19.4 < 7.8 - 4x < 19.4$$
$$-27.2 < -4x < 11.6$$
$$\frac{-27.2}{-4} > \frac{-4x}{-4} > \frac{11.6}{-4}$$
$$6.8 > x > -2.9 \quad \text{or} \quad -2.9 < x < 6.8$$

The solution set is $\{x | -2.9 < x < 6.8\}$. The solution set in interval notation is $(-2.9, 6.8)$.

▶ **Now Try Exercise 43**

4 Solve Inequalities of the Form $|x| > a$, $a > 0$

Now we look at inequalities of the form $|x| > a$. Consider $|x| > 3$. This inequality represents the set of values that are greater than 3 units from 0 on a number line (see **Fig. 10.6c** on page 632). The solution set is $\{x | x < -3 \text{ or } x > 3\}$. The solution set to $|x| > a$ is the set of values that are *greater than a units from* 0 on a number line.

EXAMPLE 6 ▶ Solve the inequality $|2x - 3| > 5$ and graph the solution on a number line.

Solution   The solution to $|2x - 3| > 5$ is the set of values such that the distance between $2x - 3$ and 0 on a number line will be greater than 5. The quantity $2x - 3$ must be either less than -5 or greater than 5 (see **Fig. 10.8**).

FIGURE 10.8

Since $2x - 3$ must be either less than -5 or greater than 5, we set up and solve the following compound inequality:

$$2x - 3 < -5 \quad \text{or} \quad 2x - 3 > 5$$
$$2x < -2 \qquad\qquad 2x > 8$$
$$x < -1 \qquad\qquad x > 4$$

The solution set to $|2x - 3| > 5$ is $\{x | x < -1 \text{ or } x > 4\}$. When x is any number less than -1 or greater than 4, the expression $2x - 3$ will represent a number that is greater than 5 units from 0 on a number line (or a number less than -5 or greater than 5).

▶ Now Try Exercise 51

To solve inequalities of the form $|x| > a$, we can use the following procedure.

> **To Solve Inequalities of the Form $|x| > a$**
>
> If $|x| > a$ and $a > 0$, then $x < -a$ or $x > a$.

EXAMPLE 7 ▶ Solve the inequality $|2x - 1| \geq 7$ and graph the solution on a number line.

Solution   Since this inequality is of the form $|x| \geq a$, we use the procedure given above.

$$2x - 1 \leq -7 \quad \text{or} \quad 2x - 1 \geq 7$$
$$2x \leq -6 \qquad\qquad 2x \geq 8$$
$$x \leq -3 \qquad\qquad x \geq 4$$

Any value of x less than or equal to -3, or greater than or equal to 4, would result in $2x - 1$ representing a number that is greater than or equal to 7 units from 0 on a number line. The solution set is $\{x | x \leq -3 \text{ or } x \geq 4\}$. In interval notation, the solution is $(-\infty, -3] \cup [4, \infty)$.

▶ Now Try Exercise 53

EXAMPLE 8 ▸ Solve the inequality $\left|\dfrac{3x-4}{2}\right| \geq 9$ and graph the solution on a number line.

Solution Since this inequality is of the form $|x| \geq a$, we write

$$\frac{3x-4}{2} \leq -9 \quad \text{or} \quad \frac{3x-4}{2} \geq 9$$

Now multiply both sides of each inequality by the least common denominator, 2. Then solve each inequality.

$$2\left(\frac{3x-4}{2}\right) \leq -9\cdot 2 \qquad \text{or} \qquad 2\left(\frac{3x-4}{2}\right) \geq 9\cdot 2$$

$$3x - 4 \leq -18 \qquad\qquad\qquad 3x - 4 \geq 18$$

$$3x \leq -14 \qquad\qquad\qquad\qquad 3x \geq 22$$

$$x \leq -\frac{14}{3} \qquad\qquad\qquad\qquad x \geq \frac{22}{3}$$

▸ **Now Try Exercise 57**

Helpful Hint

Some general information about equations and inequalities containing absolute value follows. For real numbers a, b, and c, where $a \neq 0$ and $c > 0$:

Form of Equation or Inequality	The Solution Will Be:	Solution on a Number Line:
$\lvert ax + b\rvert = c$	Two distinct numbers, p and q	
$\lvert ax + b\rvert < c$	The set of numbers between two numbers, $p < x < q$	
$\lvert ax + b\rvert > c$	The set of numbers less than one number or greater than a second number, $x < p$ or $x > q$	

5 Solve Inequalities of the Form $|x| > a$ or $|x| < a$, $a < 0$

We have solved inequalities of the form $|x| < a$ where $a > 0$. Now let us consider what happens in an absolute value inequality when $a < 0$. Consider the inequality $|x| < -3$. Since $|x|$ will always have a value greater than or equal to 0 for any real number x, this inequality can never be true, and the solution is the empty set, $\varnothing$. Whenever we have an absolute value inequality of this type, the solution will be the empty set.

EXAMPLE 9 ▸ Solve the inequality $|6x - 8| + 5 < 3$.

Solution Begin by subtracting 5 from both sides of the inequality.

$$|6x - 8| + 5 < 3$$
$$|6x - 8| < -2$$

Since $|6x - 8|$ will always be greater than or equal to 0 for any real number x, this inequality can never be true. Therefore, the solution is the empty set, $\varnothing$.

▸ **Now Try Exercise 41**

Now consider the inequality $|x| > -3$. Since $|x|$ will always have a value greater than or equal to 0 for any real number x, this inequality will always be true. Since every value of x will make this inequality a true statement, the solution is the set of all real numbers, $\mathbb{R}$. Whenever we have an absolute value inequality of this type, the solution will be the set of all real numbers, $\mathbb{R}$.

EXAMPLE 10 ▶ Solve the inequality $|5x + 3| + 4 \geq -9$.

Solution   Begin by subtracting 4 from both sides of the inequality.

$$|5x + 3| + 4 \geq -9$$
$$|5x + 3| \geq -13$$

Since $|5x + 3|$ will always be greater than or equal to 0 for any real number x, this inequality is true for all real numbers. Thus, the solution is the set of all real numbers, $\mathbb{R}$.

▶ **Now Try Exercise 59**

6 Solve Inequalities of the Form $|x| > 0$ or $|x| < 0$

Now let us discuss inequalities where one side of the inequality is 0. The only value that satisfies the equation $|x - 5| = 0$ is 5, since 5 makes the expression inside the absolute value sign 0. Now consider $|x - 5| \leq 0$. Since the absolute value can never be negative, this inequality is true only when $x = 5$. The inequality $|x - 5| < 0$ has no solution. Can you explain why? What is the solution to $|x - 5| \geq 0$? Since any value of x will result in the absolute value being greater than or equal to 0, the solution is the set of all real numbers, $\mathbb{R}$. What is the solution to $|x - 5| > 0$? The solution is every real number except 5. Can you explain why 5 is excluded from the solution?

EXAMPLE 11 ▶ Solve each inequality.   **a)** $|x + 2| > 0$ **b)** $|3x - 8| \leq 0$

Solution

a) The inequality will be true for every value of x except -2. The solution set is $\{x | x < -2 \text{ or } x > -2\}$.

b) Determine the number that makes the absolute value equal to 0 by setting the expression within the absolute value sign equal to 0 and solving for x.

$$3x - 8 = 0$$
$$3x = 8$$
$$x = \frac{8}{3}$$

The inequality will be true only when $x = \frac{8}{3}$. The solution set is $\left\{\frac{8}{3}\right\}$.

▶ **Now Try Exercise 61**

7 Solve Equations of the Form $|x| = |y|$

Now we will discuss absolute value equations where an absolute value appears on both sides of the equation. To solve equations of the form $|x| = |y|$, use the procedure that follows.

To Solve Equations of the Form $|x| = |y|$

If $|x| = |y|$, then $x = y$ or $x = -y$.

When solving an absolute value equation with an absolute value expression on each side of the equal sign, the two expressions must have the same absolute value. Therefore, *the expressions must be equal to each other or be opposites of each other.*

EXAMPLE 12 ▶ Solve the equation $|z + 3| = |2z - 7|$.

Solution   If we let $z + 3$ be x and $2z - 7$ be y, this equation is of the form $|x| = |y|$. Using the procedure given on the previous page, we obtain the two equations

$$z + 3 = 2z - 7 \quad \text{or} \quad z + 3 = -(2z - 7)$$

Now solve each equation.

$$
\begin{array}{ll}
z + 3 = 2z - 7 \quad \text{or} & z + 3 = -(2z - 7) \\
\quad\quad 3 = z - 7 & z + 3 = -2z + 7 \\
\quad\quad 10 = z & 3z + 3 = 7 \\
& 3z = 4 \\
& z = \dfrac{4}{3}
\end{array}
$$

Check $z = 10$ $|z + 3| = |2z - 7|$ $z = \dfrac{4}{3}$ $|z + 3| = |2z - 7|$

$$|10 + 3| \overset{?}{=} |2(10) - 7| \qquad\qquad \left|\dfrac{4}{3} + 3\right| \overset{?}{=} \left|2\left(\dfrac{4}{3}\right) - 7\right|$$

$$|13| \overset{?}{=} |20 - 7| \qquad\qquad \left|\dfrac{13}{3}\right| \overset{?}{=} \left|\dfrac{8}{3} - \dfrac{21}{3}\right|$$

$$|13| \overset{?}{=} |13| \qquad\qquad \left|\dfrac{13}{3}\right| \overset{?}{=} \left|-\dfrac{13}{3}\right|$$

$$13 = 13 \quad \textit{True} \qquad\qquad \dfrac{13}{3} = \dfrac{13}{3} \quad \textit{True}$$

The solution set is $\left\{10, \dfrac{4}{3}\right\}$.

▶ **Now Try Exercise 63**

EXAMPLE 13 ▶ Solve the equation $|4x - 7| = |6 - 4x|$.

Solution

$$
\begin{array}{ll}
4x - 7 = 6 - 4x \quad \text{or} & 4x - 7 = -(6 - 4x) \\
8x - 7 = 6 & 4x - 7 = -6 + 4x \\
8x = 13 & -7 = -6 \quad \textit{False} \\
x = \dfrac{13}{8} &
\end{array}
$$

Since the equation $4x - 7 = -(6 - 4x)$ results in a false statement, the absolute value equation has only one solution. A check will show that the solution set is $\left\{\dfrac{13}{8}\right\}$.

▶ **Now Try Exercise 69**

> ## Summary of Procedures for Solving Equations and Inequalities Containing Absolute Value
>
> For $a > 0$,
>
> If $|x| = a$, then $x = a$ or $x = -a$.
> If $|x| < a$, then $-a < x < a$.
> If $|x| > a$, then $x < -a$ or $x > a$.
> If $|x| = |y|$, then $x = y$ or $x = -y$.

EXERCISE SET 10.2

Concept/Writing Exercises

1. How do we solve equations of the form $|x| = a, a > 0$?

2. For each of the following equations, find the solution set and explain how you determined your answer.
   a) $|x| = -2$
   b) $|x| = 0$
   c) $|x| = 2$

3. How do we solve inequalities of the form $|x| < a, a > 0$?

4. How do you check to see whether -7 is a solution to $|2x + 3| = 11$? Is -7 a solution?

5. How do we solve inequalities of the form $|x| > a, a > 0$?

6. What is the solution to $|x| < 0$? Explain your answer.

7. What is the solution to $|x| > 0$? Explain your answer.

8. Suppose m and n $(m < n)$ are two distinct solutions to the equation $|ax + b| = c$. Indicate the solutions, using both inequality symbols and the number line, to each inequality. (See the Helpful Hint on page 636.)
   a) $|ax + b| < c$

   b) $|ax + b| > c$

9. Explain how to solve an equation of the form $|x| = |y|$.

10. How many solutions will $|ax + b| = k, a \neq 0$ have if
   a) $k < 0$,
   b) $k = 0$,
   c) $k > 0$?

11. How many solutions are there to the following equations or inequalities if $a \neq 0$ and $k > 0$?
   a) $|ax + b| = k$
   b) $|ax + b| < k$
   c) $|ax + b| > k$

12. Match each absolute value equation or inequality labeled **a)** through **e)** with the graph of its solution set, labeled A.–E.
   a) $|x| = 4$ A. (number line graph from -6 to 6)
   b) $|x| < 4$ B. (number line graph from -6 to 6)
   c) $|x| > 4$ C. (number line graph from -6 to 6)
   d) $|x| \geq 4$ D. (number line graph from -6 to 6)
   e) $|x| \leq 4$ E. (number line graph from -6 to 6)

13. Match each absolute value equation or inequality, labeled **a)** through **e)**, with its solution set labeled A.–E.
   a) $|x| = 5$ A. $\{x | x \leq -5 \text{ or } x \geq 5\}$
   b) $|x| < 5$ B. $\{x | -5 < x < 5\}$
   c) $|x| > 5$ C. $\{x | -5 \leq x \leq 5\}$
   d) $|x| \leq 5$ D. $\{-5, 5\}$
   e) $|x| \geq 5$ E. $\{x | x < -5 \text{ or } x > 5\}$

14. Suppose $|x| < |y|$ and $x < 0$ and $y < 0$.
   a) Which of the following must be true: $x < y$, $x > y$, or $x = y$?
   b) Give an example to support your answer to part **a)**.

Practice the Skills

Find the solution set for each equation.

15. $|a| = 2$

16. $|b| = 17$

17. $|c| = \dfrac{1}{2}$

18. $|x| = 0$

19. $|d| = -\dfrac{5}{6}$

20. $|l + 4| = 6$

21. $|x + 5| = 8$

22. $|3 + y| = \dfrac{3}{5}$

23. $|4.5q + 31.5| = 0$

24. $|4.7 - 1.6z| = 14.3$

25. $|5 - 3x| = \dfrac{1}{2}$

26. $|6(y + 4)| = 24$

27. $\left|\dfrac{x - 3}{4}\right| = 5$

28. $\left|\dfrac{3z + 5}{6}\right| - 2 = 7$

29. $\left|\dfrac{x - 3}{4}\right| + 8 = 8$

30. $\left|\dfrac{5x - 3}{2}\right| + 5 = 9$

Find the solution set for each inequality.

31. $|w| < 11$

32. $|p| \leq 9$

33. $|q + 5| \leq 8$

34. $|7 - x| < 6$

35. $|5b - 15| < 10$

36. $|x - 3| - 7 < -2$

37. $|2x + 3| - 5 \leq 10$

38. $|4 - 3x| - 4 < 11$

39. $|3x - 7| + 8 < 14$

40. $\left|\dfrac{2x - 1}{9}\right| \leq \dfrac{5}{9}$

41. $|2x - 6| + 5 \leq 1$

42. $|2x - 3| < -10$

43. $\left|\dfrac{1}{2}j + 4\right| < 7$

44. $\left|\dfrac{k}{4} - \dfrac{3}{8}\right| < \dfrac{7}{16}$

45. $\left|\dfrac{x - 3}{2}\right| - 4 \leq -2$

46. $\left|7x - \dfrac{1}{2}\right| < 0$

Find the solution set for each inequality.

47. $|y| > 2$

48. $|a| \geq 13$

49. $|x + 4| > 5$

50. $|2b - 7| > 3$

51. $|7 - 3b| > 5$

52. $\left|\dfrac{6 + 2z}{3}\right| > 2$

53. $|2h - 5| > 3$

54. $|2x - 1| \geq 12$

55. $|0.1x - 0.4| + 0.4 > 0.6$

56. $|3.7d + 6.9| - 2.1 > -5.4$

57. $\left|\dfrac{x}{2} + 4\right| \geq 5$

58. $\left|4 - \dfrac{3x}{5}\right| \geq 9$

59. $|7w + 3| - 12 \geq -12$ **60.** $|2.6 - x| \geq 0$

61. $|4 - 2x| > 0$

62. $|2c - 8| > 0$

Find the solution set for each equation.

63. $|3p - 5| = |2p + 10|$

64. $|6n + 3| = |4n - 13|$

65. $|6x| = |3x - 9|$

66. $|5t - 10| = |10 - 5t|$

67. $\left|\dfrac{2r}{3} + \dfrac{5}{6}\right| = \left|\dfrac{r}{2} - 3\right|$

68. $|3x - 8| = |3x + 8|$

69. $\left|-\dfrac{3}{4}m + 8\right| = \left|7 - \dfrac{3}{4}m\right|$

70. $\left|\dfrac{3}{2}r + 2\right| = \left|8 - \dfrac{3}{2}r\right|$

Find the solution set for each equation or inequality.

71. $|h| = 1$

72. $|y| \leq 8$

73. $|q + 6| > 2$

74. $|9d + 7| \leq -9$

75. $|2w - 7| \leq 9$

76. $|2z - 7| + 5 > 8$

77. $|5a - 1| = 9$

78. $|2x - 4| + 5 = 13$

79. $|5 + 2x| > 0$

80. $|7 - 3b| = |5b + 15|$

81. $|4 + 3x| \leq 9$

82. $|2.4x + 4| + 4.9 > 3.9$

83. $|3n + 8| - 4 = -10$

84. $|4 - 2x| - 3 = 7$

85. $\left|\dfrac{w + 4}{3}\right| + 5 < 9$

86. $\left|\dfrac{5t - 10}{6}\right| > \dfrac{5}{3}$

87. $\left|\dfrac{3x - 2}{4}\right| - \dfrac{1}{3} \geq -\dfrac{1}{3}$

88. $\left|\dfrac{2x - 4}{5}\right| = 14$

89. $|2x - 8| = \left|\dfrac{1}{2}x + 3\right|$

90. $\left|\dfrac{1}{3}y + 3\right| = \left|\dfrac{2}{3}y - 1\right|$

91. $|2 - 3x| = \left|4 - \dfrac{5}{3}x\right|$

92. $\left|\dfrac{-2u + 3}{7}\right| \leq 5$

Problem Solving

93. Glass Thickness Certain types of glass manufactured by PPG Industries ideally will have a thickness of 0.089 inches. However, due to limitations in the manufacturing process, the thickness is allowed to vary from the ideal thickness by up to 0.004 inch. If t represents the actual thickness of the glass, then the allowable range of thicknesses can be represented using the inequality $|t - 0.089| \leq 0.004$.

Source: www.ppg.com

a) Solve this inequality for t (use interval notation).

b) What is the smallest thickness the glass is allowed to be?

c) What is the largest thickness the glass is allowed to be?

94. Plywood Guarantee Certain plywood manufactured by Lafor International is guaranteed to be $\dfrac{5}{8}$ inch thick with a tolerance of plus or minus $\dfrac{1}{56}$ of an inch. If t represents the actual thickness of the plywood, then the allowable range of thicknesses can be represented using the inequality $\left|t - \dfrac{5}{8}\right| \leq \dfrac{1}{56}$.

Source: www.sticktrade.com

a) Solve this inequality for t (use interval notation).

b) What is the smallest thickness the plywood is allowed to be?

c) What is the largest thickness the plywood is allowed to be?

95. Submarine Depth A submarine is 160 feet below sea level. It has rock formations above and below it, and should not change its depth by more than 28 feet. Its distance below sea level, d, can be described by the inequality $|d - 160| \leq 28$.

a) Solve this inequality for d. Write your answer in interval notation.

b) Between what vertical distances, measured from sea level, may the submarine move?

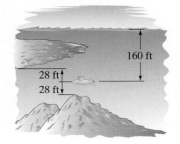

96. A Bouncing Spring A spring hanging from a ceiling is bouncing up and down so that its distance, d, above the ground satisfies the inequality $|d - 4| \leq \dfrac{1}{2}$ foot (see the figure).

a) Solve this inequality for d. Write your answer in interval notation.

b) Between what distances, measured from the ground, will the spring oscillate?

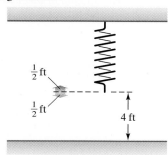

In Exercises 97–100, determine an equation or inequality that involves an absolute value that has the given solution set.

97. $\{-5, 5\}$

98. $\{x | -5 < x < 5\}$

99. $\{x | x \leq -5 \text{ or } x \geq 5\}$

100. $\{x | -5 \leq x \leq 5\}$

101. For what value of x will the inequality $|ax + b| \leq 0$ be true? Explain.

102. For what value of x will the inequality $|ax + b| > 0$ *not* be true? Explain.

103. a) Explain how to find the solution to the equation $|ax + b| = c$. (Assume that $c > 0$ and $a \neq 0$.)

b) Solve this equation for x.

104. a) Explain how to find the solution to the inequality $|ax + b| < c$. (Assume that $a > 0$ and $c > 0$.)

b) Solve this inequality for x.

105. a) Explain how to find the solution to the inequality $|ax + b| > c$. (Assume that $a > 0$ and $c > 0$.)

b) Solve this inequality for x.

106. a) What is the first step in solving the inequality $-4|3x - 5| \leq -12$?

b) Solve this inequality and give the solution in interval notation.

Determine what values of x will make each equation true. Explain your answer.

107. $|x - 4| = |4 - x|$

108. $|x - 4| = -|x - 4|$

109. $|x| = x$

110. $|x + 2| = x + 2$

Solve. Explain how you determined your answer.

111. $|x + 1| = 2x - 1$

112. $|3x + 1| = x - 3$

113. $|x - 4| = -(x - 4)$

Challenge Problems

Solve by considering the possible signs for x.

114. $|x| + x = 8$

115. $x + |-x| = 8$

116. $|x| - x = 8$

117. $x - |x| = 8$

Group Activity

Discuss and answer Exercise 118 as a group.

118. Consider the equation $|x + y| = |y + x|$.

a) Have each group member select an x value and a y value and determine whether the equation holds. Repeat for two other pairs of x- and y-values.

b) As a group, determine for what values of x and y the equation is true. Explain your answer.

c) Now consider $|x - y| = -|y - x|$. Under what conditions will this equation be true?

Cumulative Review Exercises

Evaluate.

[1.9] **119.** $\dfrac{1}{3} + \dfrac{1}{4} \div \dfrac{2}{5}\left(\dfrac{1}{3}\right)^2$

120. $4(x + 3y) - 5xy$ when $x = 1, y = 3$

[3.4] **121. Swimming** Terry Chong swims across a lake averaging 2 miles an hour. Then he turns around and swims back across the lake, averaging 1.6 miles per hour. If his total swimming time is 1.5 hours, what is the width of the lake?

[10.1] **122.** Find the solution set to the inequality $3(x - 2) - 4(x - 3) > 2$.

Mid-Chapter Test: 10.1–10.2

To find out how well you understand the chapter material to this point, take this brief test. The answers, and the section where the material was initially discussed, are given in the back of the book. Review any questions that you answered incorrectly.

1. Express $x > \dfrac{5}{2}$ on a number line.

2. Express $-4 \le w < 5$ in interval notation.

Solve each inequality and gruph the solution on a number line.

3. $6x - 5 \le 4x + 61$

4. $5y - 9y + 8 \le 2(-2y + 1)$

5. $2.4w - 3.2 < 3.6w - 0.8$

Solve each inequality and give the solution in interval notation.

6. $3 - 2x < 6 + 2x + 1$

7. $-8 < p - 3 \le 4$

8. $\dfrac{1}{3} \le 4x + 1 < 5$

Solve each inequality and indicate the solution set.

9. $-3 \le \dfrac{2x + 1}{5} \le 1$

10. $m \le 7$ and $m > -6$

11. $9 - x < 3$ or $7x - 5 \le -5$

12. Find $A \cup B$ and $A \cap B$ for $A = \{7, 8, 9, 10\}$ and $B = \{8, 10, 12, 15\}$.

Find the solution set for each equation or inequality.

13. $|t - 4| = 6$

14. $\left| \dfrac{w - 5}{3} \right| \le 7$

15. $|z| \ge 8$

16. $|5 - 2x| \ge 0$

17. $|4 - 2x| = 5$

18. $|-3m + 4| < 7$

19. $|6y + 9| = |4y - 1|$

20. $\left| \dfrac{1}{2}z - 3 \right| = \left| \dfrac{3}{2}z + 2 \right|$

10.3 Graphing Linear Inequalities in Two Variables and Systems of Linear Inequalities

1 Graph linear inequalities in two variables.

2 Solve systems of linear inequalities.

3 Solve linear programming problems.

4 Solve systems of linear inequalities containing absolute value.

1 Graph Linear Inequalities in Two Variables

A **linear inequality** results when the equal sign in a linear equation is replaced with an inequality sign.

Examples of Linear Inequalities in Two Variables

$$2x + 3y > 2 \qquad\qquad 3y < 4x - 9$$
$$-x - 2y \le 3 \qquad\qquad 5x \ge 2y - 7$$

A line divides a plane into three regions: the line itself and the two **half-planes** on either side of the line. The line is called the **boundary**. Consider the linear equation $2x + 3y = 6$. The graph of this line, the boundary line, divides the plane into the set of points that satisfy the inequality $2x + 3y < 6$ from the set of points that satisfy the inequality $2x + 3y > 6$. An inequality may or may not include the boundary line. Since the inequality $2x + 3y \le 6$ means $2x + 3y < 6$ or $2x + 3y = 6$, the inequality $2x + 3y \le 6$ contains the boundary line. Similarly, the inequality $2x + 3y \ge 6$ contains the boundary line. The graph of the inequalities $2x + 3y < 6$ and $2x + 3y > 6$ do not contain the boundary line. Now let's discuss how to graph linear inequalities.

> ### To Graph a Linear Inequality in Two Variables
>
> 1. Replace the inequality symbol with an equal sign.
> 2. Draw the graph of the equation in step 1. If the original inequality contains a $\geq$ or $\leq$ symbol, draw the graph using a solid line. If the original inequality contains a $>$ or $<$ symbol, draw the graph using a dashed line.
> 3. Select any point not on the line and determine if this point is a solution to the original inequality. If the point selected is a solution, shade the region on the side of the line containing this point. If the selected point does not satisfy the inequality, shade the region on the side of the line not containing this point.

In step 3 we are deciding which set of points satisfies the given inequality.

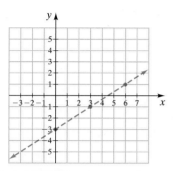

FIGURE 10.9

EXAMPLE 1 ▶ Graph the inequality $y < \frac{2}{3}x - 3$.

Solution First graph the equation $y = \frac{2}{3}x - 3$. Since the original inequality contains a less than sign, $<$, use a dashed line when drawing the graph (**Fig. 10.9**). The dashed line indicates that the points on this line are not solutions to the inequality $y < \frac{2}{3}x - 3$. Select a point not on the line and determine if this point satisfies the inequality. Often the easiest point to use is the origin, $(0, 0)$.

Checkpoint $(0, 0)$

$$y < \frac{2}{3}x - 3$$

$$0 \overset{?}{<} \frac{2}{3}(0) - 3$$

$$0 \overset{?}{<} 0 - 3$$

$$0 < -3 \quad \textit{False}$$

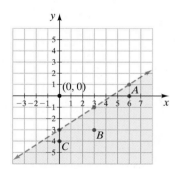

FIGURE 10.10

Since 0 is not less than -3, the point $(0, 0)$ does not satisfy the inequality. The solution will be all points on the other side of the line from the point $(0, 0)$. Shade in this region (**Fig. 10.10**). Every point in the shaded area satisfies the given inequality. Let's check a few selected points A, B, and C.

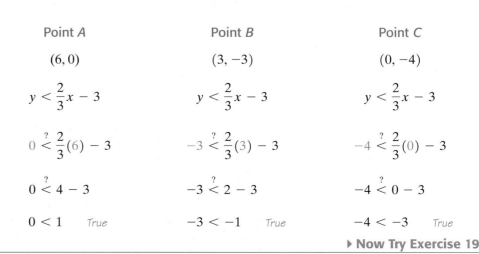

Point A	Point B	Point C
$(6, 0)$	$(3, -3)$	$(0, -4)$
$y < \frac{2}{3}x - 3$	$y < \frac{2}{3}x - 3$	$y < \frac{2}{3}x - 3$
$0 \overset{?}{<} \frac{2}{3}(6) - 3$	$-3 \overset{?}{<} \frac{2}{3}(3) - 3$	$-4 \overset{?}{<} \frac{2}{3}(0) - 3$
$0 \overset{?}{<} 4 - 3$	$-3 \overset{?}{<} 2 - 3$	$-4 \overset{?}{<} 0 - 3$
$0 < 1 \quad$ *True*	$-3 < -1 \quad$ *True*	$-4 < -3 \quad$ *True*

▶ **Now Try Exercise 19**

EXAMPLE 2 ▶ Graph the inequality $y \geq -\frac{1}{2}x$.

Solution   First, we graph the equation $y = -\frac{1}{2}x$. Since the inequality is $\geq$, we use a solid line to indicate that the points on the line are solutions to the inequality (**Fig. 10.11**). Since the point $(0, 0)$ is on the line, we cannot select that point to find the solution. Let's arbitrarily select the point $(3, 1)$.

Checkpoint $(3, 1)$

$$y \geq -\frac{1}{2}x$$

$$1 \overset{?}{\geq} -\frac{1}{2}(3)$$

$$1 \geq -\frac{3}{2} \qquad \textit{True}$$

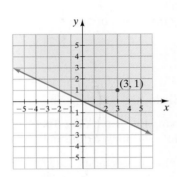

FIGURE 10.11

Since the point $(3, 1)$ satisfies the inequality, every point on the same side of the line as $(3, 1)$ will also satisfy the inequality $y \geq -\frac{1}{2}x$. Shade this region as indicated. Every point in the shaded region, as well as every point on the line, satisfies the inequality.

▶ **Now Try Exercise 13**

EXAMPLE 3 ▶ Graph the inequality $3x - 2y < -6$.

Solution   First, we graph the equation $3x - 2y = -6$. Since the inequality is $<$, we use a dashed line when drawing the graph **Fig. 10.12**. Substituting the checkpoint $(0, 0)$ into the inequality results in a false statement.

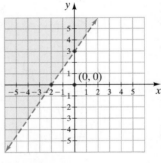

Checkpoint $(0, 0)$

$$3x - 2y < -6$$

$$3(0) - 2(0) \overset{?}{<} -6$$

$$0 < -6 \qquad \textit{False}$$

FIGURE 10.12

The solution is, therefore, that part of the plane that does not contain the origin.

▶ **Now Try Exercise 27**

USING YOUR GRAPHING CALCULATOR

A graphing calculator can also display graphs of inequalities. The procedure to display the graphs varies from calculator to calculator. In **Figure 10.13** we show the graph of $y > 2x + 3$. Read your graphing calculator manual and learn how to display graphs of inequalities.

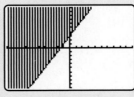

FIGURE 10.13

2 Solve Systems of Linear Inequalities

In Section 9.1 we learned how to solve systems of linear equations graphically. In this section we show how to solve **systems of linear inequalities** graphically.

> **To Solve a System of Linear Inequalities**
>
> Graph each inequality on the same axes. The solution is the set of points whose coordinates satisfy all the inequalities in the system.

EXAMPLE 4 ▶ Determine the solution to the following system of inequalities.

$$y < -\frac{1}{2}x + 2$$
$$x - y \le 4$$

Solution   First graph the inequality $y < -\frac{1}{2}x + 2$ (**Fig. 10.14**). Now on the same axes graph the inequality $x - y \le 4$ (**Fig. 10.15**). The solution is the set of points common to the graphs of both inequalities. It is the part of the graph that contains both shadings. The dashed line is not part of the solution, but the part of the solid line that satisfies both inequalities is part of the solution.

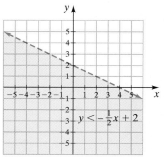

FIGURE 10.14

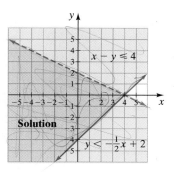

FIGURE 10.15

▶ **Now Try Exercise 29**

EXAMPLE 5 ▶ Determine the solution to the following system of inequalities.

$$3x - y < 6$$
$$2x + 2y \ge 5$$

Solution   Graph $3x - y < 6$ (see **Fig. 10.16**). Graph $2x + 2y \ge 5$ on the same axes (**Fig. 10.17**). The solution is the part of the graph that contains both shadings and the part of the solid line that satisfies both inequalities.

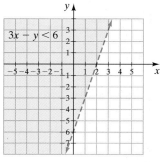

FIGURE 10.16

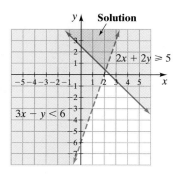

FIGURE 10.17

▶ **Now Try Exercise 31**

EXAMPLE 6 ▶ Determine the solution to the following system of inequalities.

$$y > -1$$
$$x \leq 4$$

Solution   The solution is illustrated in **Figure 10.18**.

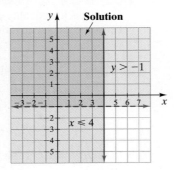

FIGURE 10.18

▶ **Now Try Exercise 39**

3 Solve Linear Programming Problems

There is a mathematical process called **linear programming** for which you often have to graph more than two linear inequalities on the same axes. These inequalities are called **constraints**. The following two examples illustrate how to determine the solution to a system of more than two inequalities.

EXAMPLE 7 ▶ Determine the solution to the following system of inequalities.

$$x \geq 0$$
$$y \geq 0$$
$$2x + 3y \leq 12$$
$$2x + \ y \leq 8$$

Solution   The first two inequalities, $x \geq 0$ and $y \geq 0$, indicate that the solution must be in the first quadrant because that is the only quadrant where both x and y are positive. **Figure 10.19** illustrates the graphs of the four inequalities.

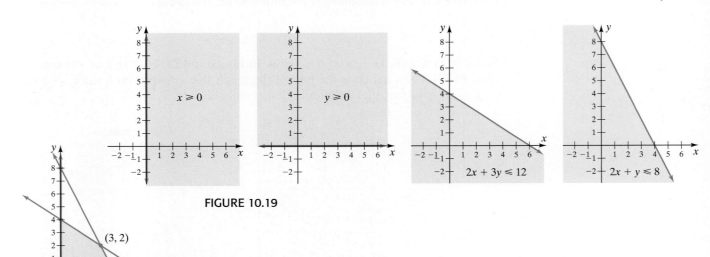

FIGURE 10.19

Figure 10.20 illustrates the four graphs on the same axes and the solution to the system of inequalities. Note that every point in the shaded area and every point on the lines that form the polygonal region is part of the answer.

FIGURE 10.20

▶ **Now Try Exercise 47**

EXAMPLE 8 ▶ Determine the solution to the following system of inequalities.

$$x \geq 0$$
$$y \geq 0$$
$$x \leq 15$$
$$8x + 8y \leq 160$$
$$4x + 12y \leq 180$$

Solution The first two inequalities indicate that the solution must be in the first quadrant. The third inequality indicates that *x* must be a value less than or equal to 15. **Figure 10.21a** indicates the graphs of the five corresponding equations and shows the region that satisfies all the inequalities in the system. **Figure 10.21b** indicates the solution to the system of inequalities.

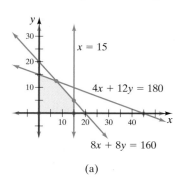

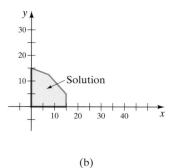

FIGURE 10.21

(a)

(b)

▶ **Now Try Exercise 53**

4 Solve Systems of Linear Inequalities Containing Absolute Value

Now we will graph *systems of linear inequalities containing absolute value* in the Cartesian coordinate system. Before we do some examples, let us recall the rules for absolute value inequalities that we learned in Section 10.2. Recall the following:

Solving Absoute Value Inequalities

If $|x| < a$ and $a > 0$, then $-a < x < a$.

If $|x| > a$ and $a > 0$, then $x < -a$ or $x > a$.

EXAMPLE 9 ▶ Graph $|x| < 3$ in the Cartesian coordinate system.

Solution From the rules given, we know that $|x| < 3$ means $-3 < x < 3$. We draw dashed vertical lines through -3 and 3 and shade the area between the two (**Fig. 10.22**).

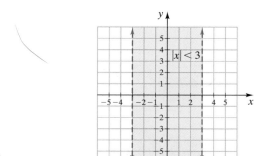

FIGURE 10.22

▶ **Now Try Exercise 57**

EXAMPLE 10 ▸ Graph $|y + 1| > 3$ in the Cartesian coordinate system.

Solution   From the rules given, we know that $|y + 1| > 3$ means $y + 1 < -3$ or $y + 1 > 3$. First we solve each inequality.

$$y + 1 < -3 \quad \text{or} \quad y + 1 > 3$$
$$y < -4 \qquad\qquad y > 2$$

Now we graph both inequalities and take the *union* of the two graphs. The solution is the shaded area in **Figure 10.23**.

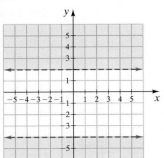

FIGURE 10.23

▸ **Now Try Exercise 59**

EXAMPLE 11 ▸ Determine the solution to following system of inequalities.

$$|x| < 3$$
$$|y + 1| > 3$$

Solution   We draw both inequalities on the same axes. Therefore, we combine the graph drawn in Example 9 with the graph drawn in Example 10 (see **Fig. 10.24**). The points common to both inequalities form the solution to the system.

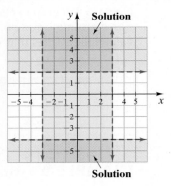

FIGURE 10.24

▸ **Now Try Exercise 65**

EXERCISE SET 10.3   *Math XL*   **MyMathLab**
MathXL®   MyMathLab

Concept/Writing Exercises

1. When graphing an inequality containing $>$ or $<$, why are points on the line not solutions to the inequality?

2. When graphing an inequality containing $\geq$ or $\leq$, why are points on the line solutions to the inequality?

3. When graphing a linear inequality, when can $(0, 0)$ not be used as a checkpoint?

4. When graphing a linear inequality of the form $y > ax + b$ where a and b are real numbers, will the solution always be above the line? Explain.

5. Explain how to find the solution to a system of linear inequalities graphically.

6. If in a system of two inequalities, one inequality contains $<$ and the other inequality contains $\geq$, is the point of intersection of the two boundary lines of the inequalities in the solution set? Explain.

7. If in a system of two inequalities, one inequality contains $\leq$ and the other inequality contains $\geq$, is the point of intersec-

tion of the two boundary lines of the inequalities in the solution set? Explain.

8. If in a system of two inequalities, one inequality contains $<$ and the other inequality contains $>$, is the point of intersection of the two boundary lines of the inequalities in the solution set? Explain.

Practice the Skills

Graph each inequality.

9. $x > 1$

10. $x \geq 4$

11. $y < -2$

12. $y < x$

13. $y \geq -\dfrac{1}{2}x$

14. $y < \dfrac{1}{2}x$

15. $y < 2x + 1$

16. $y \geq 3x - 1$

17. $y > 2x - 1$

18. $y \leq -x + 4$

19. $y \geq \dfrac{1}{2}x - 3$

20. $y < 3x + 2$

21. $2x + 3y > 6$

22. $2x - 3y \geq 12$

23. $y \leq -3x + 5$

24. $y \leq \dfrac{2}{3}x + 3$

25. $2x + y < 4$

26. $3x - 4y \leq 12$

27. $10 \geq 5x - 2y$

28. $-x - 2y > 4$

Determine the solution to each system of inequalities.

29. $2x - y < 4$
    $y \geq -x + 2$

30. $y \leq -2x + 1$
    $y > -3x$

31. $y < 3x - 2$
    $y \leq -2x + 3$

32. $y \geq 2x - 5$
    $y > -3x + 5$

33. $y < x$
    $y \geq 3x + 2$

34. $-3x + 2y \geq -5$
    $y \leq -4x + 7$

35. $-2x + 3y < -5$
    $3x - 8y > 4$

36. $-4x + 3y \geq -4$
    $y > -3x + 3$

37. $-4x + 5y < 20$
    $x \geq -3$

38. $y \geq -\dfrac{2}{3}x + 1$
    $y > -4$

39. $x \leq 4$
    $y \geq -2$

40. $x \geq 0$
    $x - 3y < 6$

41. $5x + 2y > 10$
    $3x - y > 3$

42. $3x + 2y > 8$
    $x - 5y < 5$

43. $-2x > y + 4$
    $-x < \dfrac{1}{2}y - 1$

44. $y \leq 3x - 2$
    $\dfrac{1}{3}y < x + 1$

45. $y < 3x - 4$
    $6x \geq 2y + 8$

46. $\dfrac{1}{2}x + \dfrac{1}{2}y \geq 2$
    $2x - 3y \leq -6$

Determine the solution to each system of inequalities. Use the method discussed in Examples 7 and 8.

47. $x \geq 0$
    $y \geq 0$
    $2x + 3y \leq 6$
    $4x + y \leq 4$

48. $x \geq 0$
    $y \geq 0$
    $x + y \leq 6$
    $7x + 4y \leq 28$

49. $x \geq 0$
    $y \geq 0$
    $2x + 3y \leq 8$
    $4x + 2y \leq 8$

50. $x \geq 0$
    $y \geq 0$
    $3x + 2y \leq 18$
    $2x + 4y \leq 20$

51. $x \geq 0$
    $y \geq 0$
    $3x + y \leq 9$
    $2x + 5y \leq 10$

52. $x \geq 0$
    $y \geq 0$
    $5x + 4y \leq 16$
    $x + 6y \leq 18$

53. $x \geq 0$
    $y \geq 0$
    $x \leq 4$
    $x + y \leq 6$
    $x + 2y \leq 8$

54. $x \geq 0$
    $y \geq 0$
    $x \leq 4$
    $2x + 3y \leq 18$
    $4x + 2y \leq 20$

55. $x \geq 0$
    $y \geq 0$
    $x \leq 15$
    $30x + 25y \leq 750$
    $10x + 40y \leq 800$

56. $x \geq 0$
    $y \geq 0$
    $x \leq 15$
    $40x + 25y \leq 1000$
    $5x + 30y \leq 900$

Graph the solution to each inequality.

57. $|x| < 2$

58. $|x| > 1$

59. $|y - 2| \le 4$

60. $|y| \ge 2$

Determine the solution to each system of inequalities.

61. $|y| > 2$
$y \le x + 3$

62. $|x| > 1$
$y \le 3x + 2$

63. $|y| < 4$
$y \ge -2x + 2$

64. $|x - 2| \le 3$
$x - y > 2$

65. $|x + 2| < 3$
$|y| > 4$

66. $|x - 2| > 1$
$y > -2$

67. $|x - 3| \le 4$
$|y + 2| \le 1$

68. $|x + 1| \le 2$
$|y - 3| \le 1$

Problem Solving

69. Life Insurance The monthly rates for $100,000 of life insurance for women from the American General Financial Group increases approximately linearly from age 35 through age 50. The rate for a 35-year-old woman is $10.15 per month and the rate for a 50-year-old woman is $16.45 per month.

  a) Draw a graph that fits this data.

  b) On the graph, darken the part of the graph where the rate is less than or equal to $15 per month.

  c) Estimate the age at which the rate first exceeds $15 per month.

70. Consumer Price Index The consumer price index (CPI) is a measure of inflation. Since 1990, the CPI has been increasing approximately linearly. The CPI in 1990 was 130.7 and in 2005 the CPI was 190.7.

  Source: U.S. Bureau of the Census

  a) Draw a graph that fits this data.

  b) On the graph, darken the part of the graph where the CPI is greater than or equal to 171.

  c) Estimate the first year in which the CPI was greater than or equal to 171.

71. Fewer Smokers The percentage of Americans 18 and over who smoke has been decreasing approximately linearly since 1997. In 1997, approximately 29.2% of Americans 18 and older smoked. In 2004, approximately 20.9% of those 18 and older smoked.

  Source: Centers for Disease Control and Prevention

  a) Draw a graph that fits these data.

  b) On the graph, darken the part of the graph where the percentage of Americans 18 and older is less than or equal to 25%.

  c) Estimate the first year that the percentage of Americans 18 and older was less than 23%.

72. China Travel The number of Chinese travelers has increased approximately linearly from 1993 through 2005. In 1993, there were about 3.7 million Chinese travelers. In 2005, there were about 17.9 million Chinese travelers.

  Source: Travel Industry Association of America.

  a) Draw a graph that fits these data.

  b) On the graph, darken the part of the graph where the number of Chinese travelers is greater than or equal to 10 million.

  c) Estimate the first year in which the number of Chinese travelers is greater than or equal to 12 million.

73. Tax Returns The following graph shows the percent of all U.S. federal tax returns filed electronically and by paper for the years 2001–2005 and projected to the year 2009. The information for the graph was obtained from the IRS Web site.

Federal Tax Return Method

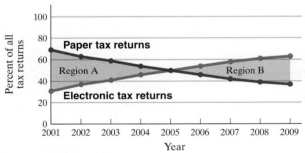

Source: Internal Revenue Service: www.irs.gov/pubs

Let $P(r)$ represent the paper tax returns (purple line) and $E(t)$ represent the electronic tax returns (green line). The regions between the two curves are identified as either Region A, shaded in blue, or Region B shaded in orange.

  a) Which region, A or B, is a solution to the system of inequalities?
$$y \le P(t)$$
$$y \ge E(t)$$

  b) Which region, A or B, is a solution to the system of inequalities?
$$y \ge P(t)$$
$$y \le E(t)$$

74. Income from Department Stores The following graph shows the annual net income, in millions of dollars, for the Federated Department Stores (Macy's, Bloomingdale's, Burdines) and the May Department Stores (Hecht's, Lord & Taylor, Filene's) for the years 1999–2003.

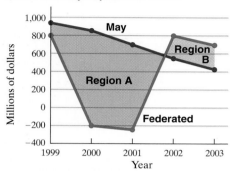

Annual Net Income for Federated Department Stores and May Department Stores

Source: Thomson Financial, the companies, *The Washington Post* (1/21/05)

Let $P(t)$ represent the income for the May Department Stores (purple line) and $Q(t)$ represent the income for the Federated Department stores (green line). The regions between the two curves are identified as either Region A, shaded in blue, or Region B, shaded in orange.

a) Which region, A or B, is a solution to the system of inequalities?
$$y \geq P(t)$$
$$y \leq Q(t)$$

b) Which region, A or B, is a solution to the system of inequalities?
$$y \leq P(t)$$
$$y \geq Q(t)$$

75. a) Graph $f(x) = 2x - 4$.

b) On the graph, shade the region bounded by $f(x)$, $x = 2$, $x = 4$, and the x-axis.

76. a) Graph $g(x) = -x + 4$.

b) On the graph, shade the region bounded by $g(x)$, $x = 1$, and the x- and y-axes.

77. Is it possible for a system of linear inequalities to have no solution? Explain. Make up an example to support your answer.

78. Is it possible for a system of two linear inequalities to have exactly one solution? Explain. If you answer yes, make up an example to support your answer.

Without graphing, determine the number of solutions in each indicated system of inequalities. Explain your answers.

79. $3x - y \leq 4$
$3x - y > 4$

80. $2x + y < 6$
$2x + y > 6$

81. $5x - 2y \leq 3$
$5x - 2y \geq 3$

82. $5x - 3y > 5$
$5x - 3y > -1$

83. $2x - y < 7$
$3x - y < -2$

84. $x + y \leq 0$
$x - y \geq 0$

Challenge Problems

Determine the solution to each system of inequalities.

85. $y \geq x^2$
$y \leq 4$

86. $y < 4 - x^2$
$y > -5$

87. $y < |x|$
$y < 4$

88. $y \geq |x - 2|$
$y \leq -|x - 2|$

Cumulative Review Exercises

[2.6] **89.** A formula for levers in physics is $f_1 d_1 + f_2 d_2 = f_3 d_3$. Solve this formula for f_2.

[8.2] *State the domain and range of each function.*

90. $\{(4, 3), (5, -2), (-1, 2), (0, -5)\}$

91. $f(x) = \dfrac{2}{3}x - 4$

92.

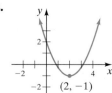

Chapter 10 Summary

IMPORTANT FACTS AND CONCEPTS	EXAMPLES

Section 10.1

Properties Used to Solve Inequalities

1. If $a > b$, then $a + c > b + c$.
2. If $a > b$, then $a - c > b - c$.
3. If $a > b$, and $c > 0$, then $ac > bc$.
4. If $a > b$, and $c > 0$, then $\dfrac{a}{c} > \dfrac{b}{c}$.
5. If $a > b$, and $c < 0$, then $ac < bc$.
6. If $a > b$, and $c < 0$, then $\dfrac{a}{c} < \dfrac{b}{c}$.

1. If $6 > 5$, then $6 + 3 > 5 + 3$.
2. If $6 > 5$, then $6 - 3 > 5 - 3$.
3. If $7 > 3$, then $7 \cdot 4 > 3 \cdot 4$.
4. If $7 > 3$, then $\dfrac{7}{4} > \dfrac{3}{4}$.
5. If $9 > 2$, then $9(-3) < 2(-3)$.
6. If $9 > 2$, then $\dfrac{9}{-3} < \dfrac{2}{-3}$.

A **compound inequality** is formed by joining two inequalities with the word *and* or *or*.

To find the solution set of an inequality containing the word *and* take the **intersection** of the solution sets of the two inequalities.

To find the solution set of an inequality containing the word *or*, take the **union** of the solution sets of the two inequalities.

$x \le 7$ and $x > 5$ is a compound inequality.
$x < -1$ or $x \ge 4$ is a compound inequality.

Solve $x \le 7$ and $x > 5$.
The intersection of $\{x | x \le 7\}$ and $\{x | x > 5\}$ is $\{x | 5 < x \le 7\}$ or $(5, 7]$.

Solve $x < -1$ or $x \ge 4$.
The union of $\{x | x < -1\}$ or $\{x | x \ge 4\}$ is $\{x | x < -1$ or $x \ge 4\}$ or $(-\infty, -1) \cup [4, \infty)$.

Section 10.2

To Solve Equations of the Form $|x| = a$
If $|x| = a$ and $a > 0$, then $x = a$ or $x = -a$.

Solve $|x| = 6$.
$$|x| = 6 \text{ gives } x = 6 \text{ or } x = -6.$$

To Solve Inequalities of the Form $|x| < a$
If $|x| < a$ and $a > 0$, then $-a < x < a$.

Solve $|3x + 1| < 13$.
$$-13 < 3x + 1 < 13$$
$$-14 < 3x < 12$$
$$-\frac{14}{3} < x < 4$$
$$\left\{ x \middle| -\frac{14}{3} < x < 4 \right\} \text{ or } \left(-\frac{14}{3}, 4 \right)$$

To Solve Inequalities of the Form $|x| > a$
If $|x| > a$ and $a > 0$, then $x < -a$ or $x > a$.

Solve $|2x - 3| \ge 5$.
$$2x - 3 \le -5 \quad \text{or} \quad 2x - 3 \ge 5$$
$$2x \le -2 \qquad\qquad 2x \ge 8$$
$$x \le -1 \qquad\qquad x \ge 4$$
$$\{x | x \le -1 \text{ or } x \ge 4\} \quad \text{or} \quad (-\infty, -1] \cup [4, \infty)$$

If $|x| > a$ and $a < 0$, the solution set is $\mathbb{R}$.
If $|x| < a$ and $a < 0$, the solution set is $\varnothing$.

$|x| > -7$, the solution set is $\mathbb{R}$.
$|x| < -7$, the solution set is $\varnothing$.

To Solve Equations of the Form $|x| = |y|$.
If $|x| = |y|$, then $x = y$ or $x = -y$.

Solve $|x| = |3|$.
$$x = 3 \text{ or } x = -3.$$

Section 10.3

A **linear inequality** results when the equal sign of a linear equation is replaced with an inequality sign.

$3x - 4y > 1$ and $2x + 5y \le -4$ are both linear inequalities in two variables.

IMPORTANT FACTS AND CONCEPTS	EXAMPLES

Section 10.3 (continued)

To Graph a Linear Inequality in Two Variables

1. Replace the inequality symbol with an equal sign.

2. Draw the graph of the equation in step 1. If the original inequality is $\geq$ or $\leq$ draw a solid line. If the original inequality is a $>$ or $<$ draw a dashed line.

3. Select any point not on the line. If the point selected is a solution, shade the region on the side of the line containing this point. If the selected point does not satisfy the inequality, shade the region on the side of the line not containing this point.

Graph $y > -x + 1$.

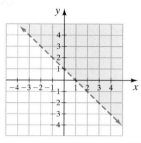

To solve a system of linear inequalities, graph each inequality on the same axes. The solution is the set of points whose coordinates satisfy all the inequalities in the system.

Determine the solution to the system of inequalities.

$$y < -\frac{1}{3}x + 1$$
$$x - y \leq 2$$

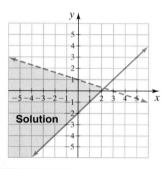

Linear programming is a process where more than two linear inequalities are graphed on the same axes.

Determine the solution to the system of inequalities.

$$x \geq 0$$
$$y \geq 0$$
$$x \leq 8$$
$$x + y \leq 10$$
$$x + 2y \leq 16$$

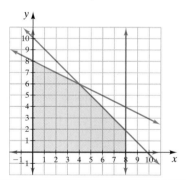

For systems of linear inequalities with absolute values:

If $|x| < a$ and $a > 0$, then $-a < x < a$.

If $|x| > a$ and $a > 0$, then $x < -a$ or $x > a$.

Determine the solution to the system of inequalities.

$$|x| < 2$$
$$|y - 1| > 3$$

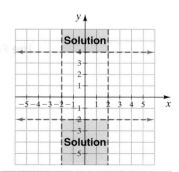

Chapter 10 Review Exercises

[10.1] *Solve the inequality. Graph the solution on a real number line.*

1. $3z + 9 \leq 15$

2. $8 - 2w > -4$

3. $2x + 1 > 6$

4. $26 \leq 4x + 5$

5. $\dfrac{4x + 3}{3} > -5$

6. $2(x - 1) > 3x + 8$

7. $-4(x - 2) \geq 6x + 8 - 10x$

8. $\dfrac{x}{2} + \dfrac{3}{4} > x - \dfrac{x}{2} + 1$

In Exercises 9–12, write an inequality that can be used to solve each problem. Solve the inequality and answer the question.

9. Weight Limit A canoe can safely carry a total weight of 560 pounds. If Bob and Kathie together weigh a total of 300 pounds, what is the maximum number of 40-pound boxes they can carry in their canoe?

10. Phone Booth Call Michael Lamb, a telephone operator, informs a customer in a phone booth that the charge for calling Omaha, Nebraska, is $4.50 for the first 3 minutes and 95¢ for each additional minute or any part thereof. How long can the customer talk if he has $8.65?

11. Fitness Center A fitness center guarantees that customers will lose a minimum of 5 pounds the first week and $1\frac{1}{2}$ pounds each additional week. Find the maximum amount of time needed to lose 27 pounds.

12. Exam Scores Patrice Lee's first four exam scores are 94, 73, 72, and 80. If a final average greater than or equal to 80 and less than 90 is needed to receive a final grade of B in the course, what range of scores on the fifth and last exam will result in Patrice's receiving a B in the course? Assume a maximum score of 100.

Find $A \cup B$ and $A \cap B$ for each set A and B.

13. $A = \{1, 2, 3, 4, 5\}, B = \{2, 3, 4, 5\}$

14. $A = \{3, 5, 7, 9\}, B = \{2, 4, 6, 8\}$

15. $A = \{1, 2, 3, 4, \ldots\}, B = \{2, 4, 6, \ldots\}$

16. $A = \{4, 6, 9, 10, 11\}, B = \{3, 5, 9, 10, 12\}$

Solve each inequality. Write the solution in interval notation.

17. $1 < x - 4 < 7$

18. $8 < p + 11 \leq 16$

19. $3 < 2x - 4 < 12$

20. $-12 < 6 - 3x < -2$

21. $-1 < \dfrac{5}{9}x + \dfrac{2}{3} \leq \dfrac{11}{9}$

22. $-8 < \dfrac{4 - 2x}{3} < 0$

Find the solution set to each compound inequality.

23. $h \leq 1$ and $7h - 4 > -25$

24. $2x - 1 > 5$ or $3x - 2 \leq 10$

25. $4x - 5 < 11$ and $-3x - 4 \geq 8$

26. $\dfrac{7 - 2g}{3} \leq -5$ or $\dfrac{3 - g}{9} > 1$

[10.2] *Find the solution set to each equation or inequality.*

27. $|a| = 2$

28. $|x| < 8$

29. $|x| \geq 9$

30. $|l + 5| = 13$

31. $|x - 2| \geq 5$

32. $|4 - 2x| = 5$

33. $|-2q + 9| < 7$

34. $\left|\dfrac{2x - 3}{5}\right| = 1$

35. $\left|\dfrac{x - 4}{3}\right| < 6$

36. $|4d - 1| = |6d + 9|$

37. $|2x - 3| + 4 \geq -17$

[10.1, 10.2] *Solve each inequality. Give the solution in interval notation.*

38. $|3c + 8| - 6 \leq 1$

39. $3 < 2x - 5 \leq 11$

40. $-6 \le \dfrac{3 - 2x}{4} < 5$

41. $2p - 5 < 7$ and $9 - 3p \le 15$

42. $x - 3 \le 4$ or $2x - 5 > 7$

43. $-10 < 3(x - 4) \le 18$

[10.3] *Graph each inequality.*

44. $y \ge -5$

45. $x < 4$

46. $y \le 4x - 3$

47. $y < \dfrac{1}{3}x - 2$

Determine the solution to each system of inequalities.

48. $-x + 3y > 6$
$2x - y \le 2$

49. $5x - 2y \le 10$
$3x + 2y > 6$

50. $y > 2x + 3$
$y < -x + 4$

51. $x > -2y + 4$
$y < -\dfrac{1}{2}x - \dfrac{3}{2}$

Determine the solution to the system of inequalities.

52. $x \ge 0$
$y \ge 0$
$x + y \le 6$
$4x + y \le 8$

53. $x \ge 0$
$y \ge 0$
$2x + y \le 6$
$4x + 5y \le 20$

54. $|x| \le 3$
$|y| > 2$

55. $|x| > 4$
$|y - 2| \le 3$

Chapter 10 Practice Test

 To find out how well you understand the chapter material, take this practice test. The answers, and the section where the material was initially discussed, are given in the back of the book. Each problem is also fully worked out on the **Chapter Test Prep Video CD.** *Review any questions that you answered incorrectly.*

Find $A \cup B$ and $A \cap B$ for sets A and B.

1. $A = \{8, 10, 11, 14\}$, $B = \{5, 7, 8, 9, 10\}$

2. $A = \{1, 3, 5, 7, \ldots\}$, $B = \{3, 5, 7, 9, 11\}$

Solve each inequality and graph the solution on a number line.

3. $3(2q + 4) < 5(q - 1) + 7$

4. $\dfrac{6 - 2x}{5} \ge -12$

Solve each inequality and write the solution in interval notation.

5. $x - 3 \le 4$ and $2x + 1 > 10$

6. $7 \le \dfrac{2u - 5}{3} < 9$

Find the solution set to the following equations.

7. $|2b + 5| = 9$

8. $|2x - 3| = \left|\dfrac{1}{2}x - 10\right|$

Find the solution set to the following inequalities.

9. $|4z + 12| = 0$

10. $|2x - 3| + 6 > 11$

11. $\left|\dfrac{2x - 3}{8}\right| \le \dfrac{1}{4}$

12. Graph $y < 3x - 2$.

Determine the solution to each system of inequalities.

13. $3x + 2y < 9$
$-2x + 5y \le 10$

14. $|x| > 3$
$|y| \le 1$

Cumulative Review Test

Take the following test and check your answers with those given in the back of the book. Review any questions that you answered incorrectly. The section where the material was covered is indicated after the answer.

1. Evaluate $-40 - 3(4 - 8)^2$.

2. Evaluate $2(x + 2y) + 4x - 5y$ when $x = 2$ and $y = 3$.

3. Simplify $\dfrac{1}{2}(x + 3) + \dfrac{1}{3}(3x + 6)$.

4. Solve $\dfrac{a - 7}{3} = \dfrac{a + 5}{2} - \dfrac{7a - 1}{6}$.

5. **Spreading Fertilizer** If a 40-pound bag of fertilizer covers 5000 square feet, how many pounds of fertilizer are needed to cover an area of 26,000 square feet?

6. **Financial Planning** Belen Poltorade, a financial planner, is offering her customers two financial plans for managing their assets. With plan 1 she charges a planning fee of $1000 plus 1% of the assets she will manage for the customers. With plan 2 she charges a planning fee of $500 plus 2% of the assets she will manage. How much in customer assets would result in both plans having the same total fee?

7. **Bird Food** At Agway Gardens, bird food is sold in bulk. In one barrel are sunflower seeds that sell for $1.80 per pound. In a second barrel is cracked corn that sells for $1.40 per pound. If the store makes bags of a mixture of the two by mixing 2.5 pounds of the sunflower seeds with 1 pound of the cracked corn, what should be the cost per pound of the mixture?

8. Simplify $\left(\dfrac{4a^3b^{-2}}{2a^{-2}b^{-3}} \right)^{-2}$.

9. Multiply $(0.003)(0.00015)$ and write the answer in scientific notation.

10. Factor $24p^3q + 16p^2q - 30pq$.

11. Solve $\dfrac{2b}{b + 1} = 2 - \dfrac{5}{2b}$.

12. Solve $\dfrac{4}{r + 5} + \dfrac{1}{r + 3} = \dfrac{2}{r^2 + 8r + 15}$.

13. Graph $y = x^2 - 2$.

14. Indicate which of the following relations are functions.

a) $\{(4, 2), (-5, 3), (6, 3), (5, 3)\}$

b)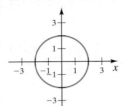

c)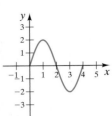

15. Find the equation of the line whose graph has a y-intercept of $(0, -2)$ and is parallel to the graph of $3x - 5y = 7$.

16. Find the equation of the line whose graph passes through $(3, -2)$ and is perpendicular to the graph of $4x - 3y = 12$.

17. Given $f(x) = x^2 - 11x + 30$ and $g(x) = x - 5$, find

a) $(f \cdot g)(x)$.

b) $(f/g)(x)$.

c) the domain of $(f/g)(x)$.

18. Solve the system of equations.
$$2a - b - 2c = -1$$
$$a - 2b - c = 1$$
$$a + b + c = 4$$

19. Solve $\dfrac{1}{2} < 3x + 4 < 6$ and give the solution in interval notation.

20. Graph $y < 2x - 3$.

11 Roots, Radicals, and Complex Numbers

GOALS OF THIS CHAPTER

In this chapter we explain how to add, subtract, multiply, and divide radical expressions. We also graph radical functions, solve radical equations, and introduce imaginary numbers and complex numbers.

Make sure that you understand the three requirements for simplifying radical expressions as discussed in Section 11.5.

MANY SCIENTIFIC FORMULAS, INCLUDING many that pertain to real-life situations, contain radical expressions. In Exercise 135 on page 696, we will see how a radical is used to determine the relationship between the illumination on an object and the distance the object is from a light source.

11.1 Roots and Radicals

1 Find square roots.

2 Find cube roots.

3 Understand odd and even roots.

4 Evaluate radicals using absolute value.

In this section we introduce the concept of radicals. In the expression $\sqrt{x}$, the $\sqrt{\ }$ is called the **radical sign**. The expression within the radical sign is called the **radicand**.

Radical sign

$\sqrt{x}$

Radicand

The entire expression, including the radical sign and radicand, is called the **radical expression**. Another part of the radical expression is its index. The **index** (plural *indices* or *indexes*) gives the "root" of the expression. Square roots have an index of 2. The index of a square root is generally not written. Thus,

$$\sqrt{x} \quad \text{means} \quad \sqrt[2]{x}$$

1 Find Square Roots

Every positive number has two square roots, a principal or positive square root and a negative square root. For any positive number x, the positive square root is written $\sqrt{x}$, and the negative square root is written $-\sqrt{x}$.

Number	Principal or Positive Square Root	Negative Square Root
25	$\sqrt{25}$	$-\sqrt{25}$
19	$\sqrt{19}$	$-\sqrt{19}$

> **Principal Square Root**
>
> The **principal square root** of a positive number a, written $\sqrt{a}$, is the *positive* number b such that $b^2 = a$.

Examples

$\sqrt{25} = 5$ since $5^2 = 5 \cdot 5 = 25$

$\sqrt{0.49} = 0.7$ since $(0.7)^2 = (0.7)(0.7) = 0.49$

$\sqrt{\dfrac{4}{9}} = \dfrac{2}{3}$ since $\left(\dfrac{2}{3}\right)^2 = \left(\dfrac{2}{3}\right)\left(\dfrac{2}{3}\right) = \dfrac{4}{9}$

Remember that $-\sqrt{25}$ means the opposite of $\sqrt{25}$. Because $\sqrt{25} = 5$, $-\sqrt{25} = -5$.

In this book whenever we use the words *square root* we will be referring to the principal or positive square root. If you are asked to find the value of $\sqrt{25}$, your answer will be 5.

In Chapter 1, we indicated that a rational number is one that can be represented as either a terminating or repeating decimal number. If you use the square root key on your calculator, $\boxed{\sqrt{\ }}$, to evaluate the three examples above, you will find they are all terminating or repeating decimal numbers. Thus, they are all *rational numbers*. Many radicals, such as $\sqrt{2}$ and $\sqrt{19}$, are not rational numbers. When evaluating $\sqrt{2}$ and $\sqrt{19}$ on a calculator, the results are nonterminating, nonrepeating decimal numbers. Thus, $\sqrt{2}$ and $\sqrt{19}$ are *irrational numbers*.

Radical	Calculator Results	
$\sqrt{2}$	1.414213562	*Nonterminating, nonrepeating decimal*
$\sqrt{19}$	4.35889894	*Nonterminating, nonrepeating decimal*

Now let's consider $\sqrt{-25}$. Since the square of any real number will always be greater than or equal to 0, there is no real number that when squared equals -25. For this reason, $\sqrt{-25}$ is *not a real number*. Since the square of any real number cannot be negative, *the square root of a negative number is not a real number*. If you evaluate $\sqrt{-25}$ on a calculator, you will get an error message. We will discuss numbers like $\sqrt{-25}$ later in this chapter.

Radical	Calculator Results	
$\sqrt{-25}$	Error	$\sqrt{-25}$ is not a real number.
$\sqrt{-3}$	Error	$\sqrt{-3}$ is not a real number.

Helpful Hint

Do not confuse $-\sqrt{36}$ with $\sqrt{-36}$. Because $\sqrt{36} = 6$, $-\sqrt{36} = -6$. However, $\sqrt{-36}$ is not a real number. The square root of a negative number is not a real number.

$$\sqrt{36} = 6$$
$$-\sqrt{36} = -6$$
$$\sqrt{-36} \text{ is not a real number.}$$

The Square Root Function

When graphing square root functions, functions of the form $f(x) = \sqrt{x}$, we must always remember that the radicand, x, cannot be negative. Thus, the domain of $f(x) = \sqrt{x}$ is $\{x | x \geq 0\}$, or $[0, \infty)$ in interval notation. To graph $f(x) = \sqrt{x}$, we can select some convenient values of x and find the corresponding values of $f(x)$ or y and then plot the ordered pairs, as shown in **Figure 11.1**.

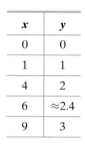

x	y
0	0
1	1
4	2
6	≈ 2.4
9	3

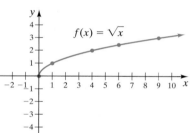

FIGURE 11.1

Since the value of $f(x)$ can never be negative, the range of $f(x) = \sqrt{x}$ is $\{y | y \geq 0\}$, or $[0, \infty)$ in interval notation.

By studying **Figure 11.1**, do you think you can graph $g(x) = -\sqrt{x}$? The graph of $g(x) = -\sqrt{x}$ would be similar to the graph in **Figure 11.1**, but the graph would be drawn below the x-axis. Can you explain why? What about the graph of $h(x) = \sqrt{x-4}$? For the graph of $h(x) = \sqrt{x-4}$, you would only select values of $x \geq 4$ since the radicand cannot be negative. The domain of $h(x) = \sqrt{x-4}$ is $\{x | x \geq 4\}$ or $[4, \infty)$.

To evaluate radical functions, it may be necessary to use a calculator.

EXAMPLE 1 ▶ For each function, find the indicated value(s).

a) $f(x) = \sqrt{11x - 2}$, $f(6)$ **b)** $g(r) = -\sqrt{-3r + 1}$, $g(-5)$ and $g(7)$

Solution

a) $f(6) = \sqrt{11(6) - 2}$ *Substitute 6 for x.*
$= \sqrt{64}$
$= 8$

b) $g(-5) = -\sqrt{-3(-5) + 1}$ *Substitute −5 for r.*
$= -\sqrt{16}$
$= -4$

$g(7) = -\sqrt{-3(7) + 1}$ *Substitute 7 for r.*
$= -\sqrt{-20}$ *Not a real number.*

Thus, $g(7)$ is not a real number.

▶ **Now Try Exercise 77**

2 Find Cube Roots

The concept used to explain square roots can be expanded to explain cube roots. The index of a cube root is 3. The cube root of a number a is written $\sqrt[3]{a}$.

> **Cube Root**
>
> The **cube root** of a number a, written $\sqrt[3]{a}$, is the number b such that $b^3 = a$.

Examples

$$\sqrt[3]{8} = 2 \qquad \text{since } 2^3 = 8$$
$$\sqrt[3]{-27} = -3 \qquad \text{since } (-3)^3 = -27$$

For each real number, there is only one cube root. The cube root of a positive number is positive and the cube root of a negative number is negative. The cube root function, $f(x) = \sqrt[3]{x}$, has all real numbers as its domain.

EXAMPLE 2 ▶ For each function, find the indicated value(s).

a) $f(x) = \sqrt[3]{10x + 34}$, $f(3)$ **b)** $g(r) = \sqrt[3]{12r - 20}$, $g(-4)$ and $g(1)$

Solution

a) $f(3) = \sqrt[3]{10(3) + 34}$ *Substitute 3 for x.*

$$= \sqrt[3]{64} = 4$$

b) $g(-4) = \sqrt[3]{12(-4) - 20}$ *Substitute −4 for r.*

$$= \sqrt[3]{-68}$$
$$\approx -4.081655102 \qquad \textit{From a calculator}$$

$$g(1) = \sqrt[3]{12(1) - 20} \qquad \textit{Substitute 1 for r.}$$
$$= \sqrt[3]{-8}$$
$$= -2$$

▶ **Now Try Exercise 83**

The Cube Root Function

Figure 11.2 shows the graph of $y = \sqrt[3]{x}$. To obtain this graph, we substituted values for x and found the corresponding values of $f(x)$ or y.

x	y
-8	-2
-1	-1
0	0
1	1
8	2

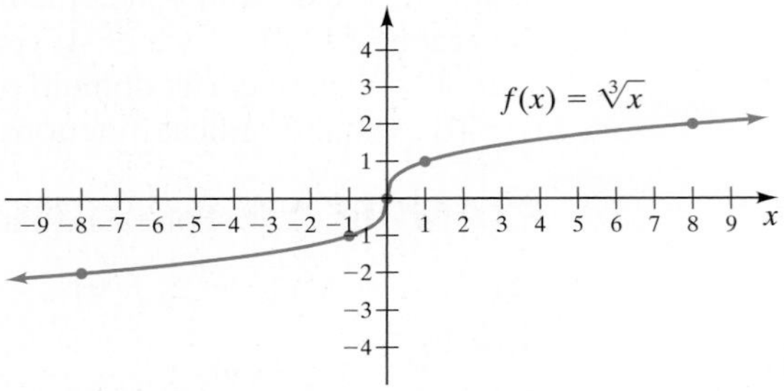

FIGURE 11.2

Notice that both the domain and range are all real numbers, $\mathbb{R}$. We will ask you to graph cube root functions on your graphing calculator in the exercise set.

3 Understand Odd and Even Roots

Up to this point we have discussed square and cube roots. Other radical expressions have different indices. For example, in the expression $\sqrt[5]{xy}$, (read "the fifth root of xy") the index is 5 and the radicand is xy.

Radical expressions that have indices of $2, 4, 6, \ldots$, or any even integer are **even roots**. Square roots are even roots since their index is 2. Radical expressions that have indices of $3, 5, 7, \ldots$, or any odd integer are **odd roots**.

Even Indices

The nth root of a, $\sqrt[n]{a}$, where n is an *even index* and a is a nonnegative real number, is the nonnegative real number b such that $b^n = a$.

Examples of Even Roots

$\sqrt{9} = 3$ \qquad since $3^2 = 3 \cdot 3 = 9$

$\sqrt[4]{16} = 2$ \qquad since $2^4 = 2 \cdot 2 \cdot 2 \cdot 2 = 16$

$\sqrt[6]{729} = 3$ \qquad since $3^6 = 3 \cdot 3 \cdot 3 \cdot 3 \cdot 3 \cdot 3 = 729$

$\sqrt[4]{\dfrac{1}{256}} = \dfrac{1}{4}$ \qquad since $\left(\dfrac{1}{4}\right)^4 = \left(\dfrac{1}{4}\right)\left(\dfrac{1}{4}\right)\left(\dfrac{1}{4}\right)\left(\dfrac{1}{4}\right) = \dfrac{1}{256}$

Any real number when raised to an even power results in a positive real number. Thus, *when the index of a radical is even, the radicand must be nonnegative for the radical to be a real number.*

Helpful Hint

There is an important difference between $-\sqrt[4]{16}$ and $\sqrt[4]{-16}$. The number $-\sqrt[4]{16}$ is the opposite of $\sqrt[4]{16}$. Because $\sqrt[4]{16} = 2$, $-\sqrt[4]{16} = -2$. However, $\sqrt[4]{-16}$ is not a real number since no real number when raised to the fourth power equals -16.

$$-\sqrt[4]{16} = -(\sqrt[4]{16}) = -2$$

$\sqrt[4]{-16}$ is not a real number.

Odd Indices

The nth root of a, $\sqrt[n]{a}$, where n is an *odd index* and a is *any real number*, is the real number b such that $b^n = a$.

Examples of Odd Roots

$\sqrt[3]{8} = 2$ \qquad since $2^3 = 2 \cdot 2 \cdot 2 = 8$

$\sqrt[3]{-8} = -2$ \qquad since $(-2)^3 = (-2)(-2)(-2) = -8$

$\sqrt[5]{243} = 3$ \qquad since $3^5 = 3 \cdot 3 \cdot 3 \cdot 3 \cdot 3 = 243$

$\sqrt[5]{-243} = -3$ \qquad since $(-3)^5 = (-3)(-3)(-3)(-3)(-3) = -243$

An odd root of a positive number is a positive number, and an odd root of a negative number is a negative number.

It is important to realize that a radical with an even index must have a nonnegative radicand if it is to be a real number. A radical with an odd index will be a real number with any real number as its radicand. Note that $\sqrt[n]{0} = 0$, regardless of whether n is an odd or even index.

EXAMPLE 3 ▶ Indicate whether or not each radical expression is a real number. If the expression is a real number, find its value.

a) $\sqrt[4]{-81}$ \qquad **b)** $-\sqrt[4]{81}$ \qquad **c)** $\sqrt[5]{-32}$ \qquad **d)** $-\sqrt[5]{-32}$

Solution

a) Not a real number. Even roots of negative numbers are not real numbers.

b) Real number, $-\sqrt[4]{81} = -(\sqrt[4]{81}) = -(3) = -3$

c) Real number, $\sqrt[5]{-32} = -2$ since $(-2)^5 = -32$

d) Real number, $-\sqrt[5]{-32} = -(-2) = 2$

▶ Now Try Exercise 21

Table 11.1 summarizes the information about even and odd roots.

Table 11.1		
	n Is Even	**n Is Odd**
$a > 0$	$\sqrt[n]{a}$ is a positive real number.	$\sqrt[n]{a}$ is a positive real number.
$a < 0$	$\sqrt[n]{a}$ is not a real number.	$\sqrt[n]{a}$ is a negative real number.
$a = 0$	$\sqrt[n]{0} = 0$	$\sqrt[n]{0} = 0$

4 Evaluate Radicals Using Absolute Value

You may think that $\sqrt{a^2} = a$, but this is not necessarily true. Below we evaluate $\sqrt{a^2}$ for $a = 2$ and $a = -2$. You will see that when $a = -2$, $\sqrt{a^2} \neq a$.

$a = 2$: $\sqrt{a^2} = \sqrt{2^2} = \sqrt{4} = 2$ *Note that $\sqrt{2^2} = 2$.*

$a = -2$: $\sqrt{a^2} = \sqrt{(-2)^2} = \sqrt{4} = 2$ *Note that $\sqrt{(-2)^2} \neq -2$.*

By examining these examples and other examples we could make up, we can reason that $\sqrt{a^2}$ *will always be a positive real number* for any nonzero real number a. Recall from Section 1.5 that the *absolute value* of any real number a, or $|a|$, is also a positive number for any nonzero number. We use these facts to reason that

Radicals and Absolute Value

For any real number a,

$$\sqrt{a^2} = |a|$$

This indicates that the principal square root of a^2 is the absolute value of a.

EXAMPLE 4 ▶ Use absolute value to evaluate.

a) $\sqrt{9^2}$ **b)** $\sqrt{0^2}$ **c)** $\sqrt{(15.7)^2}$

Solution

a) $\sqrt{9^2} = |9| = 9$ **b)** $\sqrt{0^2} = |0| = 0$ **c)** $\sqrt{(15.7)^2} = |15.7| = 15.7$

▶ **Now Try Exercise 41**

When simplifying a square root, if the radicand contains a variable and we are not sure that the radicand is positive, we need to use absolute value signs when simplifying.

EXAMPLE 5 ▶ Simplify.

a) $\sqrt{(x + 8)^2}$ **b)** $\sqrt{16x^2}$ **c)** $\sqrt{25y^6}$ **d)** $\sqrt{a^2 - 6a + 9}$

Solution   Each square root has a radicand that contains a variable. Since we do not know the value of the variable, we do not know whether it is positive or negative. Therefore, we must use absolute value signs when simplifying.

a) $\sqrt{(x + 8)^2} = |x + 8|$

b) Write $16x^2$ as $(4x)^2$, then simplify.

$$\sqrt{16x^2} = \sqrt{(4x)^2} = |4x|$$

c) Write $25y^6$ as $(5y^3)^2$, then simplify.

$$\sqrt{25y^6} = \sqrt{(5y^3)^2} = |5y^3|$$

d) Notice that $a^2 - 6a + 9$ is a perfect square trinomial. Write the trinomial as the square of a binomial, then simplify.

$$\sqrt{a^2 - 6a + 9} = \sqrt{(a - 3)^2} = |a - 3|$$

▶ **Now Try Exercise 63**

If you have a square root whose radicand contains a variable and are given instructions like "Assume all variables represent positive values and the radicand is nonnegative," then it is not necessary to use the absolute value sign when simplifying.

EXAMPLE 6 ▶ Simplify. Assume all variables represent positive values and the radicand is nonnegative.

a) $\sqrt{64x^2}$ **b)** $\sqrt{81p^4}$ **c)** $\sqrt{49x^6}$ **d)** $\sqrt{4x^2 - 12xy + 9y^2}$

Solution

a) $\sqrt{64x^2} = \sqrt{(8x)^2} = 8x$ *Write $64x^2$ as $(8x)^2$.*

b) $\sqrt{81p^4} = \sqrt{(9p^2)^2} = 9p^2$ *Write $81p^4$ as $(9p^2)^2$.*

c) $\sqrt{49x^6} = \sqrt{(7x^3)^2} = 7x^3$ *Write $49x^6$ as $(7x^3)^2$.*

d) $\sqrt{4x^2 - 12xy + 9y^2} = \sqrt{(2x - 3y)^2}$ *Write $4x^2 - 12xy + 9y^2$ as $(2x - 3y)^2$.*
$= 2x - 3y$

▶ **Now Try Exercise 67**

We only need to be concerned about adding absolute value signs when discussing square (and other even) roots. We do not need to use absolute value signs when the index is odd.

EXERCISE SET 11.1 Math XL MyMathLab
MathXL®   MyMathLab

Concept/Writing Exercises

1. a) How many square roots does every positive real number have? Name them.

b) Find all square roots of the number 49.

c) In this text, when we refer to "the square root," which square root are we referring to?

d) Find the square root of 49.

2. a) What are even roots? Give an example of an even root.

b) What are odd roots? Give an example of an odd root.

3. Explain why $\sqrt{-81}$ is not a real number.

4. Will a radical expression with an odd index and a real number as the radicand always be a real number? Explain your answer.

5. Will a radical expression with an even index and a real number as the radicand always be a real number? Explain your answer.

6. a) To what is $\sqrt{a^2}$ equal?

b) To what is $\sqrt{a^2}$ equal if we know $a \geq 0$?

7. a) Evaluate $\sqrt{a^2}$ for $a = 1.3$.

b) Evaluate $\sqrt{a^2}$ for $a = -1.3$.

8. a) Evaluate $\sqrt{a^2}$ for $a = 5.72$.

b) Evaluate $\sqrt{a^2}$ for $a = -5.72$.

9. a) Evaluate $\sqrt[3]{27}$.

b) Evaluate $-\sqrt[3]{27}$.

c) Evaluate $\sqrt[3]{-27}$.

10. a) Evaluate $\sqrt[4]{16}$

b) Evaluate $-\sqrt[4]{16}$

c) Evaluate $\sqrt[4]{-16}$.

Practice the Skills

Evaluate each radical expression if it is a real number. Use a calculator to approximate irrational numbers to the nearest hundredth. If the expression is not a real number, so state.

11. $\sqrt{36}$

12. $-\sqrt{36}$

13. $\sqrt[3]{-64}$

14. $\sqrt[3]{125}$

15. $\sqrt[3]{-125}$

16. $-\sqrt[3]{-125}$

17. $\sqrt[5]{-1}$

18. $-\sqrt[5]{-1}$

19. $\sqrt[5]{1}$

20. $\sqrt[6]{64}$

21. $\sqrt[6]{-64}$

22. $\sqrt[4]{-81}$

23. $\sqrt[3]{-343}$

24. $\sqrt{121}$

25. $\sqrt{-36}$

26. $\sqrt{45.3}$

27. $\sqrt{-45.3}$

28. $\sqrt{53.9}$

29. $\sqrt{\dfrac{1}{25}}$

30. $\sqrt{-\dfrac{1}{25}}$

31. $\sqrt[3]{\dfrac{1}{8}}$

32. $\sqrt[3]{-\dfrac{1}{8}}$

33. $\sqrt{\dfrac{4}{49}}$

34. $\sqrt[3]{\dfrac{8}{27}}$

35. $\sqrt[3]{-\dfrac{8}{27}}$

36. $\sqrt[4]{-8.9}$

37. $-\sqrt[4]{18.2}$

38. $\sqrt[5]{93}$

Use absolute value to evaluate.

39. $\sqrt{7^2}$

40. $\sqrt{(-7)^2}$

41. $\sqrt{19^2}$

42. $\sqrt{(-19)^2}$

43. $\sqrt{119^2}$

44. $\sqrt{(-119)^2}$

45. $\sqrt{(235.23)^2}$

46. $\sqrt{(-201.5)^2}$

47. $\sqrt{(0.06)^2}$

48. $\sqrt{(-0.19)^2}$

49. $\sqrt{\left(\dfrac{12}{13}\right)^2}$

50. $\sqrt{\left(-\dfrac{101}{319}\right)^2}$

Write as an absolute value.

51. $\sqrt{(x-4)^2}$

52. $\sqrt{(a+10)^2}$

53. $\sqrt{(x-3)^2}$

54. $\sqrt{(7a-11b)^2}$

55. $\sqrt{(3x^2-1)^2}$

56. $\sqrt{(7y^2-3y)^2}$

57. $\sqrt{(6a^3-5b^4)^2}$

58. $\sqrt{(9y^4-2z^3)^2}$

Use absolute value to simplify. You may need to factor first.

59. $\sqrt{a^{14}}$

60. $\sqrt{y^{22}}$

61. $\sqrt{z^{32}}$

62. $\sqrt{x^{200}}$

63. $\sqrt{a^2-8a+16}$

64. $\sqrt{x^2-12x+36}$

65. $\sqrt{9a^2+12ab+4b^2}$

66. $\sqrt{4x^2+20xy+25y^2}$

Simplify. Assume that all variables represent positive values and that the radicand is nonnegative.

67. $\sqrt{49x^2}$

68. $\sqrt{100a^4}$

69. $\sqrt{16c^6}$

70. $\sqrt{121z^8}$

71. $\sqrt{x^2+4x+4}$

72. $\sqrt{9a^2-6a+1}$

73. $\sqrt{4x^2+4xy+y^2}$

74. $\sqrt{16b^2-40bc+25c^2}$

Find the indicated value of each function. Use your calculator to approximate irrational numbers. Round irrational numbers to the nearest thousandth.

75. $f(x)=\sqrt{5x-6}, f(2)$

76. $f(c)=\sqrt{7c+1}, f(5)$

77. $q(x)=\sqrt{76-3x}, q(4)$

78. $q(b)=\sqrt{9b+34}, q(-1)$

79. $t(a)=\sqrt{-15a-9}, t(-6)$

80. $f(a)=\sqrt{14a-36}, f(4)$

81. $g(x)=\sqrt{64-8x}, g(-3)$

82. $p(x)=\sqrt[3]{8x+9}, p(2)$

83. $h(x)=\sqrt[3]{9x^2+4}, h(4)$

84. $k(c)=\sqrt[4]{16c-5}, k(6)$

85. $f(x)=\sqrt[3]{-2x^2+x-6}, f(-3)$

86. $t(x)=\sqrt[4]{2x^3-3x^2+6x}, t(2)$

Problem Solving

87. Find $f(81)$ if $f(x)=x+\sqrt{x}+7$.

88. Find $g(25)$ if $g(x)=x^2+\sqrt{x}-13$.

89. Find $t(18)$ if $t(x)=\dfrac{x}{2}+\sqrt{2x}-4$.

90. Find $m(36)$ if $m(x)=\dfrac{x}{3}+\sqrt{4x}+10$.

91. Find $k(8)$ if $k(x)=x^2+\sqrt{\dfrac{x}{2}}-21$.

92. Find $r(45)$ if $r(x)=\dfrac{x}{9}+\sqrt{\dfrac{x}{5}}+13$.

93. Select a value for x for which $\sqrt{(2x+1)^2}\neq 2x+1$.

94. Select a value for x for which $\sqrt{(5x-3)^2}\neq 5x-3$.

95. For what values of x will $\sqrt{(x-1)^2}=x-1$? Explain how you determined your answer.

96. For what values of x will $\sqrt{(x+3)^2}=x+3$? Explain how you determined your answer.

97. For what values of x will $\sqrt{(2x-6)^2}=2x-6$? Explain how you determined your answer.

98. For what values of x will $\sqrt{(3x-8)^2}=3x-8$? Explain how you determined your answer.

99. a) For what values of a is $\sqrt{a^2}=|a|$?
b) For what values of a is $\sqrt{a^2}=a$?
c) For what values of a is $\sqrt[3]{a^3}=a$?

100. Under what circumstances is the expression $\sqrt[n]{x}$ not a real number?

101. Explain why the expression $\sqrt[n]{x^n}$ is a real number for any real number x.

102. Under what circumstances is the expression $\sqrt[n]{x^m}$ not a real number?

103. Find the domain of $\dfrac{\sqrt{x+5}}{\sqrt[3]{x+5}}$. Explain how you determined your answer.

104. Find the domain of $\dfrac{\sqrt[3]{x-2}}{\sqrt[6]{x+1}}$. Explain how you determined your answer.

*By considering the domains of the functions in Exercises 105 through 108, match each function with its graph, labeled **a)** through **d)**.*

105. $f(x)=\sqrt{x}$

106. $f(x)=\sqrt{x^2}$

107. $f(x)=\sqrt{x-5}$

108. $f(x)=\sqrt{x+5}$

a)

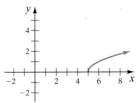

b)

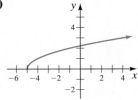

c)

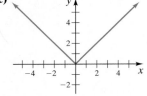

d)

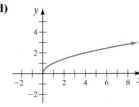

109. Give a radical function whose domain is $\{x \mid x \geq 8\}$.

110. Give a radical function whose domain is $\{x \mid x \leq 5\}$.

111. If $f(x) = -\sqrt{x}$, can $f(x)$ ever be
  a) greater than 0,
  b) equal to 0,
  c) less than 0?
  Explain your answers.

112. If $f(x) = \sqrt{x + 5}$, can $f(x)$ ever be
  a) less than 0,
  b) equal to 0,
  c) greater than 0?
  Explain your answers.

113. Velocity of an Object The velocity, V, of an object, in feet per second, after it has fallen a distance, h, in feet, can be found by the formula $V = \sqrt{64.4h}$. A pile driver is a large mass that is used as a hammer to drive pilings into soft earth to support a building or other structure.

With what velocity will the hammer hit the piling if it falls from

  a) 20 feet above the top of the piling?
  b) 40 feet above the top of the piling?

114. Wave Action Scripps Institute of Oceanography in La Jolla, California, developed the formula for relating wind speed, u, in knots, with the height, H, in feet, of the waves the wind produces in certain areas of the ocean. This formula is

$$u = \sqrt{\frac{H}{0.026}}$$

If waves produced by a storm have a height of 15 feet, what is the wind speed producing the waves?

115. Graph $f(x) = \sqrt{x + 1}$.

116. Graph $g(x) = -\sqrt{x}$.

117. Graph $g(x) = \sqrt{x} + 1$.

118. Graph $f(x) = \sqrt{x - 2}$.

For Exercises 119–124, use your graphing calculator.

119. Check the graph drawn in Exercise 115.

120. Check the graph drawn in Exercise 117.

121. Determine whether the domain you gave in Exercise 103 is correct.

122. Determine whether the domain you gave in Exercise 104 is correct.

123. Graph $y = \sqrt[3]{x + 4}$.

124. Graph $f(x) = \sqrt[3]{2x - 3}$.

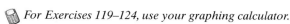

Group Activity

In this activity, you will determine the conditions under which certain properties of radicals are true. We will discuss these properties later in this chapter. Discuss and answer these exercises as a group.

125. The property $\sqrt[n]{a} \cdot \sqrt[n]{b} = \sqrt[n]{ab}$, called the *multiplication property for radicals*, is true for certain real numbers a and b. By substituting values for a and b, determine under what conditions this property is true.

126. The property $\dfrac{\sqrt[n]{a}}{\sqrt[n]{b}} = \sqrt[n]{\dfrac{a}{b}}$, called the *division property for radicals*, is true for certain real numbers a and b. By substituting values for a and b, determine under what conditions this property is true.

Cumulative Review Exercises

Factor.

[6.2] **127.** $9ax - 3bx + 12ay - 4by$

[6.4] **128.** $3x^3 - 18x^2 + 24x$

 129. $8x^4 + 10x^2 - 3$

[6.5] **130.** $x^3 - \dfrac{8}{27}y^3$

11.2 Rational Exponents

1 Change a radical expression to an exponential expression.

2 Simplify radical expressions.

3 Apply the rules of exponents to rational and negative exponents.

4 Factor expressions with rational exponents.

1 Change a Radical Expression to an Exponential Expression

In this section we discuss changing radical expressions to exponential expressions and vice versa. When you see a rational exponent, you should realize that the expression can be written as a radical expression by using the following procedure.

> **Exponential Form of $\sqrt[n]{a}$**
>
> $$\sqrt[n]{a} = a^{1/n}$$
>
> When a is nonnegative, n can be any index.
> When a is negative, n must be odd.

For the remainder of this chapter, unless you are instructed otherwise, assume that all variables in radicands represent nonnegative real numbers and that the radicand is nonnegative. With this assumption, we will not need to state that the variable is nonnegative whenever we have a radical with an even index. This will allow us to write many answers without absolute value signs.

EXAMPLE 1 ▶ Write each expression in exponential form (with rational exponents).

a) $\sqrt{7}$ b) $\sqrt[3]{15ab}$ c) $\sqrt[7]{-4x^2y^5}$ d) $\sqrt[8]{\dfrac{5x^7}{2z^{11}}}$

Solution

a) $\sqrt{7} = 7^{1/2}$ *Recall that the index of a square root is 2.*

b) $\sqrt[3]{15ab} = (15ab)^{1/3}$ c) $\sqrt[7]{-4x^2y^5} = (-4x^2y^5)^{1/7}$ d) $\sqrt[8]{\dfrac{5x^7}{2z^{11}}} = \left(\dfrac{5x^7}{2z^{11}}\right)^{1/8}$

▶ **Now Try Exercise 19**

Exponential expressions can be converted to radical expressions by reversing the procedure.

EXAMPLE 2 ▶ Write each expression in radical form (without rational exponents).

a) $9^{1/2}$ b) $(-8)^{1/3}$ c) $y^{1/4}$ d) $(10x^2y)^{1/7}$ e) $5rs^{1/2}$

Solution

a) $9^{1/2} = \sqrt{9} = 3$ b) $(-8)^{1/3} = \sqrt[3]{-8} = -2$

c) $y^{1/4} = \sqrt[4]{y}$ d) $(10x^2y)^{1/7} = \sqrt[7]{10x^2y}$ e) $5rs^{1/2} = 5r\sqrt{s}$

▶ **Now Try Exercise 33**

2 Simplify Radical Expressions

We can expand the preceding rule so that radicals of the form $\sqrt[n]{a^m}$ can be written as exponential expressions. Consider $a^{2/3}$. We can write $a^{2/3}$ as $(a^{1/3})^2$ or $(a^2)^{1/3}$. This suggests $a^{2/3} = (\sqrt[3]{a})^2 = \sqrt[3]{a^2}$.

> **Exponential Form of $\sqrt[n]{a^m}$**
>
> For any nonnegative number a, and integers m and n,
>
> $$\sqrt[n]{a^m} = \left(\sqrt[n]{a}\right)^m = a^{m/n} \; \leftarrow \text{Index}$$
>
> *Power*

This rule can be used to change an expression from radical form to exponential form and vice versa. When changing a radical expression to exponential form, the *power* is placed in the *numerator*, and the *index or root* is placed in the *denominator* of the rational exponent. Thus, for example, $\sqrt[3]{x^4}$ can be written $x^{4/3}$. Also $(\sqrt[5]{y})^2$ can be written $y^{2/5}$. Additional examples follow.

Examples

$$\sqrt{y^3} = y^{3/2} \qquad \sqrt[3]{z^2} = z^{2/3} \qquad \sqrt[5]{2^8} = 2^{8/5}$$
$$(\sqrt{p})^3 = p^{3/2} \qquad (\sqrt[4]{x})^3 = x^{3/4} \qquad (\sqrt[4]{7})^3 = 7^{3/4}$$

By this rule, for nonnegative values of the variable we can write

$$\sqrt{x^5} = (\sqrt{x})^5 \qquad (\sqrt[4]{p})^3 = \sqrt[4]{p^3}$$

EXAMPLE 3 ▶ Write each expression in exponential form (with rational exponents) and then simplify.

a) $\sqrt[4]{x^{12}}$ b) $(\sqrt[3]{y})^{15}$ c) $(\sqrt[6]{x})^{12}$

Solution

a) $\sqrt[4]{x^{12}} = x^{12/4} = x^3$ b) $(\sqrt[3]{y})^{15} = y^{15/3} = y^5$

c) $(\sqrt[6]{x})^{12} = x^{12/6} = x^2$

▶ **Now Try Exercise 45**

Exponential expressions with rational exponents can be converted to radical expressions by reversing the procedure. The *numerator* of the rational exponent is the *power*, and the *denominator* of the rational exponent is the *index or root* of the radical expression. Here are some examples.

Examples

$$x^{1/2} = \sqrt{x} \qquad\qquad\qquad 5^{1/3} = \sqrt[3]{5}$$
$$7^{2/3} = \sqrt[3]{7^2} \text{ or } (\sqrt[3]{7})^2 \qquad\qquad y^{3/10} = \sqrt[10]{y^3} \text{ or } (\sqrt[10]{y})^3$$
$$x^{9/5} = \sqrt[5]{x^9} \text{ or } (\sqrt[5]{x})^9 \qquad\qquad z^{10/3} = \sqrt[3]{z^{10}} \text{ or } (\sqrt[3]{z})^{10}$$

Notice that you may choose, for example, to write $6^{2/3}$ as either $\sqrt[3]{6^2}$ or $(\sqrt[3]{6})^2$.

EXAMPLE 4 ▶ Write each expression in radical form (without rational exponents).

a) $x^{2/5}$ b) $(3ab)^{5/4}$

Solution

a) $x^{2/5} = \sqrt[5]{x^2} \text{ or } (\sqrt[5]{x})^2$ b) $(3ab)^{5/4} = \sqrt[4]{(3ab)^5} \text{ or } (\sqrt[4]{3ab})^5$

▶ **Now Try Exercise 35**

EXAMPLE 5 ▶ Simplify.

a) $4^{3/2}$ b) $\sqrt[6]{(49)^3}$ c) $\sqrt[4]{(xy)^{20}}$ d) $(\sqrt[15]{z})^5$

Solution

a) Sometimes an expression with a rational exponent can be simplified more easily by writing the expression as a radical, as illustrated.

$$4^{3/2} = (\sqrt{4})^3 \qquad \textit{Write as a radical.}$$
$$= (2)^3$$
$$= 8$$

b) Sometimes a radical expression can be simplified more easily by writing the expression with rational exponents, as illustrated in parts **b)** through **d)**.

$$\sqrt[6]{(49)^3} = 49^{3/6} \qquad \textit{Write with a rational exponent.}$$
$$= 49^{1/2} \qquad \textit{Reduce exponent.}$$
$$= \sqrt{49} \qquad \textit{Write as a radical.}$$
$$= 7 \qquad \textit{Simplify.}$$

c) $\sqrt[4]{(xy)^{20}} = (xy)^{20/4} = (xy)^5$

d) $(\sqrt[15]{z})^5 = z^{5/15} = z^{1/3}$ or $\sqrt[3]{z}$

▶ **Now Try Exercise 51**

Now let's consider $\sqrt[5]{x^5}$. When written in exponential form, this is $x^{5/5} = x^1 = x$. This leads to the following rule.

> ### Exponential form of $\sqrt[n]{a^n}$
> For any nonnegative real number a,
> $$\sqrt[n]{a^n} = (\sqrt[n]{a})^n = a^{n/n} = a$$

In the preceding box, we specified that a was nonnegative. If n is an even index and a is a negative real number, $\sqrt[n]{a^n} = |a|$ and not a. For example, $\sqrt[6]{(-5)^6} = |-5| = 5$. *Since we are assuming, except where noted otherwise, that variables in radicands represent nonnegative real numbers*, we may write $\sqrt[6]{x^6} = x$ and not $|x|$. This assumption also lets us write $\sqrt{x^2} = x$ and $(\sqrt[4]{z})^4 = z$.

Examples

$$\sqrt{3^2} = 3 \qquad\qquad \sqrt[4]{y^4} = y$$
$$\sqrt[6]{(xy)^6} = xy \qquad\qquad (\sqrt[5]{z})^5 = z$$

3 Apply the Rules of Exponents to Rational and Negative Exponents

In Section 5.1, we introduced and discussed the rules of exponents. In that section we only used exponents that were whole numbers. The rules still apply when the exponents are rational numbers. Let's review those rules now.

> ### Rules of Exponents
> For all real numbers a and b and all rational numbers m and n,
>
> | | |
> |---|---|
> | Product rule | $a^m \cdot a^n = a^{m+n}$ |
> | Quotient rule | $\dfrac{a^m}{a^n} = a^{m-n}, \quad a \neq 0$ |
> | Negative exponent rule | $a^{-m} = \dfrac{1}{a^m}, \quad a \neq 0$ |
> | Zero exponent rule | $a^0 = 1, \quad a \neq 0$ |
> | Raising a power to a power | $(a^m)^n = a^{m \cdot n}$ |
> | Raising a product to a power | $(ab)^m = a^m b^m$ |
> | Raising a quotient to a power | $\left(\dfrac{a}{b}\right)^m = \dfrac{a^m}{b^m}, \quad b \neq 0$ |

Using these rules, we will now work some problems in which the exponents are rational numbers.

EXAMPLE 6 ▸ Evaluate. **a)** $8^{-2/3}$ **b)** $(-27)^{-5/3}$ **c)** $(-32)^{-6/5}$

Solution

a) Begin by using the negative exponent rule.

$$8^{-2/3} = \frac{1}{8^{2/3}} \qquad \textit{Negative exponent rule}$$

$$= \frac{1}{(\sqrt[3]{8})^2} \qquad \textit{Write the denominator as a radical.}$$

$$= \frac{1}{2^2} \qquad \textit{Simplify the denominator.}$$

$$= \frac{1}{4}$$

b) $(-27)^{-5/3} = \dfrac{1}{(-27)^{5/3}} = \dfrac{1}{(\sqrt[3]{-27})^5} = \dfrac{1}{(-3)^5} = -\dfrac{1}{243}$

c) $(-32)^{-6/5} = \dfrac{1}{(-32)^{6/5}} = \dfrac{1}{(\sqrt[5]{-32})^6} = \dfrac{1}{(-2)^6} = \dfrac{1}{64}$

▸ **Now Try Exercise 81**

Note that Example 6 **a)** could have been evaluated as follows:

$$8^{-2/3} = \frac{1}{8^{2/3}} = \frac{1}{\sqrt[3]{8^2}} = \frac{1}{\sqrt[3]{64}} = \frac{1}{4}$$

However, it is generally easier to evaluate the root before applying the power.

Consider the expression $(-16)^{3/4}$. This can be rewritten as $(\sqrt[4]{-16})^3$. Since $(\sqrt[4]{-16})^3$ is not a real number, the expression $(-16)^{3/4}$ is not a real number.

In Chapter 5, we indicated that

$$\left(\frac{a}{b}\right)^{-n} = \left(\frac{b}{a}\right)^{n}$$

We use this fact in the following example.

EXAMPLE 7 ▸ Evaluate. **a)** $\left(\dfrac{9}{25}\right)^{-1/2}$ **b)** $\left(\dfrac{27}{8}\right)^{-1/3}$

Solution

a) $\left(\dfrac{9}{25}\right)^{-1/2} = \left(\dfrac{25}{9}\right)^{1/2} = \sqrt{\dfrac{25}{9}} = \dfrac{5}{3}$

b) $\left(\dfrac{27}{8}\right)^{-1/3} = \left(\dfrac{8}{27}\right)^{1/3} = \sqrt[3]{\dfrac{8}{27}} = \dfrac{2}{3}$

▸ **Now Try Exercise 83**

Helpful Hint

How do the expressions $-25^{1/2}$ and $(-25)^{1/2}$ differ?
Recall that $-x^2$ means $-(x^2)$. The same principle applies here.

$$-25^{1/2} = -(25)^{1/2} = -\sqrt{25} = -5$$

$$(-25)^{1/2} = \sqrt{-25}, \text{ which is not a real number.}$$

EXAMPLE 8 ▶ Simplify each expression and write the answer without negative exponents.

a) $a^{1/2} \cdot a^{-2/3}$ **b)** $(6x^2y^{-4})^{-1/2}$ **c)** $3.2x^{1/3}(2.4x^{1/2} + x^{-1/4})$ **d)** $\left(\dfrac{9x^{-4}z^{2/5}}{z^{-3/5}}\right)^{1/8}$

Solution

a) $a^{1/2} \cdot a^{-2/3} = a^{(1/2)-(2/3)}$ *Product rule*

$\qquad\qquad\quad = a^{-1/6}$ *Find the LCD and subtract the exponents.*

$\qquad\qquad\quad = \dfrac{1}{a^{1/6}}$ *Negative exponent rule*

b) $(6x^2y^{-4})^{-1/2} = 6^{-1/2}x^{2(-1/2)}y^{-4(-1/2)}$ *Raise the product to a power.*

$\qquad\qquad\qquad = 6^{-1/2}x^{-1}y^2$ *Multiply the exponents.*

$\qquad\qquad\qquad = \dfrac{y^2}{6^{1/2}x}\left(\text{or } \dfrac{y^2}{x\sqrt{6}}\right)$ *Negative exponent rule*

c) Begin by using the distributive property.

$3.2x^{1/3}(2.4x^{1/2} + x^{-1/4}) = (3.2x^{1/3})(2.4x^{1/2}) + (3.2x^{1/3})(x^{-1/4})$ *Distributive property*

$\qquad\qquad\qquad\qquad\quad = (3.2)(2.4)(x^{(1/3)+(1/2)}) + 3.2x^{(1/3)-(1/4)}$ *Product rule*

$\qquad\qquad\qquad\qquad\quad = 7.68x^{5/6} + 3.2x^{1/12}$

d) $\left(\dfrac{9x^{-4}z^{2/5}}{z^{-3/5}}\right)^{1/8} = (9x^{-4}z^{(2/5)-(-3/5)})^{1/8}$ *Quotient rule*

$\qquad\qquad\qquad\quad = (9x^{-4}z)^{1/8}$ *Subtract the exponents.*

$\qquad\qquad\qquad\quad = 9^{1/8}x^{-4(1/8)}z^{1/8}$ *Raise the product to a power.*

$\qquad\qquad\qquad\quad = 9^{1/8}x^{-4/8}z^{1/8}$ *Multiply the exponents.*

$\qquad\qquad\qquad\quad = 9^{1/8}x^{-1/2}z^{1/8}$ *Simplify exponent.*

$\qquad\qquad\qquad\quad = \dfrac{9^{1/8}z^{1/8}}{x^{1/2}}$ *Negative exponent rule*

▶ **Now Try Exercise 105**

EXAMPLE 9 ▶ Simplify. **a)** $\sqrt[15]{(7y)^5}$ **b)** $\left(\sqrt[4]{a^2b^3c}\right)^{20}$ **c)** $\sqrt[4]{\sqrt[3]{x}}$

Solution

a) $\sqrt[15]{(7y)^5} = (7y)^{5/15}$ *Write with a rational exponent.*

$\qquad\qquad\; = (7y)^{1/3}$ *Simplify the exponent.*

$\qquad\qquad\; = \sqrt[3]{7y}$ *Write as a radical.*

b) $\left(\sqrt[4]{a^2b^3c}\right)^{20} = (a^2b^3c)^{20/4}$ *Write with a rational exponent.*

$\qquad\qquad\qquad = (a^2b^3c)^5$

$\qquad\qquad\qquad = a^{10}b^{15}c^5$ *Raise the product to a power.*

c) $\sqrt[4]{\sqrt[3]{x}} = \sqrt[4]{x^{1/3}}$ *Write $\sqrt[3]{x}$ as $x^{1/3}$.*

$\qquad\quad = (x^{1/3})^{1/4}$ *Write with a rational exponent.*

$\qquad\quad = x^{1/12}$ *Raise the power to a power.*

$\qquad\quad = \sqrt[12]{x}$ *Write as a radical.*

▶ **Now Try Exercise 53**

Finding Roots or Expressions with Rational Exponents on a Scientific or Graphing Calculator

There are often many ways to evaluate an expression like $(\sqrt[5]{845})^3$ or $845^{3/5}$ on a calculator. The procedure to use varies from calculator to calculator. One general method is to write the expression with a rational exponent and use the $\boxed{y^x}$ or $\boxed{a^x}$ or $\boxed{\wedge}$ key with parentheses keys, as shown below.*

Scientific Calculator

To evaluate $845^{3/5}$, press

Answer displayed

845 $\boxed{y^x}$ $\boxed{(}$ 3 $\boxed{\div}$ 5 $\boxed{)}$ $\boxed{=}$ 57.03139903

To evaluate $845^{-3/5}$, press

Answer displayed

845 $\boxed{y^x}$ $\boxed{(}$ 3 $\boxed{+/_-}$ $\boxed{\div}$ 5 $\boxed{)}$ $\boxed{=}$ 0.017534201

Graphing Calculator

To evaluate $845^{3/5}$, press the following keys.

Answer displayed

845 $\boxed{\wedge}$ $\boxed{(}$ 3 $\boxed{\div}$ 5 $\boxed{)}$ $\boxed{\text{ENTER}}$ 57.03139903

To evaluate $845^{-3/5}$, press the following keys.

Answer displayed

845 $\boxed{\wedge}$ $\boxed{(}$ $\boxed{(-)}$ 3 $\boxed{\div}$ 5 $\boxed{)}$ $\boxed{\text{ENTER}}$ .0175342008

*Keystrokes to use vary from calculator to calculator. Read your calculator manual to learn how to evaluate exponential expressions.

4   Factor Expressions with Rational Exponents

In higher-level math courses, you may have to factor out a variable with a rational exponent. To factor a rational expression, factor out the term with the smallest (or most negative) exponent.

EXAMPLE 10 ▶ Factor $x^{2/5} + x^{-3/5}$.

Solution   The smallest of the two exponents is $-3/5$. Therefore, we will factor $x^{-3/5}$ from both terms. To find the new exponent on the variable that originally had the greater exponent, we subtract the exponent that was factored out from the original exponent.

Original exponent

Exponent factored out

$$x^{2/5} + x^{-3/5} = x^{-3/5}\left(x^{2/5-(-3/5)} + 1\right)$$

$$= x^{-3/5}(x^1 + 1)$$

$$= x^{-3/5}(x + 1)$$

$$= \frac{x + 1}{x^{3/5}}$$

We can check our factoring by multiplying.

$$x^{-3/5}(x + 1) = x^{-3/5} \cdot x + x^{-3/5} \cdot 1$$

$$= x^{(-3/5)+1} + x^{-3/5}$$

$$= x^{2/5} + x^{-3/5}$$

Since we obtained the original expression, the factoring is correct.

▶ **Now Try Exercise 135**

EXERCISE SET 11.2

MathXL®   MyMathLab

Concept/Writing Exercises

1. a) Under what conditions is $\sqrt[n]{a}$ a real number?

   b) When $\sqrt[n]{a}$ is a real number, how can it be expressed with rational exponents?

2. a) Under what conditions is $\sqrt[n]{a^m}$ a real number?

   b) Under what conditions is $(\sqrt[n]{a})^m$ a real number?

   c) When $\sqrt[n]{a^m}$ is a real number, how can it be expressed with rational exponents?

3. a) Under what conditions is $\sqrt[n]{a^n}$ a real number?

   b) When n is even and $a \geq 0$, what is $\sqrt[n]{a^n}$ equal to?

   c) When n is odd, what is $\sqrt[n]{a^n}$ equal to?

d) When n is even and a may be any real number, what is $\sqrt[n]{a^n}$ equal to?

4. a) Explain the difference between $-16^{1/2}$ and $(-16)^{1/2}$.

   b) Evaluate each expression in part **a)** if possible.

5. a) Is $(xy)^{1/2} = xy^{1/2}$? Explain.

   b) Is $(xy)^{-1/2} = \dfrac{x^{1/2}}{y^{-1/2}}$? Explain.

6. a) Is $\sqrt[6]{(3y)^3} = (3y)^{6/3}$? Explain.

   b) Is $\sqrt{(ab)^4} = (ab)^2$? Explain.

Practice the Skills

In this exercise set, assume that all variables represent positive real numbers. Write each expression in exponential form.

7. $\sqrt{a^3}$
8. $\sqrt{y^7}$
9. $\sqrt{9^5}$
10. $\sqrt[3]{y}$

11. $\sqrt[3]{z^5}$
12. $\sqrt[3]{x^{11}}$
13. $\sqrt[3]{7^{10}}$
14. $\sqrt[5]{9^{11}}$

15. $\sqrt[4]{9^7}$
16. $(\sqrt{x})^9$
17. $(\sqrt[3]{y})^{14}$
18. $\sqrt{ab^5}$

19. $\sqrt[4]{a^3b}$
20. $\sqrt[3]{x^4y}$
21. $\sqrt[4]{x^9z^5}$
22. $\sqrt[6]{y^{11}z}$

23. $\sqrt[6]{3a + 8b}$
24. $\sqrt[9]{3x + 5z^4}$
25. $\sqrt[5]{\dfrac{2x^6}{11y^7}}$
26. $\sqrt[4]{\dfrac{3a^8}{11b^5}}$

Write each expression in radical form.

27. $a^{1/2}$
28. $b^{2/3}$
29. $c^{5/2}$
30. $19^{1/2}$

31. $18^{5/3}$
32. $y^{17/6}$
33. $(24x^3)^{1/2}$
34. $(85a^3)^{5/2}$

35. $(11b^2c)^{3/5}$
36. $(8x^3y^2)^{7/4}$
37. $(6a + 5b)^{1/5}$
38. $(8x^2 + 9y)^{7/3}$

39. $(b^3 - d)^{-1/3}$
40. $(7x^2 - 2y^3)^{-1/6}$

Simplify each radical expression by changing the expression to exponential form. Write the answer in radical form when appropriate.

41. $\sqrt{a^6}$
42. $\sqrt[4]{a^8}$
43. $\sqrt[3]{x^9}$
44. $\sqrt[4]{x^{12}}$

45. $\sqrt[6]{y^2}$
46. $\sqrt[8]{b^4}$
47. $\sqrt[6]{y^3}$
48. $\sqrt[12]{z^4}$

49. $(\sqrt{19.3})^2$
50. $\sqrt[4]{(6.83)^4}$
51. $\left(\sqrt[3]{xy^2}\right)^{15}$
52. $\left(\sqrt[4]{a^4bc^3}\right)^{40}$

53. $(\sqrt[8]{xyz})^4$
54. $\left(\sqrt[9]{a^2bc^4}\right)^3$
55. $\sqrt{\sqrt{x}}$
56. $\sqrt{\sqrt[3]{a}}$

57. $\sqrt{\sqrt[4]{y}}$
58. $\sqrt[3]{\sqrt[4]{b}}$
59. $\sqrt[3]{\sqrt[3]{x^2y}}$
60. $\sqrt[4]{\sqrt[3]{7y}}$

61. $\sqrt{\sqrt[5]{a^9}}$
62. $\sqrt[5]{\sqrt[4]{ab}}$

Evaluate if possible. If the expression is not a real number, so state.

63. $25^{1/2}$
64. $121^{1/2}$
65. $64^{1/3}$
66. $81^{1/4}$

67. $64^{2/3}$
68. $27^{2/3}$
69. $(-49)^{1/2}$
70. $(-64)^{1/4}$

71. $\left(\dfrac{25}{9}\right)^{1/2}$
72. $\left(\dfrac{100}{49}\right)^{1/2}$
73. $\left(\dfrac{1}{8}\right)^{1/3}$
74. $\left(\dfrac{1}{32}\right)^{1/5}$

75. $-81^{1/2}$
76. $(-81)^{1/2}$
77. $-64^{1/3}$
78. $(-64)^{1/3}$

79. $64^{-1/3}$
80. $49^{-1/2}$
81. $16^{-3/2}$
82. $64^{-2/3}$

83. $\left(\dfrac{64}{27}\right)^{-1/3}$
84. $(-81)^{3/4}$
85. $(-100)^{3/2}$
86. $-\left(\dfrac{25}{49}\right)^{-1/2}$

87. $121^{1/2} + 169^{1/2}$
88. $49^{-1/2} + 36^{-1/2}$
89. $343^{-1/3} + 16^{-1/2}$
90. $16^{-1/2} - 256^{-3/4}$

Simplify. Write the answer in exponential form without negative exponents.

91. $x^4 \cdot x^{1/2}$

92. $x^6 \cdot x^{1/2}$

93. $\dfrac{x^{1/2}}{x^{1/3}}$

94. $x^{-6/5}$

95. $(x^{1/2})^{-2}$

96. $(a^{-1/3})^{-1/2}$

97. $(9^{-1/3})^0$

98. $\dfrac{x^4}{x^{-1/2}}$

99. $\dfrac{5y^{-1/3}}{60y^{-2}}$

100. $x^{-1/2}x^{-2/5}$

101. $4x^{5/3}3x^{-7/2}$

102. $(x^{-4/5})^{1/3}$

103. $\left(\dfrac{3}{24x}\right)^{1/3}$

104. $\left(\dfrac{54}{2x^4}\right)^{1/3}$

105. $\left(\dfrac{22x^{3/7}}{2x^{1/2}}\right)^2$

106. $\left(\dfrac{x^{-1/3}}{x^{-2}}\right)^2$

107. $\left(\dfrac{a^4}{4a^{-2/5}}\right)^{-3}$

108. $\left(\dfrac{27z^{1/4}y^3}{3z^{1/4}}\right)^{1/2}$

109. $\left(\dfrac{x^{3/4}y^{-3}}{x^{1/2}y^2}\right)^4$

110. $\left(\dfrac{250a^{-3/4}b^5}{2a^{-2}b^2}\right)^{2/3}$

Multiply.

111. $4z^{-1/2}(2z^4 - z^{1/2})$

112. $-3a^{-4/9}(5a^{1/9} - a^2)$

113. $5x^{-1}(x^{-4} + 4x^{-1/2})$

114. $-9z^{3/2}(z^{3/2} - z^{-3/2})$

115. $-6x^{5/3}(-2x^{1/2} + 3x^{1/3})$

116. $\dfrac{1}{2}x^{-2}(10x^{4/3} - 38x^{-1/2})$

Use a calculator to evaluate each expression. Give the answer to the nearest hundredth.

117. $\sqrt{180}$

118. $\sqrt[3]{168}$

119. $\sqrt[5]{402.83}$

120. $\sqrt[4]{1096}$

121. $93^{2/3}$

122. $38.2^{3/2}$

123. $1000^{-1/2}$

124. $8060^{-3/2}$

Problem Solving

125. Under what conditions will $\sqrt[n]{a^n} = (\sqrt[n]{a})^n = a$?

126. By selecting values for a and b, show that $(a^2 + b^2)^{1/2}$ *is not equal to* $a + b$.

127. By selecting values for a and b, show that $(a^{1/2} + b^{1/2})^2$ *is not equal to* $a + b$.

128. By selecting values for a and b, show that $(a^3 + b^3)^{1/3}$ is not equal to $a + b$.

129. By selecting values for a and b, show that $(a^{1/3} + b^{1/3})^3$ is not equal to $a + b$.

130. Determine whether $\sqrt[3]{\sqrt{x}} = \sqrt{\sqrt[3]{x}}$, $x \geq 0$.

Factor. Write the answer without negative exponents.

131. $x^{3/2} + x^{1/2}$

132. $x^{1/4} - x^{5/4}$

133. $y^{1/3} - y^{7/3}$

134. $x^{-1/2} + x^{1/2}$

135. $y^{-2/5} + y^{8/5}$

136. $a^{6/5} + a^{-4/5}$

In Exercises 137 through 142, use a calculator where appropriate.

137. Growing Bacteria The function, $B(t) = 2^{10} \cdot 2^t$, approximates the number of bacteria in a certain culture after t hours.

   a) The initial number of bacteria is determined when $t = 0$. What is the initial number of bacteria?

   b) How many bacteria are there after $\dfrac{1}{2}$ hour?

138. Carbon Dating Carbon dating is used by scientists to find the age of fossils, bones, and other items. The formula used in carbon dating is $P = P_0 2^{-t/5600}$, where P_0 represents the original amount of carbon 14 (C_{14}) present and P represents the amount of C_{14} present after t years. If 10 milligrams (mg) of C_{14} is present in an animal bone recently excavated, how many milligrams will be present in 5000 years?

139. Retirement Plans Each year more and more people contribute to their companies' 401(k) retirement plans. The total assets, $A(t)$, in U.S. 401(k) plans, in billions of dollars, can be approximated by the function $A(t) = 2.69t^{3/2}$, where t is years since 1993 and $1 \leq t \leq 16$. (Therefore, this function holds from 1994 through 2009.) Estimate the total assets in U.S. 401(k) plans in **a)** 2000 and **b)** 2009.

140. Internet Sales Retail Internet sales are increasing annually. The total amount, $I(t)$, in billions of dollars, of Internet sales can be approximated by the function $I(t) = 0.25t^{5/3}$, where t is years since 1999 and $1 \leq t \leq 9$. Find the total Internet sales in **a)** 2000 and **b)** 2008.

141. Evaluate $(3^{\sqrt{2}})^{\sqrt{2}}$. Explain how you determined your answer.

142. a) On your calculator, evaluate 3^π.

   b) Explain why your value from part **a)** does or does not make sense.

143. Find the domain of $f(x) = (x - 7)^{1/2}(x + 3)^{-1/2}$.

144. Find the domain of $f(x) = (x + 4)^{1/2}(x - 3)^{-1/2}$.

145. Assume that x can be any real number. Simplify $\sqrt[n]{(x - 6)^{2n}}$ if

   a) n is even.

   b) n is odd.

Determine the index to be placed in the shaded area to make the statement true. Explain how you determined your answer.

146. $\sqrt[4]{\sqrt[\square]{\sqrt{x}}} = x^{1/24}$

147. $\sqrt[4]{\sqrt[5]{\sqrt[\square]{\sqrt[3]{z}}}} = z^{1/120}$

148. a) Write $f(x) = \sqrt{2x + 3}$ in exponential form.

   b) On your grapher, check that the answer you gave in part **a)** is correct by graphing both $f(x)$ as given and the function you gave in exponential form.

Cumulative Review Exercises

[7.5] **149.** Simplify $\dfrac{a^{-2} + ab^{-1}}{ab^{-2} - a^{-2}b^{-1}}$.

[7.6] **150.** Solve $\dfrac{3x - 2}{x + 4} = \dfrac{2x + 1}{3x - 2}$.

[7.7] **151. Flying a Plane** Amy Mayfield can fly her plane 500 miles against the wind in the same time it takes her to fly 560 miles with the wind. If the wind blows at 25 miles per hour, find the speed of the plane in still air.

[8.2] **152.** Determine which of the following relations are also functions.

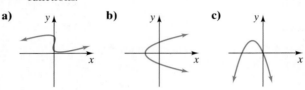

a)   **b)**   **c)**

11.3 Simplifying Radicals

1 Understand perfect powers.

2 Simplify radicals using the product rule for radicals.

3 Simplify radicals using the quotient rule for radicals.

1 Understand Perfect Powers

In this section, we will simplify radicals using the **product rule for radicals** and the **quotient rule for radicals**. However, before we introduce these rules let's look at **perfect powers**, which will help us with our discussion.

A number or expression is a **perfect square** if it is the square of an expression. Examples of perfect squares are illustrated below.

Perfect squares $1,\quad 4,\quad 9,\quad 16,\quad 25,\quad 36, \ldots$
                        $\downarrow\quad \downarrow\quad \downarrow\quad \downarrow\quad \downarrow\quad \downarrow$
Square of a number $1^2,\quad 2^2,\quad 3^2,\quad 4^2,\quad 5^2,\quad 6^2, \ldots$

Variables with exponents may also be perfect squares, as illustrated below.

Perfect squares $x^2,\quad x^4,\quad x^6,\quad x^8,\quad x^{10}, \ldots$
                        $\downarrow\quad\quad \downarrow\quad\quad \downarrow\quad\quad \downarrow\quad\quad \downarrow$
Square of an expression $(x)^2,\quad (x^2)^2,\quad (x^3)^2,\quad (x^4)^2,\quad (x^5)^2, \ldots$

Notice that the exponents on the variables in the perfect squares are all multiples of 2.

Just as there are perfect squares, there are perfect cubes. A number or expression is a **perfect cube** if it can be written as the cube of an expression. Examples of perfect cubes are illustrated below.

Perfect cubes $1,\quad 8,\quad 27,\quad 64,\quad 125,\quad 216, \ldots$
                        $\downarrow\quad \downarrow\quad \downarrow\quad \downarrow\quad \downarrow\quad \downarrow$
Cube of a number $1^3,\quad 2^3,\quad 3^3,\quad 4^3,\quad 5^3,\quad 6^3, \ldots$

Perfect cubes $x^3,\quad x^6,\quad x^9,\quad x^{12},\quad x^{15}, \ldots$
                        $\downarrow\quad\quad \downarrow\quad\quad \downarrow\quad\quad \downarrow\quad\quad \downarrow$
Cube of an expression $(x)^3,\quad (x^2)^3,\quad (x^3)^3,\quad (x^4)^3,\quad (x^5)^3, \ldots$

Notice that the exponents on the variables in the perfect cubes are all multiples of 3.

We can expand our discussion to perfect powers of a variable for any radicand. In general, the radicand x^n is a perfect power *when n is a multiple of the index* of the radicand (or where *n* is divisible by the index).

Example

Perfect powers of x^n for index n x^n, x^{2n}, x^{3n}, x^{4n}, x^{5n}, ...

For example, if the index of a radical expression is 5, then x^5, x^{10}, x^{15}, x^{20}, and so on, are perfect powers of the index.

Helpful Hint

A quick way to determine if a radicand x^n is a perfect power for an index is to determine if the exponent n is divisible by the index of the radical. For example, consider $\sqrt[5]{x^{20}}$. Since the exponent, 20, is divisible by the index, 5, x^{20} is a perfect fifth power. Now consider $\sqrt[6]{x^{20}}$. Since the exponent, 20, is not divisible by the index, 6, x^{20} is not a perfect sixth power. However, x^{18} and x^{24} are both perfect sixth powers since 6 divides both 18 and 24.

Notice that the square root of a perfect square simplifies to an expression without a radical sign, the cube root of a perfect cube simplifies to an expression without a radical sign, and so on.

Examples

$$\sqrt{36} = \sqrt{6^2} = 6^{2/2} = 6$$
$$\sqrt[3]{27} = \sqrt[3]{3^3} = 3^{3/3} = 3$$
$$\sqrt{x^6} = x^{6/2} = x^3$$
$$\sqrt[3]{z^{12}} = z^{12/3} = z^4$$
$$\sqrt[5]{n^{35}} = n^{35/5} = n^7$$

Now we are ready to discuss the product rule for radicals.

2 Simplify Radicals Using the Product Rule for Radicals

To introduce the **product rule for radicals**, observe that $\sqrt{4} \cdot \sqrt{9} = 2 \cdot 3 = 6$. Also, $\sqrt{4 \cdot 9} = \sqrt{36} = 6$. We see that $\sqrt{4} \cdot \sqrt{9} = \sqrt{4 \cdot 9}$. This is one example of the product rule for radicals.

Product Rule for Radicals

For nonnegative real numbers a and b,

$$\sqrt[n]{a} \cdot \sqrt[n]{b} = \sqrt[n]{ab}$$

Examples of the Product Rule for Radicals

$$\sqrt{20} = \begin{cases} \sqrt{1} \cdot \sqrt{20} \\ \sqrt{2} \cdot \sqrt{10} \\ \sqrt{4} \cdot \sqrt{5} \end{cases}$$

$\sqrt{20}$ can be factored into any of these forms.

$$\sqrt[3]{20} = \begin{cases} \sqrt[3]{1} \cdot \sqrt[3]{20} \\ \sqrt[3]{2} \cdot \sqrt[3]{10} \\ \sqrt[3]{4} \cdot \sqrt[3]{5} \end{cases}$$

$\sqrt[3]{20}$ can be factored into any of these forms.

$$\sqrt{x^7} = \begin{cases} \sqrt{x} \cdot \sqrt{x^6} \\ \sqrt{x^2} \cdot \sqrt{x^5} \\ \sqrt{x^3} \cdot \sqrt{x^4} \end{cases}$$

$\sqrt{x^7}$ can be factored into any of these forms.

$$\sqrt[3]{x^7} = \begin{cases} \sqrt[3]{x} \cdot \sqrt[3]{x^6} \\ \sqrt[3]{x^2} \cdot \sqrt[3]{x^5} \\ \sqrt[3]{x^3} \cdot \sqrt[3]{x^4} \end{cases}$$

$\sqrt[3]{x^7}$ can be factored into any of these forms.

Now that we have introduced the product rule for radicals, we will use this rule to simplify radicals. Here is a general procedure that can be used to simplify radicals using the product rule.

To Simplify Radicals Using the Product Rule

1. If the radicand contains a coefficient other than 1, write it as a product of two numbers, one of which is the largest perfect power for the index.
2. Write each variable factor as a product of two factors, one of which is the largest perfect power of the variable for the index.
3. Use the product rule to write the radical expression as a product of radicals. Place all the perfect powers (numbers and variables) under the same radical.
4. Simplify the radical containing the perfect powers.

If we are simplifying a *square* root, we will write the radicand as the product of the largest *perfect square* and another number. If we are simplifying a *cube* root, we will write the radicand as the product of the largest *perfect cube* and another number, and so on.

EXAMPLE 1 ▶ Simplify.   **a)** $\sqrt{32}$ **b)** $\sqrt{60}$ **c)** $\sqrt[3]{54}$ **d)** $\sqrt[4]{96}$

Solution   The radicands in this example contain no variables. We will follow step 1 of the procedure.

a) Since we are evaluating a square root, we look for the largest perfect square that divides 32. The largest perfect square that divides, or is a factor of, 32 is 16.
$$\sqrt{32} = \sqrt{16 \cdot 2} = \sqrt{16}\,\sqrt{2} = 4\sqrt{2}$$

b) The largest perfect square that is a factor of 60 is 4.
$$\sqrt{60} = \sqrt{4 \cdot 15} = \sqrt{4}\,\sqrt{15} = 2\sqrt{15}$$

c) The largest perfect cube that is a factor of 54 is 27.
$$\sqrt[3]{54} = \sqrt[3]{27 \cdot 2} = \sqrt[3]{27}\sqrt[3]{2} = 3\sqrt[3]{2}$$

d) The largest perfect fourth power that is a factor of 96 is 16.
$$\sqrt[4]{96} = \sqrt[4]{16 \cdot 6} = \sqrt[4]{16}\,\sqrt[4]{6} = 2\sqrt[4]{6}$$

▶ **Now Try Exercise 19**

Helpful Hint

In Example 1 **a)**, if you first thought that 4 was the largest perfect square that divided 32, you could proceed as follows:
$$\sqrt{32} = \sqrt{4 \cdot 8} = \sqrt{4}\,\sqrt{8} = 2\sqrt{8}$$
$$= 2\sqrt{4 \cdot 2} = 2\sqrt{4}\,\sqrt{2} = 2 \cdot 2\sqrt{2} = 4\sqrt{2}$$

Note that the final result is the same, but you must perform more steps. The lists of perfect squares and perfect cubes on page 674 can help you determine the largest perfect square or perfect cube that is a factor of a radicand.

In Example 1 **b)**, $\sqrt{15}$ can be factored as $\sqrt{5 \cdot 3}$; however, since neither 5 nor 3 is a perfect square, $\sqrt{15}$ cannot be simplified.

When the radicand is a perfect power for the index, the radical can be simplified by writing it in exponential form, as in Example 2.

EXAMPLE 2 ▶ Simplify.   **a)** $\sqrt{x^4}$ **b)** $\sqrt[3]{x^{12}}$ **c)** $\sqrt[5]{z^{40}}$

Solution

a) $\sqrt{x^4} = x^{4/2} = x^2$ **b)** $\sqrt[3]{x^{12}} = x^{12/3} = x^4$ **c)** $\sqrt[5]{z^{40}} = z^{40/5} = z^8$

▶ **Now Try Exercise 33**

EXAMPLE 3 ▸ Simplify.   **a)** $\sqrt{x^9}$ **b)** $\sqrt[5]{x^{23}}$ **c)** $\sqrt[4]{y^{33}}$

Solution   Because the radicands have coefficients of 1, we start with step 2 of the procedure.

a) The largest perfect square less than or equal to x^9 is x^8.

$$\sqrt{x^9} = \sqrt{x^8 \cdot x} = \sqrt{x^8} \cdot \sqrt{x} = x^{8/2}\sqrt{x} = x^4\sqrt{x}$$

b) The largest perfect fifth power less than or equal to x^{23} is x^{20}.

$$\sqrt[5]{x^{23}} = \sqrt[5]{x^{20} \cdot x^3} = \sqrt[5]{x^{20}}\,\sqrt[5]{x^3} = x^{20/5}\sqrt[5]{x^3} = x^4\sqrt[5]{x^3}$$

c) The largest perfect fourth power less than or equal to y^{33} is y^{32}.

$$\sqrt[4]{y^{33}} = \sqrt[4]{y^{32} \cdot y} = \sqrt[4]{y^{32}}\,\sqrt[4]{y} = y^{32/4}\sqrt[4]{y} = y^8\sqrt[4]{y}$$

▸ **Now Try Exercise 39**

If you observe the answers to Example 3, you will see that the exponent on the variable in the radicand is always less than the index. **When a radical is simplified, the radicand does not have a variable with an exponent greater than or equal to the index.**
In Example 3 **b)**, we simplified $\sqrt[5]{x^{23}}$. If we divide 23, the exponent in the radicand, by 5, the index, we obtain

$$\begin{array}{r} 4 \quad\longleftarrow \text{\textit{Quotient}} \\ 5\overline{)23} \\ \underline{20} \\ 3 \quad\longleftarrow \text{\textit{Remainder}} \end{array}$$

Notice that $\sqrt[5]{x^{23}}$ simplifies to $x^4\sqrt[5]{x^3}$ and

$$\textit{Quotient} \longrightarrow x^4\sqrt[5]{x^3} \longleftarrow \textit{Remainder}$$

When simplifying a radical, if you divide the exponent within the radical by the index, the quotient will be the exponent on the variable outside the radical sign and the remainder will be the exponent on the variable within the radical sign. Simplify Example 3 **c)** using this technique now.

EXAMPLE 4 ▸ Simplify.   **a)** $\sqrt{x^{12}y^{17}}$ **b)** $\sqrt[4]{x^6 y^{23}}$

Solution

a) x^{12} is a perfect square. The largest perfect square that is a factor of y^{17} is y^{16}. Write y^{17} as $y^{16} \cdot y$.

$$\sqrt{x^{12}y^{17}} = \sqrt{x^{12} \cdot y^{16} \cdot y} = \sqrt{x^{12}y^{16}}\,\sqrt{y}$$
$$= \sqrt{x^{12}}\sqrt{y^{16}}\,\sqrt{y}$$
$$= x^{12/2}y^{16/2}\sqrt{y}$$
$$= x^6 y^8 \sqrt{y}$$

b) We begin by finding the largest perfect fourth power factors of x^6 and y^{23}. For an index of 4, the largest perfect power that is a factor of x^6 is x^4. The largest perfect power that is a factor of y^{23} is y^{20}.

$$\sqrt[4]{x^6 y^{23}} = \sqrt[4]{x^4 \cdot x^2 \cdot y^{20} \cdot y^3}$$
$$= \sqrt[4]{x^4 y^{20} \cdot x^2 y^3}$$
$$= \sqrt[4]{x^4 y^{20}}\,\sqrt[4]{x^2 y^3}$$
$$= xy^5\sqrt[4]{x^2 y^3}$$

▸ **Now Try Exercise 51**

Often the steps where we change the radical expression to exponential form are done mentally, and those steps are not illustrated. For instance, in Example 4 **b)**, we changed $\sqrt[4]{x^4 y^{20}}$ to xy^5 mentally and did not show the intermediate steps.

EXAMPLE 5 ▶ Simplify. **a)** $\sqrt{80x^5y^{12}z^3}$ **b)** $\sqrt[3]{54x^{17}y^{25}}$

Solution

a) The largest perfect square that is a factor of 80 is 16. The largest perfect square that is a factor of x^5 is x^4. The expression y^{12} is a perfect square. The largest perfect square that is a factor of z^3 is z^2. Place all the perfect squares under the same radical, and then simplify.

$$\sqrt{80x^5y^{12}z^3} = \sqrt{16 \cdot 5 \cdot x^4 \cdot x \cdot y^{12} \cdot z^2 \cdot z}$$
$$= \sqrt{16x^4y^{12}z^2 \cdot 5xz}$$
$$= \sqrt{16x^4y^{12}z^2} \cdot \sqrt{5xz}$$
$$= 4x^2y^6z\sqrt{5xz}$$

b) The largest perfect cube that is a factor of 54 is 27. The largest perfect cube that is a factor of x^{17} is x^{15}. The largest perfect cube that is a factor of y^{25} is y^{24}.

$$\sqrt[3]{54x^{17}y^{25}} = \sqrt[3]{27 \cdot 2 \cdot x^{15} \cdot x^2 \cdot y^{24} \cdot y}$$
$$= \sqrt[3]{27x^{15}y^{24} \cdot 2x^2y}$$
$$= \sqrt[3]{27x^{15}y^{24}} \cdot \sqrt[3]{2x^2y}$$
$$= 3x^5y^8\sqrt[3]{2x^2y}$$

▶ **Now Try Exercise 57**

Helpful Hint

In Example 4 **b)**, we showed that

$$\sqrt[4]{x^6y^{23}} = xy^5\sqrt[4]{x^2y^3}$$

As mentioned on page 677, this radical can also be simplified by dividing the exponents on the variables in the radicand, 6 and 23, by the index, 4, and observing the quotients and remainders.

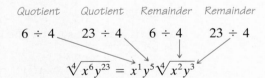

Can you explain why this procedure works? You may wish to use this procedure to work or check certain problems.

Now let's introduce the quotient rule for radicals.

3 Simplify Radicals Using the Quotient Rule for Radicals

In mathematics we sometimes need to simplify a quotient of two radicals. To do so we use the **quotient rule for radicals**.

Quotient Rule for Radicals

For nonnegative real numbers a and b,

$$\frac{\sqrt[n]{a}}{\sqrt[n]{b}} = \sqrt[n]{\frac{a}{b}}, \quad b \neq 0$$

Examples of the Quotient Rule for Radicals

$$\frac{\sqrt{18}}{\sqrt{3}} = \sqrt{\frac{18}{3}} \qquad\qquad \sqrt{\frac{9}{25}} = \frac{\sqrt{9}}{\sqrt{25}}$$

$$\frac{\sqrt{x^3}}{\sqrt{x}} = \sqrt{\frac{x^3}{x}} \qquad\qquad \sqrt{\frac{x^4}{y^2}} = \frac{\sqrt{x^4}}{\sqrt{y^2}}$$

$$\frac{\sqrt[3]{y^5}}{\sqrt[3]{y^2}} = \sqrt[3]{\frac{y^5}{y^2}} \qquad\qquad \sqrt[3]{\frac{z^9}{27}} = \frac{\sqrt[3]{z^9}}{\sqrt[3]{27}}$$

Examples 6 and 7 illustrate how to use the quotient rule to simplify radical expressions.

EXAMPLE 6 ▸ Simplify. **a)** $\dfrac{\sqrt{75}}{\sqrt{3}}$ **b)** $\dfrac{\sqrt[3]{24x}}{\sqrt[3]{3x}}$ **c)** $\dfrac{\sqrt[3]{x^4 y^7}}{\sqrt[3]{xy^{-5}}}$

Solution   In each part we use the quotient rule to write the quotient of radicals as a single radical. Then we simplify.

a) $\dfrac{\sqrt{75}}{\sqrt{3}} = \sqrt{\dfrac{75}{3}} = \sqrt{25} = 5$

b) $\dfrac{\sqrt[3]{24x}}{\sqrt[3]{3x}} = \sqrt[3]{\dfrac{24x}{3x}} = \sqrt[3]{8} = 2$

c) $\dfrac{\sqrt[3]{x^4 y^7}}{\sqrt[3]{xy^{-5}}} = \sqrt[3]{\dfrac{x^4 y^7}{xy^{-5}}}$ *Quotient rule for radicals*

$\qquad\qquad = \sqrt[3]{x^3 y^{12}}$ *Simplify the radicand.*

$\qquad\qquad = xy^4$

▸ **Now Try Exercise 93**

In Section 11.1, when we introduced radicals we indicated that $\sqrt{\dfrac{4}{9}} = \dfrac{2}{3}$ since $\dfrac{2}{3} \cdot \dfrac{2}{3} = \dfrac{4}{9}$.

The quotient rule may be helpful in evaluating square roots containing fractions as illustrated in Example 7 **a)**.

EXAMPLE 7 ▸ Simplify. **a)** $\sqrt{\dfrac{121}{25}}$ **b)** $\sqrt[3]{\dfrac{8x^4 y}{27xy^{10}}}$ **c)** $\sqrt[4]{\dfrac{18xy^5}{3x^9 y}}$

Solution   In each part we first simplify the radicand, if possible. Then we use the quotient rule to write the given radical as a quotient of radicals.

a) $\sqrt{\dfrac{121}{25}} = \dfrac{\sqrt{121}}{\sqrt{25}} = \dfrac{11}{5}$

b) $\sqrt[3]{\dfrac{8x^4 y}{27xy^{10}}} = \sqrt[3]{\dfrac{8x^3}{27y^9}} = \dfrac{\sqrt[3]{8x^3}}{\sqrt[3]{27y^9}} = \dfrac{2x}{3y^3}$

c) $\sqrt[4]{\dfrac{18xy^5}{3x^9 y}} = \sqrt[4]{\dfrac{6y^4}{x^8}} = \dfrac{\sqrt[4]{6y^4}}{\sqrt[4]{x^8}} = \dfrac{\sqrt[4]{y^4}\sqrt[4]{6}}{x^2} = \dfrac{y\sqrt[4]{6}}{x^2}$

▸ **Now Try Exercise 97**

Avoiding Common Errors

The following simplifications are correct because the numbers and variables divided out are not within square roots.

CORRECT

$$\frac{\overset{2}{\cancel{6}}\sqrt{2}}{\underset{1}{\cancel{3}}} = 2\sqrt{2}$$

CORRECT

$$\frac{\cancel{x}\sqrt{2}}{\cancel{x}} = \sqrt{2}$$

An expression within a square root cannot be divided by an expression not within the square root.

CORRECT

$$\frac{\sqrt{2}}{2} \quad \textit{Cannot be simplified further}$$

INCORRECT

$$\frac{\sqrt{\cancel{2}^{1}}}{\underset{1}{\cancel{2}}} = \sqrt{1} = 1$$

$$\frac{\sqrt{x^3}}{x} = \frac{\sqrt{x^2}\sqrt{x}}{x} = \frac{\cancel{x}\sqrt{x}}{\cancel{x}} = \sqrt{x}$$

$$\frac{\sqrt{x^{3}}}{\cancel{x}} = \sqrt{x^2} = x$$

EXERCISE SET 11.3   MathXL   MyMathLab

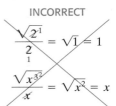

Concept/Writing Exercises

1. a) How do you obtain the numbers that are perfect squares?

   b) List the first six perfect squares.

2. a) How do you obtain the numbers that are perfect cubes?

   b) List the first six perfect cube numbers.

3. a) How do you obtain numbers that are perfect fifth powers?

   b) List the first five perfect fifth-power numbers.

4. State the product rule for radicals.

5. When we gave the product rule, we stated that for nonnegative real numbers a and b, $\sqrt[n]{a} \cdot \sqrt[n]{b} = \sqrt[n]{ab}$. Why is it necessary to specify that both a and b are nonnegative real numbers?

6. State the quotient rule for radicals.

7. When we gave the quotient rule, we stated that for nonnegative real numbers a and b, $\dfrac{\sqrt[n]{a}}{\sqrt[n]{b}} = \sqrt[n]{\dfrac{a}{b}}, b \neq 0$. Why is it necessary to state that both a and b are nonnegative real numbers?

8. In the quotient rule discussed in Exercise 7, why can the denominator never equal 0?

Practice the Skills

In this exercise set, assume that all variables represent positive real numbers.

Simplify.

9. $\sqrt{8}$

10. $\sqrt{28}$

11. $\sqrt{24}$

12. $\sqrt{18}$

13. $\sqrt{32}$

14. $\sqrt{12}$

15. $\sqrt{50}$

16. $\sqrt{72}$

17. $\sqrt{75}$

18. $\sqrt{300}$

19. $\sqrt{40}$

20. $\sqrt{600}$

21. $\sqrt[3]{16}$

22. $\sqrt[3]{24}$

23. $\sqrt[3]{54}$

24. $\sqrt[3]{81}$

25. $\sqrt[3]{32}$

26. $\sqrt[3]{108}$

27. $\sqrt[3]{40}$

28. $\sqrt[4]{80}$

29. $\sqrt[4]{48}$

30. $\sqrt[4]{162}$

31. $-\sqrt[5]{64}$

32. $-\sqrt[5]{243}$

33. $\sqrt[3]{b^9}$

34. $6\sqrt{y^{12}}$

35. $\sqrt[3]{x^6}$

36. $\sqrt[5]{y^{20}}$

37. $\sqrt{x^3}$

38. $-\sqrt{x^5}$

39. $\sqrt{a^{11}}$

40. $\sqrt[3]{b^{13}}$

41. $8\sqrt[3]{z^{32}}$

42. $\sqrt[3]{a^7}$

43. $\sqrt[4]{b^{23}}$

44. $\sqrt[5]{z^7}$

45. $\sqrt[6]{x^9}$

46. $\sqrt[7]{y^{15}}$

47. $3\sqrt[5]{y^{23}}$

48. $\sqrt{24x^3}$

49. $2\sqrt{50y^9}$

50. $\sqrt{75a^7b^{11}}$

51. $\sqrt[3]{x^3y^7}$

52. $\sqrt{x^5y^9}$

53. $\sqrt[5]{a^6b^{23}}$

54. $-\sqrt{20x^6y^7z^{12}}$

55. $\sqrt{24x^{15}y^{20}z^{27}}$

56. $\sqrt[3]{16x^3y^6}$

57. $\sqrt[3]{81a^6b^8}$

58. $\sqrt[3]{128a^{10}b^{11}c^{12}}$

59. $\sqrt[4]{32x^8y^9z^{19}}$

60. $\sqrt[4]{48x^{11}y^{21}}$

61. $\sqrt[4]{81a^8b^9}$

62. $-\sqrt[4]{32x^{18}y^{31}}$

63. $\sqrt[5]{32a^{10}b^{12}}$

64. $\sqrt[6]{64x^{12}y^{23}z^{50}}$

Simplify.

65. $\sqrt{\dfrac{75}{3}}$ **66.** $\sqrt{\dfrac{36}{4}}$ **67.** $\sqrt{\dfrac{81}{100}}$ **68.** $\sqrt{\dfrac{8}{50}}$

69. $\dfrac{\sqrt{27}}{\sqrt{3}}$ **70.** $\dfrac{\sqrt{72}}{\sqrt{2}}$ **71.** $\dfrac{\sqrt{3}}{\sqrt{48}}$ **72.** $\dfrac{\sqrt{15}}{\sqrt{60}}$

73. $\sqrt[3]{\dfrac{3}{24}}$ **74.** $\sqrt[3]{\dfrac{2}{54}}$ **75.** $\dfrac{\sqrt[3]{3}}{\sqrt[3]{81}}$ **76.** $\dfrac{\sqrt[3]{32}}{\sqrt[3]{4}}$

77. $\sqrt[4]{\dfrac{3}{48}}$ **78.** $\dfrac{\sqrt[4]{243}}{\sqrt[4]{3}}$ **79.** $\sqrt[5]{\dfrac{96}{3}}$ **80.** $\dfrac{\sqrt[5]{2}}{\sqrt[5]{64}}$

81. $\sqrt{\dfrac{r^4}{4}}$ **82.** $\sqrt{\dfrac{100a^8}{49b^6}}$ **83.** $\sqrt{\dfrac{16x^4}{25y^{10}}}$ **84.** $\sqrt{\dfrac{49a^8b^{10}}{121c^{14}}}$

85. $\sqrt[3]{\dfrac{c^6}{64}}$ **86.** $\sqrt[3]{\dfrac{27x^6}{y^{12}}}$ **87.** $\sqrt[3]{\dfrac{a^8b^{12}}{b^{-8}}}$ **88.** $\sqrt[4]{\dfrac{16x^{16}y^{32}}{81x^{-4}}}$

89. $\dfrac{\sqrt{24}}{\sqrt{3}}$ **90.** $\dfrac{\sqrt{64x^5}}{\sqrt{2x^3}}$ **91.** $\dfrac{\sqrt{27x^6}}{\sqrt{3x^2}}$ **92.** $\dfrac{\sqrt{72x^3y^5}}{\sqrt{8x^3y^7}}$

93. $\dfrac{\sqrt{48x^6y^9}}{\sqrt{6x^2y^6}}$ **94.** $\dfrac{\sqrt{300a^{10}b^{11}}}{\sqrt{2ab^4}}$ **95.** $\sqrt[3]{\dfrac{5xy}{8x^{13}}}$ **96.** $\sqrt[3]{\dfrac{64a^5b^{12}}{27a^{14}b^5}}$

97. $\sqrt[3]{\dfrac{25x^2y^9}{5x^8y^2}}$ **98.** $\sqrt[3]{\dfrac{54x^4y^4z^{17}}{18x^{13}z^4}}$ **99.** $\sqrt[4]{\dfrac{10x^4y}{81x^{-8}}}$ **100.** $\sqrt[4]{\dfrac{3a^6b^5}{16a^{-6}b^{13}}}$

Problem Solving

101. Prove $\sqrt{a\cdot b} = \sqrt{a}\,\sqrt{b}$ by converting $\sqrt{a\cdot b}$ to exponential form.

102. Will the product of two radicals always be a radical? Give an example to support your answer.

103. Will the quotient of two radicals always be a radical? Give an example to support your answer.

104. Prove $\sqrt[n]{\dfrac{a}{b}} = \dfrac{\sqrt[n]{a}}{\sqrt[n]{b}}$ by converting $\sqrt[n]{\dfrac{a}{b}}$ to exponential form.

105. a) Will $\dfrac{\sqrt[n]{x}}{\sqrt[n]{x}}$ always equal 1?

    b) If your answer to part **a)** was no, under what conditions does $\dfrac{\sqrt[n]{x}}{\sqrt[n]{x}}$ equal 1?

Cumulative Review Exercises

[2.6] **106.** Solve the formula $F = \dfrac{9}{5}C + 32$ for C.

[5.6] **107.** Divide $\dfrac{15x^{12} - 5x^9 + 20x^6}{5x^6}$.

[6.5] **108.** Factor $(x-3)^3 + 8$.

[6.6] **109.** Solve $(2x-3)(x-2) = 4x - 6$.

[10.2] **110.** Solve $\left|\dfrac{2x-4}{5}\right| = 12$.

11.4 Adding, Subtracting, and Multiplying Radicals

1 Add and subtract radicals.

2 Multiply radicals.

1 Add and Subtract Radicals

Like radicals are radicals having the same radicand and index. **Unlike radicals** are radicals differing in either the radicand or the index.

Examples of Like Radicals	Examples of Unlike Radicals	
$\sqrt{5}, 3\sqrt{5}$	$\sqrt{5}, \sqrt[3]{5}$	*Indices differ.*
$6\sqrt{7}, -2\sqrt{7}$	$\sqrt{6}, \sqrt{7}$	*Radicands differ.*
$\sqrt{x}, 5\sqrt{x}$	$\sqrt{x}, \sqrt{2x}$	*Radicands differ.*
$\sqrt[3]{2x}, -4\sqrt[3]{2x}$	$\sqrt{x}, \sqrt[3]{x}$	*Indices differ.*
$\sqrt[4]{x^2y^5}, -\sqrt[4]{x^2y^5}$	$\sqrt[3]{xy}, \sqrt[3]{x^2y}$	*Radicands differ.*

Like radicals are added and subtracted in exactly the same way that like terms are added or subtracted. To add or subtract like radicals, add or subtract their numerical coefficients and multiply this sum or difference by the like radical.

Examples of Adding and Subtracting Like Radicals

$$3\sqrt{6} + 2\sqrt{6} = (3 + 2)\sqrt{6} = 5\sqrt{6}$$
$$5\sqrt{x} - 7\sqrt{x} = (5 - 7)\sqrt{x} = -2\sqrt{x}$$
$$\sqrt[3]{4x^2} + 5\sqrt[3]{4x^2} = (1 + 5)\sqrt[3]{4x^2} = 6\sqrt[3]{4x^2}$$
$$4\sqrt{5x} - y\sqrt{5x} = (4 - y)\sqrt{5x}$$

EXAMPLE 1 ▶ Simplify.

a) $6 + 4\sqrt{2} - \sqrt{2} + 7$ **b)** $2\sqrt[3]{x} + 8x + 4\sqrt[3]{x} - 3$

Solution

a) $6 + 4\sqrt{2} - \sqrt{2} + 7 = 6 + 7 + 4\sqrt{2} - \sqrt{2}$ *Place like terms together.*
$$= 13 + (4 - 1)\sqrt{2}$$
$$= 13 + 3\sqrt{2} \quad (\text{or } 3\sqrt{2} + 13)$$

b) $2\sqrt[3]{x} + 8x + 4\sqrt[3]{x} - 3 = 6\sqrt[3]{x} + 8x - 3$

▶ **Now Try Exercise 15**

It is sometimes possible to convert unlike radicals into like radicals by simplifying one or more of the radicals, as was discussed in Section 11.3.

EXAMPLE 2 ▶ Simplify $\sqrt{3} + \sqrt{27}$.

Solution   Since $\sqrt{3}$ and $\sqrt{27}$ are unlike radicals, they cannot be added in their present form. We can simplify $\sqrt{27}$ to obtain like radicals.

$$\sqrt{3} + \sqrt{27} = \sqrt{3} + \sqrt{9}\sqrt{3}$$
$$= \sqrt{3} + 3\sqrt{3} = 4\sqrt{3}$$

▶ **Now Try Exercise 19**

To Add or Subtract Radicals

1. Simplify each radical expression.
2. Combine like radicals (if there are any).

EXAMPLE 3 ▶ Simplify.

a) $5\sqrt{24} + \sqrt{54}$ **b)** $2\sqrt{45} - \sqrt{80} + \sqrt{20}$ **c)** $\sqrt[3]{27} + \sqrt[3]{81} - 7\sqrt[3]{3}$

Solution

a) $5\sqrt{24} + \sqrt{54} = 5 \cdot \sqrt{4} \cdot \sqrt{6} + \sqrt{9} \cdot \sqrt{6}$
$$= 5 \cdot 2\sqrt{6} + 3\sqrt{6}$$
$$= 10\sqrt{6} + 3\sqrt{6} = 13\sqrt{6}$$

b) $2\sqrt{45} - \sqrt{80} + \sqrt{20} = 2 \cdot \sqrt{9} \cdot \sqrt{5} - \sqrt{16} \cdot \sqrt{5} + \sqrt{4} \cdot \sqrt{5}$
$$= 2 \cdot 3\sqrt{5} - 4\sqrt{5} + 2\sqrt{5}$$
$$= 6\sqrt{5} - 4\sqrt{5} + 2\sqrt{5} = 4\sqrt{5}$$

c) $\sqrt[3]{27} + \sqrt[3]{81} - 7\sqrt[3]{3} = 3 + \sqrt[3]{27} \cdot \sqrt[3]{3} - 7\sqrt[3]{3}$
$$= 3 + 3\sqrt[3]{3} - 7\sqrt[3]{3} = 3 - 4\sqrt[3]{3}$$

▶ **Now Try Exercise 23**

EXAMPLE 4 ▶ Simplify.   **a)** $\sqrt{x^2} - \sqrt{x^2 y} + x\sqrt{y}$ **b)** $\sqrt[3]{x^{13}y^2} - \sqrt[3]{x^4 y^8}$

Solution

a) $\sqrt{x^2} - \sqrt{x^2 y} + x\sqrt{y} = x - \sqrt{x^2} \cdot \sqrt{y} + x\sqrt{y}$

$$= x - x\sqrt{y} + x\sqrt{y}$$

$$= x$$

b) $\sqrt[3]{x^{13}y^2} - \sqrt[3]{x^4 y^8} = \sqrt[3]{x^{12}} \cdot \sqrt[3]{xy^2} - \sqrt[3]{x^3 y^6} \cdot \sqrt[3]{xy^2}$

$$= x^4 \sqrt[3]{xy^2} - xy^2 \sqrt[3]{xy^2}$$

Now factor out the common factor, $\sqrt[3]{xy^2}$.

$$= (x^4 - xy^2)\sqrt[3]{xy^2}$$

▶ **Now Try Exercise 35**

Helpful Hint

The product rule and quotient rule for radicals presented in Section 11.3 are

$$\sqrt[n]{a} \cdot \sqrt[n]{b} = \sqrt[n]{ab} \qquad \frac{\sqrt[n]{a}}{\sqrt[n]{b}} = \sqrt[n]{\frac{a}{b}}$$

Students often incorrectly assume similar properties exist for addition and subtraction. They do not. To illustrate this, let n be a square root (index 2), $a = 9$, and $b = 16$.

$$\sqrt[n]{a} + \sqrt[n]{b} \ne \sqrt[n]{a + b}$$

$$\sqrt{9} + \sqrt{16} \ne \sqrt{9 + 16}$$

$$3 + 4 \ne \sqrt{25}$$

$$7 \ne 5$$

We will now discuss multiplication of radicals.

2 Multiply Radicals

To multiply radicals, we use the product rule given earlier. After multiplying, we can often simplify the new radical (see Examples 5 and 6).

EXAMPLE 5 ▶ Multiply and simplify.

a) $\sqrt{6x^3}\sqrt{8x^6}$ **b)** $\sqrt[3]{2x}\sqrt[3]{4x^2}$ **c)** $\sqrt[4]{4x^{11}y}\sqrt[4]{16x^6 y^{22}}$

Solution

a) $\sqrt{6x^3}\sqrt{8x^6} = \sqrt{6x^3 \cdot 8x^6}$ *Product rule for radicals*

$$= \sqrt{48x^9}$$

$$= \sqrt{16x^8}\sqrt{3x}$$ *$16x^8$ is a perfect square.*

$$= 4x^4\sqrt{3x}$$

b) $\sqrt[3]{2x}\sqrt[3]{4x^2} = \sqrt[3]{2x \cdot 4x^2}$ *Product rule for radicals*

$$= \sqrt[3]{8x^3}$$ *$8x^3$ is a perfect cube.*

$$= 2x$$

c) $\sqrt[4]{4x^{11}y}\sqrt[4]{16x^6 y^{22}} = \sqrt[4]{4x^{11}y \cdot 16x^6 y^{22}}$ *Product rule for radicals*

$$= \sqrt[4]{64x^{17}y^{23}}$$

$$= \sqrt[4]{16x^{16}y^{20}}\sqrt[4]{4xy^3}$$ *The largest perfect fourth root factors are 16, x^{16}, and y^{20}.*

$$= 2x^4 y^5 \sqrt[4]{4xy^3}$$

▶ **Now Try Exercise 47**

Remember, as stated earlier, when a radical is simplified, the radicand does not have any variable with an exponent greater than or equal to the index.

EXAMPLE 6 ▶ Multiply and simplify $\sqrt{2x}(\sqrt{8x} - \sqrt{50})$.

Solution   Begin by using the distributive property.

$$\sqrt{2x}(\sqrt{8x} - \sqrt{50}) = (\sqrt{2x})(\sqrt{8x}) + (\sqrt{2x})(-\sqrt{50})$$
$$= \sqrt{16x^2} - \sqrt{100x}$$
$$= 4x - \sqrt{100}\sqrt{x}$$
$$= 4x - 10\sqrt{x}$$

▶ **Now Try Exercise 53**

Note in Example 6 that the same result could be obtained by first simplifying $\sqrt{8x}$ and $\sqrt{50}$ and then multiplying. You may wish to try this now.

Now we will multiply two binomial factors. To multiply binomial factors, each term in one factor must be multiplied by each term in the other factor. This can be accomplished using the FOIL method that was discussed earlier.

EXAMPLE 7 ▶ Multiply $(\sqrt{x} - \sqrt{y})(\sqrt{x} - y)$.

Solution   We will multiply using the FOIL method.

$$\begin{array}{cccc} \text{F} & \text{O} & \text{I} & \text{L} \\ \downarrow & \downarrow & \downarrow & \downarrow \\ (\sqrt{x})(\sqrt{x}) \ + & (\sqrt{x})(-y) \ + & (-\sqrt{y})(\sqrt{x}) \ + & (-\sqrt{y})(-y) \end{array}$$

$$= \ \sqrt{x^2} \ - \ y\sqrt{x} \ - \ \sqrt{xy} \ + \ y\sqrt{y}$$
$$= x - y\sqrt{x} - \sqrt{xy} + y\sqrt{y}$$

▶ **Now Try Exercise 63**

EXAMPLE 8 ▶ Simplify.   **a)** $(2\sqrt{6} - \sqrt{3})^2$ **b)** $\left(\sqrt[3]{x} - \sqrt[3]{2y^2}\right)\left(\sqrt[3]{x^2} - \sqrt[3]{8y}\right)$

Solution

a) $(2\sqrt{6} - \sqrt{3})^2 = (2\sqrt{6} - \sqrt{3})(2\sqrt{6} - \sqrt{3})$

Now multiply the factors using the FOIL method.

$$\begin{array}{cccc} \text{F} & \text{O} & \text{I} & \text{L} \end{array}$$
$$(2\sqrt{6})(2\sqrt{6}) + (2\sqrt{6})(-\sqrt{3}) + (-\sqrt{3})(2\sqrt{6}) + (-\sqrt{3})(-\sqrt{3})$$
$$= 4(6) - 2\sqrt{18} - 2\sqrt{18} + 3$$
$$= 24 - 2\sqrt{18} - 2\sqrt{18} + 3$$
$$= 27 - 4\sqrt{18}$$
$$= 27 - 4\sqrt{9}\sqrt{2}$$
$$= 27 - 12\sqrt{2}$$

b) Multiply the factors using the FOIL method.

$$\begin{array}{cccc} & \text{F} & \text{O} & \text{I} & \text{L} \end{array}$$
$$\left(\sqrt[3]{x} - \sqrt[3]{2y^2}\right)\left(\sqrt[3]{x^2} - \sqrt[3]{8y}\right) = (\sqrt[3]{x})(\sqrt[3]{x^2}) + (\sqrt[3]{x})(-\sqrt[3]{8y}) + \left(-\sqrt[3]{2y^2}\right)(\sqrt[3]{x^2}) + \left(-\sqrt[3]{2y^2}\right)(-\sqrt[3]{8y})$$
$$= \sqrt[3]{x^3} - \sqrt[3]{8xy} - \sqrt[3]{2x^2y^2} + \sqrt[3]{16y^3}$$
$$= \sqrt[3]{x^3} - \sqrt[3]{8}\sqrt[3]{xy} - \sqrt[3]{2x^2y^2} + \sqrt[3]{8y^3}\sqrt[3]{2}$$
$$= x - 2\sqrt[3]{xy} - \sqrt[3]{2x^2y^2} + 2y\sqrt[3]{2}$$

▶ **Now Try Exercise 99**

EXAMPLE 9 ▶ Multiply $(3 + \sqrt{6})(3 - \sqrt{6})$.

Solution   We can multiply using the FOIL method.

$$
\begin{array}{cccc}
\text{F} & \text{O} & \text{I} & \text{L}
\end{array}
$$
$$(3 + \sqrt{6})(3 - \sqrt{6}) = 3(3) + 3(-\sqrt{6}) + (\sqrt{6})(3) + (\sqrt{6})(-\sqrt{6}).$$
$$= 9 \quad - \quad 3\sqrt{6} \quad + \quad 3\sqrt{6} \quad - \quad \sqrt{36}$$
$$= 9 - \sqrt{36}$$
$$= 9 - 6 = 3$$

▶ **Now Try Exercise 59**

Note that in Example 9, we multiplied *the sum and difference of the same two terms*. Recall from Section 6.5 that $(a + b)(a - b) = a^2 - b^2$. If we let $a = 3$ and $b = \sqrt{6}$, then we can multiply as follows.

$$(a + b)(a - b) = a^2 - b^2$$
$$(3 + \sqrt{6})(3 - \sqrt{6}) = 3^2 - (\sqrt{6})^2$$
$$= 9 - 6$$
$$= 3$$

When multiplying the sum and difference of the same two terms, you may obtain the answer using the difference of the squares of the two terms. We will see additional multiplications of this type in Section 11.5.

EXAMPLE 10 ▶ If $f(x) = \sqrt[3]{x^2}$ and $g(x) = \sqrt[3]{x^4} + \sqrt[3]{x^2}$, find **a)** $(f \cdot g)(x)$ and **b)** $(f \cdot g)(6)$.

Solution

a) From Section 8.5, we know that $(f \cdot g)(x) = f(x) \cdot g(x)$.

$$(f \cdot g)(x) = f(x) \cdot g(x)$$
$$= \sqrt[3]{x^2}\left(\sqrt[3]{x^4} + \sqrt[3]{x^2}\right) \qquad \textit{Substitute given values.}$$
$$= \sqrt[3]{x^2}\,\sqrt[3]{x^4} + \sqrt[3]{x^2}\,\sqrt[3]{x^2} \qquad \textit{Distributive property}$$
$$= \sqrt[3]{x^6} + \sqrt[3]{x^4} \qquad \textit{Product rule for radicals}$$
$$= x^2 + x\sqrt[3]{x} \qquad \textit{Simplify radicals.}$$

b) To compute $(f \cdot g)(6)$, substitute 6 for x in the answer obtained in part **a)**.

$$(f \cdot g)(x) = x^2 + x\sqrt[3]{x}$$
$$(f \cdot g)(6) = 6^2 + 6\sqrt[3]{6} \qquad \textit{Substitute 6 for x.}$$
$$= 36 + 6\sqrt[3]{6}$$

▶ **Now Try Exercise 77**

EXAMPLE 11 ▶ Simplify $f(x)$ if **a)** $f(x) = \sqrt{x + 3}\,\sqrt{x + 3}$, $x \geq -3$ and **b)** $f(x) = \sqrt{3x^2 - 30x + 75}$; assume that the variable may be any real number.

Solution

a) $f(x) = \sqrt{x + 3}\,\sqrt{x + 3}$
$$= \sqrt{(x + 3)(x + 3)} \qquad \textit{Product rule for radicals}$$
$$= \sqrt{(x + 3)^2}$$
$$= x + 3$$

Since we are told that $x \geq -3$, we can use the product rule. Note that the radicand will be a nonnegative number for any $x \geq -3$ and we can write the answer as $x + 3$ rather than $|x + 3|$.

b) $f(x) = \sqrt{3x^2 - 30x + 75}$

$= \sqrt{3(x^2 - 10x + 25)}$ *Factor out 3.*

$= \sqrt{3(x - 5)^2}$ *Write as the square of a binomial.*

$= \sqrt{3}\sqrt{(x - 5)^2}$ *Product rule for radicals*

$= \sqrt{3}|x - 5|$

Since the variable could be any real number, we write our answer with absolute value signs. If we had been told that $x - 5$ was nonnegative, then we could have written our answer as $\sqrt{3}(x - 5)$.

▶ **Now Try Exercise 105**

EXERCISE SET 11.4   Math XP   MyMathLab

MathXL®   MyMathLab

Concept/Writing Exercises

1. What are like radicals?

2. a) Explain how to add like radicals.

   b) Using the procedure in part **a)**, add $\frac{3}{5}\sqrt{5} + \frac{5}{4}\sqrt{5}$.

3. Use a calculator to estimate $\sqrt{3} + 3\sqrt{2}$.

4. Use a calculator to estimate $2\sqrt{3} + \sqrt{5}$.

5. Does $\sqrt{a} + \sqrt{b} = \sqrt{a + b}$? Explain your answer and give an example supporting your answer.

6. Since $64 + 36 = 100$, does $\sqrt{64} + \sqrt{36} = \sqrt{100}$? Explain your answer.

Practice the Skills

In this exercise set, assume that all variables represent positive real numbers.

Simplify.

7. $\sqrt{3} - \sqrt{3}$

8. $2\sqrt{6} - \sqrt{6}$

9. $6\sqrt{5} - 2\sqrt{5}$

10. $3\sqrt{2} + 7\sqrt{2} - 11$

11. $2\sqrt{3} - 2\sqrt{3} - 4\sqrt{3} + 5$

12. $6\sqrt[3]{7} - 8\sqrt[3]{7}$

13. $2\sqrt[4]{y} - 9\sqrt[4]{y}$

14. $3\sqrt[5]{a} + 7 + 5\sqrt[5]{a} - 2$

15. $3\sqrt{5} - \sqrt[3]{x} + 6\sqrt{5} + 3\sqrt[3]{x}$

16. $9 + 4\sqrt[4]{a} - 7\sqrt[4]{a} + 5$

17. $5\sqrt{x} - 8\sqrt{y} + 3\sqrt{x} + 2\sqrt{y} - \sqrt{x}$

18. $8\sqrt{a} + 4\sqrt[3]{b} + 7\sqrt{a} - 12\sqrt[3]{b}$

Simplify.

19. $\sqrt{5} + \sqrt{20}$

20. $\sqrt{75} + \sqrt{108}$

21. $-6\sqrt{75} + 5\sqrt{125}$

22. $3\sqrt{250} + 4\sqrt{160}$

23. $-4\sqrt{90} + 3\sqrt{40} + 2\sqrt{10}$

24. $3\sqrt{40x^2y} + 2x\sqrt{490y}$

25. $\sqrt{500xy^2} + y\sqrt{320x}$

26. $5\sqrt{8} + 2\sqrt{50} - 3\sqrt{72}$

27. $2\sqrt{5x} - 3\sqrt{20x} - 4\sqrt{45x}$

28. $3\sqrt{27c^2} - 2\sqrt{108c^2} - \sqrt{48c^2}$

29. $3\sqrt{50a^2} - 3\sqrt{72a^2} - 8a\sqrt{18}$

30. $4\sqrt[3]{5} - 5\sqrt[3]{40}$

31. $\sqrt[3]{108} + \sqrt[3]{32}$

32. $3\sqrt[3]{16} + \sqrt[3]{54}$

33. $\sqrt[3]{27} - 5\sqrt[3]{8}$

34. $3\sqrt{45x^3} + \sqrt{5x}$

35. $2\sqrt[3]{a^4b^2} + 4a\sqrt[3]{ab^2}$

36. $5y\sqrt[4]{48x^5} - x\sqrt[4]{3x^5y^4}$

37. $\sqrt{4r^7s^5} + 3r^2\sqrt{r^3s^5} - 2rs\sqrt{r^5s^3}$

38. $x\sqrt[3]{27x^5y^2} - x^2\sqrt[3]{x^2y^2} + 4\sqrt[3]{x^8y^2}$

39. $\sqrt[3]{128x^8y^{10}} - 2x^2y\sqrt[3]{16x^2y^7}$

40. $5\sqrt[3]{320x^5y^8} + 3x\sqrt[3]{135x^2y^8}$

Simplify.

41. $\sqrt{3}\sqrt{27}$

42. $\sqrt[3]{2}\sqrt[3]{4}$

43. $\sqrt[4]{4}\sqrt[4]{14}$

44. $\sqrt[3]{3}\sqrt[3]{54}$

45. $\sqrt{9m^3n^7}\sqrt{3mn^4}$

46. $\sqrt[3]{5ab^2}\sqrt[3]{25a^4b^{12}}$

47. $\sqrt[3]{9x^7y^{10}}\sqrt[3]{6x^4y^3}$

48. $\sqrt[4]{3x^9y^{12}}\sqrt[4]{54x^4y^7}$

49. $\sqrt[5]{x^{24}y^{30}z^9}\sqrt[5]{x^{13}y^8z^7}$

50. $\sqrt[4]{8x^4yz^3}\sqrt[4]{2x^2y^3z^7}$

51. $\left(\sqrt[3]{2x^3y^4}\right)^2$

52. $\sqrt{2}\left(\sqrt{6} + \sqrt{18}\right)$

53. $\sqrt{5}\left(\sqrt{5} - \sqrt{3}\right)$

54. $\sqrt{3}\left(\sqrt{12} + \sqrt{8}\right)$

55. $\sqrt[3]{y}\left(2\sqrt[3]{y} - \sqrt[3]{y^8}\right)$

56. $\sqrt{3y}\left(\sqrt{27y^2} - \sqrt{y}\right)$

57. $2\sqrt[3]{x^4y^5}\left(\sqrt[3]{8x^{12}y^4} + \sqrt[3]{16xy^9}\right)$

58. $\sqrt[5]{16x^7y^6}\left(\sqrt[5]{2x^6y^9} - \sqrt[5]{10x^3y^7}\right)$

59. $(8 + \sqrt{5})(8 - \sqrt{5})$

60. $(9 - \sqrt{5})(9 + \sqrt{5})$

61. $(\sqrt{6} + x)(\sqrt{6} - x)$

62. $(\sqrt{x} + y)(\sqrt{x} - y)$

63. $(\sqrt{7} - \sqrt{z})(\sqrt{7} + \sqrt{z})$

64. $(3\sqrt{a} - 5\sqrt{b})(3\sqrt{a} + 5\sqrt{b})$

65. $(\sqrt{3} + 4)(\sqrt{3} + 5)$

66. $(1 + \sqrt{5})(8 + \sqrt{5})$

67. $(3 - \sqrt{2})(4 - \sqrt{8})$

68. $(5\sqrt{6} + 3)(4\sqrt{6} - 1)$

69. $(4\sqrt{3} + \sqrt{2})(\sqrt{3} - \sqrt{2})$

70. $(\sqrt{3} + 7)^2$

71. $(2\sqrt{5} - 3)^2$

72. $(\sqrt{y} + \sqrt{6z})(\sqrt{2z} - \sqrt{8y})$

73. $(2\sqrt{3x} - \sqrt{y})(3\sqrt{3x} + \sqrt{y})$

74. $(\sqrt[3]{9} + \sqrt[3]{2})(\sqrt[3]{3} + \sqrt[3]{4})$

75. $(\sqrt[3]{4} - \sqrt[3]{6})(\sqrt[3]{2} - \sqrt[3]{36})$

76. $(\sqrt[3]{4x} - \sqrt[3]{2y})(\sqrt[3]{4x} + \sqrt[3]{10})$

In Exercises 77–82, f(x) and g(x) are given. Find $(f \cdot g)(x)$.

77. $f(x) = \sqrt{2x}, g(x) = \sqrt{8x} - \sqrt{32}$

78. $f(x) = \sqrt{6x}, g(x) = \sqrt{6x} - \sqrt{10x}$

79. $f(x) = \sqrt[3]{x}, g(x) = \sqrt[3]{x^5} + \sqrt[3]{x^4}$

80. $f(x) = \sqrt[3]{2x^2}, g(x) = \sqrt[3]{4x} + \sqrt[3]{32x^2}$

81. $f(x) = \sqrt[4]{3x^2}, g(x) = \sqrt[4]{9x^4} - \sqrt[4]{x^7}$

82. $f(x) = \sqrt[4]{2x^3}, g(x) = \sqrt[4]{8x^5} - \sqrt[4]{5x^6}$

Simplify. These exercises are a combination of the types of exercises presented earlier in this exercise set.

83. $\sqrt{24}$

84. $\sqrt{300}$

85. $\sqrt{125} - \sqrt{20}$

86. $4\sqrt{7} + 2\sqrt{63} - 2\sqrt{28}$

87. $(3\sqrt{2} - 4)(\sqrt{2} + 5)$

88. $(\sqrt{5} + \sqrt{2})(\sqrt{2} + \sqrt{20})$

89. $\sqrt{6}(5 - \sqrt{2})$

90. $3\sqrt[3]{81} + 4\sqrt[3]{24}$

91. $\sqrt{150}\sqrt{3}$

92. $\sqrt[4]{2}\sqrt[4]{40}$

93. $\sqrt[3]{80x^{11}}$

94. $\sqrt[3]{x^9 y^{11} z}$

95. $\sqrt[6]{128ab^{17}c^9}$

96. $\sqrt[5]{14x^4y^2}\sqrt[5]{3x^4y^3}$

97. $2b\sqrt[4]{a^4b} + ab\sqrt[4]{16b}$

98. $2\sqrt[3]{24a^3y^4} + 4a\sqrt[3]{81y^4}$

99. $(\sqrt[3]{x^2} - \sqrt[3]{y})(\sqrt[3]{x} - 2\sqrt[3]{y^2})$

100. $(\sqrt[3]{a} + 5)(\sqrt[3]{a^2} - 6)$

101. $\sqrt[3]{3ab^2}(\sqrt[3]{4a^4b^3} - \sqrt[3]{8a^5b^4})$

102. $\sqrt[4]{4st^2}(\sqrt[4]{2s^5t^6} + \sqrt[4]{5s^9t^2})$

Simplify the following. In Exercises 105 and 106, assume the variable can be any real number. See Example 11.

103. $f(x) = \sqrt{2x - 5}\sqrt{2x - 5}, x \geq \dfrac{5}{2}$

104. $g(a) = \sqrt{3a + 7}\sqrt{3a + 7}, a \geq -\dfrac{7}{3}$

105. $h(r) = \sqrt{4r^2 - 32r + 64}$

106. $f(b) = \sqrt{20b^2 + 60b + 45}$

Problem Solving

Find the perimeter and area of the following figures. Write the perimeter and area in radical form with the radicals simplified.

107. 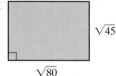 $\sqrt{45}$ $\sqrt{80}$

108. 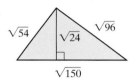 $\sqrt{54}$ $\sqrt{24}$ $\sqrt{96}$ $\sqrt{150}$

109. 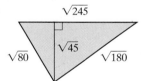 $\sqrt{245}$ $\sqrt{80}$ $\sqrt{45}$ $\sqrt{180}$

110. 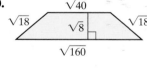 $\sqrt{40}$ $\sqrt{18}$ $\sqrt{8}$ $\sqrt{18}$ $\sqrt{160}$

111. Will the sum of two radicals always be a radical? Give an example to support your answer.

112. Will the difference of two radicals always be a radical? Give an example to support your answer.

113. Skid Marks Law enforcement officials sometimes use the formula $s = \sqrt{30FB}$ to determine a car's speed, s, in miles per hour, from a car's skid marks. The F in the formula represents the "road factor," which is determined by the road's surface, and the B represents the braking distance, in feet. Officer Jenkins is investigating an accident. Find the car's speed if the skid marks are 80 feet long and **a)** the road was dry asphalt, whose road factor is 0.85, **b)** the road was wet gravel, whose road factor is 0.52.

114. Water through a Fire Hose The rate at which water flows through a particular fire hose, R, in gallons per minute, can be approximated by the formula $R = 28d^2\sqrt{P}$, where d is the diameter of the nozzle, in inches, and P is the nozzle pressure, in pounds per square inch. If a nozzle has a diameter of 2.5 inches and the nozzle pressure is 80 pounds per square inch, find the flow rate.

115. Height of Girls The function $f(t) = 3\sqrt{t} + 19$ can be used to approximate the median height, $f(t)$, in inches, for U.S. girls of age t, in months, where $1 \le t \le 60$. Estimate the median height of girls at age **a)** 36 months and **b)** 40 months.

116. Standard Deviation In statistics, the standard deviation of the population, σ, read "sigma," is a measure of the spread of a set of data about the mean of the data. The greater the spread, the greater the standard deviation. One formula used to determine sigma is $\sigma = \sqrt{npq}$, where n represents the sample size, p represents the percent chance (or probability) that something specific happens, and q represents the percent chance (or probability) that the specific thing does not happen. In a sample of 600 people who purchase airline tickets, the percent that showed up for their flight, p, was 0.93, and the percent that did not show up for their flight, q, was 0.07. Use this information to find σ.

117. The graph of $f(x) = \sqrt{x}$ is shown.

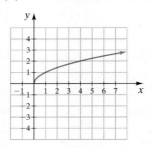

a) If $g(x) = 2$, sketch the graph of $(f + g)(x)$.

b) What effect does adding 2 have to the graph of $f(x)$?

118. The graph of $f(x) = -\sqrt{x}$ is shown.

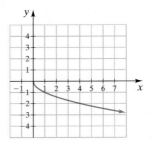

a) If $g(x) = 3$, sketch the graph of $(f + g)(x)$.

b) What effect does adding 3 have to the graph of $f(x)$?

119. You are given that $f(x) = \sqrt{x}$ and $g(x) = \sqrt{x} - 2$.

a) Sketch the graph of $(f - g)(x)$. Explain how you determined your answer.

b) What is the domain of $(f - g)(x)$?

120. You are given that $f(x) = \sqrt{x}$ and $g(x) = -\sqrt{x} - 3$.

a) Sketch the graph of $(f + g)(x)$. Explain how you determined your answer.

b) What is the domain of $(f + g)(x)$?

121. Graph the function $f(x) = \sqrt{x^2}$.

122. Graph the function $f(x) = \sqrt{x^2} - 4$.

Cumulative Review Exercises

[1.4] **123.** What is a rational number?

124. What is a real number?

125. What is an irrational number?

[1.5] **126.** What is the definition of $|a|$?

[2.6] **127.** Solve the formula $E = \frac{1}{2}mv^2$ for m.

[10.1] **128.** Solve the inequality $-4 < 2x - 3 \le 7$ and indicate the solution **a)** on the number line; **b)** in internal notation; **c)** in set builder notation.

Mid-Chapter Test: 11.1–11.4

To find out how well you understand the chapter material to this point, take this brief test. The answers, and the section where the material was initially discussed, are given in the back of the book. Review any questions that you answered incorrectly.

Find the indicated root.

1. $\sqrt{121}$

2. $\sqrt[3]{-\dfrac{27}{64}}$

Use absolute value to evaluate.

3. $\sqrt{(-16.3)^2}$

4. $\sqrt{(3a^2 - 4b^3)^2}$

5. Find $g(16)$ if $g(x) = \dfrac{x}{8} + \sqrt{4x} - 7$.

6. Write $\sqrt[5]{7a^4b^3}$ in exponential form.

7. Evaluate $-49^{1/2} + 81^{3/4}$.

Simplify each expression.

8. $\left(\sqrt[4]{a^2b^3c}\right)^{20}$

9. $7x^{-5/2} \cdot 2x^{3/2}$

10. Multiply $8x^{-2}(x^3 + 2x^{-1/2})$.

Simplify each radical.

11. $\sqrt{32x^4y^9}$

12. $\sqrt[6]{64a^{13}b^{23}c^{15}}$

13. $\dfrac{\sqrt[3]{3}}{\sqrt[3]{81}}$

14. $\dfrac{\sqrt{20x^5y^{12}}}{\sqrt{180x^{15}y^7}}$

Simplify.

15. $2\sqrt{x} - 3\sqrt{y} + 9\sqrt{x} + 15\sqrt{y}$

16. $2\sqrt{90x^2y} + 3x\sqrt{490y}$

17. $(x + \sqrt{5})(2x - 3\sqrt{5})$

18. $2\sqrt{3a}\left(\sqrt{27a^2} - 5\sqrt{4a}\right)$

19. $3b\sqrt[4]{a^5b} + 2ab\sqrt[4]{16ab}$

20. When simplifying the following square roots, in which parts will the answer contain an absolute value? Explain your answer and simplify parts **a)** and **b)**.

  a) $\sqrt{(x - 3)^2}$

  b) $\sqrt{64x^2}$, $x \geq 0$

11.5 Dividing Radicals

1 Rationalize denominators.

2 Rationalize a denominator using the conjugate.

3 Understand when a radical is simplified.

4 Use rationalizing the denominator in an addition problem.

5 Divide radical expressions with different indices.

1 Rationalize Denominators

We introduced the quotient rule for radicals in Section 11.3. Now we will use the quotient rule to work other division problems and to rationalize denominators.

When the denominator of a fraction contains a radical, we generally simplify the expression by **rationalizing the denominator**. To rationalize a denominator is to remove all radicals from the denominator. When adding radicals, it may be necessary to rationalize denominators, as will be illustrated in Example 6.

> **To Rationalize a Denominator**
>
> Multiply both the numerator and the denominator of the fraction by a radical that will result in the radicand in the denominator becoming a perfect power.

When both the numerator and denominator are multiplied by the same radical expression, you are in effect multiplying the fraction by 1, which does not change its value.

EXAMPLE 1 ▶ Simplify. **a)** $\dfrac{1}{\sqrt{5}}$ **b)** $\dfrac{x}{4\sqrt{3}}$ **c)** $\dfrac{11}{\sqrt{2x}}$ **d)** $\dfrac{\sqrt[3]{16a^4}}{\sqrt[3]{b}}$

Solution To simplify each expression, we must rationalize the denominators. We do so by multiplying both the numerator and denominator by a radical that will result in the denominator becoming a perfect power for the given index.

a) $\dfrac{1}{\sqrt{5}} = \dfrac{1}{\sqrt{5}} \cdot \dfrac{\sqrt{5}}{\sqrt{5}} = \dfrac{\sqrt{5}}{\sqrt{25}} = \dfrac{\sqrt{5}}{5}$

b) $\dfrac{x}{4\sqrt{3}} = \dfrac{x}{4\sqrt{3}} \cdot \dfrac{\sqrt{3}}{\sqrt{3}} = \dfrac{x\sqrt{3}}{4 \cdot 3} = \dfrac{x\sqrt{3}}{12}$

c) There are two factors in the radicand, 2 and x. We must make each factor a perfect square. Since 2^2 or 4 is a perfect square and x^2 is a perfect square, we multiply both numerator and denominator by $\sqrt{2x}$.

$$\dfrac{11}{\sqrt{2x}} = \dfrac{11}{\sqrt{2x}} \cdot \dfrac{\sqrt{2x}}{\sqrt{2x}}$$

$$= \dfrac{11\sqrt{2x}}{\sqrt{4x^2}}$$

$$= \dfrac{11\sqrt{2x}}{2x}$$

d) There are no common factors in the numerator and denominator. Before we rationalize the denominator, let's simplify the numerator.

$$\frac{\sqrt[3]{16a^4}}{\sqrt[3]{b}} = \frac{\sqrt[3]{8a^3}\,\sqrt[3]{2a}}{\sqrt[3]{b}} \qquad \textit{Product rule for radicals}$$

$$= \frac{2a\sqrt[3]{2a}}{\sqrt[3]{b}} \qquad \textit{Simplify the numerator.}$$

Now we rationalize the denominator. Since the denominator is a cube root, we need to make the radicand a perfect cube. Since the denominator contains b and we want b^3, we need two more factors of b, or b^2. We therefore multiply both numerator and denominator by $\sqrt[3]{b^2}$.

$$= \frac{2a\sqrt[3]{2a}}{\sqrt[3]{b}} \cdot \frac{\sqrt[3]{b^2}}{\sqrt[3]{b^2}}$$

$$= \frac{2a\sqrt[3]{2ab^2}}{\sqrt[3]{b^3}}$$

$$= \frac{2a\sqrt[3]{2ab^2}}{b}$$

▶ **Now Try Exercise 15**

EXAMPLE 2 ▶ Simplify.   **a)** $\sqrt{\dfrac{5}{7}}$ **b)** $\sqrt[3]{\dfrac{x}{2y^2}}$ **c)** $\sqrt[4]{\dfrac{32x^9y^6}{3z^2}}$

Solution   In each part, we will use the quotient rule to write the radical as a quotient of two radicals.

a) $\sqrt{\dfrac{5}{7}} = \dfrac{\sqrt{5}}{\sqrt{7}} \cdot \dfrac{\sqrt{7}}{\sqrt{7}} = \dfrac{\sqrt{35}}{\sqrt{49}} = \dfrac{\sqrt{35}}{7}$

b) $\sqrt[3]{\dfrac{x}{2y^2}} = \dfrac{\sqrt[3]{x}}{\sqrt[3]{2y^2}}$

The denominator is $\sqrt[3]{2y^2}$ and we want to change it to $\sqrt[3]{2^3y^3}$. We now multiply both the numerator and denominator by the cube root of an expression that will make the radicand in the denominator $\sqrt[3]{2^3y^3}$. Since $2 \cdot 2^2 = 2^3$ and $y^2 \cdot y = y^3$, we multiply both numerator and denominator by $\sqrt[3]{2^2y}$.

$$\frac{\sqrt[3]{x}}{\sqrt[3]{2y^2}} = \frac{\sqrt[3]{x}}{\sqrt[3]{2y^2}} \cdot \frac{\sqrt[3]{2^2y}}{\sqrt[3]{2^2y}}$$

$$= \frac{\sqrt[3]{x}\,\sqrt[3]{4y}}{\sqrt[3]{2^3y^3}}$$

$$= \frac{\sqrt[3]{4xy}}{2y}$$

c) After using the quotient rule, we simplify the numerator.

$$\sqrt[4]{\frac{32x^9y^6}{3z^2}} = \frac{\sqrt[4]{32x^9y^6}}{\sqrt[4]{3z^2}} \qquad \textit{Quotient rule for radicals}$$

$$= \frac{\sqrt[4]{16x^8y^4}\,\sqrt[4]{2xy^2}}{\sqrt[4]{3z^2}} \qquad \textit{Product rule for radicals}$$

$$= \frac{2x^2y\sqrt[4]{2xy^2}}{\sqrt[4]{3z^2}} \qquad \textit{Simplify the numerator.}$$

Now we rationalize the denominator. To make the radicand in the denominator a perfect fourth power, we need to get each factor to a power of 4. Since the denominator contains one factor of 3, we need 3 more factors of 3, or 3^3. Since there are two factors of z, we need 2 more factors of z, or z^2. Thus we will multiply both numerator and denominator by $\sqrt[4]{3^3 z^2}$.

$$= \frac{2x^2 y \sqrt[4]{2xy^2}}{\sqrt[4]{3z^2}} \cdot \frac{\sqrt[4]{3^3 z^2}}{\sqrt[4]{3^3 z^2}}$$

$$= \frac{2x^2 y \sqrt[4]{2xy^2} \sqrt[4]{27 z^2}}{\sqrt[4]{3z^2} \sqrt[4]{3^3 z^2}}$$

$$= \frac{2x^2 y \sqrt[4]{54 x y^2 z^2}}{\sqrt[4]{3^4 z^4}} \qquad \textit{Product rule for radicals}$$

$$= \frac{2x^2 y \sqrt[4]{54 x y^2 z^2}}{3z}$$

Note: There are no perfect fourth power factors of 54, and each exponent in the radicand is less than the index.

▶ **Now Try Exercise 53**

2 Rationalize a Denominator Using the Conjugate

When the denominator of a rational expression is a binomial that contains a radical, we rationalize the denominator. We do this by multiplying both the numerator and the denominator of the fraction by the **conjugate** of the denominator. The conjugate of a binomial is a binomial having the same two terms with the sign of the second term changed.

Expression	Conjugate
$9 + \sqrt{2}$	$9 - \sqrt{2}$
$8\sqrt{3} - \sqrt{5}$	$8\sqrt{3} + \sqrt{5}$
$\sqrt{x} + \sqrt{y}$	$\sqrt{x} - \sqrt{y}$
$6a - \sqrt{b}$	$6a + \sqrt{b}$

When a binomial is multiplied by its conjugate, the outer and inner products will sum to 0. We multiplied radicals containing binomial factors in Section 11.4. We will work one more example of multiplication of radical expressions in Example 3.

EXAMPLE 3 ▶ Multiply $(6 + \sqrt{3})(6 - \sqrt{3})$.

Solution Multiply using the FOIL method.

$$\begin{array}{cccc} & F & O & I & L \\ (6 + \sqrt{3})(6 - \sqrt{3}) = & 6(6) & + 6(-\sqrt{3}) & + 6(\sqrt{3}) & + \sqrt{3}(-\sqrt{3}) \end{array}$$

$$= 36 \; \boxed{-6\sqrt{3} + 6\sqrt{3}} \; - \sqrt{9}$$

$$= 36 - \sqrt{9}$$

$$= 36 - 3$$

$$= 33$$

▶ **Now Try Exercise 57**

In Example 3, we would get the same result using the formula for the product of the sum and difference of the same two terms. The product results in the difference of two squares, $(a + b)(a - b) = a^2 - b^2$. In Example 3, if we let $a = 6$ and $b = \sqrt{3}$, then using the formula we get the following.

$$(a + b)(a - b) = a^2 - b^2$$
$$\downarrow \downarrow \downarrow \downarrow \downarrow \downarrow$$
$$(6 + \sqrt{3})(6 - \sqrt{3}) = 6^2 - (\sqrt{3})^2$$
$$= 36 - 3$$
$$= 33$$

Now let's work an example where we rationalize a denominator containing two terms.

EXAMPLE 4 ▶ Simplify.   **a)** $\dfrac{13}{4 + \sqrt{3}}$ **b)** $\dfrac{6}{\sqrt{5} - \sqrt{2}}$ **c)** $\dfrac{a - \sqrt{b}}{a + \sqrt{b}}$

Solution   In each part, we rationalize the denominator by multiplying the numerator and the denominator by the conjugate of the denominator.

a)
$$\frac{13}{4 + \sqrt{3}} = \frac{13}{4 + \sqrt{3}} \cdot \frac{4 - \sqrt{3}}{4 - \sqrt{3}}$$
$$= \frac{13(4 - \sqrt{3})}{(4 + \sqrt{3})(4 - \sqrt{3})}$$
$$= \frac{13(4 - \sqrt{3})}{16 - 3}$$
$$= \frac{\overset{1}{\cancel{13}}(4 - \sqrt{3})}{\underset{1}{\cancel{13}}} \text{ or } 4 - \sqrt{3}$$

b)
$$\frac{6}{\sqrt{5} - \sqrt{2}} = \frac{6}{\sqrt{5} - \sqrt{2}} \cdot \frac{\sqrt{5} + \sqrt{2}}{\sqrt{5} + \sqrt{2}}$$
$$= \frac{6(\sqrt{5} + \sqrt{2})}{5 - 2}$$
$$= \frac{\overset{2}{\cancel{6}}(\sqrt{5} + \sqrt{2})}{\underset{1}{\cancel{3}}}$$
$$= 2(\sqrt{5} + \sqrt{2}) \quad \text{or} \quad 2\sqrt{5} + 2\sqrt{2}$$

c)
$$\frac{a - \sqrt{b}}{a + \sqrt{b}} = \frac{a - \sqrt{b}}{a + \sqrt{b}} \cdot \frac{a - \sqrt{b}}{a - \sqrt{b}}$$
$$= \frac{a^2 - a\sqrt{b} - a\sqrt{b} + \sqrt{b^2}}{a^2 - b}$$
$$= \frac{a^2 - 2a\sqrt{b} + b}{a^2 - b}$$

Remember that you cannot divide out a^2 or b because they are terms, not factors.

▶ **Now Try Exercise 75**

Now that we have illustrated how to rationalize denominators, let's discuss the criteria a radical must meet to be considered simplified.

3 Understand when a Radical Is Simplified

After you have simplified a radical expression, you should check it to make sure that it is simplified as far as possible.

> **A Radical Expression Is Simplified When the Following Are All True**
> 1. No perfect powers are factors of the radicand and all exponents in the radicand are less than the index.
> 2. No radicand contains a fraction.
> 3. No denominator contains a radical.

EXAMPLE 5 ▶ Determine whether the following expressions are simplified. If not, explain why. Simplify the expressions if not simplified.

a) $\sqrt{48x^5}$ **b)** $\sqrt{\dfrac{1}{2}}$ **c)** $\dfrac{1}{\sqrt{6}}$

Solution

a) This expression is not simplified because 16 is a perfect square factor of 48 and because x^4 is a perfect square factor of x^5. Notice that the exponent on the variable in the radicand, 5, is greater than the index, 2. Whenever the exponent on the variable in the radicand is greater than the index, the radicand has a perfect power factor of the variable, and the radical needs to be simplified further. Let's simplify the radical.

$$\sqrt{48x^5} = \sqrt{16x^4 \cdot 3x} = \sqrt{16x^4} \cdot \sqrt{3x} = 4x^2\sqrt{3x}$$

b) This expression is not simplified since the radicand contains the fraction $\dfrac{1}{2}$. This violates item 2. We simplify by first using the quotient rule, and then we rationalize the denominator, as follows.

$$\sqrt{\frac{1}{2}} = \frac{\sqrt{1}}{\sqrt{2}} \cdot \frac{\sqrt{2}}{\sqrt{2}} = \frac{\sqrt{2}}{2}$$

c) This expression is not simplified since the denominator, $\sqrt{6}$, contains a radical. This violates item 3. We simplify by rationalizing the denominator, as follows.

$$\frac{1}{\sqrt{6}} = \frac{1}{\sqrt{6}} \cdot \frac{\sqrt{6}}{\sqrt{6}} = \frac{\sqrt{6}}{6}$$

▶ **Now Try Exercise 7**

4 Use Rationalizing the Denominator in an Addition Problem

Now let's work an addition problem that requires rationalizing the denominator. This example makes use of the methods discussed in Sections 11.3 and 11.4 to add and subtract radicals.

EXAMPLE 6 ▶ Simplify $4\sqrt{2} - \dfrac{3}{\sqrt{8}} + \sqrt{32}$.

Solution   Begin by rationalizing the denominator and by simplifying $\sqrt{32}$.

$$4\sqrt{2} - \frac{3}{\sqrt{8}} + \sqrt{32} = 4\sqrt{2} - \frac{3}{\sqrt{8}} \cdot \frac{\sqrt{2}}{\sqrt{2}} + \sqrt{16}\,\sqrt{2} \qquad \textit{Rationalize denominator.}$$

$$= 4\sqrt{2} - \frac{3\sqrt{2}}{\sqrt{16}} + 4\sqrt{2} \qquad \textit{Product rule}$$

$$= 4\sqrt{2} - \frac{3}{4}\sqrt{2} + 4\sqrt{2} \qquad \textit{Write } \frac{3\sqrt{2}}{\sqrt{16}} \textit{ as } \frac{3}{4}\sqrt{2}.$$

$$= \left(4 - \frac{3}{4} + 4\right)\sqrt{2} \qquad \textit{Simplify.}$$

$$= \frac{29\sqrt{2}}{4}$$

▶ **Now Try Exercise 115**

5   Divide Radical Expressions with Different Indices

Now we will divide radical expressions where the radicals have different indices. To divide such problems, write each radical in exponential form. Then use the rules of exponents, with the rational exponents, as was discussed in Section 11.2, to simplify the expression. Example 7 illustrates this procedure.

EXAMPLE 7 ▸ Simplify.   **a)** $\dfrac{\sqrt[5]{(m + n)^7}}{\sqrt[3]{(m + n)^4}}$ **b)** $\dfrac{\sqrt[3]{a^5 b^4}}{\sqrt{a^2 b}}$

Solution   Begin by writing the numerator and denominator with rational exponents.

a) $\dfrac{\sqrt[5]{(m + n)^7}}{\sqrt[3]{(m + n)^4}} = \dfrac{(m + n)^{7/5}}{(m + n)^{4/3}}$ *Write with rational exponents.*

$\qquad\qquad = (m + n)^{(7/5) - (4/3)}$ *Quotient rule for exponents*

$\qquad\qquad = (m + n)^{1/15}$

$\qquad\qquad = \sqrt[15]{m + n}$ *Write as a radical.*

b) $\dfrac{\sqrt[3]{a^5 b^4}}{\sqrt{a^2 b}} = \dfrac{(a^5 b^4)^{1/3}}{(a^2 b)^{1/2}}$ *Write with rational exponents.*

$\qquad\quad = \dfrac{a^{5/3} b^{4/3}}{a b^{1/2}}$ *Raise the product to a power.*

$\qquad\quad = a^{(5/3) - 1} b^{(4/3) - (1/2)}$ *Quotient rule for exponents*

$\qquad\quad = a^{2/3} b^{5/6}$

$\qquad\quad = a^{4/6} b^{5/6}$ *Write the fractions with denominator 6.*

$\qquad\quad = (a^4 b^5)^{1/6}$ *Rewrite using the laws of exponents.*

$\qquad\quad = \sqrt[6]{a^4 b^5}$ *Write as a radical.*

▸ **Now Try Exercise 133**

EXERCISE SET 11.5   Math XL   MyMathLab

MathXL®   MyMathLab

Concept/Writing Exercises

1. a) What is the conjugate of a binomial?
   b) What is the conjugate of $x - \sqrt{3}$?

2. What does it mean to rationalize a denominator?

3. a) Explain how to rationalize a denominator that contains a radical expression of one term.
   b) Rationalize $\dfrac{4}{\sqrt{3y}}$ using the procedure you specified in part **a)**.

4. a) Explain how to rationalize a denominator that contains a binomial in which one or both terms is a radical expression.

b) Rationalize $\dfrac{\sqrt{2} + \sqrt{5}}{\sqrt{2} - \sqrt{5}}$ using the procedure you specified in part **a)**.

5. What are the three conditions that must be met for a radical expression to be simplified?

6. Explain why each of the following is not simplified.

   a) $\sqrt{x^5}$ **b)** $\sqrt{\dfrac{1}{2}}$ **c)** $\dfrac{1}{\sqrt{3}}$

Simplify. In this exercise set, assume all variables represent positive real numbers.

7. $\dfrac{1}{\sqrt{3}}$ **8.** $\dfrac{1}{\sqrt{6}}$ **9.** $\dfrac{4}{\sqrt{5}}$ **10.** $\dfrac{3}{\sqrt{7}}$

11. $\dfrac{6}{\sqrt{6}}$ **12.** $\dfrac{17}{\sqrt{17}}$ **13.** $\dfrac{1}{\sqrt{z}}$ **14.** $\dfrac{y}{\sqrt{y}}$

15. $\dfrac{p}{\sqrt{2}}$

16. $\dfrac{m}{\sqrt{13}}$

17. $\dfrac{\sqrt{y}}{\sqrt{7}}$

18. $\dfrac{\sqrt{19}}{\sqrt{q}}$

19. $\dfrac{6\sqrt{3}}{\sqrt{6}}$

20. $\dfrac{15x}{\sqrt{x}}$

21. $\dfrac{\sqrt{x}}{\sqrt{y}}$

22. $\dfrac{2\sqrt{3}}{\sqrt{a}}$

23. $\sqrt{\dfrac{5m}{8}}$

24. $\dfrac{9\sqrt{3}}{\sqrt{y^3}}$

25. $\dfrac{2n}{\sqrt{18n}}$

26. $\sqrt{\dfrac{120x}{4y^3}}$

27. $\sqrt{\dfrac{18x^4y^3}{2z^3}}$

28. $\sqrt{\dfrac{7pq^4}{2r}}$

29. $\sqrt{\dfrac{20y^4z^3}{3xy^{-4}}}$

30. $\sqrt{\dfrac{5xy^6}{3z}}$

31. $\sqrt{\dfrac{48x^6y^5}{3z^3}}$

32. $\sqrt{\dfrac{45y^{12}z^{10}}{2x}}$

Simplify.

33. $\dfrac{1}{\sqrt[3]{2}}$

34. $\dfrac{1}{\sqrt[3]{4}}$

35. $\dfrac{8}{\sqrt[3]{y}}$

36. $\dfrac{2}{\sqrt[3]{a^2}}$

37. $\dfrac{1}{\sqrt[4]{3}}$

38. $\dfrac{z}{\sqrt[4]{4}}$

39. $\dfrac{a}{\sqrt[4]{8}}$

40. $\dfrac{8}{\sqrt[4]{z}}$

41. $\dfrac{5}{\sqrt[4]{z^2}}$

42. $\dfrac{13}{\sqrt[4]{z^3}}$

43. $\dfrac{10}{\sqrt[5]{y^3}}$

44. $\dfrac{x}{\sqrt[5]{y^4}}$

45. $\dfrac{2}{\sqrt[7]{a^4}}$

46. $\sqrt[3]{\dfrac{4x}{y}}$

47. $\sqrt[3]{\dfrac{1}{2x}}$

48. $\sqrt[3]{\dfrac{7c}{9y^2}}$

49. $\dfrac{5m}{\sqrt[4]{2}}$

50. $\dfrac{3}{\sqrt[4]{a}}$

51. $\sqrt[4]{\dfrac{5}{3x^3}}$

52. $\sqrt[4]{\dfrac{2x^3}{4y^2}}$

53. $\sqrt[3]{\dfrac{3x^2}{2y^2}}$

54. $\sqrt[3]{\dfrac{15x^6y^7}{2z^2}}$

55. $\sqrt[3]{\dfrac{14xy^2}{2z^2}}$

56. $\sqrt[6]{\dfrac{r^4s^9}{2r^5}}$

Multiply.

57. $(5-\sqrt{6})(5+\sqrt{6})$

58. $(7+\sqrt{3})(7-\sqrt{3})$

59. $(8+\sqrt{2})(8-\sqrt{2})$

60. $(6-\sqrt{7})(6+\sqrt{7})$

61. $(2-\sqrt{10})(2+\sqrt{10})$

62. $(3+\sqrt{17})(3-\sqrt{17})$

63. $(\sqrt{a}-\sqrt{b})(\sqrt{a}+\sqrt{b})$

64. $(\sqrt{x}-\sqrt{y})(\sqrt{x}+\sqrt{y})$

65. $(2\sqrt{x}-3\sqrt{y})(2\sqrt{x}+3\sqrt{y})$

66. $(5\sqrt{c}-4\sqrt{d})(5\sqrt{c}+4\sqrt{d})$

Simplify by rationalizing the denominator.

67. $\dfrac{2}{\sqrt{3}+1}$

68. $\dfrac{4}{\sqrt{2}+1}$

69. $\dfrac{1}{2+\sqrt{3}}$

70. $\dfrac{3}{5-\sqrt{7}}$

71. $\dfrac{5}{\sqrt{2}-7}$

72. $\dfrac{6}{\sqrt{2}+\sqrt{3}}$

73. $\dfrac{\sqrt{5}}{2\sqrt{5}-\sqrt{6}}$

74. $\dfrac{1}{\sqrt{17}-\sqrt{8}}$

75. $\dfrac{3}{6+\sqrt{x}}$

76. $\dfrac{4\sqrt{5}}{\sqrt{a}-3}$

77. $\dfrac{4\sqrt{x}}{\sqrt{x}-y}$

78. $\dfrac{\sqrt{8x}}{x+\sqrt{y}}$

79. $\dfrac{\sqrt{2}-2\sqrt{3}}{\sqrt{2}+4\sqrt{3}}$

80. $\dfrac{\sqrt{c}-\sqrt{2d}}{\sqrt{c}-\sqrt{d}}$

81. $\dfrac{\sqrt{a^3}+\sqrt{a^7}}{\sqrt{a}}$

82. $\dfrac{2\sqrt{xy}-\sqrt{xy}}{\sqrt{x}+\sqrt{y}}$

83. $\dfrac{4}{\sqrt{x+2}-3}$

84. $\dfrac{8}{\sqrt{y-3}+6}$

Simplify. These exercises are a combination of the types of exercises presented earlier in this exercise set.

85. $\sqrt{\dfrac{x}{16}}$

86. $\sqrt[4]{\dfrac{x^4}{16}}$

87. $\sqrt{\dfrac{2}{9}}$

88. $\sqrt{\dfrac{a}{b}}$

89. $(\sqrt{7}+\sqrt{6})(\sqrt{7}-\sqrt{6})$

90. $\sqrt[3]{\dfrac{1}{16}}$

91. $\sqrt{\dfrac{24x^3y^6}{5z}}$

92. $\dfrac{5}{4-\sqrt{y}}$

93. $\sqrt{\dfrac{28xy^4}{2x^3y^4}}$

94. $\dfrac{8x}{\sqrt[3]{5y}}$

95. $\dfrac{1}{\sqrt{a}+7}$

96. $\dfrac{\sqrt{x}}{\sqrt{x}+6\sqrt{y}}$

97. $-\dfrac{7\sqrt{x}}{\sqrt{98}}$

98. $\sqrt{\dfrac{2xy^4}{50xy^2}}$

99. $\sqrt[4]{\dfrac{3y^2}{2x}}$

100. $\sqrt{\dfrac{49x^2y^5}{3z}}$

101. $\sqrt[3]{\dfrac{32y^{12}z^{10}}{2x}}$ **102.** $\dfrac{\sqrt{3}+2}{\sqrt{2}+\sqrt{3}}$ **103.** $\dfrac{\sqrt{ar}}{\sqrt{a}-2\sqrt{r}}$ **104.** $\sqrt[4]{\dfrac{2}{9x}}$

105. $\dfrac{\sqrt[3]{6x}}{\sqrt[3]{5xy}}$ **106.** $\dfrac{\sqrt[3]{16m^{2}n}}{\sqrt[3]{2mn^{2}}}$ **107.** $\sqrt[4]{\dfrac{2x^{7}y^{12}z^{4}}{3x^{9}}}$ **108.** $\dfrac{9}{\sqrt{y+9}-\sqrt{y}}$

Simplify.

109. $\dfrac{1}{\sqrt{2}}+\dfrac{\sqrt{2}}{2}$ **110.** $\dfrac{1}{\sqrt{3}}+\dfrac{\sqrt{3}}{3}$ **111.** $\sqrt{5}-\dfrac{2}{\sqrt{5}}$

112. $\dfrac{\sqrt{6}}{2}-\dfrac{2}{\sqrt{6}}$ **113.** $4\sqrt{\dfrac{1}{6}}+\sqrt{24}$ **114.** $5\sqrt{3}-\dfrac{3}{\sqrt{3}}+2\sqrt{18}$

115. $5\sqrt{2}-\dfrac{2}{\sqrt{8}}+\sqrt{50}$ **116.** $\dfrac{2}{3}+\dfrac{1}{\sqrt{3}}+\sqrt{75}$ **117.** $\sqrt{\dfrac{1}{2}}+7\sqrt{2}+\sqrt{18}$

118. $\dfrac{1}{2}\sqrt{18}-\dfrac{3}{\sqrt{2}}-9\sqrt{50}$ **119.** $\dfrac{2}{\sqrt{50}}-3\sqrt{50}-\dfrac{1}{\sqrt{8}}$ **120.** $\dfrac{\sqrt{3}}{3}+\dfrac{5}{\sqrt{3}}+\sqrt{12}$

121. $\sqrt{\dfrac{3}{8}}+\sqrt{\dfrac{3}{2}}$ **122.** $2\sqrt{\dfrac{8}{3}}-4\sqrt{\dfrac{100}{6}}$ ⦿ **123.** $-2\sqrt{\dfrac{x}{y}}+3\sqrt{\dfrac{y}{x}}$

124. $-5x\sqrt{\dfrac{y}{y^{2}}}+9x\sqrt{\dfrac{1}{y}}$ **125.** $\dfrac{3}{\sqrt{a}}-\sqrt{\dfrac{9}{a}}+2\sqrt{a}$ **126.** $6\sqrt{x}+\dfrac{1}{\sqrt{x}}+\sqrt{\dfrac{1}{x}}$

Simplify.

⦿ **127.** $\dfrac{\sqrt{(a+b)^{4}}}{\sqrt[3]{a+b}}$ **128.** $\dfrac{\sqrt[3]{c+2}}{\sqrt[4]{(c+2)^{3}}}$ **129.** $\dfrac{\sqrt[5]{(a+2b)^{4}}}{\sqrt[3]{(a+2b)^{2}}}$ **130.** $\dfrac{\sqrt[6]{(r+3)^{5}}}{\sqrt[3]{(r+3)^{5}}}$

131. $\dfrac{\sqrt[3]{r^{2}s^{4}}}{\sqrt{rs}}$ **132.** $\dfrac{\sqrt{a^{2}b^{4}}}{\sqrt[3]{ab^{2}}}$ **133.** $\dfrac{\sqrt[5]{x^{4}y^{6}}}{\sqrt[3]{(xy)^{2}}}$ **134.** $\dfrac{\sqrt[6]{4m^{8}n^{4}}}{\sqrt[4]{m^{4}n^{2}}}$

Problem Solving

135. Illumination of a Light Under certain conditions the formula

$$d=\sqrt{\dfrac{72}{I}}$$

is used to show the relationship between the illumination on an object, I, in lumens per meter, and the distance, d, in meters, the object is from the light source. If the illumination on a person standing near a light source is 5.3 lumens per meter, how far is the person from the light source?

136. Strength of a Board When sufficient pressure is applied to a particular particle board, the particle board will break (or rupture). The thicker the particle board the greater will be the pressure that will need to be applied before the board breaks. The formula

$$T=\sqrt{\dfrac{0.05\,LB}{M}}$$

relates the thickness of a specific particle board, T, in inches, the board's length, L, in inches, the board's load that will cause the board to rupture, B, in pounds, and the modulus of rupture, M, in pounds per square inch. The modulus of rupture is a constant determined by sample tests on the specific type of particle board.

Find the thickness of a 36-inch-long particle board if the modulus of rupture is 2560 pounds per square inch and the board ruptures when 800 pounds are applied.

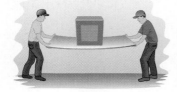

137. Volume of a Fish Tank A new restaurant wants to have a spherical fish tank in its lobby. The radius, r, in inches, of a spherical tank can be found by the formula

$$r=\sqrt[3]{\dfrac{3V}{4\pi}}$$

where V is the volume of the tank in cubic inches. Find the radius of a spherical tank whose volume is 7238.23 cubic inches.

138. Consecutive Numbers If we consider the set of consecutive natural numbers $1, 2, 3, 4, \ldots, n$ to be the population, the standard deviation, σ, which is a measure of the spread of the data from the mean, can be calculated by the formula

$$\sigma=\sqrt{\dfrac{n^{2}-1}{12}}$$

where n represents the number of natural numbers in the population. Find the standard deviation for the first 100 consecutive natural numbers.

139. U.S. Farms The number of farms in the United States is declining annually (however, the size of the remaining farms is increasing). A function that can be used to estimate the number of farms, $N(t)$, in millions, is

$$N(t) = \frac{6.21}{\sqrt[4]{t}}$$

where t is years since 1959 and $1 \le t \le 50$. Estimate the number of farms in the United States in **a)** 1960 and **b)** 2008.

140. Infant Mortality Rate The U.S. infant mortality rate has been declining steadily. The infant mortality rate, $N(t)$, defined as deaths per 1000 live births, can be estimated by the function

$$N(t) = \frac{28.46}{\sqrt[3]{t^2}}$$

where t is years since 1969 and $1 \le t \le 37$. Estimate the infant mortality rate in **a)** 1970 and **b)** 2006.

141. Which is greater, $\dfrac{2}{\sqrt{2}}$ or $\dfrac{3}{\sqrt{3}}$? Explain.

142. Which is greater, $\dfrac{\sqrt{3}}{2}$ or $\dfrac{2}{\sqrt{3}}$? Explain.

143. Which is greater, $\dfrac{1}{\sqrt{3}+2}$ or $2 + \sqrt{3}$? (Do not use a calculator.) Explain how you determined your answer.

144. Which is greater, $\dfrac{1}{\sqrt{3}} + \sqrt{75}$ or $\dfrac{2}{\sqrt{12}} + \sqrt{48} + 2\sqrt{3}$? (Do not use a calculator.) Explain how you determined your answer.

145. Consider the functions $f(x) = x^{a/2}$ and $g(x) = x^{b/3}$.

    a) List three values for a that will result in $x^{a/2}$ being a perfect square.

    b) List three values for b that will result in $x^{b/3}$ being a perfect cube.

    c) If $x \ge 0$, find $(f \cdot g)(x)$.

    d) If $x \ge 0$, find $(f/g)(x)$.

Rationalize each denominator.

146. $\dfrac{1}{\sqrt{a+b}}$

147. $\dfrac{3}{\sqrt{2a-3b}}$

In higher math courses, it may be necessary to rationalize the numerators of radical expressions. Rationalize the numerators of the following expressions. (Your answers will contain radicals in the denominators.)

148. $\dfrac{\sqrt{7}}{3}$

149. $\dfrac{5-\sqrt{5}}{6}$

150. $\dfrac{6\sqrt{x}-\sqrt{3}}{x}$

151. $\dfrac{\sqrt{x+h}-\sqrt{x}}{h}$

Group Activity

Similar Figures *The following two exercises will reinforce many of the concepts presented in this chapter. Work each problem as a group. Make sure each member of the group understands each step in obtaining the solution. The figures in each exercise are similar. For each exercise, use a proportion to find the length of side x. Write the answer in radical form with a rationalized denominator.*

152.

153.

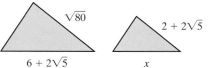

Cumulative Review Exercises

[2.6] **154.** Solve the equation $A = \dfrac{1}{2}h(b_1 + b_2)$ for b_2.

[3.4] **155. Moving Vehicles** Two cars leave from West Point at the same time traveling in opposite directions. One travels 10 miles per hour faster than the other. If the two cars are 270 miles apart after 3 hours, find the speed of each car.

[5.5] **156.** Multiply $(x - 2)(4x^2 + 9x - 2)$.

[7.6] **157.** Solve $\dfrac{x}{2} - \dfrac{4}{x} = -\dfrac{7}{2}$.

11.6 Solving Radical Equations

1 Solve equations containing one radical.

2 Solve equations containing two radicals.

3 Solve equations containing two radical terms and a nonradical term.

4 Solve applications using radical equations.

5 Solve for a variable in a radicand.

1 Solve Equations Containing One Radical

A **radical equation** is an equation that contains a variable in a radicand.

Examples of Radical Equations

$\sqrt{x} = 5$, $\sqrt[3]{y + 4} = 9$, $\sqrt{x - 2} = 7 + \sqrt{x + 8}$

To Solve Radical Equations

1. Rewrite the equation so that one radical containing a variable is by itself (isolated) on one side of the equation.

2. Raise each side of the equation to a power equal to the index of the radical.

3. Combine like terms.

4. If the equation still contains a term with a variable in a radicand, repeat steps 1 through 3.

5. Solve the resulting equation for the variable.

6. Check all solutions in the original equation for extraneous solutions.

Recall from Section 7.6 that an extraneous solution is a number obtained when solving an equation that is not a solution to the original equation.

The following examples illustrate the procedure for solving radical equations.

EXAMPLE 1 ▶ Solve the equation $\sqrt{x} = 5$.

Solution The square root containing the variable is already by itself on one side of the equation. Square both sides of the equation.

$$\sqrt{x} = 5$$
$$(\sqrt{x})^2 = (5)^2$$
$$x = 25$$

Check $\sqrt{x} = 5$

$$\sqrt{25} \overset{?}{=} 5$$
$$5 = 5 \quad \textit{True}$$

▶ **Now Try Exercise 11**

EXAMPLE 2 ▶ Solve.

a) $\sqrt{x - 4} - 6 = 0$ **b)** $\sqrt[3]{x} + 10 = 8$ **c)** $\sqrt{x} + 3 = 0$

Solution The first step in each part will be to isolate the term containing the radical.

a)

$$\sqrt{x - 4} - 6 = 0$$
$$\sqrt{x - 4} = 6 \qquad \textit{Isolate the radical containing the variable.}$$
$$(\sqrt{x - 4})^2 = 6^2 \qquad \textit{Square both sides.}$$
$$x - 4 = 36 \qquad \textit{Solve for the variable.}$$
$$x = 40$$

A check will show that 40 is the solution.

b)

$$\sqrt[3]{x} + 10 = 8$$
$$\sqrt[3]{x} = -2 \qquad \textit{Isolate the radical containing the variable.}$$
$$(\sqrt[3]{x})^3 = (-2)^3 \qquad \textit{Cube both sides.}$$
$$x = -8$$

A check will show that −8 is the solution.

c)
$$\sqrt{x} + 3 = 0$$
$$\sqrt{x} = -3 \qquad \textit{Isolate the radical containing the variable.}$$
$$(\sqrt{x})^2 = (-3)^2 \qquad \textit{Square both sides.}$$
$$x = 9$$

Check $\sqrt{x} + 3 = 0$
$$\sqrt{9} + 3 \overset{?}{=} 0$$
$$3 + 3 \overset{?}{=} 0$$
$$6 = 0 \qquad \textit{False}$$

A check shows that 9 is not a solution. The answer to part **c)** is "no real solution." You may have realized there was no real solution to the problem when you obtained the equation $\sqrt{x} = -3$, because $\sqrt{x}$ cannot equal a negative real number.

▶ **Now Try Exercise 17**

Helpful Hint

Don't forget to check your solutions in the original equation. When you raise both sides of an equation to a power you may introduce extraneous solutions.

Consider the equation $x = 2$. Note what happens when you square both sides of the equation.

$$x = 2$$
$$x^2 = 2^2$$
$$x^2 = 4$$

Note that the equation $x^2 = 4$ has two solutions, $+2$ and -2. Since the original equation $x = 2$ has only one solution, 2, we introduced the extraneous solution, -2.

EXAMPLE 3 ▶ Solve $\sqrt{2x - 3} = x - 3$.

Solution   Since the radical is already isolated, we square both sides of the equation. Then we solve the resulting quadratic equation.

$$(\sqrt{2x - 3})^2 = (x - 3)^2$$
$$2x - 3 = (x - 3)(x - 3)$$
$$2x - 3 = x^2 - 6x + 9$$
$$0 = x^2 - 8x + 12$$

Now we factor and use the zero-factor property.

$$x^2 - 8x + 12 = 0$$
$$(x - 6)(x - 2) = 0$$
$$x - 6 = 0 \quad \text{or} \quad x - 2 = 0$$
$$x = 6 \qquad\qquad x = 2$$

Check
$$x = 6 \qquad\qquad\qquad x = 2$$
$$\sqrt{2x - 3} = x - 3 \qquad\qquad \sqrt{2x - 3} = x - 3$$
$$\sqrt{2(6) - 3} \overset{?}{=} 6 - 3 \qquad\qquad \sqrt{2(2) - 3} \overset{?}{=} 2 - 3$$
$$\sqrt{9} \overset{?}{=} 3 \qquad\qquad\qquad \sqrt{1} \overset{?}{=} -1$$
$$3 = 3 \quad \textit{True} \qquad\qquad 1 = -1 \quad \textit{False}$$

Thus, 6 is a solution, but 2 is not a solution to the equation. The 2 is an extraneous solution because 2 satisfies the equation $(\sqrt{2x - 3})^2 = (x - 3)^2$, but not the original equation, $\sqrt{2x - 3} = x - 3$.

▶ **Now Try Exercise 43**

USING YOUR GRAPHING CALCULATOR

In Example 3, we found the solution to $\sqrt{2x-3} = x-3$ to be 6. If we let $Y_1 = \sqrt{2x-3}$ and $Y_2 = x-3$ and graph Y_1 and Y_2 on a graphing calculator, we get **Figure 11.3**. Notice the graphs appear to intersect at $x = 6$, which is what we expect.

The table of values in **Figure 11.4** shows that the y-coordinate at the point of intersection is 3. In the table, ERROR appears under Y_1 for the values of x of 0 and 1. For any values less than $\frac{3}{2}$, the value of $2x-3$ is negative and therefore $\sqrt{2x-3}$ is not a real number. The domain of function Y_1 is $\left\{ x \,\middle|\, x \geq \frac{3}{2} \right\}$, which may be found by solving the inequality $2x-3 \geq 0$.

You can use your graphing calculator to either solve or check radical equations.

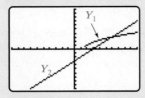

FIGURE 11.3

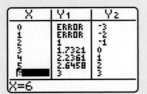

FIGURE 11.4

EXERCISES

Use your graphing calculator to determine whether the indicated value is the solution to the radical equation. If it is not the solution, use your grapher to determine the solution.

  1. $\sqrt{2x+9} = 5(x-7), 8$ **2.** $\sqrt{3x+4} = \sqrt{x+12}, 6$

EXAMPLE 4 ▸ Solve $x - 2\sqrt{x} - 3 = 0$.

Solution   First, isolate the radical term by writing the radical term by itself on one side of the equation.

$$x - 2\sqrt{x} - 3 = 0$$
$$-2\sqrt{x} = -x + 3$$
$$2\sqrt{x} = x - 3$$

Now square both sides of the equation.

$$(2\sqrt{x})^2 = (x-3)^2$$
$$4x = x^2 - 6x + 9$$
$$0 = x^2 - 10x + 9$$
$$0 = (x-1)(x-9)$$
$$x - 1 = 0 \quad \text{or} \quad x - 9 = 0$$
$$x = 1 \qquad\qquad\qquad x = 9$$

Check $x = 1$ $x = 9$

$$x - 2\sqrt{x} - 3 = 0 \qquad\qquad\qquad x - 2\sqrt{x} - 3 = 0$$
$$1 - 2\sqrt{1} - 3 \overset{?}{=} 0 \qquad\qquad\qquad 9 - 2\sqrt{9} - 3 \overset{?}{=} 0$$
$$1 - 2(1) - 3 \overset{?}{=} 0 \qquad\qquad\qquad 9 - 2(3) - 3 \overset{?}{=} 0$$
$$1 - 2 - 3 \overset{?}{=} 0 \qquad\qquad\qquad\quad 9 - 6 - 3 \overset{?}{=} 0$$
$$-4 = 0 \quad \textit{False} \qquad\qquad\qquad\quad 3 - 3 \overset{?}{=} 0$$
$$0 = 0 \quad \textit{True}$$

The solution is 9. The value 1 is an extraneous solution.

▸ **Now Try Exercise 41**

2 Solve Equations Containing Two Radicals

Now we will look at some equations that contain two radicals.

EXAMPLE 5 ▸ Solve $\sqrt{9x^2 + 6} = 3\sqrt{x^2 + x - 2}$.

Solution   Since the two radicals appear on different sides of the equation, we square both sides of the equation.

$$\left(\sqrt{9x^2 + 6}\right)^2 = \left(3\sqrt{x^2 + x - 2}\right)^2 \qquad \textit{Square both sides.}$$

$$9x^2 + 6 = 9(x^2 + x - 2)$$

$$9x^2 + 6 = 9x^2 + 9x - 18 \qquad \textit{Distributive property.}$$

$$6 = 9x - 18 \qquad \textit{9x}^2 \textit{ was subtracted from both sides.}$$

$$24 = 9x$$

$$\frac{8}{3} = x$$

A check will show that $\frac{8}{3}$ is the solution.

▸ **Now Try Exercise 27**

In higher mathematics courses, equations are sometimes given using exponents rather than radicals. Example 6 illustrates such an equation.

EXAMPLE 6 ▸ For $f(x) = 3(x - 2)^{1/3}$ and $g(x) = (17x - 14)^{1/3}$, find all values of x for which $f(x) = g(x)$.

Solution   You should realize that alternate ways of writing $f(x)$ and $g(x)$ are $f(x) = 3\sqrt[3]{x - 2}$ and $g(x) = \sqrt[3]{17x - 14}$. We could therefore work this example using radicals, but we will work instead with rational exponents. We set the two functions equal to each other and solve for x.

$$f(x) = g(x)$$

$$3(x - 2)^{1/3} = (17x - 14)^{1/3}$$

$$[3(x - 2)^{1/3}]^3 = [(17x - 14)^{1/3}]^3 \qquad \textit{Cube both sides.}$$

$$3^3(x - 2) = 17x - 14$$

$$27(x - 2) = 17x - 14$$

$$27x - 54 = 17x - 14$$

$$10x - 54 = -14$$

$$10x = 40$$

$$x = 4$$

A check will show that the solution is 4. If you substitute 4 into both $f(x)$ and $g(x)$, you will find they both simplify to $3\sqrt[3]{2}$. Check this now.

▸ **Now Try Exercise 69**

In Example 6, if you solve the equation $3\sqrt[3]{x - 2} = \sqrt[3]{17x - 14}$ you will obtain the solution 4. For additional practice, do this now.

3 Solve Equations Containing Two Radical Terms and a Nonradical Term

When a radical equation contains two radical terms and a third nonradical term, you will sometimes need to raise both sides of the equation to a given power twice to obtain the solution. First, isolate one radical term. Then raise both sides of the equation to the given power. This will eliminate one of the radicals. Next, isolate the remaining radical on one side of the equation. Then raise both sides of the equation to the given power a second time. This procedure is illustrated in Example 7.

EXAMPLE 7 ▸ Solve $\sqrt{5x - 1} - \sqrt{3x - 2} = 1$.

Solution We must isolate one radical term on one side of the equation. We will begin by adding $\sqrt{3x - 2}$ to both sides of the equation to isolate $\sqrt{5x - 1}$. Then we will square both sides of the equation and combine like terms.

$$\sqrt{5x - 1} = 1 + \sqrt{3x - 2}$$ Isolate $\sqrt{5x - 1}$.
$$(\sqrt{5x - 1})^2 = (1 + \sqrt{3x - 2})^2$$ Square both sides.
$$5x - 1 = (1 + \sqrt{3x - 2})(1 + \sqrt{3x - 2})$$ Write as a product.
$$5x - 1 = 1 + \sqrt{3x - 2} + \sqrt{3x - 2} + (\sqrt{3x - 2})^2$$ Multiply.
$$5x - 1 = 1 + 2\sqrt{3x - 2} + 3x - 2$$ Combine like terms; simplify.
$$5x - 1 = 3x - 1 + 2\sqrt{3x - 2}$$ Combine like terms.
$$2x = 2\sqrt{3x - 2}$$ Isolate the radical term.
$$x = \sqrt{3x - 2}$$ Both sides were divided by 2.

We have isolated the remaining radical term. We now square both sides of the equation again and solve for x.

$$x = \sqrt{3x - 2}$$
$$x^2 = (\sqrt{3x - 2})^2$$ Square both sides.
$$x^2 = 3x - 2$$
$$x^2 - 3x + 2 = 0$$
$$(x - 2)(x - 1) = 0$$
$$x - 2 = 0 \quad \text{or} \quad x - 1 = 0$$
$$x = 2 \qquad\qquad x = 1$$

A check will show that both 2 and 1 are solutions of the equation.

▸ **Now Try Exercise 61**

EXAMPLE 8 ▸ For $f(x) = \sqrt{5x - 1} - \sqrt{3x - 2}$, find all values of x for which $f(x) = 1$.

Solution Substitute 1 for $f(x)$. This gives

$$1 = \sqrt{5x - 1} - \sqrt{3x - 2}$$

Since this is the same equation we solved in Example 7, the answers are $x = 2$ and $x = 1$. Verify for yourself that $f(2) = 1$ and $f(1) = 1$.

▸ **Now Try Exercise 121**

Avoiding Common Errors

In Chapter 5, we stated that $(a + b)^2 \neq a^2 + b^2$. Be careful when squaring a binomial like $1 + \sqrt{x}$. Look at the following computations carefully so that you do not make the mistake shown on the right.

CORRECT	INCORRECT
$(1 + \sqrt{x})^2 = (1 + \sqrt{x})(1 + \sqrt{x})$	$(1 + \sqrt{x})^2 = 1^2 + (\sqrt{x})^2$
$\qquad\quad$ F $\quad$ O $\quad$ I $\quad$ L	
$= 1 + \sqrt{x} + \sqrt{x} + \sqrt{x}\,\sqrt{x}$	$= 1 + x$
$= 1 + 2\sqrt{x} + x$	

4 Solve Applications Using Radical Equations

Now we will look at a few of the many applications of radicals.

EXAMPLE 9 ▶ The Green Monster In Fenway Park, where the Boston Red Sox play baseball, the distance from home plate down the third base line to the bottom of the wall in left field is 310 feet. In left field at the end of the baseline there is a green wall perpendicular to the ground that is 37 feet tall. This green wall is commonly known as *the Green Monster* (see photo). Determine the distance from home plate to the top of the Green Monster along the third base line.

Solution Understand **Figure 11.5** illustrates the problem. We need to find the distance from home plate to the top of the wall in left field.

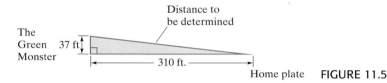

The Green Monster 37 ft

Distance to be determined

310 ft.

Home plate   FIGURE 11.5

Translate To solve the problem we use the Pythagorean Theorem, which was discussed earlier, $\text{leg}^2 + \text{leg}^2 = \text{hyp}^2$, or $a^2 + b^2 = c^2$.

$$310^2 + 37^2 = c^2 \qquad \textit{Substitute known values.}$$

Carry Out

$$96{,}100 + 1369 = c^2$$

$$97{,}469 = c^2$$

$$\sqrt{97{,}469} = \sqrt{c^2} \qquad \textit{Take the square root of both sides.}$$

$$\sqrt{97{,}469} = c \qquad \textit{* See footnote.}$$

$$312.20 \approx c$$

Answer The distance from home plate to the top of the wall is about 312.20 feet.

▶ **Now Try Exercise 99**

EXAMPLE 10 ▶ Period of a Pendulum The length of time it takes for a pendulum to make one complete swing back and forth is called the *period* of the pendulum. See **Figure 11.6**. The period of a pendulum, T, in seconds, can be found by the formula $T = 2\pi\sqrt{\dfrac{L}{32}}$, where L is the length of the pendulum in feet. Find the period of a pendulum if its length is 5 feet.

Solution Substitute 5 for L and 3.14 for π in the formula. If you have a calculator that has a $\boxed{\pi}$ key, use it to enter π.

$$T = 2\pi\sqrt{\frac{L}{32}}$$

$$\approx 2(3.14)\sqrt{\frac{5}{32}}$$

$$\approx 2(3.14)\sqrt{0.15625} \approx 2.48$$

Thus, the period is about 2.48 seconds. If you have a grandfather clock with a 5-foot pendulum, it will take about 2.48 seconds for it to swing back and forth.

▶ **Now Try Exercise 103**

FIGURE 11.6

---

*$c^2 = 97{,}469$ has two solutions: $c = \sqrt{97{,}469}$ and $c = -\sqrt{97{,}469}$. Since we are solving for a length, which must be a positive quantity, we use the positive root.

5   Solve for a Variable in a Radicand

You may be given a formula and be asked to solve for a variable in a radicand. To do so, follow the same general procedure used to solve a radical equation. Begin by isolating the radical expression. Then raise both sides of the equation to the same power as the index of the radical. This procedure is illustrated in Example 11 **b)**.

EXAMPLE 11 ▸ **Error of Estimation**   A formula in statistics for finding the maximum error of estimation is $E = Z\dfrac{\sigma}{\sqrt{n}}$.

a) Find E if $Z = 1.28$, $\sigma = 10$, and $n = 36$.

b) Solve this equation for n.

Solution

a) $E = Z\dfrac{\sigma}{\sqrt{n}} = 1.28\left(\dfrac{10}{\sqrt{36}}\right) = 1.28\left(\dfrac{10}{6}\right) \approx 2.13$

b) First multiply both sides of the equation by $\sqrt{n}$ to eliminate fractions. Then isolate $\sqrt{n}$. Finally, solve for n by squaring both sides of the equation.

$$E = Z\dfrac{\sigma}{\sqrt{n}}$$

$$\sqrt{n}\,(E) = \left(Z\dfrac{\sigma}{\sqrt{n}}\right)\sqrt{n} \qquad \textit{Eliminate fractions.}$$

$$\sqrt{n}\,(E) = Z\sigma$$

$$\sqrt{n} = \dfrac{Z\sigma}{E} \qquad \textit{Isolate the radical term.}$$

$$(\sqrt{n})^2 = \left(\dfrac{Z\sigma}{E}\right)^2 \qquad \textit{Square both sides.}$$

$$n = \left(\dfrac{Z\sigma}{E}\right)^2 \quad \text{or} \quad n = \dfrac{Z^2\sigma^2}{E^2}$$

▸ **Now Try Exercise 75**

EXERCISE SET 11.6   *Math XP* **MyMathLab**
MathXL®   MyMathLab

Concept/Writing Exercises

1. a) Explain how to solve a radical equation.
   b) Solve $\sqrt{2x + 26} - 2 = 4$ using the procedure you gave in part **a)**.

2. Consider the equation $\sqrt{x + 3} = -\sqrt{2x - 1}$. Explain why this equation can have no real solution.

3. Consider the equation $-\sqrt{x^2} = \sqrt{(-x)^2}$. By studying the equation, can you determine its solution? Explain.

4. Consider the equation $\sqrt[3]{x^2} = -\sqrt[3]{x^2}$. By studying the equation, can you determine its solution? Explain.

5. Explain without solving the equation how you can tell that $\sqrt{x - 3} + 4 = 0$ has no solution.

6. Why is it necessary to check solutions to radical equations?

7. Does the equation $\sqrt{x} = 5$ have one or two solutions? Explain.

8. Does the equation $x^2 = 9$ have one or two solutions? Explain.

Practice the Skills

Solve and check your solution(s). If the equation has no real solution, so state.

9. $\sqrt{x} = 4$

10. $\sqrt{x} = 13$

11. $\sqrt{x} = -9$

12. $\sqrt[3]{x} = 4$

13. $\sqrt[3]{x} = -4$

14. $\sqrt{a} + 5 = 0$

15. $\sqrt{2x + 3} = 5$

16. $\sqrt[3]{7x - 6} = 4$

17. $\sqrt[3]{3x} + 4 = 7$

18. $2\sqrt{4x + 5} = 14$

19. $\sqrt[3]{2x + 29} = 3$

20. $\sqrt[3]{6x + 2} = -4$

21. $\sqrt[4]{x} = 3$

22. $\sqrt[4]{x} = -3$

23. $\sqrt[4]{x + 10} = 3$

24. $\sqrt[4]{3x - 2} = 2$

25. $\sqrt[4]{2x + 1} + 6 = 2$

26. $\sqrt{2x + 7} = 13$

27. $\sqrt{x + 8} = \sqrt{x - 8}$

28. $\sqrt{r + 5} + 7 = 10$

29. $2\sqrt[3]{x - 1} = \sqrt[3]{x^2 + 2x}$

30. $\sqrt[3]{6t - 1} = \sqrt[3]{2t + 3}$

31. $\sqrt[4]{x + 8} = \sqrt[4]{2x}$

32. $\sqrt[4]{3x - 1} + 4 = 0$

33. $\sqrt{5x + 1} - 6 = 0$

34. $\sqrt{x^2 + 12x + 3} = -x$

35. $\sqrt{m^2 + 6m - 4} = m$

36. $\sqrt{x^2 + 3x + 12} = x$

37. $\sqrt{5c + 1} - 9 = 0$

38. $\sqrt{b^2 - 2} = b + 4$

39. $\sqrt{z^2 + 5} = z + 1$

40. $\sqrt{x} + 6x = 1$

41. $\sqrt{2y + 5} + 5 - y = 0$

42. $\sqrt{4x + 1} = \dfrac{1}{2}x + 2$

43. $\sqrt{5x + 6} = 2x - 6$

44. $\sqrt{4b + 5} + b = 10$

45. $(2a + 9)^{1/2} - a + 3 = 0$

46. $(3x + 4)^{1/2} - x = -2$

47. $(2x^2 + 4x + 9)^{1/2} = (2x^2 + 9)^{1/2}$

48. $(2x + 1)^{1/2} + 7 = x$

49. $(r + 4)^{1/3} = (3r + 10)^{1/3}$

50. $(7x + 6)^{1/3} + 4 = 0$

51. $(5x + 7)^{1/4} = (9x + 1)^{1/4}$

52. $(5b + 3)^{1/4} = (2b + 17)^{1/4}$

53. $\sqrt[4]{x + 5} = -2$

54. $\sqrt{x^2 + x - 1} = -\sqrt{x + 3}$

Solve. You will have to square both sides of the equation twice to eliminate all radicals.

55. $\sqrt{4x + 1} = \sqrt{2x} + 1$

56. $3\sqrt{b} - 1 = \sqrt{b + 21}$

57. $\sqrt{3a + 1} = \sqrt{a - 4} + 3$

58. $\sqrt{x + 1} = 2 - \sqrt{x}$

59. $\sqrt{x + 3} = \sqrt{x} - 3$

60. $\sqrt{y + 1} = 2 + \sqrt{y - 7}$

61. $\sqrt{x + 7} = 6 - \sqrt{x - 5}$

62. $\sqrt{b - 3} = 4 - \sqrt{b + 5}$

63. $\sqrt{4x - 3} = 2 + \sqrt{2x - 5}$

64. $\sqrt{r + 10} + 2 + \sqrt{r - 5} = 0$

65. $\sqrt{y + 1} = \sqrt{y + 10} - 3$

66. $3 + \sqrt{x + 1} = \sqrt{3x + 12}$

For each pair of functions, find all real values of x where $f(x) = g(x)$.

67. $f(x) = \sqrt{x + 8},\ g(x) = \sqrt{2x + 1}$

68. $f(x) = \sqrt{x^2 - 6x + 10},\ g(x) = \sqrt{x - 2}$

69. $f(x) = \sqrt[3]{5x - 19},\ g(x) = \sqrt[3]{6x - 23}$

70. $f(x) = (14x - 8)^{1/2},\ g(x) = 2(3x + 2)^{1/2}$

71. $f(x) = 2(8x + 24)^{1/3},\ g(x) = 4(2x - 2)^{1/3}$

72. $f(x) = 2\sqrt{x + 2},\ g(x) = 8 - \sqrt{x + 14}$

Solve each formula for the indicated variable.

73. $p = \sqrt{2v}$, for v

74. $l = \sqrt{4r}$, for r

75. $v = \sqrt{2gh}$, for g

76. $v = \sqrt{\dfrac{2E}{m}}$, for E

77. $v = \sqrt{\dfrac{FR}{M}}$, for F

78. $\omega = \sqrt{\dfrac{a_0}{b_0}}$, for b_0

79. $x = \sqrt{\dfrac{m}{k}}V_0$, for m

80. $T = 2\pi\sqrt{\dfrac{L}{32}}$, for L

81. $r = \sqrt{\dfrac{A}{\pi}}$, for A

82. $r = \sqrt[3]{\dfrac{3V}{4\pi}}$, for V

Problem Solving

Use the Pythagorean Theorem to find the length of the unknown side of each triangle. Write the answer as a radical in simplified form.

83.
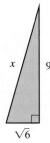

84.

85.
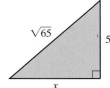

86.
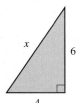

Solve. You will need to square both sides of the equation twice.

87. $\sqrt{x + 5} - \sqrt{x} = \sqrt{x - 3}$

88. $\sqrt{2x} - \sqrt{x - 4} = \sqrt{12 - x}$

89. $\sqrt{4y + 6} + \sqrt{y + 5} = \sqrt{y + 1}$

90. $\sqrt{2b - 2} + \sqrt{b - 5} = \sqrt{4b}$

91. $\sqrt{c + 1} + \sqrt{c - 2} = \sqrt{3c}$

92. $\sqrt{2t - 1} + \sqrt{t - 4} = \sqrt{3t + 1}$

93. $\sqrt{a + 2} - \sqrt{a - 3} = \sqrt{a - 6}$

94. $\sqrt{r - 1} - \sqrt{r + 6} = \sqrt{r - 9}$

Solve. You will need to square both sides of the equation twice.

95. $\sqrt{2 - \sqrt{x}} = \sqrt{x}$

96. $\sqrt{6 + \sqrt{x + 4}} = \sqrt{2x - 1}$

97. $\sqrt{2 + \sqrt{x + 1}} = \sqrt{7 - x}$

98. $\sqrt{1 + \sqrt{x - 1}} = \sqrt{x - 6}$

99. Baseball Diamond A regulation baseball diamond is a square with 90 feet between bases. How far is second base from home plate?

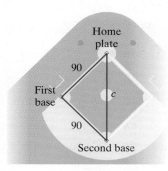

100. Wire from Telephone Pole A telephone pole is at a right, or 90°, angle with the ground as shown in the figure. Find the length of the wire that connects to the pole 40 feet above the ground and is anchored to the ground 20 feet from the base of the pole.

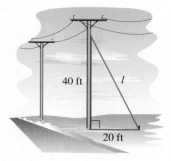

101. Side of a Garden When you are given the area of a square, the length of a side can be found by the formula $s = \sqrt{A}$. Find the side of Tom Kim's square garden if it has an area of 169 square feet.

102. Radius of Basketball Hoop When you are given the area of a circle, its radius can be found by the formula $r = \sqrt{A/\pi}$.

a) Find the radius of a basketball hoop if the area enclosed by the hoop is 254.47 square inches.

b) If the diameter of a basketball is 9 inches, what is the minimum distance possible between the hoop and the ball when the center of the ball is in the center of the hoop?

103. Period of a Pendulum The formula for the period of a pendulum is

$$T = 2\pi\sqrt{\frac{l}{g}}$$

where T is the period in seconds, l is its length in feet, and g is the acceleration of gravity. On Earth, gravity is 32 ft/sec². The formula when used on Earth becomes

$$T = 2\pi\sqrt{\frac{l}{32}}$$

a) Find the period of a pendulum whose length is 8 feet.

b) If the length of a pendulum is doubled, what effect will this have on the period? Explain.

c) The gravity on the Moon is 1/6 that on Earth. If a pendulum has a period of 2 seconds on Earth, what will be the period of the same pendulum on the Moon?

104. Diagonal of a Suitcase A formula for the length of a diagonal from the upper corner of a box to the opposite lower corner is $d = \sqrt{L^2 + W^2 + H^2}$, where L, W, and H are the length, width, and height, respectively.

a) Find the length of the diagonal of a suitcase of length 22 inches, width 15 inches, and height 12 inches.

b) If the length, width, and height are all doubled, how will the diagonal change?

c) Solve the formula for W.

105. Blood Flowing in an Artery The formula

$$r = \sqrt[4]{\frac{8\mu l}{\pi R}}$$

is used in determining movement of blood through arteries. In the formula, R represents the resistance to blood flow, μ is the viscosity of blood, l is the length of the artery, and r is the radius of the artery. Solve this equation for R.

106. Falling Object The formula

$$t = \frac{\sqrt{19.6s}}{9.8}$$

can be used to tell the time, t, in seconds, that an object has been falling if it has fallen s meters. Suppose an object has been dropped from a helicopter and has fallen 100 meters. How long has it been in free fall?

107. Earth Days For any planet in our solar system, its "year" is the time it takes for the planet to revolve once around the Sun. The number of Earth days in a given planet's year, N, is approximated by the formula $N = 0.2(\sqrt{R})^3$, where R is the mean distance of the planet to the Sun in millions of kilometers. Find the number of Earth days in the year of the planet Earth, whose mean distance to the Sun is 149.4 million kilometers.

108. Earth Days Find the number of Earth days in the year of the planet Mercury, whose mean distance to the Sun is 58 million kilometers. See Exercise 107.

109. Forces on a Car When two forces, F_1 and F_2, pull at right angles to each other as illustrated below, the resultant, or the effective force, R, can be found by the formula $R = \sqrt{F_1^2 + F_2^2}$. Two cars are trying to pull a third out of the mud, as illustrated. If car A is exerting a force of 60 pounds and car B is exerting a force of 80 pounds, find the resulting force on the car stuck in the mud.

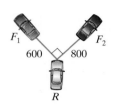
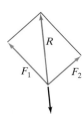

110. Escape Velocity The escape velocity, or the velocity needed for a spacecraft to escape a planet's gravitational field, is found by the formula $v_e = \sqrt{2gR}$, where g is the force of gravity of the planet and R is the radius of the planet. Find the escape velocity for Earth, in meters per second, where $g = 9.75$ meters per second squared and $R = 6{,}370{,}000$ meters.

111. Motion of a Wave A formula used in the study of shallow water wave motion is $c = \sqrt{gH}$, in which c is wave velocity, H is water depth, and g is the acceleration due to gravity. Find the wave velocity if the water's depth is 10 feet. (Use $g = 32$ ft/sec^2.)

112. Diagonal of a Box The top of a rectangular box measures 20 inches by 32 inches. Find the length of the diagonal for the top of the box.

113. Flower Garden A rectangular flower garden measures 25 meters by 32 meters. Find the length of the diagonal for the garden.

114. Speed of Sound When sound travels through air (or any gas), the velocity of the sound wave is dependent on the air (or gas) temperature. The velocity, v, in meters per second, at air temperature, t, in degrees Celsius, can be found by the formula

$$v = 331.3\sqrt{1 + \frac{t}{273}}$$

Find the speed of sound in air whose temperature is 20°C (equivalent to 68°F).

A formula that we have already mentioned and that we will be discussing in more detail shortly is the quadratic formula

$$x = \frac{-b \pm \sqrt{b^2 - 4ac}}{2a}$$

115. Find x when $a = 1, b = 0, c = -4$.

116. Find x when $a = 1, b = 1, c = -12$.

117. Find x when $a = -1, b = 4, c = 5$.

118. Find x when $a = 2, b = 5, c = -12$.

Given $f(x)$, find all values of x for which $f(x)$ is the indicated value.

119. $f(x) = \sqrt{x - 5}, f(x) = 5$

120. $f(x) = \sqrt[3]{2x + 3}, f(x) = 3$

121. $f(x) = \sqrt{3x^2 - 11} + 7, f(x) = 15$

122. $f(x) = 8 + \sqrt[3]{x^2 + 152}, f(x) = 14$

123. a) Consider the equation $\sqrt{4x - 12} = x - 3$. Setting each side of the equation equal to y yields the following system of equations.

$$y = \sqrt{4x - 12}$$
$$y = x - 3$$

The graphs of the equations in the system are illustrated in the figure.

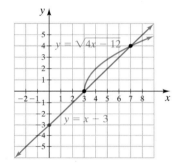

From the graphs, determine the values that appear to be solutions to the equation $\sqrt{4x - 12} = x - 3$. Explain how you determined your answer.

b) Substitute the values found in part **a)** into the original equation and determine whether they are the solutions to the equation.

c) Solve the equation $\sqrt{4x - 12} = x - 3$ algebraically and see if your solution agrees with the values obtained in part **a)**.

124. If the graph of a radical function, $f(x)$, does not intersect the x-axis, then the equation $f(x) = 0$ has no real solutions. Explain why.

125. Suppose we are given a rational function $g(x)$. If $g(4) = 0$, then the graph of $g(x)$ must intersect the x-axis at 4. Explain why.

126. The graph of the equation $y = \sqrt{x - 3} + 2$ is illustrated in the figure.

a) What is the domain of the function?

b) How many real solutions does the equation $\sqrt{x - 3} + 2 = 0$ have? List all the real solutions. Explain how you determined your answer.

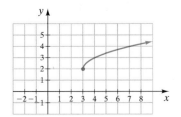

127. Confidence Interval In statistics, a "confidence interval" is a range of values that is likely to contain the true value of the population. For a "95% confidence interval," the lower limit of the range, L_1, and the upper limit of the range, L_2, can be found by the formulas

$$L_1 = p - 1.96\sqrt{\frac{p(1-p)}{n}}$$

$$L_2 = p + 1.96\sqrt{\frac{p(1-p)}{n}}$$

where p represents the percent obtained from a sample and n is the size of the sample. Francesco, a statistician, takes a sample of 36 families and finds that 60% of those surveyed use an answering machine in their home. He can be 95% certain that the true percent of families that use an answering machine in their home is between L_1 and L_2. Find the values of L_1 and L_2. Use $p = 0.60$ and $n = 36$ in the formulas.

128. Quadratic Mean The *quadratic mean* (or *root mean square*, *RMS*) is often used in physical applications. In power distribution systems, for example, voltages and currents are usually referred to in terms of their RMS values. The quadratic mean of a set of scores is obtained by squaring each score and adding the results (signified by Σx^2), then dividing the value obtained by the number of scores, n, and then taking the square root of this value. We may express the formula as

$$\text{quadratic mean} = \sqrt{\frac{\Sigma x^2}{n}}$$

Find the quadratic mean of the numbers 2, 4, and 10.

In Exercises 129 and 130, solve the equation.

129. $\sqrt{x^2 + 49} = (x^2 + 49)^{1/2}$

130. $\sqrt{x^2 - 16} = (x^2 - 16)^{1/2}$

In Exercises 131–134, use your graphing calculator to solve the equations. Round your solutions to the nearest tenth.

131. $\sqrt{x + 8} = \sqrt{3x + 5}$

132. $\sqrt{10x - 16} - 15 = 0$

133. $\sqrt[3]{5x^2 - 6} - 4 = 0$

134. $\sqrt[3]{5x^2 - 22} = \sqrt[3]{4x + 83}$

Challenge Problems

Solve.

135. $\sqrt{\sqrt{x + 25} - \sqrt{x}} = 5$

136. $\sqrt{\sqrt{x + 9} + \sqrt{x}} = 3$

Solve each equation for n.

137. $z = \dfrac{\bar{x} - \mu}{\dfrac{\sigma}{\sqrt{n}}}$

138. $z = \dfrac{p' - p}{\sqrt{\dfrac{pq}{n}}}$

Group Activity

Discuss and answer Exercise 139 as a group.

139. Heron's Formula The area of a triangle is $A = \frac{1}{2}bh$. If the height is not known but we know the lengths of the three sides, we can use Heron's formula to find the area, A. Heron's formula is

$$A = \sqrt{S(S - a)(S - b)(S - c)}$$

where a, b, and c are the lengths of the three sides and

$$S = \frac{a + b + c}{2}$$

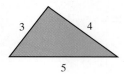

a) Have each group member use Heron's formula to find the area of a triangle whose sides are 3 inches, 4 inches, and 5 inches.

b) Compare your answers for part **a)**. If any member of the group did not get the correct answer, make sure they understand their error.

c) Have each member of the group do the following.

   1. Draw a triangle on the following grid. Place each vertex of the triangle at the intersection of two grid lines.

   2. Measure with a ruler the length of each side of your triangle.

   3. Use Heron's formula to find the area of your triangle.

   4. Compare and discuss your work from part **c)**.

Cumulative Review Exercises

[2.6] **140.** Solve the formula $P_1P_2 - P_1P_3 = P_2P_3$ for P_2.

[7.1] **141.** Simplify $\dfrac{x(x-5) + x(x-2)}{2x-7}$.

Perform each indicated operation.

[7.2] **142.** $\dfrac{4a^2 - 9b^2}{4a^2 + 12ab + 9b^2} \cdot \dfrac{6a^2b}{8a^2b^2 - 12ab^3}$

143. $(t^2 - 2t - 15) \div \dfrac{t^2 - 9}{t^2 - 3t}$

[7.4] **144.** $\dfrac{2}{x+3} - \dfrac{1}{x-3} + \dfrac{2x}{x^2-9}$

[7.6] **145.** Solve $2 + \dfrac{3x}{x-1} = \dfrac{8}{x-1}$.

11.7 Complex Numbers

1 Recognize a complex number.

2 Add and subtract complex numbers.

3 Multiply complex numbers.

4 Divide complex numbers.

5 Find powers of i.

1 Recognize a Complex Number

In Section 11.1 we mentioned that the square roots of negative numbers, such as $\sqrt{-4}$, are not real numbers. Numbers like $\sqrt{-4}$ are called **imaginary numbers**. Such numbers are called imaginary because when they were introduced many mathematicians refused to believe that they existed. Although they do not belong to the set of real numbers, the imaginary numbers, by definition, do exist and are very useful in mathematics and science.

Every imaginary number has $\sqrt{-1}$ as a factor. The $\sqrt{-1}$, called the **imaginary unit**, is often denoted by the letter i.

Imaginary Unit

$$i = \sqrt{-1}$$

To write the square root of a negative number in terms of i, use the following property.

Square Root of a Negative Number

For any positive real number n,
$$\sqrt{-n} = \sqrt{-1}\,\sqrt{n} = i\sqrt{n}$$

Therefore, we can write

$$\sqrt{-4} = \sqrt{-1}\,\sqrt{4} = i2 \quad \text{or} \quad 2i$$
$$\sqrt{-9} = \sqrt{-1}\,\sqrt{9} = i3 \quad \text{or} \quad 3i$$
$$\sqrt{-7} = \sqrt{-1}\,\sqrt{7} = i\sqrt{7}$$

In this book we will generally write $i\sqrt{7}$ rather than $\sqrt{7}i$ to avoid confusion with $\sqrt{7i}$. Also, $3\sqrt{5}i$ is written as $3i\sqrt{5}$.

Examples

$$\sqrt{-81} = 9i \qquad\qquad \sqrt{-6} = i\sqrt{6}$$
$$\sqrt{-49} = 7i \qquad\qquad \sqrt{-10} = i\sqrt{10}$$

The real number system is a part of a larger number system, called the *complex number system*. Now we will discuss **complex numbers**.

Complex Number

Every number of the form
$$a + bi$$
where a and b are real numbers, is a **complex number**.

Every real number and every imaginary number are also complex numbers. A complex number has two parts: a real part, a, and an imaginary part, b.

Real part ⟶   ⟵ Imaginary part

$$a + b\,i$$

If $b = 0$, the complex number is a real number. If $a = 0$, the complex number is a *pure imaginary number*.

Examples of Complex Numbers

$3 + 2i$	$a = 3, b = 2$	
$5 - i\sqrt{6}$	$a = 5, b = -\sqrt{6}$	
4	$a = 4, b = 0$	(real number, $b = 0$)
$8i$	$a = 0, b = 8$	(imaginary number, $a = 0$)
$-i\sqrt{7}$	$a = 0, b = -\sqrt{7}$	(imaginary number, $a = 0$)

We stated that all real numbers and imaginary numbers are also complex numbers. The relationship between the various sets of numbers is illustrated in **Figure 11.7**.

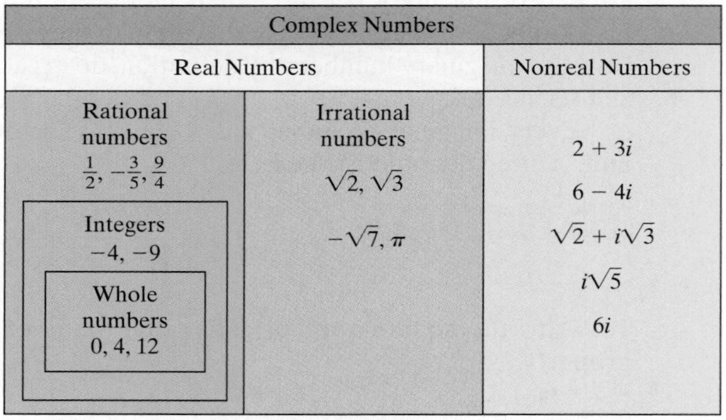

Complex Numbers		
Real Numbers		Nonreal Numbers
Rational numbers $\frac{1}{2}, -\frac{3}{5}, \frac{9}{4}$ Integers $-4, -9$ Whole numbers $0, 4, 12$	Irrational numbers $\sqrt{2}, \sqrt{3}$ $-\sqrt{7}, \pi$	$2 + 3i$ $6 - 4i$ $\sqrt{2} + i\sqrt{3}$ $i\sqrt{5}$ $6i$

FIGURE 11.7

EXAMPLE 1 ▶ Write each complex number in the form $a + bi$.

a) $7 + \sqrt{-36}$ **b)** $4 - \sqrt{-12}$ **c)** 19 **d)** $\sqrt{-50}$ **e)** $6 + \sqrt{10}$

Solution

a) $7 + \sqrt{-36} = 7 + \sqrt{-1}\,\sqrt{36}$
$$= 7 + i6 \quad \text{or} \quad 7 + 6i$$

b) $4 - \sqrt{-12} = 4 - \sqrt{-1}\,\sqrt{12}$
$$= 4 - \sqrt{-1}\,\sqrt{4}\,\sqrt{3}$$
$$= 4 - i(2)\sqrt{3} \quad \text{or} \quad 4 - 2i\sqrt{3}$$

c) $19 = 19 + 0i$

d) $\sqrt{-50} = 0 + \sqrt{-50}$
$$= 0 + \sqrt{-1}\,\sqrt{25}\,\sqrt{2}$$
$$= 0 + i(5)\sqrt{2} \quad \text{or} \quad 0 + 5i\sqrt{2}$$

e) Both 6 and $\sqrt{10}$ are real numbers. Written as a complex number, the answer is $(6 + \sqrt{10}) + 0i$.

▶ Now Try Exercise 23

Complex numbers can be added, subtracted, multiplied, and divided. To perform these operations, we use the definitions that $i = \sqrt{-1}$ and

Definition of i^2

$$i^2 = -1$$

2 Add and Subtract Complex Numbers

We now explain how to add or subtract complex numbers.

> **To Add or Subtract Complex Numbers**
> 1. Change all imaginary numbers to bi form.
> 2. Add (or subtract) the real parts of the complex numbers.
> 3. Add (or subtract) the imaginary parts of the complex numbers.
> 4. Write the answer in the form $a + bi$.

EXAMPLE 2 ▸ Add $(9 + 15i) + (-6 - 2i) + 18$.

Solution $(9 + 15i) + (-6 - 2i) + 18 = 9 + 15i - 6 - 2i + 18$
$$= 9 - 6 + 18 + 15i - 2i \qquad \textit{Rearrange terms.}$$
$$= 21 + 13i \qquad \textit{Combine like terms.}$$

▸ **Now Try Exercise 27**

EXAMPLE 3 ▸ Subtract $(8 - \sqrt{-27}) - (-3 + \sqrt{-48})$.
Solution
$$(8 - \sqrt{-27}) - (-3 + \sqrt{-48}) = (8 - \sqrt{-1}\sqrt{27}) - (-3 + \sqrt{-1}\sqrt{48})$$
$$= (8 - \sqrt{-1}\sqrt{9}\sqrt{3}) - (-3 + \sqrt{-1}\sqrt{16}\sqrt{3})$$
$$= (8 - 3i\sqrt{3}) - (-3 + 4i\sqrt{3})$$
$$= 8 - 3i\sqrt{3} + 3 - 4i\sqrt{3}$$
$$= 8 + 3 - 3i\sqrt{3} - 4i\sqrt{3}$$
$$= 11 - 7i\sqrt{3}$$

▸ **Now Try Exercise 35**

3 Multiply Complex Numbers

Now let's discuss how to multiply complex numbers.

> **To Multiply Complex Numbers**
> 1. Change all imaginary numbers to bi form.
> 2. Multiply the complex numbers as you would multiply polynomials.
> 3. Substitute -1 for each i^2.
> 4. Combine the real parts and the imaginary parts. Write the answer in $a + bi$ form.

EXAMPLE 4 ▸ Multiply.

a) $5i(6 - 2i)$ **b)** $\sqrt{-9}(\sqrt{-3} + 8)$ **c)** $(2 - \sqrt{-18})(\sqrt{-2} + 5)$

Solution

a) $5i(6 - 2i) = 5i(6) + 5i(-2i)$ *Distributive property*
$$= 30i - 10i^2$$
$$= 30i - 10(-1) \qquad \textit{Replace } i^2 \textit{ with } -1.$$
$$= 30i + 10 \quad \text{or} \quad 10 + 30i$$

b) $\sqrt{-9}(\sqrt{-3} + 8) = 3i(i\sqrt{3} + 8)$ *Change imaginary numbers to bi form.*
$$= 3i(i\sqrt{3}) + 3i(8) \qquad \textit{Distributive property}$$
$$= 3i^2\sqrt{3} + 24i$$
$$= 3(-1)\sqrt{3} + 24i \qquad \textit{Replace } i^2 \textit{ with } -1.$$
$$= -3\sqrt{3} + 24i$$

c) $(2 - \sqrt{-18})(\sqrt{-2} + 5) = (2 - \sqrt{-1}\,\sqrt{18})(\sqrt{-1}\,\sqrt{2} + 5)$

$\qquad\qquad\qquad\qquad\quad = (2 - \sqrt{-1}\,\sqrt{9}\,\sqrt{2})(\sqrt{-1}\,\sqrt{2} + 5)$

$\qquad\qquad\qquad\qquad\quad = (2 - 3i\sqrt{2})(i\sqrt{2} + 5)$

Now use the FOIL method to multiply.

$(2 - 3i\sqrt{2})(i\sqrt{2} + 5) = (2)(i\sqrt{2}) + (2)(5) + (-3i\sqrt{2})(i\sqrt{2}) + (-3i\sqrt{2})(5)$

$\qquad\qquad\qquad\qquad\quad = 2i\sqrt{2} + 10 - 3i^2(2) - 15i\sqrt{2}$

$\qquad\qquad\qquad\qquad\quad = 2i\sqrt{2} + 10 - 3(-1)(2) - 15i\sqrt{2}$

$\qquad\qquad\qquad\qquad\quad = 2i\sqrt{2} + 10 + 6 - 15i\sqrt{2}$

$\qquad\qquad\qquad\qquad\quad = 16 - 13i\sqrt{2}$

▶ **Now Try Exercise 45**

Avoiding Common Errors

What is $\sqrt{-4} \cdot \sqrt{-2}$?

CORRECT	INCORRECT
$\sqrt{-4} \cdot \sqrt{-2} = 2i \cdot i\sqrt{2}$	$\sqrt{-4} \cdot \sqrt{-2} = \sqrt{8}$
$\qquad\qquad\quad = 2i^2\sqrt{2}$	$\qquad\qquad\quad = \sqrt{4} \cdot \sqrt{2}$
$\qquad\qquad\quad = 2(-1)\sqrt{2}$	$\qquad\qquad\quad = 2\sqrt{2}$
$\qquad\qquad\quad = -2\sqrt{2}$	

Recall that $\sqrt{a} \cdot \sqrt{b} = \sqrt{ab}$ only for *nonnegative* real numbers a and b.

---

4 Divide Complex Numbers

The **conjugate of a complex number** $a + bi$ is $a - bi$. For example,

Complex Number	Conjugate
$3 + 7i$	$3 - 7i$
$1 - i\sqrt{3}$	$1 + i\sqrt{3}$
$2i$ (or $0 + 2i$)	$-2i$ (or $0 - 2i$)

When a complex number is multiplied by its conjugate using the FOIL method, the inner and outer products will sum to 0, and the result is a real number. For example,

$$(5 + 3i)(5 - 3i) = 25 - 15i + 15i - 9i^2$$

$$= 25 - 9i^2$$

$$= 25 - 9(-1)$$

$$= 25 + 9 = 34$$

Now we explain how to divide complex numbers.

To Divide Complex Numbers

1. Change all imaginary numbers to *bi* form.
2. Rationalize the denominator by multiplying both the numerator and denominator by the conjugate of the denominator.
3. Write the answer in $a + bi$ form.

EXAMPLE 5 ▶ Divide $\dfrac{9+i}{i}$.

Solution   Begin by multiplying both numerator and denominator by $-i$, the conjugate of i.

$$\frac{9+i}{i} \cdot \frac{-i}{-i} = \frac{(9+i)(-i)}{-i^2}$$

$$= \frac{-9i - i^2}{-i^2} \qquad \textit{Distributive property}$$

$$= \frac{-9i - (-1)}{-(-1)} \qquad \textit{Replace } i^2 \textit{ with } -1.$$

$$= \frac{-9i + 1}{1}$$

$$= 1 - 9i$$

▶ **Now Try Exercise 59**

EXAMPLE 6 ▶ Divide $\dfrac{3+2i}{4-i}$.

Solution   Multiply both numerator and denominator by $4+i$, the conjugate of $4-i$.

$$\frac{3+2i}{4-i} \cdot \frac{4+i}{4+i} = \frac{12 + 3i + 8i + 2i^2}{16 - i^2}$$

$$= \frac{12 + 11i + 2(-1)}{16 - (-1)}$$

$$= \frac{10 + 11i}{17} \quad \text{or} \quad \frac{10}{17} + \frac{11}{17}i$$

▶ **Now Try Exercise 65**

EXAMPLE 7 ▶ **Impedance** A concept needed for the study of electronics is *impedance*. Impedance affects the current in a circuit. The impedance, Z, in a circuit is found by the formula $Z = \dfrac{V}{I}$, where V is voltage and I is current. Find Z when $V = 1.6 - 0.3i$ and $I = -0.2i$, where $i = \sqrt{-1}$.

Solution $Z = \dfrac{V}{I} = \dfrac{1.6 - 0.3i}{-0.2i}$. Now multiply both numerator and denominator by the conjugate of the denominator, $0.2i$.

$$Z = \frac{1.6 - 0.3i}{-0.2i} \cdot \frac{0.2i}{0.2i} = \frac{0.32i - 0.06i^2}{-0.04i^2}$$

$$= \frac{0.32i + 0.06}{0.04}$$

$$= \frac{0.32i}{0.04} + \frac{0.06}{0.04}$$

$$= 8i + 1.5 \quad \text{or} \quad 1.5 + 8i$$

▶ **Now Try Exercise 127**

Most algebra books use i as the imaginary unit. However, most electronics books use j as the imaginary unit because i is often used to represent current.

5 Find Powers of i

Using $i = \sqrt{-1}$ and $i^2 = -1$, we can find other **powers of i**. For example,

$$i^3 = i^2 \cdot i = -1 \cdot i = -i \qquad\qquad i^6 = i^4 \cdot i^2 = 1(-1) = -1$$
$$i^4 = i^2 \cdot i^2 = (-1)(-1) = 1 \qquad i^7 = i^4 \cdot i^3 = 1(-i) = -i$$
$$i^5 = i^4 \cdot i^1 = 1 \cdot i = i \qquad\qquad i^8 = i^4 \cdot i^4 = (1)(1) = 1$$

Note that successive powers of i rotate through the four values i, -1, $-i$, and 1 (see **Fig. 11.8**).

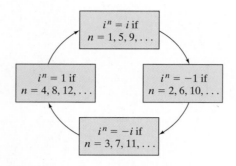

FIGURE 11.8

EXAMPLE 8 ▶ Evaluate.   **a)** i^{35} **b)** i^{101}

Solution   Write each expression as a product of factors such that the exponent of one factor is the largest multiple of 4 less than or equal to the given exponent. Then write this factor as i^4 raised to some power. Since i^4 has a value of 1, the expression i^4 raised to a power will also have a value of 1.

a) $i^{35} = i^{32} \cdot i^3 = (i^4)^8 \cdot i^3 = 1 \cdot i^3 = 1(-i) = -i$

b) $i^{101} = i^{100} \cdot i^1 = (i^4)^{25} \cdot i = 1 \cdot i = i$

▶ **Now Try Exercise 101**

Helpful Hint

A quick way of evaluating i^n is to divide the exponent by 4 and observe the remainder.

If the remainder is 0, the value is 1.   If the remainder is 2, the value is -1.
If the remainder is 1, the value is i.   If the remainder is 3, the value is $-i$.

For Example 8 **a)**
$$\begin{array}{r} 8 \\ 4\overline{)35} \\ \underline{32} \\ 3 \end{array}$$ *Answer is $-i$.*

For 8 **b)**
$$\begin{array}{r} 25 \\ 4\overline{)101} \\ \underline{8} \\ 21 \\ \underline{20} \\ 1 \end{array}$$ ← *Answer is i.*

$$i^{35} = (i^4)^8 \cdot i^3 = (1)^8 \cdot i^3 = 1 \cdot i^3 = i^3 = -i$$

EXAMPLE 9 ▶ Let $f(x) = x^2$. Find   **a)** $f(6i)$ **b)** $f(3 + 7i)$.

Solution

a) $f(x) = x^2$

$f(6i) = (6i)^2 = 36i^2 = 36(-1) = -36$

b) $f(x) = x^2$

$f(3 + 7i) = (3 + 7i)^2 = (3)^2 + 2(3)(7i) + (7i)^2$
$$= 9 + 42i + 49i^2$$
$$= 9 + 42i + 49(-1)$$
$$= 9 + 42i - 49$$
$$= -40 + 42i$$

▶ **Now Try Exercise 117**

EXERCISE SET 11.7   Math_{XL}   MyMathLab

MathXL®   *MyMathLab*

1. a) What does i equal?
  b) What does i^2 equal?
2. Write $\sqrt{-n}$ using i.
3. Are all of the following complex numbers? If any are not complex numbers, explain why.

  a) 9   **b)** $-\dfrac{1}{2}$ **c)** $4 - \sqrt{-2}$
  d) $7 - 3i$ **e)** $4.2i$ **f)** $11 + \sqrt{3}$

4. What does i^4 equal?
5. Is every real and every imaginary number a complex number?
6. Is every complex number a real number?
7. What is the conjugate of $a + bi$?

8. a) Is $i \cdot i$ a real number? Explain.
  b) Is $i \cdot i \cdot i$ a real number? Explain.
9. List, if possible, a number that is *not*
  a) a rational number.
  b) an irrational number.
  c) a real number.
  d) an imaginary number.
  e) a complex number.
10. Write a paragraph or two explaining the relationship between the real numbers, imaginary numbers, and complex numbers. Include in your discussion how the various sets of numbers relate to each other.

Practice the Skills

Write each expression as a complex number in the form $a + bi$.

11. 7   **12.** $3i$ **13.** $\sqrt{25}$ **14.** $\sqrt{-100}$
15. $21 - \sqrt{-36}$ **16.** $\sqrt{3} + \sqrt{-3}$ **17.** $\sqrt{-24}$ **18.** $\sqrt{49} - \sqrt{-49}$
19. $8 - \sqrt{-12}$ **20.** $\sqrt{-9} + \sqrt{-81}$ **21.** $3 + \sqrt{-98}$ **22.** $\sqrt{-9} + 7i$
23. $12 - \sqrt{-25}$ **24.** $10 + \sqrt{-32}$ **25.** $7i - \sqrt{-45}$ **26.** $\sqrt{144} + \sqrt{-96}$

Add or subtract.

27. $(19 - i) + (2 + 9i)$
28. $(22 + i) - 5(11 - 3i) + 4$
29. $(8 - 3i) + (-8 + 3i)$
30. $(7 - \sqrt{-4}) - (-1 - \sqrt{-16})$
31. $(1 + \sqrt{-1}) + (-18 - \sqrt{-169})$
32. $(16 - i\sqrt{3}) + (17 - \sqrt{-3})$
33. $(\sqrt{3} + \sqrt{2}) + (3\sqrt{2} - \sqrt{-8})$
34. $(8 - \sqrt{2}) - (5 + \sqrt{-15})$
35. $(5 - \sqrt{-72}) + (6 + \sqrt{-8})$
36. $(29 + \sqrt{-75}) + (\sqrt{-147})$
37. $(\sqrt{4} - \sqrt{-45}) + (-\sqrt{25} + \sqrt{-5})$
38. $(\sqrt{20} - \sqrt{-12}) + (2\sqrt{5} + \sqrt{-75})$

Multiply.

39. $2(3 - i)$
40. $-7(5 + 3i\sqrt{5})$
41. $i(4 + 9i)$
42. $3i(6 - i)$
43. $\sqrt{-9}(6 + 11i)$
44. $\dfrac{1}{2}i\left(\dfrac{1}{3} - 18i\right)$
45. $\sqrt{-16}(\sqrt{3} - 7i)$
46. $-\sqrt{-24}(\sqrt{6} - \sqrt{-3})$
47. $\sqrt{-27}(\sqrt{3} - \sqrt{-3})$
48. $\sqrt{-32}(\sqrt{2} + \sqrt{-8})$
49. $(3 + 2i)(1 + i)$
50. $(6 - 2i)(3 + i)$
51. $(10 - 3i)(10 + 3i)$
52. $(-4 + 3i)(2 - 5i)$
53. $(7 + \sqrt{-2})(5 - \sqrt{-8})$
54. $(\sqrt{4} - 3i)(4 + \sqrt{-4})$
55. $\left(\dfrac{1}{2} - \dfrac{1}{3}i\right)\left(\dfrac{1}{4} + \dfrac{2}{3}i\right)$
56. $\left(\dfrac{3}{5} - \dfrac{1}{4}i\right)\left(\dfrac{2}{3} + \dfrac{2}{5}i\right)$

Divide.

57. $\dfrac{8}{3i}$
58. $\dfrac{5}{4i}$
59. $\dfrac{2 + 3i}{2i}$
60. $\dfrac{7 - 3i}{2i}$
61. $\dfrac{6}{2 - i}$
62. $\dfrac{9}{5 + i}$
63. $\dfrac{3}{1 - 2i}$
64. $\dfrac{13}{-3 - 4i}$
65. $\dfrac{6 - 3i}{4 + 2i}$
66. $\dfrac{4 - 3i}{4 + 3i}$
67. $\dfrac{4}{6 - \sqrt{-4}}$
68. $\dfrac{2}{3 + \sqrt{-5}}$
69. $\dfrac{\sqrt{2}}{5 + \sqrt{-12}}$
70. $\dfrac{\sqrt{6}}{\sqrt{3} - \sqrt{-9}}$
71. $\dfrac{\sqrt{10} + \sqrt{-3}}{5 - \sqrt{-20}}$
72. $\dfrac{12 - \sqrt{-12}}{\sqrt{3} + \sqrt{-5}}$
73. $\dfrac{\sqrt{-75}}{\sqrt{-3}}$
74. $\dfrac{\sqrt{-30}}{\sqrt{-2}}$
75. $\dfrac{\sqrt{-32}}{\sqrt{-18}\sqrt{8}}$
76. $\dfrac{\sqrt{-40}\sqrt{-20}}{\sqrt{-4}}$

Perform each indicated operation. These exercises are a combination of the types of exercises presented earlier in this exercise set.

77. $(9 - 2i) + (3 - 5i)$

78. $\left(\dfrac{1}{2} + 2i\right) - \left(\dfrac{3}{5} - \dfrac{2}{3}i\right)$

79. $(\sqrt{50} - \sqrt{2}) - (\sqrt{-12} - \sqrt{-48})$

80. $(8 - \sqrt{-6}) - (2 - \sqrt{-24})$

81. $5.2(4 - 3.2i)$

82. $\sqrt{-6}(\sqrt{3} - \sqrt{-10})$

83. $(9 + 2i)(3 - 5i)$

84. $(\sqrt{3} + 2i)(\sqrt{6} - \sqrt{-8})$

85. $\dfrac{11 + 4i}{2i}$

86. $\dfrac{1}{4 + 3i}$

87. $\dfrac{6}{\sqrt{3} - \sqrt{-4}}$

88. $\dfrac{5 - 2i}{3 + 2i}$

89. $\left(11 - \dfrac{5}{9}i\right) - \left(4 - \dfrac{3}{5}i\right)$

90. $\dfrac{8}{7}\left(4 - \dfrac{2}{5}i\right)$

91. $\left(\dfrac{2}{3} - \dfrac{1}{5}i\right)\left(\dfrac{3}{5} - \dfrac{3}{4}i\right)$

92. $\sqrt{\dfrac{4}{9}}\left(\sqrt{\dfrac{25}{36}} - \sqrt{-\dfrac{4}{25}}\right)$

93. $\dfrac{\sqrt{-48}}{\sqrt{-12}}$

94. $\dfrac{-6 - 2i}{2 + \sqrt{-5}}$

95. $(5.23 - 6.41i) - (9.56 + 4.5i)$

96. $(\sqrt{-6} + 3)(\sqrt{-15} + 5)$

For each imaginary number, indicate whether its value is i, -1, $-i$, or 1.

97. i^6

98. i^{63}

99. i^{160}

100. i^{231}

101. i^{93}

102. i^{103}

103. i^{811}

104. i^{1213}

Problem Solving

105. Consider the complex number $2 + 3i$.

   a) Find the additive inverse.

   b) Find the multiplicative inverse. Write the answer in simplified form.

106. Consider the complex number $4 - 5i$.

   a) Find the additive inverse.

   b) Find the multiplicative inverse. Write the answer in simplified form.

In Exercises 107–110, answer true or false. Support your answer with an example.

107. The product of two pure imaginary numbers is always a real number.

108. The sum of two pure imaginary numbers is always an imaginary number.

109. The product of two complex numbers is always a real number.

110. The sum of two complex numbers is always a complex number.

111. What values of n will result in i^n being a real number? Explain.

112. What values of n will result in i^{2n} being a real number? Explain.

113. If $f(x) = x^2$, find $f(2i)$.

114. If $f(x) = x^2$, find $f(4i)$.

115. If $f(x) = x^4 - 2x$, find $f(2i)$.

116. If $f(x) = x^3 - 4x^2$, find $f(5i.)$

117. If $f(x) = x^2 + 2x$, find $f(3 + i)$.

118. If $f(x) = \dfrac{x^2}{x - 2}$, find $f(4 - i)$.

Evaluate each expression for the given value of x.

119. $x^2 - 2x + 5,\ x = 1 + 2i$

120. $x^2 - 2x + 5,\ x = 1 - 2i$

121. $x^2 + 2x + 7,\ x = -1 + i\sqrt{5}$

122. $x^2 + 2x + 9,\ x = -1 - i\sqrt{5}$

In Exercises 123–126, determine whether the given value of x is a solution to the equation.

123. $x^2 - 4x + 5 = 0,\ x = 2 - i$

124. $x^2 - 4x + 5 = 0,\ x = 2 + i$

125. $x^2 - 6x + 11 = 0,\ x = -3 + i\sqrt{3}$

126. $x^2 - 6x + 15 = 0,\ x = 3 - i\sqrt{3}$

127. **Impedance** Find the impedance, Z, using the formula $Z = \dfrac{V}{I}$ when $V = 1.8 + 0.5i$ and $I = 0.6i$. See Example 7.

128. **Impedance** Refer to Exercise 127. Find the impedance when $V = 2.4 - 0.6i$ and $I = -0.4i$.

129. Impedance Under certain conditions, the total impedance, Z_T, of a circuit is given by the formula

$$Z_T = \frac{Z_1 Z_2}{Z_1 + Z_2}$$

Find Z_T when $Z_1 = 2 - i$ and $Z_2 = 4 + i$.

130. Impedance Refer to Exercise 129. Find Z_T when $Z_1 = 3 - i$ and $Z_2 = 5 + i$.

131. Determine whether i^{-1} is equal to i, -1, $-i$, or 1. Show your work.

132. Determine whether i^{-5} is equal to i, -1, $-i$, or 1. Show your work.

In Chapter 12, we will use the quadratic formula $x = \dfrac{-b \pm \sqrt{b^2 - 4ac}}{2a}$ *to solve equations of the form* $ax^2 + bx + c = 0$. **a)** *Use the quadratic formula to solve the following quadratic equations.* **b)** *Check each of the two solutions by substituting the values found for x (one at a time) back into the original equation. In these exercises, the ± (read "plus or minus") results in two distinct complex answers.*

133. $x^2 - 2x + 6 = 0$

134. $x^2 - 4x + 6 = 0$

Given the complex numbers $a = 5 + 2i\sqrt{3}$, $b = 1 + i\sqrt{3}$, *evaluate each expression.*

135. $a + b$

136. $a - b$

137. ab

138. $\dfrac{a}{b}$

Cumulative Review Exercises

[5.6] **139.** Divide $\dfrac{8c^2 + 6c - 35}{4c + 9}$.

[7.4] **140.** Add $\dfrac{b}{a - b} + \dfrac{a + b}{b}$.

[7.6] **141.** Solve $\dfrac{x}{4} + \dfrac{1}{2} = \dfrac{x - 1}{2}$.

[9.5] **142. Mixture** Berreda Coughlin, a grocer in Dallas, has two coffees, one selling for $5.50 per pound and the other for $6.30 per pound. How many pounds of each type of coffee should he mix to make 40 pounds of coffee to sell for $6.00 per pound?

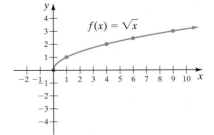

Chapter 11 Summary

IMPORTANT FACTS AND CONCEPTS	EXAMPLES
Section 11.1	
A **radical expression** has the form $\sqrt[n]{x}$, where n is the index and x is the radicand.	In the radical expression $\sqrt[3]{x}$, 3 is the index and x is the radicand.
The **principal square root** of a positive number a, written $\sqrt{a}$, is the positive number b such that $b^2 = a$.	$\sqrt{81} = 9$, since $9^2 = 81$ $\sqrt{0.36} = 0.6$ since $(0.6)^2 = 0.36$
The **square root function** is $f(x) = \sqrt{x}$. Its domain is $[0, \infty)$ and its range is $[0, \infty)$.	
The **cube root** of a number a, written $\sqrt[3]{a}$, is the number b such that $b^3 = a$.	$\sqrt[3]{27} = 3$ since $3^3 = 27$ $\sqrt[3]{-125} = -5$ since $(-5)^3 = -125$

IMPORTANT FACTS AND CONCEPTS	EXAMPLES

Section 11.1 (continued)

The **cube root function** is $f(x) = \sqrt[3]{x}$. Its domain is $(-\infty, \infty)$ or $\mathbb{R}$ and its range is $(-\infty, \infty)$ or $\mathbb{R}$.

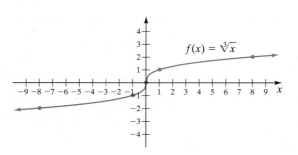

The nth root of a, $\sqrt[n]{a}$, where a is an **even index** and a is a nonnegative real number, is the nonnegative number b such that $b^n = a$.

$\sqrt{4} = 2$ since $2^2 = 2 \cdot 2 = 4$
$\sqrt[4]{81} = 3$ since $3^4 = 3 \cdot 3 \cdot 3 \cdot 3 = 81$

The nth root of a, $\sqrt[n]{a}$, where n is an **odd index** and a is any real number, is the real number b such that $b^n = a$.

$\sqrt[3]{27} = 3$ since $3^3 = 3 \cdot 3 \cdot 3 = 27$
$\sqrt[5]{-32} = -2$ since $(-2)^5 = (-2)(-2)(-2)(-2)(-2) = -32$

For any real number a, $\sqrt{a^2} = |a|$.

$\sqrt{(-6)^2} = |-6| = 6$
$\sqrt{(y+8)^2} = |y+8|$

Section 11.2

Rational Exponent

$$\sqrt[n]{a} = a^{1/n}$$

When a is nonnegative, n can be any index.
When a is negative, n must be odd.

$\sqrt{17} = 17^{1/2}$
$\sqrt[4]{21x^3y^2} = (21x^3y^2)^{1/4}$

For any nonnegative number a, and integers m and n,

$$\sqrt[n]{a^m} = \left(\sqrt[n]{a}\right)^m = a^{m/n}$$

(Power — over m; Index — over n)

$\sqrt[4]{z^9} = \left(\sqrt[4]{z}\right)^9 = z^{9/4}$

For any nonnegative real number a,

$$\sqrt[n]{a^n} = \left(\sqrt[n]{a}\right)^n = a^{n/n} = a$$

$\sqrt[4]{y^4} = y$, $\sqrt[8]{14^8} = 14$

Rules of Exponents

For all real numbers a and b and all rational numbers m and n,

Product rule $a^m \cdot a^n = a^{m+n}$

Quotient rule $\dfrac{a^m}{a^n} = a^{m-n}, \quad a \neq 0$

Negative exponent rule $a^{-m} = \dfrac{1}{a^m}, \quad a \neq 0$

Zero exponent rule $a^0 = 1, \quad a \neq 0$

Raising a power to a power $(a^m)^n = a^{m \cdot n}$

Raising a product to a power $(ab)^m = a^m b^m$

Raising a quotient to a power $\left(\dfrac{a}{b}\right)^m = \dfrac{a^m}{b^m}, \quad b \neq 0$

$\left(\dfrac{a}{b}\right)^{-n} = \left(\dfrac{b}{a}\right)^n = \dfrac{b^n}{a^n}, \quad a \neq 0, b \neq 0$

$x^{1/3} \cdot x^{4/3} = x^{(1/3)+(4/3)} = x^{5/3}$

$\dfrac{x^{4/5}}{x^{1/2}} = x^{(4/5)-(1/2)} = x^{(8/10)-(5/10)} = x^{3/10}$

$x^{-1/7} = \dfrac{1}{x^{1/7}}$

$m^0 = 1$

$(c^{1/8})^{16} = c^{(1/8) \cdot 16} = c^2$

$(p^3 q^4)^{1/8} = p^{3/8} q^{1/2}$

$\left(\dfrac{81}{49}\right)^{-1/2} = \left(\dfrac{49}{81}\right)^{1/2} = \dfrac{49^{1/2}}{81^{1/2}} = \dfrac{7}{9}$

IMPORTANT FACTS AND CONCEPTS	EXAMPLES

Section 11.3

A number or expression is a **perfect square** if it is the square of an expression.

A number or expression is a **perfect cube** if it is the cube of an expression.

Perfect squares:	49 81	x^{12}	y^{50}
	↓ ↓	↓	↓
Square of a number or expression:	7^2 9^2	$(x^6)^2$	$(y^{25})^2$
Perfect cubes:	27 −27	y^{18}	z^{30}
	↓ ↓	↓	↓
Cube of a number or expression:	3^3 $(-3)^3$	$(y^6)^3$	$(z^{10})^3$

Product Rule for Radicals

For nonnegative real numbers a and b,

$$\sqrt[n]{a} \cdot \sqrt[n]{b} = \sqrt[n]{ab}$$

$$\sqrt{2} \cdot \sqrt{8} = \sqrt{16} = 4, \quad \sqrt[3]{2x^3} = \sqrt[3]{x^3} \cdot \sqrt[3]{2} = x\sqrt[3]{2}$$

To Simplify Radicals Using the Product Rule

1. If the radicand contains a coefficient other than 1, write it as a product of two numbers, one of which is the largest perfect power for the index.
2. Write each variable factor as a product of two factors, one of which is the largest perfect power of the variable for the index.
3. Use the product rule to write the radical expression as a product of radicals. Place all the perfect powers (numbers and variables) under the same radical.
4. Simplify the radical containing the perfect powers.

$$\sqrt{24} = \sqrt{4 \cdot 6} = \sqrt{4}\,\sqrt{6} = 2\sqrt{6}$$

$$\sqrt[3]{16x^5y^9} = \sqrt[3]{8x^3y^9 \cdot 2x^2}$$
$$= \sqrt[3]{8x^3y^9}\,\sqrt[3]{2x^2}$$
$$= 2xy^3\sqrt[3]{2x^2}$$

Quotient Rule for Radicals

For nonnegative real numbers a and b,

$$\frac{\sqrt[n]{a}}{\sqrt[n]{b}} = \sqrt[n]{\frac{a}{b}}, \quad b \neq 0$$

$$\frac{\sqrt{32}}{\sqrt{2}} = \sqrt{\frac{32}{2}} = \sqrt{16} = 4, \quad \sqrt[3]{\frac{x^6}{y^{12}}} = \frac{\sqrt[3]{x^6}}{\sqrt[3]{y^{12}}} = \frac{x^2}{y^4}$$

Section 11.4

Like radicals are radicals with the same radicand and index.
Unlike radicals are radicals with a different radicand or index.

Like Radicals   Unlike Radicals
$\sqrt{3}, \quad 12\sqrt{3}$ $\sqrt{3}, \quad 7\sqrt[4]{3}$
$2\sqrt[4]{xy^3}, \quad -3\sqrt[4]{xy^3}$ $\sqrt[5]{xy^3}, \quad x\sqrt[5]{y^3}$

To Add or Subtract Radicals

1. Simplify each radical expression.
2. Combine like radicals (if there are any).

$$\sqrt{27} + \sqrt{48} - 2\sqrt{75} = \sqrt{9} \cdot \sqrt{3} + \sqrt{16} \cdot \sqrt{3} - 2 \cdot \sqrt{25} \cdot \sqrt{3}$$
$$= 3\sqrt{3} + 4\sqrt{3} - 10\sqrt{3}$$
$$= -3\sqrt{3}$$

To Multiply Radicals

Use the product rule.

$$\sqrt[n]{a} \cdot \sqrt[n]{b} = \sqrt[n]{ab}$$

$$\sqrt[4]{8c^2}\,\sqrt[4]{4c^3} = \sqrt[4]{32c^5} = \sqrt[4]{16c^4}\,\sqrt[4]{2c}$$
$$= 2c\sqrt[4]{2c}$$

Section 11.5

To **rationalize a denominator** multiply both the numerator and the denominator of the fraction by a radical that will result in the radicand in the denominator becoming a perfect power.

$$\frac{6}{\sqrt{3x}} \cdot \frac{\sqrt{3x}}{\sqrt{3x}} = \frac{6\sqrt{3x}}{\sqrt{9x^2}} = \frac{6\sqrt{3x}}{3x} = \frac{2\sqrt{3x}}{x}$$

IMPORTANT FACTS AND CONCEPTS	EXAMPLES

Section 11.5 (continued)

A Radical Expression Is Simplified When the Following Are All True

1. No perfect powers are factors of the radicand and all exponents in the radicand are less than the index.
2. No radicand contains a fraction.
3. No denominator contains a radical.

	Not Simplified	Simplified
1.	$\sqrt{x^3}$	$x\sqrt{x}$
2.	$\sqrt{\dfrac{1}{2}}$	$\dfrac{\sqrt{2}}{2}$
3.	$\dfrac{1}{\sqrt{2}}$	$\dfrac{\sqrt{2}}{2}$

Section 11.6

To Solve Radical Equations

1. Rewrite the equation so that one radical containing a variable is by itself (isolated) on one side of the equation.
2. Raise each side of the equation to a power equal to the index of the radical.
3. Combine like terms.
4. If the equation still contains a term with a variable in a radicand, repeat steps 1 through 3.
5. Solve the resulting equation for the variable.
6. Check all solutions in the original equation for extraneous solutions.

Solve $\sqrt{x} - 8 = 0$.

$$\sqrt{x} - 8 = 0$$
$$\sqrt{x} = 8$$
$$(\sqrt{x})^2 = 8^2$$
$$x = 64$$

A check shows that 64 is the solution.

Section 11.7

The **imaginary unit**, i is defined as $i = \sqrt{-1}$. (Also, $i^2 = -1$.)

$$\sqrt{-25} = \sqrt{25}\sqrt{-1} = 5i$$

Imaginary Number
For any positive number n,
$$\sqrt{-n} = i\sqrt{n}.$$

$$\sqrt{-19} = i\sqrt{19}$$

A **complex number** is a number of the form $a + bi$, where a and b are real numbers.

$3 + 2i$ and $26 - 15i$ are complex numbers.

To Add or Subtract Complex Numbers

1. Change all imaginary numbers to bi form.
2. Add (or subtract) the real parts of the complex numbers.
3. Add (or subtract) the imaginary parts of the complex numbers.
4. Write the answer in the form $a + bi$.

Add $(8 - 3i) + (12 + 5i)$.

$$(8 - 3i) + (12 + 5i)$$
$$= 8 + 12 - 3i + 5i$$
$$= 20 + 2i$$

To Multiply Complex Numbers

1. Change all imaginary numbers to bi form.
2. Multiply the complex numbers as you would multiply polynomials.
3. Substitute -1 for each i^2.
4. Combine the real parts and the imaginary parts. Write the answer in $a + bi$ form.

Multiply $(7 + 2i\sqrt{3})(5 - 4i\sqrt{3})$.

$$(7 + 2i\sqrt{3})(5 - 4i\sqrt{3})$$
$$= 35 - 28i\sqrt{3} + 10i\sqrt{3} - 8(i^2)(3)$$
$$= 35 - 28i\sqrt{3} + 10i\sqrt{3} + 24$$
$$= 59 - 18i\sqrt{3}$$

The **conjugate of a complex number** $a + bi$ is $a - bi$.

Complex Number	Conjugate
$14 + 2i$	$14 - 2i$
$17 - 8i$	$17 + 8i$

IMPORTANT FACTS AND CONCEPTS	EXAMPLES

Section 11.7 (continued)

To Divide Complex Numbers

1. Change all imaginary numbers to *bi* form.
2. Rationalize the denominator by multiplying both the numerator and denominator by the conjugate of the denominator.
3. Write the answer in $a + bi$ form.

Divide $\dfrac{2 - i}{5 + 3i}$.

$$\frac{2 - i}{5 + 3i} \cdot \frac{5 - 3i}{5 - 3i} = \frac{10 - 6i - 5i + 3i^2}{25 - 9i^2} = \frac{7 - 11i}{34}$$

Powers of *i*
$$i^2 = -1, i^3 = -i, i^4 = 1, i^5 = i$$

$$i^{38} = i^{36} \cdot i^2 = (i^4)^9 \cdot i^2 = 1^9 \cdot (-1) = -1$$
$$i^{63} = i^{60} \cdot i^3 = (i^4)^{15} \cdot i^3 = 1^{15}(-i) = -i$$

Chapter 11 Review Exercises

[11.1] *Evaluate.*

1. $\sqrt{100}$ **2.** $\sqrt[3]{-27}$ **3.** $\sqrt[3]{-125}$ **4.** $\sqrt[4]{256}$

Use absolute value to evaluate.

5. $\sqrt{(-8)^2}$ **6.** $\sqrt{(38.2)^2}$

Write as an absolute value.

7. $\sqrt{x^2}$ **8.** $\sqrt{(x - 3)^2}$ **9.** $\sqrt{(x - y)^2}$ **10.** $\sqrt{(x^2 - 4x + 12)^2}$

11. Let $f(x) = \sqrt{10x + 9}$. Find $f(4)$.   **12.** Let $k(x) = 2x + \sqrt{\dfrac{x}{3}}$. Find $k(27)$.

13. Let $g(x) = \sqrt[3]{2x + 3}$. Find $g(4)$ and round the answer to the nearest tenth.   **14. Area** The area of a square is 144 square meters. Find the length of its side.

For the remainder of these review exercises, assume that all variables represent positive real numbers.

[11.2] *Write in exponential form.*

15. $\sqrt{x^7}$ **16.** $\sqrt[3]{x^5}$ **17.** $(\sqrt[4]{y})^{13}$ **18.** $\sqrt[7]{6^{-2}}$

Write in radical form.

19. $x^{1/2}$ **20.** $a^{4/5}$ **21.** $(8m^2n)^{7/4}$ **22.** $(x + y)^{-5/3}$

Simplify each radical expression by changing the expression to exponential form. Write the answer in radical form when appropriate.

23. $\sqrt[3]{4^6}$ **24.** $\sqrt{x^{12}}$ **25.** $(\sqrt[4]{9})^8$ **26.** $\sqrt[20]{a^5}$

Evaluate if possible. If the expression is not a real number, so state.

27. $-36^{1/2}$ **28.** $(-36)^{1/2}$ **29.** $\left(\dfrac{64}{27}\right)^{-1/3}$ **30.** $64^{-1/2} + 8^{-2/3}$

Simplify. Write the answer without negative exponents.

31. $x^{3/5} \cdot x^{-1/3}$ **32.** $\left(\dfrac{64}{y^9}\right)^{1/3}$ **33.** $\left(\dfrac{a^{-6/5}}{a^{2/5}}\right)^{2/3}$ **34.** $\left(\dfrac{20x^5y^{-3}}{4y^{1/2}}\right)^2$

Multiply.

35. $a^{1/2}(5a^{3/2} - 3a^2)$ **36.** $4x^{-2/3}\left(x^{-1/2} + \dfrac{11}{4}x^{2/3}\right)$

Factor each expression. Write the answer without negative exponents.

37. $x^{2/5} + x^{7/5}$ **38.** $a^{-1/2} + a^{3/2}$

For each function, find the indicated value of the function. Use your calculator to evaluate irrational numbers. Round irrational numbers to the nearest thousandth.

39. If $f(x) = \sqrt{6x - 11}$, find $f(6)$.   **40.** If $g(x) = \sqrt[3]{9x - 17}$, find $g(4)$.

Graph the following functions.

41. $f(x) = \sqrt{x}$

42. $f(x) = \sqrt{x} - 4$

[11.2–11.5] Simplify.

43. $\sqrt{48}$

44. $\sqrt[3]{128}$

45. $\sqrt{\dfrac{49}{9}}$

46. $\sqrt[3]{\dfrac{8}{125}}$

47. $-\sqrt{\dfrac{81}{49}}$

48. $\sqrt[3]{-\dfrac{27}{125}}$

49. $\sqrt{32}\,\sqrt{2}$

50. $\sqrt[3]{32}\,\sqrt[3]{2}$

51. $\sqrt{18x^2y^3z^4}$

52. $\sqrt{75x^3y^7}$

53. $\sqrt[3]{54a^7b^{10}}$

54. $\sqrt[3]{125x^8y^9z^{16}}$

55. $\left(\sqrt[6]{x^2y^3z^5}\right)^{42}$

56. $\left(\sqrt[5]{2ab^4c^6}\right)^{15}$

57. $\sqrt{5x}\,\sqrt{8x^5}$

58. $\sqrt[3]{2x^2y}\,\sqrt[3]{4x^9y^4}$

59. $\sqrt[3]{2x^4y^5}\,\sqrt[3]{16x^4y^4}$

60. $\sqrt[4]{4x^4y^7}\,\sqrt[4]{4x^5y^9}$

61. $\sqrt{3x}\left(\sqrt{12x} - \sqrt{20}\right)$

62. $\sqrt[3]{2x^2y}\left(\sqrt[3]{4x^4y^7} + \sqrt[3]{9x}\right)$

63. $\sqrt{\sqrt{a^3b^2}}$

64. $\sqrt{\sqrt[3]{x^5y^2}}$

65. $\left(\dfrac{4r^2p^{1/3}}{r^{1/2}p^{4/3}}\right)^3$

66. $\left(\dfrac{6y^{2/5}z^{1/3}}{x^{-1}y^{3/5}}\right)^{-1}$

67. $\sqrt{\dfrac{3}{5}}$

68. $\sqrt[3]{\dfrac{7}{9}}$

69. $\sqrt[4]{\dfrac{5}{4}}$

70. $\dfrac{x}{\sqrt{10}}$

71. $\dfrac{8}{\sqrt{x}}$

72. $\dfrac{m}{\sqrt[3]{25}}$

73. $\dfrac{10}{\sqrt[3]{y^2}}$

74. $\dfrac{9}{\sqrt[4]{z}}$

75. $\sqrt[3]{\dfrac{x^3}{27}}$

76. $\dfrac{\sqrt[3]{2x^{10}}}{\sqrt[3]{16x^7}}$

77. $\sqrt{\dfrac{32x^2y^5}{2x^8y}}$

78. $\sqrt[4]{\dfrac{48x^9y^{15}}{3xy^3}}$

79. $\sqrt{\dfrac{6x^4}{y}}$

80. $\sqrt{\dfrac{12a}{7b}}$

81. $\sqrt{\dfrac{18x^4y^5}{3z}}$

82. $\sqrt{\dfrac{125x^2y^5}{3z}}$

83. $\sqrt[3]{\dfrac{108x^3y^7}{2y^3}}$

84. $\sqrt[3]{\dfrac{3x}{5y}}$

85. $\sqrt[3]{\dfrac{9x^5y^3}{x^6}}$

86. $\sqrt[3]{\dfrac{y^6}{5x^2}}$

87. $\sqrt[4]{\dfrac{2a^2b^{11}}{a^5b}}$

88. $\sqrt[4]{\dfrac{3x^2y^6}{8x^3}}$

89. $(3 - \sqrt{2})(3 + \sqrt{2})$

90. $(\sqrt{x} + y)(\sqrt{x} - y)$

91. $(x - \sqrt{y})(x + \sqrt{y})$

92. $(\sqrt{3} + 2)^2$

93. $(\sqrt{x} - \sqrt{3y})(\sqrt{x} + \sqrt{5y})$

94. $(\sqrt[3]{2x} - \sqrt[3]{3y})(\sqrt[3]{3x} - \sqrt[3]{2y})$

95. $\dfrac{6}{2 + \sqrt{5}}$

96. $\dfrac{x}{4 + \sqrt{x}}$

97. $\dfrac{a}{4 - \sqrt{b}}$

98. $\dfrac{x}{\sqrt{y} - 7}$

99. $\dfrac{\sqrt{x}}{\sqrt{x} + \sqrt{y}}$

100. $\dfrac{\sqrt{x} - 3\sqrt{y}}{\sqrt{x} - \sqrt{y}}$

101. $\dfrac{2}{\sqrt{a - 1} - 2}$

102. $\dfrac{5}{\sqrt{y + 2} - 3}$

103. $\sqrt[3]{x} + 10\sqrt[3]{x} - 2\sqrt[3]{x}$

104. $\sqrt{3} + \sqrt{27} - \sqrt{192}$

105. $\sqrt[3]{16} - 5\sqrt[3]{54} + 3\sqrt[3]{64}$

106. $\sqrt{2} - \dfrac{3}{\sqrt{32}} + \sqrt{50}$

107. $9\sqrt{x^5y^6} - \sqrt{16x^7y^8}$

108. $8\sqrt[3]{x^7y^8} - \sqrt[3]{x^4y^2} + 3\sqrt[3]{x^{10}y^2}$

In Exercises 109 and 110, $f(x)$ and $g(x)$ are given. Find $(f \cdot g)(x)$.

109. $f(x) = \sqrt{3x},\, g(x) = \sqrt{6x} - \sqrt{15}$

110. $f(x) = \sqrt[3]{2x^2},\, g(x) = \sqrt[3]{4x^4} + \sqrt[3]{16x^5}$

Simplify. In Exercise 112, assume that the variable can be any real number.

111. $f(x) = \sqrt{2x + 7}\,\sqrt{2x + 7},\, x \geq -\dfrac{7}{2}$

112. $g(a) = \sqrt{20a^2 + 100a + 125}$

Simplify.

113. $\dfrac{\sqrt[3]{(x + 5)^5}}{\sqrt{(x + 5)^3}}$

114. $\dfrac{\sqrt[3]{a^3b^2}}{\sqrt[4]{a^4b}}$

Perimeter and Area *For each figure, find* **a)** *the perimeter and* **b)** *the area. Write the perimeter and area in radical form with the radicals simplified.*

115.

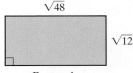

Rectangle

116.

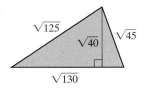

117. The graph of $f(x) = \sqrt{x} + 2$ is given.

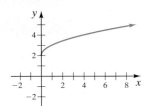

a) For $g(x) = -3$, sketch the graph of $(f + g)(x)$.

b) What is the domain of $(f + g)(x)$?

118. The graph of $f(x) = -\sqrt{x}$ is given.

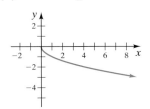

a) For $g(x) = \sqrt{x} + 2$, sketch the graph of $(f + g)(x)$.

b) What is the domain of $(f + g)(x)$?

[11.6] *Solve each equation and check your solutions.*

119. $\sqrt{x} = 9$

120. $\sqrt{x} = -4$

121. $\sqrt[3]{x} = 4$

122. $\sqrt[3]{x} = -5$

123. $7 + \sqrt{x} = 10$

124. $7 + \sqrt[3]{x} = 12$

125. $\sqrt{3x + 4} = \sqrt{5x + 14}$

126. $\sqrt{x^2 + 2x - 8} = x$

127. $\sqrt[3]{x - 9} = \sqrt[3]{5x + 3}$

128. $(x^2 + 7)^{1/2} = x + 1$

129. $\sqrt{x + 3} = \sqrt{3x + 9}$

130. $\sqrt{6x - 5} - \sqrt{2x + 6} - 1 = 0$

For each pair of functions, find all values of x for which $f(x) = g(x)$.

131. $f(x) = \sqrt{3x + 4}, g(x) = 2\sqrt{2x - 4}$

132. $f(x) = (4x + 5)^{1/3}, g(x) = (6x - 7)^{1/3}$

Solve the following for the indicated variable.

133. $V = \sqrt{\dfrac{2L}{w}}$, for L

134. $r = \sqrt{\dfrac{A}{\pi}}$, for A

Find the length of the unknown side of each right triangle. Write the answer as a radical in simplified form.

135.

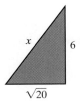

136.

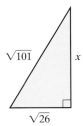

Solve.

137. Telephone Pole How long a wire does a phone company need to reach the top of a 5-meter telephone pole from a point on the ground 2 meters from the base of the pole?

138. Velocity Use the formula $v = \sqrt{2gh}$ to find the velocity of an object after it has fallen 20 feet ($g = 32$ ft/s^2).

139. Pendulum Use the formula

$$T = 2\pi\sqrt{\dfrac{L}{32}}$$

to find the period of a pendulum, T, if its length, L, is 64 feet.

140. Kinetic and Potential Energy There are two types of energy: kinetic and potential. Potential energy is due to position and kinetic energy is due to motion. For example, if you hold a billiard ball above the ground it has potential energy. If you let go of the ball the potential energy is changed to kinetic energy as the ball drops. The formula

$$V = \sqrt{\dfrac{2K}{m}}$$

can be used to determine the velocity, V, in meters per second, when a mass, m, in kilograms, has a kinetic energy, K, in joules. A 0.145-kg baseball is thrown. If the kinetic energy of the moving ball is 45 joules, at what speed is the ball moving?

141. Speed of Light Albert Einstein found that if an object at rest, with mass m_0, is made to travel close to the speed of light, its mass increases to m, where

$$m = \frac{m_0}{\sqrt{1 - \dfrac{v^2}{c^2}}}$$

In the formula, v is the velocity of the moving object and c is the speed of light.* In an accelerator used for cancer therapy, particles travel at speeds of $0.98c$, that is, at 98% the speed of light. At a speed of $0.98c$, determine a particle's mass, m, in terms of its rest mass, m_0. Use $v = 0.98c$ in the above formula.

[11.7] *Write each expression as a complex number in the form* $a + bi$.

142. 5

143. -8

144. $7 - \sqrt{-256}$

145. $9 + \sqrt{-16}$

Perform each indicated operation.

146. $(3 + 2i) + (10 - i)$

147. $(9 - 6i) - (3 - 4i)$

148. $(\sqrt{3} + \sqrt{-5}) + (11\sqrt{3} - \sqrt{-7})$

149. $\sqrt{-6}(\sqrt{6} + \sqrt{-6})$

150. $(4 + 3i)(2 - 3i)$

151. $(6 + \sqrt{-3})(4 - \sqrt{-15})$

152. $\dfrac{8}{3i}$

153. $\dfrac{2 + \sqrt{3}}{2i}$

154. $\dfrac{4}{3 + 2i}$

155. $\dfrac{\sqrt{3}}{5 - \sqrt{-6}}$

Evaluate each expression for the given value of x.

156. $x^2 - 2x + 9, x = 1 + 2i\sqrt{2}$

157. $x^2 - 2x + 12, x = 1 - 2i$

For each imaginary number, indicate whether its value is $i, -1, -i,$ *or* 1.

158. i^{33}

159. i^{59}

160. i^{404}

161. i^{802}

---

*The speed of light is 3.00×10^8 meters per second. However, you do not need this information to solve this problem.

Chapter 11 Practice Test

To find out how well you understand the chapter material, take this practice test. The answers, and the section where the material was initially discussed, are given in the back of the book. Each problem is also fully worked out on the **Chapter Test Prep Video CD.** *Review any questions that you answered incorrectly.*

1. Write $\sqrt{(5x - 3)^2}$ as an absolute value.

2. Simplify $\left(\dfrac{x^{2/5} \cdot x^{-1}}{x^{3/5}}\right)^2$.

3. Factor $x^{-2/3} + x^{4/3}$.

4. Graph $g(x) = \sqrt{x} + 1$.

In Exercises 5–14, simplify. Assume that all variables represent positive real numbers.

5. $\sqrt{54x^7 y^{10}}$

6. $\sqrt[3]{25x^5 y^2} \sqrt[3]{10x^6 y^8}$

7. $\sqrt{\dfrac{7x^6 y^3}{8z}}$

8. $\dfrac{9}{\sqrt[3]{x}}$

9. $\dfrac{\sqrt{3}}{3 + \sqrt{27}}$

10. $2\sqrt{24} - 6\sqrt{6} + 3\sqrt{54}$

11. $\sqrt[3]{8x^3 y^5} + 4\sqrt[3]{x^6 y^8}$

12. $(\sqrt{3} - 2)(6 - \sqrt{8})$

13. $\sqrt[4]{\sqrt{x^5 y^3}}$

14. $\dfrac{\sqrt[4]{(7x + 2)^5}}{\sqrt[3]{(7x + 2)^2}}$

In Exercises 15–17, solve the equation.

15. $\sqrt{2x + 19} = 3$

16. $\sqrt{x^2 - x - 12} = x + 3$

17. $\sqrt{a - 8} = \sqrt{a} - 2$

18. For $f(x) = (9x + 37)^{1/3}$ and $g(x) = 2(2x + 2)^{1/3}$, find all values of x such that $f(x) = g(x)$.

19. Solve the formula $w = \dfrac{\sqrt{2gh}}{4}$ for g.

20. Falling Object The velocity, V, in feet per second, after an object has fallen a distance, h, in feet, can be found by the formula $V = \sqrt{64.4h}$. Find the velocity of a pen after it has fallen 200 feet.

21. Ladder A ladder is placed against a house. If the base of the ladder is 5 feet from the house and the ladder rests on the house 12 feet above the ground, find the length of the ladder.

22. Springs A formula used in the study of springs is

$$T = 2\pi\sqrt{\frac{m}{k}}$$

where T is the period of a spring (the time for the spring to stretch and return to its rest point), m is the mass on the spring, in kilograms, and k is the spring's constant, in newtons/meter. A mass of 1400 kilograms rests on a spring. Find the period of the spring if the spring's constant is 65,000 newtons/meter.

23. Multiply $(6 - \sqrt{-4})(2 + \sqrt{-16})$.

24. Divide $\dfrac{5 - i}{7 + 2i}$.

25. Evaluate $x^2 + 6x + 12$ for $x = -3 + i$.

Cumulative Review Test

Take the following test and check your answers with those given in the back of the book. Review any questions that you answered incorrectly. The section where the material was covered is indicated after the answer.

1. Solve $\dfrac{1}{5}(x - 3) = \dfrac{3}{4}(x + 3) - x$.

2. Solve $3(x - 4) = 6x - (4 - 5x)$.

3. Sweater When the price of a sweater is decreased by 60%, it costs $16. Find the original price of the sweater.

4. Graph $y = \dfrac{3}{2}x - 3$.

5. Multiply $(5xy - 3)(5xy + 3)$.

6. Volume The volume of the box that follows is $6r^3 + 5r^2 + r$. Find w in terms of r.

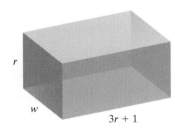

7. Factor $4x^3 - 9x^2 + 5x$.

8. Factor $(x + 1)^3 - 27$.

9. Solve $8x^2 - 3 = -10x$.

10. Multiply $\dfrac{4x + 4y}{x^2y} \cdot \dfrac{y^3}{12x}$.

11. Add $\dfrac{x - 4}{x - 5} - \dfrac{3}{x + 5} - \dfrac{10}{x^2 - 25}$.

12. Solve $\dfrac{4}{x} - \dfrac{1}{6} = \dfrac{1}{x}$.

13. Determine whether the graphs of the given equations are parallel lines, perpendicular lines, or neither.
$$y = 3x - 8$$
$$6y = 18x + 12$$

14. Find the equation of the line through $(1, -4)$ that is perpendicular to the graph of $3x - 2y = 6$.

15. Given $f(x) = x^2 - 3x + 4$ and $g(x) = 2x - 9$, find $(g - f)(x)$.

16. Solve the system of equations.
$$x + 2y = 12$$
$$4x = 8$$
$$3x - 4y + 5z = 20$$

17. Evaluate the determinant.
$$\begin{vmatrix} 3 & -6 & -1 \\ 2 & 1 & -2 \\ 1 & 3 & 1 \end{vmatrix}$$

18. Find the solution set of $|3 - 2x| < 5$.

19. Solve $\sqrt{2x^2 + 7} + 3 = 8$.

20. Falling Object The distance, d, an object drops in free fall is directly proportional to the square of the time, t. If an object falls 16 feet in 1 second, how far will the object fall in 5 seconds?

Quadratic Functions

THERE ARE MANY REAL-LIFE situations that can be represented or approximated with quadratic equations. As you read through this chapter, you will see several real-life applications of quadratic equations and quadratic functions. For instance, in Exercises 101 and 102 on page 747, we will use quadratic equations and the quadratic formula to determine the time it takes for a drop of water on top of a waterfall to reach the bottom of the waterfall.

12.1 Solving Quadratic Equations by Completing the Square

1 Use the square root property to solve equations.

2 Understand perfect square trinomials.

3 Solve quadratic equations by completing the square.

In this section, we introduce two concepts, the square root property and completing the square. The square root property will be used in several sections of this book.

In Section 6.6, we solved quadratic, or second-degree, equations by factoring. Quadratic equations that cannot be solved by factoring can be solved by completing the square, or by the quadratic formula, which is presented in Section 12.2.

1 Use the Square Root Property to Solve Equations

In Section 11.1, we stated that every positive number has two square roots. Thus far, we have been using only the positive square root. In this section, we use both the positive and negative square roots of a number.

Positive Square Root of 25   Negative Square Root of 25

$\sqrt{25} = 5$ $-\sqrt{25} = -5$

A convenient way to indicate the two square roots of a number is to use the plus or minus symbol, $\pm$. For example, the square roots of 25 can be indicated by ± 5, read "plus or minus 5." The equation $x^2 = 25$ has two solutions, the two square roots of 25, which are ± 5. If you check each root, you will see that each value satisfies the equation. The **square root property** can be used to find the solutions to equations of the form $x^2 = a$.

> **Square Root Property**
>
> If $x^2 = a$, where a is a real number, then $x = \pm\sqrt{a}$.

EXAMPLE 1 ▶ Solve the following equations.

a) $x^2 - 9 = 0$ **b)** $x^2 + 10 = 85$

Solution

a) Add 9 to both sides of the equation to isolate the variable.

$$x^2 - 9 = 0$$
$$x^2 = 9 \qquad \textit{Isolate the variable.}$$
$$x = \pm\sqrt{9} \qquad \textit{Square root property}$$
$$= \pm 3$$

Check the solutions in the original equation.

$x = 3$ $x = -3$

$x^2 - 9 = 0$ $x^2 - 9 = 0$

$3^2 - 9 \stackrel{?}{=} 0$ $(-3)^2 - 9 \stackrel{?}{=} 0$

$0 = 0$ *True* $0 = 0$ *True*

In both cases the check is true, which means that both 3 and -3 are solutions to the equation.

b)
$$x^2 + 10 = 85$$
$$x^2 = 75 \qquad \textit{Isolate the variable.}$$
$$x = \pm\sqrt{75} \qquad \textit{Square root property}$$
$$= \pm\sqrt{25}\sqrt{3} \qquad \textit{Simplify.}$$
$$= \pm 5\sqrt{3}$$

The solutions are $5\sqrt{3}$ and $-5\sqrt{3}$.

▶ **Now Try Exercise 13**

Not all quadratic equations have real solutions, as is illustrated in Example 2.

EXAMPLE 2 ▶ Solve the equation $x^2 + 7 = 0$.

Solution

$$x^2 + 7 = 0$$
$$x^2 = -7 \qquad \textit{Isolate the variable.}$$
$$x = \pm\sqrt{-7} \qquad \textit{Square root property}$$
$$= \pm i\sqrt{7}$$

The solutions are $i\sqrt{7}$ and $-i\sqrt{7}$, both of which are imaginary numbers.

▶ **Now Try Exercise 15**

EXAMPLE 3 ▶ Solve   **a)** $(a - 5)^2 = 32$ **b)** $(z + 3)^2 + 28 = 0$.

Solution

a) Since the term containing the variable is already isolated, begin by using the square root property.

$$(a - 5)^2 = 32$$
$$a - 5 = \pm\sqrt{32} \qquad \textit{Square root property}$$
$$a = 5 \pm \sqrt{32} \qquad \textit{Add 5 to both sides.}$$
$$= 5 \pm \sqrt{16}\sqrt{2} \qquad \textit{Simplify.}$$
$$= 5 \pm 4\sqrt{2}$$

The solutions are $5 + 4\sqrt{2}$ and $5 - 4\sqrt{2}$.

b) Begin by subtracting 28 from both sides of the equation to isolate the term containing the variable.

$$(z + 3)^2 + 28 = 0$$
$$(z + 3)^2 = -28$$

Now use the square root property.

$$z + 3 = \pm\sqrt{-28} \qquad \textit{Square root property.}$$
$$z = -3 \pm \sqrt{-28} \qquad \textit{Subtract 3 from both sides.}$$
$$= -3 \pm \sqrt{28}\sqrt{-1}$$
$$= -3 \pm i\sqrt{4}\sqrt{7} \qquad \textit{Simplify } \sqrt{28} \textit{ and replace } \sqrt{-1} \textit{ with i.}$$
$$= -3 \pm 2i\sqrt{7}$$

The solutions are $-3 + 2i\sqrt{7}$ and $-3 - 2i\sqrt{7}$. Note that the solutions to the equation $(z + 3)^2 + 28 = 0$ are not real numbers. The solutions are complex numbers.

▶ **Now Try Exercise 23**

2 Understand Perfect Square Trinomials

Now that we know the square root property we can focus our attention on completing the square. To understand this procedure, you need to know how to form perfect square trinomials. A *perfect square trinomial* is a trinomial that can be expressed as the square of a binomial. In a perfect square trinomial, the first and last terms are perfect squares, and the middle term is twice the product of the square root of the first term and the square root of the last term. Some examples follow.

Perfect Square Trinomials		Factors		Square of a Binomial
$x^2 + 8x + 16$	$=$	$(x + 4)(x + 4)$	$=$	$(x + 4)^2$
$x^2 - 8x + 16$	$=$	$(x - 4)(x - 4)$	$=$	$(x - 4)^2$
$x^2 + 10x + 25$	$=$	$(x + 5)(x + 5)$	$=$	$(x + 5)^2$
$x^2 - 10x + 25$	$=$	$(x - 5)(x - 5)$	$=$	$(x - 5)^2$

In a perfect square trinomial with a leading coefficient of 1, there is a relationship between the coefficient of the first-degree term and the constant term. In such trinomials the constant term is the square of one-half the coefficient of the first degree term. Let's examine some perfect square trinomials for which the leading coefficient is 1.

$$x^2 + 8x + 16 = (x + 4)^2$$
$$\left[\tfrac{1}{2}(8)\right]^2 = (4)^2$$

$$x^2 - 10x + 25 = (x - 5)^2$$
$$\left[\tfrac{1}{2}(-10)\right]^2 = (-5)^2$$

When a perfect square trinomial with a leading coefficient of 1 is written as the square of a binomial, the constant in the binomial is one-half the coefficient of the first-degree term in the trinomial. For example,

$$x^2 + 8x + 16 = (x + 4)^2$$
$$\tfrac{1}{2}(8)$$

$$x^2 - 10x + 25 = (x - 5)^2$$
$$\tfrac{1}{2}(-10)$$

3 Solve Quadratic Equations by Completing the Square

Now we introduce completing the square. To solve a quadratic equation by **completing the square**, we add a constant to both sides of the equation so that the remaining trinomial is a perfect square trinomial. Then we use the square root property to solve the resulting equation. We will now summarize the procedure.

To Solve a Quadratic Equation by Completing the Square

1. Use the multiplication (or division) property of equality if necessary to make the leading coefficient 1.
2. Rewrite the equation with the constant by itself on the right side of the equation.
3. Take one-half the numerical coefficient of the first-degree term, square it, and add this quantity to both sides of the equation.
4. Replace the trinomial with the square of a binomial.
5. Use the square root property to take the square root of both sides of the equation.
6. Solve for the variable.
7. Check your solutions in the *original* equation.

EXAMPLE 4 ▶ Solve the equation $x^2 + 6x + 5 = 0$ by completing the square.

Solution   Since the leading coefficient is 1, step 1 is not necessary.

Step 2:   Move the constant, 5, to the right side of the equation by subtracting 5 from both sides of the equation.

$$x^2 + 6x + 5 = 0$$
$$x^2 + 6x = -5$$

Step 3: Determine the square of one-half the numerical coefficient of the first-degree term, 6.

$$\frac{1}{2}(6) = 3, \qquad 3^2 = \boxed{9}$$

Add this value to both sides of the equation.

$$x^2 + 6x \boxed{+ 9} = -5 \boxed{+ 9}$$
$$x^2 + 6x + 9 = 4$$

Step 4: By following this procedure, we produce a perfect square trinomial on the left side of the equation. The expression $x^2 + 6x + 9$ is a perfect square trinomial that can be expressed as $(x + 3)^2$.

$\frac{1}{2}$ the numerical coefficient of the first-degree term is $\frac{1}{2}(6) = +3$.

$$(x + 3)^2 = 4$$

Step 5: Use the square root property.

$$x + 3 = \pm\sqrt{4}$$
$$x + 3 = \pm 2$$

Step 6: Finally, solve for x by subtracting 3 from both sides of the equation.

$$x + 3 \boxed{- 3} = \boxed{-3} \pm 2$$
$$x = -3 \pm 2$$
$$x = -3 + 2 \qquad \text{or} \qquad x = -3 - 2$$
$$x = -1 \qquad\qquad\qquad x = -5$$

Step 7: Check both solutions in the original equation.

$$x = -1 \qquad\qquad\qquad\qquad x = -5$$
$$x^2 + 6x + 5 = 0 \qquad\qquad\qquad x^2 + 6x + 5 = 0$$
$$(-1)^2 + 6(-1) + 5 \overset{?}{=} 0 \qquad\qquad (-5)^2 + 6(-5) + 5 \overset{?}{=} 0$$
$$1 - 6 + 5 \overset{?}{=} 0 \qquad\qquad\qquad 25 - 30 + 5 \overset{?}{=} 0$$
$$0 = 0 \quad \textit{True} \qquad\qquad\qquad 0 = 0 \quad \textit{True}$$

Since each number checks, both -1 and -5 are solutions to the original equation.

▶ **Now Try Exercise 49**

Helpful Hint

When solving the equation $x^2 + bx + c = 0$ by completing the square, we obtain $x^2 + bx + \left(\dfrac{b}{2}\right)^2$ on the left side of the equation and a constant on the right side of the equation. We then replace $x^2 + bx + \left(\dfrac{b}{2}\right)^2$ with $\left(x + \dfrac{b}{2}\right)^2$. In the figure on the next page we show why

$$x^2 + bx + \left(\frac{b}{2}\right)^2 = \left(x + \frac{b}{2}\right)^2$$

The figure is a square with sides of length $x + \dfrac{b}{2}$. The area is therefore $\left(x + \dfrac{b}{2}\right)^2$. The area of the square can also be determined by adding the areas of the four sections as follows:

$$x^2 + \frac{b}{2}x + \frac{b}{2}x + \left(\frac{b}{2}\right)^2 = x^2 + bx + \left(\frac{b}{2}\right)^2$$

(*continued on the next page*)

Comparing the areas, we see that $x^2 + bx + \left(\dfrac{b}{2}\right)^2 = \left(x + \dfrac{b}{2}\right)^2$.

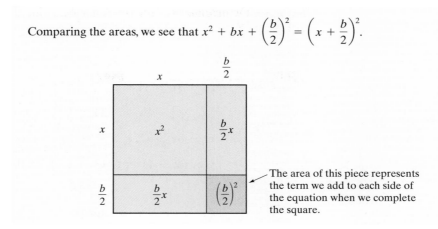

The area of this piece represents the term we add to each side of the equation when we complete the square.

EXAMPLE 5 ▶ Solve the equation $-x^2 = -3x - 18$ by completing the square.

Solution   The numerical coefficient of the squared term must be 1, not -1. Therefore, begin by multiplying both sides of the equation by -1 to make the coefficient of the squared term equal to 1.

$$-x^2 = -3x - 18$$
$$-1(-x^2) = -1(-3x - 18)$$
$$x^2 = 3x + 18$$

Now move all terms except the constant to the left side of the equation.

$$x^2 - 3x = 18$$

Take half the numerical coefficient of the x-term, square it, and add this product to both sides of the equation. Then rewrite the left side of the equation as the square of a binomial.

$$\frac{1}{2}(-3) = -\frac{3}{2} \qquad \left(-\frac{3}{2}\right)^2 = \frac{9}{4}$$

$$x^2 - 3x + \boxed{\frac{9}{4}} = 18 + \boxed{\frac{9}{4}} \qquad \textit{Complete the square.}$$

$$\left(x - \frac{3}{2}\right)^2 = 18 + \frac{9}{4} \qquad \begin{array}{l}\textit{Rewrite the trinomial as the}\\ \textit{square of a binomial.}\end{array}$$

$$\left(x - \frac{3}{2}\right)^2 = \frac{72}{4} + \frac{9}{4}$$

$$\left(x - \frac{3}{2}\right)^2 = \frac{81}{4}$$

$$x - \frac{3}{2} = \pm\sqrt{\frac{81}{4}} \qquad \textit{Square root property}$$

$$x - \frac{3}{2} = \pm\frac{9}{2} \qquad \textit{Simplify.}$$

$$x = \frac{3}{2} \pm \frac{9}{2} \qquad \textit{Add } \frac{3}{2} \textit{ to both sides.}$$

$$x = \frac{3}{2} + \frac{9}{2} \qquad \text{or} \qquad x = \frac{3}{2} - \frac{9}{2}$$

$$x = \frac{12}{2} = 6 \qquad\qquad x = -\frac{6}{2} = -3$$

The solutions are 6 and -3.

▶ **Now Try Exercise 53**

In the following examples we will not show some of the intermediate steps.

EXAMPLE 6 ▶ Solve $x^2 - 8x + 34 = 0$ by completing the square.

Solution

$$x^2 - 8x + 34 = 0$$

$$x^2 - 8x = -34 \qquad \text{\textit{Move the constant term to the right side.}}$$

$$x^2 - 8x \boxed{+\ 16} = -34 \boxed{+\ 16} \qquad \text{\textit{Complete the square.}}$$

$$(x - 4)^2 = -18 \qquad \text{\textit{Write the trinomial as the square of a binomial.}}$$

$$x - 4 = \pm\sqrt{-18} \qquad \text{\textit{Square root property}}$$

$$x - 4 = \pm 3i\sqrt{2} \qquad \text{\textit{Simplify.}}$$

$$x = 4 \pm 3i\sqrt{2} \qquad \text{\textit{Solve for x.}}$$

The solutions are $4 + 3i\sqrt{2}$ and $4 - 3i\sqrt{2}$.

▶ **Now Try Exercise 61**

EXAMPLE 7 ▶ Solve the equation $-4m^2 + 8m + 32 = 0$ by completing the square.

Solution

$$-4m^2 + 8m + 32 = 0$$

$$\boxed{-\frac{1}{4}}(-4m^2 + 8m + 32) = \boxed{-\frac{1}{4}}(0) \qquad \text{\textit{Multiply by } } -\frac{1}{4} \text{ \textit{to obtain a leading coefficient of 1.}}$$

$$m^2 - 2m - 8 = 0$$

Now proceed as before.

$$m^2 - 2m = 8 \qquad \text{\textit{Move the constant term to the right side.}}$$

$$m^2 - 2m \boxed{+\ 1} = 8 \boxed{+\ 1} \qquad \text{\textit{Complete the square.}}$$

$$(m - 1)^2 = 9 \qquad \text{\textit{Write the trinomial as the square of a binomial.}}$$

$$m - 1 = \pm 3 \qquad \text{\textit{Square root property}}$$

$$m = 1 \pm 3 \qquad \text{\textit{Solve for m.}}$$

$$m = 1 + 3 \quad \text{or} \quad m = 1 - 3$$

$$m = 4 \qquad\qquad m = -2$$

▶ **Now Try Exercise 75**

If you were asked to solve the equation $-\frac{1}{4}x^2 + 2x - 8 = 0$ by completing the square, what would you do first? If you answered, "Multiply both sides of the equation by -4 to make the leading coefficient 1," you answered correctly. To solve the equation $\frac{2}{3}x^2 + 3x - 5 = 0$, you would first multiply both sides of the equation by $\frac{3}{2}$ to obtain a leading coefficient of 1.

Generally, quadratic equations that cannot be easily solved by factoring will be solved by the *quadratic formula*, which will be presented in the next section. We introduced completing the square because we use it to derive the quadratic formula in Section 12.2. We will use completing the square later in this chapter and in a later chapter.

EXAMPLE 8 ▶ **Compound Interest** The compound interest formula $A = p\left(1 + \dfrac{r}{n}\right)^{nt}$ can be used to find the amount, A, when an initial principal, p, is invested at an annual interest rate, r, compounded n times a year for t years.

a) Josh Adams initially invested $1000 in a savings account where interest is compounded annually (once a year). If after 2 years the amount, or balance, in the account is $1102.50, find the annual interest rate, r.

b) Trisha McDowell initially invested $1000 in a savings account where interest is compounded quarterly. If after 3 years the amount in the account is $1195.62, find the annual interest rate, r.

Solution **a)** Understand   We are given the following information:

$$p = \$1000, \qquad A = \$1102.50, \qquad n = 1, \qquad t = 2$$

We are asked to find the annual rate, r. To do so, we substitute the appropriate values into the formula and solve for r.

Translate

$$A = p\left(1 + \frac{r}{n}\right)^{nt}$$

$$1102.50 = 1000\left(1 + \frac{r}{1}\right)^{1(2)}$$

Carry Out

$$1102.50 = 1000(1 + r)^2$$

$$1.10250 = (1 + r)^2 \qquad \textit{Divide both sides by 1000.}$$

$$\sqrt{1.10250} = 1 + r \qquad \textit{Square root property; use principal root since r must be positive.}$$

$$1.05 = 1 + r$$

$$0.05 = r \qquad \textit{Subtract 1 from both sides.}$$

Answer   The annual interest rate is 0.05 or 5%.

b) Understand   We are given

$$p = 1000, \qquad A = \$1195.62, \qquad n = 4, \qquad t = 3$$

To find r, we substitute the appropriate values into the formula and solve for r.

Translate

$$A = p\left(1 + \frac{r}{n}\right)^{nt}$$

$$1195.62 = 1000\left(1 + \frac{r}{4}\right)^{4(3)}$$

$$1.19562 = \left(1 + \frac{r}{4}\right)^{12} \qquad \textit{Divide both sides by 1000.}$$

Carry Out

$$\sqrt[12]{1.19562} = 1 + \frac{r}{4} \qquad \textit{Take the 12th root of both sides (or raise both sides to the 1/12 power).}$$

$$1.015 \approx 1 + \frac{r}{4} \qquad \textit{Approximate } \sqrt[12]{1.19562} \textit{ on a calculator.}$$

$$0.015 \approx \frac{r}{4} \qquad \textit{Subtract 1 from both sides.}$$

$$0.06 \approx r \qquad \textit{Multiply both sides by 4.}$$

Answer   The annual interest rate is approximately 0.06 or 6%.

▶ **Now Try Exercise 103**

Helpful Hint *Study Tip*

In this chapter, you will be working with roots and radicals. This material was discussed in Chapter 11. If you do not remember how to evaluate or simplify radicals, review Chapter 11 now.

EXERCISE SET 12.1   Math XL   MyMathLab

MathXL®   MyMathLab

Concept/Writing Exercises

1. Write the two square roots of 36.

2. Write the two square roots of 17.

3. Write the square root property.

4. What is the first step in completing the square?

5. Explain how to determine whether a trinomial is a perfect square trinomial.

6. Write a paragraph explaining how to construct a perfect square trinomial.

7. **a)** Is $x = 4$ the solution to $x - 4 = 0$? If not, what is the correct solution? Explain.

   b) Is $x = 2$ the solution to $x^2 - 4 = 0$? If not, what is the correct solution? Explain.

8. **a)** Is $x = -7$ the solution to $x + 7 = 0$? If not, what is the correct solution? Explain.

   b) Is $x = \pm\sqrt{7}$ the solution to $x^2 + 7 = 0$? If not, what is the correct solution? Explain.

9. What is the first step in solving the equation $2x^2 + 3x = 9$ by completing the square? Explain.

10. What is the first step in solving the equation $\frac{1}{7}x^2 + 12x = -4$ by completing the square? Explain.

11. When solving the equation $x^2 - 6x = 17$ by completing the square, what number do we add to both sides of the equation? Explain.

12. When solving the equation $x^2 + 10x = 39$ by completing the square, what number do we add to both sides of the equation? Explain.

Practice the Skills

Use the square root property to solve each equation.

13. $x^2 - 25 = 0$

14. $x^2 - 49 = 0$

15. $x^2 + 49 = 0$

16. $x^2 - 24 = 0$

17. $x^2 + 24 = 0$

18. $y^2 - 10 = 51$

19. $y^2 + 10 = -51$

20. $(x - 3)^2 = 49$

21. $(p - 4)^2 = 16$

22. $(x + 3)^2 = 49$

23. $(x + 3)^2 + 25 = 0$

24. $(a - 3)^2 = 45$

25. $(a - 2)^2 + 45 = 0$

26. $(a + 2)^2 + 45 = 0$

27. $\left(b + \dfrac{1}{3}\right)^2 = \dfrac{4}{9}$

28. $\left(b - \dfrac{1}{3}\right)^2 = \dfrac{4}{9}$

29. $\left(b - \dfrac{2}{3}\right)^2 + \dfrac{4}{9} = 0$

30. $(x - 0.2)^2 = 0.64$

31. $(x + 0.8)^2 = 0.81$

32. $\left(x + \dfrac{1}{2}\right)^2 = \dfrac{16}{9}$

33. $(2a - 5)^2 = 18$

34. $(4y + 1)^2 = 12$

35. $\left(2y + \dfrac{1}{2}\right)^2 = \dfrac{4}{25}$

36. $\left(3x - \dfrac{1}{4}\right)^2 = \dfrac{9}{25}$

Solve each equation by completing the square.

37. $x^2 + 3x - 4 = 0$

38. $x^2 - 3x - 4 = 0$

39. $x^2 + 8x + 15 = 0$

40. $x^2 - 8x + 15 = 0$

41. $x^2 + 6x + 8 = 0$

42. $x^2 - 6x + 8 = 0$

43. $x^2 - 7x + 6 = 0$

44. $x^2 + 9x + 18 = 0$

45. $2x^2 + x - 1 = 0$

46. $3c^2 - 4c - 4 = 0$

47. $2z^2 - 7z - 4 = 0$

48. $4a^2 + 9a = 9$

49. $x^2 - 13x + 40 = 0$

50. $x^2 + x - 12 = 0$

51. $-x^2 + 6x + 7 = 0$

52. $-a^2 - 5a + 14 = 0$

53. $-z^2 + 9z - 20 = 0$

54. $-z^2 - 4z + 12 = 0$

55. $b^2 = 3b + 28$

56. $-x^2 = 6x - 27$

57. $x^2 + 10x = 11$

58. $-x^2 + 40 = -3x$

59. $x^2 - 4x - 10 = 0$

60. $x^2 - 6x + 2 = 0$

61. $r^2 + 8r + 5 = 0$

62. $a^2 + 4a - 8 = 0$

63. $c^2 - c - 3 = 0$

64. $p^2 - 5p = 4$

65. $x^2 + 3x + 6 = 0$

66. $z^2 - 5z + 7 = 0$

67. $9x^2 - 9x = 0$

68. $4y^2 + 12y = 0$

69. $-\dfrac{3}{4}b^2 - \dfrac{1}{2}b = 0$

70. $\dfrac{1}{3}a^2 - \dfrac{5}{3}a = 0$

71. $36z^2 - 6z = 0$

72. $x^2 = \dfrac{9}{2}x$

73. $-\dfrac{1}{2}p^2 - p + \dfrac{3}{2} = 0$

74. $2x^2 + 6x = 20$

75. $2x^2 = 8x + 64$

76. $3x^2 + 33x + 72 = 0$

77. $2x^2 + 18x + 4 = 0$

78. $\dfrac{2}{3}x^2 + \dfrac{4}{3}x + 1 = 0$

79. $\dfrac{3}{4}w^2 + \dfrac{1}{2}w - \dfrac{1}{4} = 0$

80. $\dfrac{3}{4}c^2 - 2c + 1 = 0$

81. $2x^2 - x = -5$

82. $\dfrac{5}{2}x^2 + \dfrac{3}{2}x - \dfrac{5}{4} = 0$

83. $-3x^2 + 6x = 6$

84. $x^2 + 2x = -5$

Problem Solving

Area *In Exercises 85–88, the area, A, of each rectangle is given.* **a)** *Write an equation for the area.* **b)** *Solve the equation for x.*

85.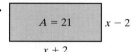
$A = 21$ $x - 2$
$x + 2$

86.
$A = 35$ $x + 3$
$x + 5$

87.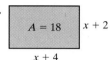
$A = 18$ $x + 2$
$x + 4$

88.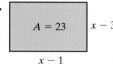
$A = 23$ $x - 3$
$x - 1$

89. Stopping Distance on Snow The formula for approximating the stopping distance, d, in feet, for a specific car on snow is $d = \dfrac{1}{6}x^2$, where x is the speed of the car, in miles per hour, before the brakes are applied. If the car's stopping distance was 150 feet, what was the car's speed before the brakes were applied?

90. Stopping Distance on Dry Pavement The formula for approximating the stopping distance, d, in feet, for a specific car on dry pavement is $d = \dfrac{1}{10}x^2$, where x is the speed of the car, in miles per hour, before the brakes are applied. If the car's stopping distance was 40 feet, what was the car's speed before the brakes were applied?

91. Integers The product of two consecutive positive odd integers is 35. Find the two odd integers.

92. Integers The larger of two integers is 2 more than twice the smaller. Find the two numbers if their product is 12.

93. Rectangular Garden Donna Simm has marked off an area in her yard where she will plant tomato plants. Find the dimensions of the rectangular area if the length is 2 feet more than twice the width and the area is 60 square feet.

94. Driveway Manuel Cortez is planning to blacktop his driveway. Find the dimensions of the rectangular driveway if its area is 381.25 square feet and its length is 18 feet greater than its width.

95. Patio Bill Justice is designing a square patio whose diagonal is 6 feet longer than the length of a side. Find the dimensions of the patio.

Use the formula $A = p\left(1 + \dfrac{r}{n}\right)^{nt}$ to answer Exercises 101–104.

101. Savings Account Frank Dipalo initially invested $500 in a savings account where interest is compounded annually. If after 2 years the amount in the account is $540.80, find the annual interest rate.

102. Savings Account Margret Chang initially invested $1000 in a savings account where interest is compounded annually. If after 2 years the amount in the account is $1102.50, find the annual interest rate.

96. Wading Pool The Lakeside Hotel is planning to build a shallow wading pool for children. If the pool is to be square and the diagonal of the square is 7 feet longer than a side, find the dimensions of the pool.

97. Inscribed Triangle When a triangle is inscribed in a semicircle where a diameter of the circle is a side of the triangle, the triangle formed is always a right triangle. If an isosceles triangle (two equal sides) is inscribed in a semicircle of radius 10 inches, find the length of the other two sides of the triangle.

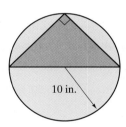
10 in.

98. Inscribed Triangle Refer to Exercise 97. Suppose a triangle is inscribed in a semicircle whose diameter is 12 meters. If one side of the inscribed triangle is 6 meters, find the third side.

99. Area of Circle The area of a circle is 24π square feet. Use the formula $A = \pi r^2$ to find the radius of the circle.

100. Area of Circle The area of a circle is 16.4π square meters. Find the radius of the circle.

103. Savings Account Steve Rodi initially invested $1200 in a savings account where interest is compounded semiannually. If after 3 years the amount in the account is $1432.86, find the annual interest rate.

104. Savings Account Angela Reyes initially invested $1500 in a savings account where interest is compounded semiannually. If after 4 years the amount in the account is $2052.85, find the annual interest rate.

105. Surface Area and Volume The surface area, S, and volume, V, of a right circular cylinder of radius, r, and height, h, are given by the formulas

$$S = 2\pi r^2 + 2\pi rh, \quad V = \pi r^2 h$$

a) Find the surface area of the cylinder if its height is 10 inches and its volume is 160 cubic inches.

b) Find the radius if the height is 10 inches and the volume is 160 cubic inches.

c) Find the radius if the height is 10 inches and the surface area is 160 square inches.

Group Activity

Discuss and answer Exercise 106 as a group.

106. On the following grid, the points $(x_1, y_1), (x_2, y_2)$, and (x_1, y_2) are plotted.

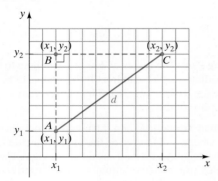

a) Explain why (x_1, y_2) is placed where it is and not somewhere else on the graph.

b) Express the length of the orange dashed line in terms of y_2 and y_1. Explain how you determined your answer.

c) Express the length of the green dashed line in terms of x_2 and x_1.

d) Using the Pythagorean Theorem and the right triangle ABC, derive a formula for the distance, d, between points (x_1, y_1) and (x_2, y_2).* Explain how you determined the formula.

e) Use the formula you determined in part **d)** to find the distance of the line segment between the points $(1, 4)$ and $(3, 7)$.

Cumulative Review Exercises

[2.5] **107.** Solve $-4(2z - 6) = -3(z - 4) + z$.

[3.2] **108. Investment** Thea Prettyman invested $10,000 for 1 year, part at 7% and part at $6\frac{1}{4}$%. If she earned a total interest of $656.50, how much was invested at each rate?

[4.3] **109.** Find the slope of the line through $(-2, 5)$ and $(0, 5)$.

[5.5] **110.** Multiply $(x - 2)(4x^2 + 9x - 3)$.

[10.2] **111.** Solve $|x + 3| = |2x - 7|$.

12.2 Solving Quadratic Equations by the Quadratic Formula

1 Derive the quadratic formula.

2 Use the quadratic formula to solve equations.

3 Determine a quadratic equation given its solutions.

4 Use the discriminant to determine the number of real solutions to a quadratic equation.

5 Study applications that use quadratic equations.

1 Derive the Quadratic Formula

The quadratic formula can be used to solve any quadratic equation. *It is the most useful and most versatile method of solving quadratic equations.* It is generally used in place of completing the square because of its efficiency.

The standard form of a quadratic equation is $ax^2 + bx + c = 0$, where a is the coefficient of the squared term, b is the coefficient of the first-degree term, and c is the constant.

Quadratic Equation in Standard Form	Values of Coefficients		
$x^2 - 3x + 4 = 0$	$a = 1,$	$b = -3,$	$c = 4$
$1.3x^2 - 7.9 = 0$	$a = 1.3,$	$b = 0,$	$c = -7.9$
$-\dfrac{5}{6}x^2 + \dfrac{3}{8}x = 0$	$a = -\dfrac{5}{6},$	$b = \dfrac{3}{8},$	$c = 0$

*The distance formula will be discussed in a later chapter.

We can derive the quadratic formula by starting with a quadratic equation in standard form and completing the square, as discussed in the preceding section.

$$ax^2 + bx + c = 0$$

$$\frac{ax^2}{a} + \frac{b}{a}x + \frac{c}{a} = 0 \qquad \text{\textit{Divide both sides by a.}}$$

$$x^2 + \frac{b}{a}x = -\frac{c}{a} \qquad \text{\textit{Subtract c/a from both sides.}}$$

$$x^2 + \frac{b}{a}x + \boxed{\frac{b^2}{4a^2}} = -\frac{c}{a} + \boxed{\frac{b^2}{4a^2}} \qquad \begin{array}{l} \textit{Take 1/2 of b/a (that is, b/2a), and} \\ \textit{square it to get } b^2/4a^2. \textit{ Then add this} \\ \textit{expression to both sides.} \end{array}$$

$$\left(x + \frac{b}{2a}\right)^2 = \frac{b^2}{4a^2} - \frac{c}{a} \qquad \begin{array}{l} \textit{Rewrite the left side of the equation} \\ \textit{as the square of a binomial.} \end{array}$$

$$\left(x + \frac{b}{2a}\right)^2 = \frac{b^2 - 4ac}{4a^2} \qquad \begin{array}{l} \textit{Write the right side with a common} \\ \textit{denominator.} \end{array}$$

$$x + \frac{b}{2a} = \pm\sqrt{\frac{b^2 - 4ac}{4a^2}} \qquad \text{\textit{Square root property}}$$

$$x + \frac{b}{2a} = \pm\frac{\sqrt{b^2 - 4ac}}{2a} \qquad \text{\textit{Quotient rule for radicals}}$$

$$x = -\frac{b}{2a} \pm \frac{\sqrt{b^2 - 4ac}}{2a} \qquad \text{\textit{Subtract b/2a from both sides.}}$$

$$x = \frac{-b \pm \sqrt{b^2 - 4ac}}{2a} \qquad \begin{array}{l} \textit{Write with a common denominator to} \\ \textit{get the quadratic formula.} \end{array}$$

2 Use the Quadratic Formula to Solve Equations

Now that we have derived the quadratic formula, we will use it to solve quadratic equations.

To Solve a Quadratic Equation by the Quadratic Formula

1. Write the quadratic equation in standard form, $ax^2 + bx + c = 0$, and determine the numerical values for $a, b,$ and c.

2. Substitute the values for $a, b,$ and c into the quadratic formula and then evaluate the formula to obtain the solution.

The Quadratic Formula

$$x = \frac{-b \pm \sqrt{b^2 - 4ac}}{2a}$$

EXAMPLE 1 ▶ Solve $x^2 + 2x - 8 = 0$ by the quadratic formula.

Solution   In this equation, $a = 1, b = 2,$ and $c = -8$.

$$x = \frac{-b \pm \sqrt{b^2 - 4ac}}{2a}$$

$$x = \frac{-2 \pm \sqrt{2^2 - 4(1)(-8)}}{2(1)}$$

$$= \frac{-2 \pm \sqrt{4 + 32}}{2}$$

$$= \frac{-2 \pm \sqrt{36}}{2}$$

$$= \frac{-2 \pm 6}{2}$$

$$x = \frac{-2 + 6}{2} \quad \text{or} \quad x = \frac{-2 - 6}{2}$$

$$x = \frac{4}{2} = 2 \qquad\qquad x = \frac{-8}{2} = -4$$

A check will show that both 2 and -4 are solutions to the equation. Note that the solutions to the equation $x^2 + 2x - 8 = 0$ are two real numbers.

▸ **Now Try Exercise 23**

The solution to Example 1 could also be obtained by factoring, as follows:

$$x^2 + 2x - 8 = 0$$
$$(x + 4)(x - 2) = 0$$
$$x + 4 = 0 \quad \text{or} \quad x - 2 = 0$$
$$x = -4 \qquad\qquad x = 2$$

When you are given a quadratic equation to solve and the method to solve it has not been specified, you may try solving by factoring first (as we discussed in Section 6.6). If the equation cannot be easily factored, use the quadratic formula.

When solving a quadratic equation using the quadratic formula, the calculations may be easier if the leading coefficient, a, is positive. Thus, if solving the quadratic equation $-x^2 + 3x = 2$, you may wish to rewrite the equation as $x^2 - 3x + 2 = 0$.

EXAMPLE 2 ▸ Solve $-9x^2 = -6x + 1$ by the quadratic formula.

Solution   Begin by adding $9x^2$ to both sides of the equation to obtain

$$0 = 9x^2 - 6x + 1$$
$$\text{or} \quad 9x^2 - 6x + 1 = 0$$
$$a = 9, \qquad b = -6, \qquad c = 1$$

$$x = \frac{-b \pm \sqrt{b^2 - 4ac}}{2a}$$

$$= \frac{-(-6) \pm \sqrt{(-6)^2 - 4(9)(1)}}{2(9)}$$

$$= \frac{6 \pm \sqrt{36 - 36}}{18} = \frac{6 \pm \sqrt{0}}{18} = \frac{6}{18} = \frac{1}{3}$$

Note that the solution to the equation $-9x^2 = -6x + 1$ is $\frac{1}{3}$, a single value. Some quadratic equations have just one value as the solution. This occurs when $b^2 - 4ac = 0$.

▸ **Now Try Exercise 39**

Avoiding Common Errors

The entire numerator of the quadratic formula must be divided by $2a$.

CORRECT	INCORRECT
$x = \dfrac{-b \pm \sqrt{b^2 - 4ac}}{2a}$	$x = -b \pm \dfrac{\sqrt{b^2 - 4ac}}{2a}$
	$x = \dfrac{-b}{2a} \pm \sqrt{b^2 - 4ac}$

EXAMPLE 3 ▸ Solve $p^2 + \frac{1}{3}p + \frac{5}{6} = 0$ by the quadratic formula.

Solution   Do not let the change in variable worry you. The quadratic formula is used exactly the same way as when x is the variable.

We could solve this equation using the quadratic formula with $a = 1, b = \frac{1}{3}$, and $c = \frac{5}{6}$. However, when a quadratic equation contains fractions, it is generally easier to begin by multiplying both sides of the equation by the least common denominator. In this example, the least common denominator is 6.

$$6\left(p^2 + \frac{1}{3}p + \frac{5}{6}\right) = 6(0)$$
$$6p^2 + 2p + 5 = 0$$

Now we can use the quadratic formula with $a = 6$, $b = 2$, and $c = 5$.

$$p = \frac{-b \pm \sqrt{b^2 - 4ac}}{2a}$$
$$= \frac{-2 \pm \sqrt{2^2 - 4(6)(5)}}{2(6)}$$
$$= \frac{-2 \pm \sqrt{-116}}{12}$$
$$= \frac{-2 \pm \sqrt{-4}\sqrt{29}}{12}$$
$$= \frac{-2 \pm 2i\sqrt{29}}{12}$$
$$= \frac{\overset{1}{2}(-1 \pm i\sqrt{29})}{\underset{6}{\cancel{12}}}$$
$$= \frac{-1 \pm i\sqrt{29}}{6}$$

The solutions are $\dfrac{-1 + i\sqrt{29}}{6}$ and $\dfrac{-1 - i\sqrt{29}}{6}$. Note that neither solution is a real number. Both solutions are complex numbers.

▶ **Now Try Exercise 53**

Avoiding Common Errors

Some students use the quadratic formula correctly until the last step, where they make an error. Below are illustrated both the correct and incorrect procedures for simplifying an answer.

When *both* terms in the numerator *and* the denominator have a common factor, that common factor may be divided out, as follows:

CORRECT

$$\frac{2 + 4\sqrt{3}}{2} = \frac{\overset{1}{\cancel{2}}(1 + 2\sqrt{3})}{\underset{1}{\cancel{2}}} = 1 + 2\sqrt{3}$$

$$\frac{6 + 3\sqrt{3}}{6} = \frac{\overset{1}{\cancel{3}}(2 + \sqrt{3})}{\underset{2}{\cancel{6}}} = \frac{2 + \sqrt{3}}{2}$$

Below are some common errors. Study them carefully so you will not make them. Can you explain why each of the following procedures is incorrect?

INCORRECT

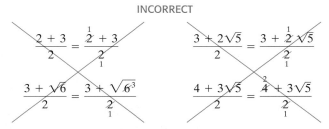

Note that $(2 + 3)/2$ simplifies to $5/2$. However, $(3 + 2\sqrt{5})/2$, $(3 + \sqrt{6})/2$, and $(4 + 3\sqrt{5})/2$ cannot be simplified any further.

EXAMPLE 4 ▶ Given $f(x) = 2x^2 + 4x$, find all real values of x for which $f(x) = 5$.

Solution   We wish to determine all real values of x for which

$$2x^2 + 4x = 5$$

We can solve this equation with the quadratic formula. First, write the equation in standard form.

$$2x^2 + 4x - 5 = 0$$

Now, use the quadratic formula with $a = 2$, $b = 4$, and $c = -5$.

$$x = \frac{-b \pm \sqrt{b^2 - 4ac}}{2a}$$

$$= \frac{-4 \pm \sqrt{4^2 - 4(2)(-5)}}{2(2)} = \frac{-4 \pm \sqrt{56}}{4} = \frac{-4 \pm 2\sqrt{14}}{4}$$

Next, factor out 2 from both terms in the numerator, and then divide out the common factor.

$$x = \frac{\overset{1}{2}(-2 \pm \sqrt{14})}{\underset{2}{4}} = \frac{-2 \pm \sqrt{14}}{2} \,^{*}$$

Thus, the solutions are $\dfrac{-2 + \sqrt{14}}{2}$ and $\dfrac{-2 - \sqrt{14}}{2}$.

Note that the expression in Example 4, $2x^2 + 4x - 5$, is not factorable. Therefore, Example 4 could not be solved by factoring.

▶ **Now Try Exercise 69**

If all the numerical coefficients in a quadratic equation have a common factor, you should factor it out before using the quadratic formula. For example, consider the equation $3x^2 + 12x + 3 = 0$. Here $a = 3$, $b = 12$, and $c = 3$. If we use the quadratic formula, we would eventually obtain $x = -2 \pm \sqrt{3}$ as solutions. By factoring the equation before using the formula, we get

$$3x^2 + 12x + 3 = 0$$
$$3(x^2 + 4x + 1) = 0$$

If we consider $x^2 + 4x + 1 = 0$, then $a = 1$, $b = 4$, and $c = 1$. If we use these new values of a, b, and c in the quadratic formula, we will obtain the identical solutions, $x = -2 \pm \sqrt{3}$. However, the calculations with these smaller values of a, b, and c are simplified. Solve both equations now using the quadratic formula to convince yourself.

3 Determine a Quadratic Equation Given Its Solutions

If we are given the solutions of an equation, we can find the equation by working backward. This procedure is illustrated in Example 5.

EXAMPLE 5 ▶ Determine an equation that has the following solutions:

a) -5 and 1 **b)** $3 + 2i$ and $3 - 2i$

Solution

a) If the solutions are -5 and 1 we write

$$x = -5 \qquad \text{or} \qquad x = 1$$
$$x + 5 = 0 \qquad\qquad x - 1 = 0 \qquad \text{\textit{Set equations equal to 0.}}$$
$$(x + 5)(x - 1) = 0 \qquad \text{\textit{Zero-factor property}}$$
$$x^2 - x + 5x - 5 = 0 \qquad \text{\textit{Multiply factors.}}$$
$$x^2 + 4x - 5 = 0 \qquad \text{\textit{Combine like terms.}}$$

---

*Solutions will be given in this form in the Answer Section.

Thus, the equation is $x^2 + 4x - 5 = 0$. Many other equations have solutions -5 and 1. In fact, any equation of the form $k(x^2 + 4x - 5) = 0$, where k is a constant, has those solutions. Can you explain why?

b)
$$x = 3 + 2i \qquad \text{or} \qquad x = 3 - 2i$$

$x - (3 + 2i) = 0 \qquad x - (3 - 2i) = 0$	*Set equations equal to 0.*
$[x - (3 + 2i)][x - (3 - 2i)] = 0$	*Zero-factor property*
$x \cdot x - x(3 - 2i) - x(3 + 2i) + (3 + 2i)(3 - 2i) = 0$	*Multiply.*
$x^2 - 3x + 2xi - 3x - 2xi + (9 - 4i^2) = 0$	*Distributive property; multiply*
$x^2 - 6x + 9 - 4i^2 = 0$	*Combine like terms.*
$x^2 - 6x + 9 - 4(-1) = 0$	*Substitute $i^2 = -1$.*
$x^2 - 6x + 13 = 0$	*Simplify.*

The equation $x^2 - 6x + 13 = 0$ has the complex solutions $3 + 2i$ and $3 - 2i$.

▶ **Now Try Exercise 75**

In Example 5 **a)**, we determined that the equation $x^2 + 4x - 5 = 0$ has solutions -5 and 1. Consider the graph of $f(x) = x^2 + 4x - 5$. The x-intercepts of the graph of $f(x)$ occur when $f(x) = 0$, or when $x^2 + 4x - 5 = 0$. Therefore, the x-intercepts of the graph of $f(x) = x^2 + 4x - 5$ are $(-5, 0)$ and $(1, 0)$, as shown in **Figure 12.1**. In Example 5 **b)**, we determined that the equation $x^2 - 6x + 13 = 0$ has no real solutions. Thus, the graph of $f(x) = x^2 - 6x + 13$ has no x-intercepts. The graph of $f(x) = x^2 - 6x + 13$ is shown in **Figure 12.2**.

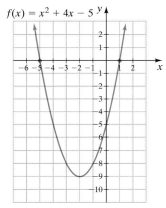

FIGURE 12.1

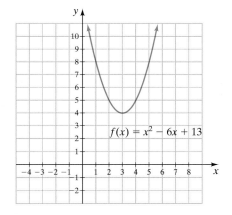

FIGURE 12.2

4 Use the Discriminant to Determine the Number of Real Solutions to a Quadratic Equation

The expression under the radical sign in the quadratic formula is called the **discriminant**.

$$\underbrace{b^2 - 4ac}_{\text{discriminant}}$$

The discriminant provides information to determine the number and kinds of solutions of a quadratic equation.

Solutions of a Quadratic Equation
For a quadratic equation of the form $ax^2 + bx + c = 0$, $a \neq 0$:
If $b^2 - 4ac > 0$, the quadratic equation has two distinct real number solutions.
If $b^2 - 4ac = 0$, the quadratic equation has a single real number solution.
If $b^2 - 4ac < 0$, the quadratic equation has no real number solution.

EXAMPLE 6 ▶

a) Find the discriminant of the equation $x^2 - 8x + 16 = 0$.

b) How many real number solutions does the given equation have?

c) Use the quadratic formula to find the solution(s).

Solution

a) $a = 1,$ $b = -8,$ $c = 16$

$$b^2 - 4ac = (-8)^2 - 4(1)(16)$$
$$= 64 - 64 = 0$$

b) Since the discriminant equals 0, there is a single real number solution.

c)
$$x = \frac{-b \pm \sqrt{b^2 - 4ac}}{2a}$$

$$= \frac{-(-8) \pm \sqrt{0}}{2(1)} = \frac{8 \pm 0}{2} = \frac{8}{2} = 4$$

The only solution is 4.

▶ **Now Try Exercise 9**

EXAMPLE 7 ▶ Without actually finding the solutions, determine whether the following equations have two distinct real number solutions, a single real number solution, or no real number solution.

a) $2x^2 - 4x + 6 = 0$ **b)** $x^2 - 5x - 3 = 0$ **c)** $4x^2 - 12x = -9$

Solution   We use the discriminant of the quadratic formula to answer these questions.

a) $b^2 - 4ac = (-4)^2 - 4(2)(6) = 16 - 48 = -32$

Since the discriminant is negative, this equation has no real number solution.

b) $b^2 - 4ac = (-5)^2 - 4(1)(-3) = 25 + 12 = 37$

Since the discriminant is positive, this equation has two distinct real number solutions.

c) First, rewrite $4x^2 - 12x = -9$ as $4x^2 - 12x + 9 = 0$.

$$b^2 - 4ac = (-12)^2 - 4(4)(9) = 144 - 144 = 0$$

Since the discriminant is 0, this equation has a single real number solution.

▶ **Now Try Exercise 15**

The discriminant can be used to find the number of real solutions to an equation of the form $ax^2 + bx + c = 0$. Since the x-intercepts of a quadratic function, $f(x) = ax^2 + bx + c$, occur where $f(x) = 0$, the discriminant can also be used to find the number of x-intercepts of a quadratic function. **Figure 12.3** shows the relationship between the discriminant and the number of x-intercepts for a function of the form $f(x) = ax^2 + bx + c$.

Graphs of $f(x) = ax^2 + bx + c$

If $b^2 - 4ac > 0$, $f(x)$ has two distinct x-intercepts.

If $b^2 - 4ac = 0$, $f(x)$ has a single x-intercept.

If $b^2 - 4ac < 0$, $f(x)$ has no x-intercepts.

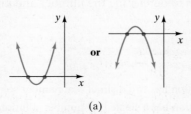

(a)

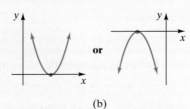

(b)

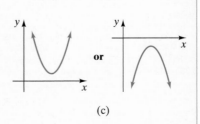

(c)

FIGURE 12.3

We will discuss graphing quadratic functions in detail in Section 12.5.

5 Study Applications That Use Quadratic Equations

We will now look at some applications of quadratic equations.

EXAMPLE 8 ▶ **Cell Phones** Mary Olson owns a business that manufactures and sells cell phones. The revenue, $R(n)$, from selling the cell phones is determined by multiplying the number of cell phones by the cost per phone. Suppose the revenue from selling n cell phones, $n \leq 50$, is

$$R(n) = n(50 - 0.2n)$$

where $(50 - 0.2n)$ is the price per cell phone, in dollars.

a) Find the revenue when 30 cell phones are sold.

b) How many cell phones must be sold to have a revenue of $480?

Solution   **a)** To find the revenue when 30 cell phones are sold, we evaluate the revenue function for $n = 30$.

$$R(n) = n(50 - 0.2n)$$
$$R(30) = 30[50 - 0.2(30)]$$
$$= 30(50 - 6)$$
$$= 30(44)$$
$$= 1320$$

The revenue from selling 30 cell phones is $1320.

b) Understand   We want to find the number of cell phones that need to be sold to have $480 in revenue. Thus, we need to let $R(n) = 480$ and solve for n.

$$R(n) = n(50 - 0.2n)$$
$$480 = n(50 - 0.2n)$$
$$480 = 50n - 0.2n^2$$
$$0.2n^2 - 50n + 480 = 0$$

Now we can use the quadratic formula to solve the equation.

Translate $a = 0.2, \qquad b = -50, \qquad c = 480$

$$n = \frac{-b \pm \sqrt{b^2 - 4ac}}{2a}$$

$$= \frac{-(-50) \pm \sqrt{(-50)^2 - 4(0.2)(480)}}{2(0.2)}$$

Carry Out $$= \frac{50 \pm \sqrt{2500 - 384}}{0.4}$$

$$= \frac{50 \pm \sqrt{2116}}{0.4}$$

$$= \frac{50 \pm 46}{0.4}$$

$$n = \frac{50 + 46}{0.4} = 240 \quad \text{or} \quad n = \frac{50 - 46}{0.4} = 10$$

Answer   Since the problem specified that $n \leq 50$, the only acceptable solution is $n = 10$. Thus, to obtain $480 in revenue, Mary must sell 10 cell phones.

▶ **Now Try Exercise 87**

An important formula in physics is $h = \dfrac{1}{2}gt^2 + v_0 t + h_0$. When an object is projected upward from an initial height, h_0, with initial velocity of v_0, this formula can be used to find the height, h, of the object above the ground at any time, t. The g in the formula is the acceleration of gravity. Since the acceleration of Earth's gravity is -32 ft/sec^2, we use -32 for g in the formula when discussing Earth. This formula can also be used in describing projectiles on the Moon and other planets, but the value of g in the formula will need to change for each planetary body. We will use this formula in Example 9.

EXAMPLE 9 ▶ Throwing a Ball Betsy Farber is standing on top of a building and throws a ball upward from a height of 60 feet with an initial velocity of 30 feet per second. Use the formula $h = \dfrac{1}{2}gt^2 + v_0 t + h_0$ to answer the following questions.

a) How long after the ball is thrown, to the nearest tenth of a second, will the ball be 25 feet above the ground?

b) How long after the ball is thrown will the ball strike the ground?

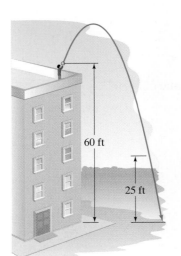

FIGURE 12.4

Solution   **a)** Understand   We will illustrate this problem with a diagram (see **Fig. 12.4**). Here $g = -32$, $v_0 = 30$, and $h_0 = 60$. We are asked to find the time, t, it takes for the ball to reach a height, h, of 25 feet above the ground. We substitute these values into the formula and then solve for t.

Translate

$$h = \frac{1}{2}gt^2 + v_0 t + h_0$$

$$25 = \frac{1}{2}(-32)t^2 + 30t + 60$$

Carry Out   Now we write the quadratic equation in standard form and solve for t by using the quadratic formula.

$$0 = -16t^2 + 30t + 35$$
$$\text{or} \quad -16t^2 + 30t + 35 = 0$$
$$a = -16, \quad b = 30, \quad c = 35$$

$$t = \frac{-b \pm \sqrt{b^2 - 4ac}}{2a}$$

$$= \frac{-30 \pm \sqrt{(30)^2 - 4(-16)(35)}}{2(-16)}$$

$$= \frac{-30 \pm \sqrt{3140}}{-32}$$

$$t = \frac{-30 + \sqrt{3140}}{-32} \quad \text{or} \quad t = \frac{-30 - \sqrt{3140}}{-32}$$
$$\approx -0.8 \qquad\qquad\qquad \approx 2.7$$

Answer   Since time cannot be negative, the only reasonable solution is 2.7 seconds. Thus, about 2.7 seconds after the ball is thrown upward, it will be 25 feet above the ground.

b) Understand   We wish to find the time at which the ball strikes the ground. When the ball strikes the ground, its distance above the ground is 0. We substitute $h = 0$ into the formula and solve for t.

Translate

$$h = \frac{1}{2}gt^2 + v_0 t + h_0$$

$$0 = \frac{1}{2}(-32)t^2 + 30t + 60$$

Carry Out

$$0 = -16t^2 + 30t + 60$$

$$a = -16, \quad b = 30, \quad c = 60$$

$$t = \frac{-b \pm \sqrt{b^2 - 4ac}}{2a}$$

$$= \frac{-30 \pm \sqrt{(30)^2 - 4(-16)(60)}}{2(-16)}$$

$$= \frac{-30 \pm \sqrt{4740}}{-32}$$

$$t = \frac{-30 + \sqrt{4740}}{-32} \quad \text{or} \quad t = \frac{-30 - \sqrt{4740}}{-32}$$

$$\approx -1.2 \qquad\qquad\qquad \approx 3.1$$

Answer   Since time cannot be negative, the only reasonable solution is 3.1 seconds. Thus, the ball will strike the ground approximately 3.1 seconds after it is thrown.

▸ **Now Try Exercise 103**

EXERCISE SET 12.2   *Math XL* **MyMathLab**
MathXL®   MyMathLab

Concept/Writing Exercises

1. Give the quadratic formula. (You should memorize this formula.)

2. To solve the equation $3x + 2x^2 - 9 = 0$ using the quadratic formula, what are the values for $a, b,$ and c?

3. To solve the equation $6x - 3x^2 + 8 = 0$ using the quadratic formula, what are the values for $a, b,$ and c?

4. To solve the equation $4x^2 - 5x = 7$ using the quadratic formula, what are the values for $a, b,$ and c?

5. Consider the two equations $-6x^2 + \frac{1}{2}x - 5 = 0$ and $6x^2 - \frac{1}{2}x + 5 = 0$. Must the solutions to these two equations be the same? Explain your answer.

6. Consider $12x^2 - 15x - 6 = 0$ and $3(4x^2 - 5x - 2) = 0$.

a) Will the solution to the two equations be the same? Explain.

b) Solve $12x^2 - 15x - 6 = 0$.

c) Solve $3(4x^2 - 5x - 2) = 0$.

7. a) Explain how to find the discriminant.

b) What is the discriminant for the equation $3x^2 - 6x + 10 = 0$?

c) Write a paragraph or two explaining the relationship between the value of the discriminant and the number of real solutions to a quadratic equation. In your paragraph, explain *why* the value of the discriminant determines the number of real solutions.

8. Write a paragraph or two explaining the relationship between the value of the discriminant and the number of x-intercepts of $f(x) = ax^2 + bx + c$. In your paragraph, explain when the function will have no, one, and two x-intercepts.

Practice the Skills

Use the discriminant to determine whether each equation has two distinct real solutions, a single real solution, or no real solution.

9. $x^2 + 3x + 1 = 0$

10. $2x^2 + x + 3 = 0$

11. $4z^2 + 6z + 5 = 0$

12. $-a^2 + 3a - 6 = 0$

 13. $5p^2 + 3p - 7 = 0$

14. $2x^2 = 16x - 32$

15. $-5x^2 + 5x - 8 = 0$

16. $4.1x^2 - 3.1x - 2.8 = 0$

17. $x^2 + 10.2x + 26.01 = 0$

18. $\frac{1}{2}x^2 + \frac{2}{3}x + 10 = 0$

19. $b^2 = -3b - \frac{9}{4}$

20. $\frac{x^2}{3} = \frac{2x}{7}$

Solve each equation by the quadratic formula.

21. $x^2 - 9x + 18 = 0$

22. $x^2 + 9x + 18 = 0$

23. $a^2 - 6a + 8 = 0$

24. $a^2 + 6a + 8 = 0$

25. $x^2 = -6x + 7$

26. $-a^2 - 9a + 10 = 0$

27. $-b^2 = 4b - 20$

28. $a^2 - 16 = 0$

29. $b^2 - 64 = 0$

30. $2x^2 = 4x + 1$

31. $3w^2 - 4w + 5 = 0$

32. $x^2 - 6x = 0$

33. $c^2 - 5c = 0$

34. $-t^2 - t - 1 = 0$

35. $4s^2 - 8s + 6 = 0$

36. $-3r^2 = 9r + 6$

37. $a^2 + 2a + 1 = 0$

38. $y^2 + 16y + 64 = 0$

39. $16x^2 - 8x + 1 = 0$

40. $100m^2 + 20m + 1 = 0$

41. $x^2 - 2x - 1 = 0$

42. $2 - 3r^2 = -4r$

43. $-n^2 = 3n + 6$

44. $-9d - 3d^2 = 5$

45. $2x^2 + 5x - 3 = 0$

46. $(r - 3)(3r + 4) = -10$

47. $(2a + 3)(3a - 1) = 2$

48. $6x^2 = 21x + 27$

49. $\frac{1}{2}t^2 + t - 12 = 0$

50. $\frac{2}{3}x^2 = 8x - 18$

51. $9r^2 + 3r - 2 = 0$

52. $2x^2 - 4x - 2 = 0$

53. $\frac{1}{2}x^2 + 2x + \frac{2}{3} = 0$

54. $x^2 - \frac{11}{3}x = \frac{10}{3}$

55. $a^2 - \frac{a}{5} - \frac{1}{3} = 0$

56. $b^2 = -\frac{b}{2} + \frac{2}{3}$

57. $c = \frac{c - 6}{4 - c}$

58. $3y = \frac{5y + 6}{2y + 3}$

59. $2x^2 - 4x + 5 = 0$

60. $3a^2 - 4a = -5$

61. $y^2 + \frac{y}{2} = -\frac{3}{2}$

62. $2b^2 - \frac{7}{3}b + \frac{4}{3} = 0$

63. $0.1x^2 + 0.6x - 1.2 = 0$

64. $2.3x^2 - 5.6x - 0.4 = 0$

For each function, determine all real values of the variable for which the function has the value indicated.

65. $f(x) = x^2 - 2x + 5, f(x) = 5$

66. $g(x) = x^2 + 3x + 8, g(x) = 8$

67. $k(x) = x^2 - x - 15, k(x) = 15$

68. $p(r) = r^2 + 17r + 81, p(r) = 9$

69. $h(t) = 2t^2 - 7t + 6, h(t) = 2$

70. $t(x) = x^2 + 5x - 4, t(x) = 3$

71. $g(a) = 2a^2 - 3a + 16, g(a) = 14$

72. $h(x) = 6x^2 + 3x + 1, h(x) = -7$

Determine an equation that has the given solutions.

73. $2, 5$

74. $-3, 4$

75. $1, -9$

76. $-2, -6$

77. $-\frac{3}{5}, \frac{2}{3}$

78. $-\frac{1}{3}, -\frac{3}{4}$

79. $\sqrt{2}, -\sqrt{2}$

80. $\sqrt{5}, -\sqrt{5}$

81. $3i, -3i$

82. $8i, -8i$

83. $3 + \sqrt{2}, 3 - \sqrt{2}$

84. $5 - \sqrt{3}, 5 + \sqrt{3}$

85. $2 + 3i, 2 - 3i$

86. $5 - 4i, 5 + 4i$

Problem Solving

In Exercises 87–90, **a)** *set up a revenue function, R(n), that can be used to solve the problem, and then,* **b)** *solve the problem. See Example 8.*

87. **Selling Lamps** A business sells n lamps, $n \leq 65$, at a price of $(10 - 0.02n)$ dollars per lamp. How many lamps must be sold to have a revenue of $450?

88. **Selling Batteries** A business sells n batteries, $n \leq 26$, at a price of $(25 - 0.1n)$ dollars per battery. How many batteries must be sold to have a revenue of $460?

89. **Selling Chairs** A business sells n chairs, $n \leq 50$, at a price of $(50 - 0.4n)$ dollars per chair. How many chairs must be sold to have a revenue of $660?

90. **Selling Watches** A business sells n watches, $n \leq 75$, at a price of $(30 - 0.15n)$ dollars per watch. How many watches must be sold to have a revenue of $1260?

91. Give your own example of a quadratic equation that can be solved by the quadratic formula but not by factoring over the set of integers.

92. Are there any quadratic equations that **a)** can be solved by the quadratic formula that cannot be solved by completing the square? **b)** can be solved by completing the square that cannot be solved by factoring over the set of integers?

93. When solving a quadratic equation by the quadratic formula, if the discriminant is a perfect square, must the equation be factorable over the set of integers?

94. When solving a quadratic equation by the quadratic formula, if the discriminant is a natural number, must the equation be factorable over the set of integers?

In Exercises 95–102, use a calculator as needed to give the solution in decimal form. Round irrational numbers to the nearest hundredth.

95. Numbers Twice the square of a positive number increased by three times the number is 27. Find the number.

96. Numbers Three times the square of a positive number decreased by twice the number is 21. Find the number.

97. Rectangular Garden The length of a rectangular garden is 1 foot less than 3 times its width. Find the length and width if the area of the garden is 24 square feet.

98. Rectangular Region Lora Wallman wishes to fence in a rectangular region along a riverbank by constructing fencing as illustrated in the diagram. If she has only 400 feet of fencing and wishes to enclose an area of 15,000 square feet, find the dimensions of the rectangular region.

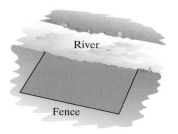

River

Fence

99. Photo John Williams, a professional photographer, has a 6-inch-by-8-inch photo. He wishes to reduce the photo by the same amount on each side so that the resulting photo will have half the area of the original photo. By how much will he have to reduce the length of each side?

100. Rectangular Garden Bart Simmons has a 12-meter-by-9-meter flower garden. He wants to build a gravel path of uniform width along the inside of the garden on each side so

that the resulting garden will have half the area of the original garden. What will be the width of the gravel path?

101. Waterfalls When a drop of water (or other object) at the top of the Lower Falls in Yellowstone National Park goes over the top of the falls, the height, h, in feet, of the drop of water above the pool of water at the bottom of the falls can be determined by the equation $h = -16t^2 + 308$. In the equation, t is the time, in seconds, after the drop goes over the falls. Determine the time it takes for the drop of water to reach the bottom of the falls (when $h = 0$).

102. Waterfalls When a drop of water (or other object) at the top of Niagara Falls goes over the top of the falls, the height, h, in feet, of the drop of water above the pool of water at the bottom of the falls can be determined by the equation $h = -16t^2 + 176$. In the equation, t is the time, in seconds, after the drop goes over the falls. Determine the time it takes for the drop of water to reach the bottom of the falls (when $h = 0$).

In Exercises 103 and 104, use the equation $h = \dfrac{1}{2}gt^2 + v_0t + h_0$ (refer to Example 9).

103. Throwing a Horseshoe A horseshoe is thrown upward from an initial height of 80 feet with an initial velocity of 60 feet per second. How long after the horseshoe is projected upward

   a) will it be 20 feet from the ground?

   b) will it strike the ground?

104. Gravity on the Moon Gravity on the Moon is about one-sixth of that on Earth. Suppose Neil Armstrong is standing on a hill on the Moon 60 feet high. If he jumps upward with a velocity of 40 feet per second, how long will it take for him to land on the ground below the hill?

Solve by the quadratic formula.

105. $x^2 - \sqrt{5}x - 10 = 0$

106. $x^2 + 5\sqrt{6}x + 36 = 0$

Challenge Problems

107. Heating a Metal Cube A metal cube expands when heated. If each edge increases 0.20 millimeter after being heated and the total volume increases by 6 cubic millimeters, find the original length of a side of the cube.

108. Six Solutions The equation $x^n = 1$ has n solutions (including the complex solutions). Find the six solutions to $x^6 = 1$. (*Hint:* Rewrite the equation as $x^6 - 1 = 0$, then factor using the formula for the difference of two squares.)

109. Throwing a Rock Travis Hawley is on the fourth floor of an eight-story building and Courtney Prenzlow is on the roof. Travis is 60 feet above the ground while Courtney is 120 feet above the ground.

   a) If Travis drops his rock out of a window, determine the time it takes for the rock to strike the ground.

   b) If Courtney drops her rock off the roof, determine the time it takes for the rock to strike the ground.

   c) If Travis throws a rock upward with an initial velocity of 100 feet per second at the same time that Courtney throws a rock upward at 60 feet per second, whose rock will strike the ground first? Explain.

   d) Will the rocks ever be at the same distance above the ground? If so, at what time?

Cumulative Review Exercises

[5.3] **110.** Evaluate $\dfrac{5.55 \times 10^3}{1.11 \times 10^1}$.

[8.2] **111.** If $f(x) = x^2 + 2x - 8$, find $f(3)$.

[9.3] **112.** Solve the system of equations.
$$3x + 4y = 2$$
$$2x = -5y - 1$$

[11.5] **113.** Simplify $\dfrac{x + \sqrt{y}}{x - \sqrt{y}}$.

[11.6] **114.** Solve $\sqrt{x^2 - 6x - 4} = x$.

12.3 Quadratic Equations: Applications and Problem Solving

1 Solve additional applications.

2 Solve for a variable in a formula.

1 Solve Additional Applications

We have already discussed a few applications of quadratic equations. In this section, we will explore several more applications. We will also discuss solving for a variable in a formula. We start by investigating a profit for a new company.

EXAMPLE 1 ▶ **Company Profits** Laserox, a start-up company, projects that its annual profits, $p(n)$, in thousands of dollars, over the first 6 years of operation can be approximated by the function $p(n) = 1.2n^2 + 4n - 8$, where n is the number of years completed.

   a) Estimate the profit (or loss) of the company after the first year.

   b) Estimate the profit (or loss) of the company after 6 years.

   c) Estimate the time needed for the company to break even.

Solution   **a)** To estimate the profit after 1 year, we evaluate the function at 1.
$$p(n) = 1.2n^2 + 4n - 8$$
$$p(1) = 1.2(1)^2 + 4(1) - 8 = -2.8$$

Thus, at the end of the first year the company projects a loss of $2.8 thousand or a loss of $2800.

b) $p(6) = 1.2(6)^2 + 4(6) - 8 = 59.2$
Thus, at the end of the sixth year the company's projected profit is $59.2 thousand, or a profit of $59,200.

c) Understand   The company will break even when the profit is 0. Thus, to find the break-even point (no profit or loss) we solve the equation
$$1.2n^2 + 4n - 8 = 0$$
We can use the quadratic formula to solve this equation.

Translate $a = 1.2, \qquad b = 4, \qquad c = -8$

$$n = \frac{-b \pm \sqrt{b^2 - 4ac}}{2a}$$

$$= \frac{-4 \pm \sqrt{4^2 - 4(1.2)(-8)}}{2(1.2)}$$

Carry Out
$$= \frac{-4 \pm \sqrt{16 + 38.4}}{2.4}$$

$$= \frac{-4 \pm \sqrt{54.4}}{2.4}$$

$$\approx \frac{-4 \pm 7.376}{2.4}$$

$$n \approx \frac{-4 + 7.376}{2.4} \approx 1.4 \quad \text{or} \quad n \approx \frac{-4 - 7.376}{2.4} \approx -4.74$$

Answer   Since time cannot be negative, the break-even time is about 1.4 years.

▶ **Now Try Exercise 29**

Now, let's consider another example that uses the quadratic formula to solve a quadratic equation.

EXAMPLE 2 ▶ **Life Expectancy** The function $N(t) = 0.0054t^2 - 1.46t + 95.11$ can be used to estimate the average number of years of life expectancy remaining for a person of age t years where $30 \le t \le 100$.

  a) Estimate the remaining life expectancy of a person of age 40.

  b) If a person has a remaining life expectancy of 14.3 years, estimate the age of the person.

Solution   **a)** Understand   We would expect that the older a person gets the shorter the remaining life expectancy. To determine the remaining life expectancy for a 40-year-old, we substitute 40 for t in the function and evaluate.

Translate
$$N(t) = 0.0054t^2 - 1.46t + 95.11$$
$$N(40) = 0.0054(40)^2 - 1.46(40) + 95.11$$

Carry Out
$$= 0.0054(1600) - 58.4 + 95.11$$
$$= 8.64 - 58.4 + 95.11$$
$$= 45.35$$

Answer and Check   The answer appears reasonable. Thus, on the average, a 40-year-old can expect to live another 45.35 years to an age of 85.35 years.

b) Understand   Here we are given the remaining life expectancy, $N(t)$, and asked to find the age of the person, t. To solve this problem, we substitute 14.3 for $N(t)$ and solve for t. To solve for t, we will use the quadratic formula.

Translate
$$N(t) = 0.0054t^2 - 1.46t + 95.11$$
$$14.3 = 0.0054t^2 - 1.46t + 95.11$$

Carry Out
$$0 = 0.0054t^2 - 1.46t + 80.81$$
$$a = 0.0054, \quad b = -1.46, \quad c = 80.81$$

$$t = \frac{-b \pm \sqrt{b^2 - 4ac}}{2a}$$

$$= \frac{-(-1.46) \pm \sqrt{(-1.46)^2 - 4(0.0054)(80.81)}}{2(0.0054)}$$

$$= \frac{1.46 \pm \sqrt{2.1316 - 1.745496}}{0.0108}$$

$$= \frac{1.46 \pm \sqrt{0.386104}}{0.0108}$$

$$\approx \frac{1.46 \pm 0.6214}{0.0108}$$

$$t \approx \frac{1.46 + 0.6214}{0.0108} \quad \text{or} \quad t \approx \frac{1.46 - 0.6214}{0.0108}$$

$$\approx 192.72 \qquad\qquad\qquad \approx 77.65$$

Answer   Since 192.72 is not a reasonable age, we can exclude that as a possibility. Thus, the average person who has a life expectancy of 14.3 years is about 77.65 years old.

▸ **Now Try Exercise 31**

Motion Problems

We first discussed motion problems in Section 3.4. The motion problem we give here is solved using the quadratic formula.

EXAMPLE 3 ▸ **Motorboat Ride**   Charles Curtis decides to go for a relaxing ride in his motorboat on the Potomac River. His trip starts in Bethesda, Maryland. He travels downstream for 12 miles with the current. He then turns around and heads back to the starting point going upstream against the current. The total time of his trip is 5 hours and the river current is 2 miles per hour. If during the entire trip he did not touch the throttle to change the speed, find the speed the boat would have traveled in still water.

Solution   Understand   We are asked to find the rate of the boat in still water. Let r = the rate of the boat in still water. We know that the total time of the trip is 5 hours. Thus, the time downriver plus the time upriver must sum to 5 hours. Since distance = rate · time, we can find the time by dividing the distance by the rate.

Direction	Distance	Rate	Time
Downriver (with current)	12	$r + 2$	$\dfrac{12}{r + 2}$
Upriver (against current)	12	$r - 2$	$\dfrac{12}{r - 2}$

Translate   time downriver + time upriver = total time

$$\frac{12}{r + 2} + \frac{12}{r - 2} = 5$$

Carry Out $(r + 2)(r - 2)\left(\dfrac{12}{r + 2} + \dfrac{12}{r - 2}\right) = (r + 2)(r - 2)(5)$ *Multiply by the LCD.*

$(r + 2)(r - 2)\left(\dfrac{12}{r + 2}\right) + (r + 2)(r - 2)\left(\dfrac{12}{r - 2}\right) = (r + 2)(r - 2)(5)$ *Distributive property*

$$12(r - 2) + 12(r + 2) = 5(r^2 - 4)$$

$$12r - 24 + 12r + 24 = 5r^2 - 20 \qquad \text{\textit{Distributive property}}$$

$$24r = 5r^2 - 20 \qquad \text{\textit{Simplify.}}$$

$$\text{or} \quad 5r^2 - 24r - 20 = 0$$

Using the quadratic formula with $a = 5$, $b = -24$, and $c = -20$, we obtain

$$r = \frac{24 \pm \sqrt{976}}{10}$$

$$r \approx 5.5 \quad \text{or} \quad r \approx -0.7$$

Answer   Since the rate cannot be negative, the rate or speed of the boat in still water is about 5.5 miles per hour.

▸ **Now Try Exercise 43**

Notice that in real-life situations most answers are not integral values.

Work Problems

Let's do an example involving a work problem. Work problems were discussed in Section 7.7. You may wish to review that section before studying the next example.

EXAMPLE 4 ▶ **Pumping Water** After a hurricane, the Durals needed to pump water from their flooded basement. They had one sump pump (used to pump out water) and borrowed a second from their local fire department. With both pumps working together, their basement would empty in about 6 hours. The fire department's pump had a higher horsepower, and it would empty their basement by itself in 2 hours less time than the Dural's pump would if it were working alone. How long would it take each pump to empty the basement if each were working alone?

Solution   Understand   Recall from Section 7.7 that the rate of work multiplied by the time worked gives the part of the task completed.

Let t = number of hours for the Durals' (slower) pump to complete the job by itself,

then $t - 2$ = number of hours for the fire department's pump to complete the job by itself.

Pump	Rate of Work	Time Worked	Part of Task Completed
Durals' pump	$\dfrac{1}{t}$	6	$\dfrac{6}{t}$
Fire department's pump	$\dfrac{1}{t-2}$	6	$\dfrac{6}{t-2}$

Translate

$$\left(\begin{array}{c}\text{part of task}\\\text{by Durals' pump}\end{array}\right) + \left(\begin{array}{c}\text{part of task}\\\text{by fire department's pump}\end{array}\right) = 1$$

$$\frac{6}{t} + \frac{6}{t-2} = 1$$

Carry Out   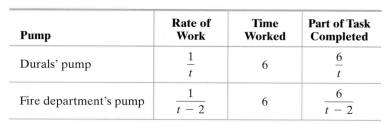 $t(t-2)\left(\dfrac{6}{t} + \dfrac{6}{t-2}\right) = t(t-2)(1)$ *Multiply both sides by the LCD, $t(t-2)$.*

$$t(t-2)\left(\frac{6}{t}\right) + t(t-2)\left(\frac{6}{t-2}\right) = t^2 - 2t \qquad \textit{Distributive property}$$

$$6(t-2) + 6t = t^2 - 2t$$

$$6t - 12 + 6t = t^2 - 2t$$

$$t^2 - 14t + 12 = 0$$

Using the quadratic formula, we obtain

$$t = \frac{14 \pm \sqrt{148}}{2}$$

$$t \approx 13.1 \quad \text{or} \quad t \approx 0.9$$

Answer   Both 13.1 and 0.9 satisfy the equation $\dfrac{6}{t} + \dfrac{6}{t-2} = 1$ (with some round-off involved). However, if we accept 0.9 as a solution, then the fire department's pump could complete the task in a negative time ($t - 2 = 0.9 - 2 = -1.1$ hours), which is not possible. Therefore, 0.9 hour is not an acceptable solution. The only solution is 13.1 hours. The Durals' pump takes approximately 13.1 hours by itself, and the fire department's pump takes approximately $13.1 - 2$ or 11.1 hours by itself to empty the basement.

▶ **Now Try Exercise 45**

2 Solve for a Variable in a Formula

When the square of a variable appears in a *formula*, you may need to use the square root property to solve for the variable. However, *when you use the square root property in most formulas, you will use only the principal or positive root*, because you are generally solving for a quantity that cannot be negative.

EXAMPLE 5 ▶

a) The formula for the area of a circle is $A = \pi r^2$. Solve this equation for the radius, r.

b) *Newton's law of universal gravity* states that every particle in the universe attracts every other particle with a force proportional to the product of their masses and inversely proportional to the square of the distance between them. We may represent Newton's law as

$$F = G\frac{m_1 m_2}{r^2}$$

Solve the equation for r.

Solution

a)
$$A = \pi r^2$$

$$\frac{A}{\pi} = r^2 \qquad \text{Isolate } r^2 \text{ by dividing both sides by } \pi.$$

$$\sqrt{\frac{A}{\pi}} = r \qquad \text{Square root property}$$

b)
$$F = G\frac{m_1 m_2}{r^2}$$

$$r^2 F = Gm_1 m_2 \qquad \text{Multiply both sides of formula by } r^2.$$

$$r^2 = \frac{Gm_1 m_2}{F} \qquad \text{Isolate } r^2 \text{ by dividing both sides by } F.$$

$$r = \sqrt{\frac{Gm_1 m_2}{F}} \qquad \text{Square root property}$$

▶ **Now Try Exercise 23**

In Example 5, since r must be greater than 0, when we used the square root property, we listed only the principal or positive square root.

EXAMPLE 6 ▶ **Diagonal of a Suitcase**   The diagonal of a box can be calculated by the formula

$$d = \sqrt{L^2 + W^2 + H^2}$$

where L is the length, W is the width, and H is the height of the box. See **Figure 12.5**.

a) Find the diagonal of a suitcase of length 30 inches, width 15 inches, and height 10 inches.

b) Solve the equation for the width, W.

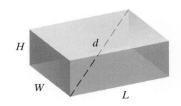

H d W L

FIGURE 12.5

Solution   **a) Understand**   To find the diagonal, we need to substitute the appropriate values into the formula and solve for the diagonal, d.

Translate
$$d = \sqrt{L^2 + W^2 + H^2}$$
$$d = \sqrt{(30)^2 + (15)^2 + (10)^2}$$

Carry Out
$$= \sqrt{900 + 225 + 100}$$
$$= \sqrt{1225}$$
$$= 35$$

Answer   Thus, the diagonal of the suitcase is 35 inches.

b) Our first step in solving for W is to square both sides of the formula.

$$d = \sqrt{L^2 + W^2 + H^2}$$
$$d^2 = \left(\sqrt{L^2 + W^2 + H^2}\right)^2 \qquad \text{Square both sides.}$$
$$d^2 = L^2 + W^2 + H^2 \qquad \text{Use } (\sqrt{a})^2 = a, a \geq 0.$$
$$d^2 - L^2 - H^2 = W^2 \qquad \text{Isolate } W^2.$$
$$\sqrt{d^2 - L^2 - H^2} = W \qquad \text{Square root property}$$

▶ **Now Try Exercise 15**

EXAMPLE 7 ▸ **Traffic Cones** The surface area of a right circular cone is

$$s = \pi r \sqrt{r^2 + h^2}$$

a) An orange traffic cone used on roads is 18 inches high with a radius of 12 inches. Find the surface area of the cone.

b) Solve the formula for h.

Solution **a)** Understand and Translate   To find the surface area, we substitute the appropriate values into the formula.

$$s = \pi r \sqrt{r^2 + h^2}$$
$$= \pi(12)\sqrt{(12)^2 + (18)^2}$$

Carry Out
$$= 12\pi\sqrt{144 + 324}$$
$$= 12\pi\sqrt{468}$$
$$\approx 815.56$$

Answer   The surface area is about 815.56 square inches.

b) To solve for h we need to isolate h on one side of the equation. There are various ways to solve the equation for h.

$$s = \pi r \sqrt{r^2 + h^2}$$

$$\frac{s}{\pi r} = \sqrt{r^2 + h^2} \qquad \text{\textit{Divide both sides by} } \pi r.$$

$$\left(\frac{s}{\pi r}\right)^2 = \left(\sqrt{r^2 + h^2}\right)^2 \qquad \text{\textit{Square both sides.}}$$

$$\frac{s^2}{\pi^2 r^2} = r^2 + h^2 \qquad \text{\textit{Use }} (\sqrt{a})^2 = a, a \geq 0.$$

$$\frac{s^2}{\pi^2 r^2} - r^2 = h^2 \qquad \text{\textit{Subtract } } r^2 \text{ \textit{from both sides.}}$$

$$\sqrt{\frac{s^2}{\pi^2 r^2} - r^2} = h \qquad \text{\textit{Square root property}}$$

Other acceptable answers are $h = \sqrt{\dfrac{s^2 - \pi^2 r^4}{\pi^2 r^2}}$ and $h = \dfrac{\sqrt{s^2 - \pi^2 r^4}}{\pi r}$. Can you explain why?

▸ **Now Try Exercise 27**

EXERCISE SET 12.3   *Math* XL   **MyMathLab**
MathXL®   MyMathLab

Concept/Writing Exercises

1. In general, when solving for a variable in a formula, whether you use the square root property or the quadratic formula, you use only the positive square root. Explain why.

2. Suppose $P = \smiley^2 + \square^2$ is a real formula. Solving for $\smiley$ gives $\smiley = \sqrt{P - \square^2}$. If $\smiley$ is to be a real number, what relationship must exist between P and $\square$?

Practice the Skills

Solve for the indicated variable. Assume the variable you are solving for must be greater than 0.

3. $A = s^2$, for s (area of a square)

4. $A = (s + 1)^2$, for s (area of a square)

5. $d = 4.9t^2$, for t (distance an object has fallen)

6. $A = S^2 - s^2$, for S (area between two squares)

7. $E = i^2 r$, for i (current in electronics)

8. $A = 4\pi r^2$, for r (surface area of a sphere)

9. $d = 16t^2$, for t (distance of a falling object)

10. $d = \dfrac{1}{9}x^2$, for x (stopping distance on pavement)

11. $E = mc^2$, for c (Einstein's famous energy formula)

12. $V = \pi r^2 h$, for r (volume of a right circular cylinder)

13. $V = \dfrac{1}{3}\pi r^2 h$, for r (volume of a right circular cone)

14. $d = \sqrt{L^2 + W^2}$, for L (diagonal of a rectangle)

15. $d = \sqrt{L^2 + W^2}$, for W (diagonal of a rectangle)

16. $a^2 + b^2 = c^2$, for a (Pythagorean Theorem)

17. $a^2 + b^2 = c^2$, for b (Pythagorean Theorem)

18. $d = \sqrt{L^2 + W^2 + H^2}$, for L (diagonal of a box)

19. $d = \sqrt{L^2 + W^2 + H^2}$, for H (diagonal of a box)

20. $A = P(1 + r)^2$, for r (compound interest formula)

21. $h = -16t^2 + s_0$, for t (height of an object)

22. $h = -4.9t^2 + s_0$, for t (height of an object)

23. $E = \dfrac{1}{2}mv^2$, for v (kinetic energy)

24. $f_x^2 + f_y^2 = f^2$, for f_x (forces acting on an object)

25. $a = \dfrac{v_2^2 - v_1^2}{2d}$, for v_1 (acceleration of a vehicle)

26. $A = 4\pi(R^2 - r^2)$, for R (surface area of two spheres)

27. $v' = \sqrt{c^2 - v^2}$, for c (relativity; v' is read "v prime")

28. $L = L_0\sqrt{1 - \dfrac{v^2}{c^2}}$, for v (art, a painting's contraction)

Problem Solving

29. Profit The profit for the Hillside Tractor Company, which sells tractors, is $P(n) = 2.7n^2 + 9n - 3$, where $P(n)$ is in hundreds of dollars.

  a) Find the profit when 5 tractors are sold.

  b) How many tractors should be sold to have a profit of $20,000?

30. Profit The profit for the Jacksons Appliances, which sells refrigerators, is $P(n) = 6.2n^2 + 6n - 3$ where $P(n)$ is in dollars.

  a) Find the profit when 7 refrigerators are sold.

  b) How many refrigerators should be sold to have a profit of $675?

31. Temperature The temperature, T, in degrees Fahrenheit, in a car's radiator during the first 4 minutes of driving is a function of time, t. The temperature can be found by the formula $T = 6.2t^2 + 12t + 32, 0 \le t \le 4$.

  a) What is the car radiator's temperature at the instant the car is turned on?

  b) What is the car radiator's temperature after the car has been driven for 2 minutes?

  c) How long after the car has begun operating will the car radiator's temperature reach 120°F?

32. School Enrollment The function $N(t) = -0.043t^2 + 1.22t + 46.0$ can be used to estimate total U.S. elementary and secondary school enrollment, in millions, between the years 1990 and 2008. In the equation t is years since 1989 and $1 \le t \le 19$.

  a) Estimate total enrollment in 1995.

  b) In what years is the total enrollment 54 million students?

33. Downloaded Songs The number of downloaded songs, in billions, from 2002 to 2006 and projected to 2008, can be estimated by the function $D = 0.04t^2 - 0.03t + 0.01$. In this function, t is the number of years since 2002 and $0 \le t \le 6$. *Source:* Price Waterhouse Coopers, LLP, RIAA, *Newsweek* (7/11/05)

  a) Estimate the number of downloaded songs in 2006.

  b) In which year is 1 billion songs projected to be downloaded?

34. Grade Point Average At a college, records show that the average person's grade point average, G, is a function of the number of hours he or she studies and does homework per week, h. The grade point average can be estimated by the equation $G = 0.01h^2 + 0.2h + 1.2, 0 \le h \le 8$.

a) What is the GPA of the average student who studies for 0 hours a week?

b) What is the GPA of the average student who studies 3 hours per week?

c) To obtain a 3.2 GPA, how many hours per week would the average student need to study?

35. Apple Production The following graph shows the annual average yield per acre, of apple trees, for the years 2000 to 2004.

Yield per Acre of Apple Trees

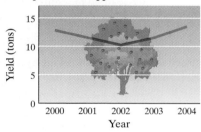

Source: National Agricultural Statistics, *USA Today* (9/15/05)

The average annual yield per acre of apple trees, in tons, can be estimated by the function $Y = 0.66t^2 - 2.49t + 12.93$. In this function, t is the number of years since 2000 and $0 \le t \le 4$.

a) Estimate the yield per acre in 2003.

b) In which year was the yield per acre 13 billion tons?

36. Drug-Free Schools The following graph summarizes data on the percent of students at various ages who say their school is not drug free.

Students Who Say Their School Is Not Drug Free

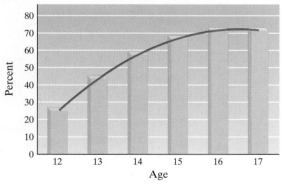

Source: National Center on Addiction and Substance Abuse

The function $f(a) = -2.32a^2 + 76.58a - 559.87$ can be used to estimate the percent of students who say their school is not drug free. In the function, a represents the student's age, where $12 \le a \le 17$. Use the function to answer the following questions.

a) Estimate the percent of 14-year-olds who say their school is not drug free.

b) At what age do 70% of the students say their school is not drug free?

37. Motorcycle Sales The following graph shows the number of new motorcycles, in millions, sold in the United States for the years 1997 to 2004.

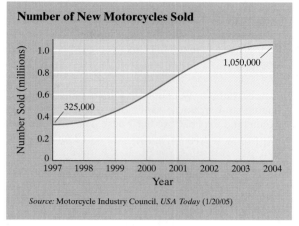

Source: Motorcycle Industry Council, *USA Today* (1/20/05)

The number of new motorcycles, $m(t)$, in millions, sold in the United States can be estimated by the function $M = -0.00434t^2 + 0.142t + 0.315$. In this function, t is the number of years since 1997.

a) If this trend continues, use this function to estimate the number of motorcycles that will be sold in the United States in 2007.

b) In what year will the number of new motorcycles sold in the United States be 1.4 million?

38. Profit A video store's weekly profit, P, in thousands of dollars, is a function of the rental price of the tapes, t. The profit equation is $P = 0.2t^2 + 1.5t - 1.2, 0 \le t \le 5$.

a) What is the store's weekly profit or loss if they charge $3 per tape?

b) What is the weekly profit if they charge $5 per tape?

c) At what tape rental price will their weekly profit be 1.4 thousand?

39. Playground The area of a children's rectangular playground is 600 square meters. The length is 10 meters longer than the width. Find the length and width of the playground.

40. Travel Hana Juarez drove for 80 miles in heavy traffic. She then reached the highway where she drove 260 miles at an average speed that was 25 miles per hour greater than the average speed she drove in heavy traffic. If the total trip took 6 hours, determine her average speed in heavy traffic and on the highway.

41. Drilling a Well Paul and Rima Jones, who live in Cedar Rapids, Iowa, want a well on their property. They hired the Ruth Cardiff Drilling Company to drill the well. The company had to drill 64 feet to hit water. The company informed the Jones that they had just ordered new drilling equipment that drills at an average of 1 foot per hour faster, and that with their new equipment, they would have hit water in 3.2 hours less time. Find the rate at which their present equipment drills.

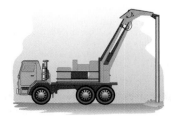

42. Car Carrier Frank Sims, a truck driver, was transporting a heavy load of new cars on a car carrier from Detroit, Michigan, to Indianapolis, Indiana. On his return trip to Detroit, since his truck was lighter, he averaged 10 miles per hour faster than on his trip to Indianapolis. If the total distance traveled each way was 300 miles and the total time he spent driving was 11 hours, find his average speed going and returning.

43. Runner Latoya Williams, a long-distance runner, starts jogging at her house. She jogs 6 miles and then turns around and jogs back to her house. The first part of her jog is mostly uphill, so her speed averages 2 miles per hour less than her returning speed. If the total time she spends jogging is $1\frac{3}{4}$ hours, find her speed going and her speed returning.

44. Red Rock Canyon Kathy Nickell traveled from the Red Rock Canyon Conservation Area, just outside Las Vegas, to Phoenix, Arizona. The total distance she traveled was 300 miles. After she got to Phoenix, she figured out that had she averaged 10 miles per hour faster, she would have arrived 1 hour earlier. Find the average speed that Kathy drove.

Red Rock Canyon

45. Build an Engine Two mechanics, Bonita Rich and Pamela Pearson, take 6 hours to rebuild an engine when they work together. If each worked alone, Bonita, the more experienced mechanic, could complete the job 1 hour faster than Pamela. How long would it take each of them to rebuild the engine working alone?

46. Riding a Bike Ricky Bullock enjoys riding his bike from Washington, D.C., to Bethesda, Maryland, and back, a total of 30 miles on the Capital Crescent path. The trip to Bethesda is uphill most of the distance. The bike's average speed going to Bethesda is 5 miles per hour slower than the average speed returning to D.C. If the round trip takes 4.5 hours, find the average speed in each direction.

47. Flying a Plane Dole Rohm flew his single-engine Cessna airplane 80 miles with the wind from Jackson Hole, Wyoming, to above Blackfoot, Idaho. He then turned around and flew back to Jackson Hole against the wind. If the wind was a constant 30 miles per hour, and the total time going

and returning was 1.3 hours, find the speed of the plane in still air.

48. Ships After a small oil spill, two cleanup ships are sent to siphon off the oil floating in Baffin Bay. The newer ship can clean up the entire spill by itself in 3 hours less time than the older ship takes by itself. Working together the two ships can clean up the oil spill in 8 hours. How long will it take the newer ship by itself to clean up the spill?

49. Janitorial Service The O'Connors own a small janitorial service. John requires $\frac{1}{2}$ hour more time to clean the Moose Club by himself than Chris does working by herself. If together they can clean the club in 6 hours, find the time required by each to clean the club.

50. Electric Heater A small electric heater requires 6 minutes longer to raise the temperature in an unheated garage to a comfortable level than does a larger electric heater. Together the two heaters can raise the garage temperature to a comfortable level in 42 minutes. How long would it take each heater by itself to raise the temperature in the garage to a comfortable level?

51. Travel Shywanda Moore drove from San Antonio, Texas, to Austin, Texas, a distance of 75 miles. She then stopped for 2 hours to see a friend in Austin before continuing her journey from Austin to Dallas, Texas, a distance of 195 miles. If she drove 10 miles per hour faster from San Antonio to Austin and the total time of the trip was 6 hours, find her average speed from San Antonio to Austin.

River Walk, San Antonio, Texas

52. Travel Lewis and his friend George are traveling from Nashville, Tennessee, to Baltimore, Maryland. Lewis travels by car while George travels by train. The train and car leave Nashville at the same time from the same point. During the trip Lewis and George speak by cellular phone, and Lewis informs George that he has just stopped for the evening after traveling for 500 miles. One and two-thirds hours later, George calls Lewis and informs him that the train had just reached Baltimore, a distance of 800 miles from Nashville. Assuming the train averaged 20 miles per hour faster than the car, find the average speed of the car and the train.

53. Widescreen TVs A widescreen television (see figure) has an aspect ratio of 16 : 9. This means that the ratio of the length to the height of the screen is 16 to 9. The figure drawn on the photo illustrates how the length and the height of the screen of a 40-inch widescreen television can be found. Determine the length and height of a 40-inch widescreen television.

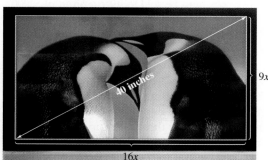

54. Standard TVs Many picture tube televisions have a screen aspect ratio of 4 : 3. Determine the length and height of a television screen that has an aspect ratio of 4 : 3 and whose diagonal is 36 inches. See Exercise 53.

55. Write your own motion problem and solve it.

56. Write your own work problem and solve it.

Challenge Problems

57. Area The area of a rectangle is 18 square meters. When the length is increased by 2 meters and the width by 3 meters, the area becomes 48 square meters. Find the dimensions of the smaller rectangle.

58. Area The area of a rectangle is 35 square inches. When the length is decreased by 1 inch and the width is increased by 1 inch, the area of the new rectangle is 36 square inches. Find the dimensions of the original rectangle.

Cumulative Review Exercises

[1.9] **59.** Evaluate $-[4(5 - 3)^3] + 2^4$.

[2.6] **60.** Solve $IR + Ir = E$ for R.

[7.4] **61.** Add $\dfrac{r}{r - 4} - \dfrac{r}{r + 4} + \dfrac{32}{r^2 - 16}$.

[11.2] **62.** Simplify $\left(\dfrac{x^{3/4} y^{-2}}{x^{1/2} y^2}\right)^8$.

[11.6] **63.** Solve $\sqrt{x^2 + 3x + 12} = x$.

Mid-Chapter Test: 12.1–12.3

To find out how well you understand the chapter material to this point, take this brief test. The answers, and the section where the material was initially discussed, are given in the back of the book. Review any questions that you answered incorrectly.

Use the square root property to solve each equation.

1. $x^2 - 12 = 86$

2. $(a - 3)^2 + 20 = 0$

3. $(2m + 7)^2 = 36$

Solve each equation by completing the square.

4. $y^2 + 4y - 12 = 0$

5. $3a^2 - 12a - 30 = 0$

6. $4c^2 + c = -9$

7. Patio The patio of a house is a square where the diagonal is 6 meters longer than a side. Find the length of one side of the patio.

8. a) Give the formula for the discriminant of a quadratic equation.

  b) Explain how to determine if a quadratic equation has two distinct real solutions, a single real solution, or no real solution.

9. Use the discriminant to determine if the equation $2b^2 - 6b - 11 = 0$ has two distinct real solutions, a single real solution, or no real solution.

Solve each equation by the quadratic formula.

10. $6n^2 + n = 15$

11. $p^2 = -4p + 8$

12. $3d^2 - 2d + 5 = 0$

In Exercises 13 and 14, determine an equation that has the given solutions.

13. $7, -2$

14. $2 + \sqrt{5}, 2 - \sqrt{5}$

15. Lamps A business sells n lamps, $n \le 20$, at a price of $(60 - 0.5n)$ dollars per lamp. How many lamps must be sold to have revenue of $550?

In Exercises 16–18, solve for the indicated variable. Assume all variables are positive.

16. $y = x^2 - r^2$, for r

17. $A = \dfrac{1}{3}kx^2$, for x

18. $D = \sqrt{x^2 + y^2}$, for y

19. Area The length of a rectangle is two feet more than twice the width. Find the dimensions if its area is 60 square feet.

20. Clocks The profit from a company selling n clocks is $p(n) = 2n^2 + n - 35$, where $p(n)$ is hundreds of dollars. How many clocks must be sold to have a profit of $2000?

12.4 Factoring Expressions and Solving Equations That Are Quadratic in Form

1 Factor trinomials that are quadratic in form.

2 Solve equations that are quadratic in form.

3 Solve equations with rational exponents.

Sometimes we need to solve an equation that is not a quadratic equation but can be rewritten in the form of a quadratic equation. We can then solve the equation in quadratic form by factoring, completing the square, or the quadratic formula. We will begin this section by factoring expressions which are quadratic in form and then we will solve equations that are quadratic in form.

1 Factor Trinomials That Are Quadratic in Form

Sometimes a higher degree trinomial can be written as a quadratic trinomial by substituting one variable for another.

> **Expression Quadratic in Form**
>
> An expression that can be written in the form $au^2 + bu + c = 0$ for $a \neq 0$, where u is an algebraic expression, is called an **expression that is quadratic in form**.

In Sections 6.3 and 6.4 we factored second degree, or quadratic, trinomials. The next two examples illustrate factoring higher degree trinomials that are quadratic in form.

EXAMPLE 1 ▶ Factor $y^4 - y^2 - 6$.

Solution   If we can rewrite this expression in the form $ax^2 + bx + c$, it will be easier to factor. Since $(y^2)^2 = y^4$, if we substitute x for y^2, the trinomial becomes

$$y^4 - y^2 - 6 = (y^2)^2 - y^2 - 6$$
$$= x^2 - x - 6 \qquad \text{\textit{Substitute x for } } y^2.$$

Now factor $x^2 - x - 6$.

$$= (x + 2)(x - 3)$$

Finally, substitute y^2 in place of x to obtain

$$= (y^2 + 2)(y^2 - 3) \qquad \text{\textit{Substitute } } y^2 \text{ \textit{for x.}}$$

Thus, $y^4 - y^2 - 6 = (y^2 + 2)(y^2 - 3)$. Note that x was substituted for y^2, and then y^2 was substituted back for x.

▶ **Now Try Exercise 7**

EXAMPLE 2 ▶ Factor $3z^4 - 17z^2 - 28$.

Solution   Let $x = z^2$. Then the trinomial can be written

$$3z^4 - 17z^2 - 28 = 3(z^2)^2 - 17z^2 - 28$$
$$= 3x^2 - 17x - 28 \qquad \text{\textit{Substitute x for } } z^2.$$
$$= (3x + 4)(x - 7) \qquad \text{\textit{Factor.}}$$

Now substitute z^2 for x.

$$= (3z^2 + 4)(z^2 - 7) \qquad \text{\textit{Substitute } } z^2 \text{ \textit{for x.}}$$

Thus, $3z^4 - 17z^2 - 28 = (3z^2 + 4)(z^2 - 7)$.

▶ **Now Try Exercise 11**

EXAMPLE 3 ▶ Factor $2(x + 5)^2 - 5(x + 5) - 12$.

Solution   We will again use a substitution, as in Examples 1 and 2. By substituting $a = x + 5$ in the equation, we obtain

$$2(x + 5)^2 - 5(x + 5) - 12$$
$$= 2a^2 - 5a - 12 \qquad \textit{Substitute a for (x + 5).}$$

Now factor $2a^2 - 5a - 12$.

$$= (2a + 3)(a - 4)$$

Finally, replace a with $x + 5$ to obtain

$$= [2(x + 5) + 3][(x + 5) - 4] \qquad \textit{Substitute (x + 5) for a.}$$
$$= [2x + 10 + 3][x + 1]$$
$$= (2x + 13)(x + 1)$$

Thus, $2(x + 5)^2 - 5(x + 5) - 12 = (2x + 13)(x + 1)$. Note that a was substituted for $x + 5$, and then $x + 5$ was substituted back for a.

▶ **Now Try Exercise 15**

In Examples 1 and 2 we used x in our substitution, whereas in Example 3 we used a. The letter selected does not affect the final answer.

2   Solve Equations That Are Quadratic in Form

Equation Quadratic in Form

An equation that can be written in the form $au^2 + bu + c = 0$ for $a \neq 0$, where u is an algebraic expression, is called **quadratic in form**.

When you are given an equation quadratic in form, make a substitution to get the equation in the form $au^2 + bu + c = 0$. In general, if the exponents are positive, the substitution to make is to let u be the middle term, without the numerical coefficient, when the expression is listed in descending order of the variable. For example,

Equation Quadratic in Form	Substitution	Equation with Substitution
$y^4 - y^2 - 6 = 0$	$u = y^2$	$u^2 - u - 6 = 0$
$2(x + 5)^2 - 5(x + 5) - 12 = 0$	$u = x + 5$	$2u^2 - 5u - 12 = 0$
$x^{2/3} + 4x^{1/3} - 3 = 0$	$u = x^{1/3}$	$u^2 + 4u - 3 = 0$

To solve equations quadratic in form, we use the following procedure. We will illustrate this procedure in Example 4.

To Solve Equations Quadratic in Form

1. Make a substitution that will result in an equation of the form $au^2 + bu + c = 0$, $a \neq 0$, where u is a function of the original variable.
2. Solve the equation $au^2 + bu + c = 0$ for u.
3. Replace u with the function of the original variable from step 1 and solve the resulting equation for the original variable.
4. Check for extraneous solutions by substituting the apparent solutions into the original equation.

EXAMPLE 4 ▶

a) Solve $x^4 - 5x^2 + 4 = 0$.

b) Find the x-intercepts of the graph of the function $f(x) = x^4 - 5x^2 + 4$.

Solution

a) To obtain an equation that is quadratic in form, write x^4 as $(x^2)^2$.

$$x^4 - 5x^2 + 4 = 0$$
$$(x^2)^2 - 5x^2 + 4 = 0 \qquad \textit{Replace x^4 with $(x^2)^2$ to obtain an equation in desired form.}$$

Now let $u = x^2$. This gives an equation that is quadratic in form.

$$u^2 - 5u + 4 = 0 \qquad \text{\textit{Substitute u for }} x^2.$$
$$(u - 4)(u - 1) = 0 \qquad \text{\textit{Solve for u.}}$$
$$u - 4 = 0 \quad \text{or} \quad u - 1 = 0$$
$$u = 4 \qquad\qquad u = 1$$
$$x^2 = 4 \qquad\qquad x^2 = 1 \qquad \text{\textit{Replace u with }} x^2.$$
$$x = \pm\sqrt{4} \qquad\quad x = \pm\sqrt{1} \qquad \text{\textit{Solve for x.}}$$
$$x = \pm 2 \qquad\qquad x = \pm 1$$

Check the four possible solutions in the original equation.

$x = 2$	$x = -2$	$x = 1$	$x = -1$
$x^4 - 5x^2 + 4 = 0$	$x^4 - 5x^2 + 4 = 0$	$x^4 - 5x^2 + 4 = 0$	$x^4 - 5x^2 + 4 = 0$
$2^4 - 5(2)^2 + 4 \overset{?}{=} 0$	$(-2)^4 - 5(-2)^2 + 4 \overset{?}{=} 0$	$1^4 - 5(1)^2 + 4 \overset{?}{=} 0$	$(-1)^4 - 5(-1)^2 + 4 \overset{?}{=} 0$
$16 - 20 + 4 \overset{?}{=} 0$	$16 - 20 + 4 \overset{?}{=} 0$	$1 - 5 + 4 \overset{?}{=} 0$	$1 - 5 + 4 \overset{?}{=} 0$
$0 = 0$	$0 = 0$	$0 = 0$	$0 = 0$
True	*True*	*True*	*True*

Thus, the solutions are 2, −2, 1, and −1.

b) The x-intercepts occur where $f(x) = 0$. Therefore, the graph will cross the x-axis at the solutions to the equation $x^4 - 5x^2 + 4 = 0$.

From part **a)**, we know the solutions are 2, −2, 1, and −1. Thus, the x-intercepts are $(2, 0), (-2, 0), (1, 0),$ and $(-1, 0)$. **Figure 12.6** is the graph of $f(x) = x^4 - 5x^2 + 4 = 0$ as illustrated on a graphing calculator. Notice that the graph crosses the x-axis at $x = 2$, $x = -2$, $x = 1$, and $x = -1$.

▶ **Now Try Exercise 31**

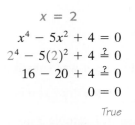

−3, 3, 1, −3, 6, 1

FIGURE 12.6

EXAMPLE 5 ▶ Solve $p^4 + 2p^2 = 8$.

Solution

$$p^4 + 2p^2 - 8 = 0 \qquad \text{\textit{Set equation equal to 0.}}$$
$$(p^2)^2 + 2p^2 - 8 = 0 \qquad \text{\textit{Write }} p^4 \text{\textit{ as }} (p^2)^2 \text{\textit{ to obtain equation in desired form.}}$$

Now let $u = p^2$. This gives an equation that is quadratic in form.

$$u^2 + 2u - 8 = 0 \qquad \text{\textit{Substitute u for }} p^2.$$
$$(u + 4)(u - 2) = 0 \qquad \text{\textit{Solve the equation for u.}}$$
$$u + 4 = 0 \quad \text{or} \quad u - 2 = 0$$
$$u = -4 \qquad\qquad u = 2$$

We are not finished. Since the variable in the original equation is p, we must solve for p, not u. Therefore, we substitute back p^2 for u and solve for p.

$$p^2 = -4 \qquad\qquad p^2 = 2 \qquad \text{\textit{Replace u with }} p^2.$$
$$p = \pm\sqrt{-4} \qquad\quad p = \pm\sqrt{2} \qquad \text{\textit{Solve for p.}}$$
$$p = \pm 2i$$

Check the four possible solutions in the *original* equation.

$p = 2i$	$p = -2i$	$p = \sqrt{2}$	$p = -\sqrt{2}$
$p^4 + 2p^2 = 8$	$p^4 + 2p^2 = 8$	$p^4 + 2p^2 = 8$	$p^4 + 2p^2 = 8$
$(2i)^4 + 2(2i)^2 \overset{?}{=} 8$	$(-2i)^4 + 2(-2i)^2 \overset{?}{=} 8$	$(\sqrt{2})^4 + 2(\sqrt{2})^2 \overset{?}{=} 8$	$(-\sqrt{2})^4 + 2(-\sqrt{2})^2 \overset{?}{=} 8$
$2^4 i^4 + 2(2^2)(i^2) \overset{?}{=} 8$	$(-2)^4 i^4 + 2(-2)^2 i^2 \overset{?}{=} 8$	$4 + 2(2) \overset{?}{=} 8$	$4 + 2(2) \overset{?}{=} 8$
$16(1) + 8(-1) \overset{?}{=} 8$	$16(1) + 8(-1) \overset{?}{=} 8$	$8 = 8$	$8 = 8$
$16 - 8 = 8$	$16 - 8 = 8$	*True*	*True*
True	*True*		

Thus, the solutions are $2i$, $-2i$, $\sqrt{2}$, and $-\sqrt{2}$.

▶ **Now Try Exercise 41**

The solutions to equations like $p^4 + 2p^2 = 8$ will always check unless a mistake has been made. In equations like this, extraneous solutions will not be introduced. However, extraneous solutions *may* be introduced when working with rational exponents, as will be shown in Example 9.

Helpful Hint

Students sometimes solve the equation for u but then forget to complete the problem by solving for the original variable. Remember that if the original equation is in x you must obtain values for x. If the original equation is in p (as in Example 5) you must obtain values for p, and so on.

EXAMPLE 6 ▶ Solve $4(2w + 1)^2 - 16(2w + 1) + 15 = 0$.

Solution　If we let $u = 2w + 1$, the equation becomes

$$4(2w + 1)^2 - 16(2w + 1) + 15 = 0$$
$$4u^2 - 16u + 15 = 0 \qquad \textit{Substitute u for 2w + 1.}$$

Now we can factor and solve.

$$(2u - 3)(2u - 5) = 0$$
$$2u - 3 = 0 \quad \text{or} \quad 2u - 5 = 0$$
$$2u = 3 \qquad\qquad 2u = 5$$
$$u = \frac{3}{2} \qquad\qquad u = \frac{5}{2}$$

We are not finished. Since the variable in the original equation is w, we must solve for w, not u. Therefore, we substitute back $2w + 1$ for u and solve for w.

$$u = \frac{3}{2} \qquad\qquad u = \frac{5}{2}$$
$$2w + 1 = \frac{3}{2} \qquad 2w + 1 = \frac{5}{2} \qquad \textit{Substitute 2w + 1 for u.}$$
$$2w = \frac{1}{2} \qquad\qquad 2w = \frac{3}{2}$$
$$w = \frac{1}{4} \qquad\qquad w = \frac{3}{4}$$

A check will show that both $\dfrac{1}{4}$ and $\dfrac{3}{4}$ are solutions to the original equation.

▶ **Now Try Exercise 53**

EXAMPLE 7 ▶ Find the x-intercepts of the graph of the function $f(x) = 2x^{-2} + x^{-1} - 1$.

Solution　The x-intercepts occur where $f(x) = 0$. Therefore, to find the x-intercepts we must solve the equation

$$2x^{-2} + x^{-1} - 1 = 0$$

This equation can be expressed as

$$2(x^{-1})^2 + x^{-1} - 1 = 0$$

When we let $u = x^{-1}$, the equation becomes

$$2u^2 + u - 1 = 0$$
$$(2u - 1)(u + 1) = 0$$
$$2u - 1 = 0 \quad \text{or} \quad u + 1 = 0$$
$$u = \frac{1}{2} \qquad\qquad u = -1$$

Now we substitute x^{-1} for u.

$$x^{-1} = \frac{1}{2} \quad \text{or} \quad x^{-1} = -1$$

$$\frac{1}{x} = \frac{1}{2} \qquad\qquad \frac{1}{x} = -1$$

$$x = 2 \qquad\qquad\quad x = -1$$

A check will show that both 2 and -1 are solutions to the original equation. Thus, the x-intercepts are $(2, 0)$ and $(-1, 0)$.

▸ **Now Try Exercise 85**

The equation in Example 7 could also be expressed as

$$\frac{2}{x^2} + \frac{1}{x} - 1 = 0$$

A second method to solve this equation is to multiply both sides of the equation by the least common denominator, x^2, then simplify.

$$x^2 \left(\frac{2}{x^2} + \frac{1}{x} - 1 \right) = x^2 \cdot 0$$

$$2 + x - x^2 = 0$$

$$x^2 - x - 2 = 0$$

$$(x - 2)(x + 1) = 0$$

$$x - 2 = 0 \quad \text{or} \quad x + 1 = 0$$

$$x = 2 \qquad\qquad\quad x = -1$$

Many of the equations solved in this section may be solved by more than one method.

3 Solve Equations with Rational Exponents

When solving equations that are quadratic in form with rational exponents, we raise both sides of the equation to some power to eliminate the rational exponents. Recall that we did this in Section 11.6 when we solved radical equations. Whenever you raise both sides of an equation to a power, you may introduce extraneous solutions. **Therefore, whenever you raise both sides of an equation to a power, you must check all apparent solutions in the original equation to make sure that none are extraneous.** We will now work two examples showing how to solve equations that contain rational exponents. We use the same procedure as used earlier.

EXAMPLE 8 ▸ Solve $x^{2/5} + x^{1/5} - 6 = 0$.

Solution This equation can be rewritten as

$$\left(x^{1/5}\right)^2 + x^{1/5} - 6 = 0$$

Let $u = x^{1/5}$. Then the equation becomes

$$u^2 + u - 6 = 0$$

$$(u + 3)(u - 2) = 0$$

$$u + 3 = 0 \quad \text{or} \quad u - 2 = 0$$

$$u = -3 \qquad\qquad u = 2$$

Now substitute $x^{1/5}$ for u and raise both sides of the equation to the fifth power to remove the rational exponents.

$$x^{1/5} = -3 \quad \text{or} \quad x^{1/5} = 2$$

$$\left(x^{1/5}\right)^5 = (-3)^5 \qquad \left(x^{1/5}\right)^5 = 2^5$$

$$x = -243 \qquad\qquad\quad x = 32$$

The two *possible* solutions are -243 and 32. Remember that whenever you raise both sides of an equation to a power, as you did here, you need to check for extraneous solutions.

Check $x = -243$ $x = 32$

$$x^{2/5} + x^{1/5} - 6 = 0$$ $$x^{2/5} + x^{1/5} - 6 = 0$$

$$(-243)^{2/5} + (-243)^{1/5} - 6 \overset{?}{=} 0$$ $$(32)^{2/5} + (32)^{1/5} - 6 \overset{?}{=} 0$$

$$(\sqrt[5]{-243})^2 + \sqrt[5]{-243} - 6 \overset{?}{=} 0$$ $$(\sqrt[5]{32})^2 + \sqrt[5]{32} - 6 \overset{?}{=} 0$$

$$(-3)^2 - 3 - 6 \overset{?}{=} 0$$ $$2^2 + 2 - 6 \overset{?}{=} 0$$

$$9 - 3 - 6 \overset{?}{=} 0$$ $$4 + 2 - 6 \overset{?}{=} 0$$

$$0 = 0 \quad \text{True}$$ $$0 = 0 \quad \text{True}$$

Since both values check, the solutions are -243 and 32.

▶ **Now Try Exercise 87**

EXAMPLE 9 ▶ Solve $2p - \sqrt{p} - 10 = 0$.

Solution   We can express this equation as

$$2p - p^{1/2} - 10 = 0$$
$$2(p^{1/2})^2 - p^{1/2} - 10 = 0$$

If we let $u = p^{1/2}$, this equation is quadratic in form.

$$2u^2 - u - 10 = 0$$
$$(2u - 5)(u + 2) = 0$$
$$2u - 5 = 0 \quad \text{or} \quad u + 2 = 0$$
$$2u = 5 \qquad\qquad\qquad u = -2$$
$$u = \frac{5}{2}$$

However, since our original equation is in the variable p, we must solve for p. We substitute $p^{1/2}$ for u.

$$p^{1/2} = \frac{5}{2} \qquad\qquad\qquad p^{1/2} = -2$$

Now we square both sides of the equation.

$$(p^{1/2})^2 = \left(\frac{5}{2}\right)^2 \qquad\qquad (p^{1/2})^2 = (-2)^2$$

$$p = \frac{25}{4} \qquad\qquad\qquad p = 4$$

We must now check both apparent solutions in the original equation.

Check $p = \dfrac{25}{4}$ $p = 4$

$$2p - \sqrt{p} - 10 = 0$$ $$2p - \sqrt{p} - 10 = 0$$

$$2\left(\frac{25}{4}\right) - \sqrt{\frac{25}{4}} - 10 \overset{?}{=} 0$$ $$2(4) - \sqrt{4} - 10 \overset{?}{=} 0$$

$$\frac{25}{2} - \frac{5}{2} - 10 \overset{?}{=} 0$$ $$8 - 2 - 10 \overset{?}{=} 0$$

$$0 = 0 \quad \text{True}$$ $$-4 = 0 \quad \text{False}$$

Since 4 does not check, it is an extraneous solution. The only solution is $\dfrac{25}{4}$.

▶ **Now Try Exercise 49**

Example 9 could also be solved by writing the equation as $\sqrt{p} = 2p - 10$ and squaring both sides of the equation. Try this now. If you have forgotten how to do this, review Section 11.6.

EXERCISE SET 12.4

Concept/Writing Exercises

1. Explain how you can determine whether a given equation can be expressed as an equation that is quadratic in form.

2. When solving an equation that is quadratic in form, when is it essential to check your answer for extraneous solutions? Explain why.

3. To solve the equation $3x^4 - 5x^2 + 1 = 0$, what is the correct choice for u to obtain an equation that is quadratic in form? Explain.

4. To solve the equation $2y^{4/3} + 9y^{2/3} - 7 = 0$, what is the correct choice for u to obtain an equation that is quadratic in form? Explain.

5. To solve the equation $z^{-2} - z^{-1} = 56$, what is the correct choice for u to obtain an equation that is quadratic in form? Explain.

6. To solve the equation $3\left(\dfrac{x+2}{x+3}\right)^2 + \left(\dfrac{x+2}{x+3}\right) - 9 = 0$, what is the correct choice for u to obtain an equation that is quadratic in form? Explain.

Practice the Skills

Factor each trinomial completely.

7. $x^4 + x^2 - 6$

8. $x^4 - 3x^2 - 10$

9. $x^4 + 5x^2 + 6$

10. $x^4 + 2x^2 - 35$

11. $6a^4 + 5a^2 - 25$

12. $(2x + 1)^2 + 2(2x + 1) - 15$

13. $4(x + 1)^2 + 8(x + 1) + 3$

14. $(2x + 3)^2 - (2x + 3) - 6$

15. $6(a + 2)^2 - 7(a + 2) - 5$

16. $6(p - 5)^2 + 11(p - 5) + 3$

17. $a^2b^2 + 8ab + 15$

18. $x^2y^2 - 10xy + 24$

19. $3x^2y^2 - 2xy - 5$

20. $3p^2q^2 + 11pq + 6$

21. $2a^2(5 - a) - 7a(5 - a) + 5(5 - a)$

22. $2y^2(y + 2) + 13y(y + 2) + 15(y + 2)$

23. $2x^2(x - 3) + 7x(x - 3) + 6(x - 3)$

24. $3x^2(x - 2) + 5x(x - 2) - 2(x - 2)$

25. $y^4 + 13y^2 + 30$

26. $3z^4 - 16z^2 + 5$

27. $x^2(x + 3) + 3x(x + 3) + 2(x + 3)$

28. $x^2(x - 1) - x(x - 1) - 30(x - 1)$

29. $5a^5b^2 - 8a^4b^3 + 3a^3b^4$

30. $2x^2y^6 + 3xy^5 - 9y^4$

Solve each equation.

31. $x^4 - 10x^2 + 9 = 0$

32. $x^4 - 37x^2 + 36 = 0$

33. $x^4 + 17x^2 + 16 = 0$

34. $x^4 + 50x^2 + 49 = 0$

35. $x^4 - 13x^2 + 36 = 0$

36. $x^4 + 13x^2 + 36 = 0$

37. $a^4 - 7a^2 + 12 = 0$

38. $b^4 + 7b^2 + 12 = 0$

39. $4x^4 - 17x^2 + 4 = 0$

40. $9d^4 - 13d^2 + 4 = 0$

41. $r^4 - 8r^2 = -15$

42. $p^4 - 8p^2 = -12$

43. $z^4 - 7z^2 = 18$

44. $a^4 + a^2 = 42$

45. $-c^4 = 4c^2 - 5$

46. $9b^4 = 57b^2 - 18$

47. $\sqrt{x} = 2x - 6$

48. $x - 2\sqrt{x} = 8$

49. $x - \sqrt{x} = 6$

50. $x - 4 = -3\sqrt{x}$

51. $9x + 3\sqrt{x} = 2$

52. $8x + 2\sqrt{x} = 1$

53. $(x + 3)^2 + 2(x + 3) = 24$

54. $(x + 1)^2 + 4(x + 1) + 3 = 0$

55. $6(a - 2)^2 = -19(a - 2) - 10$

56. $10(z + 2)^2 = 3(z + 2) + 1$

57. $(x^2 - 3)^2 - (x^2 - 3) - 6 = 0$

58. $(a^2 - 1)^2 - 5(a^2 - 1) - 14 = 0$

59. $2(b + 3)^2 + 5(b + 3) - 3 = 0$

60. $(z^2 - 6)^2 + 2(z^2 - 6) - 24 = 0$

61. $18(x^2 - 5)^2 + 27(x^2 - 5) + 10 = 0$

62. $28(x^2 - 8)^2 - 23(x^2 - 8) - 15 = 0$

63. $a^{-2} + 4a^{-1} + 4 = 0$

64. $x^{-2} + 10x^{-1} + 25 = 0$

65. $12b^{-2} - 7b^{-1} + 1 = 0$

66. $5x^{-2} + 4x^{-1} - 1 = 0$

67. $2b^{-2} = 7b^{-1} - 3$

68. $10z^{-2} - 3z^{-1} - 1 = 0$

69. $x^{-2} + 9x^{-1} = 10$

70. $6a^{-2} = a^{-1} + 12$

71. $x^{-2} = 4x^{-1} + 12$

72. $x^{2/3} - 5x^{1/3} + 6 = 0$

73. $x^{2/3} - 4x^{1/3} = -3$

74. $x^{2/3} = 3x^{1/3} + 4$

75. $b^{2/3} - 9b^{1/3} + 18 = 0$

76. $c^{2/3} - 4 = 0$

77. $-2a - 5a^{1/2} + 3 = 0$

78. $r^{2/3} - 7r^{1/3} + 10 = 0$

79. $c^{2/5} + 3c^{1/5} + 2 = 0$

80. $x^{2/5} - 5x^{1/5} + 6 = 0$

Find all x-intercepts of each function.

81. $f(x) = x - 5\sqrt{x} + 4$

82. $g(x) = x - 15\sqrt{x} + 56$

83. $h(x) = x + 14\sqrt{x} + 45$

84. $k(x) = x + 7\sqrt{x} + 12$

85. $p(x) = 4x^{-2} - 19x^{-1} - 5$

86. $g(x) = 4x^{-2} + 12x^{-1} + 9$

87. $f(x) = x^{2/3} - x^{1/3} - 6$

88 $f(x) = x^{1/2} + 6x^{1/4} - 7$

89. $g(x) = (x^2 - 3x)^2 + 2(x^2 - 3x) - 24$

90. $g(x) = (x^2 - 6x)^2 - 5(x^2 - 6x) - 24$

91. $f(x) = x^4 - 29x^2 + 100$

92. $h(x) = x^4 - 4x^2 + 3$

Problem Solving

93. Give a general procedure for solving an equation of the form $ax^4 + bx^2 + c = 0$.

94. Give a general procedure for solving an equation of the form $ax^{2n} + bx^n + c = 0$.

95. Give a general procedure for solving an equation of the form $ax^{-2} + bx^{-1} + c = 0$.

96. Give a general procedure for solving an equation of the form $a(x - r)^2 + b(x - r) - c = 0$.

97. Determine an equation of the form $ax^4 + bx^2 + c = 0$ that has solutions ± 2 and ± 1. Explain how you obtained your answer.

98. Determine an equation of the form $ax^4 + bx^2 + c = 0$ that has solutions ± 3 and $\pm 2i$. Explain how you obtained your answer.

99. Determine an equation of the form $ax^4 + bx^2 + c = 0$ that has solutions $\pm\sqrt{2}$ and $\pm\sqrt{5}$. Explain how you obtained your answer.

100. Determine an equation of the form $ax^4 + bx^2 + c = 0$ that has solutions $\pm 2i$ and $\pm 5i$. Explain how you obtained your answer.

101. Is it possible for an equation of the form $ax^4 + bx^2 + c = 0$ to have exactly one imaginary solution? Explain.

102. Is it possible for an equation of the form $ax^4 + bx^2 + c = 0$ to have exactly one real solution? Explain.

103. Solve the equation $\dfrac{3}{x^2} - \dfrac{3}{x} = 60$ by

  a) multiplying both sides of the equation by the LCD.

  b) writing the equation with negative exponents.

104. Solve the equation $1 = \dfrac{2}{x} - \dfrac{2}{x^2}$ by

  a) multiplying both sides of the equation by the LCD.

  b) writing the equation with negative exponents.

Find all real solutions to each equation.

105. $15(r + 2) + 22 = -\dfrac{8}{r + 2}$

106. $2(p + 3) + 5 = \dfrac{3}{p + 3}$

107. $4 - (x - 1)^{-1} = 3(x - 1)^{-2}$

108. $3(x - 4)^{-2} = 16(x - 4)^{-1} + 12$

109. $x^6 - 9x^3 + 8 = 0$

110. $x^6 - 28x^3 + 27 = 0$

111. $(x^2 + 2x - 2)^2 - 7(x^2 + 2x - 2) + 6 = 0$

112. $(x^2 + 3x - 2)^2 - 10(x^2 + 3x - 2) + 16 = 0$

Find all solutions to each equation.

113. $2n^4 - 6n^2 - 3 = 0$

114. $3x^4 + 8x^2 - 1 = 0$

Cumulative Review Exercises

[1.9] **115.** Evaluate $\dfrac{4}{5} - \left(\dfrac{3}{4} - \dfrac{2}{3}\right)$.

[2.4] **116.** Solve $3(x + 2) - 2(3x + 3) = -3$

[8.2] **117.** State the domain and range for $y = (x - 3)^2$.

[11.3] **118.** Simplify $\sqrt[3]{16x^3y^6}$.

[11.4] **119.** Add $\sqrt{75} + \sqrt{48}$.

12.5 Graphing Quadratic Functions

1   Identify some key characteristics of graphs of polynomial functions.

2   Find the axis of symmetry, vertex, and x-intercepts of a parabola.

3   Graph quadratic functions using the axis of symmetry, vertex, and intercepts.

4   Solve maximum and minimum problems.

5   Understand translations of parabolas.

6   Write functions in the form $f(x) = a(x - h)^2 + k$.

Quadratic functions are part of a family of functions called *polynominal functions*. In this section we begin by discussing polynomial functions and some general characteristics of their graphs. Then we look at the specific characteristics of the graphs of quadratic functions.

1   Identify Some Key Characteristics of Graphs of Polynomial Functions

In Section 5.4 we introduced polynomials. Recall that a polynomial in x is an expression containing the sum of a finite number of terms of the form ax^n for any real number a and any whole number n. The following are examples of polynomials.

$$x + 3$$
$$2x^2 - 5x + 6$$
$$5x^3 - \frac{1}{2}x^2 + 6x - 4$$

In Section 8.2 we introduced functions. Now we combine what we have learned earlier and discuss polynomial functions.

The expression $2x^3 + 6x^2 + 3$ is a polynomial. If we write $P(x) = 2x^3 + 6x^2 + 3$, then we have a polynomial function. In a **polynomial function** the expression used to describe the function is a polynomial. We often use functions that are polynomials without realizing it. All linear functions and all quadratic functions are polynomial functions. However, other functions, such as cubic (third degree) and fourth degree functions are also polynomial functions.

To evaluate a polynomial function, we use substitution, just as we did to evaluate other functions in Chapter 8.

EXAMPLE 1 ▶ For the polynomial function $P(x) = 4x^3 - 6x^2 - 2x + 8$, find

a) $P(0)$ **b)** $P(3)$ **c)** $P(-2)$

Solution

a) $P(x) = 4x^3 - 6x^2 - 2x + 8$

$P(0) = 4(0)^3 - 6(0)^2 - 2(0) + 8$
$\quad\quad = 0 - 0 - 0 + 8 = 8$

b) $P(3) = 4(3)^3 - 6(3)^2 - 2(3) + 8$
$\quad\quad\quad = 4(27) - 6(9) - 6 + 8 = 56$

c) $P(-2) = 4(-2)^3 - 6(-2)^2 - 2(-2) + 8$
$\quad\quad\quad\quad = 4(-8) - 6(4) + 4 + 8 = -44$

▶ **Now Try Exercise 15**

We will not spend time graphing polynomial functions that are of a degree higher than quadratic functions. To graph polynomial functions, whether they are linear, quadratic, cubic, or of any degree, you can plot points as we did in Section 8.1. Here we would like to give you some general guidelines as to what polynomial functions may look like. If you learn these guidelines, you will be able to tell if you pressed keys incorrectly on your graphing calculator and if you have a complete graph of the polynomial function in the window of your graphing calculator.

The graphs of all polynomial functions are smooth, continuous curves. **Figure 12.7**, on the next page, shows a graph of a quadratic polynomial function. The graphs of all quadratic polynomial functions with a *positive leading coefficient* will have the shape of the graph in **Figure 12.7**.

The graph of a cubic polynomial function with a *positive leading coefficient* may have the shape of the graph in either **Figure 12.8** or **Figure 12.9**. Notice that *whenever the leading coefficient in a polynomial function is positive, the polynomial function will increase (or move upward as x increases—the green part of the curve) to the right of some value of x.* For example, in **Figure 12.7**, the graph continues increasing to the right of $x = -1$. In **Figure 12.8** the graph is continuously increasing, and in **Figure 12.9** the graph is increasing to the right of about $x = 1.4$.

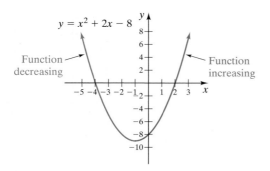

FIGURE 12.7

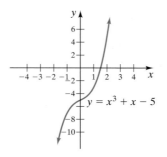

FIGURE 12.8

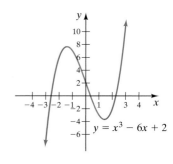

FIGURE 12.9

Polynomial functions with a negative leading coefficient will decrease (or move downward as x increases—the red part of the curve) to the right of some value of x. A quadratic polynomial function with a negative leading coefficient is shown in **Figure 12.10**, and cubic polynomial functions with negative leading coefficients are shown in **Figure 12.11** and **Figure 12.12**. In **Figure 12.10** the quadratic function is decreasing to the right of $x = 2$. In **Figure 12.11** the cubic function is continuously decreasing, and in **Figure 12.12** the cubic function is decreasing to the right of about $x = 1.2$.

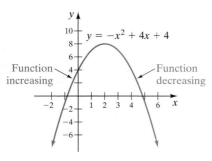

FIGURE 12.10

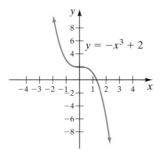

FIGURE 12.11

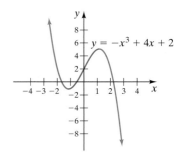

FIGURE 12.12

Why does the leading coefficient determine whether a function will increase or decrease to the right of some value of x? The leading coefficient is the coefficient of the term with the greatest exponent on the variable. As x increases, this term will eventually dominate all the other terms in the function. So if the coefficient of this term is positive, the function will *eventually* increase as x increases. If the leading coefficient is negative, the function will *eventually* decrease as x increases. This information, along with checking the y-intercept of the graph, can be useful in determining whether a graph is correct or complete. Read the Using Your Graphing Calculator box, on the next page, even if you are not using a graphing calculator. Also work Exercises 75 through 78.

Whenever you graph a polynomial function on your grapher, make sure your screen shows every change in direction of your graph. For example, suppose you graph $y = 0.1x^3 - 2x^2 + 5x - 8$ on your grapher. Using the standard window, you get the graph shown in **Figure 12.13**.

However, from our preceding discussion you should realize that since the leading coefficient, 0.1, is positive, the graph must increase to the right of some value of x. The graph in **Figure 12.13** does not show this. If you change your window as shown in **Figure 12.14**, you will get the graph shown. Now you can see how the graph increases to the right of about $x = 12$. When graphing, the y-intercept is often helpful in determining the values to use for the range. Recall that to find the y-intercept, we set $x = 0$ and solve for y. For example, if graphing

$$y = 4x^3 + 6x^2 + x - 180,$$

the y-intercept will be at -180.

EXERCISES

Use your grapher to graph each polynomial. Make sure your window shows all changes in direction of the graph.

1. $y = 0.2x^3 + 5.1x^2 - 6.2x + 9.3$
2. $y = 4.1x^3 - 19.6x^2 + 5.4x - 60.2$

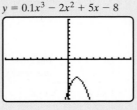
$y = 0.1x^3 - 2x^2 + 5x - 8$

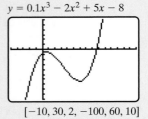
$y = 0.1x^3 - 2x^2 + 5x - 8$

$[-10, 30, 2, -100, 60, 10]$

FIGURE 12.13   **FIGURE 12.14**

2 Find the Axis of Symmetry, Vertex, and *x*-Intercepts of a Parabola

Now let's look at the graphs of quadratic functions a bit more closely. These graphs are called *parabolas*. They have a shape that resembles, but is not the same as, the letter U. As we have seen, this characteristic shape will either be increasing or decreasing to the right of some value of x. When the leading coefficient is positive, the parabola will be increasing to the right of some value of x and the parabola will open upward, see **Figure 12.15a**. When the leading coefficient is negative, the parabola will be decreasing to the right of some value of x and the parabola will open downward, see **Figure 12.15b**.

For a parabola that opens upward, the **vertex** is the lowest point on the curve. The minimum value of the function is the y-coordinate of the vertex. The minimum value is obtained when the x-coordinate of the vertex is substituted into the function. For a parabola that opens downward, the vertex is the highest point on the curve. The maximum value of the function is the y-coordinate of the vertex. The maximum value is obtained when the x-coordinate of the vertex is substituted into the function.

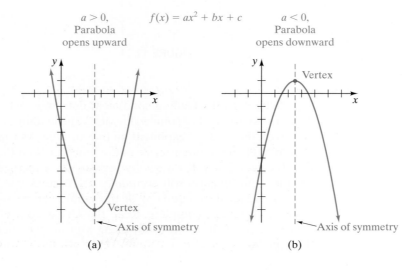

FIGURE 12.15   (a)   (b)

Graphs of quadratic functions of the form $f(x) = ax^2 + bx + c$ will have **symmetry** about a vertical line through the vertex. This means that if we fold the paper along this imaginary line, called the **axis of symmetry**, the right and left sides of the graph will coincide (see **Fig. 12.15** on page 768). We will now give the equation for finding the axis of symmetry.

To Find the Axis of Symmetry

For a function of the form $f(x) = ax^2 + bx + c$, the equation of the **axis of symmetry**, of the parabola is

$$x = -\frac{b}{2a}$$

Now we will derive the formula for the axis of symmetry, and find the coordinates of the vertex of a parabola, by beginning with a quadratic function of the form $f(x) = ax^2 + bx + c$ and completing the square on the first two terms.

$$f(x) = ax^2 + bx + c$$

$$= a\left(x^2 + \frac{b}{a}x\right) + c \quad \text{Factor out a.}$$

One half the coefficient of x is $\dfrac{b}{2a}$. Its square is $\dfrac{b^2}{4a^2}$. Add and subtract this term inside the parentheses. The sum of these two terms is zero.

$$f(x) = a\left[x^2 + \frac{b}{a}x + \frac{b^2}{4a^2} - \frac{b^2}{4a^2}\right] + c$$

Now rewrite the function as follows.

$$f(x) = a\left[x^2 + \frac{b}{a}x + \left(\frac{b^2}{4a^2}\right)\right] - a\left(\frac{b^2}{4a^2}\right) + c$$

$$= a\left(x + \frac{b}{2a}\right)^2 - \frac{b^2}{4a} + c \qquad \textit{Replace the trinomial with the square of a binomial.}$$

$$= a\left(x + \frac{b}{2a}\right)^2 - \frac{b^2}{4a} + \frac{4ac}{4a} \qquad \textit{Write fractions with a common denominator.}$$

$$= a\left(x + \frac{b}{2a}\right)^2 + \frac{4ac - b^2}{4a} \qquad \textit{Combine the last two terms; write with the variable a first.}$$

$$= a\left[x - \left(-\frac{b}{2a}\right)\right]^2 + \frac{4ac - b^2}{4a}$$

The expression $\left[x - \left(-\dfrac{b}{2a}\right)\right]^2$ will always be greater than or equal to 0. (Why?) If $a > 0$, the parabola will open upward and have a minimum value. Since $\left[x - \left(-\dfrac{b}{2a}\right)\right]^2$ will have a minimum value when $x = -\dfrac{b}{2a}$, the minimum value of the graph will occur when $x = -\dfrac{b}{2a}$. If $a < 0$, the parabola will open downward and have a maximum value. The maximum value will occur when $x = -\dfrac{b}{2a}$. To determine the lowest, or highest, point on a parabola, substitute $-\dfrac{b}{2a}$ for x in the function to find y. The resulting ordered pair will be the vertex of the parabola. Since the axis of symmetry is the vertical line through the vertex, its equation is found using the x-coordinate of the ordered pair. Thus, the equation of the axis of symmetry is $x = -\dfrac{b}{2a}$. Note that when $x = -\dfrac{b}{2a}$, the value of $f(x)$ is $\dfrac{4ac - b^2}{4a}$. Do you know why?

To Find the Vertex of a Parabola

The parabola represented by the function $f(x) = ax^2 + bx + c$ will have axis of symmetry $x = -\dfrac{b}{2a}$ and vertex

$$\left(-\frac{b}{2a}, \frac{4ac - b^2}{4a}\right)$$

Since we often find the y-coordinate of the vertex by substituting the x-coordinate of the vertex into $f(x)$, the vertex may also be designated as

$$\left(-\frac{b}{2a}, f\left(-\frac{b}{2a}\right)\right)$$

The parabola given by the function $f(x) = ax^2 + bx + c$ will open upward when a is greater than 0 and open downward when a is less than 0.

Recall that to find the x-intercept of the graph of $f(x) = ax^2 + bx + c$, we set $f(x) = 0$ and solve the equation

$$ax^2 + bx + c = 0$$

This equation may be solved by factoring, the quadratic formula, or completing the square.

As we mentioned in Section 12.2, the discriminant, $b^2 - 4ac$, may be used to determine the *number of x-intercepts*. The following table summarizes information about the discriminant.

Discriminant, $b^2 - 4ac$	Number of x-Intercepts	Possible Graphs of $f(x) = ax^2 + bx + c$
> 0	Two	
$= 0$	One	
< 0	None	

3 Graph Quadratic Functions Using the Axis of Symmetry, Vertex, and Intercepts

Now we will draw graphs of quadratic functions.

EXAMPLE 2 ▶ Consider the equation $y = -x^2 + 8x - 12$.

a) Determine whether the parabola opens upward or downward.

b) Find the y-intercept.

c) Find the vertex.

d) Find the x-intercepts, if any.

e) Draw the graph.

Solution

a) Since a is -1, which is less than 0, the parabola opens downward.

b) To find the y-intercept, set $x = 0$ and solve for y.

$$y = -(0)^2 + 8(0) - 12 = -12$$

The y-intercept is $(0, -12)$.

c) First, find the x-coordinate, then find the y-coordinate of the vertex.

$$x = -\frac{b}{2a} = -\frac{8}{2(-1)} = 4$$

$$y = \frac{4ac - b^2}{4a} = \frac{4(-1)(-12) - 8^2}{4(-1)} = \frac{48 - 64}{-4} = 4$$

The vertex is at $(4, 4)$. The y-coordinate of the vertex could also be found by substituting 4 for x in the function and finding the corresponding value of y, which is 4.

d) To find the x-intercepts, set $y = 0$.

$$0 = -x^2 + 8x - 12$$
$$\text{or}\quad x^2 - 8x + 12 = 0$$
$$(x - 6)(x - 2) = 0$$
$$x - 6 = 0 \quad \text{or} \quad x - 2 = 0$$
$$x = 6 \quad \text{or} \quad x = 2$$

Thus, the x-intercepts are $(2, 0)$ and $(6, 0)$. These values could also be found by the quadratic formula (or by completing the square).

e) Use all this information to draw the graph (**Fig. 12.16**).

▶ **Now Try Exercise 21**

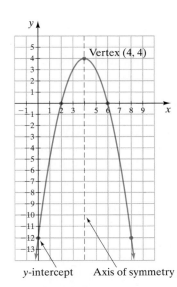

Vertex $(4, 4)$

y-intercept   Axis of symmetry

FIGURE 12.16

Notice that in Example 2, the equation is $y = -x^2 + 8x - 12$ and the y-intercept is $(0, -12)$. In general, for any equation of the form $y = ax^2 + bx + c$, the y-intercept will be $(0, c)$.

If you obtain irrational values when finding x-intercepts by the quadratic formula, use your calculator to estimate these values, and then plot these decimal values. For example, if you obtain $x = \frac{2 \pm \sqrt{10}}{2}$, you would evaluate $\frac{2 + \sqrt{10}}{2}$ and $\frac{2 - \sqrt{10}}{2}$ on your calculator and obtain 2.58 and -0.58, respectively, to the nearest hundredth. The x-intercepts would therefore be $(2.58, 0)$ and $(-0.58, 0)$.

EXAMPLE 3 ▶ Consider the function $f(x) = 2x^2 + 6x + 5$.

a) Determine whether the parabola opens upward or downward.

b) Find the y-intercept.

c) Find the vertex.

d) Find the x-intercepts, if any.

e) Draw the graph.

Solution

a) Since *a* is 2, which is greater than 0, the parabola opens upward.

b) Since $f(x)$ is the same as y, to find the y-intercept, set $x = 0$ and solve for $f(x)$, or y.

$$f(0) = 2(0)^2 + 6(0) + 5 = 5$$

The y-intercept is $(0, 5)$.

c)
$$x = -\frac{b}{2a} = -\frac{6}{2(2)} = -\frac{6}{4} = -\frac{3}{2}$$

$$y = \frac{4ac - b^2}{4a} = \frac{4(2)(5) - 6^2}{4(2)} = \frac{40 - 36}{8} = \frac{4}{8} = \frac{1}{2}$$

The vertex is $\left(-\frac{3}{2}, \frac{1}{2}\right)$. The y-coordinate of the vertex can also be found by evaluating $f\left(-\frac{3}{2}\right)$.

d) To find the x-intercepts, set $f(x) = 0$.

$$0 = 2x^2 + 6x + 5$$

This trinomial cannot be factored. To determine whether this equation has any real solutions, evaluate the discriminant.

$$b^2 - 4ac = 6^2 - 4(2)(5) = 36 - 40 = -4$$

Since the discriminant is less than 0, this equation has no real solutions. You should have expected this answer because the y-coordinate of the vertex is a positive number and therefore above the x-axis. Since the parabola opens upward, it cannot intersect the x-axis.

e) The graph is given in **Figure 12.17**.

▸ **Now Try Exercise 45**

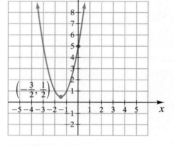

FIGURE 12.17

4 Solve Maximum and Minimum Problems

A parabola that opens upward has a **minimum value** at its vertex, as illustrated in **Figure 12.18a**. A parabola that opens downward has a **maximum value** at its vertex, as shown in **Figure 12.18b**. If you are given a function of the form $f(x) = ax^2 + bx + c$, the maximum or minimum value will occur at $-\frac{b}{2a}$, and the value will be $\frac{4ac - b^2}{4a}$.

There are many real-life problems that require finding maximum and minimum values.

$$y = ax^2 + bx + c$$

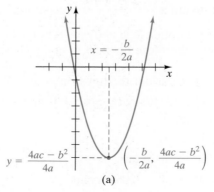

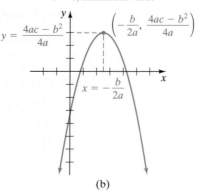

FIGURE 12.18   (a)   (b)

FIGURE 12.19

EXAMPLE 4 ▶ **Baseball** Tommy Magee plays baseball with the Yorktown Cardinals. In the seventh inning of a game against the Arlington Blue Jays, he hit the ball at a height of 3 feet above the ground (see **Fig. 12.19**). For this particular hit, the height of the ball above the ground, $f(t)$, in feet, at time, t, in seconds, can be estimated by the formula

$$f(t) = -16t^2 + 52t + 3$$

a) Find the maximum height attained by the baseball.

b) Find the time it takes for the baseball to reach its maximum height.

c) Find the time it takes for the baseball to strike the ground.

Solution **a)** Understand The baseball will follow the path of a parabola that opens downward ($a < 0$). The baseball will rise to a maximum height, then begin its fall back to the ground due to gravity. To find the maximum height, we use the formula $y = \dfrac{4ac - b^2}{4a}$.

Translate $a = -16, \qquad b = 52, \qquad c = 3$

$$y = \frac{4ac - b^2}{4a}$$

Carry Out

$$= \frac{4(-16)(3) - (52)^2}{4(-16)}$$

$$= \frac{-192 - 2704}{-64}$$

$$= \frac{-2896}{-64}$$

$$= 45.25$$

Answer The maximum height attained by the baseball is 45.25 feet.

b) The baseball reaches its maximum height at

$$t = -\frac{b}{2a} = -\frac{52}{2(-16)} = -\frac{52}{-32} = \frac{13}{8} \quad \text{or} \quad 1\frac{5}{8} \quad \text{or} \quad 1.625 \text{ seconds}$$

c) Understand and Translate When the baseball strikes the ground, its height, y, above the ground is 0. Thus, to determine when the baseball strikes the ground, we solve the equation

$$-16t^2 + 52t + 3 = 0$$

We will use the quadratic formula to solve the equation.

$$t = \frac{-b \pm \sqrt{b^2 - 4ac}}{2a}$$

Carry Out

$$= \frac{-52 \pm \sqrt{(52)^2 - 4(-16)(3)}}{2(-16)}$$

$$= \frac{-52 \pm \sqrt{2704 + 192}}{-32}$$

$$= \frac{-52 \pm \sqrt{2896}}{-32}$$

$$\approx \frac{-52 \pm 53.81}{-32}$$

$$t \approx \frac{-52 + 53.81}{-32} \quad \text{or} \quad t \approx \frac{-52 - 53.81}{-32}$$

$$\approx -0.06 \text{ second} \qquad\qquad \approx 3.31 \text{ seconds}$$

Answer   The only acceptable value is 3.31 seconds. The baseball strikes the ground in about 3.31 seconds. Notice in part **b)** that the time it takes the baseball to reach its maximum height, 1.625 seconds, is not quite half the total time the baseball was in flight, 3.31 seconds. The reason for this is that the baseball was hit from a height of 3 feet and not at ground level.

▶ **Now Try Exercise 107**

EXAMPLE 5 ▶ **Area of a Rectangle** Consider the rectangle below where the length is $x + 3$ and the width is $10 - x$.

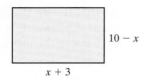

$10 - x$

$x + 3$

a) Find an equation for the area, $A(x)$.

b) Find the value for x that gives the largest (maximum) area.

c) Find the maximum area.

Solution   **a)** Area is length times width. The area function is

$$A(x) = (x + 3)(10 - x)$$
$$= -x^2 + 7x + 30$$

b) Understand and Translate   The graph of the function is a parabola that opens downward. Thus, the maximum value occurs at the vertex. Therefore, the maximum area occurs at $x = -\dfrac{b}{2a}$.

Carry Out $$x = -\frac{b}{2a} = -\frac{7}{2(-1)} = \frac{7}{2} = 3.5$$

Answer   The maximum area occurs when x is 3.5 units.

c) To find the maximum area, substitute 3.5 for each x in the equation determined in part **a)**.

$$A(x) = -x^2 + 7x + 30$$
$$A(3.5) = -(3.5)^2 + 7(3.5) + 30$$
$$= -12.25 + 24.5 + 30$$
$$= 42.25$$

Observe that for this rectangle, the length is $x + 3 = 3.5 + 3 = 6.5$ units and the width is $10 - x = 10 - 3.5 = 6.5$ units. The rectangle is actually a square, and its area is $(6.5)(6.5) = 42.25$ square units. Therefore, the maximum area is 42.25 square units.

▶ **Now Try Exercise 85**

In Example 5**c)**, the maximum area could have been determined by using the formula $y = \dfrac{4ac - b^2}{4a}$. Determine the maximum area now using this formula. You should obtain the same answer, 42.25 square units.

FIGURE 12.20

EXAMPLE 6 ▶ **Rectangular Corral** John W. Brown is building a corral for newborn calves in the shape of a rectangle (see **Fig. 12.20**). If he plans to use 160 meters of fencing, find the dimensions of the corral that will give the greatest area.

Solution Understand We are given the perimeter of the corral, 160 meters. The formula for the perimeter of a rectangle is $P = 2l + 2w$. For this problem, $160 = 2l + 2w$. We are asked to maximize the area, A, where

$$A = lw$$

We need to express the area in terms of one variable, not two. To express the area in terms of l, we solve the perimeter formula, $160 = 2l + 2w$, for w, then make a substitution.

Translate
$$160 = 2l + 2w$$
$$160 - 2l = 2w$$
$$80 - l = w$$

Carry Out Now we substitute $80 - l$ for w into $A = lw$. This gives

$$A = lw$$
$$A = l(80 - l)$$
$$A = -l^2 + 80l$$

In this quadratic equation, $a = -1$, $b = 80$, and $c = 0$. The maximum area will occur at

$$l = -\frac{b}{2a} = -\frac{80}{2(-1)} = 40$$

Answer The length that will give the largest area is 40 meters. The width, $w = 80 - l$, will also be 40 meters. Thus, a square with dimensions 40 by 40 meters will give the largest area.

The largest area can be found by substituting $l = 40$ into the formula $A = l(80 - l)$ or by using $A = \dfrac{4ac - b^2}{4a}$. In either case, we obtain an area of 1600 square meters.

▶ **Now Try Exercise 105**

In Example 6, when we obtained the equation $A = -l^2 + 80l$, we could have completed the square as follows:

$$A = -(l^2 - 80l)$$
$$= -(l^2 - 80l + 1600 - 1600)$$
$$= -(l^2 - 80l + 1600) + 1600$$
$$= -(l - 40)^2 + 1600$$

From this equation we can determine that the maximum area, 1600 square meters, occurs when the length is 40 meters.

5 Understand Translations of Parabolas

Now we will look at another method used to graph parabolas. With this method, you start with a graph of an equation of the form $f(x) = ax^2$ and **translate**, or shift, the position of the graph to obtain the graph of the function you are seeking. As a reference, **Figure 12.21a** on the next page shows the graphs of $f(x) = x^2$, $g(x) = 2x^2$, and $h(x) = \frac{1}{2}x^2$. **Figure 12.21b** on the next page shows the graphs of $f(x) = -x^2$, $g(x) = -2x^2$, and $h(x) = -\frac{1}{2}x^2$.

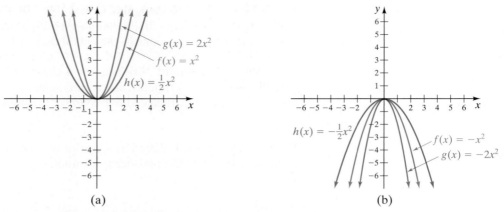

FIGURE 12.21

(a)   (b)

You can verify that each of the graphs is correct by plotting points. Notice that in **Figures 12.21a** and **b** the *value of a* in $f(x) = ax^2$ determines the width of the parabola. As $|a|$ gets larger, the parabola gets narrower and as $|a|$ gets smaller, the parabola gets wider.

Now let's consider the three functions $f(x) = x^2, g(x) = (x - 2)^2$, and $h(x) = (x + 2)^2$. These functions are graphed in **Figure 12.22**. (You can verify that these are graphs of the three functions by plotting points.)

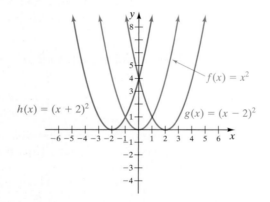

FIGURE 12.22

Notice that the graphs of $g(x)$ and $h(x)$ are identical in shape to the graph of $f(x)$ except that $g(x)$ has been translated, or shifted, 2 units to the right and $h(x)$ has been translated 2 units to the left. In general, the graph of $g(x) = a(x - h)^2$ will have the same shape as the graph of $f(x) = ax^2$. The graph of an equation of the form $g(x) = a(x - h)^2$ will be shifted horizontally from the graph of $f(x) = ax^2$. *If h is a positive real number, the graph of g(x) = a(x − h)² will be shifted h units to the right of the graph of f(x) = ax². If h is a negative real number, the graph of g(x) = a(x − h)² will be shifted |h| units to the left of the graph of f(x) = ax².*

Now consider the graphs of $f(x) = x^2, g(x) = x^2 + 3$ and $h(x) = x^2 - 3$ that are illustrated in **Figure 12.23**. You can verify that these are graphs of the three functions by plotting points.

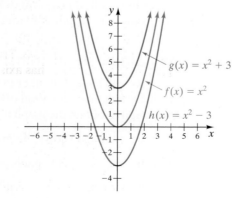

FIGURE 12.23

Notice that the graphs of $g(x)$ and $h(x)$ are identical to the graph of $f(x)$ except that $g(x)$ is translated 3 units up, and $h(x)$ is translated 3 units down. In general, *the graph of* $g(x) = ax^2 + k$ *is the graph of* $f(x) = ax^2$ *shifted k units up if k is a positive real number, and* $|k|$ *units down if k is a negative real number.*

Now consider the graphs of $f(x) = x^2$ and $g(x) = (x - 2)^2 + 3$, shown in **Figure 12.24**.

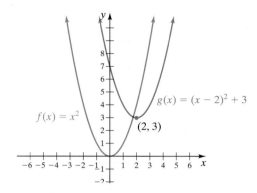

FIGURE 12.24

Notice that the graph of $g(x)$ has the same general shape as that of $f(x)$. The graph of $g(x)$ is the graph of $f(x)$ translated 2 units to the right and 3 units up. This graph and the discussion preceding it lead to the following important facts.

Parabola Shifts

For any function $f(x) = ax^2$, the graph of $g(x) = a(x - h)^2 + k$ will have the same shape as the graph of $f(x)$. The graph of $g(x)$ will be the graph of $f(x)$ shifted as follows:

- If h is a positive real number, the graph will be shifted h units to the right.
- If h is a negative real number, the graph will be shifted $|h|$ units to the left.
- If k is a positive real number, the graph will be shifted k units up.
- If k is a negative real number, the graph will be shifted $|k|$ units down.

Examine the graph of $g(x) = (x - 2)^2 + 3$ in **Figure 12.24**. Notice that its axis of symmetry is $x = 2$ and its vertex is $(2, 3)$.

Axis of Symmetry and Vertex of a Parabola

The graph of any function of the form

$$f(x) = a(x - h)^2 + k$$

will be a parabola with axis of symmetry $x = h$ and vertex at (h, k).

Example	Axis of Symmetry	Vertex	Parabola Opens
$f(x) = 2(x - 5)^2 + 7$	$x = 5$	$(5, 7)$	upward, $a > 0$
$f(x) = -\dfrac{1}{2}(x - 6)^2 - 3$	$x = 6$	$(6, -3)$	downward, $a < 0$

Now consider $f(x) = 2(x + 5)^2 + 3$. We can rewrite this as $f(x) = 2[x - (-5)]^2 + 3$. Therefore, h has a value of -5 and k has a value of 3. The graph of this function has axis of symmetry $x = -5$ and vertex at $(-5, 3)$.

Example	Axis of Symmetry	Vertex	Parabola Opens
$f(x) = 3(x + 4)^2 - 2$	$x = -4$	$(-4, -2)$	upward, $a > 0$
$f(x) = -\dfrac{1}{2}\left(x + \dfrac{1}{3}\right)^2 + \dfrac{1}{4}$	$x = -\dfrac{1}{3}$	$\left(-\dfrac{1}{3}, \dfrac{1}{4}\right)$	downward, $a < 0$

Now we are ready to graph parabolas using translations.

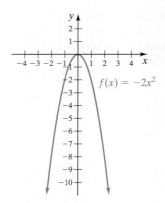

FIGURE 12.25

EXAMPLE 7 ▸ The graph of $f(x) = -2x^2$ is illustrated in **Figure 12.25**. Using this graph as a guide, graph $g(x) = -2(x + 3)^2 - 4$.

Solution   The function $g(x)$ may be written $g(x) = -2[x - (-3)]^2 - 4$. Therefore, in the function, h has a value of -3 and k has a value of -4. The graph of $g(x)$ will therefore be the graph of $f(x)$ translated 3 units to the left (because $h = -3$) and 4 units down (because $k = -4$). The graphs of $f(x)$ and $g(x)$ are illustrated in **Figure 12.26**.

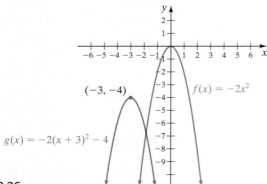

FIGURE 12.26

▸ **Now Try Exercise 55**

In objective 2, we started with a function of the form $f(x) = ax^2 + bx + c$ and completed the square to obtain

$$f(x) = a\left[x - \left(-\frac{b}{2a}\right)\right]^2 + \frac{4ac - b^2}{4a}$$

We stated that the vertex of the parabola of this function is $\left(-\dfrac{b}{2a}, \dfrac{4ac - b^2}{4a}\right)$.

Suppose we substitute h for $-\dfrac{b}{2a}$ and k for $\dfrac{4ac - b^2}{4a}$ in the function. We then get

$$f(x) = a(x - h)^2 + k$$

which we know is a parabola with vertex at (h, k). Therefore, both functions $f(x) = ax^2 + bx + c$ and $f(x) = a(x - h)^2 + k$ yield the same vertex and axis of symmetry for any given function.

6 Write Functions in the Form $f(x) = a(x - h)^2 + k$

If we wish to graph parabolas using translations, we need to change the form of a function from $f(x) = ax^2 + bx + c$ to $f(x) = a(x - h)^2 + k$. To do this, we *complete the square* as was discussed in Section 12.1. By completing the square we obtain a perfect square trinomial, which we can represent as the square of a binomial. Examples 8 and 9 explain the procedure. We will use this procedure again in Chapter 14, when we discuss conic sections.

EXAMPLE 8 ▸ Given $f(x) = x^2 - 6x + 10$,

a) Write $f(x)$ in the form $f(x) = a(x - h)^2 + k$.

b) Graph $f(x)$.

Solution

a) We use the x^2 and $-6x$ terms to obtain a perfect square trinomial.

$$f(x) = (x^2 - 6x) + 10$$

Now we take half the coefficient of the x-term and square it.

$$\left[\frac{1}{2}(-6)\right]^2 = \boxed{9}$$

We then add this value, 9, within the parentheses. Since we are adding 9 within parentheses, we add -9 outside parentheses. Adding 9 and -9 to an expression is the same as adding 0, which does not change the value of the expression.

$$f(x) = (x^2 - 6x \boxed{+ 9}) \boxed{- 9} + 10$$

By doing this we have created a perfect square trinomial within the parentheses, plus a constant outside the parentheses. We express the perfect square trinomial as the square of a binomial.

$$f(x) = (x - 3)^2 + 1$$

The function is now in the form we are seeking.

b) Since $a = 1$, which is greater than 0, the parabola opens upward. The axis of symmetry of the parabola is $x = 3$ and the vertex is at $(3, 1)$. The y-intercept can be easily obtained by substituting $x = 0$ and finding $f(x)$. When $x = 0$, $f(x) = (-3)^2 + 1 = 10$. Thus, the y-intercept is at 10. By plotting the vertex, y-intercept, and a few other points, we obtain the graph in **Figure 12.27**. The figure also shows the graph of $y = x^2$ for comparison.

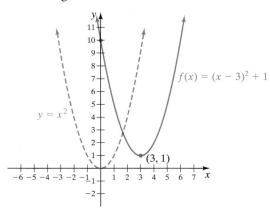

FIGURE 12.27

▶ **Now Try Exercise 65**

EXAMPLE 9 ▶ Given $f(x) = -2x^2 - 10x - 13$,

a) Write $f(x)$ in the form $f(x) = a(x - h)^2 + k$.

b) Graph $f(x)$.

Solution

a) When the leading coefficient is not 1, we factor out the leading coefficient from the terms containing the variable.

$$f(x) = -2(x^2 + 5x) - 13$$

Now we complete the square.

Half of coefficient of x-term squared
$$\downarrow$$
$$\left[\frac{1}{2}(5)\right]^2 = \boxed{\frac{25}{4}}$$

If we add $\frac{25}{4}$ within the parentheses, we are actually adding $-2\left(\frac{25}{4}\right)$ or $-\frac{25}{2}$, since each term in parentheses is multiplied by -2. Therefore, to compensate, we must add $\frac{25}{2}$ outside the parentheses.

$$f(x) = -2\left(x^2 + 5x + \boxed{\frac{25}{4}}\right) + \boxed{\frac{25}{2}} - 13$$

$$= -2\left(x + \frac{5}{2}\right)^2 - \frac{1}{2}$$

b) Since $a = -2$, the parabola opens downward. The axis of symmetry is $x = -\dfrac{5}{2}$ and the vertex is $\left(-\dfrac{5}{2}, -\dfrac{1}{2}\right)$. The y-intercept is at $f(0) = -13$. We plot a few points and draw the graph in **Figure 12.28**. In the figure, we also show the graph of $y = -2x^2$ for comparison.

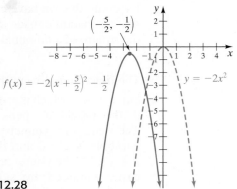

FIGURE 12.28

Notice that $f(x) = -2\left(x + \dfrac{5}{2}\right)^2 - \dfrac{1}{2}$ has no x-intercepts. Therefore, there are no real values of x for which $f(x) = 0$.

▶ **Now Try Exercise 69**

A second way to change the equation from $f(x) = ax^2 + bx + c$ to $f(x) = a(x - h)^2 + k$ form is to let $h = -\dfrac{b}{2a}$ and $k = \dfrac{4ac - b^2}{4a}$. Find the values for h and k and then substitute the values obtained into $f(x) = a(x - h)^2 + k$. For example, for the function $f(x) = -2x^2 - 10x - 13$, in Example 9, $a = -2$, $b = -10$, and $c = -13$. Then

$$h = -\frac{b}{2a} = -\frac{-10}{2(-2)} = -\frac{5}{2}$$

$$k = \frac{4ac - b^2}{4a} = \frac{4(-2)(-13) - (-10)^2}{4(-2)} = -\frac{1}{2}$$

Therefore,

$$f(x) = a(x - h)^2 + k$$

$$= -2\left[x - \left(-\frac{5}{2}\right)\right]^2 - \frac{1}{2}$$

$$= -2\left(x + \frac{5}{2}\right)^2 - \frac{1}{2}$$

This answer checks with that obtained in Example 9 **a)**.

EXERCISE SET 12.5   *Math* XL   *MyMathLab*
$\qquad\qquad\qquad\qquad\quad$ MathXL®   MyMathLab

Concept/Writing Exercises

1. What is the graph of a quadratic equation called?

2. What is the vertex of a parabola?

3. What is the axis of symmetry of a parabola?

4. What is the equation of the axis of symmetry of the graph of $f(x) = ax^2 + bx + c$?

5. What is the vertex of the graph of $f(x) = ax^2 + bx + c$?

6. How many x-intercepts does a quadratic function have if the discriminant is **a)** <0, **b)** $= 0$, **c)** >0?

7. For $f(x) = ax^2 + bx + c$, will $f(x)$ have a maximum or a minimum if **a)** $a > 0$, **b)** $a < 0$? Explain.

8. Explain how to find the x-intercepts of the graph of a quadratic function.

9. Explain how to find the y-intercept of the graph of a quadratic function.

10. Consider the graph of $f(x) = ax^2$. Explain how the shape of $f(x)$ changes as $|a|$ increases and as $|a|$ decreases.

11. Consider the graph of $f(x) = ax^2$. What is the general shape of $f(x)$ if **a)** $a > 0$, **b)** $a < 0$?

12. Will the graphs of $f(x) = ax^2$ and $g(x) = -ax^2$ have the same vertex for any nonzero real number a? Explain.

13. Does the function $f(x) = 3x^2 - 4x + 2$ have a maximum or a minimum value? Explain.

14. Does the function $g(x) = -\dfrac{1}{2}x^2 + 2x - 7$ have a maximum or a minimum value? Explain.

Practice the Skills

Evaluate each polynomial function at the given value.

15. Find $P(2)$ if $P(x) = x^2 - 6x + 4$.

16. Find $P(-1)$ if $P(x) = 4x^2 - 6x + 22$.

17. Find $P\left(\dfrac{1}{2}\right)$ if $P(x) = 2x^2 - 3x - 6$.

18. Find $P\left(\dfrac{1}{3}\right)$ if $P(x) = \dfrac{1}{2}x^3 - x^2 + 6$.

19. Find $P(0.4)$ if $P(x) = 0.2x^3 + 1.6x^2 - 2.3$.

20. Find $P(-1.2)$ if $P(x) = -1.6x^3 - 4.6x^2 - 0.1x$.

a) *Determine whether the parabola opens upward or downward.* **b)** *Find the y-intercept.* **c)** *Find the vertex.* **d)** *Find the x-intercepts (if any).* **e)** *Draw the graph.*

21. $f(x) = x^2 + 8x + 15$

22. $g(x) = x^2 + 2x - 3$

23. $f(x) = x^2 - 4x + 3$

24. $h(x) = x^2 - 2x - 8$

25. $f(x) = -x^2 - 2x + 8$

26. $p(x) = -x^2 + 8x - 15$

27. $g(x) = -x^2 + 4x + 5$

28. $n(x) = -x^2 - 2x + 24$

29. $t(x) = -x^2 + 4x - 5$

30. $g(x) = x^2 + 6x + 13$

31. $f(x) = x^2 - 4x + 4$

32. $r(x) = -x^2 + 10x - 25$

33. $r(x) = x^2 + 2$

34. $f(x) = x^2 + 4x$

35. $l(x) = -x^2 + 5$

36. $g(x) = -x^2 + 6x$

37. $f(x) = -2x^2 + 4x - 8$

38. $g(x) = -2x^2 - 6x + 4$

39. $m(x) = 3x^2 + 4x + 3$

40. $p(x) = -2x^2 + 5x + 4$

41. $y = 3x^2 + 4x - 6$

42. $y = x^2 - 6x + 4$

43. $y = 2x^2 - x - 6$

44. $g(x) = -4x^2 + 6x - 9$

45. $f(x) = -x^2 + 3x - 5$

46. $h(x) = -2x^2 + 4x - 5$

Using the graphs in **Figures 12.21** *through* **12.24** *as a guide, graph each function and label the vertex.*

47. $f(x) = (x - 3)^2$

48. $f(x) = (x - 4)^2$

49. $f(x) = (x + 1)^2$

50. $f(x) = (x + 2)^2$

51. $f(x) = x^2 + 3$

52. $f(x) = x^2 + 5$

53. $f(x) = x^2 - 1$

54. $f(x) = x^2 - 4$

55. $f(x) = (x - 2)^2 + 3$

56. $f(x) = (x - 3)^2 - 4$

57. $f(x) = (x + 4)^2 + 4$

58. $h(x) = (x + 4)^2 - 1$

59. $g(x) = -(x + 3)^2 - 2$

60. $g(x) = (x - 1)^2 + 4$

61. $y = -2(x - 2)^2 + 2$

62. $y = -2(x - 3)^2 + 1$

63. $h(x) = -2(x + 1)^2 - 3$

64. $f(x) = -(x - 5)^2 + 2$

In Exercises 65–74, **a)** *express each function in the form* $f(x) = a(x - h)^2 + k$ *and* **b)** *draw the graph of each function and label the vertex.*

65. $f(x) = x^2 - 6x + 8$

66. $g(x) = x^2 + 6x + 2$

67. $g(x) = x^2 - x - 3$

68. $f(x) = x^2 - x + 1$

69. $f(x) = -x^2 - 4x - 6$

70. $h(x) = -x^2 + 6x + 1$

71. $g(x) = x^2 - 4x - 1$

72. $p(x) = x^2 - 2x - 6$

73. $f(x) = 2x^2 + 5x - 3$

74. $k(x) = 2x^2 + 7x - 4$

Problem Solving

*In Exercises 75–78, determine which of the graphs—***a)**, **b)**, *or* **c)**—*is the graph of the given equation. Explain how you determined your answer.*

75. $y = x^2 + 3x - 4$

a)

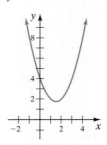

b)

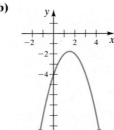

c)

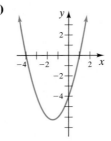

76. $y = x^3 + 2x^2 - 4$

a)

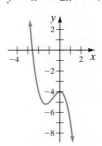

b)

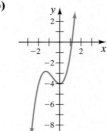

c)

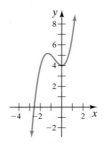

77. $y = -x^3 + 2x - 6$

a)

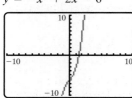

b)

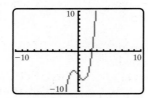

c)

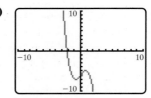

78. $y = x^3 + 4x^2 - 5$

a) 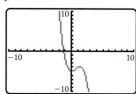 **b)** 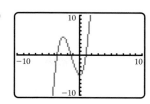 **c)**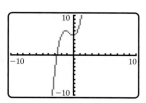

*Answer Exercises 79 and 80 using a graphing calculator if you have one. If you do not have a graphing calculator, draw the graphs in part **a)** by plotting points. Then answer parts **b)** through **e)**.*

 79. a) Graph

$$y_1 = x^3$$
$$y_2 = x^3 - 3x^2 - 3$$

b) In both graphs, for values of $x > 3$, do the functions increase or decrease as x increases?

c) When the leading term of a polynomial function is x^3, the polynomial must increase for $x > a$, where a is some real number greater than 0. Explain why this must be so.

d) In both graphs, for values of $x < -3$, do the functions increase or decrease as x decreases?

e) When the leading term of a polynomial function is x^3, the polynomial must decrease for $x < a$, where a is some real number less than 0. Explain why this must be so.

80. a) Graph

$$y_1 = x^4$$
$$y_2 = x^4 - 6x^2$$

b) In each graph, for values of $x > 3$, are the functions increasing or decreasing as x increases?

c) When the leading term of a polynomial function is x^4, the polynomial must increase for $x > a$, where a is some real number greater than 0. Explain why this must be so.

d) In each graph, for values of $x < -3$, are the functions increasing or decreasing as x decreases?

e) When the leading term of a polynomial function is x^4, the polynomial must increase for $x < a$, where a is some real number less than 0. Explain why this must be so.

*Match the functions in Exercises 81–84 with the appropriate graphs labeled **a)** through **d)**.*

a) 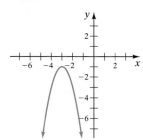 **b)** 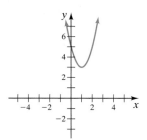 **c)** 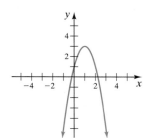 **d)**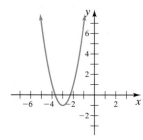

81. $f(x) = 2(x + 3)^2 - 1$ **82.** $f(x) = -2(x + 3)^2 - 1$ **83.** $f(x) = 2(x - 1)^2 + 3$ **84.** $f(x) = -2(x - 1)^2 + 3$

Area *For each rectangle, **a)** find the value for x that gives the maximum area and **b)** find the maximum area.*

85. 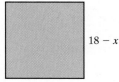 $18 - x$, $x + 4$

86. 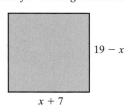 $19 - x$, $x + 7$

87. 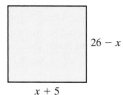 $26 - x$, $x + 5$

88. 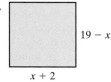 $19 - x$, $x + 2$

89. Selling Batteries The revenue function for selling n batteries is $R(n) = n(8 - 0.02n) = -0.02n^2 + 8n$. Find **a)** the number of batteries that must be sold to obtain the maximum revenue and **b)** the maximum revenue.

90. Selling Watches The revenue function for selling n watches is $R(n) = n(25 - 0.1n) = -0.1n^2 + 25n$. Find **a)** the number of watches that must be sold to obtain the maximum revenue and **b)** the maximum revenue.

91. Enrollment The enrollment in a high school in the Naplewood School District can be approximated by the function

$$N(t) = -0.043t^2 + 1.82t + 46.0$$

where t is the number of years since 1989 and $1 \le t \le 22$. In what year will the maximum enrollment be obtained?

92. Drug-Free Schools The percent of students in schools in the United States who say their school is not drug free can be approximated by the function

$$f(a) = -2.32a^2 + 76.58a - 559.87$$

where a is the student's age and $12 < a < 20$. What age has the highest percent of students who say their school is not drug free?

93. What is the distance between the vertices of the graphs of $f(x) = (x - 2)^2 + \dfrac{5}{2}$ and $g(x) = (x - 2)^2 - \dfrac{3}{2}$?

94. What is the distance between the vertices of the graphs of $f(x) = 2(x - 4)^2 - 3$ and $g(x) = -3(x - 4)^2 + 2$?

95. What is the distance between the vertices of the graphs of $f(x) = 2(x + 4)^2 - 3$ and $g(x) = -(x + 1)^2 - 3$?

96. What is the distance between the vertices of the graphs of $f(x) = -\dfrac{1}{3}(x - 3)^2 - 2$ and $g(x) = 2(x + 5)^2 - 2$?

97. Write the function whose graph has the shape of the graph of $f(x) = 2x^2$ and has a vertex at $(3, -2)$.

98. Write the function whose graph has the shape of the graph of $f(x) = -\dfrac{1}{2}x^2$ and has a vertex at $\left(\dfrac{2}{3}, -5\right)$.

99. Write the function whose graph has the shape of the graph of $f(x) = -4x^2$ and has a vertex at $\left(-\dfrac{3}{5}, -\sqrt{2}\right)$.

100. Write the function whose graph has the shape of the graph of $f(x) = \dfrac{3}{5}x^2$ and has a vertex at $(-\sqrt{3}, \sqrt{5})$.

101. Consider $f(x) = x^2 - 8x + 12$ and $g(x) = -x^2 + 8x - 12$.
   a) Without graphing, can you explain how the graphs of the two functions compare?
   b) Will the graphs have the same x-intercepts? Explain.
   c) Will the graphs have the same vertex? Explain.
   d) Graph both functions on the same axes.

102. By observing the leading coefficient in a quadratic function and by determining the coordinates of the vertex of its graph, explain how you can determine the number of x-intercepts the parabola has.

103. Selling Tickets The Johnson High School Theater Club is trying to set the price of tickets for a play. If the price is too low, they will not make enough money to cover expenses, and if the price is too high, not enough people will pay the price of a ticket. They estimate that their total income per concert, I, in hundreds of dollars, can be approximated by the formula

$$I = -x^2 + 24x - 44, \, 0 \le x \le 24$$

where x is the cost of a ticket.

   a) Draw a graph of income versus the cost of a ticket.
   b) Determine the minimum cost of a ticket for the theater club to break even.
   c) Determine the maximum cost of a ticket that the theater club can charge and break even.
   d) How much should they charge to receive the maximum income?
   e) Find the maximum income.

104. Throwing an Object An object is projected upward with an initial velocity of 192 feet per second. The object's distance above the ground, d, after t seconds may be found by the formula $d = -16t^2 + 192t$.

   a) Find the object's distance above the ground after 3 seconds.
   b) Draw a graph of distance versus time.
   c) What is the maximum height the object will reach?
   d) At what time will it reach its maximum height?
   e) At what time will the object strike the ground?

105. Profit The Fulton Bird House Company earns a weekly profit according to the function $f(x) = -0.4x^2 + 80x - 200$, where x is the number of bird feeders built and sold.

   a) Find the number of bird feeders that the company must sell in a week to obtain the maximum profit.
   b) Find the maximum profit.

106. Profit The A. B. Bronson Company earns a weekly profit according to the function $f(x) = -1.2x^2 + 180x - 280$, where x is the number of rocking chairs built and sold.

   a) Find the number of rocking chairs that the company must sell in a week to obtain the maximum profit.
   b) Find the maximum profit.

107. Firing a Cannon If a certain cannon is fired from a height of 9.8 meters above the ground, at a certain angle, the height of the cannonball above the ground, h, in meters, at time, t, in seconds, is found by the function

$$h(t) = -4.9t^2 + 24.5t + 9.8$$

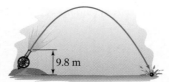

9.8 m

   a) Find the maximum height attained by the cannonball.
   b) Find the time it takes for the cannonball to reach its maximum height.
   c) Find the time it takes for the cannonball to strike the ground.

108. Throwing a Ball Ramon Loomis throws a ball into the air with an initial velocity of 32 feet per second. The height of the ball at any time t is given by the formula $h = 96t - 16t^2$. At what time does the ball reach its maximum height? What is the maximum height?

109. Rent The following graph shows the average monthly rent for an apartment in Maricopa County, Arizona (apartment complexes with 50 or more units), for the years 1994 to 2003.

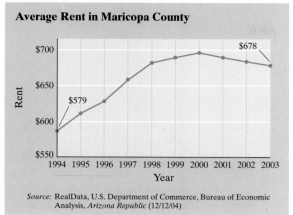

Average Rent in Maricopa County

Source: RealData, U.S. Department of Commerce, Bureau of Economic Analysis, *Arizona Republic* (12/12/04)

The function $r(t) = -2.723t^2 + 35.273t + 579$ can be used to estimate the average monthly rent for an apartment in Maricopa County, where t is the number of years since 1994.

a) If use assume the trend continues, estimate the average monthly rent for an apartment in Maricopa County in 2007.

b) In what year was the average monthly rent for an apartment a maximum?

110. Canadian Dollar The following graph shows the value of a Canadian dollar in U.S. dollars on February 22 of each year for the years from 2000 to 2005.

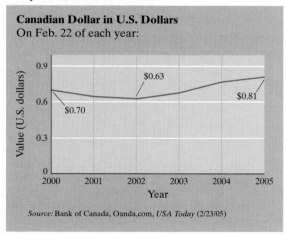

Canadian Dollar in U.S. Dollars
On Feb. 22 of each year:

Source: Bank of Canada, Oanda.com, *USA Today* (2/23/05)

The function $C(t) = 0.019t^2 - 0.074t + 0.702$ can be used to estimate the value of a Canadian dollar in U.S. dollars on February 22 of each year, where t is the number of years since 2000.

a) If we assume the trend continues, estimate the value of a Canadian dollar, in U.S. dollars, on February 22 in 2008.

b) In what year was the value of the Canadian dollar, in U.S. dollars, on February 22 at a minimum?

111. Room in a House Jake Kishner is designing plans for his house. What is the maximum possible area of a room if its perimeter is to be 80 feet?

112. Greatest Area What are the dimensions of a garden that will have the greatest area if its perimeter is to be 70 feet?

113. Minimum Product What is the minimum product of two numbers that differ by 8? What are the numbers?

114. Minimum Product What is the minimum product of two numbers that differ by 10? What are the numbers?

115. Maximum Product What is the maximum product of two numbers that add to 60? What are the numbers?

116. Maximum Product What is the maximum product of two numbers that add to 5? What are the numbers?

The profit of a company, in dollars, is the difference between the company's revenue and cost. Exercises 117 and 118 give cost, C(x), and revenue, R(x), functions for a particular company. The x represents the number of items produced and sold to distributors. Determine **a)** *the maximum profit of the company and* **b)** *the number of items that must be produced and sold to obtain the maximum profit.*

117. $C(x) = 2000 + 40x$
$R(x) = 800x - x^2$

118. $C(x) = 5000 + 12x$
$R(x) = 2000x - x^2$

Challenge Problems

*In Exercises 119 and 120, determine which of the graphs—**a)**, **b)**, or **c)**—is the graph of the given equation. Explain how you determined your answer.*

119. $y = -x^4 + 3x^3 - 5$

a)

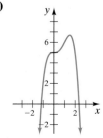

b)

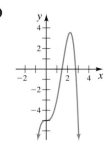

c)

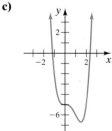

120. $y = 2x^4 + 9x^2 - 5$

a)

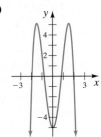

b)

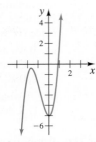

c)

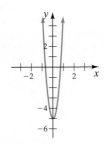

121. Baseball In Example 4 of this section, we used the function $f(t) = -16t^2 + 52t + 3$ to find that the maximum height, f, attained by a baseball hit by Tommy Magee was 45.25 feet. The ball reached this height at 1.625 seconds after the baseball was hit.
Review Example 4 now.

a) Write $f(t)$ in the form $f(t) = a(t - h)^2 + k$ by completing the square.

b) Using the function you obtained in part **a)**, determine the maximum height attained by the baseball, and the time after it was hit that the baseball attained its maximum value.

c) Are the answers you obtained in part **b)** the same answers obtained in Example 4? If not, explain why not.

Group Activity

Discuss and answer Exercises 122–124 as a group.

122. If the leading term of a polynomial function is $3x^3$, which of the following could possibly be the graph of the polynomial? Explain. Consider what happens for large positive values of x and for large negative values of x.

a)

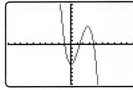

b)

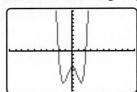

c)

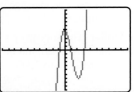

123. If the leading term of a polynomial is $-2x^4$, which of the following could possibly be the graph of the polynomial? Explain.

a)

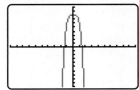

b)

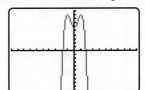

c)

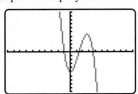

Discuss and answer Exercise 124 as a group.

124. a) Group member 1: Write two quadratic functions $f(x)$ and $g(x)$ so that the functions will not intersect.

b) Group member 2: Write two quadratic functions $f(x)$ and $g(x)$ so that neither function will have x-intercepts, and the vertices of the functions are on opposite sides of the x-axis.

c) Group member 3: Write two quadratic functions $f(x)$ and $g(x)$ so that both functions have the same vertex but one function opens upward and the other opens downward.

d) As a group, review each answer in parts **a)**–**c)** and decide whether each answer is correct. Correct any answer that is incorrect.

Cumulative Review Exercises

[3.3] **125.** Find the area shaded blue in the figure.

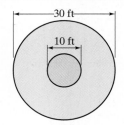

[7.2] **126.** Divide $(x - 3) \div \dfrac{x^2 + 3x - 18}{x}$.

[9.4] **127.** Solve the system of equations.
$$x - y = -5$$
$$2x + 2y - z = 0$$
$$x + y + z = 3$$

[9.7] **128.** Evaluate the determinant.
$$\begin{vmatrix} \dfrac{1}{2} & 3 \\ 2 & -4 \end{vmatrix}$$

[10.3] **129.** Graph $y \le \dfrac{2}{3}x + 3$.

12.6 Quadratic and Other Inequalities in One Variable

1 Solve quadratic inequalities.

2 Solve other polynomial inequalities.

3 Solve rational inequalities.

In Section 10.1, we discussed linear inequalities in one variable. Now we discuss quadratic inequalities in one variable.

When the equal sign in a quadratic equation of the form $ax^2 + bx + c = 0$ is replaced by an inequality sign, we get a **quadratic inequality**.

Examples of Quadratic Inequalities

$$x^2 + x - 12 > 0, \qquad 2x^2 - 9x - 5 \le 0$$

The **solution to a quadratic inequality** is the set of all values that make the inequality a true statement. For example, if we substitute 5 for x in $x^2 + x - 12 > 0$, we obtain

$$x^2 + x - 12 > 0$$
$$5^2 + 5 - 12 \overset{?}{>} 0$$
$$18 > 0 \qquad \textit{True}$$

The inequality is true when x is 5, so 5 satisfies the inequality. However, 5 is not the only solution. There are other values that satisfy (or are solutions to) the inequality. Does 4 satisfy the inequality? Does 2 satisfy the inequality?

1 Solve Quadratic Inequalities

A number of methods can be used to find the solutions to quadratic inequalities. We will begin by introducing a **sign graph**. Consider the function $f(x) = x^2 + x - 12$. Its graph is shown in **Figure 12.29a**. **Figure 12.29b** shows, in red, that when $x < -4$ or $x > 3, f(x) > 0$ or $x^2 + x - 12 > 0$. It also shows, in green, that when $-4 < x < 3, f(x) < 0$ or $x^2 + x - 12 < 0$.

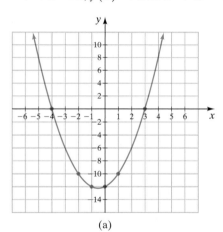   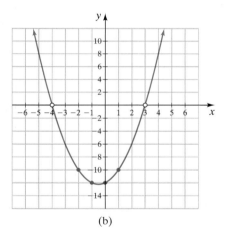

FIGURE 12.29   (a)   (b)

One way of finding the solutions to inequalities is to draw the graph and determine from the graph which values of the variable satisfy the inequality, as we just did. In many cases, it may be inconvenient or take too much time to draw the graph of a function, so we provide an alternate method to solve quadratic and other inequalities.

In Example 1, we will show how $x^2 + x - 12 > 0$ is solved using a number line. Then we will outline the procedure we used.

EXAMPLE 1 ▶ Solve the inequality $x^2 + x - 12 > 0$. Give the solution **a)** on a number line, **b)** in interval notation, and **c)** in set builder notation.

Solution   Set the inequality equal to 0 and solve the equation.

$$x^2 + x - 12 = 0$$
$$(x + 4)(x - 3) = 0$$
$$x + 4 = 0 \quad \text{or} \quad x - 3 = 0$$
$$x = -4 \qquad\qquad x = 3$$

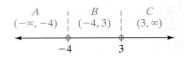

FIGURE 12.30

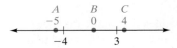

FIGURE 12.31

FIGURE 12.32

The numbers obtained are called **boundary values**. The boundary values are used to break a number line up into intervals. If the original inequality is $<$ or $>$, the boundary values are not part of the intervals. If the original inequality is $\leq$ or $\geq$, the boundary values are part of the intervals.

In **Figure 12.30**, we have labeled the intervals A, B, and C. Next, we select one test value in *each* interval. Then we substitute each of those numbers, one at a time, into either $x^2 + x - 12 > 0$ or $(x + 4)(x - 3) > 0$ and determine whether they result in a true statement. If the test value results in a true statement, all values in that interval will also satisfy the inequality. If the test value results in a false statement, no numbers in that interval will satisfy the inequality.

In this example, we will use the test values of -5 in interval A, 0 in interval B, and 4 in interval C (see **Fig. 12.31**).

Interval A	Interval B	Interval C
$(-\infty, -4)$	$(-4, 3)$	$(3, \infty)$
Test value, -5	Test value, 0	Test value, 4
Is $x^2 + x - 12 > 0$?	Is $x^2 + x - 12 > 0$?	Is $x^2 + x - 12 > 0$?
$(-5)^2 - 5 - 12 \overset{?}{>} 0$	$0^2 + 0 - 12 \overset{?}{>} 0$	$4^2 + 4 - 12 \overset{?}{>} 0$
$8 > 0$	$-12 > 0$	$8 > 0$
True	*False*	*True*

Since the test values in both intervals A and C satisfy the inequality, the solution is all real numbers in intervals A or C. Since the inequality symbol is $>$, the values -4 and 3 are not included in the solution because they make the inequality equal to 0.

The answer to parts **a)**, **b)**, and **c)** follow.

a) The solution is illustrated on a number line in **Figure 12.32**.

b) The solution in interval notation is $(-\infty, -4) \cup (3, \infty)$.

c) The solution in set builder notation is $\{x | x < -4 \text{ or } x > 3\}$.

Note that the solution, in any form, is consistent with the graph in **Figure 12.29b**.

▶ **Now Try Exercise 15**

To Solve Quadratic and Other Inequalities

1. Write the inequality as an equation and solve the equation.

2. If solving a rational inequality, determine the values that make any denominator 0.

3. Construct a number line. Mark each solution from step 1 and numbers obtained in step 2 on the number line. Mark the lowest value on the left, with values increasing from left to right.

4. Select a test value in each interval and determine whether it satisfies the inequality. Also test each boundary value.

5. Write the solution in the form requested by your instructor.

EXAMPLE 2 ▶ Solve the inequality $x^2 - 4x \geq -4$. Give the solution **a)** on a number line, **b)** in interval notation, and **c)** in set builder notation.

Solution   Write the inequality as an equation, then solve the equation.

$$x^2 - 4x = -4$$
$$x^2 - 4x + 4 = 0$$
$$(x - 2)(x - 2) = 0$$
$$x - 2 = 0 \quad \text{or} \quad x - 2 = 0$$
$$x = 2 \qquad\qquad x = 2$$

FIGURE 12.33

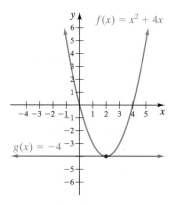

$f(x) = x^2 + 4x$

$g(x) = -4$

FIGURE 12.34

Since both factors are the same there is only one boundary value, 2 (see **Fig. 12.33**). Both test values, 1 and 3, result in true statements.

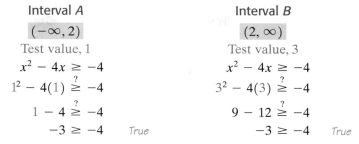

Interval A	Interval B
$(-\infty, 2)$	$(2, \infty)$
Test value, 1	Test value, 3
$x^2 - 4x \geq -4$	$x^2 - 4x \geq -4$
$1^2 - 4(1) \overset{?}{\geq} -4$	$3^2 - 4(3) \overset{?}{\geq} -4$
$1 - 4 \overset{?}{\geq} -4$	$9 - 12 \overset{?}{\geq} -4$
$-3 \geq -4$ *True*	$-3 \geq -4$ *True*

The solution set includes both intervals and the boundary value, 2. The solution set is the set of real numbers, $\mathbb{R}$. The answer to parts **a)**, **b)**, and **c)** follow.

a) (number line through 0)   **b)** $(-\infty, \infty)$ **c)** $\{x | -\infty < x < \infty\}$

▶ **Now Try Exercise 11**

We can check the solution to Example 2 using graphing. Let $f(x) = x^2 - 4x$ and $g(x) = -4$. For $x^2 - 4x \geq -4$ to be true, we want $f(x) \geq g(x)$. The graphs of $f(x)$ and $g(x)$ are given in **Figure 12.34**.

Observe that $f(x) = g(x)$ at $x = 2$ and $f(x) > g(x)$ for all other values of x. Thus, $f(x) \geq g(x)$ for all values of x, and the solution set is the set of real numbers.

In Example 2, if we rewrite the inequality $x^2 - 4x \geq -4$ as $x^2 - 4x + 4 \geq 0$ and then as $(x - 2)^2 \geq 0$ we can see that the solution must be the set of real numbers, since $(x - 2)^2$ must be greater than or equal to 0 for any real number x. The solution to $x^2 - 4x < -4$ is the empty set, $\emptyset$. Can, you explain why?

EXAMPLE 3 ▶ Solve the inequality $x^2 - 2x - 4 \leq 0$. Express the solution in interval notation.

Solution   First we need to solve the equation $x^2 - 2x - 4 = 0$. Since this equation is not factorable, we use the quadratic formula to solve.

$$x = \frac{-b \pm \sqrt{b^2 - 4ac}}{2a}$$

$$= \frac{2 \pm \sqrt{4 - 4(1)(-4)}}{2(1)} = \frac{2 \pm \sqrt{20}}{2} = \frac{2 \pm 2\sqrt{5}}{2} = 1 \pm \sqrt{5}$$

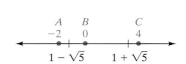

FIGURE 12.35

The boundary values are $1 - \sqrt{5}$ and $1 + \sqrt{5}$. The value of $1 - \sqrt{5}$ is about -1.24 and the value of $1 + \sqrt{5}$ is about 3.24. We will select test values of $-2, 0,$ and 4 (see **Fig. 12.35**).

Interval A	Interval B	Interval C
$(-\infty, 1 - \sqrt{5})$	$(1 - \sqrt{5}, 1 + \sqrt{5})$	$(1 + \sqrt{5}, \infty)$
Test value, -2	Test value, 0	Test value, 4
$x^2 - 2x - 4 \leq 0$	$x^2 - 2x - 4 \leq 0$	$x^2 - 2x - 4 \leq 0$
$(-2)^2 - 2(-2) - 4 \overset{?}{\leq} 0$	$0^2 - 2(0) - 4 \overset{?}{\leq} 0$	$4^2 - 2(4) - 4 \overset{?}{\leq} 0$
$4 + 4 - 4 \overset{?}{\leq} 0$	$0 - 0 - 4 \overset{?}{\leq} 0$	$16 - 8 - 4 \overset{?}{\leq} 0$
$4 \leq 0$	$-4 \leq 0$	$4 \leq 0$
False	*True*	*False*

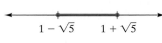

FIGURE 12.36

The boundary values are part of the solution because the inequality symbol is $\leq$ and the boundary values make the inequality equal to 0. Thus, the solution in interval notation is $[1 - \sqrt{5}, 1 + \sqrt{5}]$. The solution is illustrated on the number line in **Figure 12.36**.

▶ **Now Try Exercise 19**

Helpful Hint

If $ax^2 + bx + c = 0$, with $a > 0$, has two distinct real solutions, then:

Inequality of Form	Solution Is	Solution on Number Line
$ax^2 + bx + c \geq 0$	End intervals	←——• •——→
$ax^2 + bx + c \leq 0$	Center interval	←——•——•——→

Example 1 is an inequality of the form $ax^2 + bx + c > 0$, and Example 3 is an inequality of the form $ax^2 + bx + c \leq 0$. Example 2 does not have two distinct real solutions, so this Helpful Hint does not apply.

2 Solve Other Polynomial Inequalities

A procedure similar to the one used earlier can be used to solve other **polynomial inequalities**, as illustrated in the following examples.

EXAMPLE 4 ▶ Solve the inequality $(3x - 2)(x + 3)(x + 5) < 0$. Illustrate the solution on a number line and write the solution in both interval notation and set builder notation.

Solution We use the zero-factor property to solve the equation $(3x - 2)(x + 3)(x + 5) = 0$.

$$3x - 2 = 0 \qquad \text{or} \qquad x + 3 = 0 \qquad \text{or} \qquad x + 5 = 0$$

$$x = \frac{2}{3} \qquad\qquad x = -3 \qquad\qquad x = -5$$

The solutions -5, -3, and $\frac{2}{3}$ break the number line into four intervals (see **Fig. 12.37**). The test values we will use are -6, -4, 0, and 1. We show the results in the following table.

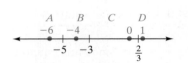

FIGURE 12.37

Interval	Test value	$(3x - 2)(x + 3)(x + 5)$	< 0
$A: (-\infty, -5)$	-6	-60	*True*
$B: (-5, -3)$	-4	14	*False*
$C: \left(-3, \dfrac{2}{3}\right)$	0	-30	*True*
$D: \left(\dfrac{2}{3}, \infty\right)$	1	24	*False*

Since the original inequality symbol is $<$, the boundary values are not part of the solution. The solution, intervals A and C, is illustrated on the number line in **Figure 12.38**.

The solution in set builder notation is $\left\{ x \middle| x < -5 \text{ or } -3 < x < \dfrac{2}{3} \right\}$.

FIGURE 12.38

The solution in interval notation is $(-\infty, -5) \cup \left(-3, \dfrac{2}{3}\right)$.

▶ **Now Try Exercise 27**

EXAMPLE 5 ▶ Given $f(x) = 3x^3 - 3x^2 - 6x$, find all values of x for which $f(x) \geq 0$. Illustrate the solution on a number line and give the solution in interval notation.

Solution   We need to solve the inequality

$$3x^3 - 3x^2 - 6x \geq 0$$

We start by solving the equation $3x^3 - 3x^2 - 6x = 0$.

$$3x(x^2 - x - 2) = 0$$

$$3x(x - 2)(x + 1) = 0$$

$$3x = 0 \quad \text{or} \quad x - 2 = 0 \quad \text{or} \quad x + 1 = 0$$

$$x = 0 \qquad\qquad x = 2 \qquad\qquad x = -1$$

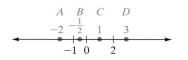

FIGURE 12.39

The solutions $-1, 0$, and 2 break the number line into four intervals (see **Fig. 12.39**). The test values that we will use are $-2, -\dfrac{1}{2}, 1$, and 3.

Interval	Test Value	$3x^3 - 3x^2 - 6x$	≥ 0
$A: (-\infty, 1)$	-2	-24	*False*
$B: (-1, 0)$	$-\dfrac{1}{2}$	$\dfrac{15}{8}$	*True*
$C: (0, 2)$	1	-6	*False*
$D: (2, \infty)$	3	36	*True*

Since the original inequality is $\geq$, the boundary values are part of the solution. The solution, intervals B and D, is illustrated on the number line in **Figure 12.40a**. The solution in interval notation is $[-1, 0] \cup [2, \infty)$. **Figure 12.40b** shows the graph of $f(x) = 3x^3 - 3x^2 - 6x$. Notice $f(x) \geq 0$ for $-1 \leq x \leq 0$ and for $x \geq 2$, which agrees with our solution.

FIGURE 12.40a

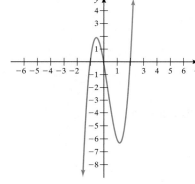

FIGURE 12.40b

▶ **Now Try Exercise 41**

In all the examples we have solved, the coefficient of the leading term has been a positive number.

Consider the inequality $-3x^3 + 3x^2 + 6x \leq 0$. Notice the coefficient of the leading term, $-3x^3$, is a negative number, -3. It is generally easier to solve an inequality where the coefficient of the leading term is a positive number. We can make the leading coefficient positive by multiplying both sides of the inequality by -1. When doing this, remember to reverse the inequality symbol.

$$-3x^3 + 3x^2 + 6x \leq 0$$

$$-1(-3x^3 + 3x^2 + 6x) \geq -1(0) \qquad \textit{Reverse inequality symbol.}$$

$$3x^3 - 3x^2 - 6x \geq 0$$

This inequality was solved in Example 5.

3 Solve Rational Inequalities

In Examples 6 and 7, we solve **rational inequalities**, which are inequalities that contain a rational expression.

EXAMPLE 6 ▶ Solve the inequality $\dfrac{x-1}{x+3} \geq 2$ and graph the solution on a number line.

Solution   Change the $\geq$ to $=$ and solve the resulting equation.

$$\frac{x-1}{x+3} = 2$$

$$\cancel{x+3} \cdot \frac{x-1}{\cancel{x+3}} = 2(x+3) \qquad \textit{Multiply both sides by x + 3.}$$

$$x - 1 = 2x + 6$$

$$-1 = x + 6$$

$$-7 = x$$

When solving rational inequalities, we also need to determine the value or values that make the denominator 0. We set the denominator equal to 0 and solve.

$$x + 3 = 0$$

$$x = -3$$

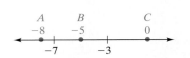

FIGURE 12.41

We use the solution to the equation, -7, and the value that makes the denominator 0, -3, to determine the intervals, shown in **Figure 12.41**. We will use -8, -5, and 0 as our test values.

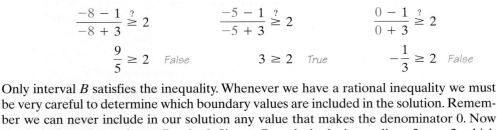

Interval A	Interval B	Interval C
$(-\infty, -7)$	$(-7, -3)$	$(-3, \infty)$
Test value, -8	Test value, -5	Test value, 0
$\dfrac{x-1}{x+3} \geq 2$	$\dfrac{x-1}{x+3} \geq 2$	$\dfrac{x-1}{x+3} \geq 2$
$\dfrac{-8-1}{-8+3} \overset{?}{\geq} 2$	$\dfrac{-5-1}{-5+3} \overset{?}{\geq} 2$	$\dfrac{0-1}{0+3} \overset{?}{\geq} 2$
$\dfrac{9}{5} \geq 2$ *False*	$3 \geq 2$ *True*	$-\dfrac{1}{3} \geq 2$ *False*

FIGURE 12.42

Only interval B satisfies the inequality. Whenever we have a rational inequality we must be very careful to determine which boundary values are included in the solution. Remember we can never include in our solution any value that makes the denominator 0. Now check the boundary values -7 and -3. Since -7 results in the inequality $-2 \geq -2$, which is true, -7 is a solution. Since division by 0 is not permitted, -3 is not a solution. Thus, the solution is $[-7, -3)$. The solution is illustrated on the number line in **Figure 12.42**.

▶ **Now Try Exercise 81**

In Example 6, we solved $\dfrac{x-1}{x+3} \geq 2$. Suppose we graphed $f(x) = \dfrac{x-1}{x+3}$.

For what values of x would $f(x) \geq 2$? If you answered $-7 \leq x < -3$ you answered correctly. **Figure 12.43** shows the graph of $f(x) = \dfrac{x-1}{x+3}$ and the graph of $y = 2$. Notice $f(x) \geq 2$ when $-7 \leq x < -3$.

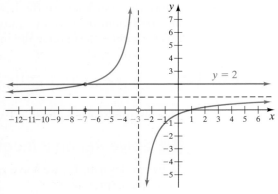

FIGURE 12.43

EXAMPLE 7 ▸ Solve the inequality $\dfrac{(x-3)(x+4)}{x+1} \geq 0$. Graph the solution on a number line and give the solution in interval notation.

Solution The solutions to the equation $\dfrac{(x-3)(x+4)}{x+1} = 0$ are 3 and -4 since these are the values that make the numerator equal to 0. The equation is not defined at -1. We therefore use the values 3, -4, and -1 to determine the intervals on the number line (see **Fig. 12.44**). Checking test values at -5, -2, 0, and 4, we find that the values in intervals B and D, $-4 < x < -1$ and $x > 3$, satisfy the inequality. Check the test values yourself to verify this. The values 3 and -4 make the inequality equal to 0 and are part of the solution. The inequality is not defined at -1, so -1 is not part of the solution. The solution is $[-4, -1) \cup [3, \infty)$. The solution is illustrated on the number line in **Figure 12.45**.

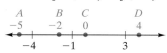

FIGURE 12.44   FIGURE 12.45

▸ **Now Try Exercise 71**

EXERCISE SET 12.6   *Math XL*   **MyMathLab**
MathXL®   MyMathLab

Concept/Writing Exercises

1. The graph of $f(x) = x^2 - 7x + 10$ is given. Find the solution to **a)** $f(x) > 0$ and **b)** $f(x) < 0$.

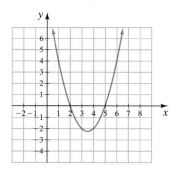

2. The graph of $f(x) = -x^2 - 4x + 5$ is given. Find the solution to **a)** $f(x) \geq 0$ and **b)** $f(x) \leq 0$.

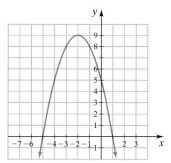

3. In solving the inequality $(x - 5)(x + 3) \geq 0$, are the boundary values 5 and -3 included in the solution set? Explain.

4. In solving the inequality $(x - 2)(x + 4) < 0$, are the boundary values 2 and -4 included in the solution set? Explain.

5. In solving the inequality $\dfrac{(x+2)(x-1)}{x+1} \leq 0$, are the boundary values -2 and 1 included in the solution set? Is the boundary value -1 included in the solution set? Explain.

6. In solving the inequality $\dfrac{(x+3)}{(x+4)(x-2)} \geq 0$, is the boundary value -3 included in the solution set? Are the boundary values -4 and 2 included in the solution set? Explain.

Practice the Skills

Solve each inequality and graph the solution on a number line.

7. $x^2 - 2x - 8 \geq 0$

8. $x^2 - 2x - 8 < 0$

9. $x^2 + 7x + 6 > 0$

10. $x^2 + 8x + 7 < 0$

11. $n^2 - 6n + 9 \geq 0$

12. $x^2 - 8x \geq 0$

13. $x^2 - 16 < 0$

14. $r^2 - 5r < 0$

15. $2x^2 + 5x - 3 \geq 0$

16. $3n^2 - 7n \leq 6$

17. $5x^2 + 6x \leq 8$

18. $3x^2 + 5x - 3 \leq 0$

19. $2x^2 - 12x + 9 \leq 0$

20. $5x^2 \leq -20x - 4$

Solve each inequality and give the solution in interval notation.

21. $(x - 2)(x + 1)(x + 5) \geq 0$

22. $(x - 2)(x + 2)(x + 5) \leq 0$

23. $(a - 3)(a + 2)(a + 4) < 0$

24. $(r - 1)(r + 2)(r + 7) < 0$

25. $(2c + 5)(3c - 6)(c + 6) > 0$

26. $(a - 4)(a - 2)(a + 8) > 0$

27. $(3x + 5)(x - 3)(x + 1) > 0$

28. $(3c - 1)(c + 4)(3c + 6) \leq 0$

29. $(x + 2)(x + 2)(3x - 8) \geq 0$

30. $(x + 3)^2(4x - 7) \leq 0$

31. $x^3 - 6x^2 + 9x < 0$

32. $x^3 + 3x^2 - 40x > 0$

For each function provided, find all values of x for which $f(x)$ satisfies the indicated conditions. Graph the solution on a number line.

33. $f(x) = x^2 - 6x, f(x) \geq 0$

34. $f(x) = x^2 - 7x, f(x) > 0$

35. $f(x) = x^2 + 4x, f(x) > 0$

36. $f(x) = x^2 + 8x, f(x) \leq 0$

37. $f(x) = x^2 - 14x + 48, f(x) < 0$

38. $f(x) = x^2 - 2x - 15, f(x) < 0$

39. $f(x) = 2x^2 + 9x - 1, f(x) \leq 5$

40. $f(x) = x^2 + 5x - 3, f(x) \leq 4$

41. $f(x) = 2x^3 + 9x^2 - 35x, f(x) \geq 0$

42. $f(x) = x^3 - 9x, f(x) \leq 0$

Solve each inequality and give the solution in set builder notation.

43. $\dfrac{x + 2}{x - 4} > 0$

44. $\dfrac{x + 2}{x - 4} \geq 0$

45. $\dfrac{x - 1}{x + 5} < 0$

46. $\dfrac{x - 1}{x + 5} \leq 0$

47. $\dfrac{x + 3}{x - 2} \geq 0$

48. $\dfrac{x - 4}{x + 6} > 0$

49. $\dfrac{a - 9}{a + 5} < 0$

50. $\dfrac{b + 7}{b + 1} \leq 0$

51. $\dfrac{c - 10}{c - 4} > 0$

52. $\dfrac{2d - 6}{d - 1} < 0$

53. $\dfrac{3y + 6}{y + 4} \leq 0$

54. $\dfrac{4z - 8}{z - 9} \geq 0$

55. $\dfrac{5a + 10}{3a - 1} \geq 0$

56. $\dfrac{x + 4}{x - 4} \leq 0$

57. $\dfrac{3x + 4}{2x - 1} < 0$

58. $\dfrac{k + 3}{k} \geq 0$

59. $\dfrac{3x + 8}{x - 2} \leq 0$

60. $\dfrac{4x - 2}{2x - 8} > 0$

Solve each inequality and give the solution in interval notation.

61. $\dfrac{(x + 1)(x - 6)}{x + 3} < 0$

62. $\dfrac{(x + 1)(x - 6)}{x + 3} \leq 0$

63. $\dfrac{(x - 2)(x + 3)}{x - 5} > 0$

64. $\dfrac{(x - 2)(x + 3)}{x - 5} \geq 0$

65. $\dfrac{(a - 1)(a - 7)}{a + 2} \geq 0$

66. $\dfrac{(b - 2)(b + 4)}{b} < 0$

67. $\dfrac{c}{(c - 3)(c + 8)} \leq 0$

68. $\dfrac{z - 5}{(z + 6)(z - 9)} \geq 0$

69. $\dfrac{x - 6}{(x + 4)(x - 1)} \leq 0$

70. $\dfrac{x + 9}{(x - 2)(x + 4)} > 0$

71. $\dfrac{(x - 3)(2x + 5)}{x - 4} \geq 0$

72. $\dfrac{r(r - 8)}{2r + 6} < 0$

Solve each inequality and graph the solution on a number line.

73. $\dfrac{2}{x - 4} \geq 1$

74. $\dfrac{2}{x - 4} > 1$

75. $\dfrac{3}{x - 1} > -1$

76. $\dfrac{3}{x + 1} \geq -1$

77. $\dfrac{5}{x + 2} \leq 1$

78. $\dfrac{5}{x + 2} < 1$

79. $\dfrac{2p - 5}{p - 4} \leq 1$

80. $\dfrac{2}{2a - 1} > 2$

81. $\dfrac{4}{x + 2} \geq 2$

82. $\dfrac{x + 6}{x + 2} > 1$

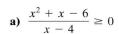 **83.** $\dfrac{w}{3w - 2} > -2$

84. $\dfrac{x - 1}{2x + 6} \leq -3$

85. The graph of $y = \dfrac{x^2 - 4x + 4}{x - 4}$ is illustrated. Determine the solution to the following inequalities.

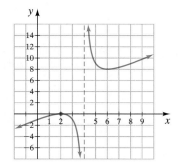

a) $\dfrac{x^2 - 4x + 4}{x - 4} > 0$

b) $\dfrac{x^2 - 4x + 4}{x - 4} < 0$

Explain how you determined your answer.

86. The graph of $y = \dfrac{x^2 + x - 6}{x - 4}$ is illustrated. Determine the solution to the following inequalities.

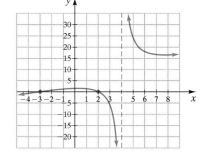

a) $\dfrac{x^2 + x - 6}{x - 4} \geq 0$

b) $\dfrac{x^2 + x - 6}{x - 4} < 0$

Explain how you determined your answer.

87. Write a quadratic inequality whose solution is

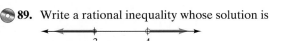

88. Write a quadratic inequality whose solution is

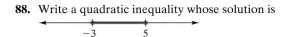

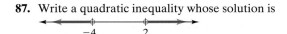 **89.** Write a rational inequality whose solution is

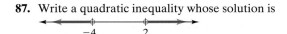

90. Write a rational inequality whose solution is

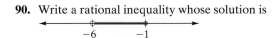

91. What is the solution to the inequality $(x + 3)^2(x - 1)^2 \geq 0$? Explain your answer.

92. What is the solution to the inequality $x^2(x - 3)^2(x + 4)^2 < 0$? Explain your answer.

93. What is the solution to the inequality $\dfrac{x^2}{(x + 2)^2} \geq 0$? Explain your answer.

94. What is the solution to the inequality $\dfrac{x^2}{(x - 3)^2} > 0$? Explain your answer.

95. If $f(x) = ax^2 + bx + c$ where $a > 0$ and the discriminant is negative, what is the solution to $f(x) < 0$? Explain.

96. If $f(x) = ax^2 + bx + c$ where $a < 0$ and the discriminant is negative, what is the solution to $f(x) > 2$? Explain.

Challenge Problems

Solve each inequality and graph the solution on the number line.

97. $(x + 1)(x - 3)(x + 5)(x + 8) \geq 0$

98. $\dfrac{(x - 4)(x + 2)}{x(x + 9)} \geq 0$

Write a quadratic inequality with the following solutions. Many answers are possible. Explain how you determined your answers.

99. $(-\infty, 0) \cup (3, \infty)$

100. $\{2\}$

101. $\varnothing$

102. $\mathbb{R}$

In Exercises 103 and 104, solve each inequality and give the solution in interval notation. Use techniques from Section 8.5 to help you find the solution.

103. $x^4 - 10x^2 + 9 > 0$

104. $x^4 - 26x^2 + 25 \le 0$

In Exercises 105 and 106, solve each inequality using factoring by grouping. Give the solution in interval notation.

105. $x^3 + x^2 - 4x - 4 \ge 0$

106. $2x^3 + x^2 - 32x - 16 < 0$

Group Activity

Discuss and answer Exercises 107 and 108 as a group.

107. Consider the number line below, where $a, b,$ and c are distinct real numbers.

   a) In which intervals will the real numbers satisfy the inequality $(x - a)(x - b)(x - c) > 0$? Explain.

   b) In which intervals will the real numbers satisfy the inequality $(x - a)(x - b)(x - c) < 0$? Explain.

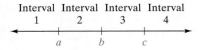

Interval Interval Interval Interval
  1   2   3   4

     a b c

108. Consider the number line below where $a, b, c,$ and d are distinct real numbers.

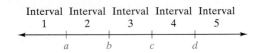

Interval Interval Interval Interval Interval
  1   2   3   4   5

     a b c d

   a) In which intervals do the real numbers satisfy the inequality $(x - a)(x - b)(x - c)(x - d) > 0$? Explain.

   b) In which interval do the real numbers satisfy the inequality $(x - a)(x - b)(x - c)(x - d) < 0$? Explain.

Cumulative Review Exercises

[3.4] **109.** **Antifreeze** How many quarts of a 100% antifreeze solution should Paul Simmons add to 10 quarts of a 20% antifreeze solution to make a 50% antifreeze solution?

[5.4] **110.** Add $(6r + 5s - t) + (-3r - 2s - 8t)$.

[7.5] **111.** Simplify $\dfrac{1 + \dfrac{x}{x + 1}}{\dfrac{2x + 1}{x - 3}}$.

[8.2] **112.** If $h(x) = \dfrac{x^2 + 4x}{x + 9}$, find $h(-3)$.

[11.7] **113.** Multiply $(3 - 4i)(6 + 5i)$.

Chapter 12 Summary

IMPORTANT FACTS AND CONCEPTS	EXAMPLES
Section 12.1	

Square Root Property If $x^2 = a$, where a is a real number, then $x = \pm\sqrt{a}$.	Solve $x^2 - 36 = 0$. $$x^2 - 36 = 0$$ $$x^2 = 36$$ $$x = \pm\sqrt{36} = \pm 6$$ The solutions are -6 and 6.
A **perfect square trinomial** is a trinomial that can be expressed as a square of a binomial.	$x^2 - 10x + 25 = (x - 5)^2$

IMPORTANT FACTS AND CONCEPTS	EXAMPLES

Section 12.1 (continued)

To Solve a Quadratic Equation by Completing the Square

1. Use the multiplication (or division) property of equality if necessary to make the leading coefficient 1.
2. Rewrite the equation with the constant by itself on the right side of the equation.
3. Take one-half the numerical coefficient of the first-degree term, square it, and add this quantity to both sides of the equation.
4. Replace the trinomial with the square of a binomial.
5. Use the square root property to take the square root of both sides of the equation.
6. Solve for the variable.
7. Check your solutions in the *original* equation.

Solve $x^2 + 4x - 12 = 0$ by completing the square.

$$x^2 + 4x - 12 = 0$$
$$x^2 + 4x = 12$$
$$x^2 + 4x + 4 = 12 + 4$$
$$(x + 2)^2 = 16$$
$$x + 2 = \pm\sqrt{16}$$
$$x + 2 = \pm 4$$
$$x = -2 \pm 4$$
$$x = -2 - 4 = -6 \quad \text{or} \quad x = -2 + 4 = 2$$

The solutions are -6 and 2.

Section 12.2

The **standard form of a quadratic equation** is $ax^2 + bx + c = 0, a \neq 0$.

$$x^2 - 5x + 17 = 0$$

To Solve a Quadratic Equation by the Quadratic Formula

1. Write the quadratic equation in standard form, $ax^2 + bx + c = 0$, and determine the numerical values for a, b, and c.
2. Substitute the values for a, b, and c into the quadratic formula and then evaluate the formula to obtain the solution.

The Quadratic Formula
$$x = \frac{-b \pm \sqrt{b^2 - 4ac}}{2a}$$

Solve $x^2 - 2x - 15 = 0$ by the quadratic formula.

$$a = 1, \quad b = -2, \quad c = -15$$
$$x = \frac{-b \pm \sqrt{b^2 - 4ac}}{2a}$$
$$= \frac{-(-2) \pm \sqrt{(-2)^2 - 4(1)(-15)}}{2(1)}$$
$$= \frac{2 \pm \sqrt{64}}{2}$$
$$= \frac{2 \pm 8}{2}$$
$$x = \frac{2 + 8}{2} = \frac{10}{2} = 5 \quad \text{or} \quad x = \frac{2 - 8}{2} = \frac{-6}{2} = -3$$

The solutions are 5 and -3.

Solutions of a Quadratic Equation

For a quadratic equation of the form $ax^2 + bx + c = 0, a \neq 0$, the **discriminant** is $b^2 - 4ac$.

If $b^2 - 4ac > 0$, the quadratic equation has two distinct real number solutions.

If $b^2 - 4ac = 0$, the quadratic equation has a single real number solution.

If $b^2 - 4ac < 0$, the quadratic equation has no real number solution.

Determine the number of solutions of $3x^2 - x + 7 = 0$.

$$a = 3, b = -1, c = 7$$
$$b^2 - 4ac = (-1)^2 - 4(3)(7)$$
$$= 1 - 84$$
$$= -83$$

Since the discriminant is negative, the equation has no real number solution.

IMPORTANT FACTS AND CONCEPTS	EXAMPLES

Section 12.4

An equation that can be written in the form $au^2 + bu + c = 0$ for $a \neq 0$, where u is an algebraic expression, is called **quadratic in form**.

To Solve Equations Quadratic in Form

1. Make a substitution that will result in an equation of the form $au^2 + bu + c = 0$, $a \neq 0$, where u is a function of the original variable.

2. Solve the equation $au^2 + bu + c = 0$ for u.

3. Replace u with the function of the original variable from step 1 and solve the resulting equation for the original variable.

4. Check for extraneous solutions by substituting the apparent solutions into the original equation.

Solve $x^4 - 17x^2 + 16 = 0$.

$$\text{Let } u = x^2.$$

Then,
$$u^2 - 17u + 16 = 0$$
$$(u - 16)(u - 1) = 0$$
$$u - 16 = 0 \quad \text{or} \quad u - 1 = 0$$
$$u = 16 \qquad\qquad u = 1$$
$$x^2 = 16 \qquad\qquad x^2 = 1$$
$$x = \pm 4 \qquad\qquad x = \pm 1$$

A check will show the solutions are 4, -4, 1, and -1.

Section 12.5

Parabola

Graphs of equations of the form $f(x) = ax^2 + bx + c$ are parabolas.

a) The parabola opens upward when $a > 0$ and downward when $a < 0$.

b) The axis of symmetry is the line $x = -\dfrac{b}{2a}$.

c) The vertex is the point $\left(-\dfrac{b}{2a}, \dfrac{4ac - b^2}{4a}\right)$ or $\left(-\dfrac{b}{2a}, f\left(-\dfrac{b}{2a}\right)\right)$.

d) The y-intercept is the point $(0, c)$.

e) To obtain the x-intercept(s), set $f(x) = 0$ and solve for x.

The graph of $f(x) = x^2 - 2x - 3$ is a parabola.

a) It opens upward since $a > 0$.

b) The axis of symmetry is $x = -\dfrac{-2}{2(1)} = 1$.

c) The vertex is $(1, -4)$.

d) The y-intercept is $(0, -3)$.

e)
$$x^2 - 2x - 3 = 0$$
$$(x - 3)(x + 1) = 0$$
$$x - 3 = 0 \quad \text{or} \quad x + 1 = 0$$
$$x = 3 \qquad\qquad x = -1$$

The x-intercepts are $(3, 0)$ and $(-1, 0)$.

The graph of $f(x) = x^2 - 2x - 3$ is shown below.

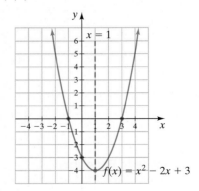

Section 12.6

A **quadratic inequality** is obtained when the equal sign in the quadratic equation $ax^2 + bx + c = 0$ is replaced by an inequality sign.

The **solution to a quadratic inequality** is the set of all values that make the inequality a true statement.

$$x^2 - 5x + 7 > 0$$

IMPORTANT FACTS AND CONCEPTS	EXAMPLES

Section 12.6 (continued)

To Solve Quadratic, Polynomial, and Rational Inequalities

1. Write the inequality as an equation and solve the equation.

2. If solving a rational inequality, determine the values that make any denominator 0.

3. Construct a number line. Mark each solution from step 1 and numbers obtained in step 2 on the number line.

4. Select a test value in each interval and determine whether it satisfies the inequality. Also test each boundary value.

5. Write the solution in the form requested by your instructor.

Solve $(2x - 1)(x - 3)(x + 1) < 0$.

$$(2x - 1)(x - 3)(x + 1) < 0$$
$$(2x - 1)(x - 3)(x + 1) = 0$$
$$2x - 1 = 0 \quad \text{or} \quad x - 3 = 0 \quad \text{or} \quad x + 1 = 0$$
$$x = \frac{1}{2} \qquad x = 3 \qquad x = -1$$

The intervals and test values selected are shown below.

Interval	Test value	$(2x - 1)(x - 3)(x + 1) < 0$
$(-\infty, -1)$	-2	-25
$\left(-1, \frac{1}{2}\right)$	0	3
$\left(\frac{1}{2}, 3\right)$	1	-4
$(3, \infty)$	5	108

The solution is $x < -1$ or $\frac{1}{2} < x < 3$.

Solution on a number line:

Solution in interval notation: $(-\infty, -1) \cup \left(\frac{1}{2}, 3\right)$

Solution in set builder notation:
$$\left\{ x \mid x < -1 \quad \text{or} \quad \frac{1}{2} < x < 3 \right\}$$

Chapter 12 Review Exercises

[12.1] *Use the square root property to solve each equation.*

1. $(x - 5)^2 = 24$ **2.** $(2x + 1)^2 = 60$ **3.** $\left(x - \frac{1}{3}\right)^2 = \frac{4}{9}$ **4.** $\left(2x - \frac{1}{2}\right)^2 = 4$

Solve each equation by completing the square.

5. $x^2 - 7x + 12 = 0$ **6.** $x^2 + 4x - 32 = 0$ **7.** $a^2 + 2a - 9 = 0$

8. $z^2 + 6z = 12$ **9.** $x^2 - 2x + 10 = 0$ **10.** $2r^2 - 8r = -64$

Area *In Exercises 11 and 12, the area, A, of each rectangle is given.* **a)** *Write an equation for the area.* **b)** *Solve the equation for x.*

11.
$A = 32$ | $x + 1$
$x + 5$

12.
$A = 63$ | $x + 2$
$x + 4$

13. Consecutive Integers The product of two consecutive positive integers is 42. Find the two integers.

14. Living Room Ronnie Sampson just moved into a new house where the living room is a square whose diagonal is 7 feet longer than the length of a side. Find the dimensions of the living room.

[12.2] *Determine whether each equation has two distinct real solutions, a single real solution, or no real solution.*

15. $2x^2 - 5x - 1 = 0$

16. $3x^2 + 2x = -6$

17. $r^2 + 16r = -64$

18. $5x^2 - x + 2 = 0$

19. $a^2 - 14a = -49$

20. $\frac{1}{2}x^2 - 3x = 8$

Solve each equation by the quadratic formula.

21. $3x^2 + 4x = 0$

22. $x^2 - 11x = -18$

23. $r^2 = 3r + 40$

24. $7x^2 = 9x$

25. $6a^2 + a - 15 = 0$

26. $4x^2 + 11x = 3$

27. $x^2 + 8x + 5 = 0$

28. $b^2 + 4b = 8$

29. $2x^2 + 4x - 3 = 0$

30. $3y^2 - 6y = 8$

31. $x^2 - x + 13 = 0$

32. $x^2 - 2x + 11 = 0$

33. $2x^2 - \frac{5}{3}x = \frac{25}{3}$

34. $4x^2 + 5x - \frac{3}{2} = 0$

For the given function, determine all real values of the variable for which the function has the value indicated.

35. $f(x) = x^2 - 4x - 35, f(x) = 25$

36. $g(x) = 6x^2 + 5x, g(x) = 6$

37. $h(r) = 5r^2 - 7r - 10, h(r) = -8$

38. $f(x) = -2x^2 + 6x + 7, f(x) = -2$

Determine an equation that has the given solutions.

39. $3, -1$

40. $\frac{2}{3}, -2$

41. $-\sqrt{11}, \sqrt{11}$

42. $3 - 2i, 3 + 2i$

[12.1–12.3]

43. Rectangular Garden Sophia Yang is designing a rectangular flower garden. If the area is to be 96 square feet and the length is to be 4 feet greater than the width, find the dimensions of the garden.

44. Triangle and Circle Find the length of side x in the figure.

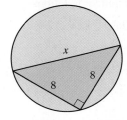

45. Savings Account Samuel Rivera invested $1000 in a savings account where the interest is compounded annually. If after 2 years the amount in the account is $1081.60, find the annual interest rate.

46. Numbers The larger of two positive numbers is 4 greater than the smaller. Find the two numbers if their product is 77.

47. Rectangle The length of a rectangle is 4 inches less than twice its width. Find the dimensions if its area is 96 square inches.

48. Wheat Crop The value, V, in dollars per acre of a wheat crop d days after planting is given by the formula $V = 12d - 0.05d^2, 20 < d < 80$. Find the value of an acre of wheat 60 days after it has been planted.

49. Expenditure by Oil Companies The expenditure, $E(t)$, in billions of dollars, by oil companies for new oil and natural gas projects can be approximated by the equation $E(t) = 7t^2 - 7.8t + 82.2$, where t is the number of years since 2001. **Source:** John S. Herald Inc., *Washington Post* (3/14/05).

a) Find the expenditure of oil companies for new oil and gas projects in 2004.

b) If this trend continues, in what year will the expenditure by oil companies be $579 billion?

50. Falling Object The distance, d, in feet, that an object is from the ground t seconds after being dropped from an airplane is given by the formula $d = -16t^2 + 784$.

a) Find the distance the object is from the ground 2 seconds after it has been dropped.

b) When will the object hit the ground?

51. Oil Leak A tractor has an oil leak. The amount of oil, $L(t)$, in milliliters per hour that leaks out is a function of the tractor's operating temperature, t, in degrees Celsius. The function is

$$L(t) = 0.0004t^2 + 0.16t + 20, 100°C \le t \le 160°C$$

a) How many milliliters of oil will leak out in 1 hour if the operating temperature of the tractor is 100°C?

b) If oil is leaking out at 53 milliliters per hour, what is the operating temperature of the tractor?

52. Molding Machines Two molding machines can complete an order in 12 hours. The larger machine can complete the order by itself in 1 hour less time than the smaller machine can by itself. How long will it take each machine to complete the order working by itself?

53. Travel Time Steve Forrester drove 25 miles at a constant speed, and then he increased his speed by 15 miles per hour for the next 65 miles. If the time required to travel 90 miles was 1.5 hours, find the speed he drove during the first 25 miles.

54. Canoe Trip Joan Banker canoed downstream going with the current for 3 miles, then turned around and canoed upstream against the current to her starting point. If the total time she spent canoeing was 4 hours and the current was 0.4 mile per hour, what is the speed she canoes in still water?

In Exercises 57–60, solve each equation for the variable indicated.

57. $a^2 + b^2 = c^2$, for a (Pythagorean Theorem)

59. $v_x^2 + v_y^2 = v^2$, for v_y (vectors)

55. Area The area of a rectangle is 80 square units. If the length is x units and the width is $x - 2$ units, find the length and the width. Round your answer to the nearest tenth of a unit.

$A = 80$	$x - 2$

x

56. Selling Tables A business sells n tables, $n \leq 40$, at a price of $(60 - 0.3n)$ dollars per table. How many tables must be sold to have a revenue of $1080?

58. $h = -4.9t^2 + c$, for t (height of an object)

60. $a = \dfrac{v_2^2 - v_1^2}{2d}$, for v_2

[12.4] Factor.

61. $x^4 - x^2 - 20$

63. $(x + 5)^2 + 10(x + 5) + 24$

62. $4x^4 + 4x^2 - 3$

64. $4(2x + 3)^2 - 12(2x + 3) + 5$

Solve each equation.

65. $x^4 - 13x^2 + 36 = 0$

67. $a^4 = 5a^2 + 24$

69. $3r + 11\sqrt{r} - 4 = 0$

71. $6(x - 2)^{-2} = -13(x - 2)^{-1} + 8$

66. $x^4 - 21x^2 + 80 = 0$

68. $3y^{-2} + 16y^{-1} = 12$

70. $2p^{2/3} - 7p^{1/3} + 6 = 0$

72. $10(r + 1) = \dfrac{12}{r + 1} - 7$

Find all x-intercepts of the given function.

73. $f(x) = x^4 - 82x^2 + 81$

75. $f(x) = x - 6\sqrt{x} + 12.$

74. $f(x) = 30x + 13\sqrt{x} - 10$

76. $f(x) = (x^2 - 6x)^2 - 5(x^2 - 6x) - 24$

[12.5] **a)** *Determine whether the parabola opens upward or downward.* **b)** *Find the y-intercept.* **c)** *Find the vertex.* **d)** *Find the x-intercepts if they exist.* **e)** *Draw the graph.*

77. $f(x) = x^2 + 5x$

79. $g(x) = -x^2 - 2$

78. $f(x) = x^2 - 2x - 8$

80. $g(x) = -2x^2 - x + 15$

81. Selling Tickets The Hamilton Outdoor Theater estimates that its total income, I, in hundreds of dollars, for its production of a play, can be approximated by the formula $I = -x^2 + 22x - 45, 2 \leq x \leq 20$, where x is the cost of a ticket.

a) How much should the theater charge to maximize its income?

b) What is the maximum income?

82. Tossing a Ball Josh Vincent tosses a ball upward from the top of a 75-foot building. The height, $s(t)$, of the ball at any time t can be determined by the function $s(t) = -16t^2 + 80t + 75$.

a) At what time will the ball attain its maximum height?

b) What is the maximum height?

Graph each function.

83. $f(x) = (x - 3)^2$ **84.** $f(x) = -(x + 2)^2 - 3$ **85.** $g(x) = -2(x + 4)^2 - 1$ **86.** $h(x) = \frac{1}{2}(x - 1)^2 + 3$

[12.6] *Graph the solution to each inequality on a number line.*

87. $x^2 + 4x + 3 \geq 0$

88. $x^2 + 3x - 10 \leq 0$

89. $x^2 \leq 11x - 20$

90. $3x^2 + 8x > 16$

91. $4x^2 - 9 \leq 0$

92. $6x^2 - 30 > 0$

Solve each inequality and give the solution in set builder notation.

93. $\frac{x + 1}{x - 5} > 0$

94. $\frac{x - 3}{x + 2} \leq 0$

95. $\frac{2x - 4}{x + 3} \geq 0$

96. $\frac{3x + 5}{x - 6} < 0$

97. $(x + 4)(x + 1)(x - 2) > 0$

98. $x(x - 3)(x - 6) \leq 0$

Solve each inequality and give the solution in interval notation.

99. $(3x + 4)(x - 1)(x - 3) \geq 0$

100. $2x(x + 2)(x + 4) < 0$

101. $\frac{x(x - 4)}{x + 2} > 0$

102. $\frac{(x - 2)(x - 8)}{x + 3} < 0$

103. $\frac{x - 3}{(x + 2)(x - 7)} \geq 0$

104. $\frac{x(x - 6)}{x + 3} \leq 0$

Solve each inequality and graph the solution on a number line.

105. $\frac{5}{x + 4} \geq -1$

106. $\frac{2x}{x - 2} \leq 1$

107. $\frac{2x + 3}{3x - 5} < 4$

Chapter 12 Practice Test

To find out how well you understand the chapter material, take this practice test. The answers, and the section where the material was initially discussed, are given in the back of the book. Each problem is also fully worked out on the **Chapter Test Prep Video CD***. Review any questions that you answered incorrectly.*

Solve by completing the square.

1. $x^2 + 2x - 15 = 0$ **2.** $a^2 + 7 = 6a$

Solve by the quadratic formula.

3. $x^2 - 6x - 16 = 0$ **4.** $x^2 - 4x = -11$

Solve by the method of your choice.

5. $3r^2 + r = 2$ **6.** $p^2 + 4 = -7p$

7. Write an equation that has *x*-intercepts $4, -\frac{2}{5}$.

8. Solve the formula $K = \frac{1}{2}mv^2$ for *v*.

9. Cost The cost, *c*, of a house in Duquoin, Illinois, is a function of the number of square feet, *s*, of the house. The cost of the house can be approximated by

$$c(s) = -0.01s^2 + 78s + 22,000, \quad 1300 \leq s \leq 3900$$

   a) Estimate the cost of a 1600-square-foot house.

   b) How large a house can Clarissa Skocy purchase if she wishes to spend $160,000 on a house?

10. Trip in a Park Tom Ficks drove his 4-wheel-drive Jeep from Anchorage, Alaska, to the Chena River State Recreation Park, a distance of 520 miles. Had he averaged 15 miles per hour faster, the trip would have taken 2.4 hours less. Find the average speed that Tom drove.

Chena River State Recreation Park

Solve.

11. $2x^4 + 15x^2 - 50 = 0$

12. $3r^{2/3} + 11r^{1/3} - 42 = 0$

13. Find all x-intercepts of $f(x) = 16x - 24\sqrt{x} + 9$.

Graph each function.

14. $f(x) = (x - 3)^2 + 2$

15. $h(x) = -\dfrac{1}{2}(x - 2)^2 - 2$

16. Determine whether $6x^2 = 2x + 3$ has two distinct real solutions, a single real solution, or no real solution. Explain your answer.

17. Consider the quadratic equation $y = x^2 + 2x - 8$.

  a) Determine whether the parabola opens upward or downward.

  b) Find the y-intercept.

  c) Find the vertex.

  d) Find the x-intercepts (if they exist).

  e) Draw the graph.

18. Write a quadratic equation whose x-intercepts are $(-7, 0), \left(\dfrac{1}{2}, 0\right)$.

Solve each inequality and graph the solution on a number line.

19. $x^2 - x \geq 42$

20. $\dfrac{(x + 5)(x - 4)}{x + 1} \geq 0$

*Solve the following inequality. Write the answer in **a)** interval notation and **b)** set builder notation.*

21. $\dfrac{x + 3}{x + 2} \leq -1$

22. **Carpet** The length of a rectangular Persian carpet is 3 feet greater than twice its width. Find the length and width of the carpet if its area is 65 square feet.

23. **Throwing a Ball** Jose Ramirez throws a ball upward from the top of a building. The distance, d, of the ball from the ground at any time, t, is $d = -16t^2 + 80t + 96$. How long will it take for the ball to strike the ground?

24. **Profit** The Leigh Ann Sims Company earns a weekly profit according to the function $f(x) = -1.4x^2 + 56x - 70$, where x is the number of wood carvings made and sold each week.

  a) Find the number of carvings the company must sell in a week to maximize its profit.

  b) What is its maximum weekly profit?

25. **Selling Brooms** A business sells n brooms, $n \leq 32$, at a price of $(10 - 0.1n)$ dollars per broom. How many brooms must be sold to have a revenue of $210?

Cumulative Review Test

Take the following test and check your answers with those given in the back of the book. Review any questions that you answered incorrectly. The section where the material was covered is indicated after the answer.

1. Evaluate $-4 \div (-2) + 18 - \sqrt{49}$.

2. Evaluate $2x^2 + 3x + 4$ when $x = 2$.

3. Simplify $6x - \{3 - [2(x - 2) - 5x]\}$.

4. Solve the equation $-\dfrac{1}{2}(4x - 6) = \dfrac{1}{3}(3 - 6x) + 2$.

5. Graph each equation.

  a) $x = -4$

  b) $y = 2$

6. Write the equation in slope-intercept form of a line passing through the points $(6, 5)$ and $(4, 3)$.

7. Express 2,540,000 in scientific notation.

8. Add $\dfrac{x + 2}{x^2 - x - 6} + \dfrac{x - 3}{x^2 - 8x + 15}$.

9. Solve the equation
$$\dfrac{1}{a - 2} = \dfrac{4a - 1}{a^2 + 5a - 14} + \dfrac{2}{a + 7}.$$

10. **Wattage Rating** The wattage rating of an appliance, w, varies jointly as the square of the current, I, and the resistance, R. If the wattage is 12 when the current is 2 amperes and the resistance is 100 ohms, find the wattage when the current is 0.8 ampere and the resistance is 600 ohms.

11. **Small Orchard** The number of baskets of apples, N, that are produced by x trees in a small orchard is given by the function $N(x) = -0.2x^2 + 40x$. How many baskets of apples are produced by 50 trees?

12. **a)** Determine whether the following graph represents a function. Explain your answer.

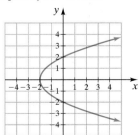

  b) Determine the domain and range of the function or relation.

13. Solve the system of equations.

$$4x - 3y = 10$$
$$2x + y = 5$$

14. Evaluate the following determinant.

$$\begin{vmatrix} 4 & 0 & -2 \\ 3 & 5 & 1 \\ 1 & -1 & 7 \end{vmatrix}$$

15. Solve the inequality $-4 < \dfrac{x + 4}{2} < 6$. Write the solution in interval notation.

16. Find the solution set for the equation $|4 - 2x| = 5$.

17. Simplify $\dfrac{3 - 4i}{2 + 5i}$.

18. Solve $4x^2 = -3x - 12$.

19. Solve the formula $V = \dfrac{1}{3}\pi r^2 h$ for r.

20. Factor $(x + 3)^2 + 10(x + 3) + 24$.

13 Exponential and Logarithmic Functions

GOALS OF THIS CHAPTER

Exponential and logarithmic functions have a wide variety of uses, some of which you will see as you read through this chapter. You often read in newspaper and magazine articles that health care spending, use of the Internet, and the world population, to list just a few, are growing exponentially. By the time you finish this chapter you should have a clear understanding of just what this means.

We also introduce two special functions, the natural exponential function and the natural logarithmic function. Many natural phenomena, such as carbon dating, radioactive decay, and the growth of savings invested in an account compounding interest continuously, can be described by natural exponential functions.

THE DEPARTMENT OF ECONOMIC and Social Affairs uses mathematical models to make estimates and projections of the world population. Currently, the world population is growing at about 1.13% per year. Since the population is growing by a percent rather than by a fixed amount, it is modeled by an exponential function rather than a linear one. In Exercise 79 on page 860, we will investigate the effect different rates have on the growth of the world population.

13.1 Composite and Inverse Functions

1 Find composite functions.

2 Understand one-to-one functions.

3 Find inverse functions.

4 Find the composition of a function and its inverse.

The focus of this chapter is logarithms. However, before we can study logarithms we discuss composite functions, one-to-one functions, and inverse functions. Let's start with composite functions.

1 Find Composite Functions

Often we come across situations in which one quantity is a function of one variable. That variable, in turn, is a function of some other variable. For example, the cost of advertising on a television show may be a function of the Nielsen rating of the show. The Nielsen rating, in turn, is a function of the number of people who watch the show. In the final outcome, the cost of advertising may be affected by the number of people who watch the show. Functions like this are called *composite functions*.

Let's consider another example. Suppose that 1 U.S. dollar can be converted into 1.20 Canadian dollars, and 1 Canadian dollar can be converted into 9.3 Mexican pesos. Using this information, we can convert 20 U.S. dollars into Mexican pesos. We have the following functions.

$$g(x) = 1.20x \text{ (U.S. dollars to Canadian dollars)}$$
$$f(x) = 9.3x \text{ (Canadian dollars to Mexican pesos)}$$

If we let $x = 20$, for \$20 U.S., then it can be converted into \$24 Canadian using function g:

$$g(x) = 1.20x$$
$$g(20) = 1.20(20) = \$24 \text{ Canadian}$$

The \$24 Canadian can, in turn, be converted into 223.20 Mexican pesos using function f:

$$f(x) = 9.3x$$
$$f(24) = 9.3(24) = 223.20 \text{ Mexican pesos}$$

Is there a way of finding this conversion without performing this string of calculations? The answer is yes. One U.S. dollar can be converted into Mexican pesos by substituting the $1.20x$ found in function $g(x)$ for the x in $f(x)$. This gives a new function, h, which converts U.S. dollars directly into Mexican pesos.

$$g(x) = 1.20x \qquad f(x) = 9.3x$$
$$h(x) = f[g(x)]$$
$$= 9.3(\boxed{1.20x}) \qquad \textit{Substitute g(x) for x in f(x).}$$
$$= 11.16x$$

Thus, for each U.S. dollar, x, we get 11.16 Mexican pesos. If we substitute \$20 for x, we get 223.20 pesos, which is what we expected.

$$h(x) = 11.16x$$
$$h(20) = 11.16(20) = 223.20$$

Function h, called a **composition of f with g**, is denoted $(f \circ g)$ and is read "f composed with g" or "f circle g." **Figure 13.1** shows how the composite function h relates to functions f and g.

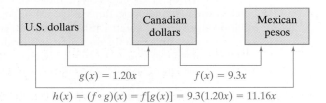

FIGURE 13.1

We now define the **composite function**.

Composite Function

The **composite function** $f \circ g$ is defined as

$$(f \circ g)(x) = f[g(x)]$$

When we are given $f(x)$ and $g(x)$, to find $(f \circ g)(x)$ we substitute $g(x)$ for x in $f(x)$ to get $f[g(x)]$.

EXAMPLE 1 ▶ Given $f(x) = x^2 - 2x + 3$ and $g(x) = x - 5$, find

a) $f(4)$ **b)** $f(a)$ **c)** $(f \circ g)(x)$ **d)** $(f \circ g)(3)$

Solution

a) To find $f(4)$, we substitute 4 for each x in $f(x)$.

$$f(x) = x^2 - 2x + 3$$
$$f(4) = 4^2 - 2 \cdot 4 + 3 = 16 - 8 + 3 = 11$$

b) To find $f(a)$, we substitute a for each x in $f(x)$.

$$f(x) = x^2 - 2x + 3$$
$$f(a) = a^2 - 2a + 3$$

c) $(f \circ g)(x) = f[g(x)]$. To find $(f \circ g)(x)$, we substitute $g(x)$, which is $x - 5$, for each x in $f(x)$.

$$f(x) = x^2 - 2 x + 3$$
$$f[g(x)] = [g(x)]^2 - 2[g(x)] + 3$$

Since $g(x) = x - 5$, we substitute as follows

$$f[g(x)] = (x - 5)^2 - 2(x - 5) + 3$$
$$= (x - 5)(x - 5) - 2x + 10 + 3$$
$$= x^2 - 10x + 25 - 2x + 13$$
$$= x^2 - 12x + 38$$

Therefore, the composite function of f with g is $x^2 - 12x + 38$.

$$(f \circ g)(x) = f[g(x)] = x^2 - 12x + 38$$

d) To find $(f \circ g)(3)$, we substitute 3 for x in $(f \circ g)(x)$.

$$(f \circ g)(x) = x^2 - 12x + 38$$
$$(f \circ g)(3) = 3^2 - 12(3) + 38 = 11$$

▶ **Now Try Exercise 9**

How do you think we would determine $(g \circ f)(x)$ or $g[f(x)]$? If you answered, "Substitute $f(x)$ for each x in $g(x)$," you answered correctly. Using $f(x)$ and $g(x)$ as given in Example 1, we find $(g \circ f)(x)$ as follows.

$$g(x) = x - 5, \qquad f(x) = x^2 - 2x + 3$$
$$g[f(x)] = f(x) - 5$$
$$g[f(x)] = (x^2 - 2x + 3) - 5$$
$$= x^2 - 2x + 3 - 5$$
$$= x^2 - 2x - 2$$

Therefore, the composite function of g with f is $x^2 - 2x - 2$.

$$(g \circ f)(x) = g[f(x)] = x^2 - 2x - 2$$

By comparing the illustrations above, we see that in this example, $f[g(x)] \neq g[f(x)]$.

EXAMPLE 2 ▶ Given $f(x) = x^2 + 4$ and $g(x) = \sqrt{x - 1}$, find

a) $(f \circ g)(x)$ **b)** $(g \circ f)(x)$

Solution

a) To find $(f \circ g)(x)$, we substitute $g(x)$, which is $\sqrt{x - 1}$ for each x in $f(x)$. You should realize that $\sqrt{x - 1}$ is a real number only when $x \geq 1$.

$$f(x) = \boxed{x}^2 + 4$$
$$(f \circ g)(x) = f[g(x)] = (\boxed{\sqrt{x - 1}})^2 + 4 = x - 1 + 4 = x + 3, x \geq 1$$

Since values of $x < 1$ are not in the domain of $g(x)$, values of $x < 1$ are not in the domain of $(f \circ g)(x)$.

b) To find $(g \circ f)(x)$, we substitute $f(x)$, which is $x^2 + 4$, for each x in $g(x)$.

$$g(x) = \sqrt{\boxed{x} - 1}$$
$$(g \circ f)(x) = g[f(x)] = \sqrt{\boxed{(x^2 + 4)} - 1} = \sqrt{x^2 + 3}$$

▶ Now Try Exercise 19

EXAMPLE 3 ▶ Given $f(x) = x - 3$ and $g(x) = x + 7$, find

a) $(f \circ g)(x)$ **b)** $(f \circ g)(2)$ **c)** $(g \circ f)(x)$ **d)** $(g \circ f)(2)$

Solution

a)
$$f(x) = \boxed{x} - 3$$
$$(f \circ g)(x) = f[g(x)] = (\boxed{x + 7}) - 3 = x + 4$$

b) We find $(f \circ g)(2)$ by substituting 2 for each x in $(f \circ g)(x)$.
$$(f \circ g)(x) = x + 4$$
$$(f \circ g)(2) = 2 + 4 = 6$$

c)
$$g(x) = \boxed{x} + 7$$
$$(g \circ f)(x) = g[f(x)] = (\boxed{x - 3}) + 7 = x + 4$$

d) Since $(g \circ f)(x) = x + 4$, $(g \circ f)(2) = 2 + 4 = 6$.

▶ Now Try Exercise 11

In general, $(f \circ g)(x) \neq (g \circ f)(x)$ as we saw at the end of Example 1. In Example 3, $(f \circ g)(x) = (g \circ f)(x)$, but this is only due to the specific functions used.

Helpful Hint

Do not confuse finding the product of two functions with finding a composite function.

Product of functions f and g: $(fg)(x)$ or $(f \cdot g)(x)$
Composite function of f with g: $(f \circ g)(x)$

When multiplying functions f and g, we can use a dot between the f and g. When finding the composite function of f with g, we use a small *open* circle.

2 Understand One-to-One Functions

Consider the following two sets of ordered pairs.

$$A = \{(1, 2), (3, 5), (4, 6), (-2, 1)\}$$
$$B = \{(1, 2), (3, 5), (4, 6), (-2, 5)\}$$

Both sets of ordered pairs, A and B, are functions since each value of x has a unique value of y. In set A, each value of y also has a unique value of x, as shown in **Figure 13.2.** In set B, each value of y does not have a unique value of x. In the ordered pairs $(3, 5)$ and $(-2, 5)$, the y-value 5 corresponds with two values of x, as shown in **Figure 13.3.**

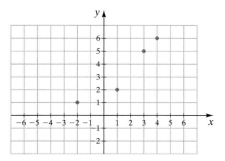

FIGURE 13.2

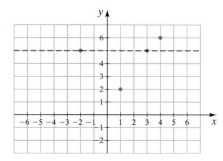

FIGURE 13.3

The set of ordered pairs in A is an example of a *one-to-one function.* The set of ordered pairs in B is not a one-to-one function. In a **one-to-one function**, each value in the range has a unique value in the domain. Thus, if y is a one-to-one function of x, in addition to each x-value having a unique y-value (the definition of a function), each y-value must also have a unique x-value.

> ### One-to-One Function
> A function is a **one-to-one function** if each value in the range corresponds with exactly one value in the domain.

For a function to be a one-to-one function, its graph must pass not only a **vertical line test** (the test to ensure that it is a function) but also a **horizontal line test** (to test the one-to-one criteria).

Consider the function $f(x) = x^2$ (see **Fig. 13.4**). Note that it is a function since its graph passes the vertical line test. For each value of x, there is a unique value of y. Does each value of y also have a unique value of x? The answer is no, as illustrated in **Figure 13.5**. Note that for the indicated value of y there are two values of x, namely x_1 and x_2. If we limit the domain of $f(x) = x^2$ to values of x greater than or equal to 0, then each x-value has a unique y-value and each y-value also has a unique x-value (see **Fig. 13.6**). The function $f(x) = x^2, x \geq 0$, **Figure 13.6**, is an example of a one-to-one function.

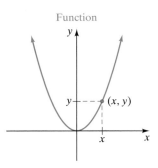

FIGURE 13.4

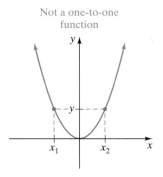

FIGURE 13.5

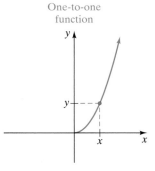

FIGURE 13.6

In **Figure 13.7**, the graphs in parts (a) through (e) are functions since they all pass the vertical line test. However, only the graphs in parts (a), (d), and (e) are one-to-one functions since they also pass the horizontal line test. The graph in part (f) is not a function; therefore, it is not a one-to-one function, even though it passes the horizontal line test.

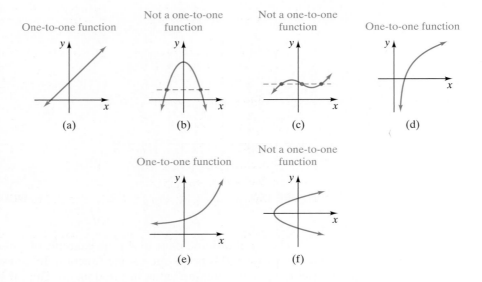

FIGURE 13.7

3 Find Inverse Functions

Now that we have discussed one-to-one functions, we can introduce inverse functions. **Only one-to-one functions have inverse functions**. If a function is one-to-one, its **inverse function** may be obtained by interchanging the first and second coordinates in each ordered pair of the function. Thus, for each ordered pair (x, y) in the function, the ordered pair (y, x) will be in the inverse function. For example,

Function: $\{(1, 4), (2, 0), (3, 7), (-2, 1), (-1, -5)\}$
Inverse function: $\{(4, 1), (0, 2), (7, 3), (1, -2), (-5, -1)\}$

Note that the domain of the function becomes the range of the inverse function, and the range of the function becomes the domain of the inverse function.

If we graph the points in the function and the points in the inverse function (**Fig. 13.8**), we see that the points are symmetric with respect to the line $y = x$.

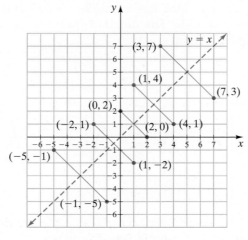

FIGURE 13.8

• Ordered pair in function
• Ordered pair in inverse function

For a function $f(x)$, the notation $f^{-1}(x)$ represents its inverse function. Note that the -1 in the notation is *not* an exponent. Therefore, $f^{-1}(x) \neq \dfrac{1}{f(x)}$.

> **Inverse Function**
>
> If $f(x)$ is a one-to-one function with ordered pairs of the form (x, y), its **inverse function**, $f^{-1}(x)$, is a one-to-one function with ordered pairs of the form (y, x).

When a function $f(x)$ and its inverse function $f^{-1}(x)$ are graphed on the same axes, $f(x)$ *and* $f^{-1}(x)$ are *symmetric about the line* $y = x$ as seen in **Figure 13.8** on page 810.

When a one-to-one function is given as an equation, its inverse function can be found by the following procedure.

> **To Find the Inverse Function of a One-to-One Function**
>
> 1. Replace $f(x)$ with y.
> 2. Interchange the two variables x and y.
> 3. Solve the equation for y.
> 4. Replace y with $f^{-1}(x)$ (this gives the inverse function using inverse function notation).

The following example will illustrate the procedure.

EXAMPLE 4 ▶

a) Find the inverse function of $f(x) = 4x + 2$.

b) On the same axes, graph both $f(x)$ and $f^{-1}(x)$.

Solution   **a)** This function is one-to-one, therefore we will follow the four-step procedure.

$$f(x) = 4x + 2 \qquad \textit{Original function}$$

Step 1 $\qquad\qquad y = 4x + 2 \qquad \textit{Replace f(x) with y.}$

Step 2 $\qquad\qquad x = 4y + 2 \qquad \textit{Interchange x and y.}$

Step 3 $\qquad\qquad x - 2 = 4y \qquad \textit{Solve for y.}$

$$\frac{x-2}{4} = y$$

$$\text{or} \qquad y = \frac{x-2}{4}$$

Step 4 $\qquad\qquad f^{-1}(x) = \dfrac{x-2}{4} \qquad \textit{Replace y with } f^{-1}(x).$

b) Below we show tables of values for $f(x)$ and $f^{-1}(x)$. The graphs of $f(x)$ and $f^{-1}(x)$ are shown in **Figure 13.9**.

x	$y = f(x)$
0	2
1	6

x	$y = f^{-1}(x)$
2	0
6	1

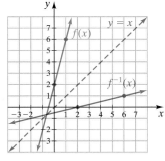

FIGURE 13.9

Note the symmetry of $f(x)$ and $f^{-1}(x)$ about the line $y = x$. Also note that both the domain and range of both $f(x)$ and $f^{-1}(x)$ are the set of real numbers, $\mathbb{R}$.

▶ **Now Try Exercise 67**

In Chapter 11 when we solved equations containing cube roots, we cubed each side of the equation. To solve cubic equations, we raise each side of the equation to the one-third power, which is equivalent to taking the cube root of each side of the equation. Recall from Chapter 11 that $\sqrt[3]{a^3} = a$ for any real number a.

EXAMPLE 5 ▶

a) Find the inverse function of $f(x) = x^3 + 2$.

b) On the same axes, graph both $f(x)$ and $f^{-1}(x)$.

Solution   **a)** This function is one-to-one; therefore we will follow the four-step procedure to find its inverse.

	$f(x) = x^3 + 2$	*Original function*
Step 1	$y = x^3 + 2$	*Replace f(x) with y.*
Step 2	$x = y^3 + 2$	*Interchange x and y.*
Step 3	$x - 2 = y^3$	*Solve for y.*
	$\sqrt[3]{x - 2} = \sqrt[3]{y^3}$	*Take the cube root of both sides.*
	$\sqrt[3]{x - 2} = y$	
	or $y = \sqrt[3]{x - 2}$	
Step 4	$f^{-1}(x) = \sqrt[3]{x - 2}$	*Replace y with f⁻¹(x).*

b) Below we show tables of values for $f(x)$ and $f^{-1}(x)$.

x	$y = f(x)$
-2	-6
-1	1
0	2
1	3
2	10

x	$y = f^{-1}(x)$
-6	-2
1	-1
2	0
3	1
10	2

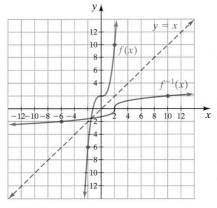

FIGURE 13.10

The graphs of $f(x)$ and $f^{-1}(x)$ are shown in **Figure 13.10**. Notice for each point (a, b) on the graph of $f(x)$, the point (b, a) appears on the graph of $f^{-1}(x)$. For example, the points $(2, 10)$ and $(-2, -6)$, indicated in blue, appear on the graph of $f(x)$, and the points $(10, 2)$ and $(-6, -2)$, indicated in red, appear on the graph of $f^{-1}(x)$.

▶ **Now Try Exercise 61**

USING YOUR GRAPHING CALCULATOR

In Example 5, we were given $f(x) = x^3 + 2$ and found that $f^{-1}(x) = \sqrt[3]{x - 2}$. The graphs of these functions are symmetric about the line $y = x$, although it may not appear that way on a graphing calculator. **Figure 13.11** on page 813 shows the two graphs using a standard calculator window.

Since the horizontal axis is longer than the vertical axis, and both axes have 10 tick marks, the graphs appear distorted. Many calculators have a feature to "square the axes." When this feature is used, the window is still rectangular, but the distance between the tick marks is equalized. To equalize the spacing between the tick marks on the vertical and horizontal axes on the

(*continued on the next page*)

TI-84 Plus calculator, press ZOOM then select option 5, ZSquare. **Figure 13.12** shows the graphs after this option is selected. A third illustration of the graphs can be obtained using ZOOM, option 4, ZDecimal. This option resets the x-axis to go from -4.7 to 4.7 and the y-axis to go from -3.1 to 3.1, as shown in **Figure 13.13**.

Standard window

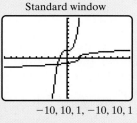

$-10, 10, 1, -10, 10, 1$

FIGURE 13.11

ZSquare window

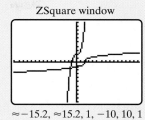

$\approx -15.2, \approx 15.2, 1, -10, 10, 1$

FIGURE 13.12

ZDecimal window

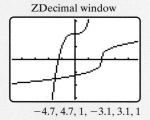

$-4.7, 4.7, 1, -3.1, 3.1, 1$

FIGURE 13.13

4 Find the Composition of a Function and Its Inverse

If two functions $f(x)$ and $f^{-1}(x)$ are inverses of each other, $(f \circ f^{-1})(x) = x$ and $(f^{-1} \circ f)(x) = x$.

EXAMPLE 6 ▸ In Example 4, we determined that for $f(x) = 4x + 2$, $f^{-1}(x) = \dfrac{x-2}{4}$. Show that

a) $(f \circ f^{-1})(x) = x$ **b)** $(f^{-1} \circ f)(x) = x$

Solution

a) To determine $(f \circ f^{-1})(x)$, substitute $f^{-1}(x)$ for each x in $f(x)$.

$$f(x) = 4\boxed{x} + 2$$
$$(f \circ f^{-1})(x) = 4\left(\boxed{\dfrac{x-2}{4}}\right) + 2$$
$$= x - 2 + 2 = x$$

b) To determine $(f^{-1} \circ f)(x)$, substitute $f(x)$ for each x in $f^{-1}(x)$.

$$f^{-1}(x) = \dfrac{\boxed{x} - 2}{4}$$
$$(f^{-1} \circ f)(x) = \dfrac{\boxed{4x + 2} - 2}{4}$$
$$= \dfrac{4x}{4} = x$$

Thus, $(f \circ f^{-1})(x) = (f^{-1} \circ f)(x) = x$.

▸ **Now Try Exercise 77**

EXAMPLE 7 ▸ In Example 5, we determined that $f(x) = x^3 + 2$ and $f^{-1}(x) = \sqrt[3]{x-2}$ are inverse functions. Show that

a) $(f \circ f^{-1})(x) = x$ **b)** $(f^{-1} \circ f)(x) = x$

Solution

a) To determine $(f \circ f^{-1})(x)$, substitute $f^{-1}(x)$ for each x in $f(x)$.

$$f(x) = \boxed{x}^3 + 2$$
$$(f \circ f^{-1})(x) = (\boxed{\sqrt[3]{x-2}})^3 + 2$$
$$= x - 2 + 2 = x$$

b) To determine $(f^{-1} \circ f)(x)$, substitute $f(x)$ for each x in $f^{-1}(x)$.

$$f^{-1}(x) = \sqrt[3]{x - 2}$$
$$(f^{-1} \circ f)(x) = \sqrt[3]{(x^3 + 2) - 2}$$
$$= \sqrt[3]{x^3} = x$$

Thus, $(f \circ f^{-1})(x) = (f^{-1} \circ f)(x) = x$.

▶ **Now Try Exercise 79**

Because a function and its inverse "undo" each other, the composite of a function with its inverse results in the given value from the domain. For example, for any function $f(x)$ and its inverse $f^{-1}(x)$, $(f^{-1} \circ f)(3) = 3$, and $(f \circ f^{-1})\left(-\dfrac{1}{2}\right) = -\dfrac{1}{2}$.

EXERCISE SET 13.1

Concept/Writing Exercises

1. Explain how to find $(f \circ g)(x)$ when you are given $f(x)$ and $g(x)$.

2. Explain how to find $(g \circ f)(x)$ when you are given $f(x)$ and $g(x)$.

3. **a)** What are one-to-one functions?

   b) Explain how you may determine whether a function is a one-to-one function.

4. Do all functions have inverse functions? If not, which functions do?

5. Consider the set of ordered pairs $\{(3, 5), (4, 2), (-1, 3), (0, -2)\}$.

   a) Is this set of ordered pairs a function? Explain.

   b) Does this function have an inverse? Explain.

   c) If this function has an inverse, give the inverse function. Explain how you determined your answer.

6. Suppose $f(x)$ and $g(x)$ are inverse functions.

   a) What is $(f \circ g)(x)$ equal to?

   b) What is $(g \circ f)(x)$ equal to?

7. What is the relationship between the domain and range of a function and the domain and range of its inverse function?

8. What is the value of $(f \circ f^{-1})(6)$? Explain.

Practice the Skills

For each pair of functions, find **a)** $(f \circ g)(x)$, **b)** $(f \circ g)(4)$, **c)** $(g \circ f)(x)$, *and* **d)** $(g \circ f)(4)$.

9. $f(x) = x^2 + 1, g(x) = x + 2$

10. $f(x) = x^2 - 3, g(x) = x + 6$

11. $f(x) = x + 3, g(x) = x^2 + x - 4$

12. $f(x) = x + 2, g(x) = x^2 + 4x - 2$

13. $f(x) = \dfrac{1}{x}, g(x) = 2x + 3$

14. $f(x) = \dfrac{2}{x}, g(x) = x^2 + 1$

15. $f(x) = 3x + 1, g(x) = \dfrac{3}{x}$

16. $f(x) = x^2 - 5, g(x) = \dfrac{4}{x}$

17. $f(x) = x^2 + 1, g(x) = x^2 + 5$

18. $f(x) = x^2 - 4, g(x) = x^2 + 3$

19. $f(x) = x - 4, g(x) = \sqrt{x + 5}, x \geq -5$

20. $f(x) = \sqrt{x + 6}, x \geq -6, g(x) = x + 7$

In Exercises 21–42, determine whether each function is a one-to-one function.

21.
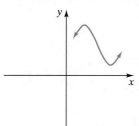

22.
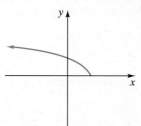

23.
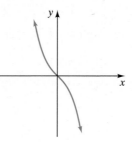

24.
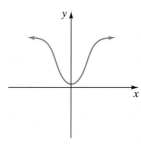

25. $\{(2, 4), (3, -7), (5, 3), (-6, 0)\}$

26. $\{(-4, 2), (2, 3), (4, 1), (0, 4)\}$

27. $\{(-4, 2), (5, 3), (0, 2), (4, 8)\}$

28. $\{(0, 5), (1, 4), (-3, 5), (4, 2)\}$

29. $y = 2x + 5$

30. $y = 3x - 8$

31. $y = x^2 - 1$

32. $y = -x^2 + 3$

33. $y = x^2 - 2x + 5$

34. $y = x^2 - 2x + 6, x \geq 1$

35. $y = x^2 - 9, x \geq 0$

36. $y = x^2 - 9, x \leq 0$

37. $y = \sqrt{x}$

38. $y = -\sqrt{x}$

39. $y = |x|$

40. $y = -|x|$

41. $y = \sqrt[3]{x}$

42. $y = x^3$

In Exercises 43–48, for the given function, find the domain and range of both $f(x)$ and $f^{-1}(x)$.

43. $\{(4, 0), (8, 9), (2, 7), (-1, 6), (-2, 4)\}$

44. $\left\{(-2, -3), (-4, 0), (5, 3), (6, 2), \left(2, \dfrac{1}{2}\right)\right\}$

45.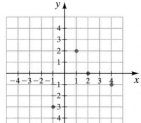

46.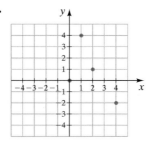

47. 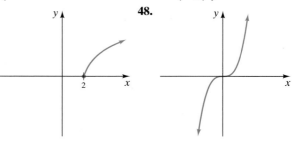 **48.**

*For each function, **a)** determine whether it is one-to-one; **b)** if it is one-to-one, find its inverse function.*

49. $f(x) = x - 2$

50. $g(x) = x + 5$

51. $h(x) = 4x$

52. $k(x) = 2x - 7$

53. $p(x) = 3x^2$

54. $r(x) = |x|$

55. $t(x) = x^2 + 3$

56. $m(x) = -x^2 + x + 8$

57. $g(x) = \dfrac{1}{x}$

58. $h(x) = \dfrac{5}{x}$

59. $f(x) = x^2 + 10$

60. $g(x) = x^3 + 9$

61. $g(x) = x^3 - 6$

62. $f(x) = \sqrt{x}, x \geq 0$

63. $g(x) = \sqrt{x + 2}, x \geq -2$

64. $f(x) = x^2 - 3, x \geq 0$

65. $h(x) = x^2 - 4, x \geq 0$

66. $h(x) = |x|$

*For each one-to-one function, **a)** find $f^{-1}(x)$ and **b)** graph $f(x)$ and $f^{-1}(x)$ on the same axes.*

67. $f(x) = 2x + 8$

68. $f(x) = -3x + 6$

69. $f(x) = \sqrt{x}, x \geq 0$

70. $f(x) = -\sqrt{x}, x \geq 0$

71. $f(x) = \sqrt{x - 1}, x \geq 1$

72. $f(x) = \sqrt{x + 4}, x \geq -4$

73. $f(x) = \sqrt[3]{x}$

74. $f(x) = \sqrt[3]{x + 3}$

75. $f(x) = \dfrac{1}{x}, x > 0$

76. $f(x) = \dfrac{1}{x}$

For each pair of inverse functions, show that $(f \circ f^{-1})(x) = x$ and $(f^{-1} \circ f)(x) = x$.

77. $f(x) = x - 8, f^{-1}(x) = x + 8$

78. $f(x) = 7x + 3, f^{-1}(x) = \dfrac{x - 3}{7}$

79. $f(x) = \dfrac{1}{2}x + 3, f^{-1}(x) = 2x - 6$

80. $f(x) = -\dfrac{1}{3}x + 2, f^{-1}(x) = -3x + 6$

81. $f(x) = \sqrt[3]{x - 2}, f^{-1}(x) = x^3 + 2$

82. $f(x) = \sqrt[3]{x + 9}, f^{-1}(x) = x^3 - 9$

83. $f(x) = \dfrac{3}{x}, f^{-1}(x) = \dfrac{3}{x}$

84. $f(x) = \sqrt{x + 5}, f^{-1}(x) = x^2 - 5, x \geq 0$

Problem Solving

85. Is $(f \circ g)(x) = (g \circ f)(x)$ for all values of x? Explain and give an example to support your answer.

86. Consider the functions $f(x) = \sqrt{x + 5}, x \geq -5$, and $g(x) = x^2 - 5, x \geq 0$.

    a) Show that $(f \circ g)(x) = (g \circ f)(x)$ for $x \geq 0$.

    b) Explain why we need to stipulate that $x \geq 0$ for part **a)** to be true.

87. Consider the functions $f(x) = x^3 + 2$ and $g(x) = \sqrt[3]{x - 2}$.

    a) Show that $(f \circ g)(x) = (g \circ f)(x)$.

    b) What are the domains of $f(x)$, $g(x)$, $(f \circ g)(x)$, and $(g \circ f)(x)$? Explain.

88. For the function $f(x) = x^3, f(2) = 2^3 = 8$. Explain why $f^{-1}(8) = 2$.

89. For the function $f(x) = x^4, x > 0, f(2) = 16$. Explain why $f^{-1}(16) = 2$.

90. The function $f(x) = 12x$ converts feet, x, into inches. Find the inverse function that converts inches into feet. In the inverse function, what do x and $f^{-1}(x)$ represent?

91. The function $f(x) = 3x$ converts yards, x, into feet. Find the inverse function that converts feet into yards. In the inverse function, what do x and $f^{-1}(x)$ represent?

92. The function $f(x) = \dfrac{22}{15}x$ converts miles per hour, x, into feet per second. Find the inverse function that converts feet per second into miles per hour.

93. The function $f(x) = \dfrac{5}{9}(x - 32)$ converts degrees Fahrenheit, x, to degrees Celsius. Find the inverse function that changes degrees Celsius into degrees Fahrenheit.

94. a) Does the function $f(x) = |x|$ have an inverse? Explain.

    b) If the domain is limited to $x \geq 0$, does the function have an inverse? Explain.

    c) Find the inverse function of $f(x) = |x|, x \geq 0$.

Composition of Functions *In Exercises 95–98, the functions $f(x)$ and $g(x)$ are given. Determine the composition $(f \circ g)(x)$. For the composition function, what does x represent and what does $(f \circ g)(x)$ represent?*

95. $f(x) = 16x$ converts pounds, x, to ounces. $g(x) = 28.35x$ converts ounces, x to grams.

96. $f(x) = 2000x$ converts tons, x, to pounds. $g(x) = 16x$ converts pounds, x, to ounces.

97. $f(x) = 3x$ converts yards, x, to feet. $g(x) = 0.305x$ converts feet, x, to meters.

98. $f(x) = 1760x$ converts miles, x, to yards. $g(x) = 0.915x$ converts yards, x, to meters.

Use your graphing calculator to determine whether the following functions are inverses.

99. $f(x) = 3x - 4, g(x) = \dfrac{x}{3} + \dfrac{4}{3}$

100. $f(x) = \sqrt{4 - x^2}, g(x) = \sqrt{4 - 2x}$

101. $f(x) = x^3 - 12, g(x) = \sqrt[3]{x + 12}$

102. $f(x) = x^5 + 5, g(x) = \sqrt[5]{x - 5}$

Challenge Problems

103. Area When a pebble is thrown into a pond, the circle formed by the pebble hitting the water expands with time. The area of the expanding circle may be found by the formula $A = \pi r^2$. The radius, r, of the circle, in feet, is a function of time, t, in seconds. Suppose that the function is $r(t) = 2t$.

    a) Find the radius of the circle at 3 seconds.

    b) Find the area of the circle at 3 seconds.

    c) Express the area as a function of time by finding $A \circ r$.

    d) Using the function found in part **c)**, find the area of the circle at 3 seconds.

    e) Do your answers in parts **b)** and **d)** agree? If not, explain.

104. Surface Area The surface area, S, of a spherical balloon of radius r, in inches, is found by $S(r) = 4\pi r^2$. If the balloon is being blown up at a constant rate by a machine, then the radius of the balloon is a function of time. Suppose that this function is $r(t) = 1.2t$, where t is in seconds.

a) Find the radius of the balloon at 2 seconds.

b) Find the surface area at 2 seconds.

c) Express the surface area as a function of time by finding $S \circ r$.

d) Using the function found in part c), find the surface area after 2 seconds.

e) Do your answers in parts b) and d) agree? If not, explain why not.

Group Activity

Discuss and answer Exercise 105 as a group.

105. Consider the function $f(x) = 2^x$. This is an example of an *exponential function*, which we will discuss in the next section.

a) Graph this function by substituting values for x and finding the corresponding values of $f(x)$.

b) Do you think this function has an inverse? Explain your answer.

c) Using the graph in part **a)**, draw the inverse function, $f^{-1}(x)$ on the same axes.

d) Explain how you obtained the graph of $f^{-1}(x)$.

Cumulative Review Exercises

[4.4] **106.** Determine the equation of a line in standard form that passes through $\left(\dfrac{1}{2}, 3\right)$ and is parallel to the graph of $2x + 3y - 9 = 0$.

[5.6] **107.** Divide using synthetic division.
$$(x^3 + 6x^2 + 6x - 8) \div (x + 2)$$

[7.5] **108.** Simplify $\dfrac{\dfrac{3}{x^2} - \dfrac{2}{x}}{\dfrac{x}{6}}$.

[11.5] **109.** Simplify $\sqrt{\dfrac{24x^3 y^2}{3xy^3}}$.

[12.1] **110.** Solve the equation $x^2 + 2x - 6 = 0$ by completing the square.

13.2 Exponential Functions

1 Graph exponential functions.

2 Solve applications of exponential functions.

1 Graph Exponential Functions

We often read about things that are growing exponentially. For example, you may read that the world population is growing exponentially or that the use of e-mail is growing exponentially. What does this indicate? The graph in **Figure 13.14** shows population growth worldwide. The graph in **Figure 13.15** also shows the shipment of smart handheld devices. Both graphs have the same general shape, as indicated by the curves. Both graphs are exponential functions and rise rapidly.

In the quadratic function $f(x) = x^2$, the variable is the base and the exponent is a constant. In the function $f(x) = 2^x$, the constant is the base and the variable is the exponent. The function $f(x) = 2^x$ is an example of an *exponential function*, which we define on page 604.

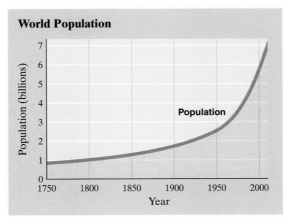

FIGURE 13.14

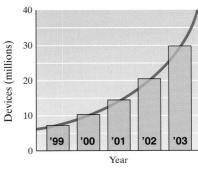

Worldwide Shipments of Smart Handheld Devices

Source: International Data Corp.; MSN MoneyCentral; CSI Inc.

FIGURE 13.15

Exponential Function

For any real number $a > 0$ and $a \neq 1$,

$$f(x) = a^x$$

is an **exponential function**.

An exponential function is a function of the form $f(x) = a^x$, where a is a positive real number not equal to 1. Notice the variable is in the exponent.

Examples of Exponential Functions

$$f(x) = 2^x, \qquad g(x) = 5^x, \qquad h(x) = \left(\frac{1}{2}\right)^x$$

Since $y = f(x)$, functions of the form $y = a^x$ are also exponential functions. Exponential functions can be graphed by selecting values for x, finding the corresponding values of y [or $f(x)$], and plotting the points.

Before we graph exponential functions, let's discuss some characteristics of the graphs of exponential functions.

Graphs of Exponential Functions

For all exponential functions of the form $y = a^x$ or $f(x) = a^x$, where $a > 0$ and $a \neq 1$,

1. The domain of the function is $(-\infty, \infty)$.

2. The range of the function is $(0, \infty)$.

3. The graph of the function passes through the points $\left(-1, \frac{1}{a}\right)$, $(0, 1)$, and $(1, a)$.

In most cases, a reasonably good exponential graph can be drawn from only the three points listed in item 3. When $a > 1$, the graph becomes almost horizontal to the left of $\left(-1, \frac{1}{a}\right)$ and somewhat vertical to the right of $(1, a)$; see Example 1. When $0 < a < 1$, the graph becomes almost horizontal to the right of $(1, a)$ and somewhat vertical to the left of $\left(-1, \frac{1}{a}\right)$; see Example 2.

EXAMPLE 1 ▶ Graph the exponential function $y = 2^x$. State the domain and range of the function.

Solution   The function is of the form $y = a^x$, where $a = 2$. First, construct a table of values. In the table, the three points listed in item 3 of the box are shown in red.

x	-4	-3	-2	-1	0	1	2	3	4
y	$\frac{1}{16}$	$\frac{1}{8}$	$\frac{1}{4}$	$\frac{1}{2}$	1	2	4	8	16

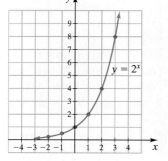

FIGURE 13.16

Now plot these points and connect them with a smooth curve (**Fig. 13.16**). The three ordered pairs in red in the table are marked in red on the graph.

$$\text{Domain: } \mathbb{R}$$
$$\text{Range: } \{y \mid y > 0\}$$

The domain of this function is the set of real numbers, $\mathbb{R}$. The range is the set of values greater than 0. If you study the equation $y = 2^x$, you should realize that y must always be positive because 2 is positive.

▶ **Now Try Exercise 7**

EXAMPLE 2 ▶ Graph $y = \left(\dfrac{1}{2}\right)^x$. State the domain and range of the function.

Solution   This function is of the form $y = a^x$, where $a = \dfrac{1}{2}$. Construct a table of values and plot the curve (**Fig. 13.17**).

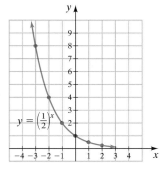

x	-4	-3	-2	-1	0	1	2	3	4
y	16	8	4	2	1	$\dfrac{1}{2}$	$\dfrac{1}{4}$	$\dfrac{1}{8}$	$\dfrac{1}{16}$

FIGURE 13.17

The domain is the set of real numbers, $\mathbb{R}$. The range is $\{y \mid y > 0\}$.

▶ **Now Try Exercise 13**

Note that the graphs in **Figures 13.16** and **13.17** are both graphs of one-to-one functions. *The graphs of exponential functions of the form* $y = a^x$ *are similar to* **Figure 13.16** *when* $a > 1$ *and similar to* **Figure 13.17** *when* $0 < a < 1$. Note that $y = 1^x$ is not a one-to-one function, so we exclude it from our discussion of exponential functions.

What will the graph of $y = 2^{-x}$ look like? Remember that 2^{-x} means $\dfrac{1}{2^x}$ *or* $\left(\dfrac{1}{2}\right)^x$. Thus, the graph of $y = 2^{-x}$ will be identical to the graph in **Figure 13.17**. Now consider the equation $y = \left(\dfrac{1}{2}\right)^{-x}$. This equation may be rewritten as $y = 2^x$ since $\left(\dfrac{1}{2}\right)^{-x} = \left(\dfrac{2}{1}\right)^x = 2^x$. Thus, the graph of $y = \left(\dfrac{1}{2}\right)^{-x}$ will be identical to the graph in **Figure 13.16**.

USING YOUR GRAPHING CALCULATOR

In **Figure 13.18**, we show the graph of the function $y = 2^x$ on the standard window of a graphing calculator. In this chapter, we will sometimes use equations like $y = 2000(1.08)^x$. If you were to graph this function on a standard calculator window, you would not see any of the graph. Can you explain why? By observing the function, can you determine the y-intercept of the graph? To determine the y-intercept, substitute 0 for x. When you do so, you find that the y-intercept is at $2000(1.08)^0 = 2000(1) = 2000$. In **Figure 13.19**, we show the graph of $y = 2000(1.08)^x$.

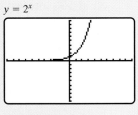

FIGURE 13.18

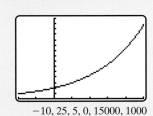

$-10, 25, 5, 0, 15000, 1000$

FIGURE 13.19

EXAMPLE 3 ▶ **Pennies Add Up** Jennifer Hewlett told her young son that if he did his chores, she would give him 2 cents the first week and double the amount each week for the next 10 weeks. The number of cents her son would receive in any given week, w, can be determined by the function $n(w) = 2^w$. Determine the number of cents Jennifer would give her son in week 8.

Solution   By evaluating 2^8 on a calculator, we determine that in week 8 Jennifer would give her son 256 cents, or \$2.56.

▶ **Now Try Exercise 35**

2 Solve Applications of Exponential Functions

Exponential functions are often used to describe the growth and decay of certain quantities. The next four examples are illustrations of exponential functions.

EXAMPLE 4 ▸ **Value of a Jeep** Ronald Yates just bought a new Jeep for $22,000. Assume the value of the Jeep depreciates at a rate of 20% per year. Therefore, the value of the Jeep is 80% of the previous year's value. One year from now, its value will be $22,000(0.80). Two years from now, its value will be $22,000(0.80)(0.80) = $22,000(0.80)^2$ and so on. Therefore, the formula for the value of the Jeep is

$$v(t) = 22{,}000(0.80)^t$$

where t is time in years. Find the value of the Jeep **a)** 1 year from now and **b)** 5 years from now.

Solution

a) To find the value 1 year from now, substitute 1 for t.

$$v(t) = 22{,}000(0.80)^t$$
$$v(1) = 22{,}000(0.80)^1 \qquad \textit{Substitute 1 for t.}$$
$$= 17{,}600$$

One year from now, the value of the Jeep will be $17,600.

b) To find the value 5 years from now, substitute 5 for t.

$$v(t) = 22{,}000(0.80)^t$$
$$v(5) = 22{,}000(0.80)^5 \qquad \textit{Substitute 5 for t.}$$
$$= 22{,}000(0.32768)$$
$$= 7208.96$$

Five years from now, the value of the Jeep will be $7208.96.

▸ **Now Try Exercise 49**

EXAMPLE 5 ▸ **Compound Interest** We have seen the *compound interest formula* $A = p\left(1 + \dfrac{r}{n}\right)^{nt}$ in earlier chapters. When interest is compounded periodically (yearly, monthly, quarterly), this formula can be used to find the amount, A.

In the formula, r is the interest rate, p is the principal, n is the number of compounding periods per year, and t is the number of years. Suppose that $10,000 is invested at 5% interest compounded quarterly for 6 years. Find the amount in the account after 6 years.

Solution **Understand** We are given that the principal, p, is $10,000. We are also given that the interest rate, r, is 5%. Because the interest is compounded quarterly, the number of compounding periods, n, is 4. The money is invested for 6 years. Therefore, t is 6.

Translate Now we substitute these values into the formula

$$A = p\left(1 + \frac{r}{n}\right)^{nt}$$
$$= 10{,}000\left(1 + \frac{0.05}{4}\right)^{4(6)}$$

Carry Out
$$= 10{,}000(1 + 0.0125)^{24}$$
$$= 10{,}000(1.0125)^{24}$$
$$\approx 10{,}000(1.347351) \qquad \textit{From a calculator}$$
$$\approx 13{,}435.51$$

Answer The original $10,000 has grown to about $13,435.51 after 6 years.

▸ **Now Try Exercise 41**

EXAMPLE 6 ▶ **Carbon 14 Dating** Carbon 14 dating is used by scientists to find the age of fossils and artifacts. The formula used in carbon dating is

$$A = A_0 \cdot 2^{-t/5600}$$

where A_0 represents the amount of carbon 14 present when the fossil was formed and A represents the amount of carbon 14 present after t years. If 500 grams of carbon 14 were present when an organism died, how many grams will be found in the fossil 2000 years later?

Solution   Understand   When the fossil died, 500 grams of carbon 14 were present. Therefore, $A_0 = 500$. To find out how many grams of carbon 14 will be present 2000 years later, we substitute 2000 for t in the formula.

Translate

$$A = A_0 \cdot 2^{-t/5600}$$
$$= 500(2)^{-2000/5600}$$

Carry Out

$$\approx 500(0.7807092) \quad \textit{From a calculator}$$
$$\approx 390.35 \text{ grams}$$

Answer   After 2000 years, about 390.35 of the original 500 grams of carbon 14 are still present.

▶ **Now Try Exercise 43**

EXAMPLE 7 ▶ **Medicare Premiums** Monthly Medicare Part B premiums have been on the rise since 2000. The graph in **Figure 13.20** shows monthly Medicare Part B premiums for the years from 2000 to 2005.

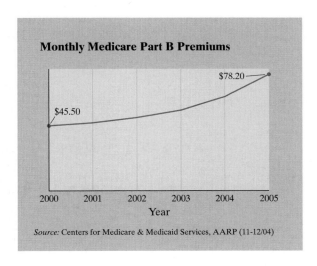

Monthly Medicare Part B Premiums

$78.20

$45.50

2000   2001   2002   2003   2004   2005
Year

Source: Centers for Medicare & Medicaid Services, AARP (11-12/04)

FIGURE 13.20

An exponential function that closely approximates this curve is $f(t) = 44.584(1.11)^t$. In this function, $f(t)$ is the monthly premium and t is the number of years since 2000. Assume that the monthly Medicare Part B premiums continue to rise as in the past. Use this function to estimate the monthly Medicare Part B premiums in **a)** 2006 and **b)** 2010.

Solution.   **a)** Understand   In this function, t is years since 2000. The year 2006 would represent $t = 6$ as 2006 is 6 years after 2000. To find the monthly premium in 2006, we need to evaluate the function for $t = 6$.

Translate and Carry Out $f(t) = 44.584(1.11)^t$

$$f(6) = 44.584(1.11)^6 \approx \$83.39 \quad \textit{From a calculator}$$

Answer   Therefore, the monthly Medicare Part B premium in 2006 would be about $83.39.

b) Since 2010 is 10 years after 2000, to find the monthly premium, we need to evaluate the function for $t = 10$.

$$f(t) = 44.584(1.11)^t$$
$$f(10) = 44.584(1.11)^{10} \approx \$126.59 \quad \textit{From a calculator}$$

Answer   Therefore, the monthly Medicare Part B premium in 2010 would be about $126.59.

▶ **Now Try Exercise 53**

EXERCISE SET 13.2

Concept/Writing Exercises

1. What are exponential functions?

2. Consider the exponential function $y = 2^x$.
   a) As x increases, what happens to y?
   b) Can y ever be 0? Explain.
   c) Can y ever be negative? Explain.

3. Consider the exponential function $y = \left(\frac{1}{2}\right)^x$.
   a) As x increases, what happens to y?
   b) Can y ever be 0? Explain.
   c) Can y ever be negative? Explain.

4. Consider the exponential function $y = 2^{-x}$. Write an equivalent exponential function that does not contain a negative sign in the exponent. Explain how you obtained your answer.

5. Consider the equations $y = 2^x$ and $y = 3^x$.
   a) Will both graphs have the same or different y-intercepts? Determine their y-intercepts.
   b) How will the graphs of the two functions compare?

6. Consider the equations $y = \left(\frac{1}{2}\right)^x$ and $y = \left(\frac{1}{3}\right)^x$.
   a) Will both graphs have the same or different y-intercepts? Determine their y-intercepts.
   b) How will the graphs of the two functions compare?

Practice the Skills

Graph each exponential function.

7. $y = 2^x$

8. $y = 3^x$

9. $y = \left(\frac{1}{2}\right)^x$

10. $y = \left(\frac{1}{3}\right)^x$

11. $y = 4^x$

12. $y = 5^x$

13. $y = \left(\frac{1}{4}\right)^x$

14. $y = \left(\frac{1}{5}\right)^x$

15. $y = 3^{-x}$

16. $y = 4^{-x}$

17. $y = \left(\frac{1}{3}\right)^{-x}$

18. $y = \left(\frac{1}{4}\right)^{-x}$

19. $y = 2^{x-1}$

20. $y = 2^{x+1}$

21. $y = \left(\frac{1}{3}\right)^{x+1}$

22. $y = \left(\frac{1}{3}\right)^{x-1}$

23. $y = 2^x + 1$

24. $y = 2^x - 1$

25. $y = 3^x - 1$

26. $y = 3^x + 2$

Problem Solving

27. We stated earlier that, for exponential functions $f(x) = a^x$, the value of a cannot equal 1.
   a) What does the graph of $f(x) = a^x$ look like when $a = 1$?
   b) Is $f(x) = a^x$ a function when $a = 1$?
   c) Does $f(x) = a^x$ have an inverse function when $a = 1$? Explain your answer.

28. How will the graphs of $y = a^x$ and $y = a^x + k, k > 0$, compare?

29. How will the graphs of $y = a^x$ and $y = a^x - k, k > 0$, compare?

30. For $a > 1$, how will the graphs of $y = a^x$ and $y = a^{x+1}$ compare?

31. For $a > 1$, how will the graphs of $y = a^x$ and $y = a^{x+2}$ compare?

32. a) Is $y = x^\pi$ an exponential function? Explain.
   b) Is $y = \pi^x$ an exponential function? Explain.

33. U.S. Population The following graph shows the growth in the U.S. population of people age 85 and older for the years from 1960 to 2000 and projected to 2050. The exponential function that closely approximates this graph is

$$f(t) = 0.592(1.042)^t$$

In this function, $f(t)$ is the population, in millions, of people age 85 and older and t is the number of years since 1960. Assuming this trend continues, use this function to estimate the number of U.S. people age 85 and older in **a)** 2060. **b)** 2100.

U.S. Population 85 and Older

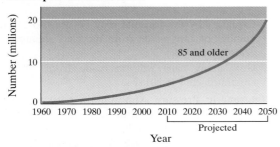

Source: U.S. Census Bureau, Decennial Census and Projections, Older Americans 2004, www.agingstats.gov

34. World Population Since about 1650 the world population has been growing exponentially. The exponential function that closely approximates the world population from 1650 and projected to 2015 is

$$f(t) = \frac{1}{2}(2.718)^{0.0072t}$$

In this function, $f(t)$ is the world population, in billions of people, and t is the number of years since 1650. If this trend continues, estimate the world population in **a)** 2010. **b)** 2015.

35. Doubling If $2 is doubled each day for 9 days, determine the amount on day 9.

36. Doubling If $2 is doubled each day for 12 days, determine the amount on day 12.

37. Simple and Compound Interest The following graph indicates linear growth of $100 invested at 7% simple interest and exponential growth at 7% interest compounded annually. In the formulas, A represents the amount in dollars and t represents the time in years.

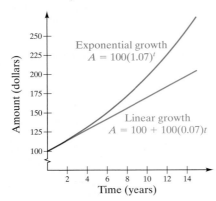

a) Use the graph to estimate the doubling time for $100 invested at 7% simple interest.

b) Estimate the doubling time for $100 invested at 7% interest compounded annually.

c) Estimate the difference in amounts after 10 years for $100 invested by each method.

d) Most banks compound interest daily instead of annually. What effect does this have on the total amount? Explain.

38. Outstanding Consumer Credit The following graph shows the outstanding consumer credit, in trillions of dollars, for the years 2001 to 2004. The exponential function that closely approximates these data is

$$f(t) = 1.841(1.045)^t$$

In this function, $f(t)$ is the outstanding consumer credit, in trillions of dollars, and t is the years since 2001. Assuming that this trend continues, use this function to estimate the outstanding consumer credit in **a)** 2007. **b)** 2011.

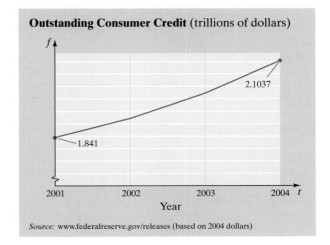

Source: www.federalreserve.gov/releases (based on 2004 dollars)

39. Bacteria in a Petri Dish Five bacteria are placed in a petri dish. The population will triple every day. The formula for the number of bacteria in the dish on day t is

$$N(t) = 5(3)^t$$

where t is the number of days after the five bacteria are placed in the dish. How many bacteria are in the dish 2 days after the four bacteria are placed in the dish?

40. Bacteria in a Petri Dish Refer to Exercise 39. How many bacteria are in the dish 6 days after the five bacteria are placed in the dish?

41. Compound Interest If Don Gecewicz invests $5000 at 6% interest compounded quarterly, find the amount after 4 years (see Example 5).

42. Compound Interest If Don Treadwell invests $8000 at 4% interest compounded quarterly, find the amount after 5 years.

43. Carbon 14 Dating If 12 grams of carbon 14 are originally present in a certain animal bone, how much will remain at the end of 1000 years? Use $A = A_0 \cdot 2^{-t/5600}$ (see Example 6).

44. Carbon 14 Dating If 60 grams of carbon 14 are originally present in the fossil Tim Jonas found at an archeological site, how much will remain after 10,000 years?

45. Radioactive Substance The amount of a radioactive substance present, in grams, at time t, in years, is given by the formula $y = 80(2)^{-0.4t}$. Find the number of grams present in **a)** 10 years. **b)** 100 years.

46. Radioactive Substance The amount of a radioactive substance present, in grams, at time t, in years, is given by the formula $y = 20(3)^{-0.6t}$. Find the number of grams present in 4 years.

47. Population The expected future population of Ackworth, which now has 2000 residents, can be approximated by the formula $y = 2000(1.2)^{0.1t}$, where t is the number of years in the future. Find the expected population of the town in **a)** 10 years. **b)** 50 years.

48. Population The expected future population of Antwerp, which currently has 6800 residents, can be approximated by the formula $y = 6800(1.4)^{-0.2t}$, where t is the number of years in the future. Find the expected population of the town 30 years in the future.

49. Value of an SUV The cost of a new SUV is $24,000. If it depreciates at a rate of 18% per year, the value of the SUV in t years can be approximated by the formula

$$V(t) = 24{,}000(0.82)^t$$

Find the value of the SUV in 4 years.

50. Value of an ATV The cost of a new all-terrain vehicle is $6200. If it depreciates at a rate of 15% per year, the value of the ATV in t years can be approximated by the formula

$$V(t) = 6200(0.85)^t$$

Find the value of the ATV in 10 years.

51. Water Use The average U.S. resident used about 580,000 gallons of water in 2005. Suppose that each year after 2005 the average resident is able to reduce the amount of water used by 5%. The amount of water used by the average resident t years after 2005 could then be found by the formula $A = 580{,}000(0.95)^t$.

a) Explain why this formula may be used to find the amount of water used.

b) What would be the average amount of water used in the year 2009?

52. Recycling Aluminum Currently, about $\frac{2}{3}$ of all aluminum cans are recycled each year, while about $\frac{1}{3}$ are disposed of in landfills. The recycled aluminum is used to make new cans. Americans used about 190,000,000 aluminum cans in 2004. The number of new cans made each year from recycled 2004 aluminum cans n years later can be estimated by the formula

$$A = 190{,}000{,}000\left(\frac{2}{3}\right)^n.$$

a) Explain why the formula may be used to estimate the number of cans made from recycled aluminum cans n years after 2004.

b) How many cans will be made from 2004 recycled aluminum cans in 2011?

53. Atmospheric Pressure Atmospheric pressure varies with altitude. The greater the altitude, the lower the pressure, as shown in the following graph.

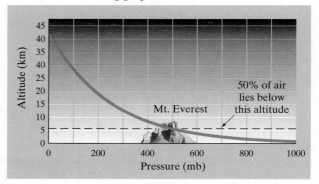

The equation $A = 41.97(0.996)^x$ can be used to estimate the altitude, A, in kilometers, for a given pressure, x, in millibars (mb). If the atmospheric pressure on top of Mt. Everest is about 389 mb, estimate the altitude of the top of Mt. Everest.

54. Centenarians Based on projections of the U.S. Census Bureau, the number of centenarians (people age 100 or older) will grow exponentially beyond 1995 (see the graph below). The function

$$f(t) = 71.24(1.045)^t$$

can be used to approximate the number of centenarians, in thousands, in the United States where t is time in years since 1995. Use this function to estimate the number of centenarians in **a)** 2060. **b)** 2070.

Number of Centenarians in the United States

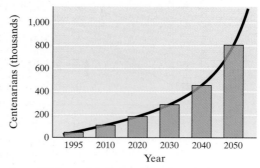

Source: U.S. Bureau of the Census; middle series projections

55. In Exercise 37, we graphed the amount for various years when $100 is invested at 7% simple interest and at 7% interest compounded annually.

   a) Use the compound interest formula given in Example 5 to determine the amount if $100 is compounded daily at 7% for 10 years (assume 365 days per year).

   b) Estimate the difference in the amount in 10 years for the $100 invested at 7% simple interest versus the 7% interest compounded daily.

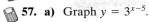 **56.** Graph $y = 2^x$ and $y = 3^x$ on the same window.

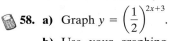 **57.** **a)** Graph $y = 3^{x-5}$.

   b) Use your graphing calculator to solve the equation $4 = 3^{x-5}$. Round your answer to the nearest hundredth.

58. **a)** Graph $y = \left(\dfrac{1}{2}\right)^{2x+3}$.

   b) Use your graphing calculator to solve the equation $-3 = \left(\dfrac{1}{2}\right)^{2x+3}$.

Challenge Problem

59. Suppose Bob Jenkins gives Carol Dantuma $1 on day 1, $2 on day 2, $4 on day 3, $8 on day 4, and continues this doubling process for 30 days.
   a) Determine how much Bob will give Carol on day 15.
   b) Determine how much Bob will give Carol on day 20.
   c) Express the amount, using exponential form, that Bob gives Carol on day n.

   d) How much, in dollars, will Bob give Carol on day 30? Write the amount in exponential form. Then use a calculator to evaluate.

   e) Express the total amount Bob gives Carol over the 30 days as a sum of exponential terms. (Do not find the actual value.)

Group Activity

60. Functions that are exponential or are approximately exponential are commonly seen.

   a) Have each member of the group individually determine a function not given in this section that may approximate an exponential function. You may use newspapers, books, or other sources.

   b) As a group, discuss one another's functions. Determine whether each function presented is an exponential function.

   c) As a group, write a paper that discusses each of the exponential functions and state why you believe each function is exponential.

Cumulative Review Exercises

[5.4] **61.** Consider the polynomial
$$2.3x^4y - 6.2x^6y^2 + 9.2x^5y^2$$

   a) Write the polynomial in descending order of the variable x.

   b) What is the degree of the polynomial?

   c) What is the leading coefficient?

[8.5] **62.** If $f(x) = x + 5$ and $g(x) = x^2 - 2x + 4$, find $(f \cdot g)(x)$.

[11.1] **63.** Write $\sqrt{a^2 - 8a + 16}$ as an absolute value.

[11.3] **64.** Simplify $\sqrt[4]{\dfrac{32x^5y^9}{2y^3z}}$.

13.3 Logarithmic Functions

1. Convert from exponential form to logarithmic form.

2. Graph logarithmic functions.

3. Compare the graphs of exponential and logarithmic functions.

4. Solve applications of logarithmic functions.

1 Convert from Exponential Form to Logarithmic Form

Now we are ready to introduce **logarithms**. Consider the exponential function $y = 2^x$. Recall from Section 13.1 that to find the inverse function, we interchange x and y and solve the resulting equation for y. Interchanging x and y gives the equation $x = 2^y$. But at this time we have no way of solving the equation $x = 2^y$ for y. To solve this equation for y, we introduce a new definition.

> **Logarithms**
>
> For all positive numbers a, where $a \neq 1$,
> $$y = \log_a x \quad \text{means} \quad x = a^y$$

By the definition of logarithm, $x = 2^y$ means $y = \log_2 x$. We can therefore reason that $y = 2^x$ and $y = \log_2 x$ are inverse functions. In general, $y = a^x$ and $y = \log_a x$ are inverse functions.

In the equation $y = \log_a x$, the word *log* is an abbreviation for the word *logarithm*; $y = \log_a x$ is read "y is the logarithm of x to the base a." The letter y represents the logarithm, the letter a represents the base, and the letter x represents the number.

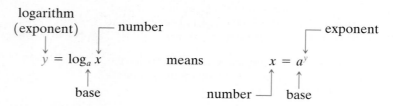

In words, the logarithm of the number x to the base a is the *exponent* to which the base a must be raised to equal the number x. In short, *a logarithm is an exponent*. For example,

$$2 = \log_{10} 100 \quad \text{means} \quad 100 = 10^2$$

In $\log_{10} 100 = 2$, the logarithm is 2, the base is 10, and the number is 100. The logarithm, 2, is the *exponent* to which the base, 10, must be raised to equal the number, 100. Note $10^2 = 100$.

Following are some examples of how an exponential expression can be converted to a logarithmic expression.

Exponential Form	Logarithmic Form
$10^0 = 1$	$\log_{10} 1 = 0$
$4^2 = 16$	$\log_4 16 = 2$
$\left(\dfrac{1}{2}\right)^5 = \dfrac{1}{32}$	$\log_{1/2} \dfrac{1}{32} = 5$
$5^{-2} = \dfrac{1}{25}$	$\log_5 \dfrac{1}{25} = -2$

Now let's do a few examples involving conversion from exponential form into logarithmic form, and vice versa.

EXAMPLE 1 ▶ Write each equation in logarithmic form.

a) $3^4 = 81$ 　　 **b)** $\left(\dfrac{1}{5}\right)^3 = \dfrac{1}{125}$ 　　 **c)** $2^{-5} = \dfrac{1}{32}$

Solution

a) $\log_3 81 = 4$ 　　 **b)** $\log_{1/5} \dfrac{1}{125} = 3$ 　　 **c)** $\log_2 \dfrac{1}{32} = -5$

▶ **Now Try Exercise 31**

EXAMPLE 2 ▶ Write each equation in exponential form.

a) $\log_7 49 = 2$ 　　 **b)** $\log_4 64 = 3$ 　　 **c)** $\log_{1/3} \dfrac{1}{81} = 4$

Solution

a) $7^2 = 49$ 　　 **b)** $4^3 = 64$ 　　 **c)** $\left(\dfrac{1}{3}\right)^4 = \dfrac{1}{81}$

▶ **Now Try Exercise 47**

EXAMPLE 3 ▶ Write each equation in exponential form; then find the unknown value.

a) $y = \log_5 25$ 　　 **b)** $2 = \log_a 16$ 　　 **c)** $3 = \log_{1/2} x$

Solution

a) $5^y = 25$. Since $5^2 = 25$, $y = 2$.

b) $a^2 = 16$. Since $4^2 = 16$, $a = 4$. Note that a must be greater than 0, so -4 is not a possible answer for a.

c) $\left(\dfrac{1}{2}\right)^3 = x$. Since $\left(\dfrac{1}{2}\right)^3 = \dfrac{1}{8}$, $x = \dfrac{1}{8}$.

▶ **Now Try Exercise 65**

2 Graph Logarithmic Functions

Now that we know how to convert from exponential form into logarithmic form and vice versa, we can graph logarithmic functions. Equations of the form $y = \log_a x$, $a > 0$, $a \neq 1$, and $x > 0$, are called **logarithmic functions**. The graphs of logarithmic functions pass the vertical line test. To graph a logarithmic function, change it to exponential form and then plot points. This procedure is illustrated in Examples 4 and 5.

Before we graph logarithmic functions, let's discuss some characteristics of the graphs of logarithmic functions.

Graphs of Logarithmic Functions

For all logarithmic functions of the form $y = \log_a x$ or $f(x) = \log_a x$, where $a > 0$, $a \neq 1$, and $x > 0$

1. The domain of the function is $(0, \infty)$.

2. The range of the function is $(-\infty, \infty)$.

3. The graph passes through the points $\left(\dfrac{1}{a}, -1\right)$, $(1, 0)$, and $(a, 1)$.

In most cases, a reasonably good logarithmic graph can be drawn from just the three points listed in item 3. When $a > 1$, the graph becomes almost vertical to the left of $\left(\dfrac{1}{a}, -1\right)$ and somewhat horizontal to the right of $(a, 1)$, see Example 4.

When $0 < a < 1$, the graph becomes almost vertical to the left of $(a, 1)$ and somewhat horizontal to the right of $\left(\dfrac{1}{a}, -1\right)$, see Example 5.

EXAMPLE 4 ▶ Graph $y = \log_2 x$. State the domain and range of the function.

Solution　This is an equation of the form $y = \log_a x$, where $a = 2$. $y = \log_2 x$ means $x = 2^y$. Using $x = 2^y$, construct a table of values. The table will be easier to develop by selecting values for y and finding the corresponding values for x. In the table, the three points listed in item 3 in the box are shown in blue.

x	$\dfrac{1}{16}$	$\dfrac{1}{8}$	$\dfrac{1}{4}$	$\dfrac{1}{2}$	1	2	4	8	16
y	-4	-3	-2	-1	0	1	2	3	4

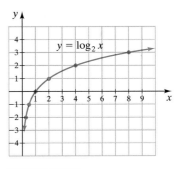

Now draw the graph (**Fig. 13.21**). The three ordered pairs in blue in the table are marked in blue on the graph. The domain, the set of x-values, is $\{x \mid x > 0\}$. The range, the set of y-values, is the set of all real numbers, $\mathbb{R}$.

FIGURE 13.21

▶ **Now Try Exercise 11**

EXAMPLE 5 ▶ Graph $y = \log_{1/2} x$. State the domain and range of the function.

Solution This is an equation of the form $y = \log_a x$, where $a = \dfrac{1}{2}$. $y = \log_{1/2} x$ means $x = \left(\dfrac{1}{2}\right)^y$. Construct a table of values by selecting values for y and finding the corresponding values of x.

x	16	8	4	2	1	$\dfrac{1}{2}$	$\dfrac{1}{4}$	$\dfrac{1}{8}$	$\dfrac{1}{16}$
y	-4	-3	-2	-1	0	1	2	3	4

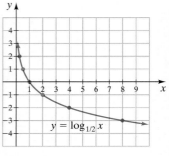

The graph is illustrated in **Figure 13.22**. The domain is $\{x \mid x > 0\}$. The range is the set of real numbers, $\mathbb{R}$.

FIGURE 13.22

▶ **Now Try Exercise 13**

If we study the domains in Examples 4 and 5, we see that the domains of both $y = \log_2 x$ and $y = \log_{1/2} x$ are $\{x \mid x > 0\}$. In fact, **for any logarithmic function $y = \log_a x$, the domain is $\{x \mid x > 0\}$.** Also note that the graphs in Examples 4 and 5 are both graphs of one-to-one functions.

3 Compare the Graphs of Exponential and Logarithmic Functions

Recall that to find inverse functions we switch x and y and solve the resulting equation for y. Consider $y = a^x$. If we switch x and y, we get $x = a^y$. By our definition of logarithm, this function may be rewritten as $y = \log_a x$, which is an equation solved for y. Therefore, $y = a^x$ *and* $y = \log_a x$ *are inverse functions.* We may therefore write: if $f(x) = a^x$, then $f^{-1}(x) = \log_a x$.

In **Figure 13.23**, on page 829, we show general graphs of $y = a^x$ and $y = \log_a x$, $a > 1$, on the same axes. Notice they are symmetric about the line $y = x$. Also, notice the following boxed information.

Graph Characteristics

	EXPONENTIAL FUNCTION $y = a^x \ (a > 0, a \neq 1)$		LOGARITHMIC FUNCTION $y = \log_a x \ (a > 0, a \neq 1)$
Domain:	$(-\infty, \infty)$		$(0, \infty)$
Range:	$(0, \infty)$		$(-\infty, \infty)$
Points on graph:	$\left(-1, \dfrac{1}{a}\right)$ $(0, 1)$ $(1, a)$	*x* becomes *y*, *y* becomes *x*	$\left(\dfrac{1}{a}, -1\right)$ $(1, 0)$ $(a, 1)$

From the information in the box, we can see that the range of the exponential function is the domain of the logarithmic function, and vice versa. We can also see that the x- and y-values in the ordered pairs are switched in the exponential and logarithmic functions.

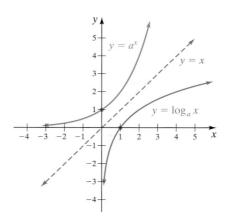

FIGURE 13.23

The graphs of $y = 2^x$ and $y = \log_2 x$ are illustrated in **Figure 13.24**. The graphs of $y = \left(\dfrac{1}{2}\right)^x$ and $y = \log_{1/2} x$ are illustrated in **Figure 13.25**. In each figure, the graphs are inverses of each other and are symmetric with respect to the line $y = x$.

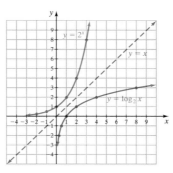

FIGURE 13.24

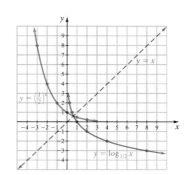

FIGURE 13.25

4 Solve Applications of Logarithmic Functions

We will see many applications of logarithms later, but let's look at one important application now.

EXAMPLE 6 ▶ Earthquakes Logarithms are used to measure the magnitude of earthquakes. The Richter scale for measuring earthquakes was developed by Charles R. Richter. The magnitude, R, of an earthquake on the Richter scale is given by the formula

$$R = \log_{10} I$$

where I represents the number of times greater (or more intense) the earthquake is than the smallest measurable activity that can be measured on a seismograph.

a) If an earthquake measures 4 on the Richter scale, how many times more intense is it than the smallest measurable activity?

b) How many times more intense is an earthquake that measures 5 on the Richter scale than an earthquake that measures 4?

Solution   **a)** Understand   The Richter number, R, is 4. To find how many times more intense the earthquake is than the smallest measurable activity, I, we substitute $R = 4$ into the formula and solve for I.

Translate $R = \log_{10} I$

$4 = \log_{10} I$

Carry Out $10^4 = I$ *Change to exponential form.*
                                             $10{,}000 = I$

Answer   Therefore, an earthquake that measures 4 on the Richter scale is 10,000 times more intense than the smallest measurable activity.

b) $5 = \log_{10} I$
                                             $10^5 = I$ *Change to exponential form.*
                                             $100{,}000 = I$

Since $(10{,}000)(10) = 100{,}000$, an earthquake that measures 5 on the Richter scale is 10 times more intense than an earthquake that measures 4 on the Richter scale .

▶ **Now Try Exercise 113**

EXERCISE SET 13.3   *Math*   **MyMathLab**
MathXL®   MyMathLab

Concept/Writing Exercises

1. Consider the logarithmic function $y = \log_a x$.

   a) What are the restrictions on a?

   b) What is the domain of the function?

   c) What is the range of the function?

2. Write $y = \log_a x$ in exponential form.

3. If some points on the graph of the exponential function

$$f(x) = a^x \text{ are } \left(-3, \frac{1}{27}\right), \left(-2, \frac{1}{9}\right), \left(-1, \frac{1}{3}\right), (0, 1), (1, 3),$$

$(2, 9)$, and $(3, 27)$, list some points on the graph of the logarithmic function $g(x) = \log_a x$. Explain how you determined your answer.

4. For the logarithmic function $y = \log_a(x - 3)$, what must be true about x? Explain.

5. Discuss the relation between the graphs of $y = a^x$ and $y = \log_a x$ for $a > 0$ and $a \neq 1$.

6. What is the x-intercept of the graph of an equation of the form $y = \log_a x$?

Practice the Skills

Graph the logarithmic function.

7. $y = \log_2 x$ **8.** $y = \log_3 x$ **9.** $y = \log_{1/2} x$ **10.** $y = \log_{1/3} x$

11. $y = \log_5 x$ **12.** $y = \log_4 x$ **13.** $y = \log_{1/5} x$ **14.** $y = \log_{1/4} x$

Graph each pair of functions on the same axes.

15. $y = 2^x, y = \log_{1/2} x$ **16.** $y = \left(\frac{1}{2}\right)^x, y = \log_2 x$ **17.** $y = 2^x, y = \log_2 x$ **18.** $y = \left(\frac{1}{2}\right)^x, y = \log_{1/2} x$

Write each equation in logarithmic form.

19. $2^3 = 8$ **20.** $3^5 = 243$ **21.** $3^2 = 9$

22. $2^6 = 64$ **23.** $16^{1/2} = 4$ **24.** $49^{1/2} = 7$

25. $8^{1/3} = 2$ **26.** $16^{1/4} = 2$ **27.** $\left(\frac{1}{2}\right)^5 = \frac{1}{32}$

28. $\left(\frac{1}{3}\right)^4 = \frac{1}{81}$ **29.** $2^{-3} = \frac{1}{8}$ **30.** $6^{-3} = \frac{1}{216}$

31. $4^{-3} = \frac{1}{64}$ **32.** $81^{1/2} = 9$ **33.** $64^{1/3} = 4$

34. $5^{-4} = \frac{1}{625}$ **35.** $8^{-1/3} = \frac{1}{2}$ **36.** $16^{-1/2} = \frac{1}{4}$

37. $81^{-1/4} = \frac{1}{3}$ **38.** $32^{-1/5} = \frac{1}{2}$ **39.** $10^{0.8451} = 7$

40. $10^{1.0792} = 12$ **41.** $e^2 = 7.3891$ **42.** $e^{-1/2} = 0.6065$

43. $a^n = b$ **44.** $c^b = w$

Write each equation in exponential form.

45. $\log_2 8 = 3$

46. $\log_5 125 = 3$

47. $\log_{1/3} \dfrac{1}{27} = 3$

48. $\log_{1/2} \dfrac{1}{64} = 6$

49. $\log_5 \dfrac{1}{25} = -2$

50. $\log_5 \dfrac{1}{625} = -4$

51. $\log_{49} 7 = \dfrac{1}{2}$

52. $\log_{64} 4 = \dfrac{1}{3}$

53. $\log_9 \dfrac{1}{81} = -2$

54. $\log_{10} \dfrac{1}{100} = -2$

55. $\log_{10} \dfrac{1}{1000} = -3$

56. $\log_{10} 1000 = 3$

57. $\log_6 216 = 3$

58. $\log_4 1024 = 5$

59. $\log_{10} 0.62 = -0.2076$

60. $\log_{10} 8 = 0.9031$

61. $\log_e 6.52 = 1.8749$

62. $\log_e 30 = 3.4012$

63. $\log_w s = -p$

64. $\log_r c = -a$

Write each equation in exponential form; then find the unknown value.

65. $\log_4 64 = y$

66. $\log_5 25 = y$

67. $\log_a 125 = 3$

68. $\log_a 81 = 4$

69. $\log_3 x = 3$

70. $\log_2 x = 5$

71. $\log_2 \dfrac{1}{16} = y$

72. $\log_8 \dfrac{1}{64} = y$

73. $\log_{1/2} x = 6$ ——

74. $\log_{1/3} x = 4$ ——

75. $\log_a \dfrac{1}{27} = -3$

76. $\log_9 \dfrac{1}{81} = y$

Evaluate the following.

77. $\log_{10} 1$

78. $\log_{10} 10$

79. $\log_{10} 100$

80. $\log_{10} 1000$

81. $\log_{10} \dfrac{1}{100}$

82. $\log_{10} \dfrac{1}{1000}$

83. $\log_{10} 10{,}000$

84. $\log_{10} 100{,}000$

85. $\log_4 256$

86. $\log_{13} 169$

87. $\log_3 \dfrac{1}{81}$

88. $\log_5 \dfrac{1}{125}$

89. $\log_8 \dfrac{1}{64}$

90. $\log_{14} \dfrac{1}{14}$

91. $\log_9 1$

92. $\log_{15} 1$

93. $\log_9 9$

94. $\log_{12} 12$

95. $\log_4 1024$

96. $\log_2 128$

Problem Solving

97. If $f(x) = 5^x$, what is $f^{-1}(x)$?

98. If $f(x) = \log_6 x$, what is $f^{-1}(x)$?

99. Between which two integers must $\log_3 62$ lie? Explain.

100. Between which two integers must $\log_{10} 0.672$ lie? Explain.

101. Between which two integers must $\log_{10} 425$ lie? Explain.

102. Between which two integers must $\log_5 0.3256$ lie? Explain.

103. For $x > 1$, which will grow faster as x increases, 2^x or $\log_{10} x$? Explain.

104. For $x > 1$, which will grow faster as x increases, x or $\log_{10} x$? Explain.

Change to exponential form, then solve for x. We will discuss rules for solving problems like this in Section 13.4.

105. $x = \log_{10} 10^6$

106. $x = \log_7 7^9$

107. $x = \log_b b^8$

108. $x = \log_e e^5$

Change to logarithmic form, then solve for x. We will discuss rules for solving problems like this in Section 13.4.

109. $x = 10^{\log_{10} 3}$

110. $x = 6^{\log_6 5}$

111. $x = b^{\log_b 9}$

112. $x = c^{\log_c 2}$

113. Earthquake If the magnitude of an earthquake is 7 on the Richter scale, how many times more intense is the earthquake than the smallest measurable activity? Use $R = \log_{10} I$ (see Example 6).

114. Earthquake If the magnitude of an earthquake is 5 on the Richter scale, how many times more intense is the earthquake than the smallest measurable activity? Use $R = \log_{10} I$.

115. Earthquake How many times more intense is an earthquake that measures 6 on the Richter scale than an earthquake that measures 2?

116. Earthquake How many times more intense is an earthquake that measures 4 on the Richter scale than an earthquake that measures 1?

117. Graph $y = \log_2 (x - 1)$.

118. Graph $y = \log_3 (x - 2)$.

Cumulative Review Exercises

[6.3–6.5] *Factor.*

119. $2x^3 - 6x^2 - 36x$ **120.** $x^4 - 16$ **121.** $40x^2 + 52x - 12$ **122.** $6r^2s^2 + rs - 1$

13.4 Properties of Logarithms

1 Use the product rule for logarithms.

2 Use the quotient rule for logarithms.

3 Use the power rule for logarithms.

4 Use additional properties of logarithms.

1 Use the Product Rule for Logarithms

When finding the logarithm of an expression, the expression is called the **argument** of the logarithm. For example, in $\log_{10} 3$, the 3 is the argument, and in $\log_{10}(2x + 4)$, the $(2x + 4)$ is the argument. When the argument contains a variable, we assume that the argument represents a positive value. *Remember, only logarithms of positive numbers exist.*

To be able to do calculations using logarithms, you must understand their properties. The first property we discuss is the product rule for logarithms.

> **Product Rule for Logarithms**
>
> For positive real numbers x, y, and $a, a \neq 1$,
> $$\log_a xy = \log_a x + \log_a y \qquad \textbf{Property 1}$$

This rule tells us that the logarithm of a product of two factors equals the sum of the logarithms of the factors.

To prove this property, we let $\log_a x = m$ and $\log_a y = n$. Remember, logarithms are exponents. Now we write each logarithm in exponential form.

$$\log_a x = m \quad \text{means} \quad a^m = x$$
$$\log_a y = n \quad \text{means} \quad a^n = y$$

By substitution and using the rules of exponents, we see that

$$xy = a^m \cdot a^n = a^{m+n}$$

We can now convert $xy = a^{m+n}$ into logarithmic form.

$$xy = a^{m+n} \quad \text{means} \quad \log_a xy = m + n$$

Finally, substituting $\log_a x$ for m and $\log_a y$ for n, we obtain

$$\log_a xy = \log_a x + \log_a y$$

which is property 1.

Examples of Property 1

$$\log_3 (6 \cdot 7) = \log_3 6 + \log_3 7$$
$$\log_4 3z = \log_4 3 + \log_4 z$$
$$\log_8 x^2 = \log_8 (x \cdot x) = \log_8 x + \log_8 x \text{ or } 2 \log_8 x$$

Property 1, the product rule, can be expanded to three or more factors, for example, $\log_a xyz = \log_a x + \log_a y + \log_a z$.

2 Use the Quotient Rule for Logarithms

Now we give the quotient rule for logarithms, which we refer to as property 2.

> **Quotient Rule for Logarithms**
>
> For positive real numbers x, y, and $a, a \neq 1$,
> $$\log_a \frac{x}{y} = \log_a x - \log_a y \qquad \textbf{Property 2}$$

This rule tells us that the logarithm of a quotient equals the difference between the logarithm of the numerator and the logarithm of the denominator.

Examples of Property 2

$$\log_3 \frac{19}{4} = \log_3 19 - \log_3 4$$

$$\log_6 \frac{x}{3} = \log_6 x - \log_6 3$$

$$\log_5 \frac{z}{z+2} = \log_5 z - \log_5 (z+2)$$

3 Use the Power Rule for Logarithms

The next property we discuss is the power rule for logarithms.

> **Power Rule for Logarithms**
>
> If x and a are positive real numbers, $a \neq 1$, and n is any real number, then
> $$\log_a x^n = n \log_a x \qquad \textbf{Property 3}$$

This rule tells us that the logarithm of a number raised to a power equals the exponent times the logarithm of the number.

Examples of Property 3

$$\log_2 4^3 = 3 \log_2 4$$
$$\log_3 x^2 = 2 \log_3 x$$

$$\log_5 \sqrt{12} = \log_5 (12)^{1/2} = \frac{1}{2} \log_5 12$$

$$\log_8 \sqrt[5]{z+3} = \log_8 (z+3)^{1/5} = \frac{1}{5} \log_8 (z+3)$$

Properties 2 and 3 can be proved in a manner similar to that given for property 1 (see Exercises 79 and 80 on page 837).

EXAMPLE 1 ▶ Use properties 1 through 3 to expand.

a) $\log_8 \dfrac{29}{43}$ **b)** $\log_4 (64 \cdot 180)$ **c)** $\log_{10} (22)^{1/5}$

Solution

a) $\log_8 \dfrac{29}{43} = \log_8 29 - \log_8 43$ *Quotient rule*

b) $\log_4 (64 \cdot 180) = \log_4 64 + \log_4 180$ *Product rule*

c) $\log_{10} (22)^{1/5} = \dfrac{1}{5} \log_{10} 22$ *Power rule*

▶ **Now Try Exercise 11**

Often we will have to use two or more of these properties in the same problem.

EXAMPLE 2 ▶ Expand.

a) $\log_{10} 4(x+2)^3$ **b)** $\log_5 \dfrac{(4-a)^2}{3}$

c) $\log_5 \left(\dfrac{4-a}{3} \right)^2$ **d)** $\log_5 \dfrac{[x(x+4)]^3}{8}$

Solution

a) $\log_{10} 4(x + 2)^3 = \log_{10} 4 + \log_{10} (x + 2)^3$ *Product rule*

$\qquad\qquad\qquad\quad = \log_{10} 4 + 3 \log_{10} (x + 2)$ *Power rule*

b) $\log_5 \dfrac{(4 - a)^2}{3} = \log_5 (4 - a)^2 - \log_5 3$ *Quotient rule*

$\qquad\qquad\quad = 2 \log_5 (4 - a) - \log_5 3$ *Power rule*

c) $\log_5 \left(\dfrac{4 - a}{3} \right)^2 = 2 \log_5 \left(\dfrac{4 - a}{3} \right)$ *Power rule*

$\qquad\qquad\qquad = 2[\log_5 (4 - a) - \log_5 3]$ *Quotient rule*

$\qquad\qquad\qquad = 2 \log_5 (4 - a) - 2 \log_5 3$ *Distributive property*

d) $\log_5 \dfrac{[x(x + 4)]^3}{8} = \log_5 [x(x + 4)]^3 - \log_5 8$ *Quotient rule*

$\qquad\qquad\qquad = 3 \log_5 x(x + 4) - \log_5 8$ *Power rule*

$\qquad\qquad\qquad = 3[\log_5 x + \log_5 (x + 4)] - \log_5 8$ *Product rule*

$\qquad\qquad\qquad = 3 \log_5 x + 3 \log_5 (x + 4) - \log_5 8$ *Distributive property*

▶ **Now Try Exercise 21**

Helpful Hint

In Example 2**b)**, when we expanded $\log_5 \dfrac{(4 - a)^2}{3}$, we first used the quotient rule. In Example 2**c)**, when we expanded $\log_5 \left(\dfrac{4 - a}{3} \right)^2$, we first used the power rule. Do you see the difference in the two problems? In $\log_5 \dfrac{(4 - a)^2}{3}$, just the numerator of the argument is squared; therefore, we use the quotient rule first. In $\log_5 \left(\dfrac{4 - a}{3} \right)^2$, the entire argument is squared, so we use the power rule first.

EXAMPLE 3 ▶ Write each of the following as the logarithm of a single expression.

a) $3 \log_8 (z + 2) - \log_8 z$

b) $\log_7 (x + 1) + 2 \log_7 (x + 4) - 3 \log_7 (x - 5)$

Solution

a) $3 \log_8 (z + 2) - \log_8 z = \log_8 (z + 2)^3 - \log_8 z$ *Power rule*

$\qquad\qquad\qquad\qquad = \log_8 \dfrac{(z + 2)^3}{z}$ *Quotient rule*

b) $\log_7 (x + 1) + 2 \log_7 (x + 4) - 3 \log_7 (x - 5)$

$\quad = \log_7 (x + 1) + \log_7 (x + 4)^2 - \log_7 (x - 5)^3$ *Power rule*

$\quad = \log_7 (x + 1)(x + 4)^2 - \log_7 (x - 5)^3$ *Product rule*

$\quad = \log_7 \dfrac{(x + 1)(x + 4)^2}{(x - 5)^3}$ *Quotient rule*

▶ **Now Try Exercise 39**

Avoiding Common Errors

THE CORRECT RULES ARE

$$\log_a xy = \log_a x + \log_a y$$

$$\log_a \frac{x}{y} = \log_a x - \log_a y$$

Note that

$$\log_a (x + y) \neq \log_a x + \log_a y \qquad \log_a xy \neq (\log_a x)(\log_a y)$$

$$\log_a (x - y) \neq \log_a x - \log_a y \qquad \log_a \frac{x}{y} \neq \frac{\log_a x}{\log_a y}$$

4 Use Additional Properties of Logarithms

The last properties we discuss in this section will be used to solve equations in Section 13.6.

Additional Properties of Logarithms

If $a > 0$, and $a \neq 1$, then

$$\log_a a^x = x \qquad \textbf{Property 4}$$

and $\qquad a^{\log_a x} = x \, (x > 0) \qquad \textbf{Property 5}$

Examples of Property 4	Examples of Property 5
$\log_6 6^5 = 5$	$3^{\log_3 7} = 7$
$\log_9 9^x = x$	$5^{\log_5 x} = x \ (x > 0)$

EXAMPLE 4 ▶ Evaluate. **a)** $\log_5 25$ **b)** $\sqrt{16}^{\,\log_4 9}$

Solution

a) $\log_5 25$ may be written as $\log_5 5^2$. By property 4,

$$\log_5 25 = \log_5 5^2 = 2$$

b) $\sqrt{16}^{\,\log_4 9}$ may be written $4^{\log_4 9}$. By property 5,

$$\sqrt{16}^{\,\log_4 9} = 4^{\log_4 9} = 9$$

▶ **Now Try Exercise 55**

EXERCISE SET 13.4   *Math XL*   *MyMathLab*
MathXL®   MyMathLab

Concept/Writing Exercises

1. Explain the product rule for logarithms.
2. Explain the quotient rule for logarithms.
3. Explain the power rule for logarithms.
4. Explain why we need to stipulate that x and y are positive real numbers when discussing the product and quotient rules.

5. Is $\log_a(xyz) = \log_a x + \log_a y + \log_a z$ a true statement? Explain.
6. Is $\log_b(x + y + z) = \log_b x + \log_b y + \log_b z$ a true statement? Explain.

Practice the Skills

Use properties 1–3 to expand.

7. $\log_4 (3 \cdot 10)$

8. $\log_5 (4 \cdot 7)$

 9. $\log_8 7(x + 3)$

10. $\log_9 x(x + 2)$

11. $\log_2 \dfrac{27}{11}$

12. $\log_5 (41 \cdot 9)$

13. $\log_{10} \dfrac{\sqrt{x}}{x-9}$

14. $\log_5 3^8$

15. $\log_6 x^7$

16. $\log_9 12(4)^6$

17. $\log_4 (r + 7)^5$

18. $\log_8 b^3(b - 2)$

19. $\log_4 \sqrt{\dfrac{a^3}{a+2}}$

20. $\log_9 (x - 6)^3 x^2$

21. $\log_3 \dfrac{d^6}{(a-8)^4}$

22. $\log_7 x^2(x - 13)$

23. $\log_8 \dfrac{y(y+4)}{y^3}$

24. $\log_{10}\left(\dfrac{z}{6}\right)^2$

25. $\log_{10} \dfrac{9m}{8n}$

26. $\log_5 \dfrac{\sqrt{a}\,\sqrt[3]{b}}{\sqrt[4]{c}}$

Write as a logarithm of a single expression.

27. $\log_5 2 + \log_5 8$

28. $\log_3 4 + \log_3 11$

29. $\log_2 9 - \log_2 5$

30. $\log_7 17 - \log_7 3$

31. $6\log_4 2$

32. $\dfrac{1}{3}\log_8 7$

33. $\log_{10} x + \log_{10} (x + 3)$

34. $\log_5 (a + 1) - \log_5 (a + 10)$

35. $2\log_9 z - \log_9 (z - 2)$

36. $3\log_8 y + 2\log_8 (y - 9)$

37. $4(\log_5 p - \log_5 3)$

38. $\dfrac{1}{2}[\log_6 (r - 1) - \log_6 r]$

39. $\log_2 n + \log_2 (n + 4) - \log_2 (n - 3)$

40. $2\log_5 t + 5\log_5 (t - 6) + \log_5 (3t + 7)$

41. $\dfrac{1}{2}[\log_5 (x - 8) - \log_5 x]$

42. $6\log_7 (a + 3) + 2\log_7 (a - 1) - \dfrac{1}{2}\log_7 a$

43. $2\log_9 4 + \dfrac{1}{3}\log_9 (r - 6) - \dfrac{1}{2}\log_9 r$

44. $5\log_6 (x + 3) - [2\log_6 (7x + 1) + 3\log_6 x]$

45. $4\log_6 3 - [2\log_6 (x + 3) + 4\log_6 x]$

46. $2\log_7 (m - 4) + 3\log_7 (m + 3) - [5\log_7 2 + 3\log_7 (m - 2)]$

Find the value by writing each argument using the numbers 2 and/or 5 and using the values $\log_a 2 = 0.3010$ *and* $\log_a 5 = 0.6990$.

47. $\log_a 10$

48. $\log_a 2.5$

49. $\log_a 0.4$

50. $\log_a \dfrac{1}{8}$

51. $\log_a 25$

52. $\log_a \sqrt[3]{5}$

Evaluate (see Example 4).

53. $5^{\log_5 10}$

54. $\log_3 3$

55. $(2^3)^{\log_8 7}$

56. $\log_8 64$

57. $\log_3 27$

58. $2\log_9 \sqrt{9}$

59. $5(\sqrt[3]{27})^{\log_3 5}$

60. $\dfrac{1}{2}\log_6 \sqrt[3]{6}$

Problem Solving

61. For $x > 0$ and $y > 0$, is $\log_a \dfrac{x}{y} = \log_a xy^{-1} = \log_a x + \log_a y^{-1} = \log_a x + \log_a \dfrac{1}{y}$?

62. Read Exercise 61. By the quotient rule, $\log_a \dfrac{x}{y} = \log_a x - \log_a y$. Can we therefore conclude that $\log_a x - \log_a y = \log_a x + \log_a \dfrac{1}{y}$?

63. Use the product rule to show that
$$\log_a \dfrac{x}{y} = \log_a x + \log_a \dfrac{1}{y}$$

64. a) Explain why
$$\log_a \dfrac{3}{xy} \neq \log_a 3 - \log_a x + \log_a y$$

  b) Expand $\log_a \dfrac{3}{xy}$ correctly.

65. Express $\log_a(x^2 - 4) - \log_a(x + 2)$ as a single logarithm and simplify.

66. Express $\log_a(x - 3) - \log_a(x^2 + 5x - 24)$ as a single logarithm and simplify.

67. Is $\log_a(x^2 + 8x + 16) = 2\log_a(x + 4)$? Explain.

68. Is $\log_a(4x^2 - 20x + 25) = 2\log_a(2x - 5)$? Explain.

If $\log_{10} x = 0.4320$, find the following.

69. $\log_{10} x^2$

70. $\log_{10} \sqrt[3]{x}$

71. $\log_{10} \sqrt[4]{x}$

72. $\log_{10} x^{11}$

If $\log_{10} x = 0.5000$ and $\log_{10} y = 0.2000$, find the following.

73. $\log_{10} xy$

74. $\log_{10}\left(\dfrac{x}{y}\right)$

75. Using the information given in the instructions for Exercises 73 and 74, is it possible to find $\log_{10}(x + y)$? Explain.

76. Are the graphs of $y = \log_b x^2$ and $y = 2\log_b x$ the same? Explain your answer by discussing the domain of each equation.

Use properties 1–3 to expand.

77. $\log_2 \dfrac{\sqrt[4]{xy}\,\sqrt[3]{a}}{\sqrt[5]{a - b}}$

78. $\log_3\left[\dfrac{(a^2 + b^2)(c^2)}{(a - b)(b + c)(c + d)}\right]^2$

79. Prove the quotient rule for logarithms.

80. Prove the power rule for logarithms.

Group Activity

Discuss and answer Exercise 81 as a group.

81. Consider $\log_a \dfrac{\sqrt{x^4 y}}{\sqrt{xy^3}}$, where $x > 0$ and $y > 0$.

  a) Group member 1: Expand the expression using the quotient rule.

  b) Group member 2: Expand the expression using the product rule.

  c) Group member 3: First simplify $\dfrac{\sqrt{x^4 y}}{\sqrt{xy^3}}$, then expand the resulting logarithm.

  d) Check each other's work and make sure all answers are correct. Can this expression be simplified by all three methods?

Cumulative Review Exercises

For Exercises 82 and 83, perform each indicated operation.

[7.2] **82.** $\dfrac{2x + 5}{x^2 - 7x + 12} \div \dfrac{x - 4}{2x^2 - x + 15}$

[7.4] **83.** $\dfrac{2x + 5}{x^2 - 7x + 12} - \dfrac{x - 4}{2x^2 - x - 15}$

[7.7] **84. Paint a House.** Mike Eisen can paint a house by himself in 4 days and Jill McGhee can paint the same house by herself in 5 days. How long would it take them to paint the house together?

[11.4] **85.** Multiply and then simplify $\sqrt[3]{4x^4 y^7} \cdot \sqrt[3]{12x^7 y^{10}}$.

[11.6] **86.** Solve $2a - 7\sqrt{a} = 30$.

Mid-Chapter Test: 13.1–13.4

To find out how well you understand the chapter material to this point, take this brief test. The answers, and the section where the material was initially discussed, are given in the back of the book. Review any questions you answered incorrectly.

1. a) Explain how to find $(f \circ g)(x)$.

  b) If $f(x) = 3x + 3$ and $g(x) = 2x + 5$, find $(f \circ g)(x)$.

2. Let $f(x) = x^2 + 5$ and $g(x) = \dfrac{6}{x}$; find

  a) $(f \circ g)(x)$

  b) $(f \circ g)(3)$

  c) $(g \circ f)(x)$

  d) $(g \circ f)(3)$

3. **a)** Explain what it means when a function is a one-to-one function.

   b) Is the function represented by the following graph a one-to-one function? Explain.

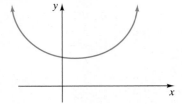

In Exercises 4–6, for each function, **a)** *determine whether it is a one-to-one function;* **b)** *if it is a one-to-one function, find its inverse function.*

4. $\{(-3, 2), (2, 3), (5, 1), (6, 8)\}$

5. $p(x) = \dfrac{1}{3}x - 5$

6. $k(x) = \sqrt{x - 4}, \quad x \geq 4$

7. Let $m(x) = -2x + 4$. Find $m^{-1}(x)$ and then graph $m(x)$ and $m^{-1}(x)$ on the same axes.

Graph each exponential function.

8. $y = 2^x$

9. $y = 3^{-x}$

10. Graph the logarithmic function $y = \log_2 x$.

11. **Bacteria** The number of bacteria in a petri-dish is $N(t) = 5(2)^t$, where t is the number of hours after the 5 original bacteria are placed in the dish. How many bacteria are in the dish
    a) 1 hour later?
    b) 6 hours later?

12. Write $27^{2/3} = 9$ in logarithmic form.

13. Write $\log_2 \dfrac{1}{64} = -6$ in exponential form.

14. Evaluate $\log_5 125$.

15. Solve the equation $\log_{1/4} \dfrac{1}{16} = x$ for x.

16. Solve the equation $\log_x 64 = 3$ for x.

Use properties 1–3 to write as a sum or difference of logarithms.

17. $\log_9 x^2(x - 5)$

18. $\log_5 \dfrac{7m}{\sqrt{n}}$

Write as a single logarithm.

19. $3 \log_2 x + \log_2 (x + 7) - 4 \log_2 (x + 1)$

20. $\dfrac{1}{2}[\log_7 (x + 2) - \log_7 x]$

13.5 Common Logarithms

1 Find common logarithms of powers of 10.

2 Find common logarithms.

3 Find antilogarithms.

1 Find Common Logarithms of Powers of 10

The properties discussed in Section 13.4 can be used with any valid base (a real number greater than 0 and not equal to 1). However, since we are used to working in base 10, we will often use the base 10 when computing with logarithms. **Base 10 logarithms** are called **common logarithms**. When we are working with common logarithms, it is not necessary to list the base. Thus, $\log x$ means $\log_{10} x$.

The properties of logarithms written as common logarithms follow. For positive real numbers x and y, and any real number n,

  1. $\log xy = \log x + \log y$

  2. $\log \dfrac{x}{y} = \log x - \log y$

  3. $\log x^n = n \log x$

The logarithms of most numbers are irrational numbers. Even the values given by calculators are usually only approximations of the actual values. *Even though we are working with approximations when evaluating most logarithms, we generally write the logarithm with an equal sign.* Thus, rather than writing $\log 6 \approx 0.77815$, we will write $\log 6 = 0.77815$. The values given for logarithms are accurate to at least four decimal places.

In Chapter 5 we learned that 1 can be expressed as 10^0 and 10 can be expressed as 10^1. Since, for example, 5 is between 1 and 10, it must also be between 10^0 and 10^1.

$$1 < 5 < 10$$
$$10^0 < 5 < 10^1$$

The number 5 can be expressed as the base 10 raised to an exponent between 0 and 1. The number 5 is approximately equal to $10^{0.69897}$. As is done with logarithms, when writing exponential expressions, we often use the equal sign, even when the values are only approximations. Thus, for example, we will generally write $10^{0.69897} = 5$ rather than $10^{0.69897} \approx 5$.

If you look up log 5 on a calculator, as will be explained shortly, it will display a value about 0.69897. Notice that

$$\log 5 = \boxed{0.69897} \quad \text{and} \quad 5 = 10^{0.69897}$$

We can see that *the common logarithm*, 0.69897, *is the exponent* on the base 10. Now we are ready to define common logarithms.

> ### Common Logarithm
>
> The **common logarithm** of a positive real number is the *exponent* to which the base 10 is raised to obtain the number.
>
> $$\text{If } \log N = L, \quad \text{then} \quad 10^L = N.$$

For example, if log 5 = 0.69897, then $10^{0.69897} = 5$.

Now consider the number 50.

$$10 < 50 < 100$$
$$10^1 < 50 < 10^2$$

The number 50 can be expressed as the base 10 raised to an exponent between 1 and 2. The number $50 = 10^{1.69897}$; thus log 50 = 1.69897.

2 Find Common Logarithms

To find common logarithms of numbers, we can use a calculator that has a logarithm key, $\boxed{\text{LOG}}$.

 USING YOUR CALCULATOR Finding Common Logarithms

Scientific Calculator

To find common logarithms, enter the number, then press the logarithm key. The answer will then be displayed.

EXAMPLE	KEYS TO PRESS	ANSWER DISPLAYED
Find log 400.	400 $\boxed{\text{LOG}}$	2.60206
Find log 0.0538.	0.0538 $\boxed{\text{LOG}}$	−1.2692177

 Graphing Calculator

On some graphing calculators, you first press the $\boxed{\text{LOG}}$ key and then you enter the number. For example, on the TI-84 Plus, you would do the following:

EXAMPLE	KEYS TO PRESS	ANSWER DISPLAYED
Find log 400.	$\boxed{\text{LOG}}$ (400 $\boxed{)}$ $\boxed{\text{ENTER}}$	2.602059991

↑
Generated by calculator

EXAMPLE 1 ▶ Find the exponent to which the base 10 must be raised to obtain the number 43,600.

Solution *We are asked to find the exponent, which is a logarithm.* We need to determine log 43,600. Using a calculator, we find

$$\log 43{,}600 = 4.6394865$$

Thus, the exponent is 4.6394865. Note that $10^{4.6394865} = 43{,}600$.

▶ **Now Try Exercise 7**

3 Find Antilogarithms

The question that should now be asked is, "If we know the common logarithm of a number, how do we find the number?" For example, if $\log N = 3.406$, what is N? To find N, the number, we need to determine the value of $10^{3.406}$. Since

$$10^{3.406} = 2546.830253$$

$N = 2546.830253$. The number 2546.830253 is the *antilogarithm* of 3.406.

When we find the value of the number from the logarithm, we say we are finding the **antilogarithm** or **inverse logarithm**. If the logarithm of N is L, then N is the antilogarithm or inverse logarithm of L.

Antilogarithm
If $\log N = L$, then $N = \text{antilog } L$.

When we are given the common logarithm, which is the exponent on the base 10, the *antilog is the number* obtained when the base 10 is raised to that exponent.

Examples

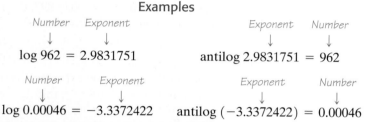

$$\log 962 = 2.9831751 \qquad \text{antilog } 2.9831751 = 962$$

$$\log 0.00046 = -3.3372422 \qquad \text{antilog } (-3.3372422) = 0.00046$$

When finding an antilog, we start with the logarithm, or the exponent, and end with the number equal to 10 raised to that logarithm or exponent. If antilog $(-3.3372422) = 0.00046$, then $10^{-3.3372422} = 0.00046$.

USING YOUR CALCULATOR Finding Antilogarithms

Scientific Calculator

To find antilogarithms on a scientific calculator, enter the logarithm, and press the $\boxed{2^{nd}}$, $\boxed{INV}$, or $\boxed{Shift}$ key, depending upon which of these keys your calculator has. Then press the $\boxed{LOG}$ key. After the $\boxed{LOG}$ key is pressed, the antilog will be displayed.

EXAMPLE	KEYS TO PRESS	ANSWER DISPLAYED
Find antilog 2.9831751.	2.9831751 $\boxed{INV}$ $\boxed{LOG}$	962.00006*
Find antilog (-3.3372422).	3.3372422 $\boxed{+/-}$ $\boxed{INV}$ $\boxed{LOG}$	0.00046**

When you are finding the antilog of a negative value, enter the value and then press the $\boxed{+/-}$ key before pressing the inverse and logarithm keys.

*Some calculators give slightly different answers, depending on their electronics.
**Some calculators may display answers in scientific notation form.

Graphing Calculator

On most graphing calculators, you press the $\boxed{2^{nd}}$ then the $\boxed{LOG}$ key before you enter the logarithm.

On the TI-84 Plus and on certain other calculators, 10^x is printed directly above the $\boxed{LOG}$ key. The antilog is actually the value of 10^x, where x is the logarithm. When you press $\boxed{2^{nd}}$ $\boxed{LOG}$, the TI-84 Plus displays $10^\wedge$ followed by a left parentheses. You then enter the logarithm followed by the $\boxed{)}$ key. After you press $\boxed{ENTER}$, the antilog is displayed.

EXAMPLE	KEYS TO PRESS	ANSWER DISPLAYED
Find antilog 2.9831751.	$\boxed{2^{nd}}$ $\boxed{LOG}$† (2.9831751 $\boxed{)}$ $\boxed{ENTER}$	962.0000619
Find antilog (-3.3372422).	$\boxed{2^{nd}}$ $\boxed{LOG}$ ((−) 3.3372422 $\boxed{)}$ $\boxed{ENTER}$	$4.599999664E^{-}4$††

†Left parenthesis is generated by the TI-84 Plus.
††Recall from scientific notation that this number is 0.0004599999664.

Since we generally do not need the accuracy given by most calculators, in the exercise set that follows we will round logarithms to four decimal places and antilogarithms to three **significant digits**. In a number written in decimal form, any zeros preceding the first nonzero digit are not significant digits. The first nonzero digit in a number, moving from left to right, is the first significant digit.

Examples

0.006 3402	First significant digit is shaded.
3.04 24080	First three significant digits are shaded.
0.0000 138483	First three significant digits are shaded.
206,435.05	First four significant digits are shaded.

Although most antilogarithms will be irrational numbers, when writing antilogarithms we will use an equal sign rather than an approximately equal to sign, just as we did when evaluating logarithms. All antilogarithms will be accurate to at least three significant digits.

EXAMPLE 2 ▸ Find the value obtained when the base 10 is raised to the -1.052 power.

Solution   We are asked to find the value of $10^{-1.052}$. Since we are given the exponent, or logarithm, we can find the value by taking the antilog of -1.052.

$$\text{antilog}\,(-1.052) = 0.0887156$$

Thus, $10^{-1.052} = 0.0887$ rounded to three significant digits.

▸ **Now Try Exercise 55**

EXAMPLE 3 ▸ Find N if $\log N = 4.192$.

Solution   We are given the logarithm and asked to find the antilog, or the number N.

$$\text{antilog}\,4.192 = 15{,}559.6563$$

Thus, $N = 15{,}559.6563$.

▸ **Now Try Exercise 33**

EXAMPLE 4 ▸ Find the following antilogs and round to three significant digits.

a) antilog 6.827   **b)** antilog (-2.35)

Solution

a) Using a calculator, we find antilog $6.827 = 6{,}714{,}288.5$. Rounding to three significant digits, we get antilog $6.827 = 6{,}710{,}000$.

b) Using a calculator, we find antilog $(-2.35) = 0.0044668$. Rounding to three significant digits, we get antilog $(-2.35) = 0.00447$.

▸ **Now Try Exercise 25**

EXAMPLE 5 ▸ **Earthquake** The magnitude of an earthquake on the Richter scale is given by the formula $R = \log I$, where I is the number of times more intense the quake is than the smallest measurable activity. How many times more intense is an earthquake measuring 6.2 on the Richter scale than the smallest measurable activity?

Solution   We want to find the value for I. We are given that $R = 6.2$. Substitute 6.2 for R in the formula $R = \log I$, and then solve for I.

$$R = \log I$$
$$6.2 = \log I \qquad \textit{Substitute 6.2 for R.}$$

To find I, we need to take the antilog of both sides of the equation.

$$\text{antilog } 6.2 = I$$
$$1,580,000 = I$$

Thus, this earthquake is about 1,580,000 times more intense than the smallest measurable activity.

▸ **Now Try Exercise 85**

EXERCISE SET 13.5   Math XL   MyMathLab

Concept/Writing Exercises

1. What are common logarithms?

2. Write $\log N = L$ in exponential form.

3. What are antilogarithms?

4. If $\log 793 = 2.8993$ what is antilog 2.8993?

Practice the Skills

Find the common logarithm of each number. Round the answer to four decimal places.

5. 86

6. 352

7. 19,200

8. 1000

9. 0.0613

10. 941,000

11. 100

12. 0.000835

13. 3.75

14. 0.375

15. 0.0173

16. 0.00872

Find the antilog of each logarithm. Round the answer to three significant digits.

17. 0.2137

18. 1.3845

19. 4.6283

20. 3.5527

21. -1.7086

22. -3.7431

23. 0.0000

24. 5.5922

25. 2.7625

26. -0.1543

27. -4.1390

28. -2.8139

Find each number N. Round N to three significant digits.

29. $\log N = 2.0000$

30. $\log N = 1.4612$

31. $\log N = 3.3817$

32. $\log N = 1.9330$

33. $\log N = 4.1409$

34. $\log N = -2.103$

35. $\log N = -1.06$

36. $\log N = -3.1469$

37. $\log N = -0.6218$

38. $\log N = 1.5177$

39. $\log N = -0.1256$

40. $\log N = -1.3206$

To what exponent must the base 10 be raised to obtain each value? Round your answer to four decimal places.

41. 3560

42. 817,000

43. 0.0727

44. 0.00612

45. 243

46. 8.16

47. 0.00592

48. 73,700,000

Find the value obtained when 10 is raised to the following exponents. Round your answer to three significant digits.

49. 2.8316

50. 3.2473

51. -0.5186

52. -3.7081

53. -1.4802

54. 4.5619

55. 1.3503

56. -2.1918

By changing the logarithm to exponential form, evaluate the common logarithm without the use of a calculator.

57. $\log 1$

58. $\log 100$

59. $\log 0.1$

60. $\log 1000$

61. $\log 0.01$

62. $\log 10$

63. $\log 0.001$

64. 0.0001

In Section 13.4, we stated that for $a > 0$, and $a \ne 1$, $\log_a a^x = x$ and $a^{\log_a x} = x \, (x > 0)$. Rewriting these properties using common logarithms $(a = 10)$, we obtain $\log 10^x = x$ and $10^{\log x} = x \, (x > 0)$, respectively. Use these properties to evaluate the following.

65. $\log 10^7$

66. $\log 10^{3.4}$

67. $10^{\log 7}$

68. $10^{\log 3.4}$

69. $4 \log 10^{5.2}$

70. $8 \log 10^{1.2}$

71. $5(10^{\log 8.3})$

72. $2.3(10^{\log 5.2})$

Problem Solving

73. On your calculator, you find log 462 and obtain the value 1.6646. Can this value be correct? Explain.

74. On your calculator, you find log 6250 and obtain the value 2.7589. Can this value be correct? Explain.

75. On your calculator, you find log 0.163 and obtain the value −2.7878. Can this value be correct? Explain.

76. On your calculator, you find log (−1.23) and obtain the value 0.08991. Can this value be correct? Explain.

77. Is $\log \dfrac{y}{4x} = \log y - \log 4 + \log x$? Explain.

78. Is $\log \dfrac{5x^2}{3} = 2(\log 5 + \log x) - \log 3$? Explain.

If log 25 = 1.3979 *and* log 5 = 0.6990, *find the answer if possible. If it is not possible to find the answer, indicate so. Do not find the logarithms on your calculator except to check answers.*

79. log 125

80. log 35

81. $\log \dfrac{1}{5}$

82. $\log \dfrac{1}{25}$

83. log 625

84. $\log \sqrt{5}$

Solve Exercises 85–88 using R = log I (see Example 5). Round your answer to three significant digits.

85. Find *I* if *R* = 3.4

86. Find *I* if *R* = 4.9

87. Find *I* if *R* = 5.7

88. Find *I* if *R* = 8.1

89. Astronomy In astronomy, a formula used to find the diameter, in kilometers, of minor planets (also called asteroids) is log *d* = 3.7 − 0.2*g*, where *g* is a quantity called the absolute magnitude of the minor planet. Find the diameter of a minor planet if its absolute magnitude is **a)** 11 and **b)** 20. **c)** Find the absolute magnitude of the minor planet whose diameter is 5.8 kilometers.

90. Standardized Test The average score on a standardized test is a function of the number of hours studied for the test. The average score, *f*(*x*), in points, can be approximated by *f*(*x*) = log 0.3*x* + 1.8, where *x* is the number of hours studied for the test. The maximum possible score on the test is 4.0. Find the score received by the average person who studied for **a)** 15 hours. **b)** 55 hours.

91. Learning Retention Sammy Barcia just finished a course in physics. The percent of the course he will remember *t* months later can be approximated by the function

$$R(t) = 94 - 46.8 \log(t + 1)$$

for 0 ≤ *t* ≤ 48. Find the percent of the course Sammy will remember **a)** 2 months later. **b)** 48 months later.

92. Learning Retention Karen Frye just finished a course in psychology. The percent of the course she will remember *t* months later can be approximated by the function

$$R(t) = 85 - 41.9 \log(t + 1)$$

for 0 ≤ *t* ≤ 48. Find the percent of the course she will remember **a)** 10 months later. **b)** 25 months later.

93. Earthquake How many times more intense is an earthquake having a Richter scale number of 3.8 than the smallest measurable activity? See Example 5.

94. Sport Utility Vehicles Since 1992, sport utility vehicle (SUV) sales in the United States have been on the rise. The number of sales each year, *f*(*t*), in millions, can be approximated by

the function *f*(*t*) = 0.98 + 1.97 log (*t* + 1), where *t* = 0 represents 1992, *t* = 1 represents 1993, and so on. If this trend continues, estimate the number of SUVs sold in **a)** 2003. **b)** 2008.

95. Energy of an Earthquake A formula sometimes used to estimate the seismic energy released by an earthquake is log *E* = 11.8 + 1.5*m*$_s$, where *E* is the seismic energy and *m*$_s$ is the surface wave magnitude.

a) Find the energy released in an earthquake whose surface wave magnitude is 6.

b) If the energy released during an earthquake is 1.2×10^{15}, what is the magnitude of the surface wave?

96. Sound Pressure The sound pressure level, s_p, is given by the formula $s_p = 20 \log \dfrac{p_r}{0.0002}$, where p_r is the sound pressure in dynes/cm^2.

a) Find the sound pressure level if the sound pressure is 0.0036 dynes/cm^2

b) If the sound pressure level is 10.0, find the sound pressure.

97. Earthquake The Richter scale, used to measure the strength of earthquakes, relates the magnitude, *M*, of the earthquake to the release of energy, *E*, in ergs, by the formula

$$M = \frac{\log E - 11.8}{1.5}$$

An earthquake releases 1.259×10^{21} ergs of energy. What is the magnitude of such an earthquake on the Richter scale?

98. pH of a Solution The pH is a measure of the acidity or alkalinity of a solution. The pH of water, for example, is 7. In general, acids have pH numbers less than 7 and alkaline solutions have pH numbers greater than 7. The pH of a solution is defined as pH = −log[H_3O^+], where H_3O^+ represents the hydronium ion concentration of the solution. Find the pH of a solution whose hydronium ion concentration is 2.8×10^{-3}.

Challenge Problems

99. Solve the formula $R = \log I$ for I.

100. Solve the formula $\log E = 11.8 + 1.5m$ for E.

101. Solve the formula $R = 26 - 41.9 \log(t + 1)$ for t.

102. Solve the formula $f = 76 - \log x$ for x.

Group Activity

103. In Section 13.7, we introduce the *change of base formula*, $\log_a x = \dfrac{\log_b x}{\log_b a}$, where a and b are bases and x is a positive number.

   a) Group member 1: Use the change of base formula to evaluate $\log_3 45$. (*Hint*: Let $b = 10$.)

   b) Group member 2: Repeat part **a)** for $\log_5 30$.

   c) Group member 3: Repeat part **a)** for $\log_6 40$.

   d) As a group, use the fact that $\log_a x = \dfrac{\log_b x}{\log_b a}$, where $b = 10$, to graph the equation $y = \log_2 x$ for $x > 0$. Use a graphing calculator if available.

Cumulative Review Exercises

[9.3] **104.** Solve the system of equations.
$$3r = -4s - 6$$
$$3s = -5r + 1$$

[9.5] **105. Cars** Two cars start at the same point in Alexandria, Virginia, and travel in opposite directions. One car travels 5 miles per hour faster than the other car. After 4 hours, the two cars are 420 miles apart. Find the speed of each car.

[12.2] **106.** Solve the following quadratic equation using the quadratic formula.
$$-3x^2 - 4x - 8 = 0$$

[12.3] **107. Motorboat** In 4 hours the Simpsons traveled 15 miles downriver in their motorboat, and then turned around and returned home. If the river current is 5 miles per hour, find the speed of their boat in still water.

[12.5] **108.** Draw the graph of $y = (x - 2)^2 + 1$.

[12.6] **109.** Graph the solution to $\dfrac{2x - 3}{5x + 10} < 0$ on a number line.

13.6 Exponential and Logarithmic Equations

1 Solve exponential and logarithmic equations.

2 Solve applications.

1 Solve Exponential and Logarithmic Equations

In Sections 13.2 and 13.3 we introduced **exponential** and **logarithmic equations**. In this section we give more examples of their use and discuss further procedures for solving such equations.

To solve exponential and logarithmic equations, we often use the following properties.

Properties for Solving Exponential and Logarithmic Equations
a. If $x = y$, then $a^x = a^y$.
b. If $a^x = a^y$, then $x = y$.
c. If $x = y$, then $\log_b x = \log_b y$ $(x > 0, y > 0)$.
d. If $\log_b x = \log_b y$, then $x = y$ $(x > 0, y > 0)$.   **Properties 6a–6d**

We will be referring to these properties when explaining the solutions to the examples in this section.

EXAMPLE 1 ▶ Solve the equation $8^x = \dfrac{1}{2}$.

Solution To solve this equation, we will write both sides of the equation with the same base, 2, and then use property 6b.

$$8^x = \frac{1}{2}$$

$$(2^3)^x = \frac{1}{2} \qquad \textit{Write 8 as } 2^3.$$

$$2^{3x} = 2^{-1} \qquad \textit{Write } \frac{1}{2} \textit{ as } 2^{-1}.$$

Using property 6b, we can write

$$3x = -1$$

$$x = -\frac{1}{3}$$

Now Try Exercise 7

When both sides of the exponential equation cannot be written as a power of the same base, we often begin by taking the logarithm of both sides of the equation, as in Example 2. In the following examples, we will round logarithms to the nearest ten-thousandth.

EXAMPLE 2 ▶ Solve the equation $5^n = 28$.

Solution   Take the logarithm of both sides of the equation and solve for n.

$$\log 5^n = \log 28$$

$$n \log 5 = \log 28 \qquad\qquad \textit{Power rule}$$

$$n = \frac{\log 28}{\log 5} \qquad\qquad \textit{Divide both sides by log 5.}$$

$$\approx \frac{1.4472}{0.6990} \approx 2.0704$$

▶ **Now Try Exercise 23**

Some logarithmic equations can be solved by expressing the equation in exponential form. **It is necessary to check logarithmic equations for extraneous solutions.** When checking a solution, if you obtain the logarithm of a nonpositive number, the solution is extraneous.

EXAMPLE 3 ▶ Solve the equation $\log_2 (x + 3)^3 = 4$.

Solution   Write the equation in exponential form.

$$(x + 3)^3 = 2^4 \qquad\qquad \textit{Write in exponential form.}$$

$$(x + 3)^3 = 16$$

$$x + 3 = \sqrt[3]{16} \qquad\qquad \textit{Take the cube root of both sides.}$$

$$x = -3 + \sqrt[3]{16} \qquad \textit{Solve for x.}$$

Check

$$\log_2 (x + 3)^3 = 4$$

$$\log_2 [(-3 + \sqrt[3]{16}) + 3]^3 \overset{?}{=} 4$$

$$\log_2 (\sqrt[3]{16})^3 \overset{?}{=} 4$$

$$\log_2 16 \overset{?}{=} 4 \qquad (\sqrt[3]{16})^3 = 16$$

$$2^4 \overset{?}{=} 16 \qquad \textit{Write in exponential form.}$$

$$16 = 16 \qquad \textit{True}$$

▶ **Now Try Exercise 43**

Other logarithmic equations can be solved using the properties of logarithms given in earlier sections.

EXAMPLE 4 ▶ Solve the equation $\log (3x + 2) + \log 9 = \log (x + 5)$.

Solution

$$\log (3x + 2) + \log 9 = \log (x + 5)$$
$$\log [(3x + 2)(9)] = \log (x + 5) \quad \textit{Product rule}$$
$$(3x + 2)(9) = (x + 5) \quad \textit{Property 6d}$$
$$27x + 18 = x + 5$$
$$26x + 18 = 5$$
$$26x = -13$$
$$x = -\frac{1}{2}$$

Check for yourself that the solution is $-\dfrac{1}{2}$.

▶ **Now Try Exercise 51**

EXAMPLE 5 ▶ Solve the equation $\log x + \log (x + 1) = \log 12$.

Solution

$$\log x + \log (x + 1) = \log 12$$
$$\log x(x + 1) = \log 12 \quad \textit{Product rule}$$
$$x(x + 1) = 12 \quad \textit{Property 6d}$$
$$x^2 + x = 12$$
$$x^2 + x - 12 = 0$$
$$(x + 4)(x - 3) = 0$$
$$x + 4 = 0 \quad \text{or} \quad x - 3 = 0$$
$$x = -4 \qquad\qquad x = 3$$

Check

$$x = -4$$
$$\log x + \log (x + 1) = \log 12$$
$$\log (-4) + \log (-4 + 1) \stackrel{?}{=} \log 12$$
$$\log (-4) + \log (-3) \stackrel{?}{=} \log 12$$

Stop. ↑ ↑
Logarithms of negative numbers are not real numbers.

$$x = 3$$
$$\log x + \log (x + 1) = \log 12$$
$$\log 3 + \log (3 + 1) \stackrel{?}{=} \log 12$$
$$\log 3 + \log 4 \stackrel{?}{=} \log 12$$
$$\log [(3)(4)] \stackrel{?}{=} \log 12$$
$$\log 12 = \log 12 \quad \textit{True}$$

Thus, -4 is an extraneous solution. The only solution is 3.

▶ **Now Try Example 65**

USING YOUR GRAPHING CALCULATOR

We have shown how equations in one variable may be solved graphically. Logarithmic and exponential equations may also be solved graphically by graphing each side of the equation and finding the x-coordinate of the point of intersection of the two graphs. In Example 5, we found that the solution to the equation $\log x + \log (x + 1) = \log 12$ was $x = 3$. **Figure 13.26** shows the graphical solution to this equation. The horizontal line is the graph of $y = \log 12$ since $\log 12$ is a constant. Notice that the x-coordinate of the point of intersection of the two graphs, 3, is the solution to the equation.

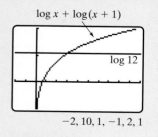

$-2, 10, 1, -1, 2, 1$

FIGURE 13.26

EXAMPLE 6 ▶ Solve the equation $\log (3x - 5) - \log 5x = 1.23$.

Solution

$$\log (3x - 5) - \log 5x = 1.23$$

$$\log \frac{3x - 5}{5x} = 1.23 \qquad \text{\textit{Quotient rule}}$$

$$\frac{3x - 5}{5x} = \text{antilog } 1.23 \qquad \text{\textit{Take the antilog of both sides.}}$$

$$\frac{3x - 5}{5x} = 17.0 \qquad \text{\textit{Rounded to three significant digits.}}$$

$$3x - 5 = 5x(17.0) \qquad \text{\textit{Multiply both sides by 5x.}}$$

$$3x - 5 = 85x$$

$$-5 = 82x$$

$$x = -\frac{5}{82} \approx -0.061$$

Check

$$\log (3x - 5) - \log 5x = 1.23$$

$$\log [3(-0.061) - 5] - \log [(5)(-0.061)] \stackrel{?}{=} 1.23$$

$$\log (-5.183) - \log (-0.305) \stackrel{?}{=} 1.23 \quad \text{\textit{Stop.}}$$

Since we have the logarithms of negative numbers, -0.061 is an extraneous solution. Thus, this equation has no solution. Its solution is the empty set, $\varnothing$.

▶ **Now Try Exercise 57**

Helpful Hint

Below we show some of the steps used in the solutions of Example 3 and Example 6 from this section.

Example 3

$$\log_2 (x + 3)^3 = 4$$

$$(x + 3)^3 = 2^4 \qquad \text{\textit{Write in exponential form.}}$$

Example 6

$$\log \frac{3x - 5}{5x} = 1.23$$

$$\frac{3x - 5}{5x} = \text{antilog } 1.23 \qquad \text{\textit{Take the antilog of both sides.}}$$

Notice the steps we used were different in Examples 3 and 6. In Example 3, we write the equation in exponential form, while in Example 6, we take the antilog of both sides of the equation. In Example 6, we could have also written the second step (line) as $10^{1.23} = \frac{3x - 5}{5x}$, and then evaluated $10^{1.23}$ on a calculator to obtain 17.0 (rounded to three significant digits). Then we could finish the problem to find the solution. However, since Example 6 is given in base 10, we decided to just take the antilog of both sides. Antilogs of base 10 numbers are easily evaluated on a calculator. You may solve problems similar to Example 6 using either method.

2 Solve Applications

Now we will look at an application that involves an exponential equation.

EXAMPLE 7 ▶ **Bacteria** If there are initially 1000 bacteria in a culture, and the number of bacteria doubles each hour, the number of bacteria after t hours can be found by the formula

$$N = 1000(2)^t$$

How long will it take for the culture to grow to 30,000 bacteria?

Solution

$$N = 1000(2)^t$$
$$30{,}000 = 1000(2)^t \qquad \textit{Substitute 30,000 for N.}$$
$$30 = (2)^t \qquad \textit{Divide both sides by 1000.}$$

We want to find the value for *t*. To accomplish this we will use logarithms. Begin by taking the logarithm of both sides of the equation.

$$\log 30 = \log (2)^t$$
$$\log 30 = t \log 2 \qquad \textit{Power rule}$$
$$\frac{\log 30}{\log 2} = t \qquad \textit{Divide both sides by log 2.}$$
$$\frac{1.4771}{0.3010} \approx t$$
$$4.91 \approx t$$

It will take about 4.91 hours for the culture to grow to 30,000 bacteria.

▶ **Now Try Exercise 69**

EXERCISE SET 13.6

Concept/Writing Exercises

1. If $\log c = \log d$, then what is the relationship between *c* and *d*?

2. If $c^r = c^s$, then what is the relationship between *r* and *s*?

3. After solving a logarithmic equation, what must you do?

4. In properties 6c and 6d, we specify that both *x* and *y* must be positive. Explain why.

5. How can you tell quickly that $\log (x + 4) = \log (-2)$ has no real solution?

6. Can $x = -1$ be a solution of the equation $\log_3 x + \log_3(x - 8) = 2$? Explain.

Practice the Skills

Solve each exponential equation without using a calculator.

7. $5^x = 125$

8. $2^x = 128$

9. $3^x = 81$

10. $4^x = 256$

11. $64^x = 8$

12. $81^x = 3$

13. $7^{-x} = \dfrac{1}{49}$

14. $6^{-x} = \dfrac{1}{216}$

15. $27^x = \dfrac{1}{3}$

16. $25^x = \dfrac{1}{5}$

17. $2^{x+2} = 64$

18. $3^{x-6} = 81$

19. $2^{3x-2} = 128$

20. $64^x = 4^{4x+1}$

21. $27^x = 3^{2x+3}$

22. $\left(\dfrac{1}{2}\right)^x = 16$

Use a calculator to solve each equation. Round your answers to the nearest hundredth.

23. $7^x = 50$

24. $1.05^x = 23$

25. $4^{x-1} = 35$

26. $2.3^{x-1} = 26.2$

27. $1.63^{x+1} = 25$

28. $4^x = 9^{x-2}$

29. $3^{x+4} = 6^x$

30. $5^x = 2^{x+5}$

Solve each logarithmic equation. Use a calculator where appropriate. If the answer is irrational, round the answer to the nearest hundredth.

31. $\log_{36} x = \dfrac{1}{2}$

32. $\log_{81} x = \dfrac{1}{2}$

33. $\log_{125} x = \dfrac{1}{3}$

34. $\log_{81} x = \dfrac{1}{4}$

35. $\log_2 x = -4$

36. $\log_7 x = -2$

37. $\log x = 2$

38. $\log x = 4$

39. $\log_2 (5 - 3x) = 3$

40. $\log_4 (3x + 7) = 3$

41. $\log_5 (x + 1)^2 = 2$

42. $\log_3 (a - 2)^2 = 2$

43. $\log_2 (r + 4)^2 = 4$

44. $\log_2 (p - 3)^2 = 6$

45. $\log (x + 8) = 2$

46. $\log (3x - 8) = 1$

47. $\log_2 x + \log_2 5 = 2$

48. $\log_3 2x + \log_3 x = 4$

49. $\log (r + 2) = \log (3r - 1)$

50. $\log 2a = \log (1 - a)$

51. $\log (2x + 1) + \log 4 = \log (7x + 8)$

52. $\log (x - 5) + \log 3 = \log (2x)$

53. $\log n + \log (3n - 5) = \log 2$

54. $\log (x + 4) - \log x = \log (x + 1)$

55. $\log 6 + \log y = 0.72$

56. $\log (x + 4) - \log x = 1.22$

57. $2 \log x - \log 9 = 2$

58. $\log 6000 - \log (x + 2) = 3.15$

59. $\log x + \log (x - 3) = 1$

60. $2 \log_2 x = 4$

61. $\log x = \dfrac{1}{3} \log 64$

62. $\log_7 x = \dfrac{3}{2} \log_7 9$

63. $\log_8 x = 4 \log_8 2 - \log_8 8$

64. $\log_4 x + \log_4 (6x - 7) = \log_4 5$

65. $\log_5 (x + 3) + \log_5 (x - 2) = \log_5 6$

66. $\log_7 (x + 6) - \log_7 (x - 3) = \log_7 4$

67. $\log_2 (x + 3) - \log_2 (x - 6) = \log_2 4$

68. $\log (x - 7) - \log (x + 3) = \log 6$

Problem Solving

Solve each problem. Round your answers to the nearest hundredth.

69. Bacteria If the initial number of bacteria in the culture in Example 7 is 4500, when will the number of bacteria in the culture reach 50,000? Use $N = 4500(2)^t$.

70. Bacteria If after 4 hours the culture in Example 7 contains 2224 bacteria, how many bacteria were present initially?

71. Radioactive Decay The amount, A, of 200 grams of a certain radioactive material remaining after t years can be found by the equation $A = 200(0.75)^t$. When will 80 grams remain?

72. Radioactive Decay The amount, A, of 70 grams of a certain radioactive material remaining after t years can be found by the equation $A = 70(0.62)^t$. When will 10 grams remain?

73. Savings Account Paul Trapper invests $2000 in a savings account earning interest at a rate of 5% compounded annually. How long will it take for the $2000 to grow to $4600? Use the compound interest formula, $A = p\left(1 + \dfrac{r}{n}\right)^{nt}$, which was discussed on page 820.

74. Savings Account If Tekar Werner invests $600 in a savings account earning interest at a rate of 6% compounded semi-annually, how long will it take for the $600 to grow to $1800?

75. Infant Mortality Rate The infant mortality rate (deaths per 1000 live births) in the United States has been decreasing since before 1959. Although it has fallen significantly, it is still higher than in many other nations. The U.S. infant mortality rate can be approximated by the function

$$f(t) = 26 - 12.1 \log (t + 1)$$

where t is the number of years since 1960 and $0 \le t \le 45$. Use this function to estimate the U.S. infant mortality rate in **a)** 1990. **b)** 2005.

76. Homicides Since 1993, the number of homicides in New York City has been on the decline. The number of homicides can be approximated by the function

$$f(t) = 1997 - 1576 \log (t + 1)$$

where t is the number of years since 1993. If this trend continues, use this function to estimate the number of homicides in New York City in 2008.

77. Depreciation A machine purchased for business use can be depreciated to reduce income tax. The value of the machine at the end of its useful life is called its *scrap value*. When the machine depreciates by a constant percentage annually, its scrap value, S, is $S = c(1 - r)^n$, where c is the original cost, r is the annual rate of depreciation as a decimal, and n is the useful life in years. Find the scrap value of a machine that costs $50,000, has a useful life of 12 years, and has an annual depreciation rate of 15%.

78. Depreciation If the machine in Exercise 77 costs $100,000, has a useful life of 15 years, and has an annual depreciation rate of 8%, find its scrap value.

79. Power Gain of an Amplifier The power gain, P, of an amplifier is defined as

$$P = 10 \log \left(\frac{P_{out}}{P_{in}}\right)$$

where P_{out} is the output power in watts and P_{in} is the input power in watts. If an amplifier has an output power of 12.6 watts and an input power of 0.146 watts, find the power gain.

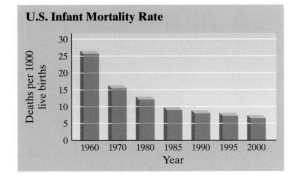

U.S. Infant Mortality Rate

80. Earthquake Measured on the Richter scale, the magnitude, R, of an earthquake of intensity I is defined by $R = \log I$, where I is the number of times more intense the earthquake is than the minimum level for comparison.

a) How many times more intense was the 1906 San Francisco earthquake, which measured 8.25 on the Richter scale, than the minimum level for comparison?

b) How many times more intense is an earthquake that measures 8.3 on the Richter scale than one that measures 4.7?

81. Magnitude of Sound The decibel scale is used to measure the magnitude of sound. The magnitude d, in decibels, of a sound is defined to be $d = 10 \log I$, where I is the number of times greater (or more intense) the sound is than the minimum intensity of audible sound.

a) An airplane engine (nearby) measures 120 decibels. How many times greater than the minimum level of audible sound is the airplane engine?

b) The intensity of the noise in a busy city street is 50 decibels. How many times greater is the intensity of the sound of the airplane engine than the sound of the city street?

82. In the following procedure, we begin with a true statement and end with a false statement. Can you find the error?

$2 < 3$	*True*
$2 \log (0.1) < 3 \log (0.1)$	*Multiply both sides by log (0.1).*
$\log (0.1)^2 < \log (0.1)^3$	*Property 3*
$(0.1)^2 < (0.1)^3$	*Property 6d*
$0.01 < 0.001$	*False*

83. Solve $8^x = 16^{x-2}$.

84. Solve $27^x = 81^{x-3}$.

85. Use equations that are quadratic in form to solve the equation $2^{2x} - 6(2^x) + 8 = 0$.

86. Use equations that are quadratic in form to solve the equation $2^{2x} - 18(2^x) + 32 = 0$.

Change the exponential or logarithmic equation to the form $ax + by = c$, and then solve the system of equations.

87. $2^x = 8^y$
$x + y = 4$

88. $3^{2x} = 9^{y+1}$
$x - 2y = -3$

89. $\log (x + y) = 2$
$x - y = 8$

90. $\log (x + y) = 3$
$2x - y = 5$

Use your calculator to estimate the solutions to the nearest tenth. If a real solution does not exist, so state.

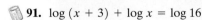 **91.** $\log (x + 3) + \log x = \log 16$

92. $\log (3x + 5) = 2.3x - 6.4$

93. $5.6 \log (5x - 12) = 2.3 \log (x - 5.4)$

94. $5.6 \log (x + 12.2) - 1.6 \log (x - 4) = 20.3 \log (2x - 6)$

Cumulative Review Exercises

[2.6] **95.** Consider the following two figures. Which has a greater volume, and by how much?

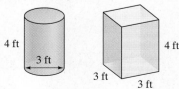

4 ft
3 ft
4 ft
3 ft
3 ft

[8.5] **96.** Let $f(x) = x^2 - x$ and $g(x) = x - 1$. Find $(g - f)(3)$.

[10.3] **97.** Determine the solution set to the system of inequalities.

$$3x - 4y \le 6$$
$$y > -x + 4$$

[11.5] **98.** Simplify $\dfrac{2\sqrt{xy} - \sqrt{xy}}{\sqrt{x} + \sqrt{y}}$.

[12.3] **99.** Solve $E = mc^2$ for c.

[12.5] **100.** Determine the function for the parabola that has the shape of $f(x) = 2x^2$ and has its vertex at $(3, -5)$.

13.7 Natural Exponential and Natural Logarithmic Functions

1 Identify the natural exponential function.

2 Identify the natural logarithmic function.

3 Find values on a calculator.

4 Find natural logarithms using the change of base formula.

5 Solve natural logarithmic and natural exponential equations.

6 Solve applications.

The **natural exponential function** and *its inverse*, the **natural logarithmic** function, are exponential functions and logarithmic functions of the type presented in the previous sections. They share all the properties of exponential functions and logarithmic functions discussed earlier. The importance of these special functions lies in the many varied applications in real life of a unique irrational number designated by the letter *e*.

1 Identify the Natural Exponential Function

In Section 13.2 we indicated that exponential functions were of the form $f(x) = a^x, a > 0$ and $a \neq 1$. Now we introduce a very special exponential function. It is called the natural exponential function, and it uses the number *e*. Like the irrational number π, the number *e* is an irrational number whose value can only be approximated by a decimal number. The number *e* plays a very important role in higher-level mathematics courses. The value of *e* is approximately 2.7183. Now we define the natural exponential function.

> **The Natural Exponential Function**
>
> The natural exponential function is
>
> $$f(x) = e^x$$
>
> where $e \approx 2.7183$.

2 Identify the Natural Logarithmic Function

We discussed common logarithms in Section 13.5. Now we will discuss natural logarithms.

> **Natural Logarithms**
>
> **Natural logarithms** are logarithms to the base *e*. Natural logarithms are indicated by the letters ln.
>
> $$\log_e x = \ln x$$

The notation ln *x* is read the "natural logarithm of *x*." The function $f(x) = \ln x$ is called the **natural logarithmic function**.

You must remember that the base of the natural logarithm is *e*. Thus, when you change a natural logarithm to exponential form, the base of the exponential expression will be *e*.

> **Natural Logarithm in Exponential Form**
>
> For $x > 0$ if $y = \ln x$, then $e^y = x$.

EXAMPLE 1 ▶ Find the value of the expression by changing the natural logarithm to exponential form.

a) ln 1   **b)** ln *e*

Solution

a) Let $y = \ln 1$; then $e^y = 1$. Since any nonzero value to the 0th power equals 1, *y* must equal 0. Thus, $\ln 1 = 0$.

b) Let $y = \ln e$; then $e^y = e$. For e^y to equal *e*, *y* must equal 1. Thus, $\ln e = 1$.

▶ **Now Try Exercise 1**

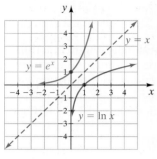

FIGURE 13.27

The functions $y = a^x$ and $y = \log_a x$ are inverse functions. Similarly, the functions $y = e^x$ and $y = \ln x$ are inverse functions. (Remember, $y = \ln x$ means $y = \log_e x$.) That is, if $f(x) = e^x$, then $f^{-1}(x) = \ln x$.

The graphs of $y = e^x$ and $y = \ln x$ are illustrated in **Figure 13.27**. Notice that the graphs are symmetric about the line $y = x$, which is what we expect of inverse functions.

Note that the graph of $y = e^x$ is similar to graphs of the form $y = a^x$, $a > 1$, and that the graph of $y = \ln x$ is similar to graphs of the form $y = \log_a x$, $a > 1$.

3 Find Values on a Calculator

Now we will learn how to find natural logarithms on a calculator.

 USING YOUR CALCULATOR Find Natural Logarithms

Natural logarithms can be found using a calculator that has a $\boxed{\text{LN}}$ key. Natural logarithms are found in the same manner that we found common logarithms on a calculator, except we use the natural log key, $\boxed{\text{LN}}$, instead of the common log key, $\boxed{\text{LOG}}$.

Scientific Calculator

EXAMPLE	KEYS TO PRESS	ANSWER DISPLAYED
Find ln 242.	242 $\boxed{\text{LN}}$	5.4889377
Find ln 0.85.	.85 $\boxed{\text{LN}}$	−0.1625189

Graphing Calculator*

On the TI-84 Plus, after the $\boxed{\text{LN}}$ key is pressed, the calculator displays ln(on the screen.

EXAMPLE	KEYS TO PRESS	ANSWER DISPLAYED
Find ln 242.	$\boxed{\text{LN}}$ (242 $\boxed{)}$ $\boxed{\text{ENTER}}$	5.488937726
Find ln 0.85.	$\boxed{\text{LN}}$ (.85 $\boxed{)}$ $\boxed{\text{ENTER}}$	−.1625189295

*The keys are for a TI-84 Plus. Read your manual for instructions on how to find natural logarithms on your calculator.

When finding the natural logarithm of a number, we are finding an exponent. The natural logarithm of a number is the exponent to which the base e must be raised to obtain that number. For example,

$$\text{If } \ln 242 = 5.4889377, \text{ then } e^{5.4889377} = 242.$$
$$\text{If } \ln 0.85 = -0.1625189, \text{ then } e^{-0.1625189} = 0.85.$$

Since $y = \ln x$ and $y = e^x$ are inverse functions, we can use the inverse key $\boxed{\text{INV}}$, in combination with the natural log key, $\boxed{\text{LN}}$, to obtain values for e^x.

 USING YOUR CALCULATOR Finding Values of e^x

Scientific Calculator

To find values of e^x, first enter the exponent on e. Then press either $\boxed{\text{shift}}$, $\boxed{2^{\text{nd}}}$, or $\boxed{\text{INV}}$, depending on your calculator. Then press the natural log key, $\boxed{\text{LN}}$. After the $\boxed{\text{LN}}$ key is pressed, the value of e^x will be displayed.

EXAMPLE	KEYS TO PRESS	ANSWER DISPLAYED
Find $e^{5.24}$.	5.24 $\boxed{\text{INV}}$ $\boxed{\text{LN}}$	188.6701
Find $e^{-1.639}$.	1.639 $\boxed{+/-}$ $\boxed{\text{INV}}$ $\boxed{\text{LN}}$	0.1941741

Graphing Calculator*

On the TI-84 Plus, after $\boxed{2^{\text{nd}}}$ $\boxed{\text{LN}}$ is pressed, the calculator displays $e^\wedge($ on the screen.

EXAMPLE	KEYS TO PRESS	ANSWER DISPLAYED
Find $e^{5.24}$.	$\boxed{2^{\text{nd}}}$ $\boxed{\text{LN}}$ (5.24 $\boxed{)}$ $\boxed{\text{ENTER}}$	188.6701024
Find $e^{-1.639}$.	$\boxed{2^{\text{nd}}}$ $\boxed{\text{LN}}$ ($\boxed{(-)}$ 1.639 $\boxed{)}$ $\boxed{\text{ENTER}}$	.1941741194

*Keys are for a TI-84 Plus. Read your manual for instructions on how to evaluate natural exponential expressions on your calculator.

Remember e is about 2.7183. When we evaluated $e^{5.24}$ or $(2.7183)^{5.24}$ in the previous calculator box we obtained a value close to 188.6701. If we found ln 188.6701 on a calculator, we would obtain a value close to 5.24. What do you think we would get if we evaluated ln 0.1941741 on a calculator? If you answered, "A value close to -1.639," you answered correctly.

EXAMPLE 2 ▶ Find N if **a)** ln $N = 5.26$ and **b)** ln $N = -0.0253$.

Solution

a) If we write ln $N = 5.26$ in exponential form, we get $e^{5.26} = N$. Thus, we simply need to evaluate $e^{5.26}$ to determine N.

$$e^{5.26} = 192.48149 \qquad \textit{From a calculator}$$

Thus, $N = 192.48149$

b) If we write ln $N = -0.0253$ in exponential form, we get $e^{-0.0253} = N$.

$$e^{-0.0253} = 0.9750174 \qquad \textit{From a calculator}$$

Thus, $N = 0.9750174$.

▶ **Now Try Exercise 19**

4 Find Natural Logarithms Using the Change of Base Formula

If you are given a logarithm in a base other than 10 or e, you will not be able to evaluate it on your calculator directly. When this occurs, you can use the **change of base formula**.

Change of Base Formula

For any logarithm bases a and b, and positive number x,

$$\log_a x = \frac{\log_b x}{\log_b a}$$

We can prove the change of base formula by beginning with $\log_a x = m$.

$$\log_a x = m$$
$$a^m = x \qquad \textit{Change to exponential form.}$$
$$\log_b a^m = \log_b x \qquad \textit{By property 6c on page 844}$$
$$m \log_b a = \log_b x \qquad \textit{Power rule}$$
$$(\log_a x)(\log_b a) = \log_b x \qquad \textit{Substitution for m}$$
$$\log_a x = \frac{\log_b x}{\log_b a} \qquad \textit{Divide both sides by } \log_b a.$$

In the change of base formula, 10 is often used in place of base b because we can find common logarithms on a calculator. Replacing base b with 10, we get

$$\log_a x = \frac{\log_{10} x}{\log_{10} a} \quad \text{or} \quad \log_a x = \frac{\log x}{\log a}$$

EXAMPLE 3 ▶ Use the change of base formula to find $\log_3 24$.

Solution   If we substitute 3 for a and 24 for x in $\log_a x = \dfrac{\log x}{\log a}$, we obtain

$$\log_3 24 = \frac{\log 24}{\log 3} \approx \frac{1.3802}{0.4771} \approx 2.8929$$

Note that $3^{2.8929} \approx 24$.

▶ **Now Try Exercise 23**

Helpful Hint:

In Example 3, if we compute the actual values in the quotient $\dfrac{\log 24}{\log 3}$ using a calculator, the value is ≈ 2.8928 instead of the value of ≈ 2.8929 we obtained in Example 3. To obtain the answer of ≈ 2.8929, we divided the *approximation* of the numerator, 1.3802, by the *approximation* of the denominator, 0.4771. When we give answers to the exercises involving the change of base formula in the answer section, we will *not* round the values in the numerator and denominator. We will only round the final answer.

We can use the same procedure as in Example 3 to find natural logarithms using the change of base formula. For example, to evaluate $\ln 20$ (or $\log_e 20$), we can substitute e for a and 20 for x in the formula $\log_a x = \dfrac{\log x}{\log a}$.

$$\log_e 20 = \frac{\log 20}{\log e} \approx \frac{1.3010}{0.4343} \approx 2.9956$$

Thus, $\ln 20 \approx 2.9956$. If you find $\ln 20$ on a calculator, you will obtain a very close value.

Since $\log e \approx 0.4343$, to evaluate natural logarithms using common logarithms, we use the formula

$$\ln x = \frac{\log x}{\log e} \approx \frac{\log x}{0.4343}$$

EXAMPLE 4 ▶ Use the change of base formula to find $\ln 95$.

Solution
$$\ln 95 = \frac{\log 95}{\log e} \approx \frac{1.9777}{0.4343} \approx 4.5538$$

If you evaluate $\ln 95$ on your calculator, you will obtain a value very close to 4.5538. Do so now and check your result.

▶ **Now Try Exercise 33**

5 Solve Natural Logarithmic and Natural Exponential Equations

The properties of logarithms discussed in Section 13.4 still hold true for natural logarithms. Following is a summary of these properties in the notation of natural logarithms.

Properties for Natural Logarithms		
$\ln xy = \ln x + \ln y$	$(x > 0 \text{ and } y > 0)$	*Product rule*
$\ln \dfrac{x}{y} = \ln x - \ln y$	$(x > 0 \text{ and } y > 0)$	*Quotient rule*
$\ln x^n = n \ln x$	$(x > 0)$	*Power rule*

Consider the expression $\ln e^x$, which means $\log_e e^x$. From property 4 on page 835, $\log_e e^x = x$. Thus, $\ln e^x = x$. Similarly, $e^{\ln x} = e^{\log_e x} = x$ by property 5. Although $\ln e^x = x$ and $e^{\ln x} = x$ are just special cases of properties 4 and 5, respectively, we will call these properties 7 and 8 so that we can make reference to them.

Additional Properties for Natural Logarithms and Natural Exponential Expressions	
$\ln e^x = x$	**Property 7**
$e^{\ln x} = x, \quad x > 0$	**Property 8**

Using property 7, $\ln e^x = x$, we can state, for example, that $\ln e^{kt} = kt$, and $\ln e^{-2.06t} = -2.06t$. Using property 8, $e^{\ln x} = x$, we can state, for example, that $e^{\ln (t+2)} = t + 2$ and $e^{\ln kt} = kt$.

EXAMPLE 5 ▸ Solve the equation $\ln y - \ln (x + 9) = t$ for y.

Solution

$$\ln y - \ln (x + 9) = t$$

$$\ln \frac{y}{x + 9} = t \qquad \text{\textit{Quotient rule}}$$

$$\frac{y}{x + 9} = e^t \qquad \text{\textit{Change to exponential form.}}$$

$$y = e^t(x + 9) \qquad \text{\textit{Solve for y.}}$$

▸ **Now Try Exercise 63**

EXAMPLE 6 ▸ Solve the equation $225 = 450e^{-0.4t}$ for t.

Solution   Begin by dividing both sides of the equation by 450 to isolate $e^{-0.4t}$.

$$\frac{225}{450} = \frac{450e^{-0.4t}}{450}$$

$$0.5 = e^{-0.4t}$$

Now take the natural logarithm of both sides of the equation to eliminate the exponential expression on the right side of the equation.

$$\ln 0.5 = \ln e^{-0.4t}$$

$$\ln 0.5 = -0.4t \qquad \text{\textit{Property 7}}$$

$$-0.6931472 = -0.4t$$

$$\frac{-0.6931472}{-0.4} = t$$

$$1.732868 = t$$

▸ **Now Try Exercise 49**

EXAMPLE 7 ▸ Solve the equation $P = P_0e^{kt}$ for t.

Solution   We can follow the same procedure as used in Example 6.

$$P = P_0e^{kt}$$

$$\frac{P}{P_0} = \frac{\cancel{P_0}e^{kt}}{\cancel{P_0}} \qquad \text{\textit{Divide both sides by } } P_0.$$

$$\frac{P}{P_0} = e^{kt}$$

$$\ln \frac{P}{P_0} = \ln e^{kt} \qquad \text{\textit{Take natural log of both sides.}}$$

$$\ln P - \ln P_0 = \ln e^{kt} \qquad \text{\textit{Quotient rule}}$$

$$\ln P - \ln P_0 = kt \qquad \text{\textit{Property 7}}$$

$$\frac{\ln P - \ln P_0}{k} = t \qquad \text{\textit{Solve for t.}}$$

▸ **Now Try Exercise 59**

6 Solve Applications

Now let's look at some applications that involve the natural exponential function and natural logarithms.

When a quantity P increases or decreases at an *exponential rate*, a formula often used to find the value of P after time t is

$$P = P_0 e^{kt}$$

where P_0 is the initial or starting value and k is the constant growth rate or decay rate compounded continuously. We will refer to this formula as the **exponential growth (or decay) formula**. In the formula, other letters may be used in place of P. When $k > 0$, P increases as t increases. When $k < 0$, P decreases and gets closer to 0 as t increases.

EXAMPLE 8 ▶ Interest Compounded Continuously Banks often credit compound interest continuously. When interest is compounded continuously, the balance, P, in the account at any time, t, can be calculated by the exponential growth formula $P = P_0 e^{kt}$, where P_0 is the principal initially invested and k is the interest rate.

a) Suppose the interest rate is 6% compounded continuously and $1000 is initially invested. Determine the balance in the account after 3 years.

b) How long will it take the account to double in value?

Solution **a) Understand and Translate** We are told that the principal initially invested, P_0, is $1000. We are also given that the time, t, is 3 years and that the interest rate, k, is 6% or 0.06. We substitute these values into the given formula and solve for P.

$$P = P_0 e^{kt}$$

$$P = 1000 e^{(0.06)(3)}$$

Carry Out
$$= 1000 e^{0.18} = 1000(1.1972174) \quad \textit{From a calculator}$$

$$\approx 1197.22$$

Answer After 3 years, the balance in the account is $\approx$ $1197.22.

b) Understand and Translate For the value of the account to double, the balance in the account would have to reach $2000. Therefore, we substitute 2000 for P and solve for t.

$$P = P_0 e^{kt}$$

$$2000 = 1000 e^{0.06t}$$

$$2 = e^{0.06t} \quad \textit{Divide both sides by 1000.}$$

Carry Out
$$\ln 2 = \ln e^{0.06t} \quad \textit{Take natural log of both sides.}$$

$$\ln 2 = 0.06t \quad \textit{Property 7}$$

$$\frac{\ln 2}{0.06} = t$$

$$\frac{0.6931472}{0.06} = t$$

$$11.552453 \approx t$$

Answer Thus, with an interest rate of 6% compounded continuously, the account will double in about 11.6 years.

▶ Now Try Exercise 69

EXAMPLE 9 ▶ **Radioactive Decay** Strontium 90 is a radioactive isotope that decays exponentially at 2.8% per year. Suppose there are initially 1000 grams of strontium 90 in a substance.

a) Find the number of grams of strontium 90 left after 50 years.

b) Find the half-life of strontium 90.

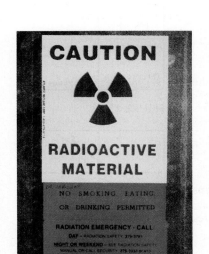

Solution **a)** Understand Since the strontium 90 is decaying over time, the value of k in the formula $P = P_0 e^{kt}$ is negative. Since the rate of decay is 2.8% per year, we use $k = -0.028$. Therefore, the formula we use is $P = P_0 e^{-0.028t}$.

Translate $P = P_0 e^{-0.028t}$

            $= 1000 e^{-0.028(50)}$

Carry Out $= 1000 e^{-1.4} = 1000(0.246597) = 246.597$

Answer   Thus, after 50 years, 246.597 grams of strontium 90 remain.

b) To find the half-life, we need to determine when 500 grams of strontium 90 are left.

$$P = P_0 e^{-0.028t}$$
$$500 = 1000 e^{-0.028t}$$
$$0.5 = e^{-0.028t} \qquad \text{\textit{Divide both sides by 1000.}}$$
$$\ln 0.5 = \ln e^{-0.028t} \qquad \text{\textit{Take natural log of both sides.}}$$
$$-0.6931472 = -0.028t \qquad \text{\textit{Property 7}}$$
$$\frac{-0.6931472}{-0.028} = t$$
$$24.755257 \approx t$$

Thus, the half-life of strontium 90 is about 24.8 years.

▶ **Now Try Exercise 71**

EXAMPLE 10 ▶ **Selling Toys** The formula for estimating the amount of money, A, spent on advertising a certain toy is $A = 350 + 650 \ln n$ where n is the expected number of toys to be sold.

a) If the company wishes to sell 2200 toys, how much money should the company expect to spend on advertising?

b) How many toys can be expected to be sold if $6000 is spent on advertising?

Solution

a) $A = 350 + 650 \ln n$

   $= 350 + 650 \ln 2200 \qquad \text{\textit{Substitute 2200 for n.}}$

   $= 350 + 650(7.6962126)$

   $= 5352.54$

Thus, $5352.54 should be expected to be spent on advertising.

b) Understand and Translate   We are asked to find the number of toys expected to be sold, n, if $6000 is spent on advertising. We substitute the given values into the equation and solve for n.

$$A = 350 + 650 \ln n$$

Carry Out $6000 = 350 + 650 \ln n \qquad \text{\textit{Substitute 6000 for A.}}$

            $5650 = 650 \ln n \qquad \text{\textit{Subtract 350 from both sides.}}$

            $\dfrac{5650}{650} = \ln n \qquad \text{\textit{Divide both sides by 650.}}$

            $8.69231 \approx \ln n$

            $e^{8.69231} \approx n \qquad \text{\textit{Change to exponential form.}}$

            $5957 \approx n \qquad \text{\textit{Obtain answer from a calculator.}}$

Answer   Thus, about 5957 toys can be expected to be sold if $6000 is spent on advertising.

▶ **Now Try Exercise 75**

USING YOUR GRAPHING CALCULATOR

Equations containing natural logarithms and natural exponential functions can be solved on your graphing calculator. For example, to solve the equation $\ln x + \ln (x + 3) = \ln 8$, we set

$$Y_1 = \ln x + \ln (x + 3)$$

$$Y_2 = \ln 8$$

and find the intersection of the graphs, as shown in **Figure 13.28**.

In **Figure 13.28**, we used the CALC, INTERSECT option to find the intersection of the graphs. The solution is the x-coordinate of the intersection. The solution to the equation is $x = 1.7016$, to the nearest ten-thousandth.

To solve the equation $4e^{0.3x} - 5 = x + 3$, we set

$$Y_1 = 4e^{0.3x} - 5$$

$$Y_2 = x + 3$$

and find the intersection of the graphs, as shown in **Figure 13.29**. This equation has two solutions since there are two intersections.

The solutions to the equation are approximately $x = -7.5896$ and $x = 3.5284$. In Exercises 93 through 97, we use the graphing calculator to check or solve equations.

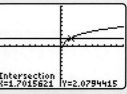

FIGURE 13.28

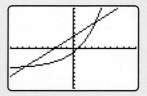

FIGURE 13.29

EXERCISE SET 13.7   *MathXL*   *MyMathLab*

MathXL®   MyMathLab

Concept/Writing Exercises

1. **a)** What is the base in the natural exponential function?
   b) What is the approximate value of e?

2. What is another way of writing $\log_e x$?

3. What is the domain of $\ln x$?

4. Under what conditions will $\ln x < 0$?

5. Give the change of base formula.

6. Is $n \log_e x = \ln x^n$? Explain.

7. To what is $\ln e^x$ equal?

8. To what is $e^{\ln x}$ equal?

9. What is the inverse of $\ln x$?

10. Under what circumstances will P in the formula $P = P_0 e^{kt}$ increase when t increases?

11. Under what conditions will P in the formula $P = P_0 e^{kt}$ decrease when t increases?

12. Is it possible to find the value of $\ln (-3.52)$? Explain.

Practice the Skills

Find the following values. Round values to four decimal places.

13. $\ln 62$

14. $\ln 791$

15. $\ln 0.813$

16. $\ln 0.000568$

Find the value of N. Round values to three significant digits.

17. $\ln N = 1.6$

18. $\ln N = 5.2$

19. $\ln N = -2.85$

20. $\ln N = 0.543$

21. $\ln N = -0.0287$

22. $\ln N = -0.674$

Use the change of base formula to find the value of the following logarithms. Do not round logarithms in the change of base formula. Write the answer rounded to the nearest ten-thousandth.

23. $\log_3 56$

24. $\log_3 198$

25. $\log_2 21$

26. $\log_2 89$

27. $\log_4 11$

28. $\log_4 316$

29. $\log_5 82$

30. $\log_5 1893$

31. $\log_6 185$

32. $\log_6 806$

33. $\ln 51$

34. $\ln 3294$

35. $\log_5 0.463$

36. $\log_3 0.0365$

Solve the following logarithmic equations.

37. $\ln x + \ln (x - 1) = \ln 12$

38. $\ln (x + 4) + \ln (x - 2) = \ln 16$

39. $\ln x + \ln (x + 4) = \ln 5$

40. $\ln (x + 3) + \ln (x - 3) = \ln 40$

41. $\ln x = 5 \ln 2 - \ln 8$

42. $\ln x = \dfrac{3}{2} \ln 16$

43. $\ln (x^2 - 4) - \ln (x + 2) = \ln 4$

44. $\ln (x + 12) - \ln (x - 4) = \ln 5$

Each of the following equations is in the form $P = P_0e^{kt}$. Solve each equation for the remaining variable. Remember, e is a constant. Write the answer rounded to the nearest ten-thousandth.

45. $P = 120e^{2.3(1.6)}$

46. $900 = P_0e^{(0.4)(3)}$

47. $50 = P_0e^{-0.5(3)}$

48. $18 = 9e^{2t}$

49. $60 = 20e^{1.4t}$

50. $29 = 58e^{-0.5t}$

51. $86 = 43e^{k(3)}$

52. $15 = 75e^{k(4)}$

53. $20 = 40e^{k(2.4)}$

54. $100 = A_0e^{-0.02(3)}$

55. $A = 6000e^{-0.08(3)}$

56. $51 = 68e^{-0.04t}$

Solve for the indicated variable.

57. $V = V_0e^{kt}$, for V_0

58. $P = P_0e^{kt}$, for P_0

59. $P = 150e^{7t}$, for t

60. $361 = P_0e^{kt}$, for t

61. $A = A_0e^{kt}$, for k

62. $167 = R_0e^{kt}$, for k

63. $\ln y - \ln x = 2.3$, for y

64. $\ln y + 9 \ln x = \ln 2$, for y

65. $\ln y - \ln (x + 6) = 5$ for y

66. $\ln (x + 2) - \ln (y - 1) = \ln 5$, for y

Problem Solving

Use a calculator to solve.

67. If $e^x = 12.183$, find the value of x. Explain how you obtained your answer.

68. To what exponent must the base e be raised to obtain the value 184.93? Explain how you obtained your answer.

69. Interest Compounded Continuously If $5000 is invested at 6% compounded continuously,

 a) determine the balance in the account after 2 years.

 b) How long would it take the value of the account to double? (See Example 8.)

70. Interest Compounded Continuously If $3000 is invested at 3% compounded continuously,

 a) determine the balance in the account after 30 years.

 b) How long would it take the value of the account to double?

71. Radioactive Decay Refer to Example 9. Determine the amount of strontium 90 remaining after 20 years if there were originally 70 grams.

72. Strontium 90 Refer to Example 9. Determine the amount of strontium 90 remaining after 40 years if there were originally 200 grams.

73. Soft Drinks For a certain soft drink, the percent of a target market, $f(t)$, that buys the soft drink is a function of the number of days, t, that the soft drink is advertised. The function that describes this relationship is $f(t) = 1 - e^{-0.04t}$.

 a) What percent of the target market buys the soft drink after 50 days of advertising?

 b) How many days of advertising are needed if 75% of the target market is to buy the soft drink?

74. Trout in a Lake In 2005, a lake had 300 trout. The growth in the number of trout is estimated by the function $g(t) = 300e^{0.07t}$ where t is the number of years after 2005. How many trout will be in the lake in **a)** 2008? **b)** 2015?

75. Walking Speed It was found in a psychological study that the average walking speed, $f(P)$, of a person living in a city is a function of the population of the city. For a city of population P, the average walking speed in feet per second is given by $f(P) = 0.37 \ln P + 0.05$. The population of Nashville, Tennessee, is 972,000.

 a) What is the average walking speed of a person living in Nashville?

 b) What is the average walking speed of a person living in New York City, population 8,567,000?

 c) If the average walking speed of the people in a certain city is 5.0 feet per second, what is the population of the city?

76. Advertising For a certain type of tie, the number of ties sold, $N(a)$, is a function of the dollar amount spent on advertising, a (in thousands of dollars). The function that describes this relationship is $N(a) = 800 + 300 \ln a$.

 a) How many ties were sold after $1500 (or $1.5 thousand) was spent on advertising?

 b) How much money must be spent on advertising to sell 1000 ties?

77. Assume that the value of the island of Manhattan has grown at an exponential rate of 8% per year since 1626 when Peter Minuet of the Dutch West India Company purchased it for $24. Then the value of Manhattan can be determined by the equation $V = 24e^{0.08t}$, where t is the number of years since 1626. Determine the value of the island of Manhattan in 2008, that is when $t = 382$ years.

78. Prescribing a Drug The percent of doctors who accept and prescribe a new drug is given by the function $P(t) = 1 - e^{-0.22t}$, where t is the time in months since the drug was placed on the market. What percent of doctors accept a new drug 2 months after it is placed on the market?

79. World Population The world population in January 2003 was estimated to be about 6.30 billion people. Let's assume that the world population will continue to grow exponentially at the current growth rate of about 1.3% per year. Then, the expected world population, in billions of people, in t years, is given by the function

$$P(t) = 6.30e^{0.013t}$$

where t is the number of years since 2003.

a) Estimate the world population in 2010.

b) In how many years will the world population double?

80. Generic Drugs Since 2001, the number of prescriptions for generic drugs used by members of the COVA Care Health Plan in the Commonwealth of Virginia has been growing exponentially (see graph). The number of prescriptions for generic drugs, $f(t)$, in 100,000s, can be approximated by the function $f(t) = 6.52e^{0.087t}$, where t is the number of years

since 2001. Assuming that this trend continues, use this function to estimate the number of prescriptions for generic drugs used by member of the COVA Health Care Plan in a) 2006. b) 2008.

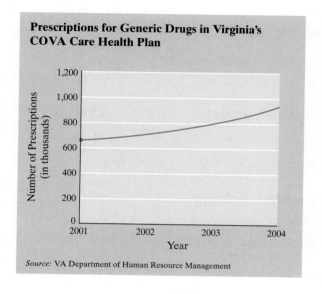

Prescriptions for Generic Drugs in Virginia's COVA Care Health Plan

Source: VA Department of Human Resource Management

81. Demand for Nurses The demand for registered nurses is expected to grow exponentially from 2005 to 2020. (See graph). The demand for registered nurses, $d(t)$, in millions, can be approximated by the function $d(t) = 2.19e^{0.0164t}$, where t is the number of years since 2005. Assuming that this trend continues, use this function to estimate the demand for registered nurses in a) 2025. b) 2040.

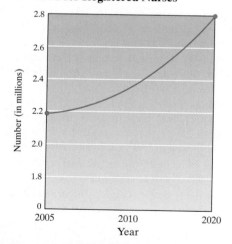

Demand for Registered Nurses

Source: Bureau of Health Professions, U.S. News & World Report (1/31/05 and 2/07/05)

82. Splenda® Products Since 2000, the number of new products using Splenda each year has been growing exponentially (see graph on the next page). The number of new products using Splenda, $N(t)$, can be approximated by the function $N(t) = 163.21e^{0.481t}$, where t is the number of years since

2000. Assuming that this trend continues, use this function to estimate the number of new products using Splenda in **a)** 2007. **b)** 2010.

New Splenda Products Introduced Each Year

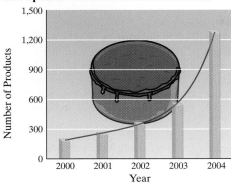

Source: Productscan Online, The New York Times (12/22/04)

83. Annual Tax Refund Since 1994, the average annual tax refund has been growing exponentially (see graph). The average annual tax refund, $r(t)$, can be approximated by the function $r(t) = 1182.3e^{0.0715t}$, where t is the number of years since 1994. Assuming that this trend continues, use this function to estimate the average annual tax refund in **a)** 2006. **b)** 2010.

Average Annual Tax Refund

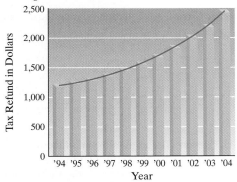

Source: Internal Revenue Service, U.S. News and World Report (4/18/05)

84. Girl's Weights The shaded area of the following graph shows the normal range (from the 5th to the 95th percentile) of weights for girls from birth up to 36 months. The median, or 50th percentile, of weights is indicated by the orange line. The function $y = 3.17 + 7.32 \ln x$ can be used to estimate the median weight of girls from 3 months to 36 months. Use this function to estimate the median weight for girls at **a)** 18 months. **b)** 30 months.

Girls

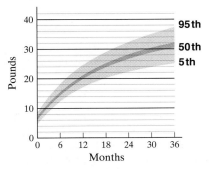

Source: *Newsweek* Special 2000 Edition, Your Child

85. Boys' Heights The shaded area of the following graph shows the normal range (from the 5th to the 95th percentile) of heights for boys from birth up to 36 months. The median, or 50th percentile, of heights is indicated by the green line. The function $y = 15.29 + 5.93 \ln x$ can be used to estimate the median height of boys from 3 months to 36 months. Use this function to estimate the median height for boys at age **a)** 18 months **b)** 30 months.

Boys

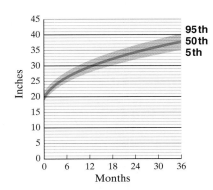

Source: *Newsweek* Special 2000 Edition, Your Child

86. Radioactive Decay Plutonium, which is commonly used in nuclear reactors, decays exponentially at a rate of 0.003% per year. The formula $A = A_0 e^{kt}$ can be used to find the amount of plutonium remaining from an initial amount, A_0, after t years. In the formula, the k is replaced with -0.00003.

a) If 1000 grams of plutonium are present in 2003, how many grams of plutonium will remain in the year 2103, 100 years later?

b) Find the half-life of plutonium.

87. Carbon Dating Carbon dating is used to estimate the age of ancient plants and objects. The radioactive element, carbon 14, is most often used for this purpose. Carbon 14 decays exponentially at a rate of 0.01205% per year. The amount of carbon 14 remaining in an object after t years can be found by the function $f(t) = v_0 e^{-0.0001205t}$, where v_0 is the initial amount present.

a) If an ancient animal bone originally had 20 grams of carbon 14, and when found it had 9 grams of carbon 14, how old is the bone?

b) How old is an item that has 50% of its original carbon 14 remaining?

88. Compound Interest At what rate, compounded continuously, must a sum of money be invested if it is to double in 13 years?

89. Compound Interest How much money must be deposited today to become $20,000 in 18 years if invested at 6% compounded continuously?

90. Radioisotope The power supply of a satellite is a radioisotope. The power P, in watts, remaining in the power supply is a function of the time the satellite is in space.

a) If there are 50 grams of the isotope originally, the power remaining after t days is $P = 50e^{-0.002t}$. Find the power remaining after 50 days.

b) When will the power remaining drop to 10 watts?

91. Radioactive Decay During the nuclear accident at Chernobyl in Ukraine in 1986, two of the radioactive materials that escaped into the atmosphere were cesium 137, with a decay rate of 2.3% and strontium 90, with a decay rate of 2.8%.

a) Which material will decompose more quickly? Explain.

b) What percentage of the cesium will remain in 2036, 50 years after the accident?

92. Radiometric Dating In the study of radiometric dating (using radioactive isotopes to determine the age of items), the formula

$$t = \frac{t_h}{0.693} \ln\left(\frac{N_0}{N}\right)$$

is often used. In the formula, t is the age of the item, t_h is the half-life of the radioactive isotope used, N_0 is the original number of radioactive atoms present, and N is the number remaining at time t. Suppose a rock originally contained 5×10^{12} atoms of uranium 238. Uranium 238 has a half-life of 4.5×10^9 years. If at present there are 4×10^{12} atoms, how old is the rock?

 In Exercises 93–97, use your graphing calculator. In Exercises 95–97, round your answers to the nearest thousandth.

93. Check your answer to Exercise 37.

94. Check your answer to Exercise 39.

95. Solve the equation $e^{x-4} = 12 \ln (x + 2)$.

96. Solve the equation $\ln (4 - x) = 2 \ln x + \ln 2.4$.

97. Solve the equation $3x - 6 = 2e^{0.2x} - 12$.

Challenge Exercises

In Exercises 98–101, when you solve for the given variable, write the answer without using the natural logarithm.

98. Intensity of Light The intensity of light as it passes through a certain medium is found by the formula $x = k(\ln I_0 - \ln I)$. Solve this equation for I_0.

99. Velocity The distance traveled by a train initially moving at velocity v_0 after the engine is shut off can be calculated by the formula $x = \frac{1}{k}\ln (kv_0 t + 1)$. Solve this equation for v_0.

100. Molecule A formula used in studying the action of a protein molecule is $\ln M = \ln Q - \ln (1 - Q)$. Solve this equation for Q.

101. Electric Circuit An equation relating the current and time in an electric circuit is $\ln i - \ln I = \frac{-t}{RC}$. Solve this equation for i.

Cumulative Review Exercises

[5.5] **102.** Multiply $(3xy^2 + y)(4x - 3xy)$.

[6.5] **103.** Find two values of b that will make $4x^2 + bx + 25$ a perfect square trinomial.

[8.2] **104.** Let $h(x) = \dfrac{x^2 + 4x}{x + 6}$. Find **a)** $h(-4)$. **b)** $h\left(\dfrac{2}{5}\right)$.

[9.5] **105. Tickets** The admission at an ice hockey game is $15 for adults and $11 for children. A total of 550 tickets were sold. Determine how many children's tickets and how many adult's tickets were sold if a total of $7290 was collected.

[11.4] **106.** Multiply $\sqrt[3]{x}\left(\sqrt[3]{x^2} + \sqrt[3]{x^5}\right)$.

Chapter 13 Summary

IMPORTANT FACTS AND CONCEPTS	EXAMPLES
Section 13.1	

The **composite function** $f \circ g$ is defined as $$(f \circ g)(x) = f[g(x)]$$	Given $f(x) = x^2 + 3x - 1$ and $g(x) = x - 4$, then $$(f \circ g)(x) = f[g(x)] = (x - 4)^2 + 3(x - 4) - 1$$ $$= x^2 - 8x + 16 + 3x - 12 - 1$$ $$= x^2 - 5x + 3$$ $$(g \circ f)(x) = g[f(x)] = (x^2 + 3x - 1) - 4$$ $$= x^2 + 3x - 5$$
A function is a **one-to-one function** if each value in the range corresponds with exactly one value in the domain.	The set $\{(1, 3), (-2, 5), (6, 2), (4, -1)\}$ is a one-to-one function since each value in the range corresponds with exactly one value in the domain.

IMPORTANT FACTS AND CONCEPTS	EXAMPLES

Section 13.1 (continued)

For a function to be a one-to-one function, its graph must pass the **vertical line test** (the test to ensure it is a function) and the **horizontal line test** (to test the one-to-one criteria).

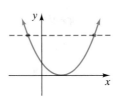

Not one-to-one function

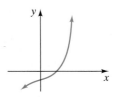

One-to-one function

If $f(x)$ is a one-to-one function with ordered pairs of the form (x, y), its **inverse function**, $f^{-1}(x)$, is a one-to-one function with ordered pairs of the form (y, x). Only one-to-one functions have inverse functions.

To Find the Inverse Function of a One-to-One Function

1. Replace $f(x)$ with y.
2. Interchange the two variables x and y.
3. Solve the equation for y.
4. Replace y with $f^{-1}(x)$ (this gives the inverse function using inverse function notation).

Find the inverse function for $f(x) = 2x + 5$. Graph $f(x)$ and $f^{-1}(x)$ on the same set of axes.

Solution:

$$f(x) = 2x + 5$$
$$y = 2x + 5$$
$$x = 2y + 5$$
$$x - 5 = 2y$$
$$\frac{1}{2}x - \frac{5}{2} = y$$

or $\quad f^{-1}(x) = \dfrac{1}{2}x - \dfrac{5}{2}$

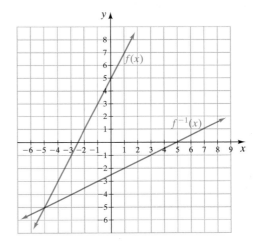

If two functions $f(x)$ and $f^{-1}(x)$ are inverses of each other, $(f \circ f^{-1})(x) = x$ and $(f^{-1} \circ f)(x) = x$.

For the previous example with $f(x) = 2x + 5$ and $f^{-1}(x) = \dfrac{1}{2}x - \dfrac{5}{2}$, then

$$(f \circ f^{-1})(x) = f[f^{-1}(x)] = 2\left(\frac{1}{2}x - \frac{5}{2}\right) + 5$$
$$= x - 5 + 5 = x$$

and

$$(f^{-1} \circ f)(x) = f^{-1}[f(x)] = \frac{1}{2}(2x + 5) - \frac{5}{2}$$
$$= x + \frac{5}{2} - \frac{5}{2} = x$$

| IMPORTANT FACTS AND CONCEPTS | EXAMPLES |

Section 13.2

For any real number $a > 0$ and $a \neq 1$,

$$f(x) = a^x$$

is an **exponential function**.

For all exponential functions of the form $y = a^x$ or $f(x) = a^x$, where $a > 0$ and $a \neq 1$,

1. The domain of the function is $(-\infty, \infty)$.
2. The range of the function is $(0, \infty)$.
3. The graph of the function passes through the points $\left(-1, \frac{1}{a}\right)$, $(0, 1)$, and $(1, a)$.

Graph $y = 3^x$.

x	y
-2	$1/9$
-1	$1/3$
0	1
1	3
2	9

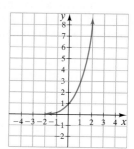

Section 13.3

Logarithms

For all positive numbers a, where $a \neq 1$,

$$y = \log_a x \quad \text{means} \quad x = a^y$$

logarithm
(exponent) ——— number   ——— exponent

$\underset{\uparrow}{y} = \log_{\underset{\uparrow}{a}} \underset{\uparrow}{x}$ means $\underset{\uparrow}{x} = a^{\underset{\uparrow}{y}}$

      base   number — base

Exponential Form	Logarithmic Form
$9^2 = 81$	$\log_9 81 = 2$
$\left(\dfrac{1}{4}\right)^3 = \dfrac{1}{64}$	$\log_{1/4} \dfrac{1}{64} = 3$

Logarithmic Functions

For all logarithmic functions of the form $y = \log_a x$ or $f(x) = \log_a x$, where $a > 0$, $a \neq 1$, and $x > 0$,

1. The domain of the function is $(0, \infty)$.
2. The range of the function is $(-\infty, \infty)$.
3. The graph of the function passes through the points $\left(\frac{1}{a}, -1\right)$, $(1, 0)$, and $(a, 1)$.

Graph $y = \log_4 x$.

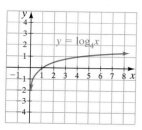

Characteristics of Exponential and Logarithmic Functions

    Exponential function   Logarithmic function
    $y = a^x \ (a > 0, a \neq 1)$ $y = \log_a x \ (a > 0, a \neq 1)$

Domain: $(-\infty, \infty)$ $(0, \infty)$
Range: $(0, \infty)$ $(-\infty, \infty)$

Points on graph:
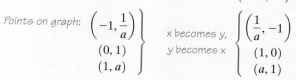

Graph $y = 3^x$ and $y = \log_3 x$ on the same set of axes.

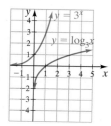

IMPORTANT FACTS AND CONCEPTS	EXAMPLES

Section 13.4

Product Rule for Logarithms

For positive real numbers x, y, and a, $a \neq 1$,

$$\log_a xy = \log_a x + \log_a y \qquad \textit{Property 1}$$

$$\log_5 (9 \cdot 13) = \log_5 9 + \log_5 13$$
$$\log_7 mn = \log_7 m + \log_7 n$$

Quotient Rule for Logarithms

For positive real numbers x, y, and a, $a \neq 1$,

$$\log_a \frac{x}{y} = \log_a x - \log_a y \qquad \textit{Property 2}$$

$$\log_3 \frac{15}{4} = \log_3 15 - \log_3 4$$
$$\log_8 \frac{z + 1}{z + 3} = \log_8 (z + 1) - \log_8 (z + 3)$$

Power Rule for Logarithms

If x and a are positive real numbers, $a \neq 1$, and n is any real number, then

$$\log_a x^n = n \log_a x \qquad \textit{Property 3}$$

$$\log_9 23^5 = 5 \log_9 23$$
$$\log_6 \sqrt[3]{x + 4} = \log_6 (x + 4)^{1/3} = \frac{1}{3} \log_6 (x + 4)$$

Additional Properties of Logarithms

If $a > 0$, and $a \neq 1$, then

$$\log_a a^x = x \qquad \textit{Property 4}$$
$$\text{and} \quad a^{\log_a x} = x \ (x > 0) \ \textit{Property 5}$$

$$\log_4 16 = \log_4 4^2 = 2$$
$$7^{\log_7 3} = 3$$

Section 13.5

Common Logarithm

Base 10 logarithms are called common logarithms.

$$\log x \text{ means } \log_{10} x$$

The **common logarithm** of a positive real number is the *exponent* to which the base 10 is raised to obtain the number.

$$\text{If } \log N = L, \quad \text{then} \quad 10^L = N.$$

To find a common logarithm, use a scientific or graphing calculator. We round the answer to four decimal places.

$\log 17$ means $\log_{10} 17$
$\log (b + c)$ means $\log_{10} (b + c)$

If $\log 14 = 1.1461$, then $10^{1.1461} = 14$.

If $\log 0.6 = -0.2218$, then $10^{-0.2218} = 0.6$.

$\log 183 = 2.2625$ (rounded to 4 places)

$\log 0.42 = -0.3768$ (rounded to 4 places)

Antilogarithm

$$\text{If } \log N = L, \quad \text{then} \quad N = \text{antilog } L.$$

To find antilogarithms, use a scientific or graphing calculator.

If $\log 1890.1662 = 3.2765$, then antilog $3.2765 = 1890.1662$.

If $\log 0.0143 = -1.8447$, then antilog $(-1.8447) = 0.0143$.

Section 13.6

Properties for Solving Exponential and Logarithmic Equations

a) If $x = y$, then $a^x = a^y$.

b) If $a^x = a^y$, then $x = y$.

c) If $x = y$, then $\log_b x = \log_b y$ $(x > 0, y > 0)$.

d) If $\log_b x = \log_b y$, then $x = y$ $(x > 0, y > 0)$.

Properties 6a–6d

a) If $x = 5$, then $3^x = 3^5$.

b) If $3^x = 3^5$, then $x = 5$.

c) If $x = 2$, then $\log x = \log 2$.

d) If $\log x = \log 2$, then $x = 2$.

IMPORTANT FACTS AND CONCEPTS	EXAMPLES

Section 13.7

The **natural exponential function** is
$$f(x) = e^x$$
where $e \approx 2.7183$.

Natural logarithms are logarithms to the base e. Natural logarithms are indicated by the letters ln.
$$\log_e x = \ln x$$
For $x > 0$, if $y = \ln x$, then $e^y = x$.

The **natural logarithmic function** is
$$g(x) = \ln x$$
where the base $e \approx 2.7183$.

To find natural exponential and natural logarithmic values, use a scientific or graphing calculator.

The natural exponential function, $f(x) = e^x$, and the natural logarithmic function, $g(x) = \ln x$, are inverses of each other.

Graph $f(x) = e^x$ and $g(x) = \ln x$ on the same set of axes.

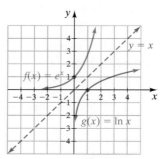

$\ln 5.83 = 1.7630$
If $\ln N = -2.09$, then $N = e^{-2.09} = 0.1237$.

Change of Base Formula
For any logarithm bases a and b, and positive number x,

$$\log_a x = \frac{\log_b x}{\log_b a}$$

$$\log_5 98 = \frac{\log 98}{\log 5} \approx \frac{1.9912}{0.6990} \approx 2.8486$$

Properties for Natural Logarithms

$\ln xy = \ln x + \ln y$ $\quad (x > 0 \text{ and } y > 0)$ *Product rule*

$\ln \dfrac{x}{y} = \ln x - \ln y$ $\quad (x > 0 \text{ and } y > 0)$ *Quotient rule*

$\ln x^n = n \ln x$ $\quad (x > 0)$ $\quad$ *Power rule*

$\ln 7 \cdot 30 = \ln 7 + \ln 30$

$\ln \dfrac{x+1}{x+8} = \ln(x+1) - \ln(x+8)$

$\ln m^5 = 5 \ln m$

Additional Properties for Natural Logarithms and Natural Exponential Expressions

$\ln e^x = x$ $\qquad$ *Property 7*

$e^{\ln x} = x, \quad x > 0$ $\qquad$ *Property 8*

$\ln e^{19} = 19$

$e^{\ln 2} = 2$

Chapter 13 Review Exercises

[13.1] *Given* $f(x) = x^2 - 3x + 4$ *and* $g(x) = 2x - 5$, *find the following.*

1. $(f \circ g)(x)$ **2.** $(f \circ g)(3)$ **3.** $(g \circ f)(x)$ **4.** $(g \circ f)(-3)$

Given $f(x) = 6x + 7$ *and* $g(x) = \sqrt{x - 3}$, $x \geq 3$, *find the following.*

5. $(f \circ g)(x)$ **6.** $(g \circ f)(x)$

Determine whether each function is a one-to-one function.

7.

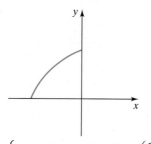

8.

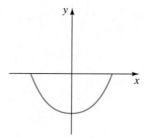

9. $\{(6, 2), (4, 0), (-5, 7), (3, 8)\}$

10. $\left\{ (0, -2), (6, 1), (3, -2), \left(\dfrac{1}{2}, 4 \right) \right\}$ **11.** $y = \sqrt{x + 8}$, $x \geq -8$ **12.** $y = x^2 - 9$

In Exercises 13 and 14, for each function, find the domain and range of both $f(x)$ and $f^{-1}(x)$.

13. $\{(5, 3), (6, 2), (-4, -3), (-1, 8)\}$

14.

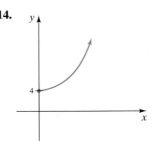

In Exercises 15 and 16, find $f^{-1}(x)$ and graph $f(x)$ and $f^{-1}(x)$ on the same axes.

15. $y = f(x) = 4x - 2$

16. $y = f(x) = \sqrt[3]{x - 1}$

17. Yards to Feet The function $f(x) = 36x$ converts yards, x, into inches. Find the inverse function that converts inches into yards. In the inverse function, what do x and $f^{-1}(x)$ represent?

18. Gallons to Quarts The function $f(x) = 4x$ converts gallons, x, into quarts. Find the inverse function that converts quarts into gallons. In the inverse function, what do x and $f^{-1}(x)$ represent?

[13.2] Graph the following functions.

19. $y = 2^x$

20. $y = \left(\dfrac{1}{2}\right)^x$

21. Smart Handheld Devices Since 1999, the number of worldwide shipments of smart handheld devices has been growing exponentially (see the graph on the right). The number of shipments, $f(t)$, in millions, can be approximated by the function $f(t) = 7.02e^{0.365t}$ where t is the number of years since 1999. Use this function to estimate the number of worldwide shipments of these devices in **a)** 2003, **b)** 2005, **c)** 2008.

Worldwide Shipments of Smart Handheld Devices

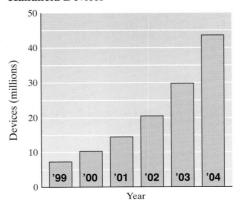

Source: International Data Corp.; MSN MoneyCentral; CSI Inc.; additional research

[13.3] Write each equation in logarithmic form.

22. $8^2 = 64$

23. $81^{1/4} = 3$

24. $5^{-3} = \dfrac{1}{125}$

Write each equation in exponential form.

25. $\log_2 32 = 5$

26. $\log_{1/4} \dfrac{1}{16} = 2$

27. $\log_6 \dfrac{1}{36} = -2$

Write each equation in exponential form and find the missing value.

28. $3 = \log_4 x$

29. $4 = \log_a 81$

30. $-3 = \log_{1/5} x$

Graph the following functions.

31. $y = \log_3 x$

32. $y = \log_{1/2} x$

[13.4] Use the properties of logarithms to expand each expression.

33. $\log_5 17^8$

34. $\log_3 \sqrt{x - 9}$

35. $\log \dfrac{6(a + 1)}{19}$

36. $\log \dfrac{x^4}{7(2x + 3)^5}$

Write the following as the logarithm of a single expression.

37. $5 \log x - 3 \log (x + 1)$

38. $4(\log 2 + \log x) - \log y$

39. $\dfrac{1}{3}[\ln x - \ln (x + 2)] - \ln 2$

40. $3 \ln x + \dfrac{1}{2}\ln (x + 1) - 6 \ln (x + 4)$

Evaluate.

41. $8^{\log_8 10}$

42. $\log_4 4^5$

43. $11^{\log_9 81}$

44. $9^{\log_8 \sqrt{8}}$

[13.5, 13.7] *Use a calculator to find each logarithm. Round your answers to the nearest ten-thousandth.*

45. $\log 819$

46. $\ln 0.0281$

Use a calculator to find the antilog of each number. Give the antilog to three significant digits.

47. 3.159

48. −3.157

Use a calculator to find N. Round your answer to three significant digits.

49. $\log N = 4.063$

50. $\log N = -1.2262$

Evaluate.

51. $\log 10^5$

52. $10^{\log 9}$

53. $7 \log 10^{3.2}$

54. $2(10^{\log 4.7})$

[13.6] *Solve without using a calculator.*

55. $625 = 5^x$

56. $49^x = \dfrac{1}{7}$

57. $2^{3x-1} = 32$

58. $27^x = 3^{2x+5}$

Solve using a calculator. Round your answers to the nearest thousandth.

59. $7^x = 152$

60. $3.1^x = 856$

61. $12.5^{x+1} = 381$

62. $3^{x+2} = 8^x$

Solve the logarithmic equation.

63. $\log_7 (2x - 3) = 2$

64. $\log x + \log (4x - 19) = \log 5$

65. $\log_3 x + \log_3 (2x + 1) = 1$

66. $\ln (x + 1) - \ln (x - 2) = \ln 4$

[13.7] *Solve each exponential equation for the remaining variable. Round your answer to the nearest thousandth.*

67. $50 = 25e^{0.6t}$

68. $100 = A_0 e^{-0.42(3)}$

Solve for the indicated variable.

69. $A = A_0 e^{kt}$, for t

70. $200 = 800 e^{kt}$, for k

71. $\ln y - \ln x = 6$, for y

72. $\ln (y + 1) - \ln (x + 8) = \ln 3$, for y

Use the change of base formula to evaluate. Write the answer rounded to the nearest ten-thousandth.

73. $\log_2 196$

74. $\log_3 47$

[13.2–13.7]

75. Compound Interest Find the amount of money accumulated if Justine Elwood puts $12,000 in a savings account yielding 6% interest per year for 8 years. Use $A = p(1 + r)^n$.

76. Interest Compounded Continuously If $6000 is placed in a savings account paying 4% interest compounded continuously, find the time needed for the account to double in value.

77. Bacteria The bacteria *Escherichia coli* are commonly found in the bladders of humans. Suppose that 2000 bacteria are present at time 0. Then the number of bacteria present t minutes later may be found by the function $N(t) = 2000(2)^{0.05t}$.

a) When will 50,000 bacteria be present?

b) Suppose that a human bladder infection is classified as a condition with 120,000 bacteria. When would a person develop a bladder infection if he or she started with 2000 bacteria?

78. Atmospheric Pressure The atmospheric pressure, P, in pounds per square inch at an elevation of x feet above sea level can be found by the formula $P = 14.7e^{-0.00004x}$. Find the atmospheric pressure at the top of the half dome in Yosemite National Park, an elevation of 8842 feet.

79. Remembering A class of history students is given a final exam at the end of the course. As part of a research project, the students are also given equivalent forms of the exam each month for n months. The average grade of the class after n months may be found by the function $A(n) = 72 - 18 \log(n + 1)$, $n \geq 0$.

a) What was the class average when the students took the original exam ($n = 0$)?

b) What was the class average for the exam given 3 months later?

c) After how many months was the class average 58.0?

Chapter 13 Practice Test

*To find out how well you understand the chapter material, take this practice test. The answers, and the section where the material was initially discussed, are given in the back of the book. Each problem is also fully worked out on the **Chapter Test Prep Video CD**. Review any questions that you answered incorrectly.*

1. a) Determine whether the following function is a one-to-one function.
$$\{(4, 2), (-3, 8), (-1, 3), (6, -7)\}$$
   b) List the set of ordered pairs in the inverse function.

2. Given $f(x) = x^2 - 3$ and $g(x) = x + 2$, find **a)** $(f \circ g)(x)$.
   b) $(f \circ g)(6)$.

3. Given $f(x) = x^2 + 8$ and $g(x) = \sqrt{x - 5}, x \geq 5$, find
   a) $(g \circ f)(x)$. **b)** $(g \circ f)(7)$.

*In Exercises 4 and 5, **a)** find $f^{-1}(x)$ and **b)** graph $f(x)$ and $f^{-1}(x)$ on the same axes.*

4. $y = f(x) = -3x - 5$

5. $y = f(x) = \sqrt{x - 1}, x \geq 1$

6. What is the domain of $y = \log_5 x$?

7. Evaluate $\log_4 \dfrac{1}{256}$.

8. Graph $y = 3^x$.

9. Graph $y = \log_2 x$.

10. Write $2^{-5} = \dfrac{1}{32}$ in logarithmic form.

11. Write $\log_5 125 = 3$ in exponential form.

Write Exercises 12 and 13 in exponential form and find the missing value.

12. $4 = \log_2(x + 3)$

13. $y = \log_{64} 16$

14. Expand $\log_2 \dfrac{x^3(x - 4)}{x + 2}$.

15. Write as the logarithm of a single expression.
$$7 \log_6(x - 4) + 2 \log_6(x + 3) - \frac{1}{2} \log_6 x.$$

16. Evaluate $10 \log_9 \sqrt{9}$.

17. a) Find $\log 4620$ rounded to 4 decimal places.
   b) Find $\ln 0.0692$ rounded to 4 decimal places.

18. Solve $3^x = 19$ for x.

19. Solve $\log 4x = \log(x + 3) + \log 2$ for x.

20. Solve $\log(x + 5) - \log(x - 2) = \log 6$ for x.

21. Find N, rounded to 4 decimal places, if $\ln N = 2.79$.

22. Evaluate $\log_6 40$, rounded to 4 decimal places, using the change of base formula.

23. Solve $100 = 250e^{-0.03t}$ for t, rounded to 4 decimal places.

24. Savings Account What amount of money accumulates if Kim Lee puts $3500 in a savings account yielding 4% interest compounded quarterly for 10 years?

25. Carbon 14 The amount of carbon 14 remaining after t years is found by the formula $v = v_0 e^{-0.0001205t}$, where v_0 is the original amount of carbon 14. If a fossil originally contained 60 grams of carbon 14, and now contains 40 grams of carbon 14, how old is the fossil?

Cumulative Review Test

Take the following test and check your answers with those given in the back of the book. Review any questions that you answered incorrectly. The section where the material is covered is indicated after the answer.

1. Evaluate $\dfrac{6 - |-18| \div 3^2 - 6}{4 - |-8| \div 2^2}$.

2. Solve $4 - (6x + 6) = -(-2x + 10)$.

3. Solve the equation $2x - 3y = 5$ for y.

4. Dinner Thomas Furgeson took his wife out to dinner at a nice restaurant. The cost of the meal before tax was $92. If the total price, including tax, was $98.90, find the tax rate.

5. Jogging Jason Sykora jogs at 4 mph while Kendra Rathbun jogs at 5 mph. Kendra starts jogging $\dfrac{1}{2}$ hour after Jason and jogs along the same path.
   a) How long after Kendra starts will the two meet?
   b) How far from the starting point will they be when they meet?

6. Find the slope of the line in the figure below. Then write the equation of the given line.

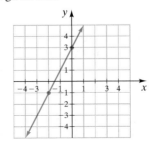

7. Simplify $\left(\dfrac{3x^4 y^{-3}}{6xy^4 z^2}\right)^{-3}$.

8. Divide $\dfrac{x^3 + 3x^2 + 5x + 4}{x + 1}$.

9. Factor $12x^2 - 5xy - 3y^2$.

10. Factor $x^2 - 2xy + y^2 - 25$.

11. Solve $\dfrac{x + 1}{x + 2} + \dfrac{x - 2}{x - 3} = \dfrac{x^2 - 4}{x^2 - x - 6}$.

12. L varies inversely as the square of P. Find L when $P = 4$ and $K = 100$.

13. Let $h(x) = \dfrac{x^2 + 4x}{x + 6}$. Find $h(-3)$.

14. Solve the system of equations.

$$0.4x + 0.6y = 3.2$$
$$1.4x - 0.3y = 1.6$$

15. Solve the system of equations using matrices.

$$x + y = 6$$
$$-2x + y = 3$$

16. Evaluate the following determinant.

$$\begin{vmatrix} 3 & 0 & -1 \\ 2 & 5 & 3 \\ -1 & 4 & 6 \end{vmatrix}$$

17. Graph $y \le \dfrac{1}{3}x + 6$.

18. Let $g(x) = x^2 - 4x - 5$.

a) Express $g(x)$ in the form

$$g(x) = a(x - h)^2 + k.$$

b) Draw the graph and label the vertex.

19. Given $f(x) = x^3 - 6x^2 + 5x$, find all values of x for which $f(x) \ge 0$. Write the answer in interval notation.

20. **Radioactive Decay** Strontium 90 is a radioactive substance that decays exponentially at 2.8% per year. Suppose there are originally 600 grams of strontium 90. **a)** Find the number of grams left after 60 years. **b)** What is the half-life of strontium 90?

14 Conic Sections

THE SHAPE OF AN ellipse gives it an unusual feature. Anything bounced off the wall of an elliptical shape from one focal point will ricochet to the other focal point. This feature has been used in architecture and medicine. One example is the National Statuary Hall in the Capitol Building, which has an elliptically shaped domed ceiling. If you whisper at one focal point, your whisper can be heard at the other focal point. Similarly, a ball hit from one focus of an elliptical billiard table will rebound to the other focal point. In Exercise 56 on page 888, you will determine the location of the foci of an elliptical billiard table.

14.1 The Parabola and the Circle

1 Identify and describe the conic sections.

2 Review parabolas.

3 Graph parabolas of the form $x = a(y - k)^2 + h$.

4 Learn the distance and midpoint formulas.

5 Graph circles with centers at the origin.

6 Graph circles with centers at (h, k).

1 Identify and Describe the Conic Sections

In previous chapters, we discussed parabolas. A **parabola** is one type of conic section. Parabolas will be discussed further in this section. Other conic sections are circles, ellipses, and hyperbolas. Each of these shapes is called a conic section because each can be made by slicing a cone and observing the shape of the slice. The methods used to slice the cone to obtain each conic section are illustrated in **Figure 14.1**.

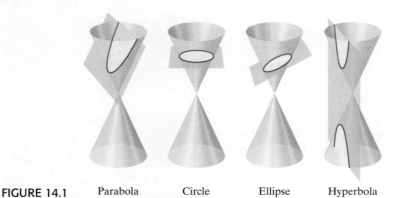

FIGURE 14.1   Parabola   Circle   Ellipse   Hyperbola

2 Review Parabolas

We discussed parabolas in Section 12.5. Example 1 will refresh your memory on how to graph parabolas in the forms $y = ax^2 + bx + c$ and $y = a(x - h)^2 + k$.

EXAMPLE 1 ▶ Consider $y = 2x^2 + 4x - 6$.

a) Write the equation in $y = a(x - h)^2 + k$ form.

b) Determine whether the parabola opens upward or downward.

c) Determine the vertex of the parabola.

d) Determine the y-intercept of the parabola.

e) Determine the x-intercepts of the parabola.

f) Graph the parabola.

Solution

a) First, factor 2 from the two terms containing the variable to make the coefficient of the squared term 1. (Do not factor 2 from the constant, -6.) Then complete the square.

$$y = 2x^2 + 4x - 6$$
$$= 2(x^2 + 2x) - 6$$
$$= 2\,(x^2 + 2x + 1) - 2 - 6 \qquad \textit{Complete the square.}$$
$$= 2(x + 1)^2 - 8$$

b) The parabola opens upward because $a = 2$, which is greater than 0.

c) The vertex of the graph of an equation in the form $y = a(x - h)^2 + k$ is (h, k). Therefore, the vertex of the graph of $y = 2(x + 1)^2 - 8$ is $(-1, -8)$. The vertex of a parabola can also be found using

$$\left(-\frac{b}{2a}, \frac{4ac - b^2}{4a}\right) \quad \text{or} \quad \left(-\frac{b}{2a}, f\left(-\frac{b}{2a}\right)\right)$$

Show that both of these procedures give $(-1, -8)$ as the vertex of the parabola now.

d) To determine the y-intercept, let $x = 0$ and solve for y.

$$y = 2(x + 1)^2 - 8$$
$$= 2(0 + 1)^2 - 8$$
$$= -6$$

The y-intercept is $(0, -6)$.

e) To determine the x-intercepts, let $y = 0$ and solve for x.

$$y = 2(x + 1)^2 - 8$$
$$0 = 2(x + 1)^2 - 8 \qquad \textit{Substitute 0 for y.}$$
$$8 = 2(x + 1)^2 \qquad \textit{Add 8 to both sides.}$$
$$4 = (x + 1)^2 \qquad \textit{Divide both sides by 2.}$$
$$\pm 2 = x + 1 \qquad \textit{Square root property}$$
$$-1 \pm 2 = x \qquad \textit{Subtract 1 from both sides.}$$
$$x = -1 - 2 \quad \text{or} \quad x = -1 + 2$$
$$x = -3 \qquad\qquad x = 1$$

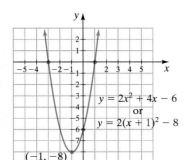

$y = 2x^2 + 4x - 6$
or
$y = 2(x + 1)^2 - 8$

$(-1, -8)$

FIGURE 14.2

The x-intercepts are $(-3, 0)$ and $(1, 0)$. The x-intercepts could also be found by substituting 0 for y in $y = 2x^2 + 4x - 6$ and solving for x using factoring or the quadratic formula. Do this now and show that you get the same x-intercepts.

f) We use the vertex and the x- and y-intercepts to draw the graph, which is shown in **Figure 14.2**.

▸ **Now Try Exercise 19**

3 Graph Parabolas of the Form $x = a(y - k)^2 + h$

Parabolas can also open to the right or left. The graph of an equation of the form $x = a(y - k)^2 + h$ will be a parabola whose vertex is at the point (h, k). If a is a positive number, the parabola will open to the right, and if a is a negative number, the parabola will open to the left. The four different forms of a parabola are shown in **Figure 14.3**.

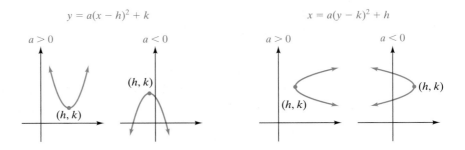

$y = a(x - h)^2 + k$

$x = a(y - k)^2 + h$

$a > 0$ $a < 0$ $a > 0$ $a < 0$

(h, k) (h, k) (h, k) (h, k)

FIGURE 14.3

Parabola with Vertex at (h, k)

1. $y = a(x - h)^2 + k, a > 0$ (opens upward)
2. $y = a(x - h)^2 + k, a < 0$ (opens downward)
3. $x = a(y - k)^2 + h, a > 0$ (opens to the right)
4. $x = a(y - k)^2 + h, a < 0$ (opens to the left)

Note that equations of the form $y = a(x - h)^2 + k$ are functions since their graphs pass the vertical line test. However, equations of the form $x = a(y - k)^2 + h$ are not functions since their graphs do not pass the vertical line test.

EXAMPLE 2 ▶ Sketch the graph of $x = -2(y + 4)^2 - 1$.

Solution The graph opens to the left since the equation is of the form $x = a(y - k)^2 + h$ and $a = -2$, which is less than 0. The equation can be expressed as $x = -2[y - (-4)]^2 - 1$. Thus, $h = -1$ and $k = -4$. The vertex of the graph is $(-1, -4)$. See **Figure 14.4**. If we set $y = 0$, we see that the x-intercept is at $-2(0 + 4)^2 - 1 = -2(16) - 1$ or -33. By substituting values for y you can find the corresponding values of x. When $y = -2$, $x = -9$, and when $y = -6$, $x = -9$. These points are marked on the graph. Notice that this graph has no y-intercept.

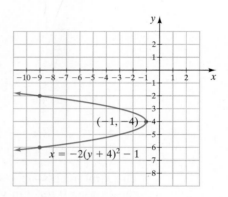

FIGURE 14.4

▶ **Now Try Exercise 31**

EXAMPLE 3 ▶

a) Write the equation $x = 2y^2 + 12y + 13$ in the form $x = a(y - k)^2 + h$.

b) Graph $x = 2y^2 + 12y + 13$.

Solution

a) First factor 2 from the first two terms. Then, complete the square on the expression within the parentheses.

$$x = 2y^2 + 12y + 13$$
$$= 2\,(y^2 + 6y) + 13$$
$$= 2\,(y^2 + 6y + 9) + (2)(-9) + 13$$
$$= 2(y^2 + 6y + 9) - 18 + 13$$
$$= 2(y + 3)^2 - 5$$

b) Since $a > 0$, the parabola opens to the right. Note that when $y = 0$, $x = 2(0)^2 + 12(0) + 13 = 13$. Thus, the x-intercept is $(13, 0)$. The vertex of the parabola is $(-5, -3)$. When $y = -6$, we find that $x = 13$. Thus, another point on the graph is $(13, -6)$. Using the quadratic formula, we can determine that the y-intercepts are about $(0, -4.6)$ and $(0, -1.4)$. The graph is shown in **Figure 14.5**.

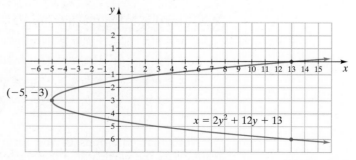

FIGURE 14.5

▶ **Now Try Exercise 45**

4 Learn the Distance and Midpoint Formulas

Now we will derive a formula to find the **distance** between two points on a line. We will use this formula shortly to develop the formula for a circle. Consider **Figure 14.6.**

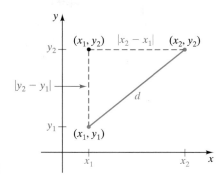

FIGURE 14.6

The horizontal distance between the two points (x_1, y_2) and (x_2, y_2), indicated by the red dashed line, is $|x_2 - x_1|$. We use the absolute value because we want the distance to be positive. If x_1 was larger than x_2, then $x_2 - x_1$ would be negative. The vertical distance between the points (x_1, y_1) and (x_1, y_2), indicated by the green dashed line, is $|y_2 - y_1|$. Using the Pythagorean Theorem where d is the distance between the two points, we get

$$d^2 = |x_2 - x_1|^2 + |y_2 - y_1|^2$$

Since any nonzero number squared is positive, we do not need absolute value signs. We can therefore write

$$d^2 = (x_2 - x_1)^2 + (y_2 - y_1)^2$$

Using the square root property, with the principal square root, we get the distance between the points (x_1, y_1) and (x_2, y_2), which is $d = \sqrt{(x_2 - x_1)^2 + (y_2 - y_1)^2}$.

Distance Formula

The distance, d, between any two points (x_1, y_1) and (x_2, y_2) can be found by the distance formula:

$$d = \sqrt{(x_2 - x_1)^2 + (y_2 - y_1)^2}$$

The distance between any two points will always be a positive number. Can you explain why? When finding the distance, it makes no difference which point we designate as point 1, (x_1, y_1), or point 2, (x_2, y_2). Note that the square of any real number will always be greater than or equal to 0. For example, $(5 - 2)^2 = (2 - 5)^2 = 9$.

EXAMPLE 4 ▶ Determine the distance between the points $(4, 5)$ and $(-2, 3)$.

Solution   As an aid, we plot the points (**Fig. 14.7**). Label $(4, 5)$ point 1 and $(-2, 3)$ point 2. Thus, (x_2, y_2) represents $(-2, 3)$ and (x_1, y_1) represents $(4, 5)$. Now use the distance formula to find the distance, d.

$$d = \sqrt{(x_2 - x_1)^2 + (y_2 - y_1)^2}$$
$$= \sqrt{(-2 - 4)^2 + (3 - 5)^2}$$
$$= \sqrt{(-6)^2 + (-2)^2}$$
$$= \sqrt{36 + 4}$$
$$= \sqrt{40} \quad \text{or} \quad \approx 6.32$$

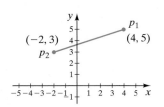

FIGURE 14.7

Thus, the distance between the points $(4, 5)$ and $(-2, 3)$ is $\sqrt{40}$ or about 6.32 units.

▶ Now Try Exercise 57

Avoiding Common Errors

Students will sometimes begin finding the distance correctly using the distance formula but will forget to take the square root of the sum $(x_2 - x_1)^2 + (y_2 - y_1)^2$ to obtain the correct answer. When taking the square root, remember that $\sqrt{a^2 + b^2} \neq a + b$.

It is often necessary to find the **midpoint** of a line segment between two given endpoints. To do this, we use the midpoint formula.

Midpoint Formula

Given any two points (x_1, y_1) and (x_2, y_2), the point halfway between the given points can be found by the midpoint formula:

$$\text{midpoint} = \left(\frac{x_1 + x_2}{2}, \frac{y_1 + y_2}{2} \right)$$

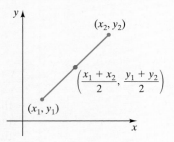

To find the midpoint, we take the average (the mean) of the x-coordinates and of the y-coordinates.

EXAMPLE 5 ▶ A line segment through the center of a circle intersects the circle at the points $(-3, 6)$ and $(4, 1)$. Find the center of the circle.

Solution   To find the center of the circle, we find the midpoint of the line segment between $(-3, 6)$ and $(4, 1)$. It makes no difference which points we label (x_1, y_1) and (x_2, y_2). We will let $(-3, 6)$ be (x_1, y_1) and $(4, 1)$ be (x_2, y_2). See **Figure 14.8.**

$$\text{midpoint} = \left(\frac{x_1 + x_2}{2}, \frac{y_1 + y_2}{2} \right)$$
$$= \left(\frac{-3 + 4}{2}, \frac{6 + 1}{2} \right) = \left(\frac{1}{2}, \frac{7}{2} \right)$$

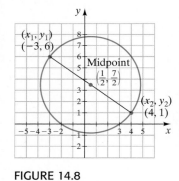

FIGURE 14.8

The point $\left(\frac{1}{2}, \frac{7}{2} \right)$ is halfway between the points $(-3, 6)$ and $(4, 1)$. It is also the center of the circle.

▶ **Now Try Exercise 69**

5 Graph Circles with Centers at the Origin

A **circle** may be defined as the set of points in a plane that are the same distance from a fixed point called its **center**.

The *standard form* of the equation of a circle whose center is at the origin may be derived using the distance formula. Let (x, y) be a point on a circle of radius r with center at $(0, 0)$. See **Figure 14.9**. Using the distance formula, we have

$$d = \sqrt{(x_2 - x_1)^2 + (y_2 - y_1)^2} \quad \text{\textit{Distance formula}}$$

$$\text{or} \quad r = \sqrt{(x - 0)^2 + (y - 0)^2} \quad \text{\textit{Substitute r for d, (x, y) for}} \atop \text{\textit{(x₂, y₂), and (O, O) for (x₁, y₁).}}$$

$$r = \sqrt{x^2 + y^2} \quad \text{\textit{Simplify the radicand.}}$$

$$r^2 = x^2 + y^2 \quad \text{\textit{Square both sides.}}$$

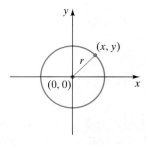

FIGURE 14.9

Circle with Its Center at the Origin and Radius *r*

$$x^2 + y^2 = r^2$$

For example, $x^2 + y^2 = 16$ is a circle with its center at the origin and radius 4, and $x^2 + y^2 = 10$ is a circle with its center at the origin and radius $\sqrt{10}$. Note that $4^2 = 16$ and $(\sqrt{10})^2 = 10$.

EXAMPLE 6 ▶ Graph the following equations.

a) $x^2 + y^2 = 64$ **b)** $y = \sqrt{64 - x^2}$ **c)** $y = -\sqrt{64 - x^2}$

Solution

a) If we rewrite the equation as

$$x^2 + y^2 = 8^2$$

we see that the radius of the circle is 8. The graph is illustrated in **Figure 14.10**.

b) If we solve the equation $x^2 + y^2 = 64$ for y, we obtain

$$y^2 = 64 - x^2$$

$$y = \pm\sqrt{64 - x^2}$$

In the equation $y = \pm\sqrt{64 - x^2}$, the equation $y = +\sqrt{64 - x^2}$ or, simply, $y = \sqrt{64 - x^2}$, represents the top half of the circle, while the equation $y = -\sqrt{64 - x^2}$ represents the bottom half of the circle. Thus, the graph of $y = \sqrt{64 - x^2}$, where y is the principal square root, lies above and on the x-axis. For any value of x in the domain of the function, the value of y must be greater than or equal to 0. Why? The graph is the semicircle shown in **Figure 14.11**.

c) The graph of $y = -\sqrt{64 - x^2}$ is also a semicircle. However, this graph lies below and on the x-axis. For any value of x in the domain of the function, the value of y must be less than or equal to 0. Why? The graph is shown in **Figure 14.12**.

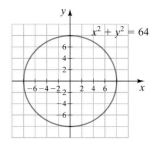

FIGURE 14.10

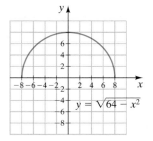

FIGURE 14.11

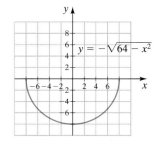

FIGURE 14.12

▶ **Now Try Exercise 101**

Consider the equations $y = \sqrt{64 - x^2}$ and $y = -\sqrt{64 - x^2}$ in Example **6 b)** and **6 c)**. If you square both sides of the equations and rearrange the terms, you will obtain $x^2 + y^2 = 64$. Try this now and see.

When using your calculator, you insert the function you wish to graph to the right of $y =$. Circles are not functions since they do not pass the vertical line test. To graph the equation $x^2 + y^2 = 64$, which is a circle of radius 8, we solve the equation for y to obtain $y = \pm\sqrt{64 - x^2}$. We then graph the two functions $Y_1 = \sqrt{64 - x^2}$ and $Y_2 = -\sqrt{64 - x^2}$ on the same axes to obtain the circle. These graphs are illustrated in **Figure 14.13**. Because of the distortion (described in the Using Your Graphing Calculator box in Section 13.1), the graph does not appear to be a circle. When you use the SQUARE feature of the calculator, the figure appears as a circle (see **Fig. 14.14**).

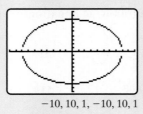

$-10, 10, 1, -10, 10, 1$

FIGURE 14.13

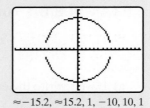

$\approx -15.2, \approx 15.2, 1, -10, 10, 1$

FIGURE 14.14

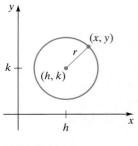

FIGURE 14.15

6 Graph Circles with Centers at (h, k)

The standard form of a circle with center at (h, k) and radius r can be derived using the distance formula. Let (h, k) be the center of the circle and let (x, y) be any point on the circle (see **Fig. 14.15**). If the radius, r, represents the distance between a point, (x, y), on the circle and its center, (h, k), then by the distance formula

$$r = \sqrt{(x - h)^2 + (y - k)^2}$$

We now square both sides of the equation to obtain the standard form of a circle with center at (h, k) and radius r.

$$r^2 = (x - h)^2 + (y - k)^2$$

Circle with Its Center at (h, k) and Radius r

$$(x - h)^2 + (y - k)^2 = r^2$$

EXAMPLE 7 ▶ Determine the equation of the circle shown in **Figure 14.16**.

Solution The center is $(-3, 2)$ and the radius is 3.

$$(x - h)^2 + (y - k)^2 = r^2$$

$$[x - (-3)]^2 + (y - 2)^2 = 3^2$$

$$(x + 3)^2 + (y - 2)^2 = 9$$

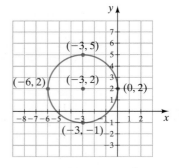

FIGURE 14.16

▶ **Now Try Exercise 87**

EXAMPLE 8 ▶

a) Show that the graph of the equation $x^2 + y^2 + 6x - 2y - 6 = 0$ is a circle.

b) Determine the center and radius of the circle and then draw the circle.

c) Find the area of the circle.

Solution

a) We will write this equation in standard form by completing the square. First we rewrite the equation, placing all the terms containing like variables together.

$$x^2 + 6x + y^2 - 2y - 6 = 0$$

Then we move the constant to the right side of the equation.

$$x^2 + 6x + y^2 - 2y = 6$$

Now we complete the square twice, once for each variable. We will first work with the variable x.

$$x^2 + 6x \boxed{+ 9} + y^2 - 2y = 6 \boxed{+ 9}$$

Now we work with the variable y.

$$x^2 + 6x + 9 + y^2 - 2y \boxed{+ 1} = 6 + 9 \boxed{+ 1}$$

or

$$\underbrace{x^2 + 6x + 9}_{(x + 3)^2} + \underbrace{y^2 - 2y + 1}_{(y - 1)^2} = 16$$

$$(x + 3)^2 + (y - 1)^2 = 16$$

$$(x + 3)^2 + (y - 1)^2 = 4^2$$

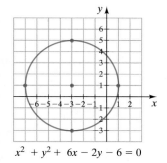

$x^2 + y^2 + 6x - 2y - 6 = 0$

FIGURE 14.17

b) The center of the circle is at $(-3, 1)$ and the radius is 4. The circle is sketched in **Figure 14.17**.

c) The area is

$$A = \pi r^2 = \pi(4)^2 = 16\pi \approx 50.3 \text{ square units}$$

▶ **Now Try Exercise 111**

EXERCISE SET 14.1   *Math XL*   **MyMathLab**
MathXL®   MyMathLab

Concept/Writing Exercises

1. Name the four conic sections. Draw a picture showing how each is formed.

2. Explain how to determine the direction a parabola will open by examining the equation.

3. Will all parabolas of the form $y = a(x - h)^2 + k, a > 0$ be functions? Explain. What will be the domain and range of $y = a(x - h)^2 + k, a > 0$?

4. Will all parabolas of the form $x = a(y - k)^2 + h, a > 0$ be functions? Explain. What will be the domain and range of $x = a(y - k)^2 + h, a > 0$?

5. How will the graphs of $y = 2(x - 3)^2 + 4$ and $y = -2(x - 3)^2 + 4$ compare?

6. Give the distance formula.

7. When the distance between two different points is found by using the distance formula, why must the distance always be a positive number?

8. Give the midpoint formula.

9. What is the definition of a circle?

10. What is the equation of a circle with center at (h, k)?

11. Is $x^2 - y^2 = 9$ an equation for a circle? Explain.

12. Is $-x^2 + y^2 = 25$ an equation for a circle? Explain.

13. Is $2x^2 + 3y^2 = 6$ an equation for a circle? Explain.

14. Is $x = y^2 - 6y + 3$ an equation for a parabola? Explain.

15. Is $x^2 = y^2 - 6y + 3$ an equation for a parabola? Explain.

16. Is $x = y + 2$ an equation for a parabola? Explain.

Practice the Skills

Graph each equation.

17. $y = (x - 2)^2 + 3$

18. $y = (x - 2)^2 - 3$

19. $y = (x + 3)^2 + 2$

20. $y = (x + 3)^2 - 2$

21. $y = (x - 2)^2 - 1$

22. $y = (x + 2)^2 + 1$

23. $y = -(x - 1)^2 + 1$

24. $y = -(x + 4)^2 - 5$

25. $y = -(x + 3)^2 + 4$

26. $y = 2(x + 1)^2 - 3$

27. $y = -3(x - 5)^2 + 3$

28. $x = (y - 1)^2 + 1$

29. $x = (y - 4)^2 - 3$

30. $x = -(y - 2)^2 + 1$

31. $x = -(y - 5)^2 + 4$

32. $x = -2(y - 4)^2 + 4$

33. $x = -5(y + 3)^2 - 6$

34. $x = 3(y + 1)^2 + 5$

35. $y = -2\left(x + \dfrac{1}{2}\right)^2 + 6$

36. $y = -\left(x - \dfrac{5}{2}\right)^2 + \dfrac{1}{2}$

In Exercises 37–50, **a)** *Write the equation in the form* $y = a(x - h)^2 + k$ *or* $x = a(y - k)^2 + h$. **b)** *Graph the equation.*

37. $y = x^2 + 2x$

38. $y = x^2 - 2x$

39. $y = x^2 + 6x$

40. $y = x^2 - 4x$

41. $x = y^2 + 4y$

42. $x = y^2 - 6y$

43. $y = x^2 + 7x + 10$

44. $y = x^2 + 2x - 7$

45. $x = -y^2 + 6y - 9$

46. $x = -y^2 - 5y - 4$

47. $y = -x^2 + 4x - 4$

48. $y = 2x^2 - 4x - 4$

49. $x = -y^2 + 3y - 4$

50. $x = 3y^2 - 12y - 36$

Determine the distance between each pair of points. Use a calculator where appropriate and round your answers to the nearest hundredth.

51. $(5, -1)$ and $(5, -6)$

52. $(-7, 2)$ and $(-3, 2)$

53. $(-1, 6)$ and $(8, 6)$

54. $(1, 8)$ and $(4, 12)$

55. $(-1, -3)$ and $(4, 9)$

56. $(-4, -5)$ and $(2, 3)$

57. $(-4, -5)$ and $(5, -2)$

58. $(6, 7)$ and $(11, 0)$

59. $(3, -1)$ and $\left(\dfrac{1}{2}, 4\right)$

60. $\left(-\dfrac{1}{4}, 2\right)$ and $\left(-\dfrac{3}{2}, 6\right)$

61. $(-1.6, 3.5)$ and $(-4.3, -1.7)$

62. $(5.2, -3.6)$ and $(-1.6, 2.3)$

63. $(\sqrt{7}, \sqrt{3})$ and $(0, 0)$

64. $(-\sqrt{2}, -\sqrt{5})$ and $(0, 0)$

Determine the midpoint of the line segment between each pair of points.

65. $(1, 3)$ and $(5, 9)$

66. $(0, 8)$ and $(4, -6)$

67. $(-7, 2)$ and $(7, -2)$

68. $(4, 7)$ and $(1, -3)$

69. $(-1, 4)$ and $(4, 6)$

70. $(-2, -9)$ and $(-6, -3)$

71. $\left(3, \dfrac{1}{2}\right)$ and $(2, -4)$

72. $\left(\dfrac{5}{2}, 3\right)$ and $\left(2, \dfrac{9}{2}\right)$

73. $(\sqrt{3}, 2)$ and $(\sqrt{2}, 7)$

74. $(-\sqrt{7}, 8)$ and $(\sqrt{5}, \sqrt{3})$

Write the equation of each circle with the given center and radius.

75. Center $(0, 0)$, radius 4

76. Center $(0, 0)$, radius 7

77. Center $(2, 0)$, radius 5

78. Center $(-3, 0)$, radius 9

79. Center $(0, 5)$, radius 1

80. Center $(0, -6)$, radius 6

81. Center $(3, 4)$, radius 8

82. Center $(-5, 2)$, radius 2

83. Center $(7, -6)$, radius 10

84. Center $(-6, -1)$, radius 7

85. Center $(1, 2)$, radius $\sqrt{5}$

86. Center $(-7, -2)$, radius $\sqrt{13}$

Write the equation of each circle. Assume the radius is a whole number.

87.

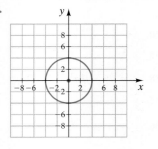

88.

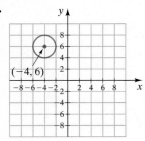

89.

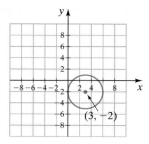

(3, −2)

90.

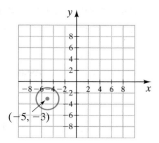

(−5, −3)

Graph each equation.

91. $x^2 + y^2 = 16$

92. $x^2 + y^2 = 5$

93. $x^2 + y^2 = 10$

94. $(x - 1)^2 + y^2 = 7$

95. $(x + 4)^2 + y^2 = 25$

96. $x^2 + (y + 1)^2 = 9$

97. $x^2 + (y - 3)^2 = 4$

98. $(x - 2)^2 + (y + 3)^2 = 16$

99. $(x + 8)^2 + (y + 2)^2 = 9$ **100.** $(x + 3)^2 + (y - 4)^2 = 36$

101. $y = \sqrt{25 - x^2}$

102. $y = \sqrt{16 - x^2}$

103. $y = -\sqrt{4 - x^2}$

104. $y = -\sqrt{49 - x^2}$

In Exercises 105–112, **a)** *use the method of completing the square to write each equation in standard form.* **b)** *Draw the graph.*

105. $x^2 + y^2 + 8x + 15 = 0$

106. $x^2 + y^2 + 4y = 0$

107. $x^2 + y^2 + 6x - 4y + 4 = 0$

108. $x^2 + y^2 + 2x - 4y - 4 = 0$

109. $x^2 + y^2 + 6x - 2y + 6 = 0$

110. $x^2 + y^2 + 4x - 6y - 3 = 0$

111. $x^2 + y^2 - 8x + 2y + 13 = 0$

112. $x^2 + y^2 - x + 3y - \dfrac{3}{2} = 0$

Problem Solving

113. Find the area of the circle in Exercise 95.

114. Find the area of the circle in Exercise 97.

In Exercises 115–118, find the x- and y-intercepts, if they exist, of the graph of each equation.

115. $x = y^2 - 6y - 7$

116. $x = -y^2 + 8y - 12$

117. $x = 2(y - 3)^2 + 6$

118. $x = -(y + 2)^2 - 8$

119. If you know the midpoint of a line segment, is it possible to determine the length of the line segment? Explain.

120. If you know one endpoint of a line segment and the length of the line segment, can you determine the other endpoint? Explain.

121. Find the length of the line segment whose midpoint is $(4, -6)$ with one endpoint at $(7, -2)$.

122. Find the length of the line segment whose midpoint is $(-2, 4)$ with one endpoint at $(3, 6)$.

123. Find the equation of a circle with center at $(-6, 2)$ that is tangent to the *x*-axis (that is, the circle touches the *x*-axis at only one point).

124. Find the equation of a circle with center at $(-3, 5)$ that is tangent to the *y*-axis.

In Exercises 125 and 126, find **a)** *the radius of the circle whose diameter is along the line shown,* **b)** *the center of the circle, and* **c)** *the equation of the circle.*

125.

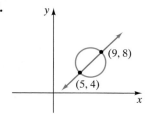

(9, 8)

(5, 4)

126.

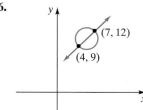

(7, 12)

(4, 9)

127. Points of Intersection What is the maximum number and the minimum number of points of intersection possible for the graphs of $y = a(x - h_1)^2 + k_1$ and $x = a(y - k_2)^2 + h_2$? Explain.

128. Inscribed Triangle Consider the figure below.

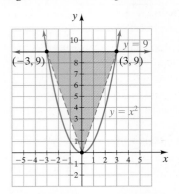

a) Find the area of the triangle outlined in green.

b) When a triangle is inscribed within a parabola, as in the figure, the area within the parabola from the base of the triangle is $\frac{4}{3}$ the area of the triangle. Find the area within the parabola from $x = -3$ to $x = 3$.

129. Ferris Wheel The Ferris wheel at Navy Pier in Chicago is 150 feet tall. The radius of the wheel itself is 68.2 feet.

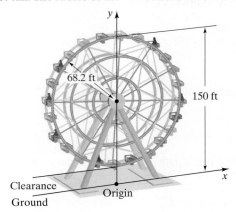

a) What is the clearance below the wheel?

b) How high is the center of the wheel from the ground?

c) Find the equation of the wheel. Assume the origin is on the ground directly below the center of the wheel.

130. Shaded Area Find the shaded area of the square in the figure. The equation of the circle is $x^2 + y^2 = 9$.

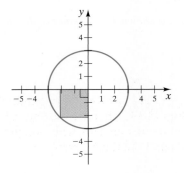

131. Shaded Area Consider the figure below. Write an equation for

a) the blue circle,

b) the red circle, and

c) the green circle.

d) Find the shaded area.

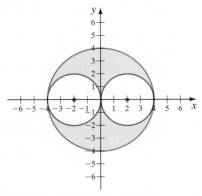

132. Points of Intersection Consider the equations $x^2 + y^2 = 16$ and $(x - 2)^2 + (y - 2)^2 = 16$. By considering the center and radius of each circle, determine the number of points of intersection of the two circles.

133. Concentric Circles Find the area between the two concentric circles whose equations are $(x - 2)^2 + (y + 4)^2 = 16$ and $(x - 2)^2 + (y + 4)^2 = 64$. *Concentric circles* are circles that have the same center.

134. Tunnel A highway department is planning to construct a semi-circular one-way tunnel through a mountain. The tunnel is to be large enough so that a truck 8 feet wide and 10 feet tall will pass through the center of the tunnel with 1 foot to spare directly above the corner of the truck when it is driving down the center of the tunnel (as shown in the figure below). Determine the minimum radius of the tunnel.

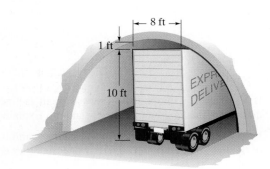

Group Activity

Discuss and answer Exercise 135 as a group.

135. Equation of a Parabola The equation of a parabola can be found if three points on the parabola are known. To do so, start with $y = ax^2 + bx + c$. Then substitute the x- and y-coordinates of the first point into the equation. This will result in an equation in a, b, and c. Repeat the procedure for the other two points. This process yields a system of three equations in three variables. Next solve the system for a, b, and c. To find the equation of the parabola, substitute the values found for a, b, and c into the equation $y = ax^2 + bx + c$.

Three points on a parabola are $(0, 12)$, $(3, -3)$, and $(-2, 32)$.

a) Individually, find a system of equations in three variables that can be used to find the equation of the parabola. Then compare your answers. If each member of the group does not have the same system, determine why.

b) Individually, solve the system and determine the values of a, b, and c. Then compare your answers.

c) Individually, write the equation of the parabola passing through $(0, 12)$, $(3, -3)$, and $(-2, 32)$. Then compare your answers.

d) Individually, write the equation in

$$y = a(x - h)^2 + k$$

form. Then compare your answers.

e) Individually, graph the equation in part **d)**. Then compare your answers.

Cumulative Review Exercises

[5.2] **136.** Simplify $\dfrac{6x^{-3}y^4}{18x^{-2}y^3}$.

[5.5] **137. a)** Write expressions to represent each of the four areas shown in the figure.

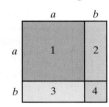

b) Express the total area shown as the square of a binomial.

[9.7] **138.** Evaluate the determinant.

$$\begin{vmatrix} 4 & 0 & 3 \\ 5 & 2 & -1 \\ 3 & 6 & 4 \end{vmatrix}$$

[10.1] **139.** Solve the inequality $-4 < 3x - 4 < 17$. Write the solution in interval notation.

[14.1] **140.** Graph $y = (x - 4)^2 + 1$.

14.2 The Ellipse

1 Graph ellipses.

2 Graph ellipses with centers at (h, k).

1 Graph Ellipses

An **ellipse** may be defined as a set of points in a plane, the sum of whose distances from two fixed points is a constant. The two fixed points are called the **foci** (each is a focus) of the ellipse (see **Fig. 14.18**). In **Figure 14.18**, F_1 and F_2 represent the two foci.

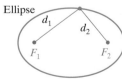
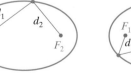
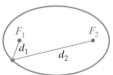

FIGURE 14.18

We can construct an ellipse using a length of string and two thumbtacks. Place the two thumbtacks fairly close together (**Fig. 14.19**). Then tie the ends of the string to the thumbtacks. With a pencil or pen pull the string taut, and, while keeping the string taut, draw the ellipse by moving the pencil around the thumbtacks.

FIGURE 14.19

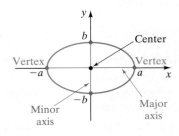

FIGURE 14.20

In **Figure 14.20**, the line segment from $-a$ to a on the x-axis is the *longer* or **major axis** and the line segment from $-b$ to b is the *shorter* or **minor axis** of the ellipse. The major axis of an ellipse may also be on the y-axis. **Figure 14.20** also shows the *center* of the ellipse and two vertices (the red dots). The vertices are the endpoints of the major axis.

The standard form of an ellipse with its center at the origin follows.

Ellipse with Its Center at the Origin

$$\frac{x^2}{a^2} + \frac{y^2}{b^2} = 1$$

where $(a, 0)$ and $(-a, 0)$ are the x-intercepts and $(0, b)$ and $(0, -b)$ are the y-intercepts.

Notice the x-intercepts are found using the constant in the denominator of the x^2-term, and the y-intercepts are found using the constant in the denominator of the y^2-term. If $a^2 > b^2$, the major axis of the ellipse will be along the x-axis. If $b^2 > a^2$, the major axis of the ellipse will be along the y-axis.

In Example 1, the major axis of the ellipse is along the x-axis.

EXAMPLE 1 ▶ Graph $\dfrac{x^2}{9} + \dfrac{y^2}{4} = 1$.

Solution   We can rewrite the equation as

$$\frac{x^2}{3^2} + \frac{y^2}{2^2} = 1$$

Thus, $a = 3$ and the x-intercepts are $(3, 0)$ and $(-3, 0)$. Since $b = 2$, the y-intercepts are $(0, 2)$ and $(0, -2)$. The ellipse is illustrated in **Figure 14.21**.

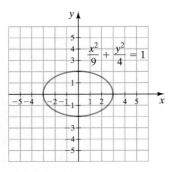

FIGURE 14.21

▶ **Now Try Exercise 15**

An equation may be written so that it may not be obvious that its graph is an ellipse. This is illustrated in Example 2.

EXAMPLE 2 ▶ Graph $20x^2 + 9y^2 = 180$.

Solution   To make the right side of the equation equal to 1, we divide both sides of the equation by 180. We then obtain an equation that we can recognize as an ellipse.

$$\frac{20x^2 + 9y^2}{180} = \frac{180}{180}$$

$$\frac{20x^2}{180} + \frac{9y^2}{180} = 1$$

$$\frac{x^2}{9} + \frac{y^2}{20} = 1$$

The equation can now be recognized as an ellipse in standard form.

$$\frac{x^2}{a^2} + \frac{y^2}{b^2} = 1$$

Since $a^2 = 9$, $a = 3$. We know that $b^2 = 20$; thus $b = \sqrt{20}$ (or approximately 4.47).

$$\frac{x^2}{3^2} + \frac{y^2}{(\sqrt{20})^2} = 1$$

The x-intercepts are $(3, 0)$ and $(-3, 0)$. The y-intercepts are $(0, -\sqrt{20})$ and $(0, \sqrt{20})$. The graph is illustrated in **Figure 14.22**. Note that the major axis lies along the y-axis.

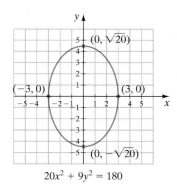

$$20x^2 + 9y^2 = 180$$

FIGURE 14.22

▶ **Now Try Exercise 19**

In Example 1, since $a^2 = 9$ and $b^2 = 4$ and $a^2 > b^2$, the major axis is along the x-axis. In Example 2, since $a^2 = 9$ and $b^2 = 20$ and $b^2 > a^2$, the major axis is along the y-axis. In the specific case where $a^2 = b^2$, the figure is a circle. Thus, the circle is a special case of an ellipse.

EXAMPLE 3 ▶ Write the equation of the ellipse shown in **Figure 14.23**.

Solution The x-intercepts are $(-\sqrt{10}, 0)$ and $(\sqrt{10}, 0)$; thus, $a = \sqrt{10}$ and $a^2 = 10$. The y-intercepts are $(0, -12)$ and $(0, 12)$; thus, $b = 12$ and $b^2 = 144$.

$$\frac{x^2}{a^2} + \frac{y^2}{b^2} = 1$$

$$\frac{x^2}{10} + \frac{y^2}{144} = 1$$

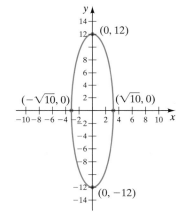

FIGURE 14.23

▶ **Now Try Exercise 45**

The formula for the **area of an ellipse** is $A = \pi ab$. In Example 1, where $a = 3$ and $b = 2$, the area is $A = \pi(3)(2) = 6\pi \approx 18.8$ square units.

In Example 2, where $a = 3$ and $b = \sqrt{20}$, the area is $A = \pi(3)(\sqrt{20}) = \pi(3)(2\sqrt{5}) = 6\pi\sqrt{5} \approx 42.1$ square units.

2 Graph Ellipses with Centers at (h, k)

Horizontal and vertical translations, similar to those used in Chapter 12, may be used to obtain the equation of an ellipse with center at (h, k).

Ellipse with Its Center at (h, k)

$$\frac{(x - h)^2}{a^2} + \frac{(y - k)^2}{b^2} = 1$$

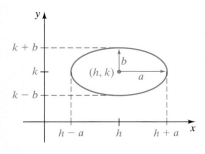

FIGURE 14.24

In the formula, the h shifts the graph left or right from the origin and k shifts the graph up or down from the origin, as shown in **Figure 14.24.**

EXAMPLE 4 ▶ Graph $\dfrac{(x-2)^2}{25} + \dfrac{(y+3)^2}{16} = 1$.

Solution This is the graph of $\dfrac{x^2}{25} + \dfrac{y^2}{16} = 1$ or $\dfrac{x^2}{5^2} + \dfrac{y^2}{4^2} = 1$ translated so that its center is at $(2, -3)$. Note that $a = 5$ and $b = 4$. The graph is shown in **Figure 14.25.**

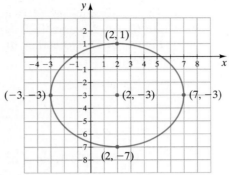

FIGURE 14.25

▶ **Now Try Exercise 33**

FIGURE 14.26

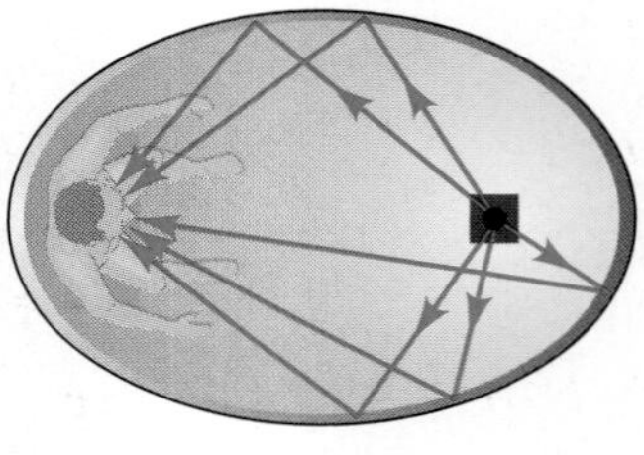

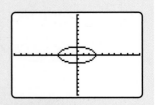

FIGURE 14.27

An understanding of ellipses is useful in many areas. Astronomers know that planets revolve in elliptical orbits around the Sun. Communications satellites move in elliptical orbits around Earth (see **Fig. 14.26**).

Ellipses are used in medicine to smash kidney stones. When a signal emerges from one focus of an ellipse, the signal is reflected to the other focus. In kidney stone machines, the person is situated so that the stone to be smashed is at one focus of an elliptically shaped chamber called a lithotripter (see **Fig. 14.27** and Exercises 57 and 58).

In certain buildings with ellipsoidal ceilings, a person standing at one focus can whisper something and a person standing at the other focus can clearly hear what the person whispered. There are many other uses for ellipses, including lamps that are made to concentrate light at a specific point.

USING YOUR GRAPHING CALCULATOR

Ellipses are not functions. To graph ellipses on a graphing calculator, we solve the equation for y. This will give the two equations that we use to graph the ellipse.

In Example 1, we graphed $\dfrac{x^2}{9} + \dfrac{y^2}{4} = 1$. Solving this equation for y, we get

$$\dfrac{x^2}{9} + \dfrac{y^2}{4} = 1$$

$$\boxed{36} \cdot \dfrac{x^2}{9} + \boxed{36} \cdot \dfrac{y^2}{4} = 1 \cdot \boxed{36} \qquad \textit{Multiply by the LCD.}$$

$$4x^2 + 9y^2 = 36$$

$$9y^2 = 36 - 4x^2$$

$$y^2 = \dfrac{36 - 4x^2}{9}$$

$$y^2 = \dfrac{4(9 - x^2)}{9} \qquad \textit{Factor 4 from the numerator.}$$

$$y = \pm\dfrac{2}{3}\sqrt{9 - x^2} \qquad \textit{Square root property}$$

To graph the ellipse, we let $Y_1 = \dfrac{2}{3}\sqrt{9 - x^2}$ and $Y_2 = -\dfrac{2}{3}\sqrt{9 - x^2}$ and graph both equations. The graphs of Y_1 and Y_2 are illustrated in **Figure 14.28**.

FIGURE 14.28

EXERCISE SET 14.2

MathXL®   MyMathLab
MathXL   MyMathLab

Concept/Writing Exercises

1. What is the definition of an ellipse?

2. What is the equation of an ellipse with its center at the origin?

3. What is the equation of an ellipse with its center at (h, k)?

4. Discuss the graphs of $\dfrac{x^2}{a^2} + \dfrac{y^2}{b^2} = 1$ when $a > b$, $a < b$, and $a = b$.

5. Explain why the circle is a special case of the ellipse.

6. In the formula $\dfrac{x^2}{a^2} + \dfrac{y^2}{b^2} = 1$, what do the a and b represent?

7. What is the first step in graphing the ellipse whose equation is $10x^2 + 36y^2 = 180$?

8. What is the first step in graphing the ellipse whose equation is $9x^2 + 25y^2 = 225$?

9. Is $\dfrac{x^2}{36} - \dfrac{y^2}{49} = 1$ an equation for an ellipse? Explain.

10. Is $-\dfrac{x^2}{49} + \dfrac{y^2}{81} = 1$ an equation for an ellipse? Explain.

Practice the Skills

Graph each equation.

11. $\dfrac{x^2}{4} + \dfrac{y^2}{1} = 1$

12. $\dfrac{x^2}{1} + \dfrac{y^2}{4} = 1$

13. $\dfrac{x^2}{4} + \dfrac{y^2}{9} = 1$

14. $\dfrac{x^2}{9} + \dfrac{y^2}{4} = 1$

15. $\dfrac{x^2}{25} + \dfrac{y^2}{9} = 1$

16. $\dfrac{x^2}{100} + \dfrac{y^2}{16} = 1$

17. $\dfrac{x^2}{16} + \dfrac{y^2}{25} = 1$

18. $\dfrac{x^2}{81} + \dfrac{y^2}{49} = 1$

19. $x^2 + 16y^2 = 16$

20. $x^2 + 25y^2 = 25$

21. $49x^2 + y^2 = 49$

22. $9x^2 + 25y^2 = 225$

23. $9x^2 + 16y^2 = 144$

24. $25x^2 + 4y^2 = 100$

25. $25x^2 + 100y^2 = 400$

26. $100x^2 + 25y^2 = 400$

27. $x^2 + 2y^2 = 8$

28. $x^2 + 36y^2 = 36$

29. $\dfrac{x^2}{16} + \dfrac{(y - 2)^2}{9} = 1$

30. $\dfrac{(x - 1)^2}{16} + \dfrac{y^2}{1} = 1$

31. $\dfrac{(x - 4)^2}{9} + \dfrac{(y + 3)^2}{25} = 1$

32. $\dfrac{(x - 3)^2}{25} + \dfrac{(y + 2)^2}{49} = 1$

33. $\dfrac{(x + 1)^2}{9} + \dfrac{(y - 2)^2}{4} = 1$

34. $\dfrac{(x - 3)^2}{16} + \dfrac{(y - 4)^2}{25} = 1$

35. $(x + 3)^2 + 9(y + 1)^2 = 81$

36. $18(x - 1)^2 + 2(y + 3)^2 = 72$

37. $(x - 5)^2 + 4(y + 4)^2 = 4$

38. $4(x - 2)^2 + 9(y + 2)^2 = 36$

39. $12(x + 4)^2 + 3(y - 1)^2 = 48$

40. $16(x - 2)^2 + 4(y + 3)^2 = 16$

Problem Solving

41. Find the area of the ellipse in Exercise 11.

42. Find the area of the ellipse in Exercise 15.

43. How many points are on the graph of $16x^2 + 25y^2 = 0$? Explain.

44. Consider the graph of the equation $\dfrac{x^2}{a^2} + \dfrac{y^2}{b^2} = 1$. Explain what will happen to the shape of the graph as the value of b gets closer to the value of a. What is the shape of the graph when $a = b$?

In Exercises 45–48, find the equation of the ellipse that has the four points as endpoints of the major and minor axes.

45. $(3, 0), (-3, 0), (0, 4), (0, -4)$

46. $(6, 0), (-6, 0), (0, 5), (0, -5)$

47. $(2, 0), (-2, 0), (0, 3), (0, -3)$

48. $(1, 0), (-1, 0), (0, 7), (0, -7)$

49. How many points of intersection will the graphs of the equations $x^2 + y^2 = 49$ and $\dfrac{x^2}{16} + \dfrac{y^2}{25} = 1$ have? Explain.

50. How many points of intersection will the graphs of the equations $y = 2(x - 2)^2 - 3$ and $\dfrac{(x - 2)^2}{4} + \dfrac{(y + 3)^2}{9} = 1$ have? Explain.

In Exercises 51 and 52, write the following equation in standard form. Determine the center of each ellipse.

51. $x^2 + 4y^2 + 6x + 16y - 11 = 0$

52. $x^2 + 4y^2 - 4x - 8y - 92 = 0$

53. Art Gallery An art gallery has an elliptical hall. The maximum distance from one focus to the wall is 90.2 feet and the minimum distance is 20.7 feet. Find the distance between the foci.

54. Communications Satellite A space shuttle transported a communications satellite to space. The satellite travels in an elliptical orbit around Earth. The maximum distance of the satellite from Earth is 23,200 miles and the minimum distance is 22,800 miles. Earth is at one focus of the ellipse. Find the distance from Earth to the other focus.

55. Tunnel through a Mountain The tunnel in the photo is the top half of an ellipse. The width of the tunnel is 20 feet and the height is 24 feet.

a) If you pictured a completed ellipse with the center of the ellipse being at the center of the road, determine the equation of the ellipse.

b) Find the area of the ellipse found in part **a)**.

c) Find the area of the opening of the tunnel.

56. Billiard Table An elliptical billiard table is 8 feet long by 5 feet wide. Determine the location of the foci. On such a table, if a ball is put at each focus and one ball is hit with enough force, it would hit the ball at the other focus no matter where it banks on the table.

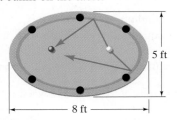

5 ft

8 ft

57. Lithotripter Machine Suppose the lithotripter machine described on page 886 is 6 feet long and 4 feet wide. Describe the location of the foci.

58. Lithotripter On page 886 we gave a brief introduction to the lithotripter, which uses ultrasound waves to shatter kidney stones. Do research and write a detailed report describing the procedure used to shatter kidney stones. Make sure that you explain how the waves are directed on the stone.

59. Whispering Gallery The National Statuary Hall in the Capitol Building in Washington, D.C., is a "whispering gallery." Do research and explain why one person standing at a certain point can whisper something and someone standing a considerable distance away can hear it.

60. Check your answer to Exercise 11 on your grapher.

61. Check your answer to Exercise 17 on your grapher.

Challenge Problems

Determine the equation of the ellipse that has the following four points as endpoints of the major and minor axes.

62. $(-7, 3), (5, 3), (-1, 5), (-1, 1)$

63. $(-3, 2), (11, 2), (4, 5), (4, -1)$

Group Activity

Work Exercise 64 individually. Then compare your answers.

64. Tunnel The photo shows an elliptical tunnel (with the bottom part of the ellipse not shown) near Rockefeller Center in New York City. The maximum width of the tunnel is 18 feet and the maximum height *from the ground to the top* is 10.5 feet.

a) If the *completed ellipse* would have a maximum height of 15 feet, how high from the ground is the center of the elliptical tunnel?

b) Consider the following graph, which could be used to represent the tunnel.

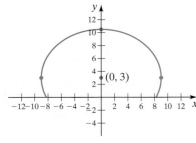

(0, 3)

If the ellipse were continued, what would be the other *y*-intercept of the graph?

c) Write the equation of the ellipse, if completed, in part **b)**.

Cumulative Review Exercises

[2.6] **65.** Solve the formula $S = \dfrac{n}{2}(f + l)$ for l.

[5.6] **66.** Divide $\dfrac{2x^2 + 2x - 7}{2x - 3}$.

[11.6] **67.** Solve $\sqrt{3b - 2} = 10 - b$.

[12.6] **68.** Solve $\dfrac{3x + 5}{x - 4} \le 0$, and give the solution in interval notation.

[13.7] **69.** Find $\log_8 321$.

Mid-Chapter Test: 14.1–14.2

To find out how well you understand the chapter material to this point, take this brief test. The answers, and the section where the material was initially discussed, are given in the back of the book. Review any questions you answered incorrectly.

Graph each equation.

 1. $y = (x - 2)^2 - 1$
 2. $y = -(x + 1)^2 + 3$
 3. $x = -(y - 4)^2 + 1$
 4. $x = 2(y + 3)^2 - 2$
 5. $y = x^2 + 6x + 10$

Find the distance between each pair of points. Where appropriate, round your answers to the nearest hundredth.

 6. $(-7, 4)$ and $(-2, -8)$
 7. $(5, -3)$ and $(2, 9)$

Find the midpoint of the line segment between each pair of points.

 8. $(9, -1)$ and $(-11, 6)$

 9. $\left(-\dfrac{5}{2}, 7\right)$ and $\left(8, \dfrac{1}{2}\right)$

 10. Write the equation of the circle with center at $(-3, 2)$ and a radius of 5 units.

Graph each equation.

 11. $x^2 + (y - 1)^2 = 16$
 12. $y = \sqrt{36 - x^2}$
 13. $x^2 + y^2 - 2x + 4y - 4 = 0$
 14. What is the definition of a circle?

Graph each equation.

 15. $\dfrac{x^2}{4} + \dfrac{y^2}{9} = 1$

 16. $\dfrac{x^2}{81} + \dfrac{y^2}{25} = 1$

 17. $\dfrac{(x - 1)^2}{49} + \dfrac{(y + 2)^2}{4} = 1$

 18. $36(x + 3)^2 + (y - 4)^2 = 36$.
 19. Find the area of the ellipse in Exercise 15.
 20. Find the equation of the ellipse that has the four points $(8, 0)$, $(-8, 0)$, $(0, 5)$, and $(0, -5)$ as the endpoints of the major and minor axes.

14.3 The Hyperbola

1 Graph hyperbolas.

2 Review conic sections.

1 Graph Hyperbolas

A **hyperbola** is the set of points in a plane, the difference of whose distances from two fixed points (called foci) is a constant. A hyperbola is illustrated in **Figure 14.29a**. In the figure, for every point on the hyperbola, the difference $M - N$ is the same constant. A hyperbola may look like a pair of parabolas. However, the shapes are actually quite different. A hyperbola has two **vertices.** The point halfway between the two vertices is the **center** of the hyperbola. The line through the vertices is called the **transverse axis.** In **Figure 14.29b** the transverse axis lies along the x-axis, and in **Figure 14.29c**, the transverse axis lies along the y-axis.

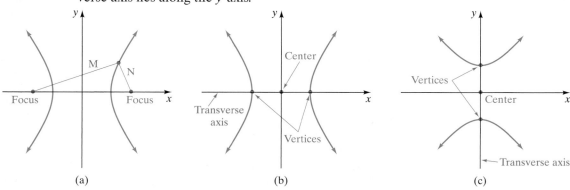

FIGURE 14.29   (a)   (b)   (c)

The dashed lines in **Figure 14.30** are called **asymptotes**. The asymptotes are not a part of the hyperbola but are used as an aid in graphing hyperbolas. (We will discuss asymptotes shortly.) Also given in **Figure 14.30** is the standard form of the equation for each hyperbola. In **Figure 14.30a**, both vertices are *a* units from the origin. In **Figure 14.30b**, both vertices are *b* units from the origin. Note that in the standard form of the equation, the denominator of the x^2 term is always a^2 and the denominator of the y^2 term is always b^2.

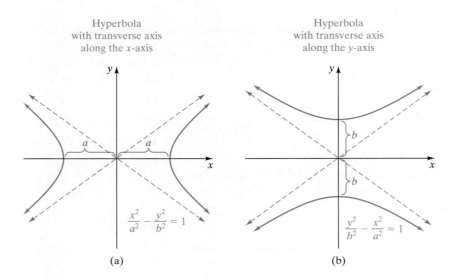

Hyperbola with transverse axis along the x-axis

$$\frac{x^2}{a^2} - \frac{y^2}{b^2} = 1$$

(a)

Hyperbola with transverse axis along the y-axis

$$\frac{y^2}{b^2} - \frac{x^2}{a^2} = 1$$

(b)

FIGURE 14.30

A hyperbola centered at the origin whose transverse axis is one of the coordinate axes has either *x*-intercepts (**Fig. 14.30a**) or *y*-intercepts (**Fig. 14.30b**), but not both. When a hyperbola is centered at the origin, the intercepts are the vertices of the hyperbola. When written in standard form, the intercepts will be on the axis indicated by the variable with the positive coefficient. The intercepts will be the positive and the negative square root of the denominator of the positive term.

Examples	Intercepts on	Intercepts
$\dfrac{x^2}{49} - \dfrac{y^2}{16} = 1$	x-axis	$(-7, 0)$ and $(7, 0)$
$\dfrac{y^2}{16} - \dfrac{x^2}{49} = 1$	y-axis	$(0, -4)$ and $(0, 4)$

Asymptotes can help you graph hyperbolas. The asymptotes are two straight lines that go through the center of the hyperbola (see **Fig. 14.30**). As the values of *x* and *y* get larger, the graph of the hyperbola approaches the asymptotes. The equations of the asymptotes of a hyperbola whose center is the origin are

$$y = \frac{b}{a}x \quad \text{and} \quad y = -\frac{b}{a}x$$

The asymptotes can be drawn quickly by plotting the four points (a, b), $(-a, b)$, $(a, -b)$, and $(-a, -b)$, and then connecting these points with dashed lines to form a rectangle. Next, draw dashed lines through the opposite corners of the rectangle to obtain the asymptotes.

Hyperbola with Its Center at the Origin

TRANSVERSE AXIS ALONG *x*-AXIS (OPENS TO THE RIGHT AND LEFT)	TRANSVERSE AXIS ALONG *y*-AXIS (OPENS UPWARD AND DOWNWARD)
$$\frac{x^2}{a^2} - \frac{y^2}{b^2} = 1$$	$$\frac{y^2}{b^2} - \frac{x^2}{a^2} = 1$$

ASYMPTOTES

$$y = \frac{b}{a}x \quad \text{and} \quad y = -\frac{b}{a}x$$

EXAMPLE 1 ▶

a) Determine the equations of the asymptotes of the hyperbola with equation

$$\frac{x^2}{9} - \frac{y^2}{16} = 1$$

b) Draw the hyperbola using the asymptotes.

Solution

a) The value of a^2 is 9; the positive square root of 9 is 3. The value of b^2 is 16; the positive square root of 16 is 4. The asymptotes are

$$y = \frac{b}{a}x \quad \text{and} \quad y = -\frac{b}{a}x$$

or

$$y = \frac{4}{3}x \quad \text{and} \quad y = -\frac{4}{3}x$$

b) To graph the hyperbola, we first graph the asymptotes. To graph the asymptotes, we can plot the points $(3, 4)$, $(-3, 4)$, $(3, -4)$, and $(-3, -4)$ and draw the rectangle as illustrated in **Figure 14.31**. The asymptotes are the dashed lines through the opposite corners of the rectangle.

Since the x-term in the original equation is positive, the graph intersects the x-axis. Since the denominator of the positive term is 9, the vertices are at $(3, 0)$ and $(-3, 0)$. Now draw the hyperbola by letting the hyperbola approach its asymptotes (**Fig. 14.32**). Note that the asymptotes are drawn using dashed lines since they are not part of the hyperbola. They are used merely to help draw the graph.

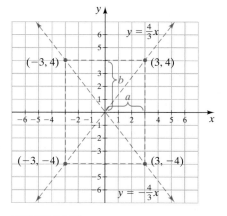

FIGURE 14.31

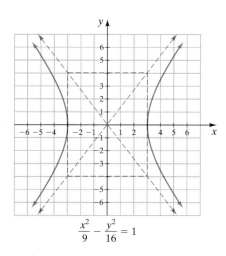

$$\frac{x^2}{9} - \frac{y^2}{16} = 1$$

FIGURE 14.32

▶ **Now Try Exercise 21**

EXAMPLE 2 ▶

a) Show that the equation $-25x^2 + 4y^2 = 100$ is a hyperbola by expressing the equation in standard form.

b) Determine the equations of the asymptotes of the graph.

c) Draw the graph.

Solution

a) We divide both sides of the equation by 100 to obtain 1 on the right side of the equation.

$$\frac{-25x^2 + 4y^2}{100} = \frac{100}{100}$$

$$\frac{-25x^2}{100} + \frac{4y^2}{100} = 1$$

$$\frac{-x^2}{4} + \frac{y^2}{25} = 1$$

Rewriting the equation in standard form (positive term first), we get

$$\frac{y^2}{25} - \frac{x^2}{4} = 1$$

b) Since $a = 2$ and $b = 5$, the equations of the asymptotes are

$$y = \frac{5}{2}x \quad \text{and} \quad y = -\frac{5}{2}x$$

c) The graph intersects the y-axis at $(0, 5)$ and $(0, -5)$. **Figure 14.33a** illustrates the asymptotes, and **Figure 14.33b** illustrates the hyperbola.

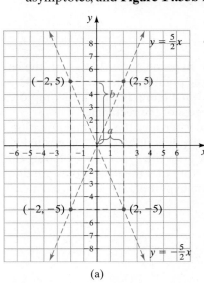

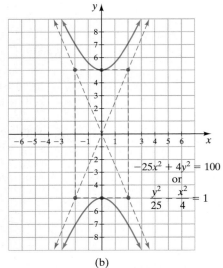

(a)   (b)

FIGURE 14.33

▶ **Now Try Exercise 29**

We have discussed hyperbolas with their centers at the origin. Hyperbolas do not have to be centered at the origin. In this book, we will not discuss such hyperbolas.

USING YOUR GRAPHING CALCULATOR

We can graph hyperbolas just as we did circles and ellipses. To graph hyperbolas on the graphing calculator, solve the equation for y and graph each part. Consider Example 1,

$$\frac{x^2}{9} - \frac{y^2}{16} = 1.$$

Show that if you solve this equation for y you get $y = \pm\frac{4}{3}\sqrt{x^2 - 9}$. Let $Y_1 = \frac{4}{3}\sqrt{x^2 - 9}$ and $Y_2 = -\frac{4}{3}\sqrt{x^2 - 9}$. **Figures 14.34a, 14.34b, 14.34c,** and **14.34d,** on the next page, give the graphs of Y_1 and Y_2 for different window settings. The window settings used are indicated above each graph.

(*continued on the next page*)

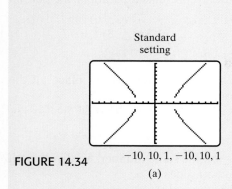

Standard setting	ZOOM: option 5 ZSquare setting	ZOOM: option 4, ZDecimal setting	Set window as shown below figure (called "friendly window settings")
$-10, 10, 1, -10, 10, 1$	$\approx-15.2, \approx15.2, 1, -10, 10, 1$	$-4.7, 4.7, 1, -3.1, 3.1, 1$	$-14.1, 14.1, 1, -9.3, 9.3, 1$
(a)	(b)	(c)	(d)

FIGURE 14.34

In part (d), the "friendly window setting," the ratio of the length of the x-axis (28.2 units) to the length of the y-axis (18.6 units) is about 1.516. This is the same ratio as the length to the width of the display window of the calculator on the TI-84 Plus.

2 Review Conic Sections

The following chart summarizes conic sections.

Parabola	Circle	Ellipse	Hyperbola
$y = a(x - h)^2 + k$ or $\quad y = ax^2 + bx + c$	$x^2 + y^2 = r^2$	$\dfrac{x^2}{a^2} + \dfrac{y^2}{b^2} = 1$	$\dfrac{x^2}{a^2} - \dfrac{y^2}{b^2} = 1$
$a > 0$			
$a < 0$	$(x - h)^2 + (y - k)^2 = r^2$	$\dfrac{(x - h)^2}{a^2} + \dfrac{(y - k)^2}{b^2} = 1$	$\dfrac{y^2}{b^2} - \dfrac{x^2}{a^2} = 1$
$x = a(y - k)^2 + h$ or $\quad x = ay^2 + by + c$			
$a > 0$			Asymptotes $y = \dfrac{b}{a}x$ and $y = -\dfrac{b}{a}x$
$a < 0$			

EXAMPLE 3 ▸ Indicate whether each equation represents a parabola, a circle, an ellipse, or a hyperbola.

a) $6x^2 = -6y^2 + 48$ **b)** $x - y^2 = 9y + 3$ **c)** $2x^2 = 8y^2 + 72$

Solution

a) This equation has an x-squared term and a y-squared term. Let's place all the squared terms on the left side of the equation.

$$6x^2 = -6y^2 + 48$$
$$6x^2 + 6y^2 = 48 \qquad \text{\textit{Add 6y}}^2 \text{ \textit{to both sides.}}$$

Since the coefficients of both squared terms are the same number, we divide both sides of the equation by this number. Divide both sides by 6.

$$\frac{6x^2 + 6y^2}{6} = \frac{48}{6}$$
$$x^2 + y^2 = 8$$

This equation is of the form $x^2 + y^2 = r^2$ where $r^2 = 8$.
   The equation $6x^2 = -6y^2 + 48$ represents a circle.

b) This equation has a y-squared term, but no x-squared term. Let's solve the equation for x.

$$x - y^2 = 9y + 3$$
$$x = y^2 + 9y + 3 \qquad \text{\textit{Add y}}^2 \text{ \textit{to both sides.}}$$

This equation is of the form $x = ay^2 + by + c$ where $a = 1$, $b = 9$, and $c = 3$.
   The equation $x - y^2 = 9y + 3$ represents a parabola that opens to the right.

c) This equation has an x-squared term and a y-squared term. Let's place all the squared terms on the left side of the equation.

$$2x^2 = 8y^2 + 72$$
$$2x^2 - 8y^2 = 72 \qquad \text{\textit{Subtract 8y}}^2 \text{ \textit{from both sides.}}$$

Since the coefficients of both squared terms are different numbers, we want to divide the equation by the constant on the right side. Divide both sides by 72.

$$\frac{2x^2 - 8y^2}{72} = \frac{72}{72}$$
$$\frac{2x^2}{72} - \frac{8y^2}{72} = 1$$
$$\frac{x^2}{36} - \frac{y^2}{9} = 1$$

This equation is of the form $\dfrac{x^2}{a^2} - \dfrac{y^2}{b^2} = 1$ where $a^2 = 36$ (or $a = 6$) and $b^2 = 9$ (or $b = 3$).

   The equation $2x^2 = 8y^2 + 72$ represents a hyperbola.

▸ **Now Try Exercise 53**

EXERCISE SET 14.3   *Math XL* **MyMathLab**
MathXL®   MyMathLab

Concept/Writing Exercises

1. What is the definition of a hyperbola?

2. What are asymptotes? How do you find the equations of the asymptotes of a hyperbola?

3. Discuss the graph of $\dfrac{x^2}{a^2} - \dfrac{y^2}{b^2} = 1$ for nonzero real numbers a and b. Include the transverse axis, vertices, and asymptotes.

4. Discuss the graph of $\dfrac{y^2}{b^2} - \dfrac{x^2}{a^2} = 1$ for nonzero real numbers a and b. Include the transverse axis, vertices, and asymptotes.

5. Is $\dfrac{x^2}{81} + \dfrac{y^2}{64} = 1$ an equation for a hyperbola? Explain.

6. Is $-\dfrac{x^2}{81} - \dfrac{y^2}{64} = 1$ an equation for a hyperbola? Explain.

7. Is $4x^2 - 25y^2 = 100$ an equation for a hyperbola? Explain.

8. Is $36x^2 - 9y^2 = -324$ an equation for a hyperbola? Explain.

9. What is the first step in graphing the hyperbola whose equation is $x^2 - 9y^2 = 81$? Explain.

10. What is the first step in graphing the hyperbola whose equation is $4x^2 - y^2 = -64$? Explain.

Practice the Skills

a) *Determine the equations of the asymptotes for each equation.* **b)** *Graph the equation.*

11. $\dfrac{x^2}{9} - \dfrac{y^2}{4} = 1$

12. $\dfrac{y^2}{4} - \dfrac{x^2}{9} = 1$

13. $\dfrac{x^2}{4} - \dfrac{y^2}{1} = 1$

14. $\dfrac{y^2}{1} - \dfrac{x^2}{4} = 1$

15. $\dfrac{x^2}{9} - \dfrac{y^2}{25} = 1$

16. $\dfrac{y^2}{25} - \dfrac{x^2}{9} = 1$

17. $\dfrac{x^2}{25} - \dfrac{y^2}{16} = 1$

18. $\dfrac{y^2}{16} - \dfrac{x^2}{25} = 1$

19. $\dfrac{y^2}{25} - \dfrac{x^2}{36} = 1$

20. $\dfrac{x^2}{36} - \dfrac{y^2}{25} = 1$

21. $\dfrac{y^2}{9} - \dfrac{x^2}{16} = 1$

22. $\dfrac{x^2}{16} - \dfrac{y^2}{9} = 1$

23. $\dfrac{y^2}{25} - \dfrac{x^2}{4} = 1$

24. $\dfrac{x^2}{4} - \dfrac{y^2}{25} = 1$

25. $\dfrac{x^2}{81} - \dfrac{y^2}{16} = 1$

26. $\dfrac{y^2}{16} - \dfrac{x^2}{81} = 1$

In Exercises 27–36, **a)** *write each equation in standard form and determine the equations of the asymptotes.* **b)** *Draw the graph.*

27. $x^2 - 25y^2 = 25$

28. $25y^2 - x^2 = 25$

29. $4y^2 - 16x^2 = 64$

30. $16x^2 - 4y^2 = 64$

31. $9y^2 - x^2 = 9$

32. $x^2 - 9y^2 = 9$

33. $25x^2 - 9y^2 = 225$

34. $9y^2 - 25x^2 = 225$

35. $4y^2 - 36x^2 = 144$

36. $64y^2 - 25x^2 = 1600$

In Exercises 37–60, indicate whether the equation represents a parabola, a circle, an ellipse, or a hyperbola. See Example 3.

37. $10x^2 + 10y^2 = 40$

38. $15x^2 - 5y^2 = 75$

39. $x^2 + 16y^2 = 64$

40. $x = 5y^2 + 15y + 1$

41. $4x^2 - 4y^2 = 29$

42. $11x^2 + 11y^2 = 99$

43. $2y = 12x^2 - 8x + 16$

44. $4y^2 - 6x^2 = 72$

45. $6x^2 + 9y^2 = 54$

46. $9.2x^2 + 9.2y^2 = 46$

47. $3x = -2y^2 + 9y - 15$

48. $12x^2 - 3y^2 = 48$

49. $6x^2 + 6y^2 = 36$

50. $9x^2 = -9y^2 + 99$

51. $14y^2 = 7x^2 + 35$

52. $9x^2 = -18y^2 + 36$

53. $x + y = 2y^2 + 6$

54. $2x^2 = -2y^2 + 32$

55. $12x^2 = 4y^2 + 48$

56. $-8x^2 = -9y^2 - 72$

57. $y - x + 4 = x^2$

58. $17x^2 = -2y^2 + 34$

59. $-3x^2 - 3y^2 = -27$

60. $x - y^2 = 15$

Problem Solving

61. Determine an equation of the hyperbola whose vertices are $(0, 2)$ and $(0, -2)$ and whose asymptotes are $y = \dfrac{1}{2}x$ and $y = -\dfrac{1}{2}x$.

62. Determine an equation of a hyperbola whose vertices are $(0, 6)$ and $(0, -6)$ and whose asymptotes are $y = \dfrac{3}{2}x$ and $y = -\dfrac{3}{2}x$.

63. Determine an equation of the hyperbola whose vertices are $(-3, 0)$ and $(3, 0)$ and whose asymptotes are $y = 2x$ and $y = -2x$.

64. Determine an equation of a hyperbola whose vertices are $(7, 0)$ and $(-7, 0)$ and whose asymptotes are $y = \dfrac{4}{7}x$ and $y = -\dfrac{4}{7}x$.

65. Determine an equation of a hyperbola whose transverse axis is along the x-axis and whose equations of the asymptotes are $y = \frac{5}{3}x$ and $y = -\frac{5}{3}x$. Is this the only possible answer? Explain.

66. Determine an equation of a hyperbola whose transverse axis is along the y-axis and whose equations of the asymptotes are $y = \frac{2}{3}x$ and $y = -\frac{2}{3}x$. Is this the only possible answer? Explain.

67. Are any hyperbolas of the form $\frac{x^2}{a^2} - \frac{y^2}{b^2} = 1$ functions? Explain.

68. Are any hyperbolas of the form $\frac{y^2}{b^2} - \frac{x^2}{a^2} = 1$ functions? Explain.

69. Considering the graph of $\frac{x^2}{25} - \frac{y^2}{4} = 1$, determine the domain and range of the relation.

70. Considering the graph of $\frac{y^2}{36} - \frac{x^2}{9} = 1$, determine the domain and range of the relation.

71. If the equation $\frac{x^2}{a^2} - \frac{y^2}{b^2} = 1$, where $a > b$, is graphed, and then the values of a and b are interchanged, and the new equation is graphed, how will the two graphs compare? Explain your answer.

72. If the equation $\frac{x^2}{a^2} - \frac{y^2}{b^2} = 1$, where $a > b$, is graphed, and then the signs of each term on the left side of the equation are changed, and the new equation is graphed, how will the two graphs compare? Explain your answer.

73. Check your answer to Exercise 15 on your grapher.

74. Check your answer to Exercise 21 on your grapher.

Cumulative Review Exercises

[4.4] **75.** Write the equation, in slope-intercept form, of the line that passes through the points $(-6, 4)$ and $(-2, 2)$.

[7.4] **76.** Add $\frac{3x}{2x - 3} + \frac{2x + 4}{2x^2 + x - 6}$.

[8.5] **77.** Let $f(x) = 3x^2 - x + 5$ and $g(x) = 6 - 4x^2$. Find $(f + g)(x)$.

[9.3] **78.** Solve the system of equations.
$$-4x + 9y = 7$$
$$5x + 6y = -3$$

[12.3] **79.** Solve the formula $E = \frac{1}{2}mv^2$ for v.

[13.6] **80.** Solve the equation $\log(x + 4) = \log 5 - \log x$.

14.4 Nonlinear Systems of Equations and Their Applications

1 Solve nonlinear systems using substitution.

2 Solve nonlinear systems using addition.

3 Solve applications.

1 Solve Nonlinear Systems Using Substitution

In Chapter 9, we discussed systems of linear equations. Here we discuss nonlinear systems of equations. A **nonlinear system of equations** is a system of equations in which at least one equation is not linear (that is, one whose graph is not a straight line).

The solution to a system of equations is the point or points that satisfy all equations in the system. Consider the system of equations

$$x^2 + y^2 = 25$$
$$3x + 4y = 0$$

Both equations are graphed on the same axes in **Figure 14.35**. Note that the graphs appear to intersect at the points $(-4, 3)$ and $(4, -3)$. The check shows that these points satisfy both equations in the system and are therefore solutions to the system.

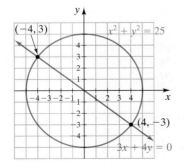

FIGURE 14.35

Check $(-4, 3)$

$$x^2 + y^2 = 25$$
$$(-4)^2 + 3^2 \overset{?}{=} 25$$
$$16 + 9 \overset{?}{=} 25$$
$$25 = 25 \quad \textit{True}$$

$$3x + 4y = 0$$
$$3(-4) + 4(3) \overset{?}{=} 0$$
$$-12 + 12 \overset{?}{=} 0$$
$$0 = 0 \quad \textit{True}$$

Check $(4, -3)$ $4^2 + (-3)^2 = 25$ $3(4) + 4(-3) = 0$

$16 + 9 \stackrel{?}{=} 25$ $12 - 12 \stackrel{?}{=} 0$

$25 = 25$ *True* $0 = 0$ *True*

The graphical procedure for solving a system of equations may be inaccurate since we have to estimate the point or points of intersection. An exact answer may be obtained algebraically.

To solve a system of equations algebraically, we often solve one or more of the equations for one of the variables and then use substitution. This procedure is illustrated in Examples 1 and 2.

EXAMPLE 1 ▶ Solve the previous system of equations algebraically using the substitution method.

$$x^2 + y^2 = 25$$
$$3x + 4y = 0$$

Solution   We first solve the linear equation $3x + 4y = 0$ for either x or y. We will solve for y.

$$3x + 4y = 0$$
$$4y = -3x$$
$$y = -\frac{3x}{4}$$

Now we substitute $-\dfrac{3x}{4}$ for y in the equation $x^2 + y^2 = 25$ and solve for the remaining variable, x.

$$x^2 + y^2 = 25$$
$$x^2 + \left(-\frac{3x}{4}\right)^2 = 25$$
$$x^2 + \frac{9x^2}{16} = 25$$
$$16\left(x^2 + \frac{9x^2}{16}\right) = 16(25)$$
$$16x^2 + 9x^2 = 400$$
$$25x^2 = 400$$
$$x^2 = \frac{400}{25} = 16$$
$$x = \pm\sqrt{16} = \pm 4$$

Next, we find the corresponding value of y for each value of x by substituting each value of x (one at a time) into the equation solved for y.

$$x = 4 \qquad\qquad\qquad x = -4$$
$$y = -\frac{3x}{4} \qquad\qquad y = -\frac{3x}{4}$$
$$= -\frac{3(4)}{4} \qquad\qquad = -\frac{3(-4)}{4}$$
$$= -3 \qquad\qquad\qquad = 3$$

The solutions are $(4, -3)$ and $(-4, 3)$. This checks with the solution obtained graphically in **Figure 14.35**.

▶ **Now Try Exercise 9**

Our objective in using substitution is to obtain a single equation containing only one variable.

Helpful Hint *Study Tip*

In this section, we will be using the substitution method and addition method to solve non-linear systems of equations. Both methods were introduced in Chapter 9 to solve linear systems of equations. If you do not remember how to use both methods to solve linear systems of equations, now is a good time to review Chapter 9.

In Examples 1 and 2, we solve systems using the substitution method, while in Examples 3 and 4, we solve systems using the addition method.

You may choose to solve a system by the substitution method if addition of the two equations will not lead to an equation that can be easily solved, as is the case with the systems in Examples 1 and 2.

EXAMPLE 2 ▶ Solve the system of equations using the substitution method.

$$y = x^2 - 3$$
$$x^2 + y^2 = 9$$

Solution   Since both equations contain x^2, we will solve one of the equations for x^2. We will choose to solve $y = x^2 - 3$ for x^2.

$$y = x^2 - 3$$
$$y + 3 = x^2$$

Now substitute $y + 3$ for x^2 in the equation $x^2 + y^2 = 9$.

$$x^2 + y^2 = 9$$
$$y + 3 + y^2 = 9$$
$$y^2 + y + 3 = 9$$
$$y^2 + y - 6 = 0$$
$$(y + 3)(y - 2) = 0$$
$$y + 3 = 0 \quad \text{or} \quad y - 2 = 0$$
$$y = -3 \qquad\qquad y = 2$$

Now find the corresponding values of x by substituting the values found for y.

$y = -3$	$y = 2$
$y = x^2 - 3$	$y = x^2 - 3$
$-3 = x^2 - 3$	$2 = x^2 - 3$
$0 = x^2$	$5 = x^2$
$0 = x$	$\pm\sqrt{5} = x$

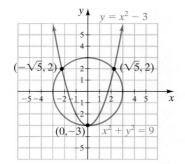

FIGURE 14.36

This system has three solutions: $(0, -3)$, $(\sqrt{5}, 2)$, and $(-\sqrt{5}, 2)$.

Note that the graph of the equation $y = x^2 - 3$ is a parabola and the graph of the equation $x^2 + y^2 = 9$ is a circle. The graphs of both equations are illustrated in **Figure 14.36**.

▶ **Now Try Exercise 19**

Helpful Hint

Students will sometimes solve for one variable and assume that they have the solution. Remember that the solution, if one exists, to a system of equations in two variables consists of one or more ordered pairs.

2 Solve Nonlinear Systems Using Addition

We can often solve systems of equations more easily using the addition method that was discussed in Section 9.3. As with the substitution method, our objective is to obtain a single equation containing only one variable.

EXAMPLE 3 ▶ Solve the system of equations using the addition method.

$$x^2 + y^2 = 9$$
$$2x^2 - y^2 = -6$$

Solution   If we add the two equations, we will obtain one equation containing only one variable.

$$
\begin{aligned}
x^2 + y^2 &= 9 \\
2x^2 - y^2 &= -6 \\
\hline
3x^2 &= 3 \\
x^2 &= 1 \\
x &= \pm 1
\end{aligned}
$$

Now solve for the variable y by substituting $x = \pm 1$ into *either* of the original equations.

$x = 1$	$x = -1$
$x^2 + y^2 = 9$	$x^2 + y^2 = 9$
$1^2 + y^2 = 9$	$(-1)^2 + y^2 = 9$
$1 + y^2 = 9$	$1 + y^2 = 9$
$y^2 = 8$	$y^2 = 8$
$y = \pm\sqrt{8}$	$y = \pm\sqrt{8}$
$= \pm 2\sqrt{2}$	$= \pm 2\sqrt{2}$

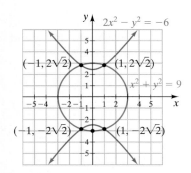

There are four solutions to this system of equations:

$$(1, 2\sqrt{2}),\ (1, -2\sqrt{2}),\ (-1, 2\sqrt{2}),\ \text{and}\ (-1, -2\sqrt{2})$$

The graphs of the equations in the system are given in **Figure 14.37**. Notice the four points of intersection of the two graphs.

FIGURE 14.37

▶ **Now Try Exercise 25**

It is possible that a system of equations has no real solution (therefore, the graphs do not intersect). Example 4 illustrates such a case.

EXAMPLE 4 ▶ Solve the system of equations using the addition method.

$$x^2 + 4y^2 = 16 \quad (eq.\,1)$$
$$x^2 + y^2 = 1 \quad (eq.\,2)$$

Solution   Multiply *(eq. 2)* by -1 and add the resulting equation to *(eq. 1)*.

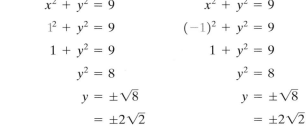

$$
\begin{aligned}
x^2 + 4y^2 &= 16 \\
-x^2 - y^2 &= -1 \qquad (eq.\,2)\quad \text{multiplied by } -1\\
\hline
3y^2 &= 15 \\
y^2 &= 5 \\
y &= \pm\sqrt{5}
\end{aligned}
$$

Now solve for x.

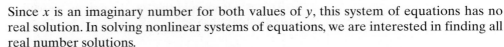

$y = \sqrt{5}$	$y = -\sqrt{5}$
$x^2 + y^2 = 1$	$x^2 + y^2 = 1$
$x^2 + (\sqrt{5})^2 = 1$	$x^2 + (-\sqrt{5})^2 = 1$
$x^2 + 5 = 1$	$x^2 + 5 = 1$
$x^2 = -4$	$x^2 = -4$
$x = \pm\sqrt{-4}$	$x = \pm\sqrt{-4}$
$x = \pm 2i$	$x = \pm 2i$

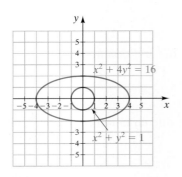

FIGURE 14.38

Since x is an imaginary number for both values of y, this system of equations has no real solution. In solving nonlinear systems of equations, we are interested in finding all real number solutions.

The graphs of the equations are shown in **Figure 14.38**. Notice that the two graphs do not intersect; therefore, there is no real solution. This agrees with the answer we obtained algebraically.

▶ **Now Try Exercise 37**

3 Solve Applications

Now we will study some applications of nonlinear systems.

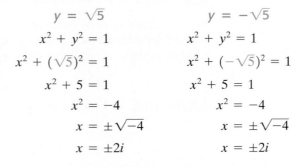

FIGURE 14.39

EXAMPLE 5 ▶ **Flower Garden** Fred and Judy Vespucci want to build a rectangular flower garden behind their house. Fred went to a local nursery and bought enough topsoil to cover 150 square meters of land. Judy went to the local hardware store and purchased 50 meters of fence for the perimeter of the garden. How should they build the garden to use all the topsoil he bought and all the fence she purchased?

Solution   Understand and Translate   We begin by drawing a sketch (see **Fig 14.39**).

$$\text{Let } x = \text{length of garden}$$
$$y = \text{width of garden.}$$

Since $A = xy$ and Fred bought topsoil to cover 150 square meters, we have

$$xy = 150$$

Since $P = 2x + 2y$ and Judy purchased 50 meters of fence for the perimeter of the garden, we have

$$2x + 2y = 50$$

The system of equations is

$$xy = 150$$
$$2x + 2y = 50$$

Carry Out   We will solve the system using substitution. The equation $2x + 2y = 50$ is a linear equation. We will solve this equation for y. (We could also solve for x.)

$$2x + 2y = 50$$
$$2y = 50 - 2x$$
$$y = \frac{50 - 2x}{2} = \frac{50}{2} - \frac{2x}{2} = 25 - x$$

Now substitute $25 - x$ for y in the equation $xy = 150$.

$$xy = 150$$
$$x(25 - x) = 150$$
$$25x - x^2 = 150$$
$$0 = x^2 - 25x + 150$$
$$0 = (x - 10)(x - 15)$$
$$x - 10 = 0 \quad \text{or} \quad x - 15 = 0$$
$$x = 10 \qquad\qquad x = 15$$

Answer If $x = 10$, then $y = 25 - 10 = 15$. And, if $x = 15$, then $y = 25 - 15 = 10$. Thus, in either case, the dimensions of the flower garden are 10 meters by 15 meters.

▶ **Now Try Exercise 43**

EXAMPLE 6 ▶ **Bicycles** Hike 'n' Bike Company produces and sells bicycles. Its weekly cost equation is $C = 50x + 400, 0 \le x \le 160$, and its weekly revenue equation is $R = 100x - 0.3x^2, 0 \le x \le 160$, where x is the number of bicycles produced and sold each week. Find the number of bicycles that must be produced and sold for Hike 'n' Bike to break even.

Solution Understand and Translate A company breaks even when its cost equals its revenue. When its cost is greater than its revenue, the company has a loss. When its revenue exceeds its cost, the company makes a profit.

The system of equations is

$$C = 50x + 400$$
$$R = 100x - 0.3x^2$$

For Hike 'n' Bike to break even, its cost must equal its revenue. Thus, we write

$$C = R$$
$$50x + 400 = 100x - 0.3x^2$$

Carry Out Writing this quadratic equation in standard form, we obtain

$$0.3x^2 - 50x + 400 = 0, \quad 0 \le x \le 160$$

We will solve this equation using the quadratic formula.

$$a = 0.3, \quad b = -50, \quad c = 400$$

$$x = \frac{-b \pm \sqrt{b^2 - 4ac}}{2a}$$

$$= \frac{-(-50) \pm \sqrt{(-50)^2 - 4(0.3)(400)}}{2(0.3)}$$

$$= \frac{50 \pm \sqrt{2020}}{0.6}$$

$$x = \frac{50 + \sqrt{2020}}{0.6} \approx 158.2 \quad \text{or} \quad x = \frac{50 - \sqrt{2020}}{0.6} \approx 8.4$$

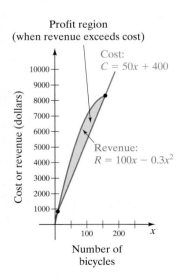

Profit region
(when revenue exceeds cost)

Cost:
$C = 50x + 400$

Cost or revenue (dollars)

Revenue:
$R = 100x - 0.3x^2$

Number of
bicycles

FIGURE 14.40

Answer The cost will equal the revenue and the company will break even when approximately 8 bicycles are sold. The cost will also equal the revenue when approximately 158 bicycles are sold. The company will make a profit when between 9 and 158 bicycles are sold. When fewer than 9 or more than 158 bicycles are sold, the company will have a loss (see **Fig. 14.40**).

▶ **Now Try Exercise 55**

USING YOUR GRAPHING CALCULATOR

To solve nonlinear systems of equations graphically, graph the equations and find the intersections of the graphs. Consider the system in Example 1, $x^2 + y^2 = 25$ and $3x + 4y = 0$. To graph $x^2 + y^2 = 25$, we use $Y_1 = \sqrt{25 - x^2}$ and $Y_2 = -\sqrt{25 - x^2}$. If we solve $3x + 4y = 0$ for y we obtain $y = -\dfrac{3}{4}x$. Thus, we use $Y_3 = -\dfrac{3}{4}x$. Therefore, to solve this system we find the intersection of

$$Y_1 = \sqrt{25 - x^2}$$
$$Y_2 = -\sqrt{25 - x^2}$$
$$Y_3 = -\frac{3}{4}x$$

The system is graphed, using the ZOOM: 5 (ZSquare) feature in **Figure 14.41a**.* In **Figure 14.41b**, we graph the same three equations using the "friendly numbers" shown below the figure. Using the calculator with either the TRACE and ZOOM features, the TABLE feature, or the INTERSECT feature, you will find that the solutions are $(4, -3)$ and $(-4, 3)$.

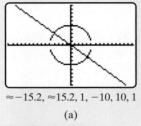

$\approx -15.2, \approx 15.2, 1, -10, 10, 1$　　　$-9.4, 9.4, 1, -6.2, 6.2, 1$

(a)　　　　　　　　　　　　　　　　(b)

FIGURE 14.41

*Start with the standard window, then select ZOOM: 5 to get this graph.

EXERCISE SET 14.4　Math XL　MyMathLab

MathXL®　　MyMathLab

Concept/Writing Exercises

1. What is a nonlinear system of equations?

2. Explain how nonlinear systems of equations may be solved graphically.

3. Can a nonlinear system of equations have exactly one real solution? If so, give an example. Explain.

4. Can a nonlinear system of equations have exactly two real solutions: If so, give an example. Explain.

5. Can a nonlinear system of equations have exactly three real solutions? If so, give an example. Explain.

6. Can a nonlinear system of equations have no real solutions? If so, give an example. Explain.

Practice the Skills

Find all real solutions to each system of equations using the substitution method.

7. $x^2 + y^2 = 18$
$x + y = 0$

8. $x^2 + y^2 = 18$
$x - y = 0$

9. $x^2 + y^2 = 9$
$x + 2y = 3$

10. $x^2 + y^2 = 4$
$x - 2y = 4$

11. $y = x^2 - 5$
$3x + 2y = 10$

12. $x + y = 4$
$x^2 - y^2 = 4$

13. $x^2 + y = 6$
$y = x^2 + 4$

14. $y - x = 2$
$x^2 - y^2 = 4$

15. $2x^2 + y^2 = 16$
$x^2 - y^2 = -4$

16. $x + y^2 = 4$
$x^2 + y^2 = 6$

17. $x^2 + y^2 = 4$
$y = x^2 - 6$

18. $x^2 - 4y^2 = 36$
$x^2 + 2y^2 = 5$

19. $x^2 + y^2 = 9$
$y = x^2 - 3$

20. $x^2 + y^2 = 16$
$y = x^2 - 4$

21. $2x^2 - y^2 = -8$
$x - y = 6$

22. $x^2 + y^2 = 1$
$y - x = 3$

Find all real solutions to each system of equations using the addition method.

23. $x^2 - y^2 = 4$
$2x^2 + y^2 = 8$

24. $x^2 + y^2 = 36$
$x^2 - y^2 = 36$

25. $x^2 + y^2 = 16$
$2x^2 - 5y^2 = 25$

26. $x^2 + y^2 = 25$
$x^2 - 2y^2 = 7$

27. $3x^2 - y^2 = 4$
$x^2 + 4y^2 = 10$

28. $3x^2 + 2y^2 = 30$
$x^2 + y^2 = 13$

29. $4x^2 + 9y^2 = 36$
$2x^2 - 9y^2 = 18$

30. $x^2 + 4y^2 = 16$
$-9x^2 + y^2 = 4$

31. $2x^2 - y^2 = 7$
$x^2 + 2y^2 = 6$

32. $5x^2 - 2y^2 = -13$
$3x^2 + 4y^2 = 39$

33. $x^2 + y^2 = 25$
$2x^2 - 3y^2 = -30$

34. $x^2 - 2y^2 = 7$
$x^2 + y^2 = 34$

35. $x^2 + y^2 = 9$
$16x^2 - 4y^2 = 64$

36. $3x^2 + 4y^2 = 35$
$2x^2 + 5y^2 = 42$

37. $x^2 + y^2 = 4$
$16x^2 + 9y^2 = 144$

38. $x^2 + y^2 = 1$
$9x^2 - 4y^2 = 36$

39. $x^2 + 4y^2 = 4$
$10y^2 - 9x^2 = 90$

40. $x^2 + y^2 = 81$
$25x^2 + 4y^2 = 100$

Problem Solving

41. Make up your own nonlinear system of equations whose solution is the empty set. Explain how you know the system has no solution.

42. If a system of equations consists of an ellipse and a hyperbola, what is the maximum number of points of intersection? Make a sketch to illustrate this.

43. Dance Floor Kris Hundley wants to build a dance floor at her gym. The dance floor is to have a perimeter of 84 meters and an area of 440 square meters. Find the dimensions of the dance floor.

44. Rectangular Region Ellen Dupree fences in a rectangular area along a riverbank as illustrated. If 20 feet of fencing encloses an area of 48 square feet, find the dimensions of the enclosed area.

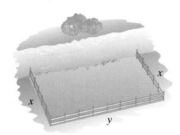

45. Vegetable Garden James Cannon is planning to build a rectangular flower garden in his yard. The garden is to have a perimeter of 78 feet and an area of 270 square feet. Find the dimensions of the vegetable garden.

46. Rectangular Region A rectangular area is to be fenced along a river as illustrated in Exercise 44. If 20 feet of fencing encloses an area of 50 square feet, find the dimensions of the enclosed area.

47. Currency A country's currency includes a bill that has an area of 112 square centimeters with a diagonal of $\sqrt{260}$ centimeters. Find the length and width of the bill.

48. Ice Rink A rectangular ice rink has an area of 3000 square feet. If the diagonal across the rink is 85 feet, find the dimensions of the rink.

Rockefeller Plaza, New York City

49. Piece of Wood Frank Samuelson, a carpenter, has a rectangular piece of plywood. When he measures the diagonal it measures 34 inches. When he cuts the wood along the diagonal, the perimeter of each triangle formed is 80 inches. Find the dimensions of the original piece of wood.

50. Sailboat A sail on a sailboat is shaped like a right triangle with a perimeter of 36 meters and a hypotenuse of 15 meters. Find the length of the legs of the triangle.

51. Baseball and Football Paul Martin throws a football upward from the ground. Its height above the ground at any time, t, is given by the formula $d = -16t^2 + 64t$. At the same time that the football is thrown, Shannon Ryan throws a baseball upward from the top of an 80-foot-tall building. Its height above the ground at any time, t, is given by the formula $d = -16t^2 + 16t + 80$. Find the time at which the two balls will be the same height above the ground. (Neglect air resistance.)

52. Tennis Ball and Snowball Robert Snell throws a tennis ball downward from a helicopter flying at a height of 950 feet. The height of the ball above the ground at any time t is found by the formula $d = -16t^2 - 10t + 950$. At the instant the ball is thrown from the helicopter, Ramon Sanchez throws a snowball upward from the top of an 750-foot-tall building. The height above the ground of the snowball at any time, t, is found by the formula $d = -16t^2 + 80t + 750$. At what time will the ball and snowball pass each other? (Neglect air resistance.)

53. Simple Interest Simple interest is calculated using the simple interest formula, interest = principal · rate · time or $i = prt$. If Seana Hayden invests a certain principal at a specific interest rate for 1 year, the interest she obtains is $7.50. If she increases the principal by $25 and the interest rate is decreased by 1%, the interest remains the same. Find the principal and the interest rate.

54. Simple Interest If Claire Brooke invests a certain principal at a specific interest rate for 1 year, the interest she obtains is $72. If she increases the principal by $120 and the interest rate is decreased by 2%, the interest remains the same. Find the principal and the interest rate. Use $i = prt$.

For the given cost and revenue equations, find the break-even point(s).

55. $C = 10x + 300, R = 30x - 0.1x^2$

56. $C = 0.6x^2 + 9, R = 12x - 0.2x^2$

57. $C = 12.6x + 150, R = 42.8x - 0.3x^2$

58. $C = 80x + 900, R = 120x - 0.2x^2$

Solve the following systems using your graphing calculator. Round your answers to the nearest hundredth.

 59. $3x - 5y = 12$
$x^2 + y^2 = 10$

60. $y = 2x^2 - x + 2$
$4x^2 + y^2 = 36$

Challenge Problems

61. Intersecting Roads The intersection of three roads forms a right triangle, as shown in the figure.

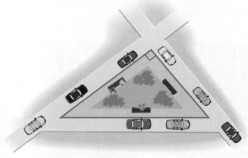

If the hypotenuse is 26 yards and the area is 120 square yards, find the length of the two legs of the triangle.

62. In the figure shown, R represents the radius of the larger orange circle and r represents the radius of the smaller orange circles. If $R = 2r$ and if the shaded area is 122.5π, find r and R.

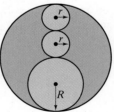

Cumulative Review Exercises

[1.9] **63.** List the order of operations we follow when evaluating an expression.

[6.5] **64.** Factor $(x + 1)^3 + 1$.

[7.8] **65.** x varies inversely as the square of P. If $x = 10$ when P is 6, find x when $P = 20$.

[11.5] **66.** Simplify $\dfrac{5}{\sqrt{x + 2} - 3}$.

[13.7] **67.** Solve $A = A_0 e^{kt}$ for k.

Chapter 14 Summary

IMPORTANT FACTS AND CONCEPTS	EXAMPLES

Section 14.1

The four **conic sections** are the parabola, circle, ellipse, and the hyperbola, which are obtained by slicing a cone.

Parabola   Circle   Ellipse   Hyperbola

The four different forms for equations of parabolas are summarized below.

Parabola with Vertex at (h, k)

1. $y = a(x - h)^2 + k, a > 0$ (opens upward)
2. $y = a(x - h)^2 + k, a < 0$ (opens downward)
3. $x = a(y - k)^2 + h, a > 0$ (opens to the right)
4. $x = a(y - k)^2 + h, a < 0$ (opens to the left)

$$y = -(x - 2)^2 + 3 \qquad x = 2(y + 1)^2 - 4$$

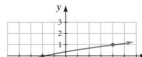

Distance Formula

The distance, d, between any two points (x_1, y_1) and (x_2, y_2) can be found by the distance formula:

$$d = \sqrt{(x_2 - x_1)^2 + (y_2 - y_1)^2}$$

The distance between $(-1, 3)$ and $(4, 15)$ is

$$d = \sqrt{[4 - (-1)]^2 + (15 - 3)^2} = \sqrt{5^2 + 12^2} = \sqrt{169} = 13$$

Midpoint Formula

Given any two points (x_1, y_1) and (x_2, y_2), the point halfway between the given points can be found by the midpoint formula:

$$\text{midpoint} = \left(\frac{x_1 + x_2}{2}, \frac{y_1 + y_2}{2} \right)$$

The midpoint of the line segment joining $(7, 6)$ and $(-11, 10)$ is

$$\text{midpoint} = \left(\frac{7 + (-11)}{2}, \frac{6 + 10}{2} \right) = \left(\frac{-4}{2}, \frac{16}{2} \right) = (-2, 8)$$

A **circle** is a set of points in a plane that are the same distance from a fixed point called its **center**.

Circle with Its Center at the Origin and Radius r

$$x^2 + y^2 = r^2$$

Sketch the graph of $x^2 + y^2 = 9$.
The graph is a circle with its center at $(0, 0)$ and radius $r = 3$.

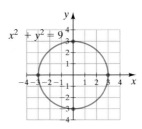

| **IMPORTANT FACTS AND CONCEPTS** | **EXAMPLES** |

Section 14.1 (continued)

Circle with Its center at (*h, k*) and Radius *r*

$$(x - h)^2 + (y - k)^2 = r^2$$

Sketch the graph of $(x - 3)^2 + (y + 5)^2 = 25$.
The graph is a circle with its center at $(3, -5)$ and radius $r = 5$.

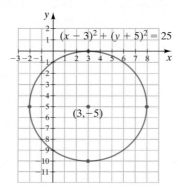

Section 14.2

An **ellipse** is a set of points in a plane, the sum of whose distances from two fixed points (called **foci**) is a constant.

Ellipse with Its Center at the Origin

$$\frac{x^2}{a^2} + \frac{y^2}{b^2} = 1$$

where $(a, 0)$ and $(-a, 0)$ are the x-intercepts and $(0, b)$ and $(0, -b)$ are the y-intercepts.

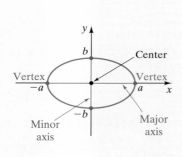 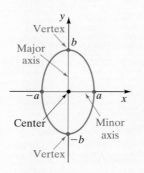

Sketch the graph of $\dfrac{x^2}{25} + \dfrac{y^2}{16} = 1$.

The graph is an ellipse. Since $a = 5$, the x-intercepts are $(-5, 0)$ and $(5, 0)$. Since $b = 4$, the y-intercepts are $(0, -4)$ and $(0, 4)$.

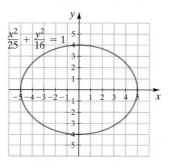

Ellipse with Its Center at (*h, k*)

$$\frac{(x - h)^2}{a^2} + \frac{(y - k)^2}{b^2} = 1$$

Sketch the graph of $\dfrac{(x - 2)^2}{9} + \dfrac{(y + 1)^2}{16} = 1$.

The graph is an ellipse with its center at $(2, -1)$, where $a = 3$ and $b = 4$.

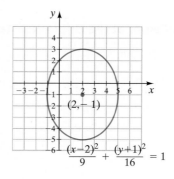

IMPORTANT FACTS AND CONCEPTS	EXAMPLES

Section 14.2 (continued)

The area, A, of an ellipse is $A = \pi ab$.

The area of the second ellipse from page 692 is

$$A = \pi ab = \pi \cdot 3 \cdot 4 = 12\pi \approx 37.70 \text{ square units.}$$

Section 14.3

A **hyperbola** is a set of points in a plane, the difference of whose distances from two fixed points (called **foci**) is a constant.

Hyperbola with Its Center at the Origin

Hyperbola
with transverse axis
along the x-axis

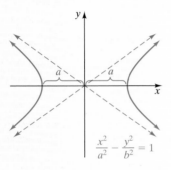

$$\frac{x^2}{a^2} - \frac{y^2}{b^2} = 1$$

Asymptotes

$$y = \frac{b}{a}x \quad \text{and} \quad y = -\frac{b}{a}x.$$

Hyperbola
with transverse axis
along the y-axis

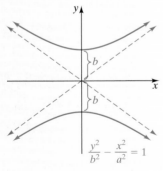

$$\frac{y^2}{b^2} - \frac{x^2}{a^2} = 1$$

Asymptotes

$$y = \frac{b}{a}x \quad \text{and} \quad y = -\frac{b}{a}x.$$

Determine the equations of the asymptotes and sketch a graph of $\frac{x^2}{4} - \frac{y^2}{9} = 1$.

The graph is a hyperbola with $a = 2$ and $b = 3$.

The equations for the asymptotes are $y = \frac{3}{2}x$ and $y = -\frac{3}{2}x$.

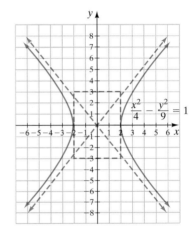

Determine the equations of the asymptotes and sketch a graph of $\frac{y^2}{25} - \frac{x^2}{16} = 1$.

The graph is a hyperbola with $a = 4$ and $b = 5$. The equations of the asymptotes are $y = \frac{5}{4}x$ and $y = -\frac{5}{4}x$.

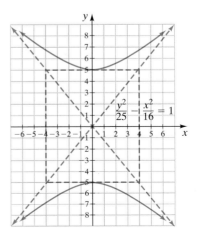

IMPORTANT FACTS AND CONCEPTS	EXAMPLES

Section 14.4

A **nonlinear system of equations** is a system of equations where at least one equation is not linear. The solution to a nonlinear system of equations is the point or points that satisfy all equations in the system.

Solve the system of equations.

$$x^2 + y^2 = 14$$
$$5x^2 - y^2 = -2$$

We will solve this system using the addition method.

$$x^2 + y^2 = 14$$
$$\underline{5x^2 - y^2 = -2}$$
$$6x^2 \qquad = 12$$
$$x^2 = 2$$
$$x = \pm\sqrt{2}$$

To obtain the value(s) for y, use the equation $x^2 + y^2 = 14$.

$x = \sqrt{2}$	$x = -\sqrt{2}$
$x^2 + y^2 = 14$	$x^2 + y^2 = 14$
$(\sqrt{2})^2 + y^2 = 14$	$(-\sqrt{2})^2 + y^2 = 14$
$2 + y^2 = 14$	$2 + y^2 = 14$
$y^2 = 12$	$y^2 = 12$
$y = \pm\sqrt{12}$	$y = \pm\sqrt{12}$
$= \pm 2\sqrt{3}$	$= \pm 2\sqrt{3}$

The system has four solutions:

$$(\sqrt{2}, 2\sqrt{3}), (\sqrt{2}, -2\sqrt{3}), (-\sqrt{2}, 2\sqrt{3}), (-\sqrt{2}, -2\sqrt{3})$$

Chapter 14 Review Exercises

[14.1] *Find the length and the midpoint of the line segment between each pair of points.*

1. $(0, 0), (5, -12)$ **2.** $(-4, 1), (-1, 5)$ **3.** $(-9, -5), (-1, 10)$ **4.** $(-4, 3), (-2, 5)$

Graph each equation.

5. $y = (x - 2)^2 + 1$ **6.** $y = (x + 3)^2 - 4$ **7.** $x = (y - 1)^2 + 4$ **8.** $x = -2(y + 4)^2 - 3$

In Exercises 9–12, **a)** *write each equation in the form* $y = a(x - h)^2 + k$ *or* $x = a(y - k)^2 + h$. **b)** *Graph the equation.*

9. $y = x^2 - 8x + 22$ **10.** $x = -y^2 - 2y + 5$ **11.** $x = y^2 + 5y + 4$ **12.** $y = 2x^2 - 8x - 24$

In Exercises 13–18, **a)** *write the equation of each circle in standard form.* **b)** *Draw the graph.*

13. Center $(0, 0)$, radius 4   **14.** Center $(-3, 4)$, radius 1   **15.** $x^2 + y^2 - 4y = 0$

16. $x^2 + y^2 - 2x + 6y + 1 = 0$ **17.** $x^2 - 8x + y^2 - 10y + 40 = 0$ **18.** $x^2 + y^2 - 4x + 10y + 17 = 0$

Graph each equation.

19. $y = \sqrt{9 - x^2}$ **20.** $y = -\sqrt{36 - x^2}$

Determine the equation of each circle.

21.

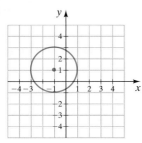

22.

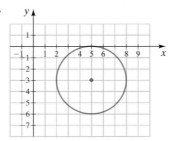

[14.2] *Graph each equation.*

23. $\dfrac{x^2}{4} + \dfrac{y^2}{9} = 1$ **24.** $\dfrac{x^2}{36} + \dfrac{y^2}{64} = 1$ **25.** $4x^2 + 9y^2 = 36$ **26.** $9x^2 + 16y^2 = 144$

27. $\dfrac{(x-3)^2}{16} + \dfrac{(y+2)^2}{4} = 1$ **28.** $\dfrac{(x+3)^2}{9} + \dfrac{y^2}{25} = 1$ **29.** $25(x-2)^2 + 9(y-1)^2 = 225$

30. For the ellipse in Exercise 23, find the area.

[14.3] *In Exercises 31–34,* **a)** *determine the equations of the asymptotes for each equation.* **b)** *Draw the graph.*

31. $\dfrac{x^2}{4} - \dfrac{y^2}{16} = 1$ **32.** $\dfrac{x^2}{4} - \dfrac{y^2}{4} = 1$ **33.** $\dfrac{y^2}{4} - \dfrac{x^2}{36} = 1$ **34.** $\dfrac{y^2}{25} - \dfrac{x^2}{16} = 1$

In Exercises 35–38, **a)** *write each equation in standard form.* **b)** *Determine the equations of the asymptotes.* **c)** *Draw the graph.*

35. $x^2 - 9y^2 = 9$ **36.** $25x^2 - 16y^2 = 400$

37. $4y^2 - 25x^2 = 100$ **38.** $49y^2 - 9x^2 = 441$

[14.1–14.3] *Identify the graph of each equation as a circle, ellipse, parabola, or hyperbola.*

39. $\dfrac{x^2}{49} - \dfrac{y^2}{16} = 1$ **40.** $4x^2 + 8y^2 = 32$ **41.** $5x^2 + 5y^2 = 125$ **42.** $4x^2 - 25y^2 = 25$

43. $\dfrac{x^2}{18} + \dfrac{y^2}{9} = 1$ **44.** $y = (x-2)^2 + 1$ **45.** $12x^2 + 9y^2 = 108$ **46.** $x = -y^2 + 8y - 9$

[14.4] *Find all real solutions to each system of equations using the substitution method.*

47. $x^2 + 2y^2 = 25$ **48.** $x^2 = y^2 + 4$ **49.** $x^2 + y^2 = 9$ **50.** $x^2 + 2y^2 = 9$
    $x^2 - 3y^2 = 25$ $x + y = 4$ $y = 3x + 9$ $x^2 - 6y^2 = 36$

Find all real solutions to each system of equations using the addition method.

51. $x^2 + y^2 = 36$ **52.** $x^2 + y^2 = 25$ **53.** $-4x^2 + y^2 = -15$ **54.** $3x^2 + 2y^2 = 6$
    $x^2 - y^2 = 36$ $x^2 - 2y^2 = -2$ $8x^2 + 3y^2 = -5$ $4x^2 + 5y^2 = 15$

55. Pool Table Jerry and Denise have a pool table in their house. It has an area of 45 square feet and a perimeter of 28 feet. Find the dimensions of the pool table.

56. Bottles of Glue The Dip and Dap Company has a cost equation of $C = 20.3x + 120$ and a revenue equation of $R = 50.2x - 0.2x^2$, where x is the number of bottles of glue sold. Find the number of bottles of glue the company must sell to break even.

57. Savings Account If Kien Kempter invests a certain principal at a specific interest rate for 1 year, the interest is $120. If he increases the principal by $2000 and the interest rate is decreased by 1%, the interest remains the same. Find the principal and interest rate. Use $i = prt$.

Chapter 14 Practice Test

*To find out how well you understand the chapter material, take this practice test. The answers, and the section where the material was initially discussed, are given in the back of the book. Each problem is also fully worked out on the **Chapter Test Prep Video CD**. Review any questions that you answered incorrectly.*

1. Why are parabolas, circles, ellipses, and hyperbolas called conic sections?

2. Determine the length of the line segment whose endpoints are $(-1, 8)$ and $(6, 7)$.

3. Determine the midpoint of the line segment whose endpoints are $(-9, 4)$ and $(7, -1)$.

4. Determine the vertex of the graph of $y = -2(x + 3)^2 + 1$, and then graph the equation.

5. Graph $x = y^2 - 2y + 4$.

6. Write the equation $x = -y^2 - 4y - 5$ in the form $x = a(y - k)^2 + h$, and then draw the graph.

7. Write the equation of a circle with center at $(2, 4)$ and radius 3 and then draw the graph of the circle.

8. Find the area of the circle whose equation is $(x + 2)^2 + (y - 8)^2 = 9$.

9. Write the equation of the circle shown.

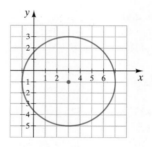

10. Graph $y = -\sqrt{16 - x^2}$.

11. Write the equation $x^2 + y^2 + 2x - 6y + 1 = 0$ in standard form, and then draw the graph.

12. Graph $4x^2 + 25y^2 = 100$.

13. Is the following graph the graph of
$$\frac{(x + 2)^2}{4} + \frac{(y + 1)^2}{16} = 1?$$ Explain your answer.

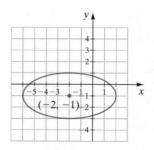

14. Graph $4(x - 4)^2 + 36(y + 2)^2 = 36$.

15. Find the center of the ellipse given by the equation $3(x - 8)^2 + 6(y + 7)^2 = 18$.

16. Explain how to determine whether the transverse axis of a hyperbola lies on the *x*- or *y*-axis.

17. What are the equations of the asymptotes of the graph of
$$\frac{x^2}{16} - \frac{y^2}{49} = 1?$$

18. Graph $\dfrac{y^2}{25} - \dfrac{x^2}{1} = 1$.

19. Graph $\dfrac{x^2}{4} - \dfrac{y^2}{9} = 1$.

In Exercises 20 and 21, determine whether the graph of the equation is a parabola, circle, ellipse, or hyperbola.

20. $4x^2 - 15y^2 = 30$

21. $25x^2 + 4y^2 = 100$

Solve each system of equations.

22. $x^2 + y^2 = 7$
    $2x^2 - 3y^2 = -1$

23. $x + y = 8$
    $x^2 + y^2 = 4$

24. **Vegetable Garden** Tom Wilson has a rectangular vegetable garden on his farm that has an area of 1500 square meters. Find the dimensions of the garden if the perimeter is 160 meters.

25. **Truck Bed** Gina Chang owns a truck. The rectangular bed of the truck has an area of 60 square feet, and the diagonal across the bed measures 13 feet. Find the dimensions of the bed of the truck.

Cumulative Review Test

Take the following test and check your answers with those given in the back of the book. Review any questions that you answered incorrectly. The section where the material was covered is indicated after the answer.

1. Solve $4x - 2(3x - 7) = 2x - 5$.

2. Find the solution set: $2(x - 5) + 2x = 4x - 7$.

3. Graph $y = -2x + 2$.

4. Simplify $(9x^2 y^5)(-3xy^4)$.

5. Factor $x^4 - x^2 - 42$.

6. A large triangular sign has a height that is 6 feet less than its base. If the area of the sign is 56 square feet, find the length of the base and the height of the sign.

7. Multiply $\dfrac{3x^2 - x - 4}{4x^2 + 7x + 3} \cdot \dfrac{2x^2 - 5x - 12}{6x^2 + x - 12}$.

8. Subtract $\dfrac{x}{x + 3} - \dfrac{x + 5}{2x^2 - 2x - 24}$.

9. Solve $\dfrac{3}{x + 3} + \dfrac{5}{x + 4} = \dfrac{12x + 19}{x^2 + 7x + 12}$.

10. If $f(x) = x^2 + 3x + 9$, find $f(10)$.

11. Solve the system of equations.

$$\frac{1}{2}x - \frac{1}{3}y = 2$$

$$\frac{1}{4}x + \frac{2}{3}y = 6$$

12. Find the solution set: $|3x + 1| > 4$.

13. Simplify $\left(\dfrac{18x^{1/2} y^3}{2x^{3/2}}\right)^{1/2}$.

14. Simplify $\dfrac{6\sqrt{x}}{\sqrt{x} - y}$.

15. Solve $3\sqrt[3]{2x + 2} = \sqrt[3]{80x - 24}$.

16. Solve $3x^2 - 4x + 5 = 0$ by the quadratic formula.

17. Solve $\log(3x - 4) + \log 4 = \log(x + 6)$.

18. Solve $35 = 70e^{-0.3t}$.

19. Graph $9x^2 + 4y^2 = 36$.

20. Graph $\dfrac{y^2}{25} - \dfrac{x^2}{16} = 1$.

Sequences, Series, and the Binomial Theorem

15

Sequences and series are discussed in this chapter. A sequence is a list of numbers in a specific order and a series is the sum of the numbers in a sequence. In this book, we discuss two types of sequences and series: arithmetic and geometric. Sequences and series can be used to solve many real-life problems as illustrated in this chapter.

In this chapter, we introduce the summation symbol, Σ, which is often used in statistics and other mathematics courses. We also discuss, in Section 15.4, the binomial theorem for expanding an expression of the form $(a + b)^n$.

IF A BALL REBOUNDS 4 feet when dropped from 6 feet, it has rebounded $66\frac{2}{3}\%$ of its original height. Theoretically, every rebound will have a rebound and the ball will never stop bouncing. In Exercise 105 on page 936, you will calculate the total distance traveled by a bouncing ball.

15.1 Sequences and Series

1 Find the terms of a sequence.

2 Write a series.

3 Find partial sums.

4 Use summation notation, Σ.

1 Find the Terms of a Sequence

Many times we see patterns in numbers. For example, suppose you are given a job offer with a starting salary of $30,000. You are given two options for your annual salary increases. One option is an annual salary increase of $2000 per year. The salary you would receive under this option is shown below.

Year	1	2	3	4	$\cdots$
	$\downarrow$	$\downarrow$	$\downarrow$	$\downarrow$	
Salary	$30,000	$32,000	$34,000	$36,000	$\cdots$

Each year the salary is $2000 greater than the previous year. The three dots on the right of the lists of numbers indicate that the list continues in the same manner.

The second option is a 5% salary increase each year. The salary you would receive under this option is shown below.

Year	1	2	3	4	$\cdots$
	$\downarrow$	$\downarrow$	$\downarrow$	$\downarrow$	
Salary	$30,000	$31,500	$33,075	$34,728.75	$\cdots$

With this option, the salary in a given year after year 1 is 5% greater than the previous year's salary.

The two lists of numbers that illustrate the salaries are examples of sequences. A **sequence** of numbers is a list of numbers arranged in a specific order. Consider the list of numbers given below, which is a sequence.

$$5, 10, 15, 20, 25, 30, \ldots$$

The first term is 5. We indicate this by writing $a_1 = 5$. Since the second term is 10, $a_2 = 10$, and so on. The three dots, called an ellipsis, indicate that the sequence continues indefinitely and is an **infinite sequence**.

Infinite Sequence

An **infinite sequence** is a function whose domain is the set of natural numbers.

Consider the infinite sequence $5, 10, 15, 20, 25, 30, 35, \ldots$

$$\text{Domain:} \quad \{1, \quad 2, \quad 3, \quad 4, \quad 5, \quad 6, \quad 7, \quad \ldots, \quad n, \quad \ldots\}$$
$$\downarrow \quad \downarrow \quad \downarrow \quad \downarrow \quad \downarrow \quad \downarrow \quad \downarrow \qquad \downarrow$$
$$\text{Range:} \quad \{5, \quad 10, \quad 15, \quad 20, \quad 25, \quad 30, \quad 35, \quad \ldots, \quad 5n, \quad \ldots\}$$

Note that the terms of the sequence $5, 10, 15, 20, \ldots$ are found by multiplying each natural number by 5. For any natural number, n, the corresponding term in the sequence is $5 \cdot n$ or $5n$. The **general term of the sequence**, a_n, which defines the sequence, is $a_n = 5n$.

$$a_n = f(n) = 5n$$

To find the twelfth term of the sequence, substitute 12 for n in the general term of the sequence: $a_{12} = 5 \cdot 12 = 60$. Thus, the twelfth term of the sequence is 60. Note that the terms in the sequence are the function values, or the numbers in the range of the function. When writing the sequence, we do not use set braces. The general form of a sequence is

$$a_1, a_2, a_3, a_4, \ldots, a_n, \ldots$$

For the infinite sequence $2, 4, 8, 16, 32, \ldots, 2^n, \ldots$ we can write

$$a_n = f(n) = 2^n$$

Notice that when $n = 1, a_1 = 2^1 = 2$; when $n = 2, a_2 = 2^2 = 4$; when $n = 3, a_3 = 2^3 = 8$; when $n = 4, a_4 = 2^4 = 16$; and so on. What is the seventh term of this sequence? The answer is $a_7 = 2^7 = 128$.

A sequence may also be **finite**.

Finite sequence

A **finite sequence** is a function whose domain includes only the first n natural numbers.

A finite sequence has only a finite number of terms.

Examples of Finite Sequences

$5, 10, 15, 20$ domain is $\{1, 2, 3, 4\}$

$2, 4, 8, 16, 32$ domain is $\{1, 2, 3, 4, 5\}$

EXAMPLE 1 ▸ Write the finite sequence defined by $a_n = 2n + 3$, for $n = 1, 2, 3, 4$.

Solution

$$a_n = 2n + 3$$
$$a_1 = 2(1) + 3 = 5$$
$$a_2 = 2(2) + 3 = 7$$
$$a_3 = 2(3) + 3 = 9$$
$$a_4 = 2(4) + 3 = 11$$

Thus, the sequence is $5, 7, 9, 11$.

▸ **Now Try Exercise 17**

Since each term of the sequence in Example 1 is larger than the preceding term, it is called an **increasing sequence**.

EXAMPLE 2 ▸ Given $a_n = \dfrac{2n + 3}{n^2}$,

a) find the first term in the sequence.

b) find the third term in the sequence.

c) find the fifth term in the sequence.

d) find the tenth term in the sequence.

Solution

a) When $n = 1, a_1 = \dfrac{2(1) + 3}{1^2} = \dfrac{5}{1} = 5$.

b) When $n = 3, a_3 = \dfrac{2(3) + 3}{3^2} = \dfrac{9}{9} = 1$.

c) When $n = 5, a_5 = \dfrac{2(5) + 3}{5^2} = \dfrac{13}{25} = 0.52$.

d) When $n = 10, a_{10} = \dfrac{2(10) + 3}{10^2} = \dfrac{23}{100} = 0.23$.

▸ **Now Try Exercise 33**

Note in Example 2 that since there is no restriction on n, a_n is the general term of an infinite sequence.

In Example 2, the first four terms of the sequence are $5, \dfrac{7}{4} = 1.75, 1, \dfrac{11}{16} = 0.6875$.

Since each term of the sequence generated by $a_n = \dfrac{2n + 3}{n^2}$ will be smaller than the preceding term, the sequence is called a **decreasing sequence**.

EXAMPLE 3 ▶ Find the first four terms of the sequence whose general term is $a_n = (-1)^n(n)$.

Solution

$$a_n = (-1)^n(n)$$
$$a_1 = (-1)^1(1) = -1$$
$$a_2 = (-1)^2(2) = 2$$
$$a_3 = (-1)^3(3) = -3$$
$$a_4 = (-1)^4(4) = 4$$

If we write the sequence, we get $-1, 2, -3, 4, \ldots, (-1)^n(n)$. Notice that each term alternates in sign. We call this an **alternating sequence**.

▶ **Now Try Exercise 25**

2 Write a Series

A **series** is the expressed sum of the terms of a sequence. A series may be finite or infinite, depending on whether the sequence it is based on is finite or infinite.

Examples

Finite Sequence

$$a_1, a_2, a_3, a_4, a_5$$

Finite Series

$$a_1 + a_2 + a_3 + a_4 + a_5$$

Infinite Sequence

$$a_1, a_2, a_3, a_4, a_5, \ldots, a_n, \ldots$$

Infinite Series

$$a_1 + a_2 + a_3 + a_4 + a_5 + \cdots + a_n + \cdots$$

EXAMPLE 4 ▶ Write the first eight terms of the sequence; then write the series that represents the sum of that sequence if

a) $a_n = \left(\dfrac{1}{2}\right)^n$ **b)** $a_n = (-2)^n$

Solution

a) We begin with $n = 1$; thus, the first eight terms of the sequence whose general term is $a_n = \left(\dfrac{1}{2}\right)^n$ are

$$\left(\frac{1}{2}\right)^1, \left(\frac{1}{2}\right)^2, \left(\frac{1}{2}\right)^3, \left(\frac{1}{2}\right)^4, \left(\frac{1}{2}\right)^5, \left(\frac{1}{2}\right)^6, \left(\frac{1}{2}\right)^7, \left(\frac{1}{2}\right)^8$$

or

$$\frac{1}{2}, \frac{1}{4}, \frac{1}{8}, \frac{1}{16}, \frac{1}{32}, \frac{1}{64}, \frac{1}{128}, \frac{1}{256}$$

The series that represents the sum of the sequence is

$$\frac{1}{2} + \frac{1}{4} + \frac{1}{8} + \frac{1}{16} + \frac{1}{32} + \frac{1}{64} + \frac{1}{128} + \frac{1}{256} = \frac{255}{256}$$

b) We again begin with $n = 1$; thus, the first eight terms of the sequence whose general term is $a_n = (-2)^n$ are

$$(-2)^1, (-2)^2, (-2)^3, (-2)^4, (-2)^5, (-2)^6, (-2)^7, (-2)^8$$

or

$$-2, 4, -8, 16, -32, 64, -128, 256$$

The series that represents the sum of this sequence is

$$-2 + 4 + (-8) + 16 + (-32) + 64 + (-128) + 256 = 170$$

▶ **Now Try Exercise 49**

3 Find Partial Sums

For an infinite sequence with the terms $a_1, a_2, a_3, \ldots, a_n, \ldots$, a **partial sum** is the sum of a finite number of consecutive terms of the sequence, beginning with the first term.

$s_1 = a_1$	*First partial sum*
$s_2 = a_1 + a_2$	*Second partial sum*
$s_3 = a_1 + a_2 + a_3$	*Third partial sum*
$\vdots$	
$s_n = a_1 + a_2 + a_3 + \cdots + a_n$	*nth partial sum*

The sum of all the terms of the infinite sequence is called an **infinite series** and is given by the following:

$$s = a_1 + a_2 + a_3 + \cdots + a_n + \cdots$$

EXAMPLE 5 ▶ Given the infinite sequence defined by $a_n = \dfrac{3 + n^2}{n}$, find the indicated partial sums.

a) s_1 and **b)** s_4

Solution

a) $s_1 = a_1 = \dfrac{3 + 1^2}{1} = \dfrac{3 + 1}{1} = 4$

b) $s_4 = a_1 + a_2 + a_3 + a_4$

$\quad = \dfrac{3 + 1^2}{1} + \dfrac{3 + 2^2}{2} + \dfrac{3 + 3^2}{3} + \dfrac{3 + 4^2}{4}$

$\quad = 4 + \dfrac{7}{2} + \dfrac{12}{3} + \dfrac{19}{4}$

$\quad = \dfrac{48}{12} + \dfrac{42}{12} + \dfrac{48}{12} + \dfrac{57}{12}$

$\quad = \dfrac{195}{12}$ or $16\dfrac{1}{4}$

▶ **Now Try Exercise 39**

4 Use Summation Notation, Σ

When the general term of a sequence is known, the Greek letter **sigma**, Σ, can be used to write a series. The sum of the first n terms of the sequence whose nth term is a_n is represented by

$$\sum_{i=1}^{n} a_i = a_1 + a_2 + a_3 + \cdots + a_n$$

where i is called the **index of summation** or simply the **index**, n is the **upper limit of summation**, and 1 is the **lower limit of summation**. In this illustration, we used i for the index; however, any letter can be used for the index.

Consider the sequence $7, 9, 11, 13, \ldots, 2n + 5, \ldots$. The sum of the first five terms can be represented using **summation notation**.

$$\sum_{i=1}^{5} (2i + 5)$$

This notation is read "the sum as i goes from 1 to 5 of $2i + 5$."

To evaluate the series represented by $\sum\limits_{i=1}^{5} (2i + 5)$, we first substitute 1 for i in $2i + 5$ and list the value obtained. Then we substitute 2 for i in $2i + 5$ and list the value. We follow this procedure for the values 1 through 5. We then sum these values to obtain the series value.

$$\sum_{i=1}^{5} (2i + 5) = (2 \cdot 1 + 5) + (2 \cdot 2 + 5) + (2 \cdot 3 + 5) + (2 \cdot 4 + 5) + (2 \cdot 5 + 5)$$

$$= 7 + 9 + 11 + 13 + 15$$

$$= 55$$

EXAMPLE 6 ▶ Write out the series $\sum\limits_{i=1}^{6} (i^2 + 1)$ and evaluate it.

Solution

$$\sum_{i=1}^{6} (i^2 + 1) = (1^2 + 1) + (2^2 + 1) + (3^2 + 1) + (4^2 + 1) + (5^2 + 1) + (6^2 + 1)$$

$$= 2 + 5 + 10 + 17 + 26 + 37$$

$$= 97$$

▶ **Now Try Exercise 61**

EXAMPLE 7 ▶ Consider the general term of a sequence $a_n = 2n^2 - 9$. Represent the third partial sum, s_3, in summation notation.

Solution The third partial sum will be the sum of the first three terms, $a_1 + a_2 + a_3$. We can represent the third partial sum as $\sum\limits_{i=1}^{3} (2i^2 - 9)$.

▶ **Now Try Exercise 69**

EXAMPLE 8 ▶ For the following set of values $x_1 = 3$, $x_2 = 4$, $x_3 = 5$, $x_4 = 6$, and $x_5 = 7$, does $\sum\limits_{i=1}^{5} (x_i)^2 = \left(\sum\limits_{i=1}^{5} x_i\right)^2$?

Solution
$$\sum_{i=1}^{5} (x_i)^2 = (x_1)^2 + (x_2)^2 + (x_3)^2 + (x_4)^2 + (x_5)^2$$

$$= 3^2 + 4^2 + 5^2 + 6^2 + 7^2$$

$$= 9 + 16 + 25 + 36 + 49 = 135$$

$$\left(\sum_{i=1}^{5} x_i\right)^2 = (x_1 + x_2 + x_3 + x_4 + x_5)^2$$

$$= (3 + 4 + 5 + 6 + 7)^2 = (25)^2 = 625$$

Since $135 \neq 625$, $\sum\limits_{i=1}^{5} (x_i)^2 \neq \left(\sum\limits_{i=1}^{5} x_i\right)^2$.

▶ **Now Try Exercise 75**

When a summation symbol is written without any upper and lower limits, it means that all the given data are to be summed.

EXAMPLE 9 ▶ A formula used to find the arithmetic mean, $\bar{x}$ (read x bar), of a set of data is $\bar{x} = \dfrac{\Sigma x}{n}$, where n is the number of pieces of data.

Joan Sally's five test scores are 70, 95, 83, 74, and 92. Find the arithmetic mean of her scores.

Solution $\bar{x} = \dfrac{\Sigma x}{n} = \dfrac{70 + 95 + 83 + 74 + 92}{5} = \dfrac{414}{5} = 82.8$

▶ **Now Try Exercise 79**

EXERCISE SET 15.1

Concept/Writing Exercises

 1. What is a sequence?

 2. What is an infinite sequence?

 3. What is a finite sequence?

 4. What is an increasing sequence?

 5. What is a decreasing sequence?

 6. What is an alternating sequence?

 7. What is a series?

 8. What is the nth partial sum of a series?

 9. Write the following notation in words: $\displaystyle\sum_{i=1}^{5} (i + 4)$.

10. Consider the summation $\displaystyle\sum_{k=1}^{5} (k + 3)$.

   a) What is the 1 called?

   b) What is the 5 called?

   c) What is the k called?

11. Let $a_n = 2n - 1$. Is this an increasing sequence or a decreasing sequence? Explain.

12. Let $a_n = -3n + 7$. Is this an increasing sequence or a decreasing sequence? Explain.

13. Let $a_n = 1 + (-2)^n$. Is this an alternating sequence? Explain.

14. Let $a_n = (-1)^{2n}$. Is this an alternating sequence? Explain.

Practice the Skills

Write the first five terms of the sequence whose nth term is shown.

15. $a_n = 6n$

16. $a_n = -5n$

17. $a_n = 4n - 1$

18. $a_n = 2n + 5$

19. $a_n = \dfrac{7}{n}$

20. $a_n = \dfrac{8}{n^2}$

21. $a_n = \dfrac{n + 2}{n + 1}$

22. $a_n = \dfrac{n - 5}{n + 6}$

23. $a_n = (-1)^n$

24. $a_n = (-1)^{2n}$

25. $a_n = (-2)^{n+1}$

26. $a_n = 3^{n-1}$

Find the indicated term of the sequence whose nth term is shown.

27. $a_n = 2n + 7$, twelfth term

28. $a_n = 3n + 2$, sixth term

29. $a_n = \dfrac{n}{4} + 8$, sixteenth term

30. $a_n = \dfrac{n}{2} - 13$ fourteenth term

31. $a_n = (-1)^n$, eighth term

32. $a_n = (-2)^n$, fourth term

33. $a_n = n(n + 2)$, ninth term

34. $a_n = (n - 1)(n + 4)$, fifth term

35. $a_n = \dfrac{n^2}{2n + 7}$, ninth term

36. $a_n = \dfrac{n(n + 6)}{n^2}$, tenth term

Find the first and third partial sums, s_1 and s_3, for each sequence.

37. $a_n = 3n - 1$

38. $a_n = 2n + 3$

39. $a_n = 2^n + 1$

40. $a_n = 3^n - 8$

41. $a_n = \dfrac{n - 1}{n + 2}$

42. $a_n = \dfrac{n}{n + 3}$

43. $a_n = (-1)^n$

44. $a_n = (-3)^n$

45. $a_n = \dfrac{n^2}{2}$

46. $a_n = \dfrac{n^2}{n + 4}$

Write the next three terms of each sequence.

47. $2, 4, 8, 16, 32, \dots$

48. $10, 15, 20, 25, 30, \dots$

49. $7, 9, 11, 13, 15, \dots$

50. $\dfrac{1}{2}, \dfrac{1}{3}, \dfrac{1}{4}, \dfrac{1}{5}, \dots$

51. $1, \dfrac{1}{2}, \dfrac{1}{3}, \dfrac{1}{4}, \dfrac{1}{5}, \dots$

52. $\dfrac{2}{3}, \dfrac{3}{4}, \dfrac{4}{5}, \dfrac{5}{6}, \dfrac{6}{7}, \dots$

53. $-1, 1, -1, 1, -1, \dots$

54. $-10, -20, -30, -40, \dots$

55. $1, \dfrac{1}{3}, \dfrac{1}{9}, \dfrac{1}{27}, \dots$

56. $\dfrac{1}{4}, \dfrac{2}{4}, \dfrac{3}{4}, \dfrac{4}{4}, \dots$

57. $1, -\dfrac{1}{2}, \dfrac{1}{4}, -\dfrac{1}{8}, \dots$

58. $\dfrac{1}{3}, \dfrac{1}{6}, \dfrac{1}{12}, \dfrac{1}{24}, \dots$

59. $37, 32, 27, 22, \dots$

60. $7, -1, -9, -17, \dots$

Write out each series, then evaluate it.

61. $\displaystyle\sum_{i=1}^{5} (3i - 1)$

62. $\displaystyle\sum_{i=1}^{4} (4i + 5)$

63. $\displaystyle\sum_{i=1}^{6} (i^2 + 1)$

64. $\displaystyle\sum_{i=1}^{5} (2i^2 - 7)$

65. $\displaystyle\sum_{i=1}^{4} \dfrac{i^2}{2}$

66. $\displaystyle\sum_{i=1}^{3} \dfrac{i^2}{5}$

67. $\displaystyle\sum_{i=4}^{9} \dfrac{i^2 + i}{i + 1}$

68. $\displaystyle\sum_{i=2}^{5} \dfrac{i^3}{i + 1}$

For the given general term a_n, write an expression using Σ to represent the indicated partial sum.

69. $a_n = n + 8$, fifth partial sum

70. $a_n = n^2 + 3$, fourth partial sum

71. $a_n = \dfrac{n^2}{4}$, third partial sum

72. $a_n = \dfrac{n^2 + 13}{n + 9}$, third partial sum

For the set of values $x_1 = 2$, $x_2 = 3$, $x_3 = 5$, $x_4 = -1$, and $x_5 = 4$, find each of the following.

73. $\displaystyle\sum_{i=1}^{5} x_i$

74. $\displaystyle\sum_{i=1}^{5} (x_i + 5)$

75. $\left(\displaystyle\sum_{i=1}^{5} x_i \right)^2$

76. $\displaystyle\sum_{i=1}^{5} 2x_i$

77. $\displaystyle\sum_{i=1}^{5} (x_i)^2$

78. $\displaystyle\sum_{i=1}^{4} (x_i^2 + 3)$

Find the arithmetic mean, $\overline{x}$, of the following sets of data.

79. $15, 20, 25, 30, 35$

80. $16, 22, 96, 18, 28$

81. $72, 83, 4, 60, 18, 20$

82. $5, 13, 9, 12, 23, 36, 70$

Problem Solving

In Exercises 83 and 84, consider the following rectangles. For the nth rectangle, the length is 2n and the width is n.

83. Perimeter

  a) Find the perimeters for the first four rectangles, and then list the perimeters in a sequence.

  b) Find the general term for the perimeter of the nth rectangle in the sequence. Use p_n for perimeter.

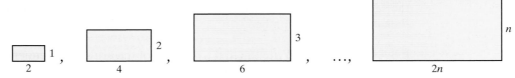

84. Area

  a) Find the areas for the four rectangles, and then list the areas in a sequence.

  b) Find the general term for the area of the nth rectangle in the sequence. Use a_n for area.

85. Create your own sequence that is an increasing sequence and list the first five terms.

86. Create your own sequence that is a decreasing sequence and list the first five terms.

87. Create your own sequence that is an alternating sequence and list the first five terms.

88. Write

  a) $\displaystyle\sum_{i=1}^{n} x_i$ as a sum of terms and

  b) $\displaystyle\sum_{j=1}^{n} x_j$ as a sum of terms.

  c) For a given set of values of x, from x_1 to x_n, will $\displaystyle\sum_{i=1}^{n} x_i = \sum_{j=1}^{n} x_j$? Explain.

89. Solve $\overline{x} = \dfrac{\Sigma x}{n}$ for Σx.

90. Solve $\bar{x} = \dfrac{\Sigma x}{n}$ for n.

91. Is $\displaystyle\sum_{i=1}^{n} 4x_i = 4\sum_{i=1}^{n} x_i$? Illustrate your answer with an example.

92. Is $\displaystyle\sum_{i=1}^{n} \frac{x_i}{3} = \frac{1}{3}\sum_{i=1}^{n} x_i$? Illustrate your answer with an example.

93. Let $x_1 = 3, x_2 = 5, x_3 = 2$, and $y_1 = 4, y_2 = 1, y_3 = 6$. Find the following. Note that $\Sigma x = x_1 + x_2 + x_3$, $\Sigma y = y_1 + y_2 + y_3$, and $\Sigma xy = x_1 y_1 + x_2 y_2 + x_3 y_3$.

a) Σx,

b) Σy,

c) $\Sigma x \cdot \Sigma y$,

d) Σxy,

e) Is $\Sigma x \cdot \Sigma y = \Sigma xy$?

Cumulative Review Exercises

[6.6]   **94.** Solve $2x^2 + 15 = 13x$ by factoring.

[12.2]   **95.** How many real solutions does the equation $6x^2 - 3x - 4 = 2$ have? Explain how you obtained your answer.

[14.2]   **96.** Graph $\dfrac{x^2}{4} + \dfrac{y^2}{1} = 1$.

[14.4]   **97.** Solve the system of equations.

$$x^2 + y^2 = 5$$
$$x = 2y$$

15.2 Arithmetic Sequences and Series

1 Find the common difference in an arithmetic sequence.

2 Find the nth term of an arithmetic sequence.

3 Find the nth partial sum of an arithmetic sequence.

1 Find the Common Difference in an Arithmetic Sequence

In the previous section, we started our discussion by assuming you got a job with a starting salary of $30,000. One option for salary increases was an increase of $2000 each year. This would result in the sequence

$$\$30{,}000, \$32{,}000, \$34{,}000, \$36{,}000, \ldots$$

This is an example of an arithmetic sequence.

> **Arithmetic Sequence**
>
> An **arithmetic sequence** is a sequence in which each term after the first differs from the preceding term by a constant amount.

The constant amount by which each pair of successive terms differs is called the **common difference**, d. The common difference can be found by subtracting any term from the term that directly follows it.

Arithmetic Sequence	Common Difference
$1, 3, 5, 7, 9, \ldots$	$d = 3 - 1 = 2$
$5, 1, -3, -7, -11, -15, \ldots$	$d = 1 - 5 = -4$
$\dfrac{7}{2}, \dfrac{2}{2}, -\dfrac{3}{2}, -\dfrac{8}{2}, -\dfrac{13}{2}, -\dfrac{18}{2}, \ldots$	$d = \dfrac{2}{2} - \dfrac{7}{2} = -\dfrac{5}{2}$

Notice that the common difference can be a positive number or a negative number. If the sequence is increasing, then d is a positive number. If the sequence is decreasing, then d is a negative number.

EXAMPLE 1 ▶ Write the first five terms of the arithmetic sequence with

a) first term 6 and common difference 4.

b) first term 3 and common difference -2.

c) first term 1 and common difference $\dfrac{1}{3}$.

Solution

a) Start with 6 and keep adding 4. The sequence is $6, 10, 14, 18, 22$.

b) $3, 1, -1, -3, -5$

c) $1, \dfrac{4}{3}, \dfrac{5}{3}, 2, \dfrac{7}{3}$

▶ Now Try Exercise 13

2 Find the *n*th Term of an Arithmetic Sequence

In general, an arithmetic sequence with first term, a_1, and common difference, d, has the following terms:

$$a_1 = a_1, \quad a_2 = a_1 + d, \quad a_3 = a_1 + 2d, \quad a_4 = a_1 + 3d, \quad \text{and so on}$$

If we continue this process, we can see that the *n*th term, a_n, can be found by the following formula:

---

nth Term of an Arithmetic Sequence

$$a_n = a_1 + (n - 1)d$$

EXAMPLE 2 ▶

a) Write an expression for the general (or *n*th) term, a_n, of the arithmetic sequence whose first term is -3 and whose common difference is 2.

b) Find the twelfth term of the sequence.

Solution

a) The *n*th term of the sequence is $a_n = a_1 + (n - 1)d$. Substituting $a_1 = -3$ and $d = 2$ we obtain

$$
\begin{aligned}
a_n &= a_1 + (n - 1)d \\
&= -3 + (n - 1)2 \\
&= -3 + 2(n - 1) \\
&= -3 + 2n - 2 \\
&= 2n - 5
\end{aligned}
$$

Thus, $a_n = 2n - 5$.

b) $a_n = 2n - 5$

$$a_{12} = 2(12) - 5 = 24 - 5 = 19$$

The twelfth term in the sequence is 19.

▶ **Now Try Exercise 11**

EXAMPLE 3 ▶ Find the number of terms in the arithmetic sequence $5, 9, 13, 17, \ldots, 41$.

Solution   The first term, a_1, is 5; the *n*th term is 41, and the common difference, d, is 4. Substitute the appropriate values into the formula for the *n*th term and solve for *n*.

$$
\begin{aligned}
a_n &= a_1 + (n - 1)d \\
41 &= 5 + (n - 1)4 \\
41 &= 5 + 4n - 4 \\
41 &= 4n + 1 \\
40 &= 4n \\
10 &= n
\end{aligned}
$$

The sequence has 10 terms.

▶ **Now Try Exercise 51**

---

3 Find the *n*th Partial Sum of an Arithmetic Sequence

An **arithmetic series** is the sum of the terms of an arithmetic sequence. A finite arithmetic series can be written

$$s_n = a_1 + (a_1 + d) + (a_1 + 2d) + (a_1 + 3d) + \cdots + (a_n - 2d) + (a_n - d) + a_n$$

If we consider the last term as a_n, the term before the last term will be $a_n - d$, the second before the last term will be $a_n - 2d$, and so on.

A formula for the nth partial sum, s_n, can be obtained by adding the reverse of s_n to itself.

$$s_n = \quad a_1 \quad + (a_1 + d) + (a_1 + 2d) + \cdots + (a_n - 2d) + (a_n - d) + \quad a_n$$
$$s_n = \quad a_n \quad + (a_n - d) + (a_n - 2d) + \cdots + (a_1 + 2d) + (a_1 + d) + \quad a_1$$
$$2s_n = (a_1 + a_n) + (a_1 + a_n) + (a_1 + a_n) + \cdots + (a_1 + a_n) + (a_1 + a_n) + (a_1 + a_n)$$

Since the right side of the equation contains n terms of $(a_1 + a_n)$, we can write

$$2s_n = n(a_1 + a_n)$$

Now divide both sides of the equation by 2 to obtain the following formula.

nth Partial Sum of an Arithmetic Sequence

$$s_n = \frac{n(a_1 + a_n)}{2}$$

EXAMPLE 4 ▶ Find the sum of the first 25 natural numbers.

Solution   The arithmetic sequence is $1, 2, 3, 4, 5, 6, \ldots, 25$. The first term, a_1, is 1; the last term, a_n, is 25. There are 25 terms; thus, $n = 25$. Using the formula for the nth partial sum, we have

$$s_n = \frac{n(a_1 + a_n)}{2} = \frac{25(1 + 25)}{2} = \frac{25(26)}{2} = 25(13) = 325$$

The sum of the first 25 natural numbers is 325. Thus, $s_{25} = 325$.

▶ **Now Try Exercise 57**

EXAMPLE 5 ▶ The first term of an arithmetic sequence is 4, and the last term is 31. If $s_n = 175$, find the number of terms in the sequence and the common difference.

Solution   We substitute the appropriate values, $a_1 = 4$, $a_n = 31$, and $s_n = 175$, into the formula for the nth partial sum and solve for n.

$$s_n = \frac{n(a_1 + a_n)}{2}$$
$$175 = \frac{n(4 + 31)}{2}$$
$$175 = \frac{35n}{2}$$
$$350 = 35n$$
$$10 = n$$

There are 10 terms in the sequence. We can now find the common difference by using the formula for the nth term of an arithmetic sequence.

$$a_n = a_1 + (n - 1)d$$
$$31 = 4 + (10 - 1)d$$
$$31 = 4 + 9d$$
$$27 = 9d$$
$$3 = d$$

The common difference is 3. The sequence is $4, 7, 10, 13, 16, 19, 22, 25, 28, 31$.

▶ **Now Try Exercise 31**

Examples 6–7 illustrate some applications of arithmetic sequences and series.

EXAMPLE 6 ▶ **Salary** Mary Tufts is given a starting salary of $35,000 and is promised a $1200 raise after each of the next 8 years. Find her salary during her eighth year of work.

Solution Understand   Her salaries during the first few years would be

$$\$35{,}000, \$36{,}200, \$37{,}400, \$38{,}600, \ldots$$

Since we are adding a constant amount each year, this is an arithmetic sequence. The general term of an arithmetic sequence is $a_n = a_1 + (n-1)d$.

Translate   In this example, $a_1 = 35{,}000$ and $d = 1200$. Thus, for $n = 8$, Mary's salary would be

$$a_8 = 35{,}000 + (8-1)1200$$

Carry Out
$$= 35{,}000 + 7(1200)$$
$$= 35{,}000 + 8400$$
$$= 43{,}400$$

Answer   During her eighth year of work, Mary's salary would be $43,400. If we listed all the salaries for the 8-year period, they would be $35,000, $36,200, $37,400, $38,600, $39,800, $41,000, $42,200, $43,400.

▶ **Now Try Exercise 83**

EXAMPLE 7 ▶ **Pendulum** Each swing of a pendulum (left to right or right to left) is 3 inches shorter than the preceding swing. The first swing is 8 feet.

a) Find the length of the twelfth swing.

b) Determine the distance traveled by the pendulum during the first 12 swings.

Solution   **a)** Understand   Since each swing is decreasing by a constant amount, this problem can be represented as an arithmetic series. Since the first swing is given in feet and the decrease in swing in inches, we will change 3 inches to 0.25 feet ($3 \div 12 = 0.25$). The twelfth swing can be considered a_{12}. The difference, d, is negative since the distance is decreasing with each swing.

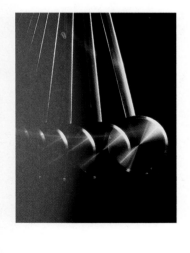

Translate
$$a_n = a_1 + (n-1)d$$
$$a_{12} = 8 + (12-1)(-0.25)$$

Carry Out
$$= 8 + 11(-0.25)$$
$$= 8 - 2.75$$
$$= 5.25 \text{ feet}$$

Answer   The twelfth swing is 5.25 feet.

b) Understand and Translate   The distance traveled during the first 12 swings can be found using the formula for the nth partial sum. The first swing, a_1, is 8 feet and the twelfth swing, a_{12}, is 5.25 feet.

$$S_n = \frac{n(a_1 + a_n)}{2}$$

$$S_{12} = \frac{12(a_1 + a_{12})}{2}$$

Carry Out
$$= \frac{12(8 + 5.25)}{2} = \frac{12(13.25)}{2} = 6(13.25) = 79.5 \text{ feet}$$

Answer   The pendulum travels 79.5 feet during its first 12 swings.

▶ **Now Try Exercise 75**

EXERCISE SET 15.2   Math XL   MyMathLab
MathXL®   MyMathLab

Concept/Writing Exercises

1. What is an arithmetic sequence?

2. What is an arithmetic series?

3. What do we call the constant amount by which each pair of successive terms in an arithmetic sequence differs?

4. How can the common difference in an arithmetic sequence be found?

5. If an arithmetic sequence is increasing, is the value for d a positive number or a negative number?

6. If an arithmetic sequence is decreasing, is the value for d a positive number or a negative number?

7. Can an arithmetic sequence consist of only negative numbers? Explain.

8. Can an arithmetic sequence consist of only odd numbers? Explain.

9. Can an arithmetic sequence consist of only even numbers? Explain.

10. Can an alternating sequence be an arithmetic sequence? Explain.

Practice the Skills

Write the first five terms of the arithmetic sequence with the given first term and common difference. Write the expression for the general (or nth) term, a_n, of the arithmetic sequence.

11. $a_1 = 4, d = 3$

12. $a_1 = -11, d = 4$

13. $a_1 = 7, d = -2$

14. $a_1 = 3, d = -5$

15. $a_1 = \frac{1}{2}, d = \frac{3}{2}$

16. $a_1 = -\frac{5}{3}, d = -\frac{1}{3}$

17. $a_1 = 100, d = -5$

18. $a_1 = \frac{7}{4}, d = -\frac{3}{4}$

Find the indicated quantity of the arithmetic sequence.

19. $a_1 = 5, d = 3$; find a_4

20. $a_1 = 10, d = -3$; find a_5

21. $a_1 = -9, d = 4$; find a_{10}

22. $a_1 = -1, d = -2$; find a_{12}

23. $a_1 = -8, d = \frac{5}{3}$; find a_{13}

24. $a_1 = 5, a_8 = -21$; find d

25. $a_1 = 11, a_9 = 27$; find d

26. $a_1 = \frac{1}{2}, a_7 = \frac{19}{2}$; find d

27. $a_1 = 4, a_n = 28, d = 3$; find n

28. $a_1 = -9, a_n = -27, d = -3$; find n

29. $a_1 = 82, a_n = 42, d = -8$; find n

30. $a_1 = -\frac{4}{3}, a_n = -\frac{14}{3}, d = -\frac{2}{3}$; find n

Find the sum, s_n, and common difference, d, of each sequence.

31. $a_1 = 1, a_{10} = 19, n = 10$

32. $a_1 = -8, a_7 = 10, n = 7$

33. $a_1 = \frac{3}{5}, a_8 = 2, n = 8$

34. $a_1 = 12, a_8 = -23, n = 8$

35. $a_1 = -5, a_6 = 13.5, n = 6$

36. $a_1 = \frac{7}{5}, a_5 = \frac{23}{5}, n = 5$

37. $a_1 = 7, a_{11} = 67, n = 11$

38. $a_1 = 14.25, a_{31} = 18.75, n = 31$

Write the first four terms of each sequence; then find a_{10} and s_{10}.

39. $a_1 = 4, d = 3$

40. $a_1 = 11, d = -6$

41. $a_1 = -6, d = 2$

42. $a_1 = -7, d = -4$

43. $a_1 = -8, d = -5$

44. $a_1 = -15, d = 4$

45. $a_1 = \frac{7}{2}, d = \frac{5}{2}$

46. $a_1 = \frac{9}{5}, d = \frac{3}{5}$

47. $a_1 = 100, d = -7$

48. $a_1 = 35, d = 6$

Find the number of terms in each sequence and find s_n.

49. $1, 4, 7, 10, \ldots, 43$

50. $-10, -8, -6, -4, \ldots, 40$

51. $-9, -5, -1, 3, \ldots, 31$

52. $6, 13, 20, 27, \ldots, 62$

53. $\dfrac{1}{2}, \dfrac{2}{2}, \dfrac{3}{2}, \dfrac{4}{2}, \dfrac{5}{2}, \ldots, \dfrac{17}{2}$

54. $-\dfrac{5}{6}, -\dfrac{7}{6}, -\dfrac{9}{6}, -\dfrac{11}{6}, \ldots, -\dfrac{21}{6}$

55. $7, 10, 13, 16, \ldots, 91$

56. $-11, -15, -19, \ldots, -51$

Problem Solving

57. Find the sum of the first 50 natural numbers.

58. Find the sum of the first 50 even numbers.

59. Find the sum of the first 50 odd numbers.

60. Find the sum of the first 40 multiples of 5.

61. Find the sum of the first 30 multiples of 3.

62. Find the sum of the numbers between 50 and 150, inclusive.

63. Determine how many numbers between 7 and 1610 are divisible by 6.

64. Determine how many numbers between 14 and 1470 are divisible by 8.

Pyramids occur everywhere. At athletic events, cheerleaders may form a pyramid where the people above stand on the shoulders of the people below. The illustration on the right shows a pyramid with 1 cheerleader on the top row, 2 cheerleaders in the middle row, and 3 cheerleaders in the bottom row. Notice that $a_1 = 1$, $a_2 = 2$, and $a_3 = 3$. Also, observe that $d = 1$, $n = 3$, and $s_3 = 6$.

At a bowling alley, the pins at the end of the bowling lane form a pyramid. The first row has 1 pin, the second row has 2 pins, the third row has 3 pins, and the fourth row has 4 pins. Thus, $a_1 = 1$, $d = 1$, $n = 4$, and $s_4 = 10$.

Use the idea of a pyramid to solve Exercises 65–70.

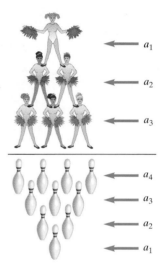

65. Auditorium An auditorium has 20 seats in the first row. Each successive row has two more seats than the previous row. How many seats are in the twelfth row? How many seats are in the first 12 rows?

66. Auditorium An auditorium has 22 seats in the first row. Each successive row has four more seats than the previous row. How many seats are in the ninth row? How many seats are in the first nine rows?

67. Logs Wolfgang Schmidt stacks logs so that there are 26 logs in the bottom layer, and each layer contains one log less than the layer below it. How many logs are in the pile?

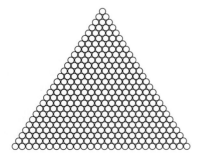

68. Logs Suppose Wolfgang, in Exercise 67, stopped stacking the logs after completing the layer containing eight logs. How many logs are in the pile?

69. Glasses in a Stack At their fiftieth wedding anniversary, Mr. and Mrs. Carlson are about to pour champagne into the top glass in the photo shown below. The top row has 1 glass, the second row has 3 glasses, the third row has 5 glasses, and so on. Each row has 2 more glasses than the row above it. This pyramid has 14 rows.

a) How many glasses are in the fourteenth row (bottom row)?

b) How many glasses are there total?

70. Candies in a Stack Individually wrapped candies are stacked in rows such that the top row has 1 candy, the second row has 3 candies, the third row has 5 candies, and so on. Each row has 2 more candies than the row above it. There are 7 rows of candies.

a) How many candies are in the seventh row (bottom row)?

b) How many candies are there total?

71. Sum of Numbers Karl Friedrich Gauss (1777–1855), a famous mathematician, as a child found the sum of the first 100 natural numbers quickly in his head $(1 + 2 + 3 + \cdots + 100)$. Explain how he might have done this and find the sum of the first 100 natural numbers as you think Gauss might have. (*Hint:* $1 + 100 = 101, 2 + 99 = 101$, etc.)

72. Sum of Numbers Use the same process from Exercise 71 to find the sum of the numbers from 101 to 150.

73. Sum of Numbers Find a formula for the sum of the first n consecutive odd numbers starting with 1.

$$1 + 3 + 5 + \cdots + (2n - 1)$$

74. Sum of Even Numbers Find a formula for the sum of the first n consecutive even numbers starting with 2.

$$2 + 4 + 6 + 8 + \cdots + 2n$$

75. Swinging on a Vine A long vine is attached to the branch of a tree. Sally Wynn swings from the vine, and each swing (left to right or right to left) is $\frac{1}{2}$ foot less than the previous swing. If her first swing is 22 feet, find

a) the length of the seventh swing, and

b) the distance traveled during her seven swings.

76. Pendulum Each swing of a pendulum is 2 inches shorter than the previous swing (left to right or right to left). The first swing is 6 feet. Find

a) the length of the eighth swing, and

b) the distance traveled by the pendulum during the eight swings.

77. Bouncing Ball Frank Holyton drops a ball from a second-story window. Each time the ball bounces, the height reached is 6 inches less than on the previous bounce. If the first bounce reaches a height of 6 feet, find the height attained on the ninth bounce.

78. Ping-Pong Ball A Ping-Pong ball falls from the table and bounces to a height of 3 feet. If each successive bounce is 3 inches less than the previous bounce, find the height attained on the tenth bounce.

79. Packages On Monday, March 17, Brian Nguyen started a new job at a packing company. On that day, he was able to prepare 105 packages for shipment. His boss expects Brian to be more productive with experience. Each day for six days, Brian is expected to prepare 10 more packages than the previous day's total.

a) How many packages is Brian expected to prepare on March 22?

b) How many packages is Brian expected to prepare during his first six days of employment?

80. Salary Marion Nickelson is making an annual salary of $37,500 at the Thompson Frozen Food Factory. Her boss has promised her an increase of $1500 in her salary each year over the next 10 years.

a) What will be Marion's salary 10 years from now?

b) What will be her total salary for these 11 years?

81. Money If Craig Campanella saves $1 on day 1, $2 on day 2, $3 on day 3, and so on, how much money, in total, will he have saved on day 31?

82. Money If Dan Currier saves 50¢ on day 1, $1.00 on day 2, $1.50 on day 3, and so on, how much, in total, will he have saved by the end of 1 year (365 days)?

83. Money Carrie Dereshi recently retired and met with her financial planner. She arranged to receive $42,000 the first year. Because of inflation, each year she will get $400 more than she received the previous year.

a) What income will she receive in her tenth year of retirement?

b) How much money will she have received in total during her first 10 years of retirement?

84. Salary Susan Forman is given a starting salary of $23,000 and is told she will receive a $1000 raise at the end of each year.

a) Find her salary during year 12.

b) How much will she receive in total during her first 12 years?

85. Angles The sum of the interior angles of a triangle, a quadrilateral, a pentagon, and a hexagon are 180°, 360°, 540°, and 720°, respectively. Use the pattern here to find the formula for the sum of the interior angles of a polygon with n sides.

86. Another formula that may be used to find the nth partial sum of an arithmetic series is

$$s_n = \frac{n}{2}[2a_1 + (n - 1)d]$$

Derive this formula using the two formulas presented in this section.

Group Activity

In calculus, a topic of importance is limits. Consider $a_n = \dfrac{1}{n}$. *The first five terms of this sequence are* $\dfrac{1}{1}, \dfrac{1}{2}, \dfrac{1}{3}, \dfrac{1}{4}, \dfrac{1}{5}$. *Since the value of* $\dfrac{1}{n}$

gets closer and closer to 0 as n gets larger and larger, we say that the limit of $\dfrac{1}{n}$ *as n approaches infinity is 0. We write this as* $\displaystyle \lim_{n \to +\infty} \dfrac{1}{n} = 0$

or $\displaystyle \lim_{n \to +\infty} a_n = 0$. *Notice that* $\dfrac{1}{n}$ *can never equal 0, but its value approaches 0 as n gets larger and larger.*

a) *Group member 1: Find* $\displaystyle \lim_{n \to +\infty} a_n$ *for Exercises 87 and 88.*   **b)** *Group member 2: Find* $\displaystyle \lim_{n \to +\infty} a_n$ *for Exercises 89 and 90.*

c) *Group member 3: Find* $\displaystyle \lim_{n \to +\infty} a_n$ *for Exercises 91 and 92.*   **d)** *Exchange work and check each other's answers.*

87. $a_n = \dfrac{1}{n-2}$

88. $a_n = \dfrac{n}{n+1}$

89. $a_n = \dfrac{1}{n^2 + 2}$

90. $a_n = \dfrac{2n+1}{n}$

91. $a_n = \dfrac{4n-3}{3n+1}$

92. $a_n = \dfrac{n^2}{n+1}$

Cumulative Review Exercises

[2.6] **93.** Solve $A = P + Prt$ for r.

[6.2] **94.** Factor $12n^2 - 6n - 30n + 15$.

[9.2] **95.** Solve the system of equations.
$$y = 2x + 1$$
$$3x - 2y = 1$$

[14.1] **96.** Graph $(x+4)^2 + y^2 = 25$.

15.3 Geometric Sequences and Series

1 Find the common ratio in a geometric sequence.

2 Find the nth term of a geometric sequence.

3 Find the nth partial sum of a geometric sequence.

4 Identify infinite geometric series.

5 Find the sum of an infinite geometric series.

6 Study applications of geometric series.

1 Find the Common Ratio in a Geometric Sequence

In Section 15.1, we assumed you got a job with a starting salary of $30,000. We also mentioned that an option for salary increases was a 5% salary increase each year. This would result in the following sequence.

$$\$30{,}000, \quad \$31{,}500, \quad \$33{,}075, \quad \$34{,}728.75, \ldots$$

This is an example of a geometric sequence.

Geometric Sequence

A **geometric sequence** is a sequence in which each term after the first is a multiple of the preceding term.

The common multiple is called the **common ratio**.

The common ratio, r, in any geometric sequence can be found by dividing any term, except the first, by the preceding term. The common ratio of the previous geometric sequence is $\dfrac{31{,}500}{30{,}000} = 1.05$ (or 105%).

Consider the geometric sequence

$$1, 3, 9, 27, 81, \ldots, 3^{n-1}, \ldots$$

The common ratio is 3 since $3 \div 1 = 3$ (or $9 \div 3 = 3$, and so on).

Geometric Sequence	Common Ratio
$4, 8, 16, 32, 64, \ldots, 4(2^{n-1}), \ldots$	2
$3, 12, 48, 192, 768, \ldots, 3(4^{n-1}), \ldots$	4
$7, \dfrac{7}{2}, \dfrac{7}{4}, \dfrac{7}{8}, \dfrac{7}{16}, \ldots, 7\left(\dfrac{1}{2}\right)^{n-1}, \ldots$	$\dfrac{1}{2}$
$5, -\dfrac{5}{3}, \dfrac{5}{9}, -\dfrac{5}{27}, \dfrac{5}{81}, \ldots, 5\left(-\dfrac{1}{3}\right)^{n-1}, \ldots$	$-\dfrac{1}{3}$

EXAMPLE 1 ▶ Determine the first five terms of the geometric sequence if $a_1 = 6$ and $r = \dfrac{1}{2}$.

Solution $a_1 = 6, \quad a_2 = 6 \cdot \dfrac{1}{2} = 3, \quad a_3 = 3 \cdot \dfrac{1}{2} = \dfrac{3}{2}, \quad a_4 = \dfrac{3}{2} \cdot \dfrac{1}{2} = \dfrac{3}{4}, \quad a_5 = \dfrac{3}{4} \cdot \dfrac{1}{2} = \dfrac{3}{8}$

Thus, the first five terms of the geometric sequence are

$$6, 3, \frac{3}{2}, \frac{3}{4}, \frac{3}{8}$$

▶ **Now Try Exercise 15**

2 Find the *n*th Term of a Geometric Sequence

In general, a geometric sequence with first term, a_1, and common ratio, r, has the following terms:

$$a_1, \qquad a_1 r, \qquad a_1 r^2, \qquad a_1 r^3, \qquad a_1 r^4, \ldots, \quad a_1 r^{n-1}, \ldots$$

↑	↑	↑	↑	↑	↑
1st	2nd	3rd	4th	5th	*n*th
term, a_1	term, a_2	term, a_3	term, a_4	term, a_5	term, a_n

Thus, we can see that the *n*th term of a geometric sequence is given by the following formula:

> **nth Term of a Geometric Sequence**
>
> $$a_n = a_1 r^{n-1}$$

EXAMPLE 2 ▶

a) Write an expression for the general (or *n*th) term, a_n, of the geometric sequence with $a_1 = 3$ and $r = -2$.

b) Find the twelfth term of this sequence.

Solution

a) The *n*th term of the sequence is $a_n = a_1 r^{n-1}$. Substituting $a_1 = 3$ and $r = -2$, we obtain

$$a_n = a_1 r^{n-1} = 3(-2)^{n-1}$$

Thus, $a_n = 3(-2)^{n-1}$.

b) $a_n = 3(-2)^{n-1}$
$$a_{12} = 3(-2)^{12-1} = 3(-2)^{11} = 3(-2048) = -6144$$

The twelfth term of the sequence is -6144. The first 12 terms of the sequence are $3, -6, 12, -24, 48, -96, 192, -384, 768, -1536, 3072, -6144$.

▶ **Now Try Exercise 35**

Helpful Hint *Study Tip*

In this chapter, you will be working with exponents and using rules for exponents. The rules for exponents were discussed in Section 5.1 and again in Chapter 11. If you do not remember the rules for exponents, now is a good time to review Section 5.1.

EXAMPLE 3 ▶ Find r and a_1 for the geometric sequence with $a_2 = 12$ and $a_5 = 324$.

Solution   The sequence can be represented with blanks for the missing terms.

$$\underline{\quad}, 12, \underline{\quad}, \underline{\quad}, 324$$

$$\uparrow \qquad\qquad\qquad \uparrow$$
$$a_2 \qquad\qquad\qquad\; a_5$$

If we assume that a_2 is the first term of a sequence with the same common ratio, we obtain

$$12, \underline{}, \underline{}, 324$$

$$\uparrow \qquad\qquad \uparrow$$

1st   4th

term   term

Now we use the formula for the nth term of a geometric sequence to find r. We let the first term, a_1, be 12 and the number of terms, n, be 4.

$$a_n = a_1 r^{n-1}$$
$$324 = 12 r^{4-1}$$
$$324 = 12 r^3$$
$$\frac{324}{12} = r^3$$
$$27 = r^3$$
$$3 = r$$

Thus, the common ratio is 3.

    The first term of the original sequence must be $12 \div 3$ or 4. Thus, $a_1 = 4$. The first term, a_1, could also be found using the formula with $a_n = 324$, $r = 3$, and $n = 5$. Find a_1 by the formula now.

▶ **Now Try Exercise 83**

---

3 Find the nth Partial Sum of a Geometric Sequence

A **geometric series** is the sum of the terms of a geometric sequence. The sum of the first n terms, s_n, of a geometric sequence can be expressed as

$$s_n = a_1 + a_1 r + a_1 r^2 + a_1 r^3 + \cdots + a_1 r^{n-2} + a_1 r^{n-1} \qquad (eq.\,1)$$

If we multiply both sides of the equation by r, we obtain

$$r s_n = a_1 r + a_1 r^2 + a_1 r^3 + \cdots + a_1 r^{n-1} + a_1 r^n \qquad (eq.\,2)$$

Now we subtract the corresponding sides of $(eq.\,2)$ from $(eq.\,1)$. The red-colored terms drop out, leaving

$$s_n - r s_n = a_1 - a_1 r^n$$

Now we solve the equation for s_n.

$$s_n(1 - r) = a_1(1 - r^n) \qquad \textit{Factor.}$$
$$s_n = \frac{a_1(1 - r^n)}{1 - r} \qquad \textit{Divide both sides by } 1 - r.$$

Thus, we have the following formula for the nth partial sum of a geometric sequence.

> **nth Partial Sum of a Geometric Sequence**
>
> $$s_n = \frac{a_1(1 - r^n)}{1 - r}, \qquad r \neq 1$$

EXAMPLE 4 ▶ Find the seventh partial sum of a geometric sequence whose first term is 16 and whose common ratio is $-\frac{1}{2}$.

Solution   Substitute the appropriate value for a, r, and n.

$$s_n = \frac{a_1(1 - r^n)}{1 - r}$$

$$s_7 = \frac{16\left[1 - \left(-\frac{1}{2}\right)^7\right]}{1 - \left(-\frac{1}{2}\right)} = \frac{16\left(1 + \frac{1}{128}\right)}{\frac{3}{2}} = \frac{16\left(\frac{129}{128}\right)}{\frac{3}{2}} = \frac{\frac{129}{8}}{\frac{3}{2}} = \frac{129}{8} \cdot \frac{2}{3} = \frac{43}{4}$$

Thus, $s_7 = \frac{43}{4}$.

▶ **Now Try Exercise 41**

EXAMPLE 5 ▶ Given $s_n = 93$, $a_1 = 3$, and $r = 2$, find n.

Solution

$$s_n = \frac{a_1(1 - r^n)}{1 - r}$$

$$93 = \frac{3(1 - 2^n)}{1 - 2} \qquad \text{\textit{Substitute values for } s_n, a_1, \text{ and } r.}$$

$$93 = \frac{3(1 - 2^n)}{-1}$$

$$-93 = 3(1 - 2^n) \qquad \text{\textit{Both sides were multiplied by } -1.}$$

$$-31 = 1 - 2^n \qquad \text{\textit{Both sides were divided by 3.}}$$

$$-32 = -2^n \qquad \text{\textit{1 was subtracted from both sides.}}$$

$$32 = 2^n \qquad \text{\textit{Both sides were divided by } -1.}$$

$$2^5 = 2^n \qquad \text{\textit{Write 32 as } 2^5.}$$

Therefore, $n = 5$.

▶ **Now Try Exercise 65**

When working with a geometric series, r can be a positive number as we saw in Example 5 or a negative number as we saw in Example 4.

4 Identify Infinite Geometric Series

All the geometric sequences that we have examined thus far have been finite since they have had a last term. The following sequence is an example of an infinite geometric sequence.

$$1, \frac{1}{2}, \frac{1}{4}, \frac{1}{8}, \frac{1}{16}, \ldots, \left(\frac{1}{2}\right)^{n-1}, \ldots$$

Note that the three dots at the end of the sequence indicate that the sequence continues indefinitely. The sum of the terms in an infinite geometric sequence forms an **infinite geometric series**. For example,

$$1 + \frac{1}{2} + \frac{1}{4} + \frac{1}{8} + \frac{1}{16} + \cdots + \left(\frac{1}{2}\right)^{n-1} + \cdots$$

is an infinite geometric series. Let's find some partial sums.

Partial Sum	Series	Sum
Second	$1 + \frac{1}{2}$	1.5
Third	$1 + \frac{1}{2} + \frac{1}{4}$	1.75
Fourth	$1 + \frac{1}{2} + \frac{1}{4} + \frac{1}{8}$	1.875
Fifth	$1 + \frac{1}{2} + \frac{1}{4} + \frac{1}{8} + \frac{1}{16}$	1.9375
Sixth	$1 + \frac{1}{2} + \frac{1}{4} + \frac{1}{8} + \frac{1}{16} + \frac{1}{32}$	1.96875

With each successive partial sum, the amount being added is less than with the previous partial sum. Also, the sum seems to be getting closer and closer to 2. In Example 6, we will show that the sum of this infinite geometric series is indeed 2.

5 Find the Sum of an Infinite Geometric Series

Consider the formula for the sum of the first n terms of an infinite geometric series:

$$s_n = \frac{a_1(1 - r^n)}{1 - r}, \qquad r \neq 1$$

What happens to r^n if $|r| < 1$ and n gets larger and larger? Suppose that $r = \frac{1}{2}$; then

$$\left(\frac{1}{2}\right)^1 = 0.5, \quad \left(\frac{1}{2}\right)^2 = 0.25, \quad \left(\frac{1}{2}\right)^3 = 0.125, \quad \left(\frac{1}{2}\right)^{20} \approx 0.000001$$

We can see that when $|r| < 1$, the value of r^n gets exceedingly close to 0 as n gets larger and larger. Thus, when considering the sum of an infinite geometric series, symbolized s_∞, the expression r^n approaches 0 when $|r| < 1$. Therefore, replacing r^n with 0 in the formula $s_n = \dfrac{a_1(1 - r^n)}{1 - r}$ leads to the following formula.

> ### Sum of an Infinite Geometric Series
>
> $$s_\infty = \frac{a_1}{1 - r} \quad \text{where} \quad |r| < 1$$

EXAMPLE 6 ▶ Find the sum of the infinite geometric series

$$1 + \frac{1}{2} + \frac{1}{4} + \frac{1}{8} + \cdots + \left(\frac{1}{2}\right)^{n-1} + \cdots.$$

Solution $a_1 = 1$ and $r = \dfrac{1}{2}$. Note that $\left|\dfrac{1}{2}\right| < 1$.

$$s_\infty = \frac{a_1}{1 - r} = \frac{1}{1 - \dfrac{1}{2}} = \frac{1}{\dfrac{1}{2}} = 2$$

Thus, $1 + \dfrac{1}{2} + \dfrac{1}{4} + \dfrac{1}{8} + \dfrac{1}{16} + \cdots + \left(\dfrac{1}{2}\right)^{n-1} + \cdots = 2.$

▶ **Now Try Exercise 69**

EXAMPLE 7 ▶ Find the sum of the infinite geometric series

$$3 - \frac{6}{5} + \frac{12}{25} - \frac{24}{125} + \frac{48}{625} + \cdots$$

Solution The terms of the corresponding sequence are $3, -\dfrac{6}{5}, \dfrac{12}{25}, -\dfrac{24}{125}, \ldots$ Note that $a_1 = 3$. To find the common ratio, r, we can divide the second term, $-\dfrac{6}{5}$, by the first term, 3.

$$r = -\frac{6}{5} \div 3 = -\frac{6}{5} \cdot \frac{1}{3} = -\frac{2}{5}.$$

Since $\left|-\dfrac{2}{5}\right| < 1$,

$$s_\infty = \frac{a_1}{1 - r}$$

$$= \frac{3}{1 - \left(-\dfrac{2}{5}\right)} = \frac{3}{1 + \dfrac{2}{5}} = \frac{3}{\dfrac{7}{5}} = \frac{15}{7}$$

▶ **Now Try Exercise 71**

EXAMPLE 8 ▶ Write $0.343434\ldots$ as a ratio of integers.

Solution We can write this decimal as

$$0.34 + 0.0034 + 0.000034 + \cdots + (0.34)(0.01)^{n-1} + \cdots$$

This is an infinite geometric series with $r = 0.01$. Since $|r| < 1$,

$$s_\infty = \frac{a_1}{1 - r} = \frac{0.34}{1 - 0.01} = \frac{0.34}{0.99} = \frac{34}{99}$$

If you divide 34 by 99 on a calculator, you will see .34343434 displayed.

▶ **Now Try Exercise 81**

What is the sum of a geometric series when $|r| > 1$? Consider the geometric sequence in which $a_1 = 1$ and $r = 2$.

$$1, 2, 4, 8, 16, 32, \ldots, 2^{n-1}, \ldots$$

The sum of its terms is

$$1 + 2 + 4 + 8 + 16 + 32 + \cdots + 2^{n-1} + \cdots$$

What is the sum of this series? As n gets larger and larger, the sum gets larger and larger. We therefore say that the sum "does not exist." For $|r| > 1$, the sum of an infinite geometric series does not exist.

6 Study Applications of Geometric Series

Now let's look at some applications of geometric sequences and series.

EXAMPLE 9 ▶ **Savings Account** Jean Simmons invests $1000 at 5% interest compounded annually in a savings account. Determine the amount in his account and the amount of interest earned at the end of 10 years.

Solution **Understand** Suppose we let P represent any principal invested. At the beginning of the second year, the amount grows to $P + 0.05P$ or $1.05P$. This amount will be the principal invested for year 2. At the beginning of the third year the second year's principal will grow by 5% to $(1.05P)(1.05)$, or $(1.05)^2P$. The amount in Jean's account at the beginning of successive years is

Savings account
5% interest annually

Year 1	Year 2	Year 3	Year 4
P	$1.05P$	$(1.05)^2P$	$(1.05)^3P$

and so on. This is a geometric series with $r = 1.05$. The amount in his account at the end of 10 years will be the same as the amount in his account at the beginning of year 11. We will therefore use the formula

$$a_n = a_1 r^{n-1}, \quad \text{with} \quad r = 1.05 \quad \text{and} \quad n = 11$$

Translate We have a geometric sequence with $a_1 = 1000, r = 1.05,$ and $n = 11$. Substituting these values into the formula, we obtain the following.

Carry Out

$$a_n = a_1 r^{n-1}$$

$$a_{11} = 1000(1.05)^{11-1}$$

$$= 1000(1.05)^{10}$$

$$\approx 1000(1.62889)$$

$$\approx 1628.89$$

Answer After 10 years, the amount in the account is about $1628.89. The amount of interest is $1628.89 - $1000 = $628.89.

▶ **Now Try Exercise 95**

EXAMPLE 10 ▶ **Money** Suppose someone offered you $1000 a day for each day of a 30-day month. Or, you could elect to take a penny on day 1, 2¢ on day 2, 4¢ on day 3, 8¢ on day 4, and so on. The amount would continue to double each day for 30 days.

a) Without doing any calculations, take a guess at which of the two offerings would provide the greater total return for 30 days.

b) Calculate the total amount you would receive by selecting $1000 a day for 30 days.

c) Calculate the amount you would receive on day 30 by selecting 1¢ on day 1 and doubling the amount each day for 30 days.

d) Calculate the total amount you would receive for 30 days by selecting 1¢ on day 1 and doubling the amount each day for 30 days.

Solution

a) Each of you will have your own answer to part **a)**.

b) If you received $1000 a day for 30 days, you would receive 30($1000) = $30,000.

c) Understand   Since the amount is doubled each day, this represents a geometric sequence with $r = 2$. The chart that follows shows the amount you would receive in each of the first 7 days. We also show the amounts written with base 2, the common ratio.

Day	1	2	3	4	5	6	7
Amount (cents)	1	2	4	8	16	32	64
Amount (cents)	2^0	2^1	2^2	2^3	2^4	2^5	2^6

Notice that for any given day, the exponent on 2 is 1 less than the given day. For example, on day 7, the amount is 2^6. In general, the amount on day n is 2^{n-1}.

Translate   To find the amount received on day 30, we evaluate $a_n = a_1 r^{n-1}$ for $n = 30$.

$$a_n = a_1 r^{n-1}$$

$$a_{30} = 1(2)^{30-1}$$

Carry Out

$$a_{30} = 1(2)^{29}$$

$$= 1(536,870,912)$$

$$= 536,870,912$$

Answer   On day 30, the amount that you would receive is 536,870,912 cents or $5,368,709.12

d) Understand and Translate   To find the total amount received over the 30 days, we find the thirtieth partial sum.

$$s_n = \frac{a_1(1 - r^n)}{1 - r}$$

$$s_{30} = \frac{1(1 - 2^{30})}{1 - 2}$$

Carry Out

$$= \frac{1(1 - 1,073,741,824)}{-1}$$

$$= 1,073,741,823$$

Answer   Therefore, over 30 days the total amount you would receive by this method would be 1,073,741,823 cents or $10,737,418.23. The amount received by this method greatly surpasses the $30,000 received by selecting $1000 a day for 30 days.

▶ **Now Try Exercise 87**

EXAMPLE 11 ▶ **Pendulum** On each swing (left to right or right to left), a certain pendulum travels 90% as far as on its previous swing. For example, if the swing to the right is 10 feet, the swing back to the left is $0.9 \times 10 = 9$ feet (see **Fig. 15.1**). If the first swing is 10 feet, determine the total distance traveled by the pendulum by the time it comes to rest.

Solution Understand This problem may be considered an infinite geometric series with $a_1 = 10$ and $r = 0.9$. We can therefore use the formula $s_\infty = \dfrac{a_1}{1 - r}$ to find the total distance traveled by the pendulum.

Translate and Carry Out

$$s_\infty = \frac{a_1}{1 - r} = \frac{10}{1 - 0.9} = \frac{10}{0.1} = 100 \text{ feet}$$

Answer   By the time the pendulum comes to rest, it has traveled 100 feet.

▶ **Now Try Exercise 99**

FIGURE 15.1

EXERCISE SET 15.3   *Math* XL   **MyMathLab**

Concept/Writing Exercises

1. What is a geometric sequence?
2. What is a geometric series?
3. Explain how to find the common ratio in a geometric sequence.
4. What is an infinite geometric series?
5. In a geometric series, if $|r| < 1$, what does r^n approach as n gets larger and larger?
6. Does the sum of an infinite geometric series when $|r| > 1$ exist?

7. In a geometric sequence, can the value of r be a negative number?
8. In a geometric sequence, can the value of r be a positive number?
9. In a geometric series, if $a_1 = 6$ and $r = 1/4$, does s_∞ exist? If so, what is its value? Explain.
10. In a geometric series, if $a_1 = 6$ and $r = -2$, does s_∞ exist? If so, what is its value? Explain.

Practice the Skills

Determine the first five terms of each geometric sequence.

11. $a_1 = 2, r = 3$
12. $a_1 = 2, r = -3$
13. $a_1 = 6, r = -\dfrac{1}{2}$

14. $a_1 = 6, r = \dfrac{1}{2}$
15. $a_1 = 72, r = \dfrac{1}{3}$
16. $a_1 = \dfrac{1}{8}, r = 4$

17. $a_1 = 90, r = -\dfrac{1}{3}$
18. $a_1 = 32, r = -\dfrac{1}{4}$
19. $a_1 = -1, r = 3$

20. $a_1 = -1, r = -3$
21. $a_1 = 5, r = -2$
22. $a_1 = -13, r = -1$

23. $a_1 = \dfrac{1}{3}, r = \dfrac{1}{2}$
24. $a_1 = \dfrac{1}{2}, r = -\dfrac{1}{3}$
25. $a_1 = 3, r = \dfrac{3}{2}$

26. $a_1 = 60, r = -\dfrac{2}{5}$

Find the indicated term of each geometric sequence.

27. $a_1 = 4, r = 2$; find a_6
28. $a_1 = 4, r = -2$; find a_6
29. $a_1 = -12, r = \dfrac{1}{2}$; find a_9

30. $a_1 = 27, r = \dfrac{1}{3}$; find a_7
31. $a_1 = \dfrac{1}{4}, r = 2$; find a_{10}
32. $a_1 = 3, r = 3$; find a_6

33. $a_1 = -3, r = -2$; find a_{12}
34. $a_1 = -10, r = -2$; find a_{10}
35. $a_1 = 2, r = \dfrac{1}{2}$; find a_8

36. $a_1 = 5, r = \dfrac{2}{3}$; find a_9

37. $a_1 = 50, r = \dfrac{1}{3}$; find a_7

38. $a_1 = -7, r = -\dfrac{3}{4}$; find a_7

Find the indicated sum.

39. $a_1 = 5, r = 2$; find s_5

40. $a_1 = 7, r = -3$; find s_5

41. $a_1 = 2, r = 5$; find s_6

42. $a_1 = 9, r = \dfrac{1}{2}$; find s_6

43. $a_1 = 80, r = 2$; find s_7

44. $a_1 = 2, r = -2$; find s_{12}

45. $a_1 = -15, r = -\dfrac{1}{2}$; find s_9

46. $a_1 = \dfrac{3}{4}, r = 3$; find s_7

47. $a_1 = -9, r = \dfrac{2}{5}$; find s_5

48. $a_1 = 35, r = \dfrac{1}{5}$; find s_{12}

For each geometric sequence, find the common ratio, r, and then write an expression for the general (or nth) term, a_n.

49. $3, \dfrac{3}{2}, \dfrac{3}{4}, \dfrac{3}{8}, \ldots$

50. $3, -\dfrac{3}{2}, \dfrac{3}{4}, -\dfrac{3}{8}, \ldots$

51. $9, 18, 36, 72, \ldots$

52. $2, 6, 18, 54, \ldots$

53. $2, -6, 18, -54, \ldots$

54. $-1, -3, -9, -18, \ldots$

55. $\dfrac{3}{4}, \dfrac{1}{2}, \dfrac{1}{3}, \dfrac{2}{9}$

56. $\dfrac{4}{3}, \dfrac{8}{3}, \dfrac{16}{3}, \dfrac{32}{3}, \ldots$

Find the sum of the terms in each geometric sequence.

57. $1, \dfrac{1}{2}, \dfrac{1}{4}, \dfrac{1}{8}, \dfrac{1}{16}, \ldots$

58. $1, -\dfrac{1}{2}, \dfrac{1}{4}, -\dfrac{1}{8}, \dfrac{1}{16}, \ldots$

59. $1, \dfrac{1}{5}, \dfrac{1}{25}, \dfrac{1}{125}, \dfrac{1}{625}, \ldots$

60. $1, -\dfrac{1}{5}, \dfrac{1}{25}, -\dfrac{1}{125}, \dfrac{1}{625}, \ldots$

61. $6, 3, \dfrac{3}{2}, \dfrac{3}{4}, \dfrac{3}{8}, \ldots$

62. $\dfrac{1}{3}, \dfrac{1}{9}, \dfrac{1}{27}, \dfrac{1}{81}, \ldots$

63. $5, 2, \dfrac{4}{5}, \dfrac{8}{25}, \ldots$

64. $-\dfrac{4}{3}, -\dfrac{4}{9}, -\dfrac{4}{27}, -\dfrac{4}{81}, \ldots$

Given s_n, a_1, and r, find n in each geometric series.

65. $s_n = 93, a_1 = 3$, and $r = 2$

66. $s_n = 80, a_1 = 2$, and $r = 3$

67. $s_n = \dfrac{189}{32}, a_1 = 3$, and $r = \dfrac{1}{2}$

68. $s_n = \dfrac{121}{9}, a_1 = 9$, and $r = \dfrac{1}{3}$

Find the sum of each infinite geometric series.

69. $2 + 1 + \dfrac{1}{2} + \dfrac{1}{4} + \dfrac{1}{8} + \cdots$

70. $8 + 4 + 2 + 1 + \cdots$

71. $8 + \dfrac{16}{3} + \dfrac{32}{9} + \dfrac{64}{27} + \cdots$

72. $6 - 2 + \dfrac{2}{3} - \dfrac{4}{9} + \cdots$

73. $-60 + 20 - \dfrac{20}{3} + \dfrac{20}{9} - \cdots$

74. $2 + \dfrac{4}{3} + \dfrac{8}{9} + \dfrac{16}{27} + \cdots$

75. $-12 - \dfrac{12}{5} - \dfrac{12}{25} - \dfrac{12}{125} - \cdots$

76. $5 - 1 + \dfrac{1}{5} - \dfrac{1}{25} + \cdots$

Write each repeating decimal as a ratio of integers.

77. $0.242424\ldots$

78. $0.454545\ldots$

79. $0.8888\ldots$

80. $0.375375\ldots$

81. $0.515151\ldots$

82. $0.742742\ldots$

Problem Solving

83. In a geometric sequence, $a_2 = 15$ and $a_5 = 405$; find r and a_1.

84. In a geometric sequence, $a_2 = 27$ and $a_5 = 1$; find r and a_1.

85. In a geometric sequence, $a_3 = 28$ and $a_5 = 112$, find r and a_1.

86. In a geometric sequence, $a_2 = 12$ and $a_5 = -324$; find r and a_1.

87. Loaf of Bread A loaf of bread currently costs $1.40. Determine the cost of a loaf of bread after 8 years (the start of the 9th year) if inflation were to grow at a constant rate of 3% per year. *Hint*: After year 1 (the start of year 2), the cost of a loaf of bread is $1.40(1.03). After year 2 (the start of year 3), the would be $1.40(1.03)^2$, and so on.

88. Bicycle A specific bicycle currently costs $400. Determine the cost of the bicycle after 12 years if inflation were to grow at a constant rate of 4% per year.

89. Mass A substance loses half its mass each day. If there are initially 600 grams of the substance, find

a) the number of days after which only 37.5 grams of the substance remain.

b) the amount of the substance remaining after 9 days.

90. Bacteria The number of a certain type of bacteria doubles every hour. If there are initially 1000 bacteria, after how many hours will the number of bacteria reach 64,000?

91. Population On July 1, 2005, the population of the United States was about 296.5 million people. If the population grows at a rate of 1.1% per year, find

   a) the population after 10 years.

   b) the number of years for the population to double.

92. Farm Equipment A piece of farm equipment that costs $105,000 decreases in value by 15% per year. Find the value of the equipment after 4 years.

93. Filtered Light The amount of light filtering through a lake diminishes by one-half for each meter of depth.

   a) Write a sequence indicating the amount of light remaining at depths of 1, 2, 3, 4, and 5 meters.

   b) What is the general term for this sequence?

   c) What is the remaining light at a depth of 7 meters?

94. Pendulum On each swing (left to right or right to left), a pendulum travels 80% as far as on its previous swing. If the first swing is 10 feet, determine the total distance traveled by the pendulum by the time it comes to rest.

95. Investment You invest $10,000 in a savings account paying 6% interest annually. Find the amount in your account at the end of 8 years.

96. Injected Dye A tracer dye is injected into Mark Damion for medical reasons. After each hour, two-thirds of the previous hour's dye remains. How much dye remains in Mark's system after 10 hours?

97. Bungee Jumping Shawna Kelly goes bungee jumping off a bridge above water. On the initial jump, the bungee cord stretches to 220 feet. Assume the first bounce reaches a height of 60% of the original jump and that each additional bounce reaches a height of 60% of the previous bounce.

   a) What will be the height of the fourth bounce?

   b) Theoretically, Shawna would never stop bouncing, but realistically, she will. Use the infinite geometric series to estimate the total distance Shawna travels in a *downward* direction.

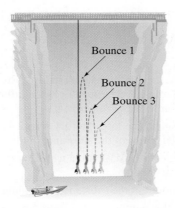

Bounce 1
Bounce 2
Bounce 3

98. Bungee Jumping Repeat Exercise 97 **b)**, but this time find the total distance traveled in an *upward* direction.

99. Ping-Pong Ball A Ping-Pong ball falls off a table 30 inches high. Assume that the first bounce reaches a height of 70% of the distance the ball fell and each additional bounce reaches a height of 70% of the previous bounce.

   a) How high will the ball bounce on the third bounce?

   b) Theoretically, the ball would never stop bouncing, but realistically, it will. Estimate the total distance the ball travels in the *downward* direction.

100. Ping-Pong Ball Repeat Exercise 99 **b)**, but this time find the total distance traveled in the *upward* direction.

101. Stack of Chips Suppose that you form stacks of blue chips such that there is one blue chip in the first stack and in each successive stack you double the number of chips. Thus, you have stacks of 1, 2, 4, 8, and so on, of blue chips. You also form stacks of red chips, starting with one red chip and then tripling the number in each successive stack. Thus the stacks will contain 1, 3, 9, 27 and so on, red chips. How many more would the sixth stack of red chips have than the sixth stack of blue chips?

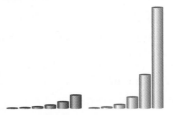

102. Stack of Money If you start with $1 and double your money each day, how many days will it take to surpass $1,000,000?

103. Depreciation One method of depreciating an item on an income tax return is the declining balance method. With this method, a given percent of the cost of the item is depreciated each year. Suppose that an item has a 5-year life and is depreciated using the declining balance method. Then, at the end of its first year, it loses $\frac{1}{5}$ of its value and $\frac{4}{5}$ of its value remains. At the end of the second year it loses $\frac{1}{5}$ of the remaining $\frac{4}{5}$ of its value, and so on. A car has a 5-year life expectancy and costs $15,000.

   a) Write a sequence showing the value of the car remaining for each of the first 3 years.

   b) What is the general term of this sequence?

   c) Find the value of the car at the end of 5 years.

104. Scrap Value On page 849, Exercise Set 13.6, Exercise 77, a formula for scrap value was given. The scrap value, S, is found by $S = c(1 - r)^n$ where c is the original cost, r is the annual depreciation rate and n is the number of years the object is depreciated.

   a) If you have not already done so, do Exercise 103 above to find the value of the car remaining at the end of 5 years.

   b) Use the formula given to find the scrap value of the car at the end of 5 years and compare this answer with the answer found in part **a)**.

105. Bouncing Ball A ball is dropped from a height of 10 feet. The ball bounces to a height of 9 feet. On each successive bounce, the ball rises to 90% of its previous height. Find the *total vertical distance* traveled by the ball when it comes to rest.

106. Wave Action A particle follows the path indicated by the wave shown. Find the *total vertical distance* traveled by the particle.

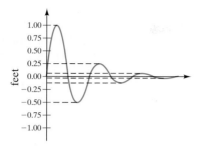

107. The formula for the nth term of a geometric sequence is $a_n = a_1 r^{n-1}$. If $a_1 = 1$, $a_n = r^{n-1}$.

   a) How do you think the graphs of $y_1 = 2^{n-1}$ and $y_2 = 3^{n-1}$ will compare?

   b) Graph both y_1 and y_2 and determine whether your answer to part **a)** was correct.

108. Use your grapher to decide the value of n to the nearest hundredth, where $100 = 3 \cdot 2^{n-1}$.

Challenge Problem

109. Find the sum of the sequence $1, 2, 4, 8, \ldots, 1{,}048{,}576$ and the number of terms in the sequence.

Cumulative Review Exercises

[5.5] **110.** Multiply $(4x^2 - 3x + 6)(2x - 3)$.

[5.6] **111.** Divide $(16x^2 + 10x - 18) \div (2x + 5)$.

[7.7] **112. Loading a Truck** It takes Mrs. Donovan twice as long to load a truck as it takes Mr. Donovan. If together they can load the truck in 8 hours, how long would it take Mr. Donovan to load the truck by himself?

[11.2] **113.** Evaluate $\left(\dfrac{9}{100}\right)^{-1/2}$.

[11.4] **114.** Simplify $\sqrt[3]{9x^2 y}\left(\sqrt[3]{3x^4 y^6} - \sqrt[3]{8xy^4}\right)$.

115. Simplify $x\sqrt{y} - 2\sqrt{x^2 y} + \sqrt{4x^2 y}$.

[11.6] **116.** Solve $\sqrt{a^2 + 9a + 3} = -a$.

Mid-Chapter Test: 15.1–15.3

To find out how well you understand the chapter material to this point, take this brief test. The answers, and the section where the material was initially discussed, are given in the back of the book. Review any questions you answered incorrectly.

1. Write the first five terms of the sequence whose nth term is $a_n = -3n + 5$.

2. If $a_n = n(n + 6)$, find the seventh term.

3. Find the first and third partial sums, s_1 and s_3, for the sequence whose nth term is $a_n = 2^n - 1$.

4. Write the next three terms of the sequence $5, 1, -3, -7, -11, \ldots$.

5. Evaluate the series $\displaystyle\sum_{i=1}^{5}(4i - 3)$.

6. If the general term of a sequence is $a_n = \dfrac{1}{3}n + 7$, write an expression using Σ to represent the fifth partial sum.

7. Write the first four terms of the arithmetic sequence with $a_1 = -6$ and $d = 5$. Find an expression for the general term a_n.

8. Find d for the arithmetic sequence with $a_1 = \dfrac{11}{2}$ and $a_7 = -\dfrac{1}{2}$.

9. Find n for the arithmetic sequence with $a_1 = 22$, $a_n = -3$, and $d = -5$.

10. Find the common difference, d, and the sum, s_6, for the arithmetic sequence with $a_1 = -8$ and $a_6 = 7$.

11. Find s_{10} for the arithmetic sequence with $a_1 = \dfrac{5}{2}$ and $d = \dfrac{1}{2}$.

12. Find the number of terms in the arithmetic sequence $-7, 0, 7, 14, \ldots, 63$.

13. Logs are stacked in a pile with 16 logs on the bottom row, 15 on the next row, 14 on the next row, and so on to the top, with the top row having one log. Each row has one log less than the row below it. How many logs are on the pile?

14. Write the first five terms of the geometric series with $a_1 = 80$ and $r = -\dfrac{1}{2}$.

15. Find a_7 for the geometric sequence with $a_1 = 81$ and $r = \dfrac{1}{3}$.

16. Find s_6 for the geometric sequence with $a_1 = 5$ and $r = 2$.

17. For the geometric sequence $8, -\dfrac{16}{3}, \dfrac{32}{9}, -\dfrac{64}{27}, \ldots$, find r.

18. Find the sum of the infinite series $12, 4, \dfrac{4}{3}, \dfrac{4}{9}, \ldots$.

19. Write the repeating decimal $0.878787\ldots$ as a ratio of two integers.

20. a) What is a sequence?
   b) What is an arithmetic sequence?
   c) What is geometric sequence?
   d) What is a series?

15.4 The Binomial Theorem

1 Evaluate factorials.

2 Use Pascal's triangle.

3 Use the binomial theorem.

1 Evaluate Factorials

To understand the binomial theorem, you must have an understanding of what **factorials** are. The notation $n!$ is read "n factorial." Its definition follows.

> ### *n* Factorial
>
> $$n! = n(n-1)(n-2)(n-3)\cdots(1)$$
>
> for any positive integer n.

Examples

$$6! = 6 \cdot 5 \cdot 4 \cdot 3 \cdot 2 \cdot 1 = 720$$
$$7! = 7 \cdot 6 \cdot 5 \cdot 4 \cdot 3 \cdot 2 \cdot 1 = 5040$$

Note that by definition **0! is 1.**

Below, we explain how to find factorials on a calculator.

USING YOUR CALCULATOR

Scientific Calculator

Factorials can be found on calculators that contain an $\boxed{n!}$ or $\boxed{x!}$ key. Often, the factorial key is a second function key. In the following examples, the answers appear after $\boxed{n!}$.

Evaluate 6!   6 $\boxed{2^{nd}}$ $\boxed{n!}$ 720

Evaluate 9!   9 $\boxed{2^{nd}}$ $\boxed{n!}$ 362880

Graphing Calculator

Graphing calculators do not have a factorial key. On some graphing calculators, factorials are found under the $\boxed{MATH}$, Probability function menu.

On the TI-84 Plus calculator, to get the probability function, PRB, menu, you press $\boxed{MATH}$, and then scroll to the right using the right arrow key, $\boxed{\blacktriangleright}$, three times, until you get to PRB. The $n!$ (or !) is the fourth item on the menu.

To find 5! or 6!, the keystrokes are as follows.

KEYSTROKES	ANSWER
5 $\boxed{MATH}$ $\boxed{\blacktriangleright}$ $\boxed{\blacktriangleright}$ $\boxed{\blacktriangleright}$ 4 $\boxed{ENTER}$	120
6 $\boxed{MATH}$ $\boxed{\blacktriangleright}$ $\boxed{\blacktriangleright}$ $\boxed{\blacktriangleright}$ 4 $\boxed{ENTER}$	720

2 Use Pascal's Triangle

Using polynomial multiplication, we can obtain the following expansions of the binomial $a + b$:

$$(a + b)^0 = 1$$
$$(a + b)^1 = a + b$$
$$(a + b)^2 = a^2 + 2ab + b^2$$
$$(a + b)^3 = a^3 + 3a^2b + 3ab^2 + b^3$$
$$(a + b)^4 = a^4 + 4a^3b + 6a^2b^2 + 4ab^3 + b^4$$
$$(a + b)^5 = a^5 + 5a^4b + 10a^3b^2 + 10a^2b^3 + 5ab^4 + b^5$$
$$(a + b)^6 = a^6 + 6a^5b + 15a^4b^2 + 20a^3b^3 + 15a^2b^4 + 6ab^5 + b^6$$

Blaise Pascal

Note that when expanding a binomial of the form $(a + b)^n$,

1. There are $n + 1$ terms in the expansion.

2. The first term is a^n and the last term is b^n.

3. Reading from left to right, the exponents on a decrease by 1 from term to term, while the exponents on b increase by 1 from term to term.

4. The sum of the exponents on the variables in each term is n.

5. The coefficients of the terms equidistant from the ends are the same.

If we examine just the variables in $(a + b)^5$, we have $a^5, a^4b, a^3b^2, a^2b^3, ab^4$, and b^5.

The numerical coefficients of each term in the expansion of $(a + b)^n$ can be found by using **Pascal's triangle**, named after Blaise Pascal, a seventeenth-century French mathematician. For example, if $n = 5$, we can determine the numerical coefficients of $(a + b)^5$ as follows.

Exponent on Binomial	Pascal's Triangle
$n = 0$	1
$n = 1$	1 1
$n = 2$	1 2 1
$n = 3$	1 3 3 1
$n = 4$	1 4 6 4 1
$n = 5$	1 5 10 10 5 1
$n = 6$	1 6 15 20 15 6 1

Examine row 5 ($n = 4$) and row 6 ($n = 5$).

$$1 + 4 + 6 + 4 + 1$$

$$1 \quad 5 \quad 10 \quad 10 \quad 5 \quad 1$$

Notice that the first and last numbers in each row are 1, and the inner numbers are obtained by adding the two numbers in the row above (to the right and left). The numerical coefficients of $(a + b)^5$ are 1, 5, 10, 10, 5, and 1. Thus, we can write the expansion of $(a + b)^5$ by using the information in 1–5 above for the variables and their exponents, and by using Pascal's triangle for the coefficients.

$$(a + b)^5 = a^5 + 5a^4b + 10a^3b^2 + 10a^2b^3 + 5ab^4 + b^5$$

This method of expanding a binomial is not practical when n is large.

3 Use the Binomial Theorem

We will shortly introduce a more practical method, called the binomial theorem, to expand expressions of the form $(a + b)^n$. However, before we introduce this formula, we need to explain how to find *binomial coefficients* of the form $\binom{n}{r}$.

Binomial Coefficients

For n and r nonnegative integers, $n \geq r$,

$$\binom{n}{r} = \frac{n!}{r! \cdot (n - r)!}$$

The binomial coefficient $\binom{n}{r}$ is read "the number of *combinations* of n items taken r at a time." Combinations are used in many areas of mathematics, including the study of probability.

EXAMPLE 1 ▸ Evaluate $\binom{6}{2}$.

Solution   Using the definition, if we substitute 6 for n and 2 for r, we obtain

$$\binom{6}{2} = \frac{6!}{2! \cdot (6-2)!} = \frac{6!}{2! \cdot 4!} = \frac{6 \cdot 5 \cdot \cancel{4 \cdot 3 \cdot 2 \cdot 1}}{(2 \cdot 1) \cdot (\cancel{4 \cdot 3 \cdot 2 \cdot 1})} = 15$$

Thus, $\binom{6}{2}$ equals 15.

▸ **Now Try Exercise 9**

EXAMPLE 2 ▸ Evaluate.

a) $\binom{7}{4}$ **b)** $\binom{8}{8}$ **c)** $\binom{5}{0}$

Solution

a) $\binom{7}{4} = \dfrac{7!}{4! \cdot (7-4)!} = \dfrac{7!}{4! \cdot 3!} = \dfrac{7 \cdot 6 \cdot 5 \cdot \cancel{4 \cdot 3 \cdot 2 \cdot 1}}{\cancel{(4 \cdot 3 \cdot 2 \cdot 1)}(3 \cdot 2 \cdot 1)} = 35$

b) $\binom{8}{8} = \dfrac{8!}{8! \, (8-8)!} = \dfrac{\cancel{8!}}{\cancel{8!} \cdot 0!} = \dfrac{1}{1} = 1$ *Remember that 0! = 1.*

c) $\binom{5}{0} = \dfrac{5!}{0! \cdot (5-0)!} = \dfrac{\cancel{5!}}{0! \cdot \cancel{5!}} = \dfrac{1}{1} = 1$

▸ **Now Try Exercise 17**

By studying Examples 2 **b)** and **c)**, you can reason that, for any positive integer n,

$$\binom{n}{n} = 1 \quad \text{and} \quad \binom{n}{0} = 1$$

USING YOUR GRAPHING CALCULATOR

All graphing calculators can evaluate binomial coefficients. On most graphers, the notation $_nC_r$ is used instead of $\binom{n}{r}$.
Thus, $\binom{7}{4}$ would be represented as $_7C_4$ on a grapher.

On the TI-84 Plus calculator, the notation $_nC_r$ can be found under the probability function, PRB, menu. This time it is item 3, $_nC_r$. To find $_7C_4$ or $_8C_2$ use the following keystrokes:

KEYSTROKES	ANSWER
$_7C_4$ 7 MATH ▶ ▶ ▶ 3 4 ENTER	35
$_8C_2$ 8 MATH ▶ ▶ ▶ 3 2 ENTER	28

If you are using a different graphing calculator, consult the manual to learn to evaluate combinations.

Now we introduce the binomial theorem.

Binomial Theorem

For any positive integer n,

$$(a+b)^n = \binom{n}{0}a^n b^0 + \binom{n}{1}a^{n-1}b^1 + \binom{n}{2}a^{n-2}b^2 + \binom{n}{3}a^{n-3}b^3 + \cdots + \binom{n}{n}a^0 b^n$$

Notice in the binomial theorem that the sum of the exponents on the variables in each term is n. In the combination, the top number is always n and the bottom number is always the same as the exponent on the second variable in the term.

For example, if we consider the term $\binom{n}{3}a^{n-3}b^3$, the sum of the exponents on the variables is $(n-3)+3=n$. Also, the exponent on the variable b is 3, and the bottom number in the combination is also 3.

If the variables and exponents on one term of the binomial theorem are $a^7 b^5$, then n must be $7+5$ or 12. Also, the combination preceding $a^7 b^5$ must be $\binom{12}{5}$. Thus, the term would be $\binom{12}{5}a^7 b^5$.

Now we will expand $(a+b)^5$ using the binomial theorem and see if we get the same expression as we did when we used polynomial multiplication and Pascal's triangle to obtain the expansion.

$$(a+b)^5 = \binom{5}{0}a^5 b^0 + \binom{5}{1}a^{5-1}b^1 + \binom{5}{2}a^{5-2}b^2 + \binom{5}{3}a^{5-3}b^3 + \binom{5}{4}a^{5-4}b^4 + \binom{5}{5}a^{5-5}b^5$$

$$= \binom{5}{0}a^5 b^0 + \binom{5}{1}a^4 b^1 + \binom{5}{2}a^3 b^2 + \binom{5}{3}a^2 b^3 + \binom{5}{4}a^1 b^4 + \binom{5}{5}a^0 b^5$$

$$= \frac{5!}{0! \cdot 5!}a^5 + \frac{5!}{1! \cdot 4!}a^4 b + \frac{5!}{2! \cdot 3!}a^3 b^2 + \frac{5!}{3! \cdot 2!}a^2 b^3 + \frac{5!}{4! \cdot 1!}ab^4 + \frac{5!}{5! \cdot 0!}b^5$$

$$= a^5 + 5a^4 b + 10a^3 b^2 + 10a^2 b^3 + 5ab^4 + b^5$$

This is the same expression as we obtained earlier.

In the binomial theorem, the first and last terms of the expansion contain a factor raised to the zero power. Since any nonzero number raised to the 0th power equals 1, we could have omitted those factors. These factors were included so that you could see the pattern better.

EXAMPLE 3 ▶ Use the binomial theorem to expand $(2x+3)^6$.

Solution   If we use $2x$ for a and 3 for b, we obtain

$$(2x+3)^6 = \binom{6}{0}(2x)^6(3)^0 + \binom{6}{1}(2x)^5(3)^1 + \binom{6}{2}(2x)^4(3)^2 + \binom{6}{3}(2x)^3(3)^3 + \binom{6}{4}(2x)^2(3)^4 + \binom{6}{5}(2x)^1(3)^5 + \binom{6}{6}(2x)^0(3)^6$$

$$= 1(2x)^6 + 6(2x)^5(3) + 15(2x)^4(9) + 20(2x)^3(27) + 15(2x)^2(81) + 6(2x)(243) + 1(729)$$

$$= 64x^6 + 576x^5 + 2160x^4 + 4320x^3 + 4860x^2 + 2916x + 729$$

▶ **Now Try Exercise 19**

EXAMPLE 4 ▶ Use the binomial theorem to expand $(5x-2y)^4$.

Solution   Write $(5x-2y)^4$ as $[5x+(-2y)]^4$. Use $5x$ in place of a and $-2y$ in place of b in the binomial theorem.

$$[5x+(-2y)]^4 = \binom{4}{0}(5x)^4(-2y)^0 + \binom{4}{1}(5x)^3(-2y)^1 + \binom{4}{2}(5x)^2(-2y)^2 + \binom{4}{3}(5x)^1(-2y)^3 + \binom{4}{4}(5x)^0(-2y)^4$$

$$= 1(5x)^4 + 4(5x)^3(-2y) + 6(5x)^2(-2y)^2 + 4(5x)(-2y)^3 + 1(-2y)^4$$

$$= 625x^4 - 1000x^3 y + 600x^2 y^2 - 160xy^3 + 16y^4$$

▶ **Now Try Exercise 25**

EXERCISE SET 15.4   *Math XL*   *MyMathLab*
MathXL®   MyMathLab

Concept/Writing Exercises

1. Explain how to construct Pascal's triangle. Construct the first five rows of Pascal's triangle.

2. Explain how to find $n!$ for any whole number n.

3. Give the value of 1!.

4. Give the value of 0!.

5. Can you evaluate $(-3)!$? Explain.

6. Can you evaluate $(-6)!$? Explain.

7. How many terms are there in the expansion of $(a + b)^{13}$? Explain.

8. How many terms are there in the expansion of $(x + y)^{20}$? Explain.

Practice the Skills

Evaluate each combination.

9. $\dbinom{5}{2}$

10. $\dbinom{6}{3}$

11. $\dbinom{5}{5}$

12. $\dbinom{9}{3}$

13. $\dbinom{7}{0}$

14. $\dbinom{10}{7}$

15. $\dbinom{8}{4}$

16. $\dbinom{12}{3}$

17. $\dbinom{8}{2}$

18. $\dbinom{11}{4}$

Use the binomial theorem to expand each expression.

19. $(x + 4)^3$

20. $(x - 4)^3$

21. $(2x - 3)^3$

22. $(2x + 3)^3$

23. $(a - b)^4$

24. $(2r + s^2)^4$

25. $(3a - b)^5$

26. $(x + 2y)^5$

27. $\left(2x + \dfrac{1}{2}\right)^4$

28. $\left(\dfrac{2}{3}x + \dfrac{3}{2}\right)^4$

29. $\left(\dfrac{x}{2} - 3\right)^4$

30. $(3x^2 + y)^5$

Write the first four terms of each expansion.

31. $(x + 10)^{10}$

32. $(2x + 3)^8$

33. $(3x - y)^7$

34. $(3p - 2q)^{11}$

35. $(x^2 - 3y)^8$

36. $\left(2x + \dfrac{y}{7}\right)^9$

Problem Solving

37. Is $n!$ equal to $n \cdot (n - 1)!$? Explain and give an example to support your answer.

38. Is $(n + 1)!$ equal to $(n + 1) \cdot n!$? Explain and give an example to support your answer.

39. Is $(n - 3)!$ equal to $(n - 3)(n - 4)(n - 5)!$ for $n \ge 5$? Explain and give an example to support your answer.

40. Is $(n + 2)!$ equal to $(n + 2)(n + 1)(n)(n - 1)!$ for $n \ge 1$? Explain and give an example to support your answer.

41. Under what conditions will $\dbinom{n}{m}$, where n and m are nonnegative integers, have a value of 1?

42. Can $\dbinom{n}{m}$ ever have a value of 0? Explain.

43. What are the first, second, next to last, and last terms of the expansion $(x + 3)^8$?

44. What are the first, second, next-to-last, and last terms of the expansion $(2x + 5)^6$?

45. Write the binomial theorem using summation notation.

46. Prove that $\dbinom{n}{r} = \dbinom{n}{n - r}$ for any whole numbers n and r, and $r \le n$.

Cumulative Review Exercises

[4.4] 47. Find the y-intercept for the line $2x + y = 10$

[6.6] 48. Solve $x(x - 11) = -18$.

[9.3] 49. Solve the system of equations.

$$\frac{1}{5}x + \frac{1}{2}y = 4$$

$$\frac{2}{3}x - y = \frac{8}{3}$$

[11.4] 50. Simplify $\sqrt{20xy^4}\,\sqrt{6x^5y^7}$.

[13.1] 51. Find $f^{-1}(x)$ if $f(x) = 3x + 8$.

Chapter 15 Summary

IMPORTANT FACTS AND CONCEPTS	EXAMPLES

Section 15.1

A **sequence** of numbers is a list of numbers arranged in a specific order. Each number is called a **term** of the sequence.	$2, 6, 10, 14, 18, 22, \ldots$ is a sequence $7, 14, 21, 28, 35, 42, \ldots$ is a sequence
An **infinite sequence** is a function whose domain is the set of natural numbers.	Domain: $\{1, \quad 2, \quad 3, \quad 4, \quad \ldots, \quad n, \quad \ldots\}$ $\qquad \downarrow \quad \downarrow \quad \downarrow \quad \downarrow \qquad\qquad \downarrow$ Range: $\{7, \quad 14, \quad 21, \quad 28, \quad \ldots \quad 7n, \quad \ldots\}$ The infinite sequence is $7, 14, 21, 28, \ldots$.
A **finite sequence** is a function whose domain includes only the first n natural numbers.	Domain: $\{1, \quad 2, \quad 3, \quad 4\}$ $\qquad \downarrow \quad \downarrow \quad \downarrow \quad \downarrow$ Range: $\{4, \quad 8, \quad 12, \quad 16\}$ The finite sequence is $4, 8, 12, 16$.
The **general term of a sequence**, a_n, can determine the sequence.	Let $a_n = n^2 - 3$. Write the first three terms of this sequence $$a_1 = 1^2 - 3 = -2$$ $$a_2 = 2^2 - 3 = 1$$ $$a_3 = 3^2 - 3 = 6$$ The first three terms of the sequence are $-2, 1, 6$.
An **increasing sequence** is a sequence where each term is larger than the preceding term. A **decreasing sequence** is a sequence where each term is smaller than the preceding term.	$-2, 5, 7, 11$ is an increasing sequence. $50, 48, 46, 44$ is a decreasing sequence.
A **series** is the sum of the terms of a sequence. A series may be finite or infinite.	If the sequence is $1, 3, 5, 7, 9$, then the series is $1 + 3 + 5 + 7 + 9 = 25$. If the sequence is $\dfrac{1}{3}, \dfrac{1}{9}, \dfrac{1}{27}, \cdots, \left(\dfrac{1}{3}\right)^n, \ldots$ then the series is $\dfrac{1}{3} + \dfrac{1}{9} + \dfrac{1}{27} + \cdots + \left(\dfrac{1}{3}\right)^n + \cdots$.
A **partial sum**, s_n, of an infinite sequence, $a_1, a_2, a_3, \ldots, a_n, \ldots$ is the sum of the first n terms. That is, $$s_1 = a_1$$ $$s_2 = a_1 + a_2$$ $$s_3 = a_1 + a_2 + a_3$$ $$\vdots \qquad \vdots$$ $$s_n = a_1 + a_2 + a_3 + \cdots + a_n$$	Let $a_n = \dfrac{5+n}{n^2}$. Compute s_1 and s_3. $$s_1 = a_1 = \frac{5+1}{1^2} = \frac{6}{1} = 6$$ $$s_3 = a_1 + a_2 + a_3$$ $$= \frac{5+1}{1^2} + \frac{5+2}{2^2} + \frac{5+3}{3^2}$$ $$= \frac{6}{1} + \frac{7}{4} + \frac{8}{9} = 8\frac{23}{36}$$
A series can be written using **summation notation**: $$\sum_{i=1}^{n} a_i = a_1 + a_2 + a_3 + \cdots + a_n.$$ i is the **index of summation**, n is the **upper limit of summation**, and 1 is the **lower limit of summation**.	$\displaystyle\sum_{i=1}^{4} (3i - 7) = (3 \cdot 1 - 7) + (3 \cdot 2 - 7) + (3 \cdot 3 - 7) + (3 \cdot 4 - 7)$ $\qquad\qquad\qquad = -4 - 1 + 2 + 5 = 2$ If $a_n = 6n^2 + 11$, the third partial sum, s_3, in summation notation, is written as $\displaystyle\sum_{i=1}^{3} (6i^2 + 11)$.

IMPORTANT FACTS AND CONCEPTS	EXAMPLES

Section 15.2

An **arithmetic sequence** is a sequence in which each term after the first differs from the preceding term by a **common difference**, **d**.	Arithmetic Sequence Common Difference, d $3, 8, 13, 18, 23, \ldots$ $\qquad d = 8 - 3 = 5$ $20, 14, 8, 2, -4, \ldots$ $\qquad d = 14 - 20 = -6$
The **nth term**, a_n, of an arithmetic sequence is $$a_n = a_1 + (n-1)d$$	The nth term of the arithmetic sequence with $a_1 = 7$ and $d = -5$ is $$a_n = 7 + (n-1)(-5)$$ $$= 7 - 5n + 5$$ $$= -5n + 12$$ For this sequence, the 20th term is $$a_{20} = -5(20) + 12 = -100 + 12$$ $$= -88$$
An **arithmetic series** is the sum of the terms of an arithmetic sequence. The sum of the first n terms, s_n of an arithmetic sequence, also known as the **nth partial sum** is $$s_n = a_1 + a_2 + a_3 + \cdots + a_n$$ For an arithmetic series, this sum is determined by the formula $$s_n = \frac{n(a_1 + a_n)}{2}$$	Find the sum of the first 30 natural numbers. That is, find the sum of $$1 + 2 + 3 + \cdots + 30$$ Since $a_1 = 1$, $a_{30} = 30$, and $n = 30$, the sum is $$s_{30} = \frac{30(1 + 30)}{2} = \frac{30(31)}{2} = 465$$

Section 15.3

A **geometric sequence** is a sequence in which each term after the first term is a common multiple of the preceding term. The common multiple is called the **common ratio**, r.	Geometric Sequence $\qquad$ Common Ratio, r $2, 6, 18, 54, 162, \ldots$ $\qquad r = \dfrac{6}{2} = 3$ $8, -2, \dfrac{1}{2}, -\dfrac{1}{8}, \dfrac{1}{32}, \ldots$ $\qquad r = \dfrac{-2}{8} = -\dfrac{1}{4}$
The **nth term**, a_n, of a geometric sequence is $$a_n = a_1 r^{n-1}$$	For the geometric sequence with $a_1 = 5$, $r = \dfrac{1}{2}$, and $n = 6$, a_6 is found as follows. $$a_6 = 5\left(\frac{1}{2}\right)^{6-1} = 5\left(\frac{1}{2}\right)^5 = \frac{5}{32}$$
A **geometric series** is the sum of the terms of a geometric sequence. The sum of the first n terms, s_n, of a geometric sequence also known as the **nth partial sum**, is $$s_n = a_1 + a_2 + a_3 + \cdots + a_n.$$ For a geometric series, this sum is determined by the formula $$s_n = \frac{a_1(1 - r^n)}{1 - r}, r \neq 1$$	To find the sum of the six terms of a geometric sequence with $a_1 = 12$ and $r = \dfrac{1}{3}$, use the formula with $n = 6$ to obtain $$s_6 = \frac{12\left[1 - \left(\frac{1}{3}\right)^6\right]}{1 - \frac{1}{3}} = \frac{12\left[1 - \frac{1}{729}\right]}{\frac{2}{3}} = \frac{12\left(\frac{728}{729}\right)}{\frac{1}{3}}$$ $$= 12\left(\frac{728}{729}\right)\left(\frac{3}{1}\right) = \frac{2912}{81} \text{ or } 35\frac{77}{81}$$
The sum of an infinite geometric series is $$s_\infty = \frac{a_1}{1 - r} \text{ where } |r| < 1$$	To find the sum of the infinite series $4 - 2 + 1 - \dfrac{1}{2} + \dfrac{1}{4} + \cdots$, use the formula with $a_1 = 4$ and $r = -\dfrac{1}{2}$ to obtain $$s_\infty = \frac{4}{1 - \left(-\frac{1}{2}\right)} = \frac{4}{\frac{3}{2}} = 4 \cdot \frac{2}{3} = \frac{8}{3} \text{ or } 2\frac{2}{3}$$

IMPORTANT FACTS AND CONCEPTS	EXAMPLES

Section 15.4

n Factorial

$$n! = n(n-1)(n-2)(n-3)\cdots(1)$$

for any positive integer n.
Note that 0! is defined to be 1.

$5! = 5 \cdot 4 \cdot 3 \cdot 2 \cdot 1 = 120$

$8! = 8 \cdot 7 \cdot 6 \cdot 5 \cdot 4 \cdot 3 \cdot 2 \cdot 1 = 40,320$

Binomial Coefficients

For n and r nonnegative integers, $n \geq r$,

$$\binom{n}{r} = \frac{n!}{r! \cdot (n-r)!}$$

$$\binom{n}{n} = 1 \text{ and } \binom{n}{0} = 1$$

$$\binom{7}{3} = \frac{7!}{3! \cdot (7-3)!} = \frac{7!}{3! \cdot 4!} = \frac{7 \cdot 6 \cdot 5 \cdot \cancel{4 \cdot 3 \cdot 2 \cdot 1}}{3 \cdot 2 \cdot 1 \cdot \cancel{4 \cdot 3 \cdot 2 \cdot 1}} = 35$$

$$\binom{10}{10} = 1, \quad \binom{10}{0} = 1$$

Binomial Theorem

For any positive integer n,

$$(a+b)^n = \binom{n}{0}a^n b^0 + \binom{n}{1}a^{n-1}b^1 + \binom{n}{2}a^{n-2}b^2 +$$

$$\binom{n}{3}a^{n-3}b^3 + \cdots + \binom{n}{n}a^0 b^n$$

$$(x+2y)^4 = \binom{4}{0}x^4 + \binom{4}{1}x^3(2y) + \binom{4}{2}x^2(2y)^2$$

$$+ \binom{4}{3}x(2y)^3 + \binom{4}{4}(2y)^4$$

$$= 1 \cdot x^4 + 4 \cdot x^3(2y) + 6 \cdot x^2(4y^2) + 4 \cdot x(8y^3) + 1 \cdot 16y^4$$

$$= x^4 + 8x^3 y + 24x^2 y^2 + 32xy^3 + 16y^4$$

Chapter 15 Review Exercises

[15.1] *Write the first five terms of each sequence.*

1. $a_n = n + 5$
2. $a_n = n^2 + n - 3$
3. $a_n = \dfrac{6}{n}$
4. $a_n = \dfrac{n^2}{n+4}$

Find the indicated term of each sequence.

5. $a_n = 3n - 10$, seventh term

6. $a_n = (-1)^n + 5$, seventh term

7. $a_n = \dfrac{n+17}{n^2}$, ninth term

8. $a_n = (n)(n-3)$, eleventh term

For each sequence, find the first and third partial sums, s_1 and s_3.

9. $a_n = 2n + 5$

10. $a_n = n^2 + 8$

11. $a_n = \dfrac{n+3}{n+2}$

12. $a_n = (-1)^n(n+8)$

Write the next three terms of each sequence. Then write an expression for the general term, a_n.

13. $2, 4, 8, 16, \ldots$

14. $-27, 9, -3, 1, \ldots$

15. $\dfrac{1}{7}, \dfrac{2}{7}, \dfrac{4}{7}, \dfrac{8}{7}, \ldots$

16. $13, 9, 5, 1, \ldots$

Write out each series. Then find the sum of the series.

17. $\displaystyle\sum_{i=1}^{3} i^2 + 9$

18. $\displaystyle\sum_{i=1}^{4} i(i+5)$

19. $\displaystyle\sum_{i=1}^{5} \dfrac{i^2}{6}$

20. $\displaystyle\sum_{i=1}^{4} \dfrac{i}{i+1}$

For the set of values $x_1 = 3$, $x_2 = 9$, $x_3 = 7$, $x_4 = 10$, evaluate the indicated sum.

21. $\displaystyle\sum_{i=1}^{4} x_i$

22. $\displaystyle\sum_{i=1}^{4} (x_i)^2$

23. $\displaystyle\sum_{i=2}^{3} (x_i^2 + 1)$

24. $\left(\displaystyle\sum_{i=1}^{4} x_i\right)^2$

In Exercises 25 and 26, consider the following rectangles. For the nth rectangle, the length is $n + 3$ and the width is n.

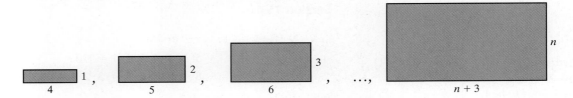

25. Perimeter

    a) Find the perimeters for the first four rectangles, and then list the perimeters in a sequence.

    b) Find the general term for the perimeter of the *n*th rectangle in the sequence. Use p_n for perimeter.

26. Area

    a) Find the areas for the first four rectangles, and then list the areas in a sequence.

    b) Find the general term for the area of the *n*th rectangle in the sequence. Use a_n for area.

[15.2] *Write the first five terms of the arithmetic sequence with the indicated first term and common difference.*

27. $a_1 = 5, d = 3$

28. $a_1 = 5, d = -\dfrac{1}{3}$

29. $a_1 = \dfrac{1}{2}, d = -2$

30. $a_1 = -100, d = \dfrac{1}{5}$

For each arithmetic sequence, find the indicated value.

31. $a_1 = 6, d = 3$; find a_9

32. $a_1 = 10, a_8 = -18$; find d

33. $a_1 = -3, a_{11} = 2$; find d

34. $a_1 = 22, a_n = -3, d = -5$; find n

Find s_n and d for each arithmetic sequence.

35. $a_1 = 7, a_8 = 21, n = 8$

36. $a_1 = -12, a_7 = -48, n = 7$

37. $a_1 = \dfrac{3}{5}, a_6 = \dfrac{13}{5}, n = 6$

38. $a_1 = -\dfrac{10}{3}, a_9 = -6, n = 9$

Write the first four terms of each arithmetic sequence. Then find a_{10} and s_{10}.

39. $a_1 = -7, d = 4$

40. $a_1 = 4, d = -3$

41. $a_1 = \dfrac{5}{6}, d = \dfrac{2}{3}$

42. $a_1 = -60, d = 5$

Find the number of terms in each arithmetic sequence. Then find s_n.

43. $4, 9, 14, \ldots, 64$

44. $-7, -4, -1, \ldots, 11$

45. $\dfrac{6}{10}, \dfrac{9}{10}, \dfrac{12}{10}, \ldots, \dfrac{36}{10}$

46. $-9, -3, 3, 9, \ldots, 45$

[15.3] *Determine the first five terms of each geometric sequence.*

47. $a_1 = 6, r = 2$

48. $a_1 = -12, r = \dfrac{1}{2}$

49. $a_1 = 20, r = -\dfrac{2}{3}$

50. $a_1 = -20, r = \dfrac{1}{5}$

Find the indicated term of each geometric sequence.

51. $a_1 = 6, r = \dfrac{1}{3}$; find a_5

52. $a_1 = 15, r = 2$; find a_6

53. $a_1 = -8, r = -3$; find a_4

54. $a_1 = \dfrac{1}{12}, r = \dfrac{2}{3}$; find a_5

Find each sum.

55. $a_1 = 7, r = 2$; find s_6

56. $a_1 = -84, r = -\dfrac{1}{4}$; find s_5

57. $a_1 = 9, r = \dfrac{3}{2}$; find s_4

58. $a_1 = 8, r = \dfrac{1}{2}$; find s_7

For each geometric sequence, find the common ratio, r, and then write an expression for the general term, a_n.

59. $6, 12, 24, \ldots$

60. $-4, -20, -100, \ldots$

61. $10, \dfrac{10}{3}, \dfrac{10}{9}, \ldots$

62. $\dfrac{9}{5}, \dfrac{18}{15}, \dfrac{36}{45}, \ldots$

Find the sum of the terms in each infinite geometric sequence.

63. $5, \dfrac{5}{2}, \dfrac{5}{4}, \dfrac{5}{8}, \ldots$

64. $\dfrac{5}{2}, 1, \dfrac{2}{5}, \dfrac{4}{25}, \ldots$

65. $-8, \dfrac{8}{3}, -\dfrac{8}{9}, \dfrac{8}{27}, \ldots$

66. $-6, -4, -\dfrac{8}{3}, -\dfrac{16}{9}, \ldots$

Find the sum of each infinite series.

67. $16 + 8 + 4 + 2 + 1 + \cdots$

68. $9 + \dfrac{9}{3} + \dfrac{9}{9} + \dfrac{9}{27} + \cdots$

69. $5 - 1 + \dfrac{1}{5} - \dfrac{1}{25} + \cdots$

70. $-4, -\dfrac{8}{3}, -\dfrac{16}{9}, -\dfrac{32}{27}, \ldots$

Write the repeating decimal as a ratio of integers.

71. $0.363636\ldots$

72. $0.621621\ldots$

[15.4] *Use the binomial theorem to expand the expression.*

73. $(3x + y)^4$

74. $(2x - 3y^2)^3$

Write the first four terms of the expansion.

75. $(x - 2y)^9$

76. $(2a^2 + 3b)^8$

[15.2]

77. Sum of Integers Find the sum of the integers between 101 and 200, inclusive.

78. Barrels of Oil Barrels of oil are stacked with 21 barrels in the bottom row, 20 barrels in the second row, 19 barrels in the third row, and so on, to the top row, which has only 1 barrel. How many barrels are there?

79. Salary Ahmed Mocanda just started a new job with an annual salary of $36,000. He has been told that his salary will increase by $1000 per year for the next 10 years.

  a) Write a sequence showing his salary for the first 4 years.

  b) Write a general term of this sequence.

  c) What will his salary be 6 years from now?

  d) How much money will he make in the first 11 years?

[15.3]

80. Money You begin with $100, double that to get $200, double that again to get $400, and so on. How much will you have after you perform this process 10 times?

81. Salary Gertude Dibble started a new job on January 1, 2006 with a monthly salary of $1600. Her boss has agreed to give her a 4% raise each month for the remainder of the year.

  a) What is Gertude's salary in July?

  b) What is Gertude's salary in December?

  c) How much money does Gertude make in 2006?

82. Inflation If the inflation rate was a constant 8% per year (each year the cost of living is 8% greater than the previous year), how much would a product that costs $200 now cost after 12 years?

83. Pendulum On each swing (left to right or right to left), a pendulum travels 92% as far as on its previous swing. If the first swing is 12 feet, find the distance traveled by the pendulum by the time it comes to rest.

Chapter 15 Practice Test

To find out how well you understand the chapter material, take this practice test. The answers, and the section where the material was initially discussed, are given in the back of the book. Each problem is also fully worked-out on the **Chapter Test Prep Video CD**. *Review any questions that you answered incorrectly.*

1. What is a series?

2. a) What is an arithmetic sequence?

    b) What is a geometric sequence?

3. Write the first five terms of the sequence if $a_n = \dfrac{n-2}{3n}$.

4. Find the first and third partial sums if $a_n = \dfrac{2n+1}{n^2}$.

5. Write out the following series and find the sum of the series.

$$\sum_{i=1}^{5} (2i^2 + 3)$$

6. For $x_1 = 4$, $x_2 = 2$, $x_3 = 8$, and $x_4 = 10$ find $\displaystyle\sum_{i=1}^{4} (x_i)^2$.

7. Write the general term for the following arithmetic sequence.

$$\frac{1}{3}, \frac{2}{3}, \frac{3}{3}, \frac{4}{3}, \ldots$$

8. Write the general term for the following geometric sequence.

$$5, 10, 20, 40, \ldots$$

In Exercises 9 and 10, write the first four terms of each sequence.

9. $a_1 = 15, d = -6$

10. $a_1 = \dfrac{5}{12}, r = \dfrac{2}{3}$

11. Find a_{11} when $a_1 = 40$ and $d = -8$.

12. Find s_8 for the arithmetic sequence with $a_1 = 7$ and $a_8 = -12$.

13. Find the number of terms in the arithmetic sequence $-4, -16, -28, \ldots, -136$.

14. Find a_6 when $a_1 = 8$ and $r = \dfrac{2}{3}$.

15. Find s_7 when $a_1 = \dfrac{3}{5}$ and $r = -5$.

16. Find the common ratio and write an expression for the general term of the sequence $15, 5, \dfrac{5}{3}, \dfrac{5}{9}, \ldots$.

17. Find the sum of the following infinite geometric series.

$$4 + \frac{8}{3} + \frac{16}{9} + \frac{32}{27} + \cdots$$

18. Write $0.3939\ldots$ as a ratio of integers.

19. Evaluate $\dbinom{8}{3}$.

20. Use the binomial theorem to expand $(x + 2y)^4$.

21. Arithmetic Mean Paul Misselwitz's five test scores are 76, 93, 83, 87, and 71. Use $\bar{x} = \dfrac{\Sigma x}{n}$ to find the arithmetic mean of Paul's scores.

22. A Pile of Logs Logs are piled with 13 logs in the bottom row, 12 logs in the second row, 11 logs in the third row, and so on, to the top. How many logs are there?

23. Saving for Retirement To save for retirement, Jamie Monroe plans to save $1000 the first year, $2000 the second year, $3000 the third year, and to increase the amount saved by $1000 in each successive year. How much will she have saved by the end of her twentieth year of savings?

24. Earnings Yolanda Rivera makes $700 per week working at an insurance office. Her boss has guaranteed her an increase of 4% per week for the next 7 weeks. How much is she making in the sixth week?

25. Culture of Bacteria The number of bacteria in a culture is tripling every hour. If there are initially 500 bacteria in the culture, how many bacteria will be in the culture by the end of the sixth hour?

Cumulative Review Test

Take the following test and check your answers with those given in the back of the book. Review any questions that you answered incorrectly. The section where the material was covered is indicated after the answer.

1. Solve $A = \dfrac{1}{2}bh$, for b.

2. Find an equation of the line through $(4, -2)$ and $(1, 9)$. Write the equation in slope-intercept form.

3. Multiply $(5x^3 + 4x^2 - 6x + 2)(x + 5)$.

4. Factor $x^3 + 2x - 6x^2 - 12$.

5. Factor $(a + b)^2 + 8(a + b) + 16$.

6. Subtract $5 - \dfrac{x-1}{x^2 + 3x - 10}$.

7. y varies directly as the square of z. If y is 80 when z is 20, find y when z is 50.

8. Solve the system of equations.

$$x + y + z = 1$$
$$2x + 2y + 2z = 2$$
$$3x + 3y + 3z = 3$$

9. If $f(x) = 2\sqrt[3]{x - 3}$ and $g(x) = \sqrt[3]{5x - 15}$, find all values of x for which $f(x) = g(x)$.

10. Solve $\sqrt{6x - 5} - \sqrt{2x + 6} - 1 = 0$.

11. Solve by completing the square.

$$x^2 + 2x + 15 = 0$$

12. Solve by the quadratic formula.

$$x^2 - \frac{x}{5} - \frac{1}{3} = 0$$

13. Numbers Twice the square of a positive number decreased by nine times the number is 5. Find the number.

14. Graph $y = 2^x - 1$.

15. Solve $\log_a \frac{1}{64} = 6$ for a.

16. Graph $y = x^2 - 4x$ and label the vertex.

17. Find an equation of a circle with center at $(-6, 2)$ and radius 7.

18. Graph $(x + 3)^2 + (y + 1)^2 = 16$.

19. Graph $9x^2 + 16y^2 = 144$.

20. Find the sum of the infinite geometric series.

$$6 + 4 + \frac{8}{3} + \frac{16}{9} + \frac{32}{27} + \cdots$$

Appendices

A   Review of Decimals and Percent
B   Finding the Greatest Common Factor and Least Common Denominator
C   Geometry
D   Review of Exponents, Polynomials, and Factoring

Appendix A Review of Decimals and Percent

Decimals

> **To Add or Subtract Numbers Containing Decimal Points**
>
> 1. Align the numbers by the decimal points.
> 2. Add or subtract the numbers as if they were whole numbers.
> 3. Place the decimal point in the sum or difference directly below the decimal points in the numbers being added or subtracted.

EXAMPLE 1 ▶ Add 4.6 + 13.813 + 9.02.

Solution

$$
\begin{array}{r}
4.600 \\
13.813 \\
+\;\; 9.020 \\
\hline
27.433
\end{array}
$$

EXAMPLE 2 ▶ Subtract 3.062 from 34.9.

Solution

$$
\begin{array}{r}
34.900 \\
-\;\; 3.062 \\
\hline
31.838
\end{array}
$$

> **To Multiply Numbers Containing Decimal Points**
>
> 1. Multiply as if the factors were whole numbers.
> 2. Determine the total number of digits to the right of the decimal points in the factors.
> 3. Place the decimal point in the product so that the product contains the same number of digits to the right of the decimal as the total found in step 2. For example, if there are a total of three digits to the right of the decimal points in the factors, there must be three digits to the right of the decimal point in the product.

EXAMPLE 3 ▶ Multiply 2.34 × 1.9.

Solution

$$
\begin{array}{r}
2.34 \quad \longleftarrow \quad \textit{two digits to the right of the decimal point} \\
\times \quad\; 1.9 \quad \longleftarrow \quad \textit{one digit to the right of the decimal point} \\
\hline
2106 \\
234 \quad\;\;\; \\
\hline
4.446 \quad \longleftarrow \quad \textit{three digits to the right of the decimal point in the product}
\end{array}
$$

EXAMPLE 4 ▸ Multiply 2.13×0.02.

Solution

$$2.13 \longleftarrow \text{ \textit{two digits to the right of the decimal point}}$$
$$\underline{\times\ 0.02} \longleftarrow \text{ \textit{two digits to the right of the decimal point}}$$
$$0.0426 \longleftarrow \text{ \textit{four digits to the right of the decimal point in the product}}$$

Note that it was necessary to add a zero preceding the digit 4 in the answer in order to have four digits to the right of the decimal point.

To Divide Numbers Containing Decimal Points

1. Multiply both the dividend and divisor by a power of 10 that will make the divisor a whole number.
2. Divide as if working with whole numbers.
3. Place the decimal point in the quotient directly above the decimal point in the dividend.

To make the divisor a whole number, multiply *both* the dividend and divisor by 10 if the divisor is given in tenths, by 100 if the divisor is given in hundredths, by 1000 if the divisor is given in thousandths, and so on. Multiplying both the numerator and denominator by the same nonzero number is the same as multiplying the fraction by 1. Therefore, the value of the fraction is unchanged.

EXAMPLE 5 ▸ Divide $\dfrac{1.956}{0.12}$.

Solution Since the divisor, 0.12, is twelve-hundredths, we multiply both the divisor and dividend by 100.

$$\frac{1.956}{0.12} \times \frac{100}{100} = \frac{195.6}{12.}$$

Now we divide.

$$\begin{array}{r} 16.3 \\ 12\overline{)195.6} \\ \underline{12} \\ 75 \\ \underline{72} \\ 36 \\ \underline{36} \\ 0 \end{array}$$

The decimal point in the answer is placed directly above the decimal point in the dividend. Thus, $\dfrac{1.956}{0.12} = 16.3$.

EXAMPLE 6 ▸ Divide 0.26 by 10.4.

Solution First, multiply both the dividend and divisor by 10.

$$\frac{0.26}{10.4} \times \frac{10}{10} = \frac{2.6}{104.}$$

Now divide.

$$\begin{array}{r} 0.025 \\ 104\overline{)2.600} \\ \underline{2\ 08} \\ 520 \\ \underline{520} \\ 0 \end{array}$$

Note that a zero had to be placed before the digit 2 in the quotient.

$$\frac{0.26}{10.4} = 0.025$$

Rounding Decimal Numbers

Now we will explain how to round decimal numbers. The explanation of the procedure will refer to the positional values to the right of the decimal point, as illustrated here:

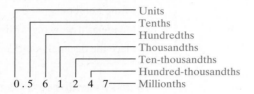

To Round a Decimal Number

1. Identify the number in the problem to the right of the positional value to which the number is to be rounded. For example, if you wish to round to tenths, you identify the number in the hundredths position. If you wish to round to hundredths, you identify the number in the thousandths position, and so on.

2. **a)** If the number you identified in step 1 is greater than or equal to 5, the number in the preceding position is increased by 1 unit, and all numbers to the right of the number that has been increased by 1 unit are eliminated.

   b) If the number you identified in step 1 is less than 5, the number you identified in step 1 and all numbers to its right are eliminated.

EXAMPLE 7 ▶ Round 4.863 to tenths.

Solution: Since we are rounding to tenths, we identify the number in the hundredths position as 6. Because the number 6 is greater than or equal to 5, we increase the number in the preceding position by 1 unit and drop the remaining digits. Therefore, 4.863 when rounded to tenths is 4.9.

EXAMPLE 8 ▶ Round 5.4738 to hundredths.

Solution: Since we are rounding to hundredths, we identify the number in the thousandths position as 3. Because the number 3 is less than 5, we drop the 3 and all numbers to the right of 3. Therefore, 5.4738 when rounded to hundredths is 5.47.

Percent

The word *percent* means "per hundred." The symbol % means percent. One percent means "one per hundred".

One Percent

$$1\% = \frac{1}{100} \quad \text{or} \quad 1\% = 0.01$$

EXAMPLE 9 ▸ Convert 16% to a decimal.

Solution Since 1% = 0.01,

$$16\% = 16(0.01) = 0.16$$

EXAMPLE 10 ▸ Convert 4.7% to a decimal.

Solution 4.7% = 4.7(0.01) = 0.047

EXAMPLE 11 ▸ Convert 1.14 to a percent.

Solution To change a decimal number to a percent, we multiply the number by 100%.

$$1.14 = 1.14 \times 100\% = 114\%$$

Often, you will need to find an amount that is a certain percent of a number. For example, when you purchase an item in a state or county that has a sales tax you must often pay a percent of the item's price as the sales tax. Examples 12 and 13 show how to find a certain percent of a number.

EXAMPLE 12 ▸ Find 32% of 300.

Solution To find a percent of a number, use multiplication. Change 32% to a decimal number, then multiply by 300.

$$(0.32)(300) = 96$$

Thus, 32% of 300 is 96.

EXAMPLE 13 ▸ Johnson County charges an 8% sales tax.

a) Find the sales tax on a stereo system that cost $580.
b) Find the total cost of the system, including tax.

Solution

a) The sales tax is 8% of 580.

$$(0.08)(580) = 46.40$$

The sales tax is $46.40.

b) The total cost is the purchase price plus the sales tax:

$$\text{total cost} = \$580 + \$46.40 = \$626.40$$

Appendix B Finding the Greatest Common Factor and Least Common Denominator

Prime Factorization

In Section 1.3, we mentioned that to simplify fractions you can divide both the numerator and denominator by the *greatest common factor* (GCF). One method to find the GCF is to use *prime factorization*. Prime factorization is the process of writing a given number as a product of prime numbers. *Prime numbers* are natural numbers, excluding

1, that can be divided by only themselves and 1. The first ten prime numbers are 2, 3, 5, 7, 11, 13, 17, 19, 23, and 29. Can you find the next prime number? If you answered 31, you answered correctly.

To write a number as a product of primes, we can use a *tree diagram*. Begin by selecting any two numbers whose product is the given number. Then continue factoring each of these numbers into prime numbers, as shown in Example 1.

EXAMPLE 1 ▶ Determine the prime factorization of the number 120.

Solution   We will use three different tree diagrams to illustrate the prime factorization of 120.

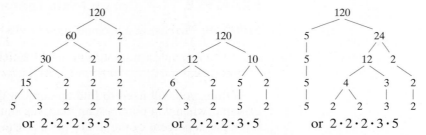

or $2 \cdot 2 \cdot 2 \cdot 3 \cdot 5$ or $2 \cdot 2 \cdot 2 \cdot 3 \cdot 5$ or $2 \cdot 2 \cdot 2 \cdot 3 \cdot 5$

Note that no matter how you start, if you do not make a mistake, you find that the prime factorization of 120 is $2 \cdot 2 \cdot 2 \cdot 3 \cdot 5$. There are other ways 120 can be factored but all will lead to the prime factorization $2 \cdot 2 \cdot 2 \cdot 3 \cdot 5$.

Greatest Common Factor

The greatest common factor (GCF) of two natural numbers is the greatest integer that is a factor of both numbers. We use the GCF when simplifying fractions.

> **To Find the Greatest Common Factor of a Given Numerator and Denominator**
>
> **1.** Write both the numerator and the denominator as a product of primes.
> **2.** Determine all the prime factors that are common to both prime factorizations.
> **3.** Multiply the prime factors found in step 2 to obtain the GCF.

EXAMPLE 2 ▶ Consider the fraction $\dfrac{108}{156}$.

a) Find the GCF of 108 and 156.   **b)** Simplify $\dfrac{108}{156}$.

Solution

a) First determine the prime factorizations of both 108 and 156.

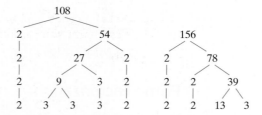

There are two 2s and one 3 common to both prime factorizations; thus

$$GCF = 2 \cdot 2 \cdot 3 = 12$$

The greatest common factor of 108 and 156 is 12. Twelve is the greatest integer that divides into both 108 and 156.

b) To simplify $\dfrac{108}{156}$, we divide both the numerator and denominator by the GCF, 12.

$$\frac{108 \div 12}{156 \div 12} = \frac{9}{13}$$

Thus, $\dfrac{108}{156}$ simplifies to $\dfrac{9}{13}$.

Least Common Denominator

When adding two or more fractions, you must write each fraction with a common denominator. The best denominator to use is the *least common denominator* (LCD). The LCD is the smallest number that each denominator divides into. Sometimes the least common denominator is referred to as the *least common multiple* of the denominators.

To Find the Least Common Denominator of Two or More Fractions

1. Write each denominator as a product of prime numbers.
2. For each prime number, determine the maximum number of times that prime number appears in any of the prime factorizations.
3. Multiply all the prime numbers found in step 2. Include each prime number the maximum number of times it appears in any of the prime factorizations. The product of all these prime numbers will be the LCD.

Example 3 illustrates the procedure to determine the LCD.

EXAMPLE 3 ▸ Consider $\dfrac{7}{108} + \dfrac{5}{156}$.

a) Determine the least common denominator.

b) Add the fractions.

Solution

a) We found in Example 2 that

$$108 = 2 \cdot 2 \cdot 3 \cdot 3 \cdot 3 \quad \text{and} \quad 156 = 2 \cdot 2 \cdot 3 \cdot 13$$

We can see that the maximum number of 2s that appear in either prime factorization is two (there are two 2s in both factorizations), the maximum number of 3s is three, and the maximum number of 13s is one. Multiply as follows:

$$2 \cdot 2 \cdot 3 \cdot 3 \cdot 3 \cdot 13 = 1404$$

Thus, the least common denominator is 1404. This is the smallest number that both 108 and 156 divide into.

b) To add the fractions, we need to write both fractions with a common denominator. The best common denominator to use is the LCD. Since $1404 \div 108 = 13$, we will multiply $\dfrac{7}{108}$ by $\dfrac{13}{13}$. Since $1404 \div 156 = 9$, we will multiply $\dfrac{5}{156}$ by $\dfrac{9}{9}$.

$$\frac{7}{108} \cdot \frac{13}{13} + \frac{5}{156} \cdot \frac{9}{9} = \frac{91}{1404} + \frac{45}{1404} = \frac{136}{1404} = \frac{34}{351}$$

Thus, $\dfrac{7}{108} + \dfrac{5}{156} = \dfrac{34}{351}$.

Appendix C Geometry

This appendix introduces or reviews important geometric concepts. **Table C.1** gives the names and descriptions of various types of angles.

Angles

Table C.1

Angle	Sketch of Angle
An **acute angle** is an angle whose measure is between 0° and 90°.	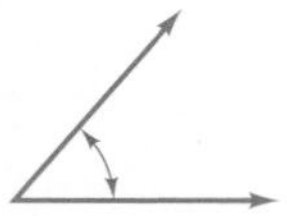
A **right angle** is an angle whose measure is 90°.	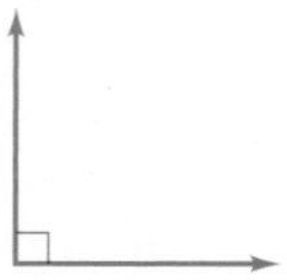
An **obtuse angle** is an angle whose measure is between 90° and 180°.	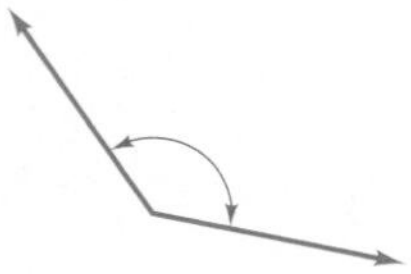
A **straight angle** is an angle whose measure is 180°.	
Two angles are **complementary angles** when the sum of their measures is 90°. Each angle is the complement of the other. Angles A and B are complementary angles.	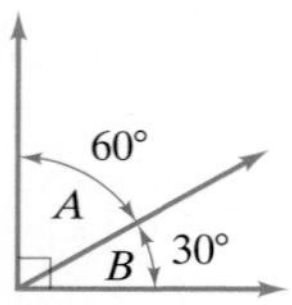 60° A B 30°
Two angles are **supplementary angles** when the sum of their measures is 180°. Each angle is the supplement of the other. Angles A and B are supplementary angles.	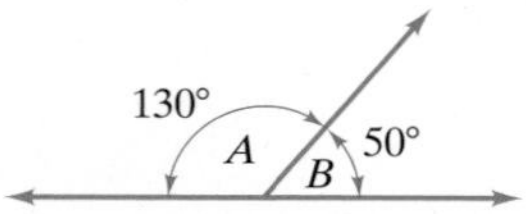 130° A B 50°

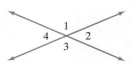

FIGURE C.1

When two lines intersect, four angles are formed as shown in **Figure C.1**. The pair of opposite angles formed by the intersecting lines are called **vertical angles**.

Angles 1 and 3 are vertical angles. Angles 2 and 4 are also vertical angles. *Vertical angles have equal measures.* Thus, angle 1, symbolized by $\angle 1$, is equal to angle 3, symbolized by $\angle 3$. We can write $\angle 1 = \angle 3$. Similarly, $\angle 2 = \angle 4$.

Parallel and Perpendicular Lines

Parallel lines are two lines in the same plane that do not intersect (**Fig. C.2**). **Perpendicular lines** are lines that intersect at right angles (**Fig. C.3**).

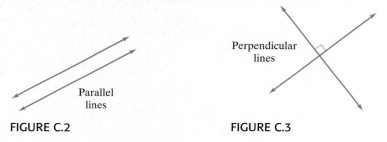

Parallel lines

FIGURE C.2

Perpendicular lines

FIGURE C.3

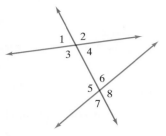

FIGURE C.4

A **transversal** is a line that intersects two or more lines at different points. When a transversal line intersects two other lines, eight angles are formed, as illustrated in **Figure C.4**. Some of these angles are given special names.

Interior angles: $3, 4, 5, 6$

Exterior angles: $1, 2, 7, 8$

Pairs of corresponding angles: 1 and 5; 2 and 6; 3 and 7; 4 and 8

Pairs of alternate interior angles: 3 and 6; 4 and 5

Pairs of alternate exterior angles: 1 and 8; 2 and 7

Parallel Lines Cut by a Transversal

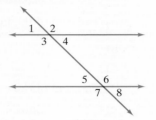

When two parallel lines are cut by a transversal,

1. Corresponding angles are equal ($\angle 1 = \angle 5, \angle 2 = \angle 6, \angle 3 = \angle 7, \angle 4 = \angle 8$).
2. Alternate interior angles are equal ($\angle 3 = \angle 6, \angle 4 = \angle 5$).
3. Alternate exterior angles are equal ($\angle 1 = \angle 8, \angle 2 = \angle 7$).

EXAMPLE 1 ▶ If line 1 and line 2 are parallel lines and $m\angle 1 = 112°$, find the measure of angles 2 through 8.

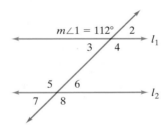

Solution   Angles 1 and 2 are supplementary, so $m\angle 2$ is $180° - 112° = 68°$. The measures of angles 1 and 4 are equal since they are vertical angles. Thus, $m\angle 4 = 112°$. Angles 1 and 5 are corresponding angles. Thus, $m\angle 5 = 112°$. It is equal to its vertical angle, $\angle 8$, so $m\angle 8 = 112°$. The measures of angles 2, 3, 6, and 7 are all equal and measure $68°$.

Polygons

A **polygon** is a closed figure in a plane determined by three or more line segments. Some polygons are illustrated in **Figure C.5**.

A **regular polygon** has sides that are all the same length, and interior angles that all have the same measure. In **Figure C.5**, (b) and (d) are regular polygons.

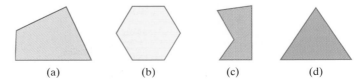

FIGURE C.5   (a)   (b)   (c)   (d)

> ## Sum of the Interior Angles of a Polygon
>
> The sum of the interior angles of a polygon can be found by the formula
>
> $$\text{Sum} = (n - 2)180°$$
>
> where n is the number of sides of the polygon.

EXAMPLE 2 ▶ Find the sum of the measures of the interior angles of **a)** a triangle; **b)** a quadrilateral (4 sides); **c)** an octagon (8 sides).

Solution

a) Since $n = 3$, we write

$$\text{Sum} = (n - 2)180°$$
$$= (3 - 2)180° = 1(180°) = 180°$$

The sum of the measures of the interior angles in a triangle is 180°.

b)
$$\text{Sum} = (n - 2)180°$$
$$= (4 - 2)180° = 2(180°) = 360°$$

The sum of the measures of the interior angles in a quadrilateral is 360°.

c)
$$\text{Sum} = (n - 2)(180°) = (8 - 2)180° = 6(180°) = 1080°$$

The sum of the measures of the interior angles in an octagon is 1080°.

Now we will briefly define several types of triangles in **Table C.2.**

Triangles

Table C.2	
Triangle	**Sketch of Triangle**
An **acute triangle** is one that has three acute angles (angles of less than 90°).	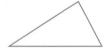
An **obtuse triangle** has one obtuse angle (an angle greater than 90°).	
A **right triangle** has one right angle (an angle equal to 90°). The longest side of a right triangle is opposite the right angle and is called the **hypotenuse**. The other two sides are called the **legs**.	
An **isosceles triangle** has two sides of equal length. The angles opposite the equal sides have the same measure.	
An **equilateral triangle** has three sides of equal length. It also has three equal angles that measure 60° each.	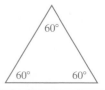

When two sides of a *right triangle* are known, the third side can be found using the **Pythagorean Theorem**, $a^2 + b^2 = c^2$, where a and b are the legs and c is the hypotenuse of the triangle. (See Section 6.7 for examples.)

Congruent and Similar Figures

If two triangles are **congruent**, it means that the two triangles are identical in size and shape. Two congruent triangles could be placed one on top of the other if we were able to move and rearrange them.

Two Triangles Are Congruent If Any One of the Following Statements Is True

1. Two angles of one triangle are equal to two corresponding angles of the other triangle, and the lengths of the sides between each pair of angles are equal. This method of showing that triangles are congruent is called the *angle, side, angle* method.

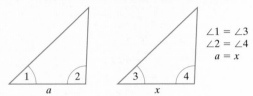

$$\angle 1 = \angle 3$$
$$\angle 2 = \angle 4$$
$$a = x$$

2. Corresponding sides of both triangles are equal. This is called the *side, side, side* method.

$$a = x$$
$$b = y$$
$$c = z$$

3. Two corresponding pairs of sides are equal, and the angle between them is equal. This is referred to as the *side, angle, side* method.

$$a = x$$
$$\angle 1 = \angle 2$$
$$b = y$$

EXAMPLE 3 ▶ Determine whether the two triangles are congruent.

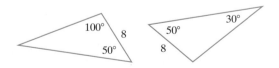

Solution   The unknown angle in the figure on the right must measure 100° since the sum of the angles of a triangle is 180°. Both triangles have the same two angles (100° and 50°), with the same length side between them, 8 units. Thus, these two triangles are congruent by the angle, side, angle method.

Two triangles are **similar** if all three pairs of corresponding angles are equal and corresponding sides are in proportion. Similar figures do not have to be the same size but must have the same general shape.

Two Triangles Are Similar If Any One of the Following Statements Is True

1. Two angles of one triangle equal two angles of the other triangle.

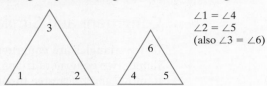

$\angle 1 = \angle 4$
$\angle 2 = \angle 5$
(also $\angle 3 = \angle 6$)

2. Corresponding sides of the two triangles are proportional.

$\dfrac{a}{x} = \dfrac{b}{y} = \dfrac{c}{z}$

3. Two pairs of corresponding sides are proportional, and the angles between them are equal.

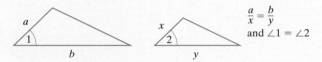

$\dfrac{a}{x} = \dfrac{b}{y}$
and $\angle 1 = \angle 2$

EXAMPLE 4 ▶ Are the triangles ABC and $AB'C'$ similar?

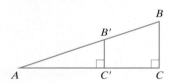

Solution   Angle A is common to both triangles. Since angle C and angle C' are equal (both $90°$), then $\angle B$ and $\angle B'$ must be equal. Since the three angles of triangle ABC equal the three angles of triangle $AB'C'$, the two triangles are similar.

Appendix D Review of Exponents, Polynomials, and Factoring

In Chapters 5 and 6 we discussed exponents, polynomials, and factoring. We review these concepts briefly in this Appendix. For a more complete discussion, please see Chapters 5 and 6.

Exponents

Exponents are shorthand for repeated multiplication. For instance, exponents can be used to rewrite the expression $5 \cdot 5 \cdot 5$ as 5^3. In the expression 5^3, the 5 is called the base and the 3 is called the exponent. The number 5^3 is read "5 cubed" or "5 to the third power" and means $5 \cdot 5 \cdot 5 = 125$. In general, an expression with n factors of b can be written as b^n.

In Chapter 5 we discussed the 6 rules of exponents. We summarize them below.

Summary of Rules of Exponents

1. $x^m \cdot x^n = x^{m+n}$ — product rule

2. $\dfrac{x^m}{x^n} = x^{m-n}, \quad x \neq 0$ — quotient rule

3. $x^0 = 1, \quad x \neq 0$ — zero exponent rule

4. $(x^m)^n = x^{m \cdot n}$ — power rule

5. $\left(\dfrac{ax}{by}\right)^m = \dfrac{a^m x^m}{b^m y^m}, \quad b \neq 0, y \neq 0$ — expanded power rule

6. $x^{-m} = \dfrac{1}{x^m}, \quad x \neq 0$ — negative exponent rule

In Example 1, we show how these properties can be used to simplify an expression.

EXAMPLE 1 ▶ Simplify $\left(\dfrac{x^4 y^{-5} z^3}{x^2 y z^3}\right)^2 x^4 y^{-8}$.

$$\left(\frac{x^4 y^{-5} z^3}{x^2 y z^3}\right)^2 x^4 y^{-8} = \left(\frac{x^{4\cdot2} y^{-5\cdot2} z^{3\cdot2}}{x^{2\cdot2} y^{1\cdot2} z^{3\cdot2}}\right) x^4 y^{-8}$$ Use expanded power rule.

$$= \left(\frac{x^8 y^{-10} z^6}{x^4 y^2 z^6}\right) x^4 y^{-8}$$ Simplify exponents.

$$= (x^{8-4} y^{-10-2} z^{6-6}) x^4 y^{-8}$$ Use quotient rule.

$$= (x^4 y^{-12} z^0) x^4 y^{-8}$$ Simplify exponents.

$$= \left(\frac{x^4 z^0}{y^{12}}\right) x^4 y^{-8}$$ Use negative exponent rule.

$$= \left(\frac{x^4 \cdot 1}{y^{12}}\right) x^4 y^{-8}$$ Use zero exponent rule.

$$= \left(\frac{x^4}{y^{12}}\right)\left(\frac{x^4}{y^8}\right)$$ Use negative exponent rule.

$$= \left(\frac{x^{4+4}}{y^{12+8}}\right)$$ Use product rule.

$$= \frac{x^8}{y^{20}}$$ Simplify exponents.

Note that this example can be simplified in many different ways but the result will always be the same.

Polynomials

A *polynomial* in x is an expression containing the sum of a finite number of terms of the form ax^n, for any real number a and any whole number n. When the exponents on the variable of a polynomial decrease from left to right, the polynomial is written in *descending order* of the variable. The parts that are added are called the *terms* of the polynomial. The *degree of a term* of a polynomial in one variable is the value of the exponent on the variable in that term. The *degree of a polynomial* in one variable is the degree of its highest-degree term. The coefficient of the highest-degree term is called the *leading coefficient*.

EXAMPLE 2 ▶ After listing the terms of the polynomial, $2x^3 + 5 - x + 3x^4$, write it in descending order, determine the degree of the polynomial, and determine the leading coefficient.

Solution   The terms of the polynomial are $2x^3$, 5, $-x$, and $3x^4$. Written in descending order we have $3x^4 + 2x^3 - x + 5$. Because the highest-degree term is $3x^4$, the degree of the polynomial is 4 and the leading coefficient is 3.

Polynomials can be added, subtracted, multiplied, and divided. When adding or subtracting polynomials, we add or subtract *like terms*. Like terms are terms that have the same variables with the same exponents.

EXAMPLE 3 ▶ Add $(4x^2 - 6x + 3)$ and $(2x^2 + 5x - 1)$.

Solution $(4x^2 - 6x + 3) + (2x^2 + 5x - 1)$

$= 4x^2 - 6x + 3 + 2x^2 + 5x - 1$ *Remove the parentheses.*

$= \underbrace{4x^2 + 2x^2} \underbrace{- 6x + 5x} \underbrace{+ 3 - 1}$ *Rearrange terms.*

$= \quad 6x^2 \qquad -x \qquad +2$ *Combine like terms.*

EXAMPLE 4 ▶ Subtract $(-x^2 - 2x + 3)$ from $(x^3 + 4x + 6)$.

Solution $(x^3 + 4x + 6) - (-x^2 - 2x + 3)$

$= (x^3 + 4x + 6) - 1(-x^2 - 2x + 3)$ *Insert 1.*

$= x^3 + 4x + 6 + x^2 + 2x - 3$ *Distributive property*

$= x^3 + x^2 + 4x + 2x + 6 - 3$ *Rearrange terms.*

$= x^3 + x^2 + 6x + 3$ *Combine like terms.*

When multiplying two polynomials, every term of the first polynomial is multiplied by every term of the second polynomial.

EXAMPLE 5 ▶ Multiply $2xy(3x^2y + 6xy^2 + 4)$.

Solution $2xy(3x^2y + 6xy^2 + 4) = (2xy)(3x^2y) + (2xy)(6xy^2) + (2xy)(4)$

$= 6x^3y^2 + 12x^2y^3 + 8xy$

When finding the product of two binomials, we use the FOIL method.

EXAMPLE 6 ▶ Multiply $(3x + 2)(x - 5)$.

Solution   F   O   I   L

$(3x + 2)(x - 5) = (3x)(x) + (3x)(-5) + (2)(x) + (2)(-5)$

$= 3x^2 - 15x + 2x - 10$

$= 3x^2 - 13x - 10$

When we multiply a binomial by a trinomial, we generally multiply vertically as illustrated in Example 7.

EXAMPLE 7 ▶ Multiply $x^2 - 3x + 2$ by $2x - 3$.

Solution

$$
\begin{array}{r}
x^2 - 3x + 2 \\
2x - 3 \\
\hline
-3x^2 + 9x - 6 \\
2x^3 - 6x^2 + 4x \\
\hline
2x^3 - 9x^2 + 13x - 6
\end{array}
$$

Multiply top expression by −3.

Multiply top expression by 2x.

Add like terms in columns.

To divide a polynomial by a monomial, we use the fact that

$$\frac{A + B}{C} = \frac{A}{C} + \frac{B}{C}$$

If the polynomial has more than two terms, we expand this procedure.

To Divide a Polynomial by a Monomial

Divide each term of the polynomial by the monomial.

EXAMPLE 8 ▶ Divide $\dfrac{4x^2 - 8x - 3}{2x}$.

Solution

$$\frac{4x^2 - 8x - 3}{2x} = \frac{4x^2}{2x} - \frac{8x}{2x} - \frac{3}{2x}$$

$$= 2x - 4 - \frac{3}{2x}$$

We divide a polynomial by a binomial in much the same way as we perform long division.

EXAMPLE 9 ▶ Divide $\dfrac{6x^2 - 5x + 5}{2x + 3}$.

Solution   In this example we will not show the change of sign in the subtractions.

$$
\begin{array}{r}
3x - 7 \\
2x + 3 \overline{)6x^2 - 5x + 5} \\
\underline{6x^2 + 9x} \qquad\qquad \longleftarrow \quad 3x(2x + 3) \\
-14x + 5 \\
\underline{-14x - 21} \quad \longleftarrow \quad -7(2x + 3) \\
26 \quad \longleftarrow \quad \text{remainder}
\end{array}
$$

Thus, $\dfrac{6x^2 - 5x + 5}{2x + 3} = 3x - 7 + \dfrac{26}{2x + 3}$.

When dividing a polynomial by a binomial, the answer may be *checked* by multiplying the divisor by the quotient and then adding the remainder. You should obtain the polynomial you began with. To check Example 9, we do the following:

$$(2x + 3)(3x - 7) + 26 = 6x^2 - 5x - 21 + 26$$

$$= 6x^2 - 5x + 5$$

Since we got the polynomial we began with, our division is correct.

   When you are dividing a polynomial by a binomial, you should list both the polynomial and binomial in descending order. If a term of any degree is missing, it is often helpful to include that term with a numerical coefficient of 0.

Factoring

Factoring is the opposite of multiplying. To factor an expression means to write it as the product of other expressions. In Chapter 6, we discussed factoring a monomial from a polynomial, factoring by grouping, factoring second degree trinomials with a leading coefficient of 1, factoring second degree trinomials with a leading coefficient not equal to 1, and factoring using special factoring formulas.

Example 10 illustrates factoring a monomial from a polynomial.

EXAMPLE 10 ▶ Factor $15x^4 - 5x^3 + 20x^2$.

Solution   The GCF is $5x^2$. Write each term as the product of the GCF and another product. Then factor out the GCF.

$$15x^4 - 5x^3 + 20x^2 = 5x^2 \cdot 3x^2 - 5x^2 \cdot x + 5x^2 \cdot 4$$
$$= 5x^2(3x^2 - x + 4)$$

Example 11 illustrates factoring by grouping.

EXAMPLE 11 ▶ Factor $x^3 - 5x^2 + 2x - 10$ by grouping.

Solution   There are no factors common to all four terms. However, x^2 is common to the first two terms and 2 is common to the last two terms.

$$x^3 - 5x^2 + 2x - 10 = x^2(x - 5) + 2(x - 5)$$
$$= (x - 5)(x^2 + 2)$$

Example 12 illustrates factoring a trinomial of the form $x^2 + bx + c$. The key to factoring these kinds of expressions is to find two numbers whose product is c and whose sum is b.

EXAMPLE 12 ▶ Factor $x^2 - x - 12$.

Solution   We must find two numbers whose product is -12 and whose sum is -1. We begin by listing the factors of -12, trying to find a pair whose sum is -1.

Factors of –12	Sum of Factors
(1)(−12)	$1 + (-12) = -11$
(2)(−6)	$2 + (-6) = -4$
(3)(−4)	$3 + (-4) = -1$

We can stop here because we found a pair whose product is -12 and whose sum is -1. Now we factor the trinomial using the 3 and -4.

$$x^2 - x - 12 = (x + 3)(x - 4)$$

Example 13 illustrates factoring a trinomial of the form $ax^2 + bx + c, a \neq 1$. This trinomial can be factored trying the various possibilities until you find the one that works.

EXAMPLE 13 ▶ Factor $3x^2 - 13x - 10$.

Solution   First we determine that the three terms have no common factor. Next we determine the possible factors of the leading coefficient, 3. The factors of 3 are 3 and 1. Therefore we write

$$3x^2 - 13x + 10 = (3x \quad)(x \quad)$$

Then we determine the possible factors of 10. Since the product of the last terms must be positive ($+10$) and the sum of the products of the outer and inner terms must be negative (-13), the two factors of 10 must both be negative. The negative factors of 10 are $(-1)(-10)$ and $(-2)(-5)$. Below is a list of the possible factors. We look for the factors that give us the correct middle term, $-13x$.

Possible Factors	Sum of Products of Outer and Inner Terms
$(3x - 1)(x - 10)$	$-31x$
$(3x - 10)(x - 1)$	$-13x$
$(3x - 2)(x - 5)$	$-17x$
$(3x - 5)(x - 2)$	$-11x$

Thus, $3x^2 - 13x + 10 = (3x - 10)(x - 1)$.

When factoring polynomials, it is also helpful to be able to recognize the special sums that have easy to remember factored forms.

Perfect Square Trinomials

$$a^2 + 2ab + b^2 = (a + b)^2$$
$$a^2 - 2ab + b^2 = (a - b)^2$$

Difference of Two Squares

$$a^2 - b^2 = (a + b)(a - b)$$

Sum of Two Cubes

$$a^3 + b^3 = (a + b)(a^2 - ab + b^2)$$

Difference of Two Cubes

$$a^3 - b^3 = (a - b)(a^2 + ab + b^2)$$

Part of the problem solving that must be done when factoring a polynomial is determining the strategy to use. Here is a general procedure.

To Factor a Polynomial

1. Determine whether all the terms in the polynomial have a greatest common factor other than 1. If so, factor out the GCF.
2. If the polynomial has two terms, determine whether it is a difference of two squares or a sum or difference of two cubes. If so, factor using the appropriate formula.
3. If the polynomial has three terms, determine whether it is a perfect square trinomial. If so, factor accordingly. If it is not, factor the trinomial using trial and error, grouping, or substitution as explained in Section 6.5.
4. If the polynomial has more than three terms, try factoring by grouping. If that does not work, see if three of the terms are the square of a binomial.
5. As a final step, examine your factored polynomial to see if any factors listed have a common factor and can be factored further. If you find a common factor, factor it out at this point.

The following examples illustrate how to use the procedure.

EXAMPLE 14 ▸ Factor $3x^2y^2 - 24xy^2 + 48y^2$.

Solution   Begin by factoring the GCF, $3y^2$, from each term.

$$3x^2y^2 - 24xy^2 + 48y^2 = 3y^2(x^2 - 8x + 16) = 3y^2(x - 4)^2$$

Note that $x^2 - 8x + 16$ is a perfect square trinomial. If you did not recognize this, you would still obtain the correct answer by factoring the trinomial into $(x - 4)(x - 4)$.

EXAMPLE 15 ▸ Factor $24x^2 - 6xy + 16xy - 4y^2$.

Solution   As always, begin by determining if all the terms in the polynomial have a common factor. In this example, 2 is common to all terms. Factor out the 2; then factor the remaining four-term polynomial by grouping.

$$\begin{aligned}
24x^2 - 6xy + 16xy - 4y^2 &= 2(12x^2 - 3xy + 8xy - 2y^2) \\
&= 2[3x(4x - y) + 2y(4x - y)] \\
&= 2(4x - y)(3x + 2y)
\end{aligned}$$

Answers

Chapter 1

Exercise Set 1.1   **1.** Answers will vary.   **3.** Answers will vary.   **5.** Answers will vary.   **7.** Answers will vary.   **9.** Answers will vary.   **11.** Do all the homework carefully and completely and preview the new material that is to be covered in class.   **13.** At least 2 hours of study and homework time for each hour of class time is generally recommended.   **15. a)** You need to do the homework in order to practice what was presented in class.   **b)** When you miss class, you miss important information; therefore, it is important that you attend class regularly.   **17.** Answers will vary.

Exercise Set 1.2   **1.** Understand, translate, calculate, check, state answer   **3.** is approximately equal to   **5.** Rank the data. The median is the value in the middle.   **7.** Divide the sum of the data by the number of pieces of data.   **9.** He actually missed a B by 10 points.   **11. a)** 76.4   **b)** 74   **13. a)** \$87.32   **b)** \$86.57   **15. a)** 11.593   **b)** 11.68   **17.** \$470   **19. a)** \$1336   **b)** \$18,036   **21.** 1,610,000,000 operations   **23. a)** 19.375 minutes   **b)** 88 minutes   **c)** 10 minutes   **25.** $\approx$18.49 miles per gallon   **27.** \$153   **29.** On the 3 on the right   **31. a)** 4106.25 gallons   **b)** $\approx$\$21.35   **33. a)** \$193   **b)** \$172   **35. a)** Hong Kong–China; 550   **b)** Mexico; 385   **c)** 165   **37. a)** 0.275 million or 275,000; 1.05 million or 1,050,000   **b)** 0.775 million or 775,000   **c)** $\approx$3.82 times greater   **39. a)** 1.394 million   **b)** 0.255 million   **c)** 0.051 million   **41. a)** 48   **b)** He cannot get a C.   **43.** Bachelor's degree   **45.** One example is 50, 60, 70, 80, 90.

Exercise Set 1.3   **1. a)** Variables are letters that represent numbers.   **b)** $x, y,$ and z **3. a)** Numerator   **b)** Denominator   **5.** Divide out the common factors.   **7. a)** The smallest number that is divisible by the two denominators   **b)** Answers will vary.   **9.** Part **b)** shows simplification.   **11.** Part **a)** is incorrect.   **13. c)**   **15.** Divide out common factors, then multiply the numerators, and multiply the denominators.   **17.** Write fractions with a common denominator, then add or subtract the numerators while keeping the common denominator.   **19.** Yes, the numerator and denominator have no common factors (other than 1).   **21.** Yes   **23.** $\dfrac{2}{3}$

25. $\dfrac{1}{4}$ **27.** $\dfrac{9}{19}$ **29.** $\dfrac{3}{7}$ **31.** Simplified   **33.** Simplified   **35.** $\dfrac{43}{15}$ **37.** $\dfrac{23}{3}$ **39.** $\dfrac{59}{18}$ **41.** $\dfrac{159}{17}$ **43.** $1\dfrac{3}{4}$ **45.** $3\dfrac{1}{4}$ **47.** $4\dfrac{4}{7}$ **49.** $6\dfrac{1}{7}$

51. $\dfrac{8}{15}$ **53.** $\dfrac{1}{9}$ **55.** $\dfrac{3}{2}$ or $1\dfrac{1}{2}$ **57.** $\dfrac{1}{2}$ **59.** 6   **61.** $\dfrac{5}{2}$ or $2\dfrac{1}{2}$ **63.** $\dfrac{43}{10}$ or $4\dfrac{3}{10}$ **65.** $\dfrac{8}{13}$ **67.** $\dfrac{5}{8}$ **69.** $\dfrac{1}{7}$ **71.** $\dfrac{6}{5}$ or $1\dfrac{1}{5}$ **73.** $\dfrac{9}{17}$

75. $\dfrac{7}{12}$ **77.** $\dfrac{13}{36}$ **79.** $\dfrac{65}{24}$ or $2\dfrac{17}{24}$ **81.** $\dfrac{17}{6}$ or $2\dfrac{5}{6}$ **83.** $\dfrac{29}{10}$ or $2\dfrac{9}{10}$ **85.** $\dfrac{277}{30}$ or $9\dfrac{7}{30}$ **87.** $\dfrac{11}{24}$ mile   **89.** $8\dfrac{15}{16}$ inches   **91.** $\dfrac{19}{50}$ **93.** $\dfrac{3}{40}$

95. 11 feet, $11\dfrac{1}{4}$ inches or $143\dfrac{1}{4}$ inches or $\approx$11.94 feet   **97.** $2\dfrac{1}{10}$ minutes   **99.** $1\dfrac{9}{16}$ inches   **101.** 5 mg   **103.** 40 times   **105.** $1\dfrac{1}{2}$ inches

107. 6 strips   **109. a)** Yes   **b)** $\dfrac{5}{8}$ inches   **c)** $52\dfrac{3}{4}$ inches   **111. a)** $\dfrac{*\,+\,?}{a}$ **b)** $\dfrac{\odot\,-\,\square}{?}$ **c)** $\dfrac{\triangle\,+\,4}{\square}$ **d)** $\dfrac{x\,-\,2}{3}$ **e)** $\dfrac{8}{x}$ **113.** 270 pills

115. Answers will vary.   **116.** 16   **117.** 15   **118.** Variables are letters used to represent numbers.

Exercise Set 1.4   **1.** A set is a collection of elements.   **3.** Answers will vary.   **5.** The set of natural numbers does not contain the number 0. The set of whole numbers contains the number 0.   **7. a)–c)** All natural numbers are whole numbers, rational numbers, and real numbers.   **9. a)** Yes   **b)** No   **c)** No   **d)** Yes   **11.** $\{\ldots, -3, -2, -1, 0, 1, 2, 3, \ldots\}$ **13.** $\{0, 1, 2, 3, \ldots\}$ **15.** $\{\ldots, -3, -2, -1\}$ **17.** True   **19.** True   **21.** False   **23.** False   **25.** True   **27.** True   **29.** False   **31.** True   **33.** True   **35.** False   **37.** True   **39.** True   **41.** False   **43.** True   **45.** False   **47.** True   **49. a)** 13   **b)** $-2, 13$ **c)** $-2, 13$ **d)** 13   **51. a)** 3, 77

b) $0, 3, 77$ **c)** $0, -2, 3, 77$ **d)** $-\dfrac{5}{7}, 0, -2, 3, 6\dfrac{1}{4}, 1.63, 77$ **e)** $\sqrt{7}, -\sqrt{3}$ **f)** $-\dfrac{5}{7}, 0, -2, 3, 6\dfrac{1}{4}, \sqrt{7}, -\sqrt{3}, 1.63, 77$

53. Answers will vary; three examples are 0, 1, and 2.   **55.** Answers will vary; three examples are $-\sqrt{2}, -\sqrt{3},$ and $-\sqrt{7}$.

57. Answers will vary; three examples are $-\dfrac{2}{3}, \dfrac{1}{2},$ and 6.3.   **59.** Answers will vary; three examples are $-13, -5,$ and -1.

61. Answers will vary; three examples are $\sqrt{2}, \sqrt{3},$ and $-\sqrt{5}$.   **63.** Answers will vary; three examples are $-7, 1,$ and 5.   **65.** 87   **67. a)** $\{1, 3, 4, 5, 8\}$ **b)** $\{2, 5, 6, 7, 8\}$ **c)** $\{5, 8\}$ **d)** $\{1, 2, 3, 4, 5, 6, 7, 8\}$ **69. a)** Set B continues beyond 4.   **b)** 4

c) An infinite number of elements   **d)** An infinite set   **71. a)** An infinite number   **b)** An infinite number   **73.** $\dfrac{27}{5}$ **74.** $5\dfrac{1}{3}$

75. $\dfrac{5}{24}$ **76.** $\dfrac{4}{45}$

Exercise Set 1.5   **1. a)**   **b)**   **c)** Greater than   **d)** $-4 < -2$ **e)** $-2 > -4$
3. a) 4 is 4 units from 0 on the number line.   **b)** -4 is 4 units from 0 on the number line.   **c)** 0 is 0 units from 0 on the number line.   **5.** Yes   **7.** No, $-4 < -3$ but $|-4| > |-3|$.   **9.** No, $|-3| < |-4|$ but $-3 > -4$.   **11.** 7   **13.** 15   **15.** 0   **17.** -5 **19.** -21 **21.** $>$ **23.** $<$ **25.** $>$ **27.** $<$ **29.** $>$ **31.** $<$ **33.** $<$ **35.** $>$ **37.** $<$ **39.** $>$ **41.** $>$ **43.** $<$ **45.** $>$ **47.** $<$ **49.** $<$

51. < **53.** > **55.** > **57.** > **59.** < **61.** < **63.** < **65.** < **67.** = **69.** = **71.** < **73.** < **75.** $-|-1|, \frac{3}{7}, \frac{4}{9}, 0.46, |-5|$

77. $\frac{5}{12}, 0.6, \frac{2}{3}, \frac{19}{25}, |-2.6|$ **79.** 4, −4 **81.** Not possible **83.** Answers will vary: one example is −3, −4, and −5. **85.** Answers will

vary: one example is 4, 5, and 6. **87.** Answers will vary: one example is 3, 4, and 5. **89. a)** Does not include the endpoints

b) Answers will vary: one example is 4.1, 5, and $5\frac{1}{2}$. **c)** No **d)** Yes **e)** True **91. a)** Dietary fiber and thiamin

b) Vitamin E, niacin, and riboflavin **93.** Greater than **95.** No **98.** $\frac{89}{15}$ or $5\frac{14}{15}$ **99.** $\{\ldots, -3, -2, -1, 0, 1, 2, 3, \ldots\}$

100. $\{0, 1, 2, 3, \ldots\}$ **101. a)** 5 **b)** 5, 0 **c)** 5, −2, 0 **d)** $5, -2, 0, \frac{1}{3}, -\frac{5}{9}, 2.3$ **e)** $\sqrt{3}$ **f)** $5, -2, 0, \frac{1}{3}, \sqrt{3}, -\frac{5}{9}, 2.3$

Mid-Chapter Test* **1.** At least two hours of study and homework for each hour of class time is generally recommended. [1.1]

2. a) $80.63 **b)** $83.81 [1.2] **3.** $824.59 [1.2] **4. a)** Natwora's **b)** $24 [1.2] **5.** $62.35 [1.2] **6.** $\frac{1}{6}$ [1.3] **7.** $\frac{39}{64}$ [1.3] **8.** $\frac{49}{40}$ or

$1\frac{9}{40}$ [1.3] **9.** $\frac{61}{20}$ or $3\frac{1}{20}$ [1.3] **10.** $54\frac{1}{3}$ feet [1.3] **11.** False [1.4] **12.** True [1.4] **13.** False [1.4]

14. True [1.4] **15.** False [1.4] **16.** $-\frac{7}{10}$ [1.5] **17.** > [1.5] **18.** > [1.5] **19.** < [1.5] **20.** = [1.5]

Exercise Set 1.6 **1.** Addition, subtraction, multiplication, and division **3. a)** No **b)** $\frac{2}{3}$ **5.** Negative **7.** Answers will vary.

9. a) Answers will vary. **b)** −77 **c)** Answers will vary. **11.** Correct **13.** −9 **15.** 28 **17.** 0 **19.** $-\frac{5}{3}$ **21.** $-2\frac{3}{5}$ **23.** −3.72

25. 11 **27.** 1 **29.** −6 **31.** 0 **33.** 0 **35.** −10 **37.** −4 **39.** −13 **41.** 0 **43.** −6 **45.** 9 **47.** −64 **49.** 16 **51.** −12 **53.** 3

55. −6 **57.** −20 **59.** −31 **61.** −39 **63.** 91 **65.** −9.9 **67.** −144.0 **69.** −141.91 **71.** −53.65 **73.** $\frac{26}{35}$ **75.** $\frac{107}{84}$ or $1\frac{23}{84}$ **77.** $\frac{4}{55}$

79. $-\frac{26}{45}$ **81.** $-\frac{16}{15}$ or $-1\frac{1}{15}$ **83.** $\frac{3}{10}$ **85.** $-\frac{13}{15}$ **87.** $\frac{19}{56}$ **89.** $-\frac{43}{60}$ **91.** $-\frac{23}{21}$ or $-1\frac{2}{21}$ **93. a)** Positive **b)** 390 **c)** Yes

95. a) Negative **b)** −373 **c)** Yes **97. a)** Negative **b)** −452 **c)** Yes **99. a)** Negative **b)** −1300 **c)** Yes **101. a)** Negative

b) −112 **c)** Yes **103. a)** Negative **b)** −3880 **c)** Yes **105. a)** Positive **b)** 1111 **c)** Yes **107. a)** Negative **b)** −2050 **c)** Yes

109. True **111.** True **113.** False **115.** $277 **117.** 21 yards **119.** 61 feet **121.** 13,796 feet **123. a)** −$12 million

b) 1999-2001: surplus, $5 million **125.** −22 **127.** 20 **129.** 0 **131.** $\frac{11}{30}$ **133.** 55 **135.** $\frac{19}{14}$ or $1\frac{5}{14}$ **136.** $\frac{43}{16}$ or $2\frac{11}{16}$ **137.** False

138. > **139.** <

Exercise Set 1.7 **1.** 2 − 7 **3.** ☺ − ? **5. a)** Answers will vary. **b)** 5 + (−14) **c)** −9 **7. a)** $a + b$ **b)** −4 + 12 **c)** 8

9. a) 3 + 6 − 5 **b)** 4 **11.** Correct **13.** 6 **15.** 7 **17.** −1 **19.** 12 **21.** −12 **23.** −7 **25.** 0 **27.** −4 **29.** −3 **31.** 9 **33.** −20

35. 9.8 **37.** 0.3 **39.** 7 **41.** 4 **43.** 22 **45.** −11 **47.** −131.0 **49.** −82 **51.** 140 **53.** −7.4 **55.** −3.93 **57.** −11 **59.** 0

61. 6.1 **63.** 18.2 **65.** −11 **67.** −18.1 **69.** $\frac{13}{72}$ **71.** $\frac{17}{45}$ **73.** $-\frac{67}{60}$ or $-1\frac{7}{60}$ **75.** $-\frac{5}{12}$ **77.** $-\frac{17}{24}$ **79.** $\frac{1}{4}$ **81.** $\frac{7}{45}$ **83.** $\frac{13}{16}$ **85.** $-\frac{13}{63}$

87. $-\frac{7}{60}$ **89. a)** Positive **b)** 99 **c)** Yes **91. a)** Negative **b)** −619 **c)** Yes **93. a)** Positive **b)** 1588 **c)** Yes

95. a) Positive **b)** 196 **c)** Yes **97. a)** Negative **b)** −448 **c)** Yes **99. a)** Positive **b)** 116.1 **c)** Yes **101. a)** Negative

b) −69 **c)** Yes **103. a)** Negative **b)** −1670 **c)** Yes **105. a)** Zero **b)** 0 **c)** Yes **107.** 4 **109.** −6 **111.** −15 **113.** −2

115. 13 **117.** −5 **119.** −32 **121.** −4 **123.** 9 **125.** −12 **127.** 12 **129.** −18 **131. a)** 43 **b)** 143 **133.** $1\frac{7}{8}$ inches

135. Dropped 100° F **137. a)** 281 **b)** 3 strokes less **139.** −5 **141. a)** 8 units **b)** −3 − (−11) **143. a)** 9 feet **b)** −3 feet

144. $\{1, 2, 3, \ldots\}$ **145.** The set of rational numbers together with the set of irrational numbers form the set of real numbers.

146. > **147.** < **148.** $-\frac{1}{24}$

Exercise Set 1.8 **1.** Like signs: product is positive. Unlike signs: product is negative. **3.** Even number of negatives: product is

positive, odd number of negatives: product is negative **5. a)** 0 **b)** Undefined **7. a)** With 3 − 5 you subtract, but with 3(−5) you

multiply. **b)** −2, −15 **9. a)** With $x - y$ you subtract, but with $x(-y)$ you multiply. **b)** 7 **c)** 10 **d)** −3 **11.** Negative

13. Positive **15.** Negative **17.** 20 **19.** −18 **21.** 80 **23.** −12.6 **25.** 42 **27.** −81 **29.** 30 **31.** 0 **33.** −84 **35.** −72 **37.** 140

39. 0 **41.** $-\frac{3}{10}$ **43.** $\frac{7}{27}$ **45.** 4 **47.** $-\frac{1}{10}$ **49.** 7 **51.** 4 **53.** 4 **55.** −18 **57.** 9.9 **59.** −10 **61.** −6 **63.** −8 **65.** 16.2

67. 0 **69.** −5.5 **71.** 6 **73.** $-\frac{2}{5}$ **75.** $\frac{5}{36}$ **77.** 1 **79.** $-\frac{144}{5}$ or $-28\frac{4}{5}$ **81.** −32 **83.** 20 **85.** −14 **87.** −9.3 **89.** −20 **91.** 1 **93.** 0

95. Undefined **97.** 0 **99.** Undefined **101. a)** Negative **b)** −3496 **c)** Yes **103. a)** Negative **b)** −16 **c)** Yes

---

* Numbers in blue brackets after the answer indicate the section where the material was discussed.

105. a) Negative **b)** -9 **c)** Yes **107. a)** Positive **b)** 6174 **c)** Yes **109. a)** Zero **b)** 0 **c)** Yes **111. a)** Undefined **b)** Undefined **c)** Yes **113. a)** Positive **b)** 3.2 **c)** Yes **115. a)** Positive **b)** 226.8 **c)** Yes **117.** False **119.** False **121.** True **123.** True **125.** False **127.** True **129.** 45 yard loss or -45 yards **131. a)** \$104 **b)** $-\$416$ **133.** \$143.85 **135. a)** 20 point loss or -20 points **b)** 80 **137. a)** 102 to 128 beats per minute **b)** Answers will vary. **139.** -125 **141.** 1 **143.** Positive **146.** $>$ **147.** $-\dfrac{41}{60}$ **148.** -2 **149.** -3 **150.** 3

Exercise Set 1.9
1. Base; exponent **3. a)** 1 **b)** 1, 3, 2, 1 **5. a)** $4y$ **b)** y^4 **7.** Parentheses, exponents, multiplication or division, then addition or subtraction **9.** No **11. a)** 13 **b)** 1 **c)** b) **13. a)** Answers will vary. **b)** -180 **15. a)** Answers will vary. **b)** -91 **17.** 25 **19.** 1 **21.** -64 **23.** 9 **25.** -1 **27.** -100 **29.** 81 **31.** 27 **33.** 256 **35.** -16 **37.** $\dfrac{9}{16}$ **39.** $-\dfrac{1}{32}$ **41.** 225 **43.** 576 **45. a)** Positive **b)** 343 **c)** Yes **47. a)** Positive **b)** 1296 **c)** Yes **49. a)** Negative **b)** -243 **c)** Yes **51. a)** Positive **b)** 625 **c)** Yes **53. a)** Negative **b)** -81 **c)** Yes **55. a)** Negative **b)** -0.140625 **c)** Yes **57.** 15 **59.** 8 **61.** 57 **63.** 0 **65.** 16 **67.** 29 **69.** -19 **71.** -77 **73.** $\dfrac{83}{100}$ **75.** 2 **77.** 10 **79.** -34 **81.** 103 **83.** 169 **85.** 13 **87.** 36.75 **89.** $\dfrac{5}{8}$ **91.** $\dfrac{1}{4}$ **93.** $\dfrac{49}{30}$ or $1\dfrac{19}{30}$ **95.** $\dfrac{5}{27}$ **97.** $\dfrac{32}{53}$ **99.** 9 **101.** -4 **103.** 1 **105. a)** 25 **b)** -25 **c)** 25 **107. a)** 4 **b)** -4 **c)** 4 **109. a)** 36 **b)** -36 **c)** 36 **111. a)** $\dfrac{1}{9}$ **b)** $-\dfrac{1}{9}$ **c)** $\dfrac{1}{9}$ **113.** 4 **115.** -45 **117.** 3 **119.** -4 **121.** -1 **123.** $\dfrac{15}{4}$ or $3\dfrac{3}{4}$ **125.** 994 **127.** -5 **129.** 193 **131.** -25 **133.** $[(6\cdot3)-4]-2; 12$ **135.** $\{[(10\cdot4)+9]-6\}\div7; \dfrac{43}{7}$ or $6\dfrac{1}{7}$ **137.** $\left(\dfrac{4}{5}+\dfrac{3}{7}\right)\cdot\dfrac{2}{3}; \dfrac{86}{105}$ **139.** All real numbers **141.** 162.5 miles **143.** 102 feet **145. a)** .08 **b)** .16 **147.** 1.71 inches **149.** $12-(4-6)+10=24$

155. a) 3 **b)**

Dogs	Number of Houses
0	4
1	5
2	3
3	1
4	1

c) 18 **d)** ≈1.29 dogs per house **156.** \$6.40 **157.** $-\dfrac{5}{36}$ **158.** $\dfrac{10}{3}$ or $3\dfrac{1}{3}$

Exercise Set 1.10
1. The commutative property of addition states that the sum of two numbers is the same regardless of the order in which they are added; $3+4=4+3$. **3.** The associative property of addition states that the sum of three numbers is the same regardless of the way the numbers are grouped; $(2+3)+4=2+(3+4)$ **5. a)** Answers will vary. **b)** 15 **c)** 44 **7.** The associative property involves changing parentheses, and uses only one operation whereas the distributive property uses two operations, multiplication and addition. **9.** 0 **11. a)** -6 **b)** $\dfrac{1}{6}$ **13. a)** 3 **b)** $-\dfrac{1}{3}$ **15. a)** $-x$ **b)** $\dfrac{1}{x}$ **17. a)** -1.6 **b)** $\dfrac{1}{1.6}$ or 0.625 **19. a)** $-\dfrac{1}{5}$ **b)** 5 **21. a)** $\dfrac{5}{6}$ **b)** $-\dfrac{6}{5}$ **23.** Distributive property **25.** Associative property of addition **27.** Commutative property of multiplication **29.** Associative property of multiplication **31.** Distributive property **33.** Identity property of multiplication **35.** Inverse property of multiplication **37.** $1+(-4)$ **39.** $(-6\cdot4)\cdot2$ **41.** $-2\cdot x+-2\cdot y$ or $-2x-2y$ **43.** $y\cdot x$ **45.** $3y+4x$ **47.** $a+(b+3)$ **49.** $3x+(4+6)$ **51.** $(m+n)3$ **53.** $4x+4y+12$ **55.** 0 **57.** $\dfrac{5}{2}n$ **59.** Yes **61.** Yes **63.** No **65.** Yes **67.** No **69.** No **71.** The $(3+4)$ is treated as one value. **73.** Commutative property of addition **75.** No; Associative property of addition **77.** $\dfrac{49}{15}$ or $3\dfrac{4}{15}$ **78.** $\dfrac{23}{16}$ or $1\dfrac{7}{16}$ **79.** -11.2 **80.** $-\dfrac{7}{8}$

Chapter 1 Review Exercises
1. 28 **2.** \$551.25 **3. a)** \$74.25 **b)** \$974.24 **4.** \$30 **5. a)** 78.4 **b)** 79 **6. a)** 11 **b)** 9.5 **7. a)** 30 minutes **b)** 27 minutes **8. a)** 1.0192 million **b)** 0.5187 million **9.** $\dfrac{1}{2}$ **10.** $\dfrac{127}{21}$ or $6\dfrac{1}{21}$ **11.** $\dfrac{25}{36}$ **12.** $\dfrac{7}{6}$ or $1\dfrac{1}{6}$ **13.** $\dfrac{23}{12}$ or $1\dfrac{11}{12}$ **14.** $\dfrac{177}{10}$ or $17\dfrac{7}{10}$ **15.** $\{1,2,3,\ldots\}$ **16.** $\{0,1,2,3,\ldots\}$ **17.** $\{\ldots,-3,-2,-1,0,1,2,3,\ldots\}$ **18.** The set of all numbers which can be expressed as the quotient of two integers, denominator not zero **19. a)** 3, 426 **b)** 3, 0, 426 **c)** 3, -5, -12, 0, 426 **d)** 3, -5, -12, 0, $\dfrac{1}{2}$, -0.62, 426, $-3\dfrac{1}{4}$ **e)** $\sqrt{7}$ **f)** 3, -5, -12, 0, $\dfrac{1}{2}$, -0.62, $\sqrt{7}$, 426, $-3\dfrac{1}{4}$ **20. a)** 1 **b)** 1 **c)** -8, -9 **d)** -8, -9, 1 **e)** -2.3, -8, -9, $1\dfrac{1}{2}$, 1, $-\dfrac{3}{17}$ **f)** -2.3, -8, -9, $1\dfrac{1}{2}$, $\sqrt{2}$, $-\sqrt{2}$, 1, $-\dfrac{3}{17}$ **21.** $<$ **22.** $>$ **23.** $<$ **24.** $>$ **25.** $<$ **26.** $>$ **27.** $=$ **28.** $>$ **29.** -14 **30.** 0 **31.** -3 **32.** -6 **33.** -6 **34.** 2 **35.** 8 **36.** 0 **37.** -5 **38.** 14 **39.** 4 **40.** -12 **41.** $\dfrac{7}{12}$ **42.** $\dfrac{11}{10}$ or $1\dfrac{1}{10}$ **43.** $-\dfrac{7}{36}$ **44.** $-\dfrac{19}{56}$ **45.** $-\dfrac{5}{4}$ or $-1\dfrac{1}{4}$ **46.** $-\dfrac{37}{84}$ **47.** $-\dfrac{7}{90}$ **48.** $\dfrac{61}{60}$ or $1\dfrac{1}{60}$ **49.** 8 **50.** -13 **51.** -12 **52.** -7 **53.** 6 **54.** 11 **55.** -63 **56.** 25.42 **57.** -120 **58.** $-\dfrac{6}{35}$ **59.** $-\dfrac{6}{11}$ **60.** $\dfrac{15}{56}$ **61.** 0 **62.** 144 **63.** -5 **64.** -6 **65.** -3.2 **66.** 4.3 **67.** 8 **68.** 9 **69.** $\dfrac{56}{27}$ or $2\dfrac{2}{27}$

70. $-\dfrac{35}{9}$ or $-3\dfrac{8}{9}$ **71.** 0   **72.** 0   **73.** Undefined   **74.** Undefined   **75.** Undefined   **76.** 0   **77.** 25   **78.** -8 **79.** 1   **80.** 3   **81.** -6

82. -32 **83.** 6   **84.** -4 **85.** 10   **86.** 1   **87.** 15   **88.** -4 **89.** -36 **90.** 36   **91.** 16   **92.** -27 **93.** -1 **94.** -32 **95.** $\dfrac{16}{25}$

96. $\dfrac{8}{125}$ **97.** 500   **98.** 4   **99.** 12   **100.** -256 **101.** 7   **102.** 32   **103.** 6.36   **104.** -17 **105.** -39 **106.** -2.3 **107.** 0   **108.** $\dfrac{9}{7}$ or $1\dfrac{2}{7}$

109. -60 **110.** 10   **111.** 20   **112.** 20   **113.** 14   **114.** 9   **115.** -4 **116.** 50   **117.** 5   **118.** 26   **119.** 45   **120.** 0   **121.** -11
122. -3 **123.** -3 **124.** 39   **125.** -215 **126.** 353.6   **127.** -2.88 **128.** 117.8   **129.** 65,536   **130.** -74.088 **131.** Associative property of addition   **132.** Distributive property   **133.** Commutative property of addition   **134.** Commutative property of multiplication   **135.** Distributive property   **136.** Associative property of addition   **137.** Identity property of multiplication   **138.** Inverse property of addition

Chapter 1 Practice Test   **1. a)** \$10.65   **b)** \$0.23   **c)** \$10.88   **d)** \$39.12 [1.2]   **2.** $\approx$2.5 times greater [1.2]
3. a) $\approx$ 13 thousand   **b)** During this specific time, half the time KFUN had more than 8.8 thousand listeners and half the time KFUN had less than 8.8 thousand listeners. [1.2]   **4. a)** 42   **b)** 42, 0   **c)** $-6, 42, 0, -7, -1$ **d)** $-6, 42, -3\dfrac{1}{2}, 0, 6.52, \dfrac{5}{9}, -7, -1$ **e)** $\sqrt{5}$

f) $-6, 42, -3\dfrac{1}{2}, 0, 6.52, \sqrt{5}, \dfrac{5}{9}, -7, -1$ [1.4]   **5.** $<$ [1.5]   **6.** $>$ [1.5]   **7.** -15 [1.6]   **8.** -11 [1.7]   **9.** -14 [1.7]   **10.** 8 [1.9]

11. -24 [1.8]   **12.** $\dfrac{16}{63}$ [1.8]   **13.** -3 [1.9]   **14.** $-\dfrac{53}{56}$ [1.7]   **15.** 12 [1.9]   **16.** $-\dfrac{32}{243}$ [1.9]   **17.** 100 [1.9]

18. $-x^2$ means $-(x^2)$ and x^2 will always be positive for any nonzero value of x. Therefore, $-x^2$ will always be negative. [1.9]   **19.** 37 [1.9]
20. 11 [1.9]   **21.** 10 [1.9]   **22.** 1 [1.9]   **23.** Commutative property of addition [1.10]   **24.** Distributive property [1.10]
25. Associative property of addition [1.10]

Chapter 2

Exercise Set 2.1   **1. a)** Terms are the parts that are added.   **b)** $3x, -4y,$ and -5 **c)** $6xy, 3x, -y,$ and -9 **3. a)** A variable
b) A constant   **c)** Coefficient   **5. a)** Yes   **b)** No, they don't have the same variable with the same exponent.   **c)** Yes   **d)** Yes
7. a) The signs of all the terms inside the parentheses change when the parentheses are removed.   **b)** $-x + 8$

9. $9x$ **11.** $3x + 6$ **13.** $5y + 3$ **15.** $\dfrac{9}{44}a$ **17.** $-6x + 7$ **19.** $-5w + 5$ **21.** $-2x$ **23.** 0   **25.** $-2x + 11$ **27.** $-2r - 8$

29. $10x^2 - 10y^2 - 7$ **31.** $2x - 3$ **33.** $b + \dfrac{23}{5}$ **35.** $0.8n + 6.42$ **37.** $\dfrac{1}{2}a + 3b + 1$ **39.** $14.6x + 8.3$ **41.** $x^2 + y$

43. $-3x - 5y$ **45.** $-3n^2 - 2n + 13$ **47.** $21.72x - 7.11$ **49.** $-\dfrac{23}{20}x - 5$ **51.** $5w^3 + 2w^2 + w + 3$ **53.** $-7z^3 - z^2 + 2z$

55. $6x^2 - 6xy + 3y^2$ **57.** $4a^2 + 3ab + b^2$ **59.** $5x + 10$ **61.** $5x + 20$ **63.** $3x - 18$ **65.** $-x + 2$ **67.** $x - 4$ **69.** $\dfrac{4}{5}s - 4$

71. $-0.9x - 1.5$ **73.** $-r + 4$ **75.** $1.4x + 0.35$ **77.** $x - y$ **79.** $-2x - 4y + 8$ **81.** $3.41x - 5.72y + 3.08$ **83.** $10x - 45y$
85. $x + 3y - 9$ **87.** $3x - 6y - 12$ **89.** $-3x + 1$ **91.** $2x + 1$ **93.** $14x + 18$ **95.** $4x - 2y + 3$ **97.** $5c$ **99.** $7x + 3$

101. $\dfrac{5}{4}x + \dfrac{1}{3}$ **103.** $\dfrac{19}{6}x - 2$ **105.** $-4s - 6$ **107.** $2x - 2$ **109.** 0   **111.** $-y - 6$ **113.** $3x - 5$ **115.** $x + 15$

117. $0.2x - 4y - 2.8$ **119.** $-6x + 7y$ **121.** $\dfrac{3}{2}x + \dfrac{7}{2}$ **123.** $2\Box + 3\ominus$ **125.** $2x + 3y + 2\triangle$ **127.** $2\triangle + 2\Box$ **129.** $1, 2, 3, 6, 9, 18$
131. $22x^2 - 25y^2 - 4x + 3$ **133.** $9x - 39$ **135.** 7   **136.** -16 **137.** -1 **138.** Answers will vary.   **139.** -12

Exercise Set 2.2   **1.** An equation is a statement that shows two algebraic expressions are equal.   **3.** Substitute the value in the equation and then determine if it results in a true statement.   **5.** Equivalent equations are two or more equations with the same solution.   **7.** Subtract 2 from both sides of the equation.   **9.** One example is $x + 2 = 1$.   **11.** They all have the same solution, 1.
13. Yes   **15.** No   **17.** Yes   **19.** Yes   **21.** No   **23.** Yes   **25.** 5   **27.** -7 **29.** -4 **31.** 43   **33.** 15   **35.** 11   **37.** -4 **39.** -5
41. -13 **43.** -30 **45.** 0   **47.** -57 **49.** -4 **51.** -20 **53.** 0   **55.** 17   **57.** -26 **59.** 28   **61.** -46.1 **63.** 46.5   **65.** -8.23
67. 5.57   **69.** No, the equation is equivalent to $1 = 2$, a false statement.   **71.** $x = \Box + \triangle$ **73.** $\Box = \odot - \triangle$
76. $\dfrac{11}{30}$ **77.** $-\dfrac{31}{24}$ **78.** $2x - 13$ **79.** $7t - 19$

Exercise Set 2.3   **1.** Answers will vary.   **3. a)** $x = -a$ **b)** $x = -5$ **c)** $x = 5$ **5.** Divide by 3 to isolate the variable.
7. Multiply both sides by 2.   **9.** 3   **11.** 21   **13.** -3 **15.** -8 **17.** 5   **19.** -3 **21.** $-\dfrac{7}{3}$ **23.** -13 **25.** 8   **27.** 39   **29.** $-\dfrac{1}{3}$ **31.** 6

33. $\dfrac{26}{43}$ **35.** 2   **37.** $\dfrac{1}{5}$ **39.** $-\dfrac{3}{40}$ **41.** -60 **43.** 240   **45.** -35 **47.** 20   **49.** -50 **51.** 0   **53.** 0   **55.** 22.5   **57.** 6   **59.** -20.2

61. 7   **63.** 9   **65.** In $5 + x = 10$, 5 is added to the variable, whereas in $5x = 10$, 5 is multiplied by the variable.   **b)** $x = 5$

c) $x = 2$ **67.** Multiply by $\frac{3}{2}$; 6   **69.** Multiply by $\frac{7}{3}$; $\frac{28}{15}$ **71. a)** $\square$ **b)** Divide both sides of the equation by $\triangle$.   **c)** $\square = \dfrac{\odot}{\triangle}$
73. -4 **74.** -30 **75.** 6   **76.** Associative property of addition   **77.** -57

Exercise Set 2.4   **1.** No, there is an x on both sides of the equation.   **3.** $x = \dfrac{1}{3}$ **5.** $x = -\dfrac{1}{2}$ **7.** $x = \dfrac{4}{9}$ **9.** Evaluate

11. a) Answers will vary.   **b)** Answers will vary.   **13. a)** Answers will vary.   **b)** $x = -2$ **15.** 5   **17.** -4 **19.** 2   **21.** $\dfrac{12}{5}$
23. 3   **25.** 3   **27.** $\dfrac{11}{3}$ **29.** $-\dfrac{19}{16}$ **31.** -10 **33.** 2   **35.** $-\dfrac{51}{5}$ **37.** 3   **39.** 6.8   **41.** 4   **43.** 12   **45.** 22   **47.** 60   **49.** -13
51. -11 **53.** -1 **55.** $\dfrac{19}{8}$ **57.** 0   **59.** -10 **61.** -1 **63.** 6   **65.** -21 **67.** $-\dfrac{19}{7}$ **69.** -4 **71.** 4   **73.** 5   **75.** 0.8   **77.** -1
79. $\dfrac{2}{7}$ **81.** -2.6 **83.** $-\dfrac{14}{5}$ **85.** 23   **87.** $-\dfrac{1}{15}$ **89.** 3   **91.** $-\dfrac{16}{21}$ **93.** 10   **95.** 5   **97.** $-\dfrac{23}{3}$ **99.** 2   **101.** $\dfrac{25}{3}$ **103.** $\dfrac{88}{135}$
105. a) You will not have to work with fractions.   **b)** $x = 3$ **107.** $\dfrac{35}{6}$ **109.** -4 **113.** False   **114.** 64   **115.** Isolate the variable on one side of the equation.   **116.** Divide both sides of the equation by -4 to isolate the variable.

Mid-Chapter Test   **1.** $-2x - 5y - 6$ [2.1]   **2.** $-\dfrac{7}{20}x - \dfrac{15}{2}$ [2.1]   **3.** $-8a + 12b - 24$ [2.1]   **4.** $3.36x - 5.44y - 8.32$ [2.1]
5. $2t - 38$ [2.1]   **6.** Yes [2.2]   **7.** No [2.2]   **8.** -4 [2.2]   **9.** -16 [2.2]   **10.** -23 [2.2] **11.** Multiply both sides by 4. [2.3]
12. $\dfrac{1}{2}$ [2.3]   **13.** 24 [2.3]   **14.** 10 [2.3]   **15.** $-\dfrac{3}{7}$ [2.3]   **16.** $\dfrac{5}{2}$ [2.4]   **17.** $-\dfrac{3}{2}$ [2.4]   **18.** $\dfrac{11}{8}$ [2.4]   **19.** -6 [2.4]   **20.** $-\dfrac{20}{9}$ [2.4]

Exercise Set 2.5   **1.** Answers will vary.   **3. a)** An identity is an equation that is true for infinitely many values of the variable.
b) All real numbers   **5.** Both sides of the equation are identical.   **7.** You will obtain a false statement.   **9. a)** Answers will vary.
b) $x = -8$ **11.** 3   **13.** 1   **15.** $\dfrac{3}{5}$ **17.** 3   **19.** 2   **21.** No Solution   **23.** 6.5   **25.** 3.2   **27.** 3   **29.** -2 **31.** No solution   **33.** $\dfrac{34}{5}$
35. $\dfrac{7}{9}$ **37.** 5   **39.** 30   **41.** $\dfrac{3}{4}$ **43.** $\dfrac{5}{2}$ **45.** 25   **47.** All real numbers   **49.** 23   **51.** 0   **53.** All real numbers   **55.** $\dfrac{21}{20}$ **57.** 14
59. $-\dfrac{35}{6}$ **61.** 5   **63.** 4   **65.** 0   **67.** 16   **69.** $-\dfrac{12}{5}$ **71.** $-\dfrac{4}{21}$ **73.** $\dfrac{10}{3}$ **75.** 30   **77.** 4   **79. a)** One example is $x + x + 1 = x + 2$.
b) It has a single solution.   **c)** For the example given in part **a)**, $x = 1$.   **81. a)** One example is $x + x + 1 = 2x + 1$.   **b)** Both sides simplify to the same expression.   **c)** All real numbers   **83. a)** One example is $x + x + 1 = 2x + 2$.   **b)** It simplifies to a false statement.   **c)** No solution   **85.** $\ast = -\dfrac{1}{4}$ **87.** All real numbers   **89.** $x = -4$ **91. a)** 4   **b)** 7   **c)** 0   **92.** ≈ 0.131687243
93. Factors are expressions that are multiplied. Terms are expressions that are added.   **94.** $7x - 10$ **95.** $\dfrac{10}{7}$ **96.** -3

Exercise Set 2.6   **1.** A formula is an equation used to express a relationship mathematically.   **3.** $i = prt$; i: simple interest, p: principal, r: rate, t: time   **5.** $d = rt$; d: distance, r: rate, t: time   **7.** No, 3.14 is an approximation for π.   **9.** When you multiply a unit by the same unit, you get a square unit.   **11.** 240   **13.** 96   **15.** 360   **17.** 26   **19.** 78.54   **21.** 82   **23.** 2   **25.** 8   **27.** 6.00   **29.** 127.03
31. ≈ 50.27 square feet   **33.** 16.5 square feet   **35.** ≈ 452.39 cubic centimeters   **37.** $C = 10°$ **39.** $F = 77°$ **41.** $P = 40$ **43.** $V = 5$
45. $w = A/l$ **47.** $t = d/r$ **49.** $t = i/(pr)$ **51.** $b = 2A/h$ **53.** $w = (P - 2l)/2$ **55.** $r = (-n + 3)/2$ **57.** $b = y - mx$
59. $b = d - a - c$ **61.** $y = (-ax - c)/b$ **63.** $h = 3V/(\pi r^2)$ **65.** $m = 2A - d$ **67.** $y = -2x + 5$ **69.** $y = x - 5$
71. $y = \dfrac{2}{3}x + \dfrac{4}{3}$ **73.** $y = \dfrac{3}{5}x - 2$ **75.** $y = \dfrac{1}{2}x - \dfrac{5}{2}$ **77.** $y = -\dfrac{1}{2}x + 4$ **79.** $y = -\dfrac{1}{3}x - \dfrac{5}{3}$ **81.** $y = 2x + \dfrac{13}{15}$
83. The distance stays the same.   **85.** The area is 4 times as large; $(2s)^2 = 4s^2$ **87.** Square;   **89.** \$1440   **91.** \$5000
93. 50 miles per hour   **95.** 7.632 miles   **97.** 48 square inches   **99.** 558 square inches
101. ≈ 75.40 feet   **103.** 3 square feet   **105.** 7 square feet   **107.** ≈ 124.1 feet
109. ≈ 381.7 cubic inches   **111. a)** $B = \dfrac{703w}{h^2}$ **b)** ≈ 23.91 **113. a)** $V = 18x^3 - 3x^2$ **b)** 6027 cubic centimeters   **c)** $S = 54x^2 - 8x$
d) 2590 square centimeters   **115.** $\dfrac{2}{15}$ **116.** -6 **117.** 0   **118.** 8

Exercise Set 2.7   **1.** A ratio is a quotient of two quantities.   **3.** c to d, $c:d$, $\dfrac{c}{d}$ **5.** Need a given ratio and one of the two parts of a second ratio   **7.** Yes, their corresponding angles must be equal and their corresponding sides must be in proportion.   **9.** Yes
11. No   **13.** 2:3   **15.** 1:2   **17.** 8:1   **19.** 7:4   **21.** 1:3   **23.** 6:1   **25.** 7:6   **27.** 8:1   **29. a)** 50:23   **b)** $\approx 2.17:1$
31. a) $5.15:3.35$ or $1.03:0.67$ **b)** $\approx 1.54:1$ **33. a)** $19.2:2.2$ or $9.6:1.1$ **b)** $6.7:3.1$ **35. a)** $40:32$ or $5:4$ **b)** $15:11$ **37.** 12
39. 45   **41.** -9 **43.** -2 **45.** -54 **47.** 6   **49.** 32 inches   **51.** 15.75 inches   **53.** 19.5 inches   **55.** 25 loads   **57.** 269.8 miles
59. 1.5 feet   **61.** 24 teaspoons   **63.** ≈ 0.43 feet   **65.** 3.75 cups   **67.** ≈ 9.49 feet   **69.** 0.55 milliliter   **71.** 570 minutes or 9 hours 30 minutes
73. ≈ 339 children   **75.** 6.5 feet   **77.** 2.9 square yards   **79.** 20 inches   **81.** ≈ 23 **83.** \$307   **85.** 2113.4 pesos
87. Yes, her ratio is $2.12:1$.   **89.** It must increase.   **91.** $\approx 41{,}667$ miles   **93.** 0.625 cubic centimeters   **96.** Commutative property of addition   **97.** Associative property of multiplication   **98.** Distributive property   **99.** All real numbers.   **100.** $m = (y - b)/x$

Chapter 2 Review Exercises   **1.** $3x + 12$ **2.** $5x - 10$ **3.** $-2x - 8$ **4.** $-x - 2$ **5.** $-m - 3$ **6.** $-16 + 4x$ **7.** $25 - 5p$
8. $24x - 30$ **9.** $-25x + 25$ **10.** $-4x + 12$ **11.** $x + 2$ **12.** $-1 - 2y$ **13.** $-x - 2y + z$ **14.** $-6a + 15b - 21$ **15.** $4x$
16. $-3y + 8$ **17.** $5x + 1$ **18.** $-3x + 3y$ **19.** $8m + 8n$ **20.** $9x + 3y + 2$ **21.** $4x + 3y + 6$ **22.** 3   **23.** $-12x^2 + 3$ **24.** 0
25. $5x + 7$ **26.** $-3b + 2$ **27.** 0   **28.** $4x - 4$ **29.** $22x - 42$ **30.** $6x^2 - 3x + y$ **31.** $-\dfrac{7}{20}d + 7$ **32.** 3   **33.** $\dfrac{1}{6}x + 2$
34. $-\dfrac{7}{12}n$ **35.** 1   **36.** -13 **37.** 11   **38.** -27 **39.** $\dfrac{11}{5}$ **40.** $\dfrac{11}{2}$ **41.** -6 **42.** -3 **43.** 12   **44.** 4   **45.** 2   **46.** -3 **47.** $\dfrac{3}{2}$ **48.** -3
49. $-\dfrac{1}{2}$ **50.** -1 **51.** All real numbers   **52.** 2   **53.** -5 **54.** -35.5 **55.** -1.125 **56.** 0.6   **57.** $-\dfrac{21}{4}$ **58.** $\dfrac{78}{7}$ **59.** $-\dfrac{3}{2}$ **60.** $-\dfrac{6}{11}$
61. 2   **62.** $\dfrac{10}{7}$ **63.** 0   **64.** -1 **65.** 10   **66.** No solution   **67.** All real numbers   **68.** -4 **69.** No solution   **70.** All real numbers
71. $\dfrac{17}{3}$ **72.** $-\dfrac{20}{7}$ **73.** No solution   **74.** 6   **75.** 52   **76.** 32   **77.** 7   **78.** -18 **79.** 3   **80.** 48   **81.** 12 square centimeters
82. ≈ 33.51 cubic inches   **83.** $l = (P - 2w)/2$ **84.** $m = \dfrac{y - y_1}{x - x_1}$ **85.** $y = \dfrac{1}{3}x + \dfrac{2}{3}$ **86.** 308.5 miles   **87.** 240 square feet
88. ≈ 25.13 cubic inches   **89.** $3:5$ **90.** $5:12$ **91.** $6:1$ **92.** 2   **93.** 20   **94.** 9   **95.** $\dfrac{135}{4}$ **96.** -10 **97.** -16 **98.** $\dfrac{108}{7}$ **99.** 90
100. 40 inches   **101.** 1 foot   **102.** 6.3 hours   **103.** 72 dishes   **104.** 440 pages   **105.** $6\dfrac{1}{3}$ inches   **106.** 15.75 feet   **107.** $\approx \$0.109$
108. 192 bottles

Chapter 2 Practice Test   **1.** $6x - 12$ [2.1]   **2.** $-x - 3y + 4$ [2.1]   **3.** $-3x + 4$ [2.1]   **4.** $-x + 10$ [2.1]   **5.** $-5x - y - 6$ [2.1]
6. $7a - 8b - 3$ [2.1]   **7.** $2x^2 + 6x - 1$ [2.1]   **8.** $x = 3$ [2.4]   **9.** $x = 8$ [2.5]   **10.** $x = -\dfrac{1}{7}$ [2.5]   **11.** No solution [2.5]
12. All real numbers [2.5]   **13.** $x = \dfrac{-by - c}{a}$ [2.6]   **14.** $y = \dfrac{6}{5}x - \dfrac{2}{5}$ [2.6]   **15.** $x = 0$ [2.5]   **16.** $x = -45$ [2.7]
17. a) Conditional equation   **b)** Contradiction   **c)** Identity [2.5]   **18.** $x = \dfrac{32}{3}$ feet or $10\dfrac{2}{3}$ feet [2.7]   **19.** 4% [2.6]
20. ≈ 28.27 inches [2.6]   **21.** 175 minutes or 2 hours 55 minutes [2.6]

Cumulative Review Test   **1.** $\dfrac{8}{3}$ [1.3]   **2.** $\dfrac{15}{16}$ [1.3]   **3.** $>$ [1.5]   **4.** 3 [1.7]   **5.** -1 [1.7]   **6.** 16 [1.9]   **7.** 3 [1.9]   **8.** 12 [1.9]
9. Distributive property [1.10]   **10.** $12x + y$ [2.1]   **11.** $\dfrac{1}{12}x + 25$ [2.1]   **12.** $3x^2 + 8x - 23$ [2.1]   **13.** -1 [2.4]   **14.** -44 [2.3]
15. $-\dfrac{5}{4}$ [2.5]   **16.** $\dfrac{12}{5}$ [2.5]   **17.** $b = 3A - a - c$ [2.6]   **18.** $\dfrac{9}{4}$ or 2.25 [2.7]   **19.** ≈ 380.13 square feet [2.6]   **20.** $\$42$ [2.7]

Chapter 3

Exercise Set 3.1   **1.** add to, more than, increased by, sum   **3.** multiplied by, product of, twice, three times   **5.** need $0.25c$
7. $25 - x$ **9.** is, was, will be, yields, gives   **11.** $h + 4$ **13.** $a - 5$ **15.** $5h$ **17.** $2d$ **19.** $\dfrac{1}{2}a$ **21.** $r - 5$ **23.** $8 - m$ **25.** $2w + 8$
27. $5a - 4$ **29.** $\dfrac{1}{3}w - 7$ **31.** $x =$ Sonya's height   **33.** $x =$ length of Jones Beach   **35.** $x =$ number of medals Finland won
37. $x =$ cost of the Chevy   **39.** $x =$ Teri's grade   **41.** $x =$ amount Kristen receives or $x =$ amount Yvonne receives
43. $x =$ Don's weight or $x =$ Angela's weight   **45.** Let $c =$ cost of chair, then $3c =$ cost of table.   **47.** Let $a =$ area of the kitchen, then $2a + 20 =$ area of the living room.   **49.** Let $w =$ width of rectangle, then $5w - 2 =$ length of rectangle.   **51.** Let $w =$ number of medals won by Sweden, then $38 - w =$ number of medals won by Brazil, or let $w =$ number of medals won by Brazil, then $38 - w =$ number of medals won by Sweden.   **53.** Let $g =$ George's age, then $\dfrac{1}{2}g + 2 =$ Mike's age.   **55.** Let $m =$ number of miles Jan walked, then $6.4 - m =$ number of miles Edward walked, or let $m =$ number of miles Edward walked, then $6.4 - m =$ number of miles Jan walked.   **57.** $n + 8$ **59.** $\dfrac{1}{2}x$ **61.** $2a - 1$ **63.** $2t - 30$ **65.** $2p - 2.3$ **67.** $80{,}000 - m$
69. $2r - 673$ **71.** $10x$ **73.** $100d$ **75.** $45 + 0.40x$ **77.** $s + 0.20s$ **79.** $e - 0.12e$ **81.** $c + 0.07c$ **83.** $m - 0.313m$
85. $f + (f + 15)$ **87.** $(2l - 1) - l$ **89.** $n - (3n - 40)$ **91.** $w + (2w - 3)$ **93.** $r + (479r + 462)$ **95.** $n + (n + 0.06n)$
97. $a - (a - 0.023a)$ **99.** $x + 4x = 20$ **101.** $x + (x + 1) = 41$ **103.** $2x - 8 = 12$ **105.** $\dfrac{1}{5}(x + 10) = 150$
107. $x + 2(x + 2) = 22$ **109.** $12.50h = 150$ **111.** $2.99x = 17.94$ **113.** $25q = 175$ **115.** $a + (2a + 1) = 52$
117. $(2s + 300) - s = 420$ **119.** $s + (2s - 4) = 890$ **121.** $m + (3m - 2) = 12.6$ **123.** $c + 0.023c = 89{,}600$
125. $p - 0.019p = 12{,}087$ **127.** $c + 0.07c = 32{,}600$ **129.** $c + 0.15c = 42.50$ **131. a)** $86{,}400d + 3600h + 60m + s$
b) $368{,}125$ seconds   **134.** 4   **135.** 15   **136.** $y = \dfrac{3x - 6}{2}$ or $y = \dfrac{3}{2}x - 3$ **137.** 2.52   **138.** 13:32

Exercise Set 3.2 **1.** Answers will vary. **3.** 5 **5.** 43, 44 **7.** 47, 49 **9.** 8, 19 **11.** 2, 6 **13.** 25, 42 **15.** 65 cards
17. 22.2 hours **19.** 13 weeks **21.** 9.6 years **23.** December: 22, June: 226 **25.** $14.74 **27.** Northern China: $0.86, Mexico $2.42
29. 15 gallons **31.** 18,100 copies **33.** 6 movies **35.** 150 miles **37.** 6 years **39.** 4000 pages **41.** 2000 newsletters **43.** $261.68
45. $23,230.77 **47.** ≈371 thousand bushels **49.** $25,000 **51.** $39,387.76 **53.** $49,867 **55.** $7500 **57.** 500 pages
59. $1071.43 **61.** $24.59 **63.** $77,777.78 **65. a)** $80 = \dfrac{74 + 88 + 76 + x}{4}$ **b)** 82 **67.** 4 **68.** Commutative property of addition
69. $h = \dfrac{2A}{b}$ **70.** 12

Mid-Chapter Test **1.** $6w$ [3.1] **2.** $3h + 5$ [3.1] **3.** $c + 0.20c$ [3.1] **4.** $40 + 0.25m$ [3.1] **5.** $50n$ [3.1] **6.** $25 - x$ [3.1]
7. $c - 0.25c$ [3.1] **8.** $x =$ length of Poison Dart Frog [3.1] **9.** Let $p =$ distance Pedro traveled, then $4p + 6 =$ distance Mary
traveled. [3.1] **10.** $v - (v - 0.18v)$ [3.1] **11.** $p + 0.12p = 38{,}619$ [3.2] **12.** $x + 3(x + 2) = 26$ [3.2] **13.** 46, 47 [3.2] **14.** 4, 11 [3.2]
15. 18 days [3.2] **16.** 15 hours [3.2] **17.** $700 [3.2] **18.** Betty: 196 clients, Anita: 404 clients [3.2] **19.** 52 miles [3.2]
20. $5000 [3.2]

Exercise Set 3.3 **1.** The area remains the same. **3.** The volume is eight times as great. **5.** The area is nine times as great.
7. An isosceles triangle is a triangle with two equal sides. **9.** 180° **11.** 46°, 46°, 88° **13.** 11.5 inches **15.** $A = 67°, B = 23°$
17. $A = 47°, B = 133°$ **19.** 88° **21.** 50°, 60°, 70° **23.** Length is 14 feet, width is 8 feet. **25.** Length is 78 feet, width is 36 feet.
27. Smaller angles are 50°; larger angles are 130° **29.** 30°, 30°, 150°, 150° **31.** 63°, 73°, 140°, 84° **33.** Width is 4 feet, height is 7 feet.
35. Width is 2.5 feet, height is 5 feet. **37.** Width is 11 feet, length is 16 feet. **39.** $ac + ad + bc + bd$ **41.** < **42.** > **43.** −10
44. $-2x - 5y + 6$ **45.** $y = -2x + 3$

Exercise Set 3.4 **1.** 1 hour **3.** 6 mph **5.** ≈0.72 hour **7.** 2.4 hours **9.** 70 mph **11.** ≈44.4 seconds **13.** 35 mph, 40 mph
15. ≈0.61 hour **17.** 1.5 miles/day, 2.25 miles/day **19.** *Apollo*: 5 mph, *Pythagoras*: 9 mph **21.** 1.5 hours **23. a)** 2.4 miles
b) 112.0 miles **c)** 26.2 miles **d)** 140.6 miles **e)** 9.01 hours **25.** $10,000 at 7%; $2000 at 5% **27.** $2400 at 6%; $3600 at 4%
29. $2000 at 4%; $8000 at 5% **31.** November **33.** 8 hours at Home Depot; 10 hours at clinic **35.** 1200 adults **37. a)** 41 shares of
Nike, 205 shares of Kellogg **b)** $37 **39.** 11.25 pounds almonds, 18.75 pounds walnuts **41.** $160: 6 cubic yards, $120: 2 cubic yards
43. $2.74 per pound **45.** ≈11.1% **47.** $1\dfrac{2}{3}$ liters **49.** 18.96% **51.** ≈11.1% **53.** 40% pure juice **55.** ≈3.77 cups Clorox,
≈2.23 cups shock treatment **57.** 11.25 pints **59.** ≈5.74 hours **62. a)** $\dfrac{22}{13}$ or $1\dfrac{9}{13}$ **b)** $\dfrac{35}{8}$ or $4\dfrac{3}{8}$ **63.** All real numbers
64. $\dfrac{3}{4}$ or 0.75 **65.** 38, 39

Chapter 3 Review Exercises **1.** $3n + 7$ **2.** $1.2g$ **3.** $d - 0.25d$ **4.** $16y$ **5.** $200 - x$ **6.** Let $d =$ Dino's age, then
$7d + 6 =$ Mario's age. **7.** $c - (c - 0.12c)$ **8.** $n - (3n - 24) = 8$ **9.** 33 and 41 **10.** 118 and 119 **11.** 38 and 7 **12.** $21,738.32
13. 19 months **14.** $2000 **15.** $618.75 **16.** 10 hours **17.** ≈$171,000 **18.** $684 **19.** 45°, 55°, 80° **20.** 30°, 40°, 150°, 140°
21. Width is 15.5 feet; length is 19.5 feet **22.** Width is 50 feet; length is 80 feet **23.** 45°, 45°, 135°, 135° **24.** Height is 2 feet; length is 4 feet
25. 2 hours **26.** 4 hours **27.** ≈8.8 feet per second **28.** $4000 at 8%; $8000 at $7\dfrac{1}{4}$% **29.** $700 at 3%; $3300 at 3.5%
30. ≈0.67 gallon **31.** 9 smaller and 21 larger **32.** 1.2 liters of 10%; 0.8 liters of 5% **33.** 103 and 105 **34.** $450 **35.** $12,000
36. 42°, 50°, 88° **37.** 8 years **38.** 70°, 70°, 110°, 110° **39.** 500 copies **40.** 0.6 mph **41.** 60 pounds of $3.50; 20 pounds of $4.10
42. Older brother: 55 mph; younger brother: 60 mph **43.** 0.4 liters **44.** Width is 16 feet; length is 24 feet **45.** 1.5 liters

Chapter 3 Practice Test **1.** $500 - n$ [3.1] **2.** $2w + 6000$ [3.1] **3.** $60t$ [3.1] **4.** $c + 0.06c$ [3.1] **5.** Let $n =$ number of packages of orange, then $7n - 105 =$ number of packages of peppermint. [3.1] **6.** Let $x =$ number of men, then $600 - x =$ number of
women, or let $x =$ number of women, then $600 - x =$ number of men. [3.1] **7.** $(2n - 1) - n$ [3.1] **8.** $n + (n + 18)$ [3.1]
9. $(c + 0.84c) - c$ [3.1] **10.** 56 and 102 [3.2] **11.** 5 and 7 [3.2] **12.** 9 and 33 [3.2] **13.** $2500 [3.2] **14.** $34.78 [3.2]
15. Peter: $40,000, Julie: $80,000 [3.2] **16.** 6 times [3.2] **17.** 7450 pages [3.2] **18.** 15 inches, 30 inches, 30 inches [3.3] **19.** Width is
6 feet, length is 8 feet [3.3] **20.** 59°, 59°, 121°, 121° [3.3] **21.** Harlene: 0.3 feet per minute, Ellis: 0.5 feet per minute [3.4] **22.** 6 mph [3.4]
23. Jelly Belly: ≈1.91 pounds, Kits: ≈1.09 pounds [3.4] **24.** 20 liters [3.4] **25.** 1 liter of 8%; 2 liters of 5% [3.4]

Cumulative Review Test **1.** $16,000 [1.2] **2. a)** 34% **b)** ≈0.06 million or ≈60,000 [1.2] **3. a)** 7.2 parts per million
b) 6 parts per million [1.2] **4.** $\dfrac{5}{9}$ [1.3] **5.** $\dfrac{13}{24}$ inch [1.3] **6. a)** $\{1, 2, 3, 4, \dots\}$ **b)** $\{0, 1, 2, 3, \dots\}$ **c)** A rational number is a quotient
of two integers where the denominator is not 0. [1.4] **7. a)** 4 **b)** $|-5|$ [1.5] **8.** −34 [1.9] **9.** $2x - 28$ [2.1] **10.** 5 [2.5]
11. $\dfrac{1}{5}$ [2.5] **12.** No solution [2.5] **13.** −8 [2.7] **14.** ≈113.10 [2.6] **15. a)** $y = -\dfrac{1}{2}x + 2$ **b)** 4 [2.6] **16.** $w = \dfrac{P - 2l}{2}$ [2.6]
17. 9 gallons [2.7] **18.** 40 minutes [3.2] **19.** 6, 23 [3.2] **20.** 40°, 45°, 90°, 185° [3.3]

Chapter 4

Using Your Graphing Calculator, 4.1

1.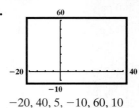
−20, 40, 5, −10, 60, 10

2.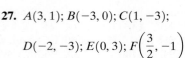
−200, 400, 100, −500, 1000, 200

3. 40
4. 100

Exercise Set 4.1   **1.** The x-coordinate   **3. a)** x-axis   **b)** y-axis   **5.** Axis is singular; axes is plural.
7. An illustration of the set of points whose coordinates satisfy the equation   **9. a)** Two   **b)** To catch errors   **11.** $ax + by = c$

13.

15. II   **17.** IV   **19.** I   **21.** III   **23.** III   **25.** II

27. $A(3, 1); B(-3, 0); C(1, -3);$

$D(-2, -3); E(0, 3); F\left(\dfrac{3}{2}, -1\right)$

29.

31.

33. The points are collinear.

35. $(-5, -3)$ is not on the line.

37. a) Point c) does not satisfy the equation.
b)

39. a) Point a) does not satisfy the equation.
b)

41. a) Point a) does not satisfy the equation.
b)

43. 2   **45.** −4   **47.** 3   **49.** $\dfrac{1}{2}$ **51.** 0   **53. a)** Latitude, 16° N; Longitude, 56° W   **b)** Latitude, 29° N; Longitude, 90.5° W
c) Latitude, 26° N; Longitude, 80.5° W   **d)** Answers will vary.   **57.** A linear equation is an equation of the form $ax + b = c$.

58. A conditional equation is a linear equation that has only one solution.   **59.** −2   **60.** $C = 6\pi \approx 18.84$ inches,
$A = 9\pi \approx 28.27$ square inches   **61.** $y = \dfrac{2x - 6}{5} = \dfrac{2}{5}x - \dfrac{6}{5}$

Using Your Graphing Calculator, 4.2

1.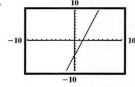
−10, 10, 1, −10, 10, 1

2.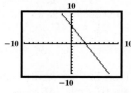
−10, 10, 1, −10, 10, 1

3.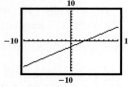
−10, 10, 1, −10, 10, 1

4.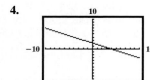
−10, 10, 1, −10, 10, 1

Using Your Graphing Calculator, 4.2   **1.** $(2, 0), (0, -8)$ **2.** $(-3, 0), (0, -6)$ **3.** $(2.5, 0), (0, -2)$ **4.** $(1.5, 0), (0, -4.5)$

Exercise Set 4.2   **1.** x-intercept: substitute 0 for y and find the corresponding value of x; y-intercept: substitute 0 for x and find the corresponding value of y **3.** A horizontal line   **5.** You may not be able to read exact answers from a graph.

7. Yes   **9.** 0   **11.** 5   **13.** 3   **15.** 2   **17.** $\dfrac{8}{3}$ **19.** −7

21.

23.

25.

27.

29.

31.

33.

35.

37.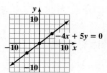

39.

41.

43.
$y = \frac{1}{2}x + 4$

45.
$y = 3x + 3$

47.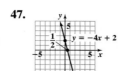
$y = -4x + 2$

49.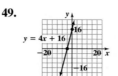
$y = 4x + 16$

51.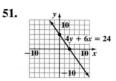
$4y + 6x = 24$

53.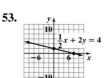
$\frac{1}{2}x + 2y = 4$

55.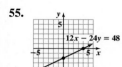
$12x - 24y = 48$

57.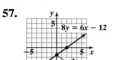
$8y = 6x - 12$

59.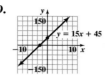
$y = 15x + 45$

61.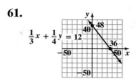
$\frac{1}{3}x + \frac{1}{4}y = 12$

63.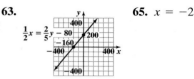
$\frac{1}{2}x = \frac{2}{5}y - 80$

65. $x = -2$

67. $y = 6$ **69.** 5   **71.** 2   **73.** Yes

75. a) $C = 0.10n + 15$

b)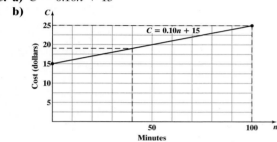

c) \$19
d) 100 minutes

77. a) $C = m + 40$

b)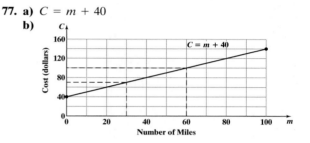

c) \$100
d) 30 miles

79. a)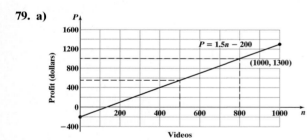

b) \$550   **c)** 800 tapes   **81.** 3,2   **83.** 6,4
85. a)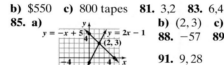
$y = -x + 5$ $y = 2x - 1$ $(2,3)$
b) $(2,3)$ **c)** Yes   **d)** No
88. -57 **89.** 18   **90.** 6.67 ounces
91. 9, 28

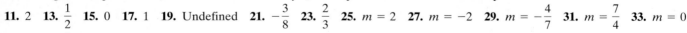

Exercise Set 4.3   **1.** The slope of a line is the ratio of the vertical change to the horizontal change between any two points on the line.   **3.** Rises from left to right   **5.** Lines that rise from the left to right have a positive slope; lines that fall from left to right have a negative slope.   **7.** No, since we cannot divide by 0, the slope is undefined.   **9.** The slopes are the same.
11. 2   **13.** $\frac{1}{2}$ **15.** 0   **17.** 1   **19.** Undefined   **21.** $-\frac{3}{8}$ **23.** $\frac{2}{3}$ **25.** $m = 2$ **27.** $m = -2$ **29.** $m = -\frac{4}{7}$ **31.** $m = \frac{7}{4}$ **33.** $m = 0$
35. $m = -\frac{2}{3}$ **37.** Undefined   **39.**   **41.**   **43.** 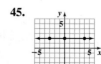   **45.**   **47.**

49. Parallel   **51.** Perpendicular   **53.** Perpendicular   **55.** Neither   **57.** Neither   **59.** Parallel   **61.** Parallel
63. Perpendicular   **65.** 3   **67.** $\frac{1}{4}$ **69.** First   **71. a)** $-\frac{23}{4}$ **b)** 11   **73.** -4 **75.** $-\frac{1}{8}$ **77.** $\frac{225}{68}$

79. a) 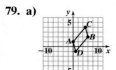   **b)** $AC, m = \frac{3}{5}$; $CB, m = -2$; $DB, m = \frac{3}{5}$; $AD, m = -2$ **c)** Yes, opposite sides are parallel.
81. a) $AB, m = 4$; $BC, m = -2$; $CD, m = 4$ **b)** $[4 + (-2) + 4]/3 = 2$
c) $AD, m = 2$ **d)** Yes   **e)** Answers will vary.
83. 0   **84. a)** $\frac{5}{2}$ **b)** 0   **85.** $\frac{84}{17}$ **86.** $c = d - a - b$ **87.** x-intercept: $(6, 0)$; y-intercept: $(0, -10)$

Mid-Chapter Test

1. IV [4.1]   **2.**   [4.1]   **3. b)** [4.1]
4. −4 [4.1]
5. 3 {4.1}
6. A graph of an equation in two variables is an illustration of a set of points whose coordinates satisfy the equation. [4.1]

7. 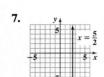 [4.2]   **8.** [4.2]   **9.** 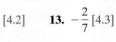 [4.2]   **10.** 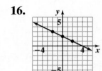 [4.2]   **11.** [4.2]

12. 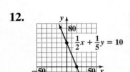 [4.2]   **13.** $-\dfrac{2}{7}$ [4.3]   **16.** [4.3]   **17.** [4.3]   **18.** Neither [4.3]
19. Perpendicular [4.3]
14. 0 [4.3]   **20.** 10 [4.3]
15. Undefined [4.3]

Using Your Graphing Calculator, 4.4

1. a) 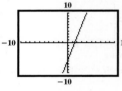   **b)** 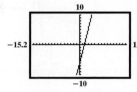   **c)**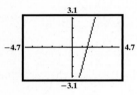
−10, 10, 1, −10, 10, 1   −15.2, 15.2, 1, −10, 10, 1   −4.7, 4.7, 1, −3.1, 3.1, 1

2. a) 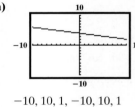   **b)** 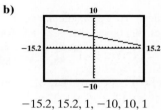   **c)**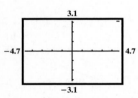
−10, 10, 1, −10, 10, 1   −15.2, 15.2, 1, −10, 10, 1   −4.7, 4.7, 1, −3.1, 3.1, 1

Exercise Set 4.4

1. $y = mx + b$ **3.** $y = 3x - 5$ **5.** Write the equations in slope-intercept form. If slopes are the same and the y-intercepts are different, the lines are parallel.   **7.** $y - y_1 = m(x - x_1)$ **9.** 2; $(0, -6)$ **11.** $\dfrac{4}{3}$; $(0, -7)$

13. 1; $(0, -3)$ **15.** 3; $(0, 2)$ **17.** 2; $(0, 0)$ **19.** 2; $(0, -3)$

21. $\dfrac{5}{2}$; $(0, -5)$ **23.** $-\dfrac{1}{2}$; $\left(0, \dfrac{3}{2}\right)$ **25.** 3; $(0, 4)$ **27.** $\dfrac{3}{2}$; $(0, 2)$

 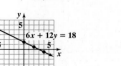   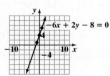

29. $y = x - 2$ **31.** $y = -\dfrac{1}{3}x + 2$ **33.** $y = -3x - 5$ **35.** $y = \dfrac{1}{3}x + 5$ **37.** Parallel   **39.** Perpendicular   **41.** Parallel

43. Neither   **45.** Perpendicular   **47.** Neither   **49.** $y = 3x + 2$ **51.** $y = -3x - 7$ **53.** $y = \dfrac{1}{2}x - \dfrac{5}{2}$ **55.** $y = \dfrac{2}{3}x + 6$

57. $y = 3x + 10$ **59.** $y = -\dfrac{3}{2}x$ **61.** $y = \dfrac{1}{2}x - 2$ **63.** $y = 7.4x - 4.5$ **65. a)** $y = 5x + 60$ **b)** \$210

67. a) Slope-intercept form   **b)** Point-slope form   **c)** Point-slope form
69. a) No   **b)** $y + 4 = 2(x + 5)$ **c)** $y - 12 = 2(x - 3)$ **d)** $y = 2x + 6$ **e)** $y = 2x + 6$ **f)** Yes

71. a) 1.465   **b)** $f = 1.465m$ **c)** ≈ 209.06 feet per second   **d)** 150 feet per second   **e)** 55 miles per hour

73. $y = -2x + 5$ **75.** $y = \dfrac{3}{4}x + 5$ **78.** $<$ **79.** True   **80.** True   **81.** False   **82.** False   **83.** 64   **84.** $r = \dfrac{i}{pt}$

Chapter 4 Review Exercises

1.

2. Not collinear
3. a), b), and **d)**
4. a) -1 **b)** -4 **c)** 6   **d)** $\dfrac{8}{3}$

5.

6.

7.
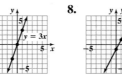
8.

9.

10.

11.
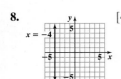
12. $5x + 2y + 10 = 0$
13.
14.
15. $-\dfrac{9}{5}$ **16.** $-\dfrac{1}{12}$
17. -2 **18.** 0
19. Undefined

20. The slope of a line is the ratio of the vertical change to the horizontal change between any two points on the line.   **21.** -2 **22.** $\dfrac{1}{4}$

23. Neither   **24.** Perpendicular   **25. a)** 214   **b)** -49 **26.** $m = -\dfrac{6}{7}; (0, 3)$ **27.** Slope is undefined; no y-intercept

28. $m = 0; (0, -3)$ **29.** $y = 3x - 3$ **30.** $y = -\dfrac{1}{2}x + 2$ **31.** Parallel   **32.** Perpendicular

33. $y = 3x + 1$ **34.** $y = -\dfrac{2}{3}x + 4$ **35.** $y = 2$ **36.** $x = 4$ **37.** $y = -\dfrac{7}{2}x - 3$ **38.** $x = -5$

Chapter 4 Practice Test
1. A graph is an illustration of the set of points whose coordinates satisfy an equation. [4.1]
2. a) IV   **b)** II [4.1]   **3. a)** $ax + by = c$ **b)** $y = mx + b$ **c)** $y - y_1 = m(x - x_1)$ [4.1-4.4]

4. b) and **d)** [4.1]   **5.** $-\dfrac{4}{3}$ [4.3]   **6.** $\dfrac{4}{9}; \left(0, -\dfrac{5}{3}\right)$ [4.4]   **7.** $y = -x - 1$ [4.4]

8.
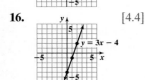
[4.2]   **9.**
[4.2]   **10.**
[4.2]   **11. a)** $y = \dfrac{1}{2}x - 2$ **b)**
[4.2]

12.
[4.2]   **13.** $y = 4x - 13$ [4.4]   **14.** $y = -\dfrac{3}{7}x + \dfrac{2}{7}$ [4.4]   **15.** The lines are parallel since they have the same slope but different y-intercepts. [4.3]

16.
[4.4]   **17.**
[4.4]   **18. a)**
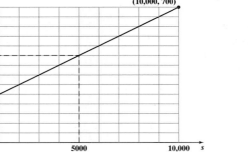
b) $450 [4.2]

Cumulative Review Test   **1. a)** $\{1, 2, 3, 4, \ldots\}$ **b)** $\{0, 1, 2, 3, \ldots\}$ [1.4]   **2.** 18 [1.7]   **3.** 35 [1.9]   **4.** 100 [1.9]
5. a) Distributive property   **b)** Commutative property of addition [1.10]   **6.** 20 [2.5]   **7.** All real numbers [2.5]   **8.** 20 [2.7]
9. $4.00 [2.7]   **10.** $w = \dfrac{v}{lh}$ [2.6]   **11.** 4 [3.2]   **12.** width: 5 feet, length: 13 feet [3.3]   **13.** 2 hours [3.4]
14. Answers will vary. [4.1]   **15.**   [4.1]   **16.**   [4.2]   **17.** 1 [4.3]   **18.** -7 [4.3]   **19.**   [4.4]

20. $y - 2 = 3(x - 5)$ [4.4]

Chapter 5

Exercise Set 5.1   **1.** In the expression t^p, t is the base, p is the exponent.   **3. a)** $\dfrac{x^m}{x^n} = x^{m-n}$, $x \neq 0$ **b)** Answers will vary.
5. a) $(x^m)^n = x^{m \cdot n}$ **b)** Answers will vary.   **7.** a^2b^5 **9.** Answers will vary.   **11.** x^9 **13.** $-z^5$ **15.** y^5 **17.** $3^5 = 243$ **19.** z^8
21. 6   **23.** x^7 **25.** $3^4 = 81$ **27.** $\dfrac{1}{y^2}$ **29.** 1   **31.** $\dfrac{1}{a^6}$ **33.** 1   **35.** 3   **37.** 4   **39.** -9 **41.** $6x^3y^2$ **43.** $-8r$ **45.** x^8 **47.** x^{25}
49. x^3 **51.** x^{12} **53.** n^{18} **55.** $-8w^6$ **57.** $-27x^9$ **59.** $64x^9y^6$ **61.** $\dfrac{x^2}{9}$ **63.** $\dfrac{y^4}{x^4}$ **65.** $-\dfrac{216}{x^3}$ **67.** $\dfrac{8x^3}{y^3}$ **69.** $\dfrac{16p^2}{25}$ **71.** $\dfrac{27x^{12}}{y^3}$ **73.** $\dfrac{a^7}{b^3}$
75. $\dfrac{x^{11}}{2y^7}$ **77.** $\dfrac{6y^4}{z^3}$ **79.** $\dfrac{7}{3x^5y^3}$ **81.** $-\dfrac{3y^2}{x^3}$ **83.** $-\dfrac{2}{x^3y^2z^5}$ **85.** $\dfrac{8}{x^6}$ **87.** $27y^9$ **89.** 1   **91.** $\dfrac{x^4}{y^4}$ **93.** $\dfrac{z^{24}}{16y^{28}}$ **95.** $\dfrac{125}{s^6t^9}$ **97.** $9x^2y^8$
99. $5ab^4$ **101.** $-6x^2y^2$ **103.** $15x^3y^6$ **105.** $-9p^6q^3$ **107.** $49r^6s^4$ **109.** x^2 **111.** x^{12} **113.** $6.25x^6$ **115.** $\dfrac{x^7}{y^4}$ **117.** $-\dfrac{m^{12}}{n^9}$
119. $-216x^9y^6$ **121.** $-8x^{12}y^6z^3$ **123.** $729r^{12}s^{15}$ **125.** $108x^5y^7$ **127.** $53.29x^4y^8$ **129.** $x^{11}y^{13}$ **131.** x^8z^8
133. Cannot be simplified   **135.** Cannot be simplified   **137.** $6z^2$ **139.** Cannot be simplified   **141.** 40   **143.** 1   **145.** The sign will
be positive because a negative number with an even exponent will be positive. This is because $(-1)^m = 1$ when m is even.
147. $8x^2$ **149.** $ab + a^2 + b^2$ **151.** $576y^8z^{13}$ **154.** 13   **155.** $x + 10$ **156.** All real numbers   **157. a)** 4 inches, 4 inches, 9 inches, 9 inches
b) $w = \dfrac{P - 2l}{2}$

Exercise Set 5.2   **1.** Answers will vary.   **3.** No, it is not simplified because of the negative exponent; $\dfrac{x^5}{y^3}$ **5.** The given simplification
is not correct since $5^{-2} = \dfrac{1}{25}$.   **7. a)** The numerator has one term, x^5y^2.   **b)** The factors of the numerator are x^5 and y^2.   **9.** The sign of
the exponent changes when a factor is moved from the numerator to the denominator of a fraction.   **11.** $\dfrac{1}{x^6}$ **13.** $\dfrac{1}{5}$ **15.** x^3 **17.** a
19. 36   **21.** $\dfrac{1}{x^6}$ **23.** $\dfrac{1}{y^{20}}$ **25.** $\dfrac{1}{x^8}$ **27.** 9   **29.** y^2 **31.** x^2 **33.** 9   **35.** $\dfrac{1}{r}$ **37.** p^3 **39.** $\dfrac{1}{x^4}$ **41.** 27   **43.** $\dfrac{1}{125}$ **45.** z^9 **47.** p^{24}
49. y^6 **51.** $\dfrac{1}{x^4}$ **53.** $\dfrac{1}{x^{15}}$ **55.** $-\dfrac{1}{16}$ **57.** $-\dfrac{1}{16}$ **59.** $-\dfrac{1}{8}$ **61.** $\dfrac{1}{36}$ **63.** $\dfrac{1}{x^{10}}$ **65.** n^2 **67.** 1   **69.** 1   **71.** 64   **73.** x^8 **75.** 1   **77.** $\dfrac{1}{4}$
79. $\dfrac{1}{49}$ **81.** x^3 **83.** $\dfrac{1}{16}$ **85.** 125   **87.** $\dfrac{1}{9}$ **89.** 1   **91.** $\dfrac{1}{36x^4}$ **93.** $\dfrac{3y^2}{x^2}$ **95.** 4   **97.** $\dfrac{64}{125}$ **99.** $\dfrac{d^4}{c^8}$ **101.** $-\dfrac{s^4}{r^{16}}$ **103.** $-\dfrac{7}{a^3b^4}$
105. $\dfrac{y^9}{64x^{15}}$ **107.** $\dfrac{18}{z^9}$ **109.** -8 **111.** $12x^5$ **113.** $-\dfrac{20z^5}{y}$ **115.** $8d^4$ **117.** $\dfrac{4}{x^2}$ **119.** $\dfrac{x^4}{2y^5}$ **121.** $\dfrac{8x^6}{y^2}$ **123.** $\dfrac{y^{14}z^6}{25x^8}$ **125.** $\dfrac{r^{20}t^{48}}{16s^{36}}$
127. $\dfrac{y^{12}}{x^{18}z^6}$ **129.** $\dfrac{1}{16p^4q^6}$ **131. a)** Yes   **b)** No   **133.** $16\dfrac{1}{16}$ **135.** $125\dfrac{1}{125}$ **137.** $\dfrac{2}{3}$ **139.** $-\dfrac{7}{8}$ **141.** $-\dfrac{5}{6}$ **143.** $\dfrac{22}{9}$ **145.** -2
147. -3 **149.** -2 **151.** $2, -3$ **153.** The product rule is $(xy)^m = x^my^m$, *not* $(x + y)^m = x^m + y^m$.   **155.** 186 miles   **156.** 94, 96
157. 12 feet by 16 feet   **158.** $3400 at 4%, $5600 at 3%   **159.** $18x^3y^9$

Exercise Set 5.3   **1.** A number in scientific notation is written as a number greater than or equal to 1 and less than 10 that is
multiplied by some power of 10.   **3. a)** Answers will vary.   **b)** 7.23×10^{-5} **5.** 6 places to the right   **7.** The exponent is positive
when the number is 10 or greater.   **9.** Negative since $0.000937 < 1$.   **11.** 1.0×10^{-6} **13.** 3.5×10^5 **15.** 7.950×10^3
17. 5.3×10^{-2} **19.** 7.26×10^{-4} **21.** 5.26×10^9 **23.** 9.14×10^{-6} **25.** 2.203×10^5 **27.** 5.104×10^{-3} **29.** 43,000
31. 0.00000932   **33.** 0.0000213   **35.** 625,000   **37.** 9,000,000   **39.** 535   **41.** 0.000000773   **43.** 10,000   **45.** 0.000008 meter
47. 125,000,000,000 watts   **49.** 15,300 meters   **51.** 0.0482 meter   **53.** 60,000,000   **55.** 0.243   **57.** 0.000064   **59.** 0.0013   **61.** 2500
63. 250,000   **65.** 4.2×10^{12} **67.** 1.28×10^{-1} **69.** 7.0×10^2 **71.** 1.75×10^2 **73.** $3.3 \times 10^{-4}, 5.3, 7.3 \times 10^2, 1.75 \times 10^6$
75. a) $\approx 6,251,000,000$ **b)** ≈ 21.9 **77.** 8,640,000,000 cubic feet   **79.** 1.6×10^7 seconds   **81. a)** 1.68×10^8 **b)** ≈ 1.39
83. a) $36,600,000,000   **b)** ≈ 4.7 **85. a)** Earth, 5.794×10^{24} metric tons; Moon, 7.34×10^{19} metric tons; Jupiter, 1.899×10^{27} metric
tons;   **b)** $\approx 7.89 \times 10^4$ **c)** $\approx 3.28 \times 10^2$ **87.** $\approx 1.316 \times 10^9$ **89. a)** 55–59: 2.0×10^5; 20–24: 5.0×10^4 **b)** 2.48×10^5 **c)** 1.52×10^5
91. 17   **93.** Answers will vary.   **95.** 1,000,000 times greater   **98.** 0   **99. a)** $\dfrac{3}{2}$ **b)** 0   **100.** 2   **101.** $-\dfrac{y^{12}}{64x^9}$

Mid-Chapter Test
1. y^{13} [5.1]   **2.** x^3 [5.1]   **3.** $6x^6y^{13}$ [5.1]   **4.** $\dfrac{2a^5b^6}{3}$ [5.1]   **5.** $-64x^6y^{12}$ [5.1]   **6.** $\dfrac{t^6}{4s^4}$ [5.1]   **7.** $63x^{11}y^{13}$ [5.1]

8. $\dfrac{1}{p^8}$ [5.2]   **9.** $\dfrac{1}{x^{10}}$ [5.2]   **10.** 1 [5.2]   **11.** $\dfrac{49}{9}$ [5.2]   **12.** $\dfrac{32x}{y}$ [5.2]   **13.** $\dfrac{3}{m^4n^6}$ [5.2]   **14.** $\dfrac{x^{12}y^{10}}{4z^4}$ [5.2]   **15. a)** and **b)** Answers will vary. [5.3]

16. 6.54×10^9 [5.3]   **17.** 0.0000327 [5.3]   **18.** 18,900 meters [5.3]   **19.** 0.0238 [5.3]   **20.** 3.0×10^{-7} [5.3]

Exercise Set 5.4
1. A polynomial is an expression containing the sum of a finite number of terms of the form ax^n, where a is a real number and n is a whole number.   **3.** No, it contains a negative exponent.   **5. a)** A monomial is a one-termed polynomial. Examples will vary.   **b)** A binomial is a two-termed polynomial. Examples will vary.   **c)** A trinomial is a three-termed polynomial. Examples will vary.   **7.** Answers will vary.   **9. b)** and **c)**   **11.** Write the polynomial with exponents on the variable decreasing from left to right.   **13.** Combine like terms.   **15.** Use the distributive property to remove parentheses and then combine like terms.
17. Fifth   **19.** Eighth   **21.** Third   **23.** Tenth   **25.** Ninth   **27.** Trinomial   **29.** Monomial   **31.** Binomial   **33.** Monomial
35. Not a polynomial   **37.** Polynomial   **39.** Trinomial   **41.** Not a polynomial   **43.** Already in descending order, 0 degree
45. $x^2 - 2x - 4$, second   **47.** $3x^2 + x - 8$, second   **49.** Already in descending order, first   **51.** Already in descending order, second   **53.** $4x^3 - 3x^2 + x - 4$, third   **55.** $-2x^4 + 3x^2 + 5x - 6$, fourth   **57.** $10x - 9$ **59.** $-x + 11$ **61.** $-2t - 1$
63. $x^2 + 6.6x + 0.8$ **65.** $5m^2 + 4$ **67.** $x^2 + 3x - 3$ **69.** $-3x^2 + x + \dfrac{17}{2}$ **71.** $8.2n^2 - 4.8n - 0.6$ **73.** $-2x^3 - 3x^2 + 4x - 3$
75. $11x^2 - 7x - y + 5$ **77.** $5x^2y - 3x + 2$ **79.** $11x - 3$ **81.** $7y^2 - 2y + 5$ **83.** $4x^2 + 2x - 4$ **85.** $2x^3 - x^2 + 6x - 2$
87. $3n^3 - 11n^2 - n + 2$ **89.** $2x - 6$ **91.** $3x + 4$ **93.** $-3r$ **95.** $-6y^2 + 1.9y - 12.7$ **97.** $8x^2 + x + 4$ **99.** $-6.4n^2 + 6n - 7.6$
101. $8x^3 - 7x^2 - 4x - 3$ **103.** $2x^3 - \dfrac{23}{5}x^2 + 2x - 2$ **105.** $x - 2$ **107.** $2x^2 - 9x + 14$ **109.** $-5c^3 - 4c^2 - 7c + 14$
111. $3x + 8$ **113.** $3a^2 - 16a + 28$ **115.** $x^2 - 3x - 3$ **117.** $-4x^2$ **119.** $4x^3 - 7x^2 + x - 2$ **121.** Answers will vary.
123. Answers will vary.   **125.** Sometimes   **127.** Sometimes   **129.** Answers will vary; one example is: $x^4 - 2x^3 + x$.
131. No, all three terms would have to be degree 4 or 1. Therefore at least two of the terms would be like terms.   **133.** $a^2 + 2ab + b^2$
135. $4x^2 + 3xy$ **137.** $-12x + 18$ **139.** $8x^2 + 28x - 24$ **141.** 4.5   **142.** $n - 5$ **143.**   **144.**

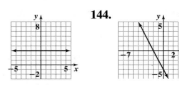

145. $\dfrac{y^3}{27x^{12}}$ **146.** 0.035

Exercise Set 5.5
1. Answers will vary.   **3.** First, Outer, Inner, Last   **5.** Yes
7. $(a + b)^2 = a^2 + 2ab + b^2;\ (a - b)^2 = a^2 - 2ab + b^2$ **9.** No, $(x - 2)^2 = x^2 - 4x + 4$ **11.** Answers will vary.
13. Answers will vary.   **15.** $-24x^6$ **17.** $20x^5y^6$ **19.** $-28x^3y^{15}$ **21.** $54x^6y^{14}$ **23.** $3x^6y$ **25.** $5.94x^8y^3$ **27.** $9x - 45$
29. $-6x^2 + 6x$ **31.** $-16y - 10$ **33.** $-2x^3 + 4x^2 - 10x$ **35.** $-20x^3 + 30x^2 - 20x$ **37.** $0.5x^5 - 3x^4 - 0.5x^2$
39. $0.6x^2y + 1.5x^2 - 1.8xy$ **41.** $x^2y^4 - 4y^7 - 3y^4$ **43.** $5x^2 + 18x - 8$ **45.** $6x^2 + 3x - 30$ **47.** $4x^2 - 16$ **49.** $-5x^2 - 22x + 48$
51. $-12x^2 + 32x - 5$ **53.** $4x^2 - 10x + 4$ **55.** $12k^2 - 30k + 12$ **57.** $x^2 - 4$ **59.** $4x^2 - 12x + 9$ **61.** $-6z^2 + 46z - 28$
63. $8x^2 - 50x + 63$ **65.** $xy - 3x + 7y - 21$ **67.** $6x^2 - 5xy - 6y^2$ **69.** $-27x^2 - 3xy + 36x + 4y$ **71.** $x^2 + 0.9x + 0.18$
73. $x^2 + \dfrac{7}{2}x - 2$ **75.** $x^2 - 36$ **77.** $9x^2 - 64$ **79.** $x^2 + 2xy + y^2$ **81.** $x^2 - 0.4x + 0.04$ **83.** $16x^2 + 40x + 25$
85. $0.16x^2 + 0.8xy + y^2$ **87.** $16c^2 - 25d^2$ **89.** $4x^2 - 36$ **91.** $49s^2 - 42st + 9t^2$ **93.** $16m^3 - 8m^2 + 9m + 18$
95. $12x^3 + 5x^2 + 13x + 10$ **97.** $-14x^3 - 22x^2 + 19x - 3$ **99.** $a^3 + b^3$ **101.** $6x^4 + 5x^3 + 5x^2 + 10x + 4$
103. $x^4 - 3x^3 + 5x^2 - 6x$ **105.** $2x^5 + 2x^4 - 23x^3 + x^2 - 12x$ **107.** $b^3 - 3b^2 + 3b - 1$ **109.** $27a^3 - 135a^2 + 225a - 125$
111. Yes   **113.** No   **115.** 6, 3, 1   **117. a)** $(x + 2)(2x + 1)$ or $2x^2 + 5x + 2$ **b)** 54 square feet   **c)** 1 foot   **119. a)** $3x^2 + 19x + 20$
b) $6x^3 + 32x^2 + 2x - 40$ **c)** 864 cubic feet   **d)** 864 cubic feet   **e)** Yes   **121.** $\dfrac{1}{3}x^2 + \dfrac{11}{45}x - \dfrac{4}{15}$ **123.** No solution
124. $C = 53°, D = 37°$ **125.** $\dfrac{x^4}{16y^8}$ **126. a)** -216 **b)** $\dfrac{1}{216}$ **127.** $-5x^2 - 2x + 14$

Exercise Set 5.6
1. To divide a polynomial by a monomial, divide each term in the polynomial by the monomial.   **3.** $x + 4$
5. The terms should be listed in descending order.   **7.** $\dfrac{x^2 + 0x - 7}{x - 2}$ **9.** $(x + 5)(x - 3) - 2 = x^2 + 2x - 17$
11. $\dfrac{x^2 - x - 42}{x - 7} = x + 6$ or $\dfrac{x^2 - x - 42}{x + 6} = x - 7$ **13.** $\dfrac{2x^2 + 5x + 3}{2x + 3} = x + 1$ or $\dfrac{2x^2 + 5x + 3}{x + 1} = 2x + 3$
15. $\dfrac{4x^2 - 9}{2x + 3} = 2x - 3$ or $\dfrac{4x^2 - 9}{2x - 3} = 2x + 3$ **17.** $x + 2$ **19.** $2n + 5$ **21.** $\dfrac{7}{3}x + 2$ **23.** $-3x + 2$ **25.** $3x + 1$ **27.** $\dfrac{1}{2}x + 4$
29. $-1 + \dfrac{5}{2}w$ **31.** $1 + \dfrac{2}{x} - \dfrac{3}{x^2}$ **33.** $-2x^3 + \dfrac{3}{x} + \dfrac{4}{x^2}$ **35.** $x^2 + 3x - \dfrac{3}{x^3}$ **37.** $3x^2 - 2x + 6 - \dfrac{5}{2x}$ **39.** $-2k^2 - \dfrac{3}{2}k + \dfrac{2}{k}$
41. $-4x^3 - x^2 + \dfrac{10}{3} + \dfrac{3}{x^2}$ **43.** $x + 3$ **45.** $5y + 1$ **47.** $2x + 4$ **49.** $x + 4$ **51.** $x + 5 - \dfrac{3}{2x - 3}$ **53.** $x + 6$
55. $3x^2 + 4x + 5 + \dfrac{4}{3x - 4}$ **57.** $4x - 3 - \dfrac{3}{2x + 3}$ **59.** $7x^2 - 5$ **61.** $2x^2 + \dfrac{12}{x - 2}$ **63.** $w^2 + 3w + 9 + \dfrac{19}{w - 3}$ **65.** $x^2 + 3x + 9$
67. $2x^2 + x - 2 - \dfrac{2}{2x - 1}$ **69.** $-m^2 - 7m - 5 - \dfrac{8}{m - 1}$ **71.** $4t^2 - 8t + 15 - \dfrac{26}{t + 2}$ **73.** No; for example $\dfrac{x + 2}{x} = 1 + \dfrac{2}{x}$ which is not a binomial.   **75.** $2x^2 + 11x + 16$ **77.** Third degree   **79.** $4x$ **81.** Since the shaded areas minus 2 must equal 3, 1, 0, and -1,

respectively, the shaded areas are 5, 3, 2, and 1, respectively.   **83.** $x^2 + \dfrac{2}{3}x + \dfrac{4}{9} - \dfrac{37}{9(3x-2)}$ **85.** $-3x + 3 + \dfrac{1}{x+3}$ **88. a)** 2 **b)** 2, 0

c) $2, -5, 0, \dfrac{2}{5}, -6.3, -\dfrac{23}{34}$ **d)** $\sqrt{7}, \sqrt{3}$ **e)** $2, -5, 0, \sqrt{7}, \dfrac{2}{5}, -6.3, \sqrt{3}, -\dfrac{23}{34}$ **89. a)** 0   **b)** Undefined   **90.** Parentheses, exponents,

multiplication or division from left to right, addition or subtraction from left to right   **91.** $-\dfrac{2}{3}$ **92.** \$39.50   **93.** x^{13}

Chapter 5 Review Exercises

1. x^7 **2.** x^6 **3.** 243   **4.** 32   **5.** x^3 **6.** 1   **7.** 25   **8.** 64   **9.** $\dfrac{1}{x^2}$ **10.** y^3 **11.** 1   **12.** 7

13. 1   **14.** 1   **15.** $25x^2$ **16.** $27a^3$ **17.** $-27x^3$ **18.** $216s^3$ **19.** $16x^8$ **20.** x^{24} **21.** p^{32} **22.** $\dfrac{4x^6}{y^2}$ **23.** $\dfrac{25y^4}{4b^2}$ **24.** $24x^5$ **25.** $\dfrac{4x}{y}$

26. $54x^4y^9$ **27.** $9x^2$ **28.** $24x^7y^7$ **29.** $16x^8y^{11}$ **30.** $6c^6d^3$ **31.** $\dfrac{27a^6}{b^{15}}$ **32.** $27x^{12}y^3$ **33.** $\dfrac{1}{b^9}$ **34.** $\dfrac{1}{27}$ **35.** $\dfrac{1}{25}$ **36.** z^2 **37.** x^7

38. 16   **39.** $\dfrac{1}{y^3}$ **40.** $\dfrac{1}{x^5}$ **41.** $\dfrac{1}{p^2}$ **42.** $\dfrac{1}{a^5}$ **43.** m^{10} **44.** x^7 **45.** $\dfrac{1}{x^6}$ **46.** $\dfrac{1}{9x^8}$ **47.** $\dfrac{x^9}{64y^3}$ **48.** $\dfrac{4n^2}{m^6}$ **49.** $12y^2$ **50.** $-\dfrac{125z^3}{y^9}$

51. $\dfrac{x^4}{16y^6}$ **52.** $\dfrac{6}{x}$ **53.** $10x^2y^2$ **54.** $\dfrac{12x^2}{y}$ **55.** $\dfrac{24y^2}{x^2}$ **56.** $3y^5$ **57.** $\dfrac{4y}{x^3}$ **58.** $\dfrac{7x^5}{y^4}$ **59.** $\dfrac{x}{2y^5}$ **60.** $\dfrac{4y^{10}}{x}$ **61.** 1.72×10^6

62. 1.53×10^{-1} **63.** 7.63×10^{-3} **64.** 4.7×10^4 **65.** 5.76×10^3 **66.** 3.14×10^{-4} **67.** 0.0075   **68.** 0.000652   **69.** 8,900,000
70. 51,200   **71.** 0.0000314   **72.** 11,030,000   **73.** 0.092 liter   **74.** 6,000,000,000 meters   **75.** 0.0000128 gram   **76.** 19,200 grams
77. 0.085   **78.** 1260   **79.** 245   **80.** 397,000,000   **81.** 0.00003   **82.** 0.0325   **83.** 3.64×10^9 **84.** 2.12×10^1 **85.** 5.0×10^9
86. 5.0×10^{-4} **87.** 3.4×10^{-3} **88.** 3.4×10^7 **89.** 50,000 gallons   **90. a)** 1,500,000,000,000   **b)** \$1.125 $\times 10^{11}$
91. Not a polynomial   **92.** Monomial, zero   **93.** $x^2 + 3x - 4$, trinomial, second   **94.** $4x^2 - x - 3$, trinomial, second
95. Not a polynomial   **96.** Binomial, third   **97.** $-4x^2 + x$, binomial, second   **98.** Not a polynomial   **99.** $2x^3 + 4x^2 - 3x - 7$,
polynomial, third   **100.** $5x - 3$ **101.** $7d + 4$ **102.** $-3x - 5$ **103.** $-4x^2 + 11x - 7$ **104.** $5m^2 - 10$ **105.** $8.1p + 2.8$
106. $-3y - 15$ **107.** $4x^2 - 12x - 15$ **108.** $3a^2 - 5a - 21$ **109.** $4x + 2$ **110.** $-5x^2 + 8x - 19$ **111.** $3x^2 + 3x$
112. $-15x^2 - 12x$ **113.** $6x^3 - 12x^2 + 21x$ **114.** $-2c^3 + 3c^2 - 5c$ **115.** $28b^3 + 21b^2 + 35b$ **116.** $x^2 + 9x + 20$
117. $-12x^2 - 21x + 6$ **118.** $25x^2 - 30x + 9$ **119.** $-6x^2 + 14x + 12$ **120.** $r^2 - 25$ **121.** $3x^3 + x^2 - 10x + 6$

122. $3x^3 + 7x^2 + 14x + 4$ **123.** $-12x^3 + 10x^2 - 30x + 14$ **124.** $x + 2$ **125.** $4y + 6$ **126.** $8x + 4$ **127.** $2x^2 + 3x - \dfrac{4}{3}$

128. $2w - \dfrac{5}{3} + \dfrac{1}{w}$ **129.** $4x^5 - 2x^4 - \dfrac{3}{4}x^2 + \dfrac{1}{4x}$ **130.** $-4m + 2$ **131.** $\dfrac{5}{2}x + \dfrac{5}{x} + \dfrac{1}{x^2}$ **132.** $\dfrac{5}{3}x - 2 + \dfrac{5}{x}$ **133.** $x + 4$

134. $5x - 2 + \dfrac{2}{x+6}$ **135.** $n + 3$ **136.** $2x^2 + 3x - 4$ **137.** $2x - 3$

Chapter 5 Practice Test

1. $15x^6$ [5.1]   **2.** $27x^3y^6$ [5.1]   **3.** $8p^5$ [5.1]   **4.** $\dfrac{x^3}{8y^6}$ [5.1]   **5.** $\dfrac{y^4}{4x^6}$ [5.2]   **6.** 4 [5.1]   **7.** $\dfrac{2x^7y}{3}$ [5.2]

8. 1.425×10^{10} [5.3]   **9.** 2.0×10^{-7} [5.3]   **10.** Monomial [5.4]   **11.** Binomial [5.4]   **12.** Not a polynomial [5.4]
13. $6x^3 - 2x^2 + 5x - 5$, third degree [5.4]   **14.** $2x^2 + x - 7$ [5.4]   **15.** $-3y^2 - 2y + 5$ [5.4]   **16.** $3x^2 - x + 3$ [5.4]
17. $15d^2 - 40d$ [5.5]   **18.** $15x^2 + 4x - 32$ [5.5]   **19.** $-12c^2 + 7c + 45$ [5.5]   **20.** $6x^3 + 2x^2 - 35x + 25$ [5.5]   **21.** $4x^2 + 2x - 1$ [5.6]

22. $4x + 2 - \dfrac{5}{3x}$ [5.6]   **23.** $4x + 5$ [5.6]   **24.** $3x - 2 - \dfrac{2}{4x+5}$ [5.6]   **25. a)** 5.73×10^3 **b)** $\approx 7.78 \times 10^5$ [5.3]

Cumulative Review Test

1. 17 [1.9]   **2.** $8x + 2$ [2.1]   **3.** -25 [1.9]   **4.** $-\dfrac{33}{20}$ [2.4]   **5.** 13 [2.5]   **6.** 0 [2.5]

7. $y = 3x + 5$ [2.6]   **8.** $-\dfrac{1}{2}$ [4.3]   **9.** No [4.4]   **10.** $\dfrac{25x^6}{y^{16}}$ [5.1]   **11.** $-7x^2 - 5x + 2$, second [5.4]   **12.** $3x^2 + 9x - 2$ [5.4]

13. $5a^2 + 6a + 5$ [5.4]   **14.** $10t^2 - 11t + 3$ [5.5]   **15.** $6x^3 - 13x^2 + 9x - 2$ [5.5]   **16.** $\dfrac{5}{2}d + 3 - \dfrac{2}{d}$ [5.6]   **17.** $2x + 5$ [5.6]

18. \$3.33 [2.7]   **19.** Bob, 56.5 mph; Nick, 63.5 mph [3.4]   **20.** $l = 10$ feet, $w = 4$ feet [3.3]

Chapter 6

Exercise Set 6.1

1. To factor an expression means to write the expression as the product of its factors.   **3.** A composite number is a positive integer other than 1 that is not prime.   **5.** The greatest common factor is the greatest number that divides into all the numbers.   **7.** $1, 2, 4, x, 2x, 4x, x^2, 2x^2, 4x^2$, and the opposites of these factors   **9.** $2^3 \cdot 7$ **11.** $2 \cdot 3^2 \cdot 5$ **13.** $2^3 \cdot 31$ **15.** 4   **17.** 14
19. 2   **21.** x **23.** $3x$ **25.** a **27.** qr **29.** x^3y^5 **31.** 1   **33.** x^2y^2 **35.** x **37.** $x - 4$ **39.** $2x - 3$ **41.** $3w + 5$ **43.** $x - 4$
45. $x - 1$ **47.** $x - 9$ **49.** $4(x - 2)$ **51.** $5(3x - 1)$ **53.** $7(q + 4)$ **55.** $3x(3x - 4)$ **57.** $x^4(7x - 9)$ **59.** $3x^2(x^3 - 4)$
61. $12x^8(3x^4 + 2)$ **63.** $9y^3(3y^{12} - 1)$ **65.** $y(1 + 6x^3)$ **67.** $a^2(7a^2 + 3)$ **69.** $4xy(4yz + x^2)$ **71.** $4x^2yz(20x^3y^2z - 9)$
73. $25x^2yz(z^2 + x)$ **75.** $x^4y^3z^9(19y^9z^4 - 8x)$ **77.** $4(2c^2 - c - 8)$ **79.** $3(3x^2 + 6x + 1)$ **81.** $4x(x^2 - 2x + 3)$
83. $8(5b^2 - 6c + 3)$ **85.** $3(5p^2 - 2p + 3)$ **87.** $3a(3a^3 - 2a^2 + b)$ **89.** $xy(8x + 12y + 5)$ **91.** $(x - 7)(x + 6)$
93. $(a - 2)(3b - 4)$ **95.** $(2x + 1)(4x + 1)$ **97.** $(2x + 1)(5x + 1)$ **99.** $(6c + 7)(3c - 2)$ **101.** $6\nabla(2 - \nabla)$

103. $4\square(3\square^2 - \square + 1)$ **105.** $2x^2(2x + 7)(3x^3 + 2x - 1)$ **107.** $(x + 2)(x + 3)$ **108.** $-3x + 17$ **109.** 2
110. $y = \dfrac{4x - 20}{5}$ or $y = \dfrac{4}{5}x - 4$ **111.** ≈ 201.06 cubic inches   **112.** $14, 27$ **113.** $\dfrac{9y^2}{4x^6}$

Exercise Set 6.2
1. The first step is to factor out a common factor, if one exists.   **3.** $x^2 - 3x - 2xy + 6y$; found by multiplying the factors.   **5.** -1 **7.** $(x + 3)(x + 2)$ **9.** $(x + 5)(x + 4)$ **11.** $(x + 2)(x + 5)$ **13.** $(c - 4)(c + 7)$ **15.** $(2x - 3)(2x + 3)$
17. $(x + 3)(3x + 1)$ **19.** $(2x + 1)(3x - 1)$ **21.** $(x + 4)(8x + 1)$ **23.** $(3t - 2)(4t - 1)$ **25.** $(x + 9)(x - 1)$
27. $(2p + 5)(3p - 2)$ **29.** $(x + 2y)(x - 3y)$ **31.** $(3x + 2y)(x - 3y)$ **33.** $(5x - 6y)(2x - 5y)$ **35.** $(x - b)(x - a)$
37. $(y + 9)(x - 5)$ **39.** $(a + 3)(a + b)$ **41.** $(y - 1)(x + 5)$ **43.** $(3 + 2y)(4 - x)$ **45.** $(z + 5)(z^2 + 1)$ **47.** $(x - 5)(x^2 + 8)$
49. $2(x - 6)(x + 4)$ **51.** $4(x + 2)(x + 2) = 4(x + 2)^2$ **53.** $x(2x + 3)(3x - 1)$ **55.** $p(p - 6q)(p + 2q)$ **57.** $(y + 5)(x + 3)$
59. $(x + 5)(y + 6)$ **61.** $(a + b)(x + y)$ **63.** $(r + 6)(s - 7)$ **65.** $(c - a)(d + 3)$ **67.** No; $xy + 2x + 5y + 10$ is factorable;
$xy + 10 + 2x + 5y$ is not factorable in this arrangement.   **69.** $(\odot + 3)(\odot - 5)$ **71. a)** $2x^2 - 5x - 6x + 15$ **b)** $(2x - 5)(x - 3)$
73. a) $2x^2 - 6x - 5x + 15$ **b)** $(x - 3)(2x - 5)$ **75. a)** $4x^2 + 3x - 20x - 15$ **b)** $(4x + 3)(x - 5)$ **77.** $(\odot + 3)(\bigstar + 2)$
79. $\dfrac{6}{5}$ **80.** 30 pounds of jelly beans; 20 pounds of gumdrops   **81.** $5x^2 - 2x - 3 + \dfrac{5}{3x}$ **82.** $a - 4$

Exercise Set 6.3
1. Since 960 is positive, both signs will be the same. Since 92 is positive, both signs will be positive.   **3.** Since -1500 is negative, one sign will be positive, the other will be negative.   **5.** Since 8000 is positive, both signs will be the same. Since -240 is negative, both signs will be negative.   **7.** The trinomial $x^2 - 11x + 24$ is obtained by multiplying the factors using the FOIL method.   **9.** The trinomial $2x^2 - 8xy - 10y^2$ is obtained by multiplying all the factors and combining like terms.   **11.** It is not completely factored. It factors to $2(x - 2)(x - 1)$.   **13.** A trinomial factoring problem can be checked by multiplying the factors.
15. $(x - 5)(x - 2)$ **17.** $(x + 2)(x + 4)$ **19.** $(x + 8)(x - 3)$ **21.** Prime   **23.** $(y - 12)(y - 1)$ **25.** $(a - 4)(a + 2)$
27. $(r - 5)(r + 3)$ **29.** $(b - 9)(b - 2)$ **31.** Prime   **33.** $(q + 9)(q - 5)$ **35.** $(x - 10)(x + 3)$ **37.** $(x + 2)^2$ **39.** $(s - 4)^2$
41. $(p - 6)^2$ **43.** $(w - 15)(w - 3)$ **45.** $(x + 13)(x - 3)$ **47.** $(x - 5)(x + 4)$ **49.** $(y + 5)(y + 8)$ **51.** $(x + 16)(x - 4)$
53. Prime   **55.** $(x - 16)(x - 4)$ **57.** $(a - 9)(a - 11)$ **59.** $(x + 2)(x + 1)$ **61.** $(w + 9)(w - 2)$ **63.** $(x - 3y)(x - 5y)$
65. $(m - 3n)^2$ **67.** $(x + 6y)(x + 2y)$ **69.** $(m + 3n)(m - 8n)$ **71.** $6(x - 4)(x - 1)$ **73.** $5(x + 3)(x + 1)$
75. $2(x - 4)(x - 5)$ **77.** $b(b - 5)(b - 2)$ **79.** $3z(z - 9)(z + 2)$ **81.** $x(x + 4)^2$ **83.** $7(a - 2b)(a - 3b)$
85. $3r(r + 4t)(r - 2t)$ **87.** $x^2(x - 7)(x + 3)$ **89.** Both negative; one positive and one negative; one positive and one negative; both positive   **91.** $x^2 - 12x + 32 = (x - 8)(x - 4)$ **93.** $x^2 - 2x - 35 = (x - 7)(x + 5)$ **95.** $(x + 0.4)(x + 0.2)$
97. $\left(x + \dfrac{1}{5}\right)\left(x + \dfrac{1}{5}\right) = \left(x + \dfrac{1}{5}\right)^2$ **99.** $(x + 8)(x - 32)$ **101.** 9   **102.** 19.6%   **103.**   **104.** $2x^3 + x^2 - 16x + 12$
105. $3x + 2 - \dfrac{2}{x - 4}$ **106.** $(5x + 2)(4x - 3)$

Using Your Graphing Calculator, 6.4
1. $(3x + 4)(2x - 7)$ **2.** $(5x - 8)(2x + 3)$ **3.** $(6x - 7)(2x + 5)$ **4.** $(9x + 4)(3x + 5)$

Exercise Set 6.4
1. Factoring trinomials is the reverse process of multiplying binomials.   **3.** The first term of the trinomial, ax^2
5. $(2x + 1)(x + 5)$ **7.** $(3x + 2)(x + 4)$ **9.** $(5x + 1)(x - 2)$ **11.** $(3r - 2)(r + 5)$ **13.** $(2z - 3)^2$ **15.** $(2z + 3)(3z - 4)$
17. Prime   **19.** $(8x + 3)(x + 2)$ **21.** Prime   **23.** $(5y - 1)(y - 3)$ **25.** $(7x + 1)(x + 6)$ **27.** $(2x + 5)(2x - 3)$ **29.** $(7t - 1)^2$
31. $(5z + 4)(z - 2)$ **33.** $(4y - 3)(y + 2)$ **35.** $(5x - 1)(2x - 5)$ **37.** $(5d + 4)(2d - 3)$ **39.** $2(4x + 1)(x - 6)$
41. $(7t + 3)(t + 1)$ **43.** $2(3x + 5)(x + 1)$ **45.** $x(2x + 1)(3x - 4)$ **47.** $4x(3x + 1)(x + 2)$ **49.** $2x(2x + 3)(x - 2)$
51. $8(2c - 1)(3c + 2)$ **53.** $4(2p + 3)(p - 1)$ **55.** $(8c + d)(c + 5d)$ **57.** $(5x + 3y)(3x - 2y)$ **59.** $2(2x - y)(3x + 4y)$
61. $(7p + 6q)(p + q)$ **63.** $(3m - 2n)(2m + n)$ **65.** $x(4x + 3y)(2x + y)$ **67.** $x^2(2x + y)(2x + 3y)$ **69.** $3x^2 - 20x - 7$; obtained by multiplying the factors.   **71.** $10x^2 + 35x + 15$; obtained by multiplying the factors.   **73.** $3t^4 + 11t^3 - 4t^2$; obtained by multiplying the factors.   **75. a)** Dividing the trinomial by the binomial gives the second factor.   **b)** $6x + 11$ **77.** $(6x - 5)(3x + 4)$
79. $(5x - 8)(3x - 20)$ **81.** $5(3a - 8)(7a + 4)$ **83.** $2x + 45$, the product of the three first terms must equal $6x^3$, and the product of the constants must equal 2250.   **85.** 49   **86.** ≈ 142.56 mph   **87.** $12xy^2(3x^3y - 1 + 2x^4y^4)$ **88.** $(b + 12)(b - 8)$

Mid-Chapter Test
1. A factoring problem may be checked by multiplying the factors. [6.1]
2. $3xy^2$ [6.1]   **3.** $4a^2b(b^2 - 6a)$ [6.1]   **4.** $(d - 6)(5c - 3)$ [6.1]   **5.** $(2x + 9)(7x + 1)$ [6.1]   **6.** $(x + 4)(x + 7)$ [6.2]
7. $(x + 5)(x - 3)$ [6.2]   **8.** $(2a + 5b)(3a - b)$ [6.2]   **9.** $(5x - 2y)(x - 9)$ [6.2]   **10.** $4x(2x + 1)(x - 6)$ [6.2]
11. $(x - 3)(x - 7)$ [6.3]   **12.** $(t + 4)(t + 5)$ [6.3]   **13.** Prime [6.3]   **14.** $(x + 8)^2$ [6.3]   **15.** $(m + 5n)(m - 9n)$ [6.3]
16. $(3x + 2)(x + 5)$ [6.4]   **17.** $(4z - 3)(z - 2)$ [6.4]   **18.** Prime [6.4]   **19.** $(3x - 1)^2$ [6.4]   **20.** $3(2a - b)(a + b)$ [6.4]

Exercise Set 6.5
1. a) $a^2 - b^2 = (a + b)(a - b)$ **b)** Answers will vary.   **3. a)** $a^3 + b^3 = (a + b)(a^2 - ab + b^2)$
b) Answers will vary.   **5.** No   **7.** Prime   **9.** $3(b^2 + 16)$ **11.** $4(4m^2 + 9n^2)$ **13.** $(y + 5)(y - 5)$ **15.** $(9 + z)(9 - z)$
17. $(x + 7)(x - 7)$ **19.** $(x + y)(x - y)$ **21.** $(3y + 5z)(3y - 5z)$ **23.** $4(4a + 3b)(4a - 3b)$ **25.** $(6 + 7x)(6 - 7x)$
27. $(z^2 + 9x)(z^2 - 9x)$ **29.** $(5x^2 + 7y^2)(5x^2 - 7y^2)$ **31.** $(6m^2 + 7n)(6m^2 - 7n)$ **33.** $2(x^2 + 5y)(x^2 - 5y)$
35. $(x^2 + 9)(x + 3)(x - 3)$ **37.** $(x + y)(x^2 - xy + y^2)$ **39.** $(x - y)(x^2 + xy + y^2)$ **41.** $(x + 4)(x^2 - 4x + 16)$
43. $(x - 3)(x^2 + 3x + 9)$ **45.** $(a + 1)(a^2 - a + 1)$ **47.** $(3x - 1)(9x^2 + 3x + 1)$ **49.** $(3a - 5)(9a^2 + 15a + 25)$
51. $(3 - 2y)(9 + 6y + 4y^2)$ **53.** $(4m + 3n)(16m^2 - 12mn + 9n^2)$ **55.** $(2a - 3b)(4a^2 + 6ab + 9b^2)$ **57.** $4(x - 3)^2$
59. $2(5x + 2)(5x - 3)$ **61.** $2(d + 4)^2$ **63.** $5(x - 3)(x + 1)$ **65.** $5(x + 2)(x - 2)$ **67.** $2(x + 5)(x - 5)$

69. $2y(x + 3)(x - 3)$ **71.** $3y^2(x + 1)(x^2 - x + 1)$ **73.** $2(x - 2)(x^2 + 2x + 4)$ **75.** $2(3x + 5)(3x - 5)$ **77.** $3r(2t^2 - 5t + 7)$
79. $2(3x - 2)(x + 4)$ **81.** $2r(s + 3)(s - 8)$ **83.** $(x + 2)(4x - 3)$ **85.** $25(b + 2)(b - 2)$ **87.** $a^3b^2(a + 2b)(a - 2b)$
89. $5x^2(x + 1)^2$ **91.** $x(x^2 + 25)$ **93.** $(y^2 + 4)(y + 2)(y - 2)$ **95.** $2(2m + 5)(4m^2 - 10m + 25)$ **97.** $(a + b)(c + 2)$
99. $9(1 + y^2)(1 + y)(1 - y)$ **101.** You cannot divide both sides of the equation by $a - b$ because it equals 0.
103. $2\blacklozenge^4(\blacklozenge^2 + 2\text{✳}^2)$ **105.** $(x^2 - 3y^3)(x^4 + 3x^2y^3 + 9y^6)$ **107.** $(x - 3 + 2y)(x - 3 - 2y)$ **109.** $(x + y + 3)(x - y + 7)$
110. $\dfrac{7}{3}$ **111.** 4 inches   **112.** -9 **113.** $\dfrac{8x^9}{27y^{12}}$ **114.** $\dfrac{1}{a^{11}}$

Using Your Graphing Calculator, 6.6   **1.** $4, -2$ **2.** $2, 5$ **3.** $2, 1$ **4.** $-1, 5$

Exercise Set 6.6   **1.** Answers will vary.   **3.** $ax^2 + bx + c = 0, a \neq 0$ **5. a)** The zero-factor property may only be used when

one side of the equation is equal to 0.   **b)** $-2, 3$ **7.** $-6, 7$ **9.** $0, 8$ **11.** $-\dfrac{7}{3}, \dfrac{11}{2}$ **13.** $4, -4$ **15.** $0, 12$ **17.** $0, -7$ **19.** 4

21. $-2, -10$ **23.** $-2, -10$ **25.** $-3, 4$ **27.** $1, -24$ **29.** $-5, -6$ **31.** $5, -3$ **33.** $30, -1$ **35.** $-3, -\dfrac{5}{4}$ **37.** $-\dfrac{1}{3}, -2$ **39.** $\dfrac{2}{3}, -5$

41. $-4, 3$ **43.** $-\dfrac{3}{4}, \dfrac{1}{2}$ **45.** $8, -8$ **47.** $0, 25$ **49.** $10, -10$ **51.** $-2, 5$ **53.** $-2, \dfrac{1}{3}$ **55.** $\dfrac{3}{2}, 6$

57. $x^2 - 2x - 24 = 0$ (other answers are possible)   **59.** $x^2 - 6x = 0$ (other answers are possible)   **61. a)** $(2x - 1)$ and $(3x + 1)$

b) $6x^2 - x - 1 = 0$ **63.** $4, 5$ **65.** $0, 3, -2$ **67.** $\dfrac{17}{45}$ **68. a)** Identity   **b)** Contradiction   **69.** ≈ 738 people   **70.** $\dfrac{9}{p^8q^2}$
71. Monomial   **72.** Binomial   **73.** Not a polynomial   **74.** Trinomial

Exercise Set 6.7   **1.** A triangle with a 90° angle   **3.** $(\text{leg})^2 + (\text{leg})^2 = (\text{hypotenuse})^2$ or $a^2 + b^2 = c^2$ **5.** 3   **7.** 13   **9.** 18
11. 39   **13.** $9, 13$ **15.** $6, 14$ **17.** $16, 18$ **19.** Width: 3 feet; length: 12 feet   **21.** Width: 10 feet; length: 15 feet   **23.** 4 meters
25. 4 seconds   **27.** Yes   **29.** Yes   **31.** 16 feet   **33.** 41 feet   **35.** 6 feet, 8 feet, 10 feet   **37.** Width: 9 inches; length: 12 inches
39. Width: 7 feet; length: 24 feet   **41.** 30 books   **43. a)** 4   **b)** 9   **45.** 432 square feet   **47.** $0, 8, -4$ **49.** $x^3 - x^2 - 6x = 0$
51. 3 and 6   **55.** $3x - 7$ **56.** $-x^2 + 7x - 4$ **57.** $6x^3 + x^2 - 10x + 4$ **58.** $2x - 3$ **59.** $2x - 3$

Chapter 6 Review Exercises   **1.** y^3 **2.** $3p$ **3.** $6c^2$ **4.** $5x^2y^2$ **5.** 1   **6.** s **7.** $x - 3$ **8.** $x + 5$ **9.** $7(x - 5)$
10. $5(7x - 1)$ **11.** $4y(6y - 1)$ **12.** $5p^2(11p - 4)$ **13.** $12ab(5a - 3b)$ **14.** $9xy(1 - 4x^2y)$ **15.** $4x^3y^2(5 + 2x^6y - 4x^2)$
16. Prime   **17.** Prime   **18.** $(5x + 3)(x - 2)$ **19.** $(x - 1)(3x + 4)$ **20.** $(4x - 3)(2x + 1)$ **21.** $(x + 6)(x + 2)$
22. $(x - 5)(x + 4)$ **23.** $(y - 6)^2$ **24.** $(y + 1)(3x + 2)$ **25.** $(a - b)(4a - 1)$ **26.** $(x + 6)(2x - 1)$ **27.** $(x + 3)(x - 2y)$
28. $(5x - y)(x + 4y)$ **29.** $(x + 3y)(4x - 5y)$ **30.** $(3a - 5b)(2a - b)$ **31.** $(p - 3)(q + 4)$ **32.** $(x - 3y)(3x + 2y)$
33. $(a + 2b)(7a - b)$ **34.** $(2x - 1)(4x + 3)$ **35.** $(x - 3)(x + 2)$ **36.** Prime   **37.** $(x + 2)(x + 9)$ **38.** $(n + 8)(n - 5)$
39. $(b + 5)(b - 4)$ **40.** $(x - 8)(x - 7)$ **41.** Prime   **42.** Prime   **43.** $x(x - 9)(x - 8)$ **44.** $t(t - 9)(t + 4)$
45. $(x + 3y)(x - 5y)$ **46.** $4x(x + 5y)(x + 3y)$ **47.** $(2x + 5)(x - 3)$ **48.** $(6x + 1)(x - 5)$ **49.** $(4x - 5)(x - 1)$
50. $(5m - 4)(m - 2)$ **51.** $(4y + 3)(4y - 1)$ **52.** $(5x - 2)(x - 6)$ **53.** Prime   **54.** $(5x - 3)(x + 8)$ **55.** $(2s + 1)(3s + 5)$
56. $(3x - 2)(2x + 5)$ **57.** $2(3x + 2)(2x - 1)$ **58.** $(5x - 3)^2$ **59.** $x(3x - 2)^2$ **60.** $2x(3x + 4)(3x - 2)$
61. $(2a - 3b)(2a - 5b)$ **62.** $(8a + b)(2a - 3b)$ **63.** $(x + 10)(x - 10)$ **64.** $(x + 6)(x - 6)$ **65.** $3(x + 4)(x - 4)$
66. $9(3x + y)(3x - y)$ **67.** $(9 + a)(9 - a)$ **68.** $(8 + x)(8 - x)$ **69.** $(4x^2 + 7y)(4x^2 - 7y)$ **70.** $(8x^3 + 7y^3)(8x^3 - 7y^3)$
71. $(a + b)(a^2 - ab + b^2)$ **72.** $(x - y)(x^2 + xy + y^2)$ **73.** $(x - 1)(x^2 + x + 1)$ **74.** $(x + 2)(x^2 - 2x + 4)$
75. $(a + 3)(a^2 - 3a + 9)$ **76.** $(b - 4)(b^2 + 4b + 16)$ **77.** $(5a + b)(25a^2 - 5ab + b^2)$ **78.** $(3 - 2y)(9 + 6y + 4y^2)$
79. $3(x - 4y)(x^2 + 4xy + 16y^2)$ **80.** $3(3x^2 + 5y)(3x^2 - 5y)$ **81.** $(x - 6)(x - 8)$ **82.** $3(x - 3)^2$ **83.** $5(q + 1)(q - 1)$
84. $8(x + 3)(x - 1)$ **85.** $4(y + 3)(y - 3)$ **86.** $(x - 9)(x + 3)$ **87.** $(3x - 1)^2$ **88.** $(7x - 3)(x + 4)$
89. $6(b - 1)(b^2 + b + 1)$ **90.** $y(x - 3)(x^2 + 3x + 9)$ **91.** $b(a + 3)(a - 5)$ **92.** $3x(2x + 3)(x + 5)$ **93.** $(x - 3y)(x - y)$
94. $(3m - 4n)(m + 2n)$ **95.** $(2x + 3y)^2$ **96.** $(5a + 7b)(5a - 7b)$ **97.** $(x + 2)(y - 7)$ **98.** $y^5(4 + 5y)(4 - 5y)$
99. $(2x - 3y)(3x + 7y)$ **100.** $2x(2x + 5y)(x + 2y)$ **101.** $x^2(4x + 1)(4x - 3)$ **102.** $(d^2 + 4)(d + 2)(d - 2)$ **103.** $0, -9$
104. $2, -6$ **105.** $-5, \dfrac{3}{4}$ **106.** $0, -7$ **107.** $0, -5$ **108.** $0, -3$ **109.** $-3, -6$ **110.** $1, 2$ **111.** $-4, 3$ **112.** $-1, -4$ **113.** $2, 4$
114. $3, -5$ **115.** $\dfrac{1}{4}, -\dfrac{3}{2}$ **116.** $-\dfrac{1}{3}, 4$ **117.** $2, -2$ **118.** $\dfrac{10}{7}, -\dfrac{10}{7}$ **119.** $\dfrac{3}{2}, \dfrac{1}{4}$ **120.** $\dfrac{3}{2}, \dfrac{5}{2}$ **121.** $a^2 + b^2 = c^2$ **122.** Hypotenuse
123. 10 feet   **124.** 12 meters   **125.** $9, 11$ **126.** $4, 14$ **127.** Width: 12 feet; length: 15 feet   **128.** 8 feet, 15 feet, 17 feet
129. 9 inches   **130.** 10 feet   **131.** 1 second   **132.** 80 dozen

Chapter 6 Practice Test   **1.** $3y^3$ [6.1]   **2.** $8p^2q^2$ [6.1]   **3.** $5x^2y^2(y - 3x^3)$ [6.1]   **4.** $4a^2b(2a - 3b + 7)$ [6.1]

5. $(x - 5)(4x + 1)$ [6.2]   **6.** $(a - 4b)(a - 5b)$ [6.2]   **7.** $(r + 8)(r - 3)$ [6.3]   **8.** $(5a - 3b)(5a + 2b)$ [6.4]   **9.** $4(x + 2)(x - 6)$ [6.4]

10. $y(2y - 3)(y + 1)$ [6.4]   **11.** $(3x + 2y)(4x - 3y)$ [6.4]   **12.** $(x + 3y)(x - 3y)$ [6.5]   **13.** $(x - 4)(x^2 + 4x + 16)$ [6.5]

14. $\dfrac{5}{6}, -3$ [6.6]   **15.** $0, 6$ [6.6]   **16.** $-8, 8$ [6.6]   **17.** -9 [6.6]   **18.** $3, 4$ [6.6]   **19.** $-2, -3$ [6.6]   **20.** 24 inches [6.7]   **21.** 34 feet [6.7]

22. $4, 9$ [6.7]   **23.** $12, 14$ [6.7]   **24.** Length: 6 meters; width: 4 meters [6.7]   **25.** 10 seconds [6.7]

Cumulative Review Test   **1.** -171 [1.9]   **2.** 9 [1.9]   **3.** $90 [3.2]   **4. a)** 7   **b)** $-6, -0.2, \frac{3}{5}, 7, 0, -\frac{5}{9}, 1.34$ **c)** $\sqrt{7}, -\sqrt{2}$

d) $-6, -0.2, \frac{3}{5}, \sqrt{7}, -\sqrt{2}, 7, 0, -\frac{5}{9}, 1.34$ [1.4]   **5.** $|-8|$ [1.5]   **6.** 13 [2.5]   **7.** 19.2 [2.7]   **8.** 7 gallons [2.7]

9. $y = -\frac{4}{3}x + \frac{7}{3}$ [2.6]   **10.** 6 liters [3.4]   **11.** $\frac{1}{4}$ hour [3.4]   **12.**   [4.4]   **13.** $y = \frac{3}{5}x + \frac{26}{5}$ [4.4]

14. $\frac{16y^6}{x^3}$ [5.2]   **15.** $-3x^3 + 2x^2 + 6x - 12$ [5.4]

$y = -\frac{3}{5}x + 1$

16. $3x^3 + 13x^2 - 28x + 12$ [5.5]   **17.** $x - 5 + \frac{21}{x + 3}$ [5.6]

18. $(r + 2)(q - 8)$ [6.2]   **19.** $(5x + 3)(x - 2)$ [6.4]   **20.** $7y(y + 3)(y - 3)$ [6.5]

Chapter 7

Using Your Graphing Calculator, 7.1   **1.** Yes   **2.** Yes   **3.** No   **4.** Yes

Exercise Set 7.1   **1. a–b)** Answers will vary.   **3.** We assume that the value of the variable does not make the denominator equal to 0.   **5.** There is no factor common in both the numerator and the denominator.   **7.** The denominator cannot be 0.   **9.** $x \ne 2$ **11.** No   **13.** All real numbers except $x = 0$.   **15.** All real numbers except $n = 4$.   **17.** All real numbers except $x = 2$ and $x = -2$.   **19.** All real numbers except $x = \frac{3}{2}$ and $x = 3$.   **21.** All real numbers   **23.** All real numbers except $p = \frac{5}{2}$ and $p = -\frac{5}{2}$

25. $\frac{x}{3y^4}$ **27.** $\frac{4}{b^5}$ **29.** $\frac{2}{1 + y}$ **31.** 5   **33.** $\frac{x^2 + 6x + 7}{2}$ **35.** $r + 1$ **37.** $\frac{x}{x + 2}$ **39.** $\frac{z - 5}{z + 5}$ **41.** $\frac{x + 1}{x + 2}$ **43.** -1

45. $-(x + 2)$ **47.** $-\frac{x + 6}{2x}$ **49.** $-(x + 3)$ **51.** $\frac{1}{4m - 5}$ **53.** $\frac{x - 5}{x + 5}$ **55.** $2x - 3$ **57.** $x - 3$ **59.** $\frac{x - 4}{x + 4}$

61. $a^2 + 2a + 4$ **63.** $3s + 4t$ **65.** $\frac{3}{x - y}$ **67.** $\frac{\smiley}{5}$ **69.** $\frac{\Delta}{2\Delta + 9}$ **71.** -1 **73.** $x + 2; (x + 2)(x - 3) = x^2 - x - 6$

75. $x^2 + 9x + 20; (x + 5)(x + 4) = x^2 + 9x + 20$ **77. a)** $x \ne -3, x \ne 2$ **b)** $\frac{1}{x + 3}$ **79. a)** $x \ne 0, x \ne -5, x \ne \frac{3}{2}$

b) $\frac{1}{x(2x - 3)}$ **81.** 1, the numerator and denominator are identical.   **84.** $y = x - 4z$ **85.** $28°, 58°,$ and $94°$ **86.** $\frac{25}{81x^4y^2}$

87. $8x^2 - 10x - 19$ **88.** $3(a + 4)(a - 6)$ **89.** 13 inches

Exercise Set 7.2   **1.** Answers will vary.   **3.** $x^2 + x - 20$; the numerator must be $(x - 4)(x + 5)$

5. $x^2 + 2x - 35$; the denominator must be $(x + 7)(x - 5)$ **7.** $\frac{6}{19}$ **9.** $-\frac{15}{18}$ **11.** $\frac{5}{16}$ **13.** $\frac{3}{5}$ **15.** $-\frac{13}{48}$ **17.** $\frac{xy}{8}$ **19.** $\frac{70x^4}{y^6}$

21. $\frac{36x^9y^2}{25z^7}$ **23.** $\frac{-3x + 2}{3x + 2}$ **25.** 1   **27.** $\frac{1}{a^2 - b^2}$ **29.** 1   **31.** $\frac{x + 2}{x + 3}$ **33.** $x + 9$ **35.** $4x^2y$ **37.** $\frac{9z}{x}$ **39.** $\frac{11}{6ab^2}$ **41.** $6r^2$ **43.** $x + 9$

45. $\frac{x - 8}{x + 2}$ **47.** $\frac{x + 3}{x - 1}$ **49.** -1 **51.** $\frac{x + 1}{x - 4}$ **53.** $4x^2y^2$ **55.** $\frac{7c}{5ab^2}$ **57.** $\frac{3y^2}{a^2}$ **59.** $\frac{8mx^7}{3y^2}$ **61.** $\frac{2(x + 3)}{x(x - 3)}$ **63.** $\frac{7x}{y}$ **65.** $\frac{4}{m^4n^{11}}$

67. $\frac{r + 2}{r - 3}$ **69.** $\frac{x - 6}{x - 3}$ **71.** $\frac{2w - 7}{w + 1}$ **73.** $\frac{q - 5}{2q + 3}$ **75.** $\frac{2n + 3}{3n - 1}$ **77.** $\frac{1}{6\Delta^3}$ **79.** $\frac{\Delta + \smiley}{9(\Delta - \smiley)}$ **81.** $x^2 + 5x + 6$ **83.** $x^2 - 4x - 12$

85. $x^2 - 3x + 2$ **87.** $\frac{x - 3}{x + 3}$ **89.** $\frac{x - 1}{x - 3}$ **91.** $x^2 - 5x + 6, x^2 - x - 20$ **94.** 1 hour   **95.** $12x^4y^5z^{11}$ **96.** $2x^2 + x - 2 - \frac{2}{2x - 1}$

97. $6(x - 5)(x + 2)$ **98.** $5, -2$

Exercise Set 7.3   **1.** Answers will vary.   **3.** Answers will vary.   **5.** $x(x + 6)$ **7.** $4x(x + 3)$ **9. a)** The negative sign in $-(2x - 7)$ was not distributed.   **b)** $\frac{4x - 3 - 2x + 9}{5x + 4}$ **11. a)** The negative sign in $-(3x^2 - 4x + 5)$ was not distributed.

b) $\frac{8x - 2 - 3x^2 + 4x - 5}{x^2 - 4x + 3}$ **13.** $\frac{6}{7}$ **15.** $\frac{5r - 1}{4}$ **17.** $\frac{x + 6}{x}$ **19.** $\frac{n + 8}{n + 1}$ **21.** $\frac{5x + 9}{x - 3}$ **23.** $\frac{t + 3}{5t^2}$ **25.** $\frac{1}{x - 4}$ **27.** $\frac{1}{m - 3}$

29. $\frac{p - 12}{p - 5}$ **31.** $x - 3$ **33.** $\frac{1}{2}$ **35.** 1   **37.** 4   **39.** $\frac{5}{x - 2}$ **41.** $\frac{3}{4}$ **43.** $\frac{x - 5}{x + 2}$ **45.** $\frac{3x + 2}{x - 4}$ **47.** $\frac{6x + 1}{x - 8}$ **49.** 5   **51.** $9n$

53. $15x$ **55.** p^3 **57.** $3m - 4$ **59.** $6x^2$ **61.** $36x^3y$ **63.** $18r^4s^7$ **65.** $m(m + 2)$ **67.** $x(x + 1)$ **69.** $4n - 1$ or $1 - 4n$

71. $4k - 5r$ or $-4k + 5r$ **73.** $18q(q + 1)$ **75.** $120x^2y^3$ **77.** $6(x + 4)(x + 2)$ **79.** $(x + 1)(x + 8)$ **81.** $(x - 8)(x + 3)(x + 8)$

83. $(a-4)^2(a-3)$ **85.** $(x+5)(x+1)(x+3)$ **87.** $(x-3)^2$ **89.** $(x-6)(x-1)$ **91.** $(3t-2)(t+4)(3t-1)$

93. $(2x+1)^2(4x+3)$ **95.** $\dfrac{19}{35}$ **97.** $\dfrac{35}{36}$ **99.** $\dfrac{1}{18}$ **101.** x^2+x-9; the sum of the numerators must be $2x^2-5x-6$

103. $x^2+9x-10$; the sum of the numerators must be $5x-7$ **105.** 5☺ **107.** $(\Delta+3)(\Delta-3)$ **109.** $\dfrac{-3x^2+12x}{x^2-25}$ **111.** $30x^{12}y^9$

113. $(x-4)(x+3)(x-2)$ **115.** $\dfrac{92}{45}$ or $2\dfrac{2}{45}$ **116.** $-\dfrac{1}{5}$ **117.** 2.25 ounces **118.** 70 hours **119.** 1 **120.** 4.0×10^{11} **121.** $\dfrac{3}{2},-1$

Exercise Set 7.4
1. For each fraction, divide the LCD by the denominator. **3. a)** Answers will vary.
b) $\dfrac{x^2+x-9}{(x+2)(x-3)(x-2)}$ **5. a)** $12z^2$ **b)** $\dfrac{3yz+10}{12z^2}$ **c)** Yes **7.** $\dfrac{3x+2y}{xy}$ **9.** $\dfrac{x+10}{2x^2}$ **11.** $\dfrac{3x+8}{x}$ **13.** $\dfrac{3x+10}{5x^2}$

15. $\dfrac{45y+12x}{20x^2y^2}$ **17.** $\dfrac{4y^2+x}{y}$ **19.** $\dfrac{9a+1}{6a}$ **21.** $\dfrac{6x^2+2y}{xy}$ **23.** $\dfrac{45a^2-4b}{5a^2b}$ **25.** $\dfrac{13x-12}{x(x-3)}$ **27.** $\dfrac{11p+6}{p(p+3)}$ **29.** $\dfrac{-d^2+14d+25}{(d+1)(3d+5)}$

31. $\dfrac{6}{p-3}$ **33.** $\dfrac{14}{x+7}$ **35.** $\dfrac{a+16}{2(a-2)}$ **37.** $\dfrac{20x}{(x-5)(x+5)}$ **39.** $\dfrac{-7n-6}{3n(2n+1)}$ **41.** $\dfrac{15w+66}{2(w+5)(w+2)}$ **43.** $\dfrac{5z-16}{(z+4)(z-4)}$

45. $\dfrac{-x+6}{(x+2)(x-2)}$ **47.** $\dfrac{r+12}{(r-4)(r-6)}$ **49.** $\dfrac{6x-11}{(x+4)(x-2)}$ **51.** $\dfrac{x^2+3x-21}{(x+5)^2}$ **53.** $\dfrac{-a+16}{(a-8)(a-1)(a+2)}$

55. $\dfrac{9x+17}{(x+3)^2(x-2)}$ **57.** $\dfrac{3x^2-12x-5}{(2x+1)(3x-2)(x+3)}$ **59.** $\dfrac{2x^2-3x-4}{(4x+3)(x+2)(2x-1)}$ **61.** $\dfrac{1}{w-3}$ **63.** $\dfrac{6}{r-3}$ **65.** $-\dfrac{x+4}{4x}$

67. All real numbers except $x=0$. **69.** All real numbers except $x=4$ and $x=-6$. **71.** $\dfrac{7}{\Delta-2}$ **73.** All real numbers except
$a=-b$ and $a=0$. **75.** 0 **77.** $\dfrac{2x-3}{2-x}$ **79.** $\dfrac{6x+5}{(x+2)(x-3)(x+1)}$ **82.** ≈1.53 hours **83.** $-\dfrac{8}{7}$ **84.** $4x-3-\dfrac{6}{2x+3}$ **85.** -1

Mid-Chapter Test
1. All real numbers except $x=\dfrac{2}{3}$ [7.1] **2.** All real numbers except $x=-2,x=7$ [7.1] **3.** 9 [7.1]
4. $\dfrac{2x+3}{3x-1}$ [7.1] **5.** $5r+6t$ [7.1] **6.** $\dfrac{6y^3}{x^3}$ [7.2] **7.** $-(m+4)$ or $-m-4$ [7.2] **8.** $x-2$ [7.2] **9.** $\dfrac{x+7}{2(x+1)}$ [7.2]
10. $\dfrac{5x+2}{7x+3}$ [7.2] **11.** $x-6$ [7.3] **12.** $x-3$ [7.3] **13.** $\dfrac{3x+2}{4x-1}$ [7.3] **14.** $3m(2m+1)$ [7.3] **15.** $(2x+3)(x-4)(x-5)$ [7.3]
16. $\dfrac{13x-1}{10x}$ [7.4] **17.** $-\dfrac{a^2+13a+23}{a^2-a-12}$ [7.4] **18.** $\dfrac{4x^2+17x-1}{2x^2+13x+6}$ [7.4] **19.** $\dfrac{x^2-7x-4}{(x+1)(x+2)(x-3)}$ [7.4]
20. Need common denominator of $x(x+1)$, $\dfrac{15x+8}{x(x+1)}$ [7.4]

Exercise Set 7.5
1. A complex fraction is a fraction whose numerator or denominator (or both) contains a fraction.
3. a) Numerator, $\dfrac{x+9}{4}$; denominator, $\dfrac{7}{x^2+5x+6}$ **b)** Numerator, $\dfrac{1}{2y}+x$; denominator, $\dfrac{3}{y}+x^2$ **5.** $\dfrac{7}{8}$ **7.** $\dfrac{57}{32}$ **9.** $\dfrac{11}{6}$ **11.** $\dfrac{x^3y^2}{21}$
13. $\dfrac{2ab^3}{21c^2}$ **15.** $\dfrac{ab-a}{3+a}$ **17.** $\dfrac{3}{x}$ **19.** $\dfrac{5x-1}{4x-1}$ **21.** $\dfrac{m-n}{m}$ **23.** $-\dfrac{a}{b}$ **25.** -1 **27.** 1 **29.** $b-a$ **31.** $\dfrac{a^2+b}{b(b+1)}$ **33.** $\dfrac{x^2y}{y-x}$
35. $\dfrac{ab^2+b^2}{a^2(b+1)}$ **37. b)–c)** $-\dfrac{224}{155}$ **39. b)–c)** $\dfrac{x-y+6}{2x+2y-7}$ **41. a)** $\dfrac{\frac{5}{12x}}{\frac{8}{x^2}-\frac{4}{3x}}$ **b)** $\dfrac{5x}{96-16x}$ **43.** $\dfrac{y+x}{3xy}$ **45.** $x+y$ **47. a)** $\dfrac{2}{7}$ **b)** $\dfrac{4}{13}$
49. $\dfrac{a^3b+a^2b^3-ab^2}{a^3-ab^3+3b^2}$ **51.** $\dfrac{17}{2}$ **52.** A polynomial is an expression containing a finite number of terms of the form ax^n where a is a
real number and n is a whole number. **53.** $(x-5)(x-8)$ **54.** $\dfrac{x^2-9x+2}{(3x-1)(x+6)(x-3)}$

Exercise Set 7.6
1. a) Answers will vary. **b)** $\dfrac{2}{3}$ **3. a)** The problem on the left is an expression to be simplified while the
problem on the right is an equation to be solved. **b)** Left: Write the fractions with the LCD, $12(x-1)$, then combine numerators;
Right: Multiply both sides of the equation by the LCD, $12(x-1)$, then solve. **c)** Left: $\dfrac{x^2-x+12}{12(x-1)}$; Right: $4,-3$ **5.** You need to
check for extraneous solutions when there is a variable in a denominator. **7.** 2 **9.** No **11.** Yes **13.** 12 **15.** -4 **17.** -20 **19.** 30
21. 4 **23.** $\dfrac{25}{6}$ **25.** 36 **27.** 8 **29.** -8 **31.** 3 **33.** 4 **35.** 2 **37.** 5 **39.** No solution **41.** 7 **43.** 15 **45.** No solution **47.** -3

49. 38   **51.** $-\frac{1}{3}, 3$ **53.** 6, -1 **55.** 4, -4 **57.** $-4, -5$ **59.** 4   **61.** $-\frac{5}{2}$ **63.** No solution   **65.** 24   **67.** -12 **69.** -3 **71.** 3

73. 5   **75.** 0   **77.** x can be any real number since the sum on the left is also $\frac{2x-4}{3}$.   **79.** 15 centimeters   **81.** -4 **83.** No, it is impossible for both sides of the equation to be equal.   **85.** More than $6\frac{1}{3}$ hours   **86.** 150 minutes   **87.** $40°, 140°$ **88.** 6.8×10^8

89. Linear equation: $ax + b = c$, $a \neq 0$; quadratic equation: $ax^2 + bx + c = 0$, $a \neq 0$

Exercise Set 7.7

1. Some examples are $A = \frac{1}{2}bh$, $A = \frac{1}{2}h(b_1 + b_2)$, $V = \frac{1}{3}\pi r^2 h$, and $V = \frac{4}{3}\pi r^3$

3. It represents 1 complete task.   **5.** length = 11 inches, width = 9 inches   **7.** base = 12 centimeters, height = 7 centimeters

9. Base: 8 feet   **11.** $\frac{8}{9}, 8$ **13.** 7   **15.** 9 miles per hour   **17.** 15 miles   **19.** 150 miles per hour, 600 miles per hour

21. No wake zone: ≈ 1.83 miles; zone to Island: ≈ 34.77 miles   **23.** 3 hours at 600 miles per hour and 2 hours at 550 miles per hour

25. 1200 feet   **27.** $3\frac{3}{7}$ hours   **29.** 4 hours   **31.** 24 minutes   **33.** 120 minutes or 2 hours.   **35.** 60 minutes or 1 hour   **37.** $8\frac{3}{4}$ days

39. $\frac{6}{5}$ hours or 1 hour, 12 minutes   **41.** 300 hours   **43.** 3 or $\frac{1}{2}$ **45.** 8 pints   **47.** $-\frac{3}{2}x - \frac{7}{2}$ **48.**

49. 1   **50.** $\dfrac{3x^2 - 9x - 15}{(2x + 3)(3x - 5)(3x + 1)}$

Exercise Set 7.8

1. a) As one quantity increases, the other increases.   **b), c)** Answers will vary.   **3.** One quantity varies as a product of two or more quantities.   **5. a)** Decrease   **b)** Inverse variation; by the definition of inverse variation   **7.** Direct
9. Inverse   **11.** Direct   **13.** Direct   **15.** Direct   **17.** Inverse   **19.** Direct   **21.** Inverse   **23.** Inverse   **25. a)** $x = ky$ **b)** 72

27. a) $y = kR$ **b)** 306   **29. a)** $R = \dfrac{k}{W}$ **b)** $\dfrac{1}{20}$ **31. a)** $A = \dfrac{kB}{C}$ **b)** 9   **33. a)** $x = ky$ **b)** 20   **35. a)** $y = kR^2$ **b)** 20

37. a) $S = \dfrac{k}{G}$ **b)** 0.96   **39. a)** $x = \dfrac{k}{P^2}$ **b)** 25   **41. a)** $F = \dfrac{kM_1 M_2}{d}$ **b)** 40   **43.** Doubled   **45.** Halved   **47.** Doubled

49. Unchanged   **51.** Doubled   **53.** $y = \dfrac{k}{x}$; $k = 5$ **55.** \$8814   **57.** 3096 milligrams   **59.** 1.05 inches   **61.** 6400 cubic centimeters

63. 3.12 hours   **65.** 45 foot-candles   **67.** 117.6 feet   **69.** 126 cubic meters   **71.** 4600 DVDs   **73.** ≈ 133.25 pounds

75. $\approx 121{,}528$ calls   **77.** $\dfrac{1}{49}$ of the light of the flash   **79. a)** $P = 14.7 + kx$ **b)** 0.43   **c)** ≈ 337.9 feet   **80.** $2x - 3 + \dfrac{6}{4x + 9}$

81. $(z - 2)(y + 8)$ **82.** $-4, 2$ **83.** $x + 8$

Chapter 7 Review Exercises

1. All real numbers except $x = 9$ **2.** All real numbers except $x = 3$ and $x = 5$

3. All real numbers except $x = \frac{1}{5}$ and $x = -1$ **4.** $\dfrac{1}{x - 3}$ **5.** $x^2 + 5x + 12$ **6.** $3x + y$ **7.** $x + 4$ **8.** $a + 9$ **9.** $-(2x + 1)$

10. $\dfrac{b - 2}{b + 2}$ **11.** $\dfrac{x - 3}{x - 2}$ **12.** $\dfrac{x - 8}{2x + 3}$ **13.** $\dfrac{5}{12b^2}$ **14.** $12xz^2$ **15.** $\dfrac{8b^3 c^4}{a^2}$ **16.** $-\dfrac{1}{9}$ **17.** $-\dfrac{2}{3}$ **18.** 1   **19.** $\dfrac{36x^2}{y}$ **20.** $\dfrac{20z}{x^3}$ **21.** $\dfrac{6}{a - b}$

22. $\dfrac{1}{8(a + 3)}$ **23.** 1   **24.** $6y(x - y)$ **25.** $\dfrac{n - 2}{n + 5}$ **26.** 4   **27.** 5   **28.** $\dfrac{4}{x + 10}$ **29.** $4h - 3$ **30.** $3x + 4$ **31.** 24   **32.** $x + 3$

33. $20x^2 y^3$ **34.** $x(x - 3)$ **35.** $(n + 5)(n - 4)$ **36.** $x(x + 2)$ **37.** $(r + s)(r - s)$ **38.** $x - 9$ **39.** $(x + 7)(x - 5)(x + 2)$

40. $\dfrac{3y^2 + 10}{6y^2}$ **41.** $\dfrac{12x + y}{4xy}$ **42.** $\dfrac{5x^2 - 18y}{3x^2 y}$ **43.** $\dfrac{7x + 12}{x + 2}$ **44.** $\dfrac{x^2 - 2xy - y^2}{xy}$ **45.** $\dfrac{9x + 8}{x(x + 4)}$ **46.** $\dfrac{-x - 4}{3x(x - 2)}$ **47.** $\dfrac{z + 14}{(z + 5)^2}$

48. $\dfrac{2x - 8}{(x - 3)(x - 5)}$ **49.** $\dfrac{5x + 38}{(x + 6)(x + 2)}$ **50.** $\dfrac{3x - 8}{x - 4}$ **51.** $\dfrac{5ab + 10b}{a - 2}$ **52.** $\dfrac{3x - 1}{(x + 3)(x - 3)}$ **53.** $\dfrac{6p^2}{q}$ **54.** $\dfrac{2x + 2}{(x + 2)(x - 3)(x - 2)}$

55. $\dfrac{8x - 29}{(x + 2)(x - 7)(x + 7)}$ **56.** $\dfrac{x}{x + y}$ **57.** $\dfrac{3(x + 3y)}{5(x - 3y)}$ **58.** $a - 3$ **59.** $\dfrac{3a^2 - 8a + 3}{(a + 1)(a - 1)(3a - 5)}$ **60.** $\dfrac{4x - 8}{x}$ **61.** $\dfrac{68}{9}$ **62.** $\dfrac{2}{3}$

63. $\dfrac{bc}{3}$ **64.** $\dfrac{8x^3 z^2}{y^3}$ **65.** $\dfrac{ab - a}{a + 1}$ **66.** $\dfrac{r^2 s + 7}{s^3}$ **67.** $\dfrac{3x + 2}{x(5x - 1)}$ **68.** $\dfrac{4}{x}$ **69.** x **70.** $\dfrac{2a + 1}{4}$ **71.** $\dfrac{-x + 1}{x + 1}$ **72.** $\dfrac{8x^2 - x^2 y}{y(y - x)}$ **73.** 15

74. 6   **75.** 12   **76.** 20   **77.** -16 **78.** $\dfrac{1}{2}$ **79.** -6 **80.** 28   **81.** No solution   **82.** 2.4 hours   **83.** $16\frac{4}{5}$ hours   **84.** $\frac{1}{6}, 1$

85. Robert: 2.1 miles per hour; Tran: 5.6 miles per hour   **86.** $\dfrac{25}{2}$ **87.** ≈ 426.7 **88.** 273 milligrams   **89.** 1.68 hours

Chapter 7 Practice Test

1. 1 [7.1]   **2.** $\dfrac{x^2 + x + 1}{x + 1}$ [7.1]   **3.** $\dfrac{8x^2z}{y}$ [7.2]   **4.** $a + 3$ [7.2]   **5.** $\dfrac{x^2 - 6x + 9}{(x + 3)(x + 2)}$ [7.2]

6. -1 [7.2]   **7.** $\dfrac{x - 2y}{5}$ [7.2]   **8.** $\dfrac{3}{y + 5}$ [7.2]   **9.** $-\dfrac{m + 6}{m - 5}$ [7.2]   **10.** $\dfrac{3x - 1}{4y}$ [7.3]   **11.** $\dfrac{7x^2 - 6x - 13}{x + 3}$ [7.3]   **12.** $\dfrac{2y^2 - 8}{xy^3}$ [7.4]

13. $-\dfrac{2z + 15}{z - 5}$ [7.4]   **14.** $\dfrac{-1}{(x + 4)(x - 4)}$ [7.4]   **15.** $\dfrac{25}{28}$ [7.5]   **16.** $\dfrac{x^2 + x^2y}{7y}$ [7.5]   **17.** $\dfrac{4x + 3}{9 - 5x}$ [7.5]   **18.** 2 [7.6]   **19.** $-\dfrac{12}{7}$ [7.6]

20. 12 [7.6]   **21.** 6 hours [7.7]   **22.** 1 [7.7]   **23.** Base: 6 inches; height: 10 inches [7.7]   **24.** 2 miles [7.7]   **25.** ≈ 1.13 feet [7.8]

Cumulative Review Test

1. 121 [1.9]   **2.** 36 [1.9]   **3.** $\dfrac{17}{8}$ [2.5]   **4.** $c + (c + 0.012c)$ [3.1]   **5.** 0 [4.3]   **6.** $y = \dfrac{3}{4}x - 3$ [4.4]

7. $8x^2 + 5x + 14$ [5.4]   **8.** $6n^3 - 23n^2 + 26n - 15$ [5.5]   **9.** $\dfrac{1}{2}x - \dfrac{19}{4}$ [5.6]   **10.** $(8a - 5)(a - 1)$ [6.2]   **11.** $13(x + 3)(x - 1)$ [6.3]

12. $4, \dfrac{3}{2}$ [6.6]   **13.** $\dfrac{x + 4}{2x + 1}$ [7.2]   **14.** $\dfrac{r^2 - 8r - 6}{(r + 2)(r - 5)}$ [7.4]   **15.** $\dfrac{10x - 18}{(x - 5)(x + 2)(x + 3)}$ [7.4]   **16.** $-\dfrac{3}{2}$ [7.6]   **17.** No solution [7.6]

18. $3000 [3.2]   **19.** 20 pounds sunflower seed, 30 pounds premixed assorted seed mix; [3.4]   **20.** First leg: 3.25 miles, second leg: 9.5 miles [3.4]

Chapter 8

Exercise Set 8.1

1. No   **3.** 0   **5.**   **7.**   **9.** 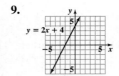   **11.**

13.   **15.** 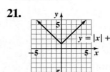   **17.**   **19.**   **21.**   **23.**

25.   **27.**   **29.** 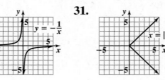   **31.**   **33.**   **35.**

37.    **39.**   **41.** Yes, the coordinates satisfy the equation

43. a)   **b)** 8 square units

45. a) 6975 yards   **b)** 7300 yards   **c)** 1990, 2000, 2005   **d)** no

47. 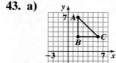   **a)** Each graph crosses the y-axis at the point corresponding to the constant term in the graph's equation.   **b)** Yes   **49.** The rate of change is 2.   **51.** The rate of change is 3.

53. $(4, -3)$, $(5, 1)$, other answers possible   **55.** c   **57.** a   **59.** d   **61.** b   **63.** b   **65.** d   **67.** b   **69.** d

71. a) 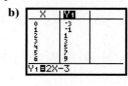   **b)** 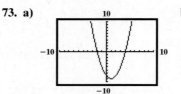   **73. a)**    **b)**

75. a)
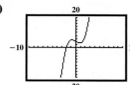

b)

77.

81. 26   **82.** $\dfrac{4}{11}$
83. -6 **84.** -1

Exercise Set 8.2   1. A correspondence where each member in the domain corresponds to exactly one member of the range
3. Yes, a relation is any set of ordered pairs.   **5.** $\{5, 6, 7, 8\}$ **7.** If a vertical line drawn through any part of the graph intersects the graph at more than one point, the graph is not a function.   **9.** The set of values for the dependent variable   **11.** Domain: $\mathbb{R}$, Range: $\mathbb{R}$ **13.** If y depends on x, then y is the dependent variable.   **15.** f of x **17. a)** Function   **b)** Domain: $\{3, 5, 11\}$, Range: $\{6, 10, 22\}$ **19. a)** Function   **b)** Domain: $\{$Cameron, Tyrone, Vishnu$\}$, Range: $\{3, 6\}$ **21. a)** Not a function   **b)** Domain: $\{1990, 2001, 2002\}$; Range: $\{20, 34, 37\}$ **23. a)** Function   **b)** Domain: $\{1, 2, 3, 4, 5\}$; Range: $\{1, 2, 3, 4, 5\}$ **25. a)** Function   **b)** Domain: $\{1, 2, 3, 4, 5, 7\}$; Range: $\{-1, 0, 2, 4, 9\}$ **27. a)** Not a function   **b)** Domain: $\{1, 2, 3\}$;   Range: $\{1, 2, 4, 5, 6\}$ **29. a)** Not a function   **b)** Domain: $\{0, 1, 2\}$; Range: $\{-7, -1, 2, 3\}$ **31.** $A = \{0\}$ **33.** $C = \{18, 20\}$ **35.** $E = \{0, 1, 2\}$ **37.** $H = \{0, 7, 14, 21, \ldots\}$

39. $J = \{1, 2, 3, 4, \ldots\}$ or $J = N$ **41. a)** Set A is the set of all x such that x is a natural number less than 7   **b)** $A = \{1, 2, 3, 4, 5, 6\}$

43.   **45.**   **47.**   **49.**   **51.**   **53.** $\{x | x \geq 1\}$
55. $\{x | x < 5 \text{ and } x \in I\}$ or $\{x | x \leq 4 \text{ and } x \in I\}$ **57.** $\{x | -3 < x \leq 5\}$ **59.** $\{x | -2.5 \leq x < 4.2\}$ **61.** $\{x | -3 \leq x \leq 1 \text{ and } x \in I\}$
63. a) Function   **b)** Domain: $\mathbb{R}$; Range: $\mathbb{R}$ **c)** 2   **65. a)** Not a function   **b)** Domain: $\{x | 0 \leq x \leq 2\}$, Range: $\{y | -3 \leq y \leq 3\}$
c) ≈ 1.5 **67. a)** Function   **b)** Domain: $\mathbb{R}$, Range: $\{y | y \geq 0\}$ **c)** $-3, -1$ **69. a)** Function   **b)** Domain: $\{-1, 0, 1, 2, 3\}$, Range:
$\{-1, 0, 1, 2, 3\}$ **c)** 2   **71. a)** Not a function   **b)** Domain: $\{x | x \geq 2\}$, Range: $\mathbb{R}$ **c)** 3   **73. a)** Function   **b)** Domain: $\{x | -2 \leq x \leq 2\}$,
Range: $\{y | -1 \leq y \leq 2\}$ **c)** $-2, 2$ **75. a)** 3   **b)** 13   **77. a)** -6 **b)** -4 **79. a)** 2   **b)** 2   **81. a)** 7   **b)** 0   **83. a)** 0   **b)** 3   **85. a)** 1

b) Undefined   **87. a)** 24 square feet   **b)** 39 square feet   **89. a)** $A(r) = \pi r^2$ **b)** ≈ 452.4 square yards   **91. a)** $C(F) = \dfrac{5}{9}(F - 32)$

b) $-35°C$ **93. a)** $18.23°C$ **b)** $27.68°C$ **95. a)** $78.32°$ **b)** $73.04°$ **97. a)** 91 oranges   **b)** 204 oranges   **99.** Answers will vary. One possible interpretation: The person warms up slowly, possibly by walking, for 5 minutes. Then the person begins jogging slowly over a period of 5 minutes. For the next 15 minutes the person jogs. For the next 5 minutes the person walks slowly and his heart rate decreases to his normal heart rate. The rate stays the same for the next 5 minutes.   **101.** Answers will vary. One possible interpretation: The man walks on level ground, about 30 feet above sea level, for 5 minutes. For the next 5 minutes he walks uphill to 45 feet above sea level. For 5 minutes he walks on level ground, then walks quickly downhill for 3 minutes to an elevation of 20 feet above sea level. For 7 minutes he walks on level ground. Then he walks quickly uphill for 5 minutes.   **103.** Answers will vary. One possible interpretation: Driver is in stop-and-go traffic, then gets on highway for about 15 minutes, then stops car for a couple of minutes, then stop-and-go traffic.   **105. a)** Yes   **b)** Year   **c)** $218,600   **d)** $865,000   **e)** $\approx 144.8\%$ **107. a)** Yes   **b)** ≈ 6.0 million   **c)** ≈ 4.4 million

d) Yes   **e)** 2006 to 2007   **109. a)**

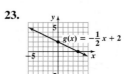

b) No. It is not a straight line.   **c)** $2,300,000

111. a)
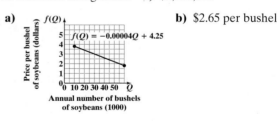

b) $2.65 per bushel

113.
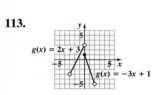

114. $\dfrac{1}{2}$ **115.** $p_2 = \dfrac{E - a_1 p_1 - a_3 p_3}{a_2}$ **116.** $y = -\dfrac{1}{2}x + \dfrac{3}{2}, m = -\dfrac{1}{2}, \left(0, \dfrac{3}{2}\right)$ **117.** $\dfrac{3x + 1}{x - 1}$

Exercise Set 8.3   1. A straight line   **3.** First replace $f(x)$ with y. To find the x-intercept, set $y = 0$ and solve for x. To find the y-intercept, set $x = 0$ and solve for y.   **5.** Graph both sides of the equation. The solution is the x-coordinate of the intersection.

7.
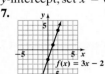

9.

11.
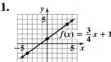

13.
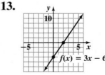

15.

17.
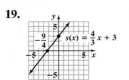

19.

21.
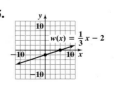

23.

25.

27.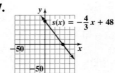

29.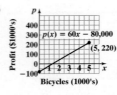

31. a) 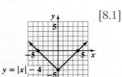   **b)** 1300 bicycles   **c)** 3800 bicycles

33. a) $s(x) = 500 + 0.15x$ **b)**   **c)** \$950   **d)** \$4000

35. a) There is only one y-value for each x-value.   **b)** Independent: length; Dependent: weight   **c)** Yes   **d)** 11.5 kilograms   **e)** 65 centimeters   **f)** 12.0–15.5 kilograms   **g)** Increases; yes, as babies get older their weights vary more.   **37.** When the graph goes through the origin, because at the origin both x and y are zero.

39. Answers will vary. One possible answer is $f(x) = 4$.   **41.** Both intercepts will be at 0.

43. a)   **b)** 2 (or -2) units   **c)** 4 (or -4) units   **d)** $\frac{1}{2}$; slope

45. 1   **47.** -3 **49.** $(-3.2, 0), (0, 6.4)$ **51.** $(-2, 0), (0, -2.5)$ **53.** $3x^2 - 12x - 88$ **54.** $3(x + 4)(x - 8)$ **55.** $\dfrac{1}{x + 2}$ **56.** 4, 6

Mid-Chapter Test

1. 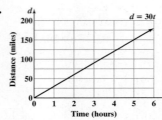   [8.1]   **2.** 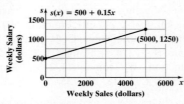   [8.1]   **3.** 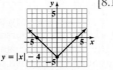   [8.1]

4.   [8.1]   **5. a)** A relation is any set of ordered pairs.   **b)** A function is a correspondence between a first set of elements, the domain, and a second set of elements, the range, such that each element of the domain corresponds to exactly one element in the range.   **c)** No   **d)** Yes [8.2]   **6.** Function; Domain: $\{1, 2, 7, -5\}$, Range: $\{5, -3, -1, 6\}$ [8.2]   **7.** Not a function; Domain: $\{x | -2 \le x \le 2\}$, Range: $\{y | -4 \le y \le 4\}$ [8.2]   **8.** Function; Domain: $\{x | -5 \le x \le 3\}$, Range: $\{y | -1 \le y \le 3\}$ [8.2]   **9.** $E = \{9, 18, 27, 36, \ldots\}$ [8.2]   **10.**   [8.2]   **11.** -21 [8.2]   **12.** 105 feet [8.2]

13.   [8.3]   **14.** 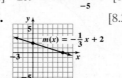   [8.3]   **15.**   [8.3]

16. a) 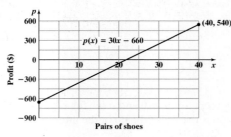   **b)** 22 pairs of shoes   **c)** 34 pairs of shoes [8.3]

Exercise Set 8.4

1. The change in y for a unit change in x **3. a)** Moved up 3 units   **b)** $(0, 1)$ **5.** Two lines are perpendicular if their slopes are negative reciprocals or if one line is vertical and the other is horizontal.

7. $m = -3; f(x) = -3x$ **9.** Slope is undefined; $x = -2$ **11.** $m = -\frac{1}{3}; f(x) = -\frac{1}{3}x + 2$ **13.** $m = -\frac{3}{2}; f(x) = -\frac{3}{2}x + 15$

15.   **17.** 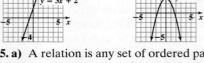   **19.** Parallel   **21.** Neither   **23.** Perpendicular   **25.** Perpendicular   **27.** Parallel   **29.** Neither   **31.** Perpendicular   **33.** Parallel   **35.** Neither   **37.** $y = 2x + 1$ **39.** $2x - 5y = 19$ **41.** $y = -\frac{5}{3}x + 5$ **43.** $f(x) = -3x + 13$ **45.** $y = -\frac{2}{3}x + 6$ **47.** 0.2   **49. a)** 11.3   **b)** Positive   **c)** 7.075

51. a–b)

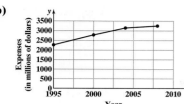

c) $123.8, 64.25, 31.75$ **d)** 1995–2000, because its line segment has the greatest slope
53. a) $h(x) = -x + 200$ **b)** 186 beats per minute
55. a) $M(t) \approx 19.34t + 159.5$ **b)** \$275.54 billion **c)** \$410.92 billion **d)** 2006
57. a) $P(t) = -1.1t + 19.4$ **b)** Negative **c)** 17.2% **d)** 9.5%
59. a) $P(t) = 8300t + 110{,}500$ **b)** \$152,000 **c)** \$235,000 **d)** 2005
61. a) $C(s) = 45.7s + 95.8$ **b)** 324.3 calories
63. a) $d(p) = -0.20p + 90$ **b)** 38 DVD players **c)** \$225
65. a) $s(p) = 95p - 60$ **b)** 206 kites **c)** \$3.00 **67. a)** $i(t) = 12.5t$ **b)** \$1500
c) 176 tickets **69. a)** $r(w) = 0.01w + 10$ **b)** \$46.13 **c)** 5000 pounds **71. a)** $y(a) = -0.865a + 79.25$ **b)** 47.2 years
c) 62.7 years old **73. a)** $w(a) \approx 0.189a + 10.6$ **b)** 14.758 kilograms **75.** $(0, 4)$ **77.** $(0, -3)$ **79. a)** $y = 3x + 1$
b) $y = 3x - 5$ **81. a)** 1 **b)** $(0, 2)$ **c)** $y = x + 2$ **83.** $y = \dfrac{3}{2}x - 7$ **85.** The y-intercept is wrong. **87.** The slope is wrong.

89. Height: 14.2 inches, width: 6.4 inches **93.** 19 **94.** 2.4 **95.** $-\dfrac{92}{5}$ **96. a)** Any set of ordered pairs **b)** A correspondence where
each member of the domain corresponds to a unique member of the range **c)** Answers will vary. **97.** Domain: $\{3, 4, 5, 6\}$;
Range: $\{-2, -1, 2, 3\}$

Exercise Set 8.5

1. Yes, this is how addition of functions is defined. **3.** $g(x) \neq 0$ since division by zero is undefined. **5.** No,
subtraction is not commutative. One example is $5 - 3 = 2$ but $3 - 5 = -2$ **7. a)** 2 **b)** -8 **c)** -15 **d)** $-\dfrac{3}{5}$ **9. a)** $x^2 + 2x + 5$
b) $a^2 + 2a + 5$ **c)** 13 **11. a)** $x^3 + x - 4$ **b)** $a^3 + a - 4$ **c)** 6 **13. a)** $4x^3 - x + 4$ **b)** $4a^3 - a + 4$ **c)** 34 **15.** -7 **17.** 29

19. -60 **21.** Undefined **23.** 13 **25.** $-\dfrac{3}{4}$ **27.** $2x^2 - 6$ **29.** 2 **31.** 18 **33.** 0 **35.** $-\dfrac{3}{7}$ **37.** $-\dfrac{1}{45}$ **39.** $-2x^2 + 2x - 6$ **41.** 3
43. -4 **45.** 1 **47.** Undefined **49.** 0 **51.** 0 **53.** -3 **55.** -2 **57. a)** 2004 **b)** \$800 **c)** \$7900 **d)** \$900
59. a) 2003, ≈ 1.8 million barrels **b)** 1998, 2001 **c)** ≈ 1.4 million barrels **d)** ≈ 4.0 million barrels **61. a)** ≈ 20 **b)** ≈ 8 **c)** ≈ 12 **d)** ≈ 23

63. a)

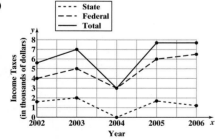

b)

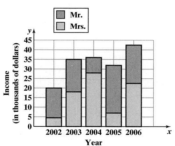

c)

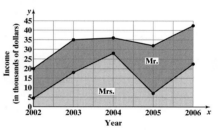

65. a)

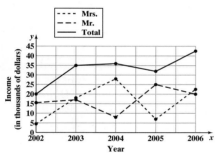

b)

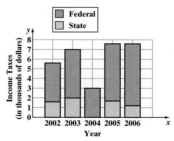

c)

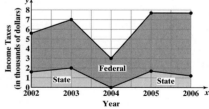

67. $f(a)$ and $g(a)$ must either be opposites or both be equal to 0. **69.** $f(a) = g(a)$ **71.** $f(a)$ and $g(a)$ must have opposite signs.

73.

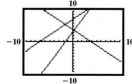

75.

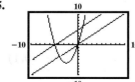

78. $-\dfrac{1}{64}$ **79.** $h = \dfrac{2A}{b}$ **80.** \$450

81.

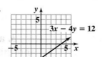

82. 2.96×10^6 **83.**

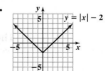

Chapter 8 Review Exercises

1.

2.

3.

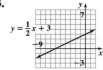

4.

5. 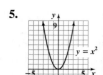   **6.**   **7.**   **8.** 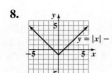   **9.**   **10.**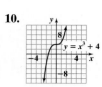

11. A function is a correspondence where each member of the domain corresponds to exactly one member of the range.
12. No, every relation is not a function. $\{(4, 2), (4, -2)\}$ is a relation but not a function. Yes, every function is a relation because it is a set of ordered pairs.   **13.** Yes, each member of the domain corresponds to exactly one member of the range.   **14.** No, the domain element 2 corresponds to more than one member of the range (5 and -2).   **15.** $\{3, 4, 5, 6\}$ **16.** $\{6, 7, 8, \ldots\}$ **17.** $(4, \infty)$
18. $(-\infty, -2]$ **19.** $(-1.5, 2.7]$ **20. a)** Yes, the relation is a function.   **b)** Domain: $\mathbb{R}$; Range: $\mathbb{R}$ **21. a)** Yes, the relation is a function.
b) Domain: $\mathbb{R}$; Range: $\{y \mid y \leq 0\}$ **22. a)** No, the relation is not a function.   **b)** Domain: $\{x \mid -3 \leq x \leq 3\}$; Range: $\{y \mid -3 \leq y \leq 3\}$
23. a) No, the relation is not a function.   **b)** Domain: $\{x \mid -2 \leq x \leq 2\}$; Range: $\{y \mid -1 \leq y \leq 1\}$ **24. a)** -2 **b)** $-h^2 + 3h - 4$
25. a) 1   **b)** $2a^3 - 3a^2 + 6$ **26.** Answers will vary. Here is one possible interpretation: The car speeds up to 50 mph. Stays at this speed for about 11 minutes. Speeds up to about 68 mph. Stays at that speed for about 5 minutes, then stops quickly. Stopped for about 5 minutes. Then in stop and go traffic for about 5 minutes.   **27. a)** 1020 baskets   **b)** 1500 baskets   **28. a)** 180 feet   **b)** 52 feet

29. 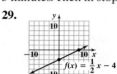   **30.** 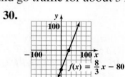   **31.**   **32. a)** 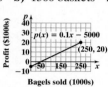   **b)** 50,000 bagels   **c)** 270,000 bagels

33. 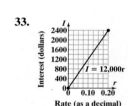   **34. a)** Yes   **b)** $f(x) = 3$ **35. a)** No   **36. a)** Yes   **b)** $f(x) = -\dfrac{1}{2}x + 2$
37. a) -2 **b)** $(0, 1)$ **c)** $y = -2x + 1$ **38.** $(0, 0)$

39. a) 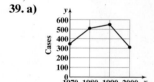   **b)** 1970–1980: 16.4; 1980–1990: 4.2; 1990–2000: -23.5 **c)** 1970–1980   **40.** $n(t) \approx 0.7t + 35.6$
41. Parallel   **42.** Perpendicular   **43.** Neither   **44.** $y = \dfrac{1}{2}x + 7$ **45.** $y = -x - 2$ **46.** $y = -\dfrac{2}{3}x + 6$
47. $y = \dfrac{5}{2}x + 3$ **48.** $y = -\dfrac{5}{3}x - 4$ **49.** $y = -\dfrac{1}{2}x + 7$ **50.** Neither   **51.** Parallel
52. Perpendicular   **53.** Neither   **54. a)** $r(a) = 0.61a - 10.59$ **b)** \$13.81
55. a) $C(r) = 1.8r + 435$ **b)** 507 calories   **c)** ≈ 91.7 yards per minute   **56.** $x^2 - x - 1$ **57.** 11
58. $-x^2 + 5x - 9$ **59.** -15 **60.** -56 **61.** 4   **62.** $-\dfrac{2}{3}$ **63.** -2 **64. a)** ≈ 4.6 billion   **b)** ≈ 2.1 billion   **c)** ≈ 0.8 billion   **d)** $\approx 33\%$
65. a) $\approx \$47,000$ **b)** $\approx \$28,000$ **c)** $\approx \$3000$

Chapter 8 Practice Test

1. 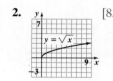 $[8.1]$ **2.** 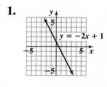 $[8.1]$ **3.** 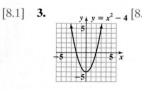 $[8.1]$ **4.** 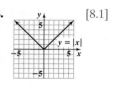 $[8.1]$

5. A function is a correspondence where each member in the domain corresponds with exactly one member in the range. $[8.2]$
6. Yes, because each member in the domain corresponds to exactly one member in the range. $[8.2]$ **7.** Yes; Domain: $\mathbb{R}$;
Range: $\{y \mid y \leq 4\}$ $[8.2]$ **8.** No; Domain: $\{x \mid -3 \leq x \leq 3\}$; Range: $\{y \mid -2 \leq y \leq 2\}$ $[8.2]$ **9.** 29 $[8.2]$ **10.**

11. a) 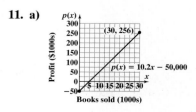   **b)** 4900 books   **c)** 14,700 books $[8.3]$
12. $p(t) = 2.9044t + 274.634$ $[8.4]$ **13.** Parallel, the slope of both lines is the same, $\dfrac{2}{3}$. $[8.4]$
14. $y = -2x + 7$ $[8.4]$ **15. a)** $r(t) = -3t + 266$
b) 248 per 100,000   **c)** 206 per 100,000 $[8.4]$ **16.** 12 $[8.5]$ **17.** $-\dfrac{3}{7}$ $[8.5]$
18. $2a^2 - a$ $[8.5]$ **19. a)** ≈ 44 million tons   **b)** ≈ 18 million tons   **c)** ≈ 26 million tons $[8.5]$

Cumulative Review Test
1. 126 [1.9]   **2.** 92 [1.9]   **3.** 0 [2.5]   **4.** $R = \dfrac{P - 2E}{3}$ [2.6]   **5.** $y = \dfrac{5}{2}x - 6$ [4.4]   **6.** $\dfrac{27y^6}{x^9}$ [5.1]

7. $8x^2 + 5x - 1$ [5.4]   **8.** $12x^3 - 38x^2 + 39x - 15$ [5.5]   **9.** $(a - 1)(6a - 5)$ [6.2]   **10.** $3, -\dfrac{5}{2}$ [6.6]   **11.** $\dfrac{x + 3}{2x + 1}$ [7.2]

12. $\dfrac{x^2 - 8x - 12}{(x + 4)(x - 5)}$ [7.4]   **13.** $\dfrac{6x + 2}{(x - 5)(x + 2)(x + 3)}$ [7.4]   **14.** $-\dfrac{3}{2}$ [7.6]   **15.** No solution [7.6]

16. a) Not a function   **b)** Domain: $\{x \mid x \le 2\}$; Range $\mathbb{R}$ [8.2]   **17.** Neither [8.4]   **18.** $x^2 + 7x - 8$ [8.5]   **19.** 260 [8.5]
20. a) 3.052×10^{12} cubic feet   **b)** 7.412×10^{12} cubic feet   **c)** 2.398×10^{13} cubic feet [5.3]

Chapter 9

Using Your Graphing Calculator, 9.1
1. $(-3, -4)$ **2.** $(-4, 3)$ **3.** $(3.1, -1.5)$ **4.** $(-0.6, 4.8)$

Exercise Set 9.1
1. The solution to a system of equations represents the ordered pairs that satisfy all the equations in the system.
3. Write the equations in slope-intercept form and compare their slopes and y-intercepts. If the slopes are different, the system has exactly one solution. If the slopes are the same and the y-intercepts are different, the system has no solution. If the slopes and y-intercepts are the same, the system has an infinite number of solutions.   **5.** The point of intersection can only be estimated.   **7.** c)   **9.** a)
11. None   **13.** a), b)   **15.** a)   **17.** Consistent; one solution   **19.** Dependent; infinite number of solutions   **21.** Consistent; one solution
23. Inconsistent; no solution   **25.** One solution   **27.** No solution   **29.** One solution   **31.** No solution   **33.** Infinite number of solutions   **35.** No solution   **37.**

37. 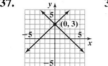   **39.**   **41.**   **43.**   **45.**
Inconsistent

47.   **49.** 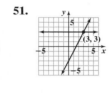 Dependent   **51.** 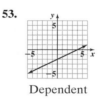   **53.** 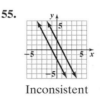 Dependent   **55.** Inconsistent   **57.**

59. 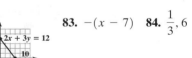   **61.** Lines are parallel; they have the same slope and different y-intercepts.
63. The system has an infinite number of solutions. If the two lines have two points in common then they must be the same line.
65. The system has no solution. Parallel lines do not intersect.
67. One; $(5, 3)$

69. 6 years   **71.** 7 hours   **79.** $-2x + 18$ **80.** 1   **81.** 0.05   **82. a)** $(6, 0), (0, 4)$ **b)**   **83.** $-(x - 7)$ **84.** $\dfrac{1}{3}, 6$

Exercise Set 9.2
1. The x in the first equation, since both 6 and 12 are divisible by 3.   **3.** You will obtain a false statement, such as $3 = 0$.   **5.** $(4, 1)$ **7.** $(-1, -1)$ **9.** No solution   **11.** $(3, -8)$ **13.** $(-2, -3)$ **15.** $(3, 5)$ **17.** Infinite number of solutions
19. $(2, 1)$ **21.** $(5, -3)$ **23.** $(-1, 0)$ **25.** $\left(-\dfrac{12}{7}, \dfrac{6}{35}\right)$ **27.** No solution   **29.** $\left(\dfrac{1}{8}, -\dfrac{1}{2}\right)$ **31.** 36, 44   **33.** Width: 8 feet, length: 17 feet
35. 1561 attendees   **37.** \$30,000   **39. a)** 16 months   **b)** Yes   **41. a)** 1.25 hours   **b)** 155 mile marker   **43. a)** $T = 180 - 10t$
b) $T = 20 + 6t$ **c)** 10 minutes   **d)** $80°F$ **45.** 23 years   **46.** 2   **47.**
48. Slope: $\dfrac{3}{5}$; y-intercept: $(0, -5)$ **49.** $y = 2x + 4$
50. $18x^2 + 9x - 14$

Exercise Set 9.3
1. Multiply the top equation by 2.   **3.** You will obtain a false statement, such as $0 = 6$.   **5.** $(5, 1)$ **7.** $(4, 5)$
9. $(5, 8)$ **11.** No solution   **13.** $(-8, -26)$ **15.** $\left(-\dfrac{3}{2}, -3\right)$ **17.** $(-2, 2)$ **19.** $(3, -1)$ **21.** Infinite number of solutions   **23.** $(1, 2)$
25. No solution   **27.** $(0, 0)$ **29.** $\left(10, \dfrac{15}{2}\right)$ **31.** $(-6, -12)$ **33.** $\left(\dfrac{20}{39}, -\dfrac{16}{39}\right)$ **35.** $\left(\dfrac{14}{5}, -\dfrac{12}{5}\right)$ **37.** No solution

39. First number: 14, second number: 6   **41.** First number: 6, second number: 4   **43.** Width: 3 inches, length: 6 inches
45. Width: 8 inches, length: 10 inches   **47.** Answers will vary.   **49. a)** $(200, 100)$ **b)** Same solution; dividing both sides of an equation by a nonzero number does not change the solution.   **51.** $(8, -1)$ **53.** $(1, 2, 3)$ **55.** 125   **56.** 7
57. $2x^2y - 9xy + 4y$ **58.** $32a^6b^9c^5$ **59.** $(y + c)(x - a)$ **60.** 14

Exercise Set 9.4   **1.** The graph will be a plane.   **3.** $(1, -2, -4)$ **5.** $\left(-7, -\dfrac{35}{4}, -3\right)$ **7.** $(0, 3, 6)$ **9.** $(1, 2, 0)$ **11.** $(-3, 15, -7)$

13. $(3, 1, -2)$ **15.** $(2, -1, 3)$ **17.** $\left(\dfrac{2}{3}, -\dfrac{1}{3}, 1\right)$ **19.** $(0, -1, 0)$ **21.** $\left(-\dfrac{11}{17}, \dfrac{7}{34}, -\dfrac{49}{17}\right)$ **23.** $(0, 0, 0)$ **25.** $(4, 6, 8)$ **27.** $\left(\dfrac{2}{3}, \dfrac{23}{15}, \dfrac{37}{15}\right)$
29. $(1, 1, 2)$ **31.** Inconsistent   **33.** Dependent   **35.** Inconsistent   **37.** No point is common to all three planes. Therefore, the system is inconsistent.   **39.** One point is common to all three planes; therefore, the system is consistent.
41. a) Yes, the 3 planes can be parallel   **b)** Yes, the 3 planes can intersect at one point   **c)** No, the 3 planes cannot intersect at exactly two points   **43.** $A = 9, B = 6, C = 2; 9x + 6y + 2z = 1$ **45.** Answers will vary. One example is $x + y + z = 10, x + 2y + z = 11, x + y + 2z = 16$ **47. a)** $a = 1, b = 2, c = -4$ **b)** $y = x^2 + 2x - 4$, Substitute 1 for a, 2 for b, and -4 for c into $y = ax^2 + bx + c$ **49.** $(1, 2, 3, 4)$ **51. a)** Commutative property of addition   **b)** Associative property of multiplication   **c)** Distributive property   **52.** $\dfrac{1}{2}$ **53. a)** $\dfrac{1}{4}$ hour or 15 minutes   **b)** 1.25 miles   **54.** Width: 3 feet, length: 8 feet

Mid-Chapter Test   **1.** b) [9.1]   **2.** a) [9.1]   **3.** One solution [9.1]   **4.** Infinite number of solutions [9.1]   **5.** No solution [9.1]

6. [9.1]   **7.** [9.1]   **8.** $(-2, 4)$ [9.2]   **9.** $\left(1, -\dfrac{1}{3}\right)$ [9.2]   **10.** No solution [9.2]
11. Length: 15 feet, width: 7 feet [9.2]   **12.** $(4, -1)$ [9.3]   **13.** $\left(-\dfrac{1}{2}, 2\right)$ [9.3]
14. Infinite number of solutions [9.3]
15. Solution must have two values, one for x and one for y. The solution is $x = 3$ and $y = 5$ or $(3, 5)$. [9.1–9.3]   **16.** $(1, 2, -1)$ [9.4]
17. $(2, 0, 3)$ [9.4]

Exercise Set 9.5   **1.** Ireland: 70,273 square kilometers, Georgia: 69,700 square kilometers   **3.** Hamburger: 21 grams, fries: 67 grams   **5.** Hot dog: $2, soda: $1   **7.** 128 MB: 72 photos, 512 MB: 288 photos   **9.** $25°, 65°$ **11.** $52°, 128°$ **13.** 12.2 miles per hour, 3.4 miles per hour   **15.** $500, 4%   **17.** 1.2 ounces of 5%, 1.8 ounces of 30%   **19.** 10 gallons concentrate, 190 gallons water

21. $17\dfrac{1}{3}$ pounds birdseed, $22\dfrac{2}{3}$ pounds sunflower seeds   **23.** Adult: $29, child: $18   **25.** $6000 at 5%, $4000 at 6%

27. 160 gallons whole, 100 gallons skim milk   **29.** 7 pounds Season's Choice, 13 pounds Garden Mix   **31.** 50 miles per hour, 55 miles per hour   **33.** Cabrina: 8 hours, Dabney: 3.4 hours   **35.** 80 grams A, 60 grams B **37.** 200 grams first alloy, 100 grams second alloy
39. 2012   **41.** Tom: 60 miles per hour, Melissa: 75 miles per hour   **43.** Personal: 3, bills and statements: 4, advertisements: 17
45. Alabama: 52, Tennessee: 45, Texas: 44   **47.** Singh: 69, Woods: 65, Mickelson: 57   **49.** Haverhill: 36.5 inches, Salem: 38 inches, Plymouth, 38 inches   **51.** Florida: 12, California: 11, Louisiana: 9   **53.** $30°, 45°, 105°$ **55.** $1500 at 3%, $3000 at 5%, $5500 at 6%
57. 4 liters of 10% solution, 2 liters of 12% solution, 2 liters of 20% solution   **59.** 10 children's chairs; 12 standard chairs; 8 executive chairs
61. $I_A = \dfrac{27}{38}; I_B = -\dfrac{15}{38}; I_C = -\dfrac{6}{19}$ **64.** $-\dfrac{35}{8}$ **65.** 4   **66.** $y = x - 10$ **67.** Use the vertical line test.

Exercise Set 9.6   **1.** It has the same number of rows and columns   **3.** Change the -2 in the second row to 1 by multiplying row 2 by $-\dfrac{1}{2}$, or $-\dfrac{1}{2}R_2$ **5.** Switch R_2 and R_3 to get a 1 in the second row, second column.   **7.** Dependent   **9.** $\begin{bmatrix} 1 & -2 & -5 \\ 3 & -7 & -4 \end{bmatrix}$

11. $\begin{bmatrix} 1 & 1 & 3 & -8 \\ 3 & 2 & 1 & -5 \\ 4 & 7 & 2 & -1 \end{bmatrix}$ **13.** $\begin{bmatrix} 1 & 3 & 12 \\ 0 & 23 & 42 \end{bmatrix}$ **15.** $\begin{bmatrix} 1 & 0 & 8 & \dfrac{1}{4} \\ 0 & 2 & -38 & -\dfrac{13}{4} \\ 6 & -3 & 1 & 0 \end{bmatrix}$ **17.** $(3, 0)$ **19.** $(-5, 1)$ **21.** $(0, 1)$ **23.** Dependent system

25. $\left(-\dfrac{1}{3}, 3\right)$ **27.** Inconsistent system   **29.** $\left(\dfrac{2}{3}, \dfrac{1}{4}\right)$ **31.** $\left(\dfrac{4}{5}, -\dfrac{7}{8}\right)$ **33.** $(2, 1, 3)$ **35.** $(3, 1, 2)$ **37.** $\left(1, -1, \dfrac{1}{2}\right)$ **39.** Dependent system

41. $\left(\dfrac{1}{2}, 2, 4\right)$ **43.** Inconsistent system   **45.** $\left(5, \dfrac{1}{3}, -\dfrac{1}{2}\right)$ **47.** No, this is the same as switching the order of the equations.
49. $\angle x = 30°, \angle y = 65°, \angle z = 85°$ **51.** 26% by Chiquita, 25% by Dole, 14% by Del Monte, 35% by other
53. $3x - 8 + \dfrac{9}{x + 4}$ **54.** $\dfrac{9}{2}, -4$ **55.** $\dfrac{-2x - 3}{(x + 2)(x - 2)}$ **56.** 30 minutes

Exercise Set 9.7 **1.** Answers will vary. **3.** If $D = 0$ and D_x, D_y, or $D_z \neq 0$, the system is inconsistent. **5.** $\left(3, -\dfrac{1}{2}\right)$ **7.** 6

9. -8 **11.** -12 **13.** 44 **15.** $(-5, 2)$ **17.** $(6, -4)$ **19.** $\left(\dfrac{1}{2}, -1\right)$ **21.** $(-7, -2)$ **23.** Infinite number of solutions **25.** $(2, -3)$

27. No solution **29.** $(2, 5)$ **31.** $(1, -1, 3)$ **33.** $\left(\dfrac{1}{2}, -\dfrac{1}{2}, 2\right)$ **35.** $\left(\dfrac{1}{2}, -\dfrac{1}{8}, 2\right)$ **37.** $(-1, 0, 2)$ **39.** Infinite number of solutions

41. $(1, -1, 2)$ **43.** No solution **45.** $(3, 4, 1)$ **47.** $(-1, 5, -2)$ **49.** It will have the opposite sign. This can be seen by comparing $a_1b_2 - a_2b_1$ to $a_2b_1 - a_1b_2$ **51.** 0 **53.** 0 **55.** Yes, it will have the opposite sign. **57.** No, same value as original value

59. Yes, value is double original value **61.** 5 **63.** 6 **65. a)** $x = \dfrac{c_1b_2 - c_2b_1}{a_1b_2 - a_2b_1}$ **b)** $y = \dfrac{a_1c_2 - a_2c_1}{a_1b_2 - a_2b_1}$ **66.** $y = \dfrac{2}{5}x - \dfrac{6}{5}$

67.   **68.**   **69.**

Chapter 9 Review Exercises **1.** c) **2.** a) **3.** Consistent, one solution **4.** Inconsistent, no solution **5.** Dependent, infinite number of solutions **6.** Consistent, one solution **7.** No solution **8.** One solution **9.** Infinite number of solutions **10.** One solution

11. 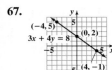 **12.** 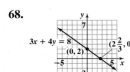 **13.**   **14.**   **15.**   **16.**   **17.**

18.   **19.** $(5, 2)$ **20.** $(-3, 2)$ **21.** $(4, 1)$ **22.** $(-15, 5)$ **23.** No solution **24.** Infinite number of solutions **25.** $\left(\dfrac{5}{2}, -1\right)$ **26.** $\left(-\dfrac{2}{7}, \dfrac{29}{7}\right)$ **27.** $(-6, -2)$ **28.** $(1, -2)$ **29.** $(-7, 19)$ **30.** $\left(\dfrac{32}{13}, \dfrac{8}{13}\right)$ **31.** $\left(-1, \dfrac{13}{3}\right)$

32. No solution **33.** Infinite number of solutions **34.** $\left(-\dfrac{78}{7}, -\dfrac{48}{7}\right)$ **35.** $(1, 2, -4)$

36. $(-1, 3, -2)$ **37.** $(-5, 1, 2)$ **38.** $(3, -2, -2)$ **39.** $\left(\dfrac{8}{3}, \dfrac{2}{3}, 3\right)$ **40.** $(0, 2, -3)$ **41.** No solution **42.** Infinite number of solutions

43. Luan: 38, Jennifer: 28 **44.** Airplane: 520 mph, wind: 40 mph **45.** Combine 2 liters of the 20% acid solution with 4 liters of the 50% acid solution. **46.** 410 adult tickets and 240 children tickets were sold. **47.** His ages were 41 years and 77 years.
48. $20,000 invested at 7%, $15,000 invested at 5%, and $5000 was invested at 3%. **49.** $(11, -2)$ **50.** $(3, 1)$ **51.** Infinite number of solutions **52.** $(2, 1, -2)$ **53.** No solution **54.** Infinite number of solutions **55.** $(2, 3)$ **56.** $(-3, 2)$ **57.** $(-1, 2)$ **58.** $(-2, 3, 4)$
59. $(1, 1, 2)$ **60.** No solution

Chapter 9 Practice Test **1.** b) [9.1] **2.** Consistent, one solution [9.1] **3.** Inconsistent, no solution [9.1] **4.** Dependent, infinite number of solutions [9.1] **5.** Consistent; one solution [9.1] **6.** Dependent; infinite number of solutions [9.1]

7. Inconsistent; no solution [9.1] **8.** 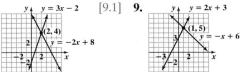 [9.1] **9.** 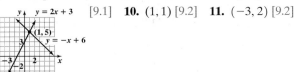 [9.1] **10.** $(1, 1)$ [9.2] **11.** $(-3, 2)$ [9.2]

12. Infinite number of solutions [9.3] **13.** $\left(\dfrac{44}{19}, \dfrac{48}{19}\right)$ [9.3] **14.** $(1, -1, 2)$ [9.4] **15.** $\begin{bmatrix} -2 & 3 & 7 & | & 5 \\ 3 & -2 & 1 & | & -2 \\ 1 & -6 & 9 & | & -13 \end{bmatrix}$ [9.6]

16. $\begin{bmatrix} 6 & -2 & 4 & | & 4 \\ 0 & 5 & -3 & | & 12 \\ 2 & -1 & 4 & | & -3 \end{bmatrix}$ [9.6] **17.** $(4, -1)$ [9.6] **18.** $(3, -1, 2)$ [9.6] **19.** -1 [9.7] **20.** 165 [9.7] **21.** $(-3, 2)$ [9.7] **22.** $(3, 1, -1)$ [9.7]

23. 8 pounds sunflower; 12 pounds bird mix [9.5] **24.** $6\dfrac{2}{3}$ liters 6% solution; $3\dfrac{1}{3}$ liters 15% solution [9.5] **25.** 4, 9, and 16 [9.5]

Cumulative Review Test **1. a)** $9, 1$ **b)** $\dfrac{1}{2}, -4, 9, 0, -4.63, 1$ **c)** $\dfrac{1}{2}, -4, 9, 0, \sqrt{3}, -4.63, 1$ [1.4] **2.** 1 [1.9] **3.** $\dfrac{29}{6}$ [2.5]

4. [4.2] **5.** $y = \dfrac{2}{5}x + \dfrac{11}{5}$ [4.4] **6.** $y = 3x - 2$ [4.4] **7.** $\dfrac{1}{7a^5b^6}$; [5.2] **8.** $(3x + 4)(x^2 + 2)$ [6.2]

9. $(x - 14)(x - 2)$ [6.3] **10.** $0, 5$ [6.6] **11.** $\dfrac{3y + 2}{8}$ [7.3] **12.** $\dfrac{41}{7}$ [7.6]

13. a) function **b)** function **c)** not a function [8.2] **14. a)** $-\dfrac{1}{7}$ **b)** $\dfrac{h + 3}{h^2 - 9}$ **c)** undefined [8.2]

15. $(1, 3)$ [9.2] **16.** $(7, -1)$ [9.3] **17.** $(2, 1, 3)$ [9.4] **18.** $10°, 80°, 90°$ [3.3] **19.** 1 hour [3.4] **20.** 600 at $20, 400 at $16 [9.5]

Chapter 10

Exercise Set 10.1   1. It is necessary to reverse the direction of the inequality symbol when multiplying or dividing both sides of the inequality by a negative number   **3. a)** When the endpoints are not included   **b)** When the endpoints are included   **c)** Answers will vary. One example is $x > 4$.   **d)** Answers will vary. One example is $x \geq 4$.   **5.** $a < x$ and $x < b$

7. a) ←—○——→ -2 **b)** $(-2, \infty)$ **c)** $\{x | x > -2\}$ **9. a)** ←——●—→ π **b)** $(-\infty, \pi]$ **c)** $\{w | w \leq \pi\}$ **11. a)** ←○———●→ -3 $\frac{4}{5}$

b) $\left(-3, \frac{4}{5}\right]$ **c)** $\left\{q \left| -3 < q \leq \frac{4}{5}\right.\right\}$ **13. a)** ←○——●→ -7 -4 **b)** $(-7, -4]$ **c)** $\{x | -7 < x \leq -4\}$ **15.** ←○———→ 3

17. ←○———→ 7 **19.** ←—●——→ 3.6 **21.** ←———+———→ 0 **23.** ←——+——→ 0 **25.** ←——+——→ 0 **27.** $\left(-\infty, \frac{3}{2}\right)$ **29.** $[2, \infty)$

31. $\left(-\infty, \frac{3}{2}\right]$ **33.** $(-\infty, \infty)$ **35.** $A \cup B = \{5, 6, 7, 8\}$; $A \cap B = \{6, 7\}$ **37.** $A \cup B = \{-1, -2, -4, -5, -6\}$; $A \cap B = \{-2, -4\}$

39. $A \cup B = \{0, 1, 2, 3\}$; $A \cap B = \emptyset$ **41.** $A \cup B = \{0, 1, 2, 3, 4, 5, 6, 7, 8\}$; $A \cap B = \emptyset$

43. $A \cup B = \{0.1, 0.2, 0.3, 0.4, \dots\}$; $A \cap B = \{0.2, 0.3\}$ **45.** $[-5, 1)$ **47.** $[-4, 5]$ **49.** $\left[4, \frac{11}{2}\right)$ **51.** $\left(-\frac{13}{3}, -4\right]$ **53.** $\{x | 3 \leq x < 7\}$

55. $\{x | 0 < x \leq 3\}$ **57.** $\left\{u \left| 4 \leq u \leq \frac{19}{3}\right.\right\}$ **59.** $\{c | -3 < c \leq 1\}$ **61.** $\emptyset$ **63.** $\{x | -5 < x < 2\}$ **65.** $(-\infty, 2) \cup [7, \infty)$ **67.** $[0, 2]$

69. $(-\infty, 0) \cup (6, \infty)$ **71.** $[0, \infty)$ **73. a)** $l + g \leq 130$ **b)** $l + 2w + 2d \leq 130$ **c)** 24.5 inches   **75.** 11 boxes   **77.** 77 minutes

79. 1881 books   **81.** 41 ounces   **83.** For sales over \$5,000 per week   **85.** 24   **87.** $76 \leq x \leq 100$ **89. a)** \$12,885.25   **b)** \$79,998.39

91. a) $[0, 3]$ **b)** $[3, 10]$ **93. a)** $[0, 5]$ **b)** $[5, 13]$ **95. a)** $[0, 8]$ **b)** None   **97.** $6.97 < x < 8.77$ **99. a)** January, February, March, May

b) March, April, May   **c)** April   **101.** Answers will vary.   **103. a)** $[17.5, 23.5]$ **b)** $[23.5, 31]$ **c)** $[27.2, 36.5]$ **105.** $84 \leq x \leq 100$

107. a) Answers will vary.   **b)** $(-3, \infty)$ **109. a)** 4   **b)** 0, 4   **c)** $-3, 4, \frac{5}{2}, 0, -\frac{29}{80}$ **d)** $-3, 4, \frac{5}{2}, \sqrt{7}, 0, -\frac{29}{80}$

110. Associative property of addition   **111.** Commutative property of addition   **112.** $V = \dfrac{R - L + Dr}{r}$

113. $A \cup B = \{1, 2, 3, 4, 5, 6, 8, 9\}$; $A \cap B = \{1, 8\}$

Exercise Set 10.2   1. Set $x = a$ or $x = -a$ **3.** $-a < x < a$ **5.** $x < -a$ or $x > a$ **7.** All real numbers except for 0; the absolute value of every real number except 0 is greater than 0.   **9.** Set $x = y$ or $x = -y$ **11. a)** Two   **b)** Infinite number

c) Infinite number   **13. a)** D   **b)** B   **c)** E   **d)** C   **e)** A   **15.** $\{-2, 2\}$ **17.** $\left\{-\frac{1}{2}, \frac{1}{2}\right\}$ **19.** $\emptyset$ **21.** $\{-13, 3\}$ **23.** $\{-7\}$ **25.** $\left\{\frac{3}{2}, \frac{11}{6}\right\}$

27. $\{-17, 23\}$ **29.** $\{3\}$ **31.** $\{w | -11 < w < 11\}$ **33.** $\{q | -13 \leq q \leq 3\}$ **35.** $\{b | 1 < b < 5\}$ **37.** $\{x | -9 \leq x \leq 6\}$

39. $\left\{x \left| \frac{1}{3} < x < \frac{13}{3}\right.\right\}$ **41.** $\emptyset$ **43.** $\{j | -22 < j < 6\}$ **45.** $\{x | -1 \leq x \leq 7\}$ **47.** $\{y | y < -2 \text{ or } y > 2\}$ **49.** $\{x | x < -9 \text{ or } x > 1\}$

51. $\left\{b \left| b < \frac{2}{3} \text{ or } b > 4\right.\right\}$ **53.** $\{h | h < 1 \text{ or } h > 4\}$ **55.** $\{x | x < 2 \text{ or } x > 6\}$ **57.** $\{x | x \leq -18 \text{ or } x \geq 2\}$ **59.** $\mathbb{R}$

61. $\{x | x < 2 \text{ or } x > 2\}$ **63.** $\{-1, 15\}$ **65.** $\{-3, 1\}$ **67.** $\left\{-23, \frac{13}{7}\right\}$ **69.** $\{10\}$ **71.** $\{-1, 1\}$ **73.** $\{q | q < -8 \text{ or } q > -4\}$

75. $\{w | -1 \leq w \leq 8\}$ **77.** $\left\{-\frac{8}{5}, 2\right\}$ **79.** $\left\{x \left| x < -\frac{5}{2} \text{ or } x > -\frac{5}{2}\right.\right\}$ **81.** $\left\{x \left| -\frac{13}{3} \leq x \leq \frac{5}{3}\right.\right\}$ **83.** $\emptyset$ **85.** $\{w | -16 < w < 8\}$

87. $\mathbb{R}$ **89.** $\left\{2, \frac{22}{3}\right\}$ **91.** $\left\{-\frac{3}{2}, \frac{9}{7}\right\}$ **93. a)** $[0.085, 0.093]$ **b)** 0.085 inch   **c)** 0.093 inch   **95. a)** $[132, 188]$ **b)** 132 to 188 feet below

sea level, inclusive   **97.** $|x| = 5$ **99.** $|x| \geq 5$ **101.** $x = -\dfrac{b}{a}$; $|ax + b|$ is never less than 0, so set $|ax + b| = 0$ and solve for x.

103. a) Set $ax + b = -c$ or $ax + b = c$ and solve each equation for x.   **b)** $x = \dfrac{-c - b}{a}$ or $x = \dfrac{c - b}{a}$ **105. a)** Write $ax + b < -c$

or $ax + b > c$ and solve each inequality for x.   **b)** $x < \dfrac{-c - b}{a}$ or $x > \dfrac{c - b}{a}$ **107.** $\mathbb{R}$; Since $3 - x = -(x - 3)$

109. $\{x | x \geq 0\}$; by definition of absolute value   **111.** $\{2\}$; set $x + 1 = 2x - 1$ or $x + 1 = -(2x - 1)$

113. $\{x | x \leq 4\}$; by definition $|x - 4| = -(x - 4)$ if $x \leq 4$ **115.** $\{4\}$ **117.** $\emptyset$ **119.** $\dfrac{29}{72}$ **120.** 25   **121.** ≈ 1.33 miles   **122.** $\{x | x < 4\}$

Mid-Chapter Test   **1.** 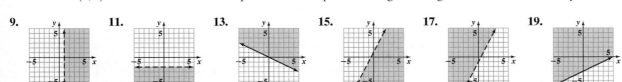 [10.1]   **2.** $[-4, 5)$ [10.1]   **3.** [10.1]   **4.** [10.1]

5. [10.1]   **6.** $(-1, \infty)$ [10.1]   **7.** $(-5, 7]$ [10.1]   **8.** $\left[-\dfrac{1}{6}, 1\right)$ [10.1]   **9.** $\{x | -8 \le x \le 2\}$ [10.1]

10. $\{m | -6 < m \le 7\}$ [10.1]   **11.** $\{x | x \le 0 \text{ or } x > 6\}$ [10.1]   **12.** $A \cup B = \{7, 8, 9, 10, 12, 15\}, A \cap B = \{8, 10\}$ [10.1]

13. $\{-2, 10\}$ [10.2]   **14.** $\{w | -16 \le w \le 26\}$ [10.2]   **15.** $\{z | z \le -8 \text{ or } z \ge 8\}$ [10.2]   **16.** $\mathbb{R}$ [10.2]   **17.** $\left\{-\dfrac{1}{2}, \dfrac{9}{2}\right\}$ [10.2]

18. $\left\{m \middle| -1 < m < \dfrac{11}{3}\right\}$ [10.2]   **19.** $\left\{-5, -\dfrac{4}{5}\right\}$ [10.2]   **20.** $\left\{-5, \dfrac{1}{2}\right\}$ [10.2]

Exercise Set 10.3   **1.** Points on the line are solutions to the corresponding equation, and are not solutions if the symbol used is $<$ or $>$.   **3.** $(0, 0)$ cannot be used as a checkpoint if the line passes through the origin.   **5.** Answers will vary.   **7.** Yes

9. **11.** **13.** **15.** **17.** **19.**

21. **23.** **25.** **27.** **29.** **31.**

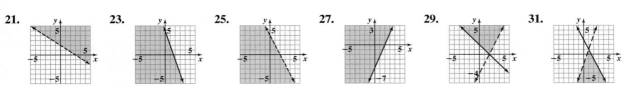

33. **35.** **37.** **39.** **41.** **43.**

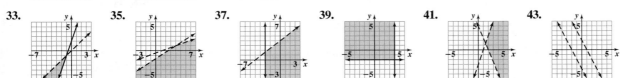

45. **47.** **49.** **51.** **53.** **55.** **57.**

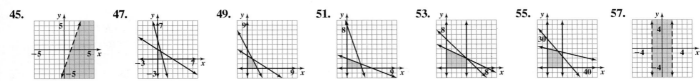

59. **61.** **63.** **65.** **67.**

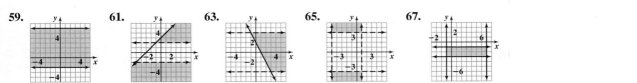

69. a)–b)   **c)** 47   **71. a)–b)**   **c)** 2003   **73. a)** Region A   **b)** Region B

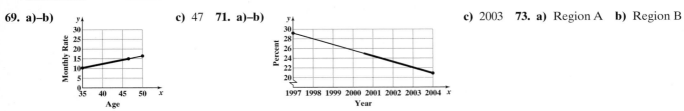

75. a) $f(x) = 2x - 4$ **b)** $f(x) = 2x - 4$ **77.** Yes. If the boundary lines are parallel, there may be no solution. One example is $y > 3x + 1; y < 3x - 2$ **79.** There is no solution. Opposite sides of the same line are being shaded and only one inequality includes the line.   **81.** There are an infinite number of solutions. Both inequalities include the line $5x - 2y = 3$.   **83.** There are an infinite number of solutions. The lines are not parallel or identical.

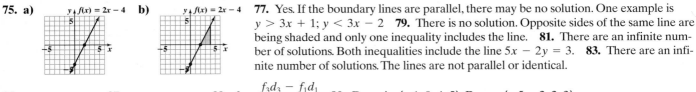

85. **87.**   **89.** $f_2 = \dfrac{f_3 d_3 - f_1 d_1}{d_2}$ **90.** Domain: $\{-1, 0, 4, 5\}$; Range: $\{-5, -2, 2, 3\}$

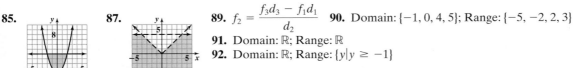

91. Domain: $\mathbb{R}$; Range: $\mathbb{R}$
92. Domain: $\mathbb{R}$; Range: $\{y | y \ge -1\}$

Chapter 10 Review Exercises

1. 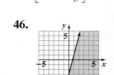   **2.**   **3.**   **4.**

5. 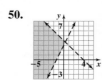   **6.**   **7.**   **8.**   **9.** 6 boxes   **10.** 7 minutes   **11.** ≈15.67 weeks

12. $\{x|81 \le x \le 100\}$ **13.** $A \cup B = \{1, 2, 3, 4, 5\}$; $A \cap B = \{2, 3, 4, 5\}$ **14.** $A \cup B = \{2, 3, 4, 5, 6, 7, 8, 9\}$; $A \cap B = \emptyset$

15. $A \cup B = \{1, 2, 3, 4, \ldots\}$; $A \cap B = \{2, 4, 6, \ldots\}$ **16.** $A \cup B = \{3, 4, 5, 6, 9, 10, 11, 12\}$; $A \cap B = \{9, 10\}$ **17.** $(5, 11)$

18. $(-3, 5]$ **19.** $\left(\dfrac{7}{2}, 8\right)$ **20.** $\left(\dfrac{8}{3}, 6\right)$ **21.** $(-3, 1]$ **22.** $(2, 14)$ **23.** $\{h|-3 < h \le 1\}$ **24.** $\mathbb{R}$ **25.** $\{x|x \le -4\}$

26. $\{g|g < -6 \text{ or } g \ge 11\}$ **27.** $\{-2, 2\}$ **28.** $\{x|-8 < x < 8\}$ **29.** $\{x|x \le -9 \text{ or } x \ge 9\}$ **30.** $\{-18, 8\}$

31. $\{x|x \le -3 \text{ or } x \ge 7\}$ **32.** $\left\{-\dfrac{1}{2}, \dfrac{9}{2}\right\}$ **33.** $\{q|1 < q < 8\}$ **34.** $\{-1, 4\}$ **35.** $\{x|-14 < x < 22\}$ **36.** $\left\{-5, -\dfrac{4}{5}\right\}$ **37.** $\mathbb{R}$

38. $\left[-5, -\dfrac{1}{3}\right]$ **39.** $(4, 8]$ **40.** $\left(-\dfrac{17}{2}, \dfrac{27}{2}\right]$ **41.** $[-2, 6)$ **42.** $(-\infty, \infty)$ **43.** $\left(\dfrac{2}{3}, 10\right]$ **44.**   **45.**

46. 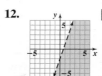   **47.**   **48.** 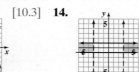   **49.**

50.   **51.** No solution   **52.**   **53.**   **54.**   **55.**

Chapter 10 Practice Test

1. $A \cup B = \{5, 7, 8, 9, 10, 11, 14\}$; $A \cap B = \{8, 10\}$ [10.1]

2. $A \cup B = \{1, 3, 5, 7, \ldots\}$; $A \cap B = \{3, 5, 7, 9, 11\}$ [10.1]   **3.** [10.1]   **4.** [10.1]   **5.** $\left(\dfrac{9}{2}, 7\right)$ [10.1]

6. $[13, 16)$ [10.1]   **7.** $\{-7, 2\}$ [10.2]   **8.** $\left\{-\dfrac{14}{3}, \dfrac{26}{5}\right\}$ [10.2]   **9.** $\{-3\}$ [10.2]   **10.** $\{x|x < -1 \text{ or } x > 4\}$ [10.2]   **11.** $\left\{x\left|\dfrac{1}{2} \le x \le \dfrac{5}{2}\right.\right\}$ [10.2]

12. [10.3]   **13.** [10.3]   **14.** [10.3]

Cumulative Review Test

1. -88 [1.9]   **2.** 9 [1.9]   **3.** $\dfrac{3}{2}x + \dfrac{7}{2}$ [2.1]   **4.** 5 [2.5]   **5.** 208 pounds [2.7]   **6.** \$50,000 [3.2]

7. \$1.69 per pound [3.4]   **8.** $\dfrac{1}{4a^{10}b^2}$ [5.2]   **9.** 4.5×10^{-7} [5.3]   **10.** $2pq(2p + 3)(6p - 5)$ [6.5]   **11.** -5 [7.6]

12. No solution [7.6]   **13.** 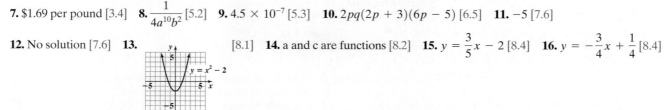 [8.1]   **14.** a and c are functions [8.2]   **15.** $y = \dfrac{3}{5}x - 2$ [8.4]   **16.** $y = -\dfrac{3}{4}x + \dfrac{1}{4}$ [8.4]

17. a) $x^3 - 16x^2 + 85x - 150$ **b)** $x - 6$ **c)** $\{x|x \ne 5\}$ [8.5]   **18.** $(2, -1, 3)$ [9.4]   **19.** $\left(-\dfrac{7}{6}, \dfrac{2}{3}\right)$ [10.1]

20. 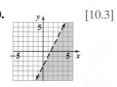 [10.3]

Chapter 11

Exercise Set 11.1   1. a) Two, positive and negative.   **b)** 7, −7   **c)** Principal square root   **d)** 7   **3.** There is no real number which, when squared gives −81.   **5.** No; if the radicand is negative, the answer is not a real number.   **7. a)** 1.3   **b)** 1.3
9. a) 3   **b)** −3   **c)** −3   **11.** 6   **13.** −4   **15.** −5   **17.** −1   **19.** 1   **21.** Not a real number   **23.** −7   **25.** Not a real number

27. Not a real number   **29.** $\dfrac{1}{5}$ **31.** $\dfrac{1}{2}$ **33.** $\dfrac{2}{7}$ **35.** $-\dfrac{2}{3}$ **37.** ≈ -2.07 **39.** 7   **41.** 19   **43.** 119   **45.** 235.23   **47.** 0.06   **49.** $\dfrac{12}{13}$

51. $|x - 4|$ **53.** $|x - 3|$ **55.** $|3x^2 - 1|$ **57.** $|6a^3 - 5b^4|$ **59.** $|a^7|$ **61.** $|z^{16}|$ **63.** $|a - 4|$ **65.** $|3a + 2b|$ **67.** $7x$ **69.** $4c^3$

71. $x + 2$ **73.** $2x + y$ **75.** 2   **77.** 8   **79.** 9   **81.** ≈ 9.381 **83.** ≈ 5.290 **85.** −3   **87.** 97   **89.** 11   **91.** 45

93. Select a value less than $-\dfrac{1}{2}$.   **95.** $x \ge 1$ **97.** $x \ge 3$ **99. a)** All real numbers   **b)** $a \ge 0$ **c)** All real numbers

101. If n is even, you are finding an even root of a positive number. If n is odd, the expression is real.   **103.** $x > -5$ **105.** d   **107.** a
109. One answer is $f(x) = \sqrt{x - 8}$ **111. a)** No   **b)** Yes, when $x = 0$ **c)** Yes   **113. a)** $\sqrt{1288} \approx 35.89$ feet per second
b) $\sqrt{2576} \approx 50.75$ feet per second.   **115.**   **117.** 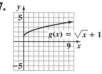   **119.**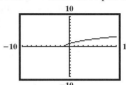

121.    **123.** 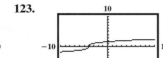   **127.** $(3a - b)(3x + 4y)$ **128.** $3x(x - 4)(x - 2)$

129. $(2x - 1)(2x + 1)(2x^2 + 3)$

130. $\left(x - \dfrac{2}{3}y\right)\left(x^2 + \dfrac{2}{3}xy + \dfrac{4}{9}y^2\right)$

Exercise Set 11.2   1. a) When n is even and $a \ge 0$, or n is odd   **b)** $a^{1/n}$ **3. a)** Always real   **b)** a **c)** a **d)** $|a|$

5. a) No; $(xy)^{1/2} = x^{1/2}y^{1/2}$ **b)** No; $(xy)^{-1/2} = x^{-1/2}y^{-1/2} = \dfrac{1}{x^{1/2}y^{1/2}}$ **7.** $a^{3/2}$ **9.** $9^{5/2}$ **11.** $z^{5/3}$ **13.** $7^{10/3}$ **15.** $9^{7/4}$ **17.** $y^{14/3}$

19. $(a^3b)^{1/4}$ **21.** $(x^9z^5)^{1/4}$ **23.** $(3a + 8b)^{1/6}$ **25.** $\left(\dfrac{2x^6}{11y^7}\right)^{1/5}$ **27.** $\sqrt{a}$ **29.** $\sqrt{c^5}$ **31.** $\sqrt[3]{18^5}$ **33.** $\sqrt{24x^3}$ **35.** $\left(\sqrt[5]{11b^2c}\right)^3$

37. $\sqrt[5]{6a + 5b}$ **39.** $\dfrac{1}{\sqrt[3]{b^3 - d}}$ **41.** a^3 **43.** x^3 **45.** $\sqrt[3]{y}$ **47.** $\sqrt{y}$ **49.** 19.3   **51.** x^5y^{10} **53.** $\sqrt{xyz}$ **55.** $\sqrt[4]{x}$ **57.** $\sqrt[8]{y}$ **59.** $\sqrt[9]{x^2y}$

61. $\sqrt[10]{a^9}$ **63.** 5   **65.** 4   **67.** 16   **69.** Not a real number   **71.** $\dfrac{5}{3}$ **73.** $\dfrac{1}{2}$ **75.** −9   **77.** −4   **79.** $\dfrac{1}{4}$ **81.** $\dfrac{1}{64}$ **83.** $\dfrac{3}{4}$

85. Not a real number   **87.** 24   **89.** $\dfrac{11}{28}$ **91.** $x^{9/2}$ **93.** $x^{1/6}$ **95.** $\dfrac{1}{x}$ **97.** 1   **99.** $\dfrac{y^{5/3}}{12}$ **101.** $\dfrac{12}{x^{11/6}}$ **103.** $\dfrac{1}{2x^{1/3}}$ **105.** $\dfrac{121}{x^{1/7}}$ **107.** $\dfrac{64}{a^{66/5}}$

109. $\dfrac{x}{y^{20}}$ **111.** $8z^{7/2} - 4$ **113.** $\dfrac{5}{x^5} + \dfrac{20}{x^{3/2}}$ **115.** $12x^{13/6} - 18x^2$ **117.** ≈ 13.42 **119.** ≈ 3.32 **121.** ≈ 20.53 **123.** ≈ 0.03

125. n is odd, or n is even and $a \ge 0$.   **127.** $(4^{1/2} + 9^{1/2})^2 \ne 4 + 9; 25 \ne 13$ **129.** $(1^{1/3} + 1^{1/3})^3 \ne 1 + 1; 8 \ne 2$ **131.** $x^{1/2}(x + 1)$

133. $y^{1/3}(1 - y)(1 + y)$ **135.** $\dfrac{1 + y^2}{y^{2/5}}$ **137. a)** $2^{10} = 1024$ bacteria   **b)** $2^{10}\sqrt{2} \approx 1448$ bacteria

139. a) $2.69\sqrt{7^3} \approx \$49.82$ billion   **b)** $2.69\sqrt{16^3} = \$172.16$ billion   **141.** 9   **143.** $\{x|x \ge 7\}$ **145. a)** $(x - 6)^2$ **b)** $(x - 6)^2$

147. $2; z^{\frac{1}{4}\frac{1}{5}\frac{1}{a}\frac{1}{3}} = z^{\frac{1}{60a}}; z^{\frac{1}{60a}} = z^{\frac{1}{120}}, 60a = 120; a = 2$ **149.** $\dfrac{b^2 + a^3b}{a^3 - b}$ **150.** 0, 3   **151.** ≈ 441.67 miles per hour   **152. c)** is a function

Exercise Set 11.3   1. a) Square the natural numbers.   **b)** $1, 4, 9, 16, 25, 36$ **3. a)** Raise the natural numbers to the fifth power.
b) $1, 32, 243, 1024, 3125$ **5.** If n is even and a or b are negative, the numbers are not real numbers.   **7.** If n is even and a or b is negative, the numbers are not real numbers;   **9.** $2\sqrt{2}$ **11.** $2\sqrt{6}$ **13.** $4\sqrt{2}$ **15.** $5\sqrt{2}$ **17.** $5\sqrt{3}$ **19.** $2\sqrt{10}$ **21.** $2\sqrt[3]{2}$ **23.** $3\sqrt[3]{2}$
25. $2\sqrt[3]{4}$ **27.** $2\sqrt[5]{5}$ **29.** $2\sqrt[3]{3}$ **31.** $-2\sqrt[5]{2}$ **33.** b^3 **35.** x^2 **37.** $x\sqrt{x}$ **39.** $a^5\sqrt{a}$ **41.** $8z^{10}\sqrt[3]{z^2}$ **43.** $b^5\sqrt[4]{b^3}$ **45.** $x\sqrt[6]{x^3}$ or $x\sqrt{x}$
47. $3y^4\sqrt[5]{y^3}$ **49.** $10y^4\sqrt{2y}$ **51.** $xy^2\sqrt[3]{y}$ **53.** $ab^4\sqrt[5]{ab^3}$ **55.** $2x^7y^{10}z^{13}\sqrt{6xz}$ **57.** $3a^2b^2\sqrt[3]{3b^2}$ **59.** $2x^2y^2z^4\sqrt[4]{2yz^3}$ **61.** $3a^2b^2\sqrt[4]{b}$
63. $2a^2b^2\sqrt[5]{b^2}$ **65.** 5   **67.** $\dfrac{9}{10}$ **69.** 3   **71.** $\dfrac{1}{4}$ **73.** $\dfrac{1}{2}$ **75.** $\dfrac{1}{3}$ **77.** $\dfrac{1}{2}$ **79.** 2   **81.** $\dfrac{r^2}{2}$ **83.** $\dfrac{4x^2}{5y^5}$ **85.** $\dfrac{c^2}{4}$ **87.** $a^2b^6\sqrt[3]{a^2b^2}$ **89.** $2\sqrt{2}$ **91.** $3x^2$
93. $2x^2y\sqrt{2y}$ **95.** $\dfrac{\sqrt[3]{5y}}{2x^4}$ **97.** $\dfrac{y^2\sqrt[3]{5y}}{x^2}$ **99.** $\dfrac{x^3\sqrt[4]{10y}}{3}$ **101.** $(a \cdot b)^{1/2} = a^{1/2}b^{1/2} = \sqrt{a}\sqrt{b}$ **103.** No; One example is $\sqrt{18}/\sqrt{2} = 3$.

105. a) No   **b)** When $\sqrt[n]{x}$ is a real number and not equal to 0.   **106.** $C = \dfrac{5}{9}(F - 32)$ **107.** $3x^6 - x^3 + 4$

108. $(x - 1)(x^2 - 8x + 19)$ **109.** $4, \dfrac{3}{2}$ **110.** $\{-28, 32\}$

Exercise Set 11.4

1. Radicals with the same radicand and index   **3.** ≈ 5.97
5. No: one example is $\sqrt{9} + \sqrt{16} \ne \sqrt{9 + 16}, 3 + 4 \ne 5, 7 \ne 5.$ **7.** 0   **9.** $4\sqrt{5}$ **11.** $-4\sqrt{3} + 5$ **13.** $-7\sqrt[4]{y}$ **15.** $2\sqrt[3]{x} + 9\sqrt{5}$
17. $7\sqrt{x} - 6\sqrt{y}$ **19.** $3\sqrt{5}$ **21.** $-30\sqrt{3} + 25\sqrt{5}$ **23.** $-4\sqrt{10}$ **25.** $18y\sqrt{5x}$ **27.** $-16\sqrt{5x}$ **29.** $-27a\sqrt{2}$ **31.** $5\sqrt[3]{4}$ **33.** -7
35. $6a\sqrt[3]{ab^2}$ **37.** $3r^3s^2\sqrt{rs}$ **39.** 0   **41.** 9   **43.** $2\sqrt[3]{7}$ **45.** $3m^2n^5\sqrt{3n}$ **47.** $3x^3y^4\sqrt[3]{2x^2y}$ **49.** $x^7y^7z^3\sqrt[5]{x^2y^3z}$ **51.** $x^2y^2\sqrt[3]{4y^2}$
53. $5 - \sqrt{15}$ **55.** $2\sqrt[3]{y^2} - y^3$ **57.** $4x^5y^3\sqrt[3]{x} + 4xy^4\sqrt[3]{2x^2y^2}$ **59.** 59   **61.** $6 - x^2$ **63.** $7 - z$ **65.** $23 + 9\sqrt{3}$ **67.** $16 - 10\sqrt{2}$
69. $10 - 3\sqrt{6}$ **71.** $29 - 12\sqrt{5}$ **73.** $18x - \sqrt{3xy} - y$ **75.** $8 - 2\sqrt[3]{18} - \sqrt[3]{12}$ **77.** $4x - 8\sqrt{x}$ **79.** $x^2 + x\sqrt[3]{x^2}$
81. $x\sqrt[4]{27x^2} - x^2\sqrt[4]{3x}$ **83.** $2\sqrt{6}$ **85.** $3\sqrt{5}$ **87.** $-14 + 11\sqrt{2}$ **89.** $5\sqrt{6} - 2\sqrt{3}$ **91.** $15\sqrt{2}$ **93.** $2x^3\sqrt[3]{10x^2}$ **95.** $2b^2c\sqrt[6]{2ab^5c^3}$
97. $4ab\sqrt[4]{b}$ **99.** $x - 2\sqrt[3]{x^2y^2} - \sqrt[3]{xy} + 2y$ **101.** $ab\sqrt[3]{12a^2b^2} - 2a^2b^2\sqrt[3]{3}$ **103.** $2x - 5$ **105.** $2|r - 4|$ **107.** $P = 14\sqrt{5}, A = 60$
109. $P = 17\sqrt{5}, A = 52.5$ **111.** No, $-\sqrt{2} + \sqrt{2} = 0$ **113. a)** ≈ 45.17 miles per hour   **b)** ≈ 35.33 miles per hour
115. a) 37 inches   **b)** ≈ 37.97 inches   **117. a)**   **119. a)**   **121.**

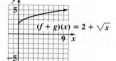

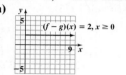

b) Raises the graph 2 units   **b)** $\{x \mid x \ge 0\}$

123. A quotient of two integers, denominator not 0.   **124.** A number that can be represented on a real number line.

125. A real number that cannot be expressed as a quotient of two integers.   **126.** $|a| = \begin{cases} a, & a \ge 0 \\ -a, & a < 0 \end{cases}$ **127.** $m = \dfrac{2E}{v^2}$

128. a)   **b)** $\left(-\dfrac{1}{2}, 5\right]$ **c)** $\left\{x \mid -\dfrac{1}{2} < x \le 5\right\}$

Mid-Chapter Test

1. 11 [11.1]   **2.** $-\dfrac{3}{4}$ [11.1]   **3.** 16.3 [11.1]   **4.** $|3a^2 - 4b^3|$ [11.1]   **5.** 3 [11.1]   **6.** $(7a^4b^3)^{1/5}$ [11.2]   **7.** 20 [11.2]

8. $a^{10}b^{15}c^5$ [11.3]   **9.** $\dfrac{14}{x}$ [11.3]   **10.** $8x + \dfrac{16}{x^{5/2}}$ [11.3]   **11.** $4x^2y^4\sqrt{2y}$ [11.3]   **12.** $2a^2b^3c^2\sqrt[6]{ab^5c^3}$ [11.3]   **13.** $\dfrac{1}{3}$ [11.3]   **14.** $\dfrac{y^2\sqrt{y}}{3x^5}$ [11.3]

15. $11\sqrt{x} + 12\sqrt{y}$ [11.4]   **16.** $27x\sqrt{10y}$ [11.4]   **17.** $2x^2 - x\sqrt{5} - 15$ [11.4]   **18.** $18a\sqrt{a} - 20a\sqrt{3}$ [11.4]   **19.** $7ab\sqrt[4]{ab}$ [11.4]

20. Part **a)** will have an absolute value.   **a)** $|x - 3|$ **b)** $8x$ [11.1]

Exercise Set 11.5

1. a) Same two terms with the sign of the second term changed.   **b)** $x + \sqrt{3}$ **3. a)** Answers will vary.
b) $\dfrac{4\sqrt{3y}}{3y}$ **5.** (1) No perfect powers are factors of any radicand. (2) No radicand contains fractions. (3) There are no radicals in any

denominator.   **7.** $\dfrac{\sqrt{3}}{3}$ **9.** $\dfrac{4\sqrt{5}}{5}$ **11.** $\sqrt{6}$ **13.** $\dfrac{\sqrt{z}}{z}$ **15.** $\dfrac{p\sqrt{2}}{2}$ **17.** $\dfrac{\sqrt{7y}}{7}$ **19.** $3\sqrt{2}$ **21.** $\dfrac{\sqrt{xy}}{y}$ **23.** $\dfrac{\sqrt{10m}}{4}$ **25.** $\dfrac{\sqrt{2n}}{3}$

27. $\dfrac{3x^2y\sqrt{yz}}{z^2}$ **29.** $\dfrac{2y^4z\sqrt{15xz}}{3x}$ **31.** $\dfrac{4x^3y^2\sqrt{yz}}{z^2}$ **33.** $\dfrac{\sqrt[3]{4}}{2}$ **35.** $\dfrac{8\sqrt[3]{y^2}}{y}$ **37.** $\dfrac{\sqrt[4]{27}}{3}$ **39.** $\dfrac{a\sqrt[4]{2}}{2}$ **41.** $\dfrac{5\sqrt[4]{z^2}}{z}$ **43.** $\dfrac{10\sqrt[5]{y^2}}{y}$ **45.** $\dfrac{2\sqrt[7]{a^3}}{a}$

47. $\dfrac{\sqrt[3]{4x^2}}{2x}$ **49.** $\dfrac{5m\sqrt[4]{8}}{2}$ **51.** $\dfrac{\sqrt[4]{135x}}{3x}$ **53.** $\dfrac{\sqrt[3]{12x^2y}}{2y}$ **55.** $\dfrac{\sqrt[5]{7xy^2z}}{z}$ **57.** 19   **59.** 62   **61.** -6 **63.** $a - b$ **65.** $4x - 9y$

67. $\sqrt{3} - 1$ **69.** $2 - \sqrt{3}$ **71.** $\dfrac{-5\sqrt{2} - 35}{47}$ **73.** $\dfrac{10 + \sqrt{30}}{14}$ **75.** $\dfrac{18 - 3\sqrt{x}}{36 - x}$ **77.** $\dfrac{4x + 4y\sqrt{x}}{x - y^2}$ **79.** $\dfrac{-13 + 3\sqrt{6}}{23}$ **81.** $a + a^3$

83. $\dfrac{4\sqrt{x + 2} + 12}{x - 7}$ **85.** $\dfrac{\sqrt{x}}{4}$ **87.** $\dfrac{\sqrt{2}}{3}$ **89.** 1   **91.** $\dfrac{2xy^3\sqrt{30xz}}{5z}$ **93.** $\dfrac{\sqrt{14}}{x}$ **95.** $\dfrac{\sqrt{a} - 7}{a - 49}$ **97.** $-\dfrac{\sqrt{2x}}{2}$ **99.** $\dfrac{\sqrt[4]{24x^3y^2}}{2x}$

101. $\dfrac{2y^4z^3\sqrt[3]{2x^2z}}{x}$ **103.** $\dfrac{a\sqrt{r} + 2r\sqrt{a}}{a - 4r}$ **105.** $\dfrac{\sqrt[3]{150y^2}}{5y}$ **107.** $\dfrac{y^3z\sqrt[4]{54x^2}}{3x}$ **109.** $\sqrt{2}$ **111.** $\dfrac{3\sqrt{5}}{5}$ **113.** $\dfrac{8\sqrt{6}}{3}$ **115.** $\dfrac{19\sqrt{2}}{2}$

117. $\dfrac{21\sqrt{2}}{2}$ **119.** $-\dfrac{301\sqrt{2}}{20}$ **121.** $\dfrac{3\sqrt{6}}{4}$ **123.** $\left(-\dfrac{2}{y} + \dfrac{3}{x}\right)\sqrt{xy}$ **125.** $2\sqrt{a}$ **127.** $\sqrt[3]{(a + b)^5}$ **129.** $\sqrt[15]{(a + 2b)^2}$ **131.** $\sqrt[6]{rs^5}$

133. $\sqrt[15]{x^2y^8}$ **135.** ≈ 3.69 meters   **137.** ≈ 12 inches   **139. a)** 6.21 million   **b)** ≈ 2.35 million   **141.** $\dfrac{3}{\sqrt{3}}; \dfrac{2}{\sqrt{2}} = \sqrt{2}, \dfrac{3}{\sqrt{3}} = \sqrt{3}$

143. $2 + \sqrt{3}$; rationalize the denominator and compare   **145. a)** $4, 8, 12$ **b)** $9, 18, 27$ **c)** $x^{(3a+2b)/6}$ **d)** $x^{(3a-2b)/6}$

147. $\dfrac{3\sqrt{2a} - 3b}{2a - 3b}$ **149.** $\dfrac{10}{15 + 3\sqrt{5}}$ **151.** $\dfrac{1}{\sqrt{x + h} + \sqrt{x}}$ **154.** $b_2 = \dfrac{2A}{h} - b_1$ **155.** 40 miles per hour, 50 miles per hour
156. $4x^3 + x^2 - 20x + 4$ **157.** $-8, 1$

Using Your Graphing Calculator, 11.6   **1.**

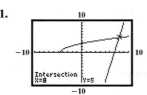

2.

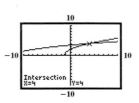

Exercise Set 11.6   **1. a)** Answers will vary.   **b)** 5   **3.** 0   **5.** Answers will vary.   **7.** 1; Answers will vary.   **9.** 16
11. No real solution   **13.** -64 **15.** 11   **17.** 9   **19.** -1 **21.** 81   **23.** 71   **25.** No real solution   **27.** No real solution   **29.** 2, 4

31. 8   **33.** 7   **35.** $\dfrac{2}{3}$ **37.** 16   **39.** 2   **41.** 10   **43.** 6   **45.** 8   **47.** 0   **49.** -3 **51.** $\dfrac{3}{2}$ **53.** No real solution   **55.** 2, 0   **57.** 5, 8

59. No real solution   **61.** 9   **63.** 3, 7   **65.** -1 **67.** 7   **69.** 4   **71.** 5   **73.** $v = \dfrac{p^2}{2}$ **75.** $g = \dfrac{v^2}{2h}$ **77.** $F = \dfrac{Mv^2}{R}$ **79.** $m = \dfrac{x^2k}{V_0^2}$

81. $A = \pi r^2$ **83.** $\sqrt{87}$ **85.** $2\sqrt{10}$ **87.** 4   **89.** No Solution   **91.** 3   **93.** 7   **95.** 1   **97.** 3   **99.** $\sqrt{16{,}200} \approx 127.28$ feet

101. 13 feet   **103. a)** ≈ 3.14 seconds   **b)** $\sqrt{2} \cdot T$; compare $\sqrt{\dfrac{l}{32}}$ with $\sqrt{\dfrac{l}{16}}$ **c)** $\sqrt{24} \approx 4.90$ seconds   **105.** $R = \dfrac{8\mu l}{\pi r^4}$

107. $0.2(\sqrt{149.4})^3 \approx 365.2$ days   **109.** $\sqrt{10{,}000} = 100$ pounds   **111.** $\sqrt{320} \approx 17.89$ feet per second   **113.** $\sqrt{1649} \approx 40.61$ meters
115. $2, -2$ **117.** $5, -1$ **119.** 30   **121.** $5, -5$ **123. a)** 3, 7; points of intersection   **b)** Yes   **c)** 3, 7; yes   **125.** At $x = 4$, $g(x)$ or
$y = 0$. Therefore the graph must have an x-intercept at 4.   **127.** $L_1 \approx 0.44, L_2 \approx 0.76$ **129.** All real numbers   **131.** 1.5

133. $\approx -3.7; \approx 3.7$ **135.** No real solution   **137.** $n = \dfrac{z^2\sigma^2}{(\overline{x} - \mu)^2}$ **140.** $P_2 = \dfrac{P_1 P_3}{P_1 - P_3}$ **141.** x **142.** $\dfrac{3a}{2b(2a + 3b)}$ **143.** $t(t - 5)$

144. $\dfrac{3}{x + 3}$ **145.** 2

Exercise Set 11.7   **1. a)** $\sqrt{-1}$ **b)** -1 **3.** Yes   **5.** Yes   **7.** $a - bi$ **9. a)** $\sqrt{2}$ **b)** 1   **c)** $\sqrt{-3}$ or $2i$ **d)** 6   **e)** Every number we have studied is a complex number.   **11.** $7 + 0i$ **13.** $5 + 0i$ **15.** $21 - 6i$ **17.** $0 + 2i\sqrt{6}$ **19.** $8 - 2i\sqrt{3}$ **21.** $3 + 7i\sqrt{2}$
23. $12 - 5i$ **25.** $0 + (7 - 3\sqrt{5})i$ **27.** $21 + 8i$ **29.** 0   **31.** $-17 - 12i$ **33.** $(4\sqrt{2} + \sqrt{3}) - 2i\sqrt{2}$ **35.** $11 - 4i\sqrt{2}$
37. $-3 - 2i\sqrt{5}$ **39.** $6 - 2i$ **41.** $-9 + 4i$ **43.** $-33 + 18i$ **45.** $28 + 4i\sqrt{3}$ **47.** $9 + 9i$ **49.** $1 + 5i$ **51.** 109   **53.** $39 - 9i\sqrt{2}$

55. $\dfrac{25}{72} + \dfrac{1}{4}i$ **57.** $-\dfrac{8}{3}i$ **59.** $\dfrac{3 - 2i}{2}$ **61.** $\dfrac{12 + 6i}{5}$ **63.** $\dfrac{3 + 6i}{5}$ **65.** $\dfrac{9 - 12i}{10}$ **67.** $\dfrac{3 + i}{5}$ **69.** $\dfrac{5\sqrt{2} - 2i\sqrt{6}}{37}$

71. $\dfrac{(5\sqrt{10} - 2\sqrt{15}) + (10\sqrt{2} + 5\sqrt{3})i}{45}$ **73.** 5   **75.** $\dfrac{\sqrt{2}}{3}$ **77.** $12 - 7i$ **79.** $4\sqrt{2} + 2i\sqrt{3}$ **81.** $20.8 - 16.64i$ **83.** $37 - 39i$

85. $\dfrac{4 - 11i}{2}$ **87.** $\dfrac{6\sqrt{3} + 12i}{7}$ **89.** $7 + \dfrac{2}{45}i$ **91.** $\dfrac{1}{4} - \dfrac{31}{50}i$ **93.** 2   **95.** $-4.33 - 10.91i$ **97.** -1 **99.** 1   **101.** i **103.** $-i$

105. a) $-2 - 3i$ **b)** $\dfrac{2 - 3i}{13}$ **107.** True; $(2i)(2i) = -4$ **109.** False; $(1 + i)(1 + 2i) = -1 + 3i$

111. Even values; i^n where n is even will either be -1 or 1   **113.** -4 **115.** $16 - 4i$ **117.** $14 + 8i$ **119.** 0   **121.** 1   **123.** Yes
125. No   **127.** $\approx 0.83 - 3i$ **129.** $\approx 1.5 - 0.33i$ **131.** $-i$ **133.** $1 + i\sqrt{5}, 1 - i\sqrt{5}$ **135.** $6 + 3i\sqrt{3}$ **137.** $-1 + 7i\sqrt{3}$

139. $2c - 3 - \dfrac{8}{4c + 9}$ **140.** $\dfrac{a^2}{b(a - b)}$ **141.** 4   **142.** 15 pounds at \$5.50, 25 pounds at \$6.30

Chapter 11 Review Exercises   **1.** 10   **2.** -3 **3.** -5 **4.** 4   **5.** 8   **6.** 38.2   **7.** $|x|$ **8.** $|x - 3|$ **9.** $|x - y|$

10. $|x^2 - 4x + 12|$ **11.** 7   **12.** 57   **13.** ≈ 2.2 **14.** 12 meters   **15.** $x^{7/2}$ **16.** $x^{5/3}$ **17.** $y^{13/4}$ **18.** $6^{-2/7}$ **19.** $\sqrt{x}$ **20.** $\sqrt[5]{a^4}$

21. $\left(\sqrt[4]{8m^2 n}\right)^7$ **22.** $\dfrac{1}{\left(\sqrt[3]{x + y}\right)^5}$ **23.** 16   **24.** x^6 **25.** 81   **26.** $\sqrt[4]{a}$ **27.** -6 **28.** Not a real number   **29.** $\dfrac{3}{4}$ **30.** $\dfrac{3}{8}$ **31.** $x^{4/15}$

32. $\dfrac{4}{y^3}$ **33.** $\dfrac{1}{a^{16/15}}$ **34.** $\dfrac{25x^{10}}{y^7}$ **35.** $5a^2 - 3a^{5/2}$ **36.** $\dfrac{4}{x^{7/6}} + 11$ **37.** $x^{2/5}(1 + x)$ **38.** $\dfrac{1 + a^2}{a^{1/2}}$

39. 5   **40.** ≈ 2.668 **41.**   **42.**   **43.** $4\sqrt{3}$ **44.** $4\sqrt[3]{2}$ **45.** $\dfrac{7}{3}$ **46.** $\dfrac{2}{5}$ **47.** $-\dfrac{9}{7}$ **48.** $-\dfrac{3}{5}$
49. 8   **50.** 4   **51.** $3xyz^2\sqrt{2y}$ **52.** $5xy^3\sqrt{3xy}$ **53.** $3a^2b^3\sqrt[3]{2ab}$

54. $5x^2y^3z^5\sqrt[3]{x^2z}$ **55.** $x^{14}y^{21}z^{35}$ **56.** $8a^3b^{12}c^{18}$ **57.** $2x^3\sqrt{10}$ **58.** $2x^3y\sqrt[3]{x^2y^2}$ **59.** $2x^2y^3\sqrt[3]{4x^2}$ **60.** $2x^2y^4\sqrt[4]{x}$ **61.** $6x - 2\sqrt{15x}$

62. $2x^2y^2\sqrt[3]{y^2} + x\sqrt[3]{18y}$ **63.** $\sqrt[4]{a^3b^2}$ **64.** $\sqrt[6]{x^5y^2}$ **65.** $\dfrac{64r^{9/2}}{p^3}$ **66.** $\dfrac{y^{1/5}}{6xz^{1/3}}$ **67.** $\dfrac{\sqrt{15}}{5}$ **68.** $\dfrac{\sqrt[3]{21}}{3}$ **69.** $\dfrac{\sqrt[4]{20}}{2}$ **70.** $\dfrac{x\sqrt{10}}{10}$ **71.** $\dfrac{8\sqrt{x}}{x}$

72. $\dfrac{m\sqrt[3]{5}}{5}$ **73.** $\dfrac{10\sqrt[3]{y}}{y}$ **74.** $\dfrac{9\sqrt[4]{z^3}}{z}$ **75.** $\dfrac{x}{3}$ **76.** $\dfrac{x}{2}$ **77.** $\dfrac{4y^2}{x^3}$ **78.** $2x^2y^3$ **79.** $\dfrac{x^2\sqrt{6y}}{y}$ **80.** $\dfrac{2\sqrt{21ab}}{7b}$ **81.** $\dfrac{x^2y^2\sqrt{6yz}}{z}$

82. $\dfrac{5xy^2\sqrt{15yz}}{3z}$ **83.** $3xy\sqrt[3]{2y}$ **84.** $\dfrac{\sqrt[3]{75xy^2}}{5y}$ **85.** $\dfrac{y\sqrt[3]{9x^2}}{x}$ **86.** $\dfrac{y^2\sqrt[3]{25x}}{5x}$ **87.** $\dfrac{b^2\sqrt[4]{2ab^2}}{a}$ **88.** $\dfrac{y\sqrt[4]{6x^3y^2}}{2x}$ **89.** 7   **90.** $x - y^2$

91. $x^2 - y$ **92.** $7 + 4\sqrt{3}$ **93.** $x + \sqrt{5xy} - \sqrt{3xy} - y\sqrt{15}$ **94.** $\sqrt[3]{6x^2} - \sqrt[3]{4xy} - \sqrt[3]{9xy} + \sqrt[3]{6y^2}$ **95.** $-12 + 6\sqrt{5}$

96. $\dfrac{4x - x\sqrt{x}}{16 - x}$ **97.** $\dfrac{4a + a\sqrt{b}}{16 - b}$ **98.** $\dfrac{x\sqrt{y} + 7x}{y - 49}$ **99.** $\dfrac{x - \sqrt{xy}}{x - y}$ **100.** $\dfrac{x - 2\sqrt{xy} - 3y}{x - y}$ **101.** $\dfrac{2\sqrt{a - 1} + 4}{a - 5}$

102. $\dfrac{5\sqrt{y + 2} + 15}{y - 7}$ **103.** $9\sqrt[3]{x}$ **104.** $-4\sqrt{3}$ **105.** $12 - 13\sqrt[3]{2}$ **106.** $\dfrac{45\sqrt{2}}{8}$ **107.** $(9x^2y^3 - 4x^3y^4)\sqrt{x}$

108. $(8x^2y^2 - x + 3x^3)\sqrt[3]{xy^2}$ **109.** $3x\sqrt{2} - 3\sqrt{5x}$ **110.** $2x^2 + 2x^2\sqrt[3]{4x}$ **111.** $2x + 7$ **112.** $\sqrt{5}|2a + 5|$ **113.** $\sqrt[6]{x + 5}$

114. $\sqrt[12]{b^5}$ **115. a)** $12\sqrt{3}$ **b)** 24   **116. a)** $8\sqrt{5} + \sqrt{130}$ **b)** $10\sqrt{13}$ **117. a)** 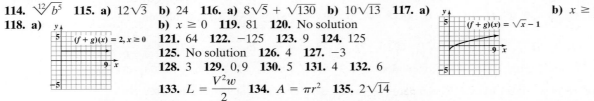   **b)** $x \geq 0$

118. a)   **b)** $x \geq 0$ **119.** 81   **120.** No solution

121. 64   **122.** -125 **123.** 9   **124.** 125

125. No solution   **126.** 4   **127.** -3

128. 3   **129.** $0, 9$ **130.** 5   **131.** 4   **132.** 6

133. $L = \dfrac{V^2 w}{2}$ **134.** $A = \pi r^2$ **135.** $2\sqrt{14}$

136. $5\sqrt{3}$ **137.** $\sqrt{29} \approx 5.39$ meters   **138.** $\sqrt{1280} \approx 35.78$ feet per second   **139.** $2\pi\sqrt{2} \approx 2.83\pi \approx 8.89$ seconds

140. $\sqrt{\dfrac{90}{0.145}} \approx 24.91$ meters per second   **141.** $m \approx 5m_0$. Thus, it is ≈ 5 times its original mass.   **142.** $5 + 0i$ **143.** $-8 + 0i$

144. $7 - 16i$ **145.** $9 + 4i$ **146.** $13 + i$ **147.** $6 - 2i$ **148.** $12\sqrt{3} + (\sqrt{5} - \sqrt{7})i$ **149.** $-6 + 6i$ **150.** $17 - 6i$

151. $(24 + 3\sqrt{5}) + (4\sqrt{3} - 6\sqrt{15})i$ **152.** $-\dfrac{8i}{3}$ **153.** $\dfrac{(-2 - \sqrt{3})i}{2}$ **154.** $\dfrac{12 - 8i}{13}$ **155.** $\dfrac{5\sqrt{3} + 3i\sqrt{2}}{31}$ **156.** 0   **157.** 7

158. i **159.** $-i$ **160.** 1   **161.** -1

Chapter 11 Practice Test

1. $|5x - 3|$ [11.1]   **2.** $\dfrac{1}{x^{12/5}}$ [11.2]   **3.** $\dfrac{1 + x^2}{x^{2/3}}$ [11.2]   **4.** 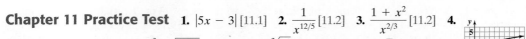 [11.1]   **5.** $3x^3y^5\sqrt{6x}$ [11.3]

6. $5x^3y^3\sqrt[3]{2x^2y}$ [11.4]   **7.** $\dfrac{x^3y\sqrt{14yz}}{4z}$ [11.5]   **8.** $\dfrac{9\sqrt[3]{x^2}}{x}$ [11.5]   **9.** $\dfrac{3 - \sqrt{3}}{6}$ [11.5]

10. $7\sqrt{6}$ [11.3]   **11.** $(2xy + 4x^2y^2)\sqrt[3]{y^2}$ [11.4]   **12.** $6\sqrt{3} - 2\sqrt{6} - 12 + 4\sqrt{2}$ [11.4]

13. $\sqrt[8]{x^5y^3}$ [11.2]   **14.** $\sqrt[12]{(7x + 2)^7}$ [11.5]   **15.** -5 [11.6]   **16.** -3 [11.6]   **17.** 9 [11.6]

18. 3 [11.6]   **19.** $g = \dfrac{8w^2}{h}$ [11.6]   **20.** $\sqrt{12{,}880} \approx 113.49$ feet per second [11.6]   **21.** 13 feet [11.6]   **22.** $2\pi\sqrt{\dfrac{1400}{65{,}000}} \approx 0.92$ second [11.6]

23. $20 + 20i$ [11.7]   **24.** $\dfrac{33 - 17i}{53}$ [11.7]   **25.** 2 [11.7]

Cumulative Review Test

1. $\dfrac{57}{9}$ [2.1]   **2.** -1 [2.1]   **3.** \$40 [2.3]   **4.** 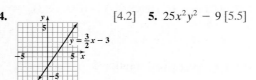 [4.2]   **5.** $25x^2y^2 - 9$ [5.5]

6. $w = 2r + 1$ [5.6]   **7.** $x(4x - 5)(x - 1)$ [6.4]

8. $(x - 2)(x^2 + 5x + 13)$ [6.5]   **9.** $\dfrac{1}{4}, -\dfrac{3}{2}$ [6.6]   **10.** $\dfrac{(x + y)y^2}{3x^3}$ [7.2]

11. $\dfrac{x + 3}{x + 5}$ [7.4]   **12.** 18 [7.6]   **13.** Parallel [8.4]   **14.** $y = -\dfrac{2}{3}x - \dfrac{10}{3}$ [8.4]   **15.** $-x^2 + 5x - 13$ [8.5]   **16.** $\left(2, 5, \dfrac{34}{5}\right)$ [9.4]

17. 40 [9.7]   **18.** $\{x | -1 < x < 4\}$ [10.2]   **19.** $3, -3$ [11.6]   **20.** 400 feet [7.8]

Chapter 12

Exercise Set 12.1   **1.** ± 6 **3.** If $x^2 = a$, then $x = \pm\sqrt{a}$.   **5.** $\left(\dfrac{b}{2}\right)^2$ must equal c.   **7. a)** Yes   **b)** No, ± 2

9. Multiply by $\dfrac{1}{2}$ to make $a = 1$.   **11.** $\left(-\dfrac{6}{2}\right)^2 = 9$ **13.** ± 5 **15.** $\pm 7i$ **17.** $\pm 2i\sqrt{6}$ **19.** $\pm i\sqrt{61}$ **21.** $8, 0$ **23.** $-3 \pm 5i$

25. $2 \pm 3i\sqrt{5}$ **27.** $-1, \dfrac{1}{3}$ **29.** $\dfrac{2 \pm 2i}{3}$ **31.** $0.1, -1.7$ **33.** $\dfrac{5 \pm 3\sqrt{2}}{2}$ **35.** $-\dfrac{1}{20}, -\dfrac{9}{20}$ **37.** $1, -4$ **39.** $-3, -5$ **41.** $-2, -4$

43. $1, 6$ **45.** $-1, \dfrac{1}{2}$ **47.** $-\dfrac{1}{2}, 4$ **49.** $5, 8$ **51.** $-1, 7$ **53.** $4, 5$ **55.** $7, -4$ **57.** $1, -11$ **59.** $2 \pm \sqrt{14}$ **61.** $-4 \pm \sqrt{11}$

63. $\dfrac{1 \pm \sqrt{13}}{2}$ **65.** $\dfrac{-3 \pm i\sqrt{15}}{2}$ **67.** $0, 1$ **69.** $0, -\dfrac{2}{3}$ **71.** $0, \dfrac{1}{6}$ **73.** $1, -3$ **75.** $8, -4$ **77.** $\dfrac{-9 \pm \sqrt{73}}{2}$ **79.** $\dfrac{1}{3}, -1$ **81.** $\dfrac{1 \pm i\sqrt{39}}{4}$

83. $1 \pm i$ **85. a)** $21 = (x + 2)(x - 2)$ **b)** 5 **87. a)** $18 = (x + 4)(x + 2)$ **b)** $-3 + \sqrt{19}$ **89.** 30 mph **91.** 5, 7

93. 5 feet by 12 feet **95.** $\dfrac{12 + \sqrt{288}}{2} \approx 14.49$ feet by 14.49 feet **97.** $\sqrt{200} \approx 14.14$ inches **99.** $\sqrt{24} \approx 4.90$ feet **101.** 4%

103. $\approx 6\%$ **105. a)** $S = 32 + 80\sqrt{\pi} \approx 173.80$ square inches **b)** $r = \dfrac{4\sqrt{\pi}}{\pi} \approx 2.26$ inches **c)** $r = -5 + \sqrt{\dfrac{80 + 25\pi}{\pi}} \approx 2.1$ inches

107. 2 **108.** \$4200 at 7%, \$5800 at $6\frac{1}{4}\%$ **109.** 0 **110.** $4x^3 + x^2 - 21x + 6$ **111.** $\left\{10, \dfrac{4}{3}\right\}$

Exercise Set 12.2 **1.** $x = \dfrac{-b \pm \sqrt{b^2 - 4ac}}{2a}$ **3.** $a = -3, b = 6, c = 8$ **5.** Yes; if you multiply both sides of one equation by -1 you get the other equation **7. a)** $b^2 - 4ac$ **b)** -84 **c)** Answers will vary. **9.** Two real solutions **11.** No real solution **13.** Two real solutions **15.** No real solution **17.** One real solution **19.** One real solution **21.** 3, 6 **23.** 2, 4 **25.** 1, -7 **27.** $-2 \pm 2\sqrt{6}$ **29.** ± 8 **31.** $\dfrac{2 \pm i\sqrt{11}}{3}$ **33.** 0, 5 **35.** $\dfrac{2 \pm i\sqrt{2}}{2}$ **37.** -1 **39.** $\dfrac{1}{4}$ **41.** $1 \pm \sqrt{2}$ **43.** $\dfrac{-3 \pm i\sqrt{15}}{2}$ **45.** $-3, \dfrac{1}{2}$ **47.** $\dfrac{1}{2}, -\dfrac{5}{3}$ **49.** 4, -6 **51.** $\dfrac{1}{3}, -\dfrac{2}{3}$ **53.** $\dfrac{-6 \pm 2\sqrt{6}}{3}$ **55.** $\dfrac{3 \pm \sqrt{309}}{30}$ **57.** $\dfrac{3 \pm \sqrt{33}}{2}$ **59.** $\dfrac{2 \pm i\sqrt{6}}{2}$ **61.** $\dfrac{-1 \pm i\sqrt{23}}{4}$ **63.** $\dfrac{-0.6 \pm \sqrt{0.84}}{0.2}$ or $-3 \pm \sqrt{21}$ **65.** 0, 2 **67.** $-5, 6$ **69.** $\dfrac{7 \pm \sqrt{17}}{4}$ **71.** No real number **73.** $x^2 - 7x + 10 = 0$ **75.** $x^2 + 8x - 9 = 0$ **77.** $15x^2 - x - 6 = 0$ **79.** $x^2 - 2 = 0$ **81.** $x^2 + 9 = 0$ **83.** $x^2 - 6x + 7 = 0$ **85.** $x^2 - 4x + 13 = 0$ **87. a)** $n(10 - 0.02n) = 450$ **b)** 50 **89. a)** $n(50 - 0.4n) = 660$ **b)** 15 **91.** Answers will vary. **93.** Yes **95.** 3 **97.** $w = 3$ feet, $l = 8$ feet **99.** 2 inches **101.** ≈ 4.39 seconds **103. a)** ≈ 4.57 seconds **b)** ≈ 4.79 seconds. **105.** $2\sqrt{5}, -\sqrt{5}$ **107.** $(-0.12 + \sqrt{14.3952})/1.2 \approx 3.0618$ millimeters **109. a)** ≈ 1.94 seconds **b)** ≈ 2.74 seconds **c)** Courtney's **d)** Yes, at 1.5 seconds **110.** 5.0×10^2 or 500 **111.** 7 **112.** $(2, -1)$ **113.** $\dfrac{x^2 + 2x\sqrt{y} + y}{x^2 - y}$ **114.** No real solution

Exercise Set 12.3 **1.** Answers will vary. **3.** $S = \sqrt{A}$ **5.** $t = \sqrt{\dfrac{d}{4.9}}$ **7.** $i = \sqrt{\dfrac{E}{r}}$ **9.** $t = \dfrac{\sqrt{d}}{4}$ **11.** $c = \sqrt{\dfrac{E}{m}}$ **13.** $r = \sqrt{\dfrac{3V}{\pi h}}$ **15.** $W = \sqrt{d^2 - L^2}$ **17.** $b = \sqrt{c^2 - a^2}$ **19.** $H = \sqrt{d^2 - L^2 - W^2}$ **21.** $t = \sqrt{\dfrac{h - s_0}{-16}}$ or $t = \dfrac{\sqrt{s_0 - h}}{4}$ **23.** $v = \sqrt{\dfrac{2E}{m}}$ **25.** $v_1 = \sqrt{v_2^2 - 2ad}$ **27.** $c = \sqrt{(v')^2 + v^2}$ **29. a)** \$10,950 **b)** ≈ 7 **31. a)** $32°F$ **b)** $80.8°F$ **c)** ≈ 2.92 minutes **33. a)** 0.53 billion **b)** 2007 **35. a)** 11.4 billion tons **b)** 2003 **37. a)** 1.301 million **b)** 2009 **39.** $l = 30$ meters, $w = 20$ meters **41.** 4 feet per hour **43.** Going 6 mph, returning 8 mph **45.** Bonita ≈ 11.52 hours; Pamela ≈ 12.52 hours **47.** 130 mph **49.** Chris ≈ 11.76 hours; John ≈ 12.26 hours **51.** 75 mph **53.** $l \approx 34.86$ inches, $h \approx 19.61$ inches **55.** Answers will vary. **57.** 6 meters by 3 meters or 2 meters by 9 meters **59.** -16 **60.** $R = \dfrac{E - Ir}{I}$ **61.** $\dfrac{8}{r - 4}$ **62.** $\dfrac{x^2}{y^{32}}$ **63.** No solution

Mid-Chapter Test **1.** $\pm 7\sqrt{2}$ [12.1] **2.** $3 \pm 2i\sqrt{5}$ [12.1] **3.** $-\dfrac{1}{2}, -\dfrac{13}{2}$ [12.1] **4.** $-6, 2$ [12.1] **5.** $2 \pm \sqrt{14}$ [12.1] **6.** $\dfrac{-1 \pm i\sqrt{143}}{8}$ [12.1] **7.** $(6 + 6\sqrt{2})$ meters [12.1] **8. a)** $b^2 - 4ac$ **b)** Two distinct real solutions: $b^2 - 4ac > 0$; single real solution: $b^2 - 4ac = 0$; no real solution: $b^2 - 4ac < 0$ [12.2] **9.** Two distinct real solutions [12.2] **10.** $-\dfrac{5}{3}, \dfrac{3}{2}$ [12.2] **11.** $-2 \pm 2\sqrt{3}$ [12.2] **12.** $\dfrac{1 \pm i\sqrt{14}}{3}$ [12.2] **13.** $x^2 - 5x - 14 = 0$ [12.2] **14.** $x^2 - 4x - 1 = 0$ [12.2] **15.** 10 lamps [12.2] **16.** $r = \sqrt{x^2 - y}$ [12.3] **17.** $x = \sqrt{\dfrac{3A}{k}}$ [12.3] **18.** $y = \sqrt{D^2 - x^2}$ [12.3] **19.** 5 feet by 12 feet [12.3] **20.** 5 clocks [12.3]

Exercise Set 12.4 **1.** Can be written in the form $au^2 + bu + c = 0$. **3.** $u = x^2$; gives equation $3u^2 - 5u + 1 = 0$. **5.** $u = z^{-1}$; gives equation $u^2 - u = 56$. **7.** $(x^2 + 3)(x^2 - 2)$ **9.** $(x^2 + 2)(x^2 + 3)$ **11.** $(2a^2 + 5)(3a^2 - 5)$ **13.** $(2x + 5)(2x + 3)$ **15.** $(3a + 1)(2a + 5)$ **17.** $(ab + 3)(ab + 5)$ **19.** $(3xy - 5)(xy + 1)$ **21.** $(2a - 5)(a - 1)(5 - a)$ **23.** $(2x + 3)(x + 2)(x - 3)$ **25.** $(y^2 + 10)(y^2 + 3)$ **27.** $(x + 2)(x + 1)(x + 3)$ **29.** $a^3 b^2 (5a - 3b)(a - b)$ **31.** $\pm 1, \pm 3$ **33.** $\pm i, \pm 4i$ **35.** $\pm 2, \pm 3$ **37.** $\pm 2, \pm \sqrt{3}$ **39.** $\pm \dfrac{1}{2}, \pm 2$ **41.** $\pm \sqrt{3}, \pm \sqrt{5}$ **43.** $\pm 3, \pm i\sqrt{2}$ **45.** $\pm 1, \pm i\sqrt{5}$ **47.** 4 **49.** 9 **51.** $\dfrac{1}{9}$ **53.** 1, -9 **55.** $\dfrac{4}{3}, -\dfrac{1}{2}$ **57.** $\pm \sqrt{6}, \pm 1$ **59.** $-6, -\dfrac{5}{2}$ **61.** $\pm \dfrac{5\sqrt{6}}{6}, \pm \dfrac{\sqrt{39}}{3}$ **63.** $-\dfrac{1}{2}$ **65.** 3, 4 **67.** $2, \dfrac{1}{3}$ **69.** $1, -\dfrac{1}{10}$ **71.** $-\dfrac{1}{2}, \dfrac{1}{6}$ **73.** 1, 27 **75.** 27, 216 **77.** $\dfrac{1}{4}$ **79.** $-32, -1$ **81.** $(1, 0), (16, 0)$ **83.** None **85.** $(-4, 0), \left(\dfrac{1}{5}, 0\right)$ **87.** $(-8, 0), (27, 0)$ **89.** $(-1, 0), (4, 0)$ **91.** $(\pm 2, 0), (\pm 5, 0)$ **93.** Let $u = x^2$ **95.** Let $u = x^{-1}$ **97.** $x^4 - 5x^2 + 4 = 0$; start with $(x - 2)(x + 2)(x - 1)(x + 1) = 0$ **99.** $x^4 - 7x^2 + 10 = 0$; start with $(x + \sqrt{2})(x - \sqrt{2})(x + \sqrt{5})(x - \sqrt{5}) = 0$ **101.** No; imaginary solution always occur in pairs.

103. a) and **b)** $\frac{1}{5}, -\frac{1}{4}$ **105.** $-\frac{14}{5}, -\frac{8}{3}$ **107.** $2, \frac{1}{4}$ **109.** $2, 1$ **111.** $-3, 1, 2, -4$ **113.** $\pm\sqrt{\dfrac{3 \pm \sqrt{15}}{2}}$ **115.** $\frac{43}{60}$ **116.** 1

117. D: $\mathbb{R}$, R: $\{y|y \geq 0\}$ **118.** $2xy^2\sqrt[3]{2}$ **119.** $9\sqrt{3}$

Using Your Graphing Calculator, 12.5   **1.**

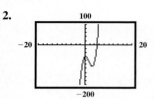   **2.**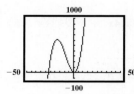

Exercise Set 12.5   **1.** The graph of a quadratic equation is called a parabola.   **3.** The axis of symmetry of a parabola is the line

where, if the graph is folded, the two sides overlap.   **5.** $\left(-\dfrac{b}{2a}, \dfrac{4ac - b^2}{4a}\right)$ **7. a)** When $a > 0$, $f(x)$ will have a minimum since the

graph opens upward.   **b)** When $a < 0$, $f(x)$ will have a maximum since the graph opens downward.   **9.** Set $x = 0$ and solve for y.

11. a) 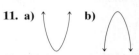   **b)**

13. Minimum value; the graph opens upward

15. -4 **17.** -7

19. -2.0312

21. a) Upward   **b)** $(0, 15)$
c) $(-4, -1)$ **d)** $(-5, 0), (-3, 0)$
e)

23. a) Upward   **b)** $(0, 3)$
c) $(2, -1)$ **d)** $(1, 0), (3, 0)$
e)

25. a) Downward   **b)** $(0, 8)$
c) $(-1, 9)$ **d)** $(-4, 0), (2, 0)$
e)

27. a) Downward   **b)** $(0, 5)$
c) $(2, 9)$ **d)** $(-1, 0), (5, 0)$
e)

29. a) Downward   **b)** $(0, -5)$
c) $(2, -1)$ **d)** No x-intercepts
e)

31. a) Upward   **b)** $(0, 4)$ **c)** $(2, 0)$
d) $(2, 0)$ **e)**

33. a) Upward   **b)** $(0, 2)$ **c)** $(0, 2)$
d) No x-intercepts
e)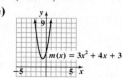

35. a) Downward   **b)** $(0, 5)$
c) $(0, 5)$ **d)** $(-\sqrt{5}, 0), (\sqrt{5}, 0)$
e)

37. a) Downward   **b)** $(0, -8)$
c) $(1, -6)$ **d)** No x-intercepts
e)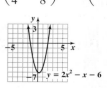

39. a) Upward   **b)** $(0, 3)$
c) $\left(-\dfrac{2}{3}, \dfrac{5}{3}\right)$ **d)** No x-intercepts
e)

41. a) Upward   **b)** $(0, -6)$
c) $\left(-\dfrac{2}{3}, -\dfrac{22}{3}\right)$
d) $\left(\dfrac{-2 + \sqrt{22}}{3}, 0\right), \left(\dfrac{-2 - \sqrt{22}}{3}, 0\right)$
e)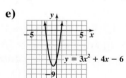

43. a) Upward   **b)** $(0, -6)$
c) $\left(\dfrac{1}{4}, -\dfrac{49}{8}\right)$ **d)** $\left(-\dfrac{3}{2}, 0\right), (2, 0)$
e)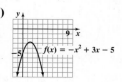

45. a) Downward   **b)** $(0, -5)$
c) $\left(\dfrac{3}{2}, -\dfrac{11}{4}\right)$ **d)** No x-intercepts
e)

47.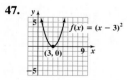

49. 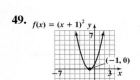 $f(x) = (x + 1)^2$ $(-1, 0)$

51. 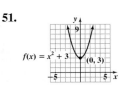 $f(x) = x^2 + 3$ $(0, 3)$

53. $f(x) = x^2 - 1$ $(0, -1)$

55. 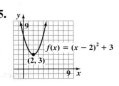 $f(x) = (x - 2)^2 + 3$ $(2, 3)$

57. 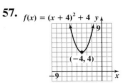 $f(x) = (x + 4)^2 + 4$ $(-4, 4)$

59. 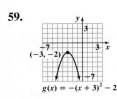 $(-3, -2)$ $g(x) = -(x + 3)^2 - 2$

61. $(2, 2)$ $y = -2(x - 2)^2 + 2$

63. 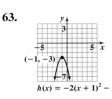 $(-1, -3)$ $h(x) = -2(x + 1)^2 - 3$

65. a) $f(x) = (x - 3)^2 - 1$
b) 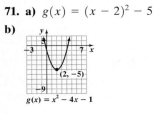 $f(x) = x^2 - 6x + 8$ $(3, -1)$

67. a) $g(x) = \left(x - \dfrac{1}{2}\right)^2 - \dfrac{13}{4}$
b) 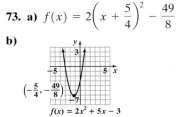 $g(x) = x^2 - x - 3$ $\left(\dfrac{1}{2}, -\dfrac{13}{4}\right)$

69. a) $f(x) = -(x + 2)^2 - 2$
b) 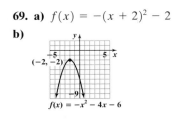 $(-2, -2)$ $f(x) = -x^2 - 4x - 6$

71. a) $g(x) = (x - 2)^2 - 5$
b) 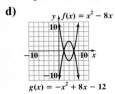 $(2, -5)$ $g(x) = x^2 - 4x - 1$

73. a) $f(x) = 2\left(x + \dfrac{5}{4}\right)^2 - \dfrac{49}{8}$
b) 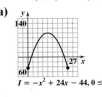 $\left(-\dfrac{5}{4}, -\dfrac{49}{8}\right)$ $f(x) = 2x^2 + 5x - 3$

75. c)   **77. c)**
79. a)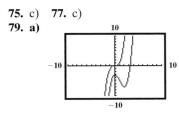
b) Increase   **c)** Answers will vary.
d) Decrease   **e)** Answers will vary.

81. d)   **83. b)**   **85. a)** $x = 7$ **b)** $A = 121$ **87. a)** $x = 10.5$ **b)** $A = 240.25$

89. a) $n = 200$ **b)** $R = \$800$ **91.** 2010   **93.** 4 units   **95.** 3 units   **97.** $f(x) = 2(x - 3)^2 - 2$ **99.** $f(x) = -4\left(x + \dfrac{3}{5}\right)^2 - \sqrt{2}$

101. a) The graphs will have the same x-intercepts but $f(x) = x^2 - 8x + 12$ will open upward and $g(x) = -x^2 + 8x - 12$ will open downward.   **b)** Yes, both at $(6, 0)$ and $(2, 0)$ **c)** No; vertex of $f(x)$ at $(4, -4)$, vertex of $g(x)$ at $(4, 4)$

d) 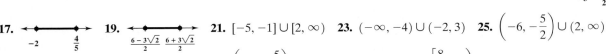 $f(x) = x^2 - 8x + 12$ $g(x) = -x^2 + 8x - 12$

103. a) $I = -x^2 + 24x - 44, \; 0 \le x \le 24$

b) $2   **c)** $22   **d)** $12   **e)** $10,000   **105. a)** 100   **b)** $3800
107. a) 40.425 meters   **b)** 2.5 seconds   **c)** $\approx$5.37 seconds
109. a) $\approx\$577$ **b)** 2000   **111.** 400 square feet   **113.** -16, 4 and -4
115. 900, 30 and 30   **117. a)** $142,400   **b)** 380   **119. b)**
121. a) $f(t) = -16(t - 1.625)^2 + 45.25$ **b)** 45.25 feet, 1.625 seconds   **c)** Same

125. 200π square feet   **126.** $\dfrac{x}{x + 6}$ **127.** $(-2, 3, 2)$ **128.** -8 **129.**

Exercise Set 12.6   **1. a)** $x < 2$ or $x > 5$ **b)** $2 < x < 5$ **3.** Yes; $\ge$ **5.** Yes, -2 and 1 make the fraction 0; No, -1 makes the fraction undefined   **7.** (-2, 4)   **9.** (-6, -1)   **11.** (0)   **13.** (-4, 4)   **15.** (-3, $\frac{1}{2}$)

17. (-2, $\frac{4}{5}$)   **19.** ($\frac{6 - 3\sqrt{2}}{2}$, $\frac{6 + 3\sqrt{2}}{2}$)   **21.** $[-5, -1] \cup [2, \infty)$ **23.** $(-\infty, -4) \cup (-2, 3)$ **25.** $\left(-6, -\dfrac{5}{2}\right) \cup (2, \infty)$

27. $\left(-1, -\dfrac{5}{3}\right) \cup (3, \infty)$ **29.** $[-2, -2] \cup \left[\dfrac{8}{3}, \infty\right)$ **31.** $(-\infty, 0)$ **33.** (0, 6)

35. (-4, 0)   **37.** (6, 8)   **39.** ($\frac{-9 - \sqrt{129}}{4}$, $\frac{-9 + \sqrt{129}}{4}$)   **41.** ($-7$, 0, $\frac{5}{2}$)   **43.** $\{x | x < -2 \text{ or } x > 4\}$ **45.** $\{x | -5 < x < 1\}$

47. $\{x | x \le -3 \text{ or } x > 2\}$ **49.** $\{a | -5 < a < 9\}$ **51.** $\{c | c < 4 \text{ or } c > 10\}$ **53.** $\{y | -4 < y \le -2\}$ **55.** $\left\{a \,\middle|\, a \le -2 \text{ or } a > \dfrac{1}{3}\right\}$

57. $\left\{ x \left| -\dfrac{4}{3} < x < \dfrac{1}{2} \right. \right\}$ **59.** $\left\{ x \left| -\dfrac{8}{3} \le x < 2 \right. \right\}$ **61.** $(-\infty, -3) \cup (-1, 6)$ **63.** $(-3, 2) \cup (5, \infty)$ **65.** $(-2, 1] \cup [7, \infty)$

67. $(-\infty, -8) \cup [0, 3)$ **69.** $(-\infty, -4) \cup (1, 6]$ **71.** $\left[-\dfrac{5}{2}, 3 \right] \cup (4, \infty)$ **73.** ⟵─○───●─⟶ **75.** ⟵───○──●───⟶ **77.** ⟵──────○──●⟶
$\qquad\qquad\qquad\qquad\qquad\qquad\qquad\qquad\qquad\qquad\quad$ 4 6 $\qquad\qquad\qquad\qquad\quad$ −2 1 $\qquad\qquad\qquad\qquad\qquad\quad$ −2 3

79. ⟵─●───○──⟶ **81.** ⟵○────●───⟶ **83.** ⟵──○──○──⟶ **85. a)** $(4, \infty)$; $y > 0$ in this interval **b)** $(-\infty, 2) \cup (2, 4)$; $y < 0$ in
$\qquad$ 1 4 $\qquad\qquad\qquad$ −2 0 $\qquad\qquad\qquad\quad \frac{4}{7}\ \frac{2}{3}$

this interval **87.** $x^2 + 2x - 8 > 0$ **89.** $\dfrac{x+3}{x-4} \ge 0$ **91.** All real numbers; for any value of x, the expression is ≥ 0. **93.** All real numbers

except -2; for any value of x except -2, the expression is ≥ 0. **95.** No solution; the graph opens upward and has no x-intercepts, so it is

always above the x-axis. **97.** ⟵●──●───●⟶ **99.** $x^2 - 3x > 0$; multiply factors containing boundary values. **101.** $x^2 < 0$; x^2 is always ≥ 0.
$\qquad\qquad\qquad\qquad\qquad\qquad\qquad$ −8 −5 −1 3

103. $(-\infty, -3) \cup (-1, 1) \cup (3, \infty)$ **105.** $[-2, -1] \cup [2, \infty)$ **109.** 6 quarts **110.** $3r + 3s - 9t$ **111.** $\dfrac{x-3}{x+1}$ **112.** $-\dfrac{1}{2}$ **113.** $38 - 9i$

Chapter 12 Review Exercises

1. $5 \pm 2\sqrt{6}$ **2.** $\dfrac{-1 \pm 2\sqrt{15}}{2}$ **3.** $-\dfrac{1}{3}, 1$ **4.** $\dfrac{5}{4}, -\dfrac{3}{4}$ **5.** $3, 4$ **6.** $4, -8$ **7.** $-1 \pm \sqrt{10}$

8. $-3 \pm \sqrt{21}$ **9.** $1 \pm 3i$ **10.** $2 \pm 2i\sqrt{7}$ **11. a)** $32 = (x+1)(x+5)$ **b)** 3 **12. a)** $63 = (x+2)(x+4)$ **b)** 5 **13.** $6, 7$

14. ≈ 16.90 ft by ≈ 16.90 ft **15.** Two real solutions **16.** No real solution **17.** One real solution **18.** No real solution

19. One real solution **20.** Two real solutions **21.** $0, -\dfrac{4}{3}$ **22.** $2, 9$ **23.** $8, -5$ **24.** $0, \dfrac{9}{7}$ **25.** $\dfrac{3}{2}, -\dfrac{5}{3}$ **26.** $\dfrac{1}{4}, -3$ **27.** $-4 \pm \sqrt{11}$

28. $-2 \pm 2\sqrt{3}$ **29.** $\dfrac{-2 \pm \sqrt{10}}{2}$ **30.** $\dfrac{3 \pm \sqrt{33}}{3}$ **31.** $\dfrac{1 \pm i\sqrt{51}}{2}$ **32.** $1 \pm i\sqrt{10}$ **33.** $\dfrac{5}{2}, -\dfrac{5}{3}$ **34.** $\dfrac{1}{4}, -\dfrac{3}{2}$ **35.** $10, -6$ **36.** $\dfrac{2}{3}, -\dfrac{3}{2}$

37. $\dfrac{7 \pm \sqrt{89}}{10}$ **38.** $\dfrac{3 \pm 3\sqrt{3}}{2}$ **39.** $x^2 - 2x - 3 = 0$ **40.** $3x^2 + 4x - 4 = 0$ **41.** $x^2 - 11 = 0$ **42.** $x^2 - 6x + 13 = 0$

43. 8 feet by 12 feet **44.** $\sqrt{128} \approx 11.31$ **45.** 4% **46.** $7, 11$ **47.** 8 inches by 12 inches **48.** \$540 **49. a)** \$121.8 billion **b)** 2010
50. a) 720 feet **b)** 7 seconds **51. a)** 40 milliliters **b)** $150°C$ **52.** larger ≈ 23.51 hours; smaller ≈ 24.51 hours **53.** 50 miles per hour
54. 1.6 miles per hour **55.** $l = 10$ units, $w = 8$ units **56.** 20 tables **57.** $a = \sqrt{c^2 - b^2}$ **58.** $t = \sqrt{\dfrac{c-h}{4.9}}$ **59.** $v_y = \sqrt{v^2 - v_x^2}$

60. $v_2 = \sqrt{v_1^2 + 2ad}$ **61.** $(x^2 + 4)(x^2 - 5)$ **62.** $(2x^2 - 1)(2x^2 + 3)$ **63.** $(x+9)(x+11)$ **64.** $(4x+1)(4x+5)$ **65.** $\pm 2, \pm 3$

66. $\pm 4, \pm \sqrt{5}$ **67.** $\pm 2\sqrt{2}, \pm i\sqrt{3}$ **68.** $\dfrac{3}{2}, -\dfrac{1}{6}$ **69.** $\dfrac{1}{9}$ **70.** $\dfrac{27}{8}, 8$ **71.** $4, \dfrac{13}{8}$ **72.** $-\dfrac{1}{5}, -\dfrac{5}{2}$ **73.** $(\pm 1, 0), (\pm 9, 0)$ **74.** $\left(\dfrac{4}{25}, 0 \right)$

75. None **76.** $(3 \pm \sqrt{17}, 0), (3 \pm \sqrt{6}, 0)$

77. a) Upward **b)** $(0, 0)$
c) $\left(-\dfrac{5}{2}, -\dfrac{25}{4} \right)$ **d)** $(0, 0), (-5, 0)$
e)

$\left(-\dfrac{5}{2}, -\dfrac{25}{4} \right)$ $\quad f(x) = x^2 + 5x$

78. a) Upward **b)** $(0, -8)$
c) $(1, -9)$ **d)** $(-2, 0), (4, 0)$
e)

$f(x) = x^2 - 2x - 8$ $\quad (1, -9)$

79. a) Downward **b)** $(0, -2)$
c) $(0, -2)$ **d)** No x-intercepts
e)

$(0, -2)$ $\quad g(x) = -x^2 - 2$

80. a) Downward **b)** $(0, 15)$
c) $\left(-\dfrac{1}{4}, \dfrac{121}{8} \right)$ **d)** $(-3, 0), \left(\dfrac{5}{2}, 0 \right)$
e)

$\left(-\dfrac{1}{4}, \dfrac{121}{8} \right)$ $\quad g(x) = -2x^2 - x + 15$

81. a) \$11 **b)** \$7600
82. a) 2.5 seconds **b)** 175 feet
83.

$f(x) = (x-3)^2$ $\quad (3, 0)$

84.

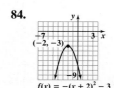

$(-2, -3)$ $\quad f(x) = -(x+2)^2 - 3$

85.

$(-4, -1)$ $\quad g(x) = -2(x+4)^2 - 1$

86.

$(1, 3)$ $\quad h(x) = \dfrac{1}{2}(x-1)^2 + 3$

87. ⟵──●──●──⟶
$\qquad$ −3 −1

88. ⟵─●────●─⟶
$\qquad$ −5 $\quad$ 2

89. ⟵─●────●─⟶
$\dfrac{11 - \sqrt{41}}{2}\quad \dfrac{11 + \sqrt{41}}{2}$

90. ⟵─○──○─⟶
$\qquad$ −4 $\quad \dfrac{4}{3}$

91. ⟵─●────●─⟶
$\quad -\dfrac{3}{2}\qquad \dfrac{3}{2}$

92. ⟵──○──○──⟶
$\quad -\sqrt{5}\ \ \sqrt{5}$

93. $\{x | x < -1 \text{ or } x > 5\}$ **94.** $\{x | -2 < x \le 3\}$ **95.** $\{x | x < -3 \text{ or } x \ge 2\}$

96. $\left\{x\left|-\dfrac{5}{3} < x < 6\right.\right\}$ **97.** $\{x|-4 < x < -1 \text{ or } x > 2\}$ **98.** $\{x|x \le 0 \text{ or } 3 \le x \le 6\}$ **99.** $\left[-\dfrac{4}{3}, 1\right] \cup [3, \infty)$

100. $(-\infty, -4) \cup (-2, 0)$ **101.** $(-2, 0) \cup (4, \infty)$ **102.** $(-\infty, -3) \cup (2, 8)$ **103.** $(-2, 3] \cup (7, \infty)$ **104.** $(-\infty, -3) \cup [0, 6]$

105.   **106.**   **107.**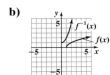

Chapter 12 Practice Test

1. $3, -5$ [12.1]   **2.** $3 \pm \sqrt{2}$ [12.1]   **3.** $8, -2$ [12.2]   **4.** $2 \pm i\sqrt{7}$ [12.2]   **5.** $\dfrac{2}{3}, -1$ [12.1–12.2]

6. $\dfrac{-7 \pm \sqrt{33}}{2}$ [12.1–12.2]   **7.** $5x^2 - 18x - 8 = 0$ [12.2]   **8.** $v = \sqrt{\dfrac{2K}{m}}$ [12.3]   **9. a)** \$121,200   **b)** ≈ 2712.57 square feet [12.1–12.3]

10. 50 mph [12.1–12.3]   **11.** $\pm\dfrac{\sqrt{10}}{2}, \pm i\sqrt{10}$ [12.4]   **12.** $\dfrac{343}{27}, -216$ [12.4]   **13.** $\left(\dfrac{9}{16}, 0\right)$ [12.4]   **14.**

15. 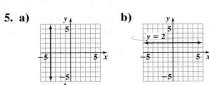   [12.5]   **16.** Two real solutions [12.5]   **17. a)** Upward   **b)** $(0, -8)$

c) $(-1, -9)$ **d)** $(-4, 0), (2, 0)$ **e)**   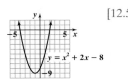   [12.5]

18. $2x^2 + 13x - 7 = 0$ [12.5]   **19.** 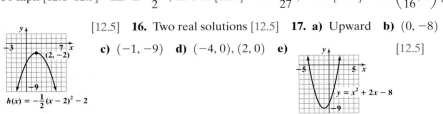 [12.6]   **20.** [12.6]   **21. a)** $\left[-\dfrac{5}{2}, -2\right)$ **b)** $\left\{x\left|-\dfrac{5}{2} \le x < -2\right.\right\}$ [12.6]

22. $w = 5$ feet, $l = 13$ feet [12.5]   **23.** 6 seconds [12.5]   **24. a)** 20   **b)** \$490 [12.5]   **25.** 30 [12.5]

Cumulative Review Test

1. 13 [1.4]   **2.** 18 [1.4]   **3.** $3x - 7$ [1.9]   **4.** All real numbers, $\mathbb{R}$ [2.5]

5. a)   **b)**   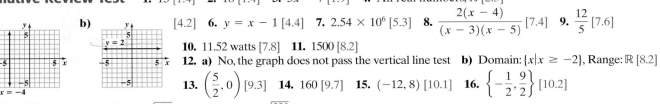   [4.2]   **6.** $y = x - 1$ [4.4]   **7.** 2.54×10^6 [5.3]   **8.** $\dfrac{2(x - 4)}{(x - 3)(x - 5)}$ [7.4]   **9.** $\dfrac{12}{5}$ [7.6]

10. 11.52 watts [7.8]   **11.** 1500 [8.2]

12. a) No, the graph does not pass the vertical line test   **b)** Domain: $\{x|x \ge -2\}$, Range: $\mathbb{R}$ [8.2]

13. $\left(\dfrac{5}{2}, 0\right)$ [9.3]   **14.** 160 [9.7]   **15.** $(-12, 8)$ [10.1]   **16.** $\left\{-\dfrac{1}{2}, \dfrac{9}{2}\right\}$ [10.2]

17. $\dfrac{-14 - 23i}{29}$ [11.7]   **18.** $\dfrac{-3 \pm i\sqrt{183}}{8}$ [12.2]   **19.** $r = \sqrt{\dfrac{3V}{\pi h}}$ [12.3]   **20.** $(x + 9)(x + 7)$ [12.4]

Chapter 13

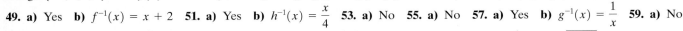

Exercise Set 13.1   **1.** To find $(f \circ g)(x)$, substitute $g(x)$ for x in $f(x)$.   **3. a)** Each y has a unique x.   **b)** Use the horizontal line test.   **5. a)** Yes; each first coordinate is paired with only one second coordinate.   **b)** Yes; each second coordinate is paired with only one first coordinate.   **c)** $\{(5, 3), (2, 4), (3, -1), (-2, 0)\}$; reverse each ordered pair.   **7.** The domain of f is the range of f^{-1} and the range of f is the domain of f^{-1}.   **9. a)** $x^2 + 4x + 5$ **b)** 37   **c)** $x^2 + 3$ **d)** 19   **11. a)** $x^2 + x - 1$ **b)** 19   **c)** $x^2 + 7x + 8$ **d)** 52

13. a) $\dfrac{1}{2x + 3}$ **b)** $\dfrac{1}{11}$ **c)** $\dfrac{2}{x} + 3$ **d)** $3\dfrac{1}{2}$ **15. a)** $\dfrac{9}{x} + 1$ **b)** $3\dfrac{1}{4}$ **c)** $\dfrac{3}{3x + 1}$ **d)** $\dfrac{3}{13}$ **17. a)** $x^4 + 10x^2 + 26$ **b)** 442

c) $x^4 + 2x^2 + 6$ **d)** 294   **19. a)** $\sqrt{x + 5} - 4$ **b)** -1 **c)** $\sqrt{x + 1}$ **d)** $\sqrt{5}$ **21.** No   **23.** Yes   **25.** Yes   **27.** No   **29.** Yes

31. No   **33.** No   **35.** Yes   **37.** Yes   **39.** No   **41.** Yes   **43.** $f(x)$: Domain: $\{-2, -1, 2, 4, 8\}$; Range: $\{0, 4, 6, 7, 9\}$; $f^{-1}(x)$: Domain: $\{0, 4, 6, 7, 9\}$; Range: $\{-2, -1, 2, 4, 8\}$ **45.** $f(x)$: Domain: $\{-1, 1, 2, 4\}$; Range: $\{-3, -1, 0, 2\}$; $f^{-1}(x)$: Domain: $\{-3, -1, 0, 2\}$; Range: $\{-1, 1, 2, 4\}$ **47.** $f(x)$: Domain: $\{x|x \ge 2\}$; Range: $\{y|y \ge 0\}$; $f^{-1}(x)$; Domain: $\{x|x \ge 0\}$; Range: $\{y|y \ge 2\}$

49. a) Yes   **b)** $f^{-1}(x) = x + 2$ **51. a)** Yes   **b)** $h^{-1}(x) = \dfrac{x}{4}$ **53. a)** No   **55. a)** No   **57. a)** Yes   **b)** $g^{-1}(x) = \dfrac{1}{x}$ **59. a)** No

61. a) Yes   **b)** $g^{-1}(x) = \sqrt[3]{x + 6}$ **63. a)** Yes   **b)** $g^{-1}(x) = x^2 - 2, x \ge 0$ **65. a)** Yes   **b)** $h^{-1}(x) = \sqrt{x + 4}, x \ge -4$

67. a) $f^{-1}(x) = \dfrac{x - 8}{2}$ **69. a)** $f^{-1}(x) = x^2, x \ge 0$ **71. a)** $f^{-1}(x) = x^2 + 1, x \ge 0$ **73. a)** $f^{-1}(x) = x^3$

b)

b)

b)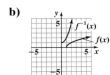

b)

75. a) $f^{-1}(x) = \dfrac{1}{x}, x > 0$ **77.** $(f \circ f^{-1})(x) = x, (f^{-1} \circ f)(x) = x$ **79.** $(f \circ f^{-1})(x) = x, (f^{-1} \circ f)(x) = x$

b)   **81.** $(f \circ f^{-1})(x) = x, (f^{-1} \circ f)(x) = x$ **83.** $(f \circ f^{-1})(x) = x, (f^{-1} \circ f)(x) = x$

85. No, composition of functions is not commutative. Let $f(x) = x^2$ and $g(x) = x + 1$.

Then $(f \circ g)(x) = x^2 + 2x + 1$, while $(g \circ f)(x) = x^2 + 1$.

87. a) $(f \circ g)(x) = x; (g \circ f)(x) = x$ **b)** The Domain is $\mathbb{R}$ for all of them.

89. The range of $f^{-1}(x)$ is the domain of $f(x)$.   **91.** $f^{-1}(x) = \dfrac{x}{3}$; x is feet and $f^{-1}(x)$ is yards

93. $f^{-1}(x) = \dfrac{9}{5}x + 32$.   **95.** $(f \circ g)(x) = 453.6x$, x is pounds, $(f \circ g)(x)$ is grams   **97.** $(f \circ g)(x) = 0.915x$, x is yards, $(f \circ g)(x)$

is meters   **99.** Yes   **101.** Yes   **103. a)** 6 feet   **b)** $36\pi \approx 113.10$ square feet   **c)** $A(t) = 4\pi t^2$ **d)** $36\pi \approx 113.10$ square feet

e) Answers should agree   **106.** $2x + 3y = 10$ **107.** $x^2 + 4x - 2 - \dfrac{4}{x + 2}$ **108.** $\dfrac{18 - 12x}{x^3}$ **109.** $\dfrac{2x\sqrt{2y}}{y}$ **110.** $-1 \pm \sqrt{7}$

Exercise Set 13.2   **1.** Exponential functions are functions of the form $f(x) = a^x, a > 0, a \neq 1$.   **3. a)** As x increases, y decreases.

b) No, $\left(\dfrac{1}{2}\right)^x$ can never be 0.   **c)** No, $\left(\dfrac{1}{2}\right)^x$ can never be negative.   **5. a)** Same; $(0, 1)$.   **b)** $y = 3^x$ will be steeper than $y = 2^x$ for $x > 0$.

7. 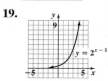   **9.**   **11.**   **13.**   **15.**   **17.**

19.   **21.** 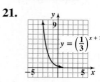   **23.** 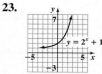   **25.** 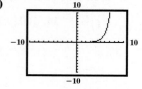   **27. a)** It is a horizontal line through $y = 1$.

b) Yes   **c)** No, not a one-to-one function

29. $y = a^x - k$ is $y = a^x$ lowered k units.

31. The graph of $y = a^{x+2}$ is the graph of $y = a^x$ shifted 2 units to the left.

33. a) ≈ 36.232 million   **b)** ≈ 187.846 million

35. \$512   **37. a)** 14 years   **b)** 10 years   **c)** \$25   **d)** Increases it   **39.** 45   **41.** $\approx$\$6344.93   **43.** $\approx$10.6 grams   **45. a)** 5 grams

b) $\approx 7.28 \times 10^{-11}$ grams   **47. a)** 2400   **b)** $\approx$4977   **49.** $\approx$\$10,850.92   **51. a)** Answers will vary.   **b)** $\approx$472,414 gallons

53. $\approx$8.83 kilometers   **55. a)** $\approx$\$201.36   **b)** $\approx$\$31.36   **57. a)**   **b)** $\approx$6.26

59. a) \$16,384   **b)** \$524,288   **c)** 2^{n-1}

d) $\$2^{29} = \$536,870,912$ **e)** $2^0 + 2^1 + 2^2 + \cdots + 2^{29}$

61. a) $-6.2x^6y^2 + 9.2x^5y^2 + 2.3x^4y$ **b)** 8   **c)** -6.2

62. $x^3 + 3x^2 - 6x + 20$ **63.** $|a - 4|$ **64.** $\dfrac{2xy\sqrt[4]{xy^2z^3}}{z}$

Exercise Set 13.3   **1. a)** $a > 0$ and $a \neq 1$ **b)** $\{x \mid x > 0\}$ **c)** $\mathbb{R}$ **3.** $\left(\dfrac{1}{27}, -3\right) \left(\dfrac{1}{9}, -2\right), \left(\dfrac{1}{3}, -1\right) (1, 0), (3, 1), (9, 2),$ and $(27, 3)$;

the functions $f(x) = a^x$ and $g(x) = \log_a x$ are inverses.   **5.** The functions $y = a^x$ and $y = \log_a x$ for $a \neq 1$ are inverses of each other,

thus the graphs are symmetric with respect to the line $y = x$. For each ordered pair (x, y) on the graph of $y = a^x$, the ordered pair

(y, x) is on the graph of $y = \log_a x$.   **7.**   **9.**   **11.**   **13.**   **15.**

17.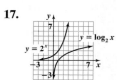

19. $\log_2 8 = 3$ **21.** $\log_3 9 = 2$ **23.** $\log_{16} 4 = \dfrac{1}{2}$ **25.** $\log_8 2 = \dfrac{1}{3}$ **27.** $\log_{1/2} \dfrac{1}{32} = 5$ **29.** $\log_2 \dfrac{1}{8} = -3$

31. $\log_4 \dfrac{1}{64} = -3$ **33.** $\log_{64} 4 = \dfrac{1}{3}$ **35.** $\log_8 \dfrac{1}{2} = -\dfrac{1}{3}$ **37.** $\log_{81} \dfrac{1}{3} = -\dfrac{1}{4}$ **39.** $\log_{10} 7 = 0.8451$ **41.** $\log_e 7.3891 = 2$

43. $\log_a b = n$ **45.** $2^3 = 8$ **47.** $\left(\dfrac{1}{3}\right)^3 = \dfrac{1}{27}$ **49.** $5^{-2} = \dfrac{1}{25}$ **51.** $49^{1/2} = 7$ **53.** $9^{-2} = \dfrac{1}{81}$ **55.** $10^{-3} = \dfrac{1}{1000}$ **57.** $6^3 = 216$

59. $10^{-0.2076} = 0.62$ **61.** $e^{1.8749} = 6.52$ **63.** $w^{-p} = s$ **65.** 3   **67.** 5   **69.** 27   **71.** -4 **73.** $\dfrac{1}{64}$ **75.** 3   **77.** 0   **79.** 2   **81.** -2 **83.** 4

85. 4   **87.** -4 **89.** -2 **91.** 0   **93.** 1   **95.** 5   **97.** $f^{-1}(x) = \log_5 x$ **99.** 3 and 4, since 62 lies between $3^3 = 27$ and $3^4 = 81$.

101. 2 and 3, since 425 lies between $10^2 = 100$ and $10^3 = 1000$.   **103.** 2^x; Note that for $x = 10, 2^x = 1024$ while $\log_{10} x = 1$.

105. 6   **107.** 8   **109.** 3   **111.** 9   **113.** 10,000,000   **115.** 10,000   **117.**   **119.** $2x(x + 3)(x - 6)$

120. $(x - 2)(x + 2)(x^2 + 4)$ **121.** $4(2x + 3)(5x - 1)$

122. $(3rs - 1)(2rs + 1)$

Exercise Set 13.4   **1.** Answers will vary.   **3.** Answers will vary.   **5.** Yes, it is an expansion of property 1.   **7.** $\log_4 3 + \log_4 10$

9. $\log_8 7 + \log_8(x + 3)$ **11.** $\log_2 27 - \log_2 11$ **13.** $\frac{1}{2}\log_{10} x - \log_{10}(x - 9)$ **15.** $7 \log_6 x$ **17.** $5 \log_4(r + 7)$

19. $\frac{3}{2}\log_4 a - \frac{1}{2}\log_4(a + 2)$ **21.** $6 \log_3 d - 4 \log_3(a - 8)$ **23.** $\log_8(y + 4) - 2 \log_8 y$ **25.** $\log_{10} 9 + \log_{10} m - \log_{10} 8 - \log_{10} n$

27. $\log_5 16$ **29.** $\log_2 \frac{9}{5}$ **31.** $\log_4 64$ **33.** $\log_{10} x(x + 3)$ **35.** $\log_9 \frac{z^2}{z - 2}$ **37.** $\log_5\left(\frac{p}{3}\right)^4$ **39.** $\log_2 \frac{n(n + 4)}{n - 3}$ **41.** $\log_5 \sqrt{\frac{x - 8}{x}}$

43. $\log_9 \frac{16\sqrt[3]{r - 6}}{\sqrt{r}}$ **45.** $\log_6 \frac{81}{(x + 3)^2 x^4}$ **47.** 1   **49.** -0.3980 **51.** 1.3980 **53.** 10   **55.** 7   **57.** 3   **59.** 25   **61.** Yes

63. $\log_a \frac{x}{y} = \log_a\left(x \cdot \frac{1}{y}\right) = \log_a x + \log_a \frac{1}{y}$ **65.** $\log_a(x - 2)$ **67.** Yes, $\log_a(x^2 + 8x + 16) = \log_a(x + 4)^2 = 2 \log_a(x + 4)$

69. 0.8640 **71.** 0.1080 **73.** 0.7000 **75.** No, there is no relationship between $\log_{10}(x + y)$ and $\log_{10} xy$ or $\log_{10}\left(\frac{x}{y}\right)$

77. $\frac{1}{4}\log_2 x + \frac{1}{4}\log_2 y + \frac{1}{3}\log_2 a - \frac{1}{5}\log_2(a - b)$ **79.** Answers will vary.   **82.** $\frac{(2x + 5)^2}{(x - 4)^2}$ **83.** $\frac{(3x + 1)(x + 9)}{(x - 4)(x - 3)(2x + 5)}$

84. $2\frac{2}{9}$ days   **85.** $2x^3 y^5 \sqrt[3]{6x^2 y^2}$ **86.** 36

Mid-Chapter Test   **1. a)** In $f(x)$, replace x by $g(x)$.   **b)** $6x + 18$ [13.1]   **2. a)** $\left(\frac{6}{x}\right)^2 + 5$ or $\frac{36}{x^2} + 5$ **b)** 9   **c)** $\frac{6}{x^2 + 5}$ **d)** $\frac{3}{7}$ [13.1]

3. a) Answers will vary.   **b)** No [13.1]   **4. a)** Yes   **b)** $\{(2, -3), (3, 2), (1, 5), (8, 6)\}$ [13.1]   **5. a)** Yes   **b)** $p^{-1}(x) = 3x + 15$ [13.1]

6. a) Yes   **b)** $k^{-1}(x) = x^2 + 4\ x \geq 0$ [13.1]   **7.** $m^{-1}(x) = -\frac{1}{2}x + 2$ [13.1]   **8.**   [13.2]

9.   [13.2]   **10.**   [13.3]

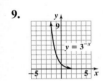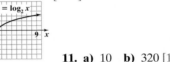

11. a) 10   **b)** 320 [13.3]   **12.** $\log_{27} 9 = \frac{2}{3}$ [13.3]   **13.** $2^{-6} = \frac{1}{64}$ [13.3]   **14.** 3 [13.3]

15. 2 [13.3]   **16.** 4 [13.3]   **17.** $2 \log_9 x + \log_9(x - 5)$ [13.4]   **18.** $\log_5 7 + \log_5 m - \frac{1}{2}\log_5 n$ [13.4]   **19.** $\log_2 \frac{x^3(x + 7)}{(x + 1)^4}$ [13.4]

20. $\log_7 \sqrt{\frac{x + 2}{x}}$ [13.4]

Exercise Set 13.5   **1.** Common logarithms are logarithms with base 10.   **3.** Antilogarithms are numbers obtained by taking 10 to the power of the logarithm.   **5.** 1.9345 **7.** 4.2833 **9.** -1.2125 **11.** 2.0000 **13.** 0.5740 **15.** -1.7620 **17.** 1.64 **19.** $42{,}500$

21. 0.0196 **23.** 1.00 **25.** 579   **27.** 0.0000726 **29.** 100   **31.** 2410   **33.** 13,800   **35.** 0.0871 **37.** 0.239 **39.** 0.749 **41.** 3.5514

43. -1.1385 **45.** 2.3856 **47.** -2.2277 **49.** 679   **51.** 0.303 **53.** 0.0331 **55.** 22.4 **57.** 0   **59.** -1 **61.** -2 **63.** -3 **65.** 7

67. 7   **69.** 20.8 **71.** 41.5 **73.** No; $10^2 = 100$ and since $462 > 100$, log 462 must be greater than 2.   **75.** No; $10^0 = 1$ and $10^{-1} = 0.1$

and since $1 > 0.163 > 0.1$, log 0.163 must be between 0 and -1.   **77.** No; $\log \frac{y}{4x} = \log y - \log 4 - \log x$ **79.** 2.0969 **81.** -0.6990

83. 2.7958 **85.** 2510   **87.** 501,000   **89. a)** ≈ 31.62 kilometers   **b)** ≈ 0.50 kilometer   **c)** ≈ 14.68 **91. a)** $\approx 72\%$ **b)** $\approx 15\%$

93. ≈ 6310 times more intense   **95. a)** $\approx 6.31 \times 10^{20}$ **b)** ≈ 2.19 **97.** ≈ 6.2 **99.** $I = \text{antilog } R$ **101.** $t = \text{antilog}\left(\frac{26 - R}{41.9}\right) - 1$

104. $(2, -3)$ **105.** 50 miles per hour, 55 miles per hour   **106.** $\frac{-2 \pm 2i\sqrt{5}}{3}$ **107.** 10 miles per hour   **108.**

109.

Exercise Set 13.6   **1.** $c = d$ **3.** Check for extraneous solutions.   **5.** $\log(-2)$ is not a real number.   **7.** 3   **9.** 4   **11.** $\frac{1}{2}$

13. 2   **15.** $-\frac{1}{3}$ **17.** 4   **19.** 3   **21.** 3   **23.** 2.01 **25.** 3.56 **27.** 5.59 **29.** 6.34 **31.** 6   **33.** 5   **35.** $\frac{1}{16}$ **37.** 100   **39.** -1 **41.** $-6, 4$

43. $0, -8$ **45.** 92   **47.** $\frac{4}{5}$ **49.** $\frac{3}{2}$ **51.** 4   **53.** 2   **55.** 0.87 **57.** 30   **59.** 5   **61.** 4   **63.** 2   **65.** 3   **67.** 9   **69.** ≈ 3.47 hours

71. ≈ 3.19 years   **73.** ≈ 17.07 years   **75. a)** ≈ 7.95 **b)** ≈ 5.88 **77.** $\approx \$7112.09$ **79.** ≈ 19.36 **81. a)** 1,000,000,000,000 times greater
b) 10,000,000 times greater   **83.** 8   **85.** $x = 1$ and $x = 2$ **87.** $(3, 1)$ **89.** $(54, 46)$ **91.** 2.8 **93.** No solution

95. The box is greater by ≈ 7.73 cubic feet   **96.** -4 **97.**

98. $\dfrac{x\sqrt{y} - y\sqrt{x}}{x - y}$ **99.** $c = \sqrt{\dfrac{E}{m}}$ **100.** $f(x) = 2(x - 3)^2 - 5$

Exercise Set 13.7

1. a) e **b)** ≈ 2.7183 **3.** $\{x \mid x > 0\}$ **5.** $\log_a x = \dfrac{\log_b x}{\log_b a}$ **7.** x **9.** e^x **11.** $k < 0$ **13.** 4.1271 **15.** -0.2070
17. 4.95 **19.** 0.0578 **21.** 0.972 **23.** 3.6640 **25.** 4.3923 **27.** 1.7297 **29.** 2.7380 **31.** 2.9135 **33.** 3.9318 **35.** -0.4784 **37.** 4
39. 1 **41.** 4 **43.** 6 **45.** $P = 4757.5673$ **47.** $P_0 = 224.0845$ **49.** $t = 0.7847$ **51.** $k = 0.2310$ **53.** $k = -0.2888$
55. $A = 4719.7672$ **57.** $V_0 = \dfrac{V}{e^{kt}}$ **59.** $t = \dfrac{\ln P - \ln 150}{7}$ **61.** $k = \dfrac{\ln A - \ln A_0}{t}$ **63.** $y = xe^{2.3}$ **65.** $y = (x + 6)e^5$
67. ≈ 2.5000; find $\ln 12.183$ **69. a)** $\approx \$5637.48$ **b)** ≈ 11.55 years   **71.** ≈ 39.98 grams   **73. a)** $\approx 86.47\%$ **b)** ≈ 34.66 days
75. a) ≈ 5.15 feet per second   **b)** ≈ 5.96 feet per second   **c)** $\approx 646{,}000$ **77.** $\approx \$449{,}004{,}412{,}200{,}000$
79. a) ≈ 6.9 billion   **b)** ≈ 53 years   **81. a)** ≈ 3.04 million   **b)** 3.89 million   **83. a)** $\approx \$2788.38$ **b)** $\approx \$3711.59$
85. a) ≈ 32.43 inches   **b)** ≈ 35.46 inches   **87. a)** ≈ 6626.62 years   **b)** ≈ 5752.26 years   **89.** $\approx \$6791.91$
91. a) Strontium 90, since it has a higher decay rate   **b)** $\approx 31.66\%$ of original amount   **93.** Answers will vary.   **95.** $-0.999, 7.286$
97. $-1.507, 16.659$ **99.** $v_0 = \dfrac{e^{xk} - 1}{kt}$ **101.** $i = Ie^{-t/RC}$ **102.** $-9x^2y^3 + 12x^2y^2 - 3xy^2 + 4xy$ **103.** $-20, 20$
104. a) 0 **b)** $\dfrac{11}{40}$ or 0.275 **105.** 240 children, 310 adults   **106.** $x + x^2$

Chapter 13 Review Exercises

1. $4x^2 - 26x + 44$ **2.** 2 **3.** $2x^2 - 6x + 3$ **4.** 39 **5.** $6\sqrt{x - 3} + 7, x \geq 3$
6. $\sqrt{6x + 4}, x \geq -\dfrac{2}{3}$ **7.** One-to-one   **8.** Not one-to-one   **9.** One-to-one   **10.** Not one-to-one   **11.** One-to-one
12. Not one-to-one   **13.** $f(x)$: Domain: $\{-4, -1, 5, 6\}$; Range: $\{-3, 2, 3, 8\}$; $f^{-1}(x)$: Domain: $\{-3, 2, 3, 8\}$; Range: $\{-4, -1, 5, 6\}$
14. $f(x)$: Domain: $\{x \mid x \geq 0\}$; Range: $\{y \mid y \geq 4\}$; $f^{-1}(x)$: Domain: $\{x \mid x \geq 4\}$; Range: $\{y \mid y \geq 0\}$
15. $f^{-1}(x) = \dfrac{x + 2}{4}$,   **16.** $f^{-1}(x) = x^3 + 1$;   **17.** $f^{-1}(x) = \dfrac{x}{36}, x$ is inches, $f^{-1}(x)$ is yards.

18. $f^{-1}(x) = \dfrac{x}{4}, x$ is quarts, $f^{-1}(x)$ is gallons   **19.**

20.

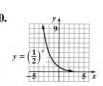

21. a) 30.23 million   **b)** 62.73 million   **c)** 187.50 million   **22.** $\log_8 64 = 2$ **23.** $\log_{81} 3 = \dfrac{1}{4}$ **24.** $\log_5 \dfrac{1}{125} = -3$ **25.** $2^5 = 32$
26. $\left(\dfrac{1}{4}\right)^2 = \dfrac{1}{16}$ **27.** $6^{-2} = \dfrac{1}{36}$ **28.** $4^3 = x; 64$ **29.** $a^4 = 81; 3$ **30.** $\left(\dfrac{1}{5}\right)^{-3} = x; 125$ **31.**

32.

33. $8 \log_5 17$ **34.** $\dfrac{1}{2} \log_3 (x - 9)$ **35.** $\log 6 + \log (a + 1) - \log 19$

36. $4 \log x - \log 7 - 5 \log (2x + 3)$ **37.** $\log \dfrac{x^5}{(x + 1)^3}$ **38.** $\log \dfrac{(2x)^4}{y}$ **39.** $\ln \dfrac{\sqrt[3]{\dfrac{x}{x + 2}}}{2}$
40. $\ln \dfrac{x^3 \sqrt{x + 1}}{(x + 4)^6}$ **41.** 10 **42.** 5 **43.** 121 **44.** 3 **45.** 2.9133 **46.** -3.5720 **47.** 1440 **48.** 0.000697 **49.** $11{,}600$ **50.** 0.0594
51. 5 **52.** 9 **53.** 22.4 **54.** 9.4 **55.** 4 **56.** $-\dfrac{1}{2}$ **57.** 2 **58.** 5 **59.** 2.582 **60.** 5.968 **61.** 1.353 **62.** 2.240 **63.** 26 **64.** 5
65. 1 **66.** 3 **67.** $t \approx 1.155$ **68.** $A_0 \approx 352.542$ **69.** $t = \dfrac{\ln A - \ln A_0}{k}$ **70.** $k = \dfrac{\ln 0.25}{t}$ **71.** $y = xe^6$ **72.** $y = 3x + 23$ **73.** 7.6147
74. 3.5046 **75.** $\approx \$19{,}126.18$ **76.** ≈ 17.3 years
77. a) ≈ 92.88 minutes   **b)** ≈ 118.14 minutes   **78.** ≈ 10.32 pounds per square inch
79. a) 72 **b)** ≈ 61.2 **c)** ≈ 5 months

Chapter 13 Practice Test   **1. a)** Yes   **b)** $\{(2, 4), (8, -3), (3, -1), (-7, 6)\}$ [13.1]   **2. a)** $x^2 + 4x + 1$ **b)** 61 [13.1]

3. a) $\sqrt{x^2 + 3}$,   **b)** $2\sqrt{13}$ [13.1]   **4. a)** $f^{-1}(x) = -\frac{1}{3}(x + 5)$ **5. a)** $f^{-1}(x) = x^2 + 1, x \geq 0$

b) [13.1]   **b)** [13.1]

6. $\{x|x > 0\}$ [13.3]   **7.** -4 [13.4]   **8.** [13.2]   **9.** [13.3]   **10.** $\log_2 \frac{1}{32} = -5$ [13.4]   **11.** $5^3 = 125$ [13.3]

12. $2^4 = x + 3, 13$ [13.3]   **13.** $64^y = 16, \frac{2}{3}$ [13.3]

14. $3 \log_2 x + \log_2 (x - 4) - \log_2(x + 2)$ [13.4]

15. $\log_6 \frac{(x - 4)^7(x + 3)^2}{\sqrt{x}}$ [13.4]   **16.** 5 [13.4]   **17. a)** 3.6646   **b)** -2.6708 [13.5]   **18.** ≈ 2.68 [13.6]   **19.** 3 [13.6]   **20.** $\frac{17}{5}$ [13.6]

21. 16.2810 [13.7]   **22.** 2.0588 [13.7]   **23.** 30.5430 [13.7]   **24.** $\approx\$5211.02$ [13.7]   **25.** ≈ 3364.86 years old [13.7]

Cumulative Review Test   **1.** -1 [1.9]   **2.** 1 [2.5]   **3.** $y = \frac{2x - 5}{3}$ [2.6]   **4.** 7.5% [3.2]   **5. a)** 2 hours   **b)** 10 miles [3.4]

6. $m = 2, y = 2x + 3$ [4.4]   **7.** $\frac{8y^{21}z^6}{x^9}$ [5.2]   **8.** $x^2 + 2x + 3 + \frac{1}{x + 1}$ [5.6]   **9.** $(4x + 3y)(3x + y)$ [6.4]

10. $(x - y + 5)(x - y - 5)$ [6.5]   **11.** -1 [7.6]   **12.** 6.25 [7.8]   **13.** -1 [8.2]   **14.** $(2, 4)$ [9.3]   **15.** $(1, 5)$ [9.6]   **16.** 41 [9.7]

17. [10.3]   **18.** $g(x) = (x - 2)^2 - 9$ **b)** [12.5] 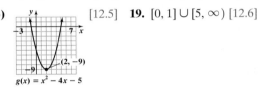   **19.** $[0, 1] \cup [5, \infty)$ [12.6]

20. a) ≈ 111.82 grams   **b)** ≈ 24.8 years [13.7]

Chapter 14

Exercise Set 14.1   **1.** Parabola, circle, ellipse, and hyperbola; see page 000 for picture.   **3.** Yes, because each value of x corresponds to only one value for y. The domain is $\mathbb{R}$, and the range is $\{y|y \geq k\}$ **5.** The graphs have the same vertex, $(3, 4)$. The first graph opens upward, and the second graph opens downward.   **7.** The distance is always a positive number because both differences are squared and we use the principal square root.   **9.** A circle is the set of all points in a plane that are the same distance from a fixed point.   **11.** No, $x^2 + y^2 = 9$ would be an equation of a circle   **13.** No, the coefficients of the x^2 and y^2 terms would need to be the same   **15.** No, if x^2 were replaced by x it would be an equation of a parabola

17. $y = (x - 2)^2 + 3$ **19.** $y = (x + 3)^2 + 2$ **21.** $y = (x - 2)^2 - 1$ **23.** $y = -(x - 1)^2 + 1$ **25.** $y = -(x + 3)^2 + 4$

27. $y = -3(x - 5)^2 + 3$ **29.** $x = (y - 4)^2 - 3$ **31.** $x = -(y - 5)^2 + 4$ **33.** $x = -5(y + 3)^2 - 6$ **35.** $y = -2\left(x + \frac{1}{2}\right)^2 + 6$

37. a) $y = (x + 1)^2 - 1$
b)
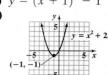

39. a) $y = (x + 3)^2 - 9$
b)

41. a) $x = (y + 2)^2 - 4$
b)
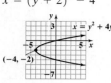

43. a) $y = \left(x + \dfrac{7}{2}\right)^2 - \dfrac{9}{4}$
b)
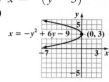

45. a) $x = -(y - 3)^2$
b)
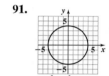

47. a) $y = -(x - 2)^2$
b)
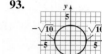

49. a) $x = -\left(y - \dfrac{3}{2}\right)^2 - \dfrac{7}{4}$
b)

51. 5   **53.** 9

55. 13   **57.** $\sqrt{90} \approx 9.49$

59. $\sqrt{\dfrac{125}{4}} \approx 5.59$

61. $\sqrt{34.33} \approx 5.86$

63. $\sqrt{10} \approx 3.16$ **65.** $(3, 6)$ **67.** $(0, 0)$ **69.** $\left(\dfrac{3}{2}, 5\right)$ **71.** $\left(\dfrac{5}{2}, -\dfrac{7}{4}\right)$ **73.** $\left(\dfrac{\sqrt{3} + \sqrt{2}}{2}, \dfrac{9}{2}\right)$ **75.** $x^2 + y^2 = 16$

77. $(x - 2)^2 + y^2 = 25$ **79.** $x^2 + (y - 5)^2 = 1$ **81.** $(x - 3)^2 + (y - 4)^2 = 64$ **83.** $(x - 7)^2 + (y + 6)^2 = 100$

85. $(x - 1)^2 + (y - 2)^2 = 5$ **87.** $x^2 + y^2 = 16$ **89.** $(x - 3)^2 + (y + 2)^2 = 9$

91.
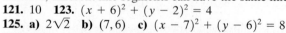

93.

95.

97.

99.

101.
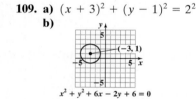

103.

105. a) $(x + 4)^2 + y^2 = 1^2$
b)

107. a) $(x + 3)^2 + (y - 2)^2 = 3^2$
b)
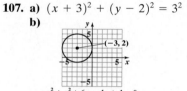

109. a) $(x + 3)^2 + (y - 1)^2 = 2^2$
b)

111. a) $(x - 4)^2 + (y + 1)^2 = 2^2$
b)

113. $25\pi \approx 78.5$ square units   **115.** x-intercept: $(-7, 0)$; y-intercepts: $(0, -1)$, $(0, 7)$

117. x-intercept: $(24, 0)$; no y-intercepts

119. No, different line segments can have the same midpoint

121. 10   **123.** $(x + 6)^2 + (y - 2)^2 = 4$

125. a) $2\sqrt{2}$ **b)** $(7, 6)$ **c)** $(x - 7)^2 + (y - 6)^2 = 8$

127. 4, 0, a parabola opening up or down and a parabola opening right or left can be drawn to have a maximum of 4 intersections, or a minimum of 0 intersections.

129. a) 13.6 feet   **b)** 81.8 feet   **c)** $x^2 + (y - 81.8)^2 = 4651.24$ **131. a)** $x^2 + y^2 = 16$ **b)** $(x - 2)^2 + y^2 = 4$

c) $(x + 2)^2 + y^2 = 4$ **d)** 8π square units   **133.** 48π square units   **136.** $\dfrac{y}{3x}$ **137. a)** 1. a^2, 2. ab, 3. ab, 4. b^2 **b)** $(a + b)^2$

138. 128   **139.** $(0, 7)$ **140.**
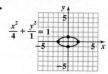

Exercise Set 14.2

1. An ellipse is a set of points in a plane, the sum of whose distances from two fixed points is constant.

3. $\dfrac{(x - h)^2}{a^2} + \dfrac{(y - k)^2}{b^2} = 1$ **5.** If $a = b$, the formula for a circle is obtained.   **7.** Divide both sides by 180.

9. No, equation for an ellipse is $\dfrac{x^2}{a^2} + \dfrac{y^2}{b^2} = 1$.   **11.**
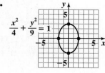

13.

15.
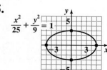

17.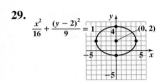
$$\frac{x^2}{16} + \frac{y^2}{25} = 1$$

19.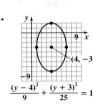
$x^2 + 16y^2 = 16$

21.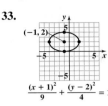
$49x^2 + y = 49$

23.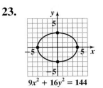
$9x^2 + 16y^2 = 144$

25.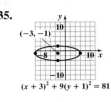
$25x^2 + 100y^2 = 400$

27.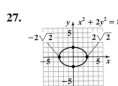
$x^2 + 2y^2 = 8$; $-2\sqrt{2}$, $2\sqrt{2}$

29.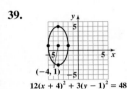
$$\frac{x^2}{16} + \frac{(y-2)^2}{9} = 1$$ (0, 2)

31.
$$\frac{(y-4)^2}{9} + \frac{(y+3)^2}{25} = 1$$ (4, −3)

33.
(−1, 2)
$$\frac{(x+1)^2}{9} + \frac{(y-2)^2}{4} = 1$$

35.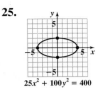
(−3, −1)
$(x+3)^2 + 9(y+1)^2 = 81$

37.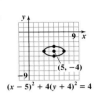
(5, −4)
$(x-5)^2 + 4(y+4)^2 = 4$

39.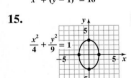
(−4, 1)
$12(x+4)^2 + 3(y-1)^2 = 48$

41. $2\pi \approx 6.3$ square units

43. one, at $(0,0)$, this is the only ordered pair that satisfies the equation   **45.** $\dfrac{x^2}{9} + \dfrac{y^2}{16} = 1$

47. $\dfrac{x^2}{4} + \dfrac{y^2}{9} = 1$

49. None, the ellipse will be within the circle   **51.** $\dfrac{(x+3)^2}{36} + \dfrac{(y+2)^2}{9} = 1; (-3, -2)$ **53.** 69.5 feet   **55. a)** $\dfrac{x^2}{100} + \dfrac{y^2}{576} = 1$

b) $240\pi \approx 753.98$ square feet   **c)** ≈ 376.99 square feet   **57.** $\sqrt{5} \approx 2.24$ feet, in both directions, from the center of the ellipse, along

the major axis.   **59.** Answers will vary.   **61.** Answers will vary.   **63.** $\dfrac{(x-4)^2}{49} + \dfrac{(y-2)^2}{9} = 1$ **65.** $l = \dfrac{2S - nf}{n}$

66. $x + \dfrac{5}{2} + \dfrac{1}{2(2x-3)}$ **67.** 6   **68.** $\left[-\dfrac{5}{3}, 4\right)$ **69.** ≈ 2.7755

Mid-Chapter Test

1.
$y = (x-2)^2 - 1$ [14.1]

2.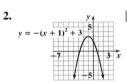
$y = -(x+1)^2 + 3$ [14.1]

3.
$x = -(y-4)^2 + 1$ [14.1]

4.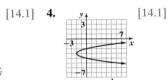
$x = 2(y+3)^2 - 2$ [14.1]

5.
$y = x^2 + 6x + 10$ [14.1]

6. 13 [14.1]   **7.** $\sqrt{153} \approx 12.37$ [14.1]   **8.** $\left(-1, \dfrac{5}{2}\right)$ [14.1]   **9.** $\left(\dfrac{11}{4}, \dfrac{15}{4}\right)$ [14.1]   **10.** $(x+3)^2 + (y-2)^2 = 25$ [14.1]

11.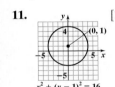
(0, 1)
$x^2 + (y-1)^2 = 16$

[14.1]   **12.**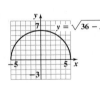
$y = \sqrt{36 - x^2}$

[14.1]   **13.**
(1, −2)
$x^2 + y^2 - 2x + 4y - 4 = 0$

[14.1]   **14.** A circle is a set of points in a plane that are the same distance from a fixed point called its center. [14.1]

15.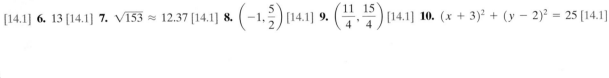
$\dfrac{x^2}{4} + \dfrac{y^2}{9} = 1$

[14.2]   **16.**
$\dfrac{x^2}{81} + \dfrac{y^2}{25} = 1$

[14.2]   **17.**
(1, −2)
$\dfrac{(x-1)^2}{49} + \dfrac{(y+2)^2}{4} = 1$

[14.2]   **18.**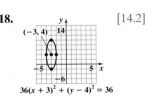
(−3, 4)
$36(x+3)^2 + (y-4)^2 = 36$

[14.2]

19. $6\pi \approx 18.85$ square units [14.2]   **20.** $\dfrac{x^2}{64} + \dfrac{y^2}{25} = 1$ [14.2]

Exercise Set 14.3

1. A hyperbola is the set of points in a plane the differences of whose distances from two fixed points is a constant.   **3.** The graph of $\dfrac{x^2}{a^2} - \dfrac{y^2}{b^2} = 1$ is a hyperbola with vertices at $(a, 0)$ and $(-a, 0)$. Its transverse axis lies along the x-axis. The asymptotes are $y = \pm\dfrac{b}{a}x$.   **5.** No, the signs of the x and y terms must differ.   **7.** Yes, divide both sides of the equation by 100 and you will see the equation is that of a hyperbola.   **9.** Divide both sides of the equation by 81.

11. a) $y = \pm\dfrac{2}{3}x$ **13. a)** $y = \pm\dfrac{1}{2}x$ **15. a)** $y = \pm\dfrac{5}{3}x$ **17. a)** $y = \pm\dfrac{4}{5}x$ **19. a)** $y = \pm\dfrac{5}{6}x$

b)   **b)**   **b)** 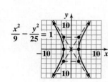   **b)**   **b)**

$\dfrac{x^2}{9} - \dfrac{y^2}{4} = 1$ $\dfrac{x^2}{4} - \dfrac{y^2}{1} = 1$ $\dfrac{x^2}{9} - \dfrac{y^2}{25} = 1$ $\dfrac{x^2}{25} - \dfrac{y^2}{16} = 1$ $\dfrac{y^2}{25} - \dfrac{x^2}{36} = 1$

21. a) $y = \pm\dfrac{3}{4}x$ **23. a)** $y = \pm\dfrac{5}{2}x$ **25. a)** $y = \pm\dfrac{4}{9}x$ **27. a)** $\dfrac{x^2}{25} - \dfrac{y^2}{1} = 1,\ y = \pm\dfrac{1}{5}x$

b)   **b)**   **b)**   **b)**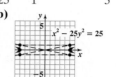

$\dfrac{y^2}{9} - \dfrac{x^2}{16} = 1$ $\dfrac{x^2}{81} - \dfrac{y^2}{16} = 1$ $\dfrac{x^2}{81} - \dfrac{y^2}{16} = 1$ $x^2 - 25y^2 = 25$

29. a) $\dfrac{y^2}{16} - \dfrac{x^2}{4} = 1,\ y = \pm2x$ **31. a)** $\dfrac{y^2}{1} - \dfrac{x^2}{9} = 1,\ y = \pm\dfrac{1}{3}x$ **33. a)** $\dfrac{x^2}{9} - \dfrac{y^2}{25} = 1,\ y = \pm\dfrac{5}{3}x$ **35. a)** $\dfrac{y^2}{36} - \dfrac{x^2}{4} = 1,\ y = \pm3x$

b)   **b)**   **b)**   **b)**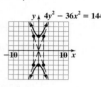

$4y^2 - 16x^2 = 64$ $9y^2 - x^2 = 9$ $25x^2 - 9y^2 = 225$ $4y^2 - 36x^2 = 144$

37. Circle   **39.** Ellipse   **41.** Hyperbola   **43.** Parabola   **45.** Ellipse   **47.** Parabola   **49.** Circle   **51.** Hyperbola   **53.** Parabola

55. Hyperbola   **57.** Parabola   **59.** Circle   **61.** $\dfrac{y^2}{4} - \dfrac{x^2}{16} = 1$ **63.** $\dfrac{x^2}{9} - \dfrac{y^2}{36} = 1$ **65.** $\dfrac{x^2}{9} - \dfrac{y^2}{25} = 1$, no, $\dfrac{x^2}{18} - \dfrac{y^2}{50} = 1$ and others will

work. The ratio of $\dfrac{b}{a}$ must be $\dfrac{5}{3}$.   **67.** No, graphs of hyperbolas of this form do not pass the vertical line test

69. D: $(-\infty, -5] \cup [5, \infty)$; R: $\mathbb{R}$ **71.** The transverse axes of both graphs are along the x-axis. The vertices of the second graph will

be closer to the origin, and the second graph will open wider.   **73.** Answers will vary.   **75.** $y = -\dfrac{1}{2}x + 1$ **76.** $\dfrac{3x + 2}{2x - 3}$

77. $-x^2 - x + 11$ **78.** $\left(-1, \dfrac{1}{3}\right)$ **79.** $v = \sqrt{\dfrac{2E}{m}}$ **80.** 1

Exercise Set 14.4   **1.** A nonlinear system of equations is a system in which at least one equation is nonlinear.

3. Yes, for example 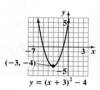   **5.** Yes, for example   **7.** $(3, -3), (-3, 3)$ **9.** $(3, 0), \left(-\dfrac{9}{5}, \dfrac{12}{5}\right)$ **11.** $(-4, 11), \left(\dfrac{5}{2}, \dfrac{5}{4}\right)$

13. $(-1, 5), (1, 5)$ **15.** $(2, 2\sqrt{2}), (2, -2\sqrt{2}), (-2, 2\sqrt{2}), (-2, -2\sqrt{2})$ **17.** No real solution   **19.** $(0, -3), (\sqrt{5}, 2)(-\sqrt{5}, 2)$
21. $(2, -4), (-14, -20)$ **23.** $(2, 0), (-2, 0)$ **25.** $(\sqrt{15}, 1)(-\sqrt{15}, 1), (\sqrt{15}, -1), (-\sqrt{15}, -1)$
27. $(\sqrt{2}, \sqrt{2}), (\sqrt{2}, -\sqrt{2}), (-\sqrt{2}, \sqrt{2}), (-\sqrt{2}, -\sqrt{2})$ **29.** $(3, 0), (-3, 0)$ **31.** $(2, 1), (2, -1), (-2, 1), (-2, -1)$
33. $(3, 4), (3, -4), (-3, 4), (-3, -4)$ **35.** $(\sqrt{5}, 2), (\sqrt{5}, -2), (-\sqrt{5}, 2), (-\sqrt{5}, -2)$ **37.** No real solution   **39.** No real solution
41. Answers will vary.   **43.** 20 meters by 22 meters   **45.** 9 feet by 30 feet   **47.** length: 14 centimeters, width: 8 centimeters
49. 16 inches by 30 inches   **51.** ≈ 1.67 seconds   **53.** $r = 6\%, p = \$125$ **55.** ≈ 16 and ≈ 184 **57.** ≈ 5 and ≈ 95
59. $(-1, -3), (3.12, -0.53)$ **61.** 10 yards, 24 yards   **63.** Parentheses, exponents, multiplication or division, addition or subtraction.

64. $(x + 2)(x^2 + x + 1)$ **65.** 0.9   **66.** $\dfrac{5\sqrt{x + 2} + 15}{x - 7}$ **67.** $k = \dfrac{\ln A - \ln A_0}{t}$

Chapter 14 Review Exercises   **1.** $13; \left(\dfrac{5}{2} - 6\right)$ **2.** $5; \left(-\dfrac{5}{2}, 3\right)$ **3.** $17; \left(-5, \dfrac{5}{2}\right)$ **4.** $\sqrt{8} \approx 2.83; (-3, 4)$

5. $y = (x - 2)^2 + 1$ (2, 1)

6. $y = (x + 3)^2 - 4$ (−3, −4)

7. $x = (y - 1)^2 + 4$ (4, 1)

8. (−3, −4) $x = -2(y + 4)^2 - 3$

9. a) $y = (x - 4)^2 + 6$ **b)** $y = x^2 - 8x + 22$ (4, 6)

10. a) $x = -(y+1)^2 + 6$

b)

$x = -y^2 - 2y + 5$

11. a) $x = \left(y + \dfrac{5}{2}\right)^2 - \dfrac{9}{4}$

b)
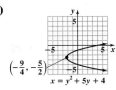
$\left(-\dfrac{9}{4}, -\dfrac{5}{2}\right)$ $x = y^2 + 5y + 4$

12. a) $y = 2(x-2)^2 - 32$

b)

$(2, -32)$ $y = 2x^2 - 8x - 24$

13. a) $x^2 + y^2 = 4^2$

b)

$(0,0)$ $x^2 + y^2 = 4^2$

14. a) $(x+3)^2 + (y-4)^2 = 1^2$

b)
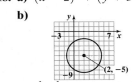
$(-3, 4)$ $(x+3)^2 + (y-4)^2 = 1^2$

15. a) $x^2 + (y-2)^2 = 2^2$

b)
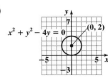
$x^2 + y^2 - 4y = 0$ $(0,2)$

16. a) $(x-1)^2 + (y+3)^2 = 3^2$

b)
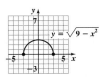
$(1,-3)$ $x^2 + y^2 - 2x + 6y + 1 = 0$

17. a) $(x-4)^2 + (y-5)^2 = 1^2$

b)
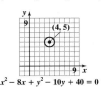
$(4,5)$ $x^2 - 8x + y^2 - 10y + 40 = 0$

18. a) $(x-2)^2 + (y+5)^2 = (\sqrt{12})^2$

b)
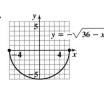
$(2,-5)$ $x^2 + y^2 - 4x + 10y + 17 = 0$

19.

$y = \sqrt{9 - x^2}$

20.

$y = -\sqrt{36 - x^2}$

21. $(x+1)^2 + (y-1)^2 = 4$

22. $(x-5)^2 + (y+3)^2 = 9$

23.

$\dfrac{x^2}{4} + \dfrac{y^2}{9} = 1$

24.
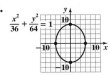
$\dfrac{x^2}{36} + \dfrac{y^2}{64} = 1$

25.

$4x^2 + 9y^2 = 36$

26.

$9x^2 + 16y^2 = 144$

27.

$(3,-2)$ $\dfrac{(x-3)^2}{16} + \dfrac{(y+2)^2}{4} = 1$

28.

$\dfrac{(x+3)^2}{9} + \dfrac{y^2}{25} = 1$

29.
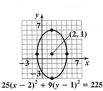
$(2,1)$ $25(x-2)^2 + 9(y-1)^2 = 225$

30. $6\pi \approx 18.85$ square units

31. a) $y = \pm 2x$

b)

$\dfrac{x^2}{4} - \dfrac{y^2}{16} = 1$

32. a) $y = \pm x$

b)
$\dfrac{x^2}{4} - \dfrac{y^2}{4} = 1$

33. a) $y = \pm \dfrac{1}{3}x$

b)
$\dfrac{y^2}{4} - \dfrac{x^2}{36} = 1$

34. a) $y = \pm \dfrac{5}{4}x$

b)

$\dfrac{y^2}{25} - \dfrac{x^2}{16} = 1$

35. a) $\dfrac{x^2}{9} - \dfrac{y^2}{1} = 1$ **b)** $y = \pm \dfrac{1}{3}x$

c)

$x^2 - 9y^2 = 9$

36. a) $\dfrac{x^2}{16} - \dfrac{y^2}{25} = 1$ **b)** $y = \pm \dfrac{5}{4}x$

c)
$25x^2 - 16y^2 = 400$

37. a) $\dfrac{y^2}{25} - \dfrac{x^2}{4} = 1$ **b)** $y = \pm \dfrac{5}{2}x$

c)

$4y^2 - 25x^2 = 100$

38. a) $\dfrac{y^2}{9} - \dfrac{x^2}{49} = 1$ **b)** $y = \pm \dfrac{3}{7}x$

c)
$49y^2 - 9x^2 = 441$

39. Hyperbola **40.** Ellipse **41.** Circle

42. Hyperbola **43.** Ellipse **44.** Parabola

45. Ellipse **46.** Parabola **47.** $(5,0), (-5,0)$

48. $\left(\dfrac{5}{2}, \dfrac{3}{2}\right)$ **49.** $(-3,0), \left(-\dfrac{12}{5}, \dfrac{9}{5}\right)$

50. No real solution **51.** $(6,0), (-6,0)$

52. $(4,3), (4,-3), (-4,3), (-4,-3)$

53. No real solution **54.** $(0, \sqrt{3}), (0, -\sqrt{3})$ **55.** 5 feet by 9 feet **56.** ≈ 4 and ≈ 145 **57.** $r = 3\%, p = \$4000$

Chapter 14 Practice Test

1. They are formed by cutting a cone or a pair of cones. [14.1]   **2.** $\sqrt{50} \approx 7.07$ [14.1]   **3.** $\left(-1, \frac{3}{2}\right)$ [14.1]

4. $(-3, 1)$, [14.1]   **5.** 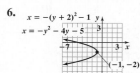 $x = y^2 - 2y + 4$ [14.1]   **6.** [14.1]   **7.** [14.1]

$y = -2(x + 3)^2 + 1$

$x = -(y + 2)^2 - 1$
$x = -y^2 - 4y - 5$

$(x - 2)^2 + (y - 4)^2 = 9$

8. $9\pi \approx 28.27$ square units [14.1]   **9.** $(x - 3)^2 + (y + 1)^2 = 16$ [14.1]   **10.** 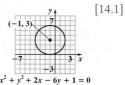 [14.1]   **11.** [14.1]

12. 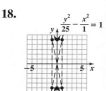 [14.2]

$4x^2 + 25y^2 = 100$

$y = -\sqrt{16 - x^2}$

$x^2 + y^2 + 2x - 6y + 1 = 0$

13. No, the major axis should be along the y-axis. [14.2]   **14.** [14.2]   **15.** $(8, -7)$ [14.2]

16. The transverse axis lies along the axis corresponding to the positive term of the equation in standard form. [14.3]

17. $y = \pm\frac{7}{4}x$ [14.3]

$4(x - 4)^2 + 36(y + 2)^2 = 36$

18. 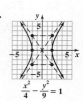 [14.3] $\frac{y^2}{25} - \frac{x^2}{1} = 1$ **19.** [14.3]   **20.** Hyperbola, divide both sides of the equation by 30. [14.3]

21. Ellipse, divide both sides of the equation by 100. [14.3]

22. $(2, \sqrt{3}), (2, -\sqrt{3}), (-2, \sqrt{3}), (-2, -\sqrt{3})$ [14.4]

23. No real solution [14.4]

24. 30 meters by 50 meters [14.4]

$\frac{x^2}{4} - \frac{y^2}{9} = 1$

25. 5 feet by 12 feet [14.4]

Cumulative Review Test

1. $\frac{19}{4}$ [2.5]   **2.** $\varnothing$ [2.5]   **3.**  [4.2]   **4.** $-27x^3y^9$ [5.1]   **5** $(x^2 + 6)(x^2 - 7)$ [6.5]

$y = -2x + 2$

6. base: 14 feet, height: 8 feet [6.7]   **7.** $\frac{x - 4}{4x + 3}$ [7.2]   **8.** $\frac{2x^2 - 9x - 5}{2(x + 3)(x - 4)}$ [7.3]   **9.** 2 [7.6]   **10.** 139 [8.2]   **11.** $(8, 6)$ [9.2]

12. $\left\{x \mid x < -\frac{5}{3} \text{ or } x > 1\right\}$ [10.2]   **13.** $\frac{3y^{3/2}}{x^{1/2}}$ [11.2]   **14.** $\frac{6x + 6y\sqrt{x}}{x - y^2}$ [11.5]   **15.** 3 [11.6]   **16.** $\frac{2 \pm i\sqrt{11}}{3}$ [12.2]   **17.** 2 [13.6]

18. ≈ 2.31 [13.7]   **19.** 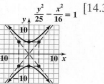 $9x^2 + 4y^2 = 36$ [14.2]   **20.** [14.3]

$\frac{y^2}{25} - \frac{x^2}{16} = 1$

Chapter 15

Exercise Set 15.1

1. A sequence is a list of numbers arranged in a specific order.   **3.** A finite sequence is a function whose domain includes only the first n natural numbers.   **5.** In a decreasing sequence, the terms decrease.   **7.** A series is the sum of the terms of a sequence.   **9.** $\sum_{i=1}^{5}(i + 4)$ is the sum as i goes from 1 to 5 of $i + 4$.   **11.** It is an increasing sequence. Each number in the sequence is greater than the preceding number   **13.** Yes, the signs of the terms alternate.   **15.** $6, 12, 18, 24, 30$

17. $3, 7, 11, 15, 19$ **19.** $7, \frac{7}{2}, \frac{7}{3}, \frac{7}{4}, \frac{7}{5}$ **21.** $\frac{3}{2}, \frac{4}{3}, \frac{5}{4}, \frac{6}{5}, \frac{7}{6}$ **23.** $-1, 1, -1, 1, -1$ **25.** $4, -8, 16, -32, 64$ **27.** 31   **29.** 12   **31.** 1

33. 99   **35.** $\frac{81}{25}$ **37.** $2, 15$ **39.** $3, 17$ **41.** $0, \frac{13}{20}$ **43.** $-1, -1$ **45.** $\frac{1}{2}, 7$ **47.** $64, 128, 256$ **49.** $17, 19, 21$ **51.** $\frac{1}{6}, \frac{1}{7}, \frac{1}{8}$ **53.** $1, -1, 1$

55. $\frac{1}{81}, \frac{1}{243}, \frac{1}{729}$ **57.** $\frac{1}{16}, -\frac{1}{32}, \frac{1}{64}$ **59.** $17, 12, 7$ **61.** $2 + 5 + 8 + 11 + 14 = 40$ **63.** $2 + 5 + 10 + 17 + 26 + 37 = 97$

65. $\frac{1}{2} + 2 + \frac{9}{2} + 8 = 15$ **67.** $4 + 5 + 6 + 7 + 8 + 9 = 39$ **69.** $\sum_{i=1}^{5}(i + 8)$ **71.** $\sum_{i=1}^{3}\frac{i^2}{4}$ **73.** 13   **75.** 169   **77.** 55   **79.** 25

81. ≈ 42.83 **83. a)** $6, 12, 18, 24$ **b)** $p_n = 6n$ **85.** Answers will vary.   **87.** Answers will vary.   **89.** $\Sigma x = n\bar{x}$

91. Yes, for example if $n = 3$, you obtain $4x_1 + 4x_2 + 4x_3 = 4(x_1 + x_2 + x_3)$ **93. a)** 10   **b)** 11   **c)** 110   **d)** 29   **e)** No

94. $5, \frac{3}{2}$ **95.** two; $b^2 - 4ac > 0$ **96.**   **97.** $(2, 1), (-2, -1)$

Exercise Set 15.2   **1.** In an arithmetic sequence, each term differs by a constant amount.   **3.** It is called the common difference.

5. Positive number   **7.** Yes, for example $-1, -2, -3, \ldots$ **9.** Yes, for example $2, 4, 6, \ldots$ **11.** $4, 7, 10, 13, 16; a_n = 3n + 1$

13. $7, 5, 3, 1, -1; a_n = -2n + 9$ **15.** $\frac{1}{2}, 2, \frac{7}{2}, 5, \frac{13}{2}; a_n = \frac{3}{2}n - 1$ **17.** $100, 95, 90, 85, 80; a_n = -5n + 105$ **19.** 14   **21.** 27   **23.** 12

25. 2   **27.** 9   **29.** 6   **31.** $s_{10} = 100; d = 2$ **33.** $s_8 = \frac{52}{5}; d = \frac{1}{5}$ **35.** $s_6 = 25.5; d = 3.7$ **37.** $s_{11} = 407; d = 6$

39. $4, 7, 10, 13; a_{10} = 31; s_{10} = 175$ **41.** $-6, -4, -2, 0; a_{10} = 12; s_{10} = 30$ **43.** $-8, -13, -18, -23; a_{10} = -53; s_{10} = -305$

45. $\frac{7}{2}, 6, \frac{17}{2}, 11; a_{10} = 26, s_{10} = 147.5$ **47.** $100, 93, 86, 79; a_{10} = 37; s_{10} = 685$ **49.** $n = 15, s_{15} = 330$ **51.** $n = 11; s_{11} = 121$

53. $n = 17; s_{17} = \frac{153}{2}$ **55.** $n = 29; s_{29} = 1421$ **57.** 1275   **59.** 2500   **61.** 1395   **63.** 267   **65.** 42,372   **67.** 351

69. a) 27   **b)** 196   **71.** $101 \cdot 50 = 5050$ **73.** $s_n = n^2$ **75. a)** 19 feet   **b)** 143.5 feet   **77.** 2 feet   **79. a)** 155   **b)** 780

81. $496   **83. a)** $45,600   **b)** $438,000   **85.** $a_n = 180°(n - 2)$ **93.** $r = \frac{A - P}{Pt}$

94. $3(2n - 5)(2n - 1)$ **95.** $(-3, -5)$ **96.**

Exercise Set 15.3   **1.** A geometric sequence is a sequence in which each term after the first is the same multiple of the preceding term.   **3.** To find the common ratio, take any term except the first and divide by the term that precedes it.   **5.** 0

7. Yes   **9.** Yes, s_∞ exists because $|r| < 1, s_\infty = 8$ **11.** $2, 6, 18, 54, 162$ **13.** $6, -3, \frac{3}{2}, -\frac{3}{4}, \frac{3}{8}$ **15.** $72, 24, 8, \frac{8}{3}, \frac{8}{9}$

17. $90, -30, 10, -\frac{10}{3}, \frac{10}{9}$ **19.** $-1, -3, -9, -27, -81$ **21.** $5, -10, 20, -40, 80$ **23.** $\frac{1}{3}, \frac{1}{6}, \frac{1}{12}, \frac{1}{24}, \frac{1}{48}$ **25.** $3, \frac{9}{2}, \frac{27}{4}, \frac{81}{8}, \frac{243}{16}$

27. 128   **29.** $-\frac{3}{64}$ **31.** 128   **33.** 6144   **35.** $\frac{1}{64}$ **37.** $\frac{50}{729}$ **39.** 155   **41.** 7812   **43.** 10,160   **45.** $-\frac{2565}{256}$ **47.** $-\frac{9279}{625}$

49. $r = \frac{1}{2}; a_n = 3\left(\frac{1}{2}\right)^{n-1}$ **51.** $r = 2; a_n = 9(2)^{n-1}$ **53.** $r = -3; a_n = 2(-3)^{n-1}$ **55.** $r = \frac{2}{3}; a_n = \frac{3}{4}\left(\frac{2}{3}\right)^{n-1}$ **57.** 2   **59.** $\frac{5}{4}$

61. 12   **63.** $\frac{25}{3}$ **65.** 5   **67.** 6   **69.** 4   **71.** 24   **73.** -45 **75.** -15 **77.** $\frac{8}{33}$ **79.** $\frac{8}{9}$ **81.** $\frac{17}{33}$ **83.** $r = 3; a_1 = 5$

85. $r = 2$ or $r = -2; a_1 = 7$ **87.** $\approx$$1.77   **89. a)** 4 days   **b)** ≈ 1.172 grams   **91. a)** ≈ 330.78 million   **b)** ≈ 63.4 years

93. a) $\frac{1}{2}, \frac{1}{4}, \frac{1}{8}, \frac{1}{16}, \frac{1}{32}$ **b)** $a_n = \frac{1}{2}\left(\frac{1}{2}\right)^{n-1} = \left(\frac{1}{2}\right)^n$ **c)** $\frac{1}{128} \approx 0.78\%$ **95.** $\approx$$15,938.48   **97. a)** 28.512 feet   **b)** 550 feet

99. a) 10.29 inches   **b)** 100 inches   **101.** 211   **103. a)** $12,000, $9600, $7680   **b)** $a_n = 12,000\left(\frac{4}{5}\right)^{n-1}$ **c)** $\approx$$4915.20

105. 190 feet   **107. a)** y_2 goes up more steeply.   **b)**

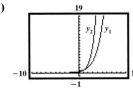

109. $n = 21; s_n = 2,097,151$
110. $8x^3 - 18x^2 + 21x - 18$
111. $8x - 15 + \frac{57}{2x + 5}$ **112.** 12 hours
113. $\frac{10}{3}$ **114.** $3x^2y^2\sqrt[3]{y} - 2xy\sqrt[3]{9y^2}$ **115.** $x\sqrt{y}$

116. $-\frac{1}{3}$

Mid-Chapter Test   **1.** $2, -1, -4, -7, -10$ [15.1]   **2.** 91 [15.1]   **3.** 1, 11 [15.1]   **4.** $-15, -19, -23$ [15.1]   **5.** 45 [15.1]

6. $\sum_{i=1}^{5} \left(\frac{1}{3}i + 7\right)$ [15.1]   **7.** $-6, -1, 4, 9; a_n = -11 + 5n$ [15.2]   **8.** -1 [15.2]   **9.** 6 [15.2]   **10.** $3, -3$ [15.2]   **11.** $47\frac{1}{2}$ [15.2]

12. 11 [15.2]   **13.** 136 [15.2]   **14.** $80, -40, 20, -10, 5$ [15.3]   **15.** $\frac{1}{9}$ [15.3]   **16.** 315 [15.3]   **17.** $-\frac{2}{3}$ [15.3]   **18.** 18 [15.3]

19. $\frac{29}{33}$ [15.3]   **20. a)** A sequence is a list of numbers arranged in a specific order.   **b)** An arithmetic sequence is a sequence where each term differs by a constant amount.   **c)** A geometric sequence is a sequence where the terms differ by a common multiple.   **d)** A series is the sum of the terms of a sequence. [15.1–15.3]

Exercise Set 15.4   **1.** Answers will vary.   **3.** 1   **5.** No, only factorials of nonnegative numbers can be found.
7. 14, the number of terms is one more than the exponent   **9.** 10   **11.** 1   **13.** 1   **15.** 70   **17.** 28   **19.** $x^3 + 12x^2 + 48x + 64$
21. $8x^3 - 36x^2 + 54x - 27$ **23.** $a^4 - 4a^3b + 6a^2b^2 - 4ab^3 + b^4$ **25.** $243a^5 - 405a^4b + 270a^3b^2 - 90a^2b^3 + 15ab^4 - b^5$

27. $16x^4 + 16x^3 + 6x^2 + x + \frac{1}{16}$ **29.** $\frac{1}{16}x^4 - \frac{3}{2}x^3 + \frac{27}{2}x^2 - 54x + 81$ **31.** $x^{10} + 100x^9 + 4500x^8 + 120{,}000x^7$
33. $2187x^7 - 5103x^6y + 5103x^5y^2 - 2835x^4y^3$ **35.** $x^{16} - 24x^{14}y + 252x^{12}y^2 - 1512x^{10}y^3$ **37.** Yes, $4! = 4 \cdot 3!$
39. Yes, $(7 - 3)! = (7 - 3)(7 - 4)(7 - 5)! = 4 \cdot 3 \cdot 2!$ **41.** $m = n$ or $m = 0$ **43.** $x^8, 24x^7, 17{,}496x, 6561$

45. $(a + b)^n = \sum_{i=0}^{n} \binom{n}{i} a^{n-i} b^i$ **47.** $(0, 10)$ **48.** $2, 9$ **49.** $(10, 4)$ **50.** $2x^3y^5\sqrt{30y}$ **51.** $f^{-1}(x) = \frac{x - 8}{3}$

Chapter 15 Review Exercises   **1.** $6, 7, 8, 9, 10$ **2.** $-1, 3, 9, 17, 27$ **3.** $6, 3, 2, \frac{3}{2}, \frac{6}{5}$ **4.** $\frac{1}{5}, \frac{2}{3}, \frac{9}{7}, 2, \frac{25}{9}$ **5.** 11   **6.** 4   **7.** $\frac{26}{81}$

8. 88   **9.** $s_1 = 7, s_3 = 27$ **10.** $s_1 = 9, s_3 = 38$ **11.** $s_1 = \frac{4}{3}, s_3 = \frac{227}{60}$ **12.** $s_1 = -9, s_3 = -10$ **13.** $32, 64, 128; a_n = 2^n$

14. $-\frac{1}{3}, \frac{1}{9}, -\frac{1}{27}; a_n = (-1)^n(3^{4-n})$ **15.** $\frac{16}{7}, \frac{32}{7}, \frac{64}{7}; a_n = \frac{2^{n-1}}{7}$ **16.** $-3, -7, -11; a_n = 17 - 4n$ **17.** $10 + 13 + 18 = 41$

18. $6 + 14 + 24 + 36 = 80$ **19.** $\frac{1}{6} + \frac{4}{6} + \frac{9}{6} + \frac{16}{6} + \frac{25}{6} = \frac{55}{6}$ **20.** $\frac{1}{2} + \frac{2}{3} + \frac{3}{4} + \frac{4}{5} = \frac{163}{60}$ **21.** 29   **22.** 239   **23.** 132   **24.** 841

25. a) $10, 14, 18, 22$ **b)** $p_n = 4n + 6$ **26. a)** $4, 10, 18, 28$ **b)** $a_n = n(n + 3) = n^2 + 3n$ **27.** $5, 8, 11, 14, 17$ **28.** $5, \frac{14}{3}, \frac{13}{3}, 4, \frac{11}{3}$

29. $\frac{1}{2}, -\frac{3}{2}, -\frac{7}{2}, -\frac{11}{2}, -\frac{15}{2}$ **30.** $-100, -\frac{499}{5}, -\frac{498}{5}, -\frac{497}{5}, -\frac{496}{5}$ **31.** 30   **32.** -4 **33.** $\frac{1}{2}$ **34.** 6   **35.** $s_8 = 112; d = 2$

36. $s_7 = -210; d = -6$ **37.** $s_7 = \frac{48}{5}; d = \frac{2}{5}$ **38.** $s_9 = -42; d = -\frac{1}{3}$ **39.** $-7, -3, 1, 5; a_{10} = 29, s_{10} = 110$

40. $4, 1, -2, -5; a_{10} = -23, s_{10} = -95$ **41.** $\frac{5}{6}, \frac{3}{2}, \frac{13}{6}, \frac{17}{6}; a_{10} = \frac{41}{6}; s_{10} = \frac{115}{3}$ **42.** $-60, -55, -50, -45; a_{10} = -15, s_{10} = -375$

43. $n = 13, s_{13} = 442$ **44.** $n = 7, s_7 = 14$ **45.** $n = 11; s_{11} = \frac{231}{10}$ **46.** $n = 10; s_{10} = 180$ **47.** $6, 12, 24, 48, 96$

48. $-12, -6, -3, -\frac{3}{2}, -\frac{3}{4}$ **49.** $20, -\frac{40}{3}, \frac{80}{9}, -\frac{160}{27}, \frac{320}{81}$ **50.** $-20, -4, -\frac{4}{5}, -\frac{4}{25}, -\frac{4}{125}$ **51.** $\frac{2}{27}$ **52.** 480   **53.** 216   **54.** $\frac{4}{243}$

55. 441   **56.** $-\frac{4305}{64}$ **57.** $\frac{585}{8}$ **58.** $\frac{127}{8}$ **59.** $r = 2; a_n = 6(2)^{n-1}$ **60.** $r = 5; a_n = -4(5)^{n-1}$ **61.** $r = \frac{1}{3}; a_n = 10\left(\frac{1}{3}\right)^{n-1}$

62. $r = \frac{2}{3}; a_n = \frac{9}{5}\left(\frac{2}{3}\right)^{n-1}$ **63.** 10   **64.** $\frac{25}{6}$ **65.** -6 **66.** -18 **67.** 32   **68.** $\frac{27}{2}$ **69.** $\frac{25}{6}$ **70.** -12 **71.** $\frac{4}{11}$ **72.** $\frac{23}{37}$
73. $81x^4 + 108x^3y + 54x^2y^2 + 12xy^3 + y^4$ **74.** $8x^3 - 36x^2y^2 + 54xy^4 - 27y^6$ **75.** $x^9 - 18x^8y + 144x^7y^2 - 672x^6y^3$
76. $256a^{16} + 3072a^{14}b + 16{,}128a^{12}b^2 + 48{,}384a^{10}b^3$ **77.** 15,050   **78.** 231   **79. a)** $\$36{,}000, \$37{,}000, \$38{,}000, \$39{,}000$
b) $a_n = \$35{,}000 + 1000n$ **c)** $\$41{,}000$ **d)** $\$451{,}000$ **80.** $\$102{,}400$ **81. a)** $\approx \$2024.51$ **b)** $\approx \$2463.13$ **c)** $\approx \$24{,}041.29$
82. $\approx \$503.63$ **83.** 150 feet

Chapter 15 Practice Test   **1.** A series is the sum of the terms of a sequence. [15.1]   **2. a)** An arithmetic sequence is one whose terms differ by a constant amount.   **b)** A geometric sequence is one whose terms differ by a common multiple. [15.2–15.3]
3. $-\frac{1}{3}, 0, \frac{1}{9}, \frac{1}{6}, \frac{1}{5}$ [15.1]   **4.** $s_1 = 3; s_3 = \frac{181}{36}$ [15.1]   **5.** $5 + 11 + 21 + 35 + 53 = 125$ [15.1]   **6.** 184 [15.1]

7. $a_n = \frac{1}{3} + \frac{1}{3}(n - 1) = \frac{1}{3}n$ [15.1]   **8.** $a_n = 5(2)^{n-1}$ [15.3]   **9.** $15, 9, 3, -3$ [15.2]   **10.** $\frac{5}{12}, \frac{5}{18}, \frac{5}{27}, \frac{10}{81}$ [15.3]   **11.** -40 [15.2]

12. -20 [15.2]   **13.** 12 [15.2]   **14.** $\frac{256}{243}$ [15.3]   **15.** $\frac{39{,}063}{5}$ [15.3]   **16.** $r = \frac{1}{3}; a_n = 15\left(\frac{1}{3}\right)^{n-1}$ [15.3]   **17.** 12 [15.3]

18. $\dfrac{13}{33}$ [15.3]   **19.** 56 [15.4]   **20.** $x^4 + 8x^3y + 24x^2y^2 + 32xy^3 + 16y^4$ [15.4]   **21.** 82 [15.2]   **22.** 91 [15.2]

23. \$210,000 [15.2]   **24.** $\approx$\$851.66 [15.3]   **25.** 364,500 [15.3]

Cumulative Review Test

1. $b = \dfrac{2A}{h}$ [2.6]   **2.** $y = -\dfrac{11}{3}x + \dfrac{38}{3}$ [4.4]   **3.** $5x^4 + 29x^3 + 14x^2 - 28x + 10$ [5.5]

4. $(x^2 + 2)(x - 6)$ [6.2]   **5.** $(a + b + 4)^2$ [6.4]   **6.** $\dfrac{5x^2 + 14x - 49}{(x + 5)(x - 2)}$ [7.3]   **7.** 500 [7.8]   **8.** Infinite number of solutions [9.4]

9. 3 [11.6]   **10.** 5 [11.6]   **11.** $-1 \pm i\sqrt{14}$ [12.1]   **12.** $\dfrac{3 \pm \sqrt{309}}{30}$ [12.2]   **13.** 5 [6.7]   **14.** [13.2]

15. $\dfrac{1}{2}$ [13.3]   **16.** [12.5]   **17.** $(x + 6)^2 + (y - 2)^2 = 49$ [14.1]

18. [14.1]   **19.** [14.2]   **20.** 18 [15.3]

Applications Index

Index

Photo Credits

Chapter 1 p. 1 © Jose Luis Pelaez/CORBIS All Rights Reserved; **p. 5** Gary Connor, PhotoEdit, Inc.; **p. 7** AP World Wide Photos; **p. 8** William Thomas Cain, Getty Images Inc.; **p. 16** Pictor, ImageState/International Stock Photography Ltd.; **p. 19** Jeopardy Productions, Getty Images Inc.; **p. 34** Allen R. Angel; **p. 40** Rhoda Peacher, R Photographs; **p. 46 (left)** Tom Pantages; **(right)** DLP ®/Courtesy of Texas Instruments Incorporated; **p. 54** Bob Daemmrich, PhotoEdit, Inc.; **p. 59** James Blank, Stock Boston; **p. 67** © Jose Luis Pelaez/CORBIS All Rights Reserved; **p. 78** Ross M. Horowitz, Getty Images Inc.; **p. 79** Doug Menuez, Getty Images, Inc.—Photodisc; **p. 85** © Zave Smith/CORBIS All Rights Reserved

Chapter 2 p. 95 Ryan McVay, Getty Images, Inc.–Photodisc; **p. 140** Getty Images – Stockbyte; **p. 143** Allen R. Angel; **p. 144** Steve Vidler, SuperStock, Inc.; **p. 149** AP Photo/Dusan Vranic; **p. 150** SuperStock, Inc.; **p. 151** Allen R. Angel; **p. 152** © Walt Disney Pictures/Pixar Animation Studios 07000; **p. 155** Johner Images, Getty Images Inc.; **p. 156** Allen R. Angel; **p. 162 (left)** Allen R. Angel; **(right)** Allen R. Angel; **p. 163 (left)** Robert Glusic, Getty Images, Inc.; **(right)** Gary Wiepert/Reuters, Landov LLC; **p. 170** Steve Mason, Getty Images, Inc.-Photodisc; **p. 171** Allen R. Angel

Chapter 3 p. 173 Dennis MacDonald, PhotoEdit, Inc.; **p. 177** Dale Pollinger; **p. 178 (top)** Allen R. Angel; **(bottom)** John A. Rizzo, Getty Images, Inc. – Photodisc; **p. 182** W. Bliss/HOA-QUI, Photo Researchers, Inc.; **p. 184** Allen R. Angel; **p. 185 (top left)** Jeff Greenberg, The Image Works; **(bottom left)** Allen R. Angel; **(right/column 1)** Allen R. Angel; **(right/column 2)** Allen R. Angel; **p. 186 (left)** Allen R. Angel; **(right)** Pictor, ImageState/International Stock Photography Ltd.; **p. 187 (left)** Allen R. Angel; **(top right)** Allen R. Angel **(bottom right)** Allen R. Angel; **p. 188** Allen R. Angel; **p. 191 (top)** © Erich Schlegel/Dallas Morning News/CORBIS All Rights Reserved; **(bottom)** Allen R. Angel; **p. 192** Dennis MacDonald, PhotoEdit, Inc.; **p. 193 (top)** Dave King © Dorling Kindersley; **(bottom)** Allen R. Angel; **p. 194** Laima Druskis, Pearson Education/PH College; **p. 196 (left)** Allen R. Angel; **(top right)** Getty Images, Inc.–Photodisc; **(bottom right)** EyeWire Collection, Getty Images, Inc.–Photodisc; **p. 197** Michael Newman, PhotoEdit, Inc.; **p. 198 (top left)** Nick Rowe, Getty Images, Inc.–Photodisc; **(bottom left)** Doug Menuez, Getty Images, Inc.–Photodisc; **(bottom right)** J. Gerard Smith, Photo Researchers, Inc.; **p. 199 (left)** Jeff Maloney, Getty Images, Inc.– Photodisc; **(right)** Jose Carrillo, PhotoEdit, Inc.; **p. 200 (top left)** Allen R. Angel; **(bottom left)** Allen R. Angel; **(right)** Karl Weatherly, Getty Images–Photodisc; **p. 203** Allen R. Angel; **p. 209** Allen R. Angel; **p. 215 (top left)** Allen R. Angel; **(bottom left)** Justin Sullivan, Getty Images, Inc.; **(right)** Allen R. Angel; **p. 216 (left)** EyeWire Collection, Getty Images, Inc.–Photodisc; **(right)** Glenn Cratty, Getty Images, Inc.; **p. 217** Peter Wilson © Dorling Kindersley Media Library, Courtesy of the Baseball Hall of Fame, Tokyo; **p. 218 (left)** © Dann Tardif/CORBIS All Rights Reserved; **(right)** © Holger Winkler/Zefa/CORBIS All Rights Reserved; **p. 222** Gary Randall, Getty Images, Inc.–Photodisc; **p. 223 (left)** © Tim McGuire/CORBIS All Rights Reserved; **(right)** © James D. Smith/Icon SMI/CORBIS All Rights Reserved

Chapter 4 p. 226 Tony Garcia, Getty Images, Inc.; **p. 227** Sheila Terry/Science Photo Library, Photo Researchers, Inc.; **p. 241** Corbis Royalty Free; **p. 244** © R. W. Jones/CORBIS; **p. 256** Sylvie Chappaz, Photo Researchers, Inc.; **p. 261** Allen R. Angel; **p. 268** Chris Oxley, Corbis/Bettmann; **p. 269** David Graham, AP Wide World Photos; **p. 273** Allen R. Angel

Chapter 5 p 275 Ryan McVay, Getty Images, Inc.–Photodisc; **p. 293** Donald Miralle, Getty Images, Inc.; **p. 298** Allen R. Angel, Disney Characters, © Disney Enterprises, Inc. Used by Permission from the Disney Enterprises, Inc.; **p. 299** Courtesy IBM; **p. 301** Canada Bureau of Travel; **p. 303** The Kobal Collection/Warner Bros/Southside Amusement Co.; **p. 327** Allen R. Angel; **p. 331** Sonda Dawes, The Image Works; **p. 333** Don Farrall, Getty Images, Inc.–Photodisc

Allen R. Angel
Chapter Test Prep Video t/a Elementary and Intermediate Algebra for College Students, 3e
0-13-159448-6
© 2008 Pearson Education, Inc.
Pearson Prentice Hall
Pearson Education, Inc.
Upper Saddle River, NJ 07458
Pearson Prentice Hall™ is a trademark of Pearson Education, Inc.

YOU SHOULD CAREFULLY READ THE TERMS AND CONDITIONS BEFORE USING THE CD-ROM PACKAGE. USING THIS CD-ROM PACKAGE INDICATES YOUR ACCEPTANCE OF THESE TERMS AND CONDITIONS.

Pearson Education, Inc. provides this program and licenses its use. You assume responsibility for the selection of the program to achieve your intended results, and for the installation, use, and results obtained from the program. This license extends only to use of the program in the United States or countries in which the program is marketed by authorized distributors.

LICENSE GRANT

You hereby accept a nonexclusive, nontransferable, permanent license to install and use the program ON A SINGLE COMPUTER at any given time. You may copy the program solely for backup or archival purposes in support of your use of the program on the single computer. You may not modify, translate, disassemble, decompile, or reverse engineer the program, in whole or in part.

TERM

The License is effective until terminated. Pearson Education, Inc. reserves the right to terminate this License automatically if any provision of the License is violated. You may terminate the License at any time. To terminate this License, you must return the program, including documentation, along with a written warranty stating that all copies in your possession have been returned or destroyed.

LIMITED WARRANTY

THE PROGRAM IS PROVIDED "AS IS" WITHOUT WARRANTY OF ANY KIND, EITHER EXPRESSED OR IMPLIED, INCLUDING, BUT NOT LIMITED TO, THE IMPLIED WARRANTIES OF MERCHANTABILITY AND FITNESS FOR A PARTICULAR PURPOSE. THE ENTIRE RISK AS TO THE QUALITY AND PERFORMANCE OF THE PROGRAM IS WITH YOU. SHOULD THE PROGRAM PROVE DEFECTIVE, YOU (AND NOT PEARSON EDUCATION, INC. OR ANY AUTHORIZED DEALER) ASSUME THE ENTIRE COST OF ALL NECESSARY SERVICING, REPAIR, OR CORRECTION. NO ORAL OR WRITTEN INFORMATION OR ADVICE GIVEN BY PEARSON EDUCATION, INC., ITS DEALERS, DISTRIBUTORS, OR AGENTS SHALL CREATE A WARRANTY OR INCREASE THE SCOPE OF THIS WARRANTY. SOME STATES DO NOT ALLOW THE EXCLUSION OF IMPLIED WARRANTIES, SO THE ABOVE EXCLUSION MAY NOT APPLY TO YOU. THIS WARRANTY GIVES YOU SPECIFIC LEGAL RIGHTS AND YOU MAY ALSO HAVE OTHER LEGAL RIGHTS THAT VARY FROM STATE TO STATE.

Pearson Education, Inc. does not warrant that the functions contained in the program will meet your requirements or that the operation of the program will be uninterrupted or error-free. However, Pearson Education, Inc. warrants the CD-ROM(s) on which the program is furnished to be free from defects in material and workmanship under normal use for a period of ninety (90) days from the date of delivery to you as evidenced by a copy of your receipt. The program should not be relied on as the sole basis to solve a problem whose incorrect solution could result in injury to person or property. If the program is employed in such a manner, it is at the user's own risk and Pearson Education, Inc. explicitly disclaims all liability for such misuse.

LIMITATION OF REMEDIES

Pearson Education, Inc.'s entire liability and your exclusive remedy shall be:

1. the replacement of any CD-ROM not meeting Pearson Education, Inc.'s "LIMITED WARRANTY" and that is returned to Pearson Education, or
2. if Pearson Education is unable to deliver a replacement CD-ROM that is free of defects in materials or workmanship, you may terminate this agreement by returning the program.

IN NO EVENT WILL PEARSON EDUCATION, INC. BE LIABLE TO YOU FOR ANY DAMAGES, INCLUDING ANY LOST PROFITS, LOST SAVINGS, OR OTHER INCIDENTAL OR CONSEQUENTIAL DAMAGES ARISING OUT OF THE USE OR INABILITY TO USE SUCH PROGRAM EVEN IF PEARSON EDUCATION, INC. OR AN AUTHORIZED DISTRIBUTOR HAS BEEN ADVISED OF THE POSSIBILITY OF SUCH DAMAGES, OR FOR ANY CLAIM BY ANY OTHER PARTY. SOME STATES DO NOT ALLOW FOR THE LIMITATION OR EXCLUSION OF LIABILITY FOR INCIDENTAL OR CONSEQUENTIAL DAMAGES, SO THE ABOVE LIMITATION OR EXCLUSION MAY NOT APPLY TO YOU.

GENERAL

You may not sublicense, assign, or transfer the license of the program. Any attempt to sublicense, assign or transfer any of the rights, duties, or obligations hereunder is void.

This Agreement will be governed by the laws of the State of New York. Should you have any questions concerning this Agreement, you may contact Pearson Education, Inc. by writing to:
ESM Media Development
Higher Education Division
Pearson Education, Inc.
1 Lake Street
Upper Saddle River, NJ 07458
Should you have any questions concerning technical support, you may write to:
New Media Production
Higher Education Division
Pearson Education, Inc.
1 Lake Street
Upper Saddle River, NJ 07458
YOU ACKNOWLEDGE THAT YOU HAVE READ THIS AGREEMENT, UNDERSTAND IT, AND AGREE TO BE BOUND BY ITS TERMS AND CONDITIONS. YOU FURTHER AGREE THAT IT IS THE COMPLETE AND EXCLUSIVE STATEMENT OF THE AGREEMENT BETWEEN US THAT SUPERSEDES ANY PROPOSAL OR PRIOR AGREEMENT, ORAL OR WRITTEN, AND ANY OTHER COMMUNICATIONS BETWEEN US RELATING TO THE SUBJECT MATTER OF THIS AGREEMENT.

System Requirements

-Windows
Pentium II 300 MHz processor
Windows 2000 (Service Pack 4) or XP
64 MB RAM (128 MB RAM required for Windows XP)
7.2 MB available hard drive space (optional—for minimum QuickTime installation)
800 x 600 resolution
8x or faster CD-ROM drive
QuickTime 7.x
Sound card

-Macintosh
PowerPC G3 233 MHz or better
Mac OS 10.x
64 MB RAM
19 MB on OS X (optional—if QuickTime installation is needed)
800 x 600 resolution
8x or faster CD-ROM drive
QuickTime 6 or 7

Support Information

If you are having problems with this software, call (800) 677-6337 between 8:00 a.m. and 8:00 p.m. EST, Monday through Friday, and 5:00 p.m. through Midnight EST on Sundays. You can also get support by filling out the web form located at: http://247.prenhall.com/mediaform

Our technical staff will need to know certain things about your system in order to help us solve your problems more quickly and efficiently. If possible, please be at your computer when you call for support. You should have the following information ready:

- Textbook ISBN
- CD-ROM ISBN
- corresponding product and title
- computer make and model
- Operating System (Windows or Macintosh) and Version
- RAM available
- hard disk space available
- Sound card? Yes or No
- printer make and model
- network connection
- detailed description of the problem, including the exact wording of any error messages.

NOTE: Pearson does not support and/or assist with the following:

- third-party software (i.e. Microsoft including Microsoft Office suite, Apple, Borland, etc.)
- homework assistance
- Textbooks and CD-ROMs purchased used are not supported and are non-replaceable. To purchase a new CD-ROM, contact Pearson Individual Order Copies at 1-800-282-0693.

Chapter 1 Real Numbers

Fractions

Addition
$$\frac{a}{c} + \frac{b}{c} = \frac{a+b}{c}$$

Subtraction
$$\frac{a}{c} - \frac{b}{c} = \frac{a-b}{c}$$

Multiplication
$$\frac{a}{b} \cdot \frac{c}{d} = \frac{a \cdot c}{b \cdot d}$$

Division
$$\frac{a}{b} \div \frac{c}{d} = \frac{a}{b} \cdot \frac{d}{c} = \frac{a \cdot d}{b \cdot c}$$

Natural numbers $\{1, 2, 3, 4, \ldots\}$

Whole numbers $\{0, 1, 2, 3, \ldots\}$

Integers $\{\ldots, -3, -2, -1, 0, 1, 2, 3, \ldots\}$

Rational numbers {quotient of two integers, denominator not 0}

The sum of two positive numbers will be a positive number.
The sum of two negative numbers will be a negative number.
The sum of a positive number and a negative number can be either a positive or negative number.

The product (or quotient) of two numbers with like signs will be a positive number.
The product (or quotient) of two numbers with unlike signs will be a negative number.

$a - b$ means $a + (-b)$

$$\frac{a}{-b} = \frac{-a}{b} = -\frac{a}{b}$$

$$b^n = \underbrace{b \cdot b \cdot b \cdot \cdots \cdot b}_{n \text{ factors of } b}$$

Order of Operations

1. Evaluate expressions within parentheses.
2. Evaluate expressions with exponents.
3. Perform multiplications or divisions moving from left to right.
4. Perform additions or subtractions moving from left to right.

Properties of the Real Numbers

Commutative: $a + b = b + a, a \cdot b = b \cdot a$

Associative: $(a + b) + c = a + (b + c), (a \cdot b) \cdot c = a \cdot (b \cdot c)$

Distributive: $a(b + c) = a \cdot b + a \cdot c$

Identity: $a + 0 = 0 + a = a, 1 \cdot a = a \cdot 1 = a$

Inverse: $a + (-a) = -a + a = 0, a \cdot \frac{1}{a} = \frac{1}{a} \cdot a = 1$

Chapter 2 Solving Linear Equations

Addition property of equality: If $a = b$, then $a + c = b + c$ for any real numbers a, b, and c.

Multiplication property of equality: If $a = b$, then $a \cdot c = b \cdot c$ for any real numbers a, b, and c.

Linear equation: $ax + b = c$, for real numbers a, b, and c.

To Solve Linear Equations with the Variable on Both Sides of the Equal Sign

1. If the equation contains fractions, multiply both sides of the equation by the LCD.
2. Use the distributive property to remove parentheses.
3. Combine like terms on the same side of the equal sign.
4. Use the addition property to rewrite the equation with all terms containing the variable on one side of the equal sign and all terms not containing the variable on the other side of the equal sign. Repeated use of the addition property will eventually result in an equation of the form $ax = b$.
5. Use the multiplication property to isolate the variable. This will give a solution of the form $x =$ some number.
6. Check the solution in the original equation.

Simple interest formula: $i = prt$

Distance formula: $d = rt$

Geometric formulas: See Section 2.6 and Appendix C.

Cross multiplication: If $\frac{a}{b} = \frac{c}{d}$, then $ad = bc$.

Similar Figures: Corresponding angles are equal and corresponding sides are in proportion.

Chapter 3 Applications of Algebra

Problem-Solving Procedure for Solving Application Problems

1. **Understand the problem.**
   Identify the quantity or quantities you are being asked to find.
2. **Translate the problem into mathematical language (express the problem as an equation).**
   a) Choose a variable to represent one quantity, *and write down exactly what it represents.* Represent any other quantity to be found in terms of this variable.
   b) Using the information from step a) write an equation that represents the application.
3. **Carry out the mathematical calculations (solve the equation).**
4. **Check the answer (using the original application).**
5. **Answer the question asked.**

Chapter 4 Graphing Linear Equations

Linear equation in two variables: $ax + by = c$

A **graph** is an illustration of the set of points whose coordinates satisfy the equation.
Every **linear equation** of the form $ax + by = c$ will be a straight line when graphed.
To find the y-intercept (where the graph crosses the y-axis) set $x = 0$ and solve for y.
To find the x-intercept (where the graph crosses the x-axis) set $y = 0$ and solve for x.

$$\text{slope } (m) = \frac{\text{change in } y}{\text{change in } x} = \frac{y_2 - y_1}{x_2 - x_1}$$

Positive slope (rises to right)

Negative slope (falls to right)

Slope is 0. (horizontal line)

Slope is undefined. (vertical line)

Linear Equations

Standard form of a linear equation: $ax + by = c$

Slope–intercept form of a linear equation: $y = mx + b$, where m is the slope and $(0, b)$ is the y-intercept.

Point–slope form of a linear equation: $y - y_1 = m(x - x_1)$, where m is the slope and (x_1, y_1) is a point on the line.

Two lines are **parallel** when they have the same slope.

Chapter 5   Exponents and Polynomials

Rules of Exponents

1. $x^m \cdot x^n = x^{m+n}$ **product rule**

2. $\dfrac{x^m}{x^n} = x^{m-n}, x \neq 0$ **quotient rule**

3. $(x^m)^n = x^{m \cdot n}$ **power rule**

4. $x^0 = 1, x \neq 0$ **zero exponent rule**

5. $x^{-m} = \dfrac{1}{x^m}, x \neq 0$ **negative exponent rule**

6. $\left(\dfrac{ax}{by}\right)^m = \dfrac{a^m x^m}{b^m y^m}, b \neq 0, y \neq 0$ **expanded power rule**

7. $\left(\dfrac{a}{b}\right)^{-m} = \left(\dfrac{b}{a}\right)^m, a \neq 0, b \neq 0$ **a fraction raised to a negative exponent rule**

FOIL method (*First, Outer, Inner, Last*) of multiplying binomials:
$(a + b)(c + d) = ac + ad + bc + bd$

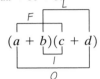

Product of the sum and difference of the same two terms:
$(a + b)(a - b) = a^2 - b^2$

Squares of binomials:
$(a + b)^2 = a^2 + 2ab + b^2$
$(a - b)^2 = a^2 - 2ab + b^2$

Chapter 6   Factoring

If $a \cdot b = c$, then a and b are **factors** of c.
Difference of two squares: $a^2 - b^2 = (a + b)(a - b)$
Sum of two cubes: $a^3 + b^3 = (a + b)(a^2 - ab + b^2)$
Difference of two cubes: $a^3 - b^3 = (a - b)(a^2 + ab + b^2)$

Quadratic equation: $ax^2 + bx + c = 0, a \neq 0$

Zero-factor property: If $ab = 0$, then $a = 0$ or $b = 0$.

Pythagorean Theorem: $a^2 + b^2 = c^2$

To Factor a Polynomial

1. If all the terms of the polynomial have a greatest common factor other than 1, factor it out.
2. If the polynomial has two terms, determine if it is a difference of two squares or a sum or a difference of two cubes. If so, factor using the appropriate formula.
3. If the polynomial has three terms, factor the trinomial using one of the procedures discussed.
4. If the polynomial has more than three terms, try factoring by grouping.
5. As a final step, examine your factored polynomial to see if the terms in any factors listed have a common factor. If you find a common factor, factor it out at this point.

To Solve a Quadratic Equation by Factoring

1. Write the equation in standard form with the squared term positive. This will result in one side of the equation being 0.
2. Factor the side of the equation that is not 0.
3. Set each factor containing a variable equal zero and solve each equation.
4. Check the solution found in step 3 in the original equation.

Chapter 7   Rational Expressions and Equations

To Simplify Rational Expressions

1. Factor both the numerator and denominator as completely as possible.
2. Divide out any factors common to both the numerator and denominator.

To Multiply Rational Expressions

1. Factor all numerators and denominators completely.
2. Divide out common factors.
3. Multiply the numerators together and multiply the denominators together.

To Add or Subtract Two Rational Expressions

1. Determine the least common denominator (LCD).
2. Rewrite each fraction as an equivalent fraction with the LCD.
3. Add or subtract numerators while maintaining the LCD.
4. When possible, factor the remaining numerator and simplify the fraction.

To Solve Rational Expressions

1. Determine the LCD of all fractions in the equation.
2. Multiply both sides of the equation by the LCD. This will result in every term in the equation being multiplied by the LCD.
3. Remove any parentheses and combine like terms on each side of the equation.
4. Solve the equation.
5. Check your solution in the original equation.

Variation

Direct Variation: $y = kx$

Inverse Variation: $y = \dfrac{k}{x}$

Joint Variation: $y = kxz$

Chapter 8   Functions and Their Graphs

A **relation** is any set of ordered pairs.

A **function** is a correspondence between a first set of elements, the domain, and a second set of elements, the range, such that each element of the domain corresponds to exactly one element in the range.

Vertical line test: If a vertical line cannot be drawn to intersect the graph at more than one point, the graph represents a function.

Two lines are **perpendicular** when their slopes are negative reciprocals.

Functions

Sum: $(f + g)(x) = f(x) + g(x)$

Difference: $(f - g)(x) = f(x) - g(x)$

Product: $(f \cdot g)(x) = f(x) \cdot g(x)$

Quotient: $\left(\dfrac{f}{g}\right)(x) = \dfrac{f(x)}{g(x)}, g(x) \neq 0$

Chapter 9   Systems of Linear Equations

Exactly 1 solution
(Nonparallel lines)

Line 1   Line 2

Consistent system

No solution
(Parallel lines)

Line 1
Line 2

Inconsistent system

Infinite number of solutions
(Same line)

Line 2
Line 1

Dependent system

A system of linear equations may be solved: (a) graphically, (b) by the substitution method, (c) by the addition or elimination method, (d) by matrices, or (e) by determinants.

$$\begin{vmatrix} a_1 & b_1 \\ a_2 & b_2 \end{vmatrix} = a_1 b_2 - a_2 b_1$$

Cramer's Rule:

Given a system of equations of the form

$$\begin{aligned} a_1 x + b_1 y &= c_1 \\ a_2 x + b_2 y &= c_2 \end{aligned}$$

then $x = \dfrac{\begin{vmatrix} c_1 & b_1 \\ c_2 & b_2 \end{vmatrix}}{\begin{vmatrix} a_1 & b_1 \\ a_2 & b_2 \end{vmatrix}}$ and $y = \dfrac{\begin{vmatrix} a_1 & c_1 \\ a_2 & c_2 \end{vmatrix}}{\begin{vmatrix} a_1 & b_1 \\ a_2 & b_2 \end{vmatrix}}$

Chapter 10   Inequalities in One and Two Variables

The **union** of set A and set B, $A \cup B$, is the set of elements that belong to either set A or set B.

The **intersection** of set A and set B, $A \cap B$, is the set of elements that belong to both set A and set B.

Inequalities

If $a > b$, then $a + c > b + c$.

If $a > b$, then $a - c > b - c$.

If $a > b$ and $c > 0$, then $a \cdot c > b \cdot c$.

If $a > b$ and $c > 0$, then $a/c > b/c$.

If $a > b$ and $c < 0$, then $a \cdot c < b \cdot c$.

If $a > b$ and $c < 0$, then $a/c < b/c$.

Absolute Value

If $|x| = a$, then $x = a$ or $x = -a$.

If $|x| < a$, then $-a < x < a$.

If $|x| > a$, then $x < -a$ or $x > a$.

If $|x| = |y|$, then $x = y$ or $x = -y$.

Chapter 11   Roots, Radicals, and Complex Numbers

If n is even and $a \geq 0$: $\sqrt[n]{a} = b$ if $b^n = a$

If n is odd: $\sqrt[n]{a} = b$ if $b^n = a$

Rules of radicals

$\sqrt{a^2} = |a|$

$\sqrt{a^2} = a, a \geq 0$

$\sqrt[n]{a^n} = a, a \geq 0$

$\sqrt[n]{a} = a^{1/n}, a \geq 0$

$\sqrt[n]{a^m} = \left(\sqrt[n]{a}\right)^m = a^{m/n}, a \geq 0$

$\sqrt[n]{a}\sqrt[n]{b} = \sqrt[n]{ab}, a \geq 0, b \geq 0$

$\dfrac{\sqrt[n]{a}}{\sqrt[n]{b}} = \sqrt[n]{\dfrac{a}{b}}, a \geq 0, b > 0$

A radical is simplified when the following are all true:

1. No perfect powers are factors of any radicand.
2. No radicand contains a fraction.
3. No denominator contains a radical.

Complex numbers: numbers of the form $a + bi$

Powers of i: $i = \sqrt{-1}, i^2 = -1, i^3 = -i, i^4 = 1$

Chapter 12   Quadratic Functions

Square Root Property:

If $x^2 = a$, where a is a real number, then $x = \pm\sqrt{a}$.

A quadratic equation may be solved by factoring, completing the square, or the quadratic formula.

Quadratic formula: $x = \dfrac{-b \pm \sqrt{b^2 - 4ac}}{2a}$

Discriminant: $b^2 - 4ac$

If $b^2 - 4ac > 0$, then equation has two distinct real number solutions.

If $b^2 - 4ac = 0$, then equation has a single real number solution.

If $b^2 - 4ac < 0$, then equation has no real number solution.

Parabolas

For $f(x) = ax^2 + bx + c$, the vertex of the parabola is

$\left(-\dfrac{b}{2a}, \dfrac{4ac - b^2}{4a}\right)$ or $\left(-\dfrac{b}{2a}, f\left(-\dfrac{b}{2a}\right)\right)$.

For $f(x) = a(x - h)^2 + k$, the vertex of the parabola is (h, k).

If $f(x) = ax^2 + bx + c, a > 0$, the function will have a minimum value of $\dfrac{4ac - b^2}{4a}$ at $x = -\dfrac{b}{2a}$.

If $f(x) = ax^2 + bx + c, a < 0$, the function will have a maximum value of $\dfrac{4ac - b^2}{4a}$ at $x = -\dfrac{b}{2a}$.